Symposium on Prospects in the Physics of Discrete Symmetries 2010

(DISCRETE 2010)

AA008916

Journal of Physics: Conference Series Volume 335

Rome, Italy
6-11 December 2010

Editors:

Antonio Di Domenico
Caterina Bloise
Riccardo Faccini
Gino Isidori
Lucio Ludovici
Cesare Bini
Fabio Bossi
Paolo Gauzzi
Paolo Lipari
Luca Silvestrini

ISBN: 978-1-62276-682-6
ISSN: 1742-6588

Printed from e-media with permission by:

Curran Associates, Inc.
57 Morehouse Lane
Red Hook, NY 12571

CURRAN ASSOCIATES INC. proceedings.com

Some format issues inherent in the e-media version may also appear in this print version.

Printed by Curran Associates, Inc. (2013)

For permission requests, please contact the Institute of Physics
at the address below.

Institute of Physics
Dirac House, Temple Back
Bristol BS1 6BE UK

Phone: 44 1 17 929 7481
Fax: 44 1 17 920 0979

techtracking@iop.org

Additional copies of this publication are available from:

Curran Associates, Inc.
57 Morehouse Lane
Red Hook, NY 12571 USA
Phone: 845-758-0400
Fax: 845-758-2634
Email: curran@proceedings.com
Web: www.proceedings.com

Plenary papers

T, C, P, CP symmetries, accidental symmetries (B, L conservation)

Neutrino mass, mixing and discrete symmetries

CPT and Lorentz symmetries, QM

Cosmology and astroparticles, dark matter

Experimental Prospects at LHC, Superflavor factories and new facilities

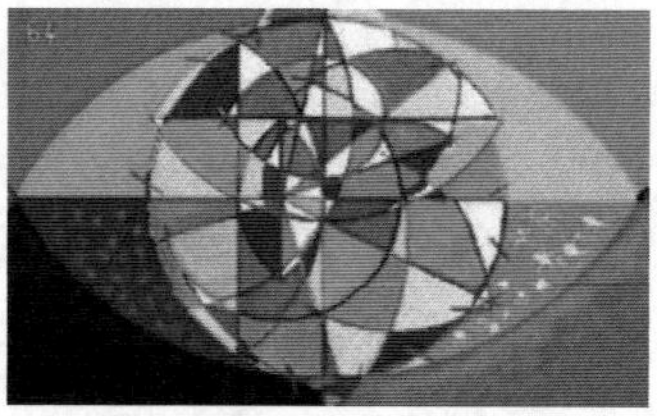

Preface

The Symposium DISCRETE2010 on Prospects in the Physics of Discrete Symmetries was held at the Sapienza Università di Roma, Italy from 6 to 11 December 2010. This second edition, after the successful one in Valencia in 2008, covered all theoretical and experimental progress in the field, and aimed at a thorough discussion on the latest developments. The topics covered included:

- T, C, P, CP symmetries;
- accidental symmetries (B, L conservation);
- CPT symmetry, decoherence, Lorentz symmetry breaking;
- neutrino mass and mixing;
- cosmology and astroparticles, dark matter searches;
- experimental prospects at LHC, Super flavor factories, and new facilities.

The Symposium was organized in plenary sessions with a total of 23 invited talks, and parallel sessions with a total of 80 talks including both invited and selected contributions from the submitted abstracts.
The speakers of the plenary sessions were: Achille Stocchi, Andreas Weiler, Kevin Pitts, Tim Gershon, Marco Sozzi, Neal Weiner, Vasiliki Mitsou, Bernard Sadoulet, Gianfranco Bertone, J. Eric Grove, Mauro Mezzetto, Alexei Yu Smirnov, Oliviero Cremonesi, Antonio Riotto, Reno Mandolesi, Brett Altschul, Jose Bernabeu, Lawrence Hall, Marco Grassi, Yannis K. Semertzidis, Riccardo Barbieri, Gigi Rolandi, Luciano Maiani.

The Symposium venue was the CNR (Consiglio Nazionale delle Ricerche) headquarter building, close to the Sapienza University. At the end of the Symposium a special open session, devoted to a wider audience, was held at the Pontifical University of the Holy Cross, in the historical center of Rome.

The symposium was attended by about 140 participants, about half coming from Italy, and the rest mainly from other European countries and United States.

Among the social events, there were a concert at the Aula Magna of the Sapienza University, and a social dinner in the historical Palazzo Pallavicini-Rospigliosi on the Quirinale Hill.

The next symposium of the series will be organised by IST, Universidade Tècnica de Lisboa, Portugal, from 3 to 7 December 2012.

Statement of Peer Review
All papers published in this volume of Journal of Physics: Conference Series have been peer reviewed through processes administered by the Editors. Reviews were conducted by expert referees to the professional and scientific standards expected of a proceedings journal published by IOP Publishing.

Roma, November 2011

The Editors

Antonio Di Domenico
Cesare Bini
Caterina Bloise
Fabio Bossi
Riccardo Faccini
Paolo Gauzzi
Gino Isidori
Paolo Lipari
Lucio Ludovici
Luca Silvestrini

International Advisory Committee

- Francisco del Aguila (Granada)
- Luis Alvarez-Gaume (CERN)
- Jose Bernabeu (Valencia)
- Andrzej J. Buras (Munich)
- Marcos Cerrada (Madrid)
- Jonathan R. Ellis (CERN)
- Domenec Espriu (Barcelona)
- Fernando Ferroni (Rome)
- Juan Fuster Verdu (Valencia)
- Marcello Giorgi (Pisa)
- Maria Jose Herrero (Madrid)
- David G. Hitlin (Caltech)
- Cecilia Jarlskog (Lund)
- Manel Martinez (Barcelona)
- Antonio Masiero (Padua)
- Nikolaos Mavromatos (London)
- Vasiliki Mitsou (Valencia)
- Hitoshi Murayama (Berkeley)
- Keith Olive (Minnesota)
- Antonio Pich (Valencia)
- Apostolos Pilaftsis (Manchester)
- Mariano Quiros (Barcelona)
- Teresa Rodrigo Anoro (Santander)
- Subir Sarkar (Oxford)
- Fabio Zwirner (Padua)

Local Organizing Committee

- Cesare Bini (Roma)
- Caterina Bloise (LNF - Frascati)
- Fabio Bossi (LNF - Frascati)
- Antonio Di Domenico (Roma) - Chairman
- Riccardo Faccini (Roma)
- Paolo Gauzzi (Roma)
- Gino Isidori (LNF - Frascati)
- Paolo Lipari (Roma)
- Lucio Ludovici (Roma)
- Guido Martinelli (Roma)
- Luca Silvestrini (Roma)

Secretariat

- Mauro Mancini
- Alba Perrotta

E-mail: discrete10@roma1.infn.it
Web page: http://www.roma1.infn.it/discrete10/

Sponsors

- Sapienza Università di Roma
- INFN - Istituto Nazionale di Fisica Nucleare
- CNR - Consiglio Nazionale delle Ricerche
- CAEN S.p.A.

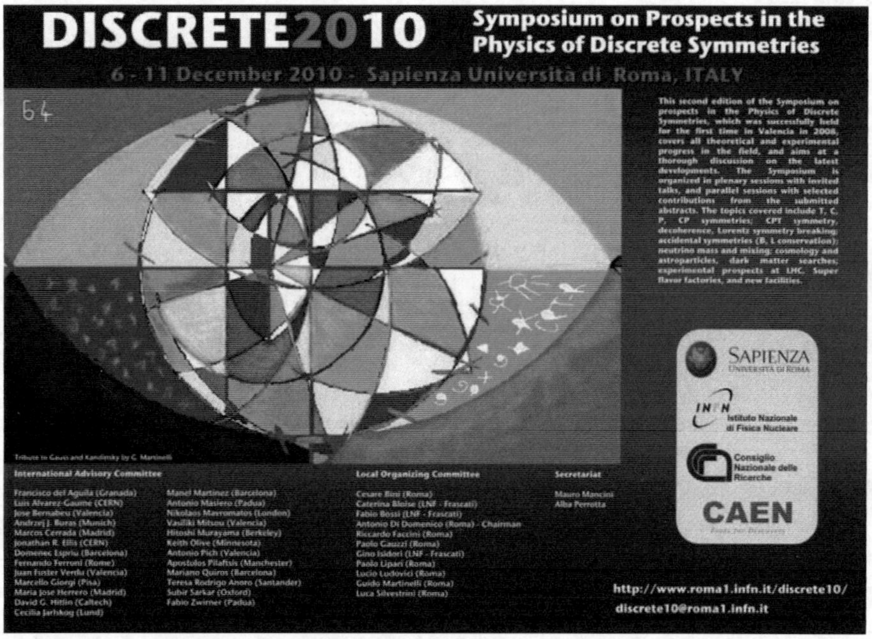

Poster

Group picture

List of Participants

ALTSCHUL, Brett	University of South Carolina, Columbia SC, USA
ALVES JUNIOR, Antonio Augusto	INFN - Sezione di Roma, ITALY
AMELINO CAMELIA, Giovanni	Sapienza Università di Roma, ITALY
ANULLI, Fabio	INFN - Sezione di Roma, ITALY
ANZIVINO, Giuseppina	Università di Perugia and INFN - Sezione di Perugia, ITALY
AZATOV, Aleksandr	Sapienza Università di Roma and INFN - Sezione di Roma, ITALY
AZZURRI, Paolo	Scuola Normale Superiore and INFN - Sezione di Pisa, ITALY
BAGNAIA, Paolo	Sapienza Università di Roma and INFN Sezione di Roma, ITALY
BARBIERI, Riccardo	Scuola Normale Superiore and INFN Sezione di Pisa, ITALY
BARZÈ, Luca	Università di Pavia and INFN - Sezione di Pavia, ITALY
BELLINI, Fabio	Sapienza Università di Roma and INFN - Sezione di Roma, ITALY
BERNABEU, Jose	Universitat de Valencia and IFIC, Valencia, SPAIN
BERTONE, Gianfranco	ITP Univertität Zürich, SWITZERLAND
BIASSONI, Pietro	Università di Milano and INFN - Sezione di Milano, ITALY
BINI, Cesare	Sapienza Università di Roma and INFN - Sezione di Roma, ITALY
BLAKE, Thomas	Imperial College, London, UNITED KINGDOM
BLOISE, Caterina	INFN - LNF, Frascati, ITALY
BOBETH, Christoph	IAS/Technische Universität München, GERMANY
BORRONI, Sara	INFN - Sezione di Roma, ITALY
BOSSI, Fabio	INFN - LNF, Frascati, ITALY
BROD, Joachim	Technische Universität München, Garching, GERMANY
CAPON, Giorgio	INFN - LNF, Frascati, ITALY
CARLUCCI, Maria Valentina	Technische Universität München, Garching, GERMANY
CASAGRANDE, Sandro	Technische Universität München, Garching, GERMANY
CASAROSA, Giulia	Università di Pisa and INFN - Sezione di Pisa, ITALY
CAVOTO, Gianluca	INFN - Sezione di Roma, ITALY
CERVELLI, Alberto	Università di Pisa and INFN - Sezione di Pisa, ITALY
CERVELLI, Franco	INFN - Sezione di Pisa, ITALY
CHRYSSOMALAKOS, Chryssomalis	Instituto de Ciencias Nucleares UNAM, México D.F., MEXICO
CIUFFOLI, Emilio	Università di Pisa and INFN Sezione di Pisa, ITALY
COLLAMATI, Francesco	Sapienza Università di Roma, ITALY
CONTINO, Roberto	Sapienza Università di Roma, ITALY
CREMONESI, Oliviero	University of Milano-Bicocca, Milano, ITALY
CZERWINSKI, Eryk	INFN - LNF, Frascati, ITALY
D'AGOSTINI, Giulio	Sapienza Università di Roma and INFN - Sezione di Roma, ITALY
D'AMBROSIO, Giancarlo	INFN - Sezione di Napoli, ITALY
DE ADELHART TOOROP, Reinier	Nikhef Theory Group, Amsterdam, NETHERLANDS
DE LUCIA, Erika	INFN - LNF, Frascati, ITALY
DE SANTIS, Antonio	Sapienza Università di Roma and INFN - Sezione di Roma, ITALY
DE SIMONE, Patrizia	INFN - LNF, Frascati, ITALY
DI DOMENICO, Antonio	Sapienza Università di Roma and INFN - Sezione di Roma, ITALY
DI LUZIO, Luca	SISSA and INFN - Sezione di Trieste, ITALY
DOMENICI, Danilo	INFN - LNF, Frascati, ITALY
DRENSKA, Nadejda Vassileva	INFN - Sezione di Roma, ITALY
DUFOUR, Fanny	University of Geneva, Genève, SWITZERLAND
FACCINI, Riccardo	Sapienza Università di Roma and INFN - Sezione di Roma, ITALY
FALCIANO, Speranza	INFN - Sezione di Roma, ITALY
FARGION, Daniele	Sapienza Università di Roma and INFN - Sezione di Roma, ITALY
FARNESE, Christian	Università di Padova and INFN - Sezione di Padova ITALY
FIORE, Salvatore	Sapienza Università di Roma and INFN Sezione di Roma, ITALY
FIRMINO DA COSTA, Joao	DESY, Hamburg, GERMANY
GADOMSKI, Szymon	University of Geneva, Genève, SWITZERLAND
GAROSI, Paola	Università di Siena and INFN - Sezione di Pisa, ITALY

GAUZZI, Paolo	Sapienza Università di Roma and INFN - Sezione di Roma, ITALY
GERSHON, Tim	University of Warwick, Coventry, UNITED KINGDOM
GERSHTEIN, Yuri	Rutgers University, Piscataway NJ, USA
GIOVANNELLA, Simona	INFN - LNF, Frascati, ITALY
GIUNTI, Carlo	Università di Torino and INFN - Sezione di Torino, ITALY
GRASSI, Marco	INFN - Sezione di Pisa, ITALY
GRASSO, Alberto	INFN - Sezione di Catania, ITALY
GROSS, Christian	Technische Universität Dortmund, GERMANY
GROVE, J. Eric	Naval Research Laboratory, Washington DC, USA
GUARDINCERRI, Elena	INFN - LNGS, L'Aquila, ITALY
GUIDO, Elisa	Università di Genova and INFN - Sezione di Genova, ITALY
HALL, Lawrence	University of California, Berkeley, USA
HAYASAKA, Kiyoshi	Nagoya University, JAPAN
HIESMAYR, Beatrix	University of Vienna, Wien, AUSTRIA
HODGKINSON, Robert	Universitat de Valencia and IFIC, Valencia, SPAIN
ISIDORI, Gino	INFN - LNF, Frascati, ITALY
IWABUCHI, Masaya	Yonsei University, Seoul, KOREA
JONES-PEREZ, Joel	INFN - LNF, Frascati, ITALY
JUHASZ, Bertalan	Stefan Meyer Institute for Subatomic Physics, Wien, AUSTRIA
KADOSH, Avihay	University of Groningen, NETHERLANDS
KOLOMENSKY, Yury	University of California Berkeley/LBNL, Berkeley, USA
KORPAR, Samo	University of Maribor, SLOVENIA
KUPSC, Andrzej	Uppsala University, SWEDEN
LALOUM, Maurice	CNRS/IN2P3/LPNHE, Paris, FRANCE
LAMANNA, Gianluca	CERN, Genève, SWITZERLAND
LAZZARO, Claudia	ETH Zürich, SWITZERLAND
LEROY, Olivier	CPPM CNRS/IN2P3 and Aix-Marseille University, Marseille, FRANCE
LIGABUE, Franco	Scuola Normale Superiore di Pisa, ITALY
LIPARI, Paolo	INFN - Sezione di Roma, ITALY
LONG, Joshua	Indiana University, Bloomington IN, USA
LOUVOT, Remi	École Polytechnique Fédérale de Lausanne, Lausanne, SWITZERLAND
LUDOVICI, Lucio	INFN - Sezione di Roma, ITALY
LUHN, Christoph	University of Southampton, UNITED KINGDOM
LUSIGNOLI, Maurizio	Sapienza Università di Roma and INFN - Sezione di Roma, ITALY
MAIANI, Luciano	CNR and Sapienza Università di Roma
MANDOLESI, Nazareno	INAF - IASF, Bologna, Italy
MARTELLOTTI, Silvia	INFN - Sezione di Roma, ITALY
MARTON, Johann	Stefan Meyer Institute for Subatomic Physics, Wien, AUSTRIA
MASINA, Isabella	Università di Ferrara and INFN - Sezione di Ferrara, ITALY
MAVROMATOS, Nikolaos	King's College, London and CERN, Genève, SWITZERLAND
MELONI, Davide	Universität Würzburg, GERMANY
MERLO, Luca	Technische Universität München, Garching, GERMANY
MEZZETTO, Mauro	INFN - Sezione di Padova, ITALY
MIGLIOZZI, Pasquale	INFN - Sezione di Napoli, ITALY
MISCETTI, Stefano	INFN - LNF, Frascati, ITALY
MITSOU, Vasiliki	IFIC, Universitat de València, SPAIN
MOLINARO, Emiliano	CFTP - Instituto Superior Técnico, Lisboa, PORTUGAL
MONDRAGON, Alfonso	Instituto de Fisica/UNAM, Mexico City, MEXICO
NEBOT, Miguel	Universitat de València and IFIC, Valencia, SPAIN
OAKES, Louise	Technische Universität München, Garching, GERMANY
ORGANTINI, Giovanni Corrado	Sapienza Università di Roma and INFN - Sezione di Roma, ITALY
PALMONARI, Federico	Università di Bologna and INFN - Sezione di Bologna, ITALY
PALOMARES, Carmen	CIEMAT, Madrid, SPAIN
PANOTOPOULOS, Grigoris	Universitat de València and IFIC, Valencia, SPAIN

PARIS, Alessio	INFN - Sezione di Padova, ITALY
PATTAVINA, Luca	INFN - Sezione di Milano-Bicocca, ITALY
PETRARCA, Silvano	INFN - Sezione di Roma, ITALY
PICCINELLI, Gabriella	Centro Tecnológico, FES Aragón, UNAM, México D.F., MEXICO
PICCINI, Mauro	INFN - Sezione di Perugia, ITALY
PINCI, Davide	INFN - Sezione di Roma, ITALY
PITTS, Kevin	University of Illinois, Urbana, USA
POULOSE, Poulose	IIT, Guwahati, INDIA
RADEMACKER, Jonas	University of Bristol, UNITED KINGDOM
REBELO, Margarida Nesbitt	CFTP - Instituto Superior Técnico, Lisboa, PORTUGAL
RIOTTO, Antonio Walter	CERN, Genève, SWITZERLAND
RODRIGUES, Eduardo	University of Glasgow, UNITED KINGDOM
ROLANDI Gigi	CERN, Genève, SWITZERLAND
RUFFINI, Fabrizio	Università di Siena and INFN - Sezione di Pisa, ITALY
SADOULET, Bernard	University of California at Berkeley, USA
SANTACESARIA, Roberta	INFN - Sezione di Roma, ITALY
SEMAEV, Igor	University of Bergen, NORWAY
SEMERTZIDIS, Yannis	Brokhaven National Laboratory, Upton NY, USA
SILVESTRINI, Luca	INFN - Sezione di Roma, ITALY
SIMOES, Catarina	CFTP - Instituto Superior Técnico, Lisboa, PORTUGAL
SIRAGUSA, Giovanni	Johannes Gutenberg Univesität, Mainz, GERMANY
SKWARNICKI, Tomasz	Syracuse University, Syracuse NY, USA
SMIRNOV, Alexei	International Centre for Theoretical Physics, Trieste, ITALY
SOLFAROLI CAMILLOCCI, Elena	INFN - Sezione di Roma, ITALY
SOZZI, Marco	Università di Pisa and INFN - Sezione di Pisa, ITALY
SPARVOLI, Roberta	Università di Roma "Tor Vergata", ITALY
STAMOU, Emmanuel	Technische Universität München, Garching, GERMANY
STOCCHI, Achille	LAL, Univ Paris-Sud, CNRS/IN2P3, Orsay, France
TIWARI, Bhupendra Nath	INFN - LNF, Frascati, ITALY
VAUDREVANGE, Patrick	LMU Munich, Mnchen, GERMANY
VERDUCCI, Monica	University of Washington, Seattle WA, USA
VERNAZZA, Leonardo	Johannes Gutenberg Universität, Mainz, GERMANY
VILLANUEVA-PÉREZ, Pablo	Universitat de València and IFIC, Valencia, SPAIN
WEILER, Andreas	CERN, Genève, SWITZERLAND
WEINER, Neal	New York University, New York, USA
WIDL, Edmund	HEPHY, Wien, AUSTRIA
WISLICKI, Wojciech	A.Soltan Institute for Nuclear Studies, Warszawa, POLAND
WONSAK, Bjoern	University of Hamburg, GERMANY

Experimental prospects for B physics and discrete symmetries at LHC and future projects

Tim Gershon

Department of Physics, University of Warwick, Coventry CV4 7AL, United Kingdom

E-mail: T.J.Gershon@warwick.ac.uk

Abstract. The experimental prospects for B physics (with brief mentions of charm and τ physics) at the Large Hadron Collider experiments are reviewed, with particular attention to the LHCb experiment. The main focus is on CP (a)symmetry, though the usefulness of rare decays to probe the symmetries of nature is also discussed. The need for upgrades and new facilities to continue the investigations beyond the first few years of LHC operations is emphasised.

1. Current status

A great deal has been learnt about heavy flavour physics over the past decade. In particular, the achievements of the B factory experiments, the Tevatron, the kaon experiments NA48 and KTeV as well as several other experiments have confirmed that the Cabibbo-Kobayashi-Maskawa (CKM) quark-mixing matrix [1, 2] is the origin of CP violation within the Standard Model (SM). This is often shown in the form of constraints on the so-called "Unitarity Triangle", as seen in Fig. 1. The strongest constraints come from the measurement of $\sin(2\beta)$ in $B^0 \to J/\psi K^0_S$ and related processes[1] from BaBar [3] and Belle [4], the value of Δm_d normalised to that obtained for Δm_s by CDF [5], the CP violation parameter in neutral kaon mixing ϵ_K and combinations of many measurements from BaBar and Belle that constrain $|V_{ub}|$, α and γ (for detailed reviews containing all original citations, see Refs. [6, 7, 8]). All measurements overlap (within a few standard deviations) at a single point, illustrating their consistency with the CKM paradigm.

However, a closer scrutiny of the situation reveals several intriguing hints of non-standard effects in flavour observables that demand further investigation. These include evidence for an inclusive same-sign dimuon asymmetry from D0 [11, 12], anomalous results for the CP violating phase in B^0_s oscillations [13, 14, 15, 16], possible deviations from the expected behaviour of the forward-backward asymmetry in $B \to K^* l^+ l^-$ decays [17, 18, 19], a larger than expected rate for $B \to \tau\nu$ [20, 21] and tension in the CKM fit [22]. Moreover, the existing constraints do not rule out the possibility of large contributions from non-SM processes in various flavour physics observables that can be tested in the next few years.

Tab. 1 presents an alternative method of summarising the current knowledge of CP violation. Definitions of the three different categories of CP violation, namely (i) CP violation in mixing, (ii) CP violation in the interference between mixing and decay, (iii) CP violation in decay (also known as direct CP violation), can be found in the review by D. Kirkby and Y. Nir

[1] The inclusion of charged conjugated processes is implied throughout.

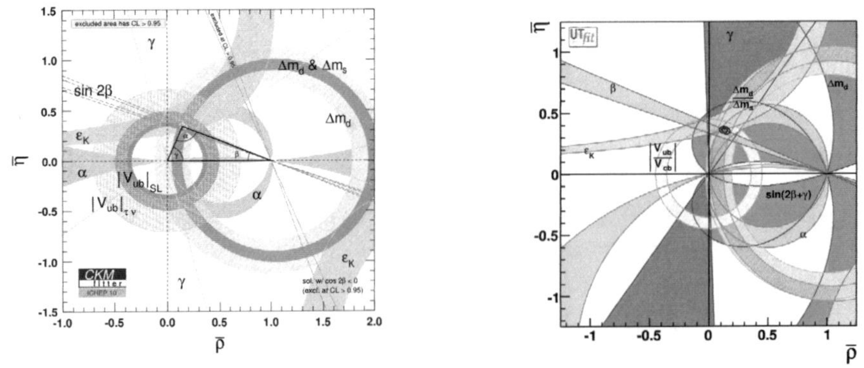

Figure 1. Constraints on the Unitarity Triangle compiled by (left) the CKMfitter collaboration [9], using frequentist methods; (right) the UTfit collaboration [10], using a Bayesian approach.

in Ref. [8]. All three categories have been observed in the neutral kaon system,[2] while some sizable effects in the B_d^0 system have been observed as expected in the SM (including direct CP violation [23, 24, 25]). Yet it is clear that despite the huge progress made by the B factory experiments, there is much that remains to be experimentally tested. Consequently, next generation experiments have enormous discovery potential. As will be discussed in the following sections, there are several sectors where large CP violation effects are expected in the SM but have not yet been investigated with sufficient precision. On the other hand, in some of the others the Standard Model prediction is for negligibly small levels of CP violation – these therefore provide sensitive "null tests" that can be used to search for the effects of physics beyond the SM.

Table 1. Summary of the systems where CP violation effects have been observed. A five standard deviation (σ) significance threshold is required for a ✓. Note that CP violation in decay is the only possible category for particles that do not undergo oscillations. Further details can be found in the review by D. Kirkby and Y. Nir in Ref. [8].

	K^0	D^0	B_d^0	B_s^0	Charged mesons	Baryons	Charged leptons
CP violation in mixing	✓	✗	✗	✗	–	–	–
CP violation in mixing/decay interference	✓	✗	✓	✗	–	–	–
CP violation in decay	✓	✗	✓	✗	✗	✗	✗

2. The Large Hadron Collider and its experiments

The year 2010 was a remarkably successful start of the LHC era. As shown in Fig. 2, the delivered luminosity per experiment approached 50 pb^{-1}, and the peak instantaneous luminosity exceeded

[2] I am grateful to Marco Sozzi for a clarification on this point.

10^2 Hz/μb $\equiv 10^{32}$ cm^{-2} s^{-1}. This roughly corresponds to the design luminosity of the LHCb experiment, and is only two orders of magnitude below the anticipated peak luminosity that will be delivered to the ATLAS and CMS experiments before a major upgrade of the machine.

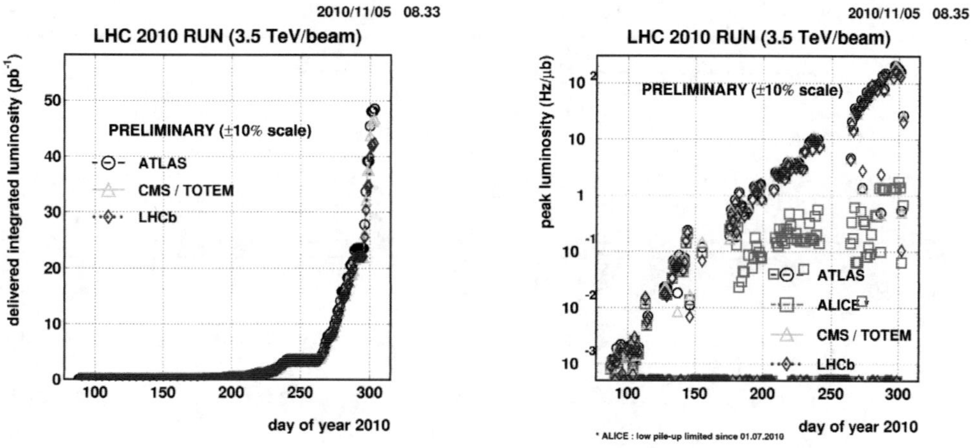

Figure 2. Performance of the LHC during 2010: (left) integrated delivered luminosity; (right) peak instantaneous luminosity.

Details of the LHCb [26], ATLAS [27] and CMS [28] experiments can be found elsewhere; the salient difference between LHCb and the general purpose detectors is that the former is instrumented to cover the forward region ($2 < \eta < 6$) where the $b\bar{b}$ production cross-section is strongly peaked. All achieved excellent operational efficiency during 2010. Among results pertinent to flavour physics, at the time of the conference, all three experiments have produced results on the J/ψ production cross-section [29, 30, 31], while LHCb has in addition measured the open charm production cross-section [32] and has published the first measurements of $b\bar{b}$ production in pp collisions at 7 TeV [33]. The first observation of the decay $B_s^0 \to D_{s2}^{*-} X \mu^+ \nu$ [34] illustrates that LHCb is already moving into a phase where it is competitive with previous B physics experiments.

3. CP violation effects expected in the Standard Model

One of the most striking features of Tab. 1 is that CP violation has not yet been observed in the decay of any charged meson. The Standard Model predicts large effects in several channels, and indeed several measurements show evidence for the predicted effects [7, 8], but none at the $5\,\sigma$ level.

One of the theoretically cleanest predictions in CP violation phenomenology is that direct CP violation will occur in $B^+ \to DK^+$ decays, when the neutral D meson is reconstructed in a final state that is accessible to both D^0 and $\bar{D}^0$ decays, at a rate that is governed by the angle of the CKM Unitarity Triangle γ. Since only tree-level decay amplitudes are involved, and all the hadronic parameters can be determined from the data, there is negligible associated theoretical uncertainty [35]. At present γ is only weakly constrained [9, 10], and therefore improving the knowledge of this quantity is one of the key goals of LHCb, as described in the recent roadmap document [36].

Already in the data accumulated in 2010, LHCb has observed signals of $B^+ \to DK^+$ and the related $B^+ \to D\pi^+$ decays, where the D meson is reconstructed in two track final states, as shown in Fig. 3. The yields are broadly in line with the Monte Carlo simulation based expectations. The yield of $B^+ \to D\pi^+$, $D \to K^+K^-$ events (1035 ± 54) exceeds that of CDF in their publication based on 1 fb^{-1} (780 ± 36 events) [37][3] but is somewhat less than in the final B factory datasets. Signals have also been seen in the related channels with the D meson reconstructed from $K^0_S\pi^+\pi^-$ or $K^0_S K^+K^-$, as also shown in Fig. 3. Measurements from the B factories have shown that these have high sensitivity to γ [38, 39].

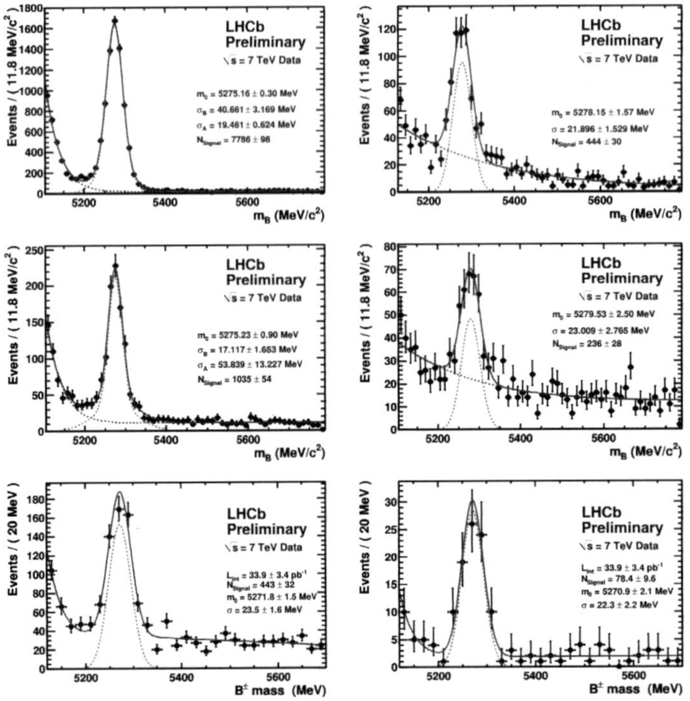

Figure 3. Signals in ~ 34 pb^{-1} of LHCb data for (top to bottom; left to right) $B^+ \to D\pi^+$, $D \to K^+\pi^-$; $B^+ \to DK^+$, $D \to K^+\pi^-$; $B^+ \to D\pi^+$, $D \to K^+K^-$; $B^+ \to D\pi^+$, $D \to \pi^+\pi^-$; $B^+ \to D\pi^+$, $D \to K^0_S\pi^+\pi^-$; $B^+ \to D\pi^+$, $D \to K^0_S K^+K^-$.

It should also be noted that several other channels are in competition for the first observation of direct CP violation in charged B decay. One notable example is $B^+ \to \rho^0 K^+$. This decay can be studied via Dalitz plot analysis of the $\pi^+\pi^-K^+$ final state, giving heightened sensitivity to CP violation effects, as demonstrated by analyses from Belle and BaBar [40, 41].

Several other decays of the type $B \to DK$ (not only those of the charged B meson) can be used to determine γ. A time-dependent analysis of $B^0_s \to D^\mp_s K^\pm$ is sensitive to the combination $2\beta_s - \gamma$. Since this quantity is not small in the SM, this channel has the potential to make a measurement of CP violation in the interference between mixing and decay in the B^0_s system. Signals for the related decay $B^0_s \to D^-_s\pi^+$ are apparent in the 2010 data, as shown in Fig. 4 (this channel will be used to measure Δm_s). As before, the yields are consistent with expectation, indicating that LHCb is on track to produce the world's best measurement of γ, with estimated sensitivity $\sim 7°$, with the data expected to be accumulated in 2011 (~ 1 fb^{-1} from collisions at $\sqrt{s} = 7$ TeV).

[3] Any such comparison must be taken "*cum granulo salis*" due to the different trigger, reconstruction and selection efficiencies.

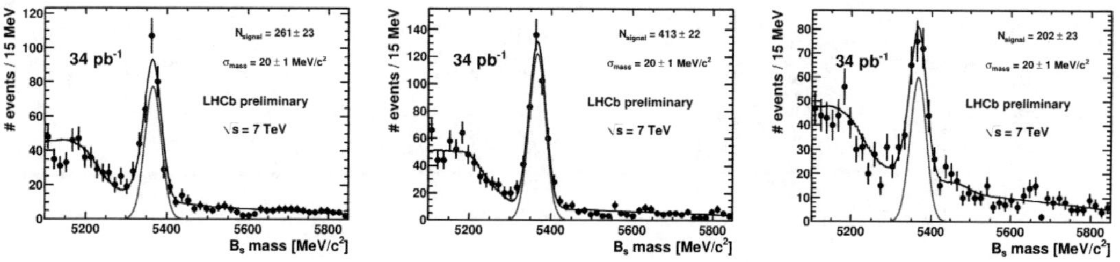

Figure 4. Signals in ~ 34 pb^{-1} of LHCb data for (left to right) $B_s^0 \to D_s^- \pi^+$, $D_s^- \to K^{*0}K^-$; $B_s^0 \to D_s^- \pi^+$, $D_s^- \to \phi\pi^-$; $B_s^0 \to D_s^- \pi^+$, $D_s^- \to K^+K^-\pi^-$ (excluding $K^{*0}K^-$ and $\phi\pi^-$).

Another channel which may provide the first observation of CP violation in the interference between mixing and decay in the B_s^0 system is $B_s^0 \to K^+K^-$. The measurement is again sensitive to γ, with somewhat larger theoretical uncertainties (compared to $B^+ \to DK^+$) that can, however, be controlled exploiting the U-spin relation to the $B_d^0 \to \pi^+\pi^-$ decay. Since the decays contain loop ("penguin") contributions, new physics particles may appear in the loops, leading to deviations from the SM predictions.

Fig. 5 shows the signals for $B_s^0 \to K^+K^-$ (254 ± 20 events) and $B_d^0 \to \pi^+\pi^-$ (229 ± 23 events) in the 2010 LHCb data.[3] These yields can be compared to those from CDF in their analysis of 1 fb^{-1}: 1307 ± 64 and 1121 ± 63 events respectively [42]. As is clear from Fig. 5, the particle identification capability provided by the RICH detectors of LHCb is a powerful tool to distinguish the different b-hadron decays to two track final states.

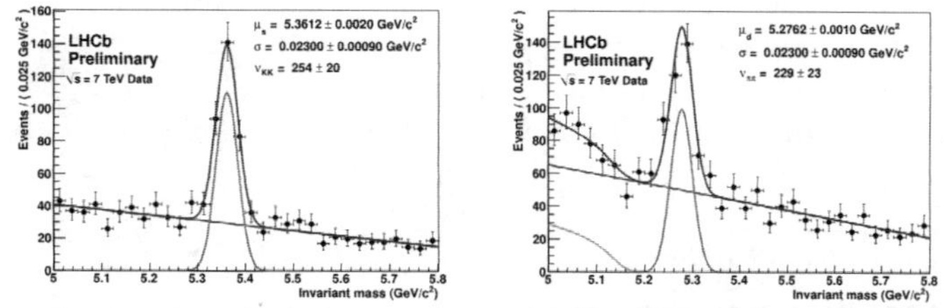

Figure 5. Signals in ~ 35 pb^{-1} of LHCb data for (left to right) $B_s^0 \to K^+K^-$; $B_d^0 \to \pi^+\pi^-$.

Two track final states offer other notable discovery possibilities. Direct CP violation in B decay was first seen in $B_d^0 \to K^+\pi^-$ decays [23, 24], and the same final state may reveal the first direct CP violation in B_s^0 decay. Fig. 6 shows the yields in the $K\pi$ final state split into the two charge conjugate modes. Raw asymmetries are clearly visible in existing data. The central values are consistent with expectations and previous measurements. Corrections due to production or detection asymmetries are being evaluated together with the other systematic uncertainties necessary to quote a meaningful quantitative measurement.[4]

Similarly, two body Λ_b decays offer excellent prospects for the first discovery of CP violation in the baryon sector. Fig. 7 shows the yields of these decays in LHCb 2010 data, illustrating that such effects are likely to become observable with the data expected to be accumulated in 2011. Charmed Λ_b decays are also being investigated.

[4] While these proceedings were in preparation, the first results including systematic uncertainties were presented [43].

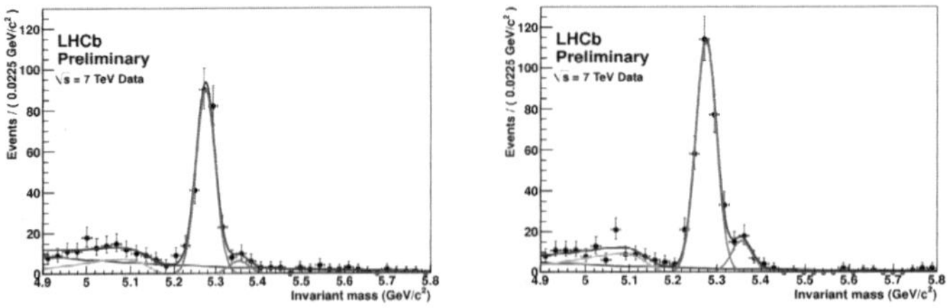

Figure 6. Signals in ~ 35 pb^{-1} of LHCb data for (left to right) $\bar{B} \to K^-\pi^+$; $B \to K^+\pi^-$.

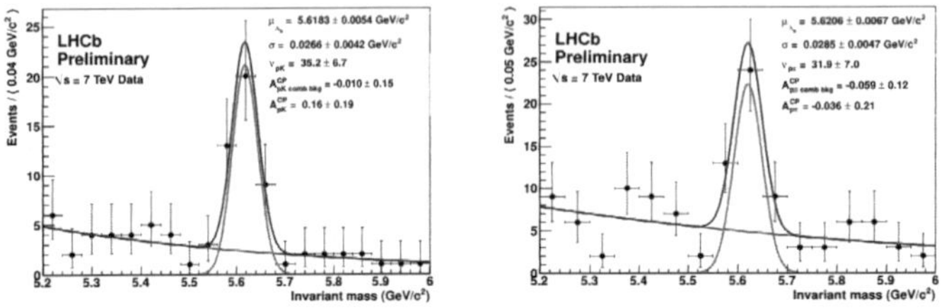

Figure 7. Signals in ~ 35 pb^{-1} of LHCb data for (left to right) $\Lambda_b^+ \to pK^+$; $\Lambda_b^+ \to p\pi^+$.

4. CP violation effects not expected in the Standard Model

While it is of undoubted importance to search for all the manifestations of CP violation that are predicted by the SM, it is perhaps even more fundamental to search for effects that would clearly indicate CP violating amplitudes that cannot be accommodated within the SM. Inconsistencies in the CKM fit that could be revealed by more precise measurements, and direct comparisons of the value of γ (for example) measured in different processes, provide exciting potential to search for "new physics" (NP) beyond the SM. These are complemented by searches for CP violation effects in processes where the SM prediction is small, and which therefore serve as "null tests" for NP.

One such process is the decay channel $B_s^0 \to J/\psi\phi$. This is the B_s^0 system counterpart to the B factory "golden mode" $B_d^0 \to J/\psi K_S^0$, and probes CP violation in the interference between mixing and decay parameterised by $2\beta_s$, which is small in the SM ($\beta_s^{\rm SM} \sim 2^\circ$). Since the final state contains two vector-mesons, an angular analysis is required to disentangle the CP-even and CP-odd components.[5] Moreover, the width difference $\Delta\Gamma_s$ cannot be neglected, unlike the case for the B_d^0 system. Consequently, the full tagged, time-dependent angular analysis is rather complicated (for full details, see Ref. [36]). First such measurements have been performed by CDF [13, 14] and D0 [15, 16] giving exciting hints of non-standard contributions, albeit with very large statistical uncertainties.

In the data collected in 2010, LHCb has already accumulated a large sample of $B_s^0 \to J/\psi\phi$ decays, as shown in Fig. 8, where the related mode $B_d^0 \to J/\psi K^{*0}$, that is used to cross-check the angular acceptance, is also shown. The number of $B_s^0 \to J/\psi\phi$ events is 877 ± 32, which can be compared to ~ 6500 in the CDF analysis based on 5.2 fb^{-1} [14] or ~ 3500 in the D0 analysis based on 6.1 fb^{-1} [16].[3] Another important aspect of the analysis is the flavour tagging, which is a serious challenge in the hadronic environment. Fig. 9 shows that the performance of the tagging algorithms was already sufficient to observe oscillations in the decay $B_d^0 \to D^{*-}X\mu^+\nu$ in the

[5] Such an angular analysis is not necessary for CP-eigenstate decays. While these proceedings were in preparation, the first observation of $B_s^0 \to J/\psi f_0(980)$ was reported by LHCb [44] and rapidly confirmed by Belle [45].

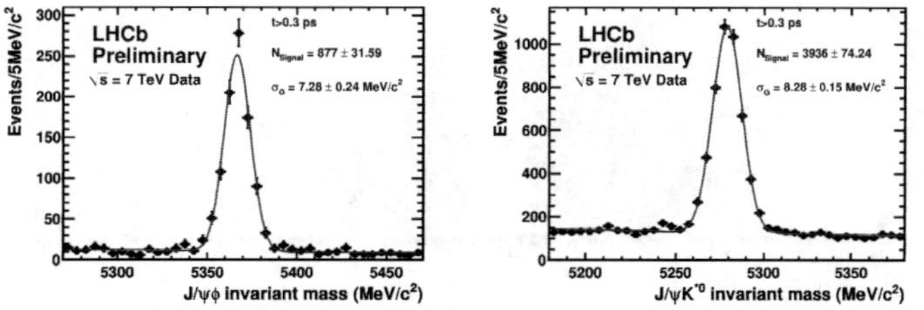

Figure 8. Signals in ~ 34 pb^{-1} of LHCb data for (left to right) $B_s^0 \to J/\psi\phi$; $B_d^0 \to J/\psi K^{*0}$.

first 1.9 pb^{-1} of data. These provide confidence that LHCb is well on the way to significantly improve the constraints on CP violation effects in B_s^0 oscillations with the data that will be accumulated in 2011. If large, non-standard effects are present in this system, this could be established with relatively little data, but in order to establish CP violation at the SM level much larger datasets (and hence an upgraded LHCb experiment) will be required.

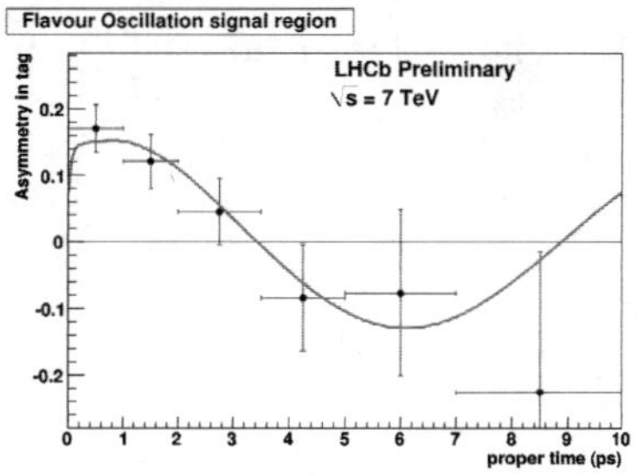

Figure 9. Raw asymmetry between N(unmixed) and N(mixed) as a function of proper time in the decay $B_d^0 \to D^{*-} X \mu^+ \nu$ in the first 1.9 pb^{-1} of LHCb 2010 data.

A complementary approach to search for New Physics effects in B_s^0 oscillations is through the measurement of CP violation that arises in the mixing amplitude itself. This can be measured using flavour-specific decays where the quantity determined is the so-called flavour-specific asymmetry, $A_{\rm fs}(B_s^0)$. Inclusive semileptonic decays provide a convenient high-statistics sample with which to search for this form of CP violation. The SM prediction is $A_{\rm fs}(B_s^0) = (2.06 \pm 0.57) \times 10^{-5}$ [46], and hence any asymmetry larger than $\mathcal{O}(10^{-4})$ could only be a consequence of NP.

The $A_{\rm fs}$ observable has received a lot of attention recently due to the $(3\,\sigma)$ evidence for an anomalous effect measured by the D0 collaboration [11, 12]. This result is based on an inclusive reconstruction, and hence probes a linear combination of the flavour-specific asymmetries in B_d^0 and B_s^0 decays. One urgent goal of the first phase of LHCb is to confirm or rule out this anomaly. Due to the high precision required, it is necessary to use methods with intrinsically low levels of systematic uncertainty. The favoured approach in early data taking is to examine the difference between the B_d^0 and B_s^0 flavour-specific asymmetries, identifying B_d^0 and B_s^0 by their decays to $D^-\mu^+X$ and $D_s^-\mu^+X$ respectively, using the identical final state $K^+K^-\pi^-$ for both D^- and D_s^- decays. Fig. 10 shows the anticipated sensitivity in the $A_{\rm fs}(B_s^0)$–$A_{\rm fs}(B_d^0)$ plane, compared to the D0 result.

The charm system provides unique opportunities to search for flavour-changing effects in flavour changing transitions of up-type quarks. All three categories of CP violation of Tab. 1 are extremely small in the SM, while current experimental constraints [7] are generally at least

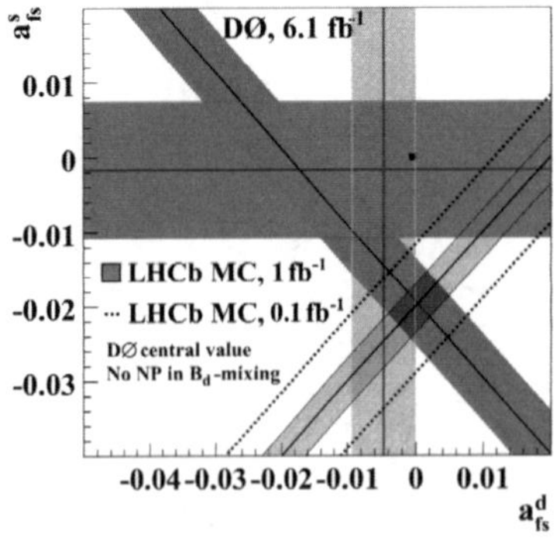

Figure 10. Anticipated sensitivity of LHCb to flavour-specific asymmetries caused by CP violation in mixing. The band shown assumes the SM value of $A_{\rm fs}(B^0_d)$. The D0 result [11, 12] is also shown.

an order of magnitude away from the interesting region.[6] As well as direct CP violation, the recent evidence for charm mixing [47, 48] has opened the door to the possibility of studying mixing-induced CP violation effects.

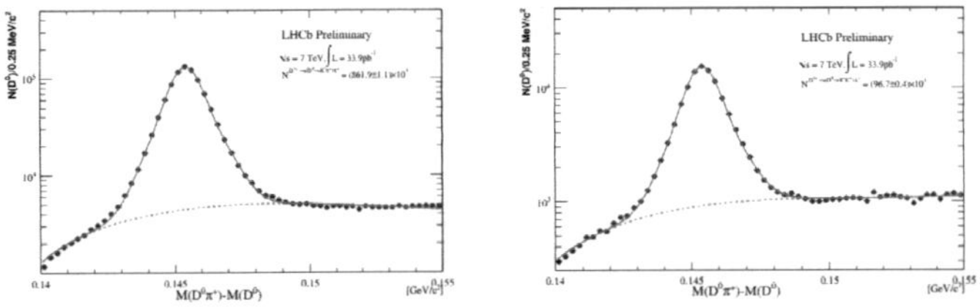

Figure 11. Signals in ~ 34 pb^{-1} of LHCb data for (left to right) $D^{*+} \to D^0\pi^+$, $D^0 \to K^-\pi^+$; $D^{*+} \to D^0\pi^+$, $D^0 \to K^+K^-$.

Owing to the large cross-section for charm production in the LHCb acceptance [32], copious samples of charm decays are already available for analysis. Fig. 11 shows the $D^0 \to K^-\pi^+$ and $D^0 \to K^+K^-$ samples in the LHCb 2010 dataset, where the charm meson is tagged using the $D^{*+} \to D^0\pi^+$ decay. These already compare well to the total samples available at previous experiments: for example there are $\sim 10^5$ $D^{*+} \to D^0\pi^+$, $D^0 \to K^+K^-$ events in Fig. 11, comparable to the number used by Belle in an analysis based on 540 fb^{-1} of e^+e^- collision data [48]. A similar picture emerges among charged charmed mesons, as shown in Fig. 12 for the decays $D^+_{(s)} \to K^+K^-\pi^+$.

Before leaving the subject of null tests of discrete symmetries, it is worthwhile to briefly consider the potential of LHCb to search for CPT violation. Although no detailed sensitivity studies have been performed, it can be foreseen that the studies of B^0_d, B^0_s and D^0 oscillations are extendable to allow searches for such effects, as has been done in B^0_d sector by the B factories [49, 50].[7] It will also be possible to test the equivalence of heavy flavoured particle and

[6] An exception is for the decays $D^+_{(s)} \to K^0_S\pi^+$ and $D^+_{(s)} \to K^0_S K^+$, where CP violation in kaon mixing leads to an apparent direct CP violation effect of $\mathcal{O}(10^{-3})$ in charm decays that is, however, still too small to be observed with current experiments.

[7] Recent results on CP violation in the B^0_s system can be interpreted in terms of constraints on CPT violation [51].

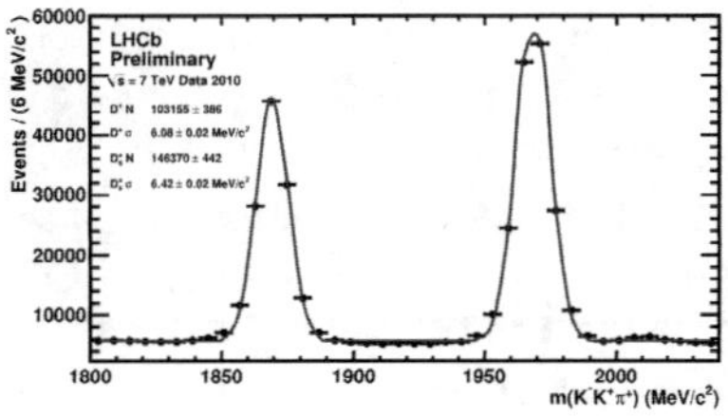

Figure 12. Signals in $\sim 10\ \mathrm{pb}^{-1}$ of LHCb data for $D^+_{(s)} \to K^+K^-\pi^+$.

antiparticle masses and lifetimes.

5. Global symmetries and rare decays

The flavour sector of the Standard Model possesses a number of accidental global symmetries that are not necessarily protected in extensions of the SM. For example, lepton universality is affected by models with extended Higgs sectors. Therefore "rare" decays (with branching fractions that are highly suppressed or even zero in the SM) that probe these symmetries provide sensitive methods to search for NP. Among the channels that can be studied in B physics, the golden channels include $B^0_s \to \mu^+\mu^-$ and $B^0_d \to K^{*0}\mu^+\mu^-$.

The prospects for the study of $B^0_s \to \mu^+\mu^-$ (and the counterpart $B^0_d \to \mu^+\mu^-$ decay) at LHCb are shown in Fig. 13. Note that this figure is based on a small subsample of the total 2010 LHCb dataset – the full dataset will be analysed in a blind manner. The signal yield will be determined from the two dimensional distribution of the dimuon invariant mass and the "geometrical likelihood" (GL), which combines information about the geometry of the event (exploiting the displaced vertex of the B candidate relative to the primary vertex). More details about the analysis strategy can be found in Ref. [36]. The background, which consists mainly of combinations of muons from semileptonic decays of different B hadrons, peaks sharply at low values of the GL and is distributed linearly in invariant mass. The signal is flat in GL (by construction of the variable) and peaks at the relevant B meson mass with an approximately Gaussian distribution.

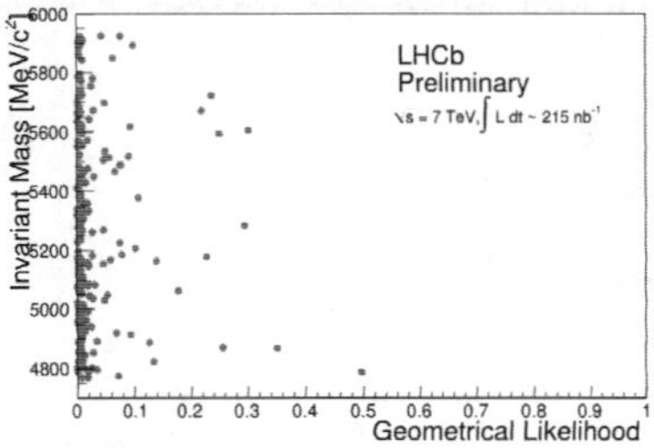

Figure 13. Invariant mass distribution of $B^0_s \to \mu^+\mu^-$ candidate invariant mass *vs.* the geometrical likelihood from 215 nb^{-1} of LHCb 2010 data.

An extrapolation of the LHCb sensitivity shows that the existing limits [52, 53] can be surpassed with less than 100 pb^{-1} of integrated luminosity. The 90% confidence level exclusion limit is predicted to approach the Standard Model prediction of $\mathcal{B}^{\rm SM}\left(B^0_s \to \mu^+\mu^-\right) = (3.2 \pm 0.2) \times 10^{-9}$ [54] with around 1 fb^{-1}.

The general purpose detectors ATLAS and CMS are also expected to have excellent sensitivity to this rare decay. A direct comparison of the sensitivity of the different experiments is difficult since it depends on many factors, such as the integrated luminosity of each experiment, the angular acceptance (and the variation of the $b\bar{b}$ production cross-section with rapidity), the trigger efficiency, the invariant mass resolution, and so on. Definitive statements on the relative sensitivities of the experiments will only be possible once the keenly anticipated first results are publicly available. Nevertheless, it is clear that the data that will be accumulated in 2011 will provide excellent discovery potential in this channel.

One of the characteristic features of the $B^0_d \to K^{*0}\mu^+\mu^-$ decay is the forward-backward asymmetry in the angular distribution of the muons in the dimuon rest-frame (relative to the B^0 direction) as a function of the dimuon invariant mass, q^2. This asymmetry arises as a consequence of interference between the contributing elecroweak diagrams. Since the relative amplitudes of these diagrams vary as a function of q^2, there is a point at which the forward-backward asymmetry crosses zero, usually denoted $q^2 = s_0$. Due to a cancellation of hadronic uncertainties, the value of s_0 is cleanly predicted in the SM to be $s_0 = \left(4.36\,^{+0.36}_{-0.33}\right)$ GeV^2 [55, 56]. First measurements of the differential distributions have been made by the B factories and CDF, with results providing an exciting hint of a deviation from the SM prediction [17, 18, 19].

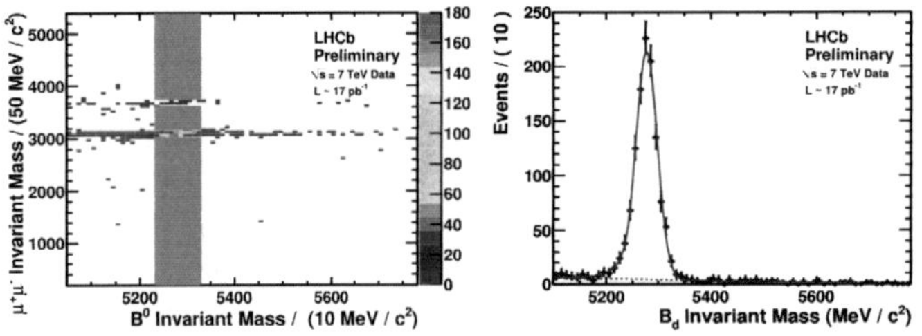

Figure 14. (Left) $K^{*0}\mu^+\mu^-$ candidates reconstructed from ~ 17 pb^{-1} of 2010 LHCb data as a function of dimuon invariant mass and the B^0_d candidate invariant mass. The horizontal bands correspond to dimuons from J/ψ and $\psi(2S)$. Note that the $B^0_d \to K^{*0}\mu^+\mu^-$ signal region is blinded. (Right) projection onto the B^0_d candidate invariant mass for events with the dimuon pair consistent with the J/ψ hypothesis. Note the much smaller background compared to Fig. 8 due to the tighter selection requirements in this analysis.

LHCb has excellent potential to improve dramatically the available sample of $B^0_d \to K^{*0}\mu^+\mu^-$ decays. The analysis is currently blinded, but as shown in Fig. 14 the backgrounds are very low, and large yields are seen in the control channel $B^0_d \to J/\psi K^{*0}$. First results on the forward-backward asymmetry can be expected within the next year. With larger data samples it will be possible to extend the analysis to include additional kinematic observables and thereby further improve the sensitivity to new physics [57, 58, 59].

The $b \to s\gamma$ transition provides another sensitive probe of the effects of virtual particles contributing to the loop amplitude. The SM predicts that the emitted photon is highly polarised, due to the V–A structure of the weak interaction, while different results are possible in NP models. Although many measurements of $b \to s\gamma$ processes exist, none have yet provided strong constraints on the photon polarisation. Several methods have been proposed to test this SM prediction, the most promising of which is based on measurements of time-dependent asymmetries of decays such as $B^0_s \to \phi\gamma$ [60, 61, 62]. While reconstruction of neutral particles such as photons presents a challenge in the hadronic environment, LHCb has already observed a signal for the decay $B^0_d \to K^{*0}\gamma$, as shown in Fig. 15.

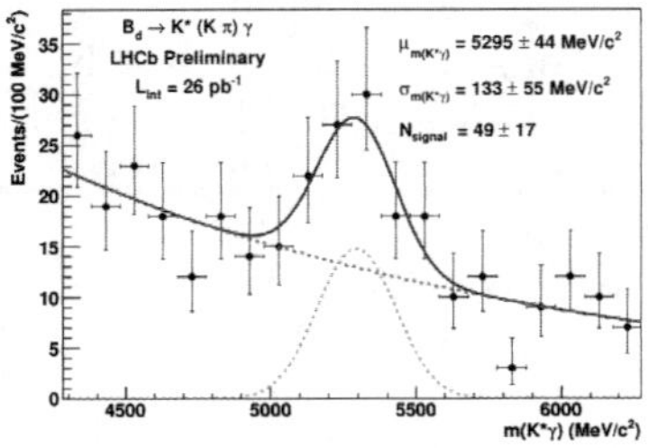

Figure 15. Signal for $B_d^0 \to K^{*0}\gamma$ in ~ 26 pb^{-1} of 2010 LHCb data.

6. Future projects

The LHCb experiment requires the instantaneous luminosity from the LHC to not exceed a value of about 2×10^{32} cm^{-2}s^{-1}. This will be achieved by displacing the LHC beams to produce only a fraction of the maximum collision rate once the available luminosity exceeds this value. The cause of the limitation is the level 0 (hardware) trigger, which has a maximum output rate of 1 MHz. Higher rates necessitate tighter cuts at level 0, particularly for hadronically triggered decay modes, so that there is no overall gain in yield. In order to overcome this, a fully flexible software trigger using the readout of the whole detector at 40 MHz is necessary.

The implementation of a software trigger is the main objective of the proposed upgrade to LHCb [63]. This will enable the full physics programme of LHCb as a general purpose detector in the forward region. The upgraded experiment has unique discovery potential that is not restricted to flavour physics – for example, several new physics models introduce new long-lived particles that will produce displaced vertices that could only be observed by LHCb. In the flavour sector, the role of the upgraded LHCb experiment will depend on the outcome of the first years of operation. In the case that NP is discovered in early data, the task will be to understand what is the correct model describing the new phenomena and to measure its parameters. In the less favourable scenario that there is no early discovery at the LHC that is inconsistent with the SM, it will be necessary to maximise the possibility of the discovery of NP, leaving no stone unturned in the search. In either case, an upgraded LHCb experiment will be essential.

However, there are certain important flavour observables that cannot be studied in the hadronic environment. Two important examples that are only accessible in e^+e^- collisions are the decay $B^+ \to \tau^+\nu$ and the lepton flavour violating decay $\tau^+ \to \mu^+\gamma$. In the context of Tab. 1, e^+e^- facilities provide unique potential to search for CP violation in charged lepton decays (for example in $\tau^+ \to K^+\pi^0\nu$ and $\tau^+ \to K_S^0\pi^+\nu$ decays). Inclusive decays, and decays with missing energy, can also be reconstructed, allowing several interesting rare decays to be studied.

Two next generation e^+e^- colliders have been proposed. Belle2/SuperKEKB is an upgrade of the Belle detector and KEKB accelerator at KEK in Japan. This project has recently been approved, and commissioning of the upgraded machine and detector is expected to commence in 2014. Super$B$ is a new Italy-based project that achieves significant cost savings by reuse of hardware from BaBar/PEP-II. Very recently, Super$B$ was approved for funding by the Italian government. The two projects share much in common in terms of machine and detector designs (for more details, see Ref. [64]), though one notable difference is the potential for beam polarisation in the Super$B$ design.

7. Summary

The LHC machine and detectors performed superbly during the 2010 run. The data accumulated has already demonstrated the potential to make groundbreaking measurements in B physics, and to test the symmetries of the Standard Model. LHCb has excellent potential for major discoveries in many sectors of flavour physics, with strong competition expected from ATLAS and CMS in some modes, most notably $B_s^0 \to \mu^+\mu^-$. The next few years promise to be extremely exciting – if large non-standard effects are present, the LHC data will point the way to progress beyond the Standard Model. Upgrades and new experiments in B physics are being planned and are necessary to ensure progress in understanding flavour physics and discrete symmetries throughout the LHC era.

Acknowledgments

I would like to thank the organisers for the invitation and for a stimulating conference. This work is supported by the Science and Technology Facilities Council (United Kingdom) and by the European Research Council under FP7.

References

[1] Cabibbo N 1963 *Phys. Rev. Lett.* **10** 531–532
[2] Kobayashi M and Maskawa T 1973 *Prog. Theor. Phys.* **49** 652–657
[3] Aubert B *et al.* (BABAR) 2009 *Phys. Rev.* **D79** 072009 (*Preprint* 0902.1708)
[4] Chen K F *et al.* (Belle) 2007 *Phys. Rev. Lett.* **98** 031802 (*Preprint* hep-ex/0608039)
[5] Abulencia A *et al.* (CDF) 2006 *Phys. Rev. Lett.* **97** 242003 (*Preprint* hep-ex/0609040)
[6] Antonelli M *et al.* 2010 *Phys. Rept.* **494** 197–414 (*Preprint* 0907.5386)
[7] Asner D *et al.* (Heavy Flavor Averaging Group) 2010 (*Preprint* 1010.1589) URL http://www.slac.stanford.edu/xorg/hfag/
[8] Nakamura K *et al.* (Particle Data Group) 2010 *J. Phys.* **G37** 075021 URL http://pdg.lbl.gov
[9] Charles J *et al.* (CKMfitter) 2005 *Eur. Phys. J.* **C41** 1–131 (*Preprint* hep-ph/0406184) URL http://ckmfitter.in2p3.fr/
[10] Bona M *et al.* (UTfit) 2006 *JHEP* **10** 081 (*Preprint* hep-ph/0606167) URL http://www.utfit.org/
[11] Abazov V M *et al.* (D0) 2010 *Phys. Rev.* **D82** 032001 (*Preprint* 1005.2757)
[12] Abazov V M *et al.* (D0) 2010 *Phys. Rev. Lett.* **105** 081801 (*Preprint* 1007.0395)
[13] Aaltonen T *et al.* (CDF) 2008 *Phys. Rev. Lett.* **100** 161802 (*Preprint* 0712.2397)
[14] 2010 CDF-Note-10206
[15] Abazov V M *et al.* (D0) 2008 *Phys. Rev. Lett.* **101** 241801 (*Preprint* 0802.2255)
[16] 2010 D0-Note-6098-CONF
[17] Aubert B *et al.* (BABAR) 2009 *Phys. Rev.* **D79** 031102 (*Preprint* 0804.4412)
[18] Wei J T *et al.* (BELLE) 2009 *Phys. Rev. Lett.* **103** 171801 (*Preprint* 0904.0770)
[19] Aaltonen T *et al.* (CDF) 2011 (*Preprint* 1101.1028)
[20] Hara K *et al.* (Belle) 2010 *Phys. Rev.* **D82** 071101 (*Preprint* 1006.4201)
[21] del Amo Sanchez P *et al.* (BABAR) 2010 (*Preprint* 1008.0104)
[22] Lunghi E and Soni A 2010 (*Preprint* 1010.6069)
[23] Lin S W *et al.* (Belle) 2008 *Nature* **452** 332–335
[24] Aubert B *et al.* (BABAR) 2008 (*Preprint* 0807.4226)
[25] Ishino H *et al.* (Belle) 2007 *Phys. Rev. Lett.* **98** 211801 (*Preprint* hep-ex/0608035)
[26] Alves A A *et al.* (LHCb) 2008 *JINST* **3** S08005
[27] Aad G *et al.* (ATLAS) 2008 *JINST* **3** S08003
[28] Adolphi R *et al.* (CMS) 2008 *JINST* **3** S08004
[29] 2010 ATLAS-CONF-2010-062
[30] Khachatryan V *et al.* (CMS) 2010 (*Preprint* 1011.4193)
[31] 2010 LHCb-CONF-2010-010
[32] 2010 LHCb-CONF-2010-013
[33] Aaij R *et al.* (LHCb) 2010 *Phys. Lett.* **B694** 209–216 (*Preprint* 1009.2731)
[34] Aaij R *et al.* (LHCb) 2011 (*Preprint* 1102.0348)
[35] Zupan J 2011 (*Preprint* 1101.0134)
[36] Adeva B *et al.* (LHCb) 2009 (*Preprint* 0912.4179)
[37] Aaltonen T *et al.* (CDF) 2010 *Phys. Rev.* **D81** 031105 (*Preprint* 0911.0425)

[38] del Amo Sanchez P *et al.* (BABAR) 2010 *Phys. Rev. Lett.* **105** 121801 (*Preprint* 1005.1096)
[39] Poluektov A *et al.* (Belle) 2010 *Phys. Rev.* **D81** 112002 (*Preprint* 1003.3360)
[40] Garmash A *et al.* (Belle) 2006 *Phys. Rev. Lett.* **96** 251803 (*Preprint* hep-ex/0512066)
[41] Aubert B *et al.* (BABAR) 2008 *Phys. Rev.* **D78** 012004 (*Preprint* 0803.4451)
[42] 2008 CDF-Note-8579
[43] 2011 CERN EP/PP/LPCC Seminar 22 February 2011 URL http://indico.cern.ch/conferenceDisplay.py?confId=128030
[44] Aaij R *et al.* (The LHCb) 2011 (*Preprint* 1102.0206)
[45] Li J and Collaboration f t B 2011 (*Preprint* 1102.2759)
[46] Lenz A and Nierste U 2007 *JHEP* **06** 072 (*Preprint* hep-ph/0612167)
[47] Aubert B *et al.* (BABAR) 2007 *Phys. Rev. Lett.* **98** 211802 (*Preprint* hep-ex/0703020)
[48] Staric M *et al.* (Belle) 2007 *Phys. Rev. Lett.* **98** 211803 (*Preprint* hep-ex/0703036)
[49] Hastings N C *et al.* (Belle) 2003 *Phys. Rev.* **D67** 052004 (*Preprint* hep-ex/0212033)
[50] Aubert B *et al.* (BABAR) 2004 *Phys. Rev.* **D70** 012007 (*Preprint* hep-ex/0403002)
[51] Kostelecky A and Van Kooten R 2010 *Phys.Rev.* **D82** 101702 (*Preprint* 1007.5312)
[52] Abazov V M *et al.* (D0) 2010 *Phys. Lett.* **B693** 539–544 (*Preprint* 1006.3469)
[53] 2010 CDF-Note-9892
[54] Buras A J, Isidori G and Paradisi P 2011 *Phys. Lett.* **B694** 402–409 (*Preprint* 1007.5291)
[55] Beneke M, Feldmann T and Seidel D 2001 *Nucl. Phys.* **B612** 25–58 (*Preprint* hep-ph/0106067)
[56] Beneke M, Feldmann T and Seidel D 2005 *Eur. Phys. J.* **C41** 173–188 (*Preprint* hep-ph/0412400)
[57] Kruger F and Matias J 2005 *Phys. Rev.* **D71** 094009 (*Preprint* hep-ph/0502060)
[58] Egede U, Hurth T, Matias J, Ramon M and Reece W 2008 *JHEP* **11** 032 (*Preprint* 0807.2589)
[59] Altmannshofer W *et al.* 2009 *JHEP* **01** 019 (*Preprint* 0811.1214)
[60] Atwood D, Gronau M and Soni A 1997 *Phys. Rev. Lett.* **79** 185–188 (*Preprint* hep-ph/9704272)
[61] Atwood D, Gershon T, Hazumi M and Soni A 2005 *Phys. Rev.* **D71** 076003 (*Preprint* hep-ph/0410036)
[62] Muheim F, Xie Y and Zwicky R 2008 *Phys. Lett.* **B664** 174–179 (*Preprint* 0802.0876)
[63] 2011 LHCb Upgrade Letter of Intent, in preparation
[64] Krizan P 2011 Proceedings of CKM2010, in preparation

Journal of Physics: Conference Series **335** (2011) 012002 doi:10.1088/1742-6596/335/1/012002

Kaon physics and discrete symmetries: status and perspectives

Marco S. Sozzi

Department of Physics, University of Pisa and INFN Pisa
Largo Pontecorvo 3, I-56127 Pisa, ITALY

E-mail: `marco.sozzi@df.unipi.it`

Abstract. A brief review of recent discrete symmetry tests with kaons is presented, and prospects for future studies are discussed.

1. From genesis to revelation

Kaons have been a remarkably effective tool to provide insight for the shaping of the Standard Model (SM) as we know it. They played the key role in the discovery that basic discrete symmetries are often broken in Nature. After leading to the fall of parity, kaons showed the existence of CP violation, thus causing the start of a large set of experimental and theoretical studies, which eventually shed a flebile light on the deep connection between discrete symmetries and flavour [1] [2].

Sixty-six years after the first published observation of a K meson in cosmic rays, one might wonder whether investigations on kaons have exhausted the information that such relatively simple but rich system can provide: it turns out that this is not yet the case, as will be discussed in the following.

2. Close to the edge

Most of the recent experimental results obtained from kaon decays stem from the pluri-decennial research endeavours which aimed at clarifying the existence of direct CP violation.

In 1980, when Jim Cronin and Val Fitch received the Nobel prize for the discovery of CP violation, achieved in 1963 [3], the situation on the most subtle violation of a discrete symmetry in Nature was still described in these words: "At present our experimental understanding of CP violation can be summarized by the statement of a single number" [4]. The above sentence referred to the long-standing lack of conclusive understanding about the existence of more straightforward CP-violation effects in kaons, such as a partial decay width difference between CP-conjugate particles.

Even more strikingly, still at the end of the '90s no other measurement of CP violation had been detected elsewhere, and the *ad-hoc* ansatz of a rather peculiar super-weak interaction [5] being responsible for CP violation required increasing uncomfortable attention.

KTeV at FNAL and NA48 at CERN were third-generation experiments aiming at the measurement of the ϵ'/ϵ parameter, representing direct CP violation in the decay amplitudes. Both experiments took data at the end of the '90s, using high-energy proton beams to produce long- and short-lived neutral kaons on a fixed target. The ingenuity involved in these experiments is well worth appreciating, but goes beyond the scope of this paper (the interested reader can find some details in [6]). Let us just recall that around 1999 the existence of direct CP violation was finally confirmed [7] [8].

The importance of such measurement is foremost a qualitative one, as it provided the first confirmation of the CKM paradigm incorporating CP violation in the Standard Model; soon after the B-factories indeed showed CP violation to be an ubiquitous effect in neutral flavoured mesons [9].

The final results on direct CP violation from the entire (1997-1999) KTeV data set have just been recently released [10]: these refer to 6 (10) million $K_L(K_S)$ decays into $\pi^0\pi^0$ and 25 (43) million $K_L(K_S)$ decays into $\pi^+\pi^-$. Improvements with respect to the previous KTeV result include a reduction of the systematic error linked to the reconstruction of the $\pi^0\pi^0$ mode, mainly through improved Monte Carlo simulations.
Together with the final NA48 result [11] this now gives the current world average:

$$\Re(\epsilon'/\epsilon) = (16.4 \pm 1.9) \cdot 10^{-4} \tag{1}$$

(see [12] and [13]) which proves the existence of direct CP violation at about 9 standard deviations, with reasonable statistical consistency ($\chi^2/\text{dof} = 5.2/3$), see figure 1.

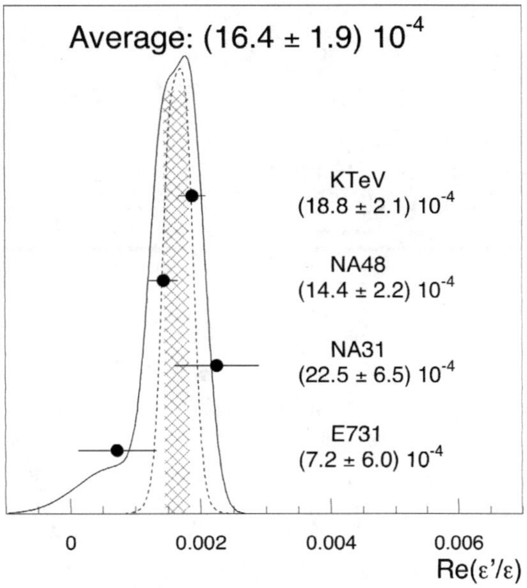

Figure 1. Recent experimental determinations of $\Re(\epsilon'/\epsilon)$.

Nature has not been too kind on this issue, hiding very well direct CP violation effects in the most accessible meson system in which it appears. Expressed in a different way, the measured value of ϵ'/ϵ describes a difference in neutral kaon partial decay widths which is indeed quite tiny:

$$\frac{\Gamma(K^0 \to \pi^+\pi^-) - \Gamma(\overline{K}^0 \to \pi^+\pi^-)}{\Gamma(K^0 \to \pi^+\pi^-) + \Gamma(\overline{K}^0 \to \pi^+\pi^-)} = (5.18 \pm 0.61) \cdot 10^{-6} \tag{2}$$

The above figure can be better compared to typical "large" measured asymmetries in the B meson system, which can reach into the ten percent range, appearing however for decay modes with branching ratios of order 10^{-5}.

The rather good experimental precision obtained calls for the possible inclusion of this measurement as a quantitative constraint on the consistency of the SM flavour structure, in the so-called unitarity triangle. Unfortunately this is not yet possible, due to the dire theoretical difficulties in computing the SM value of ϵ'/ϵ which, besides the usual uncertainties related to low-energy hadronic physics, is also plagued by large accidental cancellations. Lattice QCD holds the promise to resolve this issue, and progress in the theoretical estimation of ϵ'/ϵ has been somewhat slow but constant in time (see e.g. [14])[1].

3. More

The legacy of the KTeV and NA48 experiments is much larger than the above main result shows, as several other key measurements were obtained from the careful analysis of the huge data samples collected.

The "classic" indirect CP violation parameter ϵ must be mentioned first: its real part has been traditionally measured through semi-leptonic decay asymmetries of K_L, which are now obtained from samples of $O(10^8)$ decays per experiment, resulting in

$$\delta_L^{(e)} = \frac{\Gamma(K_L \to \pi^- e^+ \nu) - \Gamma(K_L \to \pi^+ e^- \overline{\nu})}{\Gamma(K_L \to \pi^- e^+ \nu) + \Gamma(K_L \to \pi^+ e^- \overline{\nu})} = (3.322 \pm 0.055) \cdot 10^{-3} \tag{3}$$

The KLOE experiment at the DAΦNE kaon factory was also designed for a direct CP violation measurement with a very different technique, and although it could not compete directly in such arena, it turned out to be a very important source of kaon physics measurements. With 2.5 billion correlated $K_S K_L$ pairs collected in 2000-2006, KLOE performed several precision Branching Ratio measurements, with the unique advantage of absolute flux normalization. As a significant example, KLOE performed the first measurement of the semi-leptonic decay asymmetry of K_S[15]:

$$\delta_S^{(e)} = (1.5 \pm 9.6_{\text{stat}} \pm 2.9_{\text{syst}}) \cdot 10^{-3} \tag{4}$$

While this asymmetry is still compatible with zero due to the limited statistics related to the small BR $\simeq 7 \cdot 10^{-4}$, it should be remarked that the equality of δ_S and δ_L is a test CPT symmetry in meson mixing, which could become significant if more data were available.

The straightforward way of extracting the ϵ parameter is by direct measurement of the BR of the CP-violating decay $K_L \to \pi\pi$. Recent measurements by KTeV [16], KLOE [17] and NA48 [18] resulted in a significant shift in most of the K_L branching ratios with respect to previously known values (figure 2). The main reasons for such "BR revolution" are be found in the improved treatment of radiative correction effects, and the proper accounting of systemtatic error correlations.

The value of ϵ is now known with a remarkable 0.5% precision, and while this puts a constraint on the unitarity triangle (see fig. 3), its quantitative strength as a SM test is still limited

[1] "We expect a 20% result on ϵ'/ϵ in 3 years", a lattice theoretician, whose name will go undisclosed, publicly stated in 2009.

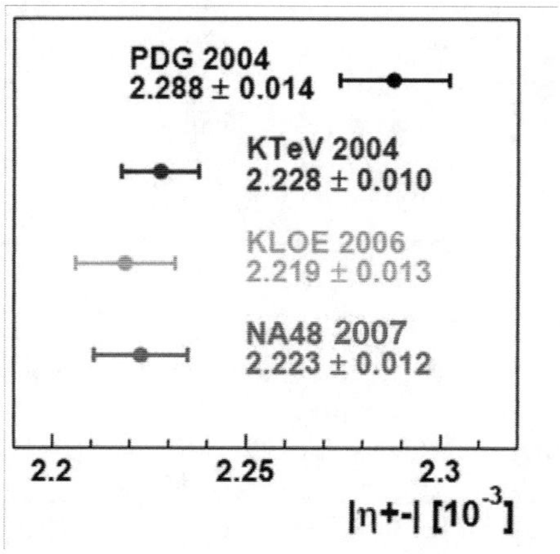

Figure 2. Recent determinations of the $|\eta_{+-}| = |A(K_L \to \pi^+\pi^-)/A(K_S \to \pi^+\pi^-)|$ CP-violation parameter.

by theory, as the uncertainties in the parameterization of hadronic effects in the computation of $K^0 - \overline{K}^0$ mixing dominate the error in the extraction of the CKM parameters from the experimental quantity.

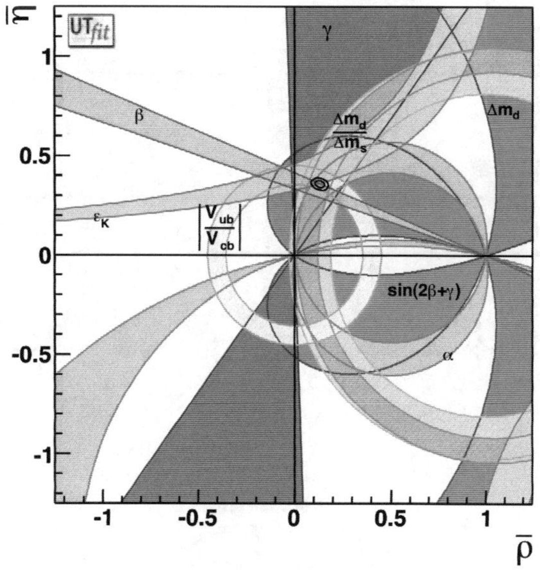

Figure 3. Recent determination of the apex $(\overline{\rho}, \overline{\eta})$ of the unitarity triangle (by the UTFIT collaboration[19]), on which the $\epsilon_K (\equiv \epsilon)$ band is visible.

Again, steady progress in lattice QCD is improving the situation, and now next-to next-to-leading order computations are becoming available, leading to theoretical errors approaching

Journal of Physics: Conference Series **335** (2011) 012002 doi:10.1088/1742-6596/335/1/012002

the 10% level (see e.g. [14]). A breakdown of the theoretical error [20] suggests that further progress is now becoming increasingly difficult, since about one third of the error is parametric, mostly due to the knowledge of the $|V_{cb}|$ CKM parameter, but many comparable sources of non-parametric theoretical uncertainty compete to build up the rest.

4. Empty spaces

Searches for CP violation in other hadronic kaon decay modes are sensitive to direct CP-violating effects, which could in principle get sizable contributions from New Physics (NP).

3π decays of K_S are the counterpart of the traditional $K_L \to 2\pi$ modes, with a large penalty due to the much larger total decay width of the K_S. They have not yet been detected, not even at the expected level driven by indirect CP violation (BR $\sim 10^{-9}$).

Experiments at hadronic machines look for such decays through $K_S - K_L$ interference searches: NA48/1 put a limit of $BR(K_S \to 3\pi^0) < 2.3 \cdot 10^{-7}$ (at 90% CL) with $O(10^8)$ kaon decays [21]. KLOE also searched for this decay, reaching a limit of $BR(K_S \to 3\pi^0) < 1.2 \cdot 10^{-7}$ with 2 signal candidates [22] (see fig. 4).

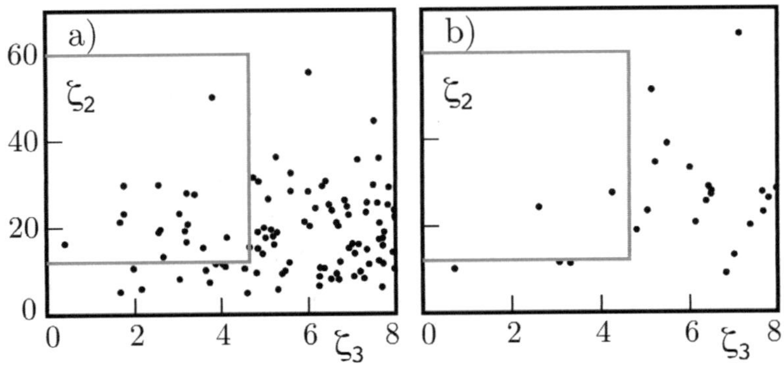

Figure 4. KLOE signal boxes for $K_S \to 3\pi^0$ on simulated (left) and real data (right); the simulated data corresponds to twice the integrated luminosity (and about 2.7 times the efficiency) of the real data.

For a long time CP violation was only detected in decays of flavoured neutral mesons, where it can exhibit a rather complex phenomenology, depending on the process (mixing, decay, or both) in which the relevant complex phases appear. Charged mesons, in contrast, can only exhibit direct CP violation in the decay, and any difference in decay properties between a positively and a negatively charged meson is the most straightforward manifestation of CP symmetry violation.

The simplest decays of $K^\pm$ which can support large CP violation are the three-pion modes; the presence of two unsuppressed isospin amplitudes in such decays allows in principle a significantly larger interference (and thus CP violation) than in the $K^0 \to 2\pi$ modes of ϵ' fame.

A high-statistics investigation of such decays was performed by the NA48/2 experiment, focusing on the difference in the linear Dalitz plot slopes g to avoid normalization issues; the pretty large BRs (1-5%) allowed to work with background-free samples of several 10^9 decays, which were collected in 2003-04. Control of systematics was the key issue in pushing the measurement to cover the allowed range of asymmetries down to the 10^{-4} level, below which the SM contribution might be expected to appear (with significant theoretical uncertainties). Using simultaneous collinear K^+ and K^- beams, and exploiting maximal cancellations of systematics,

NA48/2 closed the window for possible asymmetries significantly larger than the SM [23], as predicted by some SUSY models (see fig. 5):

$$A_g = \frac{g_+ - g_-}{g_+ + g_-} = (-1.5 \pm 2.1) \cdot 10^{-4} \quad (K^\pm \to \pi^\pm \pi^+ \pi^-) \tag{5}$$

$$A_g = \frac{g_+ - g_-}{g_+ + g_-} = (1.8 \pm 1.8) \cdot 10^{-4} \quad (K^\pm \to \pi^\pm \pi^0 \pi^0) \tag{6}$$

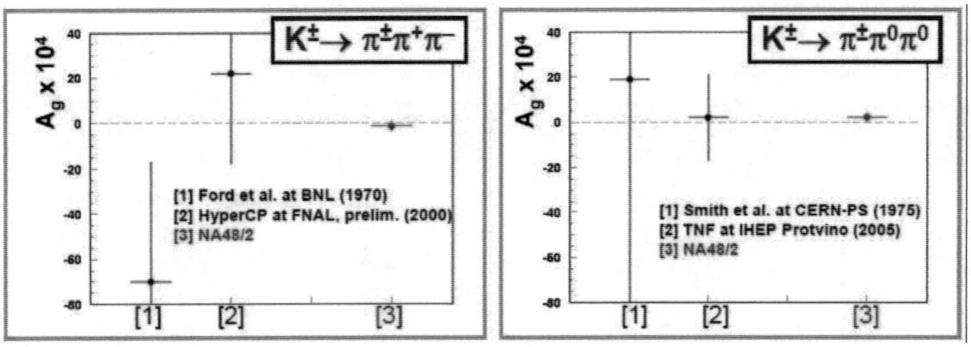

Figure 5. Experimental results on CP violation in the comparison of linear Dalitz plot slopes in $K^\pm \to \pi^\pm \pi\pi$ decays.

Radiative decay modes such as $K^\pm \to \pi^\pm \pi^0 \gamma$ also offer a chance to investigate CP violation without having to deal with flux normalization issues, through the comparison of the radiative photon spectrum. With 220 thousand events detected, NA48/2 observed for the first time the $\sim 3\%$ interference term between Bremsstrahlung and Direct Emission amplitudes (where CP violation might show up), detecting no evidence for any asymmetry [24] [25].

Radiative decays of neutral mesons, such as $K_L \to \pi^+ \pi^- \gamma$ were also investigated in the past, without detecting any sign of CP violation. From such decays, the time-honored technique of exploiting internal e^+e^- conversion to get information on photon helicity led to the measurement of a large CP asymmetry in the angle between the di-pion and di-electron planes in the rare $K_L \to \pi^+ \pi^- e^+ e^-$ decay [26]:

$$A_\phi = (13.8 \pm 2.2)\% \tag{7}$$

This result is very well consistent with the prediction, related to the usual $K_L \to \pi^+ \pi^-$ CP violation, mostly of indirect origin. No asymmetry is measured in the companion decay $K_S \to \pi^+ \pi^-$, as expected [26].

5. Any colour you like

CPT symmetry, despite being a cornerstone of all known theories[2], is nevertheless very actively tested experimentally, in many different systems.

Numerically, the most striking CPT test in particle physics is given by the measured mass difference between K^0 and $\overline{K}^0$, obtained from a combined analysis of all kaon decay modes using unitarity, in the form of the Bell-Steinberger relation.

[2] Except possibly for string theory (but which possible phenomenon is not encompassed by string theory, by the way?).

Using the recent KTeV data on $K \to 2\pi$ decays, combined with input from NA48, KLOE and CPLEAR, one obtains [10]

$$|m(K^0) - m(\overline{K}^0)| < 4.0 \cdot 10^{-19}\,\mathrm{GeV}/c^2 \tag{8}$$

at 95% CL, as a limit for validity of CPT symmetry in the meson mixing process, assuming CPT symmetry in the decays (see fig. 6).

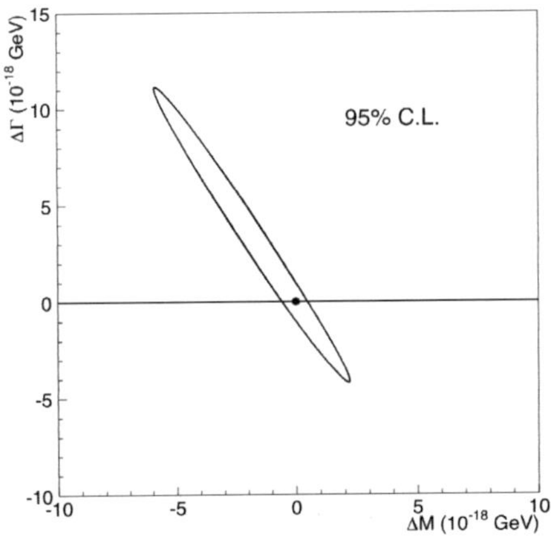

Figure 6. Determination of the mass and decay width differences between K^0 and $\overline{K}^0$.

This latter assumption is consistent with the KTeV measurement of $\Im(\epsilon'/\epsilon)$ (proportional to the phase difference of the two $K \to \pi\pi$ decays) being close to zero (see fig. 7).

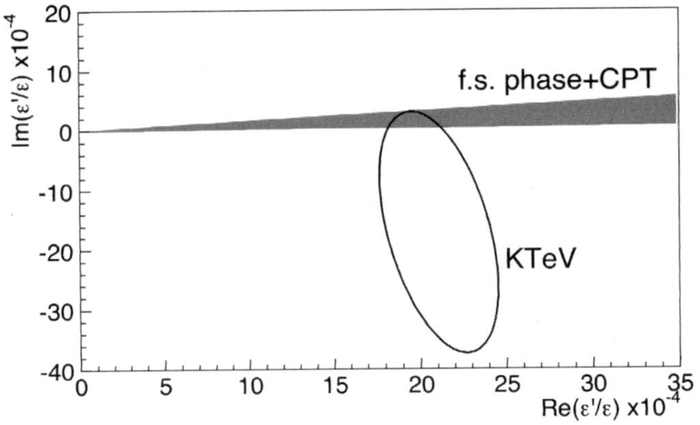

Figure 7. KTeV results on the real and imaginary parts of ϵ'/ϵ, and comparison with the expectation in case of CPT symmetry using final-state interaction phase measurements.

The only direct measurement of time reversal violation (without any assumption on CPT but still equivalent to CP) is the Kabir asymmetry measured by CPLEAR [27], between the probabilities of $K^0 \leftrightarrow \overline{K}^0$ transitions:

$$\frac{P(\overline{K}^0 \to K^0) - P(K^0 \to \overline{K}^0)}{P(\overline{K}^0 \to K^0) + P(K^0 \to \overline{K}^0)} = (6.6 \pm 1.3) \cdot 10^{-3} \tag{9}$$

While we only discussed tests of discrete symmetries, it should be pointed out that many other interesting results were obtained as byproducts of the above investigations, such as the much improved determination of the Cabibbo mixing angle (or $|V_{us}|$), to 0.23% precision, from improved measurements of $K \to \pi \ell \nu$ ($K_{\ell 3}$) decays; this resulted in a stringent test of unitarity in the first row of the CKM matrix [28].

As a second example, much improved measurements of the BR for the helicity suppressed leptonic decay $K^{\pm} \to e^{\pm}\nu$ (K_{e2}) recently allowed to perform stringent tests for deviations from the $V - A$ structure of weak interactions and the presence of lepton-flavour violating effects [29] in NP, which could significantly affect the ratio of BRs between K_{e2} and $K_{\mu 2}$.

6. High hopes

In the near future an upgrade of the KLOE experiment [30] should pursue with much higher statistics the kind of interferometry studies which were already started, allowing more stringent tests of CPT, Lorentz symmetry, and testing the validity of quantum mechanics itself (see [31]).

Other experiments are in preparation to improve the existing tests of discrete symmetries with kaons. One is the elegant TREK experiment [32], aiming at a ten-fold improvement in the measurement of the muon polarization P_T transverse to the decay plane in the $K^+ \to \pi^0 \mu^+ \nu$ decay. A non-zero value for such quantity at a level above 10^{-4} would be an indication of time-reversal violation in New Physics, still allowed by the current limit ($P_T < 5 \cdot 10^{-3}$) set by the KEK E246 experiment [33].

The next frontier in extracting physics from kaons lies in brave searches for ultra-rare decay modes: it has been long known that a few kaon decay modes are unique probes for New Physics in the flavour sector [34], due to the extraordinary precision achievable in the theoretical prediction of their BR (at the few percent level). Some features of these modes are summarized in table 1; the hard-GIM suppression in the SM which allows large NP contributions, and the extraction of the hadronic part of the amplitude from the very well measured $K_{\ell 3}$ decay, are the two key points to understand the relevance of these processes.

Mode	BR(SM)	BR(exp)
$K_L \to \pi^0 e^+ e^-$	10^{-11} ($3 \cdot 10^{-12}$ CPV$_{\rm dir}$)	$< 2.8 \cdot 10^{-10}$
$K_L \to \pi^0 \mu^+ \mu^-$	10^{-11} ($1 \cdot 10^{-12}$ CPV$_{\rm dir}$)	$< 3.8 \cdot 10^{-10}$
$K^+ \to \pi^+ \nu \overline{\nu}$	$(8.5 \pm 0.7) \cdot 10^{-11}$	$1.47^{+1.30}_{-0.89} \cdot 10^{-10}$
$K_L \to \pi^0 \nu \overline{\nu}$	$(2.6 \pm 0.4) \cdot 10^{-11}$	$< 2.6 \cdot 10^{-8}$

Table 1. Ultra-rare kaon decay modes, listing the experimental information and theoretical prediction on their BRs (CPV$_{\rm dir}$ denotes the short-distance, direct-CP violating part of the BR).

The four decay modes in table 1 probe different classes of NP; the modes with neutrinos in the final state are the cleanest from a theoretical point of view, and represent the new "holy grail" of physics with kaons and one of the most powerful probes of NP at large.

The K^+ mode was detected by the dedicated BNL experiments E787 and E949, with a total of seven candidate events [35], while only limits exist for the K_L mode [36], whose (directly) CP-violating nature contributes to reducing further the intrinsic theoretical uncertainties in the BR prediction.

Out of many proposals which were put forward for the search of these decays, at least two projects are steadily ongoing.

NA62 at CERN [37] aims to measure O(100) $K^+ \to \pi^+\nu\overline{\nu}$ events in two years of data-taking, starting in 2013, exploiting a new experimental approach based on high-energy (75 GeV/c) K^+ decay in flight (figure 8).

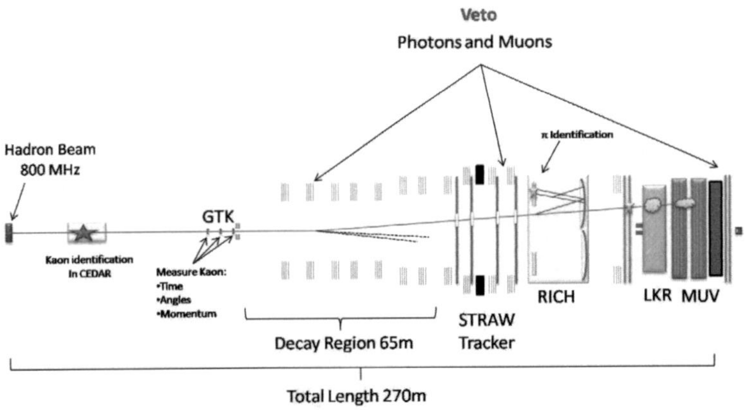

Figure 8. Layout of the NA62 experiment at CERN for the measurement of the $K^+ \to \pi^+\nu\overline{\nu}$ decay.

KOTO at J-PARC [38] is the successor of the pilot project E391a [36] at KEK, aiming at a first detection of the $K_L \to \pi^0\nu\overline{\nu}$ decay starting in 2011, with a future upgrade path to collect tens of events using a dedicated beam line. It is based on the use of a hermetic detector and a pencil-like neutral beam (figure 9).

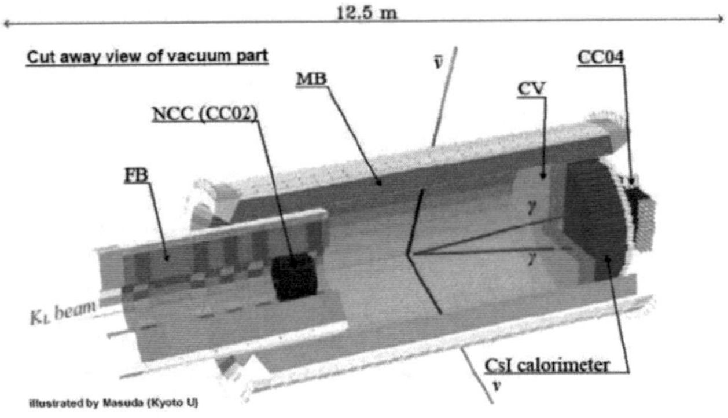

Figure 9. Layout of the KOTO experiment at J-PARC for the measurement of the $K_L \to \pi^0\nu\overline{\nu}$ decay.

Fermilab plans for physics at the intensity frontier, exploiting a future very intense proton accelerator (so-called "Project X") [39], do include ultra-rare kaon decay measurements as a central research topic. 3 MW of beam power at 3 GeV from such a machine could deliver (around 2020) enough kaons to get several hundreds decays in both the K^+ and K_L ultra-rare modes, and experiments are envisaged for both such measurements. Such a machine would also offer the possibility of performing a ultimate $K^0 \to 2\pi$ phase measurement, bringing the $K^0 - \overline{K}^0$ mass difference CPT test to a level of sensitivity corresponding to the Planck scale.

The KLOD R&D program for a K_L measurement at Protvino should also be mentioned, as well as the OKA project, in the same laboratory, using a separated charged kaon beam.

7. Los endos

The whole issue of discrete symmetries violation in fundamental particles was born with kaons, and this system - being the simplest one which exhibits the full set of weak-interaction effects linked to flavour - is still at the forefront of research on the topic which gives the title to this conference. The low mass, on the hadronic scale, is what gives kaons distinct advantages over other meson systems, both in terms of large lifetime and lifetime difference of the physical states (which allow a most varied set of experiments), of tiny mass difference (resulting in unparalleled probing power on virtual effects) and limited number of dominant decay modes (a significant experimental simplification). The theoretical limitations of dealing with lower-energy systems, on the other hand, are not such to cancel the above-mentioned advantages.

The smallest BRs ever measured were obtained from careful kaon experiments, and experimenters have now taken up the challenge of stepping up from tens to hundreds of kW of beam power.

All such considerations lead the author to believe that kaons will still be playing an important role on the stage of future investigations of Nature's broken symmetries.

It is a pleasure to thank A. Di Domenico and the Organizing Committee of Discrete 2010 for the kind invitation, and for providing such a pleasant experience at the conference.

References

[1] N. Cabibbo, Phys. Rev. Lett. **10** (1963) 531.
[2] M. Kobayashi, T. Maskawa, Prog. Theor. Phys. **49** (1973) 652.
[3] J.H. Christenson *et al* , Phys. Rev. Lett. **13** (1964) 138.
[4] J.W. Cronin, Rev. Mod. Phys. **53** (1981) 37.
[5] L. Wolfenstein, Phys. Rev. Lett. **13** (1964) 562.
[6] M.S. Sozzi, Discrete symmetries and CP violation, Oxford University Press (2008).
[7] A. Alavi-Harati *et al* (KTeV collaboration), Phys. Rev. D **67** 012005.
[8] A. Lai *et al* NA48 collaboration), Eur. Phys. Jour. C **22** (2001) 231.
[9] B. Aubert *et al* (BaBar collaboration), Phys. Rev. Lett. **87** (2001) 091801; K. Abe *et al* (Belle collaboration), Phys. Rev. Lett. **87** (2001) 091802.
[10] E. Abouzaid *et al* (KTeV collaboration), preprint `arXiV 1011.0127`, submitted to Phys. Rev. D (November 2010).
[11] J.R. Batley *et al* (NA48 collaboration), Phys. Lett. B **544** (2002) 97.
[12] K. Nakamura et al. (Particle Data Group), J. Phys. G **37** (2010) 075021.
[13] M. Sozzi, Eur. Phys. Jour. C **36** (2004) 37.
[14] `http://www.infn.it/Lattice2010`.
[15] F. Ambrosino *et al* (KLOE collaboration), Phys. Lett. B **636** (2006) 173.
[16] T. Alexopoulos *et al* (KTeV collaboration), Phys. Rev. D **70** (2004) 092006.
[17] F. Ambrosino *et al* (KLOE collaboration), Phys. Lett. B **638** (2006) 140.
[18] A. Lai *et al* (NA48 collaboration), Phys. Lett. B **645** (2007) 26.
[19] UTfit collaboration, `http://www.utfit.org`.
[20] J. Brod, these proceedings.
[21] A. Lai *et al* (NA48/1 collaboration), Phys. Lett. B **610** (2005) 165.
[22] F. Ambrosino *et al* (KLOE collaboration), Phys. Lett. B **619** (2005) 61.
[23] J.R. Batley *et al* (NA48/2 collaboration), Eur. Phys. J. C **52** (2007) 875.
[24] J.R. Batley *et al* (NA48/2 collaboration), Eur. Phys. J. C **68** (2010) 75.
[25] G. Anzivino, these proceedings.
[26] J. Adams *et al* (KTeV collaboration), Physical Review Letters **80** (1998) 4123; A. Lai *et al* (NA48 collaboration), Eur. Phys. Jour. C **30** (2003) 33.

[27] A. Angelopoulos *et al* (CPLEAR collaboration), Phys. Lett. B **444** (1998) 43.
[28] M. Antonelli *et al* , Eur. J. Phys. C **69** (2010) 399.
[29] F. Ambrosino *et al* (KLOE collaboration), Eur. Phys. J. C **64** (2009) 627; ibid. **65** (2010) 703. C. Lazzeroni *et al* (NA62 collaboration), arXiv:1008.1219, submitted to Phys. Lett. B (2011).
[30] G. Amelino-Camelia *et al* (KLOE-2 collaboration), arXiv:1003.3868 (2010).
[31] B. Hiesmayr, A. De Santis, these proceedings.
[32] J-PARC proposal P-06, at http://j-parc.jp/NuclPart/Proposal_e.html.
[33] M. Abe *et al* (KEK E246 collaboration), Phys. Rev. Lett. 93 (2004) 131601.
[34] A.J. Buras *et al* , Rev. Mod. Phys. **80** (2008) 965; E. Stamou, these proceedings.
[35] S. Adler *et al* (BNL E949 collaboration), Phys. Rev. D **77** (2008) 052003.
[36] J.K. Ahn *et al* (KEK E391a collaboration), Phys. Rev. D **81** (2010) 072004.
[37] NA62 Technical Design, CERN NA62-10-07, CERN, Geneva, December 2010.
[38] J-PARC proposal P-14, at http://j-parc.jp/NuclPart/Proposal_e.html.
[39] Project-X web pages at projectx.fnal.gov.

Dark matter searches at LHC

Vasiliki A Mitsou[1,2]

[1] Instituto de Física Corpuscular (IFIC), CSIC – Universitat de València, Parc Científic, Apartado de Correos 22085, E-46071, Valencia, Spain
[2] CERN, PH Department, CH-1211 Geneva 12, Switzerland

E-mail: vasiliki.mitsou@ific.uv.es

Abstract. Besides Standard Model measurements and other Beyond Standard Model studies, the ATLAS and CMS experiments at the LHC will search for Supersymmetry, one of the most attractive explanation for dark matter. The SUSY discovery potential with early data is presented here together with some first results obtained with 2010 collision data at 7 TeV. Emphasis is placed on measurements and parameter determination that can be performed to disentangle the possible SUSY models and SUSY look-alike and the interpretation of a possible positive supersymmetric signal as an explanation of dark matter.

1. Introduction

Unveiling the nature of dark matter (DM) [1] is a quest in both Astrophysics and Particle Physics. Among the list of well-motivated candidates, the most popular particles are cold and weakly interacting, and typically predict missing energy signals at particle colliders. Supersymmetry (SUSY) and models with extra dimensions are theoretical scenarios that inherently provide such a dark matter candidate. The Large Hadron Collider (LHC) [2] currently in operation at CERN in Geneva, Switzerland, is an ideal machine for discovering DM in colliders and exploring both phenomenological as much as purely theoretical aspects aspects of DM.

The exploration of dark matter is being complemented by other types of searches of particle dark matter: direct detection in low background underground experiments [3] and indirect detection of neutrinos, gamma-rays and antimatter with terrestrial and space-borne detectors [4]. Experiments in upcoming colliders, such as the ILC [5] and CLIC [6], are expected to further constraint such models and make a key step in understanding dark matter.

The structure of this paper is as follows. Section 2 provides a brief introduction to dark matter properties as defined by the current cosmological data. In section 3 the features of collider experiments that play a central role in exploring DM are highlighted. In section 4 we discuss the strategy for discovering supersymmetry at the LHC, some recent results and prospects for the near future. Studies on methods to constrain dark matter parameters at the LHC, such as particle masses and spins, are reviewed in section 5. Some alternative theoretical models yielding a modified DM density and its implications for SUSY searches are discussed in section 6. The paper concludes with an outlook in section 7.

2. Dark matter evidence

The nature of the dark sector of our Universe constitutes one of the major mysteries of fundamental physics. According to observations over the past two decades, most of our

Universe energy budget consists of unknown entities: $\sim 23\%$ is dark matter and $\sim 72\%$ is dark energy, a form of ground-state energy. Dark energy is believed to be responsible for the current-era acceleration of the Universe. Dark matter, on the other hand, is matter inferred to exist from gravitational effects on visible matter, but is undetectable by emitted or scattered electromagnetic radiation. A possible explanation then —other than the introduction of a new particle— is to ascribe the observed effects to modified Newtonian dynamics (MOND) [7].

The energy budget of the Cosmos (fig. 1) has been obtained by combining a variety of astrophysical data, such as type-Ia supernovae [8], cosmic microwave background (CMB) [9], baryon oscillations [10] and weak lensing data [11]. The most precise measurement comes from anisotropies of the cosmic microwave background [9, 12]. The third peak in the temperature power spectrum, shown in fig. 2, can be used to extract information about the dark matter contribution to the Universe energy budget.

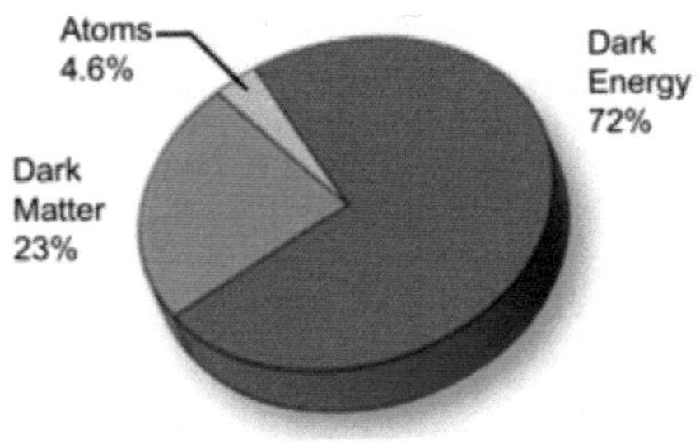

Figure 1. The energy distribution of the Universe according to recent cosmological evidence and assuming the ΛCDM model [13].

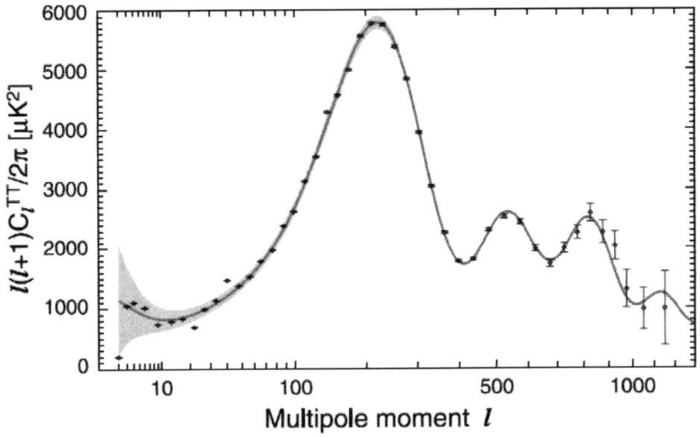

Figure 2. Temperature power spectrum from WMAP7. The third acoustic peak is sensitive to the dark matter density [12].

Evidence from the formation of large-scale structure (galaxies and their clusters) strongly favour cosmologies where non-baryonic DM is entirely composed of cold dark matter (CDM), i.e. non-relativistic particles.[1] CDM particles, in turn, may be axions [14], superheavy non-thermal relics (wimpzillas, cryptons) [15] or weakly interacting massive particles (WIMPs). The latter class of DM candidates arises naturally in models which attempt to explain the origin of electroweak symmetry breaking and this is precisely where the connection between Cosmology and Particle Physics lies. Furthermore the typical (weak-scale) cross sections characterizing these models are of the same order of magnitude as the WIMP annihilation cross section, thus establishing the so-called *WIMP miracle*.

3. WIMPs and colliders

WIMP dark matter candidates include the lightest neutralino in models with weak-scale supersymmetry [16], while Kaluza-Klein photons arise in scenarios with universal extra dimensions (UED) [17], and lightest T-odd particles are predicted in Little Higgs models [18]

[1] The interpretation of the 'raw' astrophysical data in the context of the Universe energy/matter content as presented above is based on the Standard Cosmological Model (ΛCDM) [13], involving cold DM as the dominant DM species, and a positive cosmological constant $\Lambda > 0$. Nevertheless, as discussed in section 6, the possibility for different theoretical scenarios is open, that modify the estimated DM relic abundance for given cosmological observations.

with a conserved T-parity. The common denominator in these theories is that they all predict the existence of a electrically neutral, colourless and *stable* particle, whose decay is prevented by a kind of symmetry: R-parity, connected to baryon and lepton number conservation in SUSY models; KK-parity, the four-dimensional remnant of momentum conservation in the extra dimensions; and a Z_2 discrete symmetry called T-parity in Little Higgs models. The origin of DM can be attributed to more than one particles, even within the same theoretical framework, e.g. in the degenerate scenario of the next-to-minimal supersymmetric standard model (NMSSM) [19].

In this paper, we focus on SUSY signatures, although these may be very similar to the other afore-mentioned models. R-parity is defined as: $R = (-1)^{3(B-L)+2S}$, where B, L and S are the baryon number, lepton number and spin, respectively. Hence $R = +1$ for all Standard Model particles and $R = -1$ for all SUSY particles. It is stressed that the conservation of R-parity is an *ad-hoc* assumption. The only firm restriction comes from the proton lifetime: non-conservation of both B and L leads to rapid proton decay. R-parity conservation has serious consequences in SUSY phenomenology in colliders: the SUSY particles are produced in pairs and the lightest SUSY particle (LSP) is absolutely stable.

A parenthesis is due here addressing the issue of (not necessarily cold) dark matter in SUSY models with R-parity violation (RPV) [20]. These seemingly incompatible concepts *can* be reconciled in models with a gravitino [21] or an axino [22] LSP with a lifetime exceeding the age of the Universe. In both cases, RPV is induced by bilinear terms in the superpotential that can also explain current data on neutrino masses and mixings without invoking any GUT-scale physics. Decays of the next-to-lightest superparticle occur rapidly via RPV interaction, and thus they do not upset the Big-Bang nucleosynthesis, unlike the R-parity conserving case. R-violating couplings can be sufficiently large to lead to interesting expectations for collider phenomenology; these will be the standard signatures of RPV supersymmetry, with multi-lepton or multi-jet events and the possibility of explicit lepton number violation at the final state [23]. Nevertheless determining whether R-parity is conserved or broken may not be trivial as WIMPs, whether absolutely stable or quasi-stable, cannot be detected directly in collider experiments.

Indeed weakly interacting massive particles do not interact neither electromagnetically nor hadronically with matter and thus, once produced, they traverse the various detectors layers without leaving a trace (just like neutrinos do). However by exploiting the hermeticity of the experiments, we can get a hint of the WIMP presence through the balance of the energy/momentum measured in the various detector components, the so-called *missing energy*. In hadron colliders, in particular, since the (longitudinal) momenta of the colliding partons are unknown, only the *transverse missing energy*, $E_{\mathrm{T}}^{\mathrm{miss}}$, can be reliably used to 'detect' DM particles.

4. SUSY(-like) searches

Here we focus on searches for dark matter in the two general-purpose experiments of the LHC, namely ATLAS [24] and CMS [25]. We review recent studies[2] performed on up to 300 nb^{-1} of LHC proton-proton collisions data at a centre-of-mass energy of 7 TeV. For comprehensive accounts of the envisaged analyses, based on Monte Carlo simulations, the reader is referred to the ATLAS 'CSC Book' [26] and the CMS Physics Technical Design Report [27].

At the LHC, supersymmetric particles are expected to be predominantly produced hadronically, i.e. through gluino-pair, squark-pair and squark-gluino production. Each of these (heavy) sparticles is going to decay into lighter ones in a cascade decay that finally leads to an LSP, which in most of the scenarios considered is the lightest neutralino $\tilde{\chi}_1^0$, as depicted in fig. 3. The two LSPs would escape detection giving rise to high transverse missing energy, which is

[2] As of December 2010.

rigorously defined in fig. 4. Such a simulated event as it will appear in the ATLAS detector is illustrated in fig. 5.

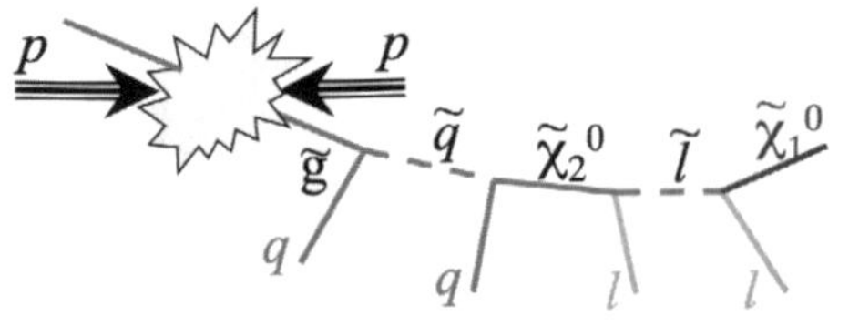

Figure 3. An example of the cascade decay of heavier to lightest sparticles. For simplicity only one branch is shown here; a second one is expected to be present. Note the high lepton and jet multiplicity.

$$E_{\mathrm{T}}^{\mathrm{miss}} \equiv \left\| -\sum_i \vec{p}_{T,i} \right\| = \left| -\sum_i p_{x,i} - \sum_i p_{y,i} \right|$$

Figure 4. $E_{\mathrm{T}}^{\mathrm{miss}}$ definition: the sums are over all reconstructed objects and calorimeter clusters/towers.

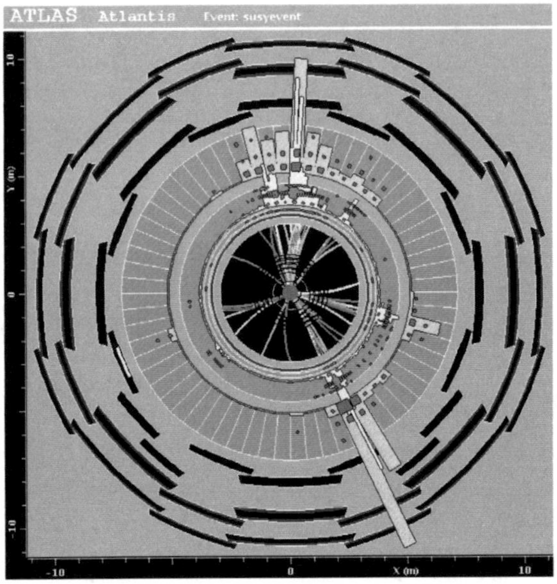

Figure 5. Transverse view of a simulated SUSY event in the ATLAS detector [28]. Note the imbalance of the deposited energy distribution towards the right-hand side. Two (stable) lightest neutralinos (not shown) escape detection towards the left-hand side.

The search strategy followed in the inclusive channels is based on the detection of high $E_{\mathrm{T}}^{\mathrm{miss}}$, many jets and possibly energetic leptons. The analyses make extensive use of data-driven Standard Model background measurements. Detailed studies have been carried out for various signatures using Monte Carlo data-sets fully simulated with Geant4 [29] for specific SUSY signal parameters and for the relevant Standard Model backgrounds. Although various SUSY-breaking mechanisms have been considered by the two collaborations, the early results and projections for higher integrated luminosity highlighted here have been performed in the context of the minimal Supergravity (mSUGRA) model. By assumption the mSUGRA model avoids both flavour-changing neutral currents and extra sources of CP violation. For masses in the TeV range, it typically predicts too much cold dark matter unless something enhances the $\tilde{\chi}_1^0$ annihilation, as discussed in section 6. The specific points mentioned here, denoted SUn for ATLAS and LMn for CMS, have been chosen to be roughly consistent with the observed CDM and represent a variety of different decay modes.

SUSY searches require careful control over backgrounds from standard model processes. Several methods for data-driven background determinations were developed and tested on early LHC pp collision data. Such a method allowed CMS to study QCD backgrounds, to control jet-energy mismeasurement, and to measure background contributions from processes producing non-prompt leptons or hadrons misidentified as leptons [30]. It is based on the α_{T} discriminating variable for the all-hadronic channel, defined for two jets j1 and j2 as:

$$\alpha_{\mathrm{T}} \equiv \frac{E_{\mathrm{T}}^{\mathrm{j2}}}{M_{\mathrm{T}}^{\mathrm{j1,j2}}}, \tag{1}$$

where $M_{\mathrm{T}}^{\mathrm{j1,j2}}$ is the transverse mass of the two jets, i.e.:

$$\alpha_{\mathrm{T}} \equiv \frac{\sqrt{E_{\mathrm{T}}^{\mathrm{j2}}/E_{\mathrm{T}}^{\mathrm{j1}}}}{\sqrt{2(1-\cos\Delta\,\phi)}}, \tag{2}$$

with$\Delta\ \phi$ the angle between the two jets. The method is roughly based on the concept of reversing a cut on a control variable ($H_{\mathrm{T}} \equiv \sum_{\mathrm{jets}\,j} p_{\mathrm{T}j}$) to check the (ideally) signal-free region of a discriminating variable (α_{T}). The MC-versus-data agreement in a control sample ($80 < H_{\mathrm{T}} < 120$ GeV) far from the signal region ($H_{\mathrm{T}} < 350$ GeV) is illustrated in fig. 6. A study of the $\alpha_{\mathrm{T}} > 0.55$ rejection power as the H_{T} threshold increases demonstrated the robust behaviour of the α_{T} cut [30]. A detailed account of recent SUSY studies performed at CMS with LHC early data is presented in ref. [31], whereas a search for long-lived ('stopped') gluinos is given in ref. [32] among other searches for New Physics.

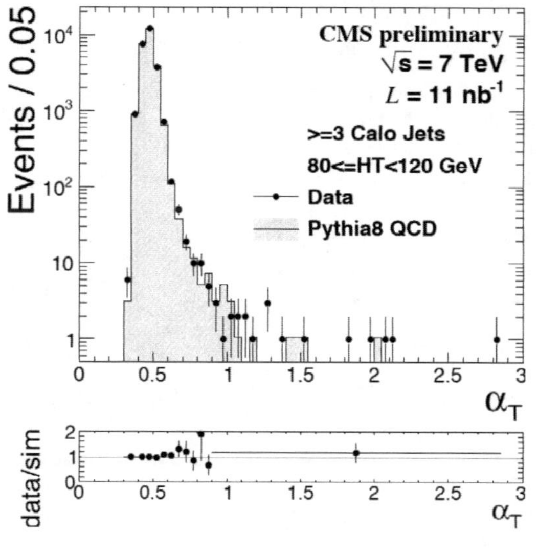

Figure 6. a_{T} distribution in CMS multijet events for $80 < H_{\mathrm{T}} < 120$ GeV. The a_{T} tail is reduced to practically zero (not shown) for $H_{\mathrm{T}} > 120$ GeV [30].

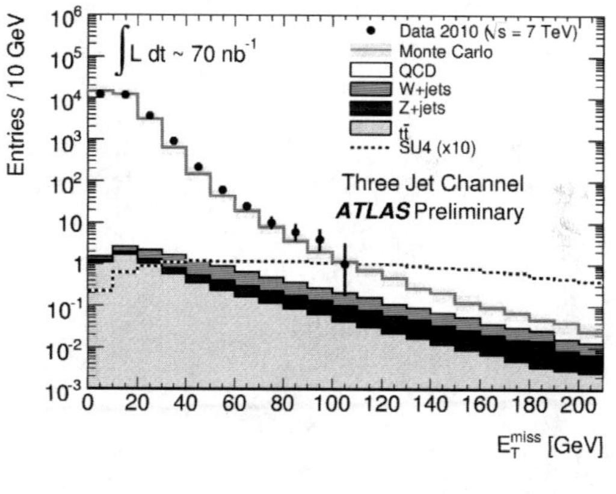

Figure 7. Distribution of the missing transverse momentum for ATLAS events in the 0-lepton, three-jet channel after basic preselection [33]. Note the long tail in the prediction for the SU4 mSUGRA point.

The first LHC collision data at 7 TeV collected by ATLAS and CMS also allowed testing the robustness of the $E_{\mathrm{T}}^{\mathrm{miss}}$-based strategy for discovering SUSY. This is illustrated in fig. 7, where the Monte Carlo $E_{\mathrm{T}}^{\mathrm{miss}}$ distributions for various SM processes that may fake a SUSY-like signal are drawn. The estimates are predominantly based on Monte Carlo simulations, however in some cases these have been scaled to data-deduced factors. The data agree well with the simulation in the 0-lepton plus three-jet selection [33]. In the same plot, the prediction for the (low mass, high cross section) mSUGRA point SU4[3] is superposed. The characteristic long,

[3] SU4: m_0=200 GeV, $m_{1/2}$=160 GeV, $\tan\beta = 10$, $A_0 = -400$ GeV, $\mu > 0$.

high-$E_T^{\rm miss}$ tail of the production of DM particles discussed in section 3 is clearly visible. More insight on current DM searches performed by ATLAS, either related to supersymmetry or to universal extra dimensions, can be found in ref. [34].

The LHC data processed and analyzed so far are $\mathcal{O}(100\ \rm nb^{-1})$, not sufficient to supersede the exclusion limits set by LEP and Tevatron. Nevertheless with the $\sim 35\ \rm pb^{-1}$ already recorded by ATLAS and CMS and the few inverse femtobarns expected to be provided in the next 1–2 years, a discovery will be within reach. This is demonstrated in fig. 8 for the CMS [35] and in fig. 9 for ATLAS [36] for integrated luminosities from 100 $\rm pb^{-1}$ to 1 $\rm fb^{-1}$. The largest coverage is achieved through the all-hadronic channel, while comparable reach is provided by the 1-lepton mode. Other channels, e.g. the 2-lepton, apart from cross checks, will play a central role —in the event of a discovery— in constraining sparticles masses, as discussed in section 5.

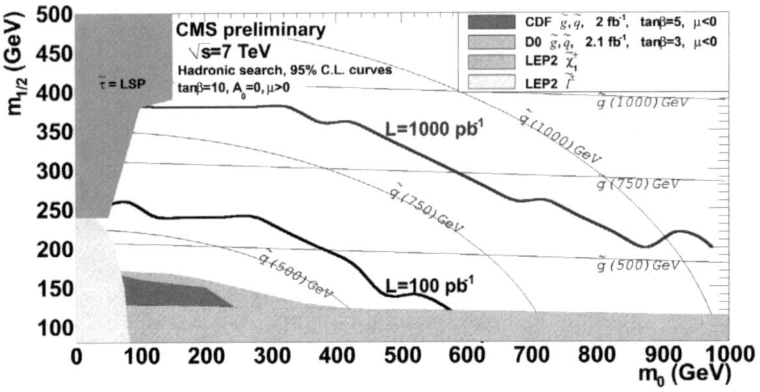

Figure 8. Estimated 95% C.L. exclusion limits for the all-hadronic SUSY search, based on simulated events, expressed in mSUGRA parameter space [35].

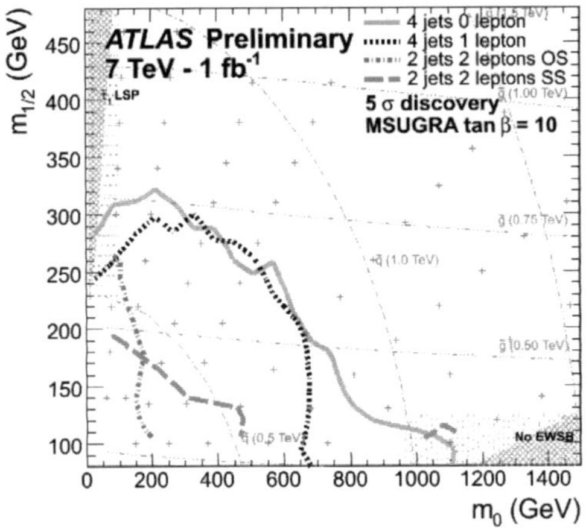

Figure 9. 5σ discovery reach as a function of m_0 and $m_{1/2}$ for a $\tan\beta = 10$ mSUGRA scan for channels with 0, 1 and 2 leptons. The assumed integrated luminosity is 1 $\rm fb^{-1}$ [36].

5. Pinning down dark matter at LHC

Once clear evidence for a possible supersymmetric signal is established, the question of whether this implies the existence of SUSY or one of its lookalike arises. In addition, if we assume that it is a SUSY signal, we need to pin down the exact SUSY breaking mechanism and measure its theoretical parameters. Both issues are discussed here, giving some examples of sparticle mass measurements and spin and parameter determination.

5.1. Constraining sparticle masses

As an example of the methods aiming at constraining sparticle-mass relations in sparticle cascade decays, we present here a dilepton analysis studied at CMS with simulated data [37, 38]. Similar analyses with ATLAS are documented in refs. [39, 40]. The analysis is targeting an integrated luminosity of 200–300 pb^{-1} at $\sqrt{s} = 10$ TeV and its objectives are: (i) to observe a significant excess of opposite-sign same-flavour leptons over the various backgrounds, and (ii) to measure the endpoint in the invariant mass distribution. The latter is directly related to sparticle-masses differences, being sensitive to opposite-sign same-flavour (OS-SF) dileptons coming from the last stages of the decay chain of sparticles:

$$\tilde{q} \to \tilde{\chi}_2^0 q \to \tilde{\ell}^{\pm}\ell^{\mp} q \to \tilde{\chi}_1^0 \ell^{\pm}\ell^{\mp} q, \ell \qquad = e \text{ or } \mu \tag{3}$$

The shape of the distribution largely depends on the exact decay chain, so various mSUGRA benchmark points have been considered. For instance at LM0,[4] the mass difference of the two lightest neutralinos is smaller than the Z boson mass and any slepton mass. Two opposite-sign same-flavour leptons come from the decay chain $\tilde{\chi}_2^0 \to \tilde{\chi}_1^0 \ell^{\pm}\ell^{\mp}$, hence the edge position represents this mass difference:

$$m_{\ell\ell,\max} = m_{\tilde{\chi}_2^0} - m_{\tilde{\chi}_1^0}. \tag{4}$$

At the LM1,[5] on the other hand, the mass difference of the two lightest neutralinos is larger than the mass of the lightest slepton, so a slepton can be an intermediate product in the neutralino decay chain $\tilde{\chi}_2^0 \to \tilde{\ell}_R \ell \to \tilde{\chi}_1^0 \ell^{\pm}\ell^{\mp}$. The equation connecting the position of the edge with sparticles masses takes the form:

$$(m_{\ell\ell,\max})^2 = \frac{(m^2_{\tilde{\chi}_2^0} - m^2_{\tilde{\ell}})(m^2_{\tilde{\ell}} - m^2_{\tilde{\chi}_1^0})}{m^2_{\tilde{\ell}}}. \tag{5}$$

The main sources of physics background are uncorrelated supersymmetric decays and SM processes: $t\bar{t}$, dibosons, and associated production of W/Z bosons and jets. A data-driven strategy has been developed to eliminate these background processes, based on different-flavour dilepton ($e\mu$) pairs in order to estimate the background in the ee and $\mu\mu$ combinations. It has been demonstrated that the background estimate is reliable for flavour-symmetric background processes.

The position of the endpoint in the invariant mass distribution of the two leptons is then extracted via a maximum-likelihood fit with components describing the background flavour-symmetric processes (see fig. 10, left) and the characteristic triangular shape of the supersymmetric signal (see fig. 10, right). The latter may be fitted by a function corresponding to a two- or a three-body decay.

The number of signal events derived from the fit agrees with the true number of signal events [38]. The theoretical endpoint value is reproduced in case of the fit with the three-body decay model, while the theoretical value is underestimated if the model is fitted for a two-body decay. At the benchmark points LM1 and LM9,[6] a higher integrated luminosity is necessary to measure the endpoint. In other studied benchmark points, an integrated luminosity needed to obtain a 5σ discovery using shape information of 250 pb^{-1} (LM1) and 350 pb^{-1} (LM9) [38].

[4] LM0: m_0=200 GeV, $m_{1/2}$=160 GeV, $\tan\beta = 10$, $A_0 = -400$ GeV, $\mu > 0$.
[5] LM1: m_0=60 GeV, $m_{1/2}$=250 GeV, $\tan\beta = 10$, $A_0 = 0$, $\mu > 0$.
[6] LM9: m_0=1450 GeV, $m_{1/2}$=175 GeV, $\tan\beta = 50$, $A_0 = 0$, $\mu > 0$.

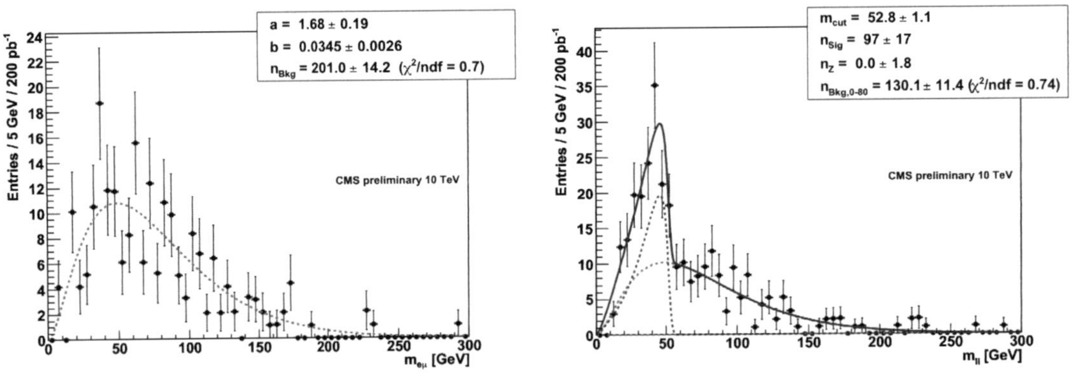

Figure 10. *Left:* The fit of the background function to the $e\mu$ invariant mass distribution [38]. *Right:* The combined fit at LM0 for 200 pb^{-1}. The green curve represents the SUSY signal model, the red curve is the background function and the light green dashed line the Z contribution. The black points represent the MC events [38].

Further constraints can be set in more complex combinations of sparticle masses if jets are added to the invariant mass calculation [26]. If sufficient endpoints are known, then the masses themselves can be deduced, e.g. by using a numerical χ^2 minimization based on the MINUIT package [41] to extract the SUSY particle masses from a combination of endpoints. A first look at sparticle masses is possible with early data, although with large uncertainties. Appropriate model assumptions and additional information will probably have to be used to constrain the fits.

5.2. Parameter determination

The next step after discovery will be to select specific supersymmetric decay chains to measure the properties of the new particles. Here we focus on how a selected set of early studies can be combined to obtain the first measurements of supersymmetric masses and of the parameters of the mSUGRA model with 1 fb^{-1} of ATLAS [26] at a centre-of-mass LHC energy of 10 TeV. Specific benchmarks in parameter space have been used to demonstrate the precision that can be expected from these measurements (such as the SU3[7] point here), but the same (or similar) techniques can be applied to a considerable portion of the SUSY parameter space accessible with LHC data.

A first glimpse of the possible parameter space can be obtained by performing a Markov-chain analysis. With this technique it is possible to efficiently explore a large-dimensional parameter spaces and check whether there are several topologically disconnected parameter regions which are favoured by a given set of measurements.

A stringent constraint on the SUSY model can be achieved by fitting theoretical calculations for a given set of parameters —performed by spectrum calculators like SPheno [42], SoftSUSY [43] or the ISASUSY [44] decay package of ISAJET [45]— to the mass combinations acquired by measuring edge points in invariant distributions. This fitting can be performed by specialized parameter-fitting packages, such as Fittino [46] or SFitter [47]. In order to estimate the expected precision for such measurements, a number of toy fits for a fixed sgn μ has been performed by ATLAS [26]. The four-dimensional distribution of parameters obtained from these toy fits is used to derive the parameter uncertainties and their correlations. The mean values

[7] SU3: m_0=100 GeV, $m_{1/2}$=300 GeV, $\tan\beta = 6$, $A_0 = -300$ GeV, $\mu > 0$.

Journal of Physics: Conference Series **335** (2011) 012003 doi:10.1088/1742-6596/335/1/012003

and uncertainties of the results for the parameters m_0, $m_{1/2}$, $\tan\beta$ and A_0 are listed in table 1. The parameters m_0 and $m_{1/2}$ can be derived reliably with uncertainties of $\mathcal{O}(10\text{ GeV})$, whereas for $\tan\beta$ and A_0 only the order of magnitude can be derived from these measurements. The χ^2 distribution of the fits can be used to evaluate the toy-fit performance. The observed mean $\chi^2 = 12.6\pm0.2$ for $\text{sgn}\,\mu = +1$ is compatible with the expected value of $N_{\text{dof}} = 11$. The solutions for the wrong assumption $\text{sgn}\,\mu = -1$, also reported in table 1, cannot however be ruled out as the observed mean $\chi^2 = 15.4 \pm 0.3$ is also acceptable.

Table 1. Results of a fit of the mSUGRA parameters to mass endpoints for the SU3 point in ATLAS [26]. The mean and RMS of the distribution of the results from toy fits is reported. The two possible assumptions for the digital parameter $\text{sgn}\,\mu = +1$, $\text{sgn}\,\mu = \pm1$ have been used, resulting in different preferred regions for the other parameters. The experimental uncertainties are also shown.

Parameter	SU3 value	Fitted value	Uncertainty
		$\text{sgn}\,\mu = +1$	
$\tan\beta$	6	7.4	4.6
m_0 [GeV]	100	98.5	9.3
$m_{1/2}$ [GeV]	300	317.7	6.9
A_0 [GeV]	−300	445	408
		$\text{sgn}\,\mu = -1$	
$\tan\beta$		13.9	2.8
m_0 [GeV]		104	18
$m_{1/2}$ [GeV]		309.6	5.9
A_0 [GeV]		489	189

Hence with 1 fb^{-1} the reconstruction of part of the supersymmetric mass spectrum will only be possible for favourable SUSY scenarios and with some assumptions about the decay chains involved. Larger integrated luminosity will help to overcome these limitations, as more measurements become possible and the precision of each increases. Furthermore the mass spectrum constraints in conjunction with precision observables, such as the $(g-2)_\mu$ and the $b \to s\gamma$, will illuminate the flavour mixing and possibly the CP properties of the supersymmetric model [48].

5.3. Spin measurement

Measurements of the number of new particles and their masses will provide us enough information to extract model parameters for one of the SUSY models. However, the mass information alone will not be enough to distinguish different new physics scenarios. For example universal extra dimensions [17] with Kaluza-Klein parity can have a mass spectrum very similar to the one of certain SUSY models. However, the spin of the new particles is different and can be used to discriminate between models [49]. Another method based on robust ratios of inclusive counts of simple physics objects has also been proposed [50].

In order to measure the spin of newly discovered particle, one possibility is to use two-body slepton decay chains as the ones described earlier in this section. In particular the cascade decay of the $\tilde{q}_L$ to $\tilde{\chi}_2^0$ which further decays to slepton (fig. 11) is very convenient for such measurements [51].

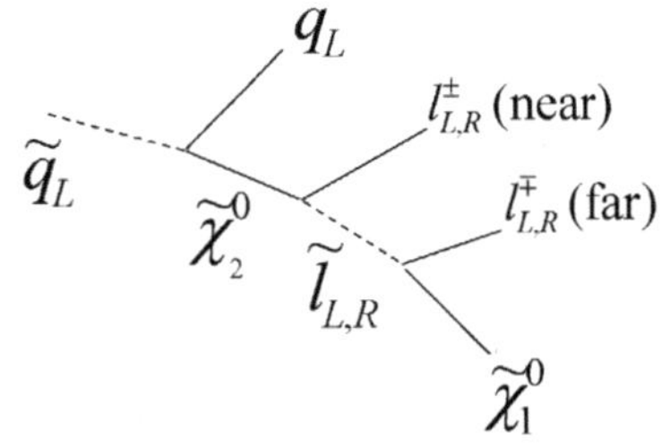

Figure 11. Schematic view of L-type squark decay. The lepton from the $\tilde{\chi}^0_2$ decay is called *near*; the lepton from $\tilde{\ell}_{L,R}$ decay is called *far*.

The charge asymmetry of ℓq pairs, for instance, can be used to measure the spin of $\tilde{\chi}^0_2$, while the shape of dilepton invariant mass spectrum measures slepton spin [52]. The first lepton in the decay chain is called the *near* lepton while the other is called the *far* lepton. The invariant masses $m(q\ell_{\text{near}})$ charge asymmetry A is trivially defined as:

$$A \equiv \frac{s^+ - s^-}{s^+ + s^-} \tag{6}$$

where $s^{\pm} = \mathrm{d}\sigma/\mathrm{d}m(q\ell^{\pm}_{\text{near}})$.

In general is not possible to distinguish between the near and the far lepton and only $m(\bar{q}\ell_{\text{near}})$ can be measured, diluting A. The expected asymmetry for the mSUGRA benchmark point SU3, as estimated by ATLAS for $\sqrt{s} = 14$ TeV, is shown in fig. 12 for a luminosity of 30 fb^{-1} [52].

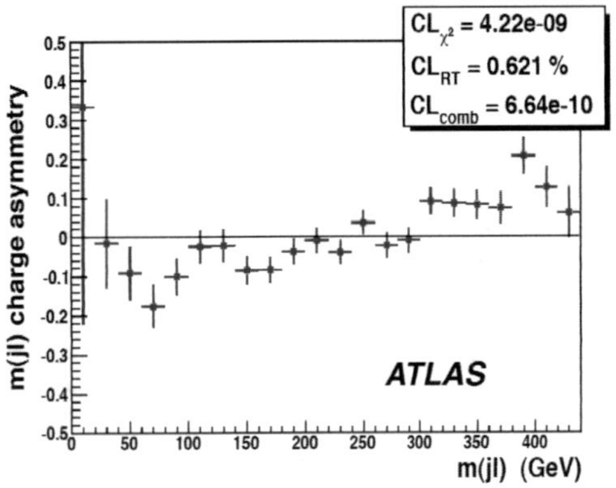

Figure 12. Charge asymmetry for lepton-jet invariant mass after SFOS-OFOS subtraction using both near and far leptons in SU3 point [52].

Results show that, in a fast simulation approach without taking into account systematic effects coming from a realistic detector description, an integrated luminosity of at least 100 fb^{-1} is needed in the case of the SU1[8] point to observe a non-zero charge asymmetry with a confidence level of about 99%, while in the more favourable case of the SU3 point 10 fb^{-1} would be sufficient [52]. It becomes therefore evident that even if the LHC experiments observe a SUSY-like signal during the next two years (2011–2012) of operation at the 'low' LHC energies of $7-8$ TeV, much more data at higher energy of 14 TeV will be necessary to establish the identity of the underlying theory.

A significant role in this context will be played by the high precision measurements expected to be performed at the ILC [5, 53]. If the determination of the properties of the DM particle by collider experiments [53] matches cosmological observations to high precision, then (and only then) we will be able to claim to have determined what DM is. Such an achievement would be

[8] SU1: m_0=70 GeV, $m_{1/2}$=350 GeV, $\tan\beta = 10$, $A_0 = 0$, $\mu > 0$.

a great success of the Particle-Physics/Cosmology connection and would give us confidence in our understanding of the Universe.

6. Interplay between the dark sector and LHC: Alternative scenarios

The simplest proposal to explain the origin of dark energy is to add a cosmological constant to Einstein's equation [13]. However, the reason why the dark matter content is comparable to the dark energy content at the present time remains a puzzle. Modifications to general relativity, braneworld scenarios, and topological defects are some of the proposals attempting to explain this fundamental issue. In string theory, the dilaton can play the role of dark energy [54]. In this section we review experimental signatures of SUSY as consequences of a rolling dilaton in the Q-cosmology scenario [54, 55] which offers an alternative framework that establishes the Supercritical (or non-critical) String Cosmology (or SSC) [54]. Such a dilaton modifies the Boltzmann equation, thus diluting the supersymmetric dark matter density (of neutralinos) by a factor $\mathcal{O}(10)$ [56, 57] and consequently the parameter-space regions excluded by the standard scenario are allowed in the SSC. Such deviating predictions in the DM relic abundance also arise in the context of space-time (D-particle) foam in string/brane-theories [58].

The mSUGRA final states at the LHC favoured by supercritical string cosmology have been studied in depth in ref. [59]. It becomes evident by inspecting the two panels in fig. 13, that the dark-matter-allowed region has larger values of m_0 compared to the standard cosmology case. Thus, the final states in the SSC scenario are different from those of the standard cosmology. For example, in the case of standard cosmology for smaller values of m_0 (also allowed by the $(g-2)_\mu$ constraint), we have low-energy taus in the final state due to the proximity of the stau to the neutralino mass in the stau-neutralino coannihilation region. On the other hand, in the SSC case the final states contain Z bosons, Higgs bosons or high-energy taus.

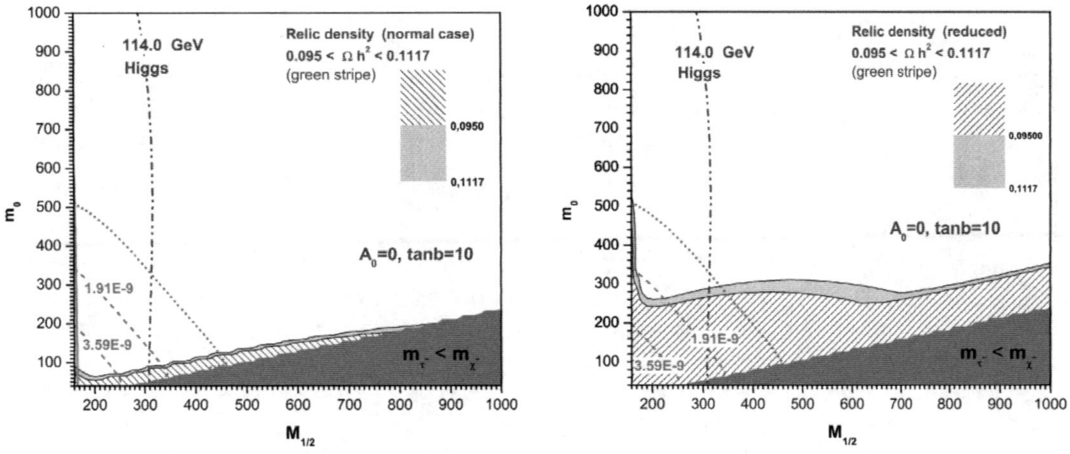

Figure 13. WMAP3-allowed parameter space in mSUGRA (m_0, $m_{1/2}$) plane for the standard cosmology (left) and the SSC (right) for $A_0 = 0$ and $\tan\beta = 10$: regions where the neutralino relic density is within the WMAP3 limits (green stripe) and where it is lower than this (hatched region) are shown. Also shown are the h^0 mass boundary (dash-dotted blue), the $(g-2)_\mu$ 1σ (dashed red) and 2σ (dotted red) boundaries and the stau-LSP region (lower solid red) [56].

In fact these final states dominate in most of the allowed SUGRA parameter space. Therefore, by analyzing the parameter space of the SSC model, most regions of the SUGRA parameter space at the LHC are explored. The following final states have been studied [59]:

– $h^0 \to b\bar{b}$ + jets + $E_{\mathrm{T}}^{\mathrm{miss}}$;
– $Z \to \ell^{\pm}\ell^{\mp}$ + jets + $E_{\mathrm{T}}^{\mathrm{miss}}$; and
– 2τ + jets + $E_{\mathrm{T}}^{\mathrm{miss}}$.

as well as constructed observables such as the endpoints of invariant mass distributions M_{bbj}, $M_{\ell\ell j}$, and $M_{\tau\tau j}$ and the peak position of $M_{\tau\tau}$. All of these analyses have been studied with ATLAS and/or CMS with simulated data at $\sqrt{s} = 10$ or 14 TeV: $h^0 \to b\bar{b}$ channel [39, 60, 26, 27], the $Z \to \ell^{\pm}\ell^{\mp}$ mode [27] and the 2-τ mode [26, 27]. In the future, when pp collision data at an energy of 14 TeV will be accumulated, searches for SUSY with these signatures will be possible.

It is remarked that the SSC scenario is consistent with the smoothly evolving dark energy at least for $0 < z < 1.6$, in accordance with the observations on supernovae [55], on the galaxy-ages-measured Hubble rate [61] and on the baryon acoustic oscillations [62]. Hence it offers a cosmologically viable solution.

7. Outlook

The origin of dark matter remains one of the most compelling mysteries in our understanding of the Universe today and the Large Hadron Collider, already delivering pp collision data at CERN at an unprecedented high-energy, is going to play a central role in constraining some of its parameters. A deviation from SM in inclusive signatures like missing energy plus jets (plus leptons) will hint a discovery of DM, however exclusive studies are required to roughly determine the new-particle properties and model parameters. Although the scheme is developed with SUSY in mind, it is applicable to other beyond-standard-model scenarios such as UED and T-parity Little Higgs.

If LHC should discover general WIMP dark matter, it will be non-trivial to prove that it has the right properties. Future e^+e^- colliders (ILC, CLIC) are expected to extend the LHC discovery potential and improve the identification of the underlying DM model. By providing more precise determination of model parameters, they will consequently put bounds on relic density, direct detection rate and WIMP annihilation processes.

The complementarity between LHC and cosmo/astroparticle experiments lies in the uncorrelated systematics and the measurement of different model parameters. In the following years we expect a continuous interplay between particle physics experiments and astrophysical/cosmological observations.

Acknowledgments

The author is grateful to the DISCRETE2010 organizers and especially Antonio Di Domenico for the kind invitation and support. Thanks to them, a warm, friendly and intellectually stimulating atmosphere was enjoyed by the speakers and participants throughout the Symposium. This work was supported in part by the Spanish Ministry of Science and Innovation (MICINN) under the project FPA2009-13234-C04-01 and by the Spanish Agency of International Cooperation for Development under the PCI projects A/023372/09 and A/030322/10. The author acknowledges support by the CERN Corresponding Associate Programme.

References

[1] For a pedagogical introduction, see e.g.: Hooper D 2009 TASI 2008 Lectures on Dark Matter *Preprint* arXiv:0901.4090 [hep-ph]
[2] Evans L and Bryant P 2008 LHC Machine *J. Instrum.* **3** S08001
[3] Akimov D 2011 Techniques and Results for the Direct Detection of Dark Matter (Review) *Nucl. Instrum. Meth.* A **628** 50–58 and references therein
[4] Morselli A 2011 Indirect detection of dark matter, current status and recent results *Prog. Part. Nucl. Phys.* **66** 208–215 and references therein
[5] Battaglia M 2009 The role of an e^+e^- linear collider in the study of cosmic dark matter *New J. Phys.* **11** 105025

Aarons G *et al* [ILC Collaboration] 2007 International Linear Collider Reference Design Report Volume 2: Physics at the ILC *Preprint* arXiv:0709.1893 [hep-ph]

[6] Accomando E *et al* [CLIC Physics Working Group] Physics at the CLIC multi-TeV linear collider *Preprint* arXiv:hep-ph/0412251

[7] For an review on MOND, see e.g.: Sanders R H and McGaugh S S 2002 Modified Newtonian Dynamics as an Alternative to Dark Matter *Ann. Rev. Astron. Astrophys.* **40** 263–317 (*Preprint* arXiv:astro-ph/0204521)

[8] Amanullah R *et al* 2010 Spectra and Light Curves of Six Type Ia Supernovae at $0.511 < z < 1.12$ and the Union2 Compilation *Astrophys. J.* **716** 712–738 (*Preprint* arXiv:1004.1711 [astro-ph.CO])

[9] Komatsu E *et al* 2011 Seven-Year Wilkinson Microwave Anisotropy Probe (WMAP) Observations: Cosmological Interpretation *Astrophys. J. Suppl.* **192** (*Preprint* arXiv:1001.4538 [astro-ph.CO])

[10] Eisenstein D J *et al* [SDSS Collaboration] 2005 Detection of the Baryon Acoustic Peak in the Large-Scale Correlation Function of SDSS Luminous Red Galaxies *Astrophys. J.* **633** 560–574 (*Preprint* astro-ph/0501171) and references therein

[11] Broadhurst T, Umetsu K, Medezinski E, Oguri M and Rephaeli Y 2008 Comparison of Cluster Lensing Profiles with Lambda CDM Predictions *Astrophys. J.* **685** L9–L12 (*Preprint* arXiv:0805.2617 [astro-ph]) and references therein

[12] Larson D *et al* 2011 Seven-Year Wilkinson Microwave Anisotropy Probe (WMAP) Observations: Power Spectra and WMAP-Derived Parameters *Astrophys. J. Suppl.* **192** 16 (*Preprint* arXiv:1001.4635 [astro-ph.CO])

[13] Carroll S M, Press W H and Turner E L 1992 The Cosmological constant *Ann. Rev. Astron. Astrophys.* **30** 499–542

[14] Zioutas K, Tsagri M, Papaevangelou T, Dafni T and Anastassopoulos V 2009 Axion Searches with Helioscopes and astrophysical signatures for axion(-like) particles *New J. Phys.* **11** 105020 (*Preprint* arXiv:0903.1807 [astro-ph.SR])

[15] Chung D J H, Kolb E W and Riotto A 1999 Superheavy dark matter *Phys. Rev.* D **59** 023501 (*Preprint* arXiv:hep-ph/9802238)

[16] Kazakov D I 2010 Supersymmetry on the Run: LHC and Dark Matter *Nucl. Phys. Proc. Suppl.* **203-204** 118–154 (*Preprint* arXiv:1010.5419 [hep-ph])

[17] Hooper D and Profumo S 2007 Dark matter and collider phenomenology of universal extra dimensions *Phys. Rept.* **453** 29–115 (*Preprint* arXiv:hep-ph/0701197)

[18] Birkedal A, Noble A, Perelstein M and Spray A 2006 Little Higgs dark matter *Phys. Rev.* D **74** 035002 (*Preprint* arXiv:hep-ph/0603077)

[19] Panotopoulos G 2011 The degenerate scenario in the NMSSM: Direct singlino-like neutralino searches with a gravitino LSP *Preprint* arXiv:1103.0140 [hep-ph] *in these proceedings*

[20] Barbier R *et al* 2005 R-parity violating supersymmetry *Phys. Rept.* **420** 1–202 (*Preprint* arXiv:hep-ph/0406039)

[21] Takayama F and Yamaguchi M 2000 Gravitino dark matter without R-parity *Phys. Lett.* B **485** 388–392 (*Preprint* arXiv:hep-ph/0005214)
Hirsch M, Porod W and Restrepo D 2005 Collider signals of gravitino dark matter in bilinearly broken R-parity *J. High Energy Phys.* JHEP03(2005)062 (*Preprint* arXiv:hep-ph/0503059)
Buchmuller W, Covi L, Hamaguchi K, Ibarra A and Yanagida T 2007 Gravitino dark matter in R-parity breaking vacua *J. High Energy Phys.* JHEP03(2007)037 (*Preprint* arXiv:hep-ph/0702184)

[22] Chun E J and Kim H B 2006 Axino Light Dark Matter and Neutrino Masses with R-parity Violation *J. High Energy Phys.* JHEP10(2006)082 (*Preprint* arXiv:hep-ph/0607076)

[23] Lola S 2009 Gravitino dark matter, neutrino masses and lepton flavor violation from broken R-parity *AIP Conf. Proc.* **1115** 318–323

[24] Aad G *et al* [ATLAS Collaboration] 2008 The ATLAS Experiment at the CERN Large Hadron Collider *J. Instrum.* **3** S08003

[25] Adolphi R *et al* [CMS Collaboration] 2008 The CMS experiment at the CERN LHC *J. Instrum.* **3** S08004

[26] Aad G *et al* [ATLAS Collaboration] 2009 Expected Performance of the ATLAS Experiment – Detector, Trigger and Physics *CERN Report* CERN-OPEN-2008-020 *Preprint* arXiv:0901.0512 [hep-ex]

[27] Bayatian G L *et al* [CMS Collaboration] 2007 CMS technical design report, volume II: Physics performance *J. Phys.* G **34** 995–1579

[28] Alan Barr, private communication

[29] Agostinelli S *et al* [GEANT4 Collaboration] 2003 GEANT4: A simulation toolkit *Nucl. Instrum. Meth.* A **506** 250–303
Allison J *et al* 2006 Geant4 developments and applications *IEEE Trans. Nucl. Sci.* **53** 270–278

[30] CMS Collaboration 2010 Performance of Methods for Data-Driven Background Estimation in SUSY Searches *CMS Physics Analysis Summary* CMS-PAS-SUS-10-001

[31] Widl E 2011 Searches for Supersymmetry with the CMS detector at the LHC *in these proceedings*
[32] Azzurri P 2011 First Results of Searches for New Physics at sqrt(s)= 7 TeV with the CMS detector *Preprint* arXiv:1103.1048 [hep-ex] *in these proceedings*
[33] ATLAS Collaboration 2010 Early supersymmetry searches in channels with jets and missing transverse momentum with the ATLAS detector *ATLAS Note* ATLAS-CONF-2010-065
[34] Siragusa G 2011 New Physics with ATLAS: experimental prospects *ATLAS Note* ATL-PHYS-PROC-2011-011 *in these proceedings*
[35] CMS Collaboration 2010 The CMS physics reach for searches at 7 TeV *CMS Note* CMS-NOTE-2010-008
[36] ATLAS Collaboration 2010 Prospects for Supersymmetry discovery based on inclusive searches at a 7 TeV centre-of-mass energy with the ATLAS detector *ATLAS Note* ATL-PHYS-PUB-2010-010
[37] CMS Collaboration 2009 Dilepton + Jets + MET channel: Observation and Measurement of $\tilde{\chi}_2^0 \rightarrow \tilde{\chi}_1^0 \ell\ell$ *CMS Physics Analysis Summary* CMS-PAS-SUS-08-001
[38] CMS Collaboration 2009 Discovery potential and measurement of a dilepton mass edge in SUSY events at $\sqrt{s} = 10$ TeV *CMS Physics Analysis Summary* CMS-PAS-SUS-09-002
[39] Airapetian A *et al* [ATLAS Collaboration] 1999 *ATLAS: Detector and physics performance technical design report* vol 1 (Geneva: CERN) 475 p (*CERN Report* CERN-LHCC-99-14)
Airapetian A *et al* [ATLAS Collaboration] 1999 *ATLAS: Detector and physics performance technical design report* vol 2 (Geneva: CERN) 519 p (*CERN Report* CERN-LHCC-99-15)
[40] ATLAS Collaboration 2009 Prospects for Supersymmetry and Univeral Extra Dimensions discovery based on inclusive searches at a 10 TeV centre-of-mass energy with the ATLAS detector *ATLAS Note* ATL-PHYS-PUB-2009-084
[41] James F and Roos M 1975 MINUIT: A System for Function Minimization and Analysis of the Parameter Errors and Correlations *Comput. Phys. Commun.* **10** 343–367
[42] Porod W 2003 SPheno, a program for calculating supersymmetric spectra, SUSY particle decays and SUSY particle production at e^+e^- colliders *Comput. Phys. Commun.* **153** 275–315 (*Preprint* arXiv:hep-ph/0301101)
[43] Allanach B C 2002 SOFTSUSY: a program for calculating supersymmetric spectra *Comput. Phys. Commun.* **143** 305–331 (*Preprint* arXiv:hep-ph/0104145)
[44] Baer H, Paige F E, Protopopescu S D and Tata X 1993 Simulating Supersymmetry With ISAJET 7.0 / ISASUSY 1.0 *Preprint* arXiv:hep-ph/9305342
[45] Paige F E, Protopopescu S D, Baer H and Tata X 2003 ISAJET 7.69: A Monte Carlo event generator for pp, $\bar{p}p$, and e^+e^- reactions *Preprint* arXiv:hep-ph/0312045
[46] Bechtle P, Desch K and Wienemann P 2006 Fittino, a program for determining MSSM parameters from collider observables using an iterative method *Comput. Phys. Commun.* **174** 47–70 (*Preprint* arXiv:hep-ph/0412012)
[47] Lafaye R, Plehn T and Zerwas D 2004 SFITTER: SUSY parameter analysis at LHC and LC *Preprint* arXiv:hep-ph/0404282
[48] Hodgkinson R N 2011 Decoding new physics at 1 fb^{-1} LHC with flavour and CP observables *in these proceedings*
[49] Smillie J M and Webber B R 2005 Distinguishing Spins in Supersymmetric and Universal Extra Dimension Models at the Large Hadron Collider *J. High Energy Phys.* JHEP10(2005)069 (*Preprint* arXiv:hep-ph/0507170)
Datta A, Kong K and Matchev K T 2005 Discrimination of supersymmetry and universal extra dimensions at hadron colliders *Phys. Rev.* D **72** 096006, Erratum-ibid. 2005 **72** 119901 (*Preprint* arXiv:hep-ph/0509246)
[50] Hubisz J, Lykken J, Pierini M and Spiropulu M 2008 Missing energy look-alikes with 100 pb^{-1} at the LHC *Phys. Rev.* D **78** 075008 (*Preprint* arXiv:0805.2398 [hep-ph])
[51] Barr A J 2004 Using lepton charge asymmetry to investigate the spin of supersymmetric particles at the LHC *Phys. Lett.* B **596** 205–212 (*Preprint* arXiv:hep-ph/0405052)
[52] Biglietti M *et al* 2007 Study of second lightest neutralino $\tilde{\chi}_2^0$ spin measurement with ATLAS detector at LHC *ATLAS Note* ATL-PHYS-PUB-2007-004
[53] Baltz E A, Battaglia M, Peskin M E and Wizansky T 2006 Determination of dark matter properties at high-energy colliders *Phys. Rev.* D **74** 103521 (*Preprint* arXiv:hep-ph/0602187)
[54] Mavromatos N E 2008 LHC Physics and Cosmology *Fundamental Interactions: Proc. Lake Louise Winter Institute 2007 (Lake Louise)* ed A Astbury *et al* (Singapore: World Scientific) pp 80–127 (*Preprint* arXiv:0708.0134 [hep-ph])
[55] Ellis J R, Mavromatos N E, Mitsou V A and Nanopoulos D V 2007 Confronting dark energy models with astrophysical data *Astropart. Phys.* **27** 185–198 (*Preprint* arXiv:astro-ph/0604272)
Mitsou V A 2008 Constraints on Dissipative Non-Equilibrium Dark Energy Models from Recent Supernova

Data *Fundamental Interactions: Proc. Lake Louise Winter Institute 2007 (Lake Louise)* ed A Astbury *et al* (Singapore: World Scientific) pp 363–367 (*Preprint* arXiv:0708.0113 [astro-ph])

[56] Lahanas A B, Mavromatos N E and Nanopoulos D V 2007 Smoothly evolving Supercritical-String Dark Energy relaxes Supersymmetric-Dark-Matter Constraints *Phys. Lett.* B **649** 83–90 (*Preprint* arXiv:hep-ph/0612152)

[57] Diamandis G A, Georgalas B C, Lahanas A B, Mavromatos N E and Nanopoulos D V 2006 Dissipative Liouville cosmology: A case study *Phys. Lett.* B **642** 179–186 (*Preprint* arXiv:hep-th/0605181)
Lahanas A B, Mavromatos N E and Nanopoulos D V 2007 Dilaton and off-shell (non-critical string) effects in Boltzmann equation for species abundances *PMC Phys.* A **1** 2 (*Preprint* arXiv:hep-ph/0608153)

[58] Mavromatos N E, Sarkar S and Vergou A 2011 Stringy Space-Time Foam, Finsler-like Metrics and Dark Matter Relics *Phys. Lett.* B **696** 300–304 (*Preprint* arXiv:1009.2880 [hep-th])
Mavromatos N E, Mitsou V A, Sarkar S and Vergou A 2010 Stochastic Finsler D-particle Space-Time Foam Enhances Dark Matter Relics *Preprint* arXiv:1012.4094 [hep-ph]

[59] Dutta B, Gurrola A, Kamon T, Krislock A, Lahanas A B, Mavromatos N E and Nanopoulos D V 2009 Supersymmetry Signals of Supercritical String Cosmology at the Large Hadron Collider *Phys. Rev.* D **79** 055002 (*Preprint* arXiv:0808.1372 [hep-ph])

[60] Mitsou V A 2002 Prospects on Higgs boson discovery and physics beyond the standard model at the LHC and development of the transition radiation tracker of the ATLAS experiment *PhD Thesis* CERN-THESIS-2002-005
Mitsou V A 1999 Observability of the $h \to b\bar{b}$ channel in cascade decay of SUSY particles within the SUGRA model *ATLAS Note* CERN-ATL-PHYS-99-017

[61] Mavromatos N E and Mitsou V A 2008 Observational Evidence for Negative-Energy Dust in Late-Times Cosmology *Astropart. Phys.* **29** 442–452 (*Preprint* arXiv:0707.4671 [astro-ph])
Mavromatos N E and Mitsou V A 2007 Relaxation dark energy in non-critical string cosmologies and astrophysical data *Proc. IDM 2006: 6th Int. Workshop on the Identification of Dark Matter (Island of Rhodes)* ed M Axenides *et al* (Singapore: World Scientific) pp 623–634 (*Preprint* arXiv:astro-ph/0611788)

[62] Mitsou V A 2010 Constraining super-critical string/brane cosmologies with astrophysical data *J. Phys. Conf. Ser.* **203** 012054 (*Preprint* arXiv:0909.5095 [astro-ph.CO])

Journal of Physics: Conference Series **335** (2011) 012004 doi:10.1088/1742-6596/335/1/012004

Identification of Dark Matter Particles

Gianfranco Bertone

Institut d'Astrophysique de Paris, UMR 7095-CNRS, Université Pierre et Marie Curie, 98 bis Boulevard Arago 75014, Paris, France & Institute für Theoretische Physik, Universität Zürich, Winterthurerstrasse 190, CH-8057 Zürich, Switzerland

E-mail: gf.bertone@gmail.com

Abstract. A reliable identification of Dark Matter particles can probably be achieved only through a combined analysis of accelerators, direct and indirect searches. I review here the status of these detection strategies, and discuss how to combine them in a self-consistent way, keeping into account all astrophysical uncertainties.

1. Introduction

Dark Matter (DM) is a fundamental ingredient of the Standard Cosmological model, which provides a satisfactory (albeit not perfect) explanation of a wide range of observations in Astrophysics and Cosmology [1, 2, 3, 4, 5, 6]. In the framework of this model, we can determine the abundance of DM in the Universe with good accuracy [7]; we have an increasingly accurate understanding of how DM is distributed in astrophysical structures, from Dwarf Galaxies up to to Clusters of galaxies, based on high-resolution numerical simulations [8], and to gravitational lensing observations [9]; we even have a rather precise idea of how the Milky Way has formed, and of what the *local* abundance of DM is [10, 11, 12, 13]. Furthermore, we know today that DM cannot be made of ordinary matter, so new particles *must* exist [14], unless we are completely misled by astrophysical and cosmological observations at all scales.

The need for new particles has prompted particle physicists to propose literally tens of possible DM candidates. Axions [15, 16], for instance, are hypothetical particles whose existence was postulated to solve the so called strong CP problem in Quantum Chromo Dynamics, and they are known to be very well motivated DM candidates. Other well-known candidates are Sterile Neutrinos [17, 18], which interact only gravitationally with ordinary matter, apart from a small mixing with the familiar neutrinos of the Standard Model. A wide array of other possibilities have been discussed in the literature, and they are currently being searched for with a variety of experimental strategies [3, 4].

The most studied class of candidates is that of WIMPs (for weakly interacting massive particles), which have the virtue of naturally achieving the correct relic abundance in the early Universe. The reason why they became so popular is that WIMP candidates arise naturally from theories that seek to extend the Standard Model of particle physics, and to embed it in a more fundamental theory. In particular, it was noticed back in 1983 that one of the most promising extensions of the Standard Model, Supersymmetry (SUSY), provides an excellent Dark Matter candidate: the neutralino [19, 20, 21, 22]. This particle fulfills all the properties of the good Dark Matter candidate, and it has become over the years a prototypical example of WIMPs. Its mass can range from about 50 GeV to a few TeV, and its interaction cross section with ordinary

Journal of Physics: Conference Series **335** (2011) 012004 doi:10.1088/1742-6596/335/1/012004

matter and with itself are such that it can account for all the Dark Matter in the Universe while still remaining consistent with all known experiments.

If DM is made of WIMPs, we should be able to detect it. We could in principle observe the interaction of DM particles with nuclei in underground detectors, as proposed back in 1985 [23] , or we may detect the products of annihilation or decay of these particles, as first discussed almost 3 decades ago [24, 25, 4]. Although all the search strategies so far devised have failed to provide incontrovertible evidence for DM particles, today a new generation of particle astrophysics experiments is about to start, or has already started taking data. Furthermore, the Large Hadron Collider at CERN has recently started operations, and it is expected to find, or to severely constrain, the most studied extensions of the Standard Model, including SUSY.

This paper is organized as follows: in Section 2 I will review the prospects to identify Dark Matter at the LHC; in Section 3 I will discuss how to combine LHC data with the results of direct detection experiments; in Section 4 I will review the complementarity of different direct detection targets; in Section 5 I will then discuss indirect searches, and conclude.

2. Accelerator Searches

What can accelerators say about DM? A lot, probably, as the discovery of new particles would mark the beginning of a new era in particle physics and cosmology, while null searches would most probaby lead to the decline of the WIMP paradigm [2]. However, even if new particles are discovered, to prove that they actually constitute the DM will be rather challenging. To illustrate this point, I will review here a specific example in the framework of a 24 parameters Minimal Supersymmetric Standard Model, recently studied in ref. [26].

The basic idea is straightforward: in case of discovery of new Physics at the LHC, the study of different kinematical variables will allow to determine the masses of several particles, with a precision that obviously depends on the properties of the specific point of the parameter space. These measurements can then be used to determine the regions of the SUSY parameter space which are consistent with such a measurement.

This can be done by applying Bayes' theorem

$$p(\mathbf{x}|\mathbf{d}) = \frac{p(\mathbf{d}|\mathbf{x})p(\mathbf{x})}{p(\mathbf{d})}, \tag{1}$$

which updates the so-called prior probability density $p(\mathbf{x})$, encapsulating the knowledge of the 24-dimensional space before taking into account the experimental constraints, $\mathbf{d}$, into the posterior probability function (pdf) $p(\mathbf{x}|\mathbf{d})$. The latter describes the probability density assigned to a generic 24-dimensional point $\mathbf{x}$ once the data have been taken into account via the likelihood function $p(\mathbf{d}|\mathbf{x})$. Furthermore, on the RHS of Eq. 1, $p(\mathbf{d})$ is the Bayesian evidence which, in our case, can be dropped since it simply plays the role of a normalization constant for the posterior in this context.

The marginal pdf of a particular subset (as e.g. only one) of the 24 parameters defining $\mathbf{x}$ can be obtained by integrating over the remaining directions:

$$p(x^i|\mathbf{d}) = \int_{[1,24]\setminus\{i\}} p(\mathbf{x}|\mathbf{d})dx^1...dx^{i-1}dx^{i+1}...dx^{24}. \tag{2}$$

The posterior encodes both the information contained in the priors and in the experimental constraints, but, ideally, it should be largely independent of the choice of priors, so that the posterior inference is dominated by the data contained in the likelihood. If some residual dependence on the prior $p(\mathbf{x})$ remains this should be considered as a sign that the experimental data employed are not constraining enough to override completely different plausible prior choices and therefore the resulting posterior should be interpreted with some care, as it might

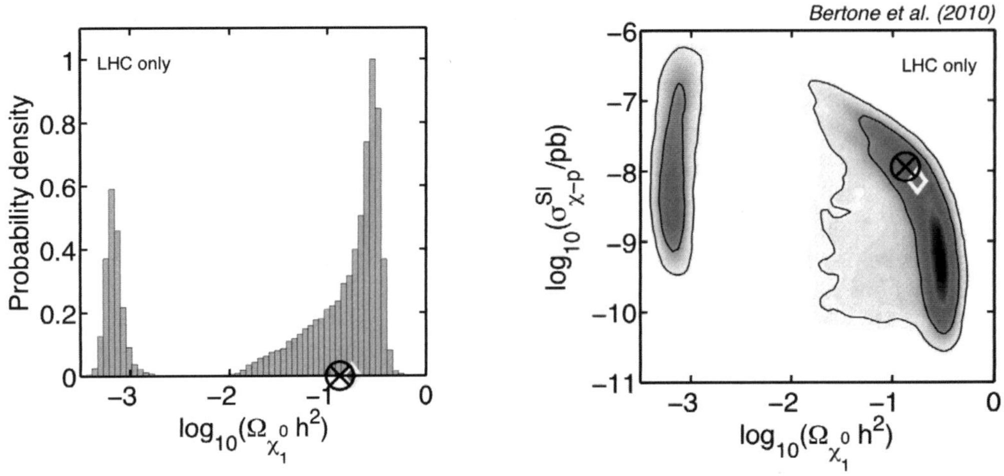

Figure 1. *Left:* Pdf of the neutralino relic density obtained from LHC data only. *Right:* As in the left panel, but in the $\sigma^{\mathrm{SI}}_{\chi-\mathrm{p}}$ vs$\Omega_{\tilde{\chi}_1^0} h^2$ plane. The inner and outer contours enclose 68% and 95% probability regions, respectively. The best fit point is shown by the encircled black cross, while the true value is given by the yellow/red diamond. From Ref. [26].

depend on the prior assumptions. The probability distribution for any observable that is a function of the 24 SUSY parameters $f(\mathbf{x})$ can also be obtained since $p(f|\mathbf{d}) = \delta(f - f(\mathbf{x}))p(\mathbf{x}|\mathbf{d})$.

In ref.[26] we have chosen a specific benchmark point in the MSSM parameter space, corresponding to the low-energy extrapolation of model LCC3 defined in Ref. [27]. This benchmark is representative of SUSY models in the co-annihilation region, where the lightest neutralino is almost degenerate in mass with the lightest stau. In this region, co-annihilation effects reduce the neutralino relic abundance down to values compatible with the results from the WMAP satellite [28], and therefore, the mass difference between the neutralino and the lightest stau is a fundamental parameter for the reconstruction of the relic density. It has been shown [27] that for this benchmark point LHC would be able to provide a measurement of the masses of a good part of the SUSY spectrum, including the two lightest neutralinos (see Ref. [29] for an extension of this analysis to the case of the ILC). However the masses of some particles (most notably the two heaviest neutralinos and both charginos) would not be measured. The set of measurements that we use as constraints in our analysis corresponds to that in Table 6 of Ref. [27] , which assumes an integrated luminosity of 300 fb^{-1}. Furthermore, as pointed out in Ref. [30], the neutralino-stau mass difference can be measured with an accuracy of 20% with 10 fb^{-1} luminosity in models where the squark masses are much larger than those of the lightest chargino and second-lightest neutralino, as is our case. We therefore also included a measurement of the neutralino-stau mass difference in our likelihood.

The reconstruction of the neutralino relic density is shown in Fig. 1. Consistently with previous analyses [27], multiple peaks can be observed, as a consequence of degeneracies in the SUSY parameters space that the LHC constraints are unable to break. In particular, the two observed peaks correspond to neutralinos with different composition: mostly Wino and mostly Bino, from left to right. This is a consequence of the fact that the LHC is assumed to be able to measure only the two lightest neutralino states, but not the two more massive ones or the charginos. The true value of the relic density for our benchmark point ($\Omega_{\tilde{\chi}_1^0} h^2 = 0.176$),

Journal of Physics: Conference Series **335** (2011) 012004 doi:10.1088/1742-6596/335/1/012004

represented by a diamond in Fig. 1, is indeed inside the peak corresponding to mostly Bino dark matter. Although this value is about 60% larger than the relic abundance measured by the WMAP satellite [28], we expect our results to remain qualitatively correct for other points in the co-annihilation region leading to the correct cosmological relic abundance.

The constraints from LHC only data are also shown, in the right panel of Fig. 1, in the plane $\sigma^{\rm SI}_{\chi-{\rm p}}$ vs$\Omega_{\tilde{\chi}^0_1} h^2$, where the true value of those quantities is given by ($\Omega_{\tilde{\chi}^0_1} h^2 = 0.176$, $\sigma^{\rm SI}_{\chi-{\rm p}} = 7.1\times 10^{-8}$ pb). The leftmost region corresponds to a neutralino which has a leading Wino component, thereby displaying a smaller relic abundance, whereas the region towards larger relic abundance corresponds to Bino-like neutralinos, for which the scattering cross section is also slightly smaller.

3. Combining Accelerator Searches with Direct Detection

In view of the results of the previous section, it is important to study complementary strategies, and to investigate whether Dark Matter direct or indirect detection can actually break the strong degeneracy in the SUSY parameter space that can be seen in Fig. 1. In ref. [26] we have simulated the response of a realistic direct detection experiment to the same benchmark point studied before.

We assumed a future signal giving a WIMP detection, namely a certain number of events N and a corresponding set of recoil energies $\{E_i\}_{i=1,\dots,N}$. The total number N of simulated events is the sum of both background events (mainly interactions of detector nuclei with neutrons from surrounding rock, from residual contaminants or from spallation of cosmic muons) and recoils due to DM. For concreteness, we exemplified the method in the case of an experiment akin to the 1-ton scale SuperCDMS experiment. We simulated the differential number of background events as in Ref. [31]. Since the capability of a simulated direct detection experiment to reconstruct the DM properties (see Refs. [32, 33, 31] for more details) is worse in the case of a constant background distribution than for an exponential one, we only consider the case of energy-independent background recoil spectrum in order to be conservative. Therefore, we adopt a constant background differential spectrum $(dN_{\rm back}/dE)$ = const which is normalized so that, when binning the spectrum in 9 bins of 10 keV width (from $E_{\rm th} = 10$ keV to $E_{\rm max} = 100$ keV) the number of background events in the first bin is the same as the number of DM signal events there.

The expected number of events λ for our benchmark model and for an exposure ϵ = 300 ton day is obtained by integrating the sum of the differential rate of WIMP and background events

$$\lambda = \epsilon \int_{E_{\rm th}}^{E_{\rm max}} \frac{dR_\chi}{dE} + \frac{dR_{\rm back}}{dE}\, dE. \tag{3}$$

The dependency of the WIMP event rate on the physical quantities in the problem becomes apparent in the following parametrization [34]

$$\frac{dR_\chi}{dE} = c_1 R_0 e^{-E/(E_0 c_2)} F^2(E)\,, \tag{4}$$

where

$$R_0 = \frac{\sigma^{\rm SI}_{\chi-{\rm p}} \rho_\chi A^2 c^2 (m_\chi + m_p)^2}{\sqrt{\pi} m^3 m_p^2 v_0}\,, \tag{5}$$

and

$$E_0 = \frac{2 m_\chi^2 v_0^2 A m_p}{(m_\chi + A m_p)^2 c^2}\,. \tag{6}$$

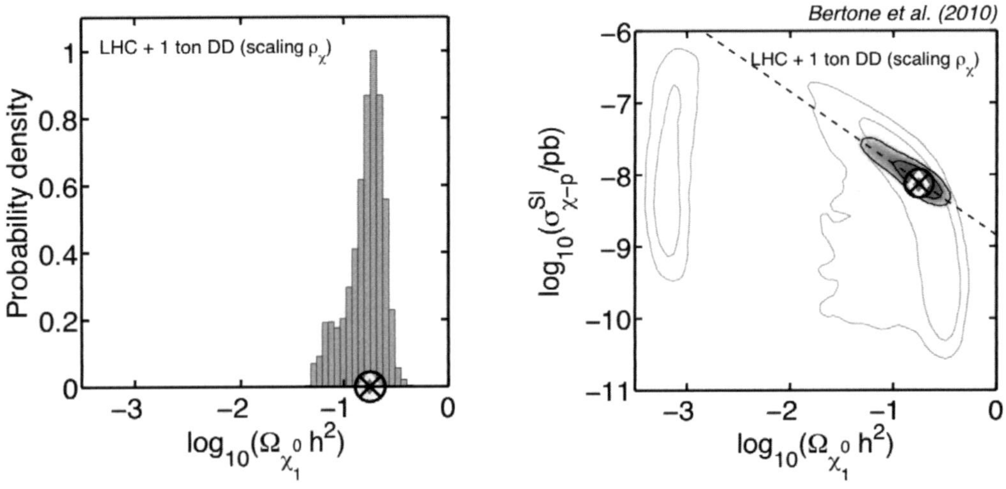

Figure 2. Pdf of the neutralino relic density obtained with a combined analysis of LHC and direct detection data, adopting a scaling Ansatz for the local density. We show for reference the LHC-only contours of the left panel (light grey), along with the direction along which our Ansatz breaks the degeneracy (dashed lines). From Ref. [26].

Here, ρ_χ is the local WIMP density, A is the mass number of the target nuclei ($A = 73$ in the case of Germanium), m_p is the proton mass, v_0 is the characteristic WIMP velocity and $F^2(E)$ denotes the nuclear form factor. A discussion on the values of the parameters c_1 and c_2 and the functional form of $F(E)$ can be found in Refs. [32, 33].

In order to combine the result of a direct detection experiment with LHC data, we run an additional scan of the SUSY parameter space including in the likelihood function an additional Poisson-distributed term that compares the number of events and their spectral shape predicted in each point in parameter space with the recoil spectrum. The overall background rate and its spectral shape are assumed to be known.

As shown by Eqs.3-5, the number of detected events is proportional to the product of the WIMP-proton cross section and the local DM density $\lambda \propto \sigma^{\rm SI}_{\chi-\rm p}\rho_\chi$. Therefore, unless one specifies the value of ρ_χ, any information on the number of events leaves the scattering cross section practically unconstrained.

In Ref.[26] we proposed the following strategy to specify ρ_χ: we assumed that the local density of the neutralino scales with the cosmological abundance. More precisely, we proposed the following *Ansatz*

$$\rho_{\tilde{\chi}^0_1}/\rho_{\rm DM} = \Omega_{\tilde{\chi}^0_1}/\Omega_{\rm DM}. \tag{7}$$

This Ansatz is strictly valid in the reasonable case where the distribution of neutralinos in large structures, and in particular in the Galaxy, traces the cosmological distribution of the DM. This is obviously true if neutralinos contribute all the DM in the Universe, but is also valid in the case where the neutralino is a subdominant component of DM, provided that DM behaves, as expected, as a cold collisionless particle. As shown below, this simple assumption is powerful tool to remove degeneracies in the parameter space.

In Fig. 2, we show the impact of adding information from direct detection experiments to the analysis based on LHC data only. These plots have been obtained by statistical posterior re-sampling of the LHC only scan, adding the relevant Ansätze and the likelihood function of a direct detection experiment as specified above. When the appropriate scaling of the local density is applied, the aforementioned Ansatz cuts the parameter space along a direction

$$\sigma^{\rm SI}_{\chi-{\rm p}} \propto \Omega^{-1}_{\tilde{\chi}^0_1}, \tag{8}$$

due to the fact that for a fixed number of events $\sigma^{\rm SI}_{\chi-{\rm p}} \propto \rho_\chi^{-1}$ and that under the scaling Ansatz $\rho_\chi \propto \Omega_{\tilde{\chi}^0_1}$. The dramatic consequences of this simple Ansatz are shown in the right-most panels of both figures. Models corresponding to a low relic density are essentially ruled out, because under the scaling Ansatz they correspond to a low local density. Given a number of observed events in direct detection searches, a low local density would require a larger scattering cross section, which is incompatible with LHC constraints. As a consequence, the parameter space region corresponding to a neutralino that is mostly Wino can now be ruled out with high confidence.

4. Combining two or more Direct Detection targets

Another interesting question is whether different detection experiments, exploiting different targets, can provide complementary information on the nature of DM particles. In ref. [35] we have discussed the reconstruction of the key phenomenological parameters of WIMPs, namely mass and scattering cross section off nuclei, in case of positive detection with one or more direct DM experiments planned for the next decade. We have in particular studied the complementarity of ton scale experiments with Xe, Ar and Ge targets, adopting experimental configurations that may realistically become available over this timescale.

target	ϵ [ton×yr]	η_{cut}	A_{NR}	ϵ_{eff} [ton×yr]	E_{thr} [keV]	$\sigma(E)$ [keV]	background events/ϵ_{eff}
Xe	5.0	0.8	0.5	2.00	10	Eq. 10	< 1
Ge	3.0	0.8	0.9	2.16	10	Eq. 9	< 1
Ar	10.0	0.8	0.8	6.40	30	Eq. 11	< 1

Table 1. Characteristics of future direct dark matter experiments using xenon, germanium and argon as target nuclei. In all cases the level of background in the fiducial mass region is negligible for the corresponding effective exposure.

Currently, the most stringent constraints on the SI WIMP-nucleon coupling are those obtained by the CDMS [36] and XENON [37] collaborations. While XENON100 should probe the cross-section region down to 5×10^{-45} cm^2 with data already in hand, the XENON1T [38] detector, whose construction is scheduled to start by mid 2011, is expected to reach another order of magnitude in sensitivity improvement. To test the σ^p_{SI} region down to 10^{-47} cm$^2 \equiv 10^{-11}$ pb and below, a new generation of detectors with larger WIMP target masses and ultra-low backgrounds is needed. Since we are interested in the prospects for detection in the next 5 to 10 years, we discuss new projects that can realistically be built on this time scale, adopting the most promising detection techniques, namely noble liquid time projection chambers (TPCs) and cryogenic detectors operated at mK temperatures.

In Europe, two large consortia, DARWIN [39] and EURECA [40], gathering the expertise of several groups working on existing DM experiments are funded for R&D and design studies to push noble liquid and cryogenic experiments to the multi-ton and ton scale, respectively. DARWIN is devoted to noble liquids, having as main goal the construction of a multi-ton liquid Xe (LXe) and/or liquid Ar (LAr) instrument [41], with data taking to start around 2016. The XENON, ArDM and WARP collaborations participate actively in the DARWIN project. EURECA is a design study dedicated to cryogenic dark matter detectors operated at mK temperatures. The proposed roadmap is to improve upon CRESST [42] and EDELWEISS [43] technologies and build a ton-scale detector by 2018, with a SI sensitivity of about 10^{-46} cm$^2 \equiv 10^{-10}$ pb. The complementarity between DARWIN and EURECA is of utmost importance for dark matter direct searches since a solid, uncontroversial discovery requires signals in distinct targets and preferentially distinct technologies. In an international context, two engineering studies (MAX [44] and LZS [45]) are funded in the US for ton to multi-ton scale LXe and LAr TPCs and the SuperCDMS/GEODM collaboration [46] plans to operate an 1.5 ton Ge cryogenic experiment at DUSEL [47]. In Japan, the XMASS experiment [48], using a total of 800 kg of liquid xenon in a single-phase detector, is under commissioning at the Kamioka underground laboratory [49], while a large single-phase liquid argon detector, DEAP-3600 [50], using 3.6 tons of LAr is under construction at SNOLab [51].

Given these developments, we will focus on the three most promising targets: Xe and Ar as examples of noble liquid detectors, and Ge as a case-study for the cryogenic technique. In the case of a Ge target, we assume an 1.5 ton detector (1 ton as fiducial target mass), 3 years of operation, an energy threshold for nuclear recoils of $E_{thr,Ge} = 10$ keV and an energy resolution given by

$$\sigma_{Ge}(E) = \sqrt{(0.3)^2 + (0.06)^2 E/\text{keV}}\ \text{keV} \quad . \tag{9}$$

For a liquid Xe detector, we assume a total mass of 8 tons (5 tons in the fiducial region), 1 year of operation, an energy threshold for nuclear recoils of $E_{thr,Xe} = 10$ keV and an energy resolution of

$$\sigma_{Xe}(E) = 0.6\ \text{keV}\sqrt{E/\text{keV}} \quad . \tag{10}$$

Finally, for a liquid Ar detector, we assume a total mass of 20 tons (10 tons in the fiducial region), 1 year of operation, an energy threshold for nuclear recoils of $E_{thr,Ar} = 30$ keV and an energy resolution of

$$\sigma_{Ar}(E) = 0.7\ \text{keV}\sqrt{E/\text{keV}} \quad . \tag{11}$$

To calculate realistic exposures, we make the following assumptions: nuclear recoils acceptances A_{NR} of 90%, 80% and 50% for Ge, Ar and Xe, respectively, and an additional, overall cut efficiency η_{cut} of 80% in all cases, which for simplicity we consider to be constant in energy. We hypothesise less than one background event per given effective exposure ϵ_{eff}, which amounts to 2.16 ton×yr in Ge, 6.4 ton×yr in Ar and 2 ton×yr in Xe, after allowing for all cuts. Such an ultra-low background will be achieved by a combination of background rejection using the ratio of charge-to-light in Ar and Xe, and charge-to-phonon in Ge, the timing characteristics of raw signals, the self-shielding properties and extreme radio-purity of detector materials, as well as minimisation of exposure to cosmic rays above ground.

The described characteristics are summarised in Table 1. We note that in the following we shall consider recoil energies below 100 keV only; to increase this maximal value may add some information but the effect is likely small given the exponential nature of WIMP-induced recoiling spectra.

We start by assuming three dark matter benchmark models with $m_\chi = 25, 50, 250$ GeV and $\sigma_{SI}^p = 10^{-9}$ pb, and fix the Galactic model parameters to their fiducial values, $\rho_0 = 0.4$ GeV/cm^3, $v_0 = 230$ km/s, $v_{esc} = 544$ km/s, $k = 1$. We then generate mock data that are in turn used to reconstruct the posterior for the DM parameters m_χ and σ_{SI}^p. The left frame of Fig. 3 presents the results for the three benchmarks and for Xe, Ge and Ar separately. Contours in the figure delimit regions of joint 68% and 95% posterior probability. Several comments are in order here. First, it is evident that the Ar configuration is less constraining than Xe or Ge ones, which can be traced back to its smaller A and larger E_{thr}. Moreover, it is also apparent that, while Ge is the most effective target for the benchmarks with $m_\chi = 25, 250$ GeV, Xe appears the best for a WIMP with $m_\chi = 50$ GeV (see below for a detailed discussion). Taking into account the differences in adopted values and procedures, our results are in qualitative agreement with Ref. [52], where a study on the supersymmetrical framework was performed. However, it is worth noticing that the contours in Ref. [52] do not extend to high masses as ours for the 250 GeV

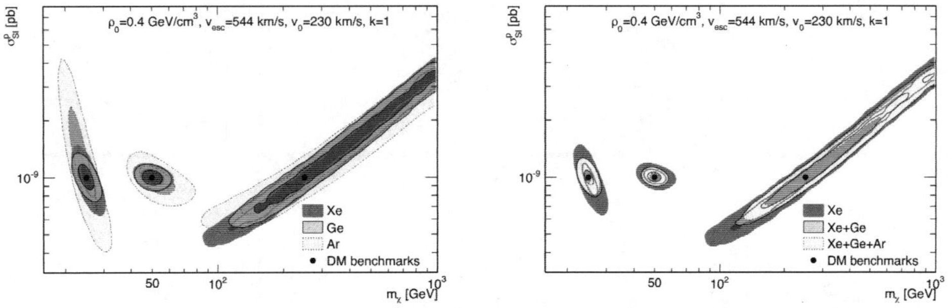

Figure 3. The joint 68% and 95% posterior probability contours in the $m_\chi - \sigma_{SI}^p$ plane for the three DM benchmarks ($m_\chi = 25, 50, 250$ GeV) with fixed astrophysical parameters. In the left frame we show the reconstruction capabilities of Xe, Ge and Ar configurations separately, whereas in the right frame the combined data sets Xe+Ge and Xe+Ge+Ar are shown.

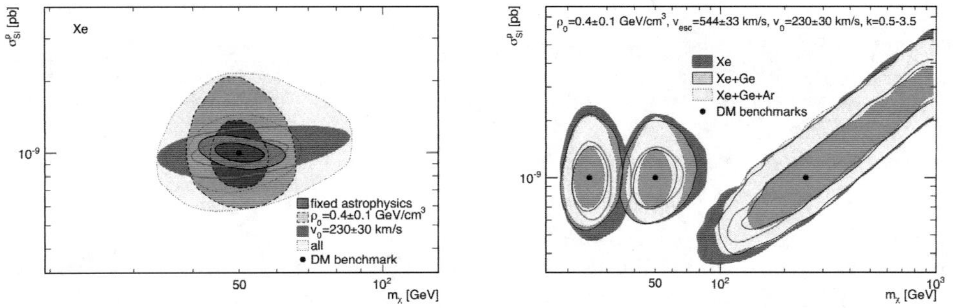

Figure 4. The joint 68% and 95% posterior probability contours in the $m_\chi - \sigma_{SI}^p$ plane for the case in which astrophysical uncertainties are taken into account. In the left frame, the effect of marginalising over ρ_0, v_0 and all four (ρ_0, v_0, v_{esc}, k) astrophysical parameters is displayed for a Xe detector and the 50 GeV benchmark WIMP. In the right frame, the combined data sets Xe+Ge and Xe+Ge+Ar are used for the three DM benchmarks ($m_\chi = 25, 50, 250$ GeV).

	Percent 1σ accuracy	
	$m_\chi = 25$ GeV	$m_\chi = 50$ GeV
Xe	6.5% (14.3%)	8.1% (20.4%)
Ge	5.5% (16.0%)	7.0% (29.6%)
Ar	12.3% (23.4%)	14.7% (86.5%)
Xe+Ge	3.9% (10.9%)	5.2% (15.2%)
Xe+Ge+Ar	3.6% (9.0%)	4.5% (10.7%)

Table 2. Marginalised percent 1σ accuracy of the DM mass reconstruction for the benchmarks $m_\chi = 25, 50$ GeV. Figures between brackets refer to scans where the astrophysical parameters were marginalised over, while the other figures refer to scans with the fiducial astrophysical setup.

benchmark – this is likely because the volume at high masses in a supersymmetrical parameter space is small.

In the right frame of Fig. 3 we show the reconstruction capabilities attained if one combines Xe and Ge data, or Xe, Ge and Ar together, *again for when the Galactic model parameters are kept fixed.* In this case, for $m_\chi = 25, 50$ GeV, the configuration Xe+Ar+Ge allows the extraction of the correct mass to better than $\mathcal{O}(10)$ GeV accuracy. For reference, the (marginalised) mass accuracy for different mock data sets is listed in Table 2. For $m_\chi = 250$ GeV, it is only possible to obtain a lower limit on m_χ.

Fig. 4 shows the results of a more realistic analysis, that keeps into account the large

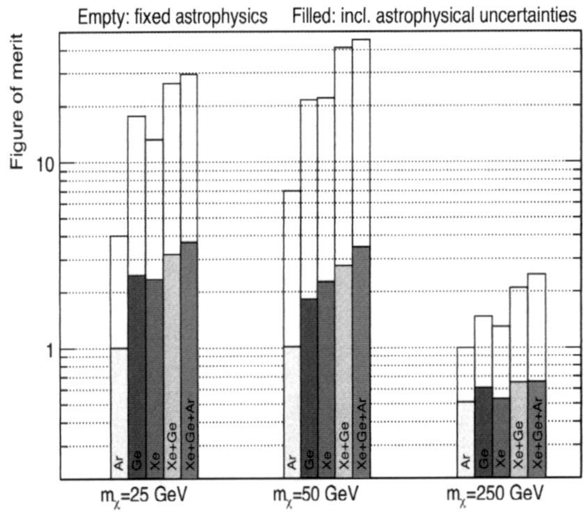

Figure 5. Figure of merit quantifying the relative information gain on Dark Matter parameters for different targets and combinations thereof. The values of the figure of merit are normalised to the Ar case at $m_\chi = 250$ GeV with fixed astrophysical parameters. Empty (filled) bars are for fixed astrophysical parameters (including astrophysical uncertainties).

uncertainties associated with Galactic model parameters. The left frame of Fig. 4 shows the effect of varying only ρ_0 (dashed lines, blue surfaces), only v_0 (solid lines, red surfaces) and all Galactic model parameters (dotted lines, yellow surfaces) for Xe and $m_\chi = 50$ GeV. The Galactic model uncertainties are dominated by ρ_0 and v_0, and, once marginalised over, they blow up the constraints obtained with fixed Galactic model parameters. This amounts to a very significant degradation of mass (cf. Table 2) and scattering cross-section reconstruction. Inevitably, the complementarity between different targets is affected – see the right frame of Fig. 4. Still, for the 50 GeV benchmark, combining Xe, Ge and Ar data improves the mass reconstruction accuracy with respect to the Xe only case, essentially by constraining the high-mass tail.

In order to be more quantitative in assessing the usefulness of different targets and their complementarity, we use as *figure of merit* the inverse area enclosed by the 95% marginalised contour in the $\log_{10}(m_\chi) - \log_{10}(\sigma_{SI}^p)$ plane. Fig. 5 displays this figure of merit for several cases, where we have normalised to the Ar target at $m_\chi = 250$ GeV with fixed Galactic model. Analyses with fixed Galactic model parameters are represented by empty bars, while the cases where all Galactic model parameters are marginalised over are represented by filled bars. Firstly, one can see that all three targets perform better for WIMP masses around 50 GeV than 25 or 250 GeV if the Galactic model is fixed. When astrophysical uncertainties are marginalised over, the constraining power of the experiments becomes very similar for benchmark WIMP masses of 25 and 50 GeV. Secondly, Fig. 5 also confirms what was already apparent from Fig. 3: Ge is the best target for $m_\chi = 25, 250$ GeV (although by a narrow margin), whereas Xe appears the most effective for a 50 GeV WIMP (again, by a narrow margin). Furthermore, the inclusion of uncertainties drastically reduces the amount of information one can extract from the data: the filled bars are systematically below the empty ones. Now, astrophysical uncertainties affect the complementarity between different targets in a non-trivial way. To understand this point, let us focus on the two rightmost bars for each benchmark in Fig. 5, corresponding to the data sets Xe+Ge and Xe+Ge+Ar. For instance, in the case of a 250 GeV WIMP, astrophysical uncertainties seem to reduce target complementarity: adding Ar to Xe+Ge leads to a significant increase in the figure of merit for analyses with fixed astrophysics (empty bars) but has a

Journal of Physics: Conference Series **335** (2011) 012004 doi:10.1088/1742-6596/335/1/012004

negligible effect for analyses with varying astrophysical parameters (filled bars). For low mass benchmarks, the effect of combining two (Xe+Ge) or three targets (Xe+Ge+Ar) is to increase the figure of merit by about a factor of 2 compared to Xe alone or Ge alone, almost independently of whether the astrophysical parameters are fixed or marginalised over. However, the overall information gain on the Dark Matter parameters (for light WIMPs) is reduced by a factor ~ 10 if astrophysical uncertainties are taken into account, compared to the case where the Galactic model is fixed.

5. Indirect detection

Indirect detection consists in the search for the annihilation or decay products of DM particles, such as photons, anti-matter and neutrinos. WIMPs in fact are expected to annihilate efficiently in regions where they accumulate, such as the center of galactic halos, or substructures such as dwarf galaxies, since the annihilation rate depends on the number density *squared*. Once they annihilate, they produce secondary particles, such as quarks and gauge bosons, which subsequently fragment and decay in the aforementioned final states. The typical energy of these final states is about a tenth of the DM particle mass, so we can search indirectly for DM by looking for an excess of photons, anti-matter or neutrinos in astrophysical data at energies between 1 GeV and 10 TeV [4].

Although in principle interesting, obtaining convincing evidence from astrophysical observations has proven a very difficult task. It is in fact easy to fit almost *any* excess in the measured energy spectrum of photons or anti-matter, at any energy, in terms of DM particles with suitable properties. One simply has to follow three easy steps: *i)* adjust the normalization of the flux by changing the distribution of DM particles and their annihilation cross section *ii)* choose a WIMP mass that provides the correct energy scale *iii)* fit the spectral features by choosing an appropriate annihilation channel and, in the case of anti-matter, by tuning the propagation parameters.

In practice, one has enough freedom to fit almost any astrophysical observation, and in fact features in the data of many experiments of the last 5-10 years, have been tentatively interpreted in terms of different DM candidates, sometimes even at the cost of making unrealistic assumptions on the nature and distribution of DM. The most recent example is the rise in the energy spectrum of the positron ratio measured by PAMELA above 10 GeV [53]. The standard WIMP model (i.e. a particle with a mass in the $10^2 - 10^3$ GeV range and a *thermal* cross section, $\sigma v \sim 10^{-26}$ cm^3 s^{-1}) can hardly account for this feature, so new ad-hoc candidates have been proposed: particles with a very large annihilation cross section (high enough to match the normalization of the positron ratio, but not too much, in order to avoid cosmological constraints [54, 55]), annihilating only to leptons (to evade anti-protons constraints [56]), and with a density profile shallower than what suggested by numerical simulations (to evade gamma-ray constraints from the Galactic center [57]). There is therefore a possible combination of parameters that can be made compatible with all observations, but this is certainly not enough to claim discovery of Dark Matter, for there are less exotic astrophysical sources that can account for the same feature without invoking new particles with ad-hoc properties.

We have shown that among the most stringent constraints on the properties of DM particle, and in particular on the DM interpretation of the PAMELA data, the most robust are provided by cosmic microwave background (CMB) observations [54, 55], since they do not depend on poorly known quantities such as the DM profile at the Galactic center, or the details of the propagation of anti-matter in the Galaxy.

Fortunately, there are actually a number of astrophysical observations that might lead to *convincing* evidence, in the sense that they could be explained *only* in terms of Dark Matter, while being incompatible with a standard astrophysical interpretation. A typical example of smoking-gun evidence is the observation of a high-energy gamma-ray line, that would point directly to the existence of new particles annihilating directly to photons. In fact, if WIMPs do not produce photons through the fragmentation and decay of secondary particles, but directly, the photons produced in the annihilation will be mono-energetic, thus producing a line in the gamma-ray spectrum at an energy equal to the mass of the Dark Matter particle. The Fermi LAT satellite has however put stringent constraints on the possibility to observe lines [58], excluding annihilation cross section to photons higher than the thermal cross section, and we can expect an improvement in sensitivity of one order of magnitude at most over the next decade (at least in the energy range where the sensitivity is limited by statistics, and not by the background, in which case the sensitivity scales with the square root of time).

Another very clean signature of Dark Matter annihilations would be the observation of high-energy neutrinos from the center of the Sun. Solar neutrinos produced in nuclear reactions have in fact energies in the MeV range, so the observation of 10^2 - 10^4 GeV neutrinos would require an explanation in terms of new physics, and the well studied process of capture and annihilation of Dark Matter particles in the Sun would provide it. The problem is that the neutrino telescope IceCube, currently under construction at the South Pole, so far has not found any evidence for an excess of neutrinos from the Sun, and in the next 5 years, the experiment will improve its sensitivity only by a factor of ~ 5, and extend the threshold down to 50 GeV, with the construction of a more densely instrumented portion of detector, called DeepCore [59]. Even with this technical improvements and longer exposure, most of the Supersymmetry parameter space will remain inaccessible, and the same holds true for the so called Kaluza–Klein Dark Matter in theory with universal extra-dimensions.

Other strategies may provide useful hints, such as the multi-wavelength approach, that consists in the combined analysis of astrophysical spectra at different wavelengths, with the aim of observing e.g. the Synchrotron and Inverse Compton emission produced by secondary electrons produced along with gamma-rays by Dark Matter annihilations [60]. Or the study of the angular power spectrum of gamma-ray anisotropies [61], that may allow to identify a Dark Matter contribution to the diffuse gamma-ray background. But even in case of detection, it would probably require a long time before these observations are considered as proof of the existence of Dark Matter, because one would have to exclude an astrophysical origin of the signal. Fortunately, although indirect searches may appear to be not particularly suited to provide incontrovertible evidence for Dark Matter, they have the big advantage of not requiring *dedicated* experiments, and that some theoretical models are indeed within the reach of current and upcoming experiments in the next 5–10 years. In absence of these (admittedly optimistic) smoking-gun observations, a convincing case for Dark Matter can be made only in case of successful searches at accelerators or direct detection experiments, in which case indirect searches may still provide useful information on the *distribution* of Dark Matter.

Acknowledgements. This paper summarizes work done in collaboration with L. Baudis, D. Cerdeno, M. Fornasa, M. Pato, R. Ruiz de Austri, L. Strigari, R. Trotta in Refs. [2, 26, 35].

[1] *Particle Dark Matter: Observations, Models and Searches*, ed. G. Bertone, 2010, Cambridge University Press
[2] G. Bertone, Nature **468** (2010) 389-393. [arXiv:1011.3532 [astro-ph.CO]].
[3] Bergstrom L., Rept. Prog. Phys. **63**, 793-841 [arXiv:hep-ph/0002126] (2000)
[4] G. Bertone, D. Hooper and J. Silk, Phys. Rept. **405**, 279 (2005) [arXiv:hep-ph/0404175].
[5] C. Munoz, Int. J. Mod. Phys. **A19**, 3093-3170 (2004). [hep-ph/0309346].
[6] D. Hooper, S. Profumo, Phys. Rept. **453**, 29-115 (2007). [hep-ph/0701197].
[7] Komatsu E. *et al.*, arXiv:1001.4538 [astro-ph.CO]. (2010)
[8] Diemand J. and Moore B., "Simulations of CDM haloes," in *Particle Dark Matter: Observations, Models and Searches*, ed. G. Bertone, p. 14-37, Cambridge University Press (2010)
[9] Mellier Y., "Gravitational lensing and dark matter," in *Particle Dark Matter: Observations, Models and Searches*, ed. G. Bertone, p.56-82, Cambridge University Press (2010)
[10] Catena R. and Ullio P., JCAP **1008** 004 arXiv:0907.0018 [astro-ph.CO]. (2010)

[11] Pato M., Agertz O., Bertone G., Moore B. and Teyssier R., Phys. Rev. D **82** 023531, arXiv:1006.1322 [astro-ph.HE] (2010)
[12] P. Salucci, F. Nesti, G. Gentile *et al.*, Astron. Astrophys. **523** (2010) A83. [arXiv:1003.3101 [astro-ph.GA]].
[13] L. E. Strigari, R. Trotta, JCAP **0911** (2009) 019. [arXiv:0906.5361 [astro-ph.HE]].
[14] Taoso M., Bertone G. and Masiero A., JCAP **0803**, 022 [arXiv:0711.4996 [astro-ph]].
[15] Sikivie P., "Axions," in *Particle Dark Matter: Observations, Models and Searches*, ed. G. Bertone, p. 204-227, Cambridge University Press (2010)
[16] Visinelli L. and Gondolo P., Phys. Rev. D **80**, 035024 [arXiv:0903.4377 [astro-ph.CO]]. (2009)
[17] Shaposhnikov M., "Sterile Neutrinos," in *Particle Dark Matter: Observations, Models and Searches*, ed. G. Bertone, p. 228-248, Cambridge University Press (2010)
[18] Boyarsky A., Ruchayskiy O. and Shaposhnikov M., Ann. Rev. Nucl. Part. Sci. **59**, 191-214 [arXiv:0901.0011 [hep-ph]]. (2009)
[19] Goldberg H., Phys. Rev. Lett. **50**, 1419-1422 (1983) [Erratum-ibid. **103**, 099905 (2009)].
[20] Ellis J. R., Hagelin J. S., Nanopoulos D. V., Olive K. A. and Srednicki M., Nucl. Phys. B **238** 453-476 (1984)
[21] Blumenthal G.R., Faber S.M., Primack J.R. & Rees M.J., Nature 311, 517-525 (1984)
[22] Jungman G. , Kamionkowski M. and Griest K., *Phys. Rept.* **267** 195-373 [arXiv:hep-ph/9506380] (1996)
[23] Goodman M.W. and Witten E., Phys. Rev., D31:3059 (1985)
[24] Silk J. and Srednicki M., Phys. Rev. Letters 53, 624-627 (1984).
[25] Silk J., Olive K.A. and Srednicki M., Phys. Rev. Lett. **55** 257-259 (1985)
[26] G. Bertone, D. G. Cerdeno, M. Fornasa *et al.*, Phys. Rev. **D82** (2010) 055008. [arXiv:1005.4280 [hep-ph]].
[27] E. A. Baltz, M. Battaglia, M. E. Peskin and T. Wizansky, Phys. Rev. D **74** (2006) 103521 [arXiv:hep-ph/0602187].
[28] E. Komatsu *et al.* [WMAP Collaboration], Astrophys. J. Suppl. **180** (2009) 330 [arXiv:0803.0547 [astro-ph]].
[29] V. Khotilovich, R. L. Arnowitt, B. Dutta and T. Kamon, Phys. Lett. B **618** (2005) 182 [arXiv:hep-ph/0503165].
[30] R. L. Arnowitt, B. Dutta, A. Gurrola, T. Kamon, A. Krislock and D. Toback, Phys. Rev. Lett. **100** (2008) 231802 [arXiv:0802.2968 [hep-ph]].
[31] N. Bernal, A. Goudelis, Y. Mambrini and C. Munoz, JCAP **0901** (2009) 046 [arXiv:0804.1976 [hep-ph]].
[32] A. M. Green, JCAP **0708** (2007) 022 [arXiv:hep-ph/0703217].
[33] A. M. Green, JCAP **0807** (2008) 005 [arXiv:0805.1704 [hep-ph]].
[34] J. D. Lewin and P. F. Smith, Astropart. Phys. **6** (1996) 87.
[35] M. Pato, L. Baudis, G. Bertone *et al.*, [arXiv:1012.3458 [astro-ph.CO]].
[36] Z. Ahmed *et al.* [The CDMS-II Collaboration], Science **327**, 1619 (2010) [arXiv:0912.3592 [astro-ph.CO]].
[37] E. Aprile *et al.* [XENON100 Collaboration], Phys. Rev. Lett. **105**, 131302 (2010) [arXiv:1005.0380 [astro-ph.CO]].
[38] E. Aprile et al. (XENON1T Collaboration), XENON1T at LNGS, Proposal, April (2010) and Technical Design Report, October (2010).
[39] `http://darwin.physik.uzh.ch/`
[40] `http://www.eureca.ox.ac.uk/`
[41] L. Baudis (for the DARWIN Consortium), in Proceedings of Science, PoS(IDM2010)122 (2010).
[42] `http://www.cresst.de/`
[43] `http://edelweiss.in2p3.fr/`
[44] `http://www.fnal.gov/pub/max/index.html`
[45] D. N. McKinsey, Journal of Physics, Conference Series 203 (2010) 012026
[46] P. Brink, Talk at DM2010, Marina del Rey (2010) `http://www.physics.ucla.edu/hep/dm10/talks/brink.pdf`
[47] `http://www.dusel.org/`
[48] Hiroyuki Sekiya (for the XMASS collaboration), in proceedings of the 1st International Workshop towards the Giant Liquid Argon Charge Imaging Experiment, arXiv:1006.1473 (2010)
[49] `http://www-sk.icrr.u-tokyo.ac.jp/index-e.html`
[50] `http://deapclean.org/`
[51] `http://www.snolab.ca/`
[52] Y. Akrami, C. Savage, P. Scott, J. Conrad and J. Edsjo, arXiv:1011.4318 [astro-ph.CO].
[53] Adriani O. *et al.* [PAMELA Collaboration], "An anomalous positron abundance in cosmic rays with energies 1.5-100 GeV," Nature **458** 607-609 [arXiv:0810.4995 [astro-ph]] (2009)
[54] Galli S., Iocco F., Bertone G. and Melchiorri A., "CMB constraints on Dark Matter models with large annihilation cross-section," Phys. Rev. D **80** 023505 [arXiv:0905.0003 [astro-ph.CO]] (2009)
[55] Slatyer T.R., Padmanabhan N. and Finkbeiner D.P., "CMB Constraints on WIMP Annihilation: Energy Absorption During the Recombination Epoch," Phys. Rev. D **80**, 043526 [arXiv:0906.1197 [astro-ph.CO]] (2009)

[56] Cirelli M., Kadastik M., Raidal M. and Strumia A., "Model-independent implications of the e+, e-, anti-proton cosmic ray spectra on properties of Dark Matter," Nucl. Phys. B **813**, 1-21 [arXiv:0809.2409 [hep-ph]]. (2009)

[57] Bertone G., Cirelli M. , Strumia A. and Taoso M., "Gamma-ray and radio tests of the e+e- excess from DM annihilations," JCAP **0903**, 009 [arXiv:0811.3744 [astro-ph]]. (2009)

[58] Abdo A. A. *et al.*,"Fermi LAT Search for Photon Lines from 30 to 200 GeV and Dark Matter Implications," Phys. Rev. Lett. **104**, 091302 [arXiv:1001.4836 [astro-ph.HE]]. (2010)

[59] Halzen F. and Hooper D., "The Indirect Search for Dark Matter with IceCube," New J. Phys. **11** 105019 [arXiv:0910.4513 [astro-ph.HE]]. (2009)

[60] Profumo S. and Ullio P., "Multi-wavelength Searches for Particle Dark Matter," arXiv:1001.4086 [astro-ph.HE], in *Particle Dark Matter: Observations, Models and Searches*, ed. G. Bertone, p. 547-564, Cambridge University Press (2010)

[61] Ando S. and Komatsu E., "Anisotropy of the cosmic gamma-ray background from dark matter Phys. Rev. D **73** 023521 [arXiv:astro-ph/0512217]. (2006)

Symposium on Prospects in the Physics of Discrete Symmetries IOP Publishing
Journal of Physics: Conference Series **335** (2011) 012005 doi:10.1088/1742-6596/335/1/012005

Search for CP and CPT violation effects in neutrino oscillations

Mauro Mezzetto

Istituto Nazionale di Fisica Nucleare, Via Marzolo 8, 35131 Padova, Italy

E-mail: mezzetto@pd.infn.it

Abstract. The present generation of long baseline neutrino experiments, both at accelerators and at reactors, is looking (or it's just going to look) for oscillation signals generated by the mixing parameter θ_{13}. The knowledge of this angle is a fundamental milestone to optimize further experiments aimed at detecting CP violation in the neutrino sector. Leptonic CP violation is a key phenomenon that has profound implications in particle physics and cosmology but it is clearly out of reach for the aforementioned experiments. Since late 90's, a world-wide activity is in progress to design facilities that can access CP violation in neutrino oscillation and perform high precision measurements of the lepton counterpart of the Cabibbo-Kobayashi-Maskawa matrix. The experimental status of searches or hints for CPT violation in the neutrino sector will be briefly summarized too.

1. Neutrino Oscillations

Neutrino oscillation experiments had been very successful in the past 15 years. Neutrino oscillations had been discovered in 1998 by the Super Kamiokande experiment [1] by analyzing atmospheric neutrinos. The measurement was confirmed by Macro[2] and Soudan II[3] . The same oscillations had been then measured by exploiting artificial neutrino beams by the K2K [3] and MINOS [4] experiments. In 2002 the SNO experiment [5] settled the solar neutrino puzzle by demonstrating, in a model independent way, that it was generated by neutrino oscillations, concluding a long standing experimental saga initiated by Ray Davis in the late sixties[6] and continued by Kamiokande[7], SuperKamiokande[8] and the two gallium experiments Gallex-GNO[9, 10] and Sage[11, 12]. Soon after SNO, the KamLAND [13] experiment detected the same kind of oscillation acting on reactor antineutrinos.

The discovery of neutrino oscillations establishes beyond doubt that neutrinos have mass and mix. This existence of neutrino masses is in fact the first solid experimental fact requiring physics beyond the Standard Model.

Neutrino oscillations are consistently described by three families ν_1, ν_2, ν_3 with mass values m_1, m_2 and m_3 that are connected to the flavor eigenstates ν_e, ν_μ and ν_τ by a mixing matrix U. The neutrino oscillation probability depends on three mixing angles, $\theta_{12}, \theta_{23}, \theta_{13}$; two mass differences, $\Delta m^2_{12} = m^2_2 - m^2_1$, $\Delta m^2_{23} = m^2_3 - m^2_2$, and a CP phase $\delta_{\rm CP}$. Additional phases are present in case neutrinos are Majorana particles, but they do not influence neutrino flavor oscillations at all.

The best-fit values and allowed range of values of the oscillation parameters at different CL, as obtained in [14], are shown in Table 1.

Table 1. Best-fit values, 2σ, and 3σ intervals (1 dof) for the three flavor neutrino oscillation parameters from global data including solar, atmospheric, reactor and accelerator experiments [14].

parameter	best fit	2σ	3σ
Δm^2_{21} $[10^{-5}eV^2]$	$7.59^{+0.23}_{-0.18}$	7.22–8.03	7.03–8.27
$\|\Delta m^2_{31}\|$ $[10^{-3}eV^2]$	$2.40^{+0.12}_{-0.11}$	2.18–2.64	2.07–2.75
$\sin^2\theta_{12}$	$0.318^{+0.019}_{-0.016}$	0.29–0.36	0.27–0.38
$\sin^2\theta_{23}$	$0.50^{+0.07}_{-0.06}$	0.39–0.63	0.36–0.67
$\sin^2\theta_{13}$	$0.013^{+0.013}_{-0.009}$	≤ 0.039	≤ 0.053

The results from the LSND experiment [15], reporting a 3.6σ evidence of $\bar{\nu}_\mu \to \bar{\nu}_e$ oscillations with a third, very distinct, neutrino mass difference: $\Delta m^2_{LSND} \sim 0.3 - 6$ eV^2, cannot be accommodated in a three neutrino scheme. They have been recently confirmed by the MiniBooNE collaboration [16] (the same collaboration did not detect any $\nu_\mu \to \nu_e$ transition [17]). To explain the whole set of data at least four different light neutrino species would have been needed (referred as "3+1" models, where the first three are active neutrinos and the fourth is a sterile neutrino). These $3+1$ models provided anyway poor overall fits to the existing global experimental data [18]. Very recently an updated full computation of the antineutrino fluxes from reactors [19] suggested that the past short baseline experiments at reactors should be interpreted as an evidence of $\overline{\nu}_e$ disappearance [20]. As a consequence the latest fits to the global neutrino data prefer a "3+2" neutrino scheme over the conventional models with three neutrinos [21].

Three parameters (out of seven) have not yet been measured in neutrino oscillations.

The mixing angle θ_{13} is the key parameter of three-neutrino oscillations and regulates at the first order all the oscillation processes that could contribute to the measurement of mass hierarchy and leptonic CP violation.

The neutrino mass hierarchy, the order by which mass eigenstates are coupled to flavor eigenstates, can be fixed by measuring the sign of Δm^2_{23}. Its value could be $+1$ (normal hierarchy), in which case ν_e would be the lightest neutrino, or -1 (inverted hierarchy), for which ν_e would be the heaviest. Its value is of great importance for double-beta decay experiments [22] and it could shed light on possible flavour symmetries.

The CP phase $\delta_{\rm CP}$ is the ultimate goal of neutrino oscillation searches. The demonstration of CP violation in the lepton sector and the knowledge of the value of this phase would be crucial to understand the origin of the baryon asymmetry in the universe, providing a strong indication, though not proof, that leptogenesis is the explanation for the observed baryon asymmetry of the Universe [23].

All these parameters can be measured via subleading $\nu_\mu \to \nu_e$ oscillations that represent the key process of any future new discovery in neutrino oscillation physics.

1.1. Sub Leading $\nu_\mu \to \nu_e$ oscillations

The $\nu_\mu \to \nu_e$ transition probability can be parametrized in different ways, see for instance [24],[25], where terms driven by θ_{13} appear together with CP-even and CP-odd terms, terms driven by the solar parameters and matter effects. A sketch of $P(\nu_\mu \to \nu_e)$ as a function of the baseline L for 1 GeV neutrinos is shown in Fig. 1.

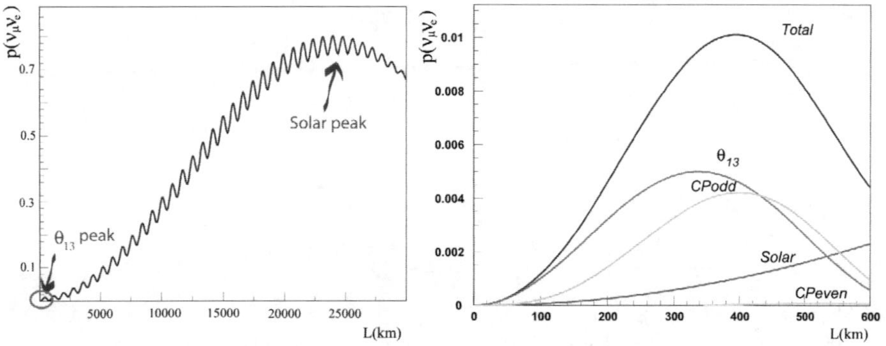

Figure 1. Sketch of $P(\nu_\mu \to \nu_e)$ as function of the baseline computed for monochromatic neutrinos of 1 GeV in the solar baseline regime for $\delta_{\rm CP}$=0 (left) and in the atmospheric baseline regime for $\delta_{\rm CP} = -\pi/2$ (right), where the different terms of $P(\nu_\mu \to \nu_e)$ are displayed. The following oscillation parameters were used in both cases: $\sin^2 2\theta_{13} = 0.01$, $\sin^2 2\theta_{12} = 0.8$, $\Delta m^2_{23} = 2.5{\cdot}10^{-3}$ eV^2, $\Delta m^2_{12} = 7{\cdot}10^{-5}$ eV^2. From Ref. [26].

1.2. Leptonic CP violation
The ultimate challenge of oscillation neutrino physics will be to determine whether CP is violated or not in neutrino oscillations. The experimental information relies on the fact that some terms of $P(\nu_\mu \to \nu_e)$ change sign by exchanging neutrinos with antineutrinos (i.e. the CP-odd term in Fig 1). In this way the probability of oscillation for neutrino will result different from the probability of antineutrinos. This allows to build a CP-violating asymmetry A_{CP}:

$$A_{CP} = \frac{P(\nu_\mu \to \nu_e) - P(\overline{\nu}_\mu \to \overline{\nu}_e)}{P(\nu_\mu \to \nu_e) + P(\overline{\nu}_\mu \to \overline{\nu}_e)} \tag{1}$$

displayed in Figure 2 as a function of θ_{13}, or the equivalent time reversal asymmetry A_T.

The richness of the $\nu_\mu \to \nu_e$ transition is also its weakness: it will be difficult to extract all the genuine parameters unambiguously. Due to the three-flavor structure of the oscillation probabilities, for a given experimental result several different disconnected regions of the multi-dimensional space of parameters could fit the experimental data, originating degenerate solutions. Traditionally these degeneracies are referred to as the intrinsic or $(\delta_{\rm CP}, \theta_{13})$ degeneracy [25]; the hierarchy or sign(Δm^2_{23})-degeneracy [28]; the octant or θ_{23}-degeneracy [29]. These lead to an eight-fold ambiguity in θ_{13} and $\delta_{\rm CP}$ [30], and hence degeneracies provide a serious limitation for the determination of θ_{13}, $\delta_{\rm CP}$, and sign(Δm^2_{23}).

2. Searches of non vanishing values of θ_{13}
The first objective of neutrino oscillation experiments is to look for non-vanishing θ_{13} values. This kind of searches can be performed by accelerator and by reactor experiments and will be briefly discussed in the following. For a comprehensive review of this subject see [31, 32].

Accelerator experiments can measure θ_{13} by detecting the appearance of ν_e neutrinos in accelerator neutrino beams.

Neutrino beams are produced by the decay of π and K mesons generated by a high energy proton beam hitting small Z, needle-shaped, segmented targets, see Fig. 3 . Positive (negative)

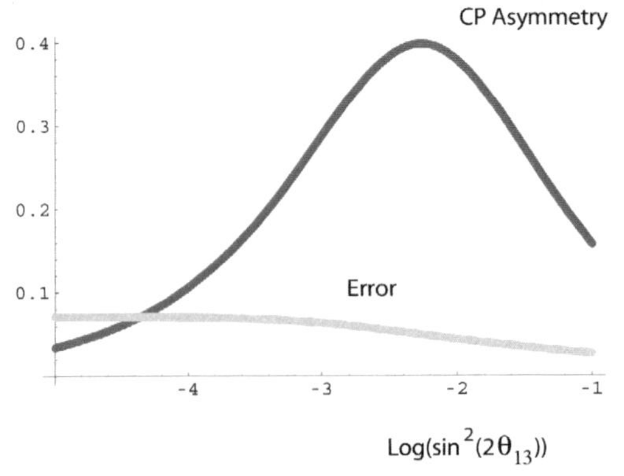

Figure 2. Magnitude of the CP asymmetry at the first oscillation maximum, for $\delta = 1$ as a function of the mixing angle $\sin^2 2\theta_{13}$. The curve marked "error" indicates the dependence of the statistical+systematic error on such a measurement. The curves have been computed for the baseline beta beam option at the fixed energy $E_\nu = 0.4$ GeV, $L = 130$ km, statistical + 2% systematic errors. From [27].

mesons are sign-selected and focused (defocused) by large acceptance magnetic lenses into a long evacuated decay tunnel where ν_μ's ($\overline{\nu}_\mu$'s) are generated. In case of positive charge selection, the ν_μ beam has typically a few percent of $\overline{\nu}_\mu$ contamination (from the decay of the residual π^-, K^- and K^0) and $\sim 1\%$ of ν_e and $\overline{\nu}_e$ coming from three-body $K^\pm$, K_0 decays and μ decays.

The precision of the evaluation of the intrinsic ν_e to ν_μ contamination is limited by the knowledge of the π and K production in the primary proton beam target requiring a devoted hadroproduction experiment. Recently the Harp experiment [33] measured both the K2K [34] and the MiniBooNE [35] targets, covering most of the useful pion phase-space, successfully improving the description of the two beam lines.

Close detectors are used to directly measure beam neutrinos and backgrounds (for a discussion about close detectors and systematic errors in future LBL experiments see [36]).

The T2K (Tokai to Kamioka) experiment [37] is aiming neutrinos from the Tokai site of J-PARC (30 GeV, 0.75 MW) to the Super-Kamiokande detector 295 km away, at an off-axis angle of 2.5 degrees from the neutrino beam, ensuring a neutrino peak energy of about 0.6 GeV. The beam line is equipped with a set of dedicated on-axis (INGRID) and off-axis (ND280) near detectors at a distance of 280 m. It is expected that the sensitivity of the experiment in a five-year ν_μ run at the full J-PARC beam intensity, will be of the order of $\sin^2 2\theta_{13} \leq 0.006$ (90% CL). The experiment started data taking at the beginning of 2010 and already released

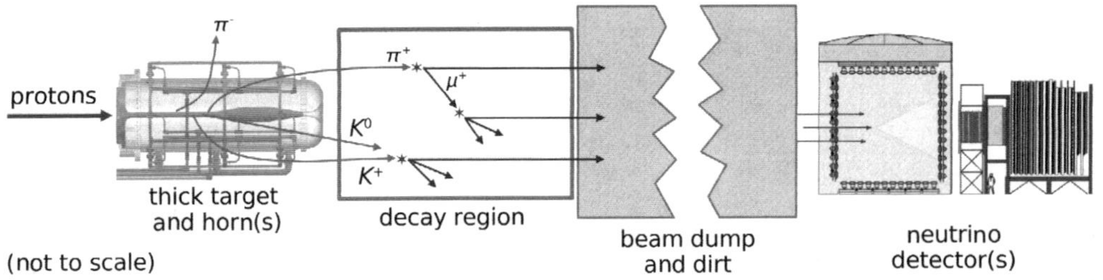

Figure 3. Sketch of a conventional accelerator neutrino beam line

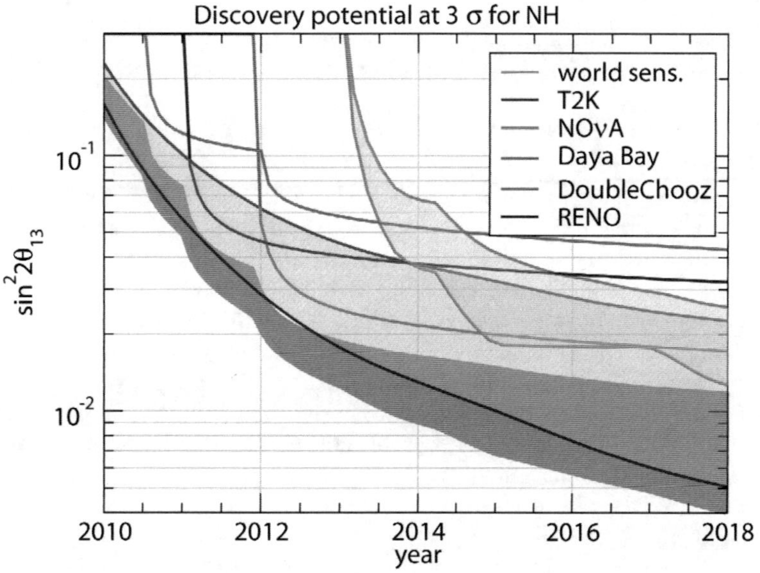

Figure 4. Evolution of the θ_{13} discovery potential as a function of time (3σ CL) for NH, showing the global sensitivity reach. The bands for the beams and the global reach reflect the (unknown) true value of δ. From [32].

the first results [38].

The NOνA experiment with an upgraded NuMI off-axis neutrino beam [39] ($E_\nu \sim 2$ GeV and a ν_e contamination lower than 0.5%), a totally active 15 kton liquid scintillator detector and with a baseline of 810 km (12 km off-axis), has been approved at FNAL with the aim to explore $\nu_\mu \to \nu_e$ oscillations with a θ_{13} sensitivity similar to T2K and with some sensitivity to sign(Δm^2_{23})thanks to the longer baseline.

Another approach to searching for non-vanishing θ_{13} is to look at $\overline{\nu}_e$ disappearance using nuclear reactors as neutrino sources. In $\overline{\nu}_e$ disappearance experiments θ_{13} is directly linked to the detected oscillation signal without any interference from $\delta_{\rm CP}$ and sign(Δm^2_{23}). Their result is truly complementary to the accelerators. On the other hand reactor experiments cannot have any role in direct searches for leptonic CP violation.

The Double Chooz [40] experiment in France will employ a far detector in the same location as the former Chooz detector as well as a near detector. The sensitivity after five years of data taking will be $\sin^2 2\theta_{13} = 0.025$ at 90% CL. The Daya Bay project in China [41] could reach a $\sin^2 2\theta_{13}$ sensitivity below 0.01, while the RENO experiment in Korea [42] should reach a sensitivity around 0.02.

A sketch of θ_{13} discovery potential of future experiments as a function of the time, following the schedule reported in the experimental proposals, is reported in Figure 4 [32].

3. A new generation of facilities for the physics of neutrino oscillations

A global fit of T2K plus NOνA plus reactors will not be able to provide firm results (3σ or better) about leptonic CP violation or sign(Δm^2_{23}) [43] whatever the value of θ_{13}.

A further generation of long-baseline neutrino experiments will be needed to address this very important search in physics. As a rule of thumb they should be at least one order of magnitude more sensitive than T2K or NOνA, a condition equivalent to an increase of two orders of magnitude on neutrino statistics, with a consequent important reduction of systematic errors.

Proposals for this very challenging task are based either on conventional neutrino beams pushed to their ultimate power, Section 3.1, or to innovative concepts about neutrino production, Section 3.2.

3.1. Neutrino Super Beams

To fulfill the needs of searches for leptonic CP violation, conventional neutrino beams must be pushed to their ultimate limits (neutrino super beams) [24] and gigantic (megaton scale) neutrino detectors must be built.

Phase II of the T2K experiment, often called T2HK [44], foresees an increase of beam power up to the maximum feasible with the accelerator and target (4 MW beam power), antineutrino runs, and a very large, 520 kt, water Čerenkov detector, HyperKamiokande or HK, to be built close to SuperKamiokande. An evolution of T2HK is the T2KK [45] project, where half of the HK detector would be installed in Japan, while the second half would be mounted in Korea, at a baseline of about 900 km, around the second oscillation maximum. Possibilities of intermediate baselines and liquid argon detectors have also been studied [46].

A wide-band beam (WBB) has been proposed at Fermilab upgrading the FNAL main injector after the end of the Tevatron programme [47]. A conventional wide-band neutrino would be sent to a megaton water Čerenkov (or liquid argon) detector at the Homestake mine at a baseline of 1290 km. It would be then displaced at the second oscillation maximum.

In Europe, the perspectives for a high intensity neutrino experiment based on Superbeams are entangled with the evolution of the CERN acceleration complex and, in particular, of the injection system of the LHC. In these directions super beams based on upgrades of the CNGS, section 3.1.1, on a high power SPL, section 3.1.2, or on a high power PS2, section 3.1.3, have been studied.

3.1.1. CNGS Upgrades The CNGS at nominal intensity can be operated to accumulate 4.5×10^{19} pot/y at an energy of 400 GeV. In the last few years, particularly in the framework of the CERN PAF ("Proton Accelerators for Future") Working Group, it has been investigated [48] the possibility of increasing the intensity of the CNGS both using present facilities and, on a longer timescale, exploiting an upgrade of the acceleration complex.

The ultimate CNGS performance are actually limited by the injection from the 50-year-old Proton Synchrotron (PS). In this scenario (CNGS as the only user of the SPS at CERN beyond the LHC), the facility could deliver up to 1×10^{20} pot/y. At a longer timescale (>2016), the replacement of the PS with a new 50 GeV synchrotron (PS2 [49]) might surpass these limitations, provided an appropriate upgrade of the SPS radio-frequency system.

It would bring CNGS to a maximum intensity (CNGS as only user of the SPS beyond the LHC) of 2×10^{20} pot/y (6.6×10^{20} NOVA pot/y).

Studies about performances of CNGS upgrades with a new setup firing a lower energy neutrino beam off-axis to a 100 kton [50] or 20 kton [51] near the LNGS show anyway that only a proton intensity one order of magnitude higher than present CNGS configuration could allow a sensitive search for leptonic CP violation. This would require a complete refurbishment of the SPS accelerator.

3.1.2. CERN-SPL In the CERN-SPL super beam project [52] the planned 4MW SPL (Superconducting Proton Linac) would deliver a 3.5 GeV/c H^- beam on a Hg target to generate a neutrino beam with an average energy of ~ 0.3 GeV.

The ν_e contamination from K will be suppressed by threshold effects and the resulting ν_e/ν_μ ratio ($\sim 0.4\%$) will be known within 2% error. The use of a near and far detector (the latter at $L = 130$ km in the Fréjus area) will allow for both ν_μ-disappearance and $\nu_\mu \to \nu_e$ appearance studies. The physics potential of the SPL super beam (SPL-SB) with a water Čerenkov far detector with a fiducial mass of 440 kt, has been extensively studied [53, 54, 55, 56]. The most updated sensitivity estimations for this setup have been published in Ref. [57].

The MEMPHYS (Megaton Mass Physics) detector [58] is a megaton-class water Čerenkov designed to be located at Fréjus, 130 km from CERN, addressing both the non-accelerator domain (nucleon decay, SuperNovae neutrino from burst event or from relic explosion, solar and atmospheric neutrinos) and the accelerator (super beam, beta beam) domain [59].

3.1.3. CERN-PS2 It has been proposed in [60] to generate a neutrino beam by a high power, 1.6 MW, version of the PS2 accelerator, a 50 GeV synchrotron designed to run at 0.4 MW to serve as a component of the new injection scheme for the LHC. Neutrinos could then be fired to a 100 kton liquid argon detector, placed at a distance of 950 km or 1544 km or 2300 km (the distances correspond to the three underground labs of Sieroszowice in Poland, Slanic in Romania and Pyhasalmi in Finland respectively, three candidates actually taken in consideration by the Laguna [61] FP7 design study).

As in the case of the WBB at Dusel, this setup would measure neutrinos at the first and at the second oscillation maximum. Liquid argon is certainly the best candidate to fulfill the requirements of this configuration.

3.2. New Concepts on Neutrino Beams

The super beam approach can be quite powerful if θ_{13} happens to be sufficiently large, in the range of values that would permit a discovery by the T2K, NOνA or the reactor experiments. For smaller values it shows evident limitations: it is not a "pure" source of neutrinos of a given flavor, being contaminated by the ν_e produced by the decay-in-flight of the kaons and muons. When seeking for sub-dominant $\nu_\mu \to \nu_e$ transitions, the systematics on the knowledge of the ν_e contamination will likely be the main limitation for a precise determination of CP violation in the leptonic sector [36]. Furthermore the ultimate precision with which the neutrino flux can be predicted is limited by the precision of the hadroproduction cross sections of the neutrino parents, that are secondary particles generated in a primary proton beam.

The intrinsic limitations of conventional neutrino beams can be overcome if the neutrino parents are fully selected, collimated and accelerated to a given energy. This can be attempted within the muon lifetime, bringing to the Neutrino Factory [62], or within beta decaying ion lifetimes, bringing to the Beta Beam [63, 64].

With this challenging approach several important improvements can be made to conventional neutrino beams:

- The neutrino fluxes would be simply derived from the knowledge of the number of parents circulating in the decay ring and from their Lorentz boost factor γ.
- The energy shape of the neutrino beam would be defined by just two parameters, the end-point energy Q_β of the beta decaying parent and its Lorentz boost factor γ.
- The intrinsic neutrino backgrounds would be suppressed (in the case of beta beam) or reduced to wrong sign muons (golden channel in neutrino factories).

The technological problems derive from the fact that the parents need to be unstable particles, requiring a fast, efficient acceleration scheme.

3.3. Neutrino Factories

Production, acceleration and stacking of high intensity muon beams for muon colliders have been envisaged since the 60's and it has been noted very early that their decays might produce useful beams of ν_μ and $\overline{\nu}_e$ (exploiting μ^- decays into $e^-\overline{\nu}_e\nu_\mu$) or $\overline{\nu}_\mu$ and ν_e (μ^+ decays into $e^+\nu_e\overline{\nu}_\mu$).

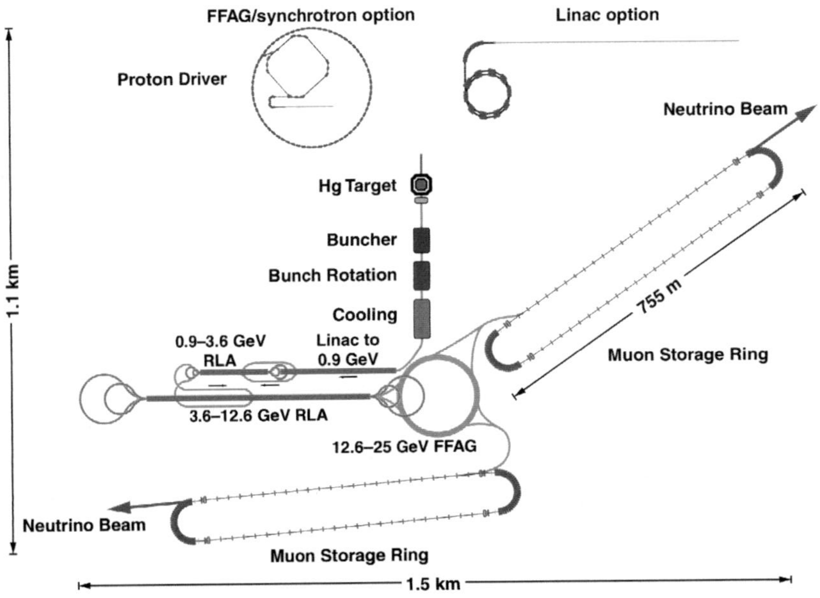

Figure 5. Layout of a Neutrino Factory facility

In the modern formulation of the "Neutrino Factory" concept, muons are created from an intense pion source at low energies, their phase space compressed to produce a bright beam, which is then accelerated to the desired energy and injected into a storage ring with long straight sections pointing to the desired direction[62] . Since μ^+ decay into $e^+\nu_e\overline{\nu}_\mu$, it is possible to investigate $\nu_e \rightarrow \nu_\mu$ oscillations seeking for the appearance of μ^- from ν_μ CC events ("wrong sign muons"), provided that these events are separated from the bulk of μ^+ ("right sign muons") coming from unoscillated $\overline{\nu}_\mu$ (the same applies to μ^- decays).

The simultaneous exploitation of μ^- and μ^+ decays would be an ideal tool to address CP violation in the leptonic sector, with outstanding performances compared with pion-based sources [65]. Moreover, the neutrino factory concept resonated with the needs of the Muon Collider accelerator community, who appreciated the possibility of a strong physics-motivated intermediate step before facing the enterprise of the Muon Collider itself [66].

The realization of the neutrino factory still represents a major accelerator challenge compared with Superbeams. It is met through a world-wide R&D programme; in Europe this programme is especially fostered by UK. Among the Neutrino Factory-oriented projects can be recalled MICE at the Rutherford Appleton Laboratories (ionization cooling), HARP at CERN (hadroproduction for the front-end proton accelerator), MERIT at CERN (targetry), EMMA at Daresbury (fixed-field alternating-gradient accelerators) and the MUCOOL R&D at Fermilab (radio-frequency and absorbers). Moreover, the Neutrino Factory has to be seeded by a very powerful low-energy proton accelerator (4 MW); its realization requires similar R&D as for the Superbeams, although its optimal energy lies in the few-GeV range (e.g. the aforementioned SPL). Current designs aim at 10^{21} muon decays per year running with a muon energy of 20 GeV.

After the work of the International Scoping Study (ISS) [67, 68, 69], there is a rather widespread consensus on the fact that the Neutrino Factory can be considered the most performing facility for the determination of θ_{13}, CP violation and the mass hierarchy. With respect to Superbeams, they profit from much smaller systematics in the knowledge of the source and much higher energies (i.e. statistics, due to the linear rise of the deep-inelastic

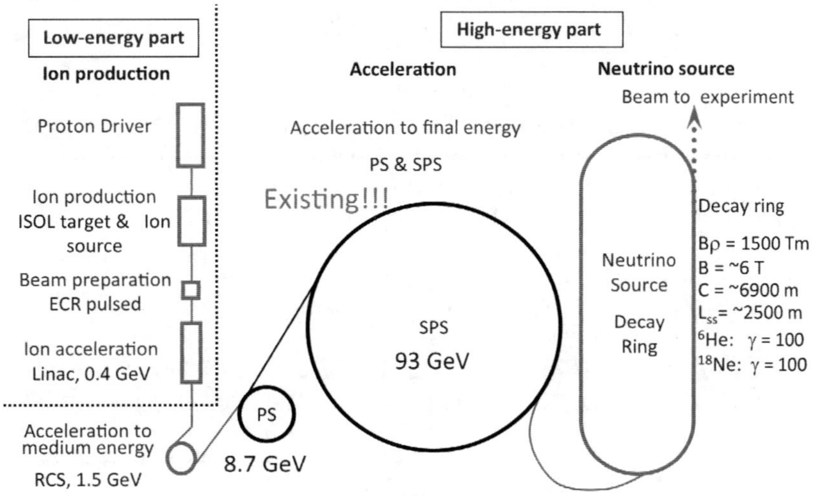

Figure 6. Layout of a Beta Beam facility

ν_μ cross section with energy). In fact, the energy is so high that for any realistic baseline ($<$ 7000 km) the ratio L/E will be off the peak of the oscillation maximum at the atmospheric scale. This condition is the main cause of the occurrence of multiple solutions when the mixing parameters are extracted from the physics observables, i.e. the rates of appearance of wrong sign muons, see the discussion of Section 1.2. The ISS suggests as an ideal solution the positioning of two detectors at baseline around 3000 and 7000 km.

Although the superior physics reach of the Neutrino Factory is nearly undisputed and no evident showstoppers have been identified, the R&D needed to build this facility remains impressive. In turn, the time schedule for its realization and the cost estimate are vague ($\sim$ 2020 after an investment of 1-2 Billion$). On the other hand, a clear indication on the size of θ_{13} will enormously boost the interest of particle physics on this technology. Neutrino Factories are virtually capable of performing real precision physics on the leptonic mixing in a way that resembles the former physics potential of the b-factories on quark mixing.

3.4. Beta Beams

The enormous progress in the technology of Radioactive Ion Beams has led P. Zucchelli [63] to the proposal of a neutrino facility based on the decay in flight of β-unstable ions (for a full review see [64]). Unlike the Neutrino Factory, these Beta Beams (BB) are pure sources of $\overline{\nu}_e$ or, in the occurrence of β^+ decays, of ν_e. Hence, they are ideal tools to study $\nu_e \to \nu_\mu$ transitions and their CP-conjugate exploiting a nearly complete absence of systematics in the knowledge of the source with the bonus of no "right sign muon" background (no ν_μ in the initial state). On the other hand, due to the very different mass-to-charge ratio between muons and β-unstable ions, the energy of the neutrinos are typically much smaller than what can be obtained at the Neutrino Factory.

The original proposal of [63] was tuned to leverage at most the present facilities of CERN - the PS and the SPS - and it was based on ^{6}He and ^{18}Ne as $\overline{\nu}_e$ and ν_e sources respectively. It goes without saying that the beta beamtriggered the interest of nuclear physics community, which was offered a stimulating synergy with the neutrino programme at CERN. As a result, such proposal [70] was studied in a systematic manner within the framework of the EURISOL

Design Study[1] (Task 12: Beta Beam aspects). The study aimed at 2.9×10^{18} antineutrinos per year from ^{6}He and 1.1×10^{18} neutrinos per year from ^{18}Ne. The outcome was extremely encouraging.

As in the case of the SPL-super beam the Eurisol Beta Beam would detect neutrino oscillation on the peak of the first oscillation maximum at a baseline that guarantees the absence of matter effects that are a source of not genuine CP violating oscillations. As discussed in [57] sensitivity on sign(Δm^2_{23})would be partially recovered by the synergic combination of beam neutrinos with atmospheric neutrinos detected by MEMPHYS [58]. On the other hand the sub-GeV energy range of the Eurisol beta beam neutrinos reflects in depleted neutrino cross sections, impacting on the overall performances of the setup.

A very interesting experimental possibility is that neutrinos created by the SPL could be fired to the same detector of the Eurisol beta beam [56]. This is possible because radioactive ion production requires a proton source of about 0.2 MW while the SPL could deliver up to 4 MW of power. Furthermore the two neutrino beams would have similar energies and so they could share the same far detector. The combination of a super beam with a beta beam in the same experiment can provide an experimental environment with very unique characteristics: the two beams can be used to separately study CP channels (like $\nu_\mu \to \nu_e$ vs $\overline{\nu}_\mu \to \overline{\nu}_e$ and $\nu_e \to \nu_\mu$ vs $\overline{\nu}_e \to \overline{\nu}_\mu$), they can be mixed to study T transitions (like $\nu_\mu \to \nu_e$ vs $\nu_e \to \nu_\mu$ and $\overline{\nu}_\mu \to \overline{\nu}_e$ vs $\overline{\nu}_e \to \overline{\nu}_\mu$) or can be mixed to study CPT transitions (like $\nu_\mu \to \nu_e$ vs $\overline{\nu}_e \to \overline{\nu}_\mu$ and $\nu_e \to \nu_\mu$ vs $\overline{\nu}_\mu \to \overline{\nu}_e$)

The addition of a super beam to a beta beam could also complement some of the weak points of the beta beam, namely the lack of sensitivity to the atmospheric parameters θ_{23} and Δm^2_{23} and the lack of ν_μ events in the close detector, useful for calibrating beta beam signal efficiency and measuring the ν_e/ν_μcross section ratio.

To improve the performances of the Eurisol Beta Beam several alternatives to the SPS have been considered: a refurbished 1 TeV SPS ("SuperSPS" [71]) envisaged for the energy and luminosity upgrade of the LHC or even the LHC itself [72, 73, 74, 75], an option that nowadays seems far in the future if not unlikely.

In 2006, C. Rubbia et al. [76] proposed the use of ^{8}Li and ^{8}B as neutrino sources noting that these isotopes could be produced in a multiturn passage of a low-energy ion beam through a low-Z target. In this case, ionization cooling techniques could increase the circulating beam lifetime and thus enhance the ion production to a level suitable for the beta beam. This option has the advantage of employing isotopes with higher Q-value than ^{18}Ne and ^{6}He, increasing correspondingly the neutrino energy (from ~ 0.5 to ~ 1.5 GeV for the SPS-based BB). This alternative approach is at focus in the framework of the EUROν Design Study[2].

A drawback with respect to the use of low-Q ions is that the flux at the far location is smaller due to the larger beam divergence and a larger amount of ions stacked in the decay ring is needed. More generally the performances of a Beta Beam in delivering neutrinos can be parametrized with a merit factor $\mathcal{M} \propto \frac{\gamma}{Q_\beta}$ [64] from which it follows that performances of a beta beam scale as the Lorentz boost factor γ and are inversely proportional to the endpoint energy Q_β. For this reason for the same baseline an high-Q beta beamrequires about an order of magnitude more ions at the source to match the performances of the Eurisol-γ beta beam.

A further option for beta beams is the possibility of creating monochromatic neutrino beams [78]. These beams are based on electron capture processes of radioactive ions, rather

[1] The EURISOL Design Study was a Project funded by the European Community within the 6th Framework Programme as a Research Infrastructures Action under the "Structuring the European Research Area Specific Programme". The Project started officially on Feb 2005, and has been completed on spring 2009.

[2] EUROν [77] is a FP7 Design Study which started on Sept 2008 and will run for 4 years. The primary aims are to study three possible future neutrino oscillation facilities for Europe (a Superbeam from CERN-to-Frejus, a RAL or CERN based Neutrino Factory and high-Q BB) and do a cost and performance comparison.

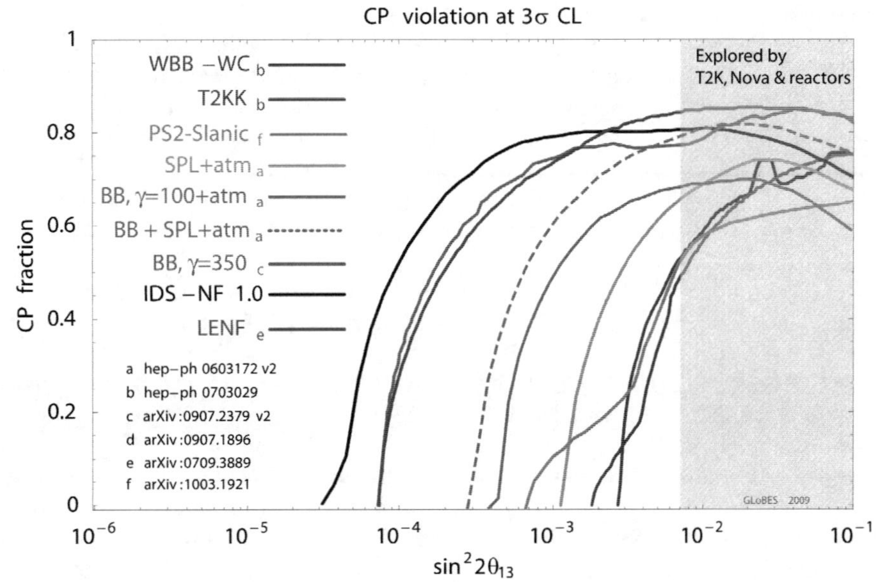

Figure 7. The discovery reach at 3 σ level for different facilities for leptonic CP violation. The discovery limits are shown as a function of the fraction of all possible values of the true value of the CP phase δ and the true value of $\sin^2 2\theta_{13}$. The curves are taken from [57], [80], [81], [43], [82], [60].

than on their beta decays, producing monochromatic neutrino beams. The main limitations of these setups are the technical difficulties of the production and acceleration schemes and the impossibility of having monochromatic antineutrino beams.

Low energy beta beams, producing neutrinos with energies in the range of few tens of MeV, could offer interesting non-oscillation neutrino physics measurements, as discussed in [79].

Concluding, beta beam performance are in between the performance of super beams and neutrino factory. The clarification of the issue of the ion production yield is considered a crucial milestone for the beta beam. Given an appropriate yield, the acceleration and stacking is viewed as less demanding than what is needed for a Neutrino Factory both from the point of view of R&D and cost. Clearly, the possibility of employing existing facilities (e.g. the CERN PS-SPS complex or its upgrades) might substantially strengthen this option.

4. Conclusions

Several different options have already been put forward to address the challenging experimental needs of future experiments looking to leptonic CP violation.

They can exploit conventional neutrino beams pushed to their ultimate performances, neutrino super beams, or innovative concepts about neutrino beam production like the neutrino factories and the beta beams.

A comparison of the sensitivities of the different facilities, see Fig. 7, shows that leptonic CP violation can be discovered provided that $\sin^2 2\theta_{13}$ is not four order of magnitudes below the present experimental limit.

Ultimate performances can be reached by the neutrino factory. However, if θ_{13} happens to be on the reach of the next generation experiments, $\sin^2 2\theta_{13} \geq 0.01$, super beams and beta beams could be very competitive being less demanding on R&D developments and costs.

5. CPT in the neutrino sector

A motivation to look for signals of CPT and Lorentz invariance violation in the neutrino sector is the possibility of a different mechanism for generating the small neutrino masses (compared to the masses of quarks and charged leptons), the so-called seesaw mechanism [83, 84, 85]. If neutrino masses are generated at energies approaching the Plank scale, they could be sensitive to new physics [86].

A consequence of CPT violation would be a difference between the parameters governing the oscillations of neutrinos and antineutrinos. A global fit to neutrino oscillations performed separating the terms coming from neutrino oscillations from those coming from antineutrino oscillations [87] (updated results have been shown in [88]) showed that at present no indication of CPT violation emerges from the global data, although the sensitivity of these comparison is quite poor.

Some renewed interest to CPT violation in the neutrino sector has been raised by the preliminary results, still not published, of the MINOS collaboration about a difference of ~ 2.4 standard deviations between the best fit values of ν_μ and $\overline{\nu}_\mu$ disappearance data [89], see also [90]. These results, if confirmed and not originated by not well controlled systematic errors, could anyway be interpreted by more conventional "non standard" scenarios, like non standard neutrino interactions in matter effects [91] or light sterile fermions plus a B-L new interaction [92].

CPT violation had been also advocated as a way out for the solution of the LSND puzzle [93] (cfr. Sec. 1), since it can provide the extra Δm^2 required by this experimental result. This possibility had been excluded by the fits of [87]. It has been also been advocated to save the sterile neutrino "3+1" model against the negative evidence of short baseline reactor neutrino experiments [94]. As discussed in Sec. 1, the bound of reactor neutrino experiments is now an evidence in favor to sterile neutrinos and steriles alone seem to be enough to describe the global data on neutrino oscillations [21].

Future neutrino facilities as those discussed in the previous sections could be a powerful tool to improve the sensitivity on CPT violation in the neutrino sector, as discussed in [95].

Finally, following the discussion of [90] there are other two kind of CPT tests that can be performed by neutrino oscillation experiments:

(i) Searches for Lorentz-violating effects in concert with CPT violation. Limits have been recently published by the LSND [96], MINOS [97] and IceCube [98] experiments.

(ii) Searches for inconsistencies in oscillation results that could point to baseline-dependent new physics with possible ramifications for CPT. They have been discussed in [99, 100, 101].

References

[1] Fukuda Y *et al.* (Super-Kamiokande) 1998 *Phys. Rev. Lett.* **81** 1562–1567 (*Preprint* hep-ex/9807003)
[2] Ambrosio M *et al.* (MACRO Collaboration) 2003 *Phys.Lett.* **B566** 35–44 (*Preprint* hep-ex/0304037)
[3] Aliu E *et al.* (K2K) 2005 *Phys. Rev. Lett.* **94** 081802 (*Preprint* hep-ex/0411038)
[4] Michael D G *et al.* (MINOS) 2006 *Phys. Rev. Lett.* **97** 191801 (*Preprint* hep-ex/0607088)
[5] Ahmad Q R *et al.* (SNO) 2002 *Phys. Rev. Lett.* **89** 011301 (*Preprint* nucl-ex/0204008)
[6] Cleveland B T *et al.* 1998 *Astrophys. J.* **496** 505–526
[7] Hirata K *et al.* (KAMIOKANDE-II Collaboration) 1989 *Phys.Rev.Lett.* **63** 16
[8] Abe K *et al.* (Super-Kamiokande Collaboration) 2010 Long author list - awaiting processing (*Preprint* 1010.0118)
[9] Anselmann P *et al.* (GALLEX Collaboration) 1992 *Phys.Lett.* **B285** 376–389
[10] Altmann M *et al.* (GNO) 2005 *Phys. Lett.* **B616** 174–190 (*Preprint* hep-ex/0504037)
[11] Abazov A, Abdurashitov D, Anosov O, Eroshkina L, Faizov E *et al.* 1991 *Nucl.Phys.Proc.Suppl.* **19** 84–93
[12] Abdurashitov J *et al.* (SAGE Collaboration) 2009 *Phys.Rev.* **C80** 015807 (*Preprint* 0901.2200)
[13] Eguchi K *et al.* (KamLAND Collaboration) 2003 *Phys.Rev.Lett.* **90** 021802 (*Preprint* hep-ex/0212021)
[14] Schwetz T, Tortola M A and Valle J W F 2008 *New J. Phys.* **10** 113011 (*Preprint* 0808.2016)
[15] Aguilar A *et al.* (LSND Collaboration) 2001 *Phys.Rev.* **D64** 112007 (*Preprint* hep-ex/0104049)
[16] Aguilar-Arevalo A *et al.* (MiniBooNE Collaboration) 2009 *Phys.Rev.Lett.* **103** 111801 (*Preprint* 0904.1958)

[17] Aguilar-Arevalo A *et al.* (The MiniBooNE Collaboration) 2007 *Phys.Rev.Lett.* **98** 231801 (*Preprint* 0704.1500)
[18] Maltoni M, Schwetz T, Tortola M A and Valle J W F 2004 *New J. Phys.* **6** 122 (*Preprint* hep-ph/0405172)
[19] Mueller T, Lhuillier D, Fallot M, Letourneau A, Cormon S *et al.* 2011 *Phys.Rev.C* (*Preprint* 1101.2663)
[20] Mention G, Fechner M, Lasserre T, Mueller T, Lhuillier D *et al.* 2011 (*Preprint* 1101.2755)
[21] Kopp J, Maltoni M and Schwetz T 2011 * Temporary entry * (*Preprint* 1103.4570)
[22] Feruglio F, Strumia A and F V 2002 *Nucl. Phys.* **B637** 345–377 (*Preprint* hep-ph/0201291)
[23] Fukugita M and Yanagida T 1986 *Phys. Lett.* **B174** 45
[24] Richter B 2000 Conventional beams or neutrino factories: The next generation of accelerator-based neutrino experiments (*Preprint* hep-ph/0008222)
[25] Burguet-Castell J, Gavela M B, Gomez-Cadenas J, Hernandez P and Mena O 2001 *Nucl. Phys.* **B608** 301–318 (*Preprint* hep-ph/0103258)
[26] Guglielmi A, Mezzetto M, Migliozzi P and Terranova F 2005 Measurement of three-family neutrino mixing and search for CP violation published in D. Bettoni and others, Phys. Rept. 434, 47 (*Preprint* hep-ph/0508034)
[27] Blondel A *et al.* 2006 *Acta Phys. Polon.* **B37** 2077–2113 (*Preprint* hep-ph/0606111)
[28] Minakata H and Nunokawa H 2001 *JHEP* **10** 001 (*Preprint* hep-ph/0108085)
[29] Fogli G L and Lisi E 1996 *Phys. Rev.* **D54** 3667–3670 (*Preprint* hep-ph/9604415)
[30] Barger V, Marfatia D and Whisnant K 2002 *Phys. Rev.* **D65** 073023 (*Preprint* hep-ph/0112119)
[31] Mezzetto M 2009 Next Challenge in Neutrino Physics: the θ_{13} Angle (*Preprint* 0905.2842)
[32] Mezzetto M and Schwetz T 2010 θ_{13}: phenomenology, present status and prospect (*Preprint* 1003.5800)
[33] Catanesi M G *et al.* (HARP) 2007 *Nucl. Instrum. Meth.* **A571** 527–561
[34] Catanesi M G *et al.* (HARP) 2006 *Nucl. Phys.* **B732** 1–45 (*Preprint* hep-ex/0510039)
[35] Catanesi M G *et al.* (HARP) 2007 *Eur. Phys. J.* **C52** 29–53 (*Preprint* hep-ex/0702024)
[36] Huber P, Mezzetto M and Schwetz T 2008 *JHEP* **03** 021 (*Preprint* 0711.2950)
[37] Itow Y *et al.* (T2K) 2001 The JHF-Kamioka neutrino project (*Preprint* hep-ex/0106019)
[38] Rubbia A T2k results presentation at the XIV Int. Workshop on Neutrino Telescopes, Venezia, March 15-18, 2011
[39] Ambats I *et al.* (NOvA) 2004 Nova proposal to build a 30-kiloton off-axis detector to study neutrino oscillations in the fermilab numi beamline (*Preprint* hep-ex/0503053)
[40] Ardellier F *et al.* (Double Chooz) 2006 Double Chooz: A search for the neutrino mixing angle θ_{13} (*Preprint* hep-ex/0606025)
[41] Guo X *et al.* (Daya-Bay) 2007 A precision measurement of the neutrino mixing angle θ_{13} using reactor antineutrinos at Daya Bay (*Preprint* hep-ex/0701029)
[42] Ahn J K *et al.* (RENO) 2010 RENO: An Experiment for Neutrino Oscillation Parameter θ_{13} Using Reactor Neutrinos at Yonggwang (*Preprint* 1003.1391)
[43] Huber P, Lindner M, Schwetz T and Winter W 2009 *JHEP* **11** 044 (*Preprint* 0907.1896)
[44] Kobayashi T 2003 *J. Phys.* **G29** 1493–1501
[45] Ishitsuka M, Kajita T, Minakata H and Nunokawa H 2005 *Phys. Rev.* **D72** 033003 (*Preprint* hep-ph/0504026)
[46] Badertscher A *et al.* 2008 A Possible Future Long Baseline Neutrino and Nucleon Decay Experiment with a 100 kton Liquid Argon TPC at Okinoshima using the J-PARC Neutrino Facility (*Preprint* 0804.2111)
[47] Diwan M *et al.* 2006 Proposal for an experimental program in neutrino physics and proton decay in the homestake laboratory (*Preprint* hep-ex/0608023)
[48] Meddahi M and Shaposhnikova E 2007 Analysis of the maximum potential proton flux to cngs cERN-AB-2007-013
[49] Information available at http://paf-ps2.web.cern.ch/paf-ps2/
[50] Meregaglia A and Rubbia A 2006 *JHEP* **11** 032 (*Preprint* hep-ph/0609106)
[51] Baibussinov B *et al.* 2008 *Astropart. Phys.* **29** 174–187 (*Preprint* 0704.1422)
[52] Autin B *et al.* Conceptual design of the SPL, a high-power superconducting H- linac at CERN cERN-2000-012
[53] Gomez-Cadenas J J *et al.* 2001 Physics potential of very intense conventional neutrino beams (*Preprint* hep-ph/0105297)
[54] Blondel A *et al.* 2001 *Nucl. Instrum. Meth.* **A503** 173–178
[55] Mezzetto M 2003 *J. Phys.* **G29** 1771–1776 (*Preprint* hep-ex/0302007)
[56] Mezzetto M 2005 *Nucl. Phys. Proc. Suppl.* **149** 179–181
[57] Campagne J E, Maltoni M, Mezzetto M and Schwetz T 2007 *JHEP* **04** 003 (*Preprint* hep-ph/0603172)
[58] de Bellefon A *et al.* 2006 MEMPHYS: A large scale water Cerenkov detector at Frejus (*Preprint* hep-ex/0607026)

[59] Autiero D *et al.* 2007 *JCAP* **0711** 011 (*Preprint* 0705.0116)
[60] Rubbia A 2010 A CERN-based high-intensity high-energy proton source for long baseline neutrino oscillation experiments with next- generation large underground detectors for proton decay searches and neutrino physics and astrophysics (*Preprint* 1003.1921)
[61] Large apparatus studying grand unification and neutrino astrophysics http://www.laguna-science.eu/
[62] Geer S 1998 *Phys. Rev.* **D57** 6989–6997 (*Preprint* hep-ph/9712290)
[63] Zucchelli P 2002 *Phys. Lett.* **B532** 166–172
[64] Lindroos M and Mezzetto M 2009 *Beta beams: Neutrino beams* london, UK: Imperial College Pr., 154 p
[65] De Rujula A, Gavela M B and Hernandez P 1999 *Nucl. Phys.* **B547** 21–38 (*Preprint* hep-ph/9811390)
[66] Geer S 2009 *Ann.Rev.Nucl.Part.Sci.* **59** 347–365
[67] Berg J S *et al.* (ISS Accelerator Working Group) 2009 *JINST* **4** P07001 (*Preprint* 0802.4023)
[68] Bandyopadhyay A *et al.* (ISS Physics Working Group) 2009 *Rept. Prog. Phys.* **72** 106201 (*Preprint* 0710.4947)
[69] Abe T *et al.* (ISS Detector Working Group) 2009 *JINST* **4** T05001 (*Preprint* 0712.4129)
[70] Bouchez J, Lindroos M and Mezzetto M 2004 *AIP Conf. Proc.* **721** 37–47 (*Preprint* hep-ex/0310059)
[71] Bruning O *et al.* Lhc luminosity and energy upgrade: A feasibility study. cERN-LHC-PROJECT-REPORT-626;
[72] Burguet-Castell J, Casper D, Gomez-Cadenas J J, Hernandez P and Sanchez F 2004 *Nucl. Phys.* **B695** 217–240 (*Preprint* hep-ph/0312068)
[73] Terranova F, Marotta A, Migliozzi P and Spinetti M 2004 *Eur. Phys. J.* **C38** 69–77 (*Preprint* hep-ph/0405081)
[74] Agarwalla S K, Raychaudhuri A and Samanta A 2005 *Phys. Lett.* **B629** 33–40 (*Preprint* hep-ph/0505015)
[75] Donini A, Fernandez-Martinez E, Migliozzi P, Rigolin S, Lavina L *et al.* 2008 *Eur.Phys.J.* **C53** 599–606 (*Preprint* hep-ph/0703209)
[76] Rubbia C, Ferrari A, Kadi Y and Vlachoudis V 2006 *Nucl. Instrum. Meth.* **A568** 475–487 (*Preprint* hep-ph/0602032)
[77] European commission fp7 design study: A high intensity neutrino oscillation facility in europe http://www.euronu.org
[78] Bernabeu J, Burguet-Castell J, Espinoza C and Lindroos M 2005 *JHEP* **12** 014 (*Preprint* hep-ph/0505054)
[79] Volpe C 2007 *J. Phys.* **G34** R1–R44 (*Preprint* hep-ph/0605033)
[80] Barger V, Huber P, Marfatia D and Winter W 2007 *Phys.Rev.* **D76** 053005 (*Preprint* hep-ph/0703029)
[81] Choubey S, Coloma P, Donini A and Fernandez-Martinez E 2009 *JHEP* **12** 020 (*Preprint* 0907.2379)
[82] Bross A D, Ellis M, Geer S, Mena O and Pascoli S 2008 *Phys.Rev.* **D77** 093012 (*Preprint* 0709.3889)
[83] Gell-Mann M, Ramond P and Slansky R 1979 *Supergravity, ed. by F. van Nieuwenhuizen and D. Freedman (Amsterdam, North Holland, 1979)* 315
[84] Yanagida T 1980 *Prog.Theor.Phys.* **64** 1103
[85] Mohapatra R N and Senjanovic G 1980 *Phys.Rev.Lett.* **44** 912
[86] Kostelecky V and Mewes M 2004 *Phys.Rev.* **D70** 031902 (*Preprint* hep-ph/0308300)
[87] Gonzalez-Garcia M C and Maltoni M 2008 *Phys. Rept.* **460** 1–129 (*Preprint* 0704.1800)
[88] Maltoni M New Physics in the Atmospheric Sector: non-Standard Interactions and CPT Violation presentation at the XIV Int. Workshop on Neutrino Telescopes, Venezia, March 15-18, 2011
[89] Danko I (MINOS Collaboration) 2009 (*Preprint* 0910.3439)
[90] Barenboim G and Lykken J D 2009 *Phys.Rev.* **D80** 113008 (*Preprint* 0908.2993)
[91] Kopp J, Machado P A and Parke S J 2010 *Phys.Rev.* **D82** 113002 (*Preprint* 1009.0014)
[92] Engelhardt N, Nelson A E and Walsh J R 2010 *Phys.Rev.* **D81** 113001 (*Preprint* 1002.4452)
[93] Murayama H and Yanagida T 2001 *Phys.Lett.* **B520** 263–268 (*Preprint* hep-ph/0010178)
[94] Giunti C and Laveder M 2010 *Phys.Rev.* **D82** 113009 (*Preprint* 1008.4750)
[95] Antusch S and Fernandez-Martinez E 2008 *Phys. Lett.* **B665** 190–196 (*Preprint* 0804.2820)
[96] Auerbach L *et al.* (LSND Collaboration) 2005 *Phys.Rev.* **D72** 076004 (*Preprint* hep-ex/0506067)
[97] Adamson P *et al.* (MINOS Collaboration) 2010 *Phys.Rev.Lett.* **105** 151601 (*Preprint* 1007.2791)
[98] Abbasi R *et al.* (IceCube Collaboration) 2010 *Phys.Rev.* **D82** 112003 (*Preprint* 1010.4096)
[99] Mavromatos N E, Meregaglia A, Rubbia A, Sakharov A and Sarkar S 2008 *Phys.Rev.* **D77** 053014 (*Preprint* 0801.0872)
[100] Sakharov A, Mavromatos N, Meregaglia A, Rubbia A and Sarkar S 2009 *J.Phys.Conf.Ser.* **171** 012038 (*Preprint* 0903.4985)
[101] Hollenberg S, Micu O, Pas H and Weiler T J 2009 *Phys.Rev.* **D80** 093005 (*Preprint* 0906.0150)

Discrete symmetries and models of flavor mixing

Alexei Yu Smirnov

International Centre for Theoretical Physics, Strada Costiera 11, 34014 Trieste, Italy

E-mail: smirnov@ictp.it

Abstract.

Evidences of a discrete symmetry behind the pattern of lepton mixing are analyzed. The program of "symmetry building" is outlined. Generic features and problems of realization of this program in consistent gauge models are formulated. The key issues include the flavor symmetry breaking, connection of mixing and masses, *ad hoc* prescription of flavor charges, "missing" representations, existence of new particles, possible accidental character of the TBM mixing. Various ways are considered to extend the leptonic symmetries to the quark sector and to reconcile them with Grand Unification. In this connection the quark-lepton complementarity could be a viable alternative to TBM. Observational consequences of the symmetries and future experimental tests of their existence are discussed.

1. Flavor of flavor models

Certain features of the leptonic mixing can be considered as an evidence of discrete symmetry. Many different realizations exist [1]. However, in spite of various interesting developments, no convincing model based on discrete symmetries has been proposed so far. The models require new extended structures, many assumptions, *ad hoc* assignments of charges and selection of the group representations for multiplets. Additional auxiliary symmetries are needed which sometimes are even more powerful than the original one. There are no natural and simple extensions of the leptonic symmetries to the quark sector. In most of the models no connection between mixing and masses exists and different physics (symmetry) is responsible for the mass hierarchies.

So, what to do? Try further using the same context [1]. Be less ambitious and explain only some features (e.g., dominant structures of the mass matrices) using the symmetry? Or modify context (some important elements can be still missed). Apply symmetry differently?

For illustration of these statements, several recently proposed models will be discussed. Generic features of the whole approach and its challenges are formulated. For details of specific models see talks [3] at this conference.

The paper is organized as follows. In sect. 2 the experimental evidences in favor of discrete symmetries will be presented. Sect. 3 is devoted to the Tri-Bimaximal (TBM) mixing and "symmetry building". In sect. 4 attempts to extend the leptonic symmetry to the quark sector are discussed. Sect. 5 is devoted to alternatives to the TBM approach, in particular, to the quark-lepton complementarity (QLC). Sect. 6 outlines perspectives in the field, and conclusions are given in Sect. 7.

[1] In attempt to further pursue this approach and to check whether something interesting is overlooked, systematic scanning (including computerized one) of all possible models within certain framework has been performed [2].

2. Data and evidences

The origin of excitement is the neutrino mass and flavor spectrum shown in fig. 1. Regularities of the flavors distribution are obvious: ν_μ and ν_τ flavors share ν_3 equally (bi-maximal mixing) and ν_e is absent in ν_3, all three flavors are presented with the same weight in ν_2 (trimaximal mixing). This can be formalized as invariance of the picture with respect to the following transformations: (1) permutation of ν_μ and ν_τ flavors in all three mass states, which also implies zero 1-3 mixing; (2) dilatation of the ν_e-flavor part by factor 2 and shrinking by 2 the rest of the state ν_2, then inverse operation in ν_1: shrinking by factor 2 the ν_e- part, and dilatation of the rest, and finally, permutation of states ν_1 and ν_2. These transformations are the basis of possible discrete symmetry.

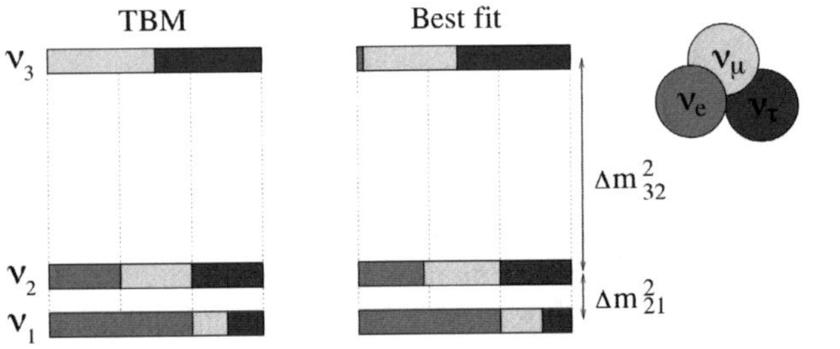

Figure 1. Mass and flavor spectrum of neutrinos in the case of TBM mixing (left) and best fit experimental values (right). The lengths of the colored parts are proportional to the moduli squared of the mixing matrix elements, $|U_{\alpha i}|$.

The regularities are described by the TBM mixing matrix [4]:

$$U_{TBM} = U_{23}(\pi/4)U_{12}(\arcsin(1/\sqrt{3})) = \begin{pmatrix} \sqrt{\frac{2}{3}} & \frac{1}{\sqrt{3}} & 0 \\ -\frac{1}{\sqrt{6}} & \frac{1}{\sqrt{3}} & -\frac{1}{\sqrt{2}} \\ -\frac{1}{\sqrt{6}} & \frac{1}{\sqrt{3}} & \frac{1}{\sqrt{2}} \end{pmatrix}. \quad (1)$$

If the length of boxes is 1, the experimental data permit deviations from the pattern shown in the left figure by amount 0.01 - 0.05. The global fit of the oscillation data [5] agrees with the TBM mixing within $(1-2)\sigma$. It shows non-zero best fit value for the 1-3 mixing, some deviation of the 2-3 mixing from maximal one with $\theta_{23} < \pi/4$, and the 1-2 mixing slightly below $\sin^2\theta_{12} = 1/3$:

$$\sin\theta_{13} = 0.118^{+0.038}_{-0.048}, \quad \sin^2\theta_{23} = 0.463^{+0.071}_{-0.048}, \quad \sin^2\theta_{12} = 0.321^{+0.016}_{-0.016} \quad (2)$$

(with 1σ errors). The deviations from the TBM values are small but robust appearing in the global analyses of different groups. Important feature is that the 1-2 and 1-3 mixing angles correlate in the data, and therefore should be considered simultaneously. The combined analyses of the solar and KamLaND results [6] [7] [8] give non-zero 1-3 mixing at the level $\sin\theta_{13} = 0.02$, $\sin^2\theta_{12} < 0.33$, and agreement with TBM is at 95% CL.. At the same time, the complete $3\nu-$ analysis of the atmospheric neutrino data from SuperKamiokande results in $\sin 2\theta_{13} = 0.0$ (< 0.04) 90%, and $\sin 2\theta_{23} = 0.5$ $(0.407 - 0.583)$ 90% [9]. That is, the best fit value of the 2-3 mixing is maximal in the case of normal mass hierarchy, in spite of the excess of $e-$like sub-GeV events which gave an indication for $\theta_{23} < \pi/4$.

Several reservations are in order.

1. Large deviations from the symmetry case are still possible and if correct (from the fundamental physics point of view) measure of mixing is $\sin\theta$, then for 1-3 mixing we have $\sin\theta_{13} = (0.10 - 0.15)$ as the best fit value which should be compared with $\sin\theta_{12} = 0.55$.

2. Mixing is not the RGE invariant, and if theory is formulated at some high mass scales (e.g., GUT scale), one may expect significant deviations from the symmetry case when running to high energies.

Journal of Physics: Conference Series **335** (2011) 012006 doi:10.1088/1742-6596/335/1/012006

3. There are several indications that new light ($m_s \sim 1$ eV^2) and almost sterile neutrino state exists which mixes substantially ($\theta_{as} \sim 0.2$) with the active neutrinos. Since $m_s \sin^2\theta_{as} \sim 0.04$ eV is bigger than the largest element of the mass matrix of active neutrinos (in the case of mass hierarchy), the presence of this state (or states) destroys the present constructions unless further complications and fine tunings are introduced.

4. Existence of the 4th generation of fermions can change the present results completely.

All this may lead to different line of developments.

Concerning the mass hierarchy, there are several important features which can be relevant for, but usually not addressed in the models under consideration: (1) The upper quarks have geometrical mass relation: $m_u m_t = m_c^2$; this hints that masses of all three generation should be considered on the same footing (and not in perturbative manner). (2) The down quark masses satisfy the Gatto-Sartori-Tonin relation $\sin\theta_C \approx \sqrt{m_d/m_s}$ [10] – explicit relation between masses and mixing. (3) The charged leptons satisfy the Koide relation [11] which indicates certain symmetry, and again, an involvement of all three generations of leptons. (4) Neutrinos have the weakest mass hierarchy among fermions which also shows connection between masses and mixing.

3. TBM and symmetry building

To a large extent relevant symmetry can be systematically derived from the data [1], [12] and this gives the main support of the approach "Mixing from discrete symmetries".

3.1. Deriving symmetry

1. In assumption that neutrinos are Majorana particles the TBM mass matrix is given by $m_{TBM} = U_{TBM} m^{diag} U_{TBM}^T$, where $m^{diag} \equiv diag(m_1, m_2, m_3)$. In the flavor basis explicitly:

$$m_{TBM} = \begin{pmatrix} a & b & b \\ \ldots & \frac{1}{2}(a+b+c) & \frac{1}{2}(a+b-c) \\ \ldots & \ldots & \frac{1}{2}(a+b+c) \end{pmatrix}, \tag{3}$$

$$a = \frac{1}{3}(2m_1 + m_2), \quad b = \frac{1}{3}(m_2 - m_1), \quad c = m_3. \tag{4}$$

Immediately one observes the S_2 symmetry of the $\nu_\mu \leftrightarrow \nu_\tau$ permutations. This symmetry plays a crucial role: it is this symmetry of the dominant structure of the mass matrix that ensures maximal 2-3 mixing and zero (small) 1-3 mixing which are robust features of the lepton mixing. It fixes two out of three angles, and probably, should be a starting point of the symmetry building.

The TBM-symmetry can be expressed in the form of the TBM-relations:

$$m_{e\mu} = m_{e\tau}, \quad m_{\mu\mu} = m_{\tau\tau}, \quad m_{ee} + m_{e\mu} = m_{\mu\mu} + m_{\mu\tau}, \tag{5}$$

or $\sum_\alpha m_{e\alpha} = \sum_\beta m_{\mu\beta}$ instead of the last equality. Notice that in general, fixing three mixing angles leads to three relations between the matrix elements. What is non-trivial is that the relations (5) are simple and of particular type which corresponds to certain symmetry. (In general relations are complicated and no symmetry can be found).

2. The TBM mass matrix (3) in the flavor basis is invariant under transformations

$$V_i m_{TBM} V_i^T = m_{TBM}, \tag{6}$$

where

$$V_1 = S = \frac{1}{3}\begin{pmatrix} -1 & 2 & 2 \\ \ldots & -1 & 2 \\ \ldots & \ldots & -1 \end{pmatrix}, \quad V_2 = U = \begin{pmatrix} 1 & 0 & 0 \\ \ldots & 0 & 1 \\ \ldots & \ldots & 0 \end{pmatrix}. \tag{7}$$

Two transformations (7) uniquely determine the form of the TBM mass matrix. They do not depend on the mass spectrum, and in fact, can be obtained immediately from the U_{TBM}. At the same time, diagonality of the mass matrix squared of the charged leptons, $m_e^\dagger m_e$, can be supported by symmetry

$$V_3^\dagger (m_e^\dagger m_e) V_3 = m_e^\dagger m_e, \quad V_3^T = diag(1, e^{i\alpha}, e^{i\beta}), \tag{8}$$

where $\alpha \neq \beta \neq \pi k$.

The main idea is that V_i are the generating elements of certain discrete symmetry group G_f which eventually determines mixing. For instance, selecting $V_3^T = T \equiv diag(1, \omega, \omega^2)$, where $\omega \equiv e^{i2\pi/3}$, one finds that S, T, U are the generating elements of the group S_4. Furthermore, it was argued that S_4 group is minimal structure which leads to TBM [12] (see discussion in [13]). (Our consideration was in the flavor basis, however neither mixing matrix nor symmetry depend on the basis once the change of basis is described by transformation V such that $VV^T = I$.)

3. The flavor symmetry G_f (which contains V_i) should be broken. (In fact, no exact flavor symmetry can be introduced in whole theory, see [14]).) Neutrino mass matrix is not invariant under T, whereas the charge lepton mass matrix is not invariant with respect to S and U. The idea is that symmetry G_f is broken *differently and partially* in the sectors which generate the neutrino masses and charged lepton masses. Namely,

$$G_f \to \text{breaking} \to \begin{cases} G_\nu & (S, U) & \text{neutrinos} \\ G_l & (T) & \text{charged leptons} \end{cases} . \tag{9}$$

The *residual* symmetries G_ν and G_l in the neutrino and charged lepton sectors are different. Clearly G_f is broken completely in whole theory. Furthermore, the two sectors communicated in high orders, and therefore G_ν and G_l are broken even in their own sectors being approximate. As a result, the TBM symmetry is broken and the TBM mixing gets corrections.

Thus, the mixing appears as a result of different ways of the flavor symmetry breaking in the neutrino and charge lepton sectors. In turn, the difference of symmetry breaking can originate (1) from different flavor assignments of the right handed (RH) components of N^c and l^c, or/and (2) from Majorana nature of the neutrino mass terms (absence of N^c): the neutrino and the charged lepton mass terms (originate from LL and Ll^c correspondingly) can have different flavor properties.

If the symmetry G_f is broken spontaneously, one should introduce different sets of the Higgs (flavon) fields for neutrinos and charged leptons. Then the form of the elements of residual symmetries, S, U and T, determines configurations of VEV's.

The items 1- 3 is a paradigm of the present day flavor model building.

4. It is possible to proceed even further in some particular way. The TBM mass matrix can be presented as the sum of three singular matrices. In the limit of small m_1 it becomes as

$$m_{TBM} \approx A \begin{pmatrix} 1 \\ 1 \\ 1 \end{pmatrix} \times (1, 1, 1) + B \begin{pmatrix} 0 \\ 1 \\ -1 \end{pmatrix} \times (0, 1, -1). \tag{10}$$

This form indicates the see-saw mechanism, for which the mass terms may have the form

$$\sum_i \frac{1}{M_i} (L f_i)(L f_i)^T, \tag{11}$$

where M_i are the masses of RH neutrinos and f_i are the triplets of flavon fields, (see e.g. [15]). The VEV's of the triplets should be $\langle f_2 \rangle (1, 1, 1)$ and $\langle f_3 \rangle (0, 1, -1)$, which can be obtained in a SUSY version of this model.

So far so good. Problems appear when we start to realize this program in consistent gauge models.

3.2. Problems

A generic problem originates from the fact that masses of fermions are given by

$$m = F(\{Y\}, \{v\}), \tag{12}$$

where F is certain functional which depends on the mechanism of neutrino mass generation. In general, it describes several different contributions and includes various corrections. $\{Y\}$ are the Yukawa couplings or coupling constants of different operators generating masses, and $\{v\}$ refer to a set of vev's of Higgs bosons. $\{Y\}$ and $\{v\}$ follow from independent sectors: from the Yukawa sector and the scalar potential. Yukawa couplings are determined by symmetry (at least partially), whereas VEV's are fixed by pattern of symmetry violation. In general, symmetry does not control how it is broken and new ingredients (dynamics, symmetries) should be introduced to fix the VEV alignment. To obtain TBM all these components should be correlated. One step constructions do not work. Essentially this means that TBM is not immediate consequence of symmetry but a result of interplay of different factors, and in this sense – accidental.

Another generic problem is related to the fact that symmetry should be broken differently in the neutrino and charged lepton Yukawa sectors. That is, different Higgs flavon multiplets (typically - several for each sector) should be used. To forbid unwanted couplings of these flavons one is forced to introduce additional symmetries with *ad hoc* charge prescription.

3.3. Flavons versus Flavored Higgses

There are two types of models with broken flavor symmetries:

1. Models with flavons f, singlets of the gauge symmetry group which have non-zero "flavor charges". In these models usual Higgs doublet(s) H are the flavor singlets, so that the electroweak symmetry and flavor symmetry breakings are separated. The Yukawa couplings and mass terms are generated by the non-renormalizable interactions:

$$\frac{1}{\Lambda_f^n} L e^c H f^n. \tag{13}$$

Here Λ_f is the scale of flavor physics which can be above the GUT scale. Clearly it is difficult to test such a scenario directly, unless Λ_f is not far above the electroweak scale.

Typical problem of this scenario is existence of high dimension operators of the type (13) which contribute to masses. Convergence of series is weak, $\langle f \rangle / \Lambda_f \sim 0.2 - 0.5$, especially if quarks (top quark) are included in consideration. Further complications (e.g. additional symmetries) are needed to control their effect.

2. Models with flavored Higgses: The Higgs doublets carry flavor charges, and usually large number of such doublets (which form flavor multiplets) should be introduced. The flavor symmetry is broken simultaneously with gauge symmetry at the EW scale. These models are testable, and in fact, strongly restricted by the FCNC, anomalous magnetic moment of muon, *etc.*. One expects to see many scalar bosons at LHC [16].

3.4. Symmetry groups

The Table 1 presents the list of small groups with irreducible representation **3**, which are used in the TBM model building. Representation **3** explains existence of three generations of fermions, however all these groups have also singlets (and some - doublets). There is no explanation why (non-trivial) singlet and doublet representations are missing. An alternative is groups, like S_3, with non-trivial irreducible representation **2**, so that the family structure appears as $\mathbf{2} + \mathbf{1}$ [17].

In the Table 1 we give the order of group (number of elements), irreducible representations and products of the representations which contain invariants. The latter determines the flavor structure of a model.

Journal of Physics: Conference Series **335** (2011) 012006 doi:10.1088/1742-6596/335/1/012006

Table 1. The simplest groups with irreducible representations **3**.

group	order	representations	invariants
A_4	12	1, 1′, 1″, 3	$3 \times 3,\ 1' \times 1''$
T'	24	1, 1′, 1″, 2, 2′, 2″, 3	$3 \times 3,\ 1' \times 1'',\ 2\times$
S_4	24	1, 1′, 2, 3, 3′	$3 \times 3,\ 3' \times 3',\ 2 \times 2,\ 1' \times 1'$
T_7	21	1, 1′, 1″, 3, 3*	$3 \times 3^*,\ 1' \times 1''$
$\Delta(27)$	27	$1_1 - 1_9$, 3, 3′	$3 \times 3',\ 1_2 \times 1_3,\ 1_4 \times 1_7,\ 1_5 \times 1_8,\ 1_6 \times 1_9$

In what follows we present structures of different models based on various discrete symmetries. The corresponding figures show explicitly *ad hoc* character of selection of the flavon multiplets and prescription of the flavor charges. An open issue is "missing" representations: not all possible low dimensional representations are used. This may create problem if discrete symmetry originates from breaking of some gauged continuous group [18]. For each model we indicate origins of mixing and shortcomings with criteria based on existence of auxiliary symmetries, presence of new fields, possibility of extension to the quarks sector and further embedding, *etc.*.

3.5. A_4 symmetry and a simplest A_4 models

The most popular group is A_4: the symmetry group of even permutations of 4 elements, or symmetry of the tetrahedron [19]. It has order 12 and two generating elements S and T which are needed to realize the TBM mixing. The presentation of the group:

$$S^2 = 1, \quad T^3 = 1, \quad (ST)^3 = 1. \tag{14}$$

The most important element, $U = A_{\mu\tau}$, is absent. So, one needs either to introduce this permutation symmetry in addition, thus extending the symmetry group, or obtain it as an accidental symmetry: as a result of particular selection of representation and the VEV alignment.

The flavor structure is determined by properties of products of representations and invariants:

$$3 \times 3 = 3 + 3 + 1 + 1' + 1'', \quad 1' \times 1'' = 1 \tag{15}$$

(and of course, by the VEV alignment).

The structure of a simplest $A4$ model for TBM mixing [20] with charge assignment is presented in fig. 2. The key features of the model include the following. There are three RH neutrinos, N^c. Four flavon multiplets participate in generation of masses. The lepton mixing appears due to different flavor assignments of the RH neutrinos and the charged leptons: N^c form triplet $N^c \sim \mathbf{3}$, whereas l^c are all singlets, $\mathbf{1}$, of A_4. The charged leptons get Dirac masses via non-renormalizable terms, whereas neutrinos - via renormalizable ones. The $U-$symmetry is accidental: due to particular selection of the flavon representations and configuration of VEV's. The auxiliary symmetry Z_4 with *ad hoc* prescribed charges is introduced; in particular, all N^c have the same whereas l^c have all different Z_4 prescriptions. The vacuum alignment is achieved by using SUSY and additional driving fields. The lepton mixing follows to a large extend from structure of mass matrix of the RH neutrinos. The model does not admit simple SO(10) embedding as well as the SU(5) embedding (if quarks also form family symmetry): l^c are singlets of the flavor symmetry group and the rest of the fermions form triplets. To resolve this problem one can introduce additional GUT matter multiplets located in the same multiplets known fermions together with new matter fields. A_4 singlet representations of L are missing; the

Journal of Physics: Conference Series **335** (2011) 012006 doi:10.1088/1742-6596/335/1/012006

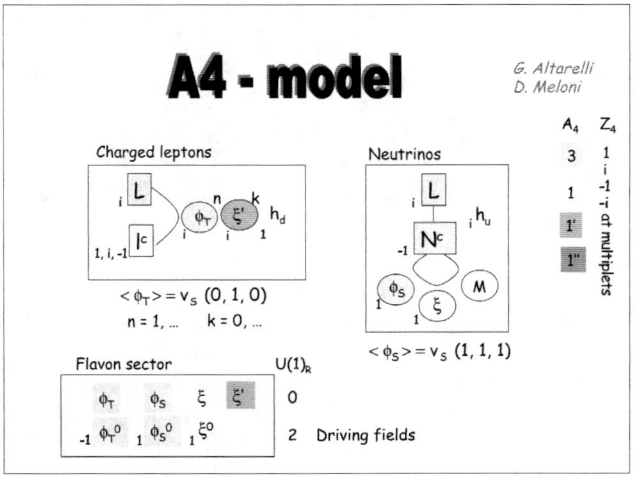

Figure 2. Scheme of the lepton mass generation in the A_4 model [20]. Colors indicate representations of the flavor symmetry group (A_4), numbers at the multiplets give transformation properties with respect to additional group (Z_4); the lines show couplings. The closed loops correspond to self-coupling of a given multiplet, e.g. $N^c N^c$.

representation $\mathbf{1}''$ is not used neither for matter fields nor for flavons. Apparently introduction of these missing multiplets will lead to new problems which could probably be cured by further complications of model.

3.6. Mixing and masses

In majority of the models mixing and masses are unrelated or have indirect relations. The latter may appear as a result of certain choice of the Higgs representation within a given symmetry context. Mixing follows from certain form invariance of the mass matrix. Usually, additional $U(1)$ (Froggatt-Nielsen type) or/and discrete symmetries are used for explanation of the mass hierarchies. Mixing is a consequence of the relations between mass matrix elements (like in eq. (5)), whereas masses depend on the absolute values of the elements.

For particular mass spectrum (set of values of masses), the mass matrix elements are fixed and in some cases this may lead to more symmetric form of the mass matrix - to additional symmetry. Then a covering symmetry group should fix both mixing and masses. (E.g., one can consider the TBM mass matrix with equality of elements $a = b$, which gives the spectrum with normal mass hierarchy and $m_1 \approx 0$, *etc.*)

4. From leptons to quarks

Do quarks need leptonic discrete symmetries? It is not accidental that in the talks devoted to flavor physics in the quark sector the leptonic symmetries proposed are not even mentioned. Although originally the first discrete symmetries have been applied for flavor in the quark sector [21]. Presently there is no clear attempts to go "from quarks to leptons" (approach which once has failed). The D_{14} symmetry has been proposed for explanation of the Cabibbo angle value without extension to leptons [22]. It is clear that the quark and lepton mixings are strongly different, and probably this difference is directly related to smallness of neutrino mass.

4.1. Extending symmetry to the quark sector

There are two different ways to extend the leptonic symmetries to the quark sector.

1) The first possibility is to use the same representations **3** and **1** for quarks as for leptons. In the lowest order one can obtain

$$V_{CKM}^0 = I, \quad U_{PMNS}^0 = U_{TBM}. \tag{16}$$

This difference of mixings can be attributed to the Majorana nature of neutrinos. As a consequence of symmetry, the Dirac mass matrices in the quark and lepton sectors are the same, both leading to zero mixing. (The Dirac matrices can be responsible for the mass hierarchies of the charged fermions.) The TBM follows via seesaw from certain structure of the Majorana mass matrix of the RH neutrinos. Then corrections from high order operators generate the CKM mixing and the deviations of lepton mixing from the TBM form. Generic problem is that corrections which would explain the Cabibbo angle lead to too large deviations from TBM, so that additional tuning is required.

2) Another way is to use different representations of the flavor symmetry group for quarks and leptons. In particular, one can choose groups which contain not only representations **3** and **1** but also **2**, and assign three generations of quarks to the representations **2** and **1** instead of **3** in lepton sector. This implies that family symmetry does not exist and leaves another question: why quarks and leptons have different symmetry properties, that is, fundamentally different.

As an example of realization of the second approach, consider model based on the symmetry T'. The T' group has order 24 being the double covering of A_4 or binary tetrahedron group. The generating elements of this group are S, T and R and presentation of the group:

$$T^3 = I, \quad S^2 = R, \quad R^2 = I, \quad (ST)^3 = 1. \tag{17}$$

Here $R = 1$ for the odd-dimensional representations and $R = -1$ for the even-dimensional representations. Again the element U is missing. Irreducible representations of the group include $\mathbf{1}$, $\mathbf{1'}$, $\mathbf{1''}$, $\mathbf{2}$, $\mathbf{2'}$,$\mathbf{2''}$, $\mathbf{3}$. The products of representations and invariants,

$$3 \times 3 = 1 + 1' + 1'' + 3 + 3, \quad 1' \times 1'' = 1, \tag{18}$$

coincide with those in the A_4 case (see Table 1). New possibilities are related to existence of the doublet representations

$$2^a \times 3 = 2 + 2' + 2'', \tag{19}$$

where $\mathbf{2}^a = \mathbf{2}$, $\mathbf{2'}$, $\mathbf{2''}$ and

$$2 \times 2 = 1 + 3, \quad 1 \times 2 = 2 \tag{20}$$

with "conservation" of primes. The singlet which appears in (20) allows one to produce new (in comparison with A_4) flavor structures. The mass generation scheme of the model [23] based on T' is shown in fig. 3.

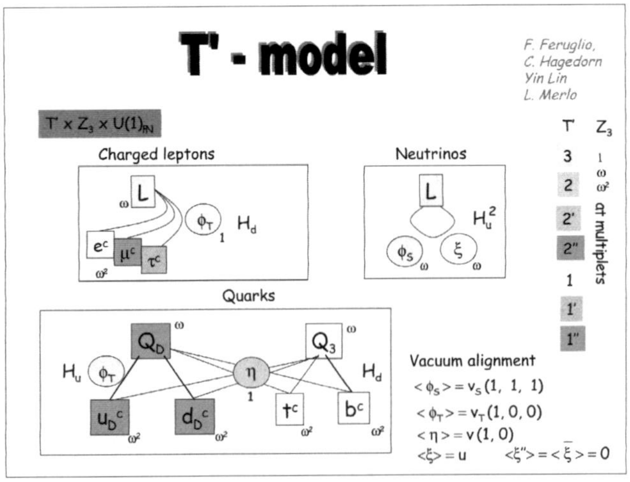

Figure 3. Scheme of the mass generation in the T' model [23]. See the caption of fig. 2 for explanation.

Features of the model include the following. The model has an auxiliary group Z_3. There is no RH neutrinos, and neutrino masses are generated by D=6 operators LLH_uH_uf, where $f = \phi_s,\ \xi$. The origin of mixing is the Majorana nature of neutrinos. In quark sector the two light generations form doublet, whereas the third one is a singlet of T'. The RH components of charged leptons are different singlets of T' but they have the same prescription of Z_3. Four different flavon fields are introduced: two triplets and one doublet $\mathbf{2}''$ (no $\mathbf{2}, \mathbf{2}'$). Only doublets $\mathbf{2}''$ are used for quarks. Z_3 prescription looks random: doublets transform with ω, all RH components have ω^2 transformation, and there are various missing representations.

4.2. TBM and GUT's

Generic problem of many models is that the flavor prescriptions required for explanation of difference of mass and mixing of quarks and leptons prevents from their embedding into Grand Unified Theories. The problem can be resolved by increasing number of matter fields and locating the known fermions with new ones in the same multiplets.

One can start immediately from the GUT structure and known matter fields:

$$GUT \times G_{flavor} + \text{new elements}, \tag{21}$$

where "new elements" may include singlet fermions and additional Higgses or/and pairs of vector-like matter fields, their mixing with usual matter, *etc.*.

As an example, structure of the model based on $SU(5) \times A_4$ [24] is shown in fig. 4. The upper quarks get masses via interactions $T_iT_jH_5\{f_{ij}\}$, where $\{f_{ij}\}$ is product of certain number (from zero to 5) of flavon fields: e.g., $f_{33} = 1$, $f_{32} = \phi_{123}\phi_3\xi^2$.

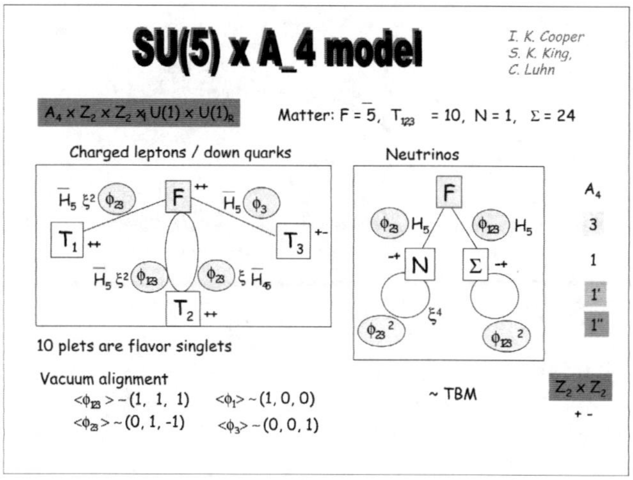

Figure 4. Scheme of generation of masses of neutrinos, charged leptons and down quarks in the $SU(5) \times A_4$ model [24]. See the caption of fig. 2 for explanation.

The following remarks are in order. Extended auxiliary symmetry $Z_2 \times Z_2 \times U(1) \times U(1)$ is imposed. The singlet N and $\Sigma = 24$ - adjoint representation of fermions are introduced apart from usual $\mathbf{10}$-plets and $\bar{\mathbf{5}}-$ plets; neutrino masses are generated by a combination of type I and type III seesaw. Only 5$-$plets form family structure, whereas 10$-$plets and RH neutrinos are singlets of A_4. Matter fields are in $\mathbf{3}$, $\mathbf{1}$, but $\mathbf{1}'$, $\mathbf{1}''$ are missing. Four different flavon multiplets with "random" $Z_2 \times Z_2$ prescriptions generate masses.

Actually, $SU(5)$ has enough flexibility: three different representations $\mathbf{10}, \mathbf{5}, \mathbf{1}$ allow one to write independent terms for the upper quarks, for down quarks and charged leptons and for neutrinos.

Models based on $SO(10)$ with all known fermions being in the same **16**-plets are more constrained. New elements should be added to the $SO(10) \times G_{flavor}$ structure. Two different ways are proposed: 1) add singlet fermions and $\mathbf{16}_H$ Higgs multiplets with couplings $\mathbf{16}S\mathbf{16}_H$ and flavons. In [25] $G_{flavor} = T_7$ and screening of the Dirac structures in the see-saw mechanism can be achieved which leads to independent structures of the mass matrices of neutrinos and charged fermions. Incomplete (partial) screening can be the origin of the TBM or bi-maximal mixings.

2). Another way of model building is to introduce **126** and $\overline{\mathbf{126}}$ plets, and thus realize the seesaw type-II mechanism which opens up a possibility to obtain (to a large extend) independent flavor mixing in the quark and lepton sectors. Realistic model proposed in [26] and based on the symmetry $S_4 \times Z_n$ requires also introduction of vector-like pairs **16**, $\overline{\mathbf{16}}$ of matter fields, additional Higgs **10**−plet, and flavons.

4.3. Is TBM accidental?

Experiment still allows relatively large deviation of the mixing parameters from the TBM values: $\Delta \sin^2\theta_{23} \sim 0.05$, $\Delta \sin^2\theta_{12} \sim 0.02$, $\Delta \sin\theta_{13} \sim 0.15$. The deviations can lead to strong (maximal) violation of the TBM-conditions (5), and consequently, to significant deviation of m_ν from the TBM form. For instance, instead of the first equality in (5), the equality with changed sign, $m_{e\mu} \sim -m_{e\tau}$, is allowed without any modification of two others. Leading structures of the mass matrix are relatively robust, whereas the sub-leading structures can change under these corrections completely. It is therefore not excluded that the approximate TBM is accidental being just an interplay of several independent factors (contributions) [27]. Alternatively it can be a manifestation of some other symmetry which differs from TBM, or other structures. This opens up new approaches to explain the data.

There are other possible applications of discrete symmetries. In the universal approach to the quark and lepton masses based on certain ansatz for the shape of the mass matrices discrete symmetries are used to get texture zeros. In this context the corrected Fritzsch ansatz has been explored in [28].

5. QLC and quark-lepton symmetries

The Quark-lepton complementarity (QLC) [29] is an alternative to description of the fermion mixings which is based on observation that

$$\theta_{12}^l + \kappa_{12}\theta_{12}^q \approx \pi/4, \quad \theta_{23}^l + \kappa_{23}\theta_{23}^q \approx \pi/4, \tag{22}$$

where $\kappa_{23}, \kappa_{12} \sim 1$, say (0.7 - 1). Qualitatively, the QLC relations mean that

- the 2-3 leptonic mixing is close to maximal because the 2-3 quark mixing is small;
- the 1-2 leptonic mixing deviates from maximal one substantially because the 1-2 quark mixing is relatively large.

In other words,

$$\text{“Lepton mixing} = \text{bi} - \text{maximal mixing} - \text{quark mixing''} \tag{23}$$

with possible implications being:

1. The quark-lepton symmetry, which, in turn, implies the quark-lepton unification, or GUT, or common family (horizontal) symmetry.
2. Existence of structure which produces the bi-maximal (BM) mixing.

The structure for BM could be related to see-saw with special properties of the RH neutrino mass matrix.

The Bi-maximal mixing, $U_{bm} = U_{23}^m U_{12}^m$, is characterized by maximal 1-2 and 2-3 rotations, and zero 1-3 rotation. There is no CP-violation. Possible scenario is that in the lowest order

$$V_{CKM}^0 = I, \quad U_{PMNS}^0 = U_{bm}, \tag{24}$$

and may be $m_1 = m_2 = 0$. If the BM structure in the lepton sector is generated by the seesaw mechanism, the corrections from the Dirac mass matrix produce (i) mass splitting, (ii) CKM and (iii) deviation from the bi-maximal mixing. The situation when the deviations and CKM mixing are related by the quark - lepton symmetry (or the same flavor symmetry for quarks and leptons) can be called "strong complementarity".

Another possibility is Cabibbo "haze" [30] [29] or the weak complementarity [31]. Deviations from BM are due to some corrections which can be of the same order in the quark and lepton sectors but not necessarily related. The corrections can be of the size of the Cabibbo angle. Possible realization is that the corrections are due to high order flavon interactions which generate simultaneously the CKM mixing and deviation from BM. In this case Grand Unification and family symmetries are not necessary.

5.1. BM-symmetry

A discrete symmetry can be behind the BM mixing as the lowest order structure (24). The bi-maximal mass relations are

$$m_{e\mu} = m_{e\tau}, \quad m_{\mu\mu} = m_{\tau\tau}, \quad m_{ee} = m_{\mu\mu} + m_{\mu\tau}. \tag{25}$$

The last equality distinguishes the bi-maximal case from TBM (see (5)). The BM mass matrix, m_{BM}, is invariant under transformations

$$V_i^T m_{BM} V_i = m_{BM}, \qquad V_i = S_{BM}, \ U, \tag{26}$$

where

$$S_{BM} = \frac{1}{2}\begin{pmatrix} 0 & \sqrt{2} & \sqrt{2} \\ \dots & -1 & 1 \\ \dots & \dots & -1 \end{pmatrix} \tag{27}$$

with property $S_{BM}^2 = I$. One can select the matrix of transformation which keeps the charged leptons mass matrix to be diagonal, in the form $T_{BM} = diag(-1, -i, i)$. In this case $T_{BM}^4 = I$, so that T and S_{BM} turn out to be the generating elements of S_4 symmetry group [32].

5.2. S_4 symmetry and model

S_4 has the order 24, it is the permutation group of 4 elements. With generating elements S_{BM} and T_{BM} it has the following presentation

$$S_{BM}^2 = T_{BM}^4 = (T_{BM} S_{BM})^3 = I \tag{28}$$

(compare with (14)). It has irreducible representations **1**, **1'**, **2**, **3**, **3'**. The products of representations read

$$3 \times 3 = 3' \times 3' = 1 + 2 + 3 + 3', \qquad 3 \times 3' = 1' + 2 + 3 + 3' \tag{29}$$

$$1' \times 1' = 1, \quad 1' \times 2 = 2 \quad 2 \times 3 = 2 \times 3' = 3 + 3' \quad 2 \times 2 = 1 + 1' + 2, \tag{30}$$

and the latter contains singlet, thus leading to new flavor structures.

Structure of the model [33] based on $S_4 \times Z_4$ and $U(1)_{FN}$ is shown in fig.5. It resembles the structure of A_4 model (see Fig. 2). The following are in order. Only S_4 representations **3**, **1**, and **1'** are used for the matter fields, the representations **3'**, **2** are missing. Flavons are in **3**, **3'**, **1** representations, and **2**, **1'** are absent. The multiplets have *ad hoc* Z_4 prescription. The Froggatt-Nielsen (FN) mechanism is introduced for the mass hierarchies and only RH components of leptons have non-zero FN-charges. Deviation from BM are due to high dimension operators with flavon fields.

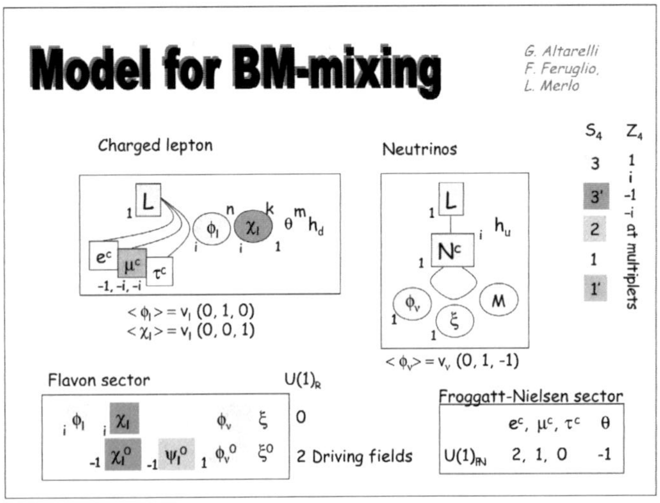

Figure 5. Scheme of the S_4 model for the weak quark-lepton complementarity [33]. See the caption of fig. 2 for explanation.

6. Perspectives and tests

Minimal and simplest models which lead to the lepton mixing of the TBM or BM type from discrete symmetries, have been systematically explored. The key problem is to check the models or at least some generic features of the whole context. Unfortunately, the majority of the models do not give specific and precise predictions which can be tested. Still one expects certain connections between the low energy observables and also probably observables at high energy accelerators under certain additional conditions.

In this connection several phenomenological directions should be mentioned.

1. Precision measurements of the neutrino parameters. Determination of the 1-3 mixing and the deviation of 2-3 mixing from maximal one are of great importance. Some models predict θ_{13}; discrimination of models with large and very small θ_{13} will be possible. Relations between θ_{13} and the deviation of θ_{23} from $\pi/2$ may reveal certain ways of realizations of the discrete symmetries.

Determination of the absolute scale of neutrino masses and mass hierarchies, establishing possible relations between mass ratios and mixings can give further insight.

2. Double beta decay. Some models lead to certain predictions for the effective Majorana mass of the electron neutrino, m_{ee}, as well as its connections to the effective electron neutrino mass, m_e, and sum of neutrino masses $\sum m_i$ [34].

3. Rare decays with lepton flavor violation: $\mu \to e+\gamma$, $\tau \to e+\gamma$, $\tau \to e+\gamma$ [35] [36]. Equality of the rates of these decays may testify for certain class of models with discrete symmetries [36]. Interesting predictions for processes like $\tau^- \to e^+\mu^-\mu^-$, $\tau^- \to \mu^+e^-e^-$ are given which depend on parameter of violation of the discrete symmetry [35].

4. LHC and high energy accelerators. Models with flavored Higgses can be directly tested in the collider experiments [37]. Even for the SM Higgs one expects modifications of the decay and production rates in the presence of horizontal symmetries![38].

5. Leptogenesis. It is affected by discrete symmetries and depends on the way the symmetries are broken [39]. In some cases suppression of the leptonic asymmetry is expected.

6. Dark matter. Particles of the dark matter (e.g. in the multi-Higgs models) can be stabilized by some discrete symmetry which is related to the flavor symmetry [40]. For instance, it may be a residual symmetry after breaking $A_4 \to Z_2$.

7. As already mentioned, some future discoveries can simply reject the described approach or require its strong modification. That includes discoveries of new fermions, like sterile neutrinos,

the 4th generation of fermions, the right handed neutrinos or W_R of the left-right symmetric models, *etc.*.

7. Conclusions

In recent years, it has been shown that the approximate TBM mixing can be consistently obtained in the context of gauge models with spontaneously broken flavor symmetries and rather extended additional structures. The considered examples of models show the price one should pay for realization of idea "mixing from discrete symmetries". There are two opposite points of view on the obtained results:

I. The features of experimental data which testify for a symmetry behind lepton mixing are actually accidental. The deviations from TBM can be significant. Realizations are too complicated with the number of assumptions being several times bigger than the number of mixing angles. This indicates that alternative approaches to explanation of the data should be pursued.

II. Some version of broken discrete symmetries give correct explanation of the data. Physics behind neutrino mass and mixing has rich extended structure and it leads to rich phenomenology. (It may happen that still some important elements of the approach are missing.

Preferable scenario? The difference of lepton and quark mixings is related *directly* to smallness of neutrino mass and probably its Majorana character. GUT's still look very appealing and there is no point to sacrifice them in favor of the present models with flavor symmetries. The observed symmetry in the lepton mixing is related to a symmetry of Hidden sector at some high mass scales. It communicates with us via the neutrino portal – mixing with neutrinos. No analogy of this in the quark sector exists. Another physics (but the same in $q-$ and $l-$ sectors) is involved in generation of CKM and deviation of PMNS from the symmetric case. Unfortunately, it is difficult, if possible, to check this possibility, but this is not the problem of Nature...

7.1. Acknowledgments

Author is grateful to D. Hernandez for numerous discussions of the material presented in this paper.

References

[1] G. Altarelli, F. Feruglio, Rev. Mod. Phys. **82** (2010) 2701. C. S. Lam, H. Ishimori, T. Kobayashi, H. Ohki, Y. Shimizu, H. Okada, M. Tanimoto, Prog. Theor. Phys. Suppl. **183** (2010) 1. R. N. Mohapatra and A. Y. Smirnov, *Ann. Rev. Nucl. Part. Sci.* **56**, 569 (2006). S. F. King, Rept. Prog. Phys. **67** (2004) 107
[2] K. M. Parattu and A. Wingerter, arXiv:1012.2842 [hep-ph]. F. Plentinger, G. Seidl and W. Winter, JHEP **0804** (2008) 077. F. Plentinger, G. Seidl and W. Winter, Nucl. Phys. B **791** (2008) 60 P. O. Ludl, arXiv:0907.5587 [hep-ph].
[3] For specific models see D. Meloni, L. Merlo, A. Toorop, C Luhn, A. Mondragon, papers in these proceedings.
[4] P. F. Harrison, D. H. Perkins and W. G. Scott, *Phys. Lett.* B **458**, 79 (1999), *Phys. Lett.* B **530**, 167 (2002); L. Wolfenstein, *Phys. Rev.* D **18**, 958 (1978).
[5] M. C. Gonzalez-Garcia, M. Maltoni, J. Salvado, JHEP **1004** (2010) 056.
[6] B. Aharmim *et al.* [SNO Collaboration], Phys. Rev. **C81** (2010) 055504.
[7] A. Gando, Y. Gando, K. Ichimura, H. Ikeda, K. Inoue, Y. Kibe, Y. Kishimoto, M. Koga *et al.*, [arXiv:1009.4771 [hep-ex]].
[8] K. Abe *et al.* [Super-Kamiokande Collaboration], [arXiv:1010.0118 [hep-ex]].
[9] R. Wendell *et al.* [Kamiokande Collaboration], Phys. Rev. **D81** (2010) 092004.
[10] R. Gatto, G. Sartori and M. Tonin, Phys. Lett. B **28** (1968) 128.
[11] Y. Koide, *Lett. Nuovo Cim.* **34** (1982) 201; *Phys. Rev.* D **28**, 252 (1983).
[12] C. S. Lam, *Phys. Rev.* D **78**, 073015 (2008). C. S. Lam, Phys. Lett. **B656** (2007) 193-198. C. S. Lam, Phys. Rev. Lett. **101** (2008) 121602.
[13] W. Grimus, L. Lavoura and P. O. Ludl, *J. Phys.* G **36**, 115007 (2009).
[14] W. Grimus and L. Lavoura, *JHEP* **0904**, 013 (2009).
[15] S. F. King, Phys. Lett. **B675** (2009) 347.

[16] R. d. A. Toorop, F. Bazzocchi, L. Merlo, A. Paris, [arXiv:1012.2091 [hep-ph]], [arXiv:1012.1791 [hep-ph]].
[17] See for instance: P. F. Harrison and W. G. Scott, Phys. Lett. B **557** (2003) 76. S. Morisi and M. Picariello, Int. J. Theor. Phys. **45** (2006) 1267. T. Teshima, Phys. Rev. D **73** (2006) 045019. R. N. Mohapatra, S. Nasri and H. B. Yu, Phys. Lett. B **639** (2006) 318. Y. Koide, Eur. Phys. J. C **50** (2007) 809. C. Y. Chen and L. Wolfenstein, Phys. Rev. D **77** (2008) 093009. Z. z. Xing, D. Yang and S. Zhou, Phys. Lett. B **690** (2010) 304.
[18] A. Adulpravitchai, A. Blum, M. Lindner, JHEP **0909** (2009) 018. C. Luhn, [arXiv:1101.2417 [hep-ph]].
[19] E. Ma and G. Rajasekaran, *Phys. Rev.* D **64**, 113012 (2001); E. Ma, *Mod. Phys. Lett.* **A 17**, 2361 (2002).
[20] G. Altarelli and D. Meloni, *J. Phys.* G **36**, 085005 (2009).
[21] S. Pakvasa and H. Sugawara, Phys. Lett. B **73** (1978) 61.
Y. Yamanaka, H. Sugawara and S. Pakvasa, Phys. Rev. D **25** (1982) 1895 [Erratum-ibid. D **29** (1984) 2135]; T. Brown, S. Pakvasa, H. Sugawara and Y. Yamanaka, Phys. Rev. D **30** (1984) 255.
[22] A. Blum, C. Hagedorn, Nucl. Phys. **B821** (2009) 327.
[23] F. Feruglio, C. Hagedorn, Y. Lin and L. Merlo, Nucl. Phys. B **775** (2007) 120 [Erratum-ibid. **836** (2010) 127].
[24] I. K. Cooper, S. F. King and C. Luhn, Phys. Lett. B **690** (2010) 396.
[25] C. Hagedorn, M. A. Schmidt and A. Y. Smirnov, *Phys. Rev.* D **79** 036002 (2009).
[26] B. Dutta, Y. Mimura, R. N. Mohapatra, JHEP **1005** (2010) 034.
[27] M. Abbas, A. Y. Smirnov, Phys. Rev. **D82** (2010) 013008.
[28] C Simoes, *talk at this conference.*
[29] A. Y. Smirnov, arXiv:hep-ph/0402264; M. Raidal, *Phys. Rev. Lett.* **93** 161801 (2004); H. Minakata and A. Y. Smirnov, *Phys. Rev.* D **70**, 073009 (2004).
[30] A. Datta, L. Everett and P. Ramond, *Phys. Lett.* B **620**, 42 (2005).
[31] G. Altarelli, arXiv:0905.2350 [hep-ph]. L. Merlo, Acta Phys. Polon. **B40** (2009) 3179.
[32] For recent studies see E. Ma, Phys. Lett. B **632** (2006) 352. C. Hagedorn, M. Lindner and R. N. Mohapatra, JHEP **0606** (2006) 042. Y. Cai and H. B. Yu, Phys. Rev. D **74** (2006) 115005. H. Zhang, Phys. Lett. B **655** (2007) 132. Y. Koide, JHEP **0708** (2007) 086. F. Bazzocchi and S. Morisi, Phys. Rev. D **80** (2009) 096005. H. Ishimori, Y. Shimizu and M. Tanimoto, Prog. Theor. Phys. **121** (2009) 769. G. J. Ding, Nucl. Phys. B **827** (2010) 82. S. Morisi, E. Peinado, Phys. Rev. **D81** (2010) 085015. F. Bazzocchi, L. Merlo, S. Morisi, Nucl. Phys. **B816** (2009) 204.
[33] G. Altarelli, F. Feruglio and L. Merlo, **JHEP 0905** 020 (2009).
[34] M. Hirsch, S. Morisi, J. W. F. Valle, Phys. Rev. **D78** (2008) 093007. J. Barry, W. Rodejohann, Nucl. Phys. **B842**, (2011) 33.
[35] F. Feruglio, C. Hagedorn, Y. Lin, L. Merlo, Nucl. Phys. **B809** (2009) 218.
[36] H. Ishimori and M. Tanimoto, arXiv:1012.2232 [hep-ph].
F. Feruglio, C. Hagedorn, Y. Lin, L. Merlo, [arXiv:0911.3874 [hep-ph]]. C. Hagedorn, E. Molinaro, S. T. Petcov, JHEP **1002** (2010) 047. F. Feruglio, A. Paris, Nucl. Phys. **B840** (2010) 405. G. -J. Ding, J. -F. Liu, JHEP **1005** (2010) 029.
[37] S. Morisi, E. Peinado, Phys. Rev. **D80**, 113011 (2009). P. H. Frampton, C. M. Ho, T. W. Kephart, S. Matsuzaki, Phys. Rev. **D82**, 113007 (2010).
[38] C. S. Lam, Phys. Lett. **B689**, 169(2010).
[39] Y. H. Ahn, S. K. Kang, C. S. Kim and T. P. Nguyen, Phys. Rev. D **82** (2010) 093005. D. Aristizabal Sierra, F. Bazzocchi, I. de Medeiros Varzielas, L. Merlo, S. Morisi, Nucl. Phys. **B827**, 34(2010). Y. H. Ahn, breaking," [arXiv:1006.2953 [hep-ph]]. Y. H. Ahn, C. -S. Chen, Phys. Rev. **D81**, 105013 (2010). C. Hagedorn, E. Molinaro, S. T. Petcov, JHEP **0909** (2009) 115. G. C. Branco, R. Gonzalez Felipe, M. N. Rebelo, H. Serodio, Phys. Rev. **D79** (2009) 093008.
[40] M. K. Parida, P. K. Sahu and K. Bora, arXiv:1011.4577 [hep-ph]. Y. Daikoku, H. Okada, T. Toma, [arXiv:1010.4963 [hep-ph]]. M. Hirsch, S. Morisi, E. Peinado, J. W. F. Valle, Phys. Rev. **D82** (2010) 116003.

Journal of Physics: Conference Series **335** (2011) 012007 doi:10.1088/1742-6596/335/1/012007

Neutrinoless Double Beta Decay Searches

Oliviero Cremonesi
INFN, Sezione di Milano Bicocca
Piazza della Scienza,3 - I-20126 Milano

E-mail: oliviero.cremonesi@mib.infn.it

Abstract. The oscillation experiments have demonstrated that neutrinos are massive particles and have provided most of the neutrino mixing matrix elements. However, two very important neutrino properties are still missing: their nature and the absolute scale of their masses. Neutrinoless double beta decay ($\beta\beta(0\nu)$) is presently the only practical tool for probing the character of neutrinos. In case neutrinos are Majorana particles $\beta\beta(0\nu)$ can moreover provide fundamental informations on their absolute mass scale. The present status of experiments searching for $\beta\beta(0\nu)$ is reviewed and the most relevant results discussed. The possibility to observe $\beta\beta(0\nu)$ at a neutrino mass scale in the range 10-50 meV is attracting a lot of interest for $\beta\beta(0\nu)$ searches. The achievement of the required experimental sensitivity is a real challenge faced by a number of new proposed projects. A review of the most relevant projects proposed for the future is given and the most relevant parameters contributing to the experimental sensitivity are finally outlined.

1. Introduction

During the past decade, the results on neutrino oscillations from atmospheric, solar, reactor and accelerator experiments have shown that neutrinos are massive particles and the three neutrino mass eigenstates are related to the three neutrino mass flavor eigenstates through the PMNS mixing matrix. This is a very strong demonstration that the Standard Model of electroweak interactions is incomplete and that new Physics beyond it must exist.

Being sensitive only to the mixing matrix parameters and to the squared mass differences between the neutrino species, the experiments on neutrino oscillations can't provide however any insight into the problems of the neutrino nature and their mass scale. Indeed, two possibile hierarchical mass arrangements (Direct and Inverted) are allowed by present data[1, 2]. The obvious quasi-degenerate option describing the situation in which the neutrino masses are all comparable and much larger than their differences is also still possible.

Two most outstanding questions are therefore still puzzling the world of neutrino Physics: the possible Majorana nature of neutrinos and their absolute mass scale. Neutrinos are the only fermions for which the Majorana formulation[3] is possible (assuming a violation of the Lepton Number). The smallness of their masses complicates however the situation and the problem of the neutrino character and absolute mass scale get mixed. Indeed the distinction between the Majorana and the Dirac formulation tends to vanish as the neutrino mass tends to zero. Although present techniques for direct measurements of the electron antineutrino mass guarantee a model-independent approach, at present they can only probe the quasi-degenerate region. On the other hand, the much more sensitive cosmological inferences and neutrinoless double-beta decay experiments could probe the inverted hierarchy but suffer from a heavy model dependance.

Published under licence by IOP Publishing Ltd

All these experimental approaches provide complementary pieces of information and a common effort is compulsory. Before the discovery that neutrinos are massive little attention was actually dedicated to the issue of Majorana neutrinos. On the other hand the situation has dramatically changed after 1998 and presently there is a common consensus that the Majorana picture is indeed the best for the physical neutrinos.

2. Neutrinoless Double Beta Decay

First suggested by M.Goeppert-Mayer in 1935[4], Double Beta Decay (DBD) is a rare nuclear process in which an initial nucleus (A,Z) decays to a member (A,Z+2) of the same isobaric multiplet with the simultaneous emission of two electrons. Given the natural trend of the nuclear masses, such a transition is possible for a number of nuclei. However, in order to avoid (or at least inhibit) the occurrence of the equivalent sequence of two single beta decays, it is generally required that both the parent and the daughter nuclei be more bound than the intermediate one. Because of the pairing term, such a condition is fulfilled in nature for a number of even-even nuclei. The decay can then proceed both to the ground state or to the excited states of the daughter nucleus. Double beta transitions accompanied by positron emission or electron capture are also possible. However they are usually characterized by lower transition energies and have correspondingly poorer experimental sensitivities. They will not be discussed in the following. We refer to the most recent reviews on $\beta\beta$ for a more complete treatment on the subject[5, 6]. Among the possible $\beta\beta$ modes two are of particular interest, the 2ν mode ($\beta\beta(2\nu)$) ${}^A_ZX \rightarrow^A_{Z+2} X + 2e^- + 2\overline{\nu}$, which observes the lepton number conservation and it is allowed in the framework of the Standard Model (SM) of electro-weak interactions, and the 0ν mode ($\beta\beta(0\nu)$) ${}^A_ZX \rightarrow^A_{Z+2} X + 2e^-$ which violates the lepton number by two units and occurs if neutrinos are their own antiparticles. A third decay mode ($\beta\beta(0\nu,\chi)$) in which one or more neutral bosons χ (Majorons) are emitted ${}^A_ZX \rightarrow^A_{Z+2} X + 2e^- + N\chi$ is also often considered. The interest in this decay is mainly related to the existence of Majorons, massless Goldstone bosons that arise upon a global breakdown of B–L symmetry and could play a significant role in the history of the early Universe and in the evolution of stars. Several models of the Majoron have been so far proposed in the framework of the supersymmetric and other extended theories[7]. Present bounds on the Majoron-neutrino coupling constant from $\beta\beta(0\nu,\chi)$ are in the range 0.35–1.9 $\cdot 10^{-4}$[8].

From the point of view of Particle Physics $\beta\beta(0\nu)$ is of course the most interesting of the $\beta\beta$ decay modes. In fact, after 70 years from its introduction by W.H. Furry[9], $\beta\beta(0\nu)$ is still one of the most powerful tools to test neutrino properties: it can exist only if neutrinos are Majorana particles and it allows then to fix important constraints on the neutrino mass scale.

When mediated by the exchange of a light virtual Majorana neutrino, the $\beta\beta(0\nu)$ rate can be expressed as

$$[T^{0\nu}_{1/2}]^{-1} = G^{0\nu}|M^{0\nu}|^2|\langle m_\nu\rangle|^2 \tag{1}$$

where $G^{0\nu}$ is the (exactly calculable) phase space integral, $|M^{0\nu}|^2$ is the nuclear matrix element and $\langle m_\nu\rangle$ is a (coherent) linear combination of the neutrino masses

$$\langle m_\nu\rangle \equiv \sum_{k=1}^{3} |U^L_{ek}|^2 m_k e^{i\phi_k} \tag{2}$$

or, explicitely for small neutrino masses

$$\langle m_\nu\rangle = c^2_{12}c^2_{13}m_1 + s^2_{12}c^2_{13}e^{i\alpha_1}m_2 + s^2_{13}e^{i\alpha_2}m_3 \tag{3}$$

where α_1 and α_2 are the neutrino Majorana phases. Unfortunately, the presence of these phases in the $\langle m_\nu\rangle$ expression implies that cancellations are possible. In particular, these cancellations are complete for a Dirac neutrino since it is equivalent to two degenerate Majorana neutrinos

Journal of Physics: Conference Series **335** (2011) 012007 doi:10.1088/1742-6596/335/1/012007

with opposite CP phases. This stresses once more the fact that $\beta\beta(0\nu)$ can occur only through the exchange of Majorana neutrinos. On the other hand $\beta\beta(0\nu)$ represents a unique possibility to measure the neutrino Majorana phases.

Altogether, the observation of $\beta\beta(0\nu)$ and the accurate determination of the $\langle m_\nu \rangle$ would definitely establish that neutrinos are Majorana particles, fixing their mass scale and providing a crucial contribution to the determination of the absolute neutrino mass scale.

It should be stressed that important constraints could be obtained even in the case that forthcoming $\beta\beta(0\nu)$ experiments would not observe any decay. Indeed, assuming that neutrinos are Majorana particles, a negative result in the 20-30 meV range for $\langle m_\nu \rangle$ would definitely rule out the inverse ordering thus fixing the neutrino hierarchy problem[1, 2].

3. $\beta\beta(0\nu)$ Nuclear Matrix Elements

As can be easily deduced from eq. (1), $\langle m_\nu \rangle$ is the only $\beta\beta(0\nu)$ measurable parameter containing direct information on the neutrino mass scale. Its derivation from the experimental $\beta\beta(0\nu)$ results requires a precise knowledge of the transition Nuclear Matrix Elements $M^{0\nu}$(NME) for which many (unfortunately often conflicting) evaluations are available in the literature. The disagreement among available calculations has lead to large uncertainty ranges for $\langle m_\nu \rangle$. This has been recognized as a critical problem by the $\beta\beta$ community and a strong effort aiming at more stable and reliable calculations has been started since few years.

The calculation of $\beta\beta(0\nu)$ nuclear matrix elements (NME) has been carried out in the last decades by many authors using different methods: Quasiparticle Random Phase Approximation (QPRPA, RQPRPA, pnQPRPA etc.), the Shell Model (SM), the Projected Hartree-Fock-Bogoliubov (PHFB) and the Interacting Boson Model (IBM2).

The different methods have complementary virtues. Indeed, QRPA calculations include many single-particle levels outside a relatively small inert core but they can hardly manage correlations. On the other hand, the shell model can include arbitrarily complicated correlations, but is limited to a few single–particle orbitals outside the inert core.

Significative improvements have been obtained recently. SM calculations are still systematically smaller than the others but NME calculations presently agree within a factor 2-3[10]. Such an agreement does not guarantee by itself the correctness of the calculations but the convergence of the results from very different methods can hardly be by chance and their comparison can help to identify the important effects responsible for the observed disagreement.

The careful check of the models in order to account for the omitted physics or the important missing informations seems the only way out of the problem. A systematic analysis of the calculation methods and their basic hypotheses have been therefore started.

A statistical analysis of the different NME calculations (comparison of different methods and model parameters) has also been recently considered[11]. Besides providing useful recipes for the comparison of the experimental results on different isotopes this approach could help in identifying systematic effects in the different calculations.

From a purely experimental point of view, the spread in the available NME calculations causes a lot of confusion in the comparison of the results and sensitivities of the different experiments. Indeed different authors tend to report $\langle m_\nu \rangle$ intervals obtained using different set of calculations thus spoiling the relevance of any comparison. Such a problem has been once more recognized recently in [6], where a practical solution consisting in referring to a Physics Motivated Average (PMA) set of NME values is suggested when comparing results or sensitivities referring to different nuclei. A different approach, consisting in disentangling the uncertainty intervals according to the different NME used calculations[12], has been also suggested recently. Such an approach has the advantage of allowing a separate comparison between different calculation methods but does not solve completely the confusion of the NME intervals.

4. Experimental approach

The only experimentally available information in $\beta\beta(0\nu)$ is carried by the daughter nucleus and the two emitted electrons. Only few experimental parameters are therefore available: sum of the electron energies, single electron energy and angular distributions, identification and/or counting of the daughter nucleus. Two main general approaches have been followed so far for $\beta\beta$ experimental investigation: i) indirect or inclusive methods, and ii) direct or counter methods. Inclusive methods are based on the measurement of anomalous concentrations of the daughter nuclei in properly selected samples, characterized by very long accumulation times. They include Geochemical and Radiochemical methods which, being completely insensitive to different $\beta\beta$ modes, can only give indirect evaluations of the $\beta\beta(0\nu)$ and $\beta\beta(2\nu)$ lifetimes. They have played a crucial role in $\beta\beta$ searches especially in the past.

Counter methods are based instead on the direct observation of the two electrons emitted in the decay. Different experimental parameters (energies and momenta of the emitted electrons, topology, etc) can then be registered according to the different capabilities of the employed detectors, which characterize also the accuracy with which the available information is collected. These methods are further classified in *inhomogeneous* (when the observed electrons originate in an external sample) and *homogeneous* experiments (when the source of $\beta\beta$'s serves also as detector).

In most cases the various $\beta\beta$ modes are separated simply on the basis of the different distribution expected for the electron sum energies: a continuous bell distribution for $\beta\beta(2\nu)$ and $\beta\beta(0\nu,\chi)$, and a sharp line at the transition energy for $\beta\beta(0\nu)$. In these cases, a good energy resolution is the most attractive experimental feature. Indeed, direct counting experiments with very good energy resolution, providing so far the best experimental results, are still the most attractive approach for $\beta\beta(0\nu)$ searches.

Table 1. Best reported results on $\beta\beta(2\nu)$ and $\beta\beta(0\nu)$ processes and most relevant $\beta\beta$ parameters. Limits are at 90% CL. $F_N^{0\nu}$ are $\beta\beta(0\nu)$ nuclear factor of merit computed using, for the NME, the PMA suggested in [6].

Isotope	$T^{2\nu}_{1/2}$[13] $(10^{19}y)$	$T^{0\nu}_{1/2}$ $(10^{24}y)$	Q (keV)	nat. ab. (%)	$F_N^{0\nu}$ $(10^{-25}y^{-1}eV^{-2})$
^{48}Ca	$(4.4^{+0.6}_{-0.5})$	> 0.0014[14]	4271	0.19	
^{76}Ge	(150 ± 10)	> 19[15]	2040	7.8	22.6
		$22.3^{+4.4}_{-3.1}$[17]			
		> 15.7[18]			
^{82}Se	(9.2 ± 0.7)	> 0.36 [19]	2995	9.2	15.1
^{96}Zr	(2.3 ± 0.2)	> 0.0092[19]	3350	2.8	
^{100}Mo	(0.71 ± 0.04)	> 1.1[19]	3034	9.6	
^{116}Cd	(2.8 ± 0.2)	> 0.17[20]	2802	7.5	
^{130}Te	(68^{+12}_{-11})	> 2.8[21]	2527	34.5	22.6
^{136}Xe	> 81[22]	> 0.45[23]	2479	8.9	15.8
^{150}Nd	$(13.3^{+4.5}_{-2.6})$	> 0.0036[24]	3367	5.6	26.7
^{238}U	(220 ± 50)	> 0.0036[24]			

Experimental evidence for several $\beta\beta(2\nu)$ decays has been provided using the measured two-electron sum energy spectra, the single electron energy distributions and the event topology[13].

On the other hand, impressive progress has been obtained during the last years also in improving $\beta\beta(0\nu)$ half-life limits for a number of isotopes (Tab.1). The best results are still maintained by the use of isotopically enriched HPGe diodes for the experimental investigation of ^{76}Ge (Heidelberg-Moscow[15] and IGEX[18]) but two other experiments have recently reached comparable sensitivities: NEMO3[19, 25] at LSM and CUORICINO at LNGS[26]. The former is a large inhomogeneous detector aiming at overcoming the intrinsic limits of the technique (relatively small active masses) by expanding the setup dimensions. The big advantage of the NEMO3 technique is the possibility to access single electron informations. CUORICINO is, on the other hand, a TeO_2 granular calorimeter based on the bolometric technique. The most attractive features of this approach are an excellent performance of the bolometers and the possibility to build large mass homogeneous detectors with any material of practical interest[27, 28]. A scan of the most interesting $\beta\beta(0\nu)$ active isotopes is thus possible. NEMO3 data taking continued until the end of 2010 while CUORICINO was stopped in June 2008 to be substituted by CUORE-0 (the first tower of CUORE) which is presently under construction at LNGS to start data taking at the end of 2011.

The evidence for a $\beta\beta(0\nu)$ signal has also been claimed[16] (and later confirmed [17]) by a small subset (KHDK) of the HDM collaboration at LNGS with $T^{0\nu}_{1/2} = 2.23^{+0.44}_{-0.31} \times 10^{25}$ y. The result is based on a sophisticated re-analysis of the HDM data heavily relying on pulse shape analysis and artificial neural network algorithms aiming to identify the $\beta\beta(0\nu)$ signal while reducing the background contributions. Such a claim has raised a lot of criticism but cannot be dismissed out of hand. On the other hand, none of the existing experiments can rule out it, and the only certain way to confirm or refute it is with additional sensitive experiments. In particular, next generation experiments should easily achieve this goal.

5. Future experiments

The performance of the different $\beta\beta(0\nu)$ experiments is usually expressed in terms of an experimental *sensitivity* or detector *factor of merit*, defined as the process half-life corresponding to the maximum signal n_B that could be hidden by the background fluctuations at a given statistical CL. At 1σ level ($n_B=\sqrt{BTM\Delta}$), one obtains:

$$F_{0\nu} = \tau_{1/2}^{Back.Fluct.} = \ln 2 \; N_{\beta\beta}\epsilon \frac{T}{n_B} = \ln 2 \times \frac{x \; \eta \; \epsilon \; N_A}{A} \sqrt{\frac{M \; T}{B \; \Delta}} \; (68\% CL)$$

where B is the background level per unit mass and energy, M is the detector mass, T is the measure time, Δ is the FWHM energy resolution, $N_{\beta\beta}$ is the number of $\beta\beta$ decaying nuclei under observation, η their isotopic abundance, N_A the Avogadro number, A the compound molecular mass, x the number of $\beta\beta$ atoms per molecule, and ϵ the detection efficiency.

Actually B never scales exactly with the detector mass but this approximation is usually reasonable and has a physical justification. Despite its simplicity, equation (4) has the unique advantage of emphasizing the role of the essential experimental parameters: mass, measuring time, isotopic abundance, background level and detection efficiency. On the other hand, it does not take into account important details like the shape of the expected signal or of the background and can't be used to analyze the case of very low statistics (discussed in the following). In these cases a more sophisticated Monte Carlo approach is needed. However, most of the criteria to be considered when optimizing the design of a new $\beta\beta(0\nu)$ experiment follow directly from eq. (4): i) a well performing detector (e.g. good energy resolution and time stability) giving the maximum number of informations (e.g. electron energies and event topology); ii) a reliable and easy to operate detector technology requiring a minimum level of maintenance (long underground running times); iii) a very large (possibly isotopically enriched) mass, of the order of one ton or larger; iv) an effective background suppression strategy. They are actually being pursued with great determination by all the next generation experiments. Unfortunately, they are often conflicting and simultaneous optimisation is rarely possible.

The case when the background level B is so low that the expected number of background events in the region of interest along the experiment life is of order of unity ($B \cdot M \cdot T \cdot \Delta \simeq O(1)$) deserves particular attention. In this case one generally speaks of "zero background" (0B) experiments, a condition met by a number of future projects. In such conditions, eq. (4) can no more be used and a good approximation to the sensitivity is given by

$$F^{0B}_{0\nu} = \ln 2 \; N_{\beta\beta}\epsilon \frac{T}{n_L} = \ln 2 \times \frac{x \; \eta \; \epsilon \; N_A}{A} \frac{M \; T}{n_L}$$

where n_L is a constant depending on the chosen CL and on the actual number of observed events. The most relevant feature of the previous equation is that $F^{0B}_{0\nu}$ does not depend on the background level or the energy resolution and scales linearly with the sensitive mass M and the measure time T. Since T is usually limited to a few years and Δ is usually fixed (meaning that for a given experimental technique is usually difficult to get sizable improvements), the 0B condition translates to $B \cdot M \simeq O(1/\Delta \cdot T)$. This means that for a given mass M there always exists a threshold for B below which no further improvement of the sensitivity is obtained or, alternatively, that it can be useless to reduce at will the background level without a corresponding increase of the experimental mass. A well designed experiment has therefore to match the condition $B \cdot M \gtrsim 1/\Delta \cdot T$. For most of the next generation high resolution calorimeters $B_T \simeq \frac{1}{10 \cdot M}$ or $B_T \simeq 10^{-4}$ for a O(1t) experiment.

A series of new proposals has been boosted in recent years by the renewed interest in $\beta\beta(0\nu)$ following neutrino oscillation results. The ultimate goal is to reach sensitivities such to allow an investigation of the inverted hierarchy (IH) of neutrino masses ($\langle m_\nu \rangle$ ~10-50 meV). From an experimental point of view this corresponds however to active masses of the order of 1 ton with background levels of the order of 1 c/keV/ton/y. A challenge that can hardly be faced by the current technology. Phased programs have been therefore proposed in USA and Europe[29, 30].

Next generation experiments are all characterized by hundred kg detectors and 1-10 c/kev/ton background rates. Their goal is to select the best technology and approach the IH region.

A restricted list of some of the most advanced forthcoming $\beta\beta(0\nu)$ projects is given in Table 2. Very different classification schemes can of course be adopted for them. They are usually based on the different strategies adopted to improve the $\beta\beta(0\nu)$ sensitivity: experimental approach, mass, energy resolution, background discrimination technique, granularity and track reconstruction, etc. In general, three broad classes can be identified: i) arrays of calorimeters with excellent energy resolution and improved background suppression methods (e.g. GERDA, MAJORANA) or based on unconventional techniques (e.g. CUORE); ii) detectors with generally poor energy resolution but topology reconstruction (e.g. EXO, SuperNEMO); iii) experiments based on suitable modifications of an existing setup aiming at a different search (e.g. SNO+, KAMLAND). In some cases technical feasibility tests are required, but the crucial issue is still the capability of each project to pursue the expected background suppression. Different estimates of the expected B levels are usually based on the extrapolation of real measurements to the final experimental conditions or on the Monte Carlo simulations based on more or less realistic expectations. The former are usually more reliable especially when based on the results of medium size detectors. The expected sensitivities are listed in Tab.2. On the other hand, the projected backgrounds are compared in Fig.1 as a function of the expected energy resolution to point out the role of the product $B \cdot \Delta$ in the comparison of the $\beta\beta(0\nu)$ sensitivities. The more realistic projections ("Reference") are here distinguished from the most optimistic expectations ("Improved"). The same holds for Fig.2 where the plotted line identifies the transition to the 0B regime. Although all proposed projects show interesting features for a second generation experiment, only few of them are characterized by a reasonable technical feasibility within the next few years.

Journal of Physics: Conference Series **335** (2011) 012007 doi:10.1088/1742-6596/335/1/012007

Table 2. List of some of the most developed $\beta\beta(0\nu)$ projects. Sensitivity at 90% CL (duration as explicitely indicated). Experimental phases are indicated as preparation (P=Progress) or development (D=R&D). $\langle m_\nu \rangle$ values are calculated using PMA NME and phase space factors from [6] and [31] respectively.

	Isotope	Mass [kg]	Lab	Status	Start	$T^{0\nu}_{1/2}$ [10^{26}y]	$\langle m_\nu \rangle$ [meV]
CUORE[32]	^{130}Te	200	LNGS	P	2014	2.1	45
GERDA[33]	^{76}Ge	18	LNGS	P	2012	2	100
		40		D		5	56
MAJORANA[34]		30	SUSEL	P	2013	5	67
EXO[35]	^{136}Xe	200	WIPP	P	2011	2.2	54
		1000		D	2015	7	30
SuperNEMO[36]	^{82}Se	100-200	LSM	D	2013-2015	0.8	90
KamLAND-Zen[37]	^{116}Cd	400	Kamioka	P	2011	3.1	45
		1000		D	2013-2015	11	24
SNO+[38]	^{150}Nd	56	Sudbury	P	2012	0.5	87
		1000		D	2015	2.2	41

5.1. High resolution calorimeters

MAJORANA and GERDA belong to the class of the high energy–resolution calorimeters and are both phased programs representing large scale extensions of past successful experiments on ^{76}Ge $\beta\beta(0\nu)$. Background control is based upon a careful choice of the setup materials and of very effective radiation shields. Active background reduction based on new detector design for single site event identification represents the new frontier and is presently gathering most of the experimental efforts. Evolved from the HM experiment, GERDA[33] aims at implementig the concept of Ge diodes immersed in a LAr bath[39] for a radical background suppression. The GERDA setup construction was completed in Gran Sasso during 2010. Since then, pre-operation with natural abundance Germanium crystals is ongoing. Two experimental phases are foreseen. GERDA-I is going to scrutinize the KHDK claim starting in 2011 with 18 kg of enriched detectors (~85%) inherited from previous experiments and an expected background level of the order of 10^{-3} c/keV/kg. 40 kg of germanium isotopically enriched in ^{76}Ge are already available for GERDA-II. One of the most attractive features of the Germanium diodes is the excellent energy resolution, of the order of 3.5 keV FWHM in the region of interest (ROI). A large part of the efforts is presently directed to develop the detectors with background reduction capability crucial for the targeted 10^{-3} c/keV/kg background level. The expected 5y sensitivity is $\sim 2.3 \times 10^{26}$ and 1.1×10^{27}y for the phase I and II respectively. Depending on the actual physics results of the two experimental phases, a third phase using 500 to 1000 kg of enriched germanium detectors is planned, merging GERDA with the US lead Majorana collaboration.

MAJORANA, a mainly USA proposal with important Canadian, Japanese, and Russian contributions, is an evolution of the IGEX experiment. The proposed configuration[34] is based on segmented n-type germanium crystals distributed inside ultra-clean electro-formed conventional cryostats (57 crystals each). The whole assembly would be enclosed in a low-background passive shield and active veto and be located deep underground. A 30 to 60 kg demonstrator (single cryostat) is presently being developed to demonstrate the viability of the technique. The detector performance is comparable to GERDA but a very low background rate, of the order of 2×10^{-4} c/keV/kg, is the distinctive and ambitious target of this project. The

Figure 1. Projected background rates as a function of the expected energy resolution for the $\beta\beta(0\nu)$ projects listed in table 2.

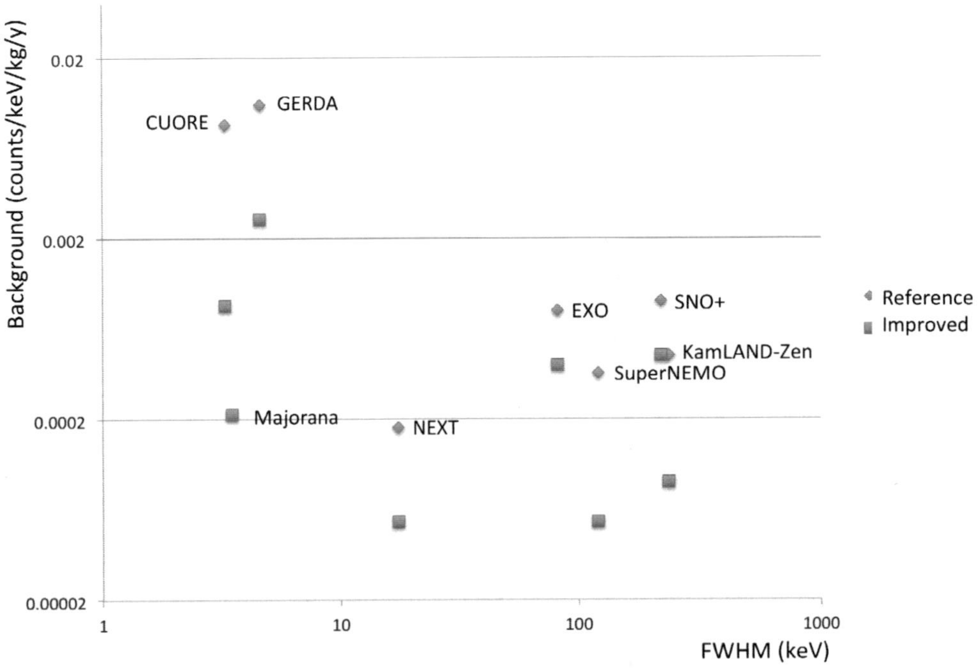

completion of this phase is expected in 2014.

CUORE[32] (*Cryogenic Underground Observatory for Rare Events*) is a very large extension of the TeO_2 bolometric array concept pioneered by the Milano group at the Gran Sasso Laboratory since the eighties. CUORE consists of a rather compact cylindrical structure of 988 cubic natural TeO_2 crystals of 5 cm side (750 g), arranged into 19 separated *towers* (13 *planes* of 4 crystals each) and operated at a temperature of 10 mK. The expected energy resolution is ~5 keV FWHM at the $\beta\beta(0\nu)$ transition energy (~2.53 MeV). A background level of the order of ~0.01 c/keV/kg/y or better is expected by extrapolating the CUORICINO background results and the dedicated CUORE R&D measurements. The expected 5y sensitivity is 2.1×10^{26}y. CUORE will therefore allow a close look at the IH region of neutrino masses. CUORE is fully funded and presently under construction at LNGS. Setup completion is expected in 2013 followed by data taking start in 2014.Thanks to the high natural abundance of ^{130}Te , CUORE is based on the use of natural tellurium even if an isotopically enriched version has been discussed as a future option. The most important limitation of this purely bolometric approach is presently represented by the difficulty to develop an active way of identify surface radioactivity contributions of the detector materials.

Thanks to the bolometer's versatility, alternative options with respect to TeO_2 are also possible. In particular, promising results have been recently obtained with scintillating bolometers which are particularly effective in recognizing the dangerous alpha background from the surface of the detector setup[40]. The use of hybrid scintillating bolometers could allow to study in the future new $\beta\beta(0\nu)$ active isotopes with improved sensitivity. A ~40 kg demonstrator

Figure 2. Projected background rates as a function of the mass and expected energy resolution product for the $\beta\beta(0\nu)$ projects listed in table 2.

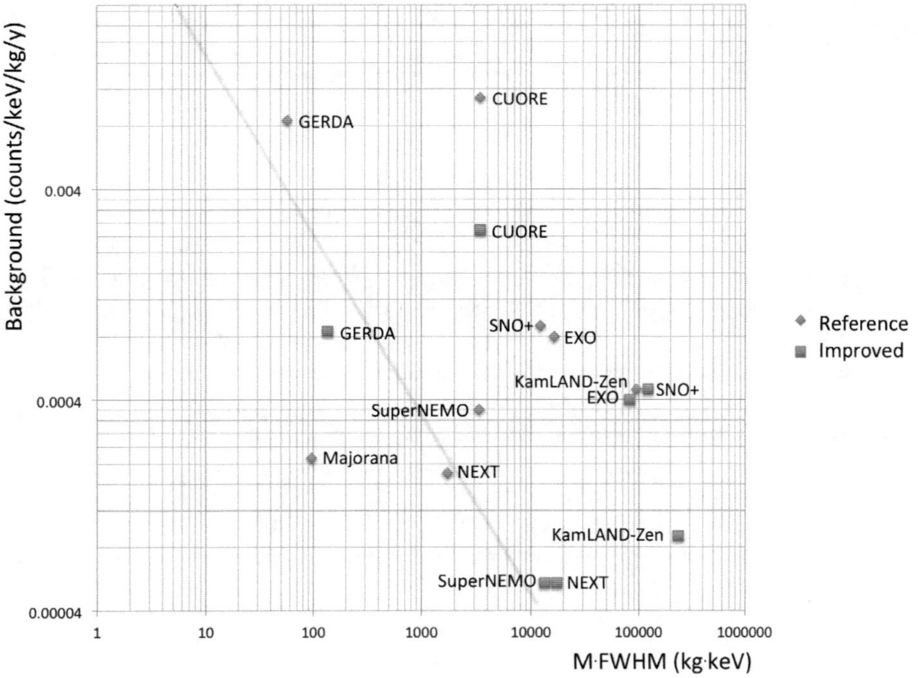

(Lucifer) aiming at applying this hybrid technique to demonstrate the feasibility of a ZnSe experiment with a background level of the order of 10^{-3} c/keV/kg was recently funded in the framework of an ERC research program. Results are expected in 2014.

5.2. Tracking detectors

Gas and liquid Time Projection Chambers (TPC's) represent another aspect of the homogeneous approach in which the limited resolution is the most relevant limitation while scalability and geometrical reconstruction are the most evident advantages. EXO[35] (*Enriched Xenon Observatory*) is a challenging project based on a large mass ($\sim$ 1–10 tons) of isotopically enriched (85% in ^{136}Xe) Xenon. An ingenious tagging of the doubly charged Ba isotope produced in the decay ($^{136}Xe \rightarrow ^{136} Ba^{++} + 2e^{-}$) would allow an excellent background suppression. The technical feasibility of such an ambitious project aiming at a complete suppression of all the backgrounds requires a hard, still ongoing R&D phase. The unavoidable $\beta\beta(2\nu)$ contribution is a serious concern due to the poor energy resolution of Xe detectors. A sizable prototype experiment with a Xe mass of 200 kg (80% ^{136}Xe), has been deployed at WIPP since summer 2009. The prototype has an expected energy resoution of $\sim$80 keV FWHM and no barium tagging. With a background level of $\sim 10^{-3}$ c/keV/kg, the primary goal is to measure ^{136}Xe $\beta\beta(2\nu)$ and to study $\beta\beta(0\nu)$ with a sensitivity of $\sim 10^{26}$ y in two years of data taking.

The use of a TPC filled with high-pressure gaseous xenon, and with capabilities for calorimetry and tracking (following the experience of the past Gotthard experiment) has been recently proposed by a mainly Spanish collaboration headed by the Valencia group [41].

Thanks to an excellent energy resolution (~1% at 2580 keV), together with a powerful background rejection provided by the distinct double-beta decay topological signature, the NEXT collaboration aims at a phased program starting with a 100 kg TPC capable of exploring the 100 meV region hence analysing the KHDH claim. Expected to operate in the Canfranc Underground Laboratory (LSC) and characterized by a projected background level of the order of few 10^{-4} c/keV/kg/y, NEXT-100 will be large enough to prove the scalability of the technology up to a 1-ton detector. Smaller scale prototypes have been already built and operated successfully.

5.3. Large mass scintillators

The idea of exploiting large mass and very low background scintillators loaded with $\beta\beta(0\nu)$ active materials dates back to the end of 90's[42]. New developments have been recently proposed in order to exploit two successful experiments on neutrino oscillation such as SNO and KamLAND. SNO+ is pursuing the goal of studying ^{150}Nd with 50 to 500 kg of isotopically enriched Neodimium depending on the results of the currently ongoing R&D program. Difficulties in reaching a significant concentration of the $\beta\beta(0\nu)$ element while maintaining a good detector performance are presently addressing most of the efforts.

A similar approach is proposed by KAMLAND-Zen, in which large masses of ^{136}Xe are dispersed in a dedicated plastic bag filled with liquid scintillator at the center of the KAMLAND detector. The program is going to start in 2011 (first phase) with 200-400 kg of isotope and continue in 2013 with 1 ton of Xenon enriched to 90% in ^{136}Xe. With a projected background of $\sim 5 \times 10^{-4}$ c/keV/kg and an energy resolution of 240 keV FWHM, the preliminary estimate of the 5y sensitivity for phase I is in the range of few 10^{26} y.

The proposed Super-NEMO experiment is the only project based on an inhomogeneous approach. It is an extension of the successful NEMO3 concept, properly scaled in order to accommodate ~100 kg of ^{82}Se foils spread among 20 detector modules. The proposed geometry is planar. The expected energy resolution is 7% FWHM (12% in NEMO-3) to improve the signal detection efficiency from 8% to 40% and reduce the $\beta\beta(2\nu)$ contribution. The projected background is $\sim 3.5 \times 10^{-4}$ c/keV/kg. The detector modules will have an active water shield to further reduce cosmic ray backgrounds. The proposed detector dimensions will require a larger hall than is currently available at Frejus and an expansion of the facility is therefore required and actively pursued. A demonstrator (single module) is presently fully funded to be completed in 2011 with a test run in the current NEMO3 site. If funded, Super-NEMO construction should immediately start.

6. Conclusions

Neutrino oscillation results have stimulated a renewed interest in the experimental study of neutrino properties. In this framework, neutrinoless $\beta\beta$ decay is recognized as a unique tool to verify the Majorana nature of the neutrino providing moreover important informations on the neutrino mass scale and intrinsic phases, unavailable to the other neutrino experiments. Present $\langle m_\nu \rangle$ sensitivities are still outside the range required to test the inverted neutrino mass hierarchy. An international effort is however supporting a phased $\beta\beta(0\nu)$ program based on a number of newly proposed experiments aiming to pursue such a goal. The success of such a program strongly depends on the true capability of the proposed projects to reach the required background levels in the $\beta\beta(0\nu)$ region. The claimed evidence for a $\beta\beta(0\nu)$ signal in the HM data could be soon verified by the presently running experiments and in any case, by the forthcoming next generation experiments.

References

[1] A. Strumia and F. Vissani, hep-ph/0606054v2.

[2] S. Pascoli, S.T. Petcov and T. Schwetz, Nucl. Phys. B 734, 24 (2006), hep-ph/0505226.
[3] E, Majorana,*Nuovo Cim.* 14 (1937) 171.
[4] M. Goeppert-Mayer, Phys. Rev. **48** (1935) 512.
[5] S.R. Elliott and P. Vogel, *Annu. Rev. Nucl. Part. Sci.* **52**, 115 (2002); hep-ph/0202264; J.D. Vergados, *Phys. Rep.* **361**, 1 (2002); J. Suhonen and O. Civitarese, *Phys. Rep.* **300**, 123 (1998); V.I. Tretyak and Y. Zdesenko, *Atomic Data and Nuclear Data Tables* **80**, 83 (2002) and **61**, 43 (1995); F.T. Avignone III,S.R. Elliott and J. Engel, *Rev. Mod. Phys* 80 (2008) 2013.
[6] J.J. Gomez-Cadenas et al, hep-ex/1010.5112v3.
[7] R.N. Mohapatra and E. Takasugi, *Phys. Lett.* B 211 (1988) 192.
[8] A.S. Barabash, nucl-ex/1002.2862
[9] W.H. Furry, *Phys. Rev.* **56**, 1184 (1939).
[10] A. Faessler, nucl-th/1104.3700v1.
[11] A.Faessler et al., hep-ph/0810.5733v2
[12] O. Cremonesi, hep-ex/1002.1437.
[13] A.S. Barabash, *Phys.Atom.Nucl.* 74 (2011) 603.
[14] I. Ogawa et al., *Nucl. Phys.* A **730** (2004) 215
[15] H.V. Klapdor-Kleingrothaus et al. (HM Coll.), *Eur. Phys. J.* A **12**, 147 (2001) and hep-ph/0103062
[16] H.V. Klapdor-Kleingrothauset al., *Mod. Phys. Lett.* A16 (2001) 2409, hep-ph/0201231.
[17] H.V. Klapdor-Kleingrothaus, *Phys. Scr.* T **127** (2006) 40; H.V. Klapdor-Kleingrothaus and I.V. Krivosheina, *Modern Physics Letters* A **21** (2006) 1547.
[18] C.E. Aalseth et al., *Phys. Rev.* D **65** (2002) 092007
[19] K. Lang (NEMO3 coll.), Proc. fo the WIN09 conference, Perugia September 2009, Italy.
[20] F.A. Danevich et al., *Phys. Rev.* C **68** (2003) 035501
[21] E. Andreotti et al., *Astropart.Phys.* 34 (2011) 822; nucl-ex/1012.3266.
[22] J.M. Gavriljuk et al., *Phys. Rev.* C **61**, 035501 (2000).
[23] R. Bernabei, *Phys. Lett.* B **546**, 23 (2002).
[24] A.S. Barabash, *Phys. At. Nucl.* **68** (2005) 414.
[25] A.S. Barabash, hep-ex/0610025.
[26] C. Arnaboldi et al., *Phys. Rev.* C **78** (2008) 035502
[27] E. Fiorini and T. Niinikoski, *Nucl. Instr. and Meth.* A **224**, 83 (1984).
[28] A. Alessandrello et al., *Nucl. Instr. and Meth.* A **440**, 397 (1998) and *Phys. Lett.* B **420**, 109 (1998)and *Phys. Lett.* B **433**, 156 (1998).
[29] S.J. Freedman, B.Kayser: "The Neutrino Matrix: DNP / DPF / DAP / DPB Joint study on the future of Neutrino Physics"; physics/0411216.
[30] Astroparticle Physics: the European strategy. http://www.aspera-eu.org.
[31] F. Boehm and P. Vogel, Physics of massive neutrinos, Cambridge Univ. Press 1992.
[32] C. Arnaboldi et al., *Nucl. Instr. Meth. A* 518 (2004) 775; C. Arnaboldi et al., hep-ex/0501010; I.C. Bandac et al., *J. Phys. Conf. Ser.* 110 (2008) 082001.
[33] I. Abt et al., hep-ex/0404039; J. Janicsko-Csathy, *Nucl. Phys. B (Proc. Suppl.)* 188 (2009) 68.
[34] C.E. Aalseth et al., hep-ex/0201021; Majorana White Paper, nucl-ex/0311013; V.E. Guiseppe, nucl-ex/0811.2446.
[35] M. Danilov et al., *Phys. Lett. B* 480 (2000) 12; R. Gornea, *J. Phys. Conf. Ser.* 179 (2009) 022004.
[36] A.S. Barabash, *Phys. At. Nucl.* 79 (2004) 10; A.S. Barabash, *Czech. J. Phys.* 52 (2002) 567, nucl-ex/0203001; A.S. Barabash, *Phys. At. Nucl.* 67 (2004) 438; E. Chauveau, *AIP Conf. Proc.* 1180 (2009) 26.
[37] K. Nakamura, Proc. Int. Conf. Neutrino2010, Athens, Greece, 13-19 June, 2010.
[38] J. Maneira, Proc. Neutrino Oscillation Workshop, Conca Specchiulla, Italy, September 4-11, 2010)
[39] H.V. Klapdor-Kleingrothaus HV, hep-ph/0103074.
[40] C. Arnaboldi et al, nucl-ex/1011.5415; C. Arnaboldi et al., *Astropart.Phys.* 34 (2011) 344, nucl-ex/1006.2721; C. Arnaboldi, *Astropart.Phys.* 34 (2010) 143, nucl-ex/1005.1239.
[41] F. Granena et al., "NEXT, Letter of Intent to the LSC Scientific Committee",hep-ex/0907.4054v1
[42] B. Caccianiga and M.G. Giammarchi, *Astropart.Phys.* 14 (2000) 15; G. Bellini et al. *Eur.Phys.J.* C 19 (2001) 43.

Symposium on Prospects in the Physics of Discrete Symmetries IOP Publishing
Journal of Physics: Conference Series **335** (2011) 012008 doi:10.1088/1742-6596/335/1/012008

Baryogenesis and Leptogenesis

Antonio Riotto

CERN PH-TH Division, Geneve 23, CH-1211, Switzerland

E-mail: antonio.riotto@cern.ch

Abstract. We summarize the state of the art of baryogenesis with particular attention to thermal leptogenesis, according to which the baryon asymmetry in the Universe is produced at very high temperatures from the decay of heavy righ-handed neutrinos, and to electroweak baryogenesis, where the baryon asymmetry is generated at the electroweak phase transition.

1. Introduction

The symmetry between particles and antiparticles, firmly established in collider physics, naturally leads to the question of why the observed universe is composed almost entirely of matter with little or no primordial antimatter.

Outside of particle accelerators, antimatter can be seen in cosmic rays in the form of a few antiprotons, present at a level of around 10^{-4} in comparison with the number of protons. However, this proportion is consistent with secondary antiproton production through accelerator-like processes, $p+p \to 3p+\bar{p}$, as the cosmic rays stream towards us. Thus there is no evidence for primordial antimatter in our galaxy. Also, if matter and antimatter galaxies were to coexist in clusters of galaxies, then we would expect there to be a detectable background of γ-radiation from nucleon-antinucleon annihilations within the clusters. This background is not observed and so we conclude that there is negligible antimatter on the scale of clusters.

More generally, if large domains of matter and antimatter exist, then annihilations would take place at the interfaces between them. If the typical size of such a domain was small enough, then the energy released by these annihilations would result in a diffuse γ-ray background and a distortion of the cosmic microwave radiation, neither of which is observed. A careful numerical analysis of this problem demonstrates that the universe must consist entirely of either matter or antimatter on all scales up to the Hubble size. It therefore seems that the universe is fundamentally matter-antimatter asymmetric.

While the above considerations put an experimental upper bound on the amount of antimatter in the universe, strict quantitative estimates of the relative abundances of baryonic matter and antimatter may also be obtained from the standard cosmology. The baryon number density does not remain constant during the evolution of the universe, instead scaling like a^{-3}, where a is the cosmological scale factor [1]. It is therefore convenient to define the baryon asymmetry of the universe in terms of the quantity

$$Y_B \equiv \frac{n_B}{s}, \tag{1}$$

where $n_B = n_b - n_{\bar{b}}$ is the difference between the number of baryons and antibaryons per unit volume and $s = (2\pi^2/45)g_\star$ is the entropy density at temperature T when the thermal plasma contained $g_\star$ relativistic degrees of freedom.

The observed baryon asymmetry of the Universe is by now accurately determined by Cosmic Microwave (CMB) Anisotropy measurements [2]

$$Y_B^{\rm CMB} = (8.75 \pm 0.23) \times 10^{-11} \,. \tag{2}$$

The physics behind the determination of Y_B and the CMB anisotropy would require another paper. It suffices to say that the baryon number plays a crucial role in determining the relative amplitude of the even peaks to the odd ones. Measuring this ratio gives a determination of the baryon asymmetry of the Universe [3]. Primordial nucleosynthesis (for a review see [4]) is also one of the most powerful predictions of the standard cosmological model. The theory allows accurate predictions of the cosmological abundances of all the light elements, H, ^{3}He, ^{4}He, D, B and ^{7}Li, while requiring only the single input parameter Y_B which has been constant since nucleosynthesis. The range of Y_B consistent with nucleosynthesis agrees with the one provided by CMB anisotropies, even though they are relevant on various ranges of temperatures, about MeV for nucleosinthesis and a fraction of eV for the CMB anisotropies.

To see that the standard cosmological model cannot explain the observed value of Y_B, suppose that initially we start with $Y_B = 0$ [5, 6]. We can compute the final number density of nucleons b that are left over after annihilations have frozen out. At temperatures $T < 1$ GeV the equilibrium abundance of nucleons and antinucleons is [1]

$$\frac{n_b}{n_\gamma} \approx \frac{n_{\bar{b}}}{n_\gamma} \approx \left(\frac{m_p}{T}\right)^{3/2} e^{-\frac{m_p}{T}} . \tag{3}$$

When the universe cools off, the number of nucleons and antinucleons decreases as long as the annihilation rate $\Gamma_{\rm ann} \approx n_b \langle \sigma_A v \rangle$ is larger than the expansion rate of the universe $H \approx 1.66 \, g_\star^{1/2} \frac{T^2}{M_{\rm P}}$. The thermally averaged annihilation cross section $\langle \sigma_A v \rangle$ is of the order of m_π^2, so at $T \approx 20$ MeV, $\Gamma_{\rm ann} \approx H$, and annihilations freeze out, nucleons and antinucleons being so rare that they cannot annihilate any longer. Therefore, from (3) we obtain

$$\frac{n_b}{n_\gamma} \approx \frac{n_{\bar{b}}}{n_\gamma} \approx 10^{-18}, \tag{4}$$

which is much smaller than the value required by nucleosynthesis. In conclusion, in the standard cosmological model there is no explanation for the value of the observed baryon asymmetry, if we start from $Y_B = 0$. An initial asymmetry may be imposed by hand as an initial condition, but this would violate any naturalness principle. Rather, the guiding principle behind *modern* cosmology is to attempt to explain the initial conditions required by the standard cosmology on the basis of quantum field theories of elementary particles in the early universe. In this context, the generation of the observed value of Y_B is referred to as *baryogenesis.*

The goal of this paper is to present the thermal leptogenesis scenario. As we shall see, it is strictly related to neutrino physics, the main subject of this School. As such, one would open, through low energy neutrino measurements, to open a window on the early Universe dynamics. For more details about the subject, the reader is forwarded to the excellent recent review of the subject [7]. Other useful reviews are [5, 6].

A note about our conventions. Throughout we use a metric with signature +2 and, unless explicitly stated otherwise, we employ units such that $\hbar = c = k = 1$.

2. The Sakharov criteria

A small baryon asymmetry Y_B may have been produced in the early universe if three necessary conditions are satisfied [8]: *i)* baryon number (B) violation; *ii)* violation of C (charge conjugation symmetry) and CP (the composition of parity and C) and *iii)* departure from thermal equilibrium. The first condition should be clear since, starting from a baryon symmetric universe with $Y_B = 0$, baryon number violation must take place in order to evolve into a universe in which Y_B does not vanish. The second Sakharov criterion is required because, if C and CP are exact symmetries, then one can prove that the total rate for any process which produces an excess of baryons is equal to the rate of the complementary process which produces an excess of antibaryons and so no net baryon number can be created. That is to say that the thermal average of the baryon number operator B, which is odd under both C and CP, is zero unless those discrete symmetries are violated. CP violation is present either if there are complex phases in the lagrangian which cannot be reabsorbed by field redifinitions (explicit breaking) or if some Higgs scalar field acquires a VEV which is not real (spontaneous breaking). We will discuss this in detail shortly.

Finally, to explain the third criterion, one can calculate the equilibrium average of B

$$\begin{aligned} \langle B \rangle_T &= \mathrm{Tr}\,(e^{-\beta H} B) = \mathrm{Tr}\,[(CPT)(CPT)^{-1} e^{-\beta H} B)] \\ &= \mathrm{Tr}\,(e^{-\beta H} (CPT)^{-1} B (CPT)] = -\mathrm{Tr}\,(e^{-\beta H} B)\ , \end{aligned} \tag{5}$$

where we have used that the Hamiltonian H commutes with CPT. Thus $\langle B \rangle_T = 0$ in equilibrium and there is no generation of net baryon number.

Of the three Sakharov conditions, baryon number violation and C and CP violation may be investigated only within a given particle physics model, while the third condition – the departure from thermal equilibrium – may be discussed in a more general way, as we shall see. Let us discuss the Sakharov criteria in more detail.

3. B-violation in the Electroweak theory.

It is well-known that the most general Lagrangian invariant under the SM gauge group and containing only color singlet Higgs fields is automatically invariant under global abelian symmetries which may be identified with the baryonic and leptonic symmetries. These, therefore, are accidental symmetries and as a result it is not possible to violate B and L at tree-level or at any order of perturbation theory. Nevertheless, in many cases the perturbative expansion does not describe all the dynamics of the theory and, indeed, in 1976 't Hooft realized that nonperturbative effects (instantons) may give rise to processes which violate the combination $(B+L)$, but not the orthogonal combination $(B-L)$. The probability of these processes occurring today is exponentially suppressed and probably irrelevant. However, in more extreme situations – like the primordial universe at very high temperatures – baryon and lepton number violating processes may be fast enough to play a significant role in baryogenesis. Let us have a closer look. At the quantum level, the baryon and the lepton symmetries are anomalous

$$\partial_\mu j^\mu_B = \partial_\mu j^\mu_L = n_f \left(\frac{g^2}{32\pi^2} W^a_{\mu\nu} \tilde{W}^{a\mu\nu} - \frac{g'^2}{32\pi^2} F_{\mu\nu} \tilde{F}^{\mu\nu} \right)\ , \tag{6}$$

where g and g' are the gauge couplings of $SU(2)_L$ and $U(1)_Y$, respectively, n_f is the number of families and $\tilde{W}^{\mu\nu} = (1/2)\epsilon^{\mu\nu\alpha\beta} W_{\alpha\beta}$ is the dual of the $SU(2)_L$ field strength tensor, with an analogous expression holding for $\tilde{F}$. To understand how the anomaly is closely related to the vacuum structure of the theory, we may compute the change in baryon number from time $t = 0$ to some arbitrary final time $t = t_f$. For transitions between vacua, the average values of the field strengths are zero at the beginning and the end of the evolution. The change in baryon number may be written as

$$\Delta B = \Delta N_{CS} \equiv n_f [N_{CS}(t_f) - N_{CS}(0)]\ . \tag{7}$$

where the Chern-Simons number is defined to be

$$N_{CS}(t) \equiv \frac{g^2}{32\pi^2}\int d^3x\,\epsilon^{ijk}\,\mathrm{Tr}\left(A_i\partial_j A_k + \frac{2}{3}igA_iA_jA_k\right)\,. \tag{8}$$

Although the Chern-Simons number is not gauge invariant, the change ΔN_{CS} is. Thus, changes in Chern-Simons number result in changes in baryon number which are integral multiples of the number of families n_f. Gauge transformations $U(x)$ which connects two degenerate vacua of the gauge theory may change the Chern-Simons number by an integer n, the winding number. If the system is able to perform a transition from the vacuum $\mathcal{G}^{(n)}_{\rm vac}$ to the closest one $\mathcal{G}^{(n\pm1)}_{\rm vac}$, the Chern-Simons number is changed by unity and $\Delta B = \Delta L = n_f$. Each transition creates 9 left-handed quarks (3 color states for each generation) and 3 left-handed leptons (one per generation). However, adjacent vacua of the electroweak theory are separated by a ridge of configurations with energies larger than that of the vacuum. The lowest energy point on this ridge is a saddle point solution to the equations of motion with a single negative eigenvalue, and is referred to as the *sphaleron*. The probability of baryon number nonconserving processes at zero temperature has been computed by 't Hooft and is highly suppressed by a factor $\exp(-4\pi/\alpha_W)$, where $\alpha_W = g^2/4\pi$. This factor may be interpreted as the probability of making a transition from one classical vacuum to the closest one by tunneling through an energy barrier of height ≈ 10 TeV corresponding to the sphaleron. On the other hand, one might think that fast baryon number violating transitions may be obtained in physical situations which involve a large number of fields. Since the sphaleron may be produced by collective and coherent excitations containing $\approx 1/\alpha_W$ quanta with wavelength of the order of $1/M_W$, one expects that at temperatures $T \gg M_W$, these modes essentially obey statistical mechanics and the transition probability may be computed via classical considerations. Analogously to the case of zero temperature and since the transition which violates the baryon number is sustained by the sphaleron configuration, the thermal rate of baryon number violation in the *broken* phase is proportional to $\exp(-S_3/T)$, where S_3 is the three-diemensional action computed along the sphaleron configuration,

$$S_3 = E_{sp}(T) \equiv (M_W(T)/\alpha_W)\mathcal{E}\,, \tag{9}$$

with the dimensionless parameter $\mathcal{E}$ lying in the range $3.1 < \mathcal{E} < 5.4$ depending on the Higgs mass. The prefactor of the thermal rate reads

$$\Gamma_{sp}(T) = \mu\left(\frac{M_W}{\alpha_W T}\right)^3 M_W^4 \exp\left(-\frac{E_{sph}(T)}{T}\right)\,, \tag{10}$$

where μ is a dimensionless constant. Although the Boltzmann suppression in (10) appears large, it is to be expected that, when the electroweak symmetry becomes restored at a temperature of around 100 GeV, there will no longer be an exponential suppression factor. Although calculation of the baryon number violating rate in the high temperature *unbroken* phase is extremely difficult, a simple estimate is possible. The only important scale in the symmetric phase is the magnetic screening length given by $\xi = (\alpha_W T)^{-1}$. Thus, on dimensional grounds, we expect the rate per unit volume of sphaleron events to be

$$\Gamma_{sp}(T) = \kappa(\alpha_W T)^4\,, \tag{11}$$

with κ another dimensionless constant. The rate of sphaleron processes can be related to the diffusion constant for Chern-Simons number by a fluctuation-dissipation theorem. In almost all numerical calculations of the sphaleron rate, this relationship is used and what is actually evaluated is the diffusion constant. The first attempts to numerically estimate κ in this way yielded $\kappa \approx 0.1 - 1$, but the approach suffered from limited statistics and large volume systematic errors. Nevertheless, more recent numerical attempts found approximately the same result. However, these approaches employ a poor definition of the Chern-Simons number which compromises their reliability.

This simple scaling argument leading to (11) is not correct however. Damping effects in the plasma suppress the rate by an extra power of α_W to give $\Gamma_{sp} \approx \alpha_W^5 T^4$. Indeed, since the transition rate involves physics at soft energies $g^2 T$ that are small compared to the typical hard energies $\approx T$ of the thermal excitations in the plasma, the simplest way of analyzing the problem is to consider an effective theory for the soft modes, where the hard modes have been integrated out, and to keep the dominant contributions, the so-called hard thermal loops. It is the resulting typical frequency ω_c of a gauge field configuration immersed in the plasma and with spatial extent $(g^2T)^{-1}$ that determines the change of baryon number per unit time and unit volume. This frequency ω_c has been estimated to be $\approx g^4 T$ when taking into account the damping effects of the hard modes. Using the effective dynamics of soft nonabelian gauge fields at finite temperature one can find that $\Gamma_{sp} \approx \alpha_W^5 T^4 \ln(1/\alpha_W)$. Lattice simulations with hard-thermal loops seem to indicate the $\Gamma_{sp} \approx 30\alpha_W^5 T^4$, which is not far from $\alpha_W^4 T^4$.

4. The standard out-of-equilibrium decay scenario

Out of the three Sakharov's conditions, the baryon number violation and C and CP violation may be investigated thoroughly only within a given particle physics model, while the third condition – the departure from thermal equilibrium –may be discussed in a more general way. Very roughly speaking, the various models of baryogenesis that have been proposed so far fall into two categories [5]:

– models where the out-of-equilibrium condition is attained thanks to the expansion of the Universe and the presence of heavy decaying particles;

– models where the departure from thermal equilibrium is attained during the phase transitions which lead to the breaking of some global and/or gauge symmetry.

In this paper we will analyse the first category –the standard out-of-equilibrium decay scenario [1].

4.1. The conditions for the out-of-equilibrium decay scenario

It is obvious that in a static Universe any particle, even very weakly interacting, will attain sooner or later thermodynamical equilibrium with the surroinding plasma. The expansion of the Universe, however, introduces a finite time-scale, $\tau_U \approx H^{-1}$. Let suppose that X is a *baryon number violating* superheavy boson field (vector or scalar) which is coupled to lighter fermionic degrees of freedom with a strength $\alpha_X^{1/2}$ (either a gauge coupling $\alpha_{\rm gauge}$ or a Yukawa coupling α_Y).

In the case in which the couplings are renormalizable, the decay rate Γ_X of the superheavy boson may be easily estimated to be

$$\Gamma_X \approx \alpha_X M_X, \tag{12}$$

where M_X is the mass of the particle X. In the opposite case in which the boson is a gauge *singlet* scalar field and it only couples to light matter through gravitational interactions – this is the case of singlets in the hidden sector of supergravity models – the decay rate is from dimensional arguments

$$\Gamma_X \approx \frac{M_X^3}{M_{\rm P}^2}. \tag{13}$$

At very large temperatures $T \gg M_X$, it is assumed that all the particles species are in thermal equilibrium, *i.e.* $n_X \approx n_{\overline{X}} \approx n_\gamma$ (up to statistical factors) and that $B = 0$. At $T < M_X$ the equilibrium abundance of X and $\overline{X}$ relative to photons is given by

$$\frac{n_X^{\rm EQ}}{n_\gamma} \approx \frac{n_{\overline{X}}^{\rm EQ}}{n_\gamma} \approx \left(\frac{M_X}{T}\right)^{3/2} e^{-\frac{M_X}{T}}, \tag{14}$$

where we have neglected the chemical potential μ_X.

For the X and $\overline{X}$ particles to mantain their equilibrium abundances, they must be able to diminish their number rapidly with respect to the Hubble rate $H(T)$. The conditions necessary for doing so are easily quantified. The superheavy X and $\overline{X}$ particles may attain equilibrium through decays with rate Γ_X, inverse decays with rate $\Gamma_X^{\rm ID}$

$$\Gamma_X^{\rm ID} \approx \Gamma_X \begin{cases} 1 & T > M_X, \\ (M_X/T)^{3/2}\exp(-M_X/T) & T < M_X, \end{cases} \tag{15}$$

and annihilation processes with rate $\Gamma_X^{\rm ann} \propto n_X$. The latter, however are "self-quenching" and therefore less important than the decay and inverse decay processes. They will be ignored from now on. Of crucial interest are the B-nonconserving scattering processes $2 \leftrightarrow 2$ mediated by the X and $\overline{X}$ particles with rate $\Gamma_X^{\rm S}$

$$\Gamma_X^{\rm S} \approx n\sigma \approx \alpha^2 T^3 \frac{T^2}{(M_X^2 + T^2)^2}, \tag{16}$$

where $\alpha \approx g^2/4\pi$ denotes the coupling strength of the X boson. At high temperatures, the $2 \leftrightarrow 2$ scatterings cross section is $\sigma \approx \alpha^2/T^2$, while at low temperatures $\sigma \approx \alpha^2 T^2/M_X^4$.

For baryogenesis, the most important rate is the decay rate, as decays (and inverse decays) are the mechanism that regulates the number of X and $\overline{X}$ particles in the plasma. It is therefore useful to define the following quantity

$$K \equiv \left.\frac{\Gamma_X}{H}\right|_{T=M_X} \tag{17}$$

which measures the effectiveness of decays at the crucial epoch ($T \approx M_X$) when the X and $\overline{X}$ particles must decrease in number if they are to stay in equilibrium. Note also that for $T < M_X$, K determines the effectiveness of inverse decays and $2 \leftrightarrow 2$ scatterings as well: $\Gamma_X^{\rm ID}/H \approx (M_X/T)^{3/2}\exp(-M_X/T)\, K$ and $\Gamma_X^{\rm S}/H \approx \alpha(T/M_X)^5\, K$.

Now, if $K \gg 1$, and therefore

$$\Gamma_X \gg H|_{T=M_X}, \tag{18}$$

then the X and $\overline{X}$ particles will adjust their abundances by decaying to their equilibrium abundances and no baryogenesis can be induced by their decays –this is simply because out-of-equilibrium conditions are not attained. Given the expression $H \sim g_\star^{1/2} T^2/M_{\rm P}$ for the expansion rate of the Universe, the condition (18) is equivalent to

$$M_X \ll g_*^{-1/2}\, \alpha_X\, M_{\rm P} \tag{19}$$

for strongly coupled scalar bosons, and to

$$M_X \gg g_\star^{1/2}\, M_{\rm P}, \tag{20}$$

for gravitationally coupled X particles. Obviously, this last condition is never satisfied for $M_X < M_{\rm P}$.

However, if the decay rate is such that $K \ll 1$, and therefore

$$\Gamma_X < H|_{T=M_X}, \tag{21}$$

then the X and $\overline{X}$ particles cannot decay on the expansion time-scale τ_U and so *they remain as abundant as photons* for $T < M_X$. In other words, at some temperature $T > M_X$, the superheavy bosons X and $\overline{X}$ are so weakly interacting that they cannot catch up with the expansion of the Universe and they decouple from the thermal bath when still *relativistic*, $n_X \approx n_{\overline{X}} \approx n_\gamma$ at the time of decoupling. Therefore, at temperature $T \approx M_X$, they will populate the Universe with an abundance which is much larger than the equilibrium one. This *overbundance* with respect to the equilibrium abundance is precisely the departure from thermal equilibrium needed to produce a final nonvanishing baryon asymmetry. Condition (21) is equivalent to

$$M_X > g_*^{-1/2}\, \alpha_X\, M_{\rm P} \tag{22}$$

for strongly coupled scalar bosons, and to

$$M_X < g_*^{1/2}\, M_{\rm P}, \tag{23}$$

for gravitationally coupled X particles. It is clear that this last condition is always satisfied, whereas the condition (22) is based on the smallness of the quantity $g_*^{-1/2}\alpha_X$. In particular, if the X particle is a gauge boson, $\alpha_X \approx \alpha_{\rm gauge}$ can span the range $(2.5 \times 10^{-2} - 10^{-1})$, while $g_\star$ is about 10^2. In this way we obtain from (22) that the condition of out-of-equilibrium can be satisfied for

$$M_X > (10^{-4} - 10^{-3})\, M_{\rm P} \approx (10^{15} - 10^{16})\,{\rm GeV}. \tag{24}$$

If X is a scalar boson, its coupling α_Y to fermions f with mass m_f is proportional to the squared mass of the fermions

$$\alpha_Y \approx \left(\frac{m_f}{m_W}\right)^2 \alpha_{\rm gauge}, \tag{25}$$

where m_W is the W-boson mass and α_Y is typically in the range $(10^{-2} - 10^{-7})$, from where

$$M_X > (10^{-8} - 10^{-3})\, M_{\rm P} \approx (10^{10} - 10^{16})\,{\rm GeV}. \tag{26}$$

Obviously, condition (26) is more easily satisfied than condition (24) and we conclude that baryogenesis is more easily produced through the decay of superheavy scalar bosons. On the other hand, as we have seen above, the condition (23) tells us that the out-of-equilibrium condition is automatically satisfied for gravitationally interacting particles.

4.2. The production of the baryon asymmetry

Let us now follow the subsequent evolution of the X and $\overline{X}$ particles. When the Universe becomes as old as the lifetime of these particles, $t \approx H^{-1} \approx \Gamma_X^{-1}$, they start decaying. This takes place at a temperature T_D defined by the condition

$$\Gamma_X \approx H|_{T=T_D}, \tag{27}$$

i.e. at

$$T_D \approx g_*^{-1/4}\, \alpha_X^{1/2}\, (M_X M_{\rm P})^{1/2} < M_X, \tag{28}$$

where the last inequality comes from (22) and is valid for particles with unsuppressed couplings. For particles with only gravitational interactions

$$T_D \approx g_*^{-1/4}\, M_X \left(\frac{M_X}{M_{\rm P}}\right)^{1/2} < M_X, \tag{29}$$

the last inequality coming from (23). At $T \approx T_D$, X and $\overline{X}$ particles start to decay and their number decrease. If their decay violate the baryon number, they will generate a net baryon number per decay.

Suppose now that the X particle may decay into two channels, let us denote them by a and b, with different baryon numbers B_a and B_b, respectively. Correspondingly, the decay channels of $\overline{X}$, $\overline{a}$ and $\overline{b}$, have baryon numbers $-B_a$ and $-B_b$, respectively. Let $r(\overline{r})$ be the branching ratio of the $X(\overline{X})$ in channel $a(\overline{a})$ and $1-r(\overline{r})$ the branching ratio of $X(\overline{X})$ in channel $b(\overline{b})$,

$$\begin{aligned} r &= \frac{\Gamma(X \to a)}{\Gamma_X}, \\ \overline{r} &= \frac{\Gamma(\overline{X} \to \overline{a})}{\Gamma_X}, \\ 1-r &= \frac{\Gamma(X \to b)}{\Gamma_X}, \\ 1-\overline{r} &= \frac{\Gamma(\overline{X} \to \overline{b})}{\Gamma_X}, \end{aligned} \tag{30}$$

where we have been using the fact that the *total* decay rates of X and $\overline{X}$ are equal because of the CPT theorem plus unitarity. The average net baryon number produced in the X decays is

$$rB_a + (1-r)B_b, \tag{31}$$

and that produced by $\overline{X}$ decays is

$$-\overline{r}B_a - (1-\overline{r})B_b. \tag{32}$$

Finally, the mean net baryon number produced in X and $\overline{X}$ decays is

$$\Delta B = (r-\overline{r})B_a + [(1-r)-(1-\overline{r})]\,B_b = (r-\overline{r})(B_a - B_b). \tag{33}$$

Equation (33) may be easily generalized to the case in which $X(\overline{X})$ may decay into a set of final states $f_n(\overline{f}_n)$ with baryon number $B_n(-B_n)$

$$\Delta B = \frac{1}{\Gamma_X}\sum_n B_n\,\left[\Gamma(X \to f_n) - \Gamma(\overline{X} \to \overline{f}_n)\right]. \tag{34}$$

At the decay temperature, $T_D < M_X$, because $K \ll 1$ both inverse decays and $2 \leftrightarrow 2$ baryon violating scatterings are impotent and can be safely ignored and thus the net baryon number produced per decay ΔB is not destroyed by the net baryon number $-\Delta B$ produced by the inverse decays and by the baryon number violating scatterings.

At $T \approx T_D$, $n_X \approx n_{\overline{X}} \approx n_\gamma$ and therefore the net baryon number density produced by the out-of-equilibrium decay is

$$n_B = \Delta B\, n_X. \tag{35}$$

The three Sakharov ingredients for producing a net baryon asymmetry can be easily traced back here:

– If B is not violated, then $B_n = 0$ and $\Delta B = 0$.

–If C and CP are not violated, then $\Gamma(X \to f_n) = \Gamma(\overline{X} \to \overline{f}_n)$, and also $\Delta B = 0$.

– In thermal equilibrium, the inverse processes are not suppressed and the net baryon number produced by decays will be erased by the inverse decays.

Since each decay produces a mean net baryon number density $n_B = \Delta B n_X \approx \Delta B n_\gamma$ and since the entropy density is $s \approx g_\star n_\gamma$, the net baryon number produced is

$$B \equiv \frac{n_B}{s} \approx \frac{\Delta B n_\gamma}{g_\star n_\gamma} \approx \frac{\Delta B}{g_\star}. \tag{36}$$

Taking $g_\star \approx 10^2$, we see that only tiny C and CP violations are required to generate $\Delta B \approx 10^{-8}$, and thus $B \approx 10^{-10}$.

To obtain (36) we have assumed that the entropy realese in X decays is negligible. However, sometimes, this is not a good approximation (especially if the X particles decay very late, at $T_D \ll M_X$, which is the case of gravitationally interacting particles). In that case, assuming that the energy density of the Universe at T_D is dominated by X particles

$$\rho_X \approx M_X\, n_X, \tag{37}$$

and that it is converted entirely into radiation at the reheating temperature T_{RH}

$$\rho = \frac{\pi^2}{30}\, g_\star\, T_{RH}^4, \tag{38}$$

we obtain

$$n_X \approx \frac{\pi^2}{30}\, g_\star\, \frac{T_{RH}^4}{M_X}. \tag{39}$$

We can therefore write the baryon number as

$$B \approx \frac{3}{4}\frac{T_{RH}}{M_X}\Delta B. \tag{40}$$

We can relate T_{RH} with the decay rate Γ_X using the decay condition

$$\Gamma_X^2 \approx H^2(T_D) \approx \frac{8\pi\rho_X}{3M_{\rm P}^2} \tag{41}$$

and so we can write

$$B \approx \left(\frac{g_\star^{-1/2}\Gamma_X M_{\rm P}}{M_X^2}\right)^{1/2} \Delta B. \tag{42}$$

For the case of strongly decaying particles (through renormalizable interactions) we obtain

$$B \approx \left(\frac{g_\star^{-1/2}\alpha M_{\rm P}}{M_X}\right)^{1/2} \Delta B. \tag{43}$$

while for the case of weakly decaying particles (through gravitational interactions) we obtain

$$B \approx \left(\frac{g_\star^{-1/2} M_X}{M_{\rm P}}\right)^{1/2} \Delta B. \tag{44}$$

In the other extreme regime $K \gg 1$, one expects the abundance of X and $\overline{X}$ bosons to track the equilibrium values as $\Gamma_X \gg H$ for $T \approx M_X$. If the equilibrium is tracked precisely enough, there will be no departure from thermal equilibrium and no baryon number may evolve. The intermediate regime, $K \approx 1$, is more interesting and to address it one has to invoke numerical analysis involving Boltzmann equations for the evolution of B. The numerical analysis essentially confirms the qualitative picture we have described so far and its discussion is beyond the scope of this paper.

Journal of Physics: Conference Series **335** (2011) 012008 doi:10.1088/1742-6596/335/1/012008

5. Baryogenesis via leptogenesis: one-flavour approximation

Baryogenesis through Leptogenesis [9] is a simple mechanism to explain this baryon asymmetry of the Universe. As we said, a lepton asymmetry is dynamically generated and then converted into a baryon asymmetry due to $(B+L)$-violating sphaleron interactions which exist in the SM. The reader is forwarded to Refs. [11] and [12] for more details.

A simple model in which this mechanism can be implemented is "Seesaw"(type I) [13], consisting of the SM plus two or three right-handed (RH) Majorana neutrinos. In this simple extension of the SM, the usual scenario that is explored (referred to as "thermal leptogenesis") consists of a hierarchical spectrum for the RH neutrinos, such that the lightest of the RH neutrinos is produced by thermal scattering after inflation, and subsequently decays out-of-equilibrium in a lepton number and CP-violating way, thus satisfying Sakharov's constraints.

This section introduces notation and reviews the calculation of the lepton asymmetry when the charged lepton Yukawa couplings are neglected. As we shall see, the commonly used formulae for the final lepton asymmetry, which we report here, may not be appropriate once flavours are considered. The reader should be patient, flavour effects will be discussed later.

Our starting point is the Lagrangian of the SM with the addition of three right-handed neutrinos N_i $(i = 1, 2, 3)$ with heavy Majorana masses $M_3 > M_2 > M_1$ and Yukawa couplings $\lambda_{\alpha i}$. Working in the basis in which the Yukawa couplings for the charged leptons are diagonal, the Lagrangian reads

$$\mathcal{L} = \mathcal{L}_{\rm SM} + \left(\frac{M_i}{2} N_i^2 + \lambda_{\alpha i} L_\alpha H N_i\right) + \text{h.c.}\,. \tag{45}$$

Here L_α is the lepton doublet with flavour $(\alpha = e, \mu, \tau)$, and H is the Higgs doublet whose neutral component has a vacuum expectation value (VEV) equal to $v = 175$ GeV. After spontaneous symmetry breaking, a Dirac mass term $m_D = \lambda\, v$, is generated by the VEV of the Higgs boson.

In the see-saw limit, $M \gg m_D$, the spectrum of neutrino mass eigenstates splits in two sets: three very heavy neutrinos, N_1, N_2 and N_3 respectively with masses $M_1 \leq M_2 \leq M_3$ almost coinciding with the eigenvalues of M, and three light neutrinos with masses $m_1 \leq m_2 \leq m_3$, the eigenvalues of the light neutrino mass matrix given by the see-saw formula [13],

$$m_\nu = -m_D \frac{1}{M} m_D^T\,. \tag{46}$$

Neutrino oscillation experiments measure two neutrino mass-squared differences. For normal schemes one has $m_3^2 - m_2^2 = \Delta m^2_{\rm atm}$ and $m_2^2 - m_1^2 = \Delta m^2_{\rm sol}$, whereas for inverted schemes one has $m_3^2 - m_2^2 = \Delta m^2_{\rm sol}$ and $m_2^2 - m_1^2 = \Delta m^2_{\rm atm}$. For $m_1 \gg m_{\rm atm} \equiv \sqrt{\Delta m^2_{\rm atm} + \Delta m^2_{\rm sol}} = (0.050 \pm 0.001)\,\text{eV}$ [14] the spectrum is quasi-degenerate, while for $m_1 \ll m_{\rm sol} \equiv \sqrt{\Delta m^2_{\rm sol}} = (0.00875 \pm 0.00012)\,\text{eV}$ [14] it is fully hierarchical (normal or inverted). Here we will restrict ourselves to the case of normal schemes. The most stringent upper bound on the absolute neutrino mass scale comes from cosmological observations. Recently, a conservative upper bound on the sum of neutrino masses, $\sum_i m_i \leq 0.61\,\text{eV}$ (95% CL), has been obtained by the WMAP collaboration combining CMB, baryon acoustic oscillations and supernovae type Ia observations [2]. Considering that it falls in the quasi-degenerate regime, it straightforwardly translates into

$$m_1 < 0.2\,\text{eV}\ \ (95\%\,\text{CL})\,. \tag{47}$$

It proves sometimes useful to adopt the bi-unitary parametrization

$$m_D = V_L^\dagger\, D_{m_D}\, U_R\,, \tag{48}$$

where V_L and U_R are two unitary matrices that diagonalize m_D and D_{m_D} is the diagonal matrix whose elements are the eigenvalues of m_D: $D_{m_D} \equiv \text{diag}(l_{D1}, l_{D2}, l_{D3})$. This shows that in the process of see-saw, 9 parameters are lost: at high energies there 18 real parameters (3 from the eigenvalues of m_D, 3 from M_i, and 12 from V_L and V_R). At low energy there are only 9 (3 from m_i and 6 from the matrix U diagonalizing m_ν). We now assume that right-handed neutrinos are hierarchical, $M_{2,3} \gg M_1$ so that studying the evolution of the number density of N_1 suffices. The final amount of $(B-L)$ asymmetry can be parametrized as $Y_{B-L} = n_{B-L}/s$, where $s = 2\pi^2 g_\star T^3/45$ is again the entropy density and $g_\star$ counts the effective number of spin-degrees of freedom in thermal equilibrium ($g_\star = 217/2$ in the SM with a single generation of right-handed neutrinos). After reprocessing by sphaleron transitions, the baryon asymmetry is related to the L asymmetry by [5]

$$Y_B = -\left(\frac{8n_G + 4n_H}{14n_G + 9n_H}\right) Y_L, \tag{49}$$

where n_H is the number of Higgs doublets, and n_G the number of fermion generations (in equilibrium). It is also useful to define an efficiency factor η which tells how efficient is the production of the baryon asymmetry, *e.g.* how much of the asymmetry per RH neutrino decay remais after wash-out processes are account for:

$$Y_B \approx 1.38 \cdot 10^{-3}\, \epsilon_1\, \eta\,, \tag{50}$$

where we have assumed $n_H = 1$. Now, the idea of thermal leptogenesis is that RH neutrinos decay in the early Universe out-of-equilibrium, thus producing a lepton asymmetry.

One defines the CP asymmetry generated by N_1 decays as

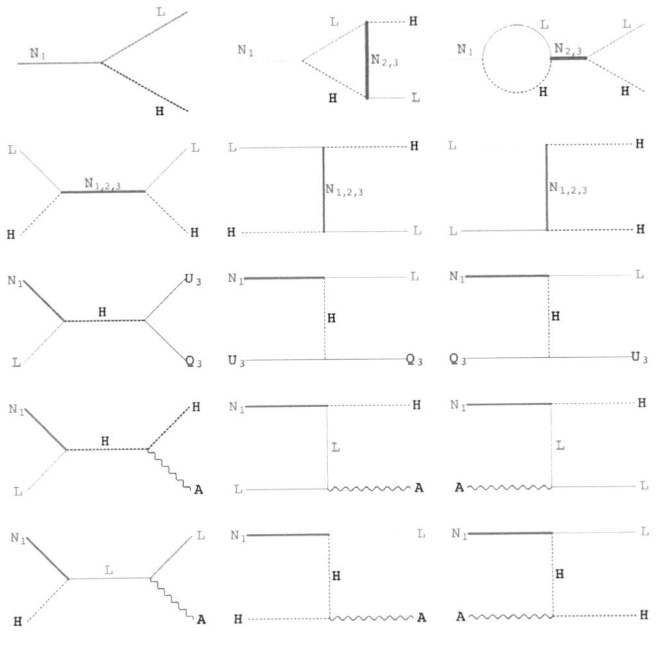

Figure 1. Feynman diagrams contributing to thermal leptogenesis.

$$\epsilon_1 \equiv \frac{\sum_\alpha[\Gamma(N_1 \to H\ell_\alpha) - \Gamma(N_1 \to \overline{H}\overline{\ell}_\alpha)]}{\sum_\alpha[\Gamma(N_1 \to H\ell_\alpha) + \Gamma(N_1 \to \overline{H}\overline{\ell}_\alpha)]} = \frac{1}{8\pi}\sum_{j\neq 1}\frac{\mathrm{Im}\left[(\lambda\lambda^\dagger)^2_{j1}\right]}{[\lambda\lambda^\dagger]_{11}}\, g\left(\frac{M_j^2}{M_1^2}\right). \tag{51}$$

It is the sum of two contributions, the vertex and the wavefunction ones. A calculation similar to the one performed in the toy model previously discussed leads to

$$g(x) = \sqrt{x}\left[\frac{1}{1-x} + 1 - (1+x)\ln\left(\frac{1+x}{x}\right)\right] \stackrel{x\gg 1}{\longrightarrow} -\frac{3}{2\sqrt{x}}\,. \tag{52}$$

Besides the CP parameter ϵ_1, the final baryon asymmetry depends on a single wash-out parameter,

$$K \equiv \frac{\sum_\alpha \Gamma(N_1 \to H\ell_\alpha)}{H(M_1)} \equiv \left(\frac{\widetilde{m}_1}{\widetilde{m}^*}\right)\,, \tag{53}$$

where $H(M_1)$ denotes the value of the Hubble rate evaluated at a temperature $T = M_1$ ($\widetilde{m}^* \approx 10^{-3}\,\mathrm{eV}$) and

$$\widetilde{m}_1 \equiv \frac{(\lambda\lambda^\dagger)_{11} v^2}{M_1} \tag{54}$$

is proportional to the total decay rate of the right-handed neutrino N_1. One could now proceed as in the previous sections by estimating the baryon asymemetry. However, we want to to a better job and we resort to the Boltzmann equations. By defining the variable $z = M_1/T$, the Boltzmann equations for the lepton asymmetry $Y_{\mathcal{L}}$, and the right-handed neutrino number density Y_{N_1} (both normalised to the entropy s), may be written in a compact form as

$$\frac{d(Y_{N_1} - Y_{N_1}^{\mathrm{EQ}})}{dz} = -\frac{z}{sH(M_1)}\left(\gamma_D + \gamma_{\Delta L=1}\right)\left(\frac{Y_{N_1}}{Y_{N_1}^{\mathrm{EQ}}} - 1\right) - \frac{dY_{N_1}^{\mathrm{EQ}}}{dz}\,, \tag{55}$$

$$\frac{dY_L}{dz} = \frac{z}{sH(M_1)}\left[\left(\frac{Y_{N_1}}{Y_{N_1}^{\mathrm{EQ}}} - 1\right)\epsilon_1\left(\gamma_D + \gamma_{\Delta L=1}\right) - \frac{Y_L}{Y_L^{\mathrm{EQ}}}\left(\gamma_D + \gamma_{\Delta L=1}\right)\right]\,. \tag{56}$$

The processes taken into account in these equations are decays and inverse decays with rate γ_D, $\Delta L = 1$ scatterings such as $(qt^c \to N\ell)$, and $\Delta L = 2$ processes mediated by heavy neutrinos. The first three modify the abundance of the lightest right-handed neutrinos. The $\Delta L = 2$ scatterings mediated by $N_{2,3}$ are neglected in our analysis for simplicity. The various γ are thermally averaged rates, including all contributions summed over flavour (s, t channel interference etc); explicit expressions can be found in the literature. Notice that in this "usual" analysis, $\Delta L = 1$ scattering contributes to the creation of N_1's and not to the production of a lepton asymmetry, only to the washout

Approximate analytic solutions for $Y_{\mathcal{L}}$ and $\Delta_{N_1} \equiv Y_{N_1} - Y_{N_1}^{\mathrm{EQ}}$ can be obtained from simplified equations. Calculating in zero temperature field theory for simplicity, one obtains

$$\gamma_D \approx sY_{N_1}^{EQ}\frac{K_1(z)}{K_2(z)}\Gamma_D, \qquad Y_{N_1}^{\mathrm{EQ}} \approx \frac{1}{4g_\star} z^2\, K_2(z)\,. \tag{57}$$

The Boltzmann equations can be approximated

$$\Delta'_{N_1} = -zK\frac{K_1(z)}{K_2(z)}\, f_1(z)\,\Delta_{N_1} - Y_{N_1}^{\mathrm{EQ}\prime}\,, \tag{58}$$

$$Y'_{\mathcal{L}} = \epsilon_1 K z\frac{K_1(z)}{K_2(z)}\Delta_{N_1} - \frac{1}{4}z^3\, K\, K_1(z)\, f_2(z)\; Y_{\mathcal{L}} \tag{59}$$

Journal of Physics: Conference Series **335** (2011) 012008 doi:10.1088/1742-6596/335/1/012008

where K_1 and K_2 are modified Bessel functions of the second kind. The function $f_1(z)$ accounts for the presence of $\Delta L = 1$ scatterings, and $f_2(z)$ accounts for scatterings in the washout term of the asymmetry. They can be approximated, in interesting limits, as

$$f_1(z) \approx \begin{cases} 1 & \text{for } z \gg 1 \\ \frac{N_c^2 m_t^2}{4\pi^2 v^2 z^2} & \text{for } z < 1 \,, \end{cases} \tag{60}$$

and

$$f_2(z) \approx \begin{cases} 1 & \text{for } z \gg 1 \\ \frac{a_K N_c^2 m_t^2}{8\pi^2 v^2 z^2} & \text{for } z < 1 \,, \end{cases} \tag{61}$$

where $\frac{N_c^2 m_t^2}{8\pi^2 v^2} \equiv K_s/K \approx 0.1$ parametrizes the strength of the $\Delta L = 1$ scatterings and $a_K = 4/3\,(2)$ for the weak (strong) wash out case. A good approximation to the rate $Kz(K_1(z)/K_2(z))f_1(z)$ is given by the function $(K_s + Kz)$ while the wash out term $-(1/4)z^3 K K_1(z) f_2(z) Y_{\mathcal{L}}$ is well approximated at small z by $-a_K K_s Y_{\mathcal{L}}$.

5.1. *Strong wash-out regime*

In the strong wash-out regime, the parameter $K \gg 1$ and the right handed neutrinos N_1's are nearly in thermal equilibrium. Under these circumstances, one can set $\Delta'_{N_1} \approx 0$ and $\Delta_{N_1} \approx (zK_2/4g_\star K)$. Exploiting a saddle-point approximation in Eq. (59) we find that the lepton asymmetry is given by

$$Y_L \approx \epsilon_1 \int_0^\infty dz\, \frac{K}{4g_\star} z^2\, e^{-\int_z^\infty dz'\,((z')^3/4) K_1(z') K} \,. \tag{62}$$

Using the steepest descent method to evaluate the integral, one finds that it gets the major contribution at $\overline{z}$ such that $\overline{z} = \log K + (5 \ln \overline{z}/2)$ when inverse decays become inefficient. The lepton asymmetry in the flavour α becomes

$$Y_L \approx 0.3\, \frac{\epsilon_1}{g_\star} \left(\frac{0.55 \times 10^{-3}\,\mathrm{eV}}{\tilde{m}_1} \right)^{1.16} \,. \tag{63}$$

5.2. *weak wash-out regime*

In this case all the $K \ll 1$. We assume that right-handed neutrinos are not initially present in the plasma, but they are generated by inverse decays and scatterings. The equation of motion for Y_{N_1} is well approximated by

$$Y'_{N_1} = -(K_s + Kz)\left(Y_{N_1} - Y_{N_1}^{\mathrm{EQ}}\right) \,, \tag{64}$$

We split the solution in two pieces. Let us define z_{EQ} the value of z at which $Y_{N_1}(z_{\mathrm{EQ}}) = Y_{N_1}^{\mathrm{EQ}}(z_{\mathrm{EQ}})$. This value has to be found a posteriori. For $z \ll z_{\mathrm{EQ}}$, we may suppose that $Y_{N_1} \ll Y_{N_1}^{\mathrm{EQ}}$ and Eq. (64) is solved by

$$\begin{aligned} Y^{-}_{N_1}(z) \approx \int_0^z dz'\,(K_s + Kz')Y_{N_1}^{\mathrm{EQ}} &= \frac{1}{4g_\star} \int_0^z dz'\,(K_s + Kz')(z')^2\, K_2(z') \\ &= \frac{K}{4g_\star}\left(\frac{K_s}{K} I_1(z) + I_2(z)\right) \end{aligned} \tag{65}$$

With I_1 and I_2 integral involving the modified Bessel functions :

$$I_1(z) = \int_0^z x^2 K_2(x)dx \approx f(z) + z^3 K_2(z) \tag{66}$$

where

$$f(z) = \frac{3\pi z^3}{((9\pi)^c + (2z^3)^c)^{1/c}}, c = 0.7 \tag{67}$$

The integral I_2 is well known, and equals

$$I_2(z) = \int_0^z x^3 K_2(x)dx = 8 - z^3 K_3(z) \tag{68}$$

Therefore

$$Y_{N_1}^-(z) \approx \frac{K}{4g_\star}\left(\frac{K_s}{K}(f(z) + z^3 K_2(z)) + 8 - z^3 K_3(z)\right) \tag{69}$$

As expected for weak washout, we find that the maximum number density of N_1 is proportional to K (recall $K_s \propto K$).

Let us now compute the value of $z_{\rm EQ}$. We expect it to be $\gg 1$ and we therefore approximate, up to $\mathcal{O}(z^{-3/2})$: $K_2(z) \approx K_3(z) \approx \sqrt{\frac{\pi}{2}} z^{-1/2} e^{-z}$. Imposing $Y_{N_1}^-(z_{\rm EQ}) = Y_{N_1}^{\rm EQ}(z_{\rm EQ})$, we find

$$z_{\rm EQ} \approx \frac{3}{2} \ln z_{\rm EQ} - \ln\left(\frac{8}{\sqrt{\pi/2}} K + 3\sqrt{\frac{\pi}{2}} K_s\right). \tag{70}$$

This solution is a good approximation to the real value for $K \ll 1$.

For $z > z_{\rm EQ}$, we have $(K_s + Kz) \approx Kz$ and

$$Y_{N_1}(z) \approx Y_{N_1}^{\rm EQ}(z_{\rm EQ}) e^{K/2(z_{\rm EQ}^2 - z^2)}. \tag{71}$$

Notice that we have included CP violation in $\Delta L = 1$ scattering, unlike the usual analysis, so we expect our solution for Y_L to have a different scaling with K than in the traditional literature: if $\not{C}\!P$ in scattering is neglected, then $Y_N \propto K$, and the N_1 decay out of equilibrium, so one expects $Y_L \propto K\epsilon_1$. However, if $\not{C}\!P$ in N_1 production ($\approx$ scattering) is included, and washout is neglected, then the equations for Y_{N_1} and Y_L are identical, so $Y_L(z \to \infty)$ vanishes. That is, for every $|1/\epsilon_1|$ N_1's that are created, be it by inverse decay or scattering, an (anti)-lepton is produced. This (anti-)asymmetry will approximately cancel against the lepton asymmetry generated later on, when the N_1 decay. However the cancellation will be imperfect, because the anti-asymmetry has more time to be washed out, so the final asymmetry should scale as K^2. After integrating by parts, this is what we find for the lepton asymmetry, which is given by

$$\begin{aligned} Y_L &\approx \epsilon_1 \int_0^\infty dz' Y_{N_1}(z') g_1(z') e^{-\int_z^\infty dz'' g_1(z'')}, \quad g_1(z) = \frac{1}{4} z^3 K K_1(z) f_2(z), \\ &\approx 1.5 \frac{\epsilon_1}{g_\star}\left(\frac{\tilde{m}_1}{3.3 \times 10^{-3}\,\text{eV}}\right)^2. \end{aligned} \tag{72}$$

Our findings hold provided that the non-resonant $\Delta L = 2$ scattering rates, in particular those mediated by the N_2 and N_3 heavy neutrinos, are slower than decays and $\Delta L = 1$ scatterings when most of the asymmetry is generated. We estimate that this applies when

$$\left(\frac{M_1}{10^{14}\,\text{GeV}}\right) < 10^{-1}. \tag{73}$$

This means that K should be larger than 10^{-4}.

Our results can be summarized with simple analytical fits for the efficiency factor

$$\frac{1}{\eta} \approx \left(\frac{3.3 \times 10^{-3}\,\text{eV}}{\tilde{m}_1}\right)^2 + \left(\frac{\tilde{m}_1}{0.55 \times 10^{-3}\,\text{eV}}\right)^{1.16} \tag{74}$$

valid for $M_{N_1} < 10^{14}$ GeV and when one starts with no RH neutrinos in the plasma. Numerical results are presented in Fig. 5.

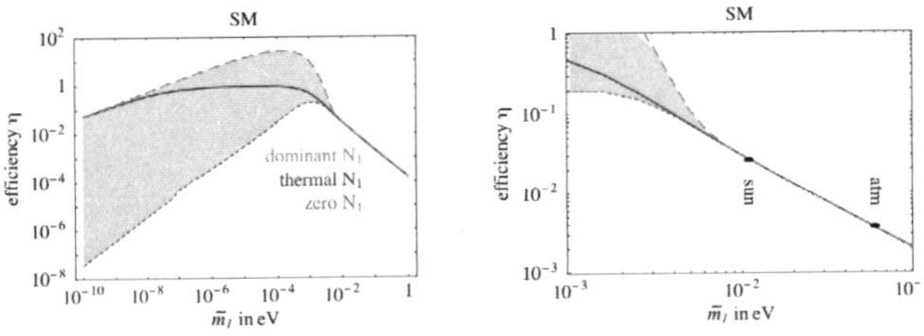

Figure 2. The efficiency factor as a function of $\tilde{m}_1$. Notice that for small values of $\tilde{m}_1$ the slop is $\tilde{m}_1^{-1}$ as the CP asymmetry from scattering is not accounted for. On the left-hand side three cases are analyzed: zero N_1 abundance, thermal abundance ($n_{N_1} \approx n_{N_1}^{\rm eq}$ and dominant N_1. On the right hand side, there is a zoom. From [11].

This enables the reader to study leptogenesis in neutrino mass models without setting up and solving the complicated Boltzmann equations.

5.3. Implications of one-flavour leptogenesis

Experiments have not yet determined the mass m_3 of the heaviest mainly left-handed neutrino. We assume $m_3 = \max(\tilde{m}_1, m_{\rm atm})$. A crucial assumption we have so far is that is that right-handed neutrinos are very hierarchical. Under this hypothesis the CP asymmetry is bounded by the expression [15] that in the hierarchical and quasi-degenerate light neutrino limit simplifies as follows

$$|\epsilon_1| \leq \frac{3}{16\pi}\frac{M_1(m_3 - m_1)}{v^2} \times \begin{cases} 1 - m_1/\tilde{m}_1 & \text{if } m_1 \ll m_3 \\ \sqrt{1 - m_1^2/\tilde{m}_1^2} & \text{if } m_1 \approx m_3 \end{cases} . \tag{75}$$

where all parameters are renormalized at the high-energy scale $\approx M_1$. The 3σ ranges of $m_{\rm atm}$ and of Y_B imply the lower bound

$$M_1 > \frac{4.5 \times 10^8\,\text{GeV}}{\eta} > 2.4 \times 10^9\,\text{GeV} \tag{76}$$

in the case in which no RH neutrinos are present in the plasma at the beginning of the dynamics and we have assumed that the efficiency is maximal (and therefore η minimal). This bound relevant because it tells us that a working model of thermal leptogenesis (in the one-flavour approximation) needs enough heavy RH neutrinos. This has implications for $SO(10)$ grand-unified theories, commonly regarded as the most attractive way to embed the see-saw mechanism. Indeed, in a traditional version of leptogenesis, where the spectrum of right-handed (RH) neutrinos is hierarchical and the asymmetry is produced from the decays of the lightest ones, there is a stringent lower bound on their mass [15], $M_1 > \mathcal{O}(10^9)$ GeV, for a sufficiently large baryon asymmetry to be produced. On the other hand, $SO(10)$ grand-unified theories typically yield, in their simplest version and for the measured values of the neutrino mixing parameters, a hierarchical spectrum with the RH neutrino masses proportional to the squared of the up-quark masses, leading to $M_1 = \mathcal{O}(10^5)$ GeV and to a final asymmetry that falls a few orders of magnitude below the observed one.

We can also work out a bound on the light neutrino masses, $\sum_i m_i < 0.15$ eV. It can be understood to arise from the lower bound on the total decay rate $m_1 \leq \tilde{m}_1$, and the upper bound on the total CP asymmetry (75). We assume the light neutrinos are degenerate, so $|m_1| \approx |m_2| \approx |m_3| \equiv \bar{m}/\sqrt{3}$ — but the masses can have different Majorana phases. Leptogenesis takes place in the strong washout regime, due to the lower bound on the total decay rate. The final baryon asymmetry can be roughly approximated as $Y_B \approx 10^{-4}\epsilon/K \propto \Delta m^2_{\rm atm}/\bar{m}^2$. As the light neutrino mass scale is increased, M_1 and the temperature of leptogenesis must increase to compensate the $\Delta m^2_{\rm atm}/\bar{m}^2$ suppression. However, this temperature is bounded from above, from the requirement of having the $\Delta L = 2$ processes out of equilibrium when leptogenesis takes place:

$$\frac{\bar{m}^2 T^3}{12\pi v^4} < \frac{10T^2}{M_{\rm P}}, \tag{77}$$

so $M_1 < 10^{10}({\rm eV}/\bar{m})^2$ GeV. There is therefore an upper bound on the baryon asymmetry which scales as $1/\bar{m}^4$, and with our rough estimates, one finds

$$\bar{m}/\sqrt{3} < 0.1\,{\rm eV}\,. \tag{78}$$

This result is confirmed by a full numerical analysis [11].

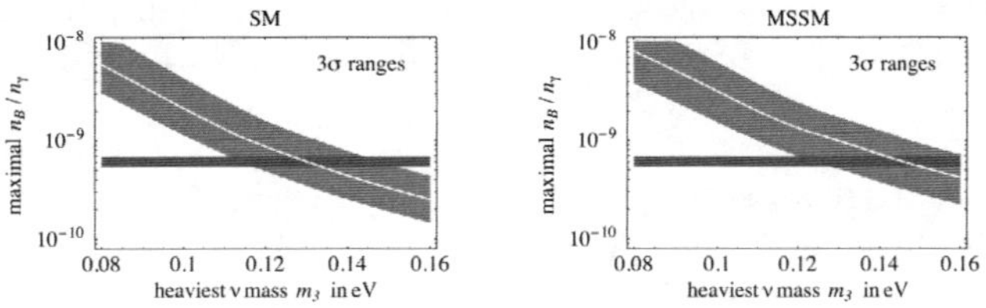

Figure 3. The bound on the heavist light neutrino mass for the SM and for the Minimal Supersymmetric Standard Model (MSSM).

6. Comments on baryogenesis via leptogenesis when flavours accounted for

In recent years, a lot of work has been devoted to a thorough analysis of thermal leptogenesis giving limited attention to the issue of lepton flavour [16]. The dynamics of leptogenesis is usually addressed within the 'one-flavour' approximation, where Boltzmann equations are written for the abundance of the lightest RH neutrino, responsible for the out of equilibrium and CP asymmetric decays, and for the total lepton asymmetry. However, this 'one-flavour' approximation is rigorously correct only when the interactions mediated by charged lepton Yukawa couplings are out of equilibrium. Flavour effects have not been included in leptogenesis calculations till very recently [17, 18, 19]. This is perhaps because perturbatively, they seem to be a small correction. For instance, if the asymmetry is a consequence of the very-out-of-equilibrium decay of an initial population of right-handed neutrinos, then the total lepton asymmetry is of order ϵ/g_*, where ϵ is the total CP asymmetry in the decay, and g_* counts for the entropy dilution factor. Clearly the small charged lepton Yukawa couplings have no effect on ϵ. However, realistic leptogenesis is a drawn-out dynamical process, involving the production and destruction of right-handed neutrinos, and of a lepton asymmetry that is distributed among *distinguishable* flavours. The processes which wash out lepton number are flavour dependent, *e.g* the inverse decays from electrons can destroy the lepton asymmetry carried by, and only by, the electrons. The asymmetries in each flavour are therefore washed out differently, and will appear with different weights in the final formula for the baryon asymmetry. This is physically inequivalent to the treatment of washout in the one-flavour approximation, where indistinguishable leptons propagate between decays and inverse decays, so inverse decays from all flavours are taken to wash out asymmetries in any flavour.

We define $Y_{\alpha\alpha}$ to be the lepton asymmetry in flavour α, where the α are the lepton mass eigenstates at the temperature of leptogenesis. The $Y_{\alpha\alpha}$ are the diagonal elements of a matrix $[Y]$ in flavour space, whose trace is the total lepton asymmetry. In this paper the off-diagonal elements are neglected. The equations of motion for the matrix $[Y]$ are more complicated than the Boltzmann equations, but at most temperatures are equivalent to Boltzmann equations written in the mass eigenstate basis of the leptons in the plasma. The off-diagonal elements of $[Y]$ could have some effect on the lepton asymmetry, if leptogenesis takes place just as a charged lepton Yukawa coupling is coming into equilibrium (so the mass eigenstate basis is changing).

The mass eigenstates for the particles in the Boltzmann equations(BE) are determined by the interactions which are fast compared to those processes included in the BE. The interaction rate for Yukawa coupling h_α can be estimated as

$$\Gamma_\alpha \approx 5 \times 10^{-3} h_\alpha^2 T, \tag{79}$$

so interactions involving the tau (mu) Yukawa coupling are out-of-equilibrium in the primeval plasma if $T > 10^{12}$ GeV ($T > 10^9$ GeV) [1]. Thermal leptogenesis takes place at temperatures on the order of M_1, and the asymmetry is generated when the rates $< H$, so we conclude that the τ (μ) lepton doublet is a distinguishable mass eigenstate, for the purposes of leptogenesis, at $T < 10^{12}(10^9)$ GeV.

Suppose therefore that M_1 is below 10^9 GeV and we forget for the time being about the bound (76). All flavours are distinguishable.

The Boltzmann equations for the flavour asymmetries $Y_{\alpha\alpha}$, are as follows. The Equation for the N_1 number density remains unchanged, and the equation for the flavoured lepton asymmetry is

$$\frac{dY^{\alpha\alpha}}{dz} = \frac{z}{sH(M_1)}\left[\left(\frac{Y_{N_1}}{Y_{N_1}^{EQ}} - 1\right)\epsilon^{\alpha\alpha}(\gamma_D + \gamma_{\Delta L=1}) - \frac{Y^{\alpha\alpha}}{Y_L^{EQ}}\left(\gamma_D^{\alpha\alpha} + \gamma_{\Delta L=1}^{\alpha\alpha}\right)\right]. \tag{80}$$

To obtain analytic solutions we could simplify this, with the approximations introduced in the previous section, to

$$Y'_{\alpha\alpha} = \epsilon_{\alpha\alpha} K z \frac{K_1(z)}{K_2(z)} f_1(z)\Delta_{N_1} - \frac{1}{4} z^3 K_1(z)\, f_2(z)\, K_{\alpha\alpha} Y_{\alpha\alpha} \tag{81}$$

$$\Delta'_{N_1} = -z\, K \frac{K_1(z)}{K_2(z)}\, f_1(z)\, \Delta_{N_1} - Y_{N_1}^{EQ\,\prime}, \tag{82}$$

where

$$K_{\alpha\alpha} Y_{\alpha\alpha} \tag{83}$$

$$\Delta'_{N_1} = -z\, K \frac{K_1(z)}{K_2(z)}\, f_1(z)\, \Delta_{N_1} - Y_{N_1}^{EQ\,\prime}, \tag{84}$$

where

$$K_{\alpha\alpha} = K \frac{\lambda_{1\alpha}\lambda_{1\alpha}^*}{\sum_\gamma |\lambda_{1\gamma}|^2} = \left(\frac{\tilde{m}_{\alpha\alpha}}{10^{-3}\,\text{eV}}\right), \quad K = \sum_\alpha K_{\alpha\alpha}\,. \tag{85}$$

$K_{\alpha\alpha}$ parametrizes the decay rate of N_1 to the α-th flavour, and the trace $\sum_\alpha K_{\alpha\alpha}$, coincides with the K parameter defined in the previous section.

[1] The electron Yukawa coupling mediates interactions relevant in the early Universe only for temperatures beneath $\approx 10^5$ GeV and can be safely disregarded.

Notice in particular that the dynamics of the right-handed neutrinos is always set by the total K.

The CP asymmetry in the α-th flavour is $\epsilon_{\alpha\alpha}$ and is normalised by the total decay rate

$$\epsilon_{\alpha\alpha} = \frac{1}{(8\pi)}\frac{1}{[\lambda\lambda^\dagger]_{11}}\sum_j \mathrm{Im}\left\{(\lambda_{1\alpha})(\lambda\lambda^\dagger)_{1j}\lambda^*_{j\alpha}\right\} g\left(\frac{M_j^2}{M_1^2}\right) \quad (86)$$

$$\rightarrow \frac{3}{(16\pi)[\lambda\lambda^\dagger]_{11}}\mathrm{Im}\left\{\lambda_{1\beta}\frac{[m^*]_{\beta\alpha}}{v^2}\lambda_{1\alpha}\right\} \quad (87)$$

where the second line is in the limit of hierarchical N_J, and $m = U^* D_m U^\dagger = v^2\lambda^T M^{-1}\lambda$ is the light neutrino mass matrix. If m_3 is the heaviest light neutrino mass ($= m_{\rm atm}$ for the non-degenerate case) and we define $\epsilon^{\rm max} = 3\Delta m^2_{\rm atm} M_1/(8\pi v^2 m_{\rm max})$ [15], then the flavour dependent CP asymmetries are bounded by

$$\epsilon_{\alpha\alpha} \leq \frac{3M_1 m_3}{16\pi v^2}\sqrt{\frac{K_{\alpha\alpha}}{K}} = \epsilon^{\rm max}\frac{m_3^2}{\Delta m^2_{\rm atm}}\sqrt{\frac{K_{\alpha\alpha}}{K}} \quad (88)$$

so the maximum CP asymmetry in a given flavour is unsuppressed for degenerate light neutrinos [17], but decreases as the square root of the branching ratio to that flavour $= K_{\alpha\alpha}/K$. The first consequence of this result is that there no upper bound on the light neutrino masses when flavours are accounted for.

The CP asymmetry $\epsilon_{\alpha\alpha}$ can be written in terms of the diagonal matrix of the light neutrino mass eigenvalues $m = \mathrm{Diag}(m_1, m_2, m_3)$, the diagonal matrix of the the right handed neutrino masses $M = \mathrm{Diag}(M_1, M_2, M_3)$ and an orthogonal complex matrix [20]

$$R = v\, M^{-1/2}\,\lambda\, U\, m^{-1/2}\,, \quad (89)$$

where the matrix U diagonalizes the light neutrino mass matrix m_ν, so that

$$U^\dagger\, m_\nu\, U^\star = -D_m \quad (90)$$

and it can be identified with the lepton mixing matrix in a basis where the charged lepton mass matrix is diagonal. The asymmetry reads

$$\epsilon_{\alpha\alpha} = -\frac{3M_1}{16\pi v^2}\frac{\mathrm{Im}\left(\sum_{\beta\rho} m_\beta^{1/2} m_\rho^{3/2} U^*_{\alpha\beta} U_{\alpha\rho} R_{\beta 1} R_{\rho 1}\right)}{\sum_\beta m_\beta\, |R_{1\beta}|^2}. \quad (91)$$

For a real R matrix, the individual CP asymmetries $\epsilon_{\alpha\alpha}$ may not vanish because of the presence of CP violation in the U matrix. On the contrary, the total CP asymmetry $\epsilon_1 = \sum_\alpha \epsilon_{\alpha\alpha}$ vanishes. This is the second implication of the flavours: leptogenesis work even if the matrix R is real. Let us say his in other words. The neutrino Yukawa coupling can be written in its singular value decomposition, $\lambda = V_L^\dagger \mathrm{Diag}(\lambda_1, \lambda_2, \lambda_3) U_R$. Hence, the CP violation in the right-handed neutrino sector is encoded in the phases in V_R, that can be extracted from diagonalizing the combination $\lambda\lambda^\dagger = U_R^\dagger \mathrm{Diag}(\lambda_1^2, \lambda_2^2, \lambda_3^2) U_R$. In the one-flavour regime, only the phases of the matrix V_R count, in the flavour regime the phase of the matrix V_L also count and the final lepton asymmetry does not vanish even if the matrix V_R is real. Implications of this result can be found in [21, 22, 23].

7. Electroweak baryogenesis

So far, we have been assuming that the departure from thermal equilibrium, necessary to generate any baryon asymmetry, is attained by late decays of heavy particles. In this section, we will focus on a different mechanism, namely the departure from equilibrium during first order phase transitions.

A first order phase transition is defined to occur if some thermodynamic quantities change discontinuously. This happens because there exist two separate thermodynamic states that are in thermal equilibrium at the time of the phase transition. The thermodynamic quantity that undergoes such a discontinuous change is generically called the order parameter ϕ. Whether a phase transition is of the first order or not depends upon the parameters of the theory and it may happen that, changing those parameters, the order parameter becomes continuous at the time of the transition. In this case, the latter is said to be of the second order at the point at which the transition becomes continuous and a continuous crossover at the other points for which all physical quantities undergo no changes. In general, we are interested in systems for which the high temperature ground state of the theory is at $\phi = 0$ and the low temperature phase is at $\phi \neq 0$ [1].

For a first order phase transition, the extremum at $\phi = 0$ becomes separated from a second local minimum of the potential by an energy barrier. At the critical temperature T_c both phases are equally favoured energetically and at later times the minimum at $\phi \neq 0$ becomes the global minimum of the theory. The phase transition proceeds by nucleation of bubbles. Initially, the bubbles are not large enough for their volume energy to overcome the competing surface tension, they shrink and disappear. However, at the nucleation temperature, critical bubbles form, *i.e.* bubbles which are just large enough to be nucleated and to grow. As the bubble walls separating the broken from the unbroken phase pass each point in space, the order parameter changes rapidly, leading to a significant departure from thermal equilibrium.

The critical bubbles of the broken (Higgs) phase have a typical profile

$$\phi(r) = \frac{\langle\phi(T_c)\rangle}{2}\left[1 + \tanh\left(\frac{r}{L_\omega}\right)\right], \tag{92}$$

where r is th spatial coordinate, L_ω is the bubble wall width and $\langle\phi(T_c)\rangle$ is the VEV of the Higgs field inside the bubble.

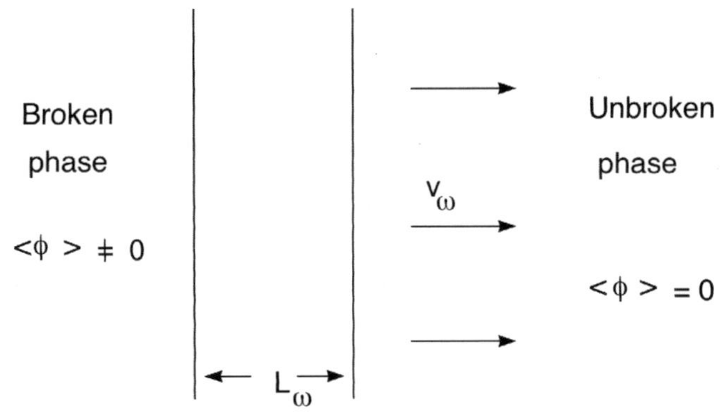

Figure 4. Schematic picture of the propagating bubble separating the broken from the unbroken phase during the electroweak phase transition.

Bubbles expand with velocity v_ω until the fill the Universe; local departure from thermal equilibrium takes place in the vicinity of the expanding bubble walls, see Fig. 10.

The fundamental idea of electroweak baryogenesis is to produce asymmetries in some local charges which are (approximately) conserved by the interactions inside the bubble walls, where local departure from thermal equilibrium is attained. These local charges will then diffuse into the unbroken phase where baryon number violation is active thanks to the unsuppressed sphaleron transitions. The latter convert the asymmetries into baryon asymmetry, because the state of minimum free energy is attained for nonvanishing baryon number. Finally, the baryon number flows into the broken phase where it remains as a remnant of the electroweak phase transition if the sphaleron transitions are suppressed in the broken phase. The recipe for electroweak baryogenesis is therefore the following:

– Look for those charges which are approximately conserved in the symmetric phase, so that they can efficiently diffuse in front of the bubble where baryon number violation is fast, and non-orthogonal to baryon number, so that the generation of a non-zero baryon charge is energetically favoured.

– Compute the CP violating currents of the plasma locally induced by the passage of the bubble wall.

– Write and solve a set of coupled differential diffusion equations for the local particle densities, including the CP violating source terms derived from the computation of the current at the previous step and the particle number changing reactions. The solution to these equations gives a net baryon number which is produced in the symmetric phase and then transmitted into the interior of the bubbles of the broken phase, where it is not wiped out if the first transition is strong enough. Since C and CP are known to be violated by the electroweak interactions, it is possible – in principle –to satisfy all Sakharov's conditions within the SM if the electroweak phase transition. CP violation is however already deadly. A naive estimate suggests that, since CP violation vanishes in the SM if any two quarks of the same charge have the same mass, the measure of CP violation should be the Jarlskog invariant

$$\begin{aligned}\frac{A_{CP}}{J} &= \left(M_t^2 - M_c^2\right)\left(M_c^2 - M_u^2\right)\left(M_u^2 - M_t^2\right) \\ &\quad \left(M_b^2 - M_s^2\right)\left(M_s^2 - M_d^2\right)\left(M_d^2 - M_b^2\right), \end{aligned} \tag{93}$$

where J is twice the area of the unitarity triangle. The quantity A_{CP} has dimension twelve. In the limit of high temperature, T much larger than the quark masses M_q, the only mass scale in the problem is the temperature itself. Therefore, the dimensionless quantity δ_{CP} is

$$\delta_{CP} \simeq \frac{A_{CP}}{T_c^{12}} \simeq 10^{-20}, \tag{94}$$

far too small for the SM to explain the observed baryon asymmetry.

7.1. Electoweak baryogenesis in the MSSM

Electroweak baryogenesis in the framework of the Minimal Supersymmetric Standard Model (MSSM) has attracted much attention in the past years, with particular emphasis on the strength of the phase transition and the mechanism of baryon number generation. A thorough analysis has shown that the electroweak phase transition is strong enough if the mass of the lightest Higgs is less than about 120 GeV and the right-handed stop is lighter than the top [25]. The new contributions to explicit violation of CP are given in the phases of the complex trilinear coupling A, the B parameter, the gaugino masses M_i ($i = 1, 2, 3$) and by the parameter μ in the superpotential. Two phases may be removed by redifining the phase of one of the two Higgs superfields, $\hat{H}_2$, in such a way that the phase of μ is opposite to that of B. The product μB is therefore real. It is also possible to remove the phase of the gaugino mass M by an R symmetry transformation. The latter leaves all the other supersymmeric couplings invariant and only modifies the trilinear ones, which get multiplied by $\exp(-\phi_M)$ where ϕ_M is the phase of M.

The phases which are left are therefore

$$\phi_A = \arg(AM) \quad \text{and} \quad \phi_\mu = -\arg(B). \tag{95}$$

The two new phases ϕ_A and ϕ_μ will be crucial for the generation of the baryon asymmetry. These new phases are essential for the generation of the baryon number since large CP violating sources may be locally induced by the passage of the bubble wall separating the broken from the unbroken phase during the electroweak phase transition. Baryogenesis is fuelled when transport properties allow the CP violating charges to efficiently diffuse in front of the advancing bubble wall where anomalous electroweak baryon violating processes are not suppressed. The new phases appear in the soft supersymmetry breaking parameters associated to the stop mixing angle and to the gaugino and neutralino mass matrices; large values of the stop mixing angle are, however, strongly restricted in order to preserve a sufficiently strong first order electroweak phase transition. Therefore, an acceptable baryon asymmetry from the stop sector may only be generated through a delicate balance between the values of the different soft supersymmetry breaking parameters contributing to the stop mixing parameter. As a result, the contribution to the final baryon asymmetry from the stop sector turns out to be negligible. On the other hand, charginos and neutralinos may be responsible for the observed baryon asymmetry if the phase of the parameter μ is large enough [24]. Yet, this is true within the MSSM. If the strength of the electroweak phase transition is enhanced by the presence of some new degrees of freedom beyond the ones contained in the MSSM, *e.g.* some extra standard model gauge singlets, light stops (predominantly the right-handed ones) and charginos/neutralinos are expected to give quantitatively the same contribution to the final baryon asymmetry.

As mentioned, a strongly first order electroweak phase transition can be achieved in the presence of a top squark lighter than the top quark [25]. In order to naturally suppress its contribution to the parameter $\Delta\rho$ and hence preserve a good agreement with the precision measurements at LEP, it should be mainly right handed. This can be achieved if the left handed stop soft supersymmetry breaking mass m_Q is much larger than M_Z. Indeed, the preservation of the baryon number asymmetry requires the order parameter $\langle\phi(T_c)\rangle/T_c$ to be larger than one (in the one Higgs limit, valid if the mass of the pseudoscalar A is high enough) . The latter is bounded from above

$$\frac{\langle\phi(T_c)\rangle}{T_c} < \left(\frac{\langle\phi(T_c)\rangle}{T_c}\right)_{\rm SM} + \frac{2\, m_t^3}{\pi\, v\, m_h^2}\,, \tag{96}$$

where $m_t = \overline{m}_t(m_t)$ is the on-shell running top quark mass in the $\overline{\rm MS}$ scheme. The first term on the right hand side of expression (96) is the Standard Model contribution

$$\left(\frac{\langle\phi(T_c)\rangle}{T_c}\right)_{\rm SM} \simeq \left(\frac{40}{m_h[{\rm GeV}]}\right)^2, \tag{97}$$

and the second term is the contribution that would be obtained if the right handed stop plasma mass vanished at the critical temperature. The difference between the SM and the MSSM is that light stops may give a large contributions to the effective potential in the MSSM.

In order to overcome the Standard Model constraints, the stop contribution must be therefore large. The stop contribution strongly depends on the value of value of the soft breaking right-handed mass squared $\widetilde{m}_U^2$, which must be small in magnitude, and negative, in order to induce a sufficiently strong first order phase transition. Indeed, large stop contributions are always associated with small values of the right handed stop plasma mass

$$m_{\widetilde{t}}^{\rm eff} = -\widetilde{m}_U^2 + \Pi_R(T), \tag{98}$$

where $\widetilde{m}_U^2 = -m_U^2$, $\Pi_R(T) \simeq 4g_3^2T^2/9 + h_t^2/3T^2$ is the finite temperature self-energy contribution to the right-handed squarks. Although large values of $\widetilde{m}_U$, of order of the critical temperature, are useful to get a strongly first order phase transition, they may also induce charge and color breaking minima. Indeed, if the effective plasma mass at the critical temperature vanished, the universe would be driven to a charge and color breaking minimum at $T \geq T_c$. Hence, the upper bound on $\langle\phi(T_c)\rangle/T_c$, Eq. (96) cannot be reached in realistic scenarios. A conservative bound on $\widetilde{m}_U$ may be obtained by demanding that the electroweak symmetry breaking minimum should be lower than any color-breaking minima induced by the presence of $\widetilde{m}_U$ at zero temperature, which yields the condition

$$\widetilde{m}_U \leq \left(\frac{m_H^2 v^2 g_3^2}{12}\right)^{1/4} . \tag{99}$$

It can be shown that this condition is sufficient to prevent dangerous color breaking minima at zero and finite temperature for any value of the mixing parameter $\widetilde{A}_t$.

In order to obtain values of $\langle\phi(T_c)\rangle/T_c$ larger than one, the Higgs mass must take small values, close to the present experimental bound. Numerically, an upper bound, of order 80 GeV, can be derived. For small mixing, the one-loop Higgs mass has a very simple form

$$m_h^2 = M_Z^2 \cos^2 2\beta + \frac{3}{4\pi^2}\frac{\overline{m}_t^4}{v^2}\log\left(\frac{m_{\widetilde{t}}^2 m_{\widetilde{T}}^2}{\overline{m}_t^4}\right), \tag{100}$$

where $m_{\widetilde{T}}^2 \simeq m_Q^2 + m_t^2$, is the heaviest stop squared mass. Hence, $\tan\beta$, which is the ratio of the vacuum expectation values of the two Higgses, must take values as small as possible. The larger the left handed stop mass, the closer to one $\tan\beta$ must be. This implies that the left handed stop effects decouple at the critical temperature and hence, different values of m_Q mainly affect the baryon asymmetry through the resulting Higgs mass. Values of the CP-odd Higgs mass $m_A < 200$ GeV are associated with a weaker first order phase transition.

Those charges which are approximately conserved in the symmetric phase, so that they can efficiently diffuse in front of the bubble where baryon number violation is fast, and non-orthogonal to baryon number, are good charges to generate a baryon number.

Charges with these characteristics in the MSSM are the axial stop charge and the Higgsino charge, which may be produced from the interactions of squarks and charginos and/or neutralinos with the bubble wall, provided a source of CP-violation is present in these sectors. This is exactly the case, since both the parameters A_t and μ may carry a physical phase [24]. The idea is that, if nonvanishing CP violating sources for the right-handed stop and higgsino numbers are induced in the bubble wall, the scattering among particles as well as diffusion will generate an asymmetry in the left-handed fermion asymmetries in the unbroken phase. The asymmetry – in turn – will fuel baryogenesis because sphaleron transitions will push the system towards the state of minimum free energy, which is the one with nonvanishing baryon asymmetry. In the next subsection, we will give some indications of how to compute the CP-violating sources.

The final baryon asymmetry turns out to be [24]

$$\frac{n_B}{s} \simeq \left(\frac{|\sin(\phi_\mu)|}{10^{-3}}\right) 4 \times 10^{-11}, \tag{101}$$

for $v_\omega \simeq 1$. A more complete analysis has shown that there exist regions of the MSSM parameter space that are consistent with both present two-loop electric dipole moment limits and the relic density and that allow for successful baryogenesisthrough resonant chargino and neutralino processes at the electroweak phase transition [26]. Under certain conditions the lightest neutralino may be simultaneously responsible for both the baryon asymmetry and relic density. Furthermore, requiring a sufficiently large asymmetry imposes lower limits on the size of the electric dipole moments which are typically on the same order, or above, the expected sensitivity of the next generation of experimental searches, implying that MSSM electroweak baryogenesis will be soon conclusively tested.

References

[1] E. W. Kolb and M. S. Turner, Front. Phys. **69**, 1 (1990).
[2] E. Komatsu *et al.* [WMAP Collaboration], arXiv:0803.0547 [astro-ph].
[3] S. Dodelson, *Amsterdam, Netherlands: Academic Pr. (2003) 440 p.*
[4] F. Iocco, G. Mangano, G. Miele, O. Pisanti and P. D. Serpico, arXiv:0809.0631 [astro-ph].
[5] A. Riotto, arXiv:hep-ph/9807454.
[6] A. Riotto and M. Trodden, Ann. Rev. Nucl. Part. Sci. **49**, 35 (1999) [arXiv:hep-ph/9901362].
[7] S. Davidson, E. Nardi and Y. Nir, arXiv:0802.2962 [hep-ph].
[8] A.D. Sakharov, Zh. Eksp. Teor. Fiz. Pis'ma **5**, 32 (1967); JETP Lett. **91B**, 24 (1967).
[9] M. Fukugita and T. Yanagida, Phys. Lett. B **174**, 45 (1986).
[10] D. H. Lyth and A. Riotto, Phys. Rept. **314**, 1 (1999) [arXiv:hep-ph/9807278].
[11] G. F. Giudice, A. Notari, M. Raidal, A. Riotto and A. Strumia, Nucl. Phys. B **685**, 89 (2004) [arXiv:hep-ph/0310123].
[12] W. Buchmuller, P. Di Bari and M. Plumacher, Annals Phys. **315**, 305 (2005).
[13] P. Minkowski, Phys. Lett. B **67**, 421 (1977); M. Gell-Mann, P. Ramond and R. Slansky, *Proceedings of the Supergravity Stony Brook Workshop*, New York 1979, eds. P. Van Nieuwenhuizen and D. Freedman; T. Yanagida, *Proceedings of the Workshop on Unified Theories and Baryon Number in the Universe*, Tsukuba, Japan 1979, ed.s A. Sawada and A. Sugamoto; R. N. Mohapatra, G. Senjanovic, Phys. Rev. Lett. **44**, 912 (1980).
[14] M. C. Gonzalez-Garcia and M. Maltoni, Phys. Rept. **460** (2008) 1; T. Schwetz, M. Tortola and J. W. F. Valle, New J.Phys. **10** (2008) 113011 [arXiv:hep-ph/0808.2016].
[15] S. Davidson and A. Ibarra, Phys. Lett. B **535**, 25 (2002) [arXiv:hep-ph/0202239].
[16] See also, R. Barbieri, P. Creminelli, A. Strumia and N. Tetradis, Nucl. Phys. B **575**, 61 (2000); T. Endoh, T. Morozumi and Z. h. Xiong, Prog. Theor. Phys. **111**, 123 (2004).
[17] A. Abada, S. Davidson, F. X. Josse-Michaux, M. Losada and A. Riotto, JCAP **0604**, 004 (2006).
[18] E. Nardi, Y. Nir, E. Roulet and J. Racker, JHEP **0601**, 164 (2006).
[19] A. Abada, S. Davidson, A. Ibarra, F. X. Josse-Michaux, M. Losada and A. Riotto, JHEP **0609**, 010 (2006).
[20] J. A. Casas and A. Ibarra, Nucl. Phys. B **618**, 171 (2001) [arXiv:hep-ph/0103065].
[21] S. Pascoli, S. T. Petcov and A. Riotto, Phys. Rev. D **75**, 083511 (2007) [arXiv:hep-ph/0609125].
[22] G. C. Branco, R. Gonzalez Felipe and F. R. Joaquim, Phys. Lett. B **645**, 432 (2007) [arXiv:hep-ph/0609297].
[23] S. Pascoli, S. T. Petcov and A. Riotto, Nucl. Phys. B **774**, 1 (2007) [arXiv:hep-ph/0611338].
[24] M. S. Carena, M. Quiros, A. Riotto, I. Vilja and C. E. M. Wagner, Nucl. Phys. B **503**, 387 (1997) [arXiv:hep-ph/9702409].
[25] M. Carena, G. Nardini, M. Quiros and C. E. M. Wagner, Nucl. Phys. B **812**, 243 (2009) [arXiv:0809.3760 [hep-ph]].
[26] V. Cirigliano, Y. Li, S. Profumo and M. J. Ramsey-Musolf, JHEP **1001**, 002 (2010) [arXiv:0910.4589 [hep-ph]]; V. Cirigliano, S. Profumo and M. J. Ramsey-Musolf, JHEP **0607**, 002 (2006) [arXiv:hep-ph/0603246].

Testing discrete symmetries with the cosmic microwave background: current constraints and Planck forecasts

N. Mandolesi[1], C. Burigana[1,2], A. Gruppuso[1] and P. Natoli[2,3,1]

[1] INAF-IASF Bologna, Istituto di Astrofisica Spaziale e Fisica Cosmica di Bologna
Istituto Nazionale di Astrofisica, via Piero Gobetti 101, I-40129 Bologna, Italy
[2] Dipartimento di Fisica, Università degli Studi di Ferrara, via Giuseppe Saragat 1, I-44100 Ferrara, Italy
[3] Agenzia Spaziale Italiana Science Data Center, c/o ESRIN, via Galileo Galilei, Frascati, Italy

E-mail: mandolesi@iasfbo.inaf.it

Abstract. Anisotropy and polarization of the CMB are probing cosmological models with unprecedented precision. The WMAP satellite data are largely consistent with concordance ΛCDM cosmology. However, intriguing flukes are known to exist that may pinpoint at physics beyond the standard model. Constraining the violations of discrete symmetries in the CMB pattern is a promising mean to investigate these discrepancies.

In this paper we constrain the Parity and $\mathcal{CPT}$ symmetries through CMB datasets. We describe the basic formalism, the relevant estimators and the overall analysis strategy. We provide marginal evidence for large scale Parity anomaly in the WMAP data that may be soon confirmed or discarded by the *Planck* satellite. *Planck* is currently measuring CMB anisotropies and their polarization with a level of precision that will remain unparalleled for many years to come. We also show how the CMB can be used to constrain fundamental symmetry violations in the photon sector through the so-called cosmological birefringence phenomenon. Finally, we provide forecasts for *Planck* and we discuss how emission from a specific diffuse foreground component arising within the Solar System needs to be kept under strict control to avoid incurring into false positive detections.

1. Introduction

The *Planck* cosmic microwave background (CMB) anisotropy probe[1], the first European and third generation mission after the COsmic Background Explorer (COBE) and Wilkinson Microwave Anisotropy Probe (WMAP) [2], represents the state-of-the-art in precision cosmology today [1, 2, 3]. *Planck* is equipped with a 1.5–m effective aperture telescope with two actively-cooled instruments observing the sky in nine frequency channels from 30 GHz to 857 GHz: the Low Frequency Instrument (LFI) operating at 20 K with pseudo-correlation radiometers, and the High Frequency Instrument (HFI) with bolometers operating at 100 mK. The coordinated

[1] *Planck* (http://www.esa.int/Planck) is a project of the European Space Agency - ESA - with instruments provided by two scientific Consortia funded by ESA member states (in particular the lead countries: France and Italy) with contributions from NASA (USA), and telescope reflectors provided in a collaboration between ESA and a scientific Consortium led and funded by Denmark.
[2] http://lambda.gsfc.nasa.gov/

use of the two different instrument technologies and analyses of their output data will allow optimal control and suppression of systematic effects, including discrimination of astrophysical sources. A summary of the LFI and HFI performance is reported in Table 1. Note that *Planck* is sensitive to linear polarization up to 353 GHz. The constraints on the thermal behaviour, required to minimize systematic effects, dictated a *Planck* cryogenic architecture that is one of the most complicated ever conceived for space. Moreover, the spacecraft has been designed to exploit the favourable thermal conditions of the orbit around the second Lagrangian point of the Sun-Earth system. The thermal system is a combination of passive and active cooling: passive radiators are used as thermal shields and pre-cooling stages, active cryocoolers are used both for instrument cooling and pre-cooling [4]. *Planck* is a spinning satellite. Thus, its receivers will observe the sky through a sequence of (almost great) circles following a scanning strategy aimed at minimizing systematic effects and achieving all-sky coverage for all receivers [5, 6]. The data analysis, its scientific exploitation, and the core cosmology programme of *Planck* are mostly carried out by two core teams (one for LFI and one for HFI), working in close connection with the Data Processing Centres (DPCs), and closely linked to the wider *Planck* scientific community, consisting of the LFI, HFI, and Telescope consortia, organized into various working groups. *Planck* is managed by the ESA *Planck* science team. *Planck* will open a new era in our understanding of the universe and of its astrophysical structures [7].

Table 1. *Planck* performance. The average sensitivity, δT/T, per FWHM^2 resolution element (FWHM is reported in arcmin) is given in CMB temperature units (i.e. equivalent thermodynamic temperature) for 28 months of integration. The white noise (per frequency channel for LFI and per detector for HFI) in 1 sec of integration (NET, in μK $\cdot\sqrt{\text{s}}$) is also given in CMB temperature units. The other used acronyms are: DT = detector technology, N of R (or B) = number of radiometers (or bolometers), EB = effective bandwidth (in GHz). Adapted from [8, 2] and [3].

LFI			
Frequency (GHz)	30	44	70
InP DT	MIC	MIC	MMIC
FWHM	33.34	26.81	13.03
N of R (or feeds)	4 (2)	6 (3)	12 (6)
EB	6	8.8	14
NET	159	197	158
δT/T [μK/K] (in T)	2.48	3.82	6.30
δT/T [μK/K] (in P)	3.51	5.40	8.91

HFI		
Frequency (GHz)	100	143
FWHM in T (P)	(9.6)	7.1 (6.9)
N of B in T (P)	(8)	4 (8)
EB in T (P)	(33)	43 (46)
NET in T (P)	100 (100)	62 (82)
δT/T [μK/K] in T (P)	2.1 (3.4)	1.6 (2.9)

Frequency (GHz)	217	353
FWHM in T (P)	4.6 (4.6)	4.7 (4.6)
N of B in T (P)	4 (8)	4 (8)
EB in T (P)	72 (63)	99 (102)
NET in T (P)	91 (132)	277 (404)
δT/T [μK/K] in T (P)	3.4 (6.4)	14.1 (26.9)

Frequency (GHz)	545	857
FWHM in T	4.7	4.3
N of B in T	4	4
EB in T	169	257
NET in T	2000	91000
δT/T [μK/K] in T	106	4243

After the successful launch in May 2009, *Planck* has already mapped the sky more than twice (at the time of writing this proceedings paper) with the expected behavior and it is planned to complete at least two further all-sky surveys.

The first scientific results have been released on January 2011 [9]. They describe the instrument performance in flight including thermal behaviour [10, 11, 12], the HFI and LFI data analysis pipelines [13, 14], the main astrophysical results about Galactic science [15, 16, 17, 18, 19, 20, 21], extragalactic sources and far-IR background [22, 23, 24, 25, 26, 27], and

Sunyaev-Zel'dovich effects and cluster properties [28, 29, 30, 31, 32], providing to the scientific community the *Planck* Early Release Compact Source Catalog (ERCSC) [33].

The first publications of the main cosmological implications are expected in early 2013.

The anisotropy pattern of the CMB, measured by WMAP, probes cosmology with unprecedented precision (see [34, 35] and references therein). WMAP data are largely consistent with the concordance Λ cold dark matter (ΛCDM) model, but there are some interesting deviations from it, in particular on the largest angular scales [36]. See also [37] for a critical point of view upon the subject.

A large number of papers dealing with these anomalies have been published in the last years. We briefly list below those that are the most studied: a) lack of power on large angular scales [38, 39]; b) hemispherical asymmetries [40, 41, 42, 43, 44, 45, 46]; c) unlikely alignments of low multipoles [47, 48, 49, 50, 51, 38, 52, 53, 54, 55, 56]; d) non-Gaussianity [46, 57, 58]; e) spots and/or excess of signal [59, 60, 45, 61], possibly linked to non-Gaussianity; f) Parity asymmetry. This anomaly represents one subject of the present paper. It has been suggested in [62] that an estimator built upon the point Parity symmetry might be used as a practical tool for detecting foregrounds. In particular these authors consider whether the observed low CMB quadrupole in temperature could more generally signal odd point-Parity, i.e. suppression of even multipoles. However they claim that WMAP dataset never supports Parity preference beyond the meagre 95% confidence level. Later, [63] found that the Parity symmetry in the temperature map of WMAP 3 and 5 year data is anomalous at the level of 4 out of 1000 in the range $\delta\ell = [2, 18]$. This analysis have been repeated in the WMAP 7 year data confirming the anomaly at same level for a slightly wider range $\delta\ell = [2, 22]$ [64]. We report in this paper that analysis and its extension to polarization [65].

In fact, the CMB polarization pattern can provide information on symmetry-violating physics beyond the standard model. In general, the breakdown of spacetime symmetries is a potential tracer of new physics [66]. Several models exist that predict non-standard $\mathcal{P}$ and $\mathcal{CP}$ violations ('$\mathcal{C}$' standing for charge conjugation), as well as $\mathcal{CPT}$ violations ('$\mathcal{T}$' being time reversal) and the related (through the anti-$\mathcal{CPT}$ theorem [67, 68]) breakdown of Lorentz invariance. A number of tests have been suggested and (in many cases) performed, either in terrestrial and orbital laboratories [69, 70] or through cosmological observations [71, 72, 73]. These violations may also be seen as anomalies in the CMB polarization pattern, since its statistical properties are constrained by the assumption of symmetry conservation.

The paper is organized as follows. In Section 2 we describe the basic formalism, the performed analysis, and the relevant symmetry estimators. Current results on symmetry estimators based on WMAP data are given in Section 3 while in Section 4 we focus on the implications for birefringence. The forecasts for *Planck* about these topics are provided in Section 5. The precise extraction of the cosmological information from microwave observations requires an extremely accurate and efficient data analysis and a careful separation of CMB and astrophysical emissions (see e.g. [74] for a discussion of this topics in the context of the *Planck* surveys). In Section 6 we focus on a specific foreground component that could be critical for the above topics [75]. Finally, our conclusions are drawn in Section 7.

2. Description of the analysis

2.1. Introduction

All-sky temperature maps, $T(\hat{n})$, are usually expanded in Spherical Harmonics $Y_{\ell m}(\hat{n})$, with $\hat{n}$ being a direction in the sky, namely depending on the couple of angles (θ, ϕ):

$$a_{T,\ell m} = \int d\Omega \, Y^{\star}_{\ell m}(\hat{n}) \, T(\hat{n}) \,, \qquad (1)$$

where $a_{T,\ell m}$ are the coefficients of the Spherical Harmonics expansion and $d\Omega = d\theta d\phi \sin\theta$. Under reflection (or Parity) symmetry ($\hat{n} \to -\hat{n}$), these coefficients behave as

$$a_{T,\ell m} \to (-1)^{\ell} a_{T,\ell m} . \tag{2}$$

Analogously for polarizations maps, taking into account the usual combination of Stokes parameters ($Q(\hat{n})$ and $U(\hat{n})$)

$$a_{\pm 2,\ell m} = \int d\Omega \, Y^{\star}_{\pm 2,\ell m}(\hat{n}) \left(Q(\hat{n}) \pm iU(\hat{n}) \right) , \tag{3}$$

where $Y_{\pm 2,\ell m}(\hat{n})$ are the Spherical Harmonics of spin 2 and $a_{\pm 2,\ell m}$ are the corresponding coefficients, it is possible to show that under Parity

$$a_{E,\ell m} \to (-1)^{\ell} a_{E,\ell m}, \tag{4}$$
$$a_{B,\ell m} \to (-1)^{\ell+1} a_{B,\ell m} , \tag{5}$$

where

$$a_{E,\ell m} = -(a_{2,\ell m} + a_{-2,\ell m})/2 , \tag{6}$$
$$a_{B,\ell m} = -(a_{2,\ell m} - a_{-2,\ell m})/2i . \tag{7}$$

Eqs. (2), (4) and (5) show cross-correlations $C_{\ell}^{TB} = C_{\ell}^{EB} = 0$.

Further details can be found for example in [76], [77] and explicit algebra is present in the Appendix of [65].

In order to evaluate the angular power spectrum (APS) we adopt the quadratic maximum likelihood (QML) estimator, introduced in [78] and extended to polarization in [79]. Further details can be found in [80].

2.2. Angular Power Spectrum Estimation, Data set and Simulations

In order to evaluate the angular power spectrum (APS) we adopt the quadratic maximum likelihood (QML) estimator, introduced in [78] and extended to polarization in [79]. Further details can be found in [80]. Now, we describe the data set that we have considered. We use the temperature ILC map smoothed at 9.8 degrees and reconstructed at HEALPix[3] [81] resolution $N_{side} = 16$, the foreground cleaned low resolution maps and the noise covariance matrix in (Q, U) publicly available at the Legacy Archive for Microwave Background Data Analysis (LAMBDA) website [4] for the frequency channels Ka, Q and V as considered by [34] for the low ℓ analysis. These frequency channels have been co-added as follows [82]

$$m_{tot} = C_{tot}(C_{Ka}^{-1} m_{Ka} + C_{Q}^{-1} m_{Q} + C_{V}^{-1} m_{V}) , \tag{8}$$

where m_i, C_i are the polarization maps and covariances (for i=Ka, Q and V) and

$$C_{tot}^{-1} = C_{Ka}^{-1} + C_{Q}^{-1} + C_{V}^{-1} . \tag{9}$$

This polarization data set has been extended to temperature considering the ILC map. We have added to the temperature map a random noise realization with variance of $1 \mu K^2$ as suggested in [83]. Consistently, the noise covariance matrix for TT is taken to be diagonal with variance equal to $1 \mu K^2$.

[3] http://healpix.jpl.nasa.gov/

[4] http://lambda.gsfc.nasa.gov/

We have also performed Monte-Carlo simulations in order to assess the significance of our results. A set of 10000 CMB + noise sky realizations has been generated: the signal extracted from the WMAP 7 years best fit model, the noise through a Cholesky decomposition of the noise covariance matrix. We have then computed the APS for each of the 10000 simulations by means of *BolPol* and built two figures of merit as explained in the next subsection.

Two masks are considered: one for T and one for Q and U. Monopole and dipole have been subtracted from the observed Internal Linear Combination (ILC) map through the HEALPix routine REMOVE-DIPOLE [81].

2.3. Estimators

We define the following quantities

$$C^X_{+/-} \equiv \frac{1}{(\ell_{max}-1)} \sum_{\ell=2,\ell_{max}}^{+/-} \frac{\ell(\ell+1)}{2\pi} \hat{C}^X_\ell \tag{10}$$

where $\hat{C}^X_\ell$ are the estimated APS obtained with the *BolPol* code [80] for the power spectrum X = TT, TE, EE and BB. The sum is meant only over the even or odd ℓ (and this is represented respectively by the symbol + or $-$) with $\ell_{max} \geq 3$.

Therefore, two estimators can be built from Eq. (10) as follows: the ratio R^X, as performed in [63] or [64],

$$R^X = C^X_+ / C^X_- \,, \tag{11}$$

and, in analogy to what performed for the hemispherical symmetry in [44], the difference D^X

$$D^X = C^X_+ - C^X_- \,, \tag{12}$$

of the two aforementioned quantities. In the following, we drop the index X for R and D specifying every time we use them which is the spectrum they refer to.

For our application to WMAP data, both estimators have been considered for the TT spectrum but only the second one for the other spectra (EE, TE and BB). This is due to the unfavorable signal-to-noise ratio of the WMAP data in polarization.

For X = TB and EB we simply use the average power

$$C^X \equiv \frac{1}{(\ell_{max}-1)} \sum_{\ell=2,\ell_{max}} \frac{\ell(\ell+1)}{2\pi} \hat{C}^X_\ell \,. \tag{13}$$

3. Results

In Fig. 1 we show the estimator R and D for TT averaged in $\delta\ell = [2,22]$ and in $\delta\ell = [2,33]$. The probability to obtain a smaller value than the WMAP one is 0.47% for R in the range $\delta\ell = [2,22]$ and 3.17% in the range $\delta\ell = [2,33]$. For the D estimator the probability is 0.63% in the range $\delta\ell = [2,22]$ and 3.17% in the range $\delta\ell = [2,33]$. The upper left panel of Fig. 1 recovers the same level of anomaly claimed in [64].

In Fig. 2 we plot the percentage related to the WMAP 7 yr Parity anomaly for TT versus ℓ_{max} in the range [10, 40] for the two considered estimators. As evident there is not a single ℓ_{max} for which the TT anomaly shows up, but rather a characteristic scale, see also [64]. For the estimator of Eq. (11) the percentage anomaly is well below 1% for almost any choice of ℓ_{max} in the range [15, 25] [5]. As also shown in Fig. 2, the estimator of Eq. (12) follows closely the other estimator although it is slightly less sensitive. Therefore, we find a whole multipole range, rather than a single ℓ_{max} value, where the WMAP 7 yr Parity anomaly holds. This dims significantly the case for posterior biasing.

[5] Only for $\ell_{max} = 21$ the estimator of Eq. (11) exhibits a percentage which is of the order of 1%.

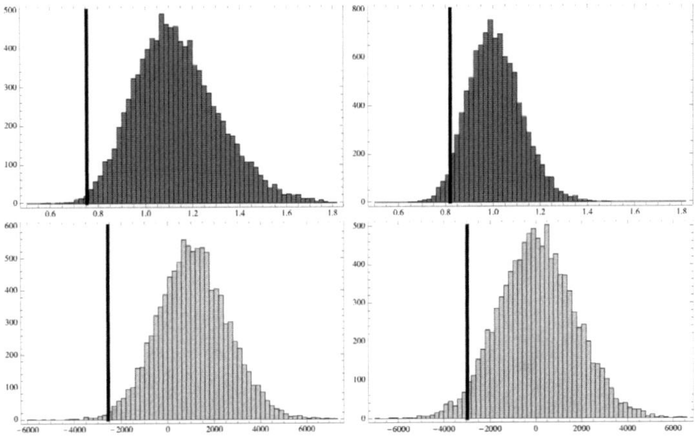

Figure 1. TT. Counts (y-axis) vs the estimator (x-axis). Upper histograms: Ratio for the range $\delta\ell = [2, 22]$ (left panel) and for the range $\delta\ell = [2, 33]$ (right panel). Lower histograms: Difference for the range $\delta\ell = [2, 22]$ (left panel) and for the range $\delta\ell = [2, 33]$ (right panel). Units for the estimator D are μK^2. The vertical line stands for the WMAP 7 year value.

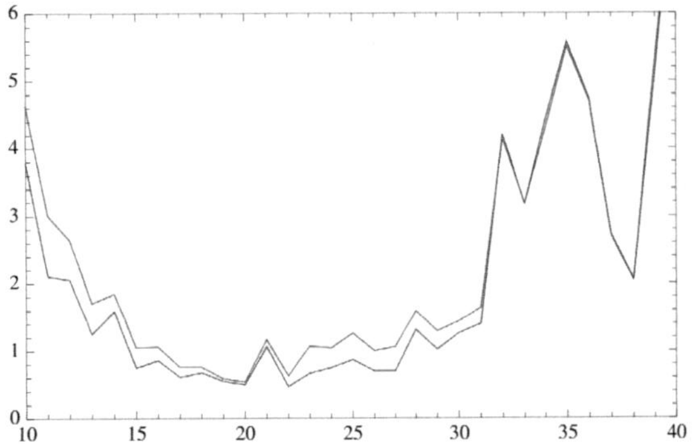

Figure 2. TT. Percentage of the WMAP 7 yr value (y-axis) vs ℓ_{max} (x-axis). Blue line is for the ratio and the red line for the difference. This analysis shows that there is no single ℓ_{max} for which the TT anomaly shows up, but rather suggests the existence of a characteristic scale, see also [64].

In Table 2 we provide the results for EE, TE and BB. As mentioned above, only D is considered and computed for the four following multipoles range $\delta\ell = [2, 4]$, $[2, 8]$, $[2, 16]$ and $[2, 22]$. No anomalies have been found and compatibility with Parity symmetry is obtained.

In Table 3 we provide the results for EB and TB where the estimator C is considered and computed for the same aforementioned four multipoles range. Both the spectra are well consistent with zero. Only the EB spectrum shows a mild anomaly in the range $\delta\ell = [2, 22]$ at the level of 97.7%. This is due to five estimates from $\ell = 18$ to $\ell = 22$ that are systematically larger than zero. When these points are excluded this mild anomaly drops. For example in the range $\delta\ell = [2, 16]$ the probability to obtain a smaller value than the WMAP one is 55.35%. The latter two estimators are shown in Fig. 3.

Table 2. Probabilities (in percentage) to obtain a smaller value than the WMAP 7 yr one.

D	$\delta\ell$ = [2,4]	$\delta\ell$ = [2,8]	$\delta\ell$ = [2,16]	$\delta\ell$ = [2,22]
EE	93.09	76.21	44.27	46.61
TE	56.35	38.88	24.79	22.77
BB	7.97	13.42	11.70	44.31

Table 3. Probabilities (in percentage) to obtain a smaller value than the WMAP 7 yr one.

C	$\delta\ell$ = [2,4]	$\delta\ell$ = [2,8]	$\delta\ell$ = [2,16]	$\delta\ell$ = [2,22]
TB	51.78	39.42	6.71	10.55
EB	62.73	69.83	55.35	97.70

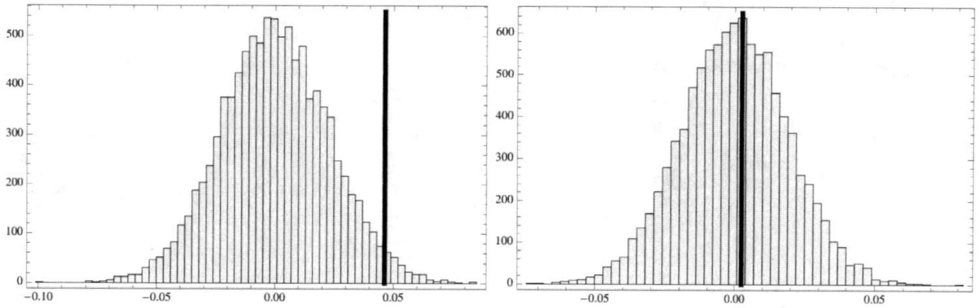

Figure 3. EB. Counts (y-axis) vs the estimator C (x-axis). Distribution of C for $\delta\ell = [2,22]$ (left panel) and $\delta\ell = [2,16]$ (right panel). Units are μK^2. The vertical line stands for the WMAP 7 yr data.

4. Birefringence

As shown above, if the physics controlling CMB fluctuations is Parity conserving then the cross spectra C_l^{TB} and C_l^{EB} must vanish due to the different handedness of the B and (T, E) harmonics. Therefore, if the standard cosmological model holds, we should expect no relevant information from TB and EB. On the other hand, detection of non-zero primordial TB and/or EB may probe fundamental physics in the early universe, such as the presence of a primordial homogeneous [84] or helical [85, 86] magnetic field which would induce Faraday rotation and non-zero TB correlations. Parity-asymmetric gravity dynamics during inflation may generate a discrepancy among left and right-handed gravitational waves, so that TB and EB are non-zero [87, 88]. Particle physics models with non-standard Parity-violating interactions also predict non-vanishing TB and EB signals [89, 90, 91].

In this section we focus on a class of models that exhibit Parity violations in the photon sector [92]. A Chern-Simons term can be introduced in the effective Lagrangian [72, 73]:

$$\Delta\mathcal{L} = -\frac{1}{4} p_\mu \epsilon^{\mu\nu\rho\sigma} F_{\rho\sigma} A_\nu \ ,$$

where $F^{\mu\nu}$ is the Maxwell tensor and A^μ the 4-potential. The 4-vector p_μ may be interpreted as the derivative of the quintessence field or the gradient of a function of the Ricci scalar [93, 94]. In either case a $\mathcal{P}$ violation always arises provided that p_0 is non-zero, while $\mathcal{C}$ and $\mathcal{T}$ remain intact. Hence, $\mathcal{CP}$ and $\mathcal{CPT}$ symmetries are also violated, as well as Lorentz invariance, since p^μ picks up a preferred direction in space-time. The net effect on a propagating photon is to

Journal of Physics: Conference Series **335** (2011) 012009 doi:10.1088/1742-6596/335/1/012009

rotate its polarization direction by an angle $\Delta\alpha$, hence the name "cosmological birefringence". Historically, the effect has being constrained by measuring polarized light from high redshift radio galaxies and quasars [72, 73, 95, 96, 97, 98, 99]. Obviously, the CMB photons would also be affected and, due to their longer journey, may get a larger rotation. A consequence for the CMB pattern is the mixing of E and B modes: the TB and EB correlations still vanish at last scattering surface, but the observable CMB spectra are distorted as [87, 88, 100]:

$$C_l'^{TB} = C_l^{TE}\sin 2\Delta\alpha \quad (14)$$

$$C_l'^{EB} = \frac{1}{2}(C_l^{EE} - C_l^{BB})\sin 4\Delta\alpha \quad (15)$$

$$C_l'^{TE} = C_l^{TE}\cos 2\Delta\alpha \quad (16)$$

$$C_l'^{EE} = C_l^{EE}\cos^2 2\Delta\alpha + C_l^{BB}\sin^2 2\Delta\alpha \quad (17)$$

$$C_l'^{BB} = C_l^{BB}\cos^2 2\Delta\alpha + C_l^{EE}\sin^2 2\Delta\alpha. \quad (18)$$

where the primed quantities are rotated. In [35] a limit $\Delta\alpha = 0.9^\circ \pm 1.4^\circ$ was derived for the multipole range $\delta\ell = [23, 800]$, whereas for $\delta\ell = [2, 23]$ they find $\Delta\alpha = -3.8^\circ \pm 5.2^\circ$. The reason for this distinction is that the low ℓ polarization pattern is only influenced by the reionization epoch, which happened at redshift $z \simeq 10$. The primary fluctuations at higher multipoles, on the other hand, can be traced to last scattering at $z \simeq 1100$ so the corresponding angular scales allow for a much longer journey of the CMB photons. A slightly more stringent limit based on QUaD[6] data has been set in [101] as $\Delta\alpha = 0.83^\circ \pm 0.94^\circ \pm 0.5^\circ$, the second error being systematic.

5. Planck Forecast

In this section we discuss how *Planck* will improve the present constraints on the symmetry violations discussed above. We first take into account the case of the low ℓ Parity anomaly. We then discuss briefly the case of birefringence.

5.1. Simulated dataset

We consider the white noise level for 143 GHz channel of *Planck*. As in [44], we consider the nominal sensitivity of the *Planck* 143 GHz channel, taken as representative of the results which can be obtained after the foreground cleaning from various frequency channels. The 143 GHz channel has an angular resolution of $7.1'$ (FWHM) and an average sensitivity of $6\,\mu K$ ($11.4\,\mu K$) per pixel - a square whose side is the FWHM size of the beam - in temperature (polarization), after 2 full sky surveys. We assume uniform uncorrelated instrumental noise and we build the corresponding diagonal covariance matrix for temperature and polarization, from which, through Cholesky decomposition we are able to extract noise realizations. For this low noise level we apply the same procedure adopted for the Monte-Carlo simulations in Subsection 2.2.

5.2. Forecasts

From the set of 10000 CMB + noise sky realizations, we find that: the T based estimators (both R and D) do not change much since at large scale the APS for T is dominated by cosmic variance and not by the noise. For EE, TE and BB it is possible to consider even the R estimator. See for example Fig. 4 where the R estimator is computed for EE in the range $\delta\ell = [2, 22]$ (left panel) and $\delta\ell = [2, 16]$ (right panel). The standard deviations for the D and C are evaluated in Table 4 for $\delta\ell = [2, 22]$ and compared to the WMAP 7 yr ones.

[6] QUaD stands for "QUEST at DASI"'. In turn, QUEST is "Q & U Extragalactic Survey Telescope" and DASI stands for "Degree Angular Scale Interferometer".

Table 4. Left Table: standard deviation for the D estimator computed in the range $\delta\ell = [2, 22]$. Right Table: standard deviation for the C estimator computed in the range $\delta\ell = [2, 22]$. Units are μK^2.

σ_D	WMAP 7 yr	*Planck*
TT	1517.17	1509.21
TE	20.19	9.08
EE	0.65	0.10
BB	0.69	0.04

σ_C	WMAP 7 yr	*Planck*
TB	0.95	0.19
EB	0.023	0.001

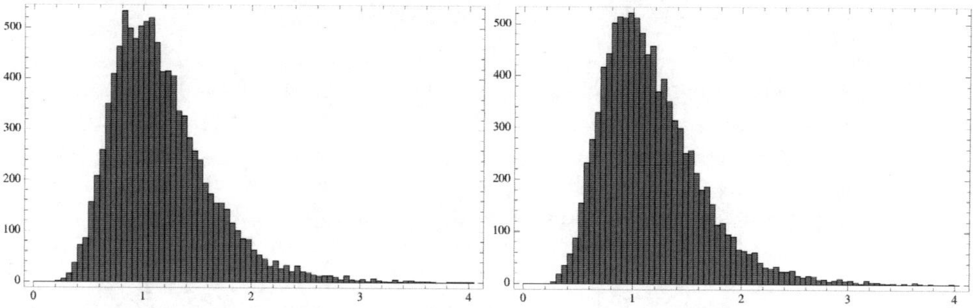

Figure 4. EE. Counts (y-axis) vs the estimator R (x-axis). Distribution of R for $\delta\ell = [2, 22]$ (left panel) and $\delta\ell = [2, 16]$ (right panel).

The case of birefringence has been investigated in [102] where the expected standard error for *Planck* in constraining the rotation angle α is given as $\Delta\alpha = 0.057°$. Note however that this error purely considers the noise level. Realistic measurements of α may be affected by systematic errors, especially arising from the uncertainties in the orientation of the polarimeters. The latter need to be properly taken into account [103].

6. Foreground impact

An open possibility is that the detected Parity anomaly in the WMAP data is due to an unknown foreground contribution relevant at large angular scales, i.e. a diffuse foreground component, or to foreground residuals left by an imperfect component separation. Foreground, by definition, is any astrophysical emission rising between us and the last scattering surface. The spectral behaviors of the most relevant Galactic diffuse foreground components (synchrotron, free-free and dust emission) are typically very different from the CMB blackbody behavior. On the other hand, it can not be excluded that diffuse Solar System emission could contribute at low ℓ's at WMAP frequencies [49, 38]. Such foreground would exhibit planar symmetry with respect to the ecliptic (or some plane slightly tilted with respect to it). According to (Far) Infra Red (IR) measures, the "classical" Zodiacal Light Emission (ZLE) is the thermal emission of sub-mm Interplanetary Dust Particles (IDPs), with orbital radii within 5 AU, characterized by a $\approx \nu^4$ scaling below 1 THz [104]. If larger IDPs exist, their emission would add to the classical ZLE, producing a signal at mm wavelengths, ranging between about $1/10 \div 10$ times the classical ZLE, with a less steep spectrum, $\approx 1/\nu$, as suggested by several models [105, 106, 107, 108, 109], see left panel of Fig. 5. Such particles can be produced by the erosion of Kuiper Belt Objects (KBOs) due to mutual collisions or by the erosion of incoming interstellar dust and they should be located beyond Jupiter. They would be then colder and possibly larger than the IDPs responsible for the classical ZLE. Impact detectors onboard interplanetary probes reveal that such particles indeed

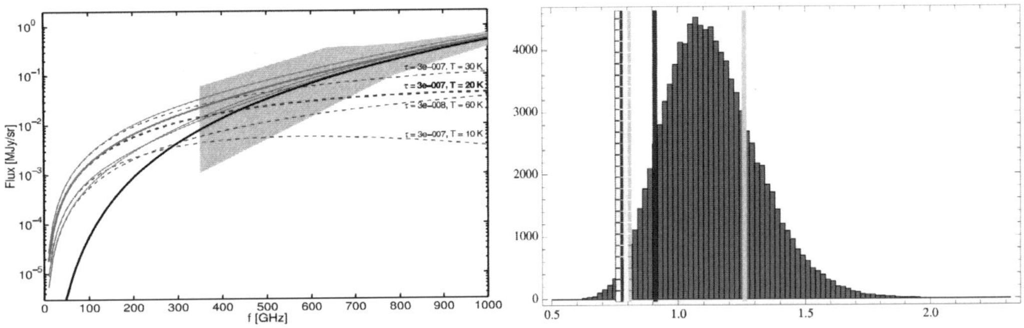

Figure 5. Left Panel: Comparison of ZLE fluxes compatible with FIRAS data [104] and a set of possible models of KBOE [106] characterized by a different dust grain temperature T and optical depth τ. The black solid line shows the ZLE derived from the best fit model to FIRAS data and extrapolated to lower frequencies. The gray band represents a sketch of the allowed region obtained from the error bars in [104]. The blue dashed lines display four different models of KBOE corresponding to different parameters values. The resulting fluxes, sum of KBOE and ZLE, are represented by the red solid lines. Note that the classical ZLE (estimated on the basis of COBE data) is negligible in practice at WMAP frequencies, whereas KBOE might not be ignored. Right Panel: Parity anomaly of the estimator r as defined in the text with $\ell_{max} = 22$. The histogram (in red) displays the distribution of r obtained from 10^5 MC realizations. Vertical lines correspond to the maps considered in this work: the black solid line (on the left) refers to the ILC 7 yr map; colored solid lines refer to the CM; colored dashed lines refer to the WM. Green, blue, and yellow lines are for $H_{KBOE} = 17.5°, 35°$, and $70°$, respectively.

exist [110], but very little is known about their nature. Motivated by these considerations, the problem of diffuse Solar System emission at cosmological frequencies has been reconsidered [75] by introducing a toy model of emission from the KBO dust particles, that we will call in this paper KBO Emission (KBOE), and evaluating its implications for CMB observations.

To explore the effect of the KBOE on CMB maps we considered a toy model in which the emission is confined within a region of width H_{wedge}, symmetric with respect to the ecliptic plane, with constant brightness inside and zero outside. Owing to the planar and cylindrical symmetry of the model, all the coefficients $a_{\ell,m}$ for the multipole expansion in ecliptic coordinates of the KBOE will be zero except for those with even ℓ and $m = 0$. Then, KBOE (and likewise ZLE) will just affect the map components with even ℓ.

Also the KBOE template we propose in this paper, as well as the standard ZLE, is highly Parity-asymmetric. Therefore it is worth to study its possible impact on the WMAP ILC 7yr map, of course under the assumption that a residual of the kind considered in the current paper, is present in the ILC map.

We consider the same estimator of eq. (11) with $\ell_{max} = 22$. We have extracted 10^5 random maps from the best fit model of WMAP 7 yr working at the HEALPix resolution $N_{side} = 64$ (i.e. all-sky maps of 49152 pixels). We have computed the R estimator for each random map and we have built the probability distribution function (pdf) for R, shown in the right panel of Fig. 5. Note that the pdf for R does not peak around the value 1, but at slightly larger values since for the chosen ℓ-range there are more terms at numerator than in the denominator. Vertical lines in right panel of Fig. 5 represent the values of R for the considered maps. Black vertical line (on left) stands for the WMAP 7 yr ILC map. The probability to have this value is as low as

0.91% [7]. When we remove our KBOE template from the maps we have the colored vertical lines. Green, blue, and yellow lines stand for $H_{KBOE} = 17.5°, 35°$, and $70°$, respectively. Colored solid lines refer to a Cold Model (CM), characterized by $T = 30$K and $\tau = 3 \times 10^{-7}$, colored dashed lines refer to a Warm Model (WM), characterized by $T = 60$K and $\tau = 3 \times 10^{-8}$ (see left panel of Fig. 5 for their spectral behaviors). Note that the probability associated to the green line (i.e. CM and $H_{KBOE} = 17.5°$) is 77.1% and the probability associated to the blue line (i.e. CM and $H_{KBOE} = 35°$) is 10.6% . The Parity anomaly is removed when these templates are properly taken into account. In the definition of the Parity estimator in the current paper we adopted $\ell_{max} = 22$, just for simplicity and since for that ℓ_{max} value the anomaly is remarkable. However, as shown in [64, 63, 65], in the WMAP TT spectrum there is a whole multipole range, rather than a single ℓ_{max} value, where the WMAP 7 yr Parity anomaly holds. This dims significantly the case for posterior biasing.

We finally remark that the inclusion of a KBOE component in the analysis could also significantly alleviate the problem of lack of power at large angular scales, reflecting itself in both the TT angular power spectrum and the two-points angular correlation function [75].

7. Conclusions

The *Planck* satellite is measuring CMB anisotropies and their polarization with a level of precision that will remain unparalleled for many years to come. The results derived from the *Planck* dataset will set a benchmark for precision cosmology. In this paper we have focused on fundamental information that the CMB may reveal about the breaking of fundamental discrete symmetries in the early universe. We have reviewed the present constraints, due to WMAP, for the cases of a hinted low resolution Parity anomaly as well as for cosmic birefringence. For the latter, the QUaD dataset has provided the most stringent limits to date. We have also presented *Planck* forecasts. *Planck* may be able to confirm or deny the existence of the low resolution Parity anomaly. Moreover, it is expected to greatly improve the knowledge of the polarization pattern of the CMB. *Planck* will also probe photon birefringence, improving the present constraints by over an order of magnitude. Through its wide frequency coverage and improved foreground cleaning *Planck* will be also capable to detect subdominant foreground components and in particular those that might impact the analyses about discrete symmetries, such as the Solar System diffuse emission considered in this review.

Acknowledgements

We acknowledge the use of the BCX and SP6 at CINECA under the agreement INAF/CINECA and the use of computing facility at NERSC. We acknowledge use of the HEALPix [81] software and analysis package for deriving the results in this paper. We acknowledge the use of the LAMBDA. Support for LAMBDA is provided by the NASA Office of Space Science. Work supported by ASI through ASI/INAF Agreement I/072/09/0 for the Planck LFI Activity of Phase E2.

References

[1] Tauber J., et al. 2010, A&A, 520, A1
[2] Bersanelli M., et al. 2010, A&A, 520, A4
[3] Lamarre J. M., et al. 2010, A&A, 520, A9
[4] Collaudin, B., Passvogel, T. 1999, Cryogenics, 39, 157
[5] Dupac, X., Tauber, J. 2005, A&A, 430, 363
[6] Maris, M., et al. 2006a, Mem. Soc. Astron. It. Suppl., 9, 460
[7] Planck Collaboration, ESA publication ESA-SCI (2005)/1 arXiv:astro-ph/0604069

[7] This value is slightly larger than what quoted in [64] that is 0.86%. This might be due to the details of the MC, like number of simulations, resolutions of maps and the considered FWHM.

[8] Mandolesi, N., et al. 2010, A&A, 520, A3
[9] Planck Collaboration, *et al.* arXiv:1101.2022 [astro-ph.IM]
[10] Planck HFI Core Team *et al.*, arXiv:1101.2039 [astro-ph.IM]
[11] Mennella, A. *et al.*, arXiv:1101.2038 [astro-ph.CO]
[12] Planck Collaboration, *et al.* arXiv:1101.2023 [astro-ph.IM]
[13] Planck HFI Core Team *et al.*, arXiv:1101.2048 [astro-ph.CO]
[14] Zacchei, A. *et al.*, arXiv:1101.2040 [astro-ph.IM]
[15] Planck Collaboration, *et al.* arXiv:1101.2037 [astro-ph.GA]
[16] Planck Collaboration, *et al.* arXiv:1101.2036 [astro-ph.GA]
[17] Planck Collaboration, *et al.* arXiv:1101.2034 [astro-ph.GA]
[18] Planck Collaboration, *et al.* arXiv:1101.2035 [astro-ph.GA]
[19] Planck Collaboration, *et al.* arXiv:1101.2032 [astro-ph.GA]
[20] Planck Collaboration, *et al.* arXiv:1101.2031 [astro-ph.GA]
[21] Planck Collaboration, *et al.* arXiv:1101.2029 [astro-ph.GA]
[22] Planck Collaboration, *et al.* arXiv:1101.2047 [astro-ph.GA]
[23] Planck Collaboration, *et al.* arXiv:1101.2046 [astro-ph.CO]
[24] Planck Collaboration, *et al.* arXiv:1101.2045 [astro-ph.CO]
[25] Planck Collaboration, *et al.* arXiv:1101.2044 [astro-ph.CO]
[26] Planck Collaboration, *et al.* arXiv:1101.1721 [astro-ph.CO]
[27] Planck Collaboration, *et al.* arXiv:1101.2028 [astro-ph.CO]
[28] Planck Collaboration, *et al.* arXiv:1101.2043 [astro-ph.CO]
[29] Planck Collaboration, *et al.* arXiv:1101.2027 [astro-ph.CO]
[30] Planck Collaboration, *et al.* arXiv:1101.2026 [astro-ph.CO]
[31] Planck Collaboration, *et al.* arXiv:1101.2025 [astro-ph.CO]
[32] Planck Collaboration, *et al.* arXiv:1101.2024 [astro-ph.CO]
[33] Planck Collaboration, *et al.* arXiv:1101.2041 [astro-ph.CO]
[34] Larson D. *et al.*, arXiv:1001.4635 [astro-ph.CO]
[35] Komatsu E. *et al.*, Astrophys. J. Suppl. **192** (2011) 18.
[36] Copi C. J., Huterer D., Schwarz D. J., Starkman G. D., arXiv:1004.5602 [astro-ph.CO]
[37] Bennett C. L. *et al.*, arXiv:1001.4758 [astro-ph.CO]
[38] Copi C. J., Huterer D., Schwarz D. J., Starkman G. D., Phys. Rev. D **75**, 023507 (2007)
[39] Copi C. J., Huterer D., Schwarz D. J., Starkman G. D., Mon. Not. Roy. Astron. Soc. **399**, 295 (2009)
[40] Eriksen H. K., Banday A. J., Gorski K. M., Hansen F. K., Lilje P. B., Astrophys. J. **660**, L81 (2007)
[41] Hansen F. K., Banday A. J., Gorski K. M., Eriksen H. K., Lilje P. B., Astrophys. J. **704**, 1448 (2009)
[42] Hoftuft J., Eriksen H. K., Banday A. J., Gorski K. M., Hansen F. K., Lilje P. B., Astrophys. J. **699**, 985 (2009)
[43] Eriksen H. K., Hansen F. K., Banday A. J., Gorski K. M., Lilje P. B., Astrophys. J. **605** (2004) 14 [Erratum-ibid. **609** (2004) 1198]
[44] Paci F., Gruppuso A., Finelli F., C., De Rosa A., Mandolesi N. , Natoli P., Mon. Not. Roy. Astron. Soc., **407**, 1, pp. 399-404 (2010)
[45] Pietrobon D., Cabella P., Balbi A., Crittenden R., de Gasperis G., Vittorio N., Mon. Not. Roy. Astron. Soc. Letters (2009), **402**, Issue 1, pp. L34-L38
[46] Vielva P., Martinez-Gonzalez E., Barreiro R. B., Sanz J. L., Cayon L., Astrophys. J. **609**, 22 (2004)
[47] Tegmark M., de Oliveira-Costa A., Hamilton A., Phys. Rev. D **68**, 123523 (2003)
[48] Copi C. J., Huterer D., Starkman G. D., Phys. Rev. D **70**, 043515 (2004)
[49] Schwarz D. J., Starkman G. D., Huterer D., Copi C. J., Phys. Rev. Lett. **93**, 221301 (2004)
[50] Weeks J. R., arXiv:astro-ph/0412231
[51] Land K., Magueijo J., Phys. Rev. Lett. **95**, 071301 (2005a)
[52] Abramo L. R., Bernui A., Ferreira I. S., Villela T., Wuensche C. A., Phys. Rev. D **74**, 063506 (2006)
[53] Wiaux Y., Vielva P., Martinez-Gonzalez E., Vandergheynst P., Phys. Rev. Lett. **96**, 151303 (2006)
[54] Vielva, P., Wiaux, Y., Martínez-González, E., Vandergheynst, P. 2007, MNRAS, 381, 932
[55] Gruppuso A., Gorski K. M., JCAP **1003**, 019 (2010)
[56] Gruppuso A., Burigana C., JCAP **0908**, 004 (2009)
[57] Bernui A. , Reboucas M. J., Phys. Rev. D **81**, 063533 (2010)
[58] Bernui A., Reboucas M. J., Teixeira A. F. F., arXiv:1005.0883 [astro-ph.CO]
[59] Cruz M., Martinez-Gonzalez E., Vielva P., Cayon L., Mon. Not. Roy. Astron. Soc. **356**, 29 (2005)
[60] Cruz M., Martinez-Gonzalez E., Vielva P., arXiv:0901.1986 [astro-ph]
[61] Diego J. M., Cruz M., González-Nuevo J., Maris M., Ascasibar Y., Burigana C., MNRAS **402**, 1213, (2010)
[62] Land K., Magueijo J., Phys. Rev. D **72**, 101302 (2005)

[63] Kim J., Naselsky P., Astrophys. J. **714**, L265 (2010a)
[64] Kim J., Naselsky P., (2010b) arXiv:1002.0148 [astro-ph.CO]
[65] Gruppuso A., Finelli F., Natoli P., Paci F., Cabella P., De Rosa A., Mandolesi N., Mon. Not. Roy. Astron. Soc., bf 411, 3, (2011)
[66] Lenhnert R., hep-ph/0611177
[67] Greenberg O.W., Phys. Rev. Lett. **89**, 231602 (2002)
[68] Found. of Phys., **36**, 1535, (2006)
[69] Bluhm R., hep-ph/0112318
[70] Mewes M., hep-ph/0307161
[71] Amelino-Camelia G. *et al.*, Nature, **393**, 763 (1998)
[72] Carroll, S.M., Field G.B., Jackiw, R. Phys. Rev. D **41**, 1231, (1990)
[73] Carroll S.M. & Field G.B., Phys. Rev. D **43**, 3789, (1991)
[74] Leach, S. M., et al. 2008, A&A, 491. 597
[75] Maris, M., Burigana, C., Gruppuso, A., Finelli, F., Diego, J. M. 2010, arXiv:1010.0830
[76] Zaldarriaga M., Astrophys. J. **503**, 1 (1998) [arXiv:astro-ph/9709271]
[77] Zaldarriaga M. , Seljak U., Phys. Rev. D **55**, 1830 (1997)
[78] Tegmark M., Phys. Rev. D **55**, 5895 (1997)
[79] Tegmark M., de Oliveira-Costa A., Phys. Rev. D **64** (2001) 063001
[80] Gruppuso A., de Rosa A., Cabella P., Paci F., Finelli F., Natoli P., de Gasperis G., Mandolesi N. Mon. Not. Roy. Astron. Soc. **400**, 1, (2009)
[81] Gorski K.M., Hivon E., Banday A.J., Wandelt B.D., Hansen F.K., Reinecke M., Bartelmann M., 2005, HEALPix: A Framework for High-resolution Discretization and Fast Analysis of Data Distributed on the Sphere, Ap.J., 622, 759-771
[82] Jarosik N. *et al.* [WMAP Collaboration], Astrophys. J. Suppl. **170**, 263 (2007)
[83] Dunkley J. et al. (WMAP), ApJ SS 180 (2009) 306
[84] Scannapieco E.S., Ferreira P.G., Phys. Rev. D **56**, 7493 (1997)
[85] Pogosian L., Vachaspati T., Winitzki S., Phys. Rev. D **65**, 083502 (2002)
[86] Giovannini M., Phys. Rev. D **71**, 021301, (2005)
[87] Lue A., Wang L., Kamionkowski M., Phys. Rev. Lett., **83**, 1506, (1999)
[88] Saito S., Ichiki K., Taruya A., arXiv:0705.3701, 2007
[89] Maity D., Majumdar P., SenGupta S., JCAP, 0406 005, (2004)
[90] Lepora N.F., gr-qc/9812077, 1998
[91] Balaji K.R.S, Brandenberger R.H., Easson D.A., JCAP, 0312, 008, (2003)
[92] Cabella P., Natoli P., Silk J., Phys. Rev. **D76**, 123014 (2007)
[93] Davoudiasl H. *et al.*, Phys. Rev. Lett., **93**, 201301, (2004)
[94] Li H., Li M., Zhang X., Phys. Rev. D **70**, 047302, (2004)
[95] Carroll S.M. , Field G.B., Phys. Rev. Lett., **79**, 2394, (1997)
[96] Carroll S.M., Phys. Rev. Lett., **81**, 3067, (1998)
[97] Nodland B., Ralston J.P., Phys. Rev. Lett., **78**, 3043, (1997)
[98] Eisenstein D.J., Bunn E.F., Phys. Rev. Lett. **79**, 1957, (1997)
[99] Leahy J.P., astro-ph/9704285, 1997
[100] Feng B., Li H., Li M., Zhang X., Phys.Lett. B **620**, 27 (2005)
[101] Wu E. Y. S.*et al.* [QUaD Collaboration], Phys. Rev. Lett. **102**, 161302 (2009)
[102] Xia J.Q., Li H., Zhao G.B. *et al.*, Int. J. Mod. Phys. **D17**, 2025-2048 (2009).
[103] Pagano L. *et al.*, Phys. Rev. D **80**, 043522 (2009)
[104] Fixsen, D.J., Dwek, E., 2002, ApJ., 578, 1009
[105] Backman, D.E., Dasgupta, A., Stencel, R.E., 1995, ApJ, 450, L35
[106] Stern, S.A., 1996, A&A, 310, 999
[107] Yamamoto, S., Mukai, T., 1998, Earth, Planets, Space, 50, 531
[108] MoroMartin, A., Malhotra, R., 2002, AJ, 124, 2305
[109] MoroMartin, A., Malhotra, R., 2003, AJ, 125, 2255
[110] Landgraf, M., Liou, J-C., Zook, H.A., Grrün, E., 2002, ApJ, 123, 2857

Journal of Physics: Conference Series **335** (2011) 012010 doi:10.1088/1742-6596/335/1/012010

Perspectives on Lorentz and CPT Symmetry Violation

Brett Altschul

Department of Physics and Astronomy, University of South Carolina, Columbia, SC 29208

E-mail: baltschu@physics.sc.edu

Abstract.

Lorentz and CPT symmetries are fundamental to our understanding of the standard model and general relativity, but there has recently been a great deal interest in the possibility that they may not be exact in nature. There is a systematically developed effective field theory, the standard model extension, that describes such symmetry violations. Many types of experiments have been used to place constraints on the parameters characterizing this theory. Some of the best constraints come from high-energy astrophysics. Observations of the radiation from ultrarelativistic electrons make it possible to map out the electron energy-momentum relation with great precision. Using observations of synchrotron and inverse Compton emissions from a variety of sources, the coefficients for spin-independent Lorentz violation in the electron sector can be constrained at the 10^{-15}–10^{-20} levels.

1. Introduction

In the past fifteen years, there has been an explosion of interest in the topic of Lorentz symmetry. So far as we know, all physical processes are invariant under rotations and Lorentz boosts. Yet an important question is whether this seemingly sacrosanct symmetry might actually be broken in nature. If a new experiment showed that Lorentz invariance does not hold exactly, that would be evidence of profoundly new physics.

Lorentz invariance is also closely related to discrete symmetries. Parity (P) and time reversal (T) are elements of the full $O(3,1)$ transformation group. Moreover, Lorentz invariance and CPT invariance are especially closely connected.

Of course, experimenters have been performing tests of Lorentz invariance since before relativity was even recognized. The famed Michelson-Morley experiment was a test of both the rotation and boost invariances that make up Lorentz invariance. The experiment's null result was a surprise at the time, but we now understand it in the framework of special relativity, as indicating that there is no preferred rest frame for the universe.

Since Einstein's introduction of special relativity in 1905, it has been subjected to a consistent barrage of experimental tests. However, until recently, such tests generally involved searches for purely *ad hoc* modifications of Einstein's framework; there was no attempt to study possible forms of Lorentz symmetry breaking systematically. More recently though, theorists and experimenters have begun looking at the problem in a much more thorough and organized fashion, as it was realized that most possible forms of Lorentz violation had been completely overlooked in earlier analyses. The awareness that there are many more ways that Lorentz

Journal of Physics: Conference Series **335** (2011) 012010 doi:10.1088/1742-6596/335/1/012010

symmetry could be broken than was previously appreciated has lead to a tremendous surge of interest in the subject.

There are two major reasons why the possibility of Lorentz invariance violation is interesting, and the two reflect differing "top-down" and "bottom-up" approaches to fundamental physics. The "top-down" reason for considering violations of Lorentz invariance is that most of the theories that are considered candidates to explain quantum gravity suggest the possibility of Lorentz symmetry violation in some regimes. These includes string theory [1], loop quantum gravity [2], noncommutative geometry theories [3], and others. However, none of these theories is yet a viable theory of four-dimensional quantum gravitation, and none of them are sufficiently well understood to be able to tell whether Lorentz violations (such as preferred frame effects) are real features of the theories—and if they are real, whether they are generic features of the theories.

The "bottom-up" approach suggests that Lorentz and CPT invariances should be tested simply because they appear to be so fundamental. Any violation of such symmetries would be a revolutionary discovery, since both the standard model of particle physics and the general theory of relativity are based on local Lorentz invariance; the entire edifice of fundamental physics is supported by this symmetry. A breaking of this symmetry would signal something profoundly new in physics and would be a critically important clue about the underlying nature of the universe. Much of the story of modern fundamental physics concerns how extremely useful and important symmetries—such as isospin, parity, and time reversal—were found not to be exact symmetries of nature. Knowing why these symmetries are violated is crucial to our understanding of the standard model.

Both Lorentz and CPT symmetries are fundamental to our understanding of the standard model, and the two are closely intertwined. It is impossible to break CPT symmetry in a local quantum theory without also breaking Lorentz invariance [4]. This means that many traditional tests of CPT can also be interpreted as tests of Lorentz invariance, and vice versa. On the other hand, many traditional phenomenalistic models of CPT violation (such as having particles and antiparticles with different masses) cannot be accommodated in a quantum field theory as we understand it.

2. The Standard Model Extension

Any realistic theory of Lorentz and CPT violation must also be capable of describing what we already know. All the fundamental physics that we current understand is described by the standard model (an effective quantum field theory) and general relativity. If we want to introduce small Lorentz-violating corrections to the standard model, we can continue to use the formalism of effective field theory, and the usual effective field theory that describes Lorentz and CPT violation is known as the standard model extension (SME) [5, 6].

The construction of the SME is straightforward. It involves only those fields (scalar, spinor, and gauge) that are part of the standard model, although the generalization to additional fields is very simple. However, in addition to the usual standard model terms, the SME Lagrange density includes all possible operators that can be constructed from these fields, subject to some appropriate restrictions. The standard model itself contains operators that are local, gauge invariant, Lorentz invariant, and renormalizable, but the SME relaxes these requirements. In particular, operators appearing in the Lagrangian do not need to be Lorentz scalars.

For example, while the usual mass term for a Dirac fermion is $\mathcal{L}_m = -m\bar{\psi}\psi$, the SME also includes mass-like terms such as $\mathcal{L}_a = -a_0\bar{\psi}\gamma^0\psi$. This operator transforms like the time component of a four-vector, and so it is not invariant under boosts. The coefficient a_0 parameterizes the size of the Lorentz violation. a_0 also represents a preferred spacetime direction. If Lorentz violation is broken spontaneously (which may be required for consistency with general relativity [7]), a_0 would be the vacuum expectation value (vev) of a vector-valued dynamical

Journal of Physics: Conference Series **335** (2011) 012010 doi:10.1088/1742-6596/335/1/012010

field. In the case of a_0, that vev is purely timelike, but models with both spacelike and timelike expectations may be viable [8]. There will also be SME coefficients with multiple indices, and each of these would correspond to the vev of an appropriate tensor-valued field [9].

Of course, the number of operators than can be constructed just from standard model fields is infinite, so some other restrictions are needed into order to have a predictive theory. The most commonly considered subset of possible operators are those that constitute the minimal SME, which contains only operators that are local, gauge invariant, and power counting renormalizable (i.e., of dimension 3 or 4). The minimal SME has become the standard framework from parameterizing the results of experimental Lorentz and CPT tests. However, there are certainly situations in which more general operators (which may be nonlocal, violate charge conservation, or have higher dimension) are of significant interest.

The minimal SME Lagrange density for a single species of fermion is

$$\mathcal{L}_\psi = \bar{\psi}(i\Gamma^\mu \partial_\mu - M)\psi, \tag{1}$$

where

$$M = m + \not{a} - \not{b}\gamma_5 + \frac{1}{2}H^{\mu\nu}\sigma_{\mu\nu} + im_5\gamma_5, \tag{2}$$

and

$$\Gamma^\mu = \gamma^\mu + c^{\nu\mu}\gamma_\nu - d^{\nu\mu}\gamma_\nu\gamma_5 + e^\mu + if^\mu\gamma_5 + \frac{1}{2}g^{\lambda\nu\mu}\sigma_{\lambda\nu}. \tag{3}$$

M and Γ contain coefficients with all possible Lorentz structures allowed by their respective dimensions. These represent all the superficially renormalizable couplings that are possible in a purely fermionic theory. With multiple species of fermions, there are many more coefficients. While each fermion species will have a full set of coefficients associated just with the behavior of that one species, there is no requirement that the Lorentz violation coefficients be diagonal in flavor space. Off-diagonal coefficients would generate effects such as neutrino oscillations.

However, some of the coefficients are more interesting than others. In particular, m_5 does not violate Lorentz symmetry, and it may be absorbed into the other coefficients by means of a field redefinition [10]. Moreover, e, f, and g are inconsistent with the coupling of the fermion field to a chiral gauge field.

As noted, CPT violation in quantum field theory implies a breaking of Lorentz symmetry as well. However, the converse does not hold; not all forms of Lorentz violation necessarily violate CPT. In fact, only those coefficients with odd numbers of indices (a, b, e, f, and g) are odd under CPT. The various operators also have different transformations under C, P, and T separately. The discrete transformation properties of the various fermion bilinears are well known, and they have important experimental consequences. Most experiments are sensitive only to a small subset of the minimal SME coefficients—those coefficients that have a certain set of symmetry properties. (For example, tests involving parity are insensitive to a_0, which is odd only under C, while tests of time reversal are insensitive to b_0, which is odd only under P.) Moreover, at nonrelativistic energies, some of the coefficients have identical effects; these means only certain special linear combinations are observable in nonrelativistic experiments.

The photon sector of the SME involves additional coefficients. The photon Lagrange density is

$$\mathcal{L}_A = -\frac{1}{4}F^{\mu\nu}F_{\mu\nu} + \frac{1}{2}k_{AF}^{\mu}\epsilon_{\mu\nu\rho\sigma}F^{\nu\rho}A^\sigma - \frac{1}{4}k_F^{\mu\nu\rho\sigma}F_{\mu\nu}F_{\rho\sigma}. \tag{4}$$

As in the fermion sector, the coefficient tensor k_{AF} with one index is CPT-odd, while k_F with four indices is CPT-even. The operator parameterized by k_{AF} has a Chern-Simons form, and its Lagrange density is not actually gauge invariant. It changes by a total derivative under a gauge transformation, but this means that the integrated action $S = \int d^4x\, \mathcal{L}$ is fully gauge invariant. However, the unusual gauge transformation properties of k_{AF} made it a subject of

Journal of Physics: Conference Series **335** (2011) 012010 doi:10.1088/1742-6596/335/1/012010

significant controversy; it was unclear whether the k_{AF} term could be generated by radiative corrections [11, 12, 13, 14]. Ultimately, it was found the the radiative corrections involved were actually undetermined.

Nonvanishing values of the various SME coefficients could have many different kinds of effects. One obvious consequence of rotation invariance would be angular momentum nonconservation. A sphere made from material with nonvanishing c_{jk} coefficients would behave as if its tensor of inertia were not $I_{jk} = I_0\delta_{jk}$ but $I_{jk} = I_0\left[\delta_{jk} + \frac{1}{2}c_{(jk)}\right]$, where $c_{(\mu\nu)} = c_{\mu\nu} + c_{\nu\mu}$. This would cause the sphere to wobble as it rotates. The absence of such wobbling in pulsar timing data makes it possible to place bounds on the c coefficients for the neutrons that compose such pulsars [15].

While the pulsar measurements are not a particularly sensitive way to constrain Lorentz violation, there are many systems and reaction processes that can be used to set strong bounds on Lorentz and CPT violation. One obvious option would be experiments with neutral mesons. CP violation has only been observed in meson oscillation experiments, which also offer a convenient way to search for CPT violation. Tests of Lorentz and CPT violation are now a standard part of meson oscillation experiments using K, D, and B mesons [16, 17, 18].

The rate of CPT violation in meson experiments is controlled by the parameter δ. In terms of the SME coefficients, the δ for kaon oscillations is

$$\delta_K = \gamma\frac{v_\mu(a_d^\mu - a_s^\mu)}{m_{K_L} - m_{K_S}}. \tag{5}$$

This depends on the a coefficients for the two quark species d and s that make up the mesons, the K_L-K_S mass difference of 3.5×10^{-12} MeV, and the velocity and Lorentz factor of the kaons. The dependence on velocity has important effects. Prior to the development of the SME, sensitivity to CPT violation in kaon experiments was generally parameterized as a sensitivity to a K^0-$\bar{K}^0$ mass difference. However, such a particle-antiparticle mass difference is forbidden in a local quantum field theory. Two experiments with the same naive mass difference sensitivity may have very different sensitivities to the SME a parameters; experiments done with more highly boosted mesons are more sensitive to local CPT violation. The best current experimental constraints on the SME coefficients coming from meson data are at the 10^{-21} GeV level for kaons and the 10^{-16} for other meson species.

While these meson experiments focus primarily on searches for CPT violation, many more coefficients can be studied by looking at violations of rotation and boost symmetries. For testing such symmetries, precision laboratory experiments are obvious options. Atomic clock and resonant cavity measurements have produced highly sensitive bounds on proton, neutron, electron and photon coefficients. Such experiments use rotating apparatuses to test whether observed resonant frequencies depend on the absolute orientation of electron, nuclear, and photons spins. The rotating setup may be implemented in two ways. The simpler way is just to utilize the Earth's rotation, which changes the orientation of a fixed device in the lab frame over the course of the planet's sidereal rotation period. More elaborate experiments also use an active rotation in the laboratory, placing the entire experiment on a turntable. The two forms of rotation complement each other; for example, by combining them, it is possible to study all the spacelike components of the various vectors b.

These tests of isotropy may also be extended to tests of boost invariance, although this requires a moving laboratory. As the Earth revolves around the sun, our velocity changes slightly, and so by making precision clock measurements at different times of year, it is possible to constrain b_0 and other boost invariance violation coefficients. However, the constraints on boost invariance violation are generally weaker; they are naturally suppressed by a factor of $v_\oplus \approx 10^{-4}$, since they involve comparisons of measurements made in rest frames that differ by that small velocity.

Atomic clock experiments and other precision laboratory experiments (such as measurements of muon magnetic moments, Penning trap frequencies, and short-distance spin-dependent forces) are extremely useful for constraining Lorentz and CPT violation. However, many of the strongest constraints actually come from astrophysical experiments that have relative poor intrinsic sensitivities. These low sensitivities can be overcome by using either of two phenomena that can be found astrophysically but not in the lab: very long distances and very high energies.

Observations of cosmologically distant objects, via photons that have propagated for 10^8–10^9 years, can be very sensitive to small alterations is the electromagnetic sector. For example, most of the k_F coefficients and all of the k_{AF} coefficients give rise to birefringence in vacuum. Photons with different polarizations propagate at different speeds, leading to gradual a rotation in the polarization direction for linearly polarized radiation. Yet no evidence of such birefringence is seen, even at extreme distances [19, 20, 21, 22, 23]. In quasar jets, for example, the polarized radiation is mostly produced by the synchrotron process, with electrons orbiting around magnetic field lines that point approximately parallel to the jet itself. The radiation from the electrons is preferentially polarized perpendicular to the field direction, and this polarization structure is exactly what is observed in the light from most quasars, independent of how far they are away.

The synchrotron processes that produce this radiation operate at extremely high energies, sometimes above 1 PeV. Just as a long line of sight can magnify a very small photon propagation effect, these immense electron energies can magnify other kinds of Lorentz-violating phenomena. In particular, observations of the synchrotron and inverse Compton radiation emitted by these incredibly energetic electrons provide the best direct tests of boost invariance violation for electrons.

3. Bounds From Synchrotron Radiation

Radiation by energetic electrons can be a sensitive probe of the electron c coefficients. It has been possible to place very strong bounds on c by looking at the relationships between energy, momentum, and velocity in the c-modified theory. At ultrarelativistic energies, the bounds on c that can be derived from observing the emissions of an electron with Lorentz factor γ scale as γ^{-2}. This strong dependence on γ is a consequence of the rapid growth in the importance of c with energy. If c is $\mathcal{O}(m/M_P)$, its effects will become dominant at scale $E \sim \sqrt{mM_P}$, and the Lorentz factor at this scale is $\gamma \sim \sqrt{M_P/m}$. So if no effects of Lorentz violation are observed up to some Lorentz factor γ, this constrains c to be smaller than $\mathcal{O}(\gamma^{-2})$.

If rotation and boost symmetries are broken, the conventional relations between energy, momentum, and velocity—$E = \gamma m$ and $\vec{\pi} = \gamma m \vec{v}$—no longer need to hold. In the presence of c, the velocity of an electron with momentum $\vec{p}$ and energy E is [24]

$$v_k = \frac{1}{E + c_{(0j)}\pi_j}\left(\pi_k - c_{kj}\pi_j - c_{jk}\pi_j + c_{jk}c_{jl}\pi_l\right) - c_{0k}. \tag{6}$$

This formula can be found either as a group velocity or using the Heisenberg equation of motion for $\vec{x}$.

From eq. (6), we can see why the effects of c become large at the scale $\sqrt{mM_P}$. According to eq. (6), the velocity might become superluminal when $|\vec{\pi}|/E \approx 1 - |c|$, where $|c|$ is a characteristic size for the Lorentz-violating coefficients. This gives us an estimate of the maximum value of γ that can be achieved before new physics must come into play if some form of causality is to be preserved. For ultrarelativistic particles, $\gamma \approx \left[2\left(1 - |\vec{\pi}|/E\right)\right]^{-1/2}$, and this diverges at an energy scale $E_{\max} \sim m/\sqrt{|c|} \sim \sqrt{mM_P}$. Above this scale, the description of the Lorentz violation through an effective field theory containing only $c^{\nu\mu}$ terms will generally break down, and higher dimension operators should become important.

We can exploit the Lorentz violation in the relationship between momentum and velocity to place bounds on c. There are two crucial and complementary effects. The first effect is that the

Journal of Physics: Conference Series **335** (2011) 012010 doi:10.1088/1742-6596/335/1/012010

maximum electron speed in a given direction $\hat{e}$ is generally different from 1. The value of this new maximum velocity can be easily calculated to first order in c,

$$(v_j\hat{e}_j)_{\max} = 1 - c_{(jk)}\hat{e}_j\hat{e}_k - c_{(0j)}\hat{e}_j, \tag{7}$$

If this is less than one, it can have readily observable consequences for the synchrotron spectrum.

The observable effects derive from the fact that the radiation by a charged particle depends on its velocity—not its momentum or its energy. This continues to be the case even in the presence of Lorentz violation, since the velocity dependence is a consequence of gauge invariance alone. The power emitted by a synchrotron electron is spread over frequencies less than the critical frequency

$$\nu_c = \frac{3}{4\pi}\gamma^3\omega_B \sin\theta = \frac{3}{4\pi m}\gamma^2 \left|e\vec{B}\right| \sin\theta, \tag{8}$$

where θ is the angle between the magnetic field and the electron velocity $\vec{v}$. Most of the power is emitted close to this frequency, and above ν_c, the radiated power falls off very rapidly.

The ν_c in (8) represents the critical frequency for the emissions of a single electron. If a source contains significant numbers of electrons with velocities up to some maximum Lorentz factor $\gamma_{\max}$, then the observed cutoff in the spectrum will be at the cutoff frequency for the most energetic electrons—$\nu_c = \frac{3}{4\pi m}\gamma_{\max}^2 \left|e\vec{B}\right| \sin\theta$.

This result for ν_c allows one to infer $\gamma_{\max}$ from observations of synchrotron spectra. However, the analysis required to find $\gamma_{\max}$ is fairly involved. From (8), we can infer that $\gamma_{\max} \propto \sqrt{\nu_c/\left|\vec{B}\right|}$. The cutoffat ν_c is an obvious feature of synchrotron spectra, and a lower bound on this cutoff frequency can easily be obtained from any spectrum that clearly has a synchrotron origin. However, the tricky part is calculating the strength of the magnetic field. The magnetic field is generally taken to be the minimum energy field which can generate the low-frequency part of the synchrotron spectrum. This turns out to depends very weakly on most of the parameters that must be fitted from the observed spectrum. Moreover, by considering only the low-frequency part of the spectrum, the value of B can be inferred independently of the measurement of ν_c that characterizes the high-energy part of the spectrum. Ultimately, $\gamma_{\max}$ depends on at most the seventh root of any fit parameter (except ν_c) that might be in error, and so its value can be quite robust.

If the existence of electrons with Lorentz factors up to some value $\gamma_{\max} = (1 - v_{\max}^2)^{-1/2}$ moving in the $\hat{e}$-direction has been observed, this implies

$$c_{jk}\hat{e}_j\hat{e}_k + c_{(0j)}\hat{e}_j < \frac{1}{2\gamma_{\max}^2}. \tag{9}$$

Radiation from relativistic sources is tightly beamed, into a narrow pencil of angles around the velocity $\vec{v}$ of the emitter. This means the relevant direction $\hat{e}$, in the bounds based on observations of a given source, is the direction from the source to the Earth. The most energetic observed synchrotron spectrum is that of the nearby Crab nebula, which contains electrons with Lorentz factors up to $\gamma_{\max} \approx 3 \times 10^9$ [25, 26, 27]. For more distant, less well studied high-energy sources, the largest inferred Lorentz factors are typically in the 10^7–10^8 range. (While many of these sources probably do have electron populations extending up into the PeV range, the small number of electrons present at the high-energy end of the spectrum makes observations of their distinctive emissions difficult.)

However, these synchrotron observations are insufficient to place two-sided bounds on the electron c coefficients. The reason is simple. If Lorentz violation decreases electron velocities, there must be a maximum electron velocity and hence Lorentz factor; this maximum γ obviously cannot be smaller than any γ that is actually observed. If the Lorentz violation increases electron velocities instead, there will be no such maximum γ; there will, however, be a complementary effect, whose effects would also be observable.

4. Bounds From Inverse Compton Scattering

The complementary effect is that there may be a maximum energy available to electrons with subluminal velocities. This will also generally depend on the direction of a particle's motion. The maximum energy for a particle moving with a speed less than 1 in the direction $\hat{e}$ is

$$\frac{E}{m} = \frac{1}{\sqrt{-2c_{jk}\hat{e}_j\hat{e}_k - 2c_{(0j)}\hat{e}_j}}, \tag{10}$$

which depends on the same combination of Lorentz violation coefficients as did the maximum value of γ. However, the maximum subluminal energy E is proportional to the inverse square root of c, which is not surprising, since for vanishing c, E must be infinite. As is obvious from (10), such a maximum value for E need not always exist. In fact, according to (7), this maximal subluminal energy does not exist precisely when the maximum speed in the relevant direction is less than or equal to one.

An obvious question is what happens to particles moving with speeds $v > 1$. The answer is that there is vacuum Cerenkov radiation, a phenomenon that is unique to Lorentz-violating theories. This process is analogous to ordinary Cerenkov radiation in matter; when charged particles move faster than the phase velocity of light, they will emit radiation. This can occur as a result of Lorentz violation in either the electromagnetic or matter sectors. The process has been studied in the context of the minimal SME [28, 29, 30, 31, 32, 33], but understanding it in detail is difficult, since new physics may appear at the same scale as the Cerenkov threshold given in eq. (10). What is clear about vacuum Cerenkov radiation is that it is an extremely efficient energy loss mechanism. (A naive estimate of the radiation rate is 10^{14} GeV$\cdot$s^{-1}.) Electrons moving superluminally would radiate almost all their energy away in times that are minuscule compared with their synchrotron lifetimes; they would have effectively no time to interact in other ways before losing almost all their energies.

So if it is possible to observe conventional radiation patterns coming from electrons with energies up to some value $E_{\rm max}$, such electrons must be moving with speeds $v < 1$. This immediately implies

$$-\frac{1}{2\left(E_{\rm max}/m\right)^2} < c_{jk}\hat{e}_j\hat{e}_k + c_{0j}\hat{e}_j, \tag{11}$$

which is the other side of the one-sided bound found in eq. (9). While synchrotron radiation provides a relatively direct probe of a charge's velocity, inverse Compton (IC) radiation provides a lower bound on $E_{\rm max}$. The highest-energy photons that are emitted by astrophysical sources arise in IC processes. Low-energy photons (which often come from synchrotron emission) scatter off ultrarelativistic electrons. An electron may transfer a substantial fraction of its own energy to a photon during such a collision, resulting in the emission of photons whose energies may range almost up to the scale of the highest electron energies.

All collisions of photons with ultrarelativistic electrons are practically all head-on when viewed in the electron's frame. Overtaking collisions are extremely rare, occurring only in a minuscule range of observer frame solid angles. In the observer frame, this means that the emitted photon is propagating in essentially the same direction as the initial electron. In the observer's rest frame, the relationship between the initial and final photon energies ϵ_i and ϵ_f in the observer frame is

$$\epsilon_f = \epsilon_i\gamma^2\frac{(1-\cos\psi_i)(1+\cos\theta^e\cos\psi_i^e)}{1+\gamma\frac{\epsilon_i}{m}(1-\cos\psi_i)(1-\cos\theta)}, \tag{12}$$

(where ψ_i, ψ^e, and θ are the scattering angles). The Lorentz factor $\gamma \gg 1$ is that of the electron, and depending on the value of $\gamma\epsilon_i/m$, the final photon energy may be $\mathcal{O}(\gamma)$ to $\mathcal{O}(\gamma^2)$ larger than its initial energy. If $\gamma\epsilon_i/m \approx 0.1$, for example, then the maximum energy that can be carried off

Journal of Physics: Conference Series **335** (2011) 012010 doi:10.1088/1742-6596/335/1/012010

by the photon is $(\epsilon_f)_{\max} \approx 0.4\gamma m$. So the IC process allows some electrons to transfer sizable fractions of their energy to photons, and observed IC photon energies range up to ~ 100 TeV.

Models of a source's structure can yield information about the energies of the electron responsible for the inverse Compton emission. In general, the models typically require maximum electron energies that are several times larger than the highest observed photon energies, because IC scattering events do not transfer all of the high-energy electrons' energies to the photons. However, to get a more robust bound, one may take $E_{\max}$ to be the highest actually observed photon energy; this conservative estimate usually differs from a model-derived bound by less than an order of magnitude. The only input required from a model is that a source's TeV γ-ray emission is well described by the IC process. This makes it easier to get reliable lower bounds on $E_{\max}$ than on $\gamma_{\max}$, and there has been a remarkable explosion in the number of TeV γ-ray sources observed in the last ten years, thanks especially to the H.E.S.S atmospheric Cerenkov telescope in Namibia.

There are nine components of c that can be bounded with the astrophysical data—the three $c_{(0j)}$ and the six $c_{(jk)}$. Each of the inequalities derived from (9) and (11) generally couples all nine of the coefficients in a nontrivial way. These bounds may be fairly awkward. However, the coupled bounds may be translated into bounds on the separate coefficients by means of linear programming, with the ultimate result that all nine of the coefficients are bounded on both sides, at the 10^{-15}–10^{-17} levels.

5. Conclusion

Other high-energy astrophysical processes can also be used to produce competitive bounds on the c coefficients. For example, if c is nonzero, the photon decay process, $\gamma \rightarrow e^+ + e^-$ may be allowed for sufficiently energetic electrons. The fact that IC photons do not decay in this fashion generates another bound, although a slightly weaker one than that discussed above. However, the bounds related to the absence of photon decay can be extended to other species of daughter particles; any $\gamma \rightarrow X^+ + X^-$ will occur rapidly if it is allowed, above a threshold $\sim m_X/\sqrt{|c_X|}$. The bounds are weaker than the electron bounds by a factor of m_X^2/m_e^2 [34, 35].

There are also generalizations of all these bounds to higher-dimension operators. However, the number of anisotropic operators increases rapidly with the operator dimension, and there has so far been no systematic study of all the possible bounds to be derived from the astrophysical observations. However, there have been analyses of isotropic Lorentz violating operators (that is, operators that violate boost invariance only) at higher dimensions.

Lorentz and CPT violation are active areas of study in physics. Understanding the possible ways that these symmetries could be violated has become a major focus of precision in recent years. There is an active experimental program testing these fundamental symmetries, and if violations are discovered, that would be of momentous importance and would provide important guidance in the future searches for new physics. And there are many interesting forms of Lorentz and CPT violation that still remain to be constrained.

References

[1] V A Kostelecký and S Samuel, *Phys. Rev.* D **39**, 683–685 (1989).
[2] R Gambini and J Pullin, *Phys. Rev.* D **59**, 124021 (1999).
[3] S M Carroll, J A Harvey, V A Kostelecký, C D Lane and T Okamoto, *Phys. Rev. Lett.* **87**, 141601 (2001).
[4] O W Greenberg, *Phys. Rev. Lett.* **89**, 231602 (2002).
[5] D Colladay and V A Kostelecký, *Phys. Rev.* D **55**, 6760–6774 (1997).
[6] D Colladay and V A Kostelecký, *Phys. Rev.* D **58**, 116002 (1998).
[7] V A Kostelecký, *Phys. Rev.* D **69**, 105009 (2004).
[8] B Altschul and V A Kostelecký, *Phys. Lett.* B **628**, 106–112 (2005).
[9] B Altschul, Q G Bailey and V A Kostelecký, *Phys. Rev.* D **81**, 065028 (2010).
[10] D Colladay and P McDonald, *J. Math. Phys.* **43**, 3554–3564 (2002).

[11] R Jackiw and V A Kostelecký, *Phys. Rev. Lett.* **82**, 3572–3575 (1999).
[12] M Pérez-Victoria, *Phys. Rev. Lett.* **83**, 2518-2521 (1999).
[13] B Altschul, *Phys. Rev.* D **69**, 125009 (2004).
[14] B Altschul, *Phys. Rev.* D **70**, 101701(R) (2004).
[15] B Altschul, *Phys. Rev.* D **75**, 023001 (2007).
[16] K Abe *et al.*, *Phys. Rev. Lett.* **86**, 3228-3232 (2001).
[17] J M Link *et al.*, *Phys. Lett.* B **556**, 7–13 (2003).
[18] B Aubert *et al.*, *Phys. Rev. Lett.* **96**, 251802 (2006).
[19] S M Carroll, G B Field and R Jackiw, *Phys. Rev.* D **41**, 1231–1240 (1990).
[20] S M Carroll and G B Field, *Phys. Rev. Lett.* **79**, 2394–2397 (1997).
[21] V A Kostelecký and M Mewes, *Phys. Rev. Lett.* **87**, 251304 (2001).
[22] V A Kostelecký and M Mewes, *Phys. Rev. Lett.* **97**, 140401 (2006).
[23] V A Kostelecký and M Mewes, *Phys. Rev. Lett.* **99**, 011601 (2007).
[24] B Altschul and D Colladay, *Phys. Rev.* D **71**, 125015 (2005).
[25] T Jacobson, S Liberati and D Mattingly, *Nature* **424**, 1019 (2003).
[26] B Altschul, *Phys. Rev. Lett.* **96**, 201101 (2006).
[27] B Altschul, *Phys. Rev.* D **74**, 083003 (2006).
[28] R Lehnert and R Potting, *Phys. Rev. Lett.* **93**, 110402 (2004).
[29] R Lehnert and R Potting, *Phys. Rev.* D **70**, 125010 (2004).
[30] C Kaufhold and F R Klinkhamer, *Nucl. Phys.* B **734**, 1–23 (2006).
[31] B Altschul, *Phys. Rev. Lett.* **98**, 041603 (2007).
[32] B Altschul, *Phys. Rev.* D **75**, 105003 (2007).
[33] B Altschul, *Nucl. Phys.* B **796**, 262–273 (2008).
[34] F W Stecker and S L Glashow, *Astropart. Phys.* **16**, 97–99 (2001).
[35] B Altschul, *Astropart. Phys.* **28**, 380–384 (2007).

T and CPT SYMMETRIES in ENTANGLED NEUTRAL MESON SYSTEMS

JOSE BERNABEU

Department of Theoretical Physics, University of Valencia, and
IFIC, Univ. Valencia-CSIC, E-46100 Burjassot, Valencia, Spain

Jose.bernabeu@uv.es

Abstract. Genuine tests of an asymmetry under T and/or CPT transformations imply the interchange between in-states and out-states. I explain a methodology to perform model-independent separate measurements of the three CP, T and CPT symmetry violations for transitions involving the decay of the neutral meson systems in B- and Φ-factories. It makes use of the quantum-mechanical entanglement only, for which the individual state of each neutral meson is not defined before the decay of its orthogonal partner. The final proof of the independence of the three asymmetries is that no other theoretical ingredient is involved and that the event sample corresponding to each case is different from the other two. The experimental analysis for the measurements of these three asymmetries as function of the time interval $\Delta t > 0$ between the first and second decays is discussed, as well as the significance of the expected results. In particular, one may advance a first observation of true, direct, evidence of Time-Reserval-Violation in B-factories by many standard deviations from zero, without any reference to, and independent of, CP-Violation.
In some quantum gravity framework the CPT-transformation is ill-defined, so there is a resulting loss of particle-antiparticle identity. This mechanism induces a breaking of the EPR correlation in the entanglement imposed by Bose statistics to the neutral meson system, the so-called ω-effect. I present results and prospects for the ω-parameter in the correlated neutral meson-antimeson states.

1. Introduction

I was asked to talk about T- and CPT-symmetries in the fundamental laws of Physics. The main point of these studies is that a genuine test of their invariance needs an interchange between in-states and out-states for a given process, a request particularly difficult to be accomplished for particles that decay during their time evolution. Can true Time Reversal Violation (TRV) be searched for in unstable systems? In this presentation I will advocate for a methodology in the neutral meson systems that makes use of the EPR-entanglement existing in B- and Φ- factories: in them, the preparation of a quantum mechanical individual state of the neutral meson is not made by measurements performed on it, but by the observation of the decay of its orthogonal correlated partner. This strategy will allow the quantum preparation of a given individual state of the (still living) neutral meson by selecting a particular decay channel of the other neutral meson.

Violation of CP invariance has been observed in the $K^0 - \overline{K}^0$ and $B^0 - \overline{B}^0$ systems. Up to now, the experimental results are in agreement [1] with the Standard Cabibbo-Kobayashi-Maskawa (CKM) mechanism in the ElectroWeak Theory. Although all present tests of CPT invariance confirm the validity of this symmetry, as imposed by any local quantum field theory with Lorentz invariance and Hermiticity [2], it would be of great interest to observe Time Reversal Violation (TRV) directly in a single experiment, independent on the question of CPT invariance. A direct evidence for true TRV would mean [3] an experiment that, considered by itself, clearly demonstrates T violation independent of, and unconnected to, the results for CP violation. There is at present no existing result that clearly shows TRV in this sense.

We are interested in Microscopic T-Symmetry Violations. Effects in particle physics odd under the change of the sign of time t ↔ -t are not necessarily T-violating. These observables can occur in theories with exact T-symmetry and are called T-odd effects, like those induced by absorptive components of the amplitude. Well known time asymmetries are the Universe t-asymmetry and the macroscopic t-asymmetry called the "arrow of time". But none of these t-asymmetries is a test of TRV. In the fundamental laws of physics, T-Violation exists in the Standard Model or any field theoretic extension of it. The observed CP-Violation in the neutral meson systems tells us that T should be violated as well. However, as emphasized above, true TRV has not been observed up to now. There are intriguing subtleties introduced by the antiunitary character of the symmetry operator and T-Violation means a non-vanishing asymmetry under the interchange of in ↔ out states.

There is no doubt that the Universe is expanding, even accelerating at present cosmological age. This natural t-asymmetry t ↔ -t is perfectly compatible with fundamental laws of physics that are Time Reversal-symmetric. It is due to the initial condition for our Universe, like Inflation. This asymmetry is similar to the fact that in our Universe we have a privileged reference frame, the one associated with the Cosmic Microwave Background (CMB) radiation at a definite temperature with fluctuations. In Figure 1 we show the WMAP result that allows fundamental measurements of cosmology, producing our New Standard Model of Cosmology.

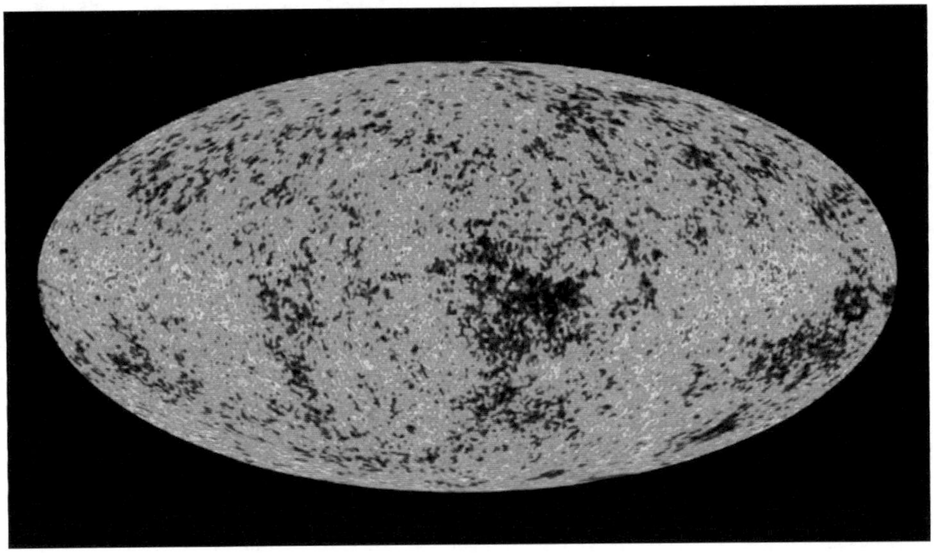

Figure 1. WMAP image of the CMB radiation

The CMB radiation has a thermal black body spectrum at a temperature of 2.725 K, as shown in Figure 2.

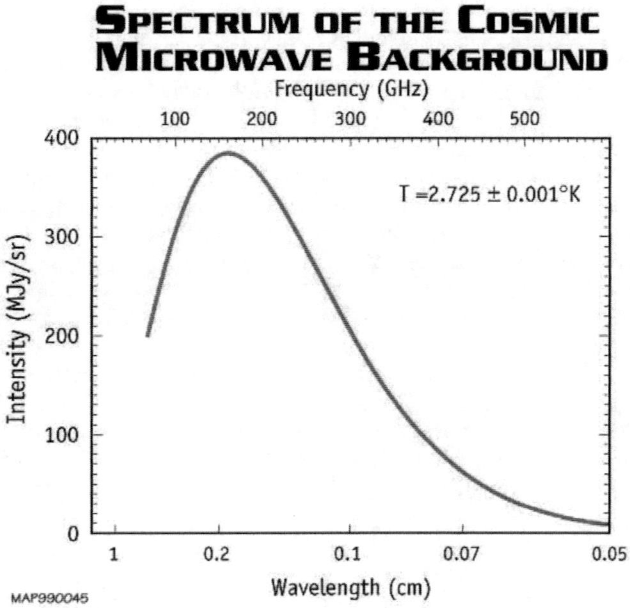

Figure 2. The CMB spectrum is the most precisely measured black body spectrum in nature

The spectrum peaks in the microwave range frequency of 160.2 GHz, corresponding to a 1.9 mm wavelength, in the intensity per unit frequency. This privileged reference frame in our Universe does not mean a violation of Lorentz invariance in the fundamental laws of physics. Similarly for the Universe time-asymmetry: it does not mean a violation of Time Reversal invariance in the fundamental laws of physics.

There is a macroscopic t-asymmetry, exemplified by the evolution of the Roman Coliseum in Figure 3, known as the "arrow of time".

Figura 3. The "arrow of time" in the Roman Coliseum

This asymmetry is in the nature of Thermodynamics. According to Eddington [4], the Time's Arrow is a property of Entropy alone: Time is asymmetric with respect to the amount of order in an isolated system. A very interesting unsolved problem in quantum epistemology is the reduction of the wave packet: Is the quantum wave function collapse related to the thermodynamic arrow of time?

In particle physics, particle decays are an example of a time-asymmetric phenomenon. If we start with an initial collection of identical unstable particles we arrive after decay at a large collection of final states. There is little, if any, chance of any collection of such final states evolving to become a set of identical unstable particle as in the initial state. This phenomenon has nothing to do with T-violation. In fact, it looks like it prevents a true test of T-symmetry in unstable systems [3, 5], test that needs an exchange between "in" and "out" states.

In Section 2 we discuss the types of experiments that can provide a direct evidence for true microscopic T-violation and the method to circumvent the problem for unstable systems. In Section 3 we identify genuine observables for TRV, not needing $\Delta\Gamma \neq 0$, in the time evolution of the system. The corresponding strategy, based on the EPR-entanglement, is applied to CP, T and CPT symmetries as independent "experimental" transformations. Section 4 presents a Monte Carlo study for TRV and the experimental significance expected, from a χ^2 test, for the TRV Asymmetries. In Section 5 we discuss the possible modifications in the EPR-entanglement induced by the ω-effect. Section 6 summarizes our conclusions.

2. Can T-symmetry be tested for unstable systems?

As emphasized before, a direct evidence for TRV would mean an experiment that, considered by itself, clearly shows TRV independent of, and unconnected to, the results for CP-violation (CPV). This evidence can be obtained from two types of experiments:

i) A non-zero expectation value of a T-odd operator for a non-degenerate stationary state, such as a non-vanishing Electric Dipole Moment (EDM), which is a P-odd, C-even, T-odd operator. The present experimental status is reviewed in [6] in this Proceedings.

The EDM can be generated by either Strong T-violation, like the θ-term $\epsilon_{\mu\nu\zeta\sigma} F^{\mu\nu} F^{\zeta\sigma}$ in the QCD Lagrangian, or Weak T-violation. The experimental small value of the θ-term needs a mechanism that protects it, the Peccei-Quinn symmetry [7]. The Standard Model theory of Electroweak Interactions or any local field theory with CP-violation, predicts T-violation effects in parallel.

ii) For a transition i→f, under the exchange in ↔ out, T-symmetry implies the connection $S_{f,i} \rightarrow S_{-i,-f}$, where –i(f) means the T-transformed state of the i(f) state.

The Kabir asymmetry $K^0 - \bar{K}^0$ vs. $\bar{K}^0 - K^0$ has been measured in CP-LEAR [8] with non-vanishing value. But $K^0 - \bar{K}^0$ is a CPT-even transition, so CP and T are experimentally identical here: a non-vanishing identically connected CP and T asymmetry needs $\Delta\Gamma \neq 0$ and the effect is time-independent.

The main question is: Is it possible to search for TRV in Mixing x Decay transitions? The origin here comes from the interference of decay amplitudes with and without mixing. We concentrate on Neutral Meson Factories. The opportunity arises [9, 10] in these Facilities from the quantum mechanical entanglement imposed by the Einstein-Podolsky-Rosen (EPR) correlation [11]. This correlation allows the quantum preparation of a physical state of a living particle by observing the decay of its partner particle: the individual state of each particle in the system is not defined before this observation. Depending on the selection of the decay channel, one can have separate tests of CP, T and CPT-symmetries!

In meson factories, the coherence between orthogonal $B^0, \bar{B}^0$ states has been used for flavour tag

$$|i\rangle = \frac{1}{\sqrt{2}}\left[B^0(t_1)\bar{B}^0(t_2) - \bar{B}^0(t_1)B^0(t_2)\right] \qquad (1)$$

where the states "1" and "2" are defined by the time of their decay with $t_1 < t_2$. The observation of $B^0 \rightarrow l^+$, for example, at time t_1, tells us that the complementary (still living) state is $\overline{B}^0$ at t_1. This is the preparation of the initial state for single state time evolution.

As said, the individual state of each neutral meson is, however, not defined before its collapse as a filter imposed by the observation of the decay of its companion. One can rewrite the same state (1) of the system $|i\rangle$ in terms of any other pair of orthogonal states of the individual neutral B-mesons: Considerer B_+ and B_- where B_- is the neutral B-state filtered by its decay to $J/\psi\ K_+$, K_+ being the neutral K-state filtered by its decay to $\pi\ \pi$, and B_+ is orthogonal to B_-, not connected to $J/\psi\ K_+$. As the final states are CP-eigenstates, we may call the filter imposed by a first observation, at time t_1, of one of these decays a "CP-tag" [12], although $B_\pm$ are not CP-eigenstates of B's necessarily.

The same entangled state of the two-body system can be rewritten

$$|i\rangle = \frac{1}{\sqrt{2}}[\ B_+(t_1)B_-(t_2) - B_-(t_1)B_+(t_2)\ ] \qquad (2)$$

Now we may proceed to a partition of the complete set of events into four categories, defined by the tag in the first decay at t_1: $B_+, B_-, B^0 or\ \overline{B}^0$, so we have eight different Decay-Intensities at our disposal, as functions of $\Delta t = t_2 - t_1 > 0$. Each one of these eight processes has an Intensity

$$I_i(\Delta t) \sim e^{\Gamma\Delta t}\{C_i \cos(\Delta m \Delta t) + S_i \sin(\Delta m \Delta t) + C_i' \cosh(\Delta\Gamma \Delta t) + S_i' \sinh(\Delta\Gamma \Delta t)\} \qquad (3)$$

where Γ is the average width.

Up to now, for CPV analyses in B-factories, BABAR & BELLE have assumed CPT invariance and $\Delta\Gamma=0$. In this case, there is a theorem which is operating [13]: Then $\Delta t \leftrightarrow -\Delta t$ exchange, i.e., the exchange of the two decay products at t_1 and t_2, which is not a T-symmetry operation, becomes equivalent to T, i.,e., the exchange of the "in" and "out" neutral B-states. In this case, only two independent Intensities remain to be compared, when CP ~ T ~ Δt are theoretically connected through their equivalence.

We notice that the Intensities (3) contain terms independent of $\Delta\Gamma$, in such a way that we will find asymmetries that, contrary to Kabir's asymmetry, do not need a non-vanishing $\Delta\Gamma$ associated to the decay properties.

3. Genuine observables not needing ΔΓ

We are now in the position of defining the genuine observables for testing the three symmetries CP, T and CPT separately.

1) Take $B_0 - B_+$ as the Reference transition and call (X, Y) the observed decay products at times t_1 and t_2, respectively. The CP, T and CPT transformed transitions are given in Table 1.

Table 1. The $B^0 \rightarrow B_+$ reference process and its CP, T and CPT transformed processes.

Transition	$B^0 \rightarrow B_+$	$\overline{B}^0 \rightarrow B_+$	$B_+ \rightarrow B^0$	$B_+ \rightarrow \overline{B}^0$
(X,Y)	(l^-,$J/\Psi K_L$)	(l^+,$J/\Psi K_L$)	($J/\Psi K_S$, l^+)	($J/\Psi K_S$, l^-)
Transformation	Reference	CP	T	CPT

The four processes are experimentally independent, as you may check, and the Δt-exchanged processes X↔Y are <u>not</u> in Table 1. This last comment is particularly important, because (X, Y) and (Y, X) are usually included in the same experimental "sample".

2) Select (Y, X) from 1) as Reference, associated with $B_{-} \rightarrow \overline{B}_0$ transition, and consider its three CP, T and CPT transformed transitions. We have thus four additional processes which are experimentally independent among themselves and with those considered in Table 1. These eight independent transitions are described by the Decay-Intensities given in equation (3).

We thus conclude that 6 independent Asymmetries, 2 for each CP, T and CPT transformations, plus the 2 Intensities for the Reference processes can be built.

The 2 genuine CPV-Asymmetries are

$$\Delta t \left[\begin{array}{l} A_{CP,1} = \dfrac{\Gamma(l^-, J/\psi\, K_L) - \Gamma(l^+, J/\psi\, K_L)}{\Gamma(l^-, J/\psi\, K_L) + \Gamma(l^+, J/\psi\, K_L)} \\ \\ A_{CP,2} = \dfrac{\Gamma(J/\psi\, K_{L,} l^-) - \Gamma(J/\psi\, K_L, l^+)}{\Gamma(J/\psi\, K_L, l^-) + \Gamma(J/\psi\, K_L, l^+)} \end{array} \right. \tag{4}$$

The 2 genuine TRV-Asymmetries are

$$\Delta t \left[\begin{array}{l} A_{T,1} = \dfrac{\Gamma(l^-, J/\psi\, K_L) - \Gamma(J/\psi\, K_S, l^+)}{\Gamma(l^-, J/\psi\, K_L) + \Gamma(J/\psi\, K_S, l^+)} \\ \\ A_{T,2} = \dfrac{\Gamma(J/\psi\, K_{L,} l^-) - \Gamma(l^+, J/\psi\, K_S)}{\Gamma(J/\psi\, K_L, l^-) + \Gamma(l^+, J/\psi\, K_S)} \end{array} \right. \tag{5}$$

The 2 genuine CPTV-Asymmetries are

$$\Delta t \left[\begin{array}{l} A_{CPT,1} = \dfrac{\Gamma(l^-, J/\psi\, K_L) - \Gamma(J/\psi\, K_S, l^-)}{\Gamma(l^-, J/\psi\, K_L) + \Gamma(J/\psi\, K_S, l^-)} \\ \\ A_{CPT,2} = \dfrac{\Gamma(J/\psi\, K_{L,} l^-) - \Gamma(l^-, J/\psi\, K_S)}{\Gamma(J/\psi\, K_L, l^-) + \Gamma(l^-, J/\psi\, K_S)} \end{array} \right. \tag{6}$$

As noticed above, the 6 Asymmetries (4), (5) and (6) are <u>independent</u>, no matter that the CPT-operation is the product of CP and T transformation. However, for a given symmetry, one can increase the statistical significance of the experimental study by considering additional processes taken as a Reference. Thus we have

3) Take $B^0 \rightarrow B_-$ as the Reference transition. The CP, T and CPT transformed transitions are

Table 2. The $B^0 \rightarrow B_-$ reference process and its CP, T and CPT transformed processes.

Transition	$B^0 \rightarrow B_-$	$\overline{B}^0 \rightarrow B_-$	$B_- \rightarrow B^0$	$B_- \rightarrow \overline{B}^0$
(X,Y)	(l^-,J/ΨK$_S$)	(l^+,J/ΨK$_S$)	(J/ΨK$_L$, l^+)	(J/ΨK$_L$, l^-)
Transformation	Reference	CP	T	CPT

Although these four processes are the same as those considered in 2), they can be combined in different ways for each of the three symmetries. For example, the CP-Asymmetry in 2) [see equation (4)] involves the decay product to J/ψ K_L only. In Table 2, however, referring to the choice of reference 3), one obtains a CP-Asymmetry involving J/ψ K_S instead. Another example: the CPT-Asymmetry in 2) [see equation (6)] involves the decay product to l^- only. In Table 2, however, one can obtain a CPT-Asymmetry involving the decay product to l^+ instead, changing the Reference.

4) Select (Y, X) from 3) as Reference, associated with $B_+ \rightarrow \overline{B}^0$ transition, and consider its three CP, T and CPT transformed transitions. The four processes are experimentally independent among themselves and with those considered in 3), i.e., in Table 2. Although these four processes in 4) are the same as those considered in 1), they can be combined in different ways for each of the three symmetries.

We thus conclude that there are another 6 Asymmetries, 2 additional ones for each CP, T and CPT transformations.

The 2 additional genuine CPV-Asymmetries are

$$\Delta t \left[\begin{array}{l} A_{CP,3} = \dfrac{\Gamma(l^-, J/\psi\, K_S) - \Gamma(l^+, J/\psi\, K_S)}{\Gamma(l^-, J/\psi\, K_S) + \Gamma(l^+, J/\psi\, K_S)} \\ \\ A_{CP,4} = \dfrac{\Gamma(J/\psi\, K_S, l^-) - \Gamma(J/\psi\, K_S, l^+)}{\Gamma(J/\psi\, K_S, l^-) + \Gamma(J/\psi\, K_S, l^+)} \end{array} \right. \tag{7}$$

The 2 additional genuine TRV-Asymmetries are

$$\Delta t \left[\begin{array}{l} A_{T,3} = \dfrac{\Gamma(l^-, J/\psi\, K_S) - \Gamma(J/\psi\, K_L, l^+)}{\Gamma(l^-, J/\psi\, K_S) + \Gamma(J/\psi\, K_L, l^+)} \\ \\ A_{T,4} = \dfrac{\Gamma(J/\psi\, K_S, l^-) - \Gamma(l^+, J/\psi\, K_L)}{\Gamma(J/\psi\, K_S, l^-) + \Gamma(l^+, J/\psi\, K_L)} \end{array} \right. \tag{8}$$

The 2 additional genuine CPTV-Asymmetries are

$$\Delta t \left[\begin{array}{l} A_{CPT,3} = \dfrac{\Gamma(l^+, J/\psi\, K_S) - \Gamma(J/\psi\, K_L, l^+)}{\Gamma(l^+, J/\psi\, K_S) + \Gamma(J/\psi\, K_L, l^+)} \\ \\ A_{CPT,4} = \dfrac{\Gamma(J/\psi\, K_S, l^+) - \Gamma(l^+, J/\psi\, K_L)}{\Gamma(J/\psi\, K_S, l^+) + \Gamma(l^+, J/\psi\, K_L)} \end{array} \right. \tag{9}$$

For a CPV test, we should use (4) plus (7) asymmetries. For a TRV test, we should use (5) plus (8) asymmetries. They are model-independent in the sense that this experimental analysis is free from any theoretical prejudice: only the EPR-correlation has been used to prepare the state of the individual neutral B. The experimental channels involved in each symmetry test are different.

And, what for genuine Asymmetries in DAPHNE ? For $K^0 \to \overline{K}^0$, contrary to $B^0 \to \overline{B}^0$, the "physical" states of definite mass and width have $\Delta\Gamma \neq 0$, so that the theorem [13] discussed before for CPT invariance, i.e., the equivalence of T (and CP) with Δt-exchange, is no longer applicable.

If K_+ is the neutral K-state filtered by the decay to 2π and K_- is its orthogonal state, the Master Table for the Φ-factory becomes

Table 3. Four asymmetries for each CP, T, CPT transformations in DAPHNE

Reference	CP	T	CPT
$K^0 \to K_+$ $(l^-, 2\pi)$	$\overline{K}^0 \to K_+$ $(l^+, 2\pi)$	$K_+ \to K_0$ $(3\pi, l^+)$	$K_+ \to \overline{K}^0$ $(3\pi, l^-)$
$K_- \to \overline{K}^0$ $(2\pi, l^-)$	$K_- \to K^0$ $(2\pi, l^+)$	$\overline{K}^0 \to K_-$ $(l^+, 3\pi)$	$K^0 \to K_-$ $(l^-, 3\pi)$
$K^0 \to K_-$ $(l^-, 3\pi)$	$\overline{K}^0 \to K_-$ $(l^+, 3\pi)$	$K_- \to K_0$ $(2\pi, l^+)$	$K_- \to \overline{K}^0$ $(2\pi, l^-)$
$K_+ \to \overline{K}^0$ $(3\pi, l^-)$	$K_+ \to K^0$ $(3\pi, l^+)$	$\overline{K}^0 \to K_+$ $(l^+, 2\pi)$	$K^0 \to K_+$ $(l^-, 2\pi)$

Experimental analyses of CP, T, CPT Asymmetries with DAΦNE data are being pursued [14]

4. Monte Carlo study for TRV in the B-system

We concentrate here on the four asymmetries for true TRV in the neutral B-system, i.e., equations (5) and (8): $B^0 \leftrightarrow B_+$, $B_- \leftrightarrow \overline{B}^0$, $B^0 \leftrightarrow B_-$, $B_+ \leftrightarrow \overline{B}^0$ as Reference.

A PDF allowing CP, T & CPT violation parameters has been developed in [15] for generating events. The true value has been taken as in the Standard Model: CPT-symmetry, $\Delta\Gamma = 0$, $C_i = 0$, $S_i = \pm\, 0.672$ ($\sin 2\beta$) in the Intensities of equation (3). For the A_T asymmetries, as function of Δt, Figure 4 gives the results including proper-time resolution, mistags and efficiency effects (as taken from BABAR published papers):

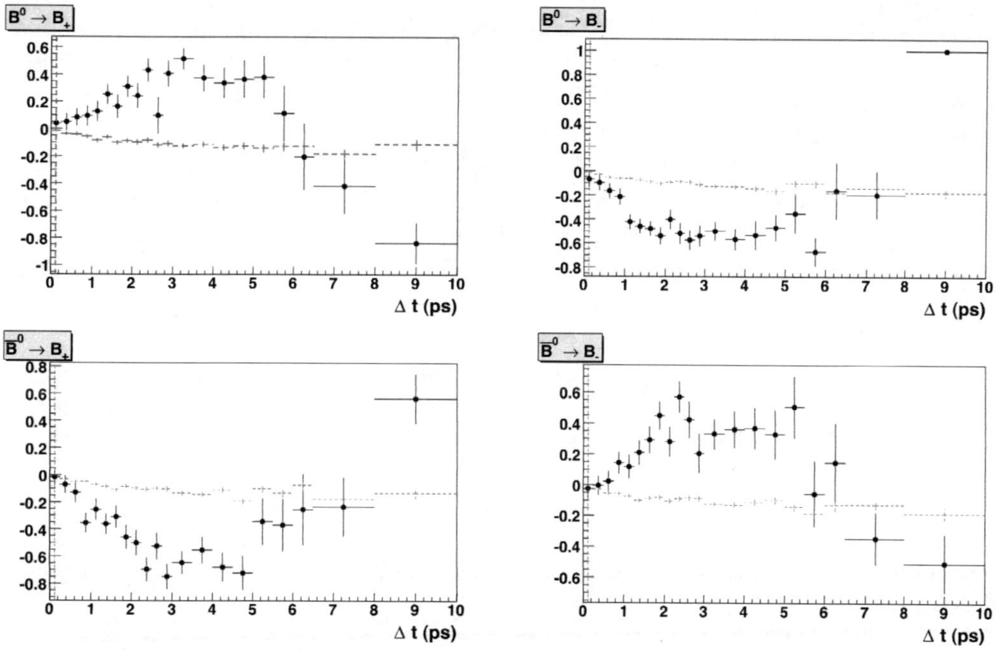

Figure 4. The four TRV-Asymmetries associated with $B_0 \leftrightarrow B_+$, $B_- \leftrightarrow \overline{B}^0$, $B_0 \leftrightarrow B_-$, $B_+ \leftrightarrow \overline{B}^0$. The red points give the expected experimental asymmetry for absence of true TRV.

Each one of the four A_T-Asymmetries has an experimental significance obtained from a χ^2 test. The main conclusion, as seen in Table 4, is that one gets much more than a 5σ-effect!

Table 4. Experimental significance for each of the 4 TRV-Asymmetries.

TRV test	$B^0 \to B_+$	$B^0 \to B_-$	$\overline{B}^0 \to B_+$	$\overline{B}^0 \to B_-$
Standard Deviations	6.70	9.84	9.42	7.34

As the four results are statistically independent, they could be combined in the χ^2-test for obtaining a global significance, assuming that the four asymmetries are theoretically related.

These simulated results provide an expected "guarantee" of a significant discovery for true TRV. A fundamental experimental result would be a first observation of true, direct evidence for genuine TRV by many standard deviations from zero, without any reference to, and independent of, CPV.

Working with the four CP-Asymmetries of Equations (4) and (7), one expects similar levels of experimental significances independently.

5. The ω-effect

The proposed tests of separate CP, T, CPT symmetries in the neutral meson systems are based on the EPR-Entanglement existing in the Meson Factories as a consequence of Particle Identity: $K^0, \overline{K}^0$ are two states of identical particles, connected by CPT.

Besides the permutation operation $\mathcal{P}$ for space-time properties, the strangeness charge connection is made by C, so that for bosons the indistinguishibility requirement is C $\mathcal{P}$= +.

In neutral meson factories, $K^0, \overline{K}^0$ are produced by Φ-decay with J=1, S=0. This implies L=1 and C= -, so that $\mathcal{P}$= -, i.e., an antisymmetric wave function. This antisymmetry is responsible for preserving $K^0\,\overline{K}^0$ terms only in the time evolution of the two-body system, including the Mixing $K^0 \leftrightarrow \overline{K}^0$. Similarly for $K_+\,K_-$ terms only. This correlation is perfect for tagging: Flavour-Tag, CP-Tag,...

The question is [16]: What if the $K^0\,\overline{K}^0$ Identity is lost? In this case, the two particle system would not satisfy the requirement C$\mathcal{P}$= +. In perturbation theory, if still J=1 with C=-, this breaking leads to a mixing of the "forbidden" $\mathcal{P}$= + symmetric state into the "allowed" $\mathcal{P}$= - antisymmetric state:

$$|i\rangle = |antisymmetric\rangle + \omega\,|symmetric\rangle \qquad (10)$$

This perturbative mixing is the ω-effect: In the time evolution of the system one finds now ω $K_0\,K_0$, ω $K_+\,K_+$ terms,..., i.e., a Demise of Tagging.

In some Quantum Gravity models, matter propagation in topologically non-trivial space-time vacua suffers a possible loss of quantum coherence or "decoherence". This effect can be originated by space-time foam backgrounds [17]. The matter quantum system is an open system, interacting with the "environment" of quantum gravitational degrees of freedom: this interaction leads to an apparent loss of unitarity for low-energy observers. As a consequence, there is not a well-defined S-matrix between asymptotic states and the CPT operator is not well defined [18].

This kind of CPT-non definition should be disentangled from the case of effective theories for Lorentz invariance violation [19], in which CPT-breaking means $[H_{eff}, CPT] \neq 0$. The CPT-Violation discussed here would be an "intrinsic" microscopic time irreversibility, so that $\overline{K}^0$ is not "well-defined" from K^0. It implies: 1) a modified single $K^0, \overline{K}^0$ evolution, leading to the (α, β, γ) parameterization [20]; 2) for entangled Kaon states in a Φ-factory, the ω-effect.

Looking for observables which are a signal of the ω-effect, consider [16] the Φ-decay amplitude described by Figure 5 and equation (11):

Figure 5. $K^0 - \overline{K}^0$ decay to (X, Y) at times t_1 and t_2, respectively

$$A(X,Y) = \langle X | K_S \rangle \langle Y | K_S \rangle N(A_1 + A_2)$$
$$A_1 = e^{-i(\lambda_L + \lambda_S)t/2} \left[\eta_X e^{-i\Delta\lambda\Delta t/2} - \eta_Y e^{i\Delta\lambda\Delta t/2} \right] \quad (11)$$
$$A_2 = \omega \left[e^{-i\lambda_S t} - \eta_X \eta_Y e^{-i\lambda_L t} \right]$$

where A_1 is the "allowed" amplitude, whereas A_2 is the "forbidden" amplitude proportional to ω.

An inspection of equation (11) tells us that the good strategy to enhance the relative effect of ω is the selection of a decay channel suppressed by the η's, the ratio between the K_L and K_S decay amplitudes. This requirement is fulfilled by the choice X = Y = $\pi^+ \pi^-$, a CP "forbidden" channel in K-physics, with η a small ratio. The relative ω-effect in equation (11) is thus $\omega / |\eta_{+-}|$.

From equation (11) we can calculate the Intensity as function of Δt

$$I(\Delta t) = \frac{1}{2} \int_{\Delta t}^{\infty} dt \, |A(X,Y)|^2 \quad (12)$$

For the ($\pi^+ \pi^-$, $\pi^+ \pi^-$) channel, and writing $\omega = |\omega| e^{i\Omega}, \eta_{+-} = |\eta_{+-}| e^{i\Phi_{+-}}$, the Intensity (12) is plotted in Figure 6 for ω = 0 (continuous line) and an ω-value comparable to η_{+-} (broken line).

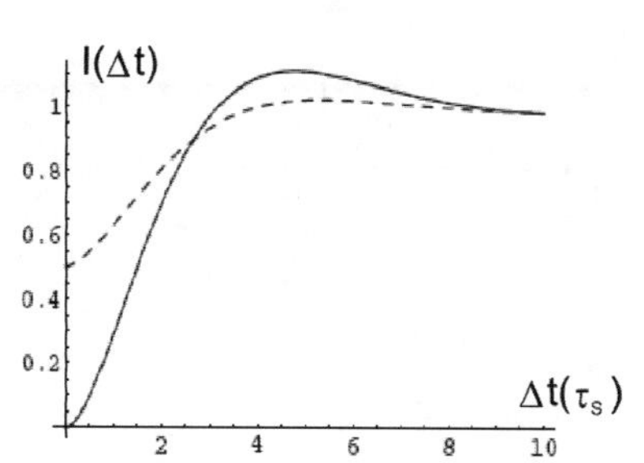

Figure 6. The Intensity I(Δt) for the decay of $K^0 - \overline{K}^0$ to ($\pi^+ \pi^-$, $\pi^+ \pi^-$), for ω=0 (—) and $|\omega| = |\eta_{+-}|, \Omega = \Phi_{+-} - 0.16\pi$ (---)

The most prominent effect is the breaking of I(Δt) ~ 0 for small values of Δt, a result that was a consequence of the particle identity anti-correlation: no identical states at t_1=t_2.

The KLOE experiment has obtained the first measurement of the ω-parameter [21]:

$$\left.\begin{aligned} \mathrm{Re}(\omega) &= \left(-1.6^{+3.0}_{-2.1\,stat} \pm 0.4_{syst}\right) \times 10^{-4} \\ \mathrm{Im}(\omega) &= \left(-1.7^{+3.3}_{-3.0\,stat} \pm 1.2_{syst}\right) \times 10^{-4} \end{aligned}\right\} \quad |\omega| < 1.0 \times 10^{-3} \, at \ 95\% \ CL \quad (13)$$

At least one order of magnitude improvement is expected with KLOE-2 at the upgraded DAΦNE.

All decoherence effects, including the ω-effect, manifest as a deviation from the quantum mechanical prediction of the EPR-correlation $I(\pi^+\pi^-, \pi^+\pi^-; \Delta t = 0) = 0$. Hence the reconstruction of events in the region near $\Delta t \sim 0$ is crucial. Experimentally, one needs vertex resolution. In Figure 7 one finds a Monte Carlo simulation of $I(\pi^+\pi^-, \pi^+\pi^-; \Delta t)$, with both the KLOE resolution $\sigma_{\Delta t} \approx \tau_S$ (black histogram) and the expected KLOE-2 resolution $\sigma_{\Delta t} \approx 0.3\ \tau_S$ (red histogram)

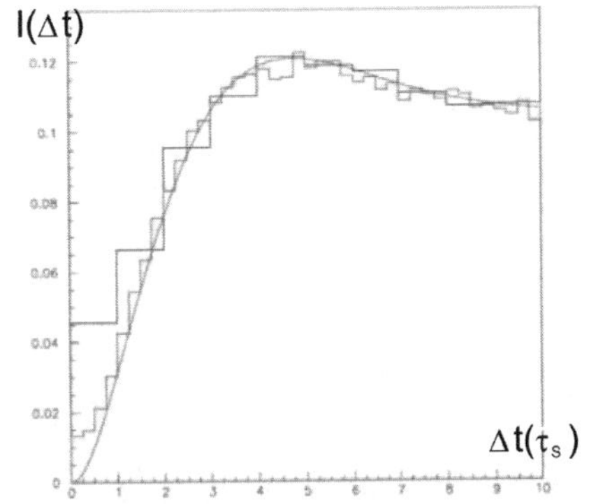

Figure 7. Comparision of the I $(\pi^+\pi^-, \pi^+\pi^-)$ simulation for KLOE Δt resolution (——) and KLOE-2 Δt-resolution (——)

In B-factories, there is no privileged channel to enhance the ω-effect. With currently available data from BABAR and BELLE, the CP-violating semileptonic charge asymmetry, in the equal sign dilepton channels $I(l^{\pm}, l^{\pm}; \Delta t)$ gives the bounds [22]

$$-0.0084 \leq \mathrm{Re}(\omega) \leq 0.0100 \text{ at } 95\%\mathrm{CL} \qquad (14)$$

The ω-effect has been described as the result of local distortions of space-time in the neighborhood of defects, interacting with matter [23]. The recoil of the Planck-mass defect leads to a metric deformation

$$g_{0i} \sim \Delta k^i / M_P = \zeta k^i / M_P \qquad (15)$$

In this framework, Lorentz invariance still holds macroscopically $< \zeta k^i > = 0$, but one has non-trivial quantum fluctuations $< \zeta^2 k^i k^j > \sim \delta_{ij} \zeta^2 \mid \vec{k} \mid^2$. The stochastic effects of the space-time foam lead to an ω-value

$$|\omega|^2 \sim \frac{\zeta^2 |\vec{k}|^4}{M_P^2 \Delta m^2} \qquad (16)$$

which is enhanced by the quasi-degeneracy of the mass eigenstates.

At the DAΦNE energy, equation (16) gives $|\omega| \sim 10^{-4}\zeta$, which lies within the sensitivity of KLOE-2 for not much small values of the momentum transfer fraction ζ.

6. Conclusions

The observed time-Asymmetries in the Universe and the macroscopic "arrow of time" are not Time-Reversal Violating. A true fundamental TRV means an Asymmetry under the exchange of in ↔ out states.

For unstable systems, there is unique opportunity for preparing the quantum states needed in a true T-symmetry test by using the EPR-Entanglement between the two neutral mesons in B and Φ factories. The Golden Channels associated with Mixing x Decay amplitudes offer 8 different Decay-Intensities.

For each one of the three symmetries CP, T and CPT, one can work with appropriate combinations of the Intensities for generating 4 Genuine Independent Asymmetries.

A Monte Carlo study for true TRV in B-factories leads to simulated results for each of the four asymmetries with more than 5 standard deviations from zero. We have thus an expectation of having experimental results leading to a first discovery of true TRV independent of, and unconnected to, CPV.

In certain quantum gravity models, the S-matrix and the CPT-operator are not well defined, leading to a breaking of the $K^0, \bar{K}^0$ particle identity and the appearance of the ω-effect. The measurement of the "forbidden" amplitude induced by the ω-parameter is better done with the privileged channel in the Φ-factory I($\pi^+ \pi^-$, $\pi^+ \pi^-$; Δt). A sensitivity to $|\omega| \sim 10^{-4}$ at KLOE-2 is not far from expectations in some models of space-time foam.

Acknowledgments
I would like to thank the conceptual discussions maintained with several colleagues, specially Antonio Di Domenico, Fernando Martinez and Pablo Villanueva. This work is supported by the Spanish Ministry MICINN and Generalitat Valenciana under Grants FPA-2008-02878 and PROMETEO-2008/004.

References

[1] Stocchi A, *talk at this conference*
[2] Luders G, *Annals Phys.***2***(1957)1;*
Pauli W in *Niels Bohr and the Development of Physics* (Pergamon Press, New York, 1955) p.30
[3] Wolfenstein L, *Int. J. Mod. Phys.* **8**(1999)501
[4] See Callender C, Thermodynamic Asymmetry in Time, The Stanford Encyclopedia of *Philosophy,* E N Zalta ed (2008)
[5] Quinn H R, *Proceedings DISCRETE'08*, *J. Phys.Conf.Ser.* 171:011001 2009
[6] Semertzidis Y, *this Proceedings*
[7] Peccei R D and Quinn H R, *Phys. Rev.Lett.* **38** (1977) 1440;
Peccei R D and Quinn H R, *Phys Rev. D* **16**(1977) 1791
[8] Angelopoulos A et al. *CP-LEAR Collaboration, Phys. Lett. B* **444**(1998)43
[9] Bañuls M C and Bernabeu J, *Phys. Lett. B* **464**(1999)117
[10] Bañuls M C and Bernabeu J, *Nucl. Phys. B* **590**(2000)19
[11] Einstein A, Podolsky B and Rosen N, *Phys Rev* **47**(1935)777
[12] Bañuls M C and Bernabeu J, *JHEP* 9906:032(1999)
[13] Bernabeu J and Bañuls M C, *TAUP99 Paris, Nucl. Phys. Proc. Suppl.* 87:315(2000);
Bernabeu J and Espinoza C, *Venice 2009, Neutrino Telescopes* p.159
[14] Bernabeu J, Di Domenico A and Villanueva P, *in preparation*
[15] Villanueva P, *this Proceedings*
[16] Bernabeu J, Mavromatos N E and Papavassiliou J, *Phys. Rev.Lett.* 92:131601(2004)
[17] For a review, see Wheeler J A and Ford K, *Geons, black holes and quantum foam: A life in physics* (New York, USA; Norton, 1998);
Ellis J R, Hagelin J S, Nanopoulos D V and Srednicki M, *Nucl. Phys. B* **241**(1984)381
[18] Wald R, *Phys. Rev.D* **21**(1980)2742
[19] Altschul B, *this Proceedings*
[20] Lindblad G, *Comm. Math. Phys.* **48**(1976)119
[21] Di Domenico A, 0904.1976 (2009)
[22] Alvarez E, Bernabeu J and Nebot M, *JHEP* 0611(2006)087
[23] Bernabeu J, Mavromatos N E and Sarkar S, *Phys. Rev. D* **74**(2006)045014

Journal of Physics: Conference Series **335** (2011) 012012 doi:10.1088/1742-6596/335/1/012012

Review of EDM experiments

Yannis K. Semertzidis

Physics Dept., Bldg 510A

Brookhaven National Laboratory, NY 11973-5000

yannis@bnl.gov

Abstract. The current limits on physics beyond the SM come in large part from the non-observation of EDMs in the sensitive electric dipole moment experiments, like the neutron, ^{199}Hg, and ^{205}Tl. New systems with enhanced EDM sensitivity are coming online and promise a resolution of the baryon asymmetry of our universe (if an EDM is observed) or a severe constraint on physics beyond the SM by the end of the current decade.

1. Introduction

Our traditional approach to physics at the frontier is dual: 1) Energy frontier with FNAL until recently and currently CERN playing that role and 2) The precision Frontier with experiments like the muon g-2, electric dipole moment experiments, B-factories, etc. Those approaches are complementary and inter-connected. The next standard model (SM) is likely to emerge with input from both; it certainly has to accommodate observations from both approaches. Currently, one of the most intriguing indications of physics beyond the SM is the discrepancy of the experimental value of the muon g-2 value from the theoretically predicted value [1]. The discrepancy is of order of 3.5 standard deviations. For those enthusiasts prescribing to super-symmetry (SUSY) it would mean a neutralino mass range of 0.1 TeV – 0.5 TeV depending on the value of $\tan\beta$, the ratio of the vacuum expectation value of the two Higgs doublets.

SUSY as well as most of the beyond the SM models (BSM) happen to predict a large electric dipole moment with the current experimental limits being embarrassingly restrictive. So, even though we may have an indication of SUSY from the muon anomalous magnetic moment measurement we also have an anti-indication of SUSY from the current experimental limits on the electric dipole moments of fundamental particles, like the electron, neutron, proton, Mercury, etc. [2]. Certainly clever theorists could accommodate the EDM non-effect for specific SUSY models. So far the fine-tuning needed for this accommodation is of order 10^{-2} so the next round of EDM experiments will probably observe a non-zero value for the first time, or SUSY is in serious trouble. The next round of EDM experiments are set to increase their sensitivity level by about one order of magnitude (Mercury), to two orders of magnitude (neutron), to three orders of magnitude (proton and Deuteron), up to four orders of magnitude for the most ambitious electron searches. The time scale for the searches range is about three years (Mercury) to 10 years (deuteron and electrons) with the neutron and proton direct searches between those two extremes. The next decade is indeed promising to be very exciting with all the new opportunities opening up for breaking the EDM mystery.

The EDM searches commenced with two pioneers who set to look for parity violation in what was called then nuclear interactions [3] more than fifty years ago. As they stated their motivation "The question of the possible existence of an electric dipole moment of a nucleus or of an elementary particle … becomes a purely experimental matter". They couldn't be more correct even today, after several orders of improvement in the EDM experimental limits. After the discovery of parity violation Landau pointed out [4] that a particle EDM requires parity as well as time violation to be allowed. Assuming CPT conservation, T-violation also means CP-violation, which is believed to play an important role in our own existence and so EDM searches are even more motivated.

Ramsey and Purcell started their EDM search with the neutron in the fifties by applying an electric field and looking for the interaction energy between the E-field and the neutron EDM. In the sixties sensitive EDM searches on electrons started using specially prepared atomic systems. In the seventies an indirect storage ring EDM method was used to put a limit on the muon EDM. In the eighties people came up with systems (polar molecules) with large enhancement factors, which could provide enough sensitivity to the electron EDM. In the nineties we saw the first attempts to utilize the new methods with large enhancement factors as well as the first dedicated Storage Ring EDM method fully developed. Since then, several storage ring EDM lattices were developed for the proton and deuteron with sensitivity at the 10^{-29} $e\cdot$cm level probing physics at the 3000 TeV scale for contact interactions and 300 TeV for SUSY-like new physics.

2. Theoretical Motivation

In quantum mechanics (QM) a non-degenerate system with spin is defined by its spin vector. If the particle has an EDM, its vector needs to be aligned with its spin vector, locked to its direction:

$$\vec{d} = d\hat{\sigma} \tag{1}$$

It needs to be either along the same direction or opposite but not both (non-degenerate) [5]. It is then and only then that the EDM of a particle with a spin violates both parity (P) and time (T) symmetries.

A permanent EDM violates both the time and parity symmetries:

$$\begin{gathered} H = -d\hat{\sigma}\cdot\vec{E} \xrightarrow{T} H = -d\left(-\hat{\sigma}\right)\cdot\vec{E} = d\hat{\sigma}\cdot\vec{E} \\ H = -d\hat{\sigma}\cdot\vec{E} \xrightarrow{P} H = -d\hat{\sigma}\cdot\left(-\vec{E}\right) = d\hat{\sigma}\cdot\vec{E} \end{gathered} \tag{2}$$

From CPT conservation we have that T-violation also means CP-violation. CP-violation is important because it is a fundamental requirement to enable a universe initially containing equal amounts of matter and antimatter to evolve into a matter-dominated universe we see today [6]. CP-violation was discovered [7] at BNL in 1964 in the K-long decays by Jim Cronin, Val Fitch *et al.*, and was confirmed in the decays of B-mesons. However, this CP-violation as was established to be described very well by a single complex phase in the CKM matrix [8], part of weak-interactions CP-violation, is way too small to account for the observed asymmetry of the matter-antimatter in our universe by several orders of magnitude. The ratio of the observed baryons to photons is

$$\frac{n_B}{n_\gamma} \approx 10^{-10} \tag{3}$$

whereas the expected level from the complex phase in the CKM matrix of the SM is only at the 10^{-18} level. A new CP-violating source is surely needed and EDMs could provide it.

EDMs are excellent probes of physics beyond the SM since most models predict values within the sensitivity of currently planned experiments: Those are SUSY, Multi-Higgs, Left-Right symmetric, etc. The SM contribution is negligible, with a caveat. There is an irony in the last statement, in that the quantum chromodynamic (QCD) part of the SM is actually predicting a much larger CP-violation than the current experimental limit from the neutron EDM. The CP-violating term in the QCD Lagrangian

$$L \approx \bar{\theta}\frac{\alpha_s}{8\pi}G\tilde{G} \tag{4}$$

violates both parity and time symmetry, with $\bar{\theta}$ (theta-QCD) an angular parameter of order one. However, the current neutron EDM limit [9] of 10^{-26} $e{\cdot}$cm sets a very severe limit on $\bar{\theta} < 10^{-10}$. Since the value of $\bar{\theta}$ was expected to be of order one and the current limit is ten orders of magnitude less than that it is all but ignored! One way out of this issue is the solution of Peccei and Quinn [10] making it into a dynamical parameter that naturally takes a zero value. The breakdown of this symmetry gives rise to a particle dubbed the axion [11]. The only problem with this solution is that no axion is found yet, so even this so-called theta value should be treated as a purely experimental matter. Therefore if an EDM found it could potentially come either from the theta-QCD term or from some physics beyond the SM. The only way to tell the source is to probe the EDM value of several particles with different combination of CP-violating contributions. The complex phase of the CKM matrix of the SM contributes an EDM at the level of 10^{-31}-10^{-32} $e{\cdot}$cm for the neutron and proton and at the level of $<10^{-38}$ $e{\cdot}$cm for the electron.

3. Experimental Approach

In the particle rest frame electric dipole moments couple only to electric fields and magnetic dipole moments couple only to magnetic fields

$$\frac{d\vec{s}}{dt} = \vec{d}\times\vec{E} + \vec{\mu}\times\vec{B} \tag{5}$$

making the first complication apparent right away: even though one can study the magnetic moment of a charged particle at rest in a magnetic field, doing the same in an electric field is challenging. For this reason the first attempt to probe the EDM of a particle was focused on the neutron, which even though is neutral it does have a magnetic moment. A small magnetic field was applied to keep the neutrons polarized and precessing at a specific frequency. Then a very strong electric field was applied along the B-field direction and the neutron precession frequency was compared for two opposite electric field directions:

$$d = \frac{\hbar(\omega_1 - \omega_2)}{4E} \tag{6}$$

which for d equal to $10^{-28}e{\cdot}$cm, E=200 kV/cm implies an angular rotation frequency of 10^{-7}rad/s or about one turn per year.

The statistical error is of the EDM value (d) is

$$\sigma_d = \frac{b\hbar}{EPA\sqrt{Nf\tau T}} \tag{7}$$

where b is a numerical factor of order one depending on the details of injection, machine/measurement cycle time, P is the polarization at the start of the experiment, A is the analyzing power of the detection mechanism, N is the number of particles per particular measurement cycle, f is the detection efficiency, τ is the polarization lifetime and T is the duration of the experiment.

3.1. Important issues in an EDM experiment

- Polarization: Preparation of the particular state, and intensity of beams
- Interaction with a strong E-field: The larger the field the better
- Analyzer: The detection efficiency and analyzing power are critical elements of the statistical sensitivity
- Scientific interpretation of the result: A simpler system makes the theoretical interpretation of any experimental result easier. A complicated system can even create disputes as to what it means and makes even harder to pin down the various CP-violating sources.

Of course any non-zero EDM observation that can satisfy Norman Ramsey's criteria for his $5K award[1] will be revolutionary and that may be well worth it.

3.2. EDM methods

One can distinguish various methods originating mainly from the specific particles those methods are applied to: neutrons, atomic and molecular systems, and storage ring EDM method for charged particles directly. In the neutron case one applies a small (holding) magnetic field, and a large E-field and probes the frequency shift with the E-field vector flipped. In the atomic and molecular systems one probes atomic and molecular transitions and compares them in different combinations of E and B-field directions. In the storage ring method one utilizes the rest-frame E-field of a moving particle in an electro-magnetic storage ring. Since the rest frame E-field is mostly in the radial direction keeping the spin along the momentum vector maximizes the sensitivity of the method. The EDM signal is the change of the vertical spin component as a function of time. Vertical here is defined as the plane perpendicular to the g-2 precession plane at the location of the polarimeter detector.

The recent advances of the various methods are: There have been improvements in reducing the effects of the stray magnetic fields, by applying spin dressing to equalize the neutron spin precession with the spin precession of the co-magnetometer, e.g. ^{3}He. Another improvement is the higher ultra-cold neutron (UCN) densities that were made possible recently using a new production method. The projected neutron production should make possible using ~10^6 neutrons per measurement cycle.

In atomic/molecular systems there has been a continuous exploration of systems capable of providing an equivalent electric field between 1-100 GV/cm, in which case even a small number of molecules could be enough to provide the needed sensitivity for up to $10^{-30}e{\cdot}$cm.

In the field of storage ring EDM systems, it turns out that 10^8 particles could be readily detected per measurement cycle. High electric fields in the lab frame are available as a result of extensive R&D for the international linear collider (ILC) and energy recovery linacs (ERL). Since polarized

[1] Norman Ramsey announced a reward for $5K for anyone who detects a non-zero EDM in any system, and it is generally accepted as such. Additional criteria include that he needs to be alive when the discovery is made, and that he understands the method.

Journal of Physics: Conference Series **335** (2011) 012012 doi:10.1088/1742-6596/335/1/012012

protons and deuterons were already utilized for other reasons at various accelerators around the world, the required tools have already been extensively developed.

The weaknesses of the various EDM methods can be summarized for the neutrons: Intensity is still an issue since eventually the experiment will be statistics limited. There is also high sensitivity to stray magnetic fields which has to be measured by other means to very high accuracy and finally the motional B-fields and geometrical phases are critical.

In atomic and molecular systems it is the intensity of the desired states that has plagued this development, and certainly in some systems the physics interpretation is the biggest obstacle.

For the storage ring EDM methods, depending on the particular method they are very sensitivity to either a net vertical E-field around the ring or a net radial B-field that could cause a problem.

Overall statistical sensitivity is critical, i.e. number of particles and spin coherence time and of course systematics originating from spurious magnetic fields, geometrical phases, etc. This is a very challenging field that a typical experimental cycle is ten to fifteen years, while the researchers tend to be intelligent with wide range of expertise and not hesitant to tackle tough problems.

3.3. Neutron EDM experiments

The current neutron EDM limit is $3\times10^{-26}e\cdot$cm (90% C.L.) comes [9] from an experiment at ILL at Grenoble. Low energy neutrons from the reactor are guided in a vertical tube against gravity; where a so-called neutron turbine knocks them down to very slow velocities, see Fig. 1.

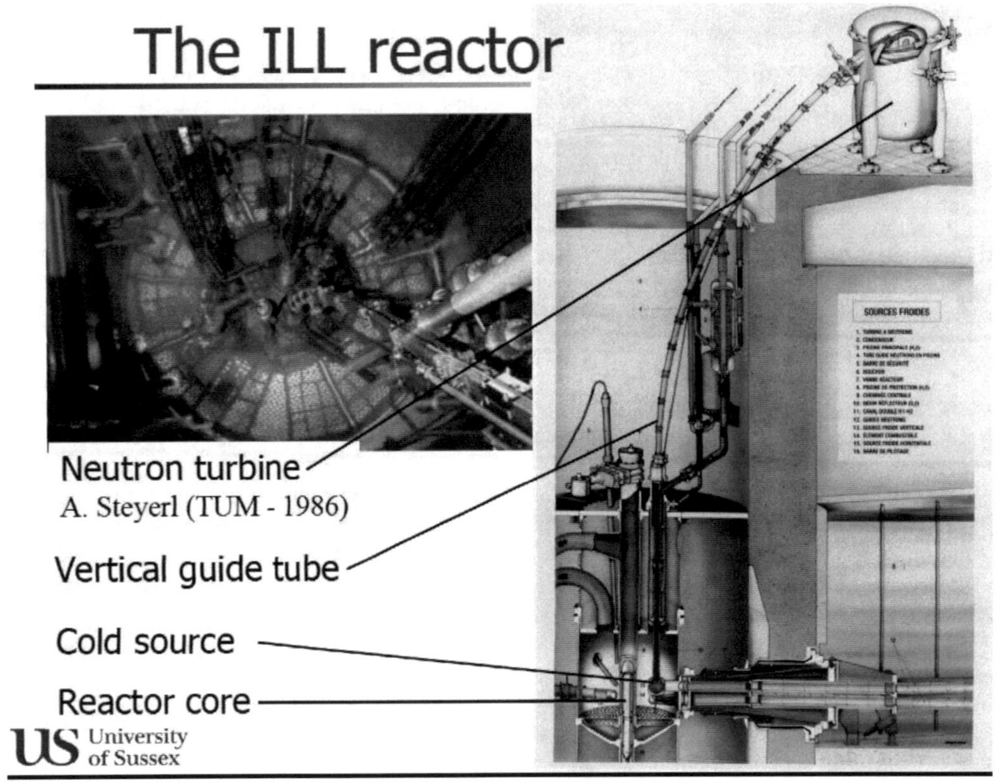

Figure 1. The ILL reactor with its neutron EDM experiment.

The new ultra-cold neutron (UCN) experiment at ILL, detail shown in Fig. 2, is going to use a superfluid helium bath of 0.5 K and improve the neutron density by a factor of 50. It also expects to improve the electric field strength by a factor of 4-6 and achieve longer spin coherence time. Overall it's expected to achieve a factor of $\sim 10^2$ better sensitivity than the present neutron EDM limit.

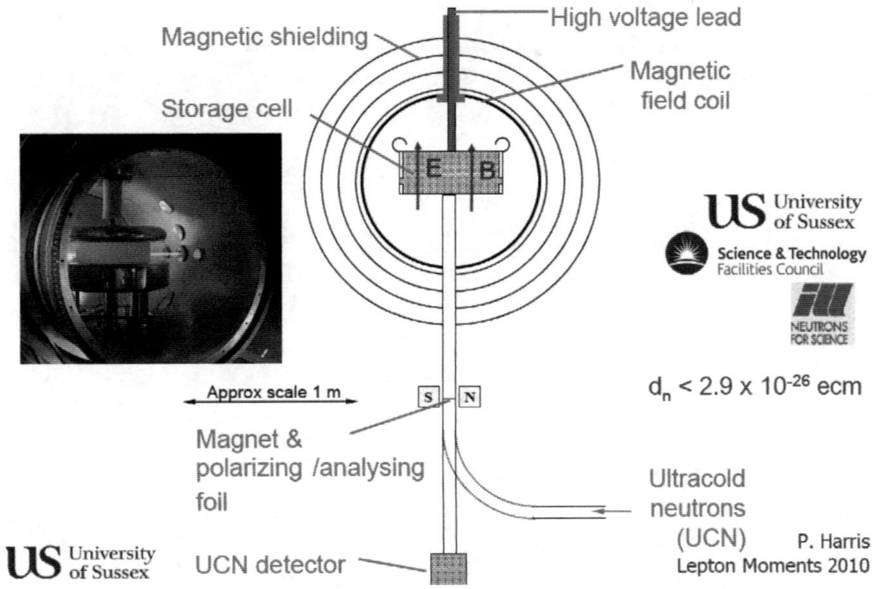

Figure 2. The new ultra-cold neutron EDM experiment at ILL with improved statistics and experimental parameters is expected to make a factor of 10^2 improvement over the present limit.

A neutron EDM experiment at PSI is underway which uses the old ILL apparatus and the neutron spallation source available at PSI with a goal of improving the sensitivity by an order of magnitude over the current limits. With further improvements in statistics and neutron production procedure they plan to improve the limit by almost another order of magnitude [12].

A neutron EDM experiment is also underway at the spallation neutron source (SNS) at Oak Ridge National Laboratory in the US, using several pioneering techniques as proposed by Golub and Lamoreaux [13]. A large helium bath volume of $\sim 10^3$ lt. at 0.4 K with a dilution refrigerator is utilized. The previous UCN were in the tail of the Maxwell-Boltzmann distribution of cold neutrons resulting to densities of 5 UCN/cm^3. The new method, suggested first by Golub and Pendlebury [14], uses neutrons that are produced by thermal down scattering in superfluid ^{4}He bath where they excite a phonon and lose ~1meV of energy. After the neutron falls in the lowest energy state, momentum and energy conservation requirements are not favored for re-exciting it into the higher energy state. The method has been applied in a small scale and it has produced the promised improvements in UCN densities. Figure 3 below shows the free neutron and ^{4}He dispersion curves.

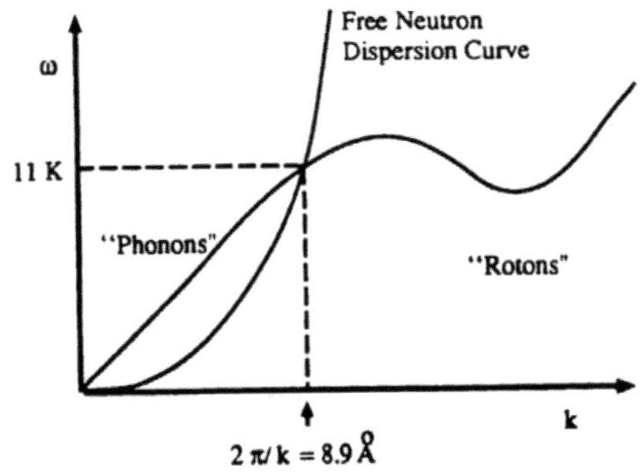

Figure 3. The dispersion of the free neutron and that of ^{4}He lattice cross at one specific point. A 1meV neutron can transfer its energy to the helium lattice with no further chance of interacting with it.

Another innovation in the method is the use of polarized ^{3}He mixed-in with the superfluid ^{4}He in a small concentration ratio of ~10^{-11}. The reason for this is that the UCN density is too small to be detected with a SQUID magnetometer. The cross section of the reaction

$$\vec{n} + {}^3\vec{He} \rightarrow t + p + 764\ \mathrm{keV} \tag{8}$$

depends very strongly on the relative direction between the spin of the neutron and that of ^{3}He. In addition, the ratio of the magnetic moments of ^{3}He and of the neutron is 1.11, i.e. they only differ by ~10% thus reducing the sensitivity to spurious magnetic fields by a factor of ~10. Special techniques, like spin dressing have also been developed by the collaboration to further reduce the background magnetic field sensitivity. The current timeline of the various neutron EDM experiments is shown in Fig. 4 below.

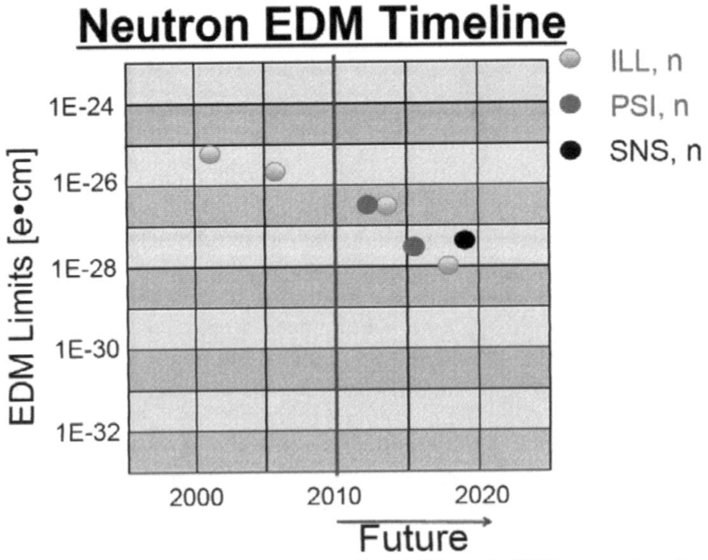

Figure 4. The neutron EDM experiments at ILL, PSI and SNS are planning to improve the experimental sensitivity by up to two orders of magnitude within the current decade.

Journal of Physics: Conference Series **335** (2011) 012012 doi:10.1088/1742-6596/335/1/012012

3.4. EDM experiments in atomic/molecular and polar systems

A fundamental difference between magnetic dipole moment experiments and electric dipole moment experiments is the fact that a uniform magnetic field would not move a stationary charged particle whereas an electric field would. To look for an EDM of a charged particle, like the electron and the proton one has to come up with ways to balance the electric field force without cancelling the effect due to the particle EDM. At the beginning Ramsey and Purcell claimed that the EDM of a charged particle is difficult to probe. A free electron for example would get lost on the walls of the electric field plates right away, whereas in an atomic system with electrons it would be shielded by the atomic electrons. The average E-field it sees should be zero, otherwise it would move. Schiff and then Sandars came up with exceptions [15] to this rule, like relativistic effects could play a role and even though the average E-field is zero, the total EDM effect is non-zero. Another exception is the effect of the finite size of a nucleus, where again the total E-field on the nucleus is zero, but the EDM effect on the atom is finite.

A review article on this effect is published recently by Commins, Jackson, and DeMille [16], where they state that a heuristic explanation of this effect is the fact that the electric dipole moment of the electron changes its length when it's Lorentz transformed from the electron to the atomic center-of-mass frame. The EDM length contraction depends on the electron's velocity and the total (integrated) interaction energy of the electron EDM with the electric field is not zero when averaged over its path.

In paramagnetic atoms (with an unpaired electron in an $s_{1/2}$ or $p_{1/2}$ orbit) it is possible to get a large enhancement factor on the atomic EDM compared to the electron EDM. The enhancement factor is

$$R \approx 10Z^3\alpha^2 \tag{9}$$

which can be larger than one for heavy atoms. For a Thallium atom it is estimated [17] to be -585. The current limit on the electron EDM, using thallium atoms [18] ^{205}Tl is

$$d_e < 1.6\times 10^{-27}\, e\cdot\text{cm, 90\% C.L.} \tag{10}$$

In diamagnetic atoms (no unpaired electrons), like ^{199}Hg there is no enhancement factor. Rather there is a so-called Schiff-moment, which is of order of 10^{-3} for ^{199}Hg, i.e. an EDM limit on the atom corresponds to a limit 10^3 larger for the nucleus EDM. The (table top) experiment utilizes a laser beam to polarize and probe the spin precession frequency to be compared for different E-field orientations. The holding B-field is 22mG, the E-field ~20 kV/cm, while the B-field drift is taken out by utilizing a number of cells in close proximity, with different combination of E and B-field orientations. The reported limit [19] is

$$\left|d\left({}^{199}\text{Hg}\right)\right| < 3.1\times 10^{-29}\, e\cdot\text{cm, 95\% C.L.} \tag{11}$$

which is a factor of 7 improvement over the previous limit. The estimated total systematic error is 60% of the statistical error, so this experiment is mostly systematics limited. This experiment also sets the current (indirect) limit on the proton EDM of $<10^{-24} e$ cm. An improvement of about a factor of 3-5 is expected at the end of the current round of experimental efforts.

There is a large effort to improve the sensitivity to the electron EDM by enhancing the effective E-field on the electron. This enhancement factor is achieved in several polar molecules: YbF, PbO, ThO, PbF, HBr, BaF, HgF, etc. Several times those fields are really impressive, e.g. up to 100 GV/cm in the combined paramagnetic molecule (radical) of Fr (paramagnetic atom) and Sr (diamagnetic atom). All these new efforts have in common the fact that the learning curve is very long like every new EDM effort has proven to be. Progress in the EDM experiments of Rn, Ra, and Xe is ongoing with large expectations.

3.5. Sensitive storage ring EDM experiments

In the face of all these EDM experiments it is natural to ask why the world needs another EDM experiment and a yet another method. It turns out the (indirect) storage ring EDM method has been with us for more than fifty years as a byproduct of the muon g-2 experiment [20]. Even in a storage ring with a pure magnetic field the particle in its own rest frame feels the presence of a very strong electric field. This is proportional to $\vec{v}\times\vec{B}$ and hence it is directed in the radial direction. For 1 T magnetic field this electric field is equivalent[2] to 300MV/m. If there is a muon EDM then, this strong electric field would contribute to the spin precession in two ways [21]: 1) it would tilt the spin precession plane, and 2) it would alter the apparent g-2 spin precession rate. The vertical beam polarization change is described by the equation

$$\Delta P_V = P\frac{\omega_d}{\Omega}\sin\left(\Omega t+\theta\right) \tag{12}$$

with

$$\Omega = \sqrt{\omega_d^2+\omega_a^2} \tag{13}$$

the total spin precession frequency, ω_d the spin precession frequency due to the muon EDM, and ω_a the spin precession frequency due to the muon anomalous magnetic moment.

What happened lately is to actually focus on the EDM itself and optimize the experimental conditions for it [22, 23, 24] utilizing the first from the two options given above. It is clear from eq. (12) that the vertical beam polarization buildup rate is maximized when ω_a is set very close to zero. In this extreme case the spin precession plane is completely vertical, instead of the horizontal we are accustomed to due to the g-2 precession alone. The ω_a=0 can be accomplished in various ways depending on the value of the anomalous magnetic moment of the particle as it will become clear from the g-2 equation in an electromagnetic field:

$$\vec{\omega}_a = -\frac{q}{m}\left\{a\vec{B}+\left[a-\left(\frac{m}{p}\right)^2\right]\frac{\vec{\beta}\times\vec{E}}{c}\right\} \tag{14}$$

with q, m the charge, and mass of the proton, and a its anomalous magnetic moment. For a positive a and in the absence of magnetic fields one can easily see that there is one momentum, the so-called magic momentum $p = m/\sqrt{a}$ where ω_a would be zero independent of the electric field value. For a negative anomalous magnetic moment (a) there is no such momentum and one needs to include a combination of magnetic and electric fields. It turns out that it is possible to achieve very high sensitivity by using polarized deuterons (a=-0.143) or protons (a=1.79). The reasons for this high sensitivity are:

1) High beam intensities (10^{10}-10^{11}) per machine cycle, with high beam polarization (>80%), and low emittance are readily available.
2) Spin coherence times of order 10^3s are possible.
3) Large E-fields for large area electrodes are possible (100-200 kV/cm)

[2] There is also another multiplying factor of γ, but this is cancelled when we estimate the spin precession rate in the lab frame.

Journal of Physics: Conference Series **335** (2011) 012012 doi:10.1088/1742-6596/335/1/012012

4) High efficiency (~1%), and high analyzing power (50%) polarimeters are available for protons and deuterons around the momentum of interest of 1GeV/c.

As a matter of fact polarized proton beams have been utilized around the world to probe spin related physics and the developed expertise is quite extensive. The storage ring EDM collaboration has developed two proposals [23, 24] both claiming a sensitivity of 10^{-29} e·cm for the deuteron and proton bare nuclei. The same method can also be applied to other particles, like ^{3}He and so on as long as there is a possibility to obtain a polarized beam and build an efficient polarimeter for it.

The main reason of the asymmetry between the B and E-fields and storage ring EDMs is the fact that in the presence of B-fields the g-2 precession frequency is velocity independent:

$$\vec{\omega}_a = -\frac{q}{m} a\vec{B} \tag{15}$$

This is in contrast to the E-field case where the g-2 precession frequency is strongly dependent on the particle's velocity:

$$\vec{\omega}_a = -\frac{q}{m}\left[a - \left(\frac{m}{p}\right)^2\right]\frac{\vec{\beta}\times\vec{E}}{c} \tag{16}$$

This is a consequence of the fact that particle EDMs, even if they are not zero, they are very small, whereas magnetic moments are large and are dominating the g-2 precession frequency. In the case of a moving particle in an electric field the particle feels the presence of a magnetic field (from Lorentz transformation), which acts on the particle magnetic moment and causes its spin to precess. The reverse, a moving particle in a magnetic field feels the presence of an electric field, which acting on a possible particle EDM it can only have a minor impact in its g-2 precession frequency.

As we mentioned above, the sensitivity to EDM is maximum when the particle spin is kept aligned to the momentum vector during storage as is shown in Figure 5.

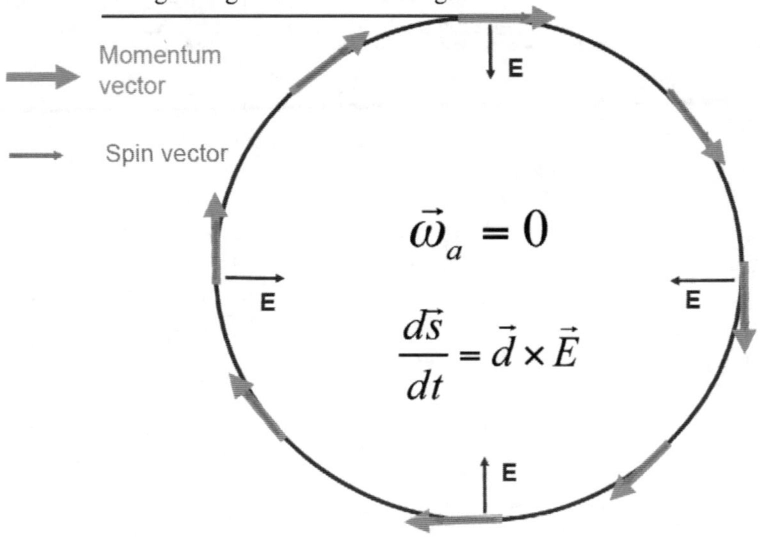

Figure 5. The optimum sensitivity to EDM is obtained when the spin is kept aligned with the momentum vector for the duration of the storage time.

The spin vector alignment with the momentum vector is going to be monitored with internal polarimeters sensitive to the transverse spin component of the beam [22, 23, 24]. If the transverse spin component is found to be larger than a pre-determined value, a feedback signal will be issued and a corrective action will take place. That means that there is a requirement to monitor the spin at all times during storage. Occasionally the spin will be let to rotate completely (multiples of 2π) without a significant loss of sensitivity. The EDM signal is any change in the vertical component of the spin as a function of storage time. The way it shows up in the polarimeter detector is as a change in the ratio (L-R)/(L+R), where L, R correspond to the counts from the left and right counters respectively, see Figure 6.

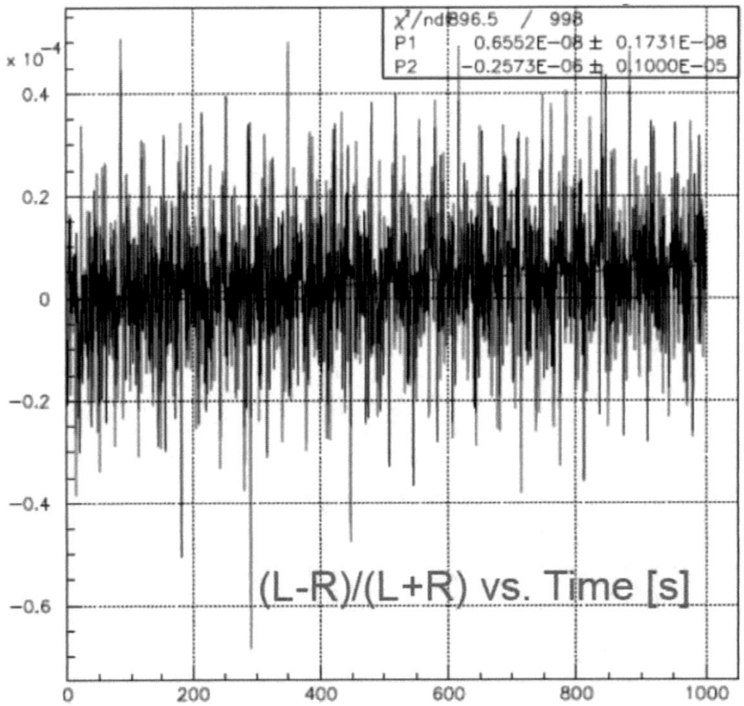

Figure 6. This is M.C. data showing how the data from the polarimeter would look as a function of time, when an appropriate ratio of the left (L) and right (R) detector counts is formed. The non-zero value of the slope of (L-R)/(L+R) as a function of time is a clear indication of the change in the vertical spin component. The slope shown above corresponds to approximate five sigma EDM signal.

The statistical sensitivity of the proton EDM experiment is given in eq. (7) with b~1.4. The values of the parameters are: E=10.5MV/m, A=0.6, P=0.8, $N=4\times10^{10}$p/cycle, $T_{tot}=10^7$s, and f=0.5%. This results to 1.4×10^{-29} e·cm/year, where 1 year is defined as 10^7s. The collaboration expects to run one year to study the systematics and set a sensitivity goal of better than $10^{-28}e$·cm, and another three years to collect more than enough statistics for $10^{-29}e$·cm sensitivity.

3.5.1 *Beam preparation.* The proton beam is to be prepared at the polarized proton source location that already exists and operates at BNL. The protons are then accelerated in the LINAC and injected into the accumulator for further acceleration (up to the kinetic energy of 233 MeV) and bunching. The beam polarization is to be kept in the vertical direction until the beam is injected into the EDM ring of ~250m circumference. In the EDM ring the beam is let to de-bunch and then re-bunched by adiabatically turning on an RF system with harmonic h=1. At this point an RF-solenoid is turned on as well to precess the beam polarization vector from the vertical direction into the horizontal direction resulting to two bunches with opposite polarization for polarimeter systematic error considerations. At this point another RF-system turns on with harmonic h=100 as needed to keep the proton spins in the forward direction for the duration of the storage. When the average spin precesses too long to the left or to the right of the momentum vector, it will be noticed by the polarimeter detector as a change in the (d-u)/(d+u), where u and d correspond to the up and down counters, and a feedback signal will be issued. Hence the longitudinal spin will be oscillating like a wind shield wiper, the opening angle of which will depend on the polarimeter statistics.

3.5.2 *Systematic errors.* The systematic errors of a typical EDM experiment include spurious magnetic fields and the so-called geometrical phase and are also present in the storage ring EDM experiments. Other effects include spin and beam dynamics resonances, which the collaboration plans to study both analytically and by using specially developed tracking software. The main systematic error for the proton at the magic momentum comes from spurious B-fields, especially a non-zero radial B-field.

A net radial B-field will induce a spin precession just like an EDM would. However, at the same time it would also split apart the counter-rotating beams. Sensitive BPMs are needed to be developed in order to probe that split. A net-radial B-field of 0.2pG could cause an EDM-like signal at the claimed sensitivity level. This magnetic field will cause a vertical split that depends on the vertical tune, i.e. the vertical betatron oscillation frequency divided by the revolution frequency as shown in equation below:

$$y=\sum_{N=0}^{\infty}\frac{\beta cR_0B_{rN}}{E_r\left(Q_y^2-N^2\right)}\cos\left(N\theta+\phi_N\right)=\frac{0.6\times 3\times 10^8(\mathrm{m/s})\times 40\mathrm{m}\times 0.2\mathrm{pG}}{10.5\mathrm{MV/m}\times\left(0.1^2\right)}=1\mathrm{pm} \quad (17)$$

where Q_y = 0.1 is the vertical tune, N is the radial magnetic field harmonic number around the ring and R_0 is the ring radius. The numerical example is an estimation for N=0, i.e. the DC component. For N=1 the vertical shift is 10^2 times less. One way to detect this split is to place magnetometers around the ring that can sense the magnetic field due to this offset. To facilitate this observation the vertical tune can be modulated by ~10% at a specific frequency, in a region where the magnetic field noise is relatively small.

The geometrical phase effect stems from the fact that spin rotations don't commute in three dimensions. When the particle spin rotates in, let's say, first quarter of the ring with respect to the vertical axis, then in the second quarter of the ring it rotates with respect to the longitudinal axis, and then continue with spin rotations with respect to the vertical and longitudinal axes but in the opposite directions than in the first half of the ring the total effect is not zero. Even though the total magnetic field, vertical and longitudinal in this case, would add to zero when integrated around the ring, their effect on the spin is a net rotation with respect to the third axis, radial in this case. The strength of this effect has been estimated by Y. Orlov to be proportional to the product of the two spin rotations [23], i.e. second order. As long as the N=1 of the average vertical and longitudinal B-field components are below 1μG, this effect is much less than the claimed sensitivity level. From eq. 17 it is clear that the sensitivity to the N=1 for the vertical B-field will be much greater than needed to eliminate this systematic error.

Other systematic errors related to the beam modulation are under consideration, but so far the collaboration hasn't found a show stopper. Special attention needs to be put on noise generated due to the RF and the quad modulation voltages around the ring, ground loops, etc. These kinds of systematic errors can be probed with the beam off and so they are easy to differentiate from the beam-induced signals. The collaboration had two technical reviews so far, one in December 2009 and one in March 2011. Both of them were great, producing a lot of good recommendations and, so far, no surprises. Every time the recommendation was to encourage the collaboration to continue this work. The collaboration is now eager to send the proposal to DOE.

4. Summary and Outlook

Clearly there have been major improvements in several fronts of searching for an EDM of a particular system. Two labs, BNL in NY and COSY in Jülich/Germany, are working towards proposals on the proton and deuteron EDM experiments respectively with high sensitivity. This decade promises to be the one that either observes or sets severe limits on popular extensions of the SM. Figure 7 shows that first we expect new results from the neutron and Hg systems. Since the CP-violating phases of physics beyond the SM are currently limited by the EDM experiments they may very well find a non-zero result already by 2015. It is expected that in the second part of the decade several new methods will start coming online and will be probing new territory. By the end of the decade the probability to discover a non-zero EDM value for one or several new systems is quite high.

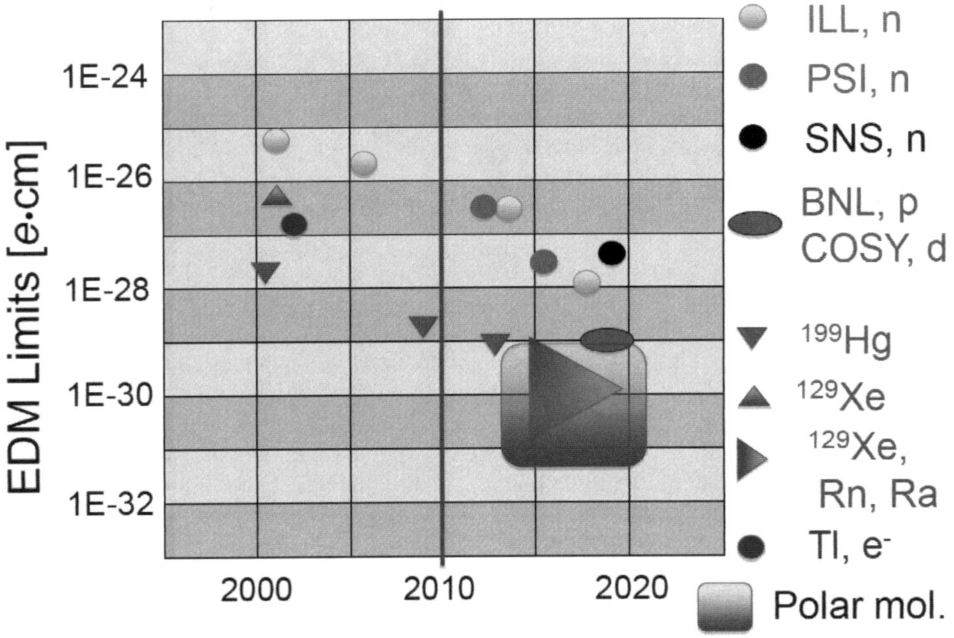

Figure 7. The past experimental sensitivity as well as the expected sensitivity of several experimental methods is shown here. The various electron EDM efforts have a similar goal as the Xe, Rn and Ra efforts and at approximately the same time period. The physics reach of the different systems varies. At present, the physics limits are dominated mostly by the ^{199}Hg, but also ^{205}Tl (at $1.6\times10^{-27}e\cdot$cm for the electron, see text) and neutron EDM results.

The physics sensitivity of Hg, and Xe suffers by a large factor of about 10^3, whereas the electron EDM is about a factor of ten less sensitive than the neutron EDM for the same nominal value due to the mass ratio of the bare quark to the electron mass. Nonetheless by the end of the decade it looks like all systems may be able to answer whether Electroweak Baryogenesis [2] plays a significant role in the baryon asymmetry of our universe (BAU). We can certainly look forward to such a noble goal.

References

[1] G.W. Bennett *et al.*, Phys. Rev. D**73**, 072003 (2006);
D. Stöckinger, Adv. Ser. On Dir. In HEP, V20, p. 393, eds. B Lee Roberts and W. Marciano, World Scientific, 2010.

[2] D.A. Demir *et al.*, Phys. Rev. D**67**, 015007 (2003);

O. Lebedev *et al.*, Phys. Rev. D**70**, 016003 (2004);
K. Blum *et al.*, JHEP **1005**, 101 (2005);
M.J. Ramsey-Musolf and S. Su, Phys. Rept. **456**, 1 (2008);
K.A. Olive *et al.*, Phys. Rev. D**72**, 075001 (2005).

[3] E.M. Purcell, and N.F. Ramsey, Phys. Rev. **78**, 807 (1950);
J.H. Smith, E.M. Purcell, and N.F. Ramsey, Phys. Rev. **108**, 120 (1957).

[4] L. Landau, Sov. Phys. JETP 5, 336 (1957).

[5] I.B. Khriplovich and S. K. Lamoreaux, *CP Violation Without Strangeness* (Springer, New York, 1997).

[6] A.D. Sakharov, JETP Lett. **5**, 24 (1967).

[7] J.H. Christenson, J. W. Cronin, V. L. Fitch, and R. Turlay, Phys. Rev. Lett. **13**, 138 (1964).

[8] M. Kobayashi and T. Maskawa, Prog. Theor. Phys. **49**, 652 (1973);
N. Cabibbo, Phys. Rev. Lett. **10**, 531 (1963).

[9] C.A. Baker *et al.*, Phys. Rev. Lett. **97**, 131801 (2006).

[10] R.D. Peccei and H.R. Quinn, Phys. Rev. Lett. **38**, 1440 (1977).

[11] S. Weinberg, Phys. Rev. Lett. **40**, 223 (1978);
F. Wilczek, Phys. Rev. Lett. **40**, 279 (1978).

[12] See http://nedm.web.psi.ch/index.htm

[13] R. Golub and S. Lamoreaux, Phys. Rep. **237**, 1 (1994).

[14] R. Golub and J.M. Pendlebury, Contemp. Phys. **13**, 519 (1972).

[15] L.I. Schiff, Phys. Rev. **132**, 2194 (1963);
P.G.H. Sandars, Phys. Lett. **14**, 194 (1965);
P.G.H. Sandars, J. Phys. B**1**, 499 (1968);
P.G.H. Sandars, J. Phys. B**1**, 511 (1968).

[16] E. Commins, J.D. Jackson, and D. DeMille, Am. J. Phys. **75**, 524 (2007).

[17] Z.W. Liu and H. P. Kelly, Phys. Rev. A **45**, R4210 (1992).

[18] B.C. Regan *et al.*, Phys. Rev. Latt. **88**, 071805 (2002).

[19] W.C. Griffith *et al.*, Phys. Rev. Lett. **102**, 101601 (2009).

[20] R.L. Gerwin and L. Lederman, Nuovo Cimento **11**, 776 (1959);
J. Bailey *et al.*, J. Phys. G**4**, 345 (1978).
D.F. Nelson *et al.*, Phys. Rev. Lett. **2**, 492 (1959).

[21] G.W. Bennett *et al.*, Phys. Rev. D**80**, 052008 (2009).

[22] Y.K. Semertzidis *et al.*, Proc. of AGS2000 workshop, BNL Spring 1996, conference web site: http://www.phy.bnl.gov/ags-2000/workshop.html, BNL report 52512, p. 67 available at http://www.osti.gov/energycitations/product.biblio.jsp?query_id=2&page=0&osti_id=383669;
Y.K. Semertzidis, Proc. of Workshop on Frontier Tests of Quant. Electrodynamics and Phys. of the Vacuum, 9-15 June 1998, Sandansky Bulgaria;
Y.K. Semertzidis *et al.*, Proc. of Intern. Workshop on High Intens. Muon Sources (HIMUS99), 1-4 Dec. 1999, KEK, Tsukuba, Japan;
Y.K. Semertzidis *et al.*, Int. J. Mod. Phys. A16S1B, 690 (2001);
F.J.M. Farley *et al.*, Phys. Rev. Lett. **93**, 052001 (2004);
Y.F. Orlov, W.M. Morse, and Y.K. Semertzidis, Phys. Rev. Lett. **96**, 214802 (2006).

[23] D. Anastassopoulos *et al.*, Search for a permanent electric dipole moment of the deuteron nucleus at the $10^{-29}e\cdot$cm level, AGS Proposal, April 2008, available from http://www.bnl.gov/edm/

[24] The proposal to DOE for a proton EDM at the $10^{-29}e\cdot$cm level is well underway and scheduled for submission within the next two months. An early version of this proposal can be found at http://www.bnl.gov/edm/review/RDplan.asp

Symposium on Prospects in the Physics of Discrete Symmetries IOP Publishing
Journal of Physics: Conference Series **335** (2011) 012013 doi:10.1088/1742-6596/335/1/012013

Beyond the standard MSSM

Riccardo Barbieri

Scuola Normale Superiore and INFN, Piazza dei Cavalieri 7, 56126 Pisa, Italy

Abstract. If supersymmetry is relevant at the Fermi scale, the lack of any direct signal so far may require going beyond the Minimal Supersymmetric Standard Model. In this talk I briefly summarize a simple and concrete extension of the MSSM that takes these issues, including a way to address the flavour and the CP problems. In its fully natural range of parameters, the expected signals for LHC, dark matter and flavour physics are clear and generally quite different from the ones of the MSSM. Gauge coupling unification may be only approximate.

1. Introduction

The Fermi scale may either find its origin in some weak coupling physics, like in the Standard Model (SM) itself with a not too heavy Higgs boson, or in some new strong force. The first case is currently preferred by indirect data, like the ElectroWeak Precision Tests (EWPT), or even by gauge coupling unification. Supersymmetry is the leading candidate for the weak coupling picture of ElectroWeak Symmetry Breaking (EWSB).

This last statement should however be contrasted with the lack of any direct signal so far. With reference with the expectations of the Minimal Supersymmetric Standard Model (MSSM), these unseen signals are: i) no Higgs boson, ii) no s-particles, iii) no clear deviation from the SM in flavour and in CP physics. This may either mean that such direct signals are there, so to say around the corner, just waiting to be discovered or, on the contrary, that weak-scale supersymmetry is not realized in nature. With LHC ready to produce data with a significant luminosity, it looks unreasonable to take now a strong position on this alternative. Rather I find more useful to give consideration to a third possibility: Does a simple and concrete extension exist of the standard MSSM that can explain the lack of signals so far? This is not a new question. My view on it is based on the following two points:

- The lack of Higgs boson and of s-particle signals may be related issues: if the Higgs boson(s) can be made heavier than in the standard MSSM, this generally relaxes in a corresponding way the naturalness bounds on all the s-particle masses;
- The suppression of any new source of flavour and CP violation may have something to do with the heaviness of all the s-fermions not coupled via the top Yukawa coupling to the Higgs system, i.e. to the relative lightness of the two stops and the left-handed sbottom only.

Although independently valid, naturalness hints at a relation between these two points. I maintain that there is a simple extension of the MSSM that realizes these points in an overall coherent way, with clear and generally quite different signatures from the ones of the standard MSSM. The key feature is summarized by the typical s-particle spectrum shown in Fig. 1, up to details to be specified below [1].

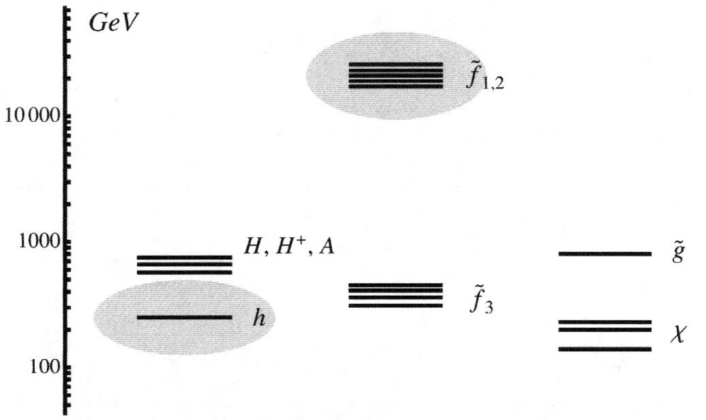

Figure 1. A representative Non Standard Supersymmetric Spectrum with $m_h = 200 \div 300$ GeV and $m_{\tilde{f}_{1,2}} \gtrsim 10$ TeV.

2. Supersymmetry without a light Higgs boson

Ways have been examined to evade the tree level upper bound on the mass of the lightest Higgs boson, valid in the MSSM, $m_h < m_Z \cos 2\beta$. They rest on a modification of the Higgs potential either though a D-term or though an F-term. The first requires an extension of the gauge group, necessarily also acting on all the matter supermultiplets via a sizable new gauge coupling [2, 3, 4]. The second may involve an extension of the Higgs system only, the simplest example being represented by the supersymmetric Yukawa coupling of the NMSSM, $\lambda \hat{S} \hat{H}_u \hat{H}_d$ [5, 6, 7, 8]. As well known, in the last case the upper bound on the mass of lightest scalar gets modified to

$$m_h \leq m_Z (\cos^2 2\beta + \frac{2\lambda^2}{g^2 + g'^2} \sin^2 2\beta)^{1/2}. \quad (1)$$

If one wants m_h increased by a significant amount, say between 200 and 300 GeV, λ cannot be too small, between 1.5 and 2. On the other hand I do not find useful to consider raising the Higgs boson mass by an effective Lagrangian approach, since the new mass scale involved has to be close to the Fermi scale, which in turn obscures the weak coupling picture of EWSB and threatens the control of the EWPT.

To me the NMSSM with a largish λ, thus dubbed λSUSY, emerges as the simplest concrete possibility to raise significantly the Higgs boson mass. It can be shown that the upper bound on the Higgs mass set in the SM by the EWPT is automatically evaded in λSUSY .Furthermore, as anticipated on general grounds, the naturalness bounds on the s-particle masses are indeed relaxed by a significant amount: with $\lambda = 2$ and a Higgs boson as heavy as 250 GeV the gluino can weight up to 1.5 TeV with less then 20% fine tuning [8].

There is a price to pay, however, which explains the general consideration of the NMSSM only for $\lambda \lesssim 0.7$, i.e. not λSUSY[1]. A coupling λ bigger than about 0.7 at the weak scale starts blowing up at a scale Λ below unification, unless some change of regime occurs before. Fig. 2 shows the upper bound on the Higgs boson mass that corresponds to $\lambda^2(\Lambda)$ being less than 4π [1]. Canonical gauge coupling unification is therefore under threat. The question then arises: Is this an unbearable step backward? I do not think so, even apart from the relevance of the issues that λSUSY addresses and potentially solves. It is conceivable that the three gauge couplings cross the threshold of a blowing interaction or even of a change in the relevant degrees

[1] See [10] and references therein

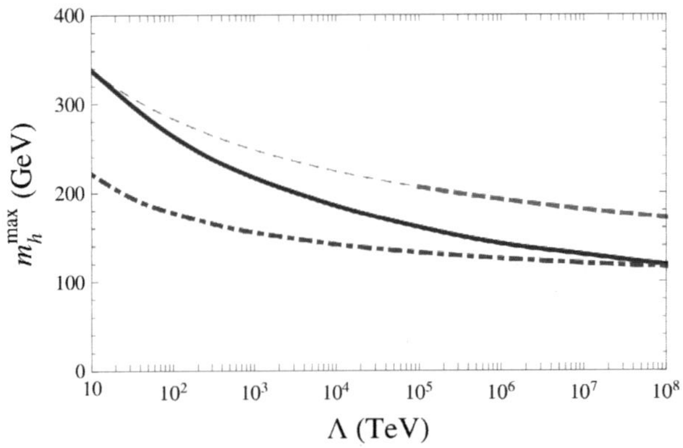

Figure 2. Upper bounds on m_h as function of the scale Λ where some coupling starts becoming semi-pertubative, $g_x^2, g_I^2, \lambda^2 = 4\pi$ for the $U(1)$ case (dotdashed, $\tan\beta >> 1$), λSUSY (solid, $\tan\beta = 1$) and $SU(2)$ (dashed, $\tan\beta >> 1$). In the $SU(2)$ case values of $m_h \gtrsim 270$ GeV are hardly compatible with naturalness and the EWPT [9].

of freedom without a significant change of their values, like it happens in the case of the fine structure constant going across the QCD scale. Furthermore I give weight to the fact that at $10^2 \div 10^3$ TeV, a scale at which λ could still be semiperturbative with a Higgs boson above 200 GeV, the three gauge couplings are already all well within a factor of 2 from each other: $g_1 \approx 0.5, g_2 \approx 0.65$ and $g_3 \approx 0.9$, achieving a kind of *approximate* uniification. *Precise* unification is definitely appealing, but it may be not the right story.

3. Addressing the flavour and the CP problems

In view of the current constraints, the idea put forward long ago [11, 12, 13, 14, 15, 16] that the heaviness of the first two generations of sfermions may explain the lack of signals from the plethora of potential new sources of flavour and CP violation present in the general MSSM is, *per se*, largely insufficient. If anything, some extra symmetry, compatible with the hierarchy in the sfermion mass spectrum, has to be operative to explain the supersymmetric flavour and CP problems.

In [17] we propose that:

- Only the sfermions that interact with the Higgs system via the top Yukawa coupling are significantly lighter than the others.
- With only the up-Yukawa couplings, Y_u, turned on, but not the down-Yukawa couplings, $Y_d = 0$, there is no flavour transition between the different families, as in the SM case.

This follows in particular from assuming for $Y_u \neq 0$ but $Y_d = 0$ that the quark sector has a flavour symmetry

$$U(1)_{\tilde{B}_1} \times U(1)_{\tilde{B}_2} \times U(1)_{\tilde{B}_3} \times U(3)_{d_R}, \tag{2}$$

where $\tilde{B}_i$ acts as baryon number but only on the supermultiplets $\hat{Q}_i$ and $\hat{u}_{R_i}$ of the i-th generation, whereas $U(3)_{d_R}$ acts on the three right-handed supermultiplets of charge 1/3. If one further assumes that, when Y_d is switched on, it behaves as a non-dynamical spurion field that leaves the standard Yukawa couplings invariant under (2), it is easy to see that the CKM matrix V controls every flavour changing transition, in particular in the gaugino interactions with matter. I do not know of any other symmetry pattern that achieves this, while still be consistent with light stops and left-handed sbottom only.

In this way all flavour and CP violating effects induced by loops of s-particle exchanges are under control. A precise examination shows that the main contraints arise from a mixed light-heavy exchange contribution to the ϵ_K parameter, controlled by the standard CKM phase, and from the one loop corrections to the electron and the neutron Electric Dipole Moments in presence of sizable flavour blind phases [18]. They require that the sfermions not coupled to the top be above $5 \div 10$ TeV, depending on the values of the other parameters. In turn such values are consistent with a low fine tuning in λSUSY, while they would not be in the MSSM[1].

With these values for the heavy sfermion masses, one remains with effective Minimal Flavour Violation and a moderate value of $\tan\beta$, as required by (1) to have a heavy enough m_h. The important residual signals are therefore related to the existence of sizable flavour blind phases and are expected to show up in EDMs at two loops or in CP asymmetries in B-physics, all mediated by the exchanges of the relatively light stops and left-handed sbottom [18]. It is noticeable that the supersymmetric flavour blind phases are precisely the sources of the unsolved CP problem in standard Minimal Flavour Violation based on $U(3)_Q \times U(3)_{u_R} \times U(3)_{d_R}$, i.e. with quasi degenerate squarks.

4. Not the standard MSSM signals

The picture outlined above and essentially represented in Fig. 2 has clear and distinctive phenomenological implications. As to the direct signals of supersymmetry, gluino pair production and subsequent decays into exclusive third generation quarks is the dominant feature. These final states may either result from chain decays with intermediate stops/sbottoms or from direct 3-body decays of the gluino. They are simply characterized by the gluino and the LSP mass, $m_{\tilde{g}}$ and m_χ, and by effective Branching Ratios for the final states

$$\tilde{g} \to t\bar{t}\chi, \ \tilde{g} \to b\bar{b}\chi, \ \tilde{g} \to t\bar{b}\chi^-(\bar{t}b\chi^+), \tag{3}$$

adding up to unity, with the last decay also followed by $\chi^\pm$ decays into χ plus a real or virtual $W^\pm$, depending on phase space[19]. With 5 fb^{-1} of integrated luminosity the LHC at 7 TeV has a significant potential for discovery of these signals with gluino masses up to 1 TeV, at least in the most favourable cases for the other parameters [20]. As mentioned above, however, the extended naturalness region of λSUSY, with gluinos above 1 TeV, may hide the direct supersymmetric signals to the LHC in its first phase.

This brings the focus on the search for the Higgs boson(s), with in mind the projected discovery potential of the SM Higgs boson at the LHC in the next year or so. Discovering a Higgs boson with mass above 200 GeV would be in contradiction with the indication of the EWPT in the SM and definitely incompatible with the MSSM. In turn, that this be possible or even necessary in λSUSY depends on the gluon-gluon-h coupling and on the h decay branching ratios, with h the lightest scalar. The studies so far show that the g-g-h coupling, hence the production cross, is very close to the SM one for the same Higgs boson mass [21, 22]. On the contrary the decay branching ratios are dependent on the composition of the h boson, which may acquire a significant $SU(2)$-singlet component. It is in particular possible that the extra singlet pseudoscalar, a, be light enough, say around 100 GeV, to permit the decay $h \to aa$, with a in turn decaying to a $b\bar{b}$ or $\tau\bar{\tau}$ pair [22]. A thorough analysis of the h branching ratios in the general parameter space of λSUSY may be worthwhile.

Finally it is worth to mention the impact of λSUSY on the interpretation of Dark Matter in terms of neutralino LSPs and the related searches. Other than the fact that the neutralino LSP can have a significant singlet component, a Higgs boson with a mass above 200 GeV produces significant differences with respect to the MSSM picture. The first obvious one is the reduction of the neutralino-nucleous cross section, inversely proportional to the forth power of m_h, of relevance to the direct searches underway. The other has to do with the relic abundance of thermally produced neutralinos in the early universe: the s-channel exchange of a h heavier than about 200 GeV in the neutralino annihilation cross section greatly influences the relic abundance of LSPs with mass in the 100 GeV range [1].

In conclusion, while the MSSM remains a crucial benchmark for the search of supersymmmetry signals, it is important to keep in mind possible alternatives like the ones discussed above.

Acknowledgements
I thank Lawrence Hall, Slava Rychkov, Yasunori Nomura, Roberto Franceschini, Duccio Pappadopulo, Enrico Bertuzzo, Paolo Lodone, Marco Farina, David Straub, Gino Isidori and Andrea Romanino for many useful discussions. This work was supported by the EU ITN "Unification in the LHC Era", contract PITN-GA-2009-237920 (UNILHC) and by MIUR under contract 2008XM9HLM.

References

[1] R. Barbieri, E. Bertuzzo, M. Farina, P. Lodone, and D. Pappadopulo, "A Non Standard Supersymmetric Spectrum," *JHEP* **08** (2010) 024, arXiv:1004.2256 [hep-ph].
[2] P. Batra, A. Delgado, D. E. Kaplan, and T. M. P. Tait, "The Higgs mass bound in gauge extensions of the minimal supersymmetric standard model," *JHEP* **02** (2004) 043, arXiv:hep-ph/0309149.
[3] P. Batra, A. Delgado, D. E. Kaplan, and T. M. P. Tait, "Running into new territory in SUSY parameter space," *JHEP* **06** (2004) 032, arXiv:hep-ph/0404251.
[4] A. Maloney, A. Pierce, and J. G. Wacker, "D-terms, unification, and the Higgs mass," *JHEP* **06** (2006) 034, arXiv:hep-ph/0409127.
[5] R. Harnik, G. D. Kribs, D. T. Larson, and H. Murayama, "The minimal supersymmetric fat Higgs model," *Phys. Rev.* **D70** (2004) 015002, arXiv:hep-ph/0311349.
[6] S. Chang, C. Kilic, and R. Mahbubani, "The new fat Higgs: Slimmer and more attractive," *Phys. Rev.* **D71** (2005) 015003, arXiv:hep-ph/0405267.
[7] A. Delgado and T. M. P. Tait, "A fat Higgs with a fat top," *JHEP* **07** (2005) 023, arXiv:hep-ph/0504224.
[8] R. Barbieri, L. J. Hall, Y. Nomura, and V. S. Rychkov, "Supersymmetry without a Light Higgs Boson," *Phys. Rev.* **D75** (2007) 035007, arXiv:hep-ph/0607332.
[9] P. Lodone, "Naturalness bounds in extensions of the MSSM without a light Higgs boson," *JHEP* **05** (2010) 068, arXiv:1004.1271 [hep-ph].
[10] U. Ellwanger, C. Hugonie, and A. M. Teixeira, "The Next-to-Minimal Supersymmetric Standard Model," *Phys. Rept.* **496** (2010) 1–77, arXiv:0910.1785 [hep-ph].
[11] M. Dine, R. G. Leigh, and A. Kagan, "Flavor symmetries and the problem of squark degeneracy," *Phys. Rev.* **D48** (1993) 4269–4274, arXiv:hep-ph/9304299.
[12] P. Pouliot and N. Seiberg, "(S)quark masses and nonAbelian horizontal symmetries," *Phys. Lett.* **B318** (1993) 169–173, arXiv:hep-ph/9308363.
[13] A. Pomarol and D. Tommasini, "Horizontal symmetries for the supersymmetric flavor problem," *Nucl. Phys.* **B466** (1996) 3–24, arXiv:hep-ph/9507462.
[14] R. Barbieri, G. R. Dvali, and L. J. Hall, "Predictions From A U(2) Flavour Symmetry In Supersymmetric Theories," *Phys. Lett.* **B377** (1996) 76–82, arXiv:hep-ph/9512388.
[15] A. G. Cohen, D. B. Kaplan, and A. E. Nelson, "The more minimal supersymmetric standard model," *Phys. Lett.* **B388** (1996) 588–598, arXiv:hep-ph/9607394.
[16] S. Dimopoulos and G. F. Giudice, "Naturalness constraints in supersymmetric theories with nonuniversal soft terms," *Phys. Lett.* **B357** (1995) 573–578, arXiv:hep-ph/9507282.
[17] R. Barbieri, E. Bertuzzo, M. Farina, P. Lodone, and D. Zhuridov, "Minimal Flavour Violation with hierarchical squark masses," arXiv:1011.0730 [hep-ph].
[18] R. Barbieri, P. Lodone, and D. M. Straub, "CP Violation in Supersymmetry with Effective Minimal Flavour Violation," arXiv:1102.0726 [hep-ph].
[19] R. Barbieri and D. Pappadopulo, "S-particles at their naturalness limits," *JHEP* **10** (2009) 061, arXiv:0906.4546 [hep-ph].
[20] G. L. Kane, E. Kuflik, R. Lu, and L.-T. Wang, "Top Channel for Early SUSY Discovery at the LHC," arXiv:1101.1963 [hep-ph].

[21] L. Cavicchia, R. Franceschini, and V. S. Rychkov, "Supersymmetry without a light Higgs boson at the LHC," *Phys. Rev.* **D77** (2008) 055006, arXiv:0710.5750 [hep-ph].
[22] R. Franceschini and S. Gori, "Solving the mu problem with a heavy Higgs boson," arXiv:1005.1070 [hep-ph].

Theory prediction of ϵ_K

Joachim Brod

Excellence Cluster Universe, Technische Universität München
85748 Garching, Germany

E-mail: joachim.brod@ph.tum.de

Abstract. The parameter ϵ_K plays an important role in the phenomenology of the standard model and its extensions. It is very sensitive to high-energy scales and can be predicted with remarkable precision. I review the theory prediction of ϵ_K in the standard model, with an emphasis on the theoretical and parametrical uncertainties. In addition, I present the results of our recent NNLO QCD calculation of η_{ct}, the charm-top-quark contribution to the effective $|\Delta S| = 2$ Hamiltonian, leading to $|\epsilon_K| = (1.90 \pm 0.26) \times 10^{-3}$.

1. Introduction

Neutral Kaon mixing proceeds via the quark-level flavour-changing neutral current (FCNC) $s - d$ transition. It is forbidden in the standard model of particle physics (SM) at tree level and induced by higher-order weak interactions, and therefore naturally sensitive to short-distance physics. The parameter ϵ_K describes indirect CP violation in the neutral Kaon system and is dominated by virtual top quarks (see Fig. 1). The coupling of the top quark to W bosons is proportional to a small Cabibbo-Kobayashi-Maskawa (CKM) matrix element, hence ϵ_K can set very stringent bounds in particular on models of "new physics" (NP) which deviate from the CKM flavour structure of the SM, i.e. which go beyond minimal flavour violation. In order to exploit this potential, a reliable theory prediction is mandatory.

After a general introduction to indirect CP violation in the neutral Kaon system in Sec. 2, we discuss the short-distance (Sec. 3) and long-distance (Sec. 4) contributions to ϵ_K in some detail. Sec. 5 contains the numerical prediction for ϵ_K using our NNLO value for η_{ct}.

2. Indirect CP violation in the neutral Kaon system

The time evolution[1] of the strangeness eigenstates $|K^0(t)\rangle$ and $|\overline{K}^0(t)\rangle$, where $|K^0(0)\rangle = |K^0\rangle$ and $|\overline{K}^0(0)\rangle = |\overline{K}^0\rangle$, is given in the Wigner-Weisskopf approximation (see Ref. [2]) by

$$i\frac{d}{dt}\begin{pmatrix}|K^0(t)\rangle \\ |\overline{K}^0(t)\rangle\end{pmatrix} = \left[\begin{pmatrix}M_{11} & M_{12} \\ M_{12}^* & M_{11}\end{pmatrix} - \frac{i}{2}\begin{pmatrix}\Gamma_{11} & \Gamma_{12} \\ \Gamma_{12}^* & \Gamma_{11}\end{pmatrix}\right]\begin{pmatrix}|K^0(t)\rangle \\ |\overline{K}^0(t)\rangle\end{pmatrix}. \tag{1}$$

The diagonal entries of the Hermitian, time-independent mass matrix M and decay matrix Γ are equal as a result of CPT invariance, and the off-diagonal entries are induced by the $|\Delta S| = 2$ transition. Eq. (1) is diagonal in the basis of mass eigenstates

$$|K_S\rangle = p|K^0\rangle + q|\overline{K}^0\rangle \quad \text{and} \quad |K_L\rangle = p|K^0\rangle - q|\overline{K}^0\rangle. \tag{2}$$

[1] The discussion in this section adopts the conventions of Ref. [1], where also some missing details can be found.

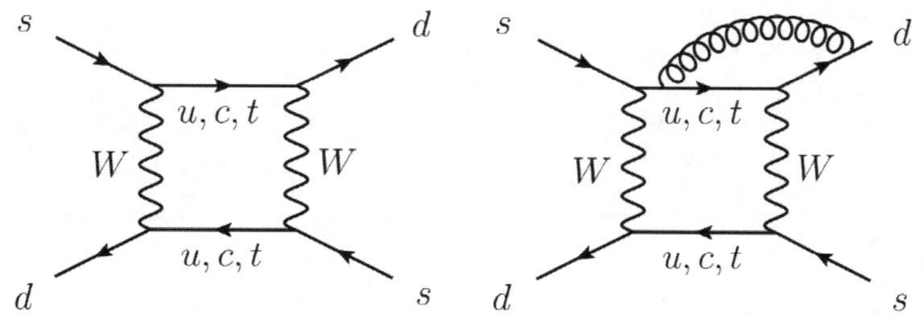

Figure 1. LO (left) and NLO (right) box diagrams inducing the $|\Delta S| = 2$ transition in the SM.

Their mass and width difference is given by $\Delta m_K = M_L - M_S$ and $\Delta\Gamma_K = \Gamma_S - \Gamma_L$, where M_L (M_S) and Γ_L (Γ_S) are the mass and the decay width of the longer-(shorter-)lived mass eigenstate.

We now define ϵ_K by

$$\epsilon_K = \frac{\eta_{00} + 2\eta_{+-}}{3} \tag{3}$$

in terms of the experimentally measured ratios $\eta_{ij} = \langle \pi^i\pi^j|K_L\rangle / \langle \pi^i\pi^j|K_S\rangle$. It can be expressed as the ratio of the decay amplitudes of a K_L and a K_S into an isospin-zero eigenstate,

$$\epsilon_K = \frac{\langle(\pi\pi)_{I=0}|K_L\rangle}{\langle(\pi\pi)_{I=0}|K_S\rangle}\left(1 + \mathcal{O}\left(\frac{A_2^2}{A_0^2}\right)\right) \tag{4}$$

up to corrections proportional to the second power of the ratio $|A_2|/|A_0|$, where $A_I = \langle(\pi\pi)_I|K^0\rangle$. The subscript I denotes the isospin of the state. The $\Delta I = 1/2$ rule implies that this ratio is small, $|A_2|/|A_0| \approx 1/22$.

The parameter ϵ_K vanishes in case of exact CP symmetry. As a consequence of Watson's theorem [3], each amplitude A_I is characterised by a single strong phase. This phase drops out in the ratio (4), hence there is no direct CP violation in ϵ_K.

Using the explicit expressions for p and q in Eq. (4) leads to [1]

$$\epsilon_K = e^{i\phi_\epsilon} \sin\phi_\epsilon \left(\frac{\mathrm{Im}M_{12}}{\Delta m_K} + \xi\right), \tag{5}$$

where $\phi_\epsilon = \arctan(2\Delta m_K/\Delta\Gamma_K) \approx 45°$, and $\xi = \mathrm{Im}A_0/\mathrm{Re}A_0$ is a correction of $\mathcal{O}(5\%)$. We have neglected terms proportional to the second power of the small quantities $\phi = \arg(-M_{12}/\Gamma_{12})$ and $|A_2|/|A_0|$, and to the first power in Γ_L/Γ_S. Taking Δm_K and ϕ_ϵ from experiment, we obtain a theory prediction for ϵ_K by computing $\mathrm{Im}M_{12}$ and ξ.

To this end, we make use of the effective field theory formalism. In this framework we can systematically separate short-distance from long-distance scales. The information about short-distance scales is contained in the Wilson coefficients of the weak effective Hamiltonian; they can in principle be calculated to any desired precision within perturbation theory. The main contributions of long-distance scales are contained in the hadronic matrix element of the effective Hamiltonian, and are governed by low-energy, non-perturbative QCD. They can be evaluated reliably by means of lattice QCD.

The matrix element M_{12} is given in terms of the effective Hamiltonian by $2\Delta m_K M_{12} = \langle \bar{K}^0|\mathcal{H}^{|\Delta S|=2}|K^0\rangle$ up to a small correction which will be discussed, together with the parameter ξ, in Sec. 4.

3. Short-distance contributions to ϵ_K

The effective Hamiltonian below the charm-quark scale, inducing the $|\Delta S| = 2$ transitions in the SM, is given by

$$\mathcal{H}^{\Delta S=2} = \frac{G_F^2}{4\pi^2} M_W^2 \left[\lambda_c^2 \eta_{cc} S(x_c) + \lambda_t^2 \eta_{tt} S(x_t) + 2\lambda_c \lambda_t \eta_{ct} S(x_c, x_t)\right] b(\mu) \tilde{Q}_{S2} + \text{H.c.} + \dots , \quad (6)$$

where G_F is the Fermi constant, M_W the W-boson mass, and $x_i = m_i^2/M_W^2$ (m_i being the quark masses). The factor $b(\mu)$ contains the remaining scale dependence, which cancels against the corresponding scale dependence of the hadronic matrix elements (see Sec. 4). $\tilde{Q}_{S2} = (\bar{s}_L \gamma_\mu d_L) \otimes (\bar{s}_L \gamma^\mu d_L)$ is the leading local four-quark operator that induces the $|\Delta S| = 2$ transition, defined in terms of the left-handed s- and d-quark fields. Higher-dimensional operators are estimated to contribute less than 1% to ImM_{12} [4].

The short-distance contributions are contained in the Wilson coefficient of $\tilde{Q}_{S2}$, which is induced in the SM at LO by the box diagram of Fig. 1 (left side). The couplings of the W bosons to the fermions are proportional to the CKM matrix elements, collected in the parameters $\lambda_i = V_{is}^* V_{id}$. Using the unitarity relation $\lambda_u = -\lambda_c - \lambda_t$, the effective Hamiltonian $\mathcal{H}^{|\Delta S|=2}$ can be split into three terms as in Eq. (6). The loop functions S, proportional to the squares of the quark masses due to the Glashow-Iliopoulos-Maiani (GIM) mechanism, denote the contribution of the LO SM box diagrams. The coefficients η_{ij} comprise the higher-order QCD corrections.

The calculation of ϵ_K involves the imaginary part of M_{12}; accordingly, the interplay between Im$\lambda_i \lambda_j$ and the loop functions S determines the relative size of the individual contributions. In contrast to B-meson mixing, where only the top-quark contribution is relevant, the smallness of λ_t^2 lets all three contributions compete in size in the case of Kaon mixing. We now discuss them in turn.

The term proportional to λ_t^2 is dominant and contributes approximately +72% to $\mathcal{H}^{|\Delta S|=2}$. The QCD corrections can be computed by a fixed-order matching calculation at the top-quark scale; there is no large logarithm. A next-to-leading-order (NLO) QCD calculation yields $\eta_{tt} = 0.5765(65)$ [5], with a small remaining theory uncertainty of about 1%.

The smallest contribution of about −14% is proportional to λ_c^2. The GIM mechanism causes the absence of a large logarithm $\log(m_c/M_W)$. A matching calculation at the charm-quark scale at NLO QCD yields $\eta_{cc} = 1.43(23)$ [6]. The relatively large remaining scale dependence is mitigated by the absolute smallness of this contribution.

The term proportional to $\lambda_c \lambda_t$ contributes roughly +42% to the effective Hamiltonian. It contains a large logarithm $\log(m_c/M_W)$, and the theoretical prediction requires a full renormalisation group analysis with double insertions of $|\Delta S| = 1$ operators (see Fig. 2). We have performed a next-to-next-to-leading-order (NNLO) QCD calculation with the result $\eta_{ct} = 0.496(47)$ [7].

In order to obtain a reliable theory prediction it is essential to estimate the theory uncertainty. The latter is a consequence of the truncation of the perturbation series. Terms proportional to powers of $\alpha_s \log(\mu/m_c)$, where μ is an arbitrary, unphysical renormalisation scale, are summed to all orders by the renormalisation group, but only incompletely cancelled in the matching calculation. η_{ct} is a scale- and scheme-independent quantity and therefore μ-independent if calculated to all orders. The residual μ-dependence gives us a handle on neglected higher-order terms. The large NNLO anomalous dimensions and an accidental cancellation of corresponding terms at leading order partly explain the still sizeable remaining scale dependence of the NNLO result.

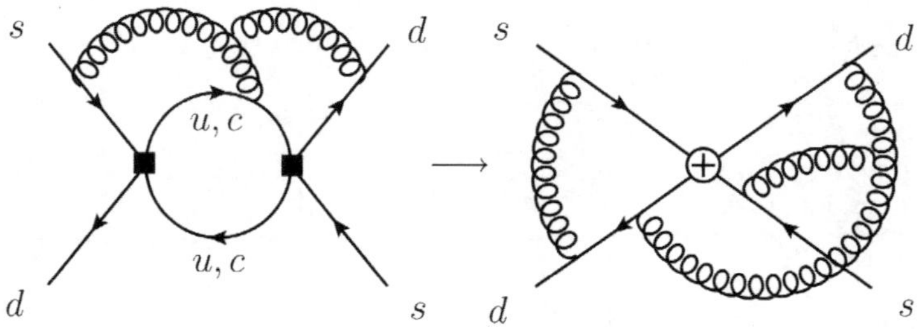

Figure 2. SM diagrams contributing to the mixing of $|\Delta S| = 1$ operators into Q_{S2} at NNLO QCD.

4. Long-distance contributions to ϵ_K

The hadronic matrix elements of the operator $\tilde{Q}_{S2}$ include the major part of the long-distance contributions to ϵ_K. They are parameterised by the bag factor

$$\hat{B}_K = \frac{3}{2} b(\mu) \frac{\langle \bar{K}^0 | \tilde{Q}_{S2} | K^0 \rangle}{f_K^2 M_K^2}, \tag{7}$$

where f_K is the Kaon decay constant, and $b(\mu)$ is factored out of Eq. (6) in such a way that $\hat{B}_K$ is a renormalisation-group invariant quantity. The determination of the bag factor was the main impediment to an accurate prediction of ϵ_K until a couple of years ago.

Nowadays, the most precise predictions for $\hat{B}_K$ come from lattice QCD calculations, where the chiral and continuum extrapolation as well as finite volume effects are now well under control [8]. A fully unquenched calculation results in $B_K = 0.725(26)$ [9], and the authors identify the transition from the non-perturbative renormalisation scheme on the lattice to the $\overline{\text{MS}}$ scheme, which is known only at one-loop, as the main source of the uncertainty. This value is consistent with the upper bound $\hat{B}_K \leq 0.75$, derived in the framework of large-N QCD [10].

There are additional long-distance contributions which are not contained in $\hat{B}_K$, proportional to the dispersive and absorptive parts of the amplitude $\int d^4x \langle \bar{K}^0 | \mathcal{H}^{\Delta S=1}(x) \mathcal{H}^{\Delta S=1}(0) | K^0 \rangle$, respectively. The parameter ξ comprises the absorptive part and has been estimated by relating it to the ratio ϵ'_K/ϵ_K [1, 11][2], yielding a -6% correction to ϵ_K. ImM_{12} receives contributions from the dispersive part of the amplitude. These have been estimated using chiral perturbation theory in [12] and enhance ϵ_K by 2.4%. The two contributions are combined together with the experimental value of ϕ_ϵ into the phenomenological parameter $\kappa_\epsilon = 0.94(2)$, which multiplies the expression (5), in which we then set $\xi = 0$ and $\phi_\epsilon = 45°$.

5. Numerics and input parameters

The parameter ϵ_K is measured with high accuracy: the value quoted by the Particle Data Group is $\epsilon_K = (2.228 \pm 0.011) \times 10^{-3} \times e^{i(43.5\pm0.7)°}$ [13]. In order to compare this with the theoretical prediction, we compute ϵ_K using the following formula[3] [18, 11],

$$|\epsilon_K| = \kappa_\epsilon C_\epsilon \hat{B}_K |V_{cb}|^2 \lambda^2 \bar{\eta} \big[|V_{cb}|^2 (1-\bar{\rho}) \eta_{tt} S(x_t) + \eta_{ct} S(x_c, x_t) - \eta_{cc} S(x_c) \big], \tag{8}$$

where

$$C_\epsilon = \frac{G_F^2 F_K^2 M_{K^0} M_W^2}{6\sqrt{2}\pi^2 \Delta M_K}. \tag{9}$$

[2] $\epsilon'_K = (\eta_{+-} - \eta_{00})/3$ measures direct CP violation in the neutral Kaon system.
[3] A term proportional to Reλ_t/Re$\lambda_c = \mathcal{O}(\lambda^4)$ has been neglected in Eq. (8) (see Ref. [18]).

Table 1. Input parameters used in our numerical analysis.

Parameter	Value	Ref.	Parameter	Value	Ref.
M_W	80.399(23) GeV	[13]	$\sin 2\beta$	0.671(23)	[13]
$m_t(m_t)$	163.7(1.1) GeV	[14]	F_K	156.1(8) MeV	[15]
$m_b(m_b)$	4.163(16) GeV	[16]	G_F	$1.166\,367(5) \times 10^{-5}$ GeV^{-2}	[13]
$m_c(m_c)$	1.279(13) GeV	[16]	λ	0.2255(7)	[15]
M_K	497.614(24) MeV	[13]	$\|V_{cb}\|$	$4.06(13) \times 10^{-2}$	[13]
κ_ϵ	0.94(2)	[12]	M_{B_d}	5.2795(3) GeV	[13]
ΔM_K	5.292(9)/ns	[13]	M_{B_s}	5.3663(6) GeV	[13]
ΔM_d	0.507(5)/ps	[13]	ΔM_s	17.77(12)/ps	[13]
ξ_s	1.243(28)	[17]	η_{tt}	0.5765(65)	[5]
$\hat{B}_K$	0.725(26)	[17]	η_{cc}	1.43(23)	[6]

We write $\bar{\eta} = R_t \sin\beta$ and $1 - \bar{\rho} = R_t \cos\beta$, where R_t is given by

$$R_t \approx \frac{\xi_s}{\lambda}\sqrt{\frac{M_{B_s}}{M_{B_d}}}\sqrt{\frac{\Delta M_d}{\Delta M_s}} \tag{10}$$

and $\xi_s = (F_{B_s}\sqrt{\hat{B}_s})/(F_{B_d}\sqrt{\hat{B}_d})$ is a ratio of B-meson decay constants and bag factors that can be computed on the lattice with high precision [17]. Using the NNLO value $\eta_{ct} = 0.496(47)$ and the numerical values given in Tab. 1, we obtain

$$|\epsilon_K| = (1.90 \pm 0.04_{\eta_{cc}} \pm 0.02_{\eta_{tt}} \pm 0.07_{\eta_{ct}} \pm 0.11_{\text{LD}} \pm 0.22_{\text{parametric}}) \times 10^{-3}\,, \tag{11}$$

where all errors have been added in quadrature.

The first three errors correspond to η_{cc}, η_{tt}, η_{ct}, respectively. While the error on η_{ct} is still dominating, a NNLO QCD calculation could reduce the error on η_{cc}.

The error indicated by LD originates from the long-distance contributions, namely ξ_s, $\hat{B}_K$, and κ_ϵ, which account for 40%, 37%, and 22% of the long-distance error, respectively. As mentioned before, the error on $\hat{B}_K$ could be reduced significantly by calculating higher order corrections to the perturbative matching to the $\overline{\text{MS}}$ scheme.

The main share of the parametric error stems from $|V_{cb}|$ (59%) and $\sin(2\beta)$ (19%), while all other contributions are well below 10%. The dominance of $|V_{cb}|$ can easily be understood from Eq. (8): we see that the dominant top-quark contribution is proportional to the fourth power of $|V_{cb}|$. At the same time, the experimental error on $|V_{cb}|$ is quite large, not least because of the two only marginally consistent values obtained from inclusive and exclusive semileptonic B-meson decays, respectively [13]. The clarification of this issue would certainly be most helpful.

6. Conclusion

The parameter ϵ_k is an important precision observable: it is very sensitive to high-energy scales and can be computed reliably. In this talk I reviewed the status of the SM prediction for ϵ_K. The recent progress in the calculation of the long- and short-distance contributions results in the numerical value $|\epsilon_K| = (1.90 \pm 0.26) \times 10^{-3}$.

Acknowledgments

I thank M. Gorbahn for the fruitful and pleasant collaboration and the organizers of DISCRETE 2010 for the interesting and inspiring conference.

References

[1] K. Anikeev *et al.*, arXiv:hep-ph/0201071.
[2] P. K. Kabir, *The CP Puzzle*, Academic Press, New York 1968
[3] K. M. Watson, Phys. Rev. **88** (1952) 1163.
[4] O. Cata and S. Peris, JHEP **0303** (2003) 060 [arXiv:hep-ph/0303162].
[5] A. J. Buras, M. Jamin and P. H. Weisz, Nucl. Phys. B **347** (1990) 491.
[6] S. Herrlich and U. Nierste, Nucl. Phys. B **476** (1996) 27 [arXiv:hep-ph/9604330].
[7] J. Brod and M. Gorbahn, Phys. Rev. D **82** (2010) 094026 [arXiv:1007.0684 [hep-ph]].
[8] G. Colangelo *et al.*, arXiv:1011.4408 [hep-lat].
[9] C. Aubin, J. Laiho and R. S. Van de Water, Phys. Rev. D **81** (2010) 014507 [arXiv:0905.3947 [hep-lat]].
[10] J. M. Gerard, arXiv:1012.2026 [hep-ph].
[11] A. J. Buras and D. Guadagnoli, Phys. Rev. D **78** (2008) 033005 [arXiv:0805.3887 [hep-ph]].
[12] A. J. Buras, D. Guadagnoli and G. Isidori, Phys. Lett. B **688** (2010) 309 [arXiv:1002.3612 [hep-ph]].
[13] K. Nakamura *et al.* (Particle Data Group), "Review of particle physics," J. Phys. G **37**, 075021 (2010).
[14] CDF Collaboration and D0 Collaboration, arXiv:1007.3178 [hep-ex].
[15] M. Antonelli *et al.* [FlaviaNet Working Group on Kaon Decays], arXiv:0801.1817 [hep-ph].
[16] K. G. Chetyrkin, J. H. Kuhn, A. Maier, P. Maierhofer, P. Marquard, M. Steinhauser and C. Sturm, Phys. Rev. D **80** (2009) 074010 [arXiv:0907.2110 [hep-ph]].
[17] J. Laiho, E. Lunghi and R. S. Van de Water, Phys. Rev. D **81** (2010) 034503 [arXiv:0910.2928 [hep-ph]].
[18] G. Buchalla, A. J. Buras and M. E. Lautenbacher, Rev. Mod. Phys. **68** (1996) 1125 [arXiv:hep-ph/9512380].

Journal of Physics: Conference Series **335** (2011) 012015 doi:10.1088/1742-6596/335/1/012015

CP Violation results from the NA48 experiments

Giuseppina Anzivino [1]

Physics Department, University of Perugia and
INFN, Sezione di Perugia
via A. Pascoli - 06123 Perugia (I)

E-mail: giuseppina.anzivino@pg.infn.it

Abstract.
The main goal of the NA48 experiments at the CERN SPS has been the search for direct CP violation (CPV) in neutral and charged kaon decays. In this paper selected results on CPV from the NA48 and NA48/2 experiments are presented. The direct CPV parameter $Re(\varepsilon'/\varepsilon)$ has been measured by NA48 from the decay rates of neutral kaons into two pions. In the charged kaon sector, NA48/2 has measured the asymmetry A_g of the linear slope parameter g in the Dalitz plot of $K^{\pm} \to 3\pi$ decays. Thanks to the high statistics collected in 2003 and 2004, further CPV studies were performed in several kaon decays. Results on CPV studies in $K^{\pm} \to \pi^{\pm}\pi^{0}\gamma$, $K^{\pm} \to \pi^{\pm}e^{+}e^{-}$ and $K^{\pm} \to \pi^{\pm}\mu^{+}\mu^{-}$ decays will be presented, as well as in the $K_S \to \pi^{+}\pi^{-}e^{+}e^{-}$ decay.

1. Introduction

The investigation of CP violation is of fundamental importance in particle physics as it addresses basic questions related to the observed matter-antimatter asymmetry in the Universe. CP violation was discovered in $K^0 \to \pi\pi$ decays in 1964 [1]. Still, kaons remain a privileged observatory for the study of the phenomenon. All three types of CP violation are observed in neutral kaons. The small asymmetry between the $K^0 \to \overline{K^0}$ and $\overline{K^0} \to K^0$ transition rates accounts for the indirect CP violation, measured with the parameter $Re(\varepsilon)$. Direct CP violation, represented by the parameter $Re(\varepsilon')$, occurs in the K^0 decays into two pions and is related to the phase difference of the two amplitudes with different $\pi\pi$ isospin values (I=0, I=2). The violation of CP symmetry takes also place in the interference between decays with and without mixing. In the Standard Model, CP violation is naturally accommodated by an irreducible complex phase in the CKM matrix, in the case of three quark families. Only direct CP violation occurs in charged kaons since mixing is not allowed.

The NA48 Collaboration has carried out over the last decade an extensive experimental programme at CERN dedicated to the study of CP violation (and rare processes) in both neutral and charged kaon decays. NA48, and its successors, was a fixed target experiment at SPS CERN. The main components were a magnetic spectrometer to measure charged particle momenta and an electromagnetic calorimeter based on liquid krypton for the measurement of electromagnetic showers. More detailed information about the detector components and performance can be found in [2].

[1] on behalf of the NA48 and NA48/2 Collaborations

Journal of Physics: Conference Series **335** (2011) 012015 doi:10.1088/1742-6596/335/1/012015

2. Highlights from the past

2.1. *Measurement of $Re(\varepsilon'/\varepsilon)$*

$Re(\varepsilon'/\varepsilon)$ is determined from the double ratio of the two pion decay rates of K_L and K_S:

$$R = \frac{\Gamma(K_L \to \pi^0\pi^0) \cdot \Gamma(K_S \to \pi^+\pi^-)}{\Gamma(K_S \to \pi^0\pi^0) \cdot \Gamma(K_L \to \pi^+\pi^-)} \approx 1 - 6Re(\varepsilon'/\varepsilon)$$

The NA48 experiment was designed to exploit cancellations of systematic effects contributing symmetrically to different components of the double ratio. The four decay modes were collected simultaneously in the same decay region, minimising sensitivity to intensity variations, detection efficiency and accidental effects. K_L and K_S acceptances become similar and approximately cancel in the ratio, despite of the large lifetime difference, by weighting K_L decays with a function of the proper time. Small differences between K_L and K_S beam intensity variations are eliminated by weighting K_S events by the K_L/K_S intensity ratio. High resolution detectors were used to achieve an efficient background rejection. A detailed description of the analysis method and event selection can be found in [3]. Data were collected in 1997, 1998-1999 and 2001. Taking into account the correlation of the systematic uncertainty between results from different periods of data taking, the combined final result is $Re(\varepsilon'/\varepsilon) = (14.7 \pm 2.2) \cdot 10^{-4}$ showing a clean evidence of direct CP violation in the neutral kaon system.

2.2. *Measurement of charged asymmetry*

Complementary with ε'/ε, the CP observable in the charged kaon sector is the asymmetry

$$A_g = (g^+ - g^-)/(g^+ + g^-) \approx \Delta g/2g$$

of the linear slope parameter g (g^+ refers to K^+, g^- to K^-). Since there is no mixing in the decays of charged kaons, any non-zero value of A_g would reflect evidence for direct CPV. SM predictions for the charge asymmetry lay in the range $10^{-6} - 10^{-5}$, while theoretical calculations involving processes beyond the SM do not exclude substantial enhancements of A_g. The goal of NA48/2 was to measure A_g with an accuracy of few 10^{-4}. From the data samples collected in 2003 and 2004, 3.11×10^9 $K^\pm \to \pi^\pm\pi^+\pi^-$ and 9.13×10^7 $K^\pm \to \pi^\pm\pi^0\pi^0$ were selected. The results for the asymmetries [4] are

$$A_g^c = (-1.5 \pm 1.5_{stat} \pm 0.9_{trig} \pm 1.1_{syst}) \times 10^{-4} = (-1.5 \pm 2.1) \times 10^{-4}$$

$$A_g^n = (1.8 \pm 1.7_{stat} \pm 0.9_{syst}) \times 10^{-4} = (1.8 \pm 1.8) \times 10^{-4}.$$

The results are compatible with the SM predictions, i.e. no evidence for direct CP violation at the order of 10^{-4} has been found. Due to the high precision achieved, the result can be used to constrain extensions of the SM predicting enhancements of the CP violating effects.

3. Recent results

3.1. *The $K^\pm \to \pi^\pm\pi^0\gamma$ decay*

The decay $K^\pm \to \pi^\pm\pi^0\gamma$ is one of the most interesting channels for studying the low energy structure of QCD. The total amplitude of the decay is the sum of two terms: the inner bremsstrahlung (IB), with a photon emitted from the outgoing charged pion, and the direct emission (DE) in which the photon is emitted at the weak vertex. The IB component of the decay is suppressed by the $\Delta I = 1/2$ rule, resulting in a relative enhancement of the DE contribution. Direct photon emission can occur through both electric and magnetic dipole transitions. The electric dipole transition can interfere with the IB amplitude giving rise to an interference term (INT), which can have CP violating contributions. The properties of the $K^\pm \to \pi^\pm\pi^0\gamma$ decay can be conveniently described using the T^*_π and W variables, where T^*_π is the kinetic energy of the charged pion in the kaon rest frame and W is a Lorentz invariant variable defined as: $W^2 = (P_K \cdot P_\gamma)(P_\pi \cdot P_\gamma)/(m_K \cdot m_\gamma)^2$ where P_K, P_π and P_γ are the 4-momenta of the kaon, the charged pion and the radiative photon respectively. Using these variables, the differential rate for the process $K^\pm \to \pi^\pm\pi^0\gamma$ can be written as [5],[6]:

$$\frac{\partial^2\Gamma^\pm_{IB}}{\partial T^*_\pi \partial W} = \frac{\partial^2\Gamma^\pm}{\partial T^*_\pi \partial W}[1 + 2\cos(\pm\phi + \delta^1_1 - \delta^2_0)m^2_\pi m^2_K X_E W^2 + m^4_\pi m^4_K (X^2_E + X^2_M)W^4]$$

where ϕ is the CP violating phase, δ^I_l are the strong pion-pion rescattering phases, and X_E and X_M are the normalized electric and magnetic amplitudes respectively. The different W dependence (the DE term is proportional to W^4 while the INT term to W^2) allows the extraction of the different decay components. From a sample of about 600k $K^\pm \to \pi^\pm\pi^0\gamma$ decay candidates the NA48/2 experiment has measured the relative amounts of DE and INT with respect to the internal bremsstrahlung (IB) contribution in the range $0 < T^*_\pi < 80$ MeV. The relative background contamination has been kept to $< 10^{-4}$ and the rate of wrong solutions for the odd photon to $< 0.1\%$. Thanks to the implementation of an algorithm to reject background from $K^\pm \to \pi^\pm\pi^0\pi^0$ decays, the cut on T^*_π could be released below the standard 55 MeV used by most of the previous experiments. Using an extended maximum likelihood technique, comparing data and simulated W distribution including non vanishing INT term, the following fractions of DE and INT with respect to IB have been obtained [7]:

$$Frac_{DE}(0 < T^*_\pi < 80MeV) = BR_{DE}/BR_{IB} = (3.32 \pm 0.15_{stat} \pm 0.14_{syst}) \cdot 10^{-2}$$

$$Frac_{INT}(0 < T^*_\pi < 80MeV) = BR_{INT}/BR_{IB} = (-2.35 \pm 0.35_{stat} \pm 0.39_{syst}) \cdot 10^{-2}$$

This measurement constitutes the first observation of an interference term in $K^\pm \to \pi^\pm\pi^0\gamma$ decays. From the above results, the electric and magnetic contributions can be extracted: $X_E = (-24 \pm 4_{stat} \pm 4_{syst})\ GeV^{-4}$ and $X_M = (254 \pm 6_{stat} \pm 6_{syst})\ GeV^{-4}$. This result is compatible with the hypothesis that the chiral anomaly, that predicts $X_M \sim 270\ GeV^{-4}$ [8], is the only source of magnetic amplitudes in the Direct Emission term, and indicates that factorization models cannot provide an appropriate description of DE and INT in the $K^\pm \to \pi^\pm\pi^0\gamma$ decay.

3.1.1. CP Violation in $K^\pm \to \pi^\pm\pi^0\gamma$ decays. The $K^\pm \to \pi^\pm\pi^0\gamma$ decay with direct photon emission is not suppressed by the $\Delta I = 1/2$ rule, therefore it is a good decay to search for CP Violation. Using a slightly modified event selection, two samples of 695k K^+ and 386k K^- have been reconstructed and used to set a limit on the CP violating asymmetry in the K^+ and K^- branching ratios. The simplest observable that can be measured is the difference in the decay rates of K^+ and K^- which can be expressed as the asymmetry on the total number of events A_N, defined as:

$$A_N = \frac{N^+ - RN^-}{N^+ + RN^-}$$

where N^+ and N^- are the number of K^+ and K^- decays to $\pi^\pm\pi^0\gamma$ in the data sample and R is the ratio of the number of K^+ to K^- in the beam. Using the $K^\pm \to \pi^\pm\pi^0\pi^0$ decay as normalization, a measurement of $R = 1.7998 \pm 0.0004$ has been performed with a high accuracy of $\delta R/R \sim 10^{-4}$. Using the full data sample, the measured asymmetry is

$$A_N = (0.0 \pm 1.0_{stat} \pm 0.6_{syst}) \cdot 10^{-3}$$

From the above value a limit for the rate asymmetry of $|A_N| < 1.5 \cdot 10^{-3}$ at 90% confidence level has been set [7]. Another CP violation observable is the asymmetry in the distribution of the Dalitz plot variable W (A_W). A fit to the distribution using the same data sample gives $A_W = (-0.6 \pm 1_{stat}) \cdot 10^{-3}$. The result is compatible with the A_N value. We can conclude that no CP violation is present at the level of 10^{-3} in $K^\pm \to \pi^\pm\pi^0\gamma$ decay, as expected by the Standard Model predictions.

3.2. The $K^\pm \to \pi^\pm e^+ e^-$ decay

The flavour-changing neutral current decays $K^\pm \to \pi^\pm l^+ l^-$ ($l = e, \mu$), induced at the one-loop level in the Standard Model (SM), are well suited to explore its structure and, possibly, its extensions. The rates of these transitions are dominated by the long-distance contributions involving one photon exchange. They have been described in the framework of Chiral Perturbation Theory (ChPT) [9] in terms of a vector interaction form factor determined by experimental measurements.

The dynamics of the $K^\pm \to \pi^\pm e^+ e^-$ decay, highly suppressed by the GIM mechanism, is completely specified by the invariant function $W(z)$, where $z = (M_{ee}/M_K)^2$ is a kinematic variable. Several models have been developed predicting the form factors that characterize the decay rate and the dilepton invariant mass distribution. In the present analysis the following parameterizations of the form factors are considered:

- Linear: $W(z) = G_F M_K^2 f_0(1 + \delta z)$, with free normalization and slope (f_0 and δ).
- Next-to-Leading Order ChPT [10]: $W(z) = G_F M_K^2(a_+ + b_+ z) + W^{\pi\pi}(z)$ with free parameters (a_+, b_+) and an explicitly calculated pion loop term $W^{\pi\pi}(z)$.
- Combined framework of ChPT and large-N_c QCD [11]: the form factor is parameterized as $W(z) \equiv W(\tilde{w}, \beta, z)$, with free parameters $(\tilde{w}, \beta)$.
- A ChPT parameterization [12] involving meson form factors $W(z) = W(M_a, M_\rho, z)$, with meson masses (M_a, M_ρ) treated as free parameters.

The aim of the analysis is to extract the form factor parameters in the framework of each of the above models and to measure the corresponding BR's in the full kinematic range and, in addition, a model-independent BR in the visible kinematic range ($z > 0.08$).

The $K^\pm \to \pi^\pm e^+ e^-$ rate is measured relatively to $K^\pm \to \pi^\pm \pi^0_D$ (with $\pi^0_D \to e^+ e^- \gamma$). Since the two decays have the same charged particles in the final state, common selection criteria have been used, resulting in cancelation of particle ID inefficiencies at first order. At the end of the selection, based on the reconstruction of three-track events, a total of 7146 $K^\pm \to \pi^\pm e^+ e^-$ candidates with 0.6% background are found in the signal region. The reconstructed $(\pi^\pm e^+ e^-)$ invariant mass spectrum is shown in Fig. 1.

The computed values of $d\Gamma_{\pi ee}/dz$ as a function of z are shown in Fig. 2 with the results of the fits to the four considered models; the measured parameters and the corresponding BR's are reported in Table 1.

Fits to all the four models are of reasonable quality, however the linear form-factor model leads to the best χ^2 (see Fig. 2). The size of the data sample is insufficient to distinguish between the models. The BR in the full kinematic range, which includes an uncertainty due to extrapolation into the inaccessible region $z < 0.08$, is

$$BR = (3.11 \pm 0.04_{stat} \pm 0.05_{syst} \pm 0.08_{ext} \pm 0.07_{mod}) \times 10^{-7}$$

$$BR(K^\pm \to \pi^\pm e^+ e^-) = (3.11 \pm 0.12) \times 10^{-7}$$

This result is in fair agreement with previous measurements. In particular, comparison to the most precise BNL E865 result [13], using the same external input, and taking into account correlation of external uncertainties between the two analyses, shows a 1.6 σ level of agreement.

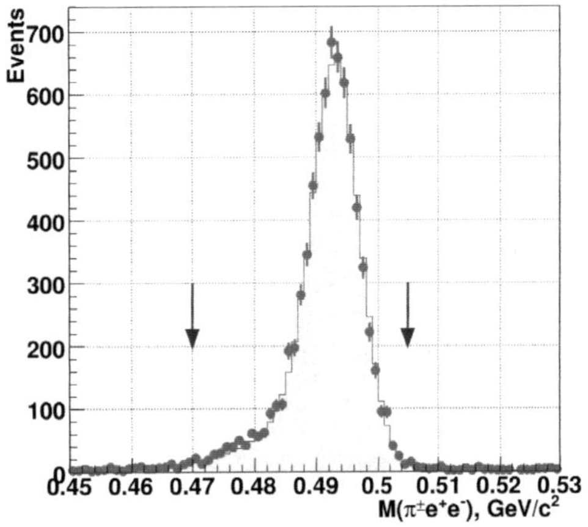

Figure 1. Reconstructed spectrum of $(\pi^{\pm}e^{+}e^{-})$ invariant mass: data (dots) and MC simulation (filled area).

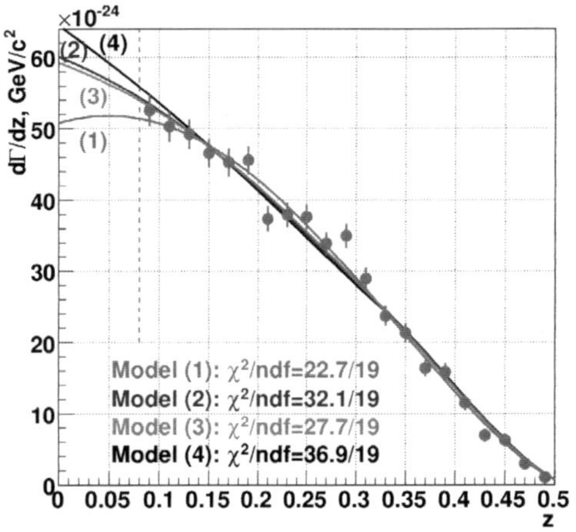

Figure 2. The computed $d\Gamma_{\pi ee}/dz$ and the results of the fits according to the four considered models.

Table 1. Results of fits to the four models and the Model-Independent BR($z > 0.08$).

Model(1)											
$\lvert f_0\rvert =$	0.531	±	$0.012_{\text{stat.}}$	±	$0.008_{\text{syst.}}$	±	$0.007_{\text{ext.}}$	=	0.531	±	0.016
$\delta =$	2.32	±	$0.15_{\text{stat.}}$	±	$0.09_{\text{syst.}}$	±	$0.00_{\text{ext.}}$	=	2.32	±	0.18
$\mathrm{BR}_1 \times 10^7 =$	3.05	±	$0.04_{\text{stat.}}$	±	$0.05_{\text{syst.}}$	±	$0.08_{\text{ext.}}$	=	3.05	±	0.10
Model(2)											
$a_+ =$	−0.578	±	$0.012_{\text{stat.}}$	±	$0.008_{\text{syst.}}$	±	$0.007_{\text{ext.}}$	=	−0.578	±	0.016
$b_+ =$	−0.779	±	$0.053_{\text{stat.}}$	±	$0.036_{\text{syst.}}$	±	$0.017_{\text{ext.}}$	=	−0.779	±	0.066
$\mathrm{BR}_2 \times 10^7 =$	3.14	±	$0.04_{\text{stat.}}$	±	$0.05_{\text{syst.}}$	±	$0.08_{\text{ext.}}$	=	3.14	±	0.10
Model(3)											
$\tilde{w} =$	0.057	±	$0.005_{\text{stat.}}$	±	$0.004_{\text{syst.}}$	±	$0.001_{\text{ext.}}$	=	0.057	±	0.007
$\beta =$	3.45	±	$0.24_{\text{stat.}}$	±	$0.17_{\text{syst.}}$	±	$0.05_{\text{ext.}}$	=	3.45	±	0.30
$\mathrm{BR}_2 \times 10^7 =$	3.13	±	$0.04_{\text{stat.}}$	±	$0.05_{\text{syst.}}$	±	$0.08_{\text{ext.}}$	=	3.13	±	0.10
Model(4)											
$M_a/\mathrm{GeV} =$	0.974	±	$0.030_{\text{stat.}}$	±	$0.019_{\text{syst.}}$	±	$0.002_{\text{ext.}}$	=	0.974	±	0.035
$M_\rho/\mathrm{GeV} =$	0.716	±	$0.011_{\text{stat.}}$	±	$0.007_{\text{syst.}}$	±	$0.002_{\text{ext.}}$	=	0.716	±	0.014
$\mathrm{BR}_3 \times 10^7 =$	3.18	±	$0.04_{\text{stat.}}$	±	$0.05_{\text{syst.}}$	±	$0.08_{\text{ext.}}$	=	3.18	±	0.10
$\mathrm{BR}_{\mathrm{MI}} \times 10^7 =$	2.28	±	$0.03_{\text{stat.}}$	±	$0.04_{\text{syst.}}$	±	$0.06_{\text{ext.}}$	=	2.28	±	0.08

Measurements of the decay parameters and BR's were performed separately for K^+ and K^- decays. In particular, the branching fractions in the full kinematic range and their statistical uncertainties were measured to be $BR^+ = (2.99 \pm 0.05_{stat}) \times 10^{-7}$, $BR^- = (3.13 \pm 0.08_{stat}) \times 10^{-7}$. This allows to perform a first measurement of the direct CP violating asymmetry of K^+ and K^- decay rates in the full kinematic range. Considering only the uncorrelated systematic uncertainties (those due to background subtraction) between K^+ and K^- samples, we found

$$\Delta(K^{\pm}_{\pi ee}) = \frac{BR^+ - BR^-}{BR^+ + BR^-} = (-2.2 \pm 1.5_{stat} \pm 0.6_{syst}) \times 10^{-2}$$

The other systematic uncertainties are correlated between the K^+ and K^- samples, and thus cancel in $\Delta(K^{\pm}_{\pi ee})$. A conservative limit for the charge asymmetry of $|\Delta(K^{\pm}_{\pi ee})| < 2.1 \times 10^{-2}$ at 90% CL can be deduced from the above value. This result is compatible with CP conservation, however the achieved precision is far from the Standard Model expectation $\Delta(K^{\pm}_{\pi ee}) \sim 10^{-5}$ [10] and even the SUSY upper limt $\Delta(K^{\pm}_{\pi ee}) \sim 10^{-3}$ [14],[15] for the CP violating asymmetry. The complete description of the study of the $K^{\pm} \to \pi^{\pm} e^+ e^-$ decay can be found in [16].

3.3. The $K^{\pm} \to \pi^{\pm}\mu^+\mu^-$ decay

The $K^{\pm} \to \pi^{\pm}\mu^+\mu^-$ rate is measured relative to the abundant $K^{\pm} \to \pi^{\pm}\pi^+\pi^-$ normalization channel. The two samples are collected concurrently using the same trigger logic. The fact that the μ and π masses are close ($m_\mu/m_\pi = 0.76$) results in similar topologies of the signal and normalization final states. This leads to first order cancellation of the systematic effects induced by imperfect kaon beam description, local detector inefficiencies and trigger inefficiency. The reconstructed $\pi^{\pm}\mu^+\mu^-$ invariant mass spectrum is shown in Fig. 3; the number of $K^{\pm} \to \pi^{\pm}\mu^+\mu^-$ candidates in the signal region is $N_{\pi\mu\mu} = 3120$. Similarly to the $K^{\pm} \to \pi^{\pm}e^+e^-$ decay, the $K^{\pm} \to \pi^{\pm}\mu^+\mu^-$ decay is described as proceeding via single virtual photon exchange, resulting in a spectrum of the $z = (M_{\mu\mu}/M_K)^2$ kinematic variable sensitive to the form factor $W(z)$. In order to extract the form factors that characterize the decay rate and the dilepton invariant mass distribution, the same four models considered in the case of the $K^{\pm} \to \pi^{\pm}e^+e^-$ decay were used. In Fig. 4 the fit to one of the models (the linear one) is shown.

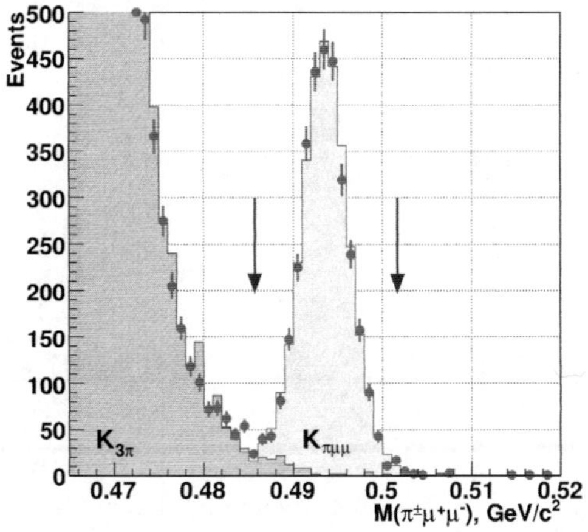

Figure 3. Reconstructed spectrum of $(\pi^{\pm}\mu^+\mu^-)$ invariant mass: data (dots) and $K_{\pi\mu\mu}$ MC simulation and $K_{3\pi}$ background estimate(filled areas).

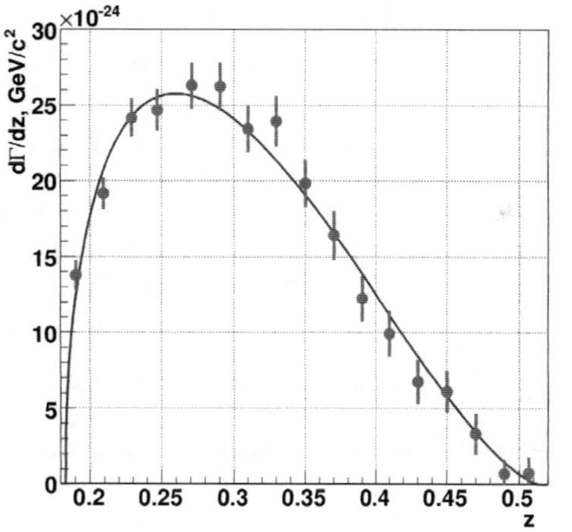

Figure 4. The reconstructed $d\Gamma_{\pi\mu\mu}/dz$ spectrum fitted to a linear form factor.

The measured model parameters and the model independent BR are reported in Table 2. The precision of our BR measurement represents a factor of ~ 3 improvement with respect to the most precise earlier measurement [17]. Each of the form factor models provides a reasonable fit to the data, but the statistical precision is insufficient to distinguish between the models.

The model-independent branching ratios measured separately for K^+ and K^- decays are

$$BR^+ = (9.70 \pm 0.26) \times 10^{-8}, \qquad BR^- = (9.49 \pm 0.35) \times 10^{-8}$$

where the quoted uncertainties are statistical only. Neglecting the systematic uncertainties of BR^+ and BR^- (which are small compared to the statistical uncertainties, and mostly common

Table 2. Results of fits to the four models and the Model-Independent Branching Ratio.

Model(1)											
$\lvert f_0 \rvert =$	0.470	$\pm$	$0.039_{\text{stat.}}$	$\pm$	$0.006_{\text{syst.}}$	$\pm$	$0.002_{\text{ext.}}$	$=$	0.470	$\pm$	0.040
$\delta =$	3.11	$\pm$	$0.56_{\text{stat.}}$	$\pm$	$0.11_{\text{syst.}}$				3.11	$\pm$	0.57
Model(2)											
$a_+ =$	-0.575	$\pm$	$0.038_{\text{stat.}}$	$\pm$	$0.006_{\text{syst.}}$	$\pm$	$0.002_{\text{ext.}}$	$=$	-0.575	$\pm$	0.039
$b_+ =$	-0.813	$\pm$	$0.142_{\text{stat.}}$	$\pm$	$0.028_{\text{syst.}}$	$\pm$	$0.013_{\text{ext.}}$	$=$	-0.813	$\pm$	0.145
Model(3)											
$\tilde{w} =$	0.064	$\pm$	$0.014_{\text{stat.}}$	$\pm$	$0.003_{\text{syst.}}$				0.064	$\pm$	0.014
$\beta =$	3.77	$\pm$	$0.61_{\text{stat.}}$	$\pm$	$0.12_{\text{syst.}}$	$\pm$	$0.02_{\text{ext.}}$	$=$	3.7	$\pm$	0.62
Model(4)											
$M_a/\text{GeV} =$	0.993	$\pm$	$0.083_{\text{stat.}}$	$\pm$	$0.016_{\text{syst.}}$	$\pm$	$0.001_{\text{ext.}}$	$=$	0.993	$\pm$	0.085
$M_\rho/\text{GeV} =$	0.721	$\pm$	$0.027_{\text{stat.}}$	$\pm$	$0.005_{\text{syst.}}$	$\pm$	$0.001_{\text{ext.}}$	$=$	0.721	$\pm$	0.028
$\text{BR}_{\text{MI}} \times 10^8 =$	9.62	$\pm$	$0.21_{\text{stat.}}$	$\pm$	$0.11_{\text{syst.}}$	$\pm$	$0.07_{\text{ext.}}$	$=$	9.62	$\pm$	0.25

to the K^+ and K^- measurements), and the possible charge asymmetry of $K_{3\pi}$ decay rates which is experimentally compatible with zero within 2×10^{-3} precision, we measure the charge asymmetry to be $\Delta(K^{\pm}_{\pi\mu\mu}) = (1.1 \pm 2.3) \times 10^{-2}$. This is a factor of ~ 5 improvement in precision with respect to the only previous measurement [18], and is compatible with CP conservation. A limit for the charge asymmetry of $|\Delta(K^{\pm}_{\pi\mu\mu}| < 2.9 \times 10^{-2}$ at 90% CL can be deduced from the above value. The experimental precision is far from the SM expectation $\Delta(K^{\pm}_{\pi\mu\mu}) \sim 10^{-4}$ [10] and even the SUSY upper limit $\Delta(K^{\pm}_{\pi\mu\mu}) \sim 10^{-3}$ [14],[15] for the CP violating asymmetry.

The forward-backward asymmetry has been measured to be $A_{FB} = (-2.4 \pm 1.8) \times 10^{-2}$, where the error is dominated by the statistical uncertainty. It corresponds to an upper limit of $|A_{FB}| < 2.3 \times 10^{-2}$ at 90% CL. The achieved precision does not reach the upper limits for the SM contribution via the two-photon intermediate state $K^{\pm} \to \pi^{\pm}\gamma^*\gamma^* \to \pi^{\pm}\mu^+\mu^-$ [19] and MSSM contribution [20], which are both of the order of 10^{-3}. The complete description of the study of the $K^{\pm} \to \pi^{\pm}\mu^+\mu^-$ decay can be found in [21].

3.4. The $K_S \to \pi^+\pi^-e^+e^-$ decay

In 2002, the NA48 experiment carried out a high intensity K_S program aiming to measure rare K_S and hyperon decays (NA48/1). Studies of the $K_S \to \pi^+\pi^-e^+e^-$ decay have recently been completed. This decay provides a testing ground for a CP non-invariance. Unlike the $K_L \to \pi^+\pi^-e^+e^-$ decay in which the CP=+1 inner bremsstrahlung (IB) process competes with the CP=-1 direct emission (DE) M1 term, the decay amplitude of $K_S \to \pi^+\pi^-e^+e^-$ is expected to be dominated by the CP-even inner bremsstrahlung transition. Therefore, the CP violating asymmetry, defined as

$$A_\phi = \frac{N_{\pi\pi ee}(\sin\phi\cos\phi > 0) - N_{\pi\pi ee}(\sin\phi\cos\phi < 0)}{N_{\pi\pi ee}(\sin\phi\cos\phi > 0) + N_{\pi\pi ee}(\sin\phi\cos\phi < 0)}$$

where ϕ is the angle between the $\pi^+\pi^-$ and the e^+e^- decay planes in the kaon center of mass, is expected to be 0.

In total 22966 $K_S \to \pi^+\pi^-e^+e^-$ candidates, with a background of 103 events were selected. The BR ratio was measured with respect to the $K_L \to \pi^+\pi^-\pi^0_D$ decay. Using the existing values for $K^{\pm} \to \pi^+\pi^-\pi^0$ and $\pi^0 \to e^+e^-\gamma$ BR's [22], we found [23]

$$BR = (4.93 \pm 0.14) \times 10^{-5}$$

Other parameters of the decay were measured, in particular the CP-violating asymmetry. The normalized differential decay rate dependence on $\sin\phi\cos\phi$ is shown in Fig. 5.

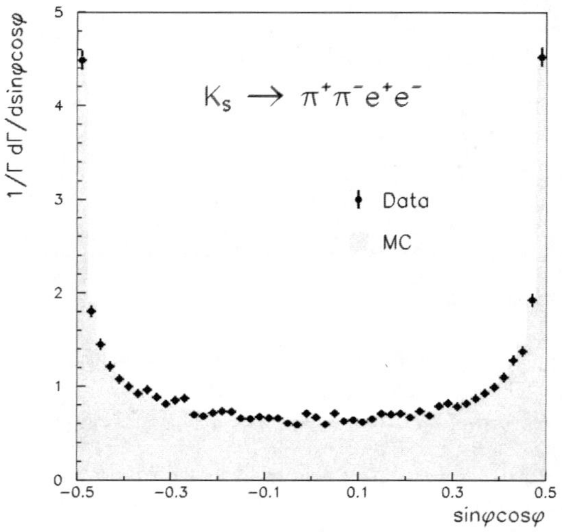

Figure 5. The $K_S \to \pi^+\pi^-e^+e^-$ normalized differential decay rate in the $\sin\phi\cos\phi$ variable.

The corresponding asymmetry parameter was found to be $A_\phi = (-0.4 \pm 0.7_{stat} \pm 0.4_{syst})\%$, consistent with zero, from which a limit of $A_\phi < 1.5\%$ at 90% CL can be set. No evidence for a CP-violating contribution in $K_S \to \pi^+\pi^-e^+e^-$ was observed.

References

[1] Christenson J H et al. 1964 *Phys. Rev. Lett.* **13** 138
[2] Fanti V et al. 2007 *Nucl. Inst. Methods A* **574** 433
[3] Batley J R et al. 2002 *Phys. Lett. B* **544** 97
[4] Batley J R et al. 2007 *Eur. Phys. J. C* **52** 875
[5] Christ N 1967 *Phys. Rev.* **159** 1292
[6] Abrams R J et al. 1972 *Phys. Rev. Lett.* **29** 1118
[7] Batley J R et al. 2010 *Eur. Phys. J. C* **68** 75
[8] Ecker G, Neufeld H, Pich A. 1992 *Phys. Lett. B* **278** 337
[9] Ecker G, Pich A and de Rafael E 1987 *Nucl. Phys. B* **291** 692
[10] D'Ambrosio G et al. 1998 JHEP 9808 4.
[11] Friot S, Greynat D and de Rafael E 2004 *Phys. Lett. B* **595** 301
[12] Dubnickova A Z et al. 2008 *Phys. Part. Nucl. Lett.* **5** 76
[13] Appel R et al. 1999 *Phys. Rev. Lett.* **83** 4482
[14] Messina A 2002 *Phys. Lett. B* **538** 130
[15] D'Ambrosio G and Dao-Neng Gao 2002 JHEP 0207 068
[16] Batley J R et al. 2009 *Phys. Lett. B* **677** 246
[17] Ma H et al. 2000 *Phys. Rev. Lett.* **84** 2580
[18] Park H K et al. 2002 *Phys. Rev. Lett.* **88** 111801
[19] Gao D N 2004 *Phys. Rev. D* **69** 094030
[20] Chen C H, Geng C Q and Ho I L 2003 *Phys. Rev. D* **67** 074029
[21] Batley J R et al. 2011 *Phys. Lett. B* **697** 107
[22] Amsler C et al. 2008 *Phys. Lett. B* **667** 1
[23] Batley J R et al. 2011 *Phys. Lett. B* **694** 301

Journal of Physics: Conference Series **335** (2011) 012016 doi:10.1088/1742-6596/335/1/012016

The Rare Decays $K \to \pi\nu\bar{\nu}$ and $K_L \to \pi^0\ell^+\ell^-$

Emmanuel Stamou

Institute for Advanced Study, Technische Universität München,
Lichtenbergstraße 2a, D-85748 Garching, Germany
Physik-Department, Technische Universität München,
James-Franck-Straße, D-85748 Garching, Germany

E-mail: estamou@ph.tum.de

Abstract. Standard-model and possible New-Physics contributions can compete in magnitude in the rare Kaon decay modes $K \to \pi\nu\bar{\nu}$ and $K_L \to \pi^0\ell^+\ell^-$. In particular the $K \to \pi\nu\bar{\nu}$ decays, the structure of short-distance physics is unclouded by non-perturbative, low-energy effects. This makes these decays excellent probes of New Physics. We review the status of the standard-model prediction of both $K \to \pi\nu\bar{\nu}$ and $K_L \to \pi^0\ell^+\ell^-$ decays and discuss the recent progress in the perturbative part of the $K \to \pi\nu\bar{\nu}$ decays.

1. Introduction

One approach to searching for deviations from the predictions of the Standard Model (SM) of particle physics is to examine low-energy observables, which are sensitive to New Physics (NP) effects and can also be predicted with a good accuracy within the SM. Four such observables are the branching ratios of the rare kaon decays $K_L \to \pi^0\nu\bar{\nu}$, $K^+ \to \pi^+\nu\bar{\nu}$, $K_L \to \pi^0\mu^+\mu^-$, and $K_L \to \pi^0e^+e^-$.

Within the SM short-distance (SD) physics accounts for approximately 90% and 40% of the branching fractions of the $K \to \pi\nu\bar{\nu}$ and $K_L \to \pi^0\ell^+\ell^-$ decays, respectively. The decays are all governed by a flavour-changing neutral current (FCNC) transition and are therefore loop induced within the SM. In loop induced processes, heavy new particles may contribute to the decay amplitudes on the same footing as SM particles. Thus, they represent ideal probes for searching and disentangling NP.

A precise SM prediction is therefore essential to disentangle NP effects from the SM background. Often, non-perturbative effects originating from low-energy dynamics spoil the SM precision. This is not the case for the two $K \to \pi\nu\bar{\nu}$ decays, which are almost completely free from non-perturbative uncertainties and thus rank among the cleanest NP probes in flavour physics. Long-distance (LD) effects are larger in the $K_L \to \pi^0\ell^+\ell^-$ decays, but useful information about the structure of NP can still be extracted from a combined analysis of both branching ratios.

In the following we briefly discuss the general structure of rare Kaon decays, we then review the SM prediction of the above-mentioned modes and also present the recent calculation of the electroweak (EW) top-quark contribution of $K \to \pi\nu\bar{\nu}$.

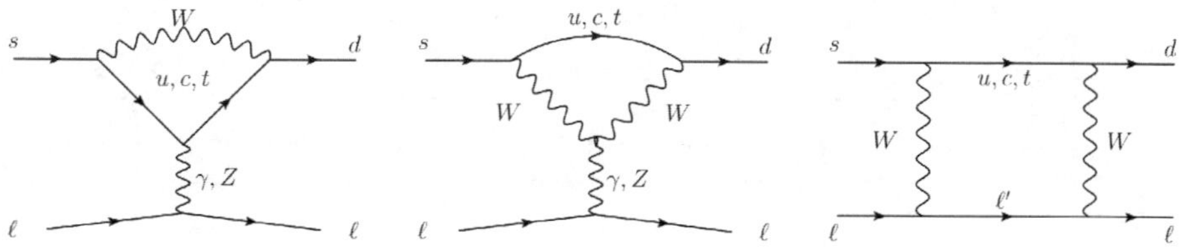

Figure 1. γ, Z penguins and W boxes contributing to the $s \to d$ transition at the one-loop level. In the case $l \equiv \nu$ the γ-penguin is absent.

2. Structure of Rare Kaon Decays

In the SM the FCNC transition $s \to d$ relevant for the rare Kaon decays is only possible at the one-loop level via the box and γ, Z-penguin diagrams of Fig. 1. There is a large energy gap between the scale in which the Feynman diagrams of Fig. 1 are to be evaluated and the scale of the outgoing physical states. The first scale is approximately the Z boson mass, while the second the mass of the Kaon, approximately 500 MeV. The gap is closed by using an appropriate effective field theory (EFT) to describe the low-energy physics, where heavy particles are no longer dynamical degrees of freedom. In this way the effect of SD physics is separated from LD effects. A master amplitude is then valid for all meson decays [1]:

$$\mathcal{A}_{\text{decay}} = \sum_i \langle Q_i \rangle \, V^i_{\text{CKM}} \, F_i. \tag{1}$$

$\langle Q_i \rangle$ are matrix elements of effective operators, that cannot be calculated perturbatively, V^i_{CKM} comprise products of CKM matrix elements and F_i are the so-called Wilson coefficients, the effective couplings incorporating the SD contributions. The sum in Eq. (1) extends over all possible operators Q_i generated by the SM or a given NP model and similarly over all possible internal particle contributions in the loop of Fig. 1. While the matrix elements of semileptonic decays can be extracted from experimental data, the Wilson coefficients depend on the high-energy model and can be calculated perturbatively within the SM.

NP contributions can enter Eq. (1) in two ways. Firstly, they can modify the Wilson coefficients. Since these are the same in rare B-, D-, and K-meson decays in the SM, the resulting correlation can be used to test the universality of the loop function in a model-independent way. Secondly, NP can generate extra operators that will contribute to the decay amplitude. This would signal a departure from Minimal Flavour Violation [2].

The master amplitude of Eq. (1) is also suited to compare the size of different contributions. The size of each contribution depends on both the CKM elements, conveniently parametrised in terms of $\lambda = |V_{us}| \approx 0.22$, and on the loop functions F_i. Within the SM, low-energy contributions are suppressed with respect to the top contribution due to the Glashow-Iliopoulos-Maiani (GIM) mechanism [3]. When SD photon penguins contribute, this suppression is only logarithmic, but in their absence (e.g. in the $K \to \pi\nu\bar{\nu}$ decays) the loop functions are proportional to m^2_{top}, m^2_{charm}, or Λ^2_{QCD} for the up-quark, and a quadratic GIM mechanism is at play.

In addition, for the Kaon decays the CKM factor of the top-quark contribution $\lambda_t = V^*_{ts} V_{td}$ is proportional to λ^5 and therefore highly suppressed. This strong suppression, absent in B- and D-decays, can imply a significant charm-contribution depending on the mode, but also renders rare Kaon decays sensitive to even very small deviations from the CKM picture of CP violation.

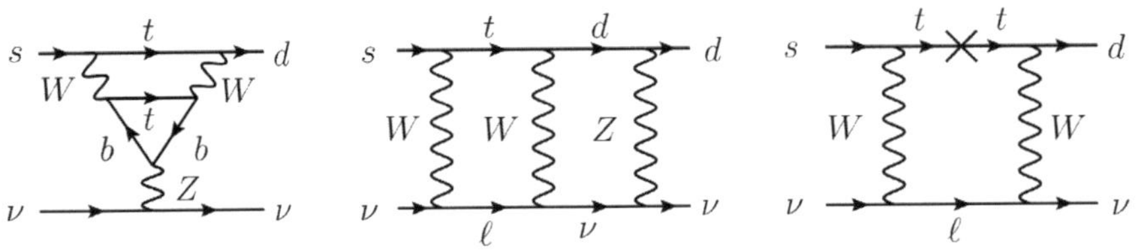

Figure 2. Sample penguin, box and counterterm diagrams contributing to X_t at NLO.

3. $K^+ \to \pi^+ \nu\bar{\nu}$ and $K_L \to \pi^0 \nu\bar{\nu}$

The weak SM effective Hamiltonian for the two $K \to \pi\nu\bar{\nu}$ decays reads [4]

$$\mathcal{H}_{\text{eff}} = \frac{4G_F}{\sqrt{2}} \frac{\alpha}{2\pi \sin^2\theta_W} \sum_{l=e,\mu,\tau} \left(\lambda_c X^l + \lambda_t X_t\right) (\bar{s}_L \gamma_\mu d_L)(\bar{\nu}_{lL} \gamma^\mu \nu_{lL}) + h.c. \tag{2}$$

and involves to a good approximation only one effective operator below $\mu = \mu_c$.

The LD part of both decays is known with high precision. The matrix elements are extracted from $K_{\ell 3}$ decays using the approximate isospin symmetry. Corrections due to isospin-breaking ($\kappa_\nu^{+,L}$) and long-distance QED effects have also been considered in [5]. At the level of accuracy reached in the SM prediction, subleading effects from light quarks and higher dimensional operators can no longer be neglected; they have been estimated using chiral perturbation theory (ChPT). Their contribution to the branching ratios is parametrised by the phenomenological parameter $\delta P_{c,u}$ [6].

The small uncertainties due to LD effects renders higher order corrections on the SD parts essential for the improvement of the SM prediction. The SD charm contribution, denoted by $P_c = \frac{1}{\lambda^4}\left(\frac{2}{3}X^e + \frac{1}{3}X^\tau\right)$, is negligible in $K_L \to \pi^0\nu\bar{\nu}$, but not in $K^+ \to \pi^+\nu\bar{\nu}$, where it amounts to approximately 30% of its branching ratio. So far P_c is known up to next-to-next-to-leading order (NNLO) in QCD [7] and also the next-to-leading order (NLO) EW corrections have been calculated [8]. NLO QCD corrections [4, 9] are known for the top-quark contribution, but until recently the NLO EW corrections were only estimated in the large-m_t limit [10]. However, these poorly approximate the full EW corrections (see Fig. 3) [10]. Therefore, the renormalisation scheme of the EW input parameters $\alpha, \sin\theta_W, M_W, M_t$, appearing at LO in Eq. (2), was far from being clear and accounted for an uncertainty of approximately 2% in X_t.

We substantially reduced the renormalisation scheme ambiguity by calculating the NLO term in the α expansion of the top-quark Wilson coefficient X_t [11]. A sample of the corresponding diagrams is given in Fig. 2. To estimate the remaining scheme- and matching-scale dependence we use both $\overline{\text{MS}}$ and on-shell scheme, as well as a mixed scheme with on-shell masses but $\overline{\text{MS}}$ couplings. With the full NLO EW corrections on X_t we reduce the renormalisation scheme uncertainty to 0.3%. The comparison of LO and NLO results in the $\overline{\text{MS}}$ and on-shell scheme is illustrated in Fig. 3 together with the remaining scale dependence.

The theoretical predictions and the current experimental values for the branching ratios then read:

$$\text{BR}^{\text{theo}}_{K^+\to\pi^+\nu\bar{\nu}} = (7.81^{+0.80}_{-0.71} \pm 0.29) \times 10^{-11} \text{ [11]} \qquad \text{BR}^{\text{exp}}_{K^+\to\pi^+\nu\bar{\nu}} = (1.73^{+1.15}_{-1.05}) \times 10^{-10} \text{ [12]}$$

$$\text{BR}^{\text{theo}}_{K_L\to\pi^0\nu\bar{\nu}} = (2.43^{+0.40}_{-0.37} \pm 0.06) \times 10^{-11} \text{ [11]} \qquad \text{BR}^{\text{exp}}_{K_L\to\pi^0\nu\bar{\nu}} < 6.7 \times 10^{-8} \text{ [13]}$$

The first error in the theory prediction is parametric, while the second summarises the remaining theory uncertainty. For the charged mode the detailed main parametric contributions are (V_{cb} : 56%, $\bar{\rho}$: 21%, m_c : 8%, m_t : 6%, $\bar{\eta}$: 4%, α_s : 3%, $\sin^2\theta_W$: 1%), while the main

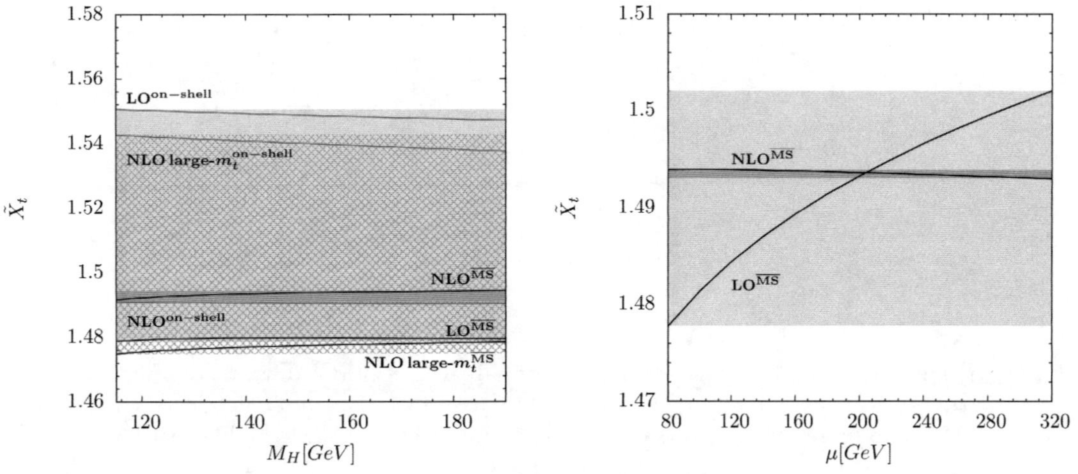

Figure 3. Left: comparison of the top-quark Wilson coefficient in the $\overline{\rm MS}$ and on-shell schemes at LO, NLO large-m_t and NLO. The grey solid band, hashed band, and solid red band represent the estimated uncertainties of LO, NLO large-m_t, and NLO result as a function of the Higgs mass, respectively. Right: the uncertainty due to EW renormalisation scale at LO and NLO. Again the grey and red band represent the estimated uncertainty before and after the NLO calculation, respectively.

theoretical contributions are ($\delta P_{c,u}$: 46%, X_t(QCD) : 24%, P_c : 20%, κ_ν^+ : 7%, X_t(EW) : 3%), respectively. Similarly for the neutral mode, the parametric uncertainties are (V_{cb} : 54%, $\bar{\eta}$: 39%, m_t : 6%) and the theoretical (X_t(QCD) : 73%, κ_ν^L : 18%, X_t(EW) : 8%, $\delta P_{c,u}$: 1%).

4. $K_L \to \pi^0\mu^+\mu^-$ and $K_L \to \pi^0 e^+e^-$

In the two $K_L \to \pi^0\ell^+\ell^-$ modes three different contributions compete in magnitude: direct-CP-violating (DCPV), indirect-CP-violating (ICPV) and CP-conserving (CPC) contribution. Long distance effects are much larger here. Still, the importance of these decay channels lies in their sensitivity to SD contributions from more than one operator. Let us now look at the three distinct contributions.

DCPV: high-energy top and charm contributions generate both vector, $Q_{7V} = (\bar{s}d)_V(\bar{\ell}\ell)_V$, and axial-vector, $Q_{7A} = (\bar{s}d)_V(\bar{\ell}\ell)_A$, dimension-six operators. Their Wilson coefficients are known to NLO QCD [2] and their matrix elements are also extracted with a few per mil precision from $K_{\ell 3}$ decays [5]. One difference between the electronic and the muonic channels is the suppression of the electronic axial-vector matrix element due to the smallness of the electron mass.

ICPV: in this case the K_L decays through its small CP even component via the γ-penguin in Fig. 4. The size of this smaller component is described by the parameter ϵ_K. We can therefore relate this contribution to the decay $K_S(\approx K_1) \to \pi^0\ell^+\ell^-$ using ChPT. The light meson loops are subleading and the amplitude is dominated by the chiral counterterm a_S [14], whose 20% uncertainty completely dominates the error for the $K_L \to \pi^0\ell^+\ell^-$ rates [15]. Although the absolute value of a_S has been extracted from NA48 measurements, its sign remains unknown. Therefore, it is still an open question whether the ICPV contribution interferes constructively or destructively with the DCPV contribution. Apart from a more precise $K_S \to \pi^0\ell^+\ell^-$ measurement, the sign of a_S can also be determined from the measurement of the integrated forward-backward asymmetry A_{FB}^ℓ in $K_L \to \pi^0\mu^+\mu^-$ [15].

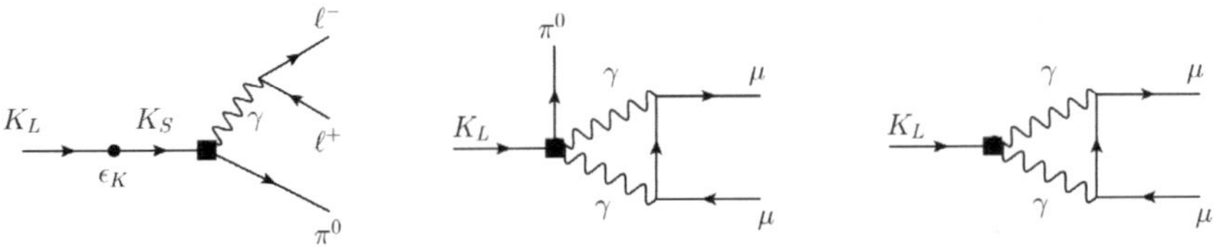

Figure 4. γ- and $\gamma\gamma$- penguins contributing to the LD part of the $K_L \to \pi^0\ell^+\ell^-$ and $K_L \to \mu^+\mu^-$ branching ratios.

CPC: is a purely LD effect described by the $\gamma\gamma$-penguin ChPT diagram in Fig. 4. The contribution is helicity suppressed for the electronic mode, a further difference between the two modes. For the muonic mode the meson loops are finite at $\mathcal{O}(p^4)$, while higher order effects of $\mathcal{O}(p^6)$ have been partially estimated from the $K_L \to \pi^0\gamma\gamma$ ratio, reducing the uncertainty to approximately 30% [16].

The current SM predictions and experimental values for the branching ratios for the two decays then read:

$$\mathrm{BR}^{\mathrm{theo}}_{K_L\to\pi^0\mu^+\mu^-} = 1.41^{+0.28}_{-0.26}\ (0.95^{+0.22}_{-0.21}) \times 10^{-11}\ [15] \qquad \mathrm{BR}^{\mathrm{exp}}_{K_L\to\pi^0\mu^+\mu^-} < 3.8 \times 10^{-10}\ [17]$$
$$\mathrm{BR}^{\mathrm{theo}}_{K_L\to\pi^0 e^+e^-} = 3.54^{+0.98}_{-0.85}\ (1.56^{+0.62}_{-0.49}) \times 10^{-11}\ [15] \qquad \mathrm{BR}^{\mathrm{exp}}_{K_L\to\pi^0 e^+e^-} < 2.8 \times 10^{-10}\ [18]$$

Theory predictions correspond to constructive interference of DCPV with ICPV contribution, while predictions in brackets to destructive interference.

5. New Physics and Outlook

The theoretical cleanness of the $K \to \pi\nu\bar{\nu}$ modes promotes both to excellent probes of NP, especially in view of the dedicated experiments, NA62 at CERN (see talk by G. Lamanna [19]) and KOTO at JPARC, which aim at measuring the charged and neutral mode, respectively, with an expected accuracy of 15%. These decays have therefore been studied extensively in models beyond the SM. A summary of predictions in different models is presented in [20]. This analysis points out the possibility of large effects in both decay modes and also illustrates the correlation between the predictions of the two decays in a large class of NP models.

NP effects on the branching ratios of the $K_L \to \pi^0\ell^+\ell^-$ modes have also been considered, in a model-independent way. A measurement of the decays could test contributions from operators not generated by the SM. Effects from scalar, pseudoscalar, tensor, and pseudotensor operators, generated by NP with or without helicity suppression, have been studied and can be disentangled by a more precise measurement of both modes [15].

To conclude, rare Kaon decays are very clean and sensitive probes of NP. The observation of deviations from the SM by the dedicated experiments would be a clear NP signal. Also, the use of both predictions and measurements as constraints of NP can shed light on NP patterns and the structure of flavour violation.

Acknowledgments

I would like to thank Marco Sozzi, convener of the first session on "T, C, P, CP symmetries, accidental symmetries" for the interesting session, the organisers of DISCRETE2010 for the hospitality in Rome, and Joachim Brod and Martin Gorbahn for their continuous support and the fruitful collaboration. This research was supported by the DFG cluster of excellence "Origin and Structure of the Universe".

Journal of Physics: Conference Series **335** (2011) 012016 doi:10.1088/1742-6596/335/1/012016

References

[1] Buchalla G, Buras A J and Lautenbacher M E 1996 *Rev. Mod. Phys.* **68** 1125–1144 (*Preprint* hep-ph/9512380)
[2] Buras A J, Gambino P, Gorbahn M, Jager S and Silvestrini L 2001 *Nucl. Phys.* **B592** 55–91 (*Preprint* hep-ph/0007313)
[3] Glashow S L, Iliopoulos J and Maiani L 1970 *Phys. Rev.* **D2** 1285–1292
[4] Buchalla G and Buras A J 1999 *Nucl. Phys.* **B548** 309–327 (*Preprint* hep-ph/9901288)
[5] Mescia F and Smith C 2007 *Phys. Rev.* **D76** 034017 (*Preprint* 0705.2025)
[6] Isidori G, Mescia F and Smith C 2005 *Nucl. Phys.* **B718** 319–338 (*Preprint* hep-ph/0503107)
[7] Buras A J, Gorbahn M, Haisch U and Nierste U 2006 *JHEP* **11** 002 (*Preprint* hep-ph/0603079)
[8] Brod J and Gorbahn M 2008 *Phys. Rev.* **D78** 034006 (*Preprint* 0805.4119)
[9] Misiak M and Urban J 1999 *Phys. Lett.* **B451** 161–169 (*Preprint* hep-ph/9901278)
[10] Buchalla G and Buras A J 1998 *Phys. Rev.* **D57** 216–223 (*Preprint* hep-ph/9707243)
[11] Brod J, Gorbahn M and Stamou E 2010 (*Preprint* 1009.0947)
[12] Artamonov A V *et al.* (E949) 2008 *Phys. Rev. Lett.* **101** 191802 (*Preprint* 0808.2459)
[13] Ahn J K *et al.* (E391a) 2010 *Phys. Rev.* **D81** 072004 (*Preprint* 0911.4789)
[14] D'Ambrosio G, Ecker G, Isidori G and Portoles J 1998 *JHEP* **08** 004 (*Preprint* hep-ph/9808289)
[15] Mescia F, Smith C and Trine S 2006 *JHEP* **08** 088 (*Preprint* hep-ph/0606081)
[16] Isidori G, Smith C and Unterdorfer R 2004 *Eur. Phys. J.* **C36** 57–66 (*Preprint* hep-ph/0404127)
[17] Alavi-Harati A *et al.* (KTeV) 2000 *Phys. Rev. Lett.* **84** 5279–5282 (*Preprint* hep-ex/0001006)
[18] Alavi-Harati A *et al.* (KTeV) 2004 *Phys. Rev. Lett.* **93** 021805 (*Preprint* hep-ex/0309072)
[19] Lamanna G 2010 *these proceedings*
[20] Mescia F and Smith C URL http://www.lnf.infn.it/wg/vus/content/Krare.html

Tests of the fundamental symmetries in η meson decays

Andrzej Kupsc[1] **and Andreas Wirzba**[2]

[1] Department of Physics and Astronomy, Uppsala University, Box 516, 75120 Uppsala, Sweden and High Energy Physics Department, The Andrzej Soltan Institute for Nuclear Studies, Hoza 69, 00-681, Warsaw, Poland

[2] Institute for Advanced Simulation and Jülich Center for Hadron Physics, Institut für Kernphysik, Forschungszentrum Jülich, D-52425 Jülich, Germany

E-mail: [1] andrzej.kupsc@physics.uu.se, [2] a.wirzba@fz-juelich.de

Abstract. Patterns of chiral symmetry violation and tests of the conservation of the fundamental C, P and CP symmetries are key physics issues in studies of the π^0, η and η' meson decays. These tests include searches for rare or forbidden decays and searches for asymmetries among the decay products in the not-so-rare decays. Some examples for the rare decays are $\eta \to 2\pi$, $\eta \to 4\pi^0$ (CP tests), decays into an odd number of photons (*e.g.*, $\eta \to 3\gamma$) and the decay $\eta \to \pi^0 e^+e^-$ (C tests). In the conversion decay $\eta \to \pi^+\pi^-e^+e^-$, the asymmetry in the distribution of the angle between the two-pion and two-lepton decay planes allows to search for CP violation in a flavor–conserving process beyond the CKM mechanism which is not constrained by the limits on the neutron dipole moment. In addition, the Dalitz decays and the decays into a lepton–antilepton pair are sensitive to contributions from a vector boson responsible for the annihilation of hypothetical light dark matter particles. The muon $g-2$ and the branching ratio for the $\pi^0 \to e^+e^-$ decay are currently the observables where hints of a deviation from the Standard Model predictions are reported. The experimental studies of the π^0, η and η' meson decays are carried out at four European accelerator research facilities: KLOE/KLOE-2 at DAFNE (Frascati), Crystal Ball at MAMI (Mainz), WASA at COSY (Jülich), Crystal Barrel at ELSA (Bonn).

1. Introduction

The π^0, η and η' mesons belong to the longest living hadrons decaying predominantly via electromagnetic or strong processes that are eigenstates of parity (P) and charge conjugation (C) operators. They therefore allow for studies of the (non)-conservation of the C, P and CP symmetries in strong and electromagnetic interactions. In the case of the π^0 and η meson all the strong and electromagnetic decays are either forbidden or at least severely suppressed. Nevertheless, there exist many open (energetically allowed) final states states for the η and η' meson decays, such that a variety of symmetry tests is possible.

The most probable decays of the η meson are [1]: the second order electromagnetic decay into two photons (BR $\approx$ 39%), which is driven by the chiral (triangle) anomaly, and the isospin-violating decays into three pions ($\eta \to \pi^0\pi^0\pi^0$ @BR $\approx$ 33%, $\eta \to \pi^+\pi^-\pi^0$ @BR $\approx$ 23%), which are mainly due to the difference between the u and d quark masses since electromagnetic contributions are suppressed. The first order radiative decay $\eta \to \pi^+\pi^-\gamma$ (BR $\approx$ 5%) is also induced by the chiral anomaly.

Journal of Physics: Conference Series **335** (2011) 012017 doi:10.1088/1742-6596/335/1/012017

Any radiative decay with one or more photons in the final state is accompanied by a process where the photon converts internally into a lepton-antilepton pair [2]. The conversion decays allow to study the *transition form factors* describing the structure of the interaction region. The form factors are also measured in the $\gamma^*\gamma^* \to \eta$ processes in e^+e^- colliders. The precise knowledge of the transition form factors of the π^0 and η mesons is needed for the calculations of the Standard Model (SM) contributions to the muon $g-2$ and to the rare π^0 and η meson decays into a lepton–antilepton pair.

2. Tests of CP symmetry

Most of the tests of the CP symmetry in η decays are modeled as flavor conserving counterparts of the corresponding K_L decay modes. The straightforward test is to search for P and CP violating η decays into two pions. The best experimental limits are from KLOE [3], $\mathrm{BR}(\eta \to \pi^+\pi^-) < 1.3 \cdot 10^{-5}$, and GAMS-$4\pi$ [4], $\mathrm{BR}(\eta \to \pi^0\pi^0) < 3.5 \cdot 10^{-4}$. The theoretical branching ratios are very small [5]: namely in the SM, the decays are G_F^2 processes additionally suppressed by a cancellation of nearly equal terms leading to the branching ratio estimate $\mathrm{BR}(\eta \to \pi\pi) \leqslant 2 \cdot 10^{-27}$. The branching ratio generated by a CP violation in the extended Higgs sector of the electroweak theory is listed in [6] as $\mathrm{BR}(\eta \to \pi^+\pi^-) = 2\mathrm{BR}(\eta \to \pi^0\pi^0) \leqslant 1.2 \cdot 10^{-15}$. If strong interaction physics via the θ term were the culprit for the CP violation, the empirical bound [1] on the electric dipole moment of the neutron would limit the pertinent branching ratio to $\mathrm{BR}(\eta \to \pi\pi)_\theta \leqslant 3 \cdot 10^{-17}$. In fact, as argued by Gorchtein [7] the bound on the electric dipole moment of the neutron can be used to impose a limit on the branching ratio $\mathrm{BR}(\eta \to \pi\pi) \leqslant 3.5 \cdot 10^{-14}$ no matter what the underlying CP violating mechanism would turn out to be.

The search for the barely (energetically) allowed η decay into $4\pi^0$, proposed by Nefkens, see Ref. [8], represents a new kind of CP test with no analog in the K_L system. If all the final pions are in relative s waves, the decay is forbidden by P and CP invariance. Due to very low excess energy of 7.9 MeV this assumption seems to be plausible. However, CP can be conserved if the final state involves higher partial waves than s or p waves. Therefore it is interesting to consider the analogue $\eta' \to 4\pi^0$ decay where the phase space is big enough for the four pions to group into two d-wave pion pairs which can couple to a combined angular momentum of $0^+, 1^+, \cdots, 4^+$. If furthermore the two pairs are then in a relative p-wave state, *i.e.* if the orbital angular momentum between the pairs is 1^-, the full four-pion system can be in a pseudoscalar $J^{PC} = 0^{-+}$ state. Since an odd number of (quasi)-Goldstone bosons is involved, the decay is driven by the chiral anomaly; *e.g.* by the triangle anomaly plus (at least) one additional pion loop/ rescattering or by the direct process $\eta^{(\prime)} \to f_2 f_2 \to (\pi^0\pi^0)(\pi^0\pi^0)$, see Figure 1. Note that the simple flavor structure of the four pions does not only exclude a contribution of the pentangle anomaly but also of the decay path $\eta^{(\prime)} \to \rho^0\rho^0 \to (\pi^0\pi^0)(\pi^0\pi^0)$ because of Bose symmetry.

Using naive phase space arguments the CP allowed 4-pion decays can be estimated – at least close to threshold – to scale as $(p\, r_0)^{1+3+3+2\cdot2+2\cdot2+2\cdot1} = (pr_0)^{17}$ where p is the averaged three-momentum modulus of a π^0 in the η, η' center-of-mass frame and $r_0 \sim 1/m_\rho$ is the effective range of the interaction between two pions expressed by the ρ-meson mass. This predicts a tiny ratio of the partial widths $\Gamma(\eta \to 4\pi^0)/\Gamma(\eta' \to 4\pi^0) \lesssim 10^{-16}$. The estimates of the branching ratio limits, however, have to involve the comparison with other decay channels (*e.g.* $\eta \to \pi^0\pi^0\pi^0$, $\eta \to \pi^+\pi^- l^+ l^-$, $\eta' \to \pi^0\pi^0\eta$ etc.) and are less secure: $\lesssim 10^{-10}$ for the η and (with more caveat because of the increased distance to the threshold) $\lesssim 10^{-8}$ for the η'. A measurement of a non-zero BR for these $\eta/\eta' \to 4\pi^0$ decays above the specified limits would probably be a signal for a CP violating process.

Experimentally the main advantage of the decay is the very low background from the direct pion production if the experiment is carried out in η meson production reactions close to threshold. In fact, the experimental limit obtained by the Crystal Ball collaboration [8], namely

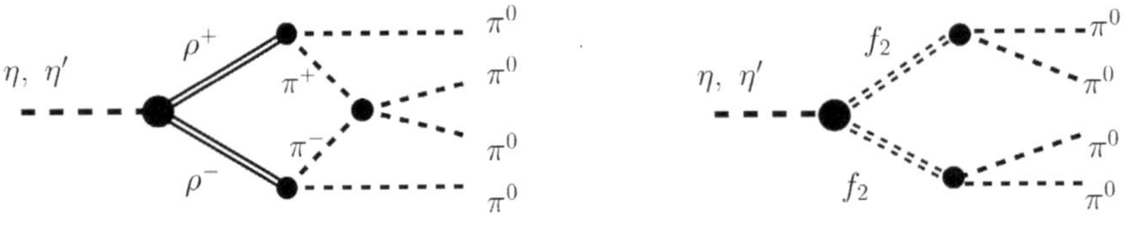

Figure 1. (Left) An example of an anomaly-driven one pion-loop/rescattering diagram contributing to the the CP allowed $\eta/\eta' \to 4\pi^0$ decay. Note that to the left of the rescattering vertex the $\pi^{\pm}\pi^0$ pairs resulting from the vector mesons are p-wave states, whereas to the right of the vertex the upper or lower $\pi^0\pi^0$ pairs have to be d-wave states. Nevertheless, the overall angular momentum is always $J^{PC} = 0^{-+}$ because of the relative p-wave orbital angular momentum between the two vector mesons and between the upper and lower $\pi^0\pi^0$ pairs.
(Right) Direct decay $\eta^{(\prime)} \to f_2 f_2 \to (\pi^0\pi^0)(\pi^0\pi^0)$ where the two virtual isoscalar d-wave bosons f_2 are emitted in a relative p-wave state.

the $BR < 6.9 \cdot 10^{-7}$, is the most sensitive result on any of the η meson decays. As mentioned above, an interesting possibility is to search for the $\eta' \to 4\pi^0$ decay. The experimental limit on the BR is $5 \cdot 10^{-4}$ from the GAMS-2000 spectrometer [9].

The search for linearly polarized photons in the $\eta \to \pi^+\pi^-\gamma$ decay was proposed by Geng and collaborators [10]. As this process is flavor-conserving and involves an extra photo-production vertex, the sensitivity of this CP test is not constrained by the experimental limits on the $\eta \to \pi\pi$ decays or the electric dipole moment of the neutron or by CP violating processes in the flavor changing kaon and B-meson sectors. The practical realization of this idea is to investigate $\eta \to \pi^+\pi^-e^+e^-$ process [11]. The violation of the CP symmetry in the $\eta \to \pi^+\pi^-e^+e^-$ decay would manifest itself as an angular asymmetry between the pionic and the lepton-antilepton decay planes. In the case of the $K_L \to \pi^+\pi^-e^+e^-$ decay such an asymmetry [12, 13, 14] is driven by standard CKM mechanism and it was observed by the KTeV [15] collaboration and the NA48 collaboration [16]. In the case of the η system a possible mechanism leading to such an asymmetry could be the interference between the usual CP allowed magnetic M_1 transition (driven by the chiral anomaly) and a CP violating flavor-conserving electric dipole operator [10] which is sensitive to the strange quark content of the decaying η meson and which is not constrained by SM physics. The experimental bound on this asymmetry, measured by the KLOE collaboration, is compatible with zero, $A_\phi = (-0.6 \pm 3.1) \cdot 10^{-2}$ [17].

3. Tests of C symmetry

There exists only limited knowledge of the C symmetry (non-)conservation in strong and electromagnetic interactions. Any η decay into *neutrals* with an *odd* number of photons in the final state will not conserve C. The simplest example of this class is the decay into three photons which, however, is heavily suppressed: each photon pair has to involve at least two units of angular momentum because of gauge invariance and Bose symmetry [18, 19]; the case of a photon pair with total angular momentum zero is excluded, since it would correspond to a radiative $0 \to 0$ transition, which is forbidden for a real photon [20], while the case of a photon pair with total angular momentum $J = 1$ is in conflict with Bose symmetry [21]. The present upper limit from the KLOE experiment is $\mathrm{BR}(\eta \to \gamma\gamma\gamma) < 1.6 \cdot 10^{-5}$ 90% CL [22].

Also the η decays into π^0 mesons and an odd number of direct photons belong to the above-mentioned class. The simplest prototype would be the $\eta \to \pi^0\gamma$ decay which, however, as a radiative $0 \to 0$ transition with a real photon is absolutely forbidden by angular momentum conservation [20]. The experimental bound on this decay from the Crystal Ball collaboration is

$\mathrm{BR}(\eta \to \pi^0\gamma) \leqslant 9 \cdot 10^{-5}$ [23]. The next simplest cases are then the decays with more than one π^0 but only one photon in the final state: $\eta \to \pi^0\pi^0\gamma$ and $\eta \to 3\pi^0\gamma$. The best branching ratio limits for these decays come again from Crystal Ball experiment and are $5 \cdot 10^{-4}$ and $6 \cdot 10^{-5}$, respectively [24].

In addition, η decays into neutrals plus an odd number of dilepton pairs in the final state as, *e.g.*, the decay $\eta \to \pi^0 e^+e^-$ may be used for tests of charge conjugation invariance. The main SM contribution to this process (shown in Figure 2) comes from C-conserving exchange of two intermediate photons ($\eta \to \pi^0\gamma^*\gamma^* \to \pi^0 e^+e^-$) with a branching ratio of about 10^{-8} [25]. The

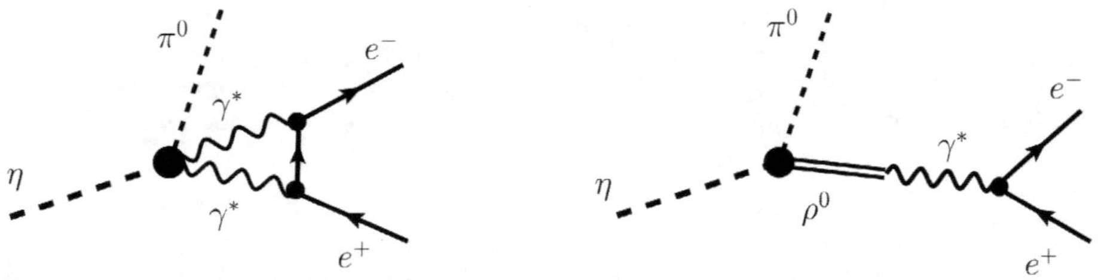

Figure 2. (Left) fourth-order electromagnetic C conserving transition: $\eta \to \pi^0\gamma^*\gamma^* \to \pi^0 e^+e^-$. (Right) Diagram for a possible C violating process: $\eta \to \pi^0\rho \to \pi^0 e^+e^-$.

process with one virtual photon γ^* in the intermediate state is forbidden by C symmetry and, moreover, has to have a vanishing transition form factor in the on-shell limit, since – as stated above – the direct $\eta \to \pi^0\gamma$ vertex is forbidden by angular momentum conservation [20].

At present, the empirical upper limit for the $BR(\eta \to \pi^0 e^+e^-)$ is equal to $4.5 \cdot 10^{-5}$ 90% CL, which comes from an 1975 experiment [26]. The data used for obtaining this limit were analyzed under assumption of a constant decay matrix element and a cut $M(e^+e^-) > 140$ MeV was applied. For the analog decay $\eta \to \pi^0\mu^+\mu^-$, there exists a similar branching ratio limit, $BR(\eta \to \pi^0\mu^+\mu^-) < 5 \cdot 10^{-6}$ [27]. Furthermore, also the single-photon contributions to the η' decays into a π^0 or η pseudoscalar and a dilepton pair (e^+e^- or $\mu^+\mu^-$) belong to the class of C-violating processes.

Finally, the C invariance can be tested also in, *e.g.*, $\eta \to \pi^+\pi^-\gamma$ and $\eta \to \pi^+\pi^-\pi^0$ decays where it can manifest itself as an asymmetry in the energy distributions for π^+ and π^- mesons in the rest frame of the η meson [28, 29]. The asymmetries in the $\eta \to \pi^+\pi^-\gamma$ and $\eta \to \pi^+\pi^-\pi^0$ decays were investigated in few experiments in 1970s [30, 31, 32, 33, 34]. For the $\eta \to \pi^+\pi^-\pi^0$ asymmetries there exist recent limits from the KLOE collaboration with sensitivity 10^{-3} [35].

4. Dark force searches

The rare decays of a neutral pseudoscalar meson P (π^0, η or η') into one dilepton pair (e^+e^- or, if kinematically allowed, $\mu^+\mu^-$) belong to the most interesting processes of low-energy hadron physics, since the SM calculation predicts minuscule tree-level contributions (via a virtual Z boson), such that a window of opportunity for the admixture of physics beyond the standard model might open up as the rates are small. In fact, the relative branching ratio $\mathrm{BR}(P \to l^+l^-)/\mathrm{BR}(P \to \gamma\gamma) \sim (\frac{\alpha}{\pi}\frac{m_l}{m_P})^2$ is not only suppressed by two powers of the fine-structure constant α resulting from the coupling of two virtual photons to the dilepton pair, but also by the helicity mismatch of the two outgoing leptons – indicated by the squared ratio of the lepton mass m_l to the pseudoscalar mass m_P. Recently the KTeV collaboration determined a new precise value of the branching ratio $\mathrm{BR}(\pi^0 \to e^+e^-) = (7.49 \pm 0.29 \pm 0.25) \cdot 10^{-8}$ which exceeds the up-to-date theoretical calculations by Dorokhov *et al.* [36, 37, 38] by 3 standard deviations.

Kahn and collaborators [39] suggested as a possible explanation of this excess the tree-level exchange of an off-shell neutral vector boson U of mass $m_U \sim (10 - 100)$ MeV. The latter is of the type proposed in the light dark matter models by Boehm *et al.* [40, 41, 42, 43] to mediate the annihilation reaction $\chi\chi \to e^+e^-$ of a neutral scalar dark matter particle χ of $(1 - 10)$ MeV mass, such that the excess positrons produced in this annihilation reaction could account for the bright 511 keV line emanating from the center of the Galaxy [39].

In order to explain the mismatch between experiment and theory for the branching ratio $\mathrm{BR}(\pi^0 \to e^+e^-)$, Kahn *et al.* [39] assumed that the neutral vector meson U would couple to the u and d quark fields of the π^0 and the e^+e^- dilepton pair via the axial-vector components g_A^u, g_A^d and g_A^e, respectively, where – for simplicity – a common axial coupling $g_A \equiv g_A^u - g_A^d \equiv g_A^e$ of the order of $g_A = (2.0 \pm 0.5) \cdot 10^{-4} \cdot m_U/(10\,\mathrm{MeV})$ was fitted. In fact, the computation of the partial width of this process is modeled according to the analog SM tree-level process $\pi^0 \to Z^* \to e^+e^-$, where the Z-boson mass m_Z is replaced by the much lighter mass m_U and where the weak coupling is replaced by $g_A^{u,d,e}$. Assuming that the octet axial-vector quark-coupling is of the same order as above, the U boson contribution to the branching ratio $\mathrm{BR}(\eta \to e^+e^-)$ is about 10^{-9}, which is of the same order as the estimates of Dorokhov [36, 37, 38] and much smaller than the experimental bound $2.7 \cdot 10^{-5}$ of the CELSIUS/WASA collaboration [44]. The same fit, however, predicts a contribution of order $2.0 \cdot 10^{-5}$ to $\mathrm{BR}(\eta \to \mu^+\mu^-)$ which is nearly an order of magnitude larger than the measured value $\mathrm{BR}(\eta \to \mu^+\mu^-) \sim (5.7 \pm 0.9) \cdot 10^{-6}$ [45], unless the axial-vector coupling of the U meson to the muon is smaller than g_A^e or the octet axial-vector quark coupling is smaller than $g_A^u - g_A^d$ or both. This is of course a limitation in the predictive power of the U boson exchange mechanism.

A different type of dark matter gauge boson, the a_μ, which is also called U-boson, was suggested by Reece and Wang [46]: here the additional $U(1)_d$ 'dark' gauge boson couples to the SM $U(1)$ gauge boson by a gauge kinetic mixing term $\mathcal{L}_{\mathrm{kin-mix}} = -2\epsilon F^{\mu\nu} F^d_{\mu\nu}$ with a strength $\epsilon \sim 10^{-3}$ or less. The dark $U(1)_d$ group is assumed to be spontaneously broken by the introduction of a dark higgs h_d field which acquires a GeV scale vacuum expectation value, such that $m_U \sim 1$ MeV – few GeV. Under a field redefinition to the standard massless photon, the U boson couples *vectorially* to all SM charged fields as $\epsilon a_\mu J^\mu_{\mathrm{EM}}$, where J^μ_{EM} is the pertinent SM electromagnetic current of the particle. In addition, the U-boson also couples to SM weak neutral currents; however, the corresponding coupling is suppressed by a factor of order m_U^2/m_Z^2. Therefore, this U boson candidate has very small axial couplings to SM matter fields and cannot serve as a candidate for fitting the missing excess in the $\pi^0 \to e^+e^-$ decay.

First searches for U bosons of the Reece and Wang type are on the way. The relevant channels are, *e.g.*, the decays $\phi \to \eta U$, $U \to e^+e^-$ and $\eta \to \gamma U$, $U \to e^+e^-$. A signature for a U boson would be a peak in the pertinent Dalitz decay $\phi \to \eta e^+e^-$ or $\eta \to \gamma e^+e^-$, while the background for such searches is the standard Dalitz decay. The searches for U bosons in the $\phi \to e^+e^-\eta$ conversion decay were reported in this conference by the KLOE collaboration[1].

Finally, as the above mention experiments will all aim at searching for a narrow peak on a large conversion background, one should also mention an attractive alternative for U boson searches: the (C)-forbidden η decays, especially the decay $\eta \to \pi^0 e^+e^-$, where both the conventional and the C-violating background are suppressed for low $M(e^+e^-)$. From the background point of view it is an ideal process to search for a low energy U boson. From the theory point of view, however, one has to cope with the additional suppression by either the violation of C-symmetry, which applies to the Reece-Wang U boson and to the Boehm *et al.* case under vector coupling, or by a violation of parity P in the Boehm *et al.* case under axial-vector coupling.

[1] S. Giovannella in these proceedings.

Journal of Physics: Conference Series **335** (2011) 012017 doi:10.1088/1742-6596/335/1/012017

5. Experiments

The experimental programs for η and η' decays are on the agenda both at e^+e^- colliders and hadro- or photo- production fixed target experiments. Currently active experiments at e^+e^- colliders are KLOE-2 at DAϕNE [47] and BESIII at BEPC [48]: the η and η' mesons then result from radiative decays of ϕ or J/ψ mesons. The π^0, η and η' mesons can copiously be produced in γp, pp or pd interactions not far from the production thresholds. The mesons originate from decays of nucleon isobars (for η mainly from $N^*(1535)$). There are two experiments which use the $\gamma p \to p\eta^{(\prime)}$ reaction close to threshold: Crystal Ball at MAMI [49] and Crystal Barrel at ELSA [50]. The $pp \to pp\eta$ reactions provides the highest useful rate of tagged η mesons enabling

Table 1. Production of η and η' mesons close to threshold; p_{thr} is the beam momentum at threshold; Q is the CMS excess energy corresponding to the maximum or optimal cross section (σ). The last column indicates the total inclusive cross section for a given initial state (σ_T). The data were extracted from references [51, 52, 53, 54, 55, 56, 57].

Reaction	p_{thr} [GeV/c]	Q [MeV]	σ	σ_T [mb]
$pp \to pp\pi^0$	0.777	122	1.3 mb	40
$pp \to pp\eta$	1.981	40	5 μb	40
$pp \to pp\eta'$	3.208	45	300 nb	40
$pd \to ^3\text{He}\eta$	1.569	2	400 nb	80
$pd \to ^3\text{He}\eta'$	2.434	60	1 nb	80
$\pi^- p \to n\eta$	0.684	36	2.6 mb	50
$\pi^- p \to n\eta'$	1.432	100	100 μb	35
$\gamma p \to p\eta$	0.706	60	16 μb	0.30
$\gamma p \to p\eta'$	1.447	40	1 μb	0.15

studies of the rare decays which have a distinct signature as, *e.g.*, the decay $\eta \to e^+e^-$ where an integrated luminosity corresponding to 10^{10} ηs is needed. The η and η' production in pp and pd interactions have the lowest ratio of the cross section to the total inclusive one. However, the signal-to-background ratio can be enhanced by working close to threshold at the price of a lower production cross section. The ideal reaction in this respect is $pd \to ^3\text{He}\eta$, where the cross section rises very quickly, reaching a plateau already at an excess energy of about 1 MeV above threshold [52]. The presence of a doubly charged, heavy ^{3}He ion in the exit channel can easily be detected and thus provides a very efficient and high purity trigger condition. Close to threshold the ions can be filtered using a magnetic spectrometer with the acceptance below one degree.

A feature of the close-to-threshold experiments is that particles from the production processes are emitted in a small forward cone. Their momenta can be identified and measured in a dedicated detector which covers a limited range of scattering angles. The meson production is identified (tagged) by the missing mass. When the beam momentum is known precisely the missing mass is kinematically constrained close to threshold and the resolution depends weakly on the precision of the kinetic energy determination of the particles. For example at light-ion storage rings, the beam-momentum resolution is of the order of 0.1% FWHM and a pp missing-mass resolution of typically a few MeV/c^2 FWHM is achieved. The light decay particles are emitted more isotropically (since the velocity of the center of mass system is not very high) and their registration requires a detector with nearly 4π sr coverage. A typical resolution for the invariant masses of the decay products is a few ten MeV/c^2. The clear separation of the phase space regions for tagging and decay particles in the close-to-threshold tagging helps, *e.g.*, in the determination of the absolute branching ratios $\Gamma_i/\Gamma_{\text{tot}}$.

There are three detectors which have started a second round of $\eta^{(\prime)}$ decay experiments: Crystal

Ball, Crystal Barrel and WASA. The Crystal Ball and Crystal Barrel detectors are now used for $\gamma p \to \eta^{(\prime)} n$ experiments. The third detector, WASA, was build at TSL Uppsala where η decay experiments were carried out in pp and pd interactions until 2005 at the CELSIUS light ion storage ring.

The design of the Crystal Ball [58], Crystal Barrel [59] and WASA [60] detectors is similar. The main part consists of a multi-segmented calorimeter of NaI, CsI(Tl) and CsI(Na), respectively. The detectors are compact enough so that they can be transported to other accelerators. In the present configuration the Crystal Ball and Crystal Barrel set-ups are extended by forward calorimeters consisting of BaF_2 crystals from the TAPS setup [61]. On the other hand they have limited capabilities for a measurement of charged decay products.

WASA is the most complicated of the three detectors. It includes a novel pellet target system allowing for low background and wide angle detection. The design of the detector was optimized for $\pi^0, \eta \to e^+e^-$ decays. In addition to the electromagnetic calorimeter, the central part of the detector includes a superconducting solenoid and a cylindrical mini drift chamber (MDC) with 17 layers of thin-walled (25μm) aluminized mylar tubes. The complete detector system was transported to the Forschungszentrum Jülich in 2005, installed at COSY storage ring, and it was operational already after one year [62]. The relocation of the detector had strengthened the experimental programme since COSY allows for higher energy beams (up to above $pp \to pp\phi$ threshold) and for polarization. The detector system was upgraded with a completely new readout system, the refurbishing of old scintillator elements, and an extension of the forward part of the detector for the higher energies. The maximum useful luminosity of the facility is about $10^{32}\text{cm}^{-2}\text{s}^{-1}$.

KLOE-2, WASA-at-COSY and Crystal Ball aim at a significant improvement of the sensitivity of the discrete symmetries tests in the decays of the η and η' mesons beyond the presently achieved limits. With an expected number of about $10^9 - 10^{10}$ η mesons tagged, a significant improvement is expected. In the nearest future, WASA-at-COSY will provide an order of magnitude improvement of the branching ratio limit for the $\eta \to e^+e^-$ decay and the first investigations of the $\eta \to \pi^0 e^+e^-$ decay in the low electron–positron invariant mass region, $M(e^+e^-) < 120$ MeV. High statistics measurements of the π^0 meson decays are also planned. The KLOE-2 experiment at DAϕNE presented at this conference[2] is starting data taking and the goal is to collect about 10^9 η and 10^7 η' events. With regard to the perspectives of other experiments, one should not forget to mention the inclusive dimuon spectrum at 7 TeV from the CMS experiment (at 40 pb^{-1}), presented by G. Rolandi at this conference. The spectrum shows a clear $\eta \to \mu^+\mu^-$ peak and thus the feasibility of a precise BR measurement using the same technique as previously the NA60 experiment [63]. The first results on η' decays from the BESIII experiment were recently published [64].

Acknowledgments

The authors gratefully acknowledge discussions with Christoph Hanhart, Bastian Kubis and Feng-Kun Guo. This work was in part supported by the European Commission under the 7th Framework Programme through the 'Research Infrastructures' action of the 'Capacities' Programme: FP7-INFRASTRUCTURES-2008-1, Grant Agreement N. 227431.

References

[1] Nakamura K *et al.* (Particle Data Group) 2010 *J. Phys.* G **37** 075021
[2] Dalitz R H 1951 *Proc. Phys. Soc.* A **64** 667–669
[3] Ambrosino F *et al.* (KLOE) 2005 *Phys. Lett.* B **606** 276–280 (*Preprint* hep-ex/0411030)
[4] Blik A M *et al.* 2007 *Phys. Atom. Nucl.* **70** 693–701
[5] Jarlskog C and Shabalin E 2002 *Phys. Scripta* T **99** 23–33

[2] S. Miscetti in these proceedings.

[6] Jarlskog C and Shabalin E 1995 *Phys. Rev.* D **52** 6327–6335
[7] Gorchtein M 2008 (*Preprint* 0803.2906)
[8] Prakhov S *et al.* (Crystal Ball) 2000 *Phys. Rev. Lett.* **84** 4802–4805
[9] Alde D *et al.* (Serpukhov-Brussels-Los Alamos-Annecy(LAPP) Collaboration) 1987 *Z. Phys.* C **36** 603
[10] Geng C Q, Ng J N and Wu T H 2002 *Mod. Phys. Lett.* A **17** 1489–1498 (*Preprint* hep-ph/0201191)
[11] Gao D N 2002 *Mod. Phys. Lett.* A **17** 1583–1588 (*Preprint* hep-ph/0202002)
[12] Cabibbo N and Maksymowicz A 1965 *Phys. Rev.* **137** B438–B443
[13] Pais A and Treiman S B 1968 *Phys. Rev.* **168** 1858–1865
[14] Bijnens J, Colangelo G, Ecker G and Gasser J 1994 (*Preprint* hep-ph/9411311)
[15] Alavi-Harati A *et al.* (KTeV Collaboration) 2000 *Phys. Rev. Lett.* **84** 408–411 (*Preprint* hep-ex/9908020)
[16] Lai A *et al.* (NA48 Collaboration) 2003 *Eur. Phys. J.* C **30** 33–49
[17] Ambrosino F *et al.* (KLOE Collaboration) 2009 *Phys. Lett.* B **675** 283–288 (*Preprint* 0812.4830)
[18] Dicus D A 1975 *Phys. Rev.* D **12** 2133–2136
[19] Grosse H and Liao Y 2001 *Phys. Lett.* B **520** 63–68 (*Preprint* hep-ph/0104260)
[20] Sakurai J J 1964 *Invariance Principles and Elementary Particles* (Princeton: University Press)
[21] Heiliger P, McKellar B and Sehgal L M 1994 *Phys. Lett.* B **327** 145–148 (*Preprint* hep-ph/9402216)
[22] Aloisio A *et al.* (KLOE Collaboration) 2004 *Phys. Lett.* B **591** 49–54 (*Preprint* hep-ex/0402011)
[23] Nefkens B M K, Prakhov S, Allgower C E, Bekrenv V, Briscoe W J *et al.* 2005 *Phys. Rev.* C **72** 035212
[24] Nefkens B M K *et al.* (Crystal Ball Collaboration) 2005 *Phys. Rev. Lett.* **94** 041601
[25] Cheng T P 1967 *Phys. Rev.* **162** 1734–1738
[26] Jane M R *et al.* 1975 *Phys. Lett.* B **59** 99–102
[27] Dzhelyadin R I *et al.* 1981 *Phys. Lett.* B **105** 239
[28] Lee T D 1965 *Phys. Rev.* **139** B1415–B1420
[29] Bernstein J, Feinberg G and Lee T D 1965 *Phys. Rev.* **139** B1650–B1659
[30] Gormley M *et al.* 1970 *Phys. Rev.* D **2** 501–505
[31] Layter J G *et al.* 1972 *Phys. Rev. Lett.* **29** 316–319
[32] Jane M R *et al.* 1974 *Phys. Lett.* B **48** 265
[33] Jane M R *et al.* 1974 *Phys. Lett.* B **48** 260
[34] Thaler J J *et al.* 1972 *Phys. Rev. Lett.* **29** 313–316
[35] Antonelli A *et al.* (KLOE) 2008 *JHEP* **05** 006 (*Preprint* 0801.2642)
[36] Dorokhov A E and Ivanov M A 2007 *Phys. Rev.* D **75** 114007 (*Preprint* 0704.3498)
[37] Dorokhov A E and Ivanov M A 2008 *JETP Lett.* **87** 531–536 (*Preprint* 0803.4493)
[38] Dorokhov A E, Ivanov M A and Kovalenko S G 2009 *Phys. Lett.* B **677** 145–149 (*Preprint* 0903.4249)
[39] Kahn Y, Schmitt M and Tait T M 2008 *Phys. Rev.* D **78** 115002 (*Preprint* 0712.0007)
[40] Boehm C and Fayet P 2004 *Nucl. Phys.* B **683** 219–263 (*Preprint* hep-ph/0305261)
[41] Boehm C, Ensslin T A and Silk J 2004 *J. Phys.* G **30** 279–286 (*Preprint* astro-ph/0208458)
[42] Boehm C *et al.* 2004 *Phys. Rev. Lett.* **92** 101301 (*Preprint* astro-ph/0309686)
[43] Boehm C and Ascasibar Y 2004 *Phys. Rev.* D **70** 115013 (*Preprint* hep-ph/0408213)
[44] Berlowski M *et al.* (CELSIUS/WASA) 2008 *Phys. Rev.* D **77** 032004
[45] Abegg R *et al.* 1994 *Phys. Rev.* D **50** 92–103
[46] Reece M and Wang L T 2009 *JHEP* **07** 051 (*Preprint* 0904.1743)
[47] Amelino-Camelia G *et al.* 2010 *Eur. Phys. J.* C **68** 619–681 (*Preprint* 1003.3868)
[48] Li H B 2009 *J. Phys.* G **36** 085009 (*Preprint* 0902.3032)
[49] Starostin A 2006 *Acta Phys. Slov.* **56** 345
[50] Schmieden H (Crystal-Barrel/TAPS) 2007 (*Preprint* arXiv:0711.4056)
[51] Prakhov S *et al.* 2005 *Phys. Rev.* C **72** 015203
[52] Mayer B *et al.* 1996 *Phys. Rev.* C **53** 2068
[53] Calen H *et al.* 1996 *Phys. Lett.* B **366** 39–43
[54] Khoukaz A *et al.* 2004 *Eur. Phys. J.* A **20** 345–350 (*Preprint* nucl-ex/0401011)
[55] Krusche B *et al.* 1995 *Phys. Rev. Lett.* **74** 3736–3739
[56] Plotzke R *et al.* (SAPHIR) 1998 *Phys. Lett.* B **444** 555–562
[57] Binnie D M *et al.* 1973 *Phys. Rev.* D **8** 2789–2813
[58] Bloom E D and Peck C 1983 *Ann. Rev. Nucl. Part. Sci.* **33** 143–197
[59] Aker E *et al.* (Crystal Barrel) 1992 *Nucl. Instrum. Meth.* A **321** 69
[60] Bargholtz C *et al.* (CELSIUS/WASA) 2008 (*Preprint* arXiv:0803.2657 [nucl-ex])
[61] Novotny R 1991 *IEEE Trans. Nucl. Sci.* **38** 379 ISSN 0018-9499
[62] Adam H H *et al.* (WASA-at-COSY) 2004 (*Preprint* nucl-ex/0411038)
[63] Arnaldi R *et al.* (NA60 Collaboration) 2009 *Phys. Lett.* B **677** 260–266 (*Preprint* 0902.2547)
[64] Ablikim M *et al.* (The BESIII Collaboration) 2011 *Phys. Rev.* D **83** 012003 (*Preprint* 1012.1117)

Journal of Physics: Conference Series **335** (2011) 012018 doi:10.1088/1742-6596/335/1/012018

Indirect tests of the Randall-Sundrum model

Sandro Casagrande

Excellence Cluster Universe, Technische Universität München
D-85748 Garching, Germany

E-mail: sandro.casagrande@ph.tum.de

Abstract. I present phenomenological implications of the Randall-Sundrum model for indirect searches, specifically a selection of flavor observables and Higgs-related collider searches. I review the interplay of constraints from CP violation in flavor physics, possible effects in rare decays, and model-specific protection mechanisms. Deviations in the Higgs couplings to fermions and, at one-loop, to gluons are unexpectedly strong and lead to strong modifications in Higgs searches.

1. Introduction

The gauge hierarchy problem is one of the driving theoretical reasons for the invention and necessity of new physics at the electroweak scale. Its strength can be quantified as a technical naturalness problem induced by the instability of the electroweak scale under radiative corrections, which gains severity given the hierarchy of 16 orders of magnitude between the electroweak and the Planck scale $M_{\rm PL}$. Models with compact extra dimensions explain this hierarchy in terms of geometry, and at the same time also the hierarchical structures observed in the fermionic masses and mixing angles via so-called geometrical sequestering [1].

This can be achieved naturally within the framework of a warped extra dimension, first proposed by Randall and Sundrum (RS) [2]. There one studies the Standard Model (SM) on a background consisting of Minkowski space, embedded in a slice of five-dimensional anti de-Sitter geometry (AdS_5) with curvature k. The fifth dimension is an S^1/Z_2 orbifold of size r, and has two branes located at orbifold fixed points, the UV and the IR brane. The geometry is given by

$$ds^2 = e^{-2\sigma(\phi)}\,\eta_{\mu\nu}\,dx^\mu dx^\nu - r^2 d\phi^2\,, \quad \sigma(\phi) = kr|\phi|\,, \tag{1}$$

where $e^{\sigma(\phi)}$ is called the warp factor. For $L \equiv kr\pi \equiv -\ln(\epsilon) \approx \ln(10^{16})$ the exponential contraction of length scales along the extra dimension mediates between a physical scale $M_{\rm KK} \equiv k' \equiv k\epsilon \sim$ TeV and a fundamental scale $k \lesssim M_{\rm Pl}$. We use the common setup of allowing the gauge and matter fields to spread in the AdS_5 bulk. This SM-like 5d gauge theory is the minimal setup from the dual conformal field theory point of view, as it implies that only the Higgs is a total TeV-scale composite of the strongly interacting sector, while SM gauge fields are fundamental and SM fermions are mostly fundamental with small mixing to the conformal sector [3]. Pragmatically thought, we only protect the scalar mass, leaving masses of higher-spin fundamentals to be protected by chiral or gauge symmetries.[1]

Fermions in the bulk admit a natural explanation of the flavor structure of the SM [5, 6, 7]. We achieve this by exponential localization of the fermion zero modes, which are the lowest

[1] For a more exhaustive review on the development of the model see [4] and the references therein.

mass harmonics of the 5d fermions. They receive their physical mass only from effective Yukawa couplings, resulting from their wave-function overlap with a Higgs boson. The latter is confined very close to the IR brane, where we may use the limit of complete localization for simplicity. The fermionic IR brane value $F(c)$ is exponentially suppressed by the volume factor L, if the normalized bulk mass parameters $c_{Q_i} = +M_{Q_i}/k$, and $c_{q_i} = -M_{q_i}/k$ are smaller than a critical value $-1/2$. Otherwise $F(c)$ is of order one. Here M_{Q_i} and M_{q_i} denote the masses of the five-dimensional $SU(2)_L$ doublet and singlet fermions. The exponential hierarchies are used to generate a hierarchical Yukawa matrix out of anarchical $\mathcal{O}(1)$ fundamental Yukawa coupling

$$\boldsymbol{Y}_q^{\text{eff}} = \text{diag}\left(F(c_{Q_i})\right)\boldsymbol{Y}_q\ \text{diag}\left(F(c_{q_i})\right), \quad F(c) \approx \exp\bigl(\min(c+1/2,0)L\bigr)\sqrt{1+2c}\,. \tag{2}$$

The model features a plethora of higher harmonics above the TeV scale, the KK excitations of gauge bosons starting at $m_{g^{(1)}} \approx 2.45\, M_{\text{KK}}$, graviton $m_{G^{(1)}} \approx 3.83\, M_{\text{KK}}$, etc. each followed by an infinite tower of states, and additional mixings with the SM fields. Yet it is determined by a rather small set of new parameters [8], *e.g.* in the quark sector we have 18 additional moduli and 9 phases. Moreover, explaining the SM fermion spectrum constrains all ratios of the values $F(c)$. E.g. in the quark sector [4, 7, 9], up to $\mathcal{O}(1)$ factors

$$m_{q_i} \sim \frac{v}{\sqrt{2}} F(c_{Q_i})F(c_{q_i})\,, \quad \lambda \sim \frac{F(c_{Q_1})}{F(c_{Q_2})}\,, \quad A\lambda^2 \sim \frac{F(c_{Q_2})}{F(c_{Q_3})}\,, \tag{3}$$

leading to the numerical distribution shown in figure 1, where we used exact formulas [4]. Thus

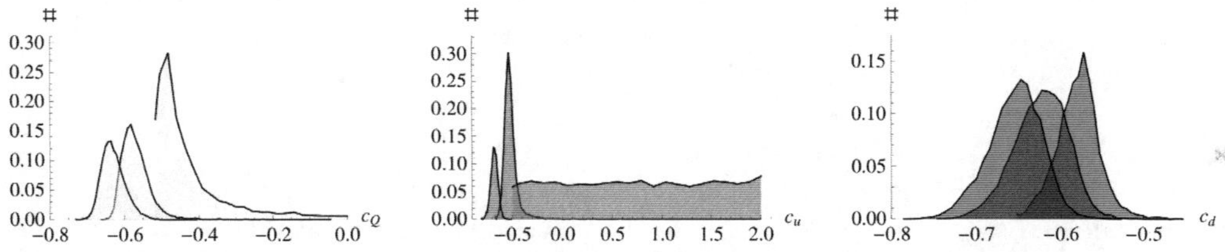

Figure 1. Distribution of bulk masses for $M_{\text{KK}} \in [1, 10]$ TeV, $\epsilon = 10^{-16}$, and flat c_{t_R} prior.

the model has well testable predictions, which could be observed in future measurements, yet we have to assure conformity with existing measurements. Most prominently, the compliance with electroweak precision data tightly restricts the minimal model to $M_{\text{KK}} > 4.0$ TeV at 99% CL [10, 4]. There exist several ways to relax this constraint, like a possible high Higgs mass of about 1 TeV, which would be natural in this model and cancels the model-dependent contribution to the T parameter, leading to a lower bound $M_{\text{KK}} > 2.6$ TeV. Also restricting the model to warp only up to an intermediate and safe UV scale can help to lower the bound (see [11] for restrictions of this approach). Alternatively one can make use of the custodial $O(4)$ symmetry of the Higgs sector, by extending the bulk hypercharge group to $SU(2)_R \times U(1)_X$, breaking it to $U(1)_Y$ on the UV brane by specific orbifold boundary conditions [3]. The Higgs mechanism on the IR brane may safely break $SU(2)_L \times SU(2)_R$ to the diagonal $SU(2)_V$ subgroup. This suppresses tree level contributions to T by three orders of magnitude and implies $M_{\text{KK}} > 2.4$ TeV, which can even be lowered further by calculable fermionic loop contributions to T.[2] A discrete left-right parity symmetry and specifically chosen fermion representations [13] furthermore relax constraints from the precisely measured $Zb\bar{b}$ vertex allowing for a wide range of right handed bottom and left handed top/bottom bulk masses, even for a low new physics scale.

[2] However, quark bi-doublets necessary to protect $Zb_L\bar{b}_L$ typically tend to give a negative contribution to T, which would even raise the bound [12].

Journal of Physics: Conference Series **335** (2011) 012018 doi:10.1088/1742-6596/335/1/012018

It is well known that strong constraints on the parameter space of new physics models can also arise from flavor observables. Depending on the flavor structure and the first quantum order of flavor changing neutral currents, this translates into a bound on the new physics scale, and might result in a tension to the resolution of the hierarchy problem. This is the so called flavor puzzle, which is partly addressed in the RS model in the way fermion hierarchies are generated. The mechanism is referred to as the RS-GIM and we show its numerical implications below.

Given the relatively high bounds on the scale $M_{\rm KK}$ and consequently on new particles, we consider it worthwhile to present possibilities which might indirectly allow for detection or exclusion of a warped extra dimension. We start at the high luminosity frontier with effects on flavor observables, emphasizing the most prominent bound coming from indirect CP violation (CPV) in K–$\bar{K}$ transitions. We show that also direct CPV provides an effective constraint on some rare decays and discuss selected correlations, based on the exhaustive survey [11]. At the high energy frontier we present a recent full one-loop calculation of the effect in Higgs production and decays [14], which shows in strong shifts in the feasibility of Higgs searching strategies.

Since the RS-GIM mechanism implies flavor dependent couplings ordered with the fermion masses, it is most reasonable to expect changes in top couplings. This could affect for instance the forward-backward asymmetry in $t\bar{t}$ production, for which the measurement by CDF has very recently been affirmed [15], and a significant deviation from the NLO QCD prediction in the high invariant mass region has been found. However one must carefully account for the explanation of the flavor structure. The latter renders the leading-order axial-vector couplings to the KK-gluon dominated contributions far too small to explain the discrepancy, and also the interference of vector couplings with the one-loop SM diagrams is outweighed by the leading-order contribution to the symmetric cross section $\sigma_{t\bar{t}}$ [16]. Another coupling of the top quark, namely to a Higgs boson and a charm quark, receives enhanced importance [17] for which we show numerics to round off this presentation.

2. Flavor related effects and constraints

Indirect CPV in $K \to \pi\pi$ decays has an outstanding constraining power in many extensions of the SM. It is induced by $K - \bar{K}$ transition and the most recent theoretical determination of the imaginary value of the corresponding transition amplitude is given by $\epsilon_K = (1.90 \pm 0.26) \cdot 10^{-3}$ [18]. In the effective-theory parametrization

$$\mathcal{H}_{\rm eff}^{\Delta S=2} = \sum_{i=1}^{5} C_i \, Q_i^{sd} + \sum_{i=1}^{3} \tilde{C}_i \, \tilde{Q}_i^{sd} \,, \tag{4}$$

helicity structures beyond the $V-A$ couplings of charged SM currents can contribute to the amplitude. Most importantly $Q_4 = (\bar{d}_R s_L)\,(\bar{d}_L s_R)$, can be generated by the exchange of new colored particles. Its coefficient enters the approximate formula for the K–$\bar{K}$ mixing amplitude

$$\left\langle K^0 \left| \mathcal{H}_{\rm eff,RS}^{\Delta S=2} \right| \bar{K}^0 \right\rangle \propto C_1^{\rm RS} + \widetilde{C}_1^{\rm RS} + 114.8 \left(1 + 0.14 \ln \left(\frac{\mu_{\rm KK}}{3\,{\rm TeV}} \right) \right) \left(C_4^{\rm RS} + \frac{C_5^{\rm RS}}{3.1} \right) . \tag{5}$$

We observe that $C_4^{\rm RS}$ has a large coefficient, which originates from the chiral enhancement of the hadronic matrix elements, and the RG evolution down to 2 GeV. Thus the strongest contribution comes from KK-gluon exchange [9, 19, 20]. The RS-GIM is at work, as each fermion attached to a flavor violating vertex contributes approximately a factor of its IR brane value $F(c)$ leading to $C_4^{\rm RS} \sim \frac{8\pi\alpha_s}{M_{\rm KK}^2} L\, F(c_{Q_1}) F(c_{Q_2}) F(c_{d_1}) F(c_{d_2})$.

In figure 2 we observe that the median value of $|\epsilon_K|$ is consistent with the measurement only for $M_{\rm KK} \gtrsim 8$ TeV. However the 5% quantile crosses the experimentally allowed range already at about 2 TeV. The slow decoupling therefore introduces a moderate amount of arbitrariness

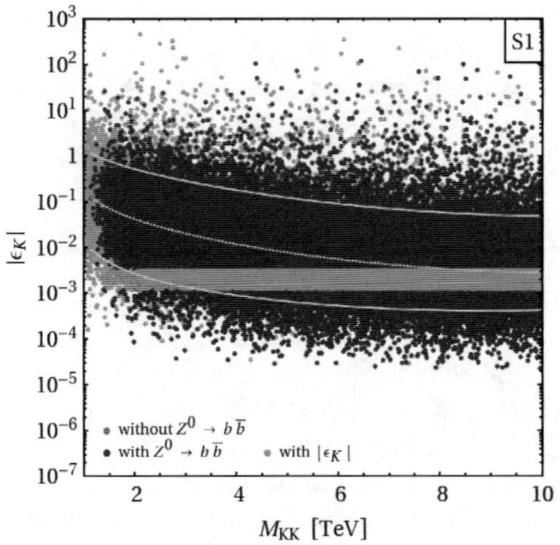

Figure 2. Prediction for $|\epsilon_K|$ in the setup with minimal field content. All points reproduce the observed fermion spectrum, and CKM matrix, are not excessively fine-tuned, and have perturbative Yukawa couplings up to the 2nd KK-level. Points colored blue are consistent with $Z^0 \to b\bar{b}$ at the 99% CL, and orange points fall inside the experimental 95% CL of $|\epsilon_K|$ combined with the theory error. The cyan lines illustrate the decoupling behavior obtained from a fit to the 5%, 50%, and 95% quantile.

in the definition of a bound on the new physics scale inferred from this observable, and it is important to be aware of the different quality compared to the aforementioned bounds from electroweak precision observables.

One possibility to protect $|\epsilon_K|$ from excessive corrections is to arrange for common bulk mass parameters in the sector of the right-handed down-type quarks [21]. In [20] we showed that the typical suppression of $C_4^{\rm RS}$ amounts to a factor of $8 \cdot 10^{-3}$ and discuss a few other resolutions.

The hadronic uncertainties involved in direct CPV in $K \to \pi\pi$ are larger compared to those in ϵ_K. Yet it turns out to be useful to look at ϵ'_K/ϵ_K, which describes the ratio of direct over indirect CPV in this decay and is defined in terms of ratios of $\Delta I = 3/2$ and $\Delta I = 1/2$ transitions

$$\frac{\epsilon'_K}{\epsilon_K} = \frac{1}{\sqrt{2}} \left(\frac{A(K_L \to (\pi\pi)_{I_2})}{A(K_L \to (\pi\pi)_{I_0})} - \frac{A(K_S \to (\pi\pi)_{I_2})}{A(K_S \to (\pi\pi)_{I_0})} \right) . \tag{6}$$

Taking into account the SM Wilson coefficients and the renormalization-group evolution from μ_W to μ_c [22, 23] we observe that contributions to the QCD penguin and the electroweak penguin

$$Q_6 = 4\,(\bar{s}_L^\alpha \gamma^\mu b_L^\beta) \sum\nolimits_q (\bar{q}_R^\beta \gamma_\mu q_R^\alpha)\,, \quad Q_8 = 6\,(\bar{s}_L^\alpha \gamma^\mu b_L^\beta) \sum\nolimits_q Q_q\,(\bar{q}_R^\beta \gamma_\mu q_R^\alpha)\,, \tag{7}$$

play the dominant role [20]. Indeed, both QCD and electroweak penguin contributions generated at a high scale are strongly enhanced by the RG evolution from μ_W down to μ_c. However, QCD

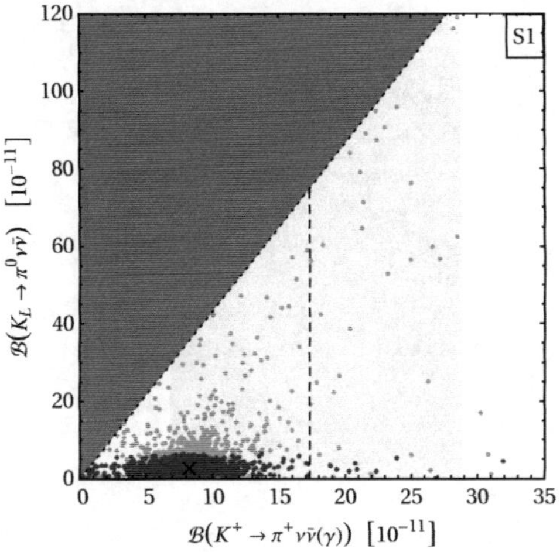

Figure 3. Impact of the ϵ'_K/ϵ_K constraint on $\mathcal{B}(K^+ \to \pi^+\nu\bar{\nu}(\gamma))$ and $\mathcal{B}(K_L \to \pi^0\nu\bar{\nu})$. All points reproduce the quark masses and mixings, and the measured value of $|\epsilon_K|$, observables in $Z^0 \to b\bar{b}$, and B_d–$\bar{B}_d$ mixing. The blue points are consistent with the measured value of ϵ'_K/ϵ_K after varying the hadronic parameters. For details see [20]. The black cross indicates the SM point, the light gray shaded area shows the region excluded by isospin conservation of QCD, and the yellow band displays the experimental 68% CL range.

corrections mainly result from the mixing of Q_6 with current-current operators $Q_{1,2}$, and are therefore mostly unaffected by the given high-scale physics. Concerning the electroweak side, in the RS model only contributions to the color-singlet operator Q_7 arise at the scale $\mu_{\rm KK}$, but they are directly fed via operator mixing into C_8 at the scale μ_W. Therefore ϵ'_K/ϵ_K in RS is mainly influenced by the electroweak penguin, similar to the decays $K \to \pi\nu\bar{\nu}$ where they are the only contributions. For those very decays of charged and neutral Kaons the SM amplitude is tiny, both because of the $V_{ts}^* V_{td}$ suppression induced by the CKM structure, and the GIM mechanism [24]. The necessary low energy matrix elements for the decays are related to precisely measured semileptonic tree-level decays K_{l3}. Therefore they offer a very clean test of physics beyond the SM from the theoretical perspective [25], but are likewise challenging to measure. Fortunately two experiments aim to perform measurements with an anticipated accuracy of 10%, namely NA62 (CERN) for the charged mode and Koto (J-PARC) for the neutral mode.

In figure 3 we observe that the RS and SM amplitudes are comparable in size. Indeed, the factors $F(c_{Q_1})F(c_{Q_2})$ associated with the flavor-violating RS vertex account for approximately the same suppression of λ^5 as in the CKM structure of the SM vertex. Moreover, the relevant RS corrections to the left-handed $s \to dZ^0$ amplitude have a free phase, in contrast to the small phase in the SM. Since the CP violating neutral mode is proportional to the imaginary part of the contributions, we see that it could in principle obtain large relative corrections, which are however excluded by the following argument. The typical ratio of RS over SM contributions in ϵ'_K/ϵ_K is even a factor of ~ 20 larger than in $K \to \pi\nu\bar{\nu}$ case. The reason is well known: The QCD and electroweak penguins cancel in the SM to a large extent, so the value of ϵ'_K/ϵ_K is accidentally small. This entails an effective constraint. The RS amplitude adds linearly and with opposite sign to the SM contribution in ϵ'_K/ϵ_K, and quadratically with the same sign in $K_L \to \pi^0\nu\bar{\nu}$. As the central value of theory prediction for ϵ'_K/ϵ_K is below the experimental one, large enhancements of $K_L \to \pi^0\nu\bar{\nu}$ are disfavored already with a conservative treatment of hadronic errors. A lattice determination of the hadronic matrix elements of $K \to \pi\pi$, taking into account the full final state, is in progress [26]. We emphasize that this would be most welcome in view of the constraining potential.

To briefly show one of the many correlations in the B-meson sector, we select here the decays $\mathcal{B}(B \to X_s l^+ l^-)^{q^2 \in [1,6]\,\mathrm{GeV}^2}$ and $\mathcal{B}(B_s \to \mu^+\mu^-)$ given in figure 4. After imposing the $Zb\bar{b}$

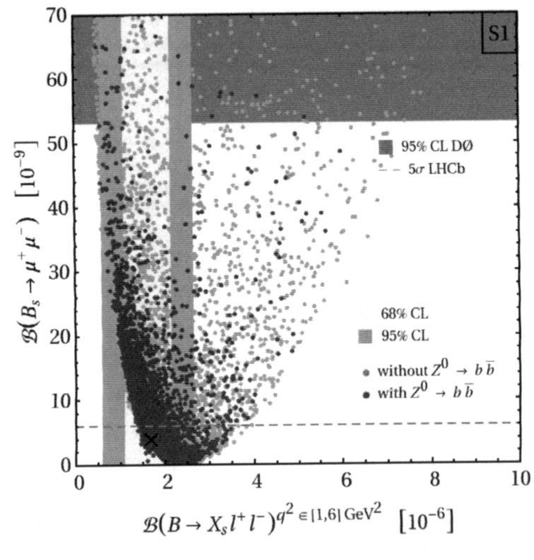

Figure 4. Correlation of the prediction for $\mathcal{B}(B \to X_s l^+ l^-)$ versus $\mathcal{B}(B_s \to \mu^+\mu^-)$ in the setup with minimal field content. The blue points are consistent with the measured $Zb\bar{b}$ couplings at the 99% CL. The black cross indicates the SM expectation. For comparison the regions of 68% (yellow) and 95% (orange) CL are also displayed. Furthermore the 95% CL exclusion of $\mathcal{B}(B_s \to \mu^+\mu^-)$ and the minimum branching fraction allowing for a discovery with 5σ at LHCb are indicated by the red band and dashed line.

constraint, enhancements in the purely leptonic mode imply values of $\mathcal{B}(B \to X_s l^+ l^-)^{q^2 \in [1,6]\,\mathrm{GeV}^2}$ safely within the experimentally allowed range in most of the parameter space. This is expected in models where third generation couplings are most strongly modified [27]. The anti-correlation can be understood by the fact that both modes receive the dominant contribution from axial vector couplings, which are aligned in flavor space, whereas the corresponding SM amplitudes have opposite sign in the two decays. Leptonic B-meson decays belong to the channels that can be studied by three LHC experiments, ATLAS, CMS, and LHCb. They will probe the branching fraction of $B_s \to \mu^+\mu^-$ down to its SM value and might reveal a signal of new physics well ahead of the direct searches.

We conclude the discussion of flavor observables with a final remark on the changes that occur when the minimal bulk gauge symmetry is extended to a custodial symmetry with left-right parity as mentioned in the introduction. The changes in the couplings have been worked out [14] and the leading effects encoded in flavor-universal rescaling factors. Custodial RS effects in $Zd_L^i\bar{d}_L^j$ (and $Zu_R^i\bar{u}_R^j$, $Wu_L^i\bar{u}_L^j$) couplings are reduced by $1/L$, but at the same time increased by a factor of approximately 9 in $Zd_R^i\bar{d}_R^j$ relative to the minimal RS model. Effects in $K \to \pi\nu\bar{\nu}$ decays are therefore typically reduced if t_R has a natural bulk mass, and large enhancements of $B_{s,d} \to \mu^+\mu^-$ like shown in figure 4 are impossible [28]. The change of the chiral nature of the Zsd vertex remarkably inverts a correlation found between the branching ratios of $K^+ \to \pi^+\nu\bar{\nu}$ and $K_L \to \mu^+\mu^-$, but an identification of the chiral structure in this way would require major theoretical progress in the prediction of this radiative decay.

Similarly one can work out the relative coupling shifts for $\Delta F = 2$ processes, which shows that Z' exchange is modified by a factor of approximately -12 in C_5. Although KK-gluon exchange is still dominant in K–$\bar{K}$ transitions, the Z' can be competitive in B_q–$\bar{B}_q$ mixing.

3. Consequences for Higgs production and decays

An estimation of the impact of the RS model on Higgs searches at hadron colliders involves first of all the production, where gluon fusion is the main channel. In the SM, the process receives the main contribution from a top-quark triangle loop. After factoring out numerical constants and $\alpha_s\sqrt{G_F}$ its contribution is encoded in the form factor $A_q^h(4m_t^2/m_h^2)$. A small correction arises from the bottom-quark, but since A_q^h vanishes proportional to its argument the lighter quarks are irrelevant. In the RS model the amplitude for the top-triangle is always reduced but still positive. The origin lies in the modified $ht\bar{t}$ coupling, which is strongly suppressed ($\sim -25\%$ for $M_{\rm KK} = 2$ TeV) because of the compositeness to the top, *i.e.* its IR localization.

Analog to the top triangle all KK modes contribute in triangle diagrams. Since $A_q^h \to 1$ in the limit of a very large quark mass, the total sum over all KK fermions appears at first sight to be divergent, as the mass splitting of the higher modes is asymptotically equidistant. However

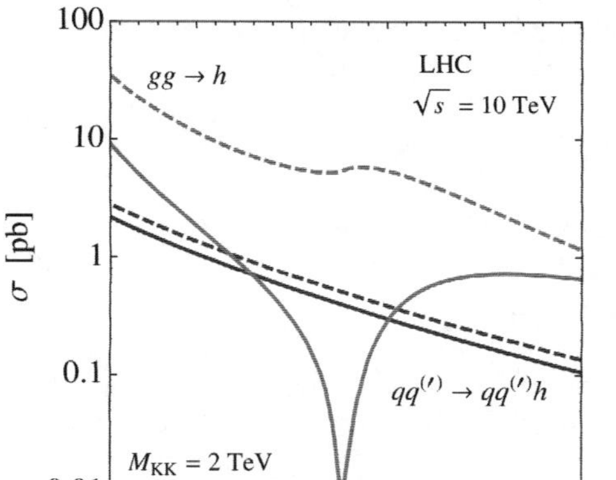

Figure 5. Cross sections for the main Higgs-boson production channels at the LHC for a center-of-mass energy of $\sqrt{s} = 10\,\text{TeV}$, and $M_{\rm KK} = 2$ TeV. The dominant channels are gluon-gluon (red) and weak gauge-boson fusion (blue). The dashed lines illustrate the SM predictions, while the solid lines indicate the results obtained in the custodial RS model.

one can show that a delicate cancellation occurs at each KK level between the two chiralities of the vector-like fermion associated to each SM fermion [14]. As a result the sum converges similar to a geometric series instead of diverging like a harmonic series.

Our numerical analysis shown in figure 5 is based on a rescaling of the SM calculation [29]. We find that the relevant amplitudes show only a small spread when scanning over the experimentally allowed RS parameter space. Therefore we display only the median values. The Higgs mass dependence can be easily understood. Up to the $t\bar{t}$ threshold the form factor increases, but it decreases above. The KK-amplitude is negative and approaches a constant in the heavy Higgs-mass limit. Where it becomes dominant, the full amplitude flips sign, what shows up as a complete destructive interference in the cross section near the $t\bar{t}$ threshold. Numerically the suppression at the LHC ranges between -45% to -100% depending on the Higgs mass. At $M_{\rm KK} = 3$ TeV the suppression is at most -90% and at 5 TeV still up to -40%. Thus we obtain a strong sensitivity of the production channel up to high new physics scales.[3] The blue curve in the plot shows the Higgs production via weak gauge-boson fusion which is affected universally by the modified WWh vertex and thus reduced by -20%. The same factor applies for associated W boson production, and associated top quark pair production is suppressed by -40%.

The calculation of the subsequent possible decay modes of the Higgs involves diagrams similar to the production channel, and furthermore gauge boson and KK-gauge boson triangle diagrams for the $\gamma\gamma$ and $Z\gamma$ final state. The WWh and ZZh couplings experience only a very small reduction. Yet, we recall the reduced production cross section, which renders a detection in the experimentally cleanest mode $h \to Z^*Z^* \to 4l$, in the region $m_h > 180$ GeV, to be more difficult. For a lower Higgs mass the decay into two photons is enhanced and can overcome the suppression of the production cross section. The product $\sigma(gg \to h)\,\mathcal{B}(h \to \gamma\gamma)$ is almost unchanged ($+3\%$) at the scale of $M_{\rm KK} = 2$ TeV and $m_h = 120$ GeV, and even enhanced for a lower new physics scale. However, for an intermediate scale of 3 TeV the cancellation in the production cross section becomes stronger in the low m_h region, and the product drops severely to 24% of the size expected in the SM.

4. Flavor changing top decays

We end this presentation with a surprising observation about the possible size of flavor changing top decays. For a Higgs coupling to fermions one usually expects factors of the corresponding fermion masses $m_{q_i}m_{q_j}/v^2$, stemming from the Yukawa mechanism. Assuming this for the RS model, one would expect the branching ratio of the flavor changing decay $t \to ch$, if kinematically allowed, to be typically two orders of magnitude smaller than for the decay $t \to cZ$.

Models with heavy vector-like quarks can behave differently, if those quarks mix with the SM quarks and additionally couple to the Higgs boson with chiralities opposite to the SM, *i.e.* if the Higgs couples to a heavy left handed singlet and a heavy right handed doublet. Under those conditions an effective operator $(H^\dagger H)(\bar{Q}_i H^c u_j)$ is induced, which contributes differently to masses and Yukawa couplings [30].

A correct treatment of the Higgs localization in RS shows that, even though heavy "wrong chirality" fermion wave-functions are zero on the IR brane by orbifold boundary conditions, the described type of misalignment exists and contributes to Higgs FCNCs [17]. Numerically this makes the decay $t \to ch$ typically even an order of magnitude larger than $t \to cZ$. We show this for a reference Higgs mass $m_h = 150$ GeV in figure 6, where we also indicate the experimental prospect of detecting the process via the subsequent decay into $b\bar{b}$ + jet [31] at later stages of the LHC running. For a lighter Higgs mass of $m_h = 120$ GeV one multiplies the shown branching ratio with a factor of 4.4. Hence we conclude that misalignment effects astoundingly might lift a detection of the htc coupling into reach of the LHC.

[3] Note that the numbers have been obtained in the model with custodial gauge symmetry, where the multiplicity of KK-fermion states is higher than in the minimal model.

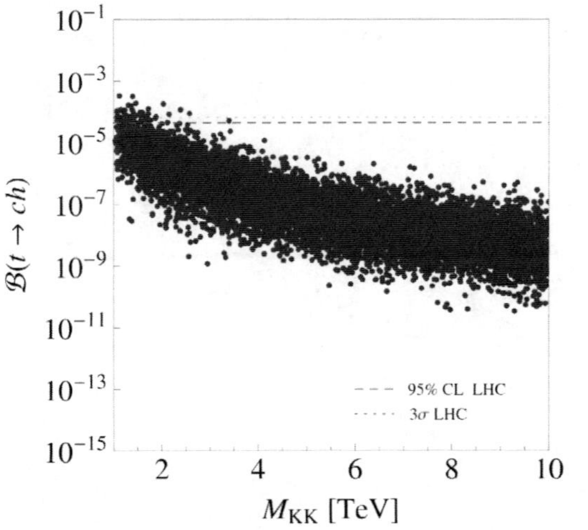

Figure 6. Prediction for the branching ratio of the rare decay $t \to ch$ in the RS model with extended custodial protection and $m_h = 150$ GeV. The red dotted (dashed) lines in the plot indicate the expected discovery (exclusion) sensitivities of LHC for $100\,\text{fb}^{-1}$ integrated luminosity. All points reproduce the correct quark masses, mixing angles, and the CKM phase. In the minimal RS model the constraints from $Zb\bar{b}$ typically eliminate points with pronounced effects.

Acknowledgments

It is a pleasure to thank Martin Bauer, Florian Goertz, Uli Haisch, Matthias Neubert and Torsten Pfoh for the fruitful collaboration and the THEP working group of the Johannes-Gutenberg-Universität Mainz for hospitality and partial support during the writeup of this text. I am also grateful to Joachim Brod and Martin Gorbahn for useful comments on the manuscript. This research was supported by the DFG cluster of excellence "Origin and Structure of the Universe".

References

[1] N. Arkani-Hamed and M. Schmaltz, Phys. Rev. D **61**, 033005 (2000)
[2] L. Randall and R. Sundrum, Phys. Rev. Lett. **83**, 3370 (1999)
[3] K. Agashe, A. Delgado, M. J. May and R. Sundrum, JHEP **0308**, 050 (2003)
[4] S. Casagrande, F. Goertz, U. Haisch, M. Neubert and T. Pfoh, JHEP **0810**, 094 (2008)
[5] Y. Grossman and M. Neubert, Phys. Lett. B **474**, 361 (2000)
[6] T. Gherghetta and A. Pomarol, Nucl. Phys. B **586**, 141 (2000)
[7] S. J. Huber, Nucl. Phys. B **666**, 269 (2003)
[8] K. Agashe, G. Perez and A. Soni, Phys. Rev. D **71**, 016002 (2005)
[9] M. Blanke, A. J. Buras, B. Duling, S. Gori and A. Weiler, JHEP **0903**, 001 (2009)
[10] M. S. Carena, A. Delgado, E. Ponton, T. M. P. Tait and C. E. M. Wagner, Phys. Rev. D **68**, 035010 (2003)
[11] M. Bauer, S. Casagrande, L. Grunder, U. Haisch and M. Neubert, Phys. Rev. D **79**, 076001 (2009)
[12] M. S. Carena, E. Ponton, J. Santiago and C. E. M. Wagner, Phys. Rev. D **76**, 035006 (2007)
[13] K. Agashe, R. Contino, L. Da Rold and A. Pomarol, Phys. Lett. B **641**, 62 (2006)
[14] S. Casagrande, F. Goertz, U. Haisch, M. Neubert and T. Pfoh, JHEP **1009**, 014 (2010)
[15] T. Aaltonen *et al.* [The CDF Collaboration], arXiv:1101.0034 [hep-ex].
[16] M. Bauer, F. Goertz, U. Haisch, T. Pfoh and S. Westhoff, JHEP **1011**, 039 (2010)
[17] A. Azatov, M. Toharia and L. Zhu, Phys. Rev. D **80**, 035016 (2009)
[18] J. Brod and M. Gorbahn, Phys. Rev. D **82**, 094026 (2010)
[19] C. Csaki, A. Falkowski and A. Weiler, JHEP **0809**, 008 (2008)
[20] M. Bauer, S. Casagrande, U. Haisch and M. Neubert, JHEP **1009**, 017 (2010)
[21] J. Santiago, JHEP **0812**, 046 (2008)
[22] A. J. Buras, M. Jamin and M. E. Lautenbacher, Nucl. Phys. B **408**, 209 (1993)
[23] S. Bosch, A. J. Buras, M. Gorbahn, S. Jäger, M. Jamin, M. E. Lautenbacher and L. Silvestrini, Nucl. Phys. B **565**, 3 (2000)
[24] G. Buchalla and A. J. Buras, Nucl. Phys. B **548**, 309 (1999)
[25] J. Brod, M. Gorbahn and E. Stamou, arXiv:1009.0947 [hep-ph].
[26] Q. Liu, RBC and U. collaborations, arXiv:1010.3768 [hep-lat].
[27] U. Haisch and A. Weiler, Phys. Rev. D **76**, 074027 (2007)
[28] M. Blanke, A. J. Buras, B. Duling, K. Gemmler and S. Gori, JHEP **0903**, 108 (2009)
[29] V. Ahrens, T. Becher, M. Neubert and L. L. Yang, Eur. Phys. J. C **62**, 333 (2009)
[30] F. del Aguila, M. Perez-Victoria and J. Santiago, JHEP **0009**, 011 (2000)
[31] J. A. Aguilar-Saavedra and G. C. Branco, Phys. Lett. B **495**, 347 (2000)

Suppression of flavor violation in an A4 warped extra dimensional model

Avihay Kadosh

Center for Theoretical Physics, University of Groningen
9747AG, Groningen, The Netherlands

E-mail: a.kadosh@rug.nl

Abstract. In an attempt to simultaneously explain the observed masses and mixing patterns of both quarks and leptons, we recently proposed a model (JHEP08(2010)115) based on the non abelian discrete flavor group A_4, implemented in a custodial RS setup with a bulk Higgs. We showed that the standard model flavor structure can be realized within the zero mode approximation (ZMA), with nearly TBM neutrino mixing and a realistic CKM matrix with rather mild assumptions. An important advantage of this framework with respect to flavor anarchic models is the vanishing of the dangerous tree level KK gluon contribution to ϵ_K and the suppression of the new physics one loop contributions to the neutron EDM, ϵ'/ϵ, $b \to s\gamma$ and Higgs mediated flavor changing neutral curent (FCNC) processes. These results are obtained beyond the ZMA, in order to account for the the full flavor structure and mixing of the zero modes and first Kaluza-Klein (KK) modes of all generations. The resulting constraints on the KK mass scale are shown to be significantly relaxed compared to the flavour anarchic case, showing explicitly the role of non abelian discrete flavor symmetries in relaxing flavor violation bounds within the RS setup. As a byproduct of our analysis we also obtain the same contributions for the custodial anarchic case with two $SU(2)_R$ doublets for each fermion generation.

1. Introduction

Recently we have proposed a model [1] based on a bulk A_4 flavor symmetry [2] in warped geometry [3], in an attempt to account for the hierarchical charged fermion masses, the hierarchical mixing pattern in the quark sector and the large mixing angles and the mild hierarchy of masses in the neutrino sector. In analogy with a previous RS realization of A_4 for the lepton sector [4], the three generations of left-handed quark doublets are unified into a triplet of A_4; this assignment forbids tree level FCNCs driven by the exchange of KK gauge bosons. The scalar sector of the RS-A_4 model consists of two bulk flavon fields, in addition to a bulk Higgs field. The bulk flavons transform as triplets of A_4, and allow for a complete "cross-talk" [5] between the $A_4 \to Z_2$ spontaneous symmetry breaking (SSB) pattern associated with the heavy neutrino sector - with scalar mediator peaked towards the UV brane - and the $A_4 \to Z_3$ SSB pattern associated with the quark and charged lepton sectors - with scalar mediator peaked towards the IR brane - and allows to obtain realistic masses and almost realistic mixing angles in the quark sector. A bulk custodial symmetry, broken differently at the two branes [6], guarantees the suppression of large contributions to electroweak precision observables [7], such as the Peskin-Takeuchi S, T parameters. However, the mixing between zero modes of

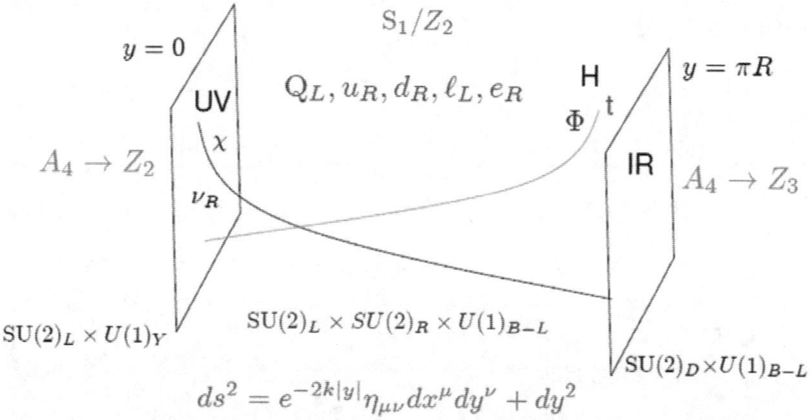

Figure 1. A pictorial description of the RS-A_4 setup. The bulk geometry is described by the metric at the bottom and $k \simeq M_{Pl}$ is the AdS_5 curvature scale. All fields propagate in the bulk and the UV(IR) peaked nature of the heavy RH neutrinos, the Higgs field, the t quark and the A_4 flavons, Φ and χ, is emphasized. The SSB patterns of the bulk symmetries on the UV and IR branes are specified on the side (for A_4) and on the bottom (for $SU(2)_L \times SU(2)_R \times U(1)_{B-L}$) of each brane.

the 5D theory and their Kaluza-Klein (KK) excitations – after 4D reduction – may still cause significant new physics (NP) contributions to SM suppressed flavor changing neutral current (FCNC) processes.

In general, when no additional flavor symmetries are present and the 5D Yukawa matrices are anarchical, FCNC processes are already generated at the tree level by a KK gauge boson exchange [8]. Stringent constraints on the KK scale come from the $K^0 - \overline{K^0}$ oscillation parameter ϵ_K, the radiative decays $b \to s(d)\gamma$ [8, 9], the direct CP violation parameter ϵ'/ϵ_K [10], and especially the neutron electric dipole moment (EDM) [8], also in the presence of an RS-GIM suppression mechanism [11, 12].

Conclusions may differ if a flavor pattern of the Yukawa couplings is assumed to hold in the 5D theory due to bulk flavor symmetries. They typically imply an increased alignment between the 4D fermion mass matrix and the Yukawa and gauge couplings, thus suppressing the amount of flavor violation induced by the interactions with KK states.

In our case, the most relevant consequence of imposing an A_4 flavor symmetry is the degeneracy of the left-handed fermion bulk profiles f_Q, i.e. $diag(f_{Q_1,Q_2,Q_3}) = f_Q \times \mathbb{1}$. In addition, the distribution of phases, CKM and Majorana-like, in the mixing matrices might induce zeros in the imaginary components of the Wilson coefficients contributing to CP violating quantities.

In the following we review the most relevant features of the RS-A_4 model [1] and its phenomenological consequences [13]. In particular, we discuss helicity flipping FCNC processes induced by dipole operators and obtain the most significant constraints coming from $b \to s\gamma$, $Re(\epsilon'/\epsilon_K)$ and the neutron EDM. As a byproduct we obtain the analogous constraints in a custodial anarchic scheme with a bulk Higgs. Implications of adopting a P_{LR} extended custodial setup [17] are then briefly discussed.

2. The RS-A_4 model

The RS-A_4 setup [1] is illustrated in Fig. 1. The bulk geometry is that of a slice of AdS_5 compactified on an orbifold S_1/Z_2 [3] and is described by the metric on the bottom of Fig. 1.

All 5D fermionic fields propagate in the bulk and transform under the following representations of $(SU(3)_c \times SU(2)_L \times SU(2)_R \times U(1)_{B-L}) \times A_4$[1]:

$$
\begin{array}{ll}
Q_L \sim \left(3,2,1,\tfrac{1}{3}\right)(\underline{\mathbf{3}}) & \ell_L \sim (1,2,1,-1)(\underline{\mathbf{3}}) \\
u_R \oplus u'_R \oplus u''_R \sim \left(3,1,2,\tfrac{1}{3}\right)(\underline{\mathbf{1}} \oplus \underline{\mathbf{1}}' \oplus \underline{\mathbf{1}}'') & \nu_R \sim (1,1,2,0)(\underline{\mathbf{3}}) \\
d_R \oplus d'_R \oplus d''_R \sim \left(3,1,2,\tfrac{1}{3}\right)(\underline{\mathbf{1}} \oplus \underline{\mathbf{1}}' \oplus \underline{\mathbf{1}}'') & e_R \oplus e'_R \oplus e''_R \sim (1,1,2,-1)(\underline{\mathbf{1}} \oplus \underline{\mathbf{1}}' \oplus \underline{\mathbf{1}}'') \,.
\end{array}
\tag{1}
$$

The SM fermions (including RH neutrinos) are identified with the zero modes of the 5D fermions above. The zero (and KK) mode profiles are determined by the bulk mass of the corresponding 5D fermion, denoted by $c_{q_L,u_i,d_i}k$ (and BC)[13]. The scalar sector contains the IR peaked Higgs field and the UV and IR peaked flavons, χ and Φ, respectively. They transform as:

$$
\Phi \sim (1,1,1,0)(\underline{\mathbf{3}}), \quad \chi \sim (1,1,1,0)(\underline{\mathbf{3}}), \quad H(1,2,2,0)(\underline{\mathbf{1}}) \,. \tag{2}
$$

The SM Higgs field is identified with the first KK mode of H. All fermionic zero modes acquire masses through Yukawa interactions with the Higgs field and the A_4 flavons after SSB. The 5D $(SU(3)_c \times SU(2)_L \times SU(2)_R \times U(1)_{B-L}) \times A_4$ invariant Yukawa Lagrangian will consist of leading order(LO) UV/IR peaked interactions and next to leading order (NLO) "cross-talk" and "cross-brane" interactions[1]. The LO interactions in the neutrino sector are shown in [1] using the see-saw I mechanism, to induce a tribimaximal (TBM)[14] pattern for neutrino mixing while NLO "cross brane" and "cross talk" interactions, induce small deviations of $\mathcal{O}(0.04)$, which are still in good agreement with the current experimental bounds [15]. Here, we focus on the quark sector and on the phenomenology relevant for flavor and CP violating processes. The relevant terms of the 5D Yukawa lagrangian are of the following form:

$$
\mathcal{L}_{5D}^{Yuk.} = \mathcal{L}_{LO} + \mathcal{L}_{NLO} = \frac{1}{k^2}\overline{Q}_L \Phi H(u_R^{(\prime,\prime\prime)}, d_R^{(\prime,\prime\prime)}) + \frac{1}{k^{3/2}}\overline{Q}_L \Phi \chi H(u_R^{(\prime,\prime\prime)}, d_R^{(\prime,\prime\prime)}) \tag{3}
$$

Notice that the LO interactions are peaked towards the IR brane while the NLO interactions mediate between the two branes due to the presence of both Φ and χ.
The VEV and physical profiles for the bulk scalars are obtained by solving the corresponding equations of motion with a UV/IR localized quartic potential term and an IR/UV localized mass term [18]. In this way one can obtain either UV or IR peaked and also flat profiles depending on the bulk mass and the choice of boundary conditions. The resulting VEV profiles of the RS-A$_4$ scalar sector are:

$$
v_{H(\Phi)}^{5D} = H_0(\phi_0) e^{(2+\beta_{H(\phi)})k(|y|-\pi R)} \qquad v_{\chi}^{5D} = \chi_0 e^{(2-\beta_\chi)k|y|}(1 - e^{(2\beta_\chi)k(|y|-\pi R)}) \,, \tag{4}
$$

where $\beta_{H\Phi,\chi} = \sqrt{4+\mu^2_{H,\Phi,\chi}}$, and $\mu_{H,\Phi,\chi}$ is the bulk mass of the corresponding scalar in units of k, the cutoff of the 5D theory. The following vacua for the Higgs and the A_4 flavons Φ and χ

$$
\langle\Phi\rangle = (v_\phi, v_\phi, v_\phi) \qquad \langle\chi\rangle = (0, v_\chi, 0) \qquad \langle H\rangle = v_H \begin{pmatrix} 1 & 0 \\ 0 & 1 \end{pmatrix}, \tag{5}
$$

provide at LO TBM neutrino mixing and zero quark mixing [2, 5]. The stability of the above vacuum alignment is discussed in [1]. The VEV of Φ induces an A$_4 \to Z_3$ SSB pattern, which in turn induces no quark mixing and is peaked towards the IR brane. Similarly, the VEV of χ induces an A$_4 \to Z_2$ SSB pattern peaked towards the UV brane and is in charge of the TBM mixing pattern in the neutrino sector. Subsequently, NLO interactions break A$_4$ completely and induce quark mixing and deviations from TBM, both in good agreement with experimental

data. The Higgs VEV is in charge of the SSB pattern $SU(2)_L \times SU(2)_R \to SU(2)_D$, which is peaked towards the IR brane. The (gauge) SSB pattern on the UV brane is driven by orbifold BC and a planckian UV localized VEV, which is effectively decoupled from the model.
To summarize the implications of the NLO interactions in the quark sector, we provide the structure of the LO+NLO quark mass matrices in the ZMA [1]:

$$\frac{1}{v}(M+\Delta M)_{u,d} = \underbrace{\begin{pmatrix} y^{4D}_{u,d} & y^{4D}_{c,s} & y^{4D}_{t,b} \\ y^{4D}_{u,d} & \omega y^{4D}_{c,s} & \omega^2 y^{4D}_{t,b} \\ y^{4D}_{u,d} & \omega^2 y^{4D}_{c,s} & \omega y^{4D}_{t,b} \end{pmatrix}}_{\sqrt{3}U(\omega)diag(y^{4D}_{u_i,d_i})} + \begin{pmatrix} f_\chi^{u,d}\tilde{x}_1^{u,d} & f_\chi^{c,s}\tilde{x}_2^{u,d} & f_\chi^{t,b}\tilde{x}_3^{u,d} \\ 0 & 0 & 0 \\ f_\chi^{u,d}\tilde{y}_1^{u,d} & f_\chi^{c,s}\tilde{y}_2^{u,d} & f_\chi^{t,b}\tilde{y}_3^{u,d} \end{pmatrix} \tag{6}$$

where $\omega = e^{2\pi i/3}$, $v = 174\text{GeV}$ is the 4D Higgs VEV, $y^{4D}_{u,c,t,d,s,b}$ are the effective 4D LO Yukawa couplings and $\tilde{x}_i^{u,d}$, $\tilde{y}_i^{u,d}$ are the coefficients of the 5D NLO Yukawa interactions. The function $f_\chi^{u_i,d_i} \simeq 2\beta_\chi C_\chi/(12 - c_{q_L} - c_{u_i d_i}) \simeq 0.05$ is the characteristic suppression of the 4D effective NLO Yukawa interactions and $C_\chi = \chi_0/M_{Pl}^{3/2} \simeq 0.155$. Finally, the unitary matrix, $U(\omega)$ is the LO left diagonalization matrix in both the up and down sectors, $(V_L^{u,d})_{LO}$, which is independent of the LO Yukawa couplings, while $(V_R^{u,d})_{LO} = \mathbb{1}$(see [1]). Using standard perturbative techniques on the matrix in Eq. (6) we obtained $(V_{L,R}^{u,d})_{NLO}$ [1, 13] at $\mathcal{O}\left(f_\chi^{u_i,d_i}(\tilde{x}_i^{u,d}, \tilde{y}_i^{u,d})\right)$ and showed that by a rather mild deviation from a universality assumption on the magnitudes and phases of the NLO coefficients, $\tilde{x}_i^{u,d}, \tilde{y}_i^{u,d}$, we are able to obtain an almost realistic CKM matrix of the characteristic form [1]:

$$V_{CKM} = \begin{pmatrix} 1 & a\lambda & b\lambda^3 \\ -a^*\lambda & 1 & c\lambda^2 \\ -b^*\lambda^3 & -c^*\lambda^2 & 1 \end{pmatrix}. \tag{7}$$

A precise matching of the CKM matrix including deviations from unity of diagonal elements, $|V_{ub}| \neq |V_{td}|$ and phase structure has to be performed at higher order in $f_\chi^{u_i,d_i}(\tilde{x}_i^{u,d}, \tilde{y}_i^{u,d})$.

3. Phenomenology of RS-A$_4$ and constraints on the KK scale

The main difference between the RS-A$_4$ setup and an anarchic RS flavor scheme [8] lies in the degeneracy of fermionic LH bulk mass parameters, which implies the universality of LH zero mode profiles and hence forbids gauge mediated FCNC processes at tree level, including the KK gluon exchange contribution to ϵ_K. The latter provides the most stringent constraint on flavor anarchic models, together with the neutron EDM [8, 10]. However, the choice of the common LH bulk mass parameter, c_q^L is strongly constrained by the matching of the top quark mass (m_t(1.8 TeV)$\approx$ 140 TeV) and the perturbativity bound of the 5D top Yukawa coupling, y_t. Most importantly, when considering the tree level corrections to the $Zb\bar{b}$ coupling against the stringent EWPM at the Z pole, we realize [1] that for an IR scale, $\Lambda_{IR} \simeq 1.8\,\text{TeV}$ and $m_h \approx 200\,\text{GeV}$, c_q^L is constrained to be larger than 0.35. Assigning $c_q^L = 0.4$ and matching with m_t we obtain $y_t < 3$, which easily satisfies the 5D Yukawa perturbativity bound. The constraint on c_q^L from $Zb\bar{b}$ has a moderate dependence on the Higgs mass, such that the constraint $\Lambda_{IR} > 1.8\,\text{TeV}$ for $c_q^L = 0.35$ and $m_h \approx 200GeV$, is relaxed to $\Lambda_{IR} > 1.3\,\text{TeV}$ for $m_h \approx 1\text{TeV}$ [13].

4. Dipole Operators and FCNC processes

FCNC processes provide us with some of the most stringent constraints for physics beyond the SM and this is also the case for the framework of flavor anarchic models in warped extra dimensions [8, 9, 10]. In the quark sector, significant bounds on the KK mass scale typically arise from the neutron electric dipole moment (EDM), the CP violation parameters, $Re(\epsilon'/\epsilon_K)$

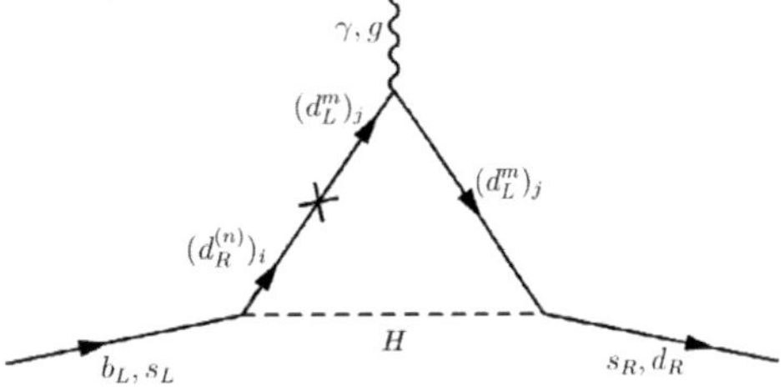

Figure 2. One-loop down-type (neutral Higgs) contribution to $b \to s\gamma$, ϵ'/ϵ_K and the neutron EDM (for external d quarks). The analogous one-loop up-type contribution (charged Higgs) contains internal up-type KK modes.

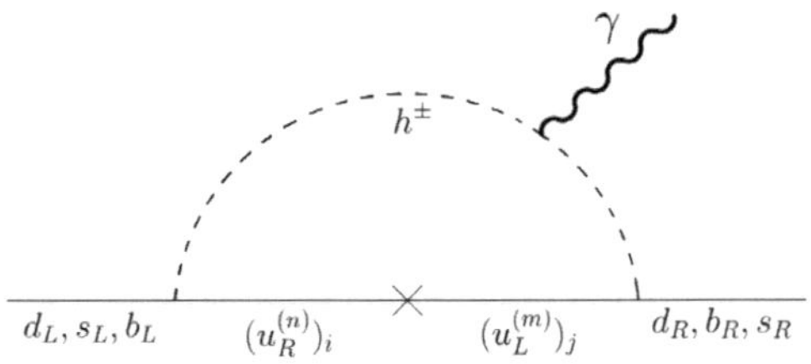

Figure 3. Charged Higgs one-loop contribution to $b \to s\gamma$ and the neutron EDM. The latter has external d quarks.

and ϵ_K, and radiative B decays such as $b \to s\gamma$. All these processes are mediated by effective dipole operators. It is also well known [19] that SM interactions only induce, to leading order, the single chirality operators $O_{7\gamma}$ and O_{8g} $(j > i)$

$$O_{7\gamma(8g)} = \bar{d}^i_L \sigma^{\mu\nu} d^j_R F_{\mu\nu}(G_{\mu\nu}). \tag{8}$$

The new RS-A_4 contributions to the FCNC processes we are interested in are generated at one-loop by the Yukawa interactions between SM fermions and their KK excitations, leading to the diagrams shown in Fig. 2 and Fig. 3.

In the mass insertion approximation, the flavor part of the corresponding amplitude can be written in terms of spurions [8]. To illustrate it, we focus on the down type contribution to the dipole operators in Fig.2, for which the corresponding Wilson coefficient is given by:

$$(C^{d-type}_{7\gamma(8g)})_{ij} = A^{1L} \frac{v}{M_{KK}} \left(F_Q \, \hat{Y}_d \hat{Y}_d^\dagger \hat{Y}_d \, F_d \right)_{ij}, \tag{9}$$

where $\hat{Y}_d$ are the 5D Yukawa matrices and the fermion profile matrices are $F_{Q,u,d} = diag(f^{-1}_{Q_i,u_j,d_j}) = (1/\sqrt{2k}) diag(\hat{\chi}_{0_{Q_i,u_i,d_i}})$ and $\hat{\chi}_{0_{Q_i,u_i,d_i}}$ is the canonically normalized zero mode wave function evaluated on the IR brane. Finally, the factor $A^{1L} = 1/(64\pi^2 M_{KK})$ comes from the one-loop integral for the diagram in Fig. 2.

To account for overlap effects and illustrate the flavor patterns generated in RS-A_4 we write the LO mass matrix for the first generation in the down-type sector following [9, 10], including the zero modes and first level KK modes and overlap effects [13]

$$\frac{\hat{\mathbf{M}}_d^{KK}}{(M_{KK})} = \begin{pmatrix} \bar{Q}_L^{d(0)} \\ \bar{d}_L^{(1^{--})} \\ \bar{Q}_L^{d(1)} \\ \bar{\tilde{d}}_L^{(1^{+-})} \end{pmatrix}^T \begin{pmatrix} \breve{y}_d f_Q^{-1} f_d^{-1} r_{00} x & 0 & \breve{y}_d f_Q^{-1} r_{01} x & \breve{y}_u f_Q^{-1} r_{101} x \\ 0 & \breve{y}_d^* r_{22} x & 1 & 0 \\ \breve{y}_d f_d^{-1} r_{10} x & 1 & \breve{y}_d r_{11} x & \breve{y}_u r_{111} x \\ 0 & \breve{y}_u^* r_{222} x & 0 & 1 \end{pmatrix} \begin{pmatrix} d_R^{(0)} \\ Q_R^{d(1^{--})} \\ d_R^{(1)} \\ \tilde{d}_R^{(1^{-+})} \end{pmatrix}, \tag{10}$$

where we factorized a common KK mass scale M_{KK}, $\breve{y}_{u,d} \equiv 2y_{u,d} v_\Phi^{4D} e^{k\pi R}/k$ and the perturbative expansion parameter is defined as $x \equiv v/M_{KK}$. In the above equation the various r's denote the ratio of the bulk and IR localized effective couplings of the modes corresponding to the matrix element in question. For simplicity, we define $r_{111} \equiv r_{11^{-+}}$, $r_{101} \equiv r_{01^{-+}}$, $r_{22} \equiv r_{1^-1^-}$, $r_{222} \equiv r_{1^-1^{+-}}$ and the notation for the rest of the overlaps is straightforward. The corresponding

Yukawa matrix, $\hat{Y}^d_{KK}$ is obtained by simply eliminating x and the 1's from the above matrix and it leads the flavor structure of the contributions of $(++)$ and $(-+)$ KK modes to the processes in Figs. 2 and 3, including overlap effects [13],

$$
\begin{aligned}
(C^{d,u}_{7\gamma(8g)})_{(++)} &\propto F_Q \hat{Y}_{d,u}\, r_{01}(c_{Q_i}, c_{d_{\ell_1},u_{\ell_1}}, \beta)\, \hat{Y}^\dagger_{d,u}\, r_{11}(c_{d_{\ell_1},u_{\ell_1}}, c_{Q_{\ell_2}}, \beta)\, \hat{Y}_{d,d}\, r_{10}(c_{Q_{\ell_2}}, c_{d_j,d_j}) F_d\,, \\
(C^{d,u}_{7\gamma(8g)})_{(-+)} &\propto F_Q \hat{Y}_{u,d}\, r_{01-+}(c_{Q_i}, c_{u_{\ell_1},d_{\ell_1}}, \beta)\, \hat{Y}^\dagger_{u,d}\, r_{1-+1}(c_{u_{\ell_1},d_{\ell_1}}, c_{Q_{\ell_2}}, \beta)\, \hat{Y}_{d,d}\, r_{10}(c_{Q_{\ell_2}}, c_{d_j,d_j}, \beta) F_d\,.
\end{aligned}
\tag{11}
$$

Considering the degeneracy of LH bulk mass parameters in the RS-A_4 setup and using the ZMA diagonalization matrices to rotate the corresponding contribution to the mass basis we obtain for the down type contribution in Fig. 2 [13]:

$$
\begin{aligned}
(C^{d-type}_{7\gamma(8g)})_{ij} &= \frac{m_{d_i} A^{1L} f_Q^2}{v^2 M_{KK}} \Big[V_R^{d\dagger} diag(f^2_{d,s,b}) (\hat{r}^d_{00})^{-1} \tilde{r}^d_{01}\, \tilde{r}^d_{11} (\hat{r}^d_{00})^{-1} V^d_R\, diag(m^2_{d,s,b}) \\
&\quad \times V_R^{d\dagger} (\hat{r}^d_{00})^{-1} \hat{r}^d_{10}\, V^d_R \Big]_{ij}\,,
\end{aligned}
\tag{12}
$$

where $\tilde{r}^{u,d}_{01} \tilde{r}^{u,d}_{11} = \hat{r}^{u,d}_{01} \hat{r}^{u,d}_{11} + \hat{r}^{u,d}_{01-+} \hat{r}^{u,d}_{1-+1}$ and all overlap matrices are real and diagonal. Since the overlap correction factors turn out to have a very mild generational dependence [13], we can take them to be degenerate and parameterize their effect by an overall multiplicative factor, B^d_P.

To improve on the mass insertion approximation we diagonalize the one generation KK-zero mass matrix in Eq. (10) to directly obtain the physical coupling between a zero mode and the three KK modes of the same generation, including the previously unaccounted $(--)$ KK modes, while generational mixing effects are still estimated using the spurion structure of Eq. (12) with all overlap matrices set to $\mathbb{1}$ [9, 10, 13].

The down-type multiplicative overlap correction factor, B^d_P, corresponding to the process in Fig. 2, will thus be extracted from

$$
(\mathcal{A}_{ij})_D^{overlap} = \left. \frac{\sum_n ((O_L^{(d_i)KK})^\dagger \hat{Y}^{KK}_{d_i} O_R^{(d_i)KK})_{1n} ((O_L^{(d_i)KK})^\dagger \hat{Y}^{KK}_{d_j} O_R^{(d_i)KK})_{n1}}{\left(M^{d_i\,(n)}_{KK}/M_{KK}\right)} \right|_{overlap} . \tag{13}
$$

We use the above procedure for obtaining the constraints on the KK scale arising from $Re(\epsilon'/\epsilon_K)$ and $b \to s\gamma$. However, by using the same procedure, the contribution to the neutron EDM turns out to be vanishing to first order in $f_\chi^{u_i,d_i}(\tilde{x}_i^{u,d}, \tilde{y}_i^{u,d})$. We can partially account for the latter by using the analog of Eq. (12) for the up type RS-A_4 neutron EDM contribution with non degenerate overlap correction factors. This yields an $\mathcal{O}(f_\chi^2 \Delta r)$ suppressed contribution, where Δr is the characteristic difference between nearly degenerate overlap correction factors and is numerically of $\mathcal{O}(0.01)$. The analytical estimations are then compared with the numerical solution of the full three generations case, where a diagonalization of the full 12×12 matrix is performed. The results of this comparison are reported in Fig. 4.

4.1. The custodial anarchic case

As a byproduct of our analysis we obtain the one generation KK estimations for a flavor anarchic setup with an additional $SU(2)_R$ doublet per generation and compare with the results of [9, 10]. The anarchic mass matrix is obtained from the matrix in Eq. (10) by eliminating all Yukawas and redefining $x = vY/M_{KK}$, where Y is the characteristic magnitude of the 5D anarchic Yukawa couplings. Generational mixing effects will simply contribute $f_{Q_i}^{-1} f_{d_j}^{-1}$ for the (ij) component of

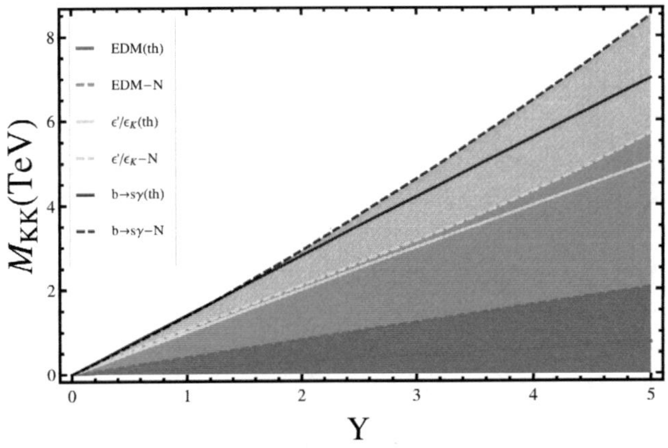

Figure 4. Bounds on the KK mass scale M_{KK} as a function of the overall Yukawa scale Y for the neutron EDM (bottom, red) ϵ'/ϵ (center, green) and $b \to s\gamma$ (top, black). The analytical (th) results (solid lines) are obtained within the one-generation approximation combined with the spurion-overlap analysis (with the exception of the EDM (see text)) and compared with the numerical (N) results (dashed lines) of the three-generation case. In both cases, predictions are obtained for the model parameters that lead to a realistic CKM matrix.

the dipole operators in Eq. (8). Together with the diagonalization of the mass matrix we obtain:

$$(C^{d}_{7\gamma(8g)})^{Cu.-An.}_{ij} = \frac{f_{Q_i}^{-1} f_{d_j}^{-1} Y^3}{64\pi^2 M_{KK}^2} \left(r_{01}r_{10}(r_{11}+r_{111}+9r_{22}+r_{222}) + r_{01}r_{101}(r_{11}+r_{111}+r_{22}+9r_{222})\right) \tag{14}$$

Using the anarchic bulk mass assignments [8] and $\beta_H = 0$, which yields the weakest bounds as in [10], we obtain the constraints on the KK mass scale from $\mathrm{Re}(\epsilon'/\epsilon_K)$ and $b \to s\gamma$ in the custodial anarchic case

$$(M_{KK})_{\epsilon'/\epsilon} \gtrsim 1.4\,\mathrm{TeV} \qquad (M_{KK})_{b\to s\gamma} \gtrsim 0.5\,\mathrm{TeV} \tag{15}$$

When combined with the constraint from ϵ_K [10], which remains unchanged in the custodial anarchic case the resulting bound on the KK scale is:

$$(M_{KK})^{Cu.-An.} \gtrsim 6.6\,\mathrm{TeV} \tag{16}$$

pushing the KK mass scale close to the limit of the LHC reach.

5. Conclusions

We have illustrated a model based on an A_4 flavor symmetry and implemented in a warped extra dimensional setup. The warped geometry supplements us, as usual, with solutions to both the gauge hierarchy problem and the fermion mass hierarchy. Simultaneously, the bulk A_4 flavor symmetry and its distinct SSB patterns towards the UV/IR branes by the A_4 flavons, Φ and χ, allows us to obtain a TBM neutrino mixing pattern and zero quark mixing as LO results, while NLO corrections induce realistic quark mixing and small deviations from TBM. While the experimentally allowed ranges for neutrino mixing angles [15] can also be explained with a bi-maximal mixing pattern at LO with large NLO corrections [21], they can be still accounted

for when considering NLO "cross-talk" and "cross-brane" interactions in the RS-A_4 model. In the quark sector, an exact matching of the CKM matrix is possible at the price of increased tuning of NLO 5D Yukawa couplings.
The most stringent constraint on the RS-A_4 model comes from the EWPM of the effective $Zb\bar{b}$ coupling. Together with the perturbativity bound of the 5D top Yukawa coupling it implies a constraint $M_{KK} \gtrsim 4\,\text{TeV}$ for $m_h \approx 200\,\text{GeV}$ and strongly constrains the common LH bulk mass parameter. On the other hand the constraints arising from the neutron EDM, ϵ'/ϵ_K and $b \to s\gamma$ were shown both analytically and numerically (see Fig. 4) to induce milder constraints, dominated by the one arising from $b \to s\gamma$.
To release the strong constraint from $Zb\bar{b}$ it is relevant to consider the implications of embedding RS-A_4 in a P_{LR} extended custodial setup [17, 22], where the dominant contributions to the anomalous $Zb\bar{b}$ coupling are strongly suppressed. However, since this embedding requires an extended non trivial fermion sector, we expect it to increase the bounds from the various FCNC processes we considered due to the presence of new KK modes and a richer flavor structure. In addition, we anticipate the typical prospects for observing $t \to cZ$ and $t \to ch$ and the reduction of Higgs production cross sections [22] to be mildly modified. A detailed study of the various phenomenological implications of a P_{LR} custodial RS-A_4 setup will be the subject of a future publication.

Acknowledgments
We thank the organizers of *DISCRETE'10 - Symposium on Prospects in the Physics of Discrete Symmetries* for the opportunity to present our work and for the lovely hospitality in Rome.

References
[1] A. Kadosh and E. Pallante, JHEP **1008**, 115 (2010) [arXiv:1004.0321 [hep-ph]].
[2] E. Ma and G. Rajasekaran, Phys. Rev. **D64**, 113012 (2001) K. S. Babu, E. Ma and J. W. F. Valle, Phys. Lett. **B552** 207 2003; E. Ma, Phys. Rev. **D70**, 031901(2004) Talk at SI2004, Fuji-Yoshida, Japan, hep-ph/0409075.
[3] L. Randall and R. Sundrum, Phys. Rev. Lett. **83**, 3370 (1999) [arXiv:hep-ph/9905221]; L. Randall and R. Sundrum, Phys. Rev. Lett. **83**, 4690 (1999) [arXiv:hep-th/9906064].
[4] C. Csaki, C. Delaunay, C. Grojean and Y. Grossman, JHEP **0810**, 055 (2008) [arXiv:0806.0356 [hep-ph]].
[5] X. G. He, Y. Y. Keum and R. R. Volkas, JHEP **0604**, 039 (2006) [arXiv:hep-ph/0601001].
[6] K. Agashe, A. Delgado, M. J. May and R. Sundrum, JHEP **0308**, 050 (2003) [arXiv:hep-ph/0308036].
[7] M. Carena, E. Ponton, J. Santiago and C.E.M. Wagner, arXiv:hep-ph/0701055.
[8] K. Agashe, G. Perez and A. Soni, Phys. Rev. D **71**, 016002 (2005) [arXiv:hep-ph/0408134];
[9] K. Agashe, A. Azatov and L. Zhu, Phys. Rev. D **79**, 056006 (2009) [arXiv:0810.1016 [hep-ph]].
[10] O. Gedalia, G. Isidori and G. Perez, Phys. Lett. **B 682**, 200 (2009) [arXiv:0905.3264 [hep-ph]].
[11] K. Agashe, G. Perez and A. Soni, Phys. Rev. Lett. **93**, 201804 (2004) [arXiv:hep-ph/0406101];
[12] G. Cacciapaglia, C. Csaki, J. Galloway, G. Marandella, J. Terning and A. Weiler, JHEP **0804**, 006 (2008)
[13] A. Kadosh and E. Pallante, [arXiv:1101.5420 [hep-ph]].
[14] P. F. Harrison, D. H. Perkins and W. G. Scott, Phys. Lett. B **458**, 79 (1999) [arXiv:hep-ph/9904297].
[15] G. L. Fogli, E. Lisi, A. Marrone, A. Palazzo and A. M. Rotunno, [arXiv:0809.2936 [hep-ph]]; T. Schwetz, M. A. Tortola, and J. W. F. Valle, New J. Phys. **10** (2008) 113011 [arXiv:0808.2016[hep-ph]].
[16] C. Amsler *et al.* [Particle Data Group], Phys. Lett. **B667** 1 (2008).
[17] K. Agashe, R. Contino, L. Da Rold and A. Pomarol, Phys. Lett. B **641**, 62 (2006) [arXiv:hep-ph/0605341].
[18] W. D. Goldberger and M. B. Wise, Phys. Rev. Lett. **83**, 4922 (1999) [arXiv:hep-ph/9907447]; G. Cacciapaglia, C. Csaki, G. Marandella and J. Terning, JHEP **0702**, 036 (2007) [arXiv:hep-ph/0611358].
[19] G. Buchalla, A. J. Buras and M. E. Lautenbacher, Rev. Mod. Phys. **68**, 1125 (1996) [arXiv:hep-ph/9512380].
[20] C. A. Baker *et al.*, Phys. Rev. Lett. **97**, 131801 (2006) [arXiv:hep-ex/0602020].
[21] L. Merlo, Talk given at DISCRETE10 - Symposium on Prospects in the Physics of Discrete Symmetries.
[22] S. Casagrande, F. Goertz, U. Haisch, M. Neubert and T. Pfoh, JHEP **1009**, 014 (2010) [arXiv:1005.4315 [hep-ph]].

Journal of Physics: Conference Series **335** (2011) 012020 doi:10.1088/1742-6596/335/1/012020

CP-violating Higgs sector of the MSSM through $gg \to H_1 \to \gamma\gamma$

S. Hesselbach[a1] **S. Moretti**[b2] **S. Munir**[c3] **P. Poulose**[d4]

[a] GSI Helmholtzzentrum fuer Schwerionenforschung GmbH, Planckstrasse 1, 64291 Darmstadt, Deutschland

[b] School of Physics & Astronomy, University of Southampton, Highfield, Southampton SO17 1BJ, UK

[c] Department of Physics, COMSATS Institute of Information Technology, Defence Road, Lahore-54000, Pakistan

[d] Physics Department, IIT Guwahati, Assam 781039, INDIA

E-mail: [1]`s.hesselbach@gsi.de`

E-mail: [2]`stefano@soton.ac.uk`

E-mail: [3]`smunir@ciitlahore.edu.pk`

E-mail: [4]`poulose@iitg.ernet.in`

Abstract. We study the effect of explicit CP violation in the Higgs sector of the MSSM in the di-photon decay of the lightest CP-mixed Higgs state. Further it is shown that the gluon fusion production mechanism along with the di-photon decay enhances CP-violating effects for a large set of suitably chosen parameter values.

1. Introduction

Phenomenology of the Minimal Supersymmetric Standard Model (MSSM) with CP-violating couplings can be very different from one of the CP-conserving MSSM. While spontaneous CP violation in the MSSM arising from complex vacuum expectation values of the Higgs fields is essentially ruled out, many parameters in the MSSM, which are absent in the Standard Model, can be complex, leading to large possibilities for CP violation. To avoid conflict with the low energy experimental constraints like those coming from measurements of the Electric Dipole Moments (EDMs) of the electron, neutron and muon, the couplings relevant to the first and second generations are considered to be real.

By building on the results of Refs. [1, 2] (for the production) and [3]–[6] (for the decay) – see also Refs. [7]–[10] – we will look here at the Large Hadron Collider (LHC) phenomenology of the $gg \to H_1 \to \gamma\gamma$ process (where H_1 labels the lightest neutral Higgs state of the CP-violating MSSM), which involves the (leading) direct effects of CP violation through couplings of the H_i ($i = 1, 2, 3$ corresponding to the three neutral Higgs bosons) to sparticles in the loops as well as the (sub-leading) indirect effects through scalar-pseudoscalar mixing yielding the CP-mixed state H_i. (See Ref. [11] for some preliminary accounts in this respect.) Here we summarize the results of [12] focusing especially on the effects of a light stop in the production of a CP-mixed H_1 state by gluon fusion and its decay into two photons.

2. CP violation in the di-photon Higgs search channel

Explicit CP violation arises in the Higgs sector of the MSSM when various related couplings become complex. One consequence is that the physical Higgs bosons are no more CP eigenstates, but a mixture of these [13]–[16]. One may then look at the production and decay of the lightest of the physical Higgs particles, hereafter labeled H_1. CP-violating effects in the combined production and decay process enter through complex H_i-$\tilde{f}$-$\tilde{f}^*$ couplings at the production and decay levels plus mixing in the propagator ($\tilde{f}$ represents a sfermion).

CP-conserving and CP-violating effects enter at *the same perturbative order* in the cross section for $gg \to$ Higgs $\to \gamma\gamma$, so that the latter is an ideal laboratory to pursue studies of the complex (or otherwise) nature of the soft supersymmetry (SUSY) breaking parameters concerned. In contrast, notice that CP-violating effects through mixing in the propagator enter only at higher order, through self-energies, as there is already a tree-level CP-conserving contribution to the propagator[1]. The dynamics of CP-violating effects in the production and decay stages are rather similar, given that the same diagrammatic topologies are involved (see Fig. 1), in particular, as shown in our previous work [3]–[6], we expect a strong impact of a light stop quark in some regions of the parameter space. The propagator is considered in the following way. A Higgs particle, H_i, produced through gluon fusion, can be converted into another mass eigenstate, H_j, through interaction of fermion or gauge boson loops and their Supersymmetric counterparts (see Fig. 1). Therefore, in the following, when talking about results for the H_1, we consider the production of any of H_i, $i = 1, 2, 3$, which, while propagating, converts into H_1. However, whenever $M_{H_{2,3}} \leq M_{H_1} + 2$ GeV (assuming that mass measurements are resolved at about 2 GeV), we consider all the degenerate Higgs particles together.

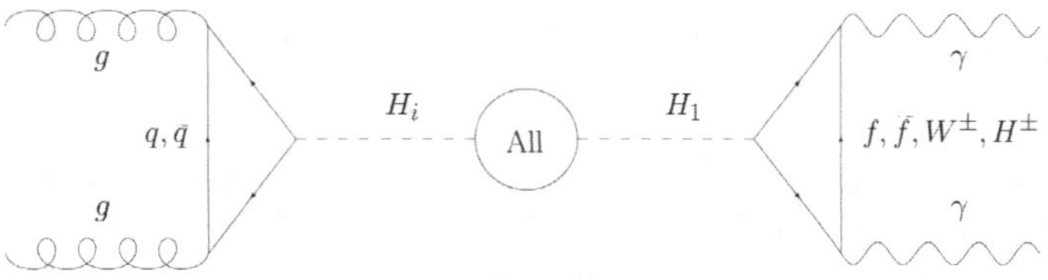

Figure 1. Feynman diagram for $gg \to H_1 \to \gamma\gamma$ including the effect of mixing in the propagator.

As intimated, the propagator matrix is obtained from the self-energy of the Higgs particles computed at one-loop level, where we used the expressions provided by [10], which include off-diagonal absorptive parts. The matrix inversion required is done numerically using the Lapack package [17]. All relevant couplings and masses are obtained from CPSuperH version 2 [18], which takes into account all applicable experimental constraints including the low energy EDMs. The cross section of the full process shown in Fig. 1 is computed numerically. The multi-dimensional integration is carried out using the CUHRE program under the CUBA package [19]. For our collider analysis we have used the CTEQ6 Parton Distribution Functions (PDFs) [20]–[23] computed at the factorization/renormalization scale $\mu = \sqrt{\hat{s}}$.

[1] Also notice that there is also an 'indirect' CP-violating contribution to the overall cross section of the process under study if one considers that the H_1 (and $H_{2,3}$) mass is subject to similar loop effects, though [3] has already shown that the consequent effects are marginal (see the left-hand side of Fig. 4 therein), so that they are included here but not dwelt upon.

3. Results

In order to illustrate the typical effects of CP violation in the MSSM, we have considered a few sample parameter space points and studied the effect of, in particular, light sparticles in the loops, chiefly, of stop squarks. We fix the following MSSM parameters which play only a minor role in CP violation studies for Higgs production and decay:

$$M_1 = 100 \text{ GeV}, \quad M_2 = M_3 = 1 \text{ TeV}, \quad M_{Q_3} = M_{D_3} = M_{L_3} = M_{E_3} = M_{\text{SUSY}} = 1 \text{ TeV}.$$

We consider the case of all the third generation trilinear couplings being unified into one single quantity, A_f. All the soft masses are taken to be the same at some unification scale, whose representative value adopted here is 1 TeV. When considering the light stop case we take a comparatively light value for $M_{U_3} \sim 250$ GeV, which corresponds to a stop mass of around 200 GeV, otherwise M_{U_3} is set to 1 TeV. We could, alternatively, consider small values for M_{Q_3} to reach light squarks, but the effects would qualitatively be the same. So, in the following we keep a fixed value of $M_{Q_3} = 1$ TeV. In the Higgs scalar-pseudoscalar mixing the product of μA_f is relevant rather than μ or A_f separately. As argued in our earlier works, the only phase that is relevant is thus the sum of the phases of μ and A_f. In our analysis we have kept $\phi_{A_f} = 0$ and studied the effect of CP violation by varying ϕ_μ. Regarding the absolute values of μ or A_f, in our numerical analysis, we have varied these parameters between 1 and 1.5 TeV. M_{H+} is instead varied between 100 and 300 GeV. The mass of the lightest Higgs particle is consequently in the range of 50–130 GeV. We then analyze cases with different values of $\tan\beta$. In particular, low $\tan\beta$ values give very small deviations from the corresponding CP-conserving cases, while large $\tan\beta$ values produce significant differences. Also, we take a representative value of $\tan\beta = 20$ to see the effect of the other parameters.

In Fig. 2 we plot the full cross section for $gg \to H_1 \to \gamma\gamma$ against M_{H^+}. We have considered $\mu = 1$ TeV, $A_f = 1$ TeV and $\tan\beta = 20$. There is appreciable variation of the cross section with ϕ_μ. Comparing the two cases of light and heavy stops, it is clear that the effect of the Higgs-stop-stop coupling is significant. Indeed, this was also noticed when we studied the di-photon decay [5].

In addition to the stop mass dependence of the $gg \to H_1 \to \gamma\gamma$ cross section, we have also studied how the latter varies with the masses of the other (s)particles entering the loops. However, in line with the results of Refs. [3]–[6] for the case of the $H_1 \to \gamma\gamma$ decay, we have found that their impact is largely negligible here too, no matter the value of the CP-violating phases.

4. Conclusions

Significant effect of CP violation in the process $gg \to H_1 \to \gamma\gamma$ is seen for the parameter values considered in this study. Varying the mass of the stop squark from about 1 TeV to around 250 GeV has also a strong quantitative impact on the cross section, possibly changing its dependence on the phases also qualitatively. The discovery channel for the mass range of the Higgs boson considered thus seem to have the potential to also disentangle its CP nature within the context of the MSSM.

Acknowledgments

SM is financially supported in part by NExT institute. PP thanks IITG for travel support.

Reference

[1] A. Dedes and S. Moretti, Phys. Rev. Lett. **84**, 22 (2000).
[2] A. Dedes and S. Moretti, Nucl. Phys. B **576**, 29 (2000).
[3] S. Moretti, S. Munir, P. Poulose, Phys. Lett. B **649**, 206 (2007).

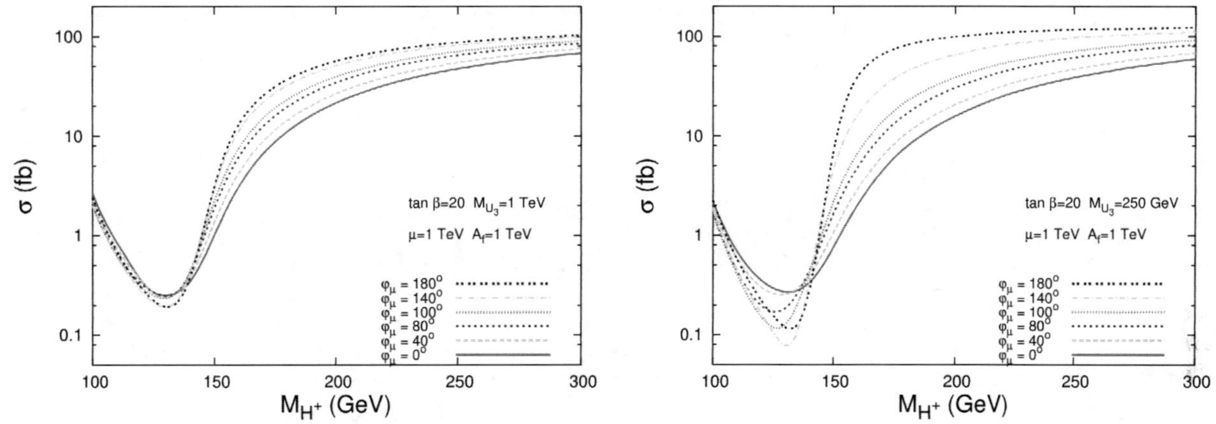

Figure 2. The cross section $\sigma(pp \to H_1 \to \gamma\gamma)$ at the LHC plotted against the charged Higgs boson mass for different ϕ_μ values. The relevant MSSM variables are set as follows: $\tan\beta = 20$, $A_f = 1$ TeV, $\mu = 1$ TeV, with $M_{U_3} = 1$ TeV (left plot) and $M_{U_3} = 250$ GeV (right plot).

[4] S. Hesselbach, S. Moretti, S. Munir, P. Poulose, Eur. Phys. J. C **54**, 129 (2008).
[5] S. Hesselbach, S. Moretti, S. Munir, P. Poulose, J. Phys. Conf. Ser. **110**, 072017 (2008).
[6] S. Hesselbach, S. Moretti, S. Munir, P. Poulose, arXiv:0710.4923 [hep-ph].
[7] A. Dedes and S. Moretti, http://www.slac.stanford.edu/spires/find/hep/www?irn=5258707,*(Prepared for Workshop on Physics at TeV Colliders, Les Houches, France, 21 May - 1 Jun 2001.)*
[8] S. Y. Choi and J. S. Lee, Phys. Rev. D **61**, 115002 (2000).
[9] S. Y. Choi, K. Hagiwara and J. S. Lee, Phys. Lett. B **529**, 212 (2002).
[10] J. R. Ellis, J. S. Lee and A. Pilaftsis, Phys. Rev. D **70**, 075010 (2004); Mod. Phys. Lett. A **21**, 1405 (2006).
[11] S. Hesselbach, S. Moretti, S. Munir and P. Poulose, AIP Conf. Proc. **1200**, 498 (2010)
[12] S. Hesselbach, S. Moretti, S. Munir and P. Poulose, Phys. Rev. D **82**, 074004 (2010)
[13] A. Pilaftsis and C. E. M. Wagner, Nucl. Phys. B **553**, 3 (1999).
[14] M. S. Carena, J. R. Ellis, A. Pilaftsis and C. E. M. Wagner, Nucl. Phys. B **586**, 92 (2000).
[15] S. Heinemeyer, Int. J. Mod. Phys. A **21**, 2659 (2006).
[16] M. Frank, T. Hahn, S. Heinemeyer, W. Hollik, H. Rzehak and G. Weiglein, JHEP **0702**, 047 (2007).
[17] See http://www.netlib.org/lapack/.
[18] J. S. Lee, M. Carena, J. Ellis, A. Pilaftsis and C. E. M. Wagner, Comput. Phys. Commun. **180**, 312 (2009).
[19] T. Hahn, Comput. Phys. Commun. **168**, 78 (2005).
[20] J. Pumplin, D. R. Stump, J. Huston, H. L. Lai, P. M. Nadolsky, W. K. Tung, JHEP **0207**, 012 (2002).
[21] D. Stump, J. Huston, J. Pumplin, W. K. Tung, H. L. Lai, S. Kuhlmann, J. F. Owens, JHEP **0310**, 046 (2003).
[22] J. F. Owens *et al.*, Phys. Rev. D **75**, 054030 (2007).
[23] H. L. Lai, P. M. Nadolsky, J. Pumplin, D. Stump, W. K. Tung, C. P. Yuan, JHEP **0704**, 089 (2007).

Symposium on Prospects in the Physics of Discrete Symmetries IOP Publishing
Journal of Physics: Conference Series **335** (2011) 012021 doi:10.1088/1742-6596/335/1/012021

Search for New Physics with Rare Heavy Flavour Decays at LHCb

Thomas Blake[1]
Imperial College London, London, UK
E-mail: thomas.blake00@imperial.ac.uk

Abstract. Rare, flavour changing neutral current processes provide an excellent probe of many new physics models. LHCb's sensitivity at $\sqrt{s} = 7\,\text{TeV}$ to three promising modes ($B_d \to K^{*0}\,\gamma$, $B_s \to \mu^+\mu^-$ and $B_d \to K^{*0}\mu^+\mu^-$) is explored based on knowledge of the first $37\,\text{pb}^{-1}$ of integrated luminosity from the LHC.

1. Introduction

The term rare decay is used to describe a class of decays that are mediated by electroweak flavour changing neutral current (FCNC) processes in the Standard Model (SM). Examples include $b \to s\ell^+\ell^-$, $b \to s\gamma$ and $B_{d,s} \to \mu^+\mu^-$. The presence of new physics (NP) particles entering in competing diagrams, in the absence of a dominant SM tree level process, can lead to comparatively large deviations from SM predictions. These deviations may be deviations in measured branching ratios (in the case of $B_s \to \mu^+\mu^-$), angular distributions (in $B_d \to K^{*0}\mu^+\mu^-$) or CP or Isospin asymmetries. Rare decay searches provide a complementary approach to direct searches at ATLAS, CMS and the TeVatron and already provide strong constraints on a wide range of NP models. A concrete example of this is shown in Figure 1.

In terms of Effective Field Theory and the operator product expansion[2], the process $b \to s\gamma$ is highly sensitive to operator $\mathcal{O}_7$ and its helicity flipped counterpart $\mathcal{O}_7'$. The process $b \to s\ell^+\ell^-$ is sensitive to $\mathcal{O}_7^{(\prime)}$ and to contributions from the vector and axial-vector operators, $\mathcal{O}_9^{(\prime)}$ and $\mathcal{O}_{10}^{(\prime)}$. $B_s \to \mu^+\mu^-$ can see large enhancements from scalar ($\mathcal{O}_S$) and pseudo-scalar operators ($\mathcal{O}_P$). In the SM the Wilson coefficients associated with the the scalar and pseudo-scalar operators are highly suppressed ($\mathcal{C}_{S,P} \propto m_\ell m_b/m_W^2 \sim 0$) as are contributions from helicity flipped operators, corresponding to right-handed currents ($\propto m_s/m_b$). Beyond the SM there can be significant contributions to both helicity flipped operators and from new operators (including $\mathcal{O}_{S,P}$).

LHCb's sensitivity to $B_d \to K^{*0}\,\gamma$, $B_d \to K^{*0}\mu^+\mu^-$ and $B_s \to \mu^+\mu^-$ is discussed below.

2. The LHCb detector

The LHCb detector is a forward-arm spectrometer at the LHC covering a rapidity range (with respect to the LHC beam axis) of $2 < \eta < 5$. The detector geometry profits from the fact that $b\bar{b}$ pairs are predominantly produced, by gluon-gluon fusion in proton-proton collisions, in

[1] On behalf of the LHCb collaboration
[2] See for example reference [2]

Published under licence by IOP Publishing Ltd

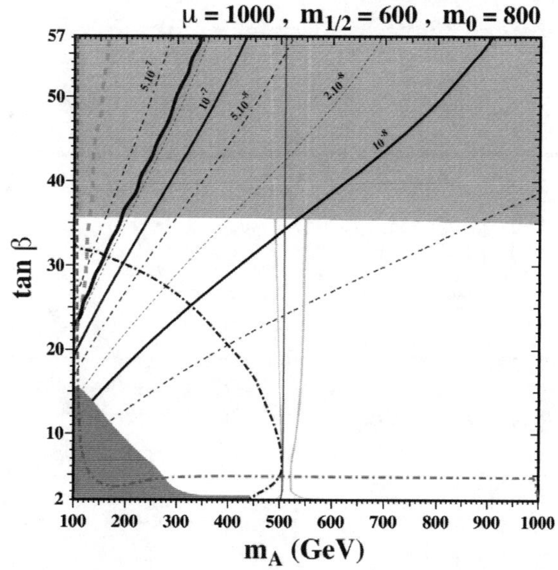

Figure 1. Experimental exclusion limits in the $m_A - \tan\beta$ plane of a MSSM NP model. The shaded region in the lower left-hand corner is the limit coming from the branching ratio of inclusive $b \to s\gamma$. The shaded exclusion region at the top of the figure comes from measurements of the muon $g-2$. Finally, the solid and dashed lines excluding from the top-left to the bottom-right of the figure come from limits on the $B_s \to \mu^+\mu^-$ branching ratio. The current experimental limit is $\sim 3.6 \times 10^{-8}$ [1]. Figure from reference [3].

the same forward (or backward) cone. This allows the interaction region to be offset in the cavern providing a long lever-arm for tracking and momentum resolutions of $\Delta p/p \sim 0.5\%$. The detector is described in detail elsewhere [4] but, briefly, it comprises: a silicon $r - \phi$ geometry vertex detector (the VELO); tracking stations with silicon strip detectors upstream of a warm dipole magnet and a combination of silicon strip detectors and straw-drift tubes downstream; a pair of ring imaging Cherenkov (RICH) detectors, to provide pion, kaon and proton separation in the momentum range $2 < p < 100\,\text{GeV}/c$; electromagnetic and hadronic calorimeters and furthest from the interaction region muon chambers.

2.1. Detector performance

Much of the emphasis in 2010 has been placed on understanding the detector performance. For the rare decay measurements discussed here, the key elements are the ability to separate cleanly primary and secondary vertices and an excellent particle identification performance. The first of these elements is provided by the VELO detector which during data taking approaches within 8 mm of the LHC beam axis and provides an impact parameter resolution of $14\,\mu$m at large transverse momenta (and typically $< 50\,\mu$m for charged tracks from B decays).

The muon identification performance has been measured to be $97 \pm 1\%$ using a tag-and-probe approach with $J/\psi \to \mu^+\mu^-$ decays (at a mis-id rate from decays in flight better than 1%). The kaon identification performance has been measured using a sample of tagged $D^* \to D(\to K\pi)\pi$ decays where the kaon and pion can be unambiguously identified. At a typical working point an average efficiency (over $2 < p < 100\,\text{GeV}/c$) of 95% can be achieved for a mis-id rate of 7%.

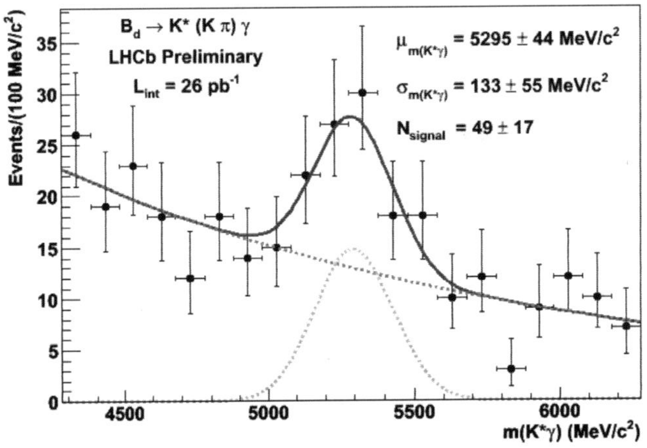

Figure 2. Reconstructed $K^{\pm}\ \pi^{\mp}\ \gamma$ candidates in $26\,\mathrm{pb}^{-1}$ integrated luminosity showing evidence for a $B_d \to K^{*0}\ \gamma$ signal at LHCb. The solid line is a fit to the data with an exponential background p.d.f. and a Gaussian signal p.d.f. The dashed lines show the individual signal and background components.

2.2. Running conditions

During 2010, the performance of the LHC has been excellent delivering $37\,\mathrm{pb}^{-1}$ of integrated luminosity to LHCb at a peak instantaneous luminosity of $2 \times 10^{32}\,\mathrm{cm}^{-2}\mathrm{s}^{-1}$. Whilst this instananeous luminosity is some way from the nominal $10^{34}\,\mathrm{cm}^{-2}\mathrm{s}^{-1}$ conditions for the general purpose detectors it is already the design luminosity of LHCb, albeit with only 400 colliding bunches. The high instantaneous luminosity, with a mean number of visible interactions that is approximately $5\times$ the design value, has consequences for the efficiency and purity of the trigger and offline selections and is reflected in the numbers shown here.

3. Radiative decays

As described in section 1, the presence of new physics particles in loop diagrams can lead to increased contributions from right-handed currents. One of LHCb's long term goals is a measurement of the decays $B_s \to \phi\gamma$ [5] and $B_d \to K^{*0}e^+e^-$ [6] that can be sensitive to the right-handed component of the photon polarisation (which is suppressed in the SM). These decay modes are not discussed here, but a logical pre-cursor to understanding LHCb's sensitivity to radiative decays is a measurement of $B_d \to K^{*0}\gamma$, a mode that has been well established at *BABAR* and Belle. Here the interest is not in a measurement of the branching ratio but of a measurement of the CP asymmetry. The most precise single measurement comes from *BABAR* [7]:

$$\mathcal{A}_{CP} = -0.016 \pm 0.022 \pm 0.007$$

based on ~ 2400 signal candidates. In $26\,\mathrm{pb}^{-1}$ LHCb observes $49 \pm 17\ B_d \to K^{*0}(\to K^+\pi^-)\gamma$ candidates with a mass resolution of $\sigma_{K\pi\gamma} = (133 \pm 55)\,\mathrm{MeV}/c^2$ (see figure 2). $B_d \to K^{*0}\gamma$ candidates have been selected offline from $K^{\pm}\pi^{\mp}\gamma$ combinations where the kaon is identified as a good kaon in LHCb's RICH detectors, the vertex of the $K\pi$ is highly displaced from the primary vertex and the photon has a large transverse energy. The full selection is described in the LHCb roadmap document [5].

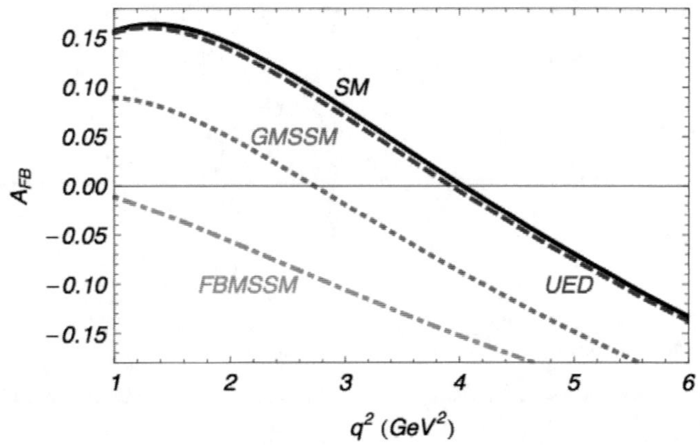

Figure 3. Predicted forward-backward asymmetry (A_{FB}) of dimuon pair as a function of $q^2 = m^2_{\mu^+\mu^-}$ in $B_d \to K^{*0}\mu^+\mu^-$ decays in the SM (solid line) and several NP models (dashed lines). Figure from [9].

The yield seen in Figure 2 is below the Monté Carlo expectation but is expected to improve with improved calorimeter calibration. Based on the observed yield, LHCb would require $\mathcal{O}(1\,\mathrm{fb}^{-1})$ of integrated luminosity to make a measurement that is competitive with the B-Factories. Such a data sample is expected in 2011.

4. $B_d \to K^{*0}\mu^+\mu^-$ at LHCb

The rare decay $B_d \to K^{*0}\mu^+\mu^-$ has particular sensitivity to new physics models that provide additional right-handed currents or introduce large contributions from scalar or pseudo-scalar operators. In many models the presence of NP particles can lead to size-able differences in angular observables from the SM predictions. To achieve ultimate sensitivity the long term goal is a full angular analysis of the decay [8], but with $< 2\,\mathrm{fb}^{-1}$ LHCb's focus is a measurement of the forward-backward asymmetry (A_{FB}) of the muon pair. This angular observable varies with the invariant mass squared of the dimuon pair (q^2) and in the standard model changes sign at well defined point, where the leading uncertainties coming from form-factors cancel. The variation of A_{FB} in the SM and several new physics models can be seen in Figure 3.

Measurements of A_{FB} from the B-Factories and CDF offer a possible hint of the effects of new physics, where the A_{FB} measured by the three experiments has the opposite sign to SM prediction at low q^2. These measurements, however, are highly limited by the available statistics. The most precise single measurement comes from Belle, based on ~ 250 signal candidates [10].

$B_d \to K^{*0}\mu^+\mu^-$ candidates in LHCb are selected by combining a track that has been identified as a good kaon by the LHCb RICH detectors, with two muon candidates and another track. The B_d vertex is then required to be highly displaced and the candidate to point back to a primary vertex. Transverse momentum cuts are avoided offline to limit possible acceptance biases on a measurement of A_{FB}.

For the early studies with the 2010 data sample the $B_d \to K^{*0}\mu^+\mu^-$ signal region has been blinded and the sample split into two $17\,\mathrm{pb}^{-1}$ sub-samples. The first $17\,\mathrm{pb}^{-1}$ was used to optimise the offline event selection using $B_d \to K^{*0}J/\psi$ candidates as a proxy for the $B_d \to K^{*0}\mu^+\mu^-$ signal and background candidates that sit outside a $\pm 50\,\mathrm{MeV}$ window around the B_d mass and a $\pm 60\,\mathrm{MeV}$ window around the J/ψ and $\psi(2S)$ masses. The second $17\,\mathrm{pb}^{-1}$

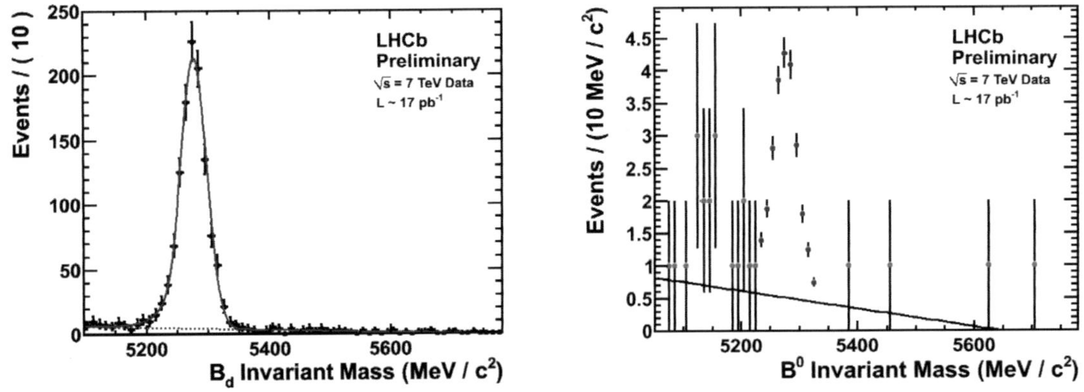

Figure 4. $B_d \to K^{*0}\ J/\psi$ candidates selected by a $B_d \to K^{*0}\mu^+\mu^-$ offline selection (left). The expected $B_d \to K^{*0}\mu^+\mu^-$ signal, scaled appropriately from $B_d \to K^{*0}\ J/\psi$, imposed on background events that lie outside a $\pm 50\,\text{MeV}$ B_d mass and $\pm 60\,\text{MeV}$ J/ψ and $\psi(2S)$ mass windows in data (right). The figure is described in the text and created from $17\,\text{pb}^{-1}$ of integrated luminosity.

was used to validate the selection. The invariant mass distribution of the offline selected B_d candidates is shown in the left-hand figure of Figure 4. It is dominated by $B_d \to K^{*0}\ J/\psi$ events that provide a useful control channel for $B_d \to K^{*0}\mu^+\mu^-$. The right-hand figure shows the distribution of background events outside J/ψ and $\psi(2S)$ mass windows. To give an indication of the expected $B_d \to K^{*0}\mu^+\mu^-$ performance, the expected $B_d \to K^{*0}\mu^+\mu^-$ signal has been extrapolated and scaled appropriately from the observed $B_d \to K^{*0}J/\psi$ yield (the mass peak itself remains blinded). This results in an expected signal yield of between 0.5 and 1 signal candidate per pb^{-1} at a signal-to-background ratio of better than one-to-one.

5. $B_s \to \mu^+\mu^-$ at LHCb

The final rare decay measurement, that completes the set of channels identified in the LHCb roadmap document, is $B_s \to \mu^+\mu^-$. As discussed in section 1 this decay is highly sensitive to new scalar and pseudo-scalar couplings and can see large enhancements in the expected branching ratio in models with an extended Higgs sector. The 'classic' example of this is the MSSM, where the presence of a second Higgs doublet can lead to enhancements in the amplitude of $C_S \propto \tan^3\beta/m_A^2$. In the SM this decay is highly supressed with an expected $\mathcal{B}(B_s \to \mu^+\mu^-) = (3.4 \pm 0.3) \times 10^{-9}$.

At LHCb, a potential $B_s \to \mu^+\mu^-$ signal is extracted by taking dimuon combinations and calculating a two-component likelihood. The likelihood is based on the mass of the dimuon system and a geometrical likelihood that discriminates signal and background using the B_s kinematics; its lifetime, impact parameter and isolation. It is driven by LHCb's excellent secondary vertex resolution and is constructed to be uniformly distributed between $0-1$ for signal events. For details please see reference [5].

The geometrical likelihood and mass resolution of signal events in the data can be characterized using $B \to hh$ decays (such as $B_{d,s} \to K\pi$, $B_d \to \pi^+\pi^-$ and $B_s \to K^+K^-$). The mass distribution of reconstructed $K^{\pm}\pi^{\mp}$ and K^+K^- events can be seen in Figure 5. The candidates have a fitted two-body mass resolution of $\sigma_{m_B} \sim 23\,\text{MeV}$.

The geometrical likelihood distribution of background events has been explored using the first $215\,\text{nb}^{-1}$ of integrated luminosity and is shown in Figure 6. Background candidates sit at

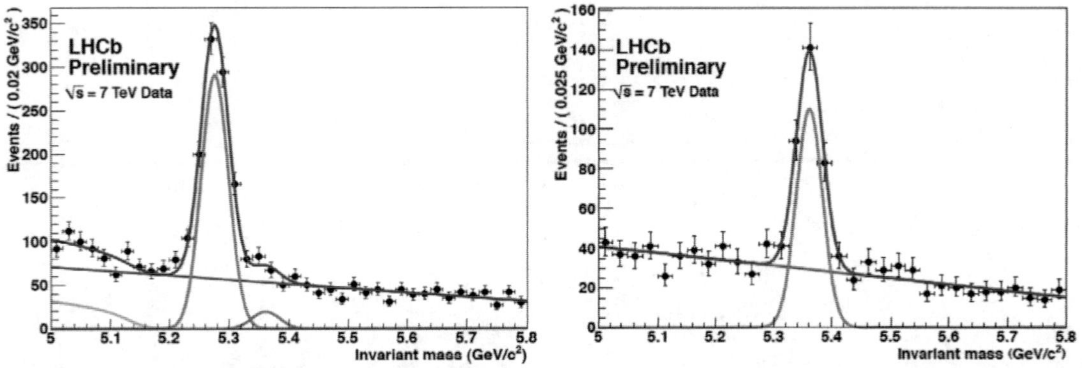

Figure 5. Reconstructed $B \rightarrow K\pi$ (left) and $B_s \rightarrow K^+K^-$ (right) decays in $26\,\text{pb}^{-1}$ of integrated luminosity from the LHC. The lines represent the result of a fit to the data with: an exponential to describe the combinatorial background component; gaussian pdf's to describe the $B_d \rightarrow K\pi$, $B_s \rightarrow K\pi$ and $B_s \rightarrow K^+K^-$ signal and a component to describe the low-mass background in $B \rightarrow K\pi$. The observed mass resolution is 23 MeV.

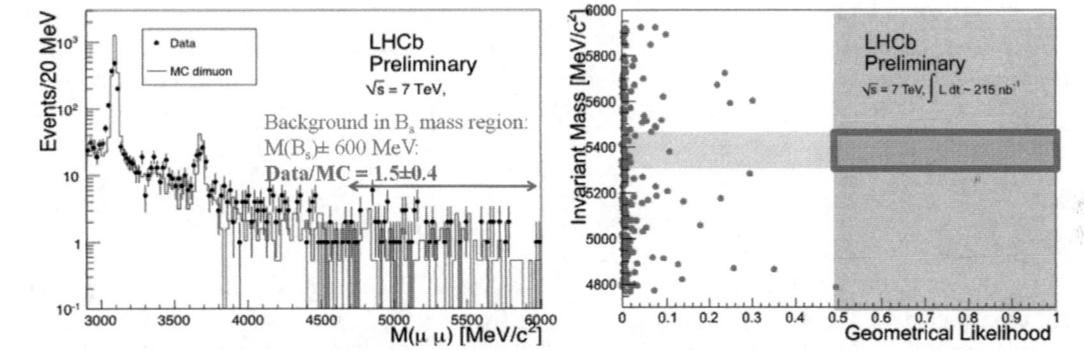

Figure 6. The invariant mass distribution of $B_s \rightarrow \mu^+\mu^-$ candidates in $215\,\text{nb}^{-1}$ of integrated luminosity (left) and the geometrical likelihood of these candidates (right). The points in the left-hand figure are the observed distribution in the data and the solid line the MC expectation.

geometrical likelihoods close to zero and there is every indication from the 2010 data that the geometrical likelihod is a good discriminating variable. Furthermore the level of background is in good agreement with MC predictions. The mass distribution of background events is shown in the left-hand plot of Figure 6. In a $\pm 600\,\text{MeV}$ window about the nominal B_s mass the ratio of data to the MC expectation is consistent with one (1.5 ± 0.4). A first measurement is now being performed on the full 2010 dataset.

Figure 7 shows the expected exclusion limit (in the background only hypothesis) on the $B_s \rightarrow \mu^+\mu^-$ branching fraction at LHCb as a function of integrated luminosity. The plot has been updated from the LHCb roadmap document [5] to reflect the knowledge of the background and the B_s mass resolution from the 2010 data. With $100\,\text{pb}^{-1}$ LHCb should be able to make a world leading measurement.

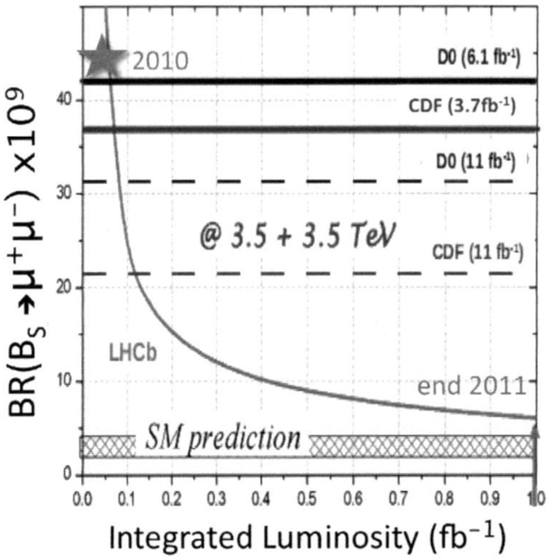

Figure 7. The expected LHCb exclusion limit on the $B_s \to \mu^+\mu^-$ BR as a function of integrated luminosity (solid curve). Superimposed on the figure are the DØ and CDF exclusion limits with 6.1 and 3.7 fb^{-1} integrated luminosity (solid horizontal lines) and the expected DØ and CDF exclusion limits with 11 fb^{-1} (dashed lines). The SM prediction for the BR is represented by the shaded region. Finally the "$\star$" highlights the expected LHCb sensitivity with 35 pb^{-1}.

6. Summary

In 2010, the performance of LHC and LHCb has been excellent, accumulating 37 pb^{-1} at a high detector efficiency. This data has been used to understand LHCb's expected performance in three rare decay channels. For $B_d \to K^{*0}\mu^+\mu^-$ early indications are that 0.5-1 signal candidates are expected per pb^{-1} allowing LHCb to make a world leading measurement with 200-400 pb^{-1}. For $B_s \to \mu^+\mu^-$ the observed background (in the first 215 nb^{-1}) indicates that limits below 3×10^{-9} could be achieved with only 100 pb^{-1} of integrated luminosity. LHCb expects to record 1 fb^{-1} in 2011.

References

[1] The CDF Collaboration, CDF Public Note 9892 (2009).
[2] A. Buras and R. Fleischer, Adv. Ser. Direct. High Energy Phys. **15** (1997).
[3] J. Ellis *et al.*, JHEP 0605:063 (2006).
[4] A. A. Alves Jr. *et al.*, JINST **3** S08005 (2008).
[5] B. Adeva *et al.*, arXiv:0912.4179v2 (2009).
[6] J. Lefrançois *et al.*, CERN-LHCb-PUB-2009-008 (2009).
[7] B. Aubert *et al.*, Phys. Rev. Lett. **103** 211802 (2009).
[8] U. Egede *et al.*, JHEP 1010:056 (2010).
[9] W. Altmannshofer *et al.*, JHEP 0901:019 (2009).
[10] J. T. Wei *et al.*, Phys. Rev. Lett. **103** 171801 (2009).
[11] M. Blanke *et al.*, JHEP 0610:003 (2006).

Journal of Physics: Conference Series **335** (2011) 012022 doi:10.1088/1742-6596/335/1/012022

First results and prospects in B physics at CMS

F Ligabue, on behalf of **The CMS Collaboration**
Scuola Normale Superiore, Pisa, Italy

E-mail: f.ligabue@sns.it

Abstract. The first results in b physics obtained by the CMS collaboration from the data collected in pp collisions at 7 TeV at the LHC. In particular, results are reported on heavy quarkoniun production (J/ψ and Υ), and on inclusive b jet production cross section. Progress on exclusive reconstructions is also mentioned.

1. Introduction

Thanks to the impressive performance of the LHC accelerator in the past months, the CMS experiment collected about 43 pb^{-1} of data from pp collisions at 7 TeV by the end of 2010. The relatively low luminosity of the first months of data-taking allowed early measurements of high-rate b-related processes such as quarkonium (J/ψ and Υ) and inclusive b-jet production. An important part in these measurements is played by muons; the loose trigger requirements allowed by the low instantaneous luminosity made it possible to extend the kinematic reach of the reconstructed muons to very low transverse momentum especially in the forward region. Muons are used to select quarkonium decays and to discriminate b jets by exploiting the semileptonic decays of the b quark. A very important role is also played by the excellent tracking and vertexing capabilities of the CMS detector, which allow a lifetime-based discrimination of b decays.
The early results presented here are obtained from just a fraction of the 2010 data statistics.

2. The CMS detector

A detailed description of the Compact Muon Solenoid (CMS) experiment can be found elsewhere[1] The central feature of the CMS apparatus is a superconducting solenoid, of 6 m internal diameter. Within the field volume are the silicon tracker, the crystal electromagnetic calorimeter (ECAL) and the brass/scintillator hadron calorimeter (HCAL). Muons are detected in the pseudorapidity window $|\eta| < 2.4$, by gaseous detectors made of three technologies: Drift Tubes (DT), Cathode Strip Chambers (CSC), and Resistive Plate Chambers (RPC), embedded in the iron return yoke.
The silicon tracker is composed of pixel detectors (three barrel layers and two forward disks in either side of the detector, made of 66 million 100×150 μm^2 pixels) followed by microstrip detectors (several layers with strips of pitch between 80 and 180 μm). Thanks to the strong magnetic field, 3.8 T, and to the high granularity of the silicon tracker, the transverse momentum, p_T , of the muons matched to reconstructed tracks is measured with a resolution better than 1.5% for p_T smaller than 100 GeV

2.1 Performance of the tracking detector and muon reconstruction

The performance of the silicon tracker has been tested with the first data collected by the CMS experiment. The left plot in figure 1 shows the measured longitudinal impact parameter resolution as a function of the track pseudorapidity; data and simulation agree very nicely. The same is true for the impact parameter significance (right plot), whose exponential tail on the right side of the peak is due to the long lived heavy quark component.

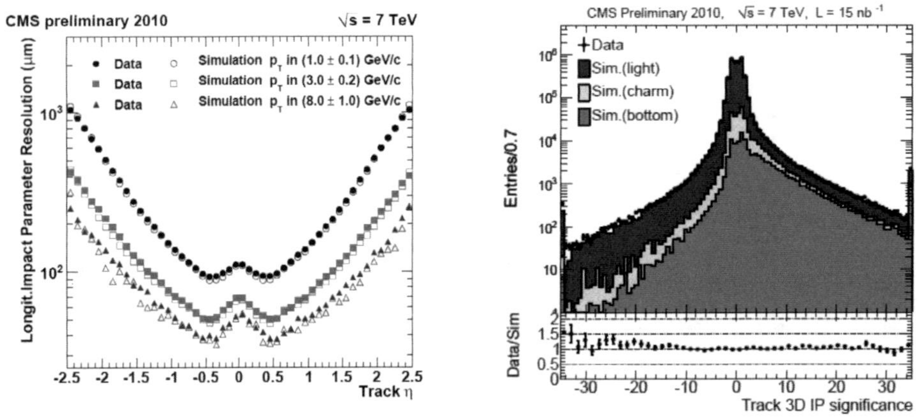

Figure 1. *Left*: measured resolution of the track longitudinal impact parameter as a function of the track pseudorapidity, for real and simulated data. *Right*: distribution of the three-dimensional track impact parameter significance for real and simulated data.

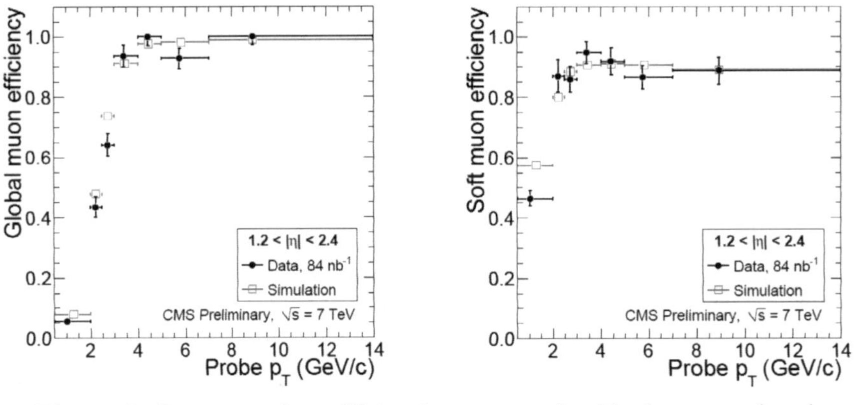

Figure 2. Reconstruction efficiencies measured with the tag-and-probe method for Global muons and for Tracker (here called «soft») muons, in real and simulated data.

Muons in CMS are reconstructed from silicon tracker and muon detector information [2]. Reconstruction algorithms use iterative Kalman-filter procedures which, starting from track «seeds», build the candidate track following either an inside-out (from the central region outwards) or an outside-in extrapolation approach. The so-called *Tracker muons* follow the inside-out approach and consist of tracks reconstructed in the silicon tracker and spatially matched to single track segments found in the muon detectors. The reconstruction of so-called *Global muons* starts from tracks reconstructed in the muon system and extrapolated from the outside in to match tracks from the silicon tracker. The latter algorithm gives a lower background contamination but also a lower efficiency, especially for low-momentum muons.

The muon reconstruction efficiency is measured in real data using the *tag and probe* technique. This is a general technique which is also used to measure other similar properties such as trigger and tracking efficiencies, and exploits the decays to muon pairs of narrow resonances such as the J/ψ, which also have the advantage of providing high statistics. Opposite-sign tracks whose invariant mass is consistent with the chosen resonance (J/ψ) are selected, where one of the two tracks (the *tag* track) is identified as a muon: the other track (the *probe*) is then used to measure the reconstruction efficiency, or other analogous properties. Figure 2 shows the measured reconstruction efficiency for Global and Tracker muons in simulated events and in minimum bias events from the first 84 nb^{-1} of CMS data.

3. Quarkonium production

Muon pairs are used to reconstruct J/ψ [3] and Y mesons, exploiting the high production rate and the use of dedicated muon triggers, made possible by the relatively low luminosity, which allowed to lower the p_T threshold and explore low transverse momentum ranges. The interest in measuring the quarkonium differential cross section lies basically in the comparison with various QCD models describing the prompt production in pp collision. In the J/ψ case, the possibility to separate the non-prompt mesons coming from b-hadron decays provides an additional test of theoretical predictions.

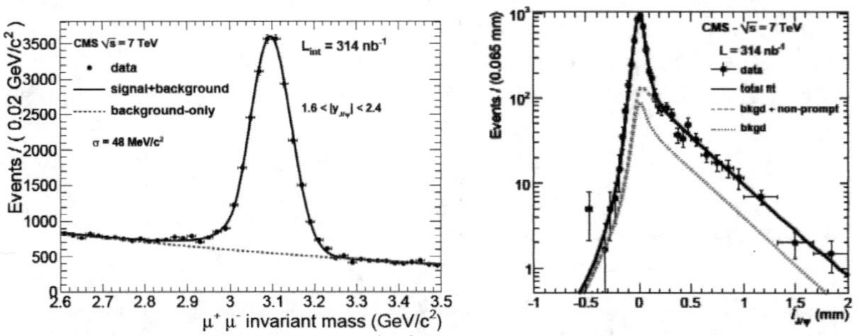

Figure 3. Left: mass fit for the selected J/ψ candidates in the rapidity range $1.6 < |y_{J/\psi}| < 2.4$. Right: lifetime fit to the measured pseudo-proper decay length of the J/ψ candidates in the same rapidity range.

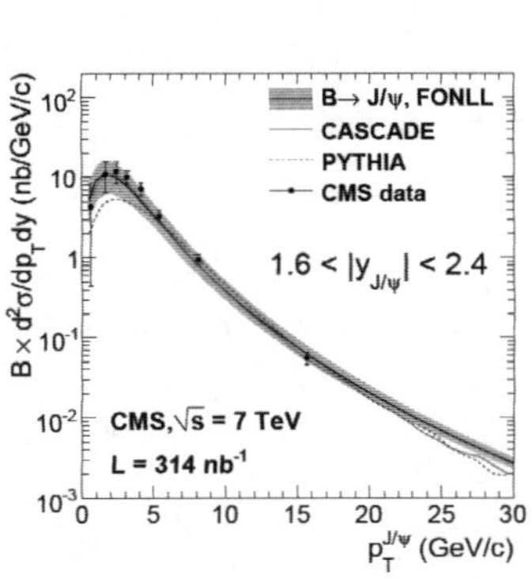

Figure 4. Measured differential cross-section as a function of pT, in the forward rapidity range, for the reconstructed J/ψ mesons coming from b hadrons, compared with different theoretical predictions.

3.1 Prompt and non-prompt J/ψ production

After some quality cuts on the reconstructed tracks and on their compatibility with coming from a common vertex, muon pairs from candidate J/ψ decays are selected by means of a fit to the invariant mass distribution (figure 3, left). The non-prompt fraction is extracted by fitting the distribution of the lifetime-related variable called pseudo-proper decay length (figure 3, right), essentially given by the projection of the decay vertex separation onto the flight direction of the decaying meson. From the first 314 nb^{-1} of data a total of about 27000 J/ψ candidates have been reconstructed. The differential cross-section for both the prompt and non-prompt production is then measured as a function of the meson's transverse momentum, in three different rapidity ranges. As shown in figure 4 (for the forward rapidity range), the results appear to be in agreement with the theoretical predictions from the models considered.

3.2 Υ production

Similar criteria are applied to select muon pairs from the three $b\bar{b}$ bound states Υ(1s), Υ(2s) and Υ(3s), where the various contributions are extracted from an invariant mass fit. The value of the measured cross section depends on the assumed polarization of the produced Υ meson, which is not known at present (this is also true for the prompt J/ψ), because the polarization determines the muon spectrum, which in turn affects the estimated acceptance. Therefore five different «extreme» polarization scenarios are assumed in the acceptance determination, which give rise to variations in the cross section up to 20%. Polarization measurements for both the Υ and the J/ψ are currently being addressed by the collaboration and will hopefully be available in the near future, already from the analysis of the full 2010 statistics.

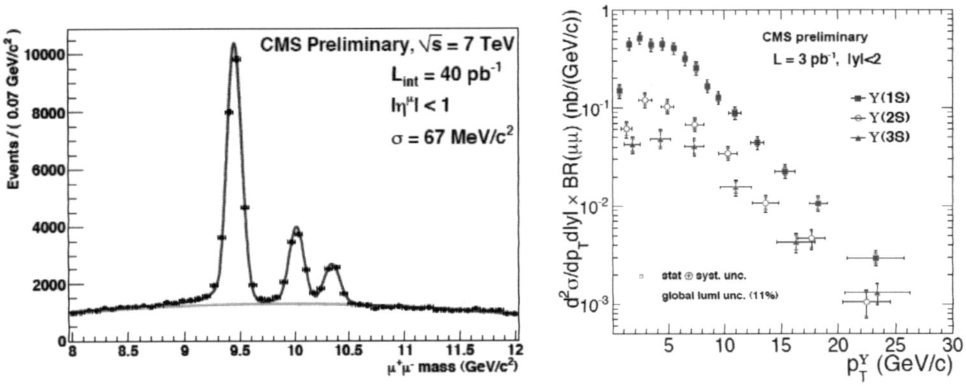

Figure 5. *Left*: invariant mass distribution for the selected muon pairs in the Υ mass range (using most of the 2010 statistics). *Right*: Measured cross section for the three Υ resonances in the central rapidity region (using the first 3pb^{-1}).

4. Exclusive B decays

J/ψ mesons are also used for exclusive B decay reconstructions such as $B \to J/\psi\ K$ and $B_s \to J/\psi\ \varphi$, for which preliminary invariant mass plots are shown in figure 6. The full 2010 statistics should already allow a differential cross-section measurement for the $B^{\pm} \to J/\psi\ K^{\pm}$ channel.

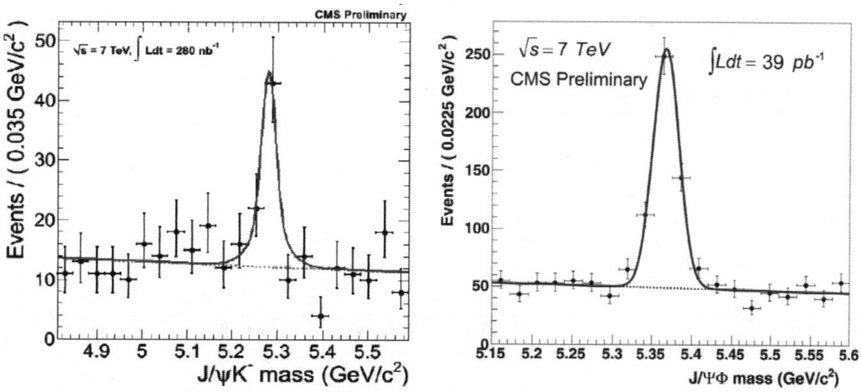

Figure 6. Preliminary invariant mass distributions for the $B^{\pm} \to J/\psi\ K^{\pm}$ (left) and $B_s \to J/\psi\ \varphi$ (right) channels, obtained with a statistics of 0.3 pb^{-1} and 39 pb^{-1} respectively.

5. Inclusive b production

Inclusive b-jet events are selected using two complementary b-tagging approaches characterized by different systematics, each of which exploits the other for consistency checks or for measuring relevant quantities such as signal efficiencies on data.

5.1 Semileptonic decays

This method exploits the high transverse momentum of leptons from b decays with respect to the jet direction, due to the comparatively high mass of B hadrons [4]. Jets containing a muon are selected, and the discriminating variable, called p_T^{rel} , is defined as the transverse component of the muon with respect to the jet axis, whose determination does not include the muon momentum. Events are selected with a single-muon trigger and the requirement of at least one well reconstructed muon; jets are formed using just the tracker information. The observed p_T^{rel} distribution in data, shown in figure 7, is fitted to a sum of templates for the expected signal and background contributions. Only the charm contribution is taken from the simulation; the shape of the light quark background, due to misidentified hadrons, is taken from the hadron track spectrum and from the misidentification probability measured on data, while the signal template is taken from simulation and validated on data using a b-jet enriched sample obtained with a vertex tagging technique similar to the one described below. The measured cross section is shown in figure 8 as a function of the muon transverse momentum and pseudorapidity. Although the figure only refers to a small fraction of the 2010 statistics, the experimental error is already comparable to the theoretical uncertainty on the next-to-leading order QCD predictions, which appear to underestimate the production cross section, especially in the central region.

5.2 Secondary vertex tagging

In the second approach [5] jets are formed starting from «particle flow» objects [6], reconstructed by a technique designed to optimize the jet energy reconstruction and resolution down to low transverse momenta of the jet. The b-jet fraction is measured using a secondary vertex tagging algorithm, based on the three-dimensional significance of the detected secondary vertex. The purity of the tagging algorithm is determined from data by means of a fit to the measured invariant mass of the secondary vertex. Similarly to what was done in the previous approach, the tagging efficiency is measured on data exploiting the properties of semileptonic decays, i.e. by fitting the p_T^{rel} distribution in a muon-enriched sample.

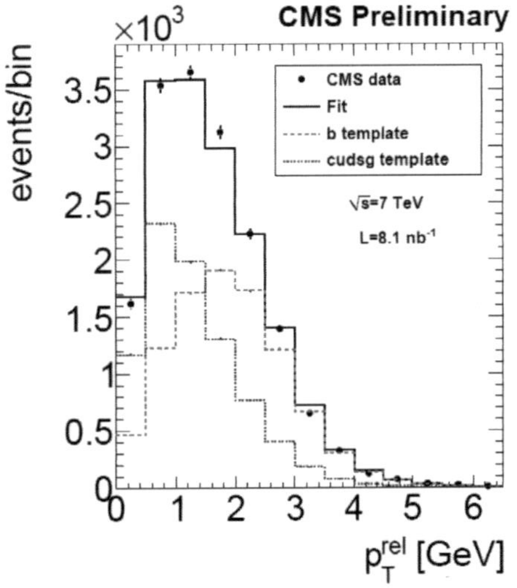

Figure 7. Measured distribution of the muon transverse momentum with respect to the track-jet axis. The signal template distribution is derived from the simulation but has been validated on a b-enriched data sample.

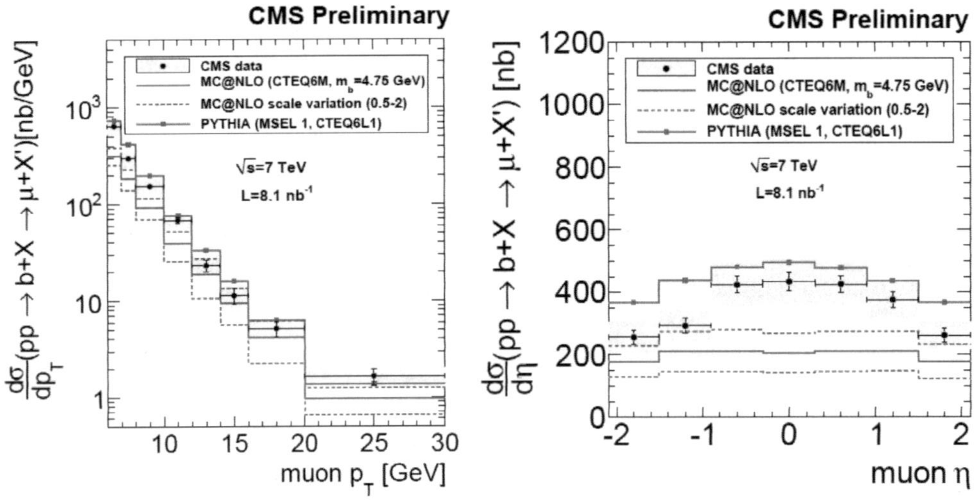

Figure 8. Preliminary cross-section for inclusive b-quark production measured at CMS using semileptonic decays, as a function of the muon transverse momentum (left) and pseudorapidity (right).

Figure 9 shows the measured double-differential cross-section as a function of the jet transverse momentum and rapidity, obtained after deconvoluting for smearing effects due to finite resolution. The experimental errors, dominated by the luminosity and tagging efficiency uncertainties, are again comparable in size to the theoretical error on the expected cross-section as computed at next-to-leading order (MC@NLO). The prediction seems to describe the data distribution fairly well, except for the tail at high transverse momentum and rapidity.

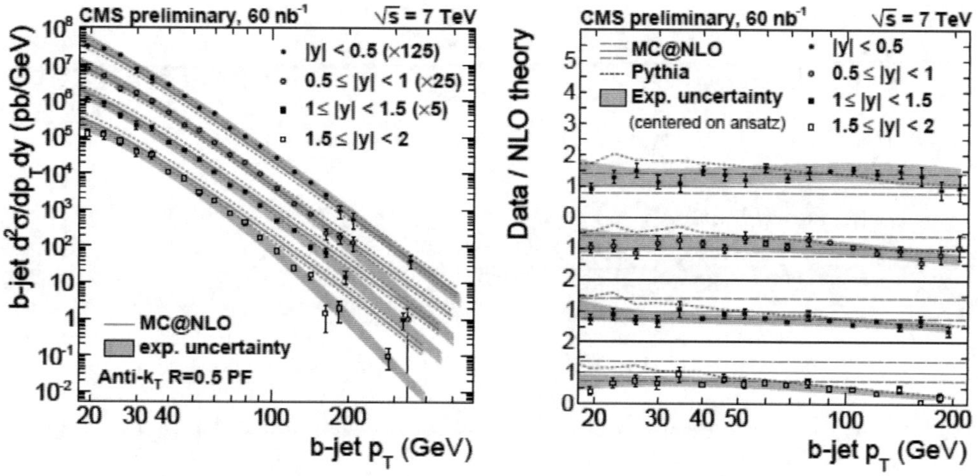

Figure 9. Preliminary cross-section for inclusive b-quark production measured at CMS using a secondary vertex tag, as a function of the muon transverse momentum and in four rapidity ranges (left). On the right the same measurements are presented as a ratio to the expected theoretical (MC@NLO) calculation.

6. Prospects for the near future and conclusions

The results presented here only refer to a (sometimes small) fraction of the statistics collected by CMS in the 2010 run. The full data sample is currently being analyzed, and aside from the updates of the above measurements, new results from a number of new interesting analyses are expected.

Among them, the measurement of the quarkonium polarization (Υ and J/ψ) will greatly reduce the corresponding uncertainty on the cross-section measurement; a study of the b-jet angular correlations will shed light on the b production mechanisms in pp collisions. Other foreseen measurements include the determination of the integrated B mixing parameter, the reconstruction of exclusive decays such as such as $B \to J/\psi\ K$ and $B_s \to J/\psi\ \varphi$, the latter being relevant for CP violation studies.

7. References

[1] CMS Collaboration 2008 The CMS experiment at the CERN LHC *JINST* **0803** doi:10.1088/1748- 0221/3/08/S08004
[2] CMS Collaboration 2010 CMS PAS MUO-10-002
[3] CMS Collaboration 2010 CMS PAS BPH-10-002
[4] CMS Collaboration 2010 CMS PAS BPH-10-007
[5] CMS Collaboration 2010 CMS PAS BPH-10-009
[6] CMS Collaboration 2009 Particle-Flow Event Reconstruction in CMS and Performance for Jets, Taus, and Missing ET, CMS PAS PFT-09-001
[7] CMS Collaboration 2010 CMS PAS BTV-10-001

Phenomenological tests of the Two-Higgs-Doublet Model with MFV and flavour-blind phases

Maria Valentina Carlucci

Physik-Department, Technische Universität München,
James-Franck-Straße, D-85748 Garching, Germany

E-mail: maria.carlucci@ph.tum.de

Abstract. In the context of a Two-Higgs-Doublet Model in which Minimal Flavour Violation (MFV) is imposed, one can allow the presence of flavour-blind CP-violating phases without obtaining electric dipole moments that overcome the experimental bounds. This choice permits to accommodate the hinted large phase in the B_s mixing and, at the same time, to soften the observed anomaly in the relation between ϵ_K and $S_{\psi K_S}$.

1. Introduction

During the last decade the B factories and the Tevatron experiments have collected a large amount of data that permits a deep analysis of precision measurements. The agreement of the experimental data of flavour-changing processes with the Standard Model (SM) predictions is verified within 20%; a similar agreement is found with the CP-violating processes, but recently a few anomalies have been detected in this sector: (i) the CP-violation in the B_s system signaled by the CP-asymmetry $S_{\psi\phi}$ in $B_s \to \psi\phi$ observed by CDF and D0 that appears to be roughly by a factor of 20 larger than the SM and MFV predictions [1, 2]; (ii) the value of $\sin 2\beta$ resulting from the UT fits tends to be significantly larger than the measured value of $S_{\psi K_S}$ [3]; (iii) the value of ϵ_K predicted in the SM by using $S_{\psi K_S}$ as the measure of the observed CP-violation is about 2σ lower than the data [3, 4]; (iv) recently D0 reported a measurement of the like-sign dimuon charge asymmetry in semileptonic b decay that is 3.2σ from the SM prediction [5]. These anomalies, if confirmed, suggest that a room for New Physics could be present in CP-violating processes, even if other CP-violating observables, such as electric dipole moments (EDMs), put tight constraints on it.

One of the most elegant and effective ways to protect the flavour in the New Physics models is the MFV hypothesis [6, 7]; it forces the flavour sector of the model to respect the same flavour symmetry and symmetry breaking as the SM, i.e. to be governed only by the same Yukawa couplings. However, while in the SM the Yukawa couplings are the only sources of both flavour breaking and CP violation, this is not necessary true in New Physics models even assuming MFV: the flavour can be broken only by Yukawa couplings, but other sources of CP violation are allowed [8]. This possibility is particularly interesting in light of the experimental results described above, in which the flavour structure of the SM seems to be confirmed, but new CP-violating mechanisms could contribute.

We have analyzed a simple extension of the SM in which two Higgs doublets are present; we have imposed MFV but allowed the presence of flavour-blind CP-violating phases in the fermion-scalar interactions; we will call this model 2HDM$_{\overline{\rm MFV}}$ [9]. First of all we have checked that the strict experimental flavour-changing neutral current (FCNCs) constraints can be naturally fulfilled, and that the bounds on CP-violation that come from EDMs are not overcome [10]. Then we have shown that a large phase in the B_s mixing can be easily accommodated and this, without extra free parameters, improves significantly in a correlated manner the issues about ϵ_K and $S_{\psi K_S}$.

2. The 2HDM$_{\overline{\rm MFV}}$

2.1. General structure of 2HDMs

The Higgs Lagrangian of a generic model with two-Higgs doublets, H_1 and H_2, with hypercharges $Y = 1/2$ and $Y = -1/2$ respectively, can be written as

$$\mathcal{L} = \sum_{i=1,2} D_\mu H_i D^\mu H_i^\dagger + \mathcal{L}_Y - V(H_1, H_2)\ , \tag{1}$$

where $D_\mu H_i = \partial_\mu H_i - ig'Y\hat{B}_\mu H_i - igT_a \hat{W}^a{}_\mu H_i$, with $T_a = \tau_a/2$.

The potential $V(H_1, H_2)$ is such that the H_i gets vacuum expectation value $\langle H^0_{1(2)}\rangle = v_{1(2)}$ with $v = \sqrt{v_1^2 + v_2^2} \approx 246$ GeV fixed by the mass of the W boson; moreover, we only consider the case in which it does not contain new sources of CP violation. The Higgs spectrum contains three Goldstone bosons $G^\pm$ and G^0, two charged Higges $H^\pm$, and three neutral Higgses h^0, H^0 (CP-even), and A^0 (CP-odd).

The most general renormalizable and gauge-invariant interaction of the two Higgs doublets with the SM quarks is

$$-\mathcal{L}_Y = \bar{Q}_L X_{d1} D_R H_1 + \bar{Q}_L X_{u1} U_R H_1^c + \bar{Q}_L X_{d2} D_R H_2^c + \bar{Q}_L X_{u2} U_R H_2 + \text{h.c.}\ , \tag{2}$$

where $H^c_{1(2)} = -i\tau_2 H^*_{1(2)}$ and the X_i are 3×3 matrices with a generic flavour structure. By performing a global rotation of angle $\beta = \arctan(v_2/v_1)$ of the Higgs fields (H_1, H_2) to the so-called Higgs basis (Φ_v, Φ_H), the mass terms and the interaction terms are separated:

$$-\mathcal{L}_Y = \bar{Q}_L \left(\frac{\sqrt{2}}{v} M_d \Phi_v + Z_d \Phi_H \right) D_R + \bar{Q}_L \left(\frac{\sqrt{2}}{v} M_u \Phi_v^c + Z_u \Phi_H^c \right) U_R + \text{h.c.}\ ; \tag{3}$$

the quark mass matrices $M_{u,d}$ and the couplings $Z_{u,d}$ are linear combinations of the X_i, weighted by the Higgs vacuum expectation values:

$$M_{u,d} = \frac{v}{\sqrt{2}} \left(\cos\beta X_{u,d\,1} + \sin\beta X_{u,d\,2} \right)\ , \qquad Z_{u,d} = \cos\beta X_{u,d\,2} - \sin\beta X_{u,d\,1}\ . \tag{4}$$

In this way it is clear that $M_{u,d}$ and $Z_{u,d}$ cannot be diagonalized simultaneously for generic X_i, and we are left with dangerous FCNC couplings to the neutral Higgses.

2.2. Minimal Flavour Violation

In order to suppress the FCNCs, some additional hypotheses on the model are needed. The oldest but still widely used method to obtain the aim is the Natural Flavour Conservation hypothesis [11], that states that renormalizable couplings contributing at the tree-level to FCNC processes are simply absent by assumption. Hence, in order to forbid these terms, one should impose some additional symmetry; however, it has been shown in [9] that this structure is not stable under quantum corrections, and sizable FCNCs operators are generated at higher energies unless a very severe fine tuning is performed.

It seems instead that a protection of the flavour symmetry and of its breaking is necessary in many New Physics models. In this frame the popular hypothesis of MFV turns out to be effective in suppressing FCNCs and stable under renormalization. Formally, MFV consists in the assumption that the $SU(3)$ quark flavour symmetry is broken only by two independent terms, Y_d and Y_u, transforming as

$$Y_u \sim (3, \bar{3}, 1)_{SU(3)^3} , \qquad Y_d \sim (3, 1, \bar{3})_{SU(3)^3} ; \tag{5}$$

in the 2HDM it implies for the X_i the structure [7]

$$X_{d1} = Y_d \tag{6a}$$
$$X_{d2} = P_{d2}(Y_u Y_u^\dagger, Y_d Y_d^\dagger) \times Y_d = \epsilon_0 Y_d + \epsilon_1 Y_d Y_d^\dagger Y_d + \epsilon_2 Y_u Y_u^\dagger Y_d + \dots \tag{6b}$$
$$X_{u1} = P_{u1}(Y_u Y_u^\dagger, Y_d Y_d^\dagger) \times Y_u = \epsilon'_0 Y_u + \epsilon'_1 Y_u Y_u^\dagger Y_u + \epsilon'_2 Y_d Y_d^\dagger Y_u + \dots \tag{6c}$$
$$X_{u2} = Y_u \tag{6d}$$

that is renormalization group invariant. Differently from the Natural Flavour Conservation case, now FCNCs are absent only at the lowest order in $Y_i Y_i^\dagger$ since the X_i are aligned [12]; they are present when one considers higher orders instead.

In order to investigate these FCNCs, one can perform an expansion in powers of suppressed off-diagonal CKM elements, so that the effective non-diagonal Z_d between the down-type quarks and the neutral Higgses assumes the form [7]

$$Z_d = W^d \lambda_d \qquad \text{with} \qquad W^d_{ij} = \bar{a}\, \delta_{ij} + \left(a_0 V^\dagger \lambda_u^2 V + a_1 V^\dagger \lambda_u^2 V \Delta + a_2 \Delta V^\dagger \lambda_u^2 V\right)_{ij} , \tag{7}$$

where $\lambda_{u,d} \propto 1/v \,\mathrm{diag}\,(m_{u,d}, m_{c,s}, m_{t,b})$, $\Delta = \mathrm{diag}\,(0, 0, 1)$, and the $\bar{a}$, a_i are parameters naturally of $\mathcal{O}(1)$; this structure already shows a large suppression due to the presence of two off-diagonal CKM elements and the down-type Yukawas. The case of Z_u is less interesting since other suppression are present, and the FCNCs are negligible.

One can check that no fine tuning is required in this case by performing a comparison with experimental FCNCs observables: constraints on the free parameters are obtained by imposing that the new physics contributions must be compatible with the experimental data within errors. We have found the bounds [9]:

$$|a_0| \tan\beta \frac{v}{M_H} < 18 \qquad \text{from } \epsilon_K , \tag{8a}$$
$$\sqrt{|(a_0^* + a_1^*)(a_0 + a_2)|} \tan\beta \frac{v}{M_H} = 10 \qquad \text{from} \Delta\, M_s , \tag{8b}$$
$$\sqrt{|a_0 + a_1|} \tan\beta \frac{v}{M_H} < 8.5 \qquad \text{from Br}\left(B_s \to \mu^+ \mu^-\right) ; \tag{8c}$$

that, as can be noted, are well compatible and perfectly natural.

2.3. Flavour-blind phases

The mechanisms of flavour and CP violation do not necessary need to be related: in MFV the Yukawa matrices are the only sources of flavour breaking, but other sources of CP violation could be present, provided that they are flavour-blind [8]. However, so far it has been assumed the effective MFV parameters a_i to be real, in order to fulfill the strong bounds on flavour-conserving CPV phases implied by the electric dipole moments. Allowing the FCNC parameters a_i to be complex, we investigate the possibility of generic CP-violating flavour-blind phases in the Higgs sector [9]; we will show that this choice implies several interesting phenomenological results for the$\Delta\ F\ = 2$ transitions, and that at the same time the electric dipole moments, although much enhanced over the SM values, are still compatible with the experimental bounds.

3. $\Delta F = 2$ amplitudes

Considering theΔ $F = 2$ FCNC transitions mediated by the neutral Higgs bosons, the leading MFV effective Hamiltonians are:

$$\mathcal{H}^{|\Delta S|=2} \propto -\frac{|a_0|^2}{M_H^2}\frac{m_s}{v}\frac{m_d}{v}\left[\left(\frac{m_t}{v}\right)^2 V_{ts}^* V_{td}\right]^2 (\bar{s}_R d_L)(\bar{s}_L d_R) + \text{h.c.}\,, \tag{9a}$$

$$\mathcal{H}_{d,s}^{|\Delta B|=2} \propto -\frac{(a_0^* + a_1^*)(a_0 + a_2)}{M_H^2}\frac{m_b}{v}\frac{m_{d,s}}{v}\left[\left(\frac{m_t}{v}\right)^2 V_{tb}^* V_{t(d,s)}\right]^2 \left[\bar{b}_R (d,s)_L\right]\left[\bar{b}_L (d,s)_R\right] + \text{h.c.}\,, \tag{9b}$$

that show two key properties:

- the impact in K^0, B_d and B_s mixing amplitudes scales with $m_s m_d$, $m_b m_d$ and $m_b m_s$ respectively, opening the possibility of sizable non-standard contributions to the B_s system without serious constraints from K^0 and B_d mixing;
- while the possible flavour-blind phases do not contribute to theΔ $S = 2$ effective Hamiltonian, they could have an impact in theΔ $B = 2$ case, offering the possibility to solve the anomaly in the B_s mixing phase.

These Hamiltonians have a direct impact on some crucial observables of the neutral mesons systems, namely the mass differences and the asymmetries in the decays; in fact, in this analysis we consider

$$\Delta M_K = \frac{1}{m_K}\text{Re}\left[\langle K^0|\,\mathcal{H}^{|\Delta S|=2}\,|\bar{K}^0\rangle\right]\,, \tag{10}$$

$$\epsilon_K \simeq e^{i\frac{\pi}{4}}\frac{1}{\sqrt{2}\,m_K \Delta M_K}\text{Im}\left[\langle K^0|\,\mathcal{H}^{|\Delta S|=2}\,|\bar{K}^0\rangle\right]\,, \tag{11}$$

$$\Delta M_{d,s} = \frac{1}{m_{B_{d,s}}}\left|\langle B^0_{d,s}|\,\mathcal{H}_{d,s}^{|\Delta B|=2}\,|\bar{B}^0_{d,s}\rangle\right|\,, \tag{12}$$

$$S_{\psi K_S} = \sin\left(\arg\langle B^0_d|\,\mathcal{H}_d^{|\Delta B|=2}\,|\bar{B}^0_d\rangle\right)\,, \tag{13}$$

$$S_{\psi\phi} = \sin\left(\arg\langle B^0_s|\,\mathcal{H}_s^{|\Delta B|=2}\,|\bar{B}^0_s\rangle\right)\,. \tag{14}$$

Using the presence of the new free phase $\arg\left[(a_0^* + a_1^*)(a_0 + a_2)\right]$ in $\mathcal{H}_s^{|\Delta B|=2}$, the hinted large value of $S_{\psi\phi}$ can be easily obtained. Moreover, MFV implies that the new phases in the B_d and the B_s systems are related by the ratio m_d/m_s, and hence a large phase in the B_s system determines an unambiguous small shift in the relation between $S_{\psi K_S}$ and the CKM phase β; as it can be seen in Fig. 1 (Left), it goes in the right direction to improve the existing tension between the experimental value of $S_{\psi K_S}$ and its SM prediction.

Due to the $m_s m_d$ factor in $\mathcal{H}^{|\Delta S|=2}$, the new physics contribution to ϵ_K is tiny and does not improve alone the agreement between data and prediction for ϵ_K. However, given the modified relation between $S_{\psi K_S}$ and the CKM phase β, the true value of β extracted in this scenario increases with respect to SM fits. As a result of this modified value of β, also the predicted value for ϵ_K increases with respect to the SM case, resulting in a better agreement with data, as shown in Fig. 1 (Right).

4. Electric dipole moments

4.1. 2HDM$_{\overline{\text{MFV}}}$ predictions

The EDM of a particle is defined by one of its electromagnetic form factors [13]. In particular, for a spin-$\frac{1}{2}$ particle f, the form-factor decomposition of the matrix element of the electromagnetic

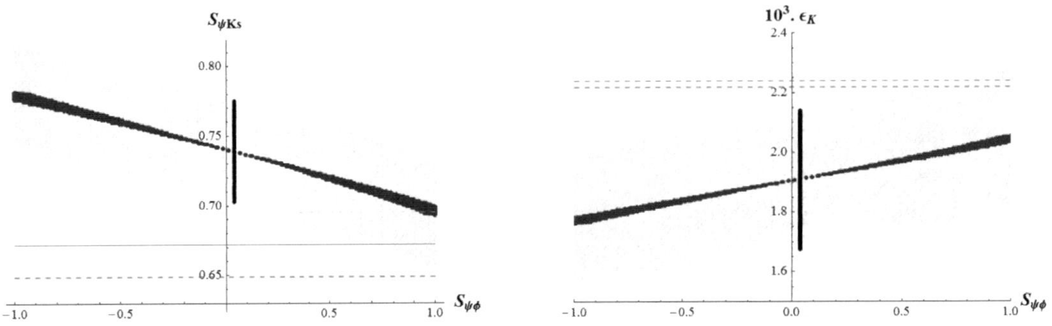

Figure 1. (Left) Correlation between $S_{\psi K_S}$ and $S_{\psi\phi}$. The dark points have been obtained with the CKM phase β fixed to its central value: the spread is determined only by the requirement of a deviation ofΔ M_s within 10% of its SM value; the light points represent the $\pm 1\sigma$ error due to the uncertainty in the extraction of β. The $\pm 1\sigma$ range of $\phi_{Bs}^{\exp}$ (light horizontal lines) and the SM prediction (black vertical line) are also shown. (Right) Correlation between ϵ_K and $S_{\psi K_S}$. Notations as before.

current J_μ is

$$\langle f(p')|\, J_\mu(0)\, |f(p)\rangle = \bar{u}(p')\Gamma_\mu(p'-p)u(p) \tag{15}$$

where

$$\Gamma_\mu(q) = F_1(q^2)\gamma_\mu + F_2(q^2) i\sigma_{\mu\nu}\frac{q^\nu}{2m} + F_A(q^2)\left(\gamma_\mu\gamma_5 q^2 - 2m\gamma_5 q_\mu\right) + F_3(q^2)\sigma_{\mu\nu}\frac{q^\nu}{2m} \,. \tag{16}$$

Since the electric dipole interaction of a particle with the electromagnetic field violates both P and T symmetry, one identifies the electric dipole moment of f with

$$d_f \equiv -\frac{F_3(0)}{2m} \;; \tag{17}$$

this corresponds to the effective electric dipole interaction

$$\mathcal{L}_{\text{eff}} = -\frac{i}{2} d_f \bar{\psi}\sigma_{\mu\nu}\gamma_5\psi F^{\mu\nu} \,, \tag{18}$$

that indeed reduces to $\mathcal{L}_{\text{eff}} = d_f\,\vec{\sigma}\cdot\vec{E}$ in the nonrelativistic limit. In an analogous way one can define the chromoelectric dipole moment. Moreover, beyond the one-loop single-particle level, there are other CP-violating operators that can have significant effects on the EDMs, like, for example, the dimension-6 Weinberg operators or the CP-odd four-fermion interactions.

The effective Lagrangian describing the fermions (C)EDMs that is relevant for our analysis reads

$$-\mathcal{L}_{\text{eff}} = \sum_f i\frac{d_f}{2}\bar{f}F_{\mu\nu}\sigma^{\mu\nu}\gamma_5 f + \sum_f i\frac{d_f^c}{2}g_s\bar{f}G_{\mu\nu}\sigma^{\mu\nu}\gamma_5 f + \sum_{f,f'} C_{ff'}(\bar{f}f)(\bar{f}'i\gamma_5 f') \,, \tag{19}$$

where $d_f^{(c)}$ stands for the quarks and leptons (C)EDMs, while $C_{ff'}$ is the coefficient of the CP-odd four fermion interactions.

Among the various atomic and hadronic EDMs, the Thallium, neutron and Mercury ones represent the most sensitive probes of CP violating effects; their EDMs can be calculated from the elementary particles EDMs using non-perturbative techniques [14]. The thallium EDM ($d_{\rm Tl}$) can be estimated as

$$d_{\rm Tl} \simeq -585 \cdot d_e - e\,(43{\rm GeV})\; C_S\;, \tag{20}$$

where

$$C_S \simeq C_{de.}\frac{29\,{\rm MeV}}{m_d} + C_{se}\frac{\kappa \times 220\,{\rm MeV}}{m_s} + C_{be}\frac{66\,{\rm MeV}(1-0.25\kappa)}{m_b}\;, \tag{21}$$

with $\kappa \simeq 0.5 \pm 0.25$; the neutron and mercury EDMs can be estimated from QCD sum rules, which leads to

$$d_n = (1 \pm 0.5)\Big[1.4(d_d - 0.25 d_u) + 1.1 e\,(d^c_d + 0.5 d^c_u)\Big] \tag{22}$$

and

$$\begin{aligned} d_{\rm Hg} \simeq\; & 7\times 10^{-3}\, e\,(d^c_u - d^c_d) + 10^{-2}\, d_e + e\,\big(3.5\times 10^{-3}\,{\rm GeV}\big)\; C_S - \\ & - e\,\big(1.4\times 10^{-5}{\rm GeV}^2\big)\left[0.5\frac{C_{dd}}{m_d} + 3.3\kappa\frac{C_{sd}}{m_s} + (1-0.25\kappa)\frac{C_{bd}}{m_b}\right] \end{aligned} \tag{23}$$

(in which the $d_f^{(c)}$ are evaluated at 1 GeV).

The values of $C_{ff'}$ and $d_f^{(c)}$ have been obtained in the context of Supersymmetry in [15, 16], and they have been extended to the case of the $2{\rm HDM}_{\overline{\rm MFV}}$ in [10]; here we present the results for the case in which CP is not violated in the scalar potential.

- For $C_{ff'}$ one has

$$C_{ff'} = \frac{m_f\, m_{f'}}{v^2}\,\frac{{\rm Im}\,\omega_{ff'}}{M_A^2}\, t_\beta^2\,, \tag{24}$$

with

$$\omega_{ff'} \simeq W^{d\star}_{ff}\, W^{d}_{f'f'}\;. \tag{25}$$

The explicit expressions for the $\omega_{ff'}$ relevant for our analysis are

$$\begin{aligned} {\rm Im}\,\omega_{ed} &= {\rm Im}\,\omega_{es} = {\rm Im}\,\sigma_d\;, & (26)\\ {\rm Im}\,\omega_{eb} &= {\rm Im}\,\xi\;, & (27)\\ {\rm Im}\,\omega_{dd} &= {\rm Im}\,\omega_{ds} = 0\;, & (28)\\ {\rm Im}\,\omega_{db} &= {\rm Im}\,(\sigma_d^\star\,\xi)\;, & (29) \end{aligned}$$

where we have defined

$$\xi = \bar a + (a_0 + a_1 + a_2)\lambda_t^2\,, \qquad \sigma_q = \bar a + |V_{tq}|^2\lambda_t^2 a_0\;. \tag{30}$$

- Concerning the fermions (C)EDMs, for the down quark they are induced already at the one-loop level by means of the exchange of the charged-Higgs boson and top quark. However, these effects are CKM suppressed by the factor $|V_{td}|^2 \approx 10^{-4}$, and even the most extreme choice of the parameters leads to predictions for d_n and $d_{\rm Hg}$ well under control [10]; hence, we can neglect these contributions in this analysis.
Instead, the two-loop Barr–Zee contributions [17] to the fermionic (C)EDMs dominate over one loop-effects since they overcome the strong CKM suppression; in the $2{\rm HDM}_{\overline{\rm MFV}}$ they read

$$\frac{d_f}{e} = -\sum_{q=t,b} \frac{N_c q_f \alpha_{\rm em}^2\, q_q^2\, m_f}{8\pi^2 s_W^2 M_W^2}\,\frac{m_q^2}{M_A^2}\,(\tan\beta)^{0,2}\left[f(\tau_q)\,{\rm Im}\,\omega_{qf} + g(\tau_q)\,{\rm Im}\,\omega_{fq}\right]\,, \tag{31}$$

$$d^c_f = -\sum_{q=t,b} \frac{\alpha_s\,\alpha_{\rm em}\, m_f}{16\pi^2 s_W^2 M_W^2}\,\frac{m_q^2}{M_A^2}\,(\tan\beta)^{0,2}\left[f(\tau_q)\,{\rm Im}\,\omega_{qf} + g(\tau_q)\,{\rm Im}\,\omega_{fq}\right]\,, \tag{32}$$

where q_ℓ is the electric charge of the fermion ℓ, $\tau_q = m_q^2/M_A^2$ and $f(\tau)$, $g(\tau)$ are the two-loop Barr–Zee functions

$$f(\tau) = \frac{\tau}{2}\int_0^1 \frac{1-2x(1-x)}{\tau - x(1-x)} \log\left[\frac{x(1-x)}{\tau}\right] , \tag{33}$$

$$g(\tau) = \frac{\tau}{2}\int_0^1 \frac{1}{\tau - x(1-x)} \log\left[\frac{x(1-x)}{\tau}\right] , \tag{34}$$

and the relevant $\omega_{ff'}$ are only

$$\mathrm{Im}\,\omega_{et} = 0 , \tag{35}$$

$$\mathrm{Im}\,\omega_{dt} = \mathrm{Im}\,\sigma_d^\star , \tag{36}$$

since, as we have already noted, the MFV structure suppresses the contributions of the u quark with a factor $(\tan\beta)^{-2}$.

A comparison of the $\mathrm{2HDM}_{\overline{\mathrm{MFV}}}$ predictions for the Thallium, neutron and Mercury EDMs with their present experimental upper bounds is shown in Tab. 1. They have been obtained with the reference values $m_A = 500$ GeV and $\tan\beta = 10$; it can be seen that for natural values of $|\bar{a}|, |a_i| = \mathcal{O}(1)$ they are compatible with the experimental limits.

Observable	Experimental bounds	$\mathrm{2HDM}_{\overline{\mathrm{MFV}}}$ predictions
$\lvert d_{Tl}\rvert$ [e cm]	$< 9.0 \times 10^{-25}$	$3 \times 10^{-25}\,\mathrm{Im}\,\xi + 5 \times 10^{-25}\,\mathrm{Im}\,\sigma_d$
$\lvert d_n\rvert$ [e cm]	$< 2.9 \times 10^{-26}$	$10^{-27}\,\mathrm{Im}\,(\sigma_d^\star \xi)$
$\lvert d_{Hg}\rvert$ [e cm]	$< 3.1 \times 10^{-29}$	$4 \times 10^{-29}\,\mathrm{Im}\,(\sigma_d^\star \xi) + 5 \times 10^{-29}\,\mathrm{Im}\,\sigma_d + 2 \times 10^{-29}\,\mathrm{Im}\,\xi$

Table 1. Predictions of the $\mathrm{2HDM}_{\overline{\mathrm{MFV}}}$ for some experimentally interesting EDMs.

4.2. Correlation between EDMs and CP violation in the B_s mixing

In the previous section we have shown that, in the context of the $\mathrm{2HDM}_{\overline{\mathrm{MFV}}}$, one can accommodate a larger B_s mixing phase obtaining at the same time some other interesting effects. However, it is necessary to check if this new phase can cause disagreements with other CP violating observables, such as the EDMs.

The correlation plot between the considered EDMs and the observable $S_{\psi\phi}$ is shown in Fig. 2. As it can be seen, the current constraints on the EDMs still allow values of $|S_{\psi\phi}|$ larger than 0.5, compatible with the highest values of the B_s mixing phase reported by the Tevatron experiments.

5. Conclusions

It is well known that the choice of introducing only one Higgs doublet in the Standard Model is just the most economical, but not the only possible one; there is a variety of New Physics models that contain more Higgs doublets, bringing interesting phenomenological features such as new sources of CP violation, dark matter candidates, axion phenomenology. In order to protect 2HDMs from FCNCs the application of MFV is natural and effective, and still allows the introduction of new flavour-blind CP-violating phases whose presence is interesting in order to address the recent experimental anomalies detected in this sector. In the $\mathrm{2HDM}_{\overline{\mathrm{MFV}}}$ in fact, once a larger phase in the B_s mixing is introduced, the tensions in the relation between ϵ_K, the CKM angle β and the asymmetry $S_{\psi K_S}$ are automatically softened, and the agreement of the predictions of flavour-violating processes (FCNCs) and flavour-conserving observables (EDMs) with experimental data is not spoiled.

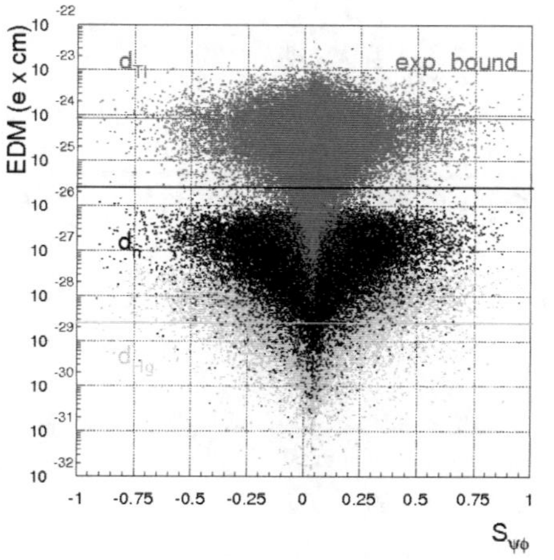

Figure 2. Correlation between EDMs of the Thallium (red dots), neutron (black dots), and Mercury (green dots), versus $S_{\psi\phi}$ [10]. It is obtained by means of the following scan: $(|\overline{a}|, |a_0|, |a_1|, |a_2|) < 2$, $0 < (\phi_{\overline{a}}, \phi_{a_0}, \phi_{a_1}, \phi_{a_2}) < 2\pi$, $\tan\beta < 60$, $M_{H^\pm} < 1.5$ TeV.

Experimental data to test the $2HDM_{\overline{MFV}}$ will probably be available in the next few years: first of all, the predicted values for the EDMs are strongly enhanced with respect to the SM, and sizable non standard values for $S_{\psi\phi}$ imply lower bounds for the aforementioned EDMs within the reach of the expected future experimental resolutions; moreover, the MFV structure of the model determines several unambiguous correlations between observables that are going to be studied by future experiments, such as the decays $B^0_{d,s} \to \mu^+\mu^-$.

Acknowledgments

I would like to thank Andrzej J. Buras, Gino Isidori and Stefania Gori for the fruitful collaboration, Paride Paradisi for valuable explanations, and Marco Bardoscia for revisions. This work has been supported in part by the Graduiertenkolleg GRK 1054 of DFG and in part by the Cluster of Excellence 'Origin and Structure of the Universe'.

References

[1] T. Aaltonen *et al.* [CDF Collaboration], Phys. Rev. Lett. **100** (2008) 161802.
[2] V. M. Abazov *et al.* [D0 Collaboration], Phys. Rev. Lett. **101** (2008) 241801.
[3] A. J. Buras and D. Guadagnoli, Phys. Rev. D **78**, 033005 (2008).
[4] E. Lunghi and A. Soni, Phys. Lett. B **666** (2008) 162.
[5] V. M. Abazov *et al.* [D0 Collaboration], Phys. Rev. D **82** (2010) 032001.
[6] A. J. Buras, P. Gambino, M. Gorbahn, S. Jager and L. Silvestrini, Phys. Lett. B **500** (2001) 161.
[7] G. D'Ambrosio, G. F. Giudice, G. Isidori and A. Strumia, Nucl. Phys. B **645** (2002) 155.
[8] A. L. Kagan, G. Perez, T. Volansky and J. Zupan, Phys. Rev. D **80** (2009) 076002.
[9] A. J. Buras, M. V. Carlucci, S. Gori and G. Isidori, JHEP **1010**, 009 (2010).
[10] A. J. Buras, G. Isidori and P. Paradisi, Phys. Lett. B **694**, 402 (2011).
[11] S. L. Glashow and S. Weinberg, Phys. Rev. D **15** (1977) 1958.
[12] A. Pich and P. Tuzon, Phys. Rev. D **80** (2009) 091702.
[13] W. Bernreuther, M. Suzuki, Rev. Mod. Phys. **63** (1991) 313-340.
[14] D. A. Demir, O. Lebedev, K. A. Olive, M. Pospelov and A. Ritz, Nucl. Phys. B **680** (2004) 339.
[15] J. R. Ellis, J. S. Lee and A. Pilaftsis, JHEP **0810** (2008) 049.
[16] J. Hisano, M. Nagai and P. Paradisi, Phys. Rev. D **80** (2009) 095014.
[17] S. M. Barr and A. Zee, Phys. Rev. Lett. **65** (1990) 21 [Erratum-ibid. **65** (1990) 2920].

Flavor Physics in SO(10) GUTs with Suppressed Proton decay Due to Gauged Discrete Symmetry

Aleksandr Azatov [1]

Dipartimento di Fisica, Università di Roma "La Sapienza" and
INFN Sezione di Roma, I-00185 Roma, Italy

E-mail: aleksandr.azatov@roma1.infn.it

Abstract. Generic $SO(10)$ GUT models suffer from the problem that Planck scale induced non-renormalizable proton decay operators require extreme suppression of their couplings to be compatible with present experimental upper limits. One way to resolve this problem is to supplement $SO(10)$ by simple gauged discrete symmetries which can also simultaneously suppress the renormalizable R-parity violating interactions when they occur and make the theory "more natural". We then present an extended $\mathbf{16}_H$ model, with three **10** and three **45**-Higgs, which is free of this problem. We propose this as a realistic and "natural" model for fermion unification and discuss the phenomenology of this model e.g. its predictions for neutrino mixings and lepton flavor violation.

1. Introduction

The neutrino observations of the past decade have put the spotlight on gauged B-L symmetry as well as unification groups such as SO(10) and $SU(2)_L \times SU(2)_R \times SU(4)_4$ containing B-L as prime candidates for theory of matter, forces and flavor. While both these groups incorporate the seesaw mechanism for neutrino masses, SO(10) has the additional attractive feature that gauge couplings unify at high scale. It is however highly nontrivial to obtain a "truly natural" SO(10) model due to such issues as doublet triplet splitting, rapid proton decay etc. In this paper we discuss how one aspect of this naturalness can be addressed i.e. how one can naturally suppress proton decay in SO(10) models while preserving our understanding neutrino masses.

We first note that SO(10) models for neutrinos discussed in recent literature can by and large be divided into two classes:

(i) One class which uses only renormalizable couplings involving the Higgs fields **10**, **120** and **126** for fermion masses and the last multiplet for breaking B-L symmetry and multiplets such as **45** and/or **210** for gauge symmetry breaking[1]. This theory could be considered as an ultraviolet complete theory by itself.

(ii)The second class uses **10** plus $\mathbf{16} \oplus \bar{\mathbf{16}}$ for fermion masses with the **16**'s breaking the B-L symmetry. Here one generally uses **45**+**54** Higgs fields for SO(10) breaking. An important feature of this class is that it has to rely on nonrenormalizable couplings to understand fermion masses and therefore has to be viewed necessarily as an effective theory at the GUT scale[2].

The first class of models leads to automatic R-parity conservation when SO(10) breaks down to MSSM so that there is a natural candidate for dark matter whereas the second class of models

[1] In collaboration with R.N.Mohapatra

suffers from R-parity breaking and hence has no stable dark matter in the absence of additional symmetries. So in principle one could argue that this class of models are not "pure" SO(10) models.

Both models have an additional naturalness problem arising from the fact that they allow R-parity conserving nonrenormalizable couplings of the form $\lambda \mathbf{16}^4_m/M_{Pl}$ which lead to rapid proton decay. Such interactions could be induced by non-perturbative Planck scale effects and it is therefore not safe to ignore them. Present proton life time limits constrain λ to be $\leq 10^{-7}$. Such a small value of λ would suggest that there is probably a symmetry responsible for its smallness. This question is particularly urgent for the class of SO(10) models with **16** Higgs since they rely on other such dimension four higher dimensional operators with coefficients of order one to understand fermion masses. This problem is generic to all non-GUT susy theories such as MSSM or left-right models as well as $SU(2)_L \times SU(2)_R \times SU(4)_4$ models and not just GUT theories. One way to understand the suppression of such operators despite the presence of non-perturbative gravitational effects, is to have an additional gauge symmetry beyond SO(10) which can forbid these unwanted terms. The simplest possibility is to have a discrete gauge symmetry[3].

The discrete gauge symmetry supplemented SO(10) models that suppress proton decay were studied for a large class of models in a recent papers[5, 6]. In particular two minimal SO(10) models- one with **16**-Higgs breaking the B-L symmetry and another with **126** breaking B-L were shown to be free of both proton decay problem as well as R-parity problem if SO(10) was supplemented by a gauged Z_6 symmetry.

In this paper we will show that within extended **16**-based models , one can achieve the desired doublet-triplet splitting and fermion masses that can match observations. This model differs from other **16**-based models in that proton decay here arises only from the gauge boson exchanges unlike other models where Planck scale induced effects as well as Higgsino exchange ones play a role[7]; (iii) we study the phenomenological implications of this model and isolate some of its tests e.g. in the domain of lepton flavor violation.

This paper is organized as follows: in sec. 2, we review the conditions discrete group has to satisfy to suppress proton decay; in sec.3 we discuss the three Higgs**16**-based model that fits fermion masses and mixings; in sec. 4, we discuss how large neutrino mixings and observed neutrino masses arise in this model; in sec. 5 we discuss bounds arising from Lepton Flavor Violating processes, and then we conclude by summarizing presented results.

2. The $SO(10) \times Z_6$ model for 16-Higgs B-L breaking

The main features of generic SO(10) models with **16**-Higgs fields breaking B-L symmetry are the following: (i) the quarks and leptons are assigned to three **16**-dimensional spinors (denoted by ψ_m, m=1,2,3); (ii) the GUT symmetry is broken down to $SU(2)_L \times SU(2)_R \times U(1)_{B-L} \times SU(3)_c$ by a $\mathbf{45} \oplus \mathbf{54}$ set of Higgs fields; (iii) $SU(2)_R \times U(1)_{B-L}$ symmetry is broken by the **16**-Higgs pair denoted by $\psi_H \oplus \bar{\psi}_H$ set to the standard model symmetry which is then broken by $SU(2)$ doublets that are linear combinations of the doublets in SO(10) **10** and **16**-Higgs fields. The standard model symmetry along with supersymmetry emerge just below the GUT sclae of 2×10^{16} GeV.

This model has several naturalness problems: it not only allows the dangerous R-parity conserving $\frac{(16_m)^4}{M_{Pl}}$ terms but also terms such as $\frac{(16_m)^3 16_H}{M_{Pl}}$ terms which on B-L symmetry breaking lead to all three types of R-parity violating operators present in general MSSM i.e. LLe^c, QLd^c and $u^c d^c d^c$ type. Thus this model for natural values of couplings will lead to extremely rapid proton decay which is unacceptable. The question addressed in Ref.[5] is to search for gauged discrete symmetries that will keep the model phenomenologically viable while keeping them "proton decay safe" and it was shown that the minimal anomaly free discrete gauge symmetry is Z_6. Similar considerations for **126** type models also led to the symmetry Z_6 and in both cases an extra **10**-Higgs field denoted by H' in addition to those considered already.

To see the discrete symmetry charges for various fields that forbid both R-parity violating terms as well as R-parity conserving baryon number violating terms, while at the same time keeping the required terms responsible for good phenomenology, we divide the superpotential terms into two classes: type I terms that must be kept for phenomenology and type II terms that must be forbidden to suppress proton decay and R-parity violating terms. They are given below:

Terms of type I: They include $\psi_m \psi_m H$, $(\psi_m \overline{\psi}_H)^2/M_{\rm P}$, $\psi_H \overline{\psi}_H$, A^2, $S^{2,3}$ and $S\,A^2$, where H, A, S are **10**-, **45**-, **54**-plets, respectively.Taking the discrete gauge symmetry to be Z_N, we can write down the constraints on the Z_N charges that are required by the type I terms:

$$2q_{\psi_m} + q_H = 0 \text{ mod}N, \quad 2q_{\psi_m} + 2q_{\overline{\psi}_H} = 0 \text{ mod } N \tag{1}$$

$$q_{\psi_H} + q_{\overline{\psi}_H} = 0 \text{ mod } N, \quad q_H + q_{H'} = 0 \text{ mod } N, \; 2q_A = 0 \text{ mod } N, \tag{2}$$

$$2q_S = 0 \text{ mod } N, \; 3q_S = 0 \text{ mod } N, \; 2q_A + q_S = 0 \text{ mod } N. \tag{3}$$

Here, we denote the Z_N charge for a field F by q_F.

Type II terms: These are the terms that must be forbidden from appearing in the superpotential and are $\psi_m \psi_m H'$, ψ_m^4, $\psi_m \overline{\psi}_H$ and $\psi_m^3 \psi_H$, $\psi_m \psi_H H$, $\psi_m \psi_H H'$ and $\psi_m \overline{\psi}_H A$. We forbid the $\psi_m \psi_m H'$ in order to avoid large Higgsino mediated contribution to proton decay since this is the very problem we are trying to solve. The necessary constraints on the Z_N charges have to be chosen such that they satisfy the inequalities

$$q_{\psi_m} + q_{H'} \neq 0 \text{ mod } N, \quad q_{\psi_m} \neq 0 \text{ mod } N, \tag{4}$$

$$q_{\psi_m} + q_{\overline{\psi}_H} \neq 0 \text{ mod } N, \; q_{\psi_m} + q_{\psi_H} \neq 0 \text{ mod } N \tag{5}$$

$$q_{\psi_m} + q_{H',H} + q_{\psi_H} \neq 0 \text{ mod } N, \; q_{\psi_m} + q_{\overline{\psi}_H} + q_A \neq 0 \text{ mod } N. \tag{6}$$

The last set of constraints come from the requirement that the discrete symmetry must be a gauge symmetry i.e. it must be anomaly free. The anomaly freedom constraints are:

$$16\,(N_g\, q_{\psi_m} + q_{\psi_H} + q_{\overline{\psi}_H}) + 10\,(q_H + q_{H'}) + 45\,q_A + 54\,q_S \;=\; 0 \text{ mod } N' \tag{7}$$

$$2N_g\, q_{\psi_m} + 2\,q_{\psi_H} + 2\,q_{\overline{\psi}_H} + q_H + q_{H'} + 8q_A + 12q_S \;=\; 0 \text{ mod } N \tag{8}$$

$$\text{where } N' = \begin{cases} N, & \text{odd } N \\ N/2, & \text{even } N \end{cases} \tag{9}$$

It was shown in [5] that the smallest symmetry allowing us to fulfill all criteria is Z_6 for number of generation $N_g = 3$. A possible charge assignments is $q_{\psi_m} = 1$, $q_{\psi_H} = -2$, $q_{\overline{\psi}_H} = +2$, $q_H = -2$, $q_{H'} = +2$, $q_{45,54} = 0$ (cf. tables 3). This charge assignment allows for seesaw couplings and the possibility of fermion masses from couplings of type $\psi_m \psi_m H$. The allowed operator $\psi_m \psi_m \overline{\psi}_H^2$ contributes to both the fermion masses as well as to the seesaw. The model also eliminates the dangerous proton decay operator $Q\,Q\,Q\,L$ or operator of type $(\psi_m)^4/M_{\rm P}$.

While the allowed set of operators provide a necessary condition for the model being phenomenologically viable, the final step where we judge whether it is acceptable first requires that we do doublet triplet splitting and see if the sub-GUT scale structure of the model can generate acceptable pattern of fermion masses or not. We address this question in the next sub-section.

3. Extended 16-Higgs model

Minimal **16** and **126** presented in [5] suffer from predicting viable phenomenology, but as was shown in [6] a simple extension of **16** can cure all these problems. The field content of the extended **16** -Higgs model and Z_6 assignments are presented in Table 1. One can see that anomaly freedom and proton decay constraints are satisfied and yet the model can lead to viable phenomenology. The superpotential is given by:

Table 1. Z_6 charge assignment of the fields

ψ_m	$\mathbf{16}_1$	ψ_H	$\mathbf{16}_{-2}$		
H_1	$\mathbf{10}_{-2}$	$\overline{\psi}_H$	$\overline{\mathbf{16}}_2$	A_3	$\mathbf{45}_2$
H_2	$\mathbf{10}_2$	A_1	$\mathbf{45}_0$	S	$\mathbf{54}_0$
H_3	$\mathbf{10}_0$	A_2	$\mathbf{45}_{-2}$		

$$\begin{aligned} W = \; & MH_1H_2 + SH_1H_2 + A_1H_1H_2 + \psi_H\psi_H H_1 + \bar{\psi}_H\bar{\psi}_H H_2 + M_\psi\bar{\psi}_H\psi_H + M_3H_3^2 + \\ & + H_2H_3A_2 + H_1H_3A_3 + \frac{H_3A_3\overline{\psi}_H\overline{\psi}_H}{M_{Pl}} + \frac{H_3A_2\psi_H\psi_H}{M_{Pl}} \end{aligned} \tag{10}$$

One can see that the vevs of the $A_{2,3}, \psi_H\overline{\psi}_H$ will break Z_6 will break Z_6 symmetry down to Z_2, so naively we lose our protection of the proton decay, but let us look at each of these vevs separately.

- vevs of the $\mathbf{16}, \overline{\mathbf{16}}$ will preserve the following combination of $2(B-L)+Z_6$, so they cannot lead to the proton decay, because it also preserves $(B-L)$.
- Potential minimization can ensure that vev $\langle A_3\rangle = 0$, so we will not have the dangerous operator $\frac{\psi_m^4 A_3}{M_{Pl}}$ which might lead to the large contribution to the proton decay operator with effective coupling $\lambda \simeq \frac{M_U}{M_{Pl}} \ll 1$ -not suppressed enough to be acceptable.
- Potential minimization can ensure that vev $\langle A_2\rangle_{\text{Triplet}} = 0$ in this case we will not have Higgsino mediated contribution to the four fermion proton decay operator, and the effective coupling of the $\frac{\psi_m^4}{M_{Pl}}$ will have a suppression $\left(\frac{\langle A_2\rangle_{\text{Doublet}}}{M_{Pl}}\right)^2 \leq 10^{-4}$ which is naively $\sim 10^{-4}$ times bigger than the present upper limit but still is a considerable improvement in naturalness.

To study the doublet triplet splitting in this model, note that the mass matrix for the **5** and $\overline{\mathbf{5}}$ of SU(5) is given by

$$\begin{pmatrix} 0 & M+A_1+S & 0 & c \\ M-A_1+S & 0 & A_2 & 0 \\ 0 & -A_2 & M_3 & 0 \\ 0 & c & \delta & M_\psi \end{pmatrix}_{ud} \tag{11}$$

Where δ in the (43) element of the matrix comes from $\frac{H_3A_2\psi_H\psi_H}{M_{Pl}}$ coupling. As before, we want the determinant of this matrix to vanish. This leads to the following constraints
Case (i)

$$M - A_1 + S = 0 \tag{12}$$

Case(ii):

$$M + A_1 + S = \frac{c^2M_3 + cA_2\delta}{M_3M_\psi} \tag{13}$$

Here we consider only the simpler of the two cases above i.e. case (i) to illustrate that our proposal leads to a realistic model. In the first case $M - A_1 + S = 0$, $D = (1,0,0,0)$ (implying that the h_d has non-zero component in the multiplet H_1) in the same way as was in the minimal model, but now due to the presence of A_2 field all the U_1 is nonvanishing, so that the "up" quarks will get masses from the $\psi_m^2 H_1$ operator. In the next sections we will discuss the detailed fit to fermion masses for this extended **16** model.

As we can see extended **16** model can solve doublet-triplet splitting problem as well as provide masses for all fermions, but now we have to check whether higgsino mediated proton decay operators are allowed. Even though quarks and leptons couple only to the H_1 field and there is no mass term $\propto H_1 H_1$, mixing between H_1, H_2, H_3, ψ_H fields can lead to the nonvanishing diagrams with higgsino exchange. The contribution of these diagrams will vanish if only the $(H_1 H_1)$ element of the inverse mass matrix (11) for the heavy triplets vanishes, thus the triplet part of the A_2 should be zero. One can show that the requirement of the $< A_3 >= 0$ combined with F flatness condition will lead to this condition(see for details [6]).

4. Fermion masses

The following couplings allowed by $Z_6 \times SO(10)$ symmetries will lead to fermion masses after symmetry breaking.

$$W = h^{10}\psi_m\psi_m H_1 + f^{10}\frac{\psi_m^2\overline{\psi}_H^2}{M_{Pl}} + f^{126}\frac{\psi_m^2\overline{\psi}_H^2}{5!M_{Pl}} + k^{120}\frac{\psi_m^2 A H_1}{3!M_{Pl}} + g^{120}\frac{\psi_m^2 A^2 H_1}{3!M_{Pl}^2} + g^{126}\frac{\psi_m^2 A^2 H_1}{5!M_{Pl}^2} \quad (14)$$

Where $h^{10}, f^{10}, f^{126}, g^{126}$ are symmetric 3×3 and k^{120}, g^{120} antisymmetric matrices. The upper index of h^{10}, f^{126}..., **10, 120, 126** shows the SO(10) structure of the fermion couplings.

$$\begin{aligned}
M_u &= v_u\left[\alpha\left(h^{10} - k^{120}\frac{a-b}{M_{Pl}} + g^{120}\frac{b(a-b)}{M_{pl}^2} + g^{126}\frac{ab+a^2}{M_{Pl}^2}\right) + \xi\left(f^{10}\frac{4c}{M_{Pl}} - f^{126}\frac{8c}{M_{Pl}}\right)\right] \\
M_\nu^D &= v_u\left[\alpha\left(h^{10} + k^{120}\frac{3a+b}{M_{Pl}} - g^{120}\frac{b(3a+b)}{M_{pl}^2} - g^{126}\frac{3(ab+a^2)}{M_{pl}^2}\right) + \xi\left(f^{10}\frac{4c}{M_{Pl}} + f^{126}\frac{24c}{M_{Pl}}\right)\right] \\
M_d &= v_d\gamma\left[h^{10} - k^{120}\frac{a+b}{M_{Pl}} - g^{120}\frac{b(a+b)}{M_{pl}^2} + g^{126}\frac{-ab+a^2}{M_{pl}^2}\right] \\
M_e &= v_d\gamma\left[h^{10} + k^{120}\frac{3a-b}{M_{Pl}} + g^{120}\frac{b(3a-b)}{M_{pl}^2} - g^{126}\frac{3(-ab+a^2)}{M_{pl}^2}\right] \\
M_\nu^M &= 16f^{126}\frac{(\xi v_u)^2}{M_{pl}} - (M_\nu^D)^T(16f^{126}\frac{c^2}{M_{Pl}})^{-1}M_\nu^D
\end{aligned} \quad (15)$$

Where the vev of the fields H^1 and $\overline{\psi}_H$ are related to the vev of the MSSM doublets h_u and h_d in the following way

$$\begin{aligned}
\langle H_{1u}\rangle &= \alpha\langle h_u\rangle \\
\langle H_{1d}\rangle &= \gamma\langle h_d\rangle \\
\langle\overline{\psi}_{Hu}\rangle &= \xi\langle h_u\rangle
\end{aligned} \quad (16)$$

Our claim is that these Yukawa coupling structure is rich enough to fit all the fermion masses. We give below an example of a scenario where correct fermion masses can arise.

We take the case where all antisymmetric couplings vanish and that down quarks and leptons are brought to the diagonal basis at the same time. Thus g^{126} and h^{10} are diagonal, and all the mixing in the quark and lepton sector arise from the couplings f^{10} and f^{126}. We now show that even under such limiting assumptions we can fit all the fermion masses. We know the quark masses at the GUT scale thus we can find corresponding h^{10}, g^{126}, but on the other hand in the case when all the quark mass matrices are symmetric the mass matrix for the up quarks is equal to

$$M_u = (CKM)^T.M_u^{diag}.CKM; \tag{17}$$

this leads to the constraint on the linear combination of f^{10} and f^{126}.On the other hand we can find f_{126} from the known neutrino masses and mixing angles [10], so this fixes f^{10} and f^{126}.

5. Constraints from the lepton flavor violation

In this section, we discuss the predictions of this model for lepton flavor violation. As is well known[11, 12], even if the slepton mass matrices are diagonal at the GUT scale the RGE running down to the scale of the righthanded neutrino will lead to the mixing in the slepton sector, which via one loop diagrams leads to lepton violation. We will assume mSUGRA boundary condition for scalar partner masses and use the renormalization group equations to run them down to the seesaw scale when the right handed neutrinos decouple.

We will work in the basis with diagonal righthanded majorana neutrino matrix, then the slepton mixing will be approximately equal to

$$(\delta^l_{ij})_{LL} = -\frac{3m_0^2 + A_0^2}{8\pi^2 m_0^2}\sum_{k=1}^{3}(Y_\nu)_{ik}(Y_\nu^*)_{jk}\ln(\frac{M_{GUT}}{M_{R_k}}) \tag{18}$$

where Y_ν are the Yukawa couplings of the Dirac neutrino. These Yukawa couplings appear to be of roughly

$$Y_\nu \sim \begin{pmatrix} 10^{-5} & 5\cdot 10^{-4} & 5\cdot 10^{-3} \\ 10^{-5} & 5\cdot 10^{-3} & 2\cdot 10^{-2} \\ 10^{-5} & 3\cdot 10^{-3} & 0.3 \end{pmatrix} \tag{19}$$

Here Y_ν is a linear combination of the Yukawa couplings h_{10}, f_{10} and f_{126} of the previous section. The slepton mixing leads to the lepton flavor violating processes $l_i \to l_j\gamma$ with the amplitude equal to

$$iM = em_{li}\epsilon^\lambda \bar{l}_j\,(iq^\mu\sigma_{\lambda\mu}(A_LP_L + A_RP_R))\,l_i \tag{20}$$

Where the q is the momentum of the photon and $P_{L,R} = \frac{1}{2}(1\mp\gamma_5)$, the exact expression for the $A_{L,R}$ can be found in [12]. The branching ratio for this processes will be equal to

$$\frac{BR(l_i\to l_j\gamma)}{BR(l_i\to l_j\nu_i\bar{\nu}_j)} = \frac{48\pi^3\alpha}{G_F^2}(|A_L^{ij}|^2)$$
$$A_l^{ij} \propto \frac{\alpha_2}{4\pi}\frac{(\delta^l_{ij})_{LL}}{\tilde{m}_0^2} \tag{21}$$

The present bounds on this processes are[13]

We will carry out our calculations for the branching ratio in the mSUGRA scenario, where there are only four parameters that will fix the low energy values of the slepton masses $M_{1/2}, m_0, A_0, tan\beta, sign(\mu)$ but our fit for the fermion masses was carried out for the $tan\beta = 55$ so we will stay with this value. In the FIG.1 one can see dependence of the branching ratios on the $M_{1/2}$ for the fixed values of $m_0, tan\beta, A_0, sign(\mu)$. We note that branching ratio for $\mu\to e+\gamma$ for almost the entire parameter range of our model is above 10^{-13} a value which is in the accessible range of the ongoing MEG experiment[13].

Table 2. Current bounds on $l_i \to l_j\gamma$ processes

$BR(\mu \to e\gamma)$	$1.2 \cdot 10^{-11}$
$BR(\tau \to \mu\gamma)$	$6.8 \cdot 10^{-8}$

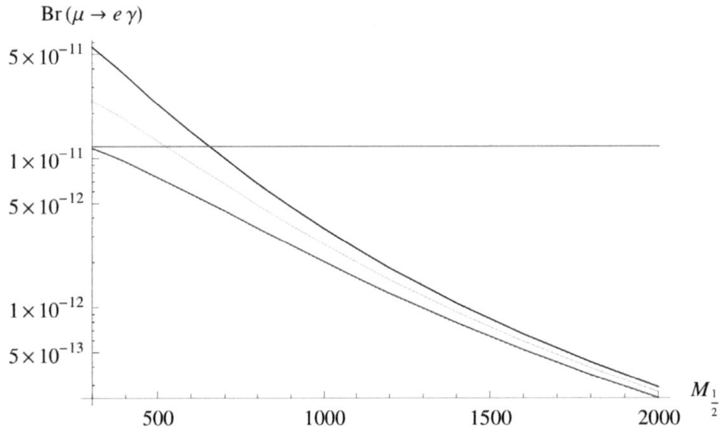

Figure 1. Branching ratios $BR(\mu \to e\gamma)$ for different values of m_0 (black 300 Gev, green 400 Gev, blue 500 Gev), $A_0 = 0, tan\beta = 55$, sign(μ)=1

6. Comments

We add a few comments on the model described before closing:

(i) In this model, the leading order proton decay operator is $\frac{\psi_m^4 A_2^2}{M_{Pl}^3}$. After GUT symmetry breaking this leads to the effective strength $\lambda \sim \frac{M_U^2}{M_{Pl}^2}$. Naively this is of order 2×10^{-5}, bigger than the present upper limit but is a considerable improvement in the naturalness. It could also be that the GUT vev could arise mainly from A_1 with $\langle A_2 \rangle$ being an order of magnitude smaller. This would then give the desired suppression to proton decay. In that case this will be the dominant graph for proton decay. Note that there are no Higgsino mediated diagrams for proton decay in this model. In addition, there is the gauge exchange diagram, present in all SO(10) GUT models.

(ii) The $\mu \to e + \gamma$ appears to be the only other low energy test of the model which is similar to such models.

(iii) For the choice of parameters used in fermion mass fitting the neutrino mixing angles and mass differences could have any values in the allowed region.

7. Conclusion

In conclusion, we have presented the minimal SO(10) **16**-Higgs model for fermion masses where the problem of extreme fine tuning of higher dimensional Planck scale induced proton decay operators has been considerably ameliorated by the presence of discrete symmetries so that in the end, we only need to tune down the coupling only by a factor of 10^{-2}. In this sense it is a more natural model. We exhibited a fit to all fermion masses and mixings including neutrinos in this model to show that it can indeed be a realistic description of nature.

Acknowledgments

Author would like to thank R.N. Mohapatra for collaboration on this project. This work was supported by the National Science Foundation Grant No. PHY-0652363. Author would like to thank organizers of the conference DISCRETE10 - Symposium on Prospects in the Physics of Discrete Symmetries for the opportunity to present this talk.

References

[1] Clark T E, Kuo T K and Nakagawa N, *Phys. Lett.* B **115**, 26 (1982); Aulakh C S and Mohapatra R N,*Phys. Rev.* **D 28**, 217 (1983); Babu K S and Mohapatra R N,*Phys. Rev. Lett.* **70**, 2845 (1993); Lee D G and Mohapatra R N, *Phys. Rev.* **D 51**, 1353 (1995); Chen M C and Mahanthappa K T,*Phys. Rev.* **D 62**, 113007 (2000); Fukuyama T and Okada N,*Preprint hep-ph/0206118*; Bajc B, Senjanović G and Vissani F, *Phys. Rev. Lett* **90**, 051802 (2003) *Preprint hep-ph/0210207.* Goh H S, Mohapatra R N and Ng S P, *Phys. Lett.* B **570**, 215 (2003) ; Fukuyama T, Ilakovac A, Kikuchi T, Meljanac S and Okada N, *JHEP* **0409**, 052 (2004);*Eur.Phys.*J.C42:191-203,2005; Dutta B, Mimura Y and Mohapatra R N, *Preprint hep-ph/0406262, Phys. Lett.* **B 603** , 35 (2004) ; *Phys. Rev. Lett.* **94**, 091804 (2005); *Phys. Rev.* **D 72**, 075009 (2005); Bajc B, Melfo A, Senjanovic G and Vissani F, *Phys. Lett.* **B 634**, 272 (2006); Aulakh C S and Garg S K, *hep-ph/0512224*; Babu K S and Macesanu C,*Phys. Rev.* **D 72**, 115003 (2005); Bertolini S, Malinsky M and Schwetz T, *Phys. Rev.* **D 73**, 115012 (2006); Grimus W and Kuhbock H, *Preprint hep-ph/0612132; arXiv:0710.1585 [hep-ph].*

[2] Barr S M and Raby S, *Phys. Rev. Lett.* **79**, 4748 (1997) Albright C, Babu K S and Barr S, *Phys. Rev. lett.* **81**, 1167 (1998); Albright C and Barr S, *Phys. Rev.* **D 58**, 013002 (1998); Babu K S, Pati J C and Wilczek F, *Preprint hep-ph/9812538*; Raby S, *Phys. Rev.* D **65**, 115004 (2002); Dermisek R and Raby S, *Phys. Lett.* B **622**, 327 (2005); Ji X, Li Y and Mohapatra R N, *Phys. Lett.* **B 633**, 755 (2006); Dermisek R, Harada M and Raby S, *Phys. Rev.* D **74**, 035011 (2006);Morisi S, Picariello M and Torrente-Lujan E, *Phys. Rev.* D **75**, 075015 (2007)

[3] Hinchliffe I and Kaeding T, *Phys. Rev.* **D47** (1993), 279; Dreiner H K, Luhn C, and Thormeier M,*Phys. Rev.* **D73** (2006), 075007, hep-ph/0512163; L. E. Ibáñez and Ross G G, *Phys. Lett.* **B260** (1991), 291; Banks T and Dine M, *Phys. Rev.* **D45** (1992), 1424; Babu K S, Gogoladze I and Wang K, *Phys. Lett.* **B570** (2003), 32; Kurosawa K, Maru N and Yanagida T, *Phys. Lett.* **B512** (2001), 203; Kajiyama Y, Itou E and Kubo J, *Nucl. Phys.* **B743** (2006), 74.

[4] For other approaches to suppressing proton decay, see Nath P and Sayed R, *Phys. Rev.* **D 77** (2008) 015015; Dreiner H and Thormeir M, *Phys. Rev.* **D 69** (2004) 053002; Sayre J, Wiesenfeldt S and Willenbrock S, *Phys. Rev.* **D 75** (2007) 037702; for a review of proton decay, see Nath P and Fileviez Perez P, *Phys.Rept.* **441** (2007)191-317.

[5] Mohapatra R N and Ratz M, *Phys. Rev.* D **76**, 095003 (2007).

[6] Azatov A and Mohapatra R N, *Phys. Rev.* D **78**, 015002 (2008)

[7] Babu K S and Barr S M, *Phys. Rev.* D **48**, 5354 (1993); Chacko Z and Mohapatra R N, *Phys. Rev.* D **59**, 011702 (1999); Dermisek R, Mafi A and Raby S, *Phys.Rev.* D **63**, 035001 (2001).

[8] Fukuyama T, Ilakovac A, Kikuchi T, Meljanac S and Okada N, *J. Math. Phys.* **46**, 033505 (2005)

[9] Das C R and Parida M K, *Eur.Phys. J. C* **20**, 121 (2001)

[10] Maltoni M, Schwetz T, Tortola M A and Valle J W F, *New J. Phys.* **6**, 122 (2004)

[11] Hisano J, Moroi T, Tobe K, Yamaguchi M and Yanagida T, *Phys. Lett.* B **357**, 579 (1995) Hisano J, Moroi T, Tobe K and Yamaguchi M, *Phys. Rev.* D **53**, 2442 (1996)

[12] Masiero A, Vempati S K and Vives O, *arXiv:0711.2903 [hep-ph]*; Ciuchini M, Masiero A, Paradisi P, Silvestrini L, Vempati S K and Vives O, *Nucl. Phys.* B **783**, 112 (2007)

[13] Bemporado et al. MEG Collaboration; PSI-R-99-05.

Symposium on Prospects in the Physics of Discrete Symmetries IOP Publishing
Journal of Physics: Conference Series **335** (2011) 012025 doi:10.1088/1742-6596/335/1/012025

Aspects of symmetry breaking in $SO(10)$ GUTs

Luca Di Luzio[1]

SISSA and INFN, Sezione di Trieste, Via Bonomea 265, 34136 Trieste, Italy

E-mail: diluzio@sissa.it

Abstract. I review some recent results on the Higgs sector of minimal $SO(10)$ grand unified theories both with and without supersymmetry. It is shown that nonsupersymmetric $SO(10)$ with just one adjoint triggering the first stage of the symmetry breaking does provide a successful gauge unification when radiative corrections are taken into account in the scalar potential, while in the supersymmetric case it is argued that the troubles in achieving a phenomenologically viable breaking with representations up to the adjoint are overcome by considering the flipped $SO(10)$ embedding of the hypercharge.

1. Introduction

While the standard model (SM) matter quantum numbers nicely fit in a few lowest-dimensional representations of the unified groups such as $SU(5)$ or $SO(10)$, this synthetic process has no counterpart in the Higgs sector where a larger set of higher-dimensional representations is usually needed in order to spontaneously break the enhanced gauge symmetry down to the SM gauge group. In this respect, establishing the minimal Higgs content needed for the grand unified theory (GUT) breaking is a basic question which has been already addressed in the early days of the GUT program. Let us remark that the quest for the simplest Higgs sector is driven not only by aesthetic criteria but it is also a phenomenologically relevant issue related to tractability and predictivity of the models. Indeed, the details of the symmetry breaking pattern, sometimes overlooked in the phenomenological analysis, give further constraints on the low-energy observables such as the proton decay and the effective SM flavor structure.

Here we will focus mainly on $SO(10)$ GUTs, both with and without supersymmetry (SUSY). Let us recall that before considering any symmetry breaking dynamics the following representations are required by the group theory in order to achieve a full breaking of $SO(10)$ to the SM:

(i) 16_H or 126_H for rank reduction. Their SM-singlet vacuum expectation value (VEV) communicates the $B-L$ breaking to neutrino masses, but preserves an $SU(5)$ little group.

(ii) 45_H or 54_H or 210_H for the further $SU(5)$ breaking. They admit for little groups different from $SU(5) \otimes U(1)$, yielding the SM when intersected with the $SU(5)$ remnant of (i).

It should be also mentioned that a one-step $SO(10) \to$ SM breaking can be achieved via only one 144_H irreducible Higgs representation [1]. However, such a setting requires an extended matter sector, including 45_F and 120_F multiplets, in order to accommodate realistic fermion masses [2].

[1] In collaboration with Stefano Bertolini (INFN and SISSA, Trieste) and Michal Malinský (IFIC, Valencia).

While the choice between 16_H or 126_H is a model dependent issue related to the details of the Yukawa sector, the simplest option in the list (ii) is certainly given by the adjoint 45_H. However, since the early 1980's, it has been observed that the vacuum dynamics aligns the adjoint along an $SU(5) \otimes U(1)$ direction, making the choice of the 45_H alone not phenomenologically viable. In the nonsupersymmetric case the alignment is only approximate [3, 4, 5], but it is such to clash with unification constraints which do not allow for any $SU(5)$-like intermediate stage, while in the supersymmetric limit the alignment is exact due to F-flatness [6, 7, 8], thus never landing to a supersymmetric SM vacuum. In the next sections we will review all these issues in more detail and provide a way out to the aforementioned problem of the vacuum alignment.

2. Adjoint breaking in nonsupersymmetric $SO(10)$

Let us consider the most general renormalizable tree level scalar potential which can be constructed out of 45_H and 16_H in nonsupersymmetric $SO(10)$

$$V_0 = V_{45_H} + V_{16_H} + V_{45_H 16_H}\,, \tag{1}$$

where, according to the notation in Ref. [9]

$$V_{45_H} = -\frac{\mu^2}{2}\,\mathrm{Tr}\,45_H^2 + \frac{a_1}{4}\,(\mathrm{Tr}\,45_H^2)^2 + \frac{a_2}{4}\,\mathrm{Tr}\,45_H^4\,, \tag{2}$$

$$V_{16_H} = -\frac{\nu^2}{2}\,16_H^\dagger 16_H + \frac{\lambda_1}{4}\,(16_H^\dagger 16_H)^2 + \frac{\lambda_2}{4}\,(16_H^T\,\Gamma\,16_H)(16_H^\dagger\,\Gamma\,16_H^*)\,, \tag{3}$$

$$V_{45_H 16_H} = \alpha\,(16_H^\dagger 16_H)\mathrm{Tr}\,45_H^2 + \beta\,16_H^\dagger 45_H^2 16_H + \tau\,16_H^\dagger 45_H 16_H\,. \tag{4}$$

There are in general three SM-singlets in the $45_H \oplus 16_H$ reducible representation of $SO(10)$. Their VEVs are defined in the following way

$$\omega_{B-L} \subset \langle(15,1,1)\rangle \subset \langle 45_H\rangle\,, \quad \omega_R \subset \langle(1,1,3)\rangle \subset \langle 45_H\rangle\,, \quad \chi_R \subset \langle(\overline{4},1,2)\rangle \subset \langle 16_H\rangle\,, \tag{5}$$

where the three submultiplets above are labeled according to the $SO(10)$ subalgebra $4_C\,2_L\,2_R$ (shorthand notation for $SU(4)_C \otimes SU(2)_L \otimes SU(2)_R$). Different VEV configurations trigger the spontaneous breakdown of the $SO(10)$ symmetry into the following subgroups. Using a self-explanatory notation, for $\chi_R = 0$ one finds

$$\begin{aligned}
\omega_R = 0,\ \omega_{B-L} \neq 0 &: & 3_c\,2_L\,2_R\,1_{B-L}\,,\\
\omega_R \neq 0,\ \omega_{B-L} = 0 &: & 4_C 2_L 1_R\,,\\
\omega_R \neq 0,\ \omega_{B-L} \neq 0 &: & 3_c\,2_L\,1_R\,1_{B-L}\,,\\
\omega_R = -\omega_{B-L} \neq 0 &: & \text{flipped}\,SU(5) \otimes U(1)_Z\,,\\
\omega_R = \omega_{B-L} \neq 0 &: & \text{standard}\,SU(5) \otimes U(1)_Z\,.
\end{aligned} \tag{6}$$

When $\chi_R \neq 0$ all intermediate symmetries are spontaneously broken down to the $3_c\,2_L\,1_Y$ of the SM with the exception of the last case which leaves the standard $SU(5)$ unbroken. In this language, the potentially viable breaking chains fulfilling the basic gauge unification constraints (labeled as VIII and XII in Ref. [10]) correspond to the settings with:

$$\omega_{B-L} \gg \omega_R > \chi_R \;:\; SO(10) \to 3_c 2_L 2_R 1_{B-L} \to 3_c 2_L 1_R 1_{B-L} \to 3_c 2_L 1_Y \qquad \text{Chain VIII}\,,$$
$$\omega_R \gg \omega_{B-L} > \chi_R \;:\; SO(10) \to 4_C 2_L 1_R \to 3_c 2_L 1_R 1_{B-L} \to 3_c 2_L 1_Y \qquad \text{Chain XII}\,.$$

The parameters (couplings and VEVs) of the scalar potential are constrained by the requirements of boundedness and the absence of tachyonic states, ensuring that the vacuum is stable and the stationary points correspond to physical minima. In particular, from the shape of the tree level masses of the $(8,1,0)$ and $(1,3,0)$ SM sub-multiplets of 45_H:

$$M^2(1,3,0) = 2a_2(\omega_{B-L}-\omega_R)(\omega_{B-L}+2\omega_R)\,, \tag{7}$$
$$M^2(8,1,0) = 2a_2(\omega_R-\omega_{B-L})(\omega_R+2\omega_{B-L})\,, \tag{8}$$

which can not be simultaneously positive unless

$$a_2<0\,, \qquad -2<\omega_{B-L}/\omega_R<-\tfrac{1}{2}\,, \tag{9}$$

one concludes that the only vacuum configurations allowed are those in the close vicinity of the flipped $SU(5)\otimes U(1)_Z$ setting. Hence, the large hierarchy (of about four orders of magnitude) between ω_R and ω_{B-L}, required by gauge coupling unification (cf. chains VIII and XII in Ref. [10]) cannot be achieved. This is the key point of the classical argument that the nonsupersymmetric $SO(10)$ GUTs with only one adjoint responsible for the first stage of the $SO(10)$ breakdown can not support the phenomenologically favoured symmetry breaking chains.

2.1. *A tree level accident*

The rationale for understanding the strong correlation among the masses of the states $(1,3,0)$ and $(8,1,0)$ can be obtained by looking at the enhancement of the global symmetry in a trivial limit of the scalar potential. When only trivial invariants of both 45_H and 16_H are considered ($a_2=\lambda_2=\beta=\tau=0$) the global symmetry of V_0 is $O(45)\otimes O(32)$. This is then broken spontaneously into $O(44)\otimes O(31)$ by the 45_H and 16_H VEVs yielding 44+31=75 Goldstone bosons (GB) in the scalar spectrum. The gauge $SO(10)$ symmetry is at the same time broken down to the SM gauge group. Therefore 75-33=42 pseudo-Goldstone bosons (PGB) are left in the spectrum and their masses should generally receive contributions from all the explicitly breaking terms a_2, λ_2, β and τ. Since the states $(1,3,0)$ and $(8,1,0)$ belong to this set of PGB, generally one would expect their mass to depend on all of a_2, λ_2, β, τ parameters. The absence of mass contributions proportional to λ_2, β, τ is just an easily understood accident of the tree level potential [9], but nothing prevents those couplings from contributing to the PGB masses at the quantum level.

2.2. *The quantum vacuum*

The relevant one-loop correction to the $(1,3,0)$ and $(8,1,0)$ PGB masses can be conveniently computed by means of the one-loop effective potential (EP). The one-loop EP can be formally written as

$$V_{\text{eff}} = V_0+\Delta V_s+\Delta V_f+\Delta V_g\,, \tag{10}$$

where $\Delta V_{s,f,g}$ denote the contributions from scalars, fermions and gauge bosons respectively. In dimensional regularization with modified minimal subtraction ($\overline{\text{MS}}$) and in the Landau gauge, they are given by

$$\Delta V_i(45_H,16_H,\mu)=\frac{\alpha_i}{64\pi^2}\text{Tr}\left[\mathcal{M}_i^4(45_H,16_H)\left(\log\frac{\mathcal{M}_i^2(45_H,16_H)}{\mu^2}-\beta_i\right)\right]\,, \tag{11}$$

where $(\alpha_s,\alpha_f,\alpha_g)=(1,-2,3)$, $(\beta_s,\beta_f,\beta_g)=(\frac{3}{2},\frac{3}{2},\frac{5}{6})$ and $\mathcal{M}_s$, $\mathcal{M}_f$ and $\mathcal{M}_g$ are the functional scalar, fermion and gauge boson mass matrices respectively, as obtained from the tree level potential. In the following we will neglect the fermionic component of the EP since there are no fermionic states at the unification scale M_G. The running masses $\overline{m}^2_{ab}$ are defined by

$$\overline{m}^2_{ab}\equiv\frac{\partial^2 V_{\text{eff}}(45_H,16_H,\mu)}{\partial\psi_a\partial\psi_b}\Big|_{\langle\psi\rangle}\,, \tag{12}$$

where ψ_a and ψ_b are generic scalar field components and the VEVs (denoted collectively $\langle\psi\rangle$) obey the one-loop stationary equations. For a given eigenvalue the physical (pole) masses M_a^2 are then obtained by

$$M_a^2 = \overline{m}_a^2 + \Delta\Sigma_a(M_a^2)\,, \tag{13}$$

where $\Delta\Sigma_a(p^2) = \Sigma_a(p^2) - \Sigma_a(0)$ and Σ_a are the $\overline{\rm MS}$ renormalized self-energies. Of particular interest is the case when M_a is substantially smaller than the mass (M_G) of the particles that contribute to Σ_a. At $\mu = M_G$ in the limit $M_a^2 \ll M_G^2$ one has

$$\Delta\Sigma_a(M_a^2) = O(M_a^4/M_G^2)\,. \tag{14}$$

In this case the running mass computed from Eq. (12) contains the leading gauge independent corrections.

The stringent tree level constraint on the ratio ω_{B-L}/ω_R, coming from the positivity of the masses of the states $(1,3,0)$ and $(8,1,0)$, that forbids non-$SU(5)$ vacua, follows from the fact that the masses depend only on the parameter a_2. On the other hand, from the discussion of the would-be global symmetries of the scalar potential we should in general expect their masses to depend on other terms in the scalar potential (in particular β, τ and gauge interactions at the one-loop level). The calculation of the EP running mass from Eq. (12) leads for the states $(1,3,0)$ and $(8,1,0)$ at $\mu = M_G$ to the mass shifts

$$\Delta M^2(1,3,0) = \frac{1}{4\pi^2}\left[\tau^2 + \beta^2(2\omega_R^2 - \omega_R\omega_{B-L} + 2\omega_{B-L}^2) + g^4\left(16\omega_R^2 + \omega_R\omega_{B-L} + 19\omega_{B-L}^2\right)\right] \tag{15}$$

$$\Delta M^2(8,1,0) = \frac{1}{4\pi^2}\left[\tau^2 + \beta^2(\omega_R^2 - \omega_R\omega_{B-L} + 3\omega_{B-L}^2) + g^4\left(13\omega_R^2 + \omega_R\omega_{B-L} + 22\omega_{B-L}^2\right)\right] \tag{16}$$

where the subleading (and gauge dependent) logarithmic terms are neglected and we have taken for simplicity $\chi_R = 0$, given that $\chi_R \ll \omega_{R,B-L}$ by unification constraints. For more details we refer the reader to Refs. [9, 11]. By comparing Eqs. (15)–(16) with Eqs. (7)–(8) it is clear that a consistent scalar mass spectrum can be obtained for the non-$SU(5)$ vacua, at variance with the tree level result. In particular, a hierarchy between ω_{B-L} and ω_R (as required by unification), while keeping the scalar states positive (minimum condition), is now possible just by taking $|a_2| \lesssim 10^{-2}$. This corresponds to keeping the masses of the PGB $(1,3,0)$ and $(8,1,0)$ one order of magnitude below M_G, which makes the EP potential computation self-consistent (since the self-energies in Eq. (13) can be neglected) and has the welcome effect of raising a little M_G. Let us also stress that this result is inherent to all the nonsupersymmetric $SO(10)$ models with one adjoint 45_H triggering the first stage of the GUT breaking: just one additional GUT representation interacting with the adjoint is needed in order to open up the non-$SU(5)$ vacua at the quantum level.

Nevertheless, the simplest scenario featuring the Higgs scalars in $10_H \oplus 16_H \oplus 45_H$ of $SO(10)$ (where the 10_H is needed for the electroweak symmetry breaking) fails when addressing the neutrino spectrum: in nonsupersymmetric models the $B-L$ breaking scale, M_{B-L}, turns out to be generally a few orders of magnitude below M_G [10]. Thus, the scale of the right-handed (RH) neutrino masses $M_R \sim M_{B-L}^2/M_P$ emerging first at the $d=5$ level from an operator of the form $16_F^2(16_H^*)^2/M_P$ (with M_P typically identified with the Planck scale) undershoots by many orders of magnitude the range of about $10^{13\div14}$ GeV naturally suggested by the seesaw mechanism. This issue can be alleviated by considering 126_H in place of 16_H in the Higgs sector, since in such a case the neutrino masses can be generated at the renormalizable level by the term $16_F^2 126_H^*$. This lifts the problematic M_{B-L}/M_P suppression factor inherent to the $d=5$ effective mass and yields $M_R \sim M_{B-L}$, that might be, at least in principle, acceptable [12]. Fermion masses and mixings in nonsupersymmetric $SO(10)$ with $10_H \oplus 126_H$ in the Yukawa sector have been recently reconsidered by the authors of Ref. [13], which show the potential predictivity of such a class of models.

Journal of Physics: Conference Series **335** (2011) 012025 doi:10.1088/1742-6596/335/1/012025

3. Adjoint breaking in supersymmetric $SO(10)$

By invoking TeV-scale SUSY, the qualitative picture changes a lot for neutrinos. Indeed, the gauge running within the minimal supersymmetric SM (MSSM) prefers M_{B-L} in the proximity of M_G and, hence, the Planck-suppressed $d = 5$ RH neutrino mass operator $16_F^2\overline{16}_H^2/M_P$, available whenever $16_H \oplus \overline{16}_H$ is present in the Higgs sector, can naturally reproduce the desired scale for M_R. Let us recall that both 16_H as well as $\overline{16}_H$ are required in order to retain SUSY below the GUT scale. It is therefore very interesting to consider the minimal Higgs setting based on the lowest-dimensional representations (up to the adjoint). On the other hand, it is well known [6, 7, 8] that the relevant superpotential does not support, at the renormalizable level, a supersymmetric breaking of the $SO(10)$ gauge group to the SM. This is due to the constraints on the vacuum manifold imposed by the F- and D-flatness conditions which, apart from linking the magnitudes of the $SU(5)$-singlet 16_H and $\overline{16}_H$ VEVs, make the the adjoint VEV $\langle 45_H \rangle$ aligned to $\langle 16_H\overline{16}_H \rangle$. As a consequence, an $SU(5)$ subgroup of the initial $SO(10)$ gauge symmetry remains unbroken.

3.1. Renormalizable vs nonrenormalizable scenarios

The alignment of the adjoint with the spinors can be broken by giving up renormalizability and allowing for effective M_P-suppressed operators in the superpotential [7]. However, this option may be rather problematic since it introduces a delicate interplay between physics at two different scales, $M_G \ll M_P$, with the consequence of spreading the GUT-scale thresholds over several orders of magnitude below M_G. In turn this may affect $d = 5$ proton decay as well as the MSSM gauge unification and it may also jeopardize the neutrino mass generation in the seesaw scheme (cf. Ref. [14] for a more detailed account of these effects). Thus, although the Planck-induced operators can provide a key to overcome the $SU(5)$ lock of the minimal SUSY $SO(10)$ Higgs model with $16_H \oplus \overline{16}_H \oplus 45_H$, such an effective scenario is prone to failure when addressing the measured proton stability and light neutrino phenomenology.

On the other hand, in the standard $SO(10)$ framework with a Higgs sector built off the lowest-dimensional representations (up to the adjoint), it is not possible to achieve a renormalizable breaking even admitting multiple copies of each type of multiplet. Firstly, with a single 45_H at play, the little group of the adjoint is $SU(5) \otimes U(1)$ regardless of the number of $16_H \oplus \overline{16}_H$ pairs. The same is true with a second 45_H added into the Higgs sector because there is no renormalizable mixing among the two 45_H's apart from the mass term that, without loss of generality, can be taken diagonal. With a third adjoint Higgs representation at play a cubic $45_1 45_2 45_3$ interaction is allowed. However, due to the total antisymmetry of the invariant and the fact that the adjoints commute on the SM vacuum, the cubic term does not contribute to the F-term equations. From this brief excursus one concludes that the $SU(5)$ lock cannot be broken at the renormalizable level by means of representations up to the adjoint.

Remarkably, all these issues are alleviated if one considers a flipped variant of the supersymmetric $SO(10)$ unification. What we have shown in Ref. [14] is that the flipped $SO(10) \otimes U(1)_X$ scenario [15, 16, 17, 18] offers an attractive option to break the gauge symmetry down to the SM (and further to $SU(3)_c \otimes U(1)_Q$) at the renormalizable level and by means of a quite simple Higgs sector, namely a couple of $SO(10)$ spinors $16_{H_{1,2}} \oplus \overline{16}_{H_{1,2}}$ and one adjoint 45_H.

3.2. Hypercharge embeddings in $SO(10) \otimes U(1)_X$

The so called flipped realization of the $SO(10)$ gauge symmetry requires an additional $U(1)_X$ gauge factor in order to provide an extra degree of freedom for the SM hypercharge identification. For a fixed embedding of the $SU(3)_c \otimes SU(2)_L$ subgroup within $SO(10)$, the SM hypercharge can be generally spanned over the three remaining Cartans generating the abelian $U(1)^3$ subgroup of the $SO(10) \otimes U(1)_X/(SU(3)_c \otimes SU(2)_L)$ coset. There are two consistent implementations of the SM hypercharge within the $SO(10)$ algebra (commonly denoted by standard and flipped $SU(5)$), while a third one becomes available due to the presence of $U(1)_X$.

In order to discuss the different embeddings we adopt the traditional left-right (LR) basis corresponding to the $SU(3)_c \otimes SU(2)_L \otimes SU(2)_R \otimes U(1)_{B-L}$ subalgebra of $SO(10)$. In full generality one can span the SM hypercharge on the generators of $U(1)_R \otimes U(1)_{B-L} \otimes U(1)_X$:

$$Y = \alpha T_R^{(3)} + \beta(B-L) + \gamma X. \tag{17}$$

The $U(1)_X$ charge is been conveniently fixed to $X_{16} = +1$ for the spinorial representation and thus $X_{10} = -2$ and also $X_1 = +4$ for the $SO(10)$ vector and singlet, respectively; this is also the minimal way to obtain an anomaly-free $U(1)_X$, that allows $SO(10) \otimes U(1)_X$ to be naturally embedded into E_6. It is a straightforward exercise to show that there are only three solutions which accommodate the SM matter quantum numbers over a reducible $16 \oplus 10 \oplus 1$ representation. On the $U(1)_R \otimes U(1)_{B-L} \otimes U(1)_X$ bases of Eq. (17) one obtains,

$$\alpha = 1, \ \beta = \tfrac{1}{2}, \ \gamma = 0\,, \tag{18}$$

which is nothing but the "standard" embedding of the SM matter into $SO(10)$. The second option is characterized by

$$\alpha = -1, \ \beta = \tfrac{1}{2}, \ \gamma = 0\,, \tag{19}$$

which is usually denoted "flipped $SU(5)$" [19, 20] embedding and corresponds to a sign flip of the $SU(2)_R$ Cartan operator $T_R^{(3)}$. A third solution corresponds to

$$\alpha = 0, \ \beta = -\tfrac{1}{4}, \ \gamma = \tfrac{1}{4}\,, \tag{20}$$

denoted as "flipped $SO(10)$" [15, 16, 17, 18] embedding of the SM hypercharge. Notice, in particular, the fundamental difference between the setting (20) with $\gamma = \frac{1}{4}$ and the two previous cases (18) and (19) where $U(1)_X$ does not play any role.

3.3. The supersymmetric flipped $SO(10) \otimes U(1)_X$ model

The active role of the $U(1)_X$ generator in the SM hypercharge identification within the flipped $SO(10)$ scenario has relevant consequences for model building. In particular, the SM decomposition of the $SO(10)$ representations changes so that there are additional SM-singlets both in $16_H \oplus \overline{16}_H$ as well as in 45_H. The presence of these additional SM-singlets provides the ground for obtaining a viable symmetry breaking with a significantly simplified renormalizable Higgs sector.

Naively, one may guess that the pair of VEVs in 16_H (plus another conjugated pair in $\overline{16}_H$ to maintain the required D-flatness) might be enough to break the GUT symmetry entirely, since one component transforms as a 10 of $SU(5) \subset SO(10)$ (cf. s_H in Table 1), while the other one (cf. n_H in Table 1) is identified with an $SU(5)$ singlet. Nevertheless, flipping is not per-se sufficient since the adjoint does not reduce the rank and the bi-spinor, in spite of the two qualitatively different SM-singlets involved, can lower it only by a single unit, leaving a residual $SU(5) \otimes U(1)$ symmetry. Only when two pairs of $16_H \oplus \overline{16}_H$ (interacting via 45_H) are introduced the two pairs of SM-singlet VEVs in the spinor multiplets may not generally be aligned and the little group is reduced to the SM [14]. Given the most general renormalizable Higgs superpotential, made of the representations $45_H \oplus 16_{H_1} \oplus \overline{16}_{H_1} \oplus 16_{H_2} \oplus \overline{16}_{H_2}$

$$W_H = \frac{\mu}{2}\,\mathrm{Tr}\,45_H^2 + \rho_{ij} 16_{H_i} \overline{16}_{H_j} + \tau_{ij} 16_{H_i} 45_H \overline{16}_{H_j}\,, \tag{21}$$

where $i, j = 1, 2$, the study of the SUSY vacuum in Ref. [14] shows that the little group is the SM for large portions of the parameter space in which the VEVs of the $16_H \oplus \overline{16}_H$ pairs are not aligned.

	$SO(10)$	$SO(10)_f$
16_F	$(D^c \oplus L)_{\overline{5}} \oplus (U^c \oplus Q \oplus E^c)_{10} \oplus (N^c)_1$	$(D^c \oplus \Lambda^c)_{\overline{5}} \oplus (\Delta^c \oplus Q \oplus S)_{10} \oplus (N^c)_1$
10_F	$(\Delta \oplus \Lambda^c)_5 \oplus (\Delta^c \oplus \Lambda)_{\overline{5}}$	$(\Delta \oplus L)_5 \oplus (U^c \oplus \Lambda)_{\overline{5}}$
1_F	$(S)_1$	$(E^c)_1$
$\langle 16_H \rangle$	$(0 \oplus \langle H_d \rangle)_{\overline{5}} \oplus (0 \oplus 0 \oplus 0)_{10} \oplus (n_H)_1$	$(0 \oplus \langle H_u \rangle)_{\overline{5}} \oplus (0 \oplus 0 \oplus s_H)_{10} \oplus (n_H)_1$
$\langle \overline{16}_H \rangle$	$(0 \oplus \langle H_u \rangle)_5 \oplus (0 \oplus 0 \oplus 0)_{\overline{10}} \oplus (n_H)_1$	$(0 \oplus \langle H_d \rangle)_5 \oplus (0 \oplus 0 \oplus s_H)_{\overline{10}} \oplus (n_H)_1$

Table 1. SM decomposition of $SO(10)$ representations in the standard (left) and flipped (right) hypercharge embedding. A self-explanatory SM notation is used, with the outer subscripts labeling the $SU(5)$ origin.

Let us stress that in the flipped embedding the spinor representations include also weak doublets H_u and H_d that may trigger the electroweak symmetry breaking and allow for renormalizable Yukawa interactions with the chiral matter fields distributed in the flipped embedding over a reducible $16_F \oplus 10_F \oplus 1_F$ representation. Notice that this matter content is needed in order to cancel the gauge anomalies of the $U(1)_X$ factor and to correctly reproduce the SM matter quantum numbers (cf. Table 1).

Considering for simplicity just one pair of spinor Higgs multiplets and imposing a Z_2 matter-parity (negative for matter and positive for Higgs superfields) the most general Yukawa superpotential (up to $d = 5$ operators) reads

$$W_Y = Y_U 16_F 10_F 16_H + \frac{1}{M_P} \left[Y_E 10_F 1_F \overline{16}_H \overline{16}_H + Y_D 16_F 16_F \overline{16}_H \overline{16}_H \right] , \qquad (22)$$

where family indexes are understood. Notice (cf. Table 1) that due to the flipped embedding the up-quarks receive mass at the renormalizable level, while all the other fermion masses need Planck-suppressed effective contributions in order to achieve a realistic texture. Thus the top/bottom hierarchy is given by an $M_G/M_P \sim 10^{-2}$ factor, which selects naturally $\mathcal{O}(1)$ values for $\tan\beta \equiv v_u/v_d$. At the end, it can be shown [14] that the Yukawa superpotential in Eq. (22) can reproduce realistic textures for the SM fermions (including neutrinos), while the exotic states are automatically kept heavy by the symmetry breaking pattern.

3.4. *Minimal E_6 embedding*

The mechanism we advocate can be embedded in an underlying nonrenormalizable E_6 Higgs model featuring a pair of $27_H \oplus \overline{27}_H$ and the adjoint 78_H. Technical similarities apart, there is, however, a crucial difference between the $SO(10) \otimes U(1)_X$ and E_6 scenarios, that is related to the fact that the Lie-algebra of E_6 is larger than that of $SO(10) \otimes U(1)_X$. It has been shown long ago [21] that the renormalizable SUSY E_6 Higgs model spanned over $27_H \oplus \overline{27}_H \oplus 78_H$ leaves an $SO(10)$ symmetry unbroken. Two pairs of $27_H \oplus \overline{27}_H$ are needed to reduce the rank by two units. In spite of the fact that the two SM-singlet directions in the 27_H are exactly those of the "flipped" 16_H, the little group of the $2 \times (27_H \oplus \overline{27}_H) \oplus 78_H$ Higgs sector remains at the renormalizable level $SU(5)$, as we explicitly show in Ref. [14]. Adding nonrenormalizable adjoint interactions allows for a misalignment of the $\langle 78_H \rangle$ from the $SU(5) \otimes U(1) \otimes U(1)$ direction, such that the little group is reduced to the SM. Since a one-step E_6 breaking with nonrenormalizable operators is phenomenologically problematic as mentioned earlier, we argue for a two-step breaking, via flipped $SO(10) \otimes U(1)_X$, with the E_6 scale near the Planck scale. Barring detailed threshold effects, it is interesting to see from Fig. 1 that the few percent mismatch observed within the two-loop MSSM gauge coupling evolution at the scale of the "one-step" grand unification is naturally accommodated in this scheme, and it is understood as an artefact of a "delayed" E_6 unification superseding the flipped $SO(10) \otimes U(1)_X$ partial unification.

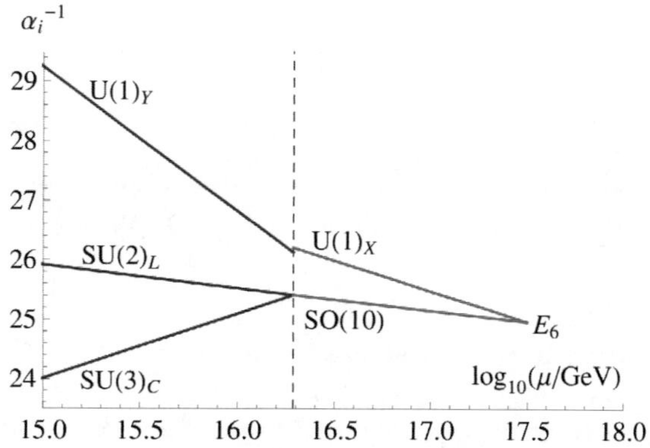

Figure 1. Sample picture of gauge unification in the E_6-embedded $SO(10) \otimes U(1)_X$ model.

4. Conclusions

Longstanding results claimed that nonsupersymmetric $SO(10)$ GUTs with just the adjoint triggering the first stage of the GUT breaking cannot provide a successful gauge unification. We argued that this conclusion is an artefact of the tree level potential and showed that quantum corrections have a dramatic impact. In particular, a model featuring $10_H \oplus 126_H \oplus 45_H$ in the Higgs sector has all the ingredients to be a viable minimal nonsupersymmetric $SO(10)$ GUT candidate [12]. Analogously, supersymmetric $SO(10)$ GUTs with representations up to the adjoint do not provide a phenomenologically viable breaking to the SM. We pointed out that the flipped $SO(10)$ embedding offers an attractive setting for breaking the gauge symmetry directly to $SU(3)_c \otimes U(1)_Q$ at the renormalizable level, by means of a quite simple Higgs sector: $2 \times (16_H \oplus \overline{16}_H) \oplus 45_H$. The case is made for a two-step breaking of a supersymmetric E_6 GUT realised in the vicinity of the Planck scale via an intermediate flipped $SO(10) \otimes U(1)_X$ stage.

References

[1] K. S. Babu, I. Gogoladze, P. Nath, R. M. Syed, Phys. Rev. **D72**, 095011 (2005). [hep-ph/0506312].
[2] P. Nath, R. M. Syed, Phys. Rev. **D81**, 037701 (2010). [arXiv:0909.2380 [hep-ph]].
[3] M. Yasue, Phys. Rev. D **24**, 1005 (1981); Phys. Lett. B **103**, 33 (1981).
[4] G. Anastaze, J. P. Derendinger and F. Buccella, Z. Phys. C **20**, 269 (1983).
[5] K. S. Babu and E. Ma, Phys. Rev. D **31**, 2316 (1985).
[6] F. Buccella, J. P. Derendinger, C. A. Savoy and S. Ferrara, Second Europhysics Study Conference on the Unication of the Fundamental Interactions, Erice, Italy, 1981, CERN Reports No. CERN-TH-3212.
[7] K. S. Babu, S. M. Barr, Phys. Rev. **D51**, 2463-2470 (1995). [hep-ph/9409285].
[8] C. S. Aulakh, B. Bajc, A. Melfo, A. Rasin and G. Senjanovic, Nucl. Phys. B **597**, 89 (2001). [hep-ph/0004031].
[9] S. Bertolini, L. Di Luzio, M. Malinsky, Phys. Rev. **D81**, 035015 (2010). [arXiv:0912.1796 [hep-ph]].
[10] S. Bertolini, L. Di Luzio, M. Malinsky, Phys. Rev. **D80**, 015013 (2009). [arXiv:0903.4049 [hep-ph]].
[11] S. Bertolini, L. Di Luzio, M. Malinsky, J. Phys. Conf. Ser. **259**, 012098 (2010). [arXiv:1010.0338 [hep-ph]].
[12] S. Bertolini, L. Di Luzio, M. Malinsky, Work in progress.
[13] B. Bajc, A. Melfo, G. Senjanovic, F. Vissani, Phys. Rev. **D73**, 055001 (2006). [hep-ph/0510139].
[14] S. Bertolini, L. Di Luzio, M. Malinsky, Phys. Rev. **D83**, 035002 (2011). [arXiv:1011.1821 [hep-ph]].
[15] T. W. Kephart, N. Nakagawa, Phys. Rev. **D30**, 1978 (1984).
[16] J. Rizos, K. Tamvakis, Phys. Lett. **B212**, 176 (1988).
[17] S. M. Barr, Phys. Rev. **D40**, 2457 (1989).
[18] N. Maekawa, T. Yamashita, Phys. Lett. **B567**, 330-338 (2003). [hep-ph/0304293].
[19] A. De Rujula, H. Georgi, S. L. Glashow, Phys. Rev. Lett. **45**, 413 (1980).
[20] S. M. Barr, Phys. Lett. **B112**, 219 (1982).
[21] F. Buccella, G. Miele, Phys. Lett. **B189**, 115-117 (1987).

Hybrid anomaly-gravity mediation with flavor

Christian Gross
Institut für Physik, Technische Universität Dortmund, 44221 Dortmund, Germany
E-mail: c.gross@physik.tu-dortmund.de

Abstract. We study models with contributions of similar size to the soft masses from anomaly- and gravity mediation, thereby curing the tachyonic slepton problem of anomaly mediation. A possible origin of this hybrid setup in a 5-dimensional brane world is briefly discussed. The absence of excessive flavor violation is explained by alignment. The gravitino can be heavy enough so that the gravitino problem of supersymmetric theories with leptogenesis is avoided. The model has a characteristic signature: It predicts the distinctive gaugino mass pattern of anomaly mediation and, at the same time, $\mathcal{O}(1)$ slepton mass splittings.

1. Introduction

While weak scale supersymmetry (SUSY) can offer solutions to some of the puzzles of the Standard Model (SM) it also introduces new challenges. For instance, sparticle-loops can induce excessive flavor- and CP violation. Apart from addressing this issue, a realistic supersymmetric model should also include a mechanism to generate tiny neutrino masses and the baryon asymmetry of the universe (BAU). An elegant way to explain neutrino masses is the seesaw mechanism. In the presence of the latter, an attractive method to explain the BAU is leptogenesis [1]. Since leptogenesis generically requires a reheating temperature of at least $\sim 10^9$ GeV (see e.g. [2]) the produced gravitini could destroy the successful predictions of Big Bang Nucleosynthesis (BBN) or overclose the universe [3]. A straightforward way to avoid this cosmological gravitino problem is to have a gravitino mass which is large enough (i.e. $m_{3/2} \gtrsim 60$ TeV, cf. e.g. [4]) that the gravitino decays before BBN.

This can be realized in models with anomaly mediated SUSY breaking (AMSB) [5, 6], where the soft terms are loop-suppressed with respect to the gravitino mass – in contrast to models with Planck-scale-mediated (a.k.a. gravity-mediated) SUSY breaking (PMSB), where $m_{3/2}$ generically is of the order of the soft terms. Apart from allowing for a heavy gravitino, AMSB has the desirable feature of avoiding the SUSY flavor problem since the soft terms are not directly sensitive to UV physics. On the other hand, AMSB has problematic aspects regarding (i) its phenomenology and (ii) its theoretical motivation: (i) AMSB leads to tachyonic sleptons. There are many proposals to cure this problem (see e.g. [5, 7]), none is agreed upon to be so compelling that this issue is considered settled, however. AMSB also suffers from the μ-problem. (ii) Being a quantum effect, AMSB generically is negligible compared to PMSB. One thus needs to justify why PMSB is suppressed.

Here, (see [8] for more details and references) we consider the possibility that the PMSB-induced soft masses are suppressed – in order for the gravitino to be much heavier than the TeV-scale –, but only so much that they still are of the same size as the AMSB contribution – so that the tachyonic slepton problem can be avoided. Of course, the flavor-blindness of AMSB is spoiled in this case since the PMSB contributions to the soft masses pick up flavor breaking from physics close to the Planck-scale. In particular, the mass splittings between sleptons of different flavor and between doublet and singlet sleptons are expected to be $\mathcal{O}(1)$. This is not necessarily a drawback, however: It is conceivable that the same flavor model which is needed in any case to explain the observed pattern of fermion masses and mixings also solves the SUSY flavor problem via alignment [9]. This case where physics beyond the SM is non-minimally flavor violating could actually open up possibilities to learn about the origin of flavor (see e.g. [10]).

Our model has an interesting phenomenology: The specific gaugino mass pattern of AMSB is preserved (as argued below) while we predict $\mathcal{O}(1)$ mass splittings for the sleptons. Both of this can lead to characteristic signals at the LHC, providing opportunities to test the model.

2. The model

The gaugino masses, A-terms and soft masses all receive contributions from AMSB. For the soft masses these read

$$m^2|_{\mathrm{AMSB}} = \frac{1}{2}|F_\Phi|^2 \mu \frac{d}{d\mu}\gamma \sim |M_\Phi|^2(\mp g^4 - g^2 Y^2 + Y^4)\,, \tag{1}$$

where γ is the chiral superfield anomalous dimension, F_Φ is the VEV of the conformal compensator F-term and $M_\Phi \equiv F_\Phi/16\pi^2$. Note that $F_\Phi \simeq m_{3/2}$. The sign in (1) depends on the sign of the beta-function of the corresponding gauge group and we omitted coefficients after the $\sim$ symbol.

By contrast, the PMSB-induced contribution to the soft masses, $m^2|_{\mathrm{PMSB}}$, is of the order F_S^2/M_*^2 in generic models. Here, F_S is the hidden sector superfield with the highest F-term VEV, F_S, and M_* is a high mass scale such as the Planck- or string scale. Since the cancellation of the vacuum energy typically requires $F_\Phi \sim F_S/M_P$, the AMSB contribution is suppressed by at least a loop factor. The desired suppression of PMSB with respect to its natural scale can be justified in 5d brane models [5], see Fig. 1. The 5d supergravity modes do not generate soft terms at tree level. The effective visible-hidden sector couplings which are induced by the exchange of heavy bulk fields are exponentially suppressed and induce an operator

$$\mathcal{L}_4 \supset \frac{e^{-M_* L}}{M_*^2}\, X_{ij}\, S\bar{S}Q_i\bar{Q}_j|_{\theta^4}\,, \tag{2}$$

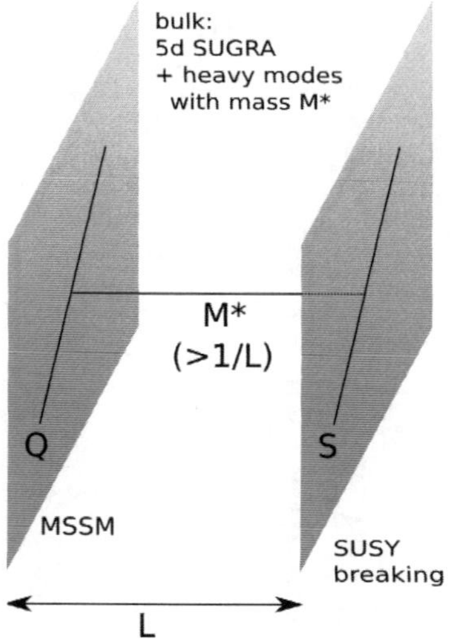

Figure 1. The matter superfields, collectively denoted by Q, as well as the gauge- and Higgs superfields are located on the MSSM brane, which is separated from the SUSY breaking brane by a distance L. S is the hidden sector superfield with the highest F-term VEV, F_S. The only light fields in the bulk come from minimal 5d supergravity. In addition to 5d supergravity, there may be other bulk modes, with a mass $M_* \gtrsim 1/L$. These induce, by the exchange of a single propagator, effective couplings of Q and S which are suppressed by $e^{-M_* L}$.

where X is a matrix in flavor-space whose diagonal entries naturally are $\mathcal{O}(1)$. The term (2) leads to

$$m^2|_{\rm PMSB} \sim \frac{e^{-M_* L}}{M_*^2}\,|F_S|^2\, X \sim |M_\Phi|^2\, rX\,, \quad \text{where} \quad r \equiv (16\pi^2 M_P/M_*)^2 e^{-M_* L}\,. \tag{3}$$

We used $F_S/M_P \sim F_\Phi$ in the last relation. Usually it is assumed that M_*L is large enough so that r is tiny and $m^2|_{\rm PMSB} \ll m^2|_{\rm AMSB}$. Here, by contrast, we assume that r is of order unity, which is the case for $M_*L \simeq 10 + \ln(M_P^2/M_*^2)$. One may view this as follows: The hierarchy between the mass scale of the sparticles and an $\mathcal{O}(100\text{ TeV})$ gravitino is realized by an exponential function with a suitable exponent.

Note that the PMSB contribution to the μ-term, gaugino masses and A-terms is negligible. This is simply because the same exponential factor suppresses a mass in these cases – in contrast to a mass squared in Eq. (3). Therefore, in hybrid anomaly-gravity mediation (i) the distinctive pattern of gaugino masses of pure AMSB is maintained and, (ii), the μ-problem of AMSB persists – we assume a viable μ-term to be induced by an unspecified mechanism.

Adding the contributions from Eqs. (1) and (3), one obtains the chirality-preserving blocks $\mathcal{M}_M^2$ (where $M = L/R$ refers to SU(2)-doublet/singlet sleptons) of the mass matrix for the charged sleptons:

$$\mathcal{M}_M^2 \simeq |M_\Phi|^2(-g_M\mathbf{1} + rX^M)\,, \tag{4}$$

where $g_L \equiv (99/50)g_1^4 + (3/2)g_2^4$ and $g_R \equiv (198/25)g_1^4$. Within our accuracy where the elements of X^M are specified up to $\mathcal{O}(1)$ coefficients only, the chirality-mixing blocks of the mass matrix as well as the flavor non-universal AMSB part and the F- and D-term contributions to $\mathcal{M}_M^2$ can all be neglected. Likewise, the effects from renormalization group (RG) running between the scale M_*, where Eq. (4) is defined, and the weak scale are not relevant. (The basic reason is that sleptons do not couple to gluinos.) We can thus employ Eq. (4) with g_M taken at the weak scale (numerically $g_{L,R} \simeq 0.3$) to deduce phenomenological consequences of our model.

3. Aligning sleptons and explaining lepton masses/mixings

The relative mass splitting between sleptons of different flavor and chirality is expected to be of order unity because we assume the diagonal elements of X^M to have arbitrary $\mathcal{O}(1)$ entries. This can lead to interesting signals for slepton spectroscopy at colliders, see Sect. 4. It also implies that FCNCs must be suppressed by alignment only, i.e. $\delta_{ij}^M \sim K_{ij}^M$, where δ_{ij}^M are the mass insertion parameters and K_{ij}^M are the couplings of the bino and neutral wino to M-chiral leptons l_i and sleptons $\tilde{l}_j$. Since both the g_M and r are of order one, we have (cf. [10])

$$\delta_{ij}^M \sim \max\left\{|X_{ij}^M|, |V_{ij}^M|, |V_{ji}^M|\right\}\,, \tag{5}$$

where V^R, V^L bring the lepton Yukawa matrix Y_E to diagonal form as $V^{R\dagger}Y_E^T V^L$.

One way to realize sufficient alignment and explain the observed fermion masses and mixings is the Froggatt-Nielsen (FN) mechanism. Consider as an illustration a $U(1)_p \times U(1)_q$ symmetry under which the Higgs superfields are neutral and the lepton doublet (singlet) superfields $L_i(\bar{E}_i)$ have the charges

$$L_1 : (3,0),\ L_2 : (1,2),\ L_3 : (0,3)\,; \qquad \bar{E}_1 : (3,1),\ \bar{E}_2 : (2,-1),\ \bar{E}_3 : (2,-3)\,. \tag{6}$$

This yields

$$Y_E \sim \lambda^2 \begin{pmatrix} \lambda^5 & 0 & 0 \\ \lambda^5 & \lambda^2 & 0 \\ \lambda^5 & \lambda^2 & 1 \end{pmatrix}, \quad X^L \sim \begin{pmatrix} 1 & \lambda^4 & \lambda^6 \\ \lambda^4 & 1 & \lambda^2 \\ \lambda^6 & \lambda^2 & 1 \end{pmatrix}, \quad X^R \sim \begin{pmatrix} 1 & \lambda^3 & \lambda^5 \\ \lambda^3 & 1 & \lambda^2 \\ \lambda^5 & \lambda^2 & 1 \end{pmatrix}, \tag{7}$$

with $\lambda_p \sim \lambda_q \sim \lambda$, where $\lambda_{p,q}$ are the ratios between the spurion VEV and the heavy messenger mass scale of the FN model. With $\lambda \sim 0.2$, this leads to realistic charged lepton masses (we assume a moderate $\tan\beta$ here).

In this example one has $|V_{ij}^M| \sim |V_{ji}^M| \lesssim |X_{ij}^M|$ so that $\delta_{ij}^M \sim X_{ij}^M$. One can now immediately check that the δ_{ij}^M's are compatible with current bounds from $l_i \to l_j\gamma$ decays (for doublet sleptons these are $\delta_{12}^L \lesssim 6\times10^{-4}$, $\delta_{13}^L \lesssim 0.08$, $\delta_{23}^L \lesssim 0.10$ [11, 8]), but within reach of upcoming FCNC tests [12]. Concerning lepton electric dipole moments (EDMs) from slepton flavor with CP violation, a rough estimate shows that a muon EDM d_μ up to $\sim 10^{-24}\,e\,\mathrm{cm}$ is possible, which is well below the current bound $d_\mu = (-0.1\pm0.9)\times10^{-19}\,e\,\mathrm{cm}$ [13], but within reach of the proposed activities to measure d_μ as low as $5\times10^{-25}\,e\,\mathrm{cm}$ [14]. The electron EDM d_e has a much stronger experimental bound but at the same time the relevant mass insertion parameters are suppressed even further so that d_e is also below the current bound.

We assume that the neutrino masses are generated by the seesaw mechanism (with neutral Majorana neutrinos N_i). The charges (6) are chosen such that the neutrino sector is anarchical,

$$(Y_N)_{ij} \sim \lambda^{n_\nu}, \qquad (m_\nu)_{ij} \sim \lambda^{2n_\nu}\langle H_u^0\rangle^2/\hat{M}_R\,, \quad \forall\, i,j\,, \tag{8}$$

with $n_\nu = 3$. One may worry that the Yukawa matrix Y_N induces excessive flavor violation via RG evolution [15]. One can show that this effect is $\delta X_{ij}^L \sim 0.1\,\lambda^{2n_\nu}$ which is negligible within our accuracy. Note also that, in order to arrive at a neutrino mass scale of ~ 0.1 eV, the N_i should have masses around $\hat{M}_R \sim 10^{10}$ GeV, which is compatible with leptogenesis (cf. e.g. [16, 2]).

Within the framework of FN symmetries, there exists a lower limit on the magnitude of the δ_{ij}^L's, for the following reason. The charges of the L_i determine both X^L and Y_N. The latter should not be suppressed too much, otherwise the seesaw scale would be too low. Therefore, the possible suppression of the X_{ij}^L's and hence δ_{ij}^L's is limited. Nevertheless, it is possible within $U(1)_p\times U(1)_q$ models to construct examples (see [8]) which produce a precise enough alignment that no signal would be seen even in planned future rare decay measurements.

4. Signatures of the model

The model has a characteristic signature: The unique pattern of gaugino masses of AMSB (see Fig. 2, 1st column) is preserved, since, as discussed above, the PMSB contribution to the gauginos is negligible. The gaugino masses thus have a ratio $|M_1| : |M_2| : |M_3|$ of $3:1:7$, yielding in particular an almost degenerate wino-like lightest neutralino $\tilde{N}_1$ and chargino $\tilde{C}_1$. A distinctive signal is the soft pion in the decay $\tilde{C}_1^\pm \to \tilde{N}_1\pi^\pm$ [17, 4].

At the same time, the model predicts $\mathcal{O}(1)$ slepton mass splittings – both between sleptons of different flavor and between singlet and doublet sleptons (as illustrated in Fig. 2, 3rd column).

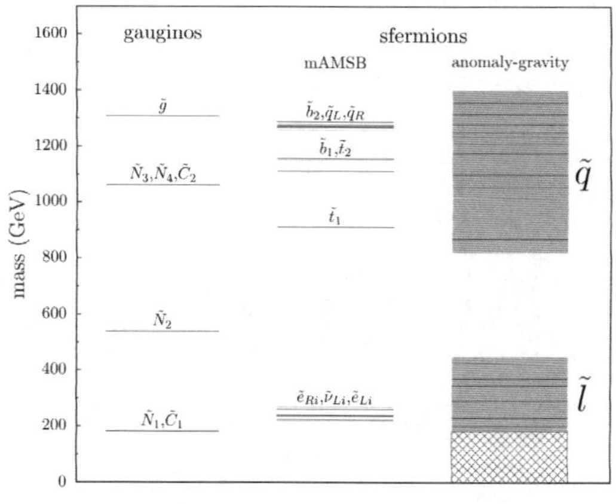

Figure 2. Schematic plot (taken from [8]) of a sample sfermion spectrum in hybrid AMSB-PMSB (3rd column), compared to an mAMSB spectrum (2nd column) with universal scalar mass uplift $m_0 = 350$ GeV and $\tan\beta = 5$, $m_{3/2} = 60$ TeV, $\mu > 0$. The gaugino masses (1st column) are as in pure AMSB (and mAMSB). The crosshatched band is disfavored phenomenologically.

This could possibly be probed at the LHC by same-flavor dilepton edge measurements with missing energy in $\tilde{N}_2 \to l\tilde{l}^*, \bar{l}\tilde{l} \to \bar{l}l\tilde{N}_1$ cascades. The position of the edge depends on the mass of the intermediate slepton. A rough estimate shows that the rates for singlets and doublets are of a similar order of magnitude, related to the bino nature of $\tilde{N}_2$. It thus seems not unlikely that the edges would be distinguishable at the LHC. The selectron-smuon mass difference could be obtained by comparing the respective kinematical edges (cf. e.g. the mSUGRA study [18]).

The squark mass squared splittings are expected to be $\mathcal{O}(10\%)$ only, due to the large flavor-universal RG-effect of the gluinos. This leads to D-$\bar{D}$ mixing close to the experimental limit [10] and a lower bound on hadronic EDMs [19].

As far as low-energy lepton flavor experiments are concerned, it is well possible that a signal is around the corner, such as is the case in the example we presented above for $\mu \to e\gamma$. A non-observation of any signal even in future FCNC experiments would however not falsify the framework of hybrid anomaly-gravity mediation.

5. Conclusions

Models where SUSY is broken by a combination of anomaly- and gravity mediation can be motivated in 5d brane worlds. The flavorful gravity mediated contribution to the slepton masses avoids the tachyonic slepton problem of AMSB. From a model-building perspective, a virtue is the possibility to have leptogenesis without the cosmological gravitino problem. From a phenomenological perspective, the model is interesting because it has a rather special signature: AMSB-like gaugino masses and – since the SUSY flavor problem is solved by alignment only – order one slepton mass splittings.

It would be worthwhile to elaborate the model further by, for instance, including a mechanism to generate a viable μ-term and a mechanism to stabilize the branes at a distance that leads to $r \sim 1$, cf. Eq. (3). Also, the phenomenological implications of the model deserve further study. In particular, it would be interesting to explore in more detail the prospects of measuring the large slepton mass splittings at the LHC.

Acknowledgments

I would like to thank Gudrun Hiller for collaboration on the work presented here. This work is supported in part by the *Bundesministerium für Bildung und Forschung (BMBF)*.

References

[1] Fukugita M and Yanagida T 1986 *Phys. Lett.* B **174** 45
[2] Giudice G F, Notari A, Raidal M, Riotto A and Strumia A 2004 *Nucl. Phys.* B **685** 89
[3] Weinberg S 1982 *Phys. Rev. Lett.* **48** 1303
[4] Gherghetta T, Giudice G F and Wells J D 1999 *Nucl. Phys.* B **559** 27
[5] Randall L and Sundrum R 1999 *Nucl. Phys.* B **557** 79
[6] Giudice G F, Luty M A, Murayama H and Rattazzi R 1998 *JHEP* **9812** 027
[7] Pomarol A and Rattazzi R 1999 *JHEP* **9905** 013
[8] Gross C and Hiller G 2011 *Preprint* arXiv:1101.5352
[9] Nir Y and Seiberg N 1993 *Phys. Lett.* B **309** 337
[10] Feng J L, Lester C G, Nir Y and Shadmi Y 2008 *Phys. Rev.* D **77** 076002
[11] Ciuchini M, Masiero A, Paradisi P, Silvestrini L, Vempati S K and Vives O 2007 *Nucl. Phys.* B **783** 112
[12] Signorelli G 2003 *J. Phys.* G **29** 2027
[13] Bennett G W *et al.* [Muon (g-2) Collaboration] 2009 *Phys. Rev.* D **80** 052008
[14] Adelmann A, Kirch K, Onderwater C J G and Schietinger T 2010 *J. Phys.* G **37** 085001
[15] Borzumati F and Masiero A 1986 *Phys. Rev. Lett.* **57** 961
[16] Davidson S and Ibarra A 2002 *Phys. Lett.* B **535** 25
[17] Feng J L, Moroi T, Randall L, Strassler M and Su S 1999 *Phys. Rev. Lett.* **83** 1731
[18] Allanach B C, Conlon J P and Lester C G 2008 *Phys. Rev.* D **77** 076006
[19] Altmannshofer W, Buras A J and Paradisi P 2010 *Phys. Lett.* B **688** 202

Journal of Physics: Conference Series **335** (2011) 012027 doi:10.1088/1742-6596/335/1/012027

Measurement of $BR(K_{e2})/BR(K_{\mu 2})$ in the NA62 experiment at CERN

Mauro Piccini

INFN - Sezione di Perugia

E-mail: Mauro.Piccini@pg.infn.it

Abstract.
The Measurement of the helicity suppressed ratio of charged kaon leptonic decay rates $R_K = \Gamma(K^\pm \to e^\pm \nu_e)/\Gamma(K^\pm \to \mu^\pm \nu_\mu)$ has long been considered as an excellent test of lepton universality and the Standard Model (SM) description of weak interactions. It was realized recently that the suppression of the SM contribution might enhance the sensitivity to SUSY-induced effects to an experimentally accessible level. The NA62 experiment at the CERN SPS has collected over 10^5 K $\to$ e ν decays during a dedicated run in 2007, aiming at achieving 0.5 % precision. Experimental strategy, details of the analysis and the final result on a partial set of data will be described. The result $R_K = (2.487 \pm 0.013) \times 10^{-5}$ is in agreement with the Standard Model expectation.

1. Introduction

In the Standard Model (SM) the decays of pseudoscalar mesons to light leptons are helicity suppressed. Although the SM predictions for such decay rates are limited by hadronic uncertainties, their specific ratios do not depend on decay constants (QCD related) and can be computed very precisely. In particular, the SM prediction for the ratio $R_K = \Gamma(K \to e\nu)/\Gamma(K \to \mu\nu)$ of kaon leptonic decay widths inclusive of internal bremsstrahlung (IB) radiation is [1]:

$$R_K^{SM} = \left(\frac{M_e}{M_\mu}\right)^2 \left(\frac{M_K^2 - M_e^2}{M_K^2 - M_\mu^2}\right)^2 (1 + \delta R_{QED}) = (2.477 \pm 0.001) \times 10^{-5} \tag{1}$$

where M_e is the electron mass, M_μ is the muon mass and M_K is the kaon mass and $R_{QED} = (-3.79 \pm 0.04)\%$ is an electromagnetic correction due to the IB and structure-dependent effects.

Within the two Higgs doublet models (2HDM), including the minimal supersymmetric model, R_K is sensitive to lepton flavour violating (LFV) effects appearing at the one-loop level via the charged Higgs boson ($H^\pm$) exchange [2, 3], representing a unique probe into mixing in the right-handed slepton sector [4]. The dominant contribution due to the LFV coupling of the $H^\pm$ is

$$R^{LVF} = R_K^{SM}\left[1 + \left(\frac{M_K}{M_H}\right)^4 \left(\frac{M_\tau}{M_e}\right)^4 |\Delta_R^{31}|^2 tan^6\beta\right] \tag{2}$$

where $tan\beta$ is the ratio of the two Higgs vacuum expectation values, and $|\Delta_R^{31}|$ is the mixing parameter between the superpartners of the right-handed leptons, which can reach $\sim 10^{-3}$.

This can enhance R_K by $\mathcal{O}(1\%)$ without contradicting any experimental constraints known at present.

The current PDG world average [5] is based on a recent result $R_K = (2.493 \pm 0.031)10^{-5}$ [6]. A new measurement of R_K based on a part of the data sample collected by the NA62 experiment at CERN in 2007 is reported here. The analyzed $K^\pm \to e^\pm \nu_e$ (from now on K_{e2}) sample is ~ 4 times larger than the total world sample, allowing a measurement of R_K with a sub-percent precision.

2. Beam, detector and data taking

The beam line and setup of the NA48/2 [7] experiment have been used for the NA62 data taking in 2007. Experimental conditions have been optimized for the $K_{e2}/K_{\mu 2}$ measurement ($K_{\mu 2}$ identifies the $K^\pm \to \mu^\pm \nu_\mu$) decay.

The beam line was originally designed to deliver simultaneous unseparated K^+ and K beams derived from the primary 400 GeV/c protons extracted from the CERN SPS. In 2007, the muon sweeping system was optimized for the positive beam, and the sample used for the present analysis was collected with the K^+ beam only with a central momentum of 74.0 GeV/c and a spread of 1.4 GeV/c (rms). The fractions of K^+ , π^+ , p , e^+ and μ^+ in the secondary beam are 0.05, 0.63, 0.21, 0.10 and 0.01, respectively. The fraction of kaons decaying in the vacuum tank at nominal momentum is 18%. The beam transverse size at the entrance to the decay volume is $\delta x = \delta y = 4$ mm (rms), and its horizontal and vertical angular divergences are about 20 μrad (rms).

Concerning the detector, charged particle properties are measured in a magnetic spectrometer, housed in a tank filled with helium at nearly atmospheric pressure, placed after the decay volume and separated from the vacuum by a thin ($\sim$0.4% radiation lengths) Kevlar window. The spectrometer comprises four drift chambers (DCHs), two upstream and two downstream of a dipole magnet which gives a horizontal transverse momentum kick of 265 MeV/c to singly-charged particles. Each DCH is composed of eight planes of sense wires, and provides a spatial resolution of 90 μm in each projection. The measured momentum resolution is $\delta_p/p = 0.48\% \pm 0.009\% \cdot p$, where p is expressed in GeV/c.

A plastic scintillator hodoscope (HOD) producing fast trigger signals and providing precise time measurements of charged particles is placed after the spectrometer. It consists of a plane of vertical strips, followed by a similar plane of horizontal strips (128 counters in total). Both planes have regular octagonal shapes and a central hole for the passage of the beam.

The HOD is followed by a quasi-homogeneous liquid krypton electromagnetic calorimeter (LKr) used for lepton identification and as a photon veto in the present analysis. The LKr is 127 cm (or 27 radiation lengths) thick along the beam, with projective readout consisting of copper/beryllium ribbons extending from the front to the back of the detector. The 13248 readout cells have a transverse size of approximately 2×2 cm^2 each and have no longitudinal segmentation. The energy resolution is $\delta_E/E = 0.032/\sqrt{E} \pm 0.09/E \pm 0.0042$ (E in GeV). The spatial resolution for the transverse coordinates x and y of an isolated electromagnetic shower is $\delta x = \delta y = 0.42/\sqrt{E} \pm 0.06$ cm (E in GeV).

An aluminium beam pipe of 158 mm outer diameter and 1.1 mm thickness traversing the centres of all detector elements allows the undecayed beam particles to continue their path in vacuum. The outer transverse sizes of the subdetectors are about 2.4 m.

During the data taking a minimum bias trigger configuration has been employed to ensure high efficiency. The K_{e2} trigger condition consists of coincidences of signals in the two HOD planes (the $Q1$ signal), loose lower and upper limits on DCH hit multiplicity (the $1-track$ signal), and LKr energy deposit (E_{LKr}) of at least 10 GeV. The $K_{\mu 2}$ trigger condition requires a coincidence of the $Q1$ and $1{-}track$ signals downscaled by a factor $D = 150$. The non-downscaled $K_{\mu 2}$ trigger rate is 0.5 MHz, and is dominated by beam halo muons; the K_{e2} trigger rate is about

10 kHz. Downscaled control samples based on trigger signals from the DCHs, HOD and LKr have been collected to monitor the performance of the main trigger signals. The data taking took place during four months starting in June 2007. About 40% of the 350k recorded good SPS spills are used for the present analysis.

3. Data analysis

The analysis strategy is based on counting the numbers of reconstructed K_{e2} and $K_{\mu 2}$ candidates collected concurrently. Therefore the analysis does not rely on an absolute beam flux measurement, and several systematic effects (due to beam simulation, accidental activity, charged track reconstruction, $Q1$ trigger efficiency, and time-dependent effects) cancel at first order. Due to the significant dependence of acceptance and background on lepton momentum, the R_K measurement is performed independently in 10 momentum bins covering a range from 13 to 65 GeV/c. The lowest momentum bin spans 7 GeV/c, while the others are 5 GeV/c wide. The selection criteria have been optimized separately in each momentum bin. The data samples in the momentum bins are statistically independent, however the systematic errors are partially correlated. The ratio R_K in each bin is computed as

$$R_K = \frac{1}{D} \cdot \frac{N(K_{e2}) - N_B(K_{e2})}{N(K_{\mu 2}) - N_B(K_{\mu 2})} \cdot \frac{A(K_{\mu 2})}{A(K_{e2})} \frac{f_\mu \times \epsilon(K_{\mu 2})}{f_e \times \epsilon(K_{e2})} \cdot \frac{1}{f_{LKr}} \quad (3)$$

where $N(K_{l2})$ are the numbers of selected K_{l2} candidates ($l = e, \mu$), $N_B(K_{l2})$ are the numbers of background events, $A(K_{\mu 2})$ and $A(K_{e2})$ are the geometric acceptances, f_l are the lepton identification efficiencies, $\epsilon(K_{l2})$ are the trigger efficiencies, f_{LKr} is the global efficiency of the LKr readout (which affects only the K_{e2} selection), and $D = 150$ is the $K_{\mu 2}$ trigger downscaling factor. To evaluate the geometric acceptance and the geometric parts of the acceptances for background processes entering the computation of $N_B(K_{l2})$, a detailed Monte Carlo (MC) simulation based on Geant3 [8] is used. It includes a description, with time variations, of the beam line optics, the full detector geometry, materials, magnetic fields, local inefficiencies of DCH wires and inactive LKR cells (0.8% of channels). Particle identification, trigger and readout efficiencies are measured directly from data.

Charged particle trajectories are reconstructed from hits and drift times in the spectrometer. Track momenta are evaluated using a detailed magnetic field map. Fine calibrations of spectrometer field integral and DCH alignment are performed by monitoring the mean reconstructed $K^+ \to \pi^+\pi^-\pi^+$ invariant mass, and the missing mass in $K_{\mu 2}$ decays. Clusters of energy deposition in the LKr are found by locating the maxima in the digitized pulses from individual cells in both space and time and accumulating the energy within a radius of approximately 11 cm. Shower energies are corrected for energy outside the cluster boundary, energy lost in inactive cells and cluster energy sharing. The energy response has been calibrated with samples of positrons from $K^+ \to \pi^0 e^+ \nu_e$ decays. Due to the topological similarity of K_{e2} and $K_{\mu 2}$ decays, a large part of the selection is common for the two decay modes, which leads to significant cancellations of the related systematic uncertainties. The main selection criteria are listed below:

- Exactly one reconstructed charged particle track (lepton candidate) geometrically consistent with originating from a kaon decay is required.
- The extrapolated track impact points in the DCHs, HOD and LKr must be within their geometrical acceptances. The LKr acceptance condition includes appropriate separations from the detector edges and inactive cells.
- The reconstructed track momentum must be in the range 13 to 65 GeV/c. The lower limit ensures high efficiency of the $E_{LKr} > 10$ GeV trigger condition. Above the upper limit, the analysis is affected by large uncertainties due to background subtraction.

- The decay vertex is reconstructed as the point of closest approach of the lepton candidate track extrapolated upstream, and the kaon beam axis. The measured stray magnetic field in the vacuum tank is taken into account. The position of the kaon beam axis is monitored with a sample of fully reconstructed $K^+ \to \pi^+\pi^-\pi^+$ decays.
- The distance from the kaon decay vertex to the beginning of the vacuum tank is required to exceed a minimum value ranging from 8 m at low lepton momentum to 43 m at high momentum.

The following two main criteria are used to distinguish K_{e2} from $K_{\mu 2}$ decays. The kinematic identification of K_{e2} ($K_{\mu 2}$) decays is based on constraining the reconstructed squared missing mass in the positron (muon) hypothesis:

$$-M_1^2 < M_{miss}^2(l) = (P_K - P_l)^2 < M_2^2 \tag{4}$$

where P_K is the average kaon four-momentum (monitored with $K^+ \to \pi^+\pi^-\pi^+$ decays), and P_l is the reconstructed lepton four-momentum (under the positron or muon mass hypothesis). The limits M_1^2 and M_2^2 have been optimized for each lepton momentum bin, taking into account the $M_{miss}^2(l)$ resolution, the radiative mass tails, and the background conditions. M_1^2 varies between 0.013 and 0.016 $(\text{GeV}/c^2)^2$ and M_2^2 between 0.010 and 0.013 $(\text{GeV}/c^2)^2$. The kinematic separation of K_{e2} and $K_{\mu 2}$ decays is illustrated in Fig. 1. The lepton identification is based on the ratio E/p of energy deposition in the LKr to momentum measured by the spectrometer. Charged particles with $(E/p)_{min} < E/p < 1.1$, where $(E/p)_{min} = 0.95$ for $p > 25$ GeV/c and $(E/p)_{min} = 0.9$ otherwise, are identified as positrons. The relaxed condition at low lepton momentum, significantly improving the identification efficiency, is applied due to the absence of backgrounds induced by particle mis-identification in that momentum range. Charged particles with $E/p < 0.85$ are classified as muons. The data E/p spectra of positrons and muons are shown in Fig2.

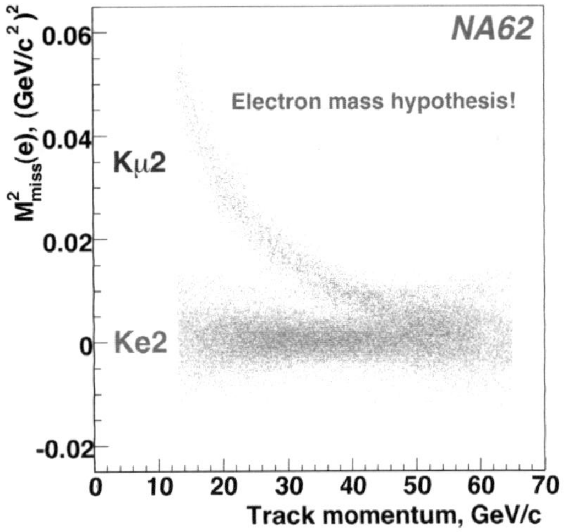

Figure 1. Squared missing mass assuming the positron mass hypothesis $M_{miss}^2(e)$ as a function of lepton momentum for reconstructed K_{e2} and $K_{\mu 2}$ decays (data).

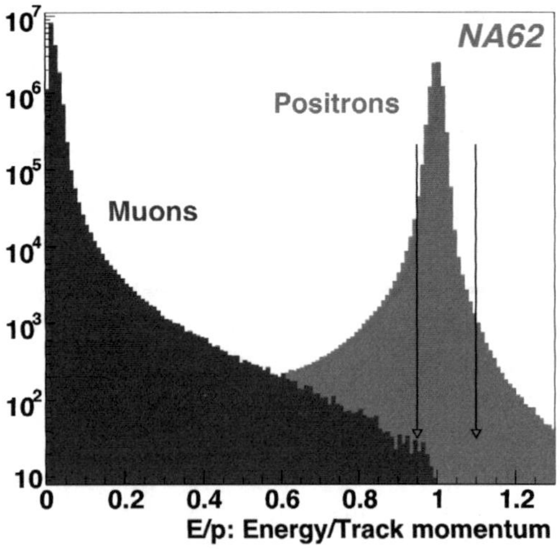

Figure 2. E/p spectra of positrons and muons (data); the positron identification limits for $p > 25$ GeV/c are indicated by arrows.

3.1. The K_{e2} sample and related background

After all the selection criteria have been applied, the number of K_{e2} candidates in the signal region is $N(K_{e2}) = 59813$. The sources of background in the K_{e2} sample is now discussed.

The main background to K_{e2} comes from $K_{\mu 2}$ decays. In fact kinematic separation of K_{e2} from $K_{\mu 2}$ decays is achievable at low lepton momentum only ($p \leq 35$ GeV/c), as shown in Fig. 1. At high lepton momentum, the $K_{\mu 2}$ decay with a mis-identified muon having $E/p > 0.95$ is the largest background source. The dominant process leading to mis-identification of the muon as a positron is "catastrophic" bremsstrahlung inside or in front of the LKr, leading to significant energy deposit in the LKr. The muon mis-identification probability $P_{\mu e}$ has been measured as a function of momentum. To collect a muon sample free from positron contamination due to μ to e decays, a 9.2 radiation lengths thick lead (Pb) wall, covering $\sim 20\%$ of the geometric acceptance, was installed approximately 1.2 m in front of the LKr calorimeter (between the two HOD planes) during a dedicated period of data taking with K^+ and K^- beams. The Pb wall reduce to a negligible level the number of electrons reaching the LKr and giving an $E/p > 0.95$, however muon passage through the Pb wall affects the measured $P_{\mu e}$ via two principal effects: ionization energy loss in Pb decreases $P_{\mu e}$ and dominates at low momentum; bremsstrahlung in Pb increases $P_{\mu e}$ and dominates at high momentum. Then to correctly evaluate $P_{\mu e}$, a dedicated MC simulation based on Geant4 (version 9.2) [9] has been developed to describe the propagation of muons downstream from the last DCH, involving all electromagnetic processes including muon bremsstrahlung. The $K_{\mu 2}$ background contamination integrated over lepton momentum has been computed to be $(6.11 \pm 0.22)\%$ using the corrected $P_{\mu e}$ measurement. A stability check of R_K with respect to variation of $(E/p)_{min}$ in the range from 0.90 to 0.97 has been performed. The observed relative stability of R_K within $\pm 0.2\%$, although the $K_{\mu 2}$ background varies from 17% to 3%, is consistent with the uncertainty assigned to the $K_{\mu 2}$ background. The $K_{\mu 2}$ decay also contributes to background via $\mu^+ \to e^+ \nu_e \overline{\nu}_\mu$ decays in flight. Energetic forward secondary positrons compatible with K_{e2} kinematics and topology are suppressed by muon polarization effects. This background contamination has been estimated to be $(0.27 \pm 0.04)\%$, where the dominant uncertainty is due to the simulated statistics.

R_K is defined to be fully inclusive of internal bremsstrahlung (IB) radiation. The structure-dependent (SD) $K^+ \to e^+ \nu \gamma$ process not suppressed by helicity represents a significant background source. The interference between the IB and SD processes is negligible. A recent measurement of the $K^+ \to e^+ \nu \gamma$ (SD) differential decay rate [6] has been used to evaluate the background contamination to be $(1.07 \pm 0.05)\%$, where the uncertainty is due to the limited precision on the $K^+ \to e^+ \nu \gamma$ (SD) decay rate, and is therefore correlated between lepton momentum bins.

Two other decays are source of background for the K_{e2} sample, $K^+ \to \pi^0 e^+ \nu$ and $K^+ \to \pi^+ \pi^0$. The $K^+ \to \pi^0 e^+ \nu$ decay produces a K_{e2} signature if the only reconstructed particle is an e^+ from K^+ or $\pi^0 \to e^+ e^- \gamma$ Dalitz (π^0_D) decays. The $K^+ \to \pi^+ \pi^0$ decay leads to a K_{e2} signature if the only reconstructed particle is a π^+ mis-identified as e^+, or an e^+ from π^0 Dalitz decay. The pion mis-identification probability ($0.95 < E/p < 1.1$) has been measured to be $(0.41 \pm 0.02)\%$ in the relevant momentum range from samples of $K^+ \to \pi^+ \pi^0$ and $K^0_L \to \pi^\pm e^\mp \nu$ decays (the latter collected during a special run). The systematic uncertainties due to subtraction of these backgrounds have been estimated as 50% of the contributions themselves, due to the limited precision of the simulation of the kaon momentum-distribution tails. The backgrounds are both at the level of $(0.05 \pm 0.03)\%$.

As no tracking is available in the beam region to tag an incoming kaon, beam halo muons can become a source of background to K_{e2} decays in case of $\mu^+ \to e^+ \nu_e \overline{\nu}_\mu$ decay or muon mis-identification as a positron. The choice of the signal region in terms of the longitudinal position of the kaon decay vertex has been dictated by the kinematic distribution of this background (which peaks in the upstream part of the vacuum volume). The halo background has been measured directly by reconstructing the K_{e2} candidates from one control data sample collected with the K^- beam transmitted by the beam line and with the K^+ beam (but not its halo) blocked. Another control data sample was collected with both K^+ and K^- beams blocked. The background contamination has been estimated to be $(1.16 \pm 0.06)\%$, where the error comes from the limited size of the control samples (uncorrelated between lepton momentum bins) and the normalization uncertainty due to decays of beam kaons and pions upstream of the decay volume (correlated between momentum bins).

The $M^2_{miss}(e)$ of K_{e2} candidates and backgrounds are shown in Fig. 3. The total background contamination is $(8.71 \pm 0.24)\%$.

3.2. The $K_{\mu 2}$ sample

The reconstructed $M^2_{miss}(\mu)$ spectrum is presented in Fig. 4. The number of $K_{\mu 2}$ candidates collected with a trigger chain involving downscaling by a factor of 150 is $N(K_{\mu 2}) = 1.803 \times 10^7$. The only significant background source in the $K_{\mu 2}$ sample is the beam halo. Its contribution is mainly at low lepton momentum, and has been measured to be $(0.38 \pm 0.01)\%$ using the same technique described for the K_{e2} sample.

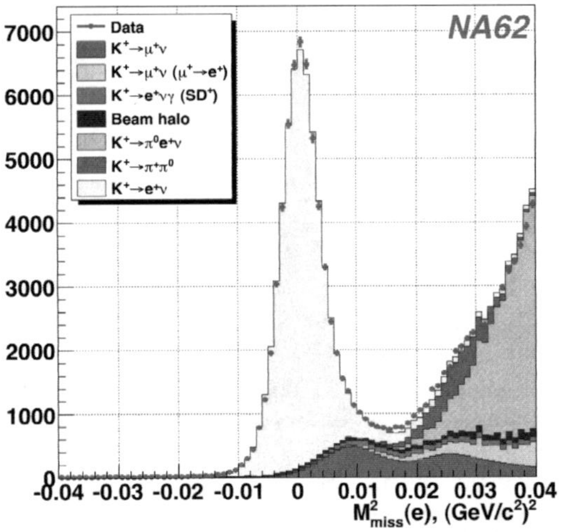

Figure 3. Reconstructed squared missing mass $M^2_{miss}(e)$ of the K_{e2} candidates compared with the sum of normalized estimated signal and background components. The small discrepancy between data and MC at low $M^2_{miss}(e)$ is due to the limited precision of MC beam description, which is taken into account by systematic uncertainty due to the acceptance correction.

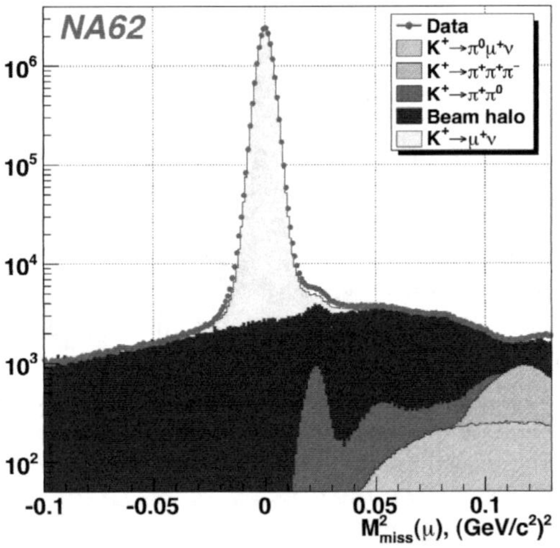

Figure 4. Reconstructed squared missing mass $M^2_{miss}(\mu)$ of the $K_{\mu 2}$ candidates compared with the sum of normalized estimated signal and background components. The deficit of reconstructed MC events in the region of the $K^+ \to \pi^+\pi^0$ peak is due to the limited precision of the beam simulation, and is mostly outside the signal region.

3.3. Corrections to R_K and systematic checks

The ratio of geometric acceptances $A(K_{\mu 2})/A(K_{e2})$ in each lepton momentum bin has been evaluated with MC simulation. The radiative $K^+ \to e^+\nu\gamma$ (IB) process, which is responsible for the loss of about 5% of the K_{e2} acceptance by increasing the reconstructed $M^2_{miss}(e)$, is taken into account following [10], including higher order corrections. The acceptance correction is strongly influenced by bremsstrahlung suffered by the positron in the material upstream of the spectrometer magnet (Kevlar window, helium, DCHs). This results in an almost momentum-independent loss of K_{e2} acceptance of about 6%, mainly by increasing the reconstructed $M^2_{miss}(e)$. The relevant material thickness has been measured by studying the spectra and rates of bremsstrahlung photons produced by low intensity 25 GeV/c and 40 GeV/c electron and positron beams steered into the DCH acceptance, using special data samples collected with the same setup by the NA48/2 experiment in 2004 and 2006. Using these measurements, the material thickness during the 2007 run has been estimated to be $(1.56\pm0.03)\%$ radiation lengths. The acceptance correction $A(K_{\mu 2})/A(K_{e2})$ is enhanced at low lepton momentum because the radial distributions of positrons from K_{e2} decays in the DCH planes are wider than those of muons from $K_{\mu 2}$ decays, and low momentum leptons are not fully contained within the geometric acceptance due to the limited transverse sizes of the DCHs.

The track reconstruction inefficiency due to interactions in spectrometer material is included into the acceptance correction. Simulation of the positron track reconstruction inefficiency (which is $\sim 10^{-3}$ in the analysis track momentum range) has been validated with a sample of $K^+ \to \pi^+\pi^0_D$ decays. The muon track reconstruction inefficiency evaluated with MC simulation is $\sim 2\times 10^{-4}$. The main sources of systematic uncertainty of the acceptance correction are the helium purity, the limited knowledge of beam profile and divergence and the simulation of soft radiative photons. A separate uncertainty has been assigned to account for the finite precision of the DCH alignment.

Concerning the lepton identification efficiencies, the E/p ratio provides powerful particle identification criteria. The momentum-dependent positron identification window $(E/p)_{min} < E/p < 1.1$ includes more than 99% of the K_{e2} events, while suppressing muons by a factor of $1/P_{\mu e} \sim 10^6$. The requirement $E/p < 0.85$ leads to a negligible inefficiency of the muon identification.

A pure sample of 4×10^7 positrons, selected kinematically from $K^+ \to \pi^0 e^+\nu$ (charged K_{e3}) decays collected with the K_{e2} trigger concurrently with the main K_{l2} data set, is used to calibrate the energy response of each LKr cell and to study f_e with respect to local position and time stability. However, the momentum range of the positrons from charged K_{e3} decays is kinematically limited, preventing a precise measurement of f_e only above 50 GeV/c. Therefore a dedicated data sample was recorded in a special 15 hour long run with a broad momentum band K^0_L beam. Electrons and positrons from the 4×10^6 collected $K^0_L \to \pi^\pm e^\mp \nu$ (neutral K_{e3}) decays allow the determination of f_e in the whole analysis momentum range. The measurements of f_e have been performed in bins of lepton momentum. The inefficiency averaged over the K_{e2} sample is $(1-f_e) = (0.73\pm 0.05)\%$, where the uncertainty takes into account the statistical precision and the small differences between charged and neutral kaon results.

The efficiency of the $Q1$ trigger condition has been measured using $K_{\mu 2}$ events triggered with a control LKr signal. The inefficiency integrated over the $K_{\mu 2}$ sample is $(1.4\pm 0.1)\%$. As a consequence of its geometric uniformity and the similarity of the K_{e2} and $K_{\mu 2}$ distributions over the HOD plane, it nearly cancels between the K_{e2} and $K_{\mu 2}$ samples, then the residual systematic bias is negligible. The inefficiency of the $1-track$ condition also largely cancels in the ratio R_K, but is anyway negligible. Thus the trigger efficiency correction $\epsilon(K_{\mu 2})/\epsilon(K_{e2})$ is determined by the efficiency $\epsilon(E_{LKr})$ of the LKr energy deposit trigger signal $E_{LKr} > 10$ GeV entering the K_{e2} trigger chain only. The inefficiency 1-$\epsilon(E_{LKr})$ is only significant in the lowest lepton momentum bin of (13, 20) GeV/c, which is close to the trigger energy threshold and is thus affected by the online energy resolution. A sample of events triggered with a control $Q1$ signal passing all K_{e2} selection criteria except the $M^2_{miss}(e)$ constraint, therefore dominated by charged K_{e3} events with two lost photons, has been used to measure 1-$\epsilon(E_{LKr})$ in the lowest momentum bin to be $(0.61\pm 0.2)\%$, once difference of positron distributions in the LKr plane between the K_{e2} sample and the control sample is taken into account. The resulting uncertainty on R_K

is negligible. Trigger efficiencies has been measured also for the main backgrounds to the K_{e2} sample in order to correctly evaluate their contributions. The global LKr readout ineffeciency, affecting the K_{e2} reconstruction only, has been measured using an independent readout system to be $1 - f_{LKr} = (0.20 \pm 0.03)\%$, stable in time.

4. Result and conclusions

A χ^2 fit to the measurements of R_K in the 10 lepton momentum bins has been performed, taking into account the bin-to-bin correlations between the systematic errors. To validate the assigned systematic uncertainties, extensive stability checks have been performed in bins of kinematic variables by varying selection criteria and analysis procedures. The fit result is

$$R_K = (2.487 \pm 0.011_{stat.} \pm 0.007_{syst.}) \times 10^{-5} = (2.487 \pm 0.013) \times 10^{-5}, \quad (5)$$

with $\chi^2/ndf = 3.6/9$. The uncertainties of the combined result are summarized in Table 1. This is the most precise R_K measurement to date. It is consistent with the KLOE measurement and with the SM expectation $R_K = (2.477 \pm 0.001) \times 10^{-5}$ and it can be used to set constraints in the phase space of relevant parameters of multi-Higgs new physics scenarios.

Table 1. Summary of the uncertainties on R_K.

Source	$\delta R_K \times 10^5$
Statistical	0.011
Backgrounds to K_{e2} sample	0.005
Helium purity	0.003
Acceptance correction	0.002
Spectrometer alignment	0.001
Positron identification efficiency	0.001
1 track trigger efficiency	0.002
LKr readout efficiency	0.001
Total systematic	0.007
Total	0.013

5. Acknowledgements

It is a pleasure to express our appreciation to the staff of the CERN laboratory and the technical staff of the participating laboratories and universities for their efforts in the operation of the experiment and data processing. This work is partially supported by Fondazione Cassa di Risparmio di Perugia.

References

[1] V. Cirigliano and I. Rosell, *Phys. Rev. Lett.* **99** (2007) 231801.
[2] A. Masiero, P. Paradisi and R. Petronzio, *Phys. Rev.* **D 74** (2006) 011701.
[3] A. Masiero, P. Paradisi and R. Petronzio, *JHEP* **0811** (2008) 042.
[4] J. Ellis, S. Lola and M. Raidal, *Nucl. Phys.* **B 812** (2009) 128.
[5] K. Nakamura *et al.*, *J. Phys.* **G 37** (2010) 075021.
[6] F. Ambrosino *et al.*, *Eur. Phys. J.* **C 64** (2009) 627; ibid. **C 65** (2010) 703.
[7] V. Fanti *et al.*, *Nucl. Instrum. Methods* **A 574** (2007) 433.
[8] GEANT description and simulation tool, CERN Program Library Long Writeup **W5013** (1994).
[9] S. Agostinelli *et al.*, *Nucl. Instrum. Methods* **A 506** (2003) 250.
[10] A. Arbuzov, A. Czarnecki and A. Gaponenko, *Phys. Rev.* **D 65** (2002) 113006.

Journal of Physics: Conference Series **335** (2011) 012028 doi:10.1088/1742-6596/335/1/012028

Lepton universality and lepton flavor violation in Υ decays

Elisa Guido (on behalf of the *BABAR* Collaboration)
Università degli Studi di Genova and INFN - Sezione di Genova, Via Dodecaneso 33, I-16146 Genova (Italy)
E-mail: elisa.guido@ge.infn.it

Abstract. The datasets collected at the $\Upsilon(3S)$ and $\Upsilon(2S)$ energies have recently allowed the *BABAR* experiment to obtain several important results in the precision tests of the Standard Model, as well as in the search for direct and indirect new physics effects. In particular, in this paper we present a test of lepton universality, measuring the ratio of decay widths $\Gamma(\Upsilon(1S) \to \tau^+\tau^-)/\Gamma(\Upsilon(1S) \to \mu^+\mu^-)$, and contributing to the search for a light pseudoscalar Higgs boson. We also show new limits on lepton flavor violating decays of the $\Upsilon(3S)$ and $\Upsilon(2S)$.

1. Introduction

The *BABAR* experiment - which is described in detail elsewhere [1, 2] - has recently obtained several important results beyond its original physics program, mainly related to CP-violation studies. In particular, results in precision tests of the standard model (SM) and in searches for new physics (NP) effects have been achieved, thanks to the large data samples collected by *BABAR* at an energy in the e^+e^- center-of-mass (CM) frame equal to the mass of the $\Upsilon(3S)$ and the $\Upsilon(2S)$. Samples of data collected 30 MeV below each Υ resonance have been used as well for background studies. The integrated luminosity of the samples employed in the analyses presented in this paper are listed in Table 1.

Table 1. Integrated luminosity recorded by *BABAR* at the energies of the $\Upsilon(3,2S)$ in the e^+e^- CM frame ("on-peak" samples), and 30 MeV below the peaks ("off-peak" samples).

$\Upsilon(nS)$	"on-peak" (fb^{-1})	"off-peak" (fb^{-1})
$\Upsilon(3S)$	28.0	2.6
$\Upsilon(2S)$	13.6	1.4

Here some recent results obtained by *BABAR* are shown: a test of lepton universality in $\Upsilon(1S)$ decays [3], which could be an indication for the existence of a light pseudoscalar Higgs boson, besides representing the most precise measurement of the ratio $\Gamma(\Upsilon(1S) \to \tau^+\tau^-)/\Gamma(\Upsilon(1S) \to \mu^+\mu^-)$; and the search for charged lepton flavor violating decays of the $\Upsilon(3S)$ and the $\Upsilon(2S)$ resonances [4].

With both these analyses the *BABAR*'s capability of performing precise measurements of the SM, as well as of constraining the NP theoretical models proposed for the different processes, is assessed.

2. Lepton universality test in $\Upsilon(1S)$ decays

A pseudoscalar Higgs boson (A^0 hereafter) is expected to exist in several extensions of the SM, such as the Next to Minimal Supersymmetric Standard Model (NMSSM) [5, 6]. A^0 could have escaped the current experimental limits [7, 8], and be assigned to a region of the parameter space which should constrain it to be light (with a mass $\sim \mathcal{O}(10\ \mathrm{GeV/c^2})$), and to have a preference for coupling to the b-type quarks (due to high $\tan\beta$ values, where $\tan\beta$ is the ratio of the vacuum expectation values of the two Higgs doublets). Therefore the B-factories, like *BABAR*, are good candidates to be the right environment where to look for A^0.

The golden channel for the search of A^0 is provided by the radiative decays of the narrow Υ resonances: $\Upsilon(nS) \to \gamma A^0$, whose predicted branching fractions range up to $\sim \mathcal{O}(10^{-4})$ [5, 6]. The following decay of the Higgs boson could lead to a set of different final states ($\mu^+\mu^-$, $\tau^+\tau^-$, or even containing only undetected, i.e. *invisible*, particles), according to the possible A^0 mass and model considered. All these possibilities - which have been recently investigated by *BABAR* [9, 10, 11, 12] - have in common the basic concept of the photon emitted in the radiative decay of the Υ resonance being energetic and detectable, and therefore the attempt of measuring its energy spectrum.

But the presence of A^0 could also be investigated by looking for an ostensible lepton universality violation in Υ decays. This measurement, besides being a precision test of the SM, can be therefore seen as a search for a light CP-odd boson, with an approach complementary to and independent of the previously mentioned analyses.

In the SM, the couplings of the gauge bosons to leptons are independent of the lepton flavor. Aside from small lepton-mass effects, the expression for the decay width $\Upsilon(1S) \to l^+l^-$ should be identical for all leptons. In the SM, one expects the quantity:

$$R_{\tau\mu}(\Upsilon(1S)) = \frac{\Gamma_{\Upsilon(1S)\to\tau^+\tau^-}}{\Gamma_{\Upsilon(1S)\to\mu^+\mu^-}} \tag{1}$$

to be very close to one (in particular, $R_{\tau\mu}(\Upsilon(1S)) \sim 0.992$ [13]).

The existence of the NMSSM pseudoscalar Higgs boson A^0 [5, 6] could perturbate this ratio, mediating the decay chain through the processes [14, 15, 16]:

$$\Upsilon(1S) \to A^0\gamma \to l^+l^-\gamma \tag{2}$$

or:

$$\Upsilon(1S) \to \eta_b(1S)\gamma, \eta_b(1S) \Leftrightarrow A^0 \to l^+l^-. \tag{3}$$

If the photon remained undetected the lepton pair would be ascribed to the $\Upsilon(1S)$. The proportionality of the coupling of the Higgs boson to the lepton mass would lead to an apparent violation of lepton universality. The deviation of $R_{\tau\mu}$ from the expected SM value depends on $X_d = \cos\theta_A \tan\beta$ (where θ_A measures the coupling of the $\Upsilon(1S)$ to the A^0) and on the mass difference between A^0 and $\eta_b(1S)$. Assuming $X_d = 12$, $\Gamma(\eta_b(1S)) = 5$ MeV, and the measured $M_{\eta_b(1S)}$ [17], the deviation of $R_{\tau\mu}(\Upsilon(1S))$ may be as large as $\sim 4\%$, depending on the A^0 mass [14, 15, 16].

A measurement of this ratio has been performed by the CLEO Collaboration, with the result $R_{\tau\mu}(\Upsilon(1S)) = 1.02 \pm 0.02(\text{stat.}) \pm 0.05(\text{syst.})$ [18].

This analysis focuses on the measurement of $R_{\tau\mu}(\Upsilon(1S))$ in the decays $\Upsilon(3S) \to \Upsilon(1S)\pi^+\pi^-$ with $\Upsilon(1S) \to l^+l^-$ and $l = \mu, \tau$ of the sample of $\Upsilon(3S)$ collected by *BABAR*. Only τ decays to a single charged particle (plus unreconstructed neutral particles) are considered, resulting in final states of exactly four detected particles for both the $\mu^+\mu^-$ and the $\tau^+\tau^-$ samples.

The event selection is optimized using Monte Carlo (MC) simulated events. Different selection criteria are used for the $\Upsilon(1S) \to \mu^+\mu^-$ decays (D_μ) and the $\Upsilon(1S) \to \tau^+\tau^-$ decays (D_τ), because the presence of neutrinos in the final state of the latter leads to a larger contamination from the background. The main sources of background are given by non-leptonic $\Upsilon(1S)$ decays and $e^+e^- \to \tau^+\tau^-$ events. The final selection efficiency for the reconstructed decay chains, estimated using a sample of MC simulated events, are $\epsilon_{\mu\mu} \sim 45\%$ and $\epsilon_{\tau\tau} \sim 17\%$ for the $\mu^+\mu^-$ and the $\tau^+\tau^-$ final states, respectively.

An extended unbinned maximum likelihood fit, performed simultaneously on the two disjoint datasets D_μ and D_τ, is used to extract $R_{\tau\mu} = \frac{N_{sig\tau}}{\epsilon_{\tau\tau}} \cdot \frac{\epsilon_{\mu\mu}}{N_{sig\mu}}$, where $N_{sig\mu}$ ($N_{sig\tau}$) indicates the number of signal events in the D_μ (D_τ) sample. For the D_μ sample, a two-dimensional probability density function (PDF) is used, based on the invariant dimuon mass $M_{\mu^+\mu^-}$ and $M^{reco}_{\pi^+\pi^-}$, the invariant mass of the system recoiling against the π-pair, defined as:

$$M^{reco}_{\pi^+\pi^-} = \sqrt{s + M^2_{\pi\pi} - 2 \cdot \sqrt{s} \cdot E^*_{\pi\pi}}, \tag{4}$$

where $\sqrt{s}$ is the e^+e^- CM energy and $E^*_{\pi\pi}$ indicates the π-pair energy calculated in the CM frame. For the D_τ sample, a one-dimensional PDF is used, based on $M^{reco}_{\pi^+\pi^-}$. The functional forms of the PDFs describing the signal components are modeled from a dedicated sub-sample consisting approximately of one tenth of the D_μ sample, and then applied only to the remaining data, in order to avoid a possible bias. The "off-peak" data, i.e. the data collected 30 MeV below the $\Upsilon(3S)$ resonance, are used to model the background shapes. Figure 1 shows the one-dimensional projections of the fit results for the three variables.

Several systematic errors cancel in the ratio. The main systematic uncertainties are related to the differences between data and simulation in the efficiency of event selection, the muon identification, and the trigger and background filters. There is also a systematic uncertainty on the signal and background yields due to the imperfect knowledge of the PDFs used in the fit. The total systematic uncertainty, obtained by summing in quadrature all the contributions, is estimated to be 2.2%.

Including all the systematic corrections, the ratio $R_{\tau\mu}$ is found to be [3]:

$$R_{\tau\mu}(\Upsilon(1S)) = 1.005 \pm 0.013(\text{stat.}) \pm 0.022(\text{syst.}).$$

No significant deviation of the ratio $R_{\tau\mu}$ from the SM expectation is observed. This result improves both the statistical and systematic precision with respect to the previous measurement [18]. Assuming values for X_d, $\Gamma(\eta_b(1S))$ and $M_{\eta_b(1S)}$ as previously stated [14, 15, 16], the present measurement excludes an A^0 with mass lower than 9 GeV/c^2 at 90% CL.

3. Lepton flavor violating Υ(3,2S) decays

Lepton flavor violation (LFV) can occur via neutrino oscillation, but it has never been observed in charged processes, such as $\Upsilon(nS) \to l^+l'^-$ with $l' \neq l$, because the tree-level contributions

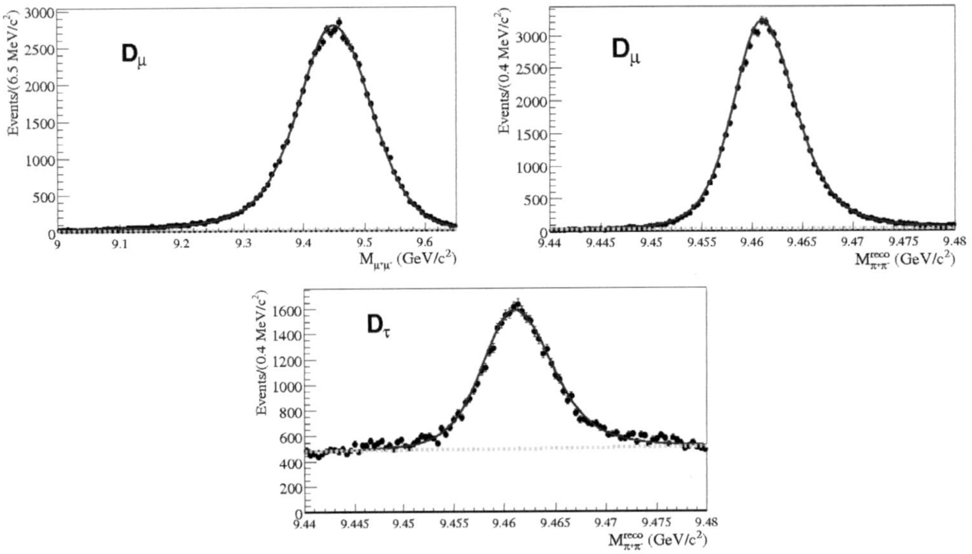

Figure 1. One-dimensional fit projections for $M_{\mu^+\mu^-}$ (top left) and for $M_{\pi\pi}^{reco}$ (top right) in the D_μ sample, and for $M_{\pi\pi}^{reco}$ (bottom) in the D_τ sample. In each plot the dashed line represents the background shape, while the solid line is the sum of signal and background contributions to the fit, and the points are the data.

are suppressed by a factor $(\Delta m_\nu^2/M_W^2)^2 < \mathcal{O}(10^{-48})$ to rates not achievable by the current experimental sensitivity. In many extensions of the SM, enhancements of these rates are possible, up to a detectable level, with expected branching fractions of $\mathcal{O}(10^{-8})$. An observation of LFV in charged decays would be a clear signature of NP, and improved limits on the branching fractions of such processes would further constrain the theoretical models proposed.

BABAR can search for charged LFV in several typologies of decays. Here we focus on searches for the charged lepton flavor violating decays $\Upsilon(nS) \to l^\pm\tau^\mp$, with $l = e, \mu$ and $n = 2, 3$, using the $\Upsilon(3S)$ and $\Upsilon(2S)$ resonances collected by BABAR.

The τ lepton is required to decay to one charged track (either a lepton or a pion) plus a number of not reconstructed neutral particles (either π^0 mesons or neutrinos). Thus the signature of the signal events consists of exactly two oppositely charged particles: a primary lepton, identified as an electron or a muon, with momentum close to the beam energy, and a secondary charged lepton or pion from the τ decay. In order to suppress quantum electrodynamics background events, such as Bhabha ($e^+e^- \to e^+e^-$) and μ-pairs events ($e^+e^- \to \mu^+\mu^-$), if the τ decays leptonically, the primary lepton and the τ-daughter are required to have different flavors. Thus, for each value of n, four signal channels are defined, consisting of leptonic and hadronic τ decay modes, and with an electron or a muon as primary lepton. The main sources of background, besides the already mentioned Bhabha and μ-pair events, come from τ-pair production and from events containing multiple pions and possible additional photons, where the pion is misidentified as a muon, and the remaining particles satisfy the selection criteria for the τ decay products (for this reason, this typology of background is referred to as "π-hadron" background).

The event selection consists of several requirements related to the particle identification and to the kinematics of the τ-daughter, which are different in each signal channel. The final selection efficiencies, estimated on samples of MC simulated events, vary in the range (4-6)% depending on the decay mode considered.

An extended unbinned maximum likelihood fit is performed to the distribution of the variable $x = |p_1|/E_B$, that is, the momentum of the primary lepton (p_1) normalized to the beam energy (E_B). The signal distribution is expected to peak at $x \sim 0.97$, while the τ-pair background x distribution is smooth and approaches zero as $x \to 0.97$, which is the effective kinematic endpoint for the lepton momentum in the decay $\tau^\pm \to l^\pm \bar{\nu}_l \nu_\tau$, boosted in the Υ rest frame. Similarly, the "π-hadron" background is expected to have a distribution which sharply falls off near the endpoint $x = 0.97$. Instead, the Bhabha and μ-pair events have a peaking behavior at $x \sim 1$. PDFs are chosen for each of these components, using samples of data and of MC simulated events. The fit is repeated in each signal channel. In Figure 2 an example of fit is shown, for the $\Upsilon(3S) \to e\tau \to e\mu(\bar{\nu}_\mu \nu_\tau)$ channel.

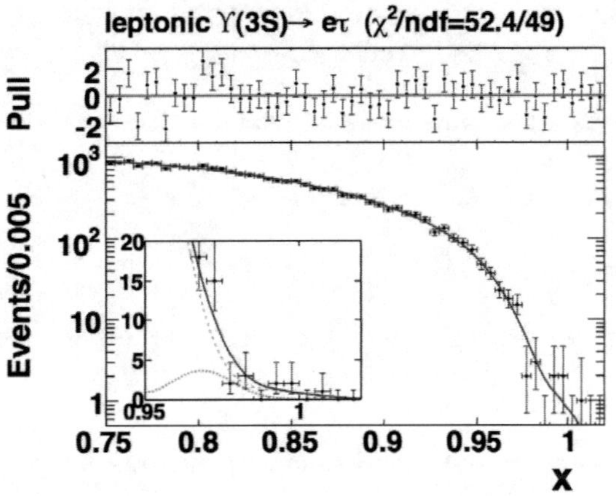

Figure 2. Fit results for the leptonic $e\tau$ channel in $\Upsilon(3S)$ data. The inset is a zoom of the region $0.95 < x < 1.02$. The dotted line represents the signal component, the dashed line represents the sum of all background components, and the solid line shows the sum of signal and background, while the points are the data. The top plot shows the normalized residuals of the fit.

The signal yield is extracted and found consistent with the no signal-hypothesis within $\pm 1.8\sigma$ in all the signal channels. Since no statistically significant signal is observed, the 90% of confidence level (CL) upper limit (UL) on the branching fraction $\mathcal{B}$ of each decay is determined, using a Bayesian technique, in which the prior likelihood is uniform in $\mathcal{B}$ and assumes that $\mathcal{B} > 0$.

In the UL calculation, the systematic uncertainties affecting the measurement are also taken into account. The dominant contribution to the systematic uncertainty comes from an imperfect knowledge of the PDF shapes. The resulting ULs [4] are summarized in Table 2 and are of $\mathcal{O}(10^{-6})$, representing the first constraints on $\mathcal{B}(\Upsilon(nS) \to e^\pm \tau^\mp)$, while improving the sensitivity with respect to the previous ULs [19] on $\mathcal{B}(\Upsilon(nS) \to \mu^\pm \tau^\mp)$.

The physics of a charged lepton flavor violating Υ decay can be described by an effective field theory, as a $b\bar{b}l^\pm\tau^\mp$ contact interaction parameterized by a NP coupling constant ($\alpha_{l\tau}$) and by a mass scale ($\Lambda^2_{l\tau}$) [20, 21]. In such a scenario, Figure 3 shows the constraint to the allowed parameter space supplied by the *BABAR* result. Assuming $\alpha_{e\tau} = \alpha_{\mu\tau} = 1$, these result translate in the 90% CL lower limits $\Lambda_{e\tau} > 1.6$ TeV and $\Lambda_{\mu\tau} > 1.7$ TeV on the mass scale of NP contributing to the lepton flavor violating decays of the Υ resonances. These constraints improve upon the previous limits [19].

Table 2. 90% CL ULs on the branching fractions $\mathcal{B}$ for signal decays $\Upsilon(nS) \to l^{\pm}\tau^{\mp}$.

Mode	UL ($\times 10^{-6}$)
$\mathcal{B}(\Upsilon(2S) \to e^{\pm}\tau^{\mp})$	< 3.2
$\mathcal{B}(\Upsilon(2S) \to \mu^{\pm}\tau^{\mp})$	< 3.3
$\mathcal{B}(\Upsilon(3S) \to e^{\pm}\tau^{\mp})$	< 4.2
$\mathcal{B}(\Upsilon(3S) \to \mu^{\pm}\tau^{\mp})$	< 3.1

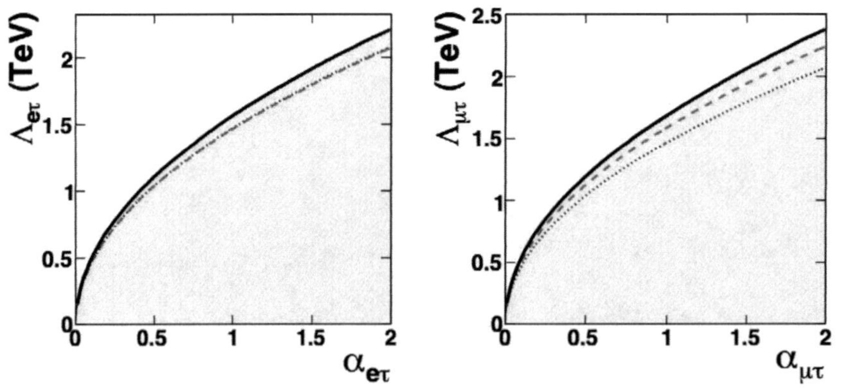

Figure 3. Excluded regions of effective field theory parameter space given by mass scale $\Lambda_{l\tau}$ versus coupling constant $\alpha_{l\tau}$. The dotted line is derived from $\Upsilon(2S)$ results only, the dashed line from $\Upsilon(3S)$ results only, and the solid line indicates the combined results, for $l = e$ (left), and $l = \mu$ (right). The shaded regions are excluded at 90% CL.

4. Conclusions

Thanks to the large amount of data collected at the $\Upsilon(2, 3S)$ resonances, *BABAR* has been able to achieve several important results, like those presented here [3, 4]. In particular, *BABAR* has performed a test of the SM, improving the precision of the lepton universality measurement in $\Upsilon(1S)$ decays. The results presented in this paper have also demonstrated that *BABAR* is able to put important constraints on theoretical predictions beyond the SM: the NP effects investigated here are the existence of the NMSSM light pseudoscalar Higgs boson, and the observation of a lepton flavor violating process involving charged leptons.

References

[1] Aubert B *et al.* (*BABAR* Collaboration) 2002 *Nucl. Instrum. Methods Phys. Res. A* **479** 1
[2] Menges W 2006 *IEEE Nucl. Sci. Symp. Conf. Rec.* **5** 1470
[3] del Amo Sanchez P *et al.* (*BABAR* Collaboration) 2010 *Phys. Rev. Lett.* **104** 191801
[4] Lees J P *et al.* (*BABAR* Collaboration) 2010 *Phys. Rev. Lett.* **104** 151802
[5] Dermisek R and Gunion J F 2005 *Phys. Rev. Lett.* **95** 041801
[6] Dermisek R and Gunion J F 2010 *Phys. Rev. D* **81** 075003
[7] Kraml S*et al.* 2006 CERN 2006-009
[8] Schael S *et al.* (ALEPH, DELPHI, L3 and OPAL Collaborations) 2006 *Eur. Phys. J. C* **47** 547
[9] Aubert B *et al.* (*BABAR* Collaboration) 2009 *Phys. Rev. Lett.* **103** 081803
[10] Aubert B *et al.* (*BABAR* Collaboration) 2009 *Phys. Rev. Lett.* **103** 181801
[11] Aubert B *et al.* (*BABAR* Collaboration) 2009 *Phys. Rev. Lett.* **103** 251801
[12] del Amo Sanchez P *et al.* (*BABAR* Collaboration) *Preprint* arXiv:1007.4646

[13] Nakamura K *et al.* (Particle Data Group Collaboration) 2010 *J. Phys. G* **37** 075021
[14] Sanchis-Lozano M A 2004 *Int. J. Mod. Phys. A* **19** 2183
[15] Fullana E and Sanchis-Lozano M A 2007 *Phys. Lett. B* **653** 67
[16] Domingo F, Ellwanger U and Sanchis-Lozano M A 2009 *Phys. Rev. Lett.* **103** 111802
[17] Aubert B *et al.* (*BABAR* Collaboration) 2008 *Phys. Rev. Lett.* **101** 071801
[18] Besson D *et al.* (CLEO Collaboration) 2007 *Phys. Rev. Lett.* **98** 052002
[19] Love W *et al.* (CLEO Collaboration) 2008 *Phys. Rev. Lett.* **101** 201601
[20] Silagadze Z K 2001 *Phys. Scripta* **64** 128-33
[21] Black D, Han T, He H-J and Sher M 2002 *Phys. Rev. D* **66** 053002

Tau lepton physics at Belle

Kiyoshi Hayasaka (KMI, Nagoya U.)
Furo-cho, Chikusa-ku, Nagoya, Japan
E-mail: hayasaka@hepl.phys.nagoya-u.ac.jp

Abstract. We report the recent results of a search for lepton-flavor-violating τ decays and a search for CP violation in $\tau \to \nu K_S^0 \pi$ using a large data sample accumulated with the Belle detector at the KEKB asymmetric-energy e^+e^- collider. The sensitivity to these modes is significantly improved compared to previous experiments.

1. Introduction

The success of the B-factory provides us the world-largest data sample of $\tau^+\tau^-$-pairs, that is useful for a search for New Physics (NP). Here, we discuss new phenomena relating to τ decays, in particular, lepton-flavor-violating τ decays and CP violation in a τ decay using an around 1 ab^{-1} data sample accumulated with the Belle detector at the KEKB asymmetric-energy e^+e^- collider,

An observation of the charged-lepton-flavor violation (cLFV) would be a clear signature of NP since the cLFV has a negligibly small probability in the Standard Model (SM) even taking into account neutrino oscillations and many of models for NP naturally predict cLFV. In particular, lepton-flavor-violating (LFV) τ decays are allowed to have various kinds of final states and some of them are expected to have a measurably large branching fraction, depending on the models. Therefore, not only to find the NP phenomenon but to distinguish the models, it is important to search for the LFV τ decays.

Similarly to the LFV τ decays, an observation of CP violation (CPV) in τ decays would be also a clear signature of NP. Here, we focus on the CPV in a τ decay into $\nu K_S^0 \pi$ induced by a new scalar particle such as a charged Higgs in a multi-Higgs-doublet model. This kind of CPV can be observed as a difference in the $\tau^\pm$ decay angular distributions by the interference effect between the scalar and (axial-)vector components of the intermediate state in the decay.

2. $\tau \to \ell M^0$ ($\ell = e, \mu$, $M^0 = \pi^0, \eta, \eta', \rho^0, K^{*0}, \bar{K}^{*0}, \omega$ and ϕ)

Generally, $\tau \to \mu\gamma$ is the most promising among various LFV τ decays in many models for NP. However, in some models, such as non-universal-Higgs-mass model [1], $\tau \to \mu\eta$ can have the largest branching fraction. In addition, a magnitude relation among the branching fractions for the LFV τ decays will help to distinguish the NP models. Here, we show the results for the two-body τ decays into a lepton and a neutral meson (pseudoscalar and vector).

2.1. Analysis Method

In the LFV τ analysis, in order to evaluate the number of signal events, two independent variables are defined, that are signal-reconstructed mass and energy in the center-of-mass (CM) frame from energies and momenta for the signal τ daughters. In the $\tau \to \mu\eta$ case, they are defined as

$$M_{\mu\eta} = \sqrt{E_{\mu\eta}^2 - P_{\mu\eta}}, \tag{1}$$

$$\Delta E = E_{\mu\eta}^{\rm CM} - E_{\rm beam}^{\rm CM}, \tag{2}$$

where $E_{\mu\eta}$ ($P_{\mu\eta}$) is a sum of the energies (a magnitude of a vector sum of the momenta) for μ and η, the superscript CM indicates that the variable is defined in the CM frame and the $E_{\rm beam}^{\rm CM}$ means the initial beam energy in the CM frame. In signal events $M_{\mu\eta}$ and ΔE are expected to be distributed around m_τ ($\sim$ 1.78 GeV/c^2) and 0 (GeV), respectively, while $M_{\mu\eta}$ and ΔE will smoothly vary without any special structure in the background (BG) events. The signal MC distribution of $\tau \to \mu\eta$ on the $M_{\mu\eta}$ – ΔE plane is shown in Fig. 1. Due to the resolution, the

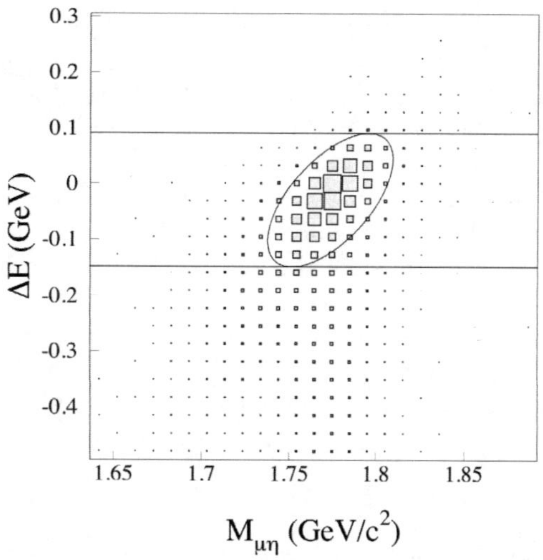

Figure 1. Distribution for the $\tau \to \mu\eta$ signal events. An elliptical region is defined as a signal region, that corresponds to $\pm 3\sigma$ and a region between two parallel lines excluding the signal region is used for the BG estimation as a sideband region, where the resolution σ is evaluated by a asymmetric-Gaussian fit to the signal MC distribution.

signal events are distributed around $M_{\mu\eta} \sim m_\tau$ and $\Delta E \sim 0$ (GeV). Taking into account the resolution, we set the elliptic signal region. (See Fig. 1.) Since the resolution for the photon energy is not so good as that for the momentum of the charged track, the signal region of the mode having some photons in the final state can be large.

To avoid any bias for our analyses, we perform the blind analysis, i.e., we do not observe the data events in the signal region before fixing the selection criteria, the evaluation of the systematic uncertainties and the expected number of the BG events. Finally, we observe some signal events or evaluate an upper limit of the number of the signal event at 90% confidence level (CL) using the expected number of the BG event and the number of the observed events in the signal region by Feldman-Cousins approach. [2]

Table 1. Summary for the efficiency (Eff.), the expected number of the BG events, ($N_{BG}^{\rm exp}$) and the upper limit on the branching fraction (UL) for each mode, where "comb." means the combined result from subdecay modes.

Mode ($\tau \to$)	Eff.(%)	$N_{BG}^{\rm exp}$	UL ($\times 10^{-8}$)	Mode ($\tau \to$)	Eff.(%)	$N_{BG}^{\rm exp}$	UL ($\times 10^{-8}$)
$\mu\eta(\to\gamma\gamma)$	8.2	0.63 ± 0.37	3.6	$e\eta(\to\gamma\gamma)$	7.0	0.66 ± 0.38	8.2
$\mu\eta(\to\pi\pi\pi^0)$	6.9	0.23 ± 0.23	8.6	$e\eta(\to\pi\pi\pi^0)$	6.3	0.69 ± 0.40	8.1
$\mu\eta$(comb.)			2.3	$e\eta$(comb.)			4.4
$\mu\eta'(\to\pi\pi\eta)$	8.1	$0.00^{+0.16}_{-0.00}$	10.0	$e\eta'(\to\pi\pi\eta)$	7.3	0.63 ± 0.45	9.4
$\mu\eta'(\to\gamma\rho^0)$	6.2	0.59 ± 0.41	6.6	$e\eta'(\to\gamma\rho^0)$	7.5	0.29 ± 0.29	6.8
$\mu\eta'$(comb.)			3.8	$e\eta'$(comb.)			3.6
$\mu\pi^0$	4.2	0.64 ± 0.32	2.7	$e\pi^0$	4.7	0.89 ± 0.40	2.2

2.2. $\tau \to \ell P^0$ ($\ell = e, \mu$, $P^0 = \pi^0, \eta, \eta'$)

We perform a new search for the τ decay into a lepton (e or μ) and a neutral pseudoscalar (π^0, η or η') with a 901 fb^{-1} data sample, that corresponds to 8.2×10^8 $\tau^+\tau^-$-pairs. A neutral pion is reconstructed from 2 photons while an η (η') is reconstructed from $\gamma\gamma$ ($\rho^0\gamma$) as well as $\pi^+\pi^-\pi^0$ ($\pi^+\pi^-\eta$) to increase the detection efficiency. When the neutral pseudoscalars are reconstructed, their four-momentum is evaluated by a mass-constrained fit to obtain a better resolution for the signal region. Because we have modified the selection criteria applied to the previous analysis, we obtain an about 1.5 times better detection efficiency while a similar background level is kept. Here, the number of the BG events in the signal region is evaluated by an interpolation from the number of the observed events in the sideband region, assuming a flat distribution for the BG events in the signal and sideband region.

As a result, we observe one event in the $\tau \to e\eta(\to\gamma\gamma)$ mode while no events are found in other modes as shown in Fig. 2. Since these results are consistent with the background estimation, we set upper limits on the following branching fractions, as shown in Table 1. Consequently, we

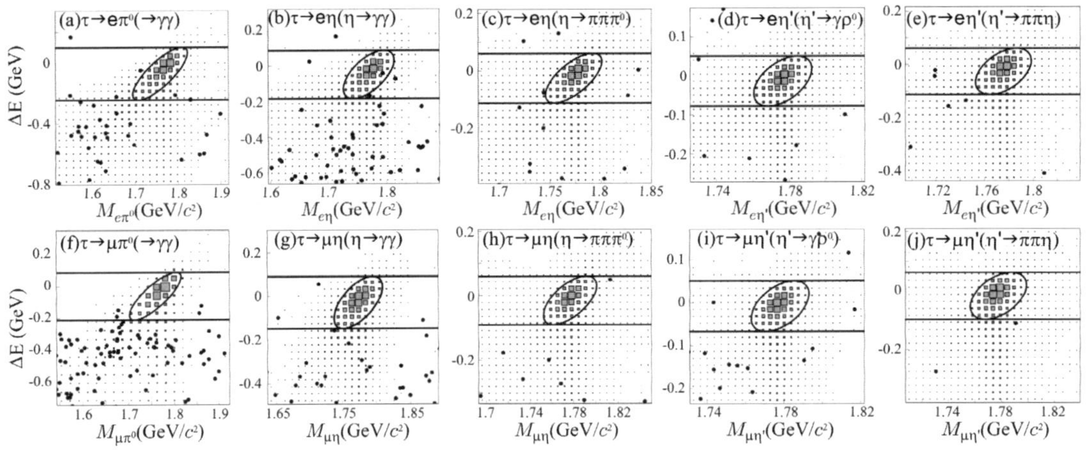

Figure 2. Resulting 2D plots for $\tau \to e\pi^0$ (a), $e\eta(\to\gamma\gamma)$ (b), $e\eta(\to\pi\pi\pi^0)$ (c), $e\eta'(\to\gamma\rho^0(\to\pi\pi))$ (d), $e\eta'(\to\pi\pi\eta(\to\gamma\gamma))$ (e), $\mu\pi^0$ (f), $\mu\eta(\to\gamma\gamma)$ (g), $\mu\eta(\to\pi\pi\pi^0)$ (h), $\mu\eta'(\to\gamma\rho^0(\to\pi\pi))$ (i) and $\mu\eta'(\to\pi\pi\eta(\to\gamma\gamma))$ (j), on the $M_{\ell P^0}$ – ΔE plane. Here, black dots (shaded boxes) express the data (signal MC), the region bounded by two lines is defined as a 3σ band for the BG estimation and the elliptic region is the signal one which corresponds to 3σ in each plot. One event is found in the signal region for $\tau \to e\eta$ while no events appear in any other modes.

Table 2. Summary for the efficiency (Eff.), the expected number of the BG events, ($N_{BG}^{\rm exp}$) and the upper limit on the branching fraction (UL) for each mode.

Mode ($\tau \to$)	Eff.(%)	$N_{BG}^{\rm exp}$	UL ($\times 10^{-8}$)	Mode ($\tau \to$)	Eff.(%)	$N_{BG}^{\rm exp}$	UL ($\times 10^{-8}$)
$\mu\rho^0$	7.1	1.48 ± 0.35	1.2	$e\rho^0$	7.6	0.29 ± 0.15	1.8
μK^{*0}	3.4	0.53 ± 0.20	7.2	eK^{*0}	4.4	0.39 ± 0.14	3.2
μK^{*0}	3.6	0.45 ± 0.17	7.0	eK^{*0}	4.4	0.08 ± 0.08	3.4
$\mu\omega$	2.4	0.72 ± 0.18	4.7	$e\omega$	2.9	0.30 ± 0.14	4.8
$\mu\phi$	3.2	0.06 ± 0.06	8.4	$e\phi$	4.2	0.47 ± 0.19	3.1

obtain a result (2.1 – 4.4) times more sensitive than previous ones.

2.3. $\tau \to \ell V^0$ ($\ell = e, \mu$, $V^0 = \rho^0, K^{*0}, \bar{K}^{*0}, \omega, \phi$)

Similarly to $\tau^- \to \ell^- P^0$, we update our results of the search for $\tau^- \to \ell^- V^0$, where $\ell = e, \mu$, $V^0 = \rho^0, K^{*0}, \bar{K}^{*0}, \omega, \phi$ with an 854 fb^{-1} data sample, that corresponds to 7.8×10^8 $\tau^+\tau^-$-pairs [3]. Here, $\rho^0, K^{*0}, \bar{K}^{*0}, \omega$ and ϕ are reconstructed from $\pi^+\pi^-$, π^-K^+, π^+K^-, $\pi^+\pi^-\pi^0$ and K^+K^-, respectively. By performing a detailed background study, we obtain a 1.2 times better efficiency in average while keeping similar level backgrounds. For the modes including a μ, BG events mainly come from $\tau\tau$ events while some two-photon process or radiative Bhabha process with a gamma conversion becomes main background in the modes having an electron. Finally, one event is found in the signal region for $\tau^- \to \mu^- K^{*0}$, $\mu^- \bar{K}^{*0}$ and $\mu^-\phi$ while no events appear in any other modes. (See Fig. 3.) They are consistent with the expected number of the backgrounds. Consequently, we set the 90% confidence level upper limits on the branching fractions as shown in Table 2. At the present, the upper limit for $\tau \to \mu\rho^0$ is the most sensitive among the all LFV τ decays. (See Fig. 4, where all of the current upper limits are shown.)

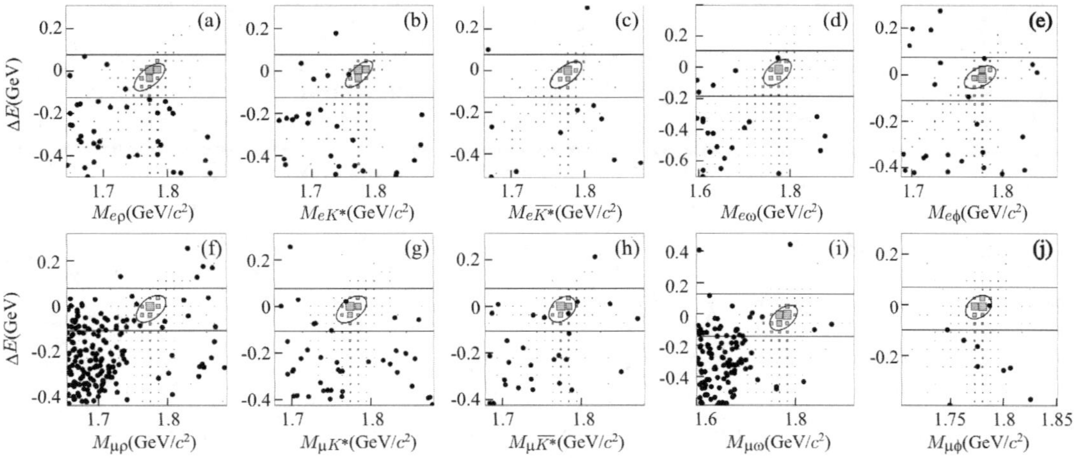

Figure 3. Resulting 2D plots for $\tau \to e\rho^0$ (a), eK^{*0} (b), $e\bar{K}^{*0}$ (c), $e\omega$ (d), $e\phi$ (e), $\mu\rho^0$ (f), μK^{*0} (g), $\mu\bar{K}^{*0}$ (h), $\mu\omega$ (i) and $\mu\phi$ (j) on the $M_{\ell V^0} - \Delta E$ plane. Here, black dots (shaded boxes) express the data (signal MC), the region bounded by two lines is defined as a 5σ band for the BG estimation and the elliptic region is the signal one which corresponds to 3σ in each plot. One event is found in the signal region for $\tau \to \mu K^{*0}$, $\mu\bar{K}^{*0}$ and $\mu\phi$ while no events appear in any other modes.

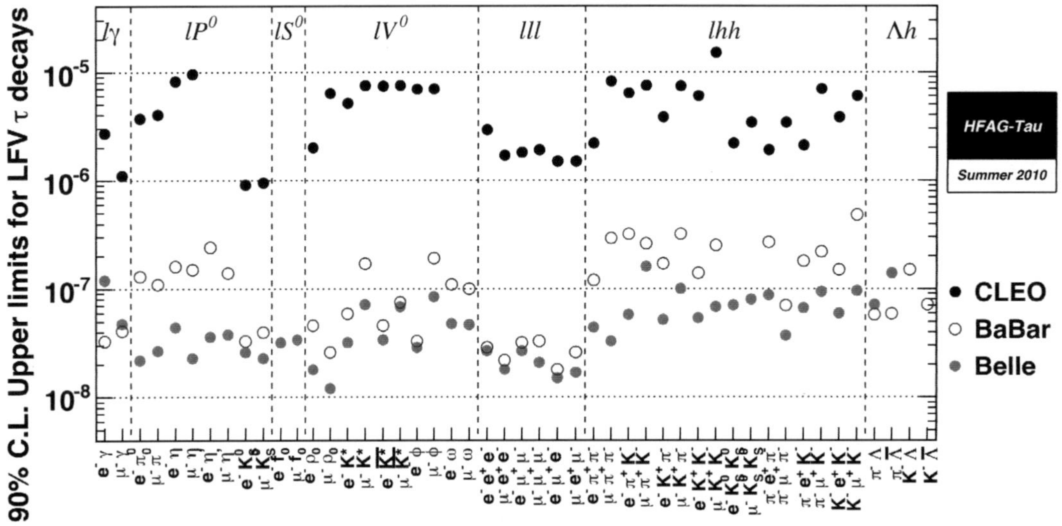

Figure 4. Upper limits on the branching fractions for the LFV τ decays. Here, red, blue and black circles express the results obtained by Belle, BaBar and CLEO, respectively [4].

3. CPV measurement with the τ decay into $\nu K_S^0 \pi$

In SM, CP in the charged lepton sector is conserved. However, by introducing a new scalar field with a complex coupling constant, CPV can be induced. A charged Higgs in the multi-Higgs-doublet model [5] is one of the candidates for such a scalar field. Here, we discuss the CPV appearing in the $\tau \to \nu K_S^0 \pi$ decay as an interference between the SM and NP processes [6].

3.1. Analysis method

A hadronic current into $K^0 \pi$ can be written as

$$\begin{aligned} J_\mu &= \langle K^0(q_1)\pi(q_2)|\bar{u}\gamma_\mu s|0\rangle \\ &= (q_1 - q_2)^\nu \left(g_{\mu\nu} - \frac{Q_\mu Q_\nu}{Q^2} \right) F_V(Q^2) + Q_\mu F_S(Q^2), \end{aligned} \tag{3}$$

where u and s are spinor fields for a u and s quark, respectively, q_1 and q_2 are 4-momenta for K^0 and π, respectively, F_V (F_S) is a vector (scalar) form factor and $Q^\mu = q_1^\mu + q_2^\mu$. To make the current induce CPV, instead of F_S, we introduce $\tilde{F}_S$, defined as

$$\tilde{F}_S = F_S + \frac{\eta_S}{m_\tau} F_H, \tag{4}$$

where F_H is a form factor for the charged Higgs, η_S is a complex coupling constant [7]. The size of CPV is proportional to an imaginary part of η_S. From the current, in the hadronic rest frame ($\vec{Q} = \vec{0}$), we can calculate the differential partial width for $\tau \to \nu K_S^0 \pi$ as

$$\begin{aligned} \frac{d\Gamma(\tau \to \nu K_S^0 \pi)}{d\omega} &= \text{(CP even terms)} \\ &\quad -4 \frac{m_\tau}{\sqrt{Q^2}} |\vec{q_1}| \Im(F_V F_H^*) \Im(\eta_S) \cos\psi \cos\beta, \end{aligned} \tag{5}$$

where $d\omega = dQ^2 d\cos\theta d\beta$, θ (β) is an opening angle between the direction opposite to the one of the CM system and the one of the hadronic system in the τ rest frame (between the direction of K_S^0 and that for the CM system in the hadronic rest frame), and ψ denotes the angle between the direction of the CM frame and the direction of the τ as seen from the hadronic rest frame.

To extract the CP violating term, we define an asymmetry as

$$\begin{aligned} A_{\psi\beta}^{CP} &= \frac{\int \cos\beta\cos\theta\left(\frac{d\Gamma_{\tau^-}}{d\omega} - \frac{d\Gamma_{\tau^+}}{d\omega}\right)d\omega}{\frac{1}{2}\left(\frac{d\Gamma_{\tau^-}}{dQ^2} + \frac{d\Gamma_{\tau^+}}{dQ^2}\right)dQ^2} \\ &\simeq \langle\cos\beta\cos\psi\rangle_{\tau^-} - \langle\cos\beta\cos\psi\rangle_{\tau^+}. \end{aligned} \tag{6}$$

Therefore, by measuring the second line, we can evaluate $A_{\psi\beta}^{CP}$ experimentally.

3.2. Result and Discussion

With a 699 fb^{-1} data sample, we obtain around 3×10^5 $\tau \to \nu K_S^0 \pi$ candidates, with a 23% contamination from BG events, as estimated by means of MC. (See Fig. 5(a).). On this sample $A_{\psi\beta}^{CP}$ is measured in bins of $W(\equiv \sqrt{Q^2})$. (See Table 3.2 and Figs. 5(b) and (c).) In the final result, $\gamma - Z$ interference and $\pi^{\pm}$ detection asymmetry effects are corrected for and the background subtraction is also performed.

Table 3. CP asymmetry in bins of the hadronic mass W. The first and second errors correspond to statistical and systematic ones, respectively.

	W (GeV/c^2)			
	0.625–0.890	0.890–1.110	1.110–1.420	1.420–1.775
$A_{\beta\psi}^{CP}$	$7.9 \pm 3.9 \pm 2.8$	$1.8 \pm 2.1 \pm 1.4$	$-4.6 \pm 7.2 \pm 1.7$	$-2.3 \pm 19.1 \pm 5.5$

Consequently, no CPV is found. From this result, we can evaluate the 90% CL upper limit for $|\Im(\eta_S)|$ varying a constant-strong-interaction-phase difference between F_V and F_S:

$$|\Im(\eta_S)| < (0.012 - 0.026). \tag{7}$$

This result is almost one order of magnitude more restrictive than that previously obtained by CLEO [8], i.e., $|\Im(\eta_S)| < 0.19$.

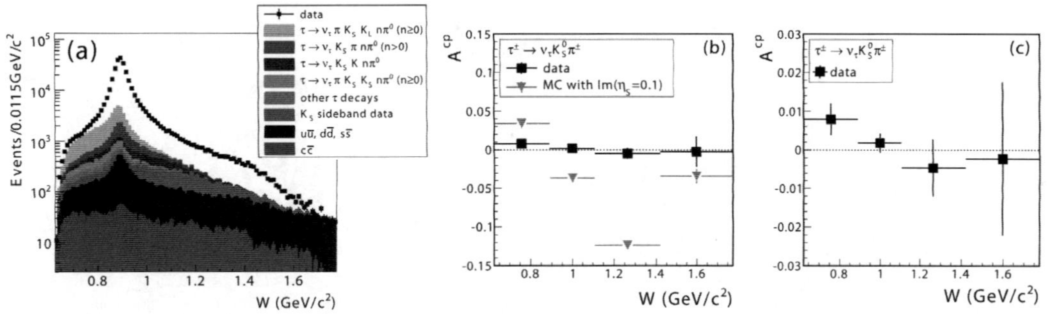

Figure 5. (a) Distribution of W for the $\tau \to \nu K_S^0 \pi$ candidate, (b) measured $A_{\beta\psi}^{CP}$ in bins of W and (c) zoom of (b).

According to a theoretical prediction for $\Im(\eta_S)$ in the multi-Higgs-doublet model, η_S is written as

$$\eta_S \simeq \frac{m_\tau m_s}{M_{H^\pm}^2} X^* Z, \tag{8}$$

where m_s is an s-quark mass, $M_{H^\pm}$ is the lightest charged Higgs mass, and X and Z are complex constants describing the coupling of the Higgs to the u and s quarks and the τ and ν, respectively. Using the limit $\Im(\eta_S) < 0.026$, we can obtain an upper limit:

$$|\Im(XZ^*)| < 0.15 \frac{M_{H^\pm}^2}{1\mathrm{GeV}^2/c^4}. \tag{9}$$

References

[1] E. Arganda, M. J. Herrero and J. Portoles, JHEP **0806**, 079 (2008).
[2] G. J. Feldman and R. D. Cousins, Phys. Rev. D **57**, 3873 (1998).
[3] Y. Miyazaki *et al.* (Belle Collaboration), arXiv:1101.0755 [hep-ex], submitted to Phys. Lett. B.
[4] It is drawn by HFAG-tau subgroup. You can find it at
http://www.slac.stanford.edu/xorg/hfag/tau/HFAG-TAU-LFV.htm
[5] S. Y. Choi, K. Hagiwara, and M. Tanabashi, Phys. Rev. D **52**, 1614 (1995).
[6] M. Bischofberger *et al.* (Belle Collaboration), arXiv:1101.0349 [hep-ex] submitted to Phys. Rev. Lett.
[7] J. H. Kühn and E. Mirkes, Phys. Lett. B **398**, 407 (1997).
[8] G. Bonvicini *et al.* (CLEO Collaboration), Phys. Rev. Lett. **88**, 111803 (2002).

Journal of Physics: Conference Series **335** (2011) 012030 doi:10.1088/1742-6596/335/1/012030

Flavour symmetries at the EW scale

Reinier de Adelhart Toorop
Nikhef Theory Group, Sciencepark 105, 1098 XG Amsterdam, The Netherlands
E-mail: reinier.de.adelhart.toorop@nikhef.nl

Abstract. We study the possibility of extending the Standard Model symmetries with a flavour symmetry at the electroweak scale. This requires the existence of a flavour-triplet of Higgses. We study the related scalar potential and its possible minima. It turns out that the new Higgses lead to extra contributions to the oblique corrections, Z and W decay and flavour violating processes. These can be used to constrain the viability of flavour models.

1. Introduction

After the discovery of neutrino oscillations and the determination of the related mixing angles, it was observed that the data may show interesting structures. The observations show two large mixing angles and one angle that is either very small or vanishing. Particularly interesting is that the measured values are reasonably close to the so-called tri-bimaximal mixing pattern, specified by $\sin^2\theta_{13} = 0$, $\sin^2\theta_{23} = 1/2$ and $\sin^2\theta_{12} = 1/3$. See figures 1 and 2 for a comparison of the data and the tri-bimaximal mixing pattern.

Figure 1. Pie charts showing the flavor content of the neutrino mass eigenstates according to the central values of the mixing angles [1].

Figure 2. Pie charts showing the flavor content of the neutrino mass eigenstates according to the tri-bimaximal prediction.

Obviously, the similarity of the data and this mixing pattern can be purely coincidental. Most of the parameters in the Standard Model (SM) are related to the fermion mass sector and it is well possible that their values are not restricted to satisfy some underlying principle, but just happen to be close to some special values. However, it might be more interesting to see if the patterns can really be explained. The answer is that they can, but the price to pay is the introduction of quite a large sector of new physics.

2. The pros and cons of flavons

Many models have been constructed to reproduce the tri-bimaximal mixing pattern; for a review, see e.g. [2]. In these models, it is assumed that there is a horizontal symmetry next to the 'vertical' gauge symmetry group of the Standard Model. Often discrete, finite groups are used and in particular the group A_4, the symmetry group of the tetrahedron, is popular.

The typical idea is to 'charge' the fermions of the different families under the flavour symmetry. The usual mass terms (Yukawa couplings and/or Majorana mass terms) are not invariant and invariance can only be restored by adding new Higgs-like fields to the theory. These scalars are called flavons; they transform trivially under (vertical) gauge transformations, but are in extended representations of the flavour symmetry. The flavons acquire vacuum expectation values (vevs) and the structure of the vevs translates to structures in the fermion mass matrices.

The idea of using flavons is quite attractive and it is possible to build models that reproduce the tri-bimaximal (or equivalent) mixing patterns, without having to tune parameters. However, three weaker points of the approach should also be stressed. Firstly, models can be quite baroque. In some set-ups, up to 10 new flavon fields have to be introduced. Some couple to neutrinos, others mostly to charged leptons or quarks, while still others don't even couple to the SM fermions directly, but are needed to help the other flavons obtain the correct vevs. Secondly, all the known techniques for flavon alignment in case of more than one flavon require additional new physics like supersymmetry or extra dimensions. There are good arguments to use these, but it would be preferable not to be dependent on them and be to able to write down more minimal models. Thirdly, the scale of the physics of the flavons is typically quite high (10^{10} GeV or higher). This has the advantage that it is possible to build models that combine flavour and grand unified symmetries that are typically also assumed to be related to that scale. The downside is that the theories are hard to test and few direct predictions can be made.

In the rest of this paper, we will discuss an alternative possibility [3, 4]. We assume that there are no separate flavons, but that the Standard Model Higgs field takes their role. In this set-up, there are three copies of the SM Higgs that are in a triplet of the flavour group. Once the Higgses get their vevs, the resulting 'vector of vevs' acts as the flavons. In this set-up, the three weak points described above are ameliorated. Although there is a substantial amount of new fields (two extra copies of the Higgs field), this is far less than in the other models. The alignment of the Higgses is now possible in a non-supersymmetric context without extra dimensions. The scale of new physics is obviously the electroweak scale and the set-up is thus testable at the LHC. The disadvantage is that we only have one flavon. This limits the amount of structure we can impose on the fermions. For instance, to reproduce the exact tri-bimaximal mixing pattern, we need to tune one parameter (versus three in the Standard Model).

3. The Higgs potential

As mentioned in the previous section, we consider three copies of the SM Higgs in the triplet representation of the flavour group, which we choose to be A_4.

$$\vec{\Phi} = (\Phi_1, \Phi_2, \Phi_3) \tag{1}$$

In this case, the potential for the Higgses is relatively simpel and depends on one dimensionful parameter (μ^2), four dimensionless parameters (λ_i) and one phase (ϵ).

$$\begin{aligned} V &= \mu^2(\Phi_1^\dagger\Phi_1 + \Phi_2^\dagger\Phi_2 + \Phi_3^\dagger\Phi_3) + \lambda_1(\Phi_1^\dagger\Phi_1 + \Phi_2^\dagger\Phi_2 + \Phi_3^\dagger\Phi_3)^2 \\ &+ \lambda_3(\Phi_1^\dagger\Phi_1\Phi_2^\dagger\Phi_2 + \Phi_1^\dagger\Phi_1\Phi_3^\dagger\Phi_3 + \Phi_2^\dagger\Phi_2\Phi_3^\dagger\Phi_3) + \lambda_4(\Phi_1^\dagger\Phi_2\Phi_2^\dagger\Phi_1 + \Phi_1^\dagger\Phi_3\Phi_3^\dagger\Phi_1 + \Phi_2^\dagger\Phi_3\Phi_3^\dagger\Phi_2) \\ &+ \frac{\lambda_5}{2}\left[e^{i\epsilon}[(\Phi_1^\dagger\Phi_2)^2 + (\Phi_2^\dagger\Phi_3)^2 + (\Phi_3^\dagger\Phi_1)^2] + e^{-i\epsilon}[(\Phi_2^\dagger\Phi_1)^2 + (\Phi_3^\dagger\Phi_2)^2 + (\Phi_1^\dagger\Phi_3)^2]\right]. \end{aligned} \tag{2}$$

In [3] we performed a systematic analysis of the possible minima of this potential. We found four stable structures. Two of them have all vacuum expectation values real:

- (v, v, v),

Journal of Physics: Conference Series **335** (2011) 012030 doi:10.1088/1742-6596/335/1/012030

- $(v, 0, 0)$.

Two others have complex vevs and thus provide an extra possibility for CP violation

- $(ve^{i\omega}, rv, 0)$
- $(ve^{i\omega}, ve^{-i\omega}, rv)$.

4. The physical Higgs fields

Our set-up is a three Higgs doublet model. This implies that there are more Higgs bosons than the one of the Standard Model. Of the three charged components of the Higgs fields, one linear combination becomes the (complex) Goldstone boson responsible for the $W^{\pm}$ masses, while two others are charged Higgs fields $H^{\pm}_{1,2}$. In the neutral sector, one linear combination of the six real components is a Goldstone boson related to the mass of the Z boson. Five others are physical Higgs fields $H^{(0)}_{1-5}$. In the CP conserving case, three of them are scalars and the two others are pseudoscalars. In the CP violating case, it is not possible to identify separate scalars and pseudoscalars.

5. Constraining minima of the potential

Not all choices of parameters that lead to the vev structures of the previous section give realistic models. In [3] we consider a number of 'tests' for the viability of a model that are model independent in the sense that they depend on the A_4 representation of the Higgs fields (namely a triplet), but not on the representations of the fermions in the theory.

Firstly, we demand m^2 to be positive for all two charged and five neutral Higgses. Secondly, we consider the tree level unitarity constraints coming from the additional scalars present in the theory. Thirdly, we study the decay of Z and $W^{\pm}$ bosons. Additional Higgses open up new decay channels that cannot give too large contributions. Lastly, we consider the effect of the extra Higgses on the oblique parameters.

In [4] we continue the analysis in a model dependent way. Once the A_4 representations of the fermions are given, it is possible to study rare decays. The additional Higgses need not be flavour conserving and thus rare decays like $\mu^- \to e^- e^- e^+$ or $\tau^- \to \mu^- \mu^- e^+$ are possible at tree level, while they are very constrained in the Standard Model. The strong experimental bounds on these processes can be used to restrict the parameter space. In the case of quarks, the additional Higgs fields also mediate meson-antimeson oscillations via tree diagrams instead of box diagrams as in the SM. See also figure 3.

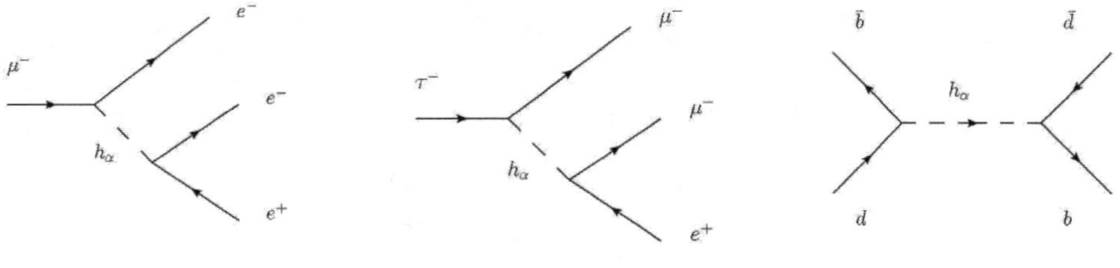

Figure 3. Feynman diagrams for charged lepton decays (left: $\mu^- \to e^- e^- e^+$; middle: $\tau^- \to \mu^- \mu^- e^+$) and meson oscillations (right: $B \leftrightarrow \bar{B}$) that are possible at tree level due to the extra Higgs bosons.

In the next two sections, we will consider two of the allowed minima in more detail. These relate to three models in the literature, [5] for the vacuum (v, v, v) and [6] (focussing on leptons) and [7] (focussing on quarks) for the vacuum $(ve^{i\omega}, ve^{-i\omega}, rv)$ and permutations.

6. The vacuum (v, v, v)

We first study the CP conserving vacuum (v, v, v) [5], in which case A_4 is not completely broken, but leaves a residual Z_3 symmetry. In the neutral Higgs sector, there is one boson that is like the SM Higgs in the sense that it is flavour conserving. The four other neutral bosons come as a pair of degenerate scalars and a pair of degenerate pseudoscalars. As can be seen in figure 4 there are quite a lot of points in parameter space that satisfy the model independent bounds. Yellow points indicate that all masses are positive, but that subsequent tests fail. If also unitarity constraints are met, points are blue. Green points indicate that also the Z- and W decay tests are passed and points that satisfy all constraints, including those from the oblique parameters are red.

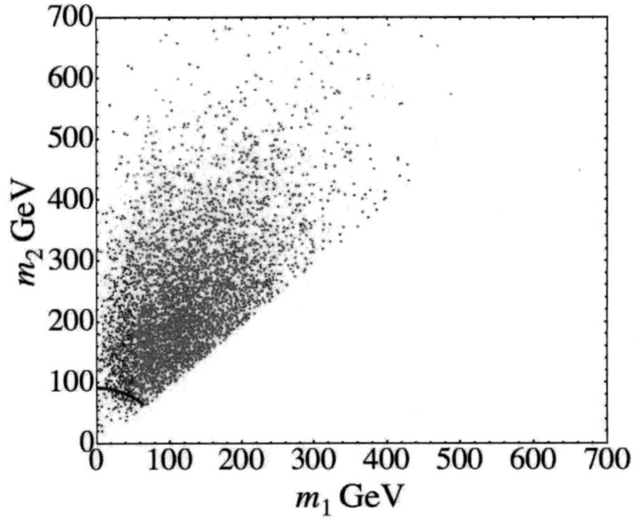

Figure 4. A slice of parameter space for the model of section 6. m_1 and m_2 are respectively the lightest and the second lightest Higgs masses. All masses ≥ 0: yellow points; unitarity OK: blue points; Z and W decay OK: green points; oblique parameters OK: red points.

Due to the residual symmetry, many rare decays are still forbidden in this model. Of the decays we studied, only $\tau^- \to \mu^-\mu^-e^+$ is allowed. Its branching ratio is below experimental bounds for most values of the lightest Higgs mass as can be seen in figure 5

We conclude that this set-up provides some interesting physics and is not too much constrained by low-energy observables.

7. The vacuum $(ve^{i\omega}, ve^{-i\omega}, rv)$

We continue to study the CP violating vacuum solution $(ve^{i\omega}, ve^{-i\omega}, rv)$, that was used in [6] and [7] to model respectively the lepton sector and the quark sector. As opposed to the previous case, there is no residual symmetry as A_4 is fully broken. Furthermore, the fact that two of the vevs are complex, shows that CP is broken in the Higgs sector.

We find that, although the minimum is a genuine minimum of the potential (2), there are very few points in parameter space that have all masses non-negative. See figure 6, where we

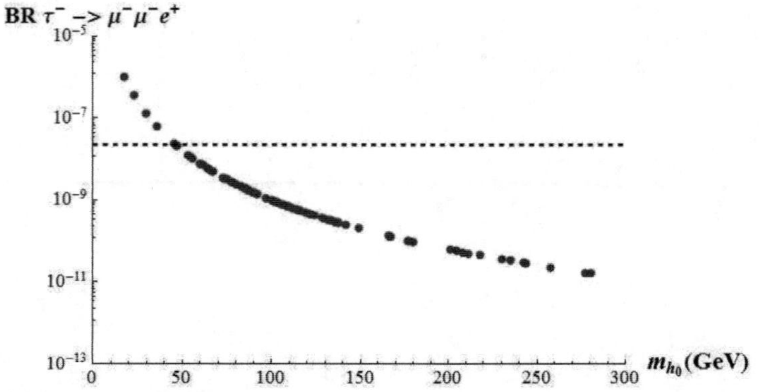

Figure 5. The branching ratio for the tau decay of figure 3 as a function of the lightest Higgs mass. The dashed line gives the experimental upper bound.

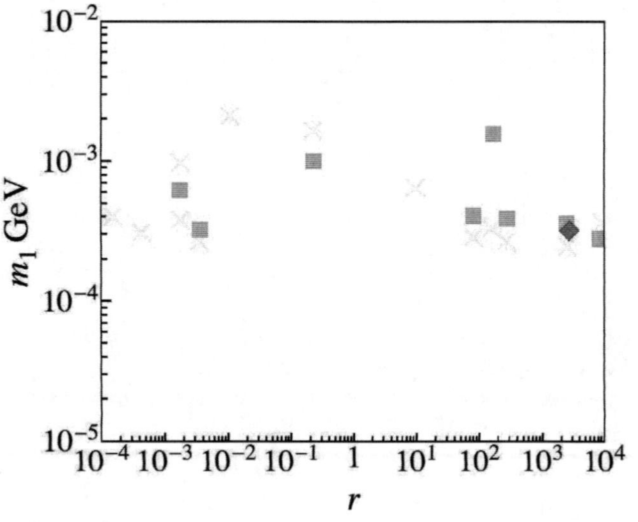

Figure 6. A slice of parameter space for the model 7 showing the r parameter versus the lightest Higgs mass. The colour coding is the same as in figure 4.

plot the r parameter that sets the ratio between the magnitude of one of the vevs and the other two against the lightest Higgs mass. We see that there are very few points that pass even the first test and even those points have unreasonably light Higgses.

We can partially solve this by adding a small term to the potential that softly breaks the A_4 symmetry

$$V_{A_4soft} = v_w^2 \frac{m}{2}(\phi_1^\dagger \phi_2 + \phi_2^\dagger \phi_1) + v_w^2 \frac{n}{2}(\phi_2^\dagger \phi_3 + \phi_3^\dagger \phi_2) + v_w^2 \frac{k}{2}(\phi_1^\dagger \phi_3 + \phi_3^\dagger \phi_1)\,, \tag{3}$$

With these terms added, more points are in the phenomenologically viable range as shown in figures 7, although they remain rather few.

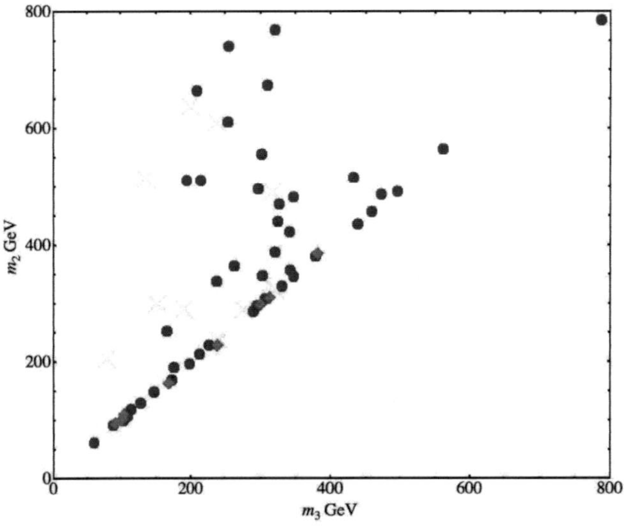

Figure 7. A slice of parameter space for the model of section 7 showing the lightest Higgs mass versus the next-to lightest Higgs mass. The colour coding is as in figure 4

As there are no remaining symmetries, many fermion number violating processes can indeed occur at tree level. We consider for instance the process $\mu^- \to e^- e^- e^+$ in the model of [6] or ΔM_{B_d} in $B_d - \bar{B}_d$ oscillations in the model of [7]. As is shown in figures 8 and 9, the current experimental bounds strongly limit the validity of these models.

8. Conclusions

In conclusion, we state that the use of flavour symmetries is an interesting tool to explain the apparent structures in the fermion mass sector. The flavour symmetries can be realized with or without the use of separate flavon fields. If we assume these to be absent, relatively minimal models can be constructed. These models have multiple copies of the Standard Model Higgs field, where the copies are related by the flavour symmetry. We analysed the case where the flavour symmetry is A_4 and only a few minima configurations are allowed by the potential. We showed that low energy observables can strongly restrict some of the possible models.

References

[1] M. C. Gonzalez-Garcia, Nucl. Phys. **A827** (2009) 5C-14C. [arXiv:0901.2505 [hep-ph]].
[2] G. Altarelli and F. Feruglio, Rev. Mod. Phys. **82** (2010) 2701 [arXiv:1002.0211 [hep-ph]].
[3] R. de Adelhart Toorop, F. Bazzocchi, L. Merlo and A. Paris, [arXiv:1012.1791 [hep-ph]].
[4] R. de Adelhart Toorop, F. Bazzocchi, L. Merlo and A. Paris, [arXiv:1012.2091 [hep-ph]].
[5] E. Ma, G. Rajasekaran, Phys. Rev. **D64** (2001) 113012. [hep-ph/0106291].
[6] S. Morisi, E. Peinado, Phys. Rev. **D80** (2009) 113011. [arXiv:0910.4389 [hep-ph]].
[7] L. Lavoura, H. Kuhbock, Eur. Phys. J. **C55** (2008) 303-308. [arXiv:0711.0670 [hep-ph]].

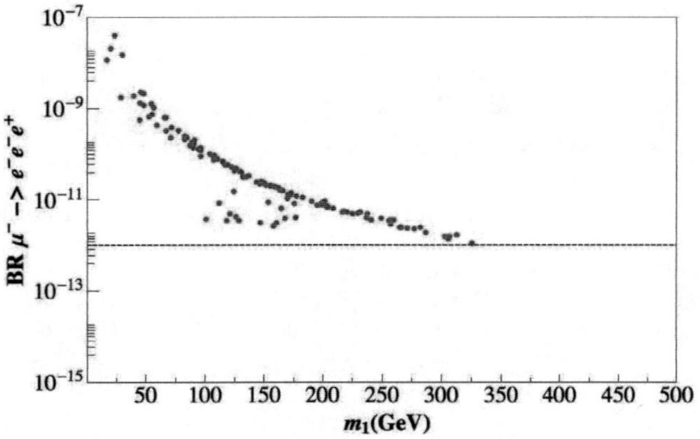

Figure 8. The branching ratio for the muon decay into two electrons and a positron in the model of [6] versus the lightest Higgs mass. The dashed line shows the experimental upper bound.

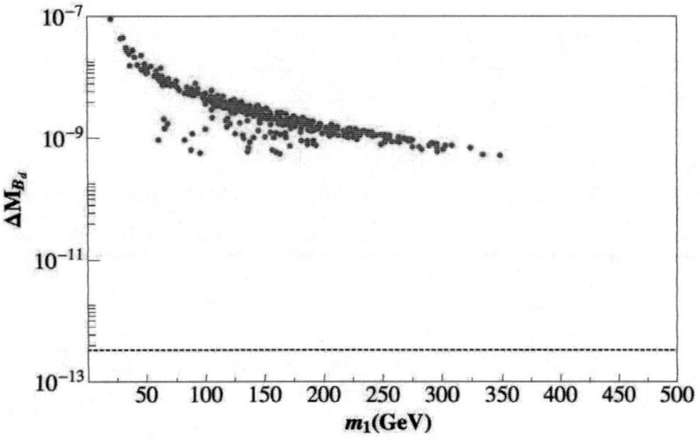

Figure 9. ΔM_{B_d} in the model of [7] versus the lightest Higgs mass. The dashed line shows the experimental value.

Nearest Neighbour Interactions and Symmetries

C Simões

CFTP,
Departamento de Física, Instituto Superior Técnico,
Avenida Rovisco Pais 1, 1049-001 Lisboa, Portugal

E-mail: csimoes@cftp.ist.utl.pt

Abstract. We study the minimal possible deviations from the Hermitic Nearest-Neighbour Interations (NNI) texture in the quark sector such that it is possible to accommodate the actual experimental data. We also show that the NNI structure can be obtained through the implementation of an Abelian discrete flavour symmetry at the Lagrangian level, where the minimal realisation is Z_4, requiring at least the presence of two Higgs doublets. Finally, we explore the consequences on the leptonic sector of this Z_4 flavour symmetry, in the context of $SU(5)$ Grand Unification with the standard fermionic content plus three right-handed neutrinos and two Higgs quintets.

1. Introduction

A huge effort has been made in the last decades in understanding the pattern of fermion masses and mixings. In the Standard Model, the Yukawa interaction terms that describe the fermion masses are unexplained. Many attempts have been made to find a framework where the fermion masses and mixings can be explained. One possible approach is to impose some zeroes on the mass matrix elements, for instance the well known Fritzsch Ansatz [1]. Such Ansatz combine the Nearest-Neighbour Interaction (NNI) [2] form with the Hermiticity condition. The NNI structure, where entries $(1,1)$, $(1,3)$, $(2,2)$ and $(3,1)$ vanish, can always be achieved for quark mass matrices M_u, M_d in the SM [2] through a weak basis transformation. However, it is no longer true in other contexts as is the case of two-Higgs doublet model. One can find many works in literature where a symmetry leads to the NNI structure, for example [6, 7] or [3] where a Z_4 flavour symmetry was implemented in the context of two Higgs doublets. Some of those models are based on Grand Unification Theories (GUT), which are very appealing to implement flavour symmetries once the SM fermions are unified in large multiplets. For example [8, 5] models based on the $SU(5)$ gauge group [9].

This work is organised as follows. In section 2 we study the minimal deviations from the Hermitic NNI structure such that both up- and down-quark mass matrices accommodate the actual quark data. In section 3 we implemet an Abelian discrete flavour symmetry that leads to quark mass matrices in the NNI form, where the minimal realisation in the context of two Higgs doublets is Z_4. This Z_4 flavour symmetry is implemented in the context of the $SU(5)$ GUT model in section 4 where the leptonic sector is analysed.

2. Minimal Deviations from Hermiticity: Quark Sector

In this section, we study the minimal deviations from the Hermitic NNI structure such that it is possible to accommodate the experimental data on both up- and down-quark mass matrices, M_u, M_d.

In the NNI basis, the quark mass matrices can be written as,

$$M_u = \begin{pmatrix} 0 & A_u & 0 \\ A'_u & 0 & B_u \\ 0 & B'_u & C_u \end{pmatrix}, \qquad M_d = K_q \begin{pmatrix} 0 & A_d & 0 \\ A'_d & 0 & B_d \\ 0 & B'_d & C_d \end{pmatrix}, \tag{1}$$

where $(A,\, A',\, B,\, B',\, C)_{u,d}$ are taken real without loss of generality and K is a diagonal phase matrix, $K_q = \mathrm{diag}(e^{i\kappa_1}, e^{i\kappa_2}, 1)$.

Since we are interested in determining the minimal possible deviations from the Hermitic NNI structure, we introduce the parameters ϵ_a^u, ϵ_b^u, ϵ_a^d and ϵ_b^d,

$$\epsilon_a^{u,d} \equiv \frac{A'_{u,d} - A_{u,d}}{A'_{u,d} + A_{u,d}}, \qquad \epsilon_b^{u,d} \equiv \frac{B'_{u,d} - B_{u,d}}{B'_{u,d} + B_{u,d}}. \tag{2}$$

To better determine the minimal values of $\epsilon_{a,b}^{u,d}$ we will work with the Hermitian matrices, H_u, H_d, defined as $H_{u,d} = M_{u,d}\, M_{u,d}^\dagger$. The complex H_d matrix can be written as,

$$H_d = K\, H_d^0\, K^\dagger\,, \tag{3}$$

where H_d^0 is a real matrix and the phases κ_1, κ_2 are given by

$$\kappa_1 = arg(H_{d_{13}})\,, \qquad \kappa_1 = arg(H_{d_{23}})\,. \tag{4}$$

With H_u and H_d^0 being Hermitian real matrices they can be diagonalised by real orthogonal matrices O_u and O_d respectively as,

$$H_u = O_u\; \mathrm{diag}(m_u^2, m_c^2, m_t^2)\, O_u^\top\,, \qquad H_d^0 = O_d\; \mathrm{diag}(m_d^2, m_s^2, m_b^2)\, O_d^\top\,, \tag{5}$$

then the quark mixing matrix, the so-called Cabibbo-Kobayashi-Maskawa (CKM) matrix [10, 11], V, is given by,

$$V = O_u^\top\, K\, O_d\,. \tag{6}$$

The deviations $\epsilon_{a,b}^{u,d}$ of Eq. (2) can be related to the quark masses through the invariants of H_u and H_d (see Ref. [3]), hence if the deviations $\epsilon_{a,b}^{u,d}$, the quark masses and the phases κ_1, κ_2 are given, the matrices M_u, M_d and consequently H_u, H_d can be fully reconstructed.

The CKM matrix is computed by Eq. (6) after diagonalization of H_u and H_d^0.

In our numerical calculations, we have performed a scan of the deviations $\epsilon_{a,b}^{u,d}$, the phases κ_1, κ_2 and the running quark masses at the electroweak scale within their errors. We have accepted solutions that lead to a CKM matrix in agreement with the actual data [12] of the CKM moduli, the angles of the unitary triangle and the strength of CP violation I. We sketched on Table 1 these input data together with the running quark masses.

To measure the global deviation from the Hermiticity of the pair of quark mass matrices, M_u, M_d, we defined the parameter ε_q,

$$\varepsilon_q \equiv \frac{1}{2}\sqrt{(\epsilon_a^u)^2 + (\epsilon_b^u)^2 + (\epsilon_a^d)^2 + (\epsilon_b^d)^2}\,, \tag{7}$$

Table 1. Values of the running up- and down-quark masses, CKM element moduli $|V_{us}|$, $|V_{cb}|$, $|V_{ub}|$ and the angles β and γ of the unitarity triangle and the strength of CP violation I, at the scale $M_Z = 91.1876 \pm 0.0021$ GeV [12]. The quark masses are calculated [13] to M_Z scale through the renormalisation group equations for QCD in the $\overline{MS}$ [14, 15, 16, 17, 18] scheme, at 4-loop level.

$m_u = 1.4 \pm 0.5\,\text{MeV}$	$m_c = 0.62^{+0.06}_{-0.07}\,\text{GeV}$	$m_t = 170.2 \pm 1.0\,\text{GeV}$
$m_d = 2.9 \pm 0.5\,\text{MeV}$	$m_s = 58^{+16}_{-12}\,\text{MeV}$	$m_b = 2.86^{+0.16}_{-0.06}\,\text{GeV}$
$\lvert V_{us}\rvert = 0.2253 \pm 0.0007$	$\lvert V_{ub}\rvert = \left(3.47^{+0.16}_{-0.12}\right) \times 10^{-3}$	
$\lvert V_{cb}\rvert = \left(41.0^{+1.1}_{-0.7}\right) \times 10^{-3}$	$\sin 2\beta = 0.673 \pm 0.023$	
$\gamma = \left(73^{+22}_{-25}\right)^\circ$	$I = \left(2.91^{+0.19}_{-0.11}\right) \times 10^{-5}$	

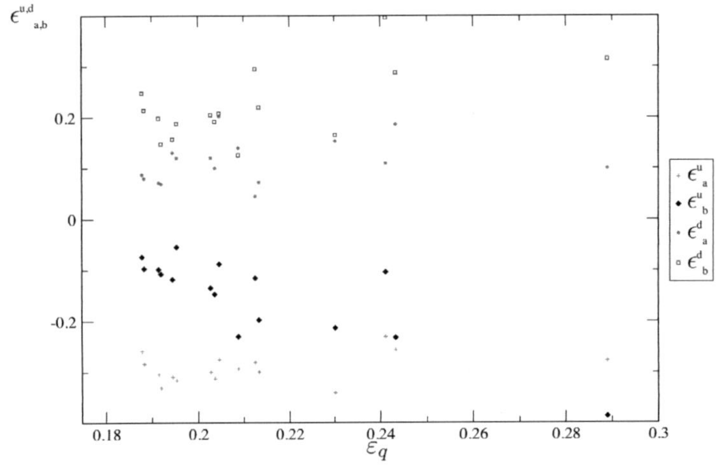

Figure 1. Set of solutions for $\epsilon^{u,d}_{a,b}$ corresponding to the constraint $\varepsilon_q \leq 0.3$.

The numerical results are shown in Fig. 1 where the deviations $\epsilon^{u,d}_{a,b}$ are plotted as a function of the parameter ε_q. From our search one sees that very small values of $\epsilon^{u,d}_{a,b}$ are not possible. Furthermore, if we consider $\epsilon_{a,b} = 0$ in one sector the deviations in the other become very large.

As the minimal global deviation is $\varepsilon_q = 0.188$ we conclude that within 20% it is possible to accomodate the actual experimental quark data on Hermitian NNI mass matrices. The values of a numerical example can be found in Ref. [3].

3. NNI from a Discrete Flavour Symmetry

In this section we show that it is possible to obtain the up- and down-quark mass matrices M_u, M_d in the NNI form through the implementation of an Abelian discrete flavour symmetry at the Lagrangian level. We shall show that the minimal realisation on cyclic groups is Z_4, requiring the presence of at least two Higgs doublets.

As we are interested in a minimal flavour symmetry, such that the NNI structure appears in the quark mass matrices, we restrict our search to Abelian and discrete symmetries and thus avoid the presence of Nambu-Goldstone bosons. We will search for symmetries of Z_n type. Under such symmetries a field, Ψ_j, with charge $\mathcal{Q}(\Psi_j)$ transforms as

$$\Psi_j \longrightarrow \Psi'_j = e^{i\,\frac{2\pi}{n}\mathcal{Q}(\Psi_j)}\,\Psi_j\,. \tag{8}$$

Let us start with the most general quark Yukawa Lagrangian,

$$-\mathcal{L}_Y = Y_u \overline{Q}_{L_i} u_{R_j} \Phi^u_{ij} + Y_d \overline{Q}_{L_i} d_{R_j} \Phi^d_{ij} + \text{H.c.}. \tag{9}$$

whereΦ^u_{ij} andΦ^d_{ij} correspond generically to the Higgs fields that couple to the (i,j)-entries of the up- and down-quark sector. To obtain the NNI form in the quark mass matrices, we must require zero charges for the trilinear matrix elements corresponding to non-zero mass matrix entries

$$\mathcal{Q}(\overline{Q}_{L_i} u_{R_j} \Phi^u_{ij}) = 0\,, \qquad \mathcal{Q}(\overline{Q}_{L_i} d_{R_j} \Phi^d_{ij}) = 0\,, \tag{10}$$

and non-zero charges for the trilinear matrix elements corresponding to zero mass matrix entries

$$\mathcal{Q}(\overline{Q}_{L_i} u_{R_j} \Phi^u_{ij}) \neq 0\,, \qquad \mathcal{Q}(\overline{Q}_{L_i} d_{R_j} \Phi^d_{ij}) \neq 0\,. \tag{11}$$

Writing down Eqs. (10) and (11) for all entries and solving for the charges of the Higgs doubletsΦ^u_{ij}, Φ^d_{ij}, we find that at least two Higgs doublets are requiredΦ_1 andΦ_2 with charges ϕ_1 and ϕ_2 respectively (see details in Ref. [3]). The quark field charges are then given by the following assignment,

$$\begin{aligned}
(q_1, q_2) &= (q_3 + \phi_1 - \phi_2, q_3 - \phi_1 + \phi_2)\,,\\
(u_1,\, u_2,\, u_3) &= (q_3 - \phi_1 + 2\phi_2, q_3 + \phi_1, q_3 + \phi_2)\,,\\
(d_1,\, d_2,\, d_3) &= (q_3 - 2\phi_1 + \phi_2, q_3 - \phi_2, q_3 - \phi_1)\,,
\end{aligned} \tag{12}$$

where $\mathcal{Q}(Q_{Li}) \equiv q_i$, $\mathcal{Q}(u_{Ri}) \equiv u_i$, $\mathcal{Q}(d_{Ri}) \equiv d_i$ with Q_{Li} the left-handed quark doublets.

Hence, the quark Yukawa Lagrangian in Eq. (9) becomes

$$-\mathcal{L}_Y = Y^1_u\, \overline{Q}_L \widetilde{\Phi}_1\, u_R + Y^2_u \overline{Q}_L \widetilde{\Phi}_2\, u_R + Y^1_d\, \overline{Q}_L \Phi_1\, d_R + Y^2_d\, \overline{Q}_L \Phi_2\, d_R + \text{H.c.}\,, \tag{13}$$

where $\widetilde{\Phi}_j \equiv i\,\sigma_2\,\Phi^*_j$ and $j = 1, 2$. After spontaneous symmetry breaking, the quark mass matrices are given by,

$$M_u = v_1^*\, Y_u^1 + v_2^*\, Y_u^2\,, \qquad M_d = v_1\, Y_d^1 + v_2\, Y_d^2\,, \tag{14}$$

taking precisely the NNI form, with $\langle \Phi_i \rangle \equiv v_i$ and $\langle \widetilde{\Phi_i} \rangle \equiv v_i^*$ and $Y^{1,2}_{u,d}$ the Yukawa matrices.

At this point we have found that two Higgs doublets are enough to achieve the NNI structure for both up- and down-quark mass matrices and the general Z_n quark charges are given by Eq. (12). Let us now find the minimal Z_n symmetry in which this implementation is possible.

From the bilinears, $\overline{Q}_{L_i} u_{R_j}$,

$$\begin{pmatrix} -2\phi_1 + 3\phi_2 & \phi_2 & -\phi_1 + 2\phi_2 \\ \phi_2 & 2\phi_1 - \phi_2 & \phi_1 \\ -\phi_1 + 2\phi_2 & \phi_1 & \phi_2 \end{pmatrix}, \tag{15}$$

and $\overline{Q}_{L_i} d_{R_j}$,

$$\begin{pmatrix} -3\phi_1 + 2\phi_2 & -\phi_1 & -2\phi_1 + \phi_2 \\ -\phi_1 & \phi_1 - 2\phi_2 & -\phi_2 \\ -2\phi_1 + \phi_2 & -\phi_2 & -\phi_1 \end{pmatrix} \tag{16}$$

one sees that neither Z_2 nor Z_3 are compatible with the NNI form for both up- and down-quark sectors. For instance, taking $(1,1)$-element of Eq. (15) one gets,

$$-2\phi_1 + 3\phi_2 = \phi_2 \pmod 2\,, \qquad -2\phi_1 + 3\phi_2 = \phi_1 \pmod 3\,, \tag{17}$$

as the up-quark sector couples with $\widetilde{\Phi}_1$ and $\widetilde{\Phi}_2$, where $\mathcal{Q}(\widetilde{\Phi}_i) = -\phi_i$; this entry becomes non-zero in the mass matrix, which is not desired. Hence, in the context of two Higgs doublets, the minimal symmetry that makes the NNI structure achievable is Z_4. In fact this conclusion is true even if the number of Higgs doublets is three or more.

4. NNI in the context of $SU(5)$

In this section we extend the Z_4 flavour symmetry, discussed in the last section, in the context of $SU(5)$ Grand Unification.

In the $SU(5)$ model the three generations of 10, 5^* fermionic multiplets are completely filled by the left-handed SM fields $(q, u^c, d^c, \ell, e^c)_i$, where $i = 1, 2, 3$ is the generation index,

$$\mathbf{10}_i = (q, u^c, e^c)_i \,, \quad \mathbf{5}^*_i = (\ell, d^c)_i \,, \tag{18}$$

Beyond the standard field content we have added, to this $SU(5) \times Z_4$ model, three right-handed neutrinos, ν_{Ri}, and two Higgs quintets. The $SU(5)$ fields assignment stated in Eq. (18) implies the follow charge relations

$$\mathcal{Q}(\mathbf{10}_i) = \mathcal{Q}(Q_{L_i}) = -\mathcal{Q}(u_{R_i}) = -\mathcal{Q}(e_{R_i}) \,, \tag{19a}$$

$$\mathcal{Q}(\mathbf{5}^*_i) = \mathcal{Q}(\ell_{L_i}) = -\mathcal{Q}(d_{R_i}) \,, \tag{19b}$$

from which one gets $\phi_2 = -2q_3$. The Z_4 charges of the $SU(5)$ multiplets are then given as:

$$\mathcal{Q}(\mathbf{10}) = (3q_3 + \phi_1,\, -q_3 - \phi_1,\, q_3) \,, \qquad \mathcal{Q}(\mathbf{5}^*) = (q_3 + 2\phi_1,\, -3q_3,\, -q_3 + \phi_1) \,. \tag{20}$$

Writting down the bilinear $\mathbf{10}_i\mathbf{10}_j$ and $\mathbf{10}_i\mathbf{5}^*_j$ one verifies that the NNI structure is achieved.

Let us now explore the leptonic sector. Due to the $SU(5)$ field assignments, Eq. (18), the charged lepton mass matrix, m_ℓ, gets the NNI form. However, since the right-handed neutrino fields, ν_{R_i}, are singlets under $SU(5)$, their Z_4 charges are free. Such freedom leads to a non parallel structure in the leptonic sector.

In this model, neutrinos acquire Majorana masses through type-I seesaw [19, 20, 21, 22]. The effective neutrino mass matrix, m_ν, can be computed from the seesaw formula, well approximated by:

$$m_\nu = -m_D\, M_R^{-1}\, m_D^\top \,, \tag{21}$$

where the symmetric Majorana mass matrix, M_R, is determined directly by the right-handed neutrino charges, ν_i, while the Dirac mass matrix, m_D, is determined in a similar way to the quark mass matrices M_u, M_d in Eq. (14). As mentioned before, the charged lepton mass matrix gets the NNI form and can be written as,

$$m_\ell = K_\ell^\dagger \begin{pmatrix} 0 & A_\ell & 0 \\ A'_\ell & 0 & B_\ell \\ 0 & B'_\ell & C_\ell \end{pmatrix} \,, \tag{22}$$

where the constants $(A,\, A',\, B,\, B',\, C)_\ell$ are taken to be real and positive and the diagonal phase matrix is $K_\ell = \mathrm{diag}(e^{i\sigma_1}, e^{i\sigma_2}, 1)$.

Performing a scan of the charges ϕ_1, q_3 and ν_i one finds that only six effective neutrino mass matrix textures are possible. From the counting of the number of parameters one finds that only two of those textures are physically viable. The remaining ones lead to small mixing angles, which is not acceptable. The two textures zeroes left,

$$a)\ \begin{pmatrix} 0 & A_\nu & 0 \\ A_\nu & B_\nu & C_\nu \\ 0 & C_\nu & D_\nu\, e^{i\varphi} \end{pmatrix} \,, \qquad b)\ \begin{pmatrix} A_\nu & B_\nu & C_\nu \\ B_\nu & 0 & 0 \\ C_\nu & 0 & D_\nu\, e^{i\varphi} \end{pmatrix} \,, \tag{23}$$

have a total of twelve parameters which are sufficient to explain the twelve physical ones (the six lepton masses, the three mixing angles, one Dirac phase and two Majorana phases).

The lepton mixing matrix or Pontecorvo-Maki-Nakagawa-Sakata (PMNS) matrix [23, 24, 25], U, is given by

$$U = O_\ell^\top \, K_\ell^\dagger \, U_\nu \,, \tag{24}$$

where U_ν is the matrix that diagonalise the effective neutrino mass matrix, a) and b) of Eq. (23). To determine the matrix O_ℓ that diagonalises the charged lepton mass matrix it is convenient to introduce the Hermitian mass matrix h_ℓ,

$$h_\ell = m_\ell \, m_\ell^\dagger \,. \tag{25}$$

Since the charged lepton mass matrix has NNI form, one can parameterise it in the same way as is done for quark sector, Eq. (2),

$$\epsilon_a^\ell \equiv \frac{A'_\ell - A_\ell}{A'_\ell + A_\ell}\,, \qquad \epsilon_b^\ell \equiv \frac{B'_\ell - B_\ell}{B'_\ell + B_\ell}, \tag{26}$$

and also define the global deviation parameter for the charged lepton mass matrix, ε_ℓ,

$$\varepsilon_\ell \equiv \frac{1}{2}\sqrt{\left(\epsilon_a^\ell\right)^2 + \left(\epsilon_b^\ell\right)^2}\,, \tag{27}$$

In our numeric procedure we have computed the PMNS matrix and taken the solutions

Table 2. Values [26] of the charged lepton masses, mixing angles θ_{13}, θ_{12} and θ_{23} and neutrino mass squared differences,$\Delta\ m_{21}^2$, $|\Delta m_{31}^2|$. The charged lepton masses are calculated [17, 18] at M_Z scale, $M_Z = 91.1876 \pm 0.0021\,\mathrm{GeV}$ [12], through the renormalisation group equations for QED in the $\overline{MS}$ scheme at 1-loop level.

$$m_e(M_Z) = 0.486661305 \pm 0.000000056 \text{ MeV}$$

$$m_\mu(M_Z) = 102.728989 \pm 0.000013 \text{ MeV}$$

$$m_\tau(M_Z) = 1746.28 \pm 0.16 \text{ MeV}$$

$$\Delta m_{21}^2 = \left(7.59_{-0.18}^{+0.23}\right) \times 10^{-5}\,\mathrm{eV}^2 \qquad \left|\Delta m_{31}^2\right| = \left(2.40_{-0.11}^{+0.12}\right) \times 10^{-3}\,\mathrm{eV}^2$$

$$\sin^2\theta_{12} = 0.318_{-0.016}^{+0.019} \qquad \sin^2\theta_{23} = 0.50_{-0.06}^{+0.07} \qquad \sin^2\theta_{13} < 0.035 \text{ at } 90\% \text{ C.L.}$$

that are in aggreement with the neutrino oscillation data [26]. We made a scan of all input parameters within their allowed range, namely the neutrino mass squared differences, $m_{21}^2\,, |\Delta m_{31}^2|$ ($\Delta m_{ij}^2 \equiv m_i^2 - m_j^2$), the lightest neutrino mass (m_1 for normal hierarchy (NH) or m_3 for inverted hierarchy (IH)), the phases ϕ,σ $_1$, σ_2, the charged lepton masses, the parameters ϵ_a^ℓ, ϵ_b^ℓ and D_ν is taken as a free parameter. The lightest neutrino mass matrix was scanned for different magnitudes below 2 eV. We sketch in Table 2 the experimental data used in the numerical procedure.

In addition to the restriction imposed by the neutrino oscillation data, one has to further consider the constraints on the effective Majorana mass, m_{ee}, proportional to the neutrinoless double beta decay amplitude [27], the constraint from Tritium β decay [12], m_{ν_e}, and the bound on the sum of light neutrino masses from cosmological and astrophysical data [28]. We found that an effective neutrino mass matrix of the form a) is only compatible with experimental data in the case of normal neutrino mass spectrum and b) only in the case of inverted neutrino mass

Table 3. Summary of the predictions for texture a) and b).

a)	b)
NH	NH
—	$\sin^2\theta_{13} > 0.010$
$0.0013\,\text{eV} \leq m_1 \leq 0.016\,\text{eV}$	$0.0042\,\text{eV} \leq m_3 \leq 0.011\,\text{eV}$
$6.4 \times 10^{-4}\,\text{eV} < \lvert m_{ee}\rvert < 2.2 \times 10^{-3}\,\text{eV}$	$0.015\,\text{eV} < \lvert m_{ee}\rvert < 0.022\,\text{eV}$
$\varepsilon_\ell > 0.0011$	$\varepsilon_\ell > 0.0013$

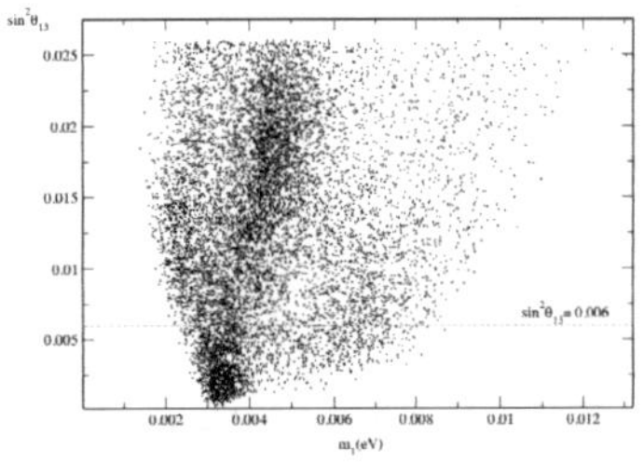

Figure 2. Plot of $\sin^2\theta_{13}$ over m_1 for texture a) and normal hierarchy.

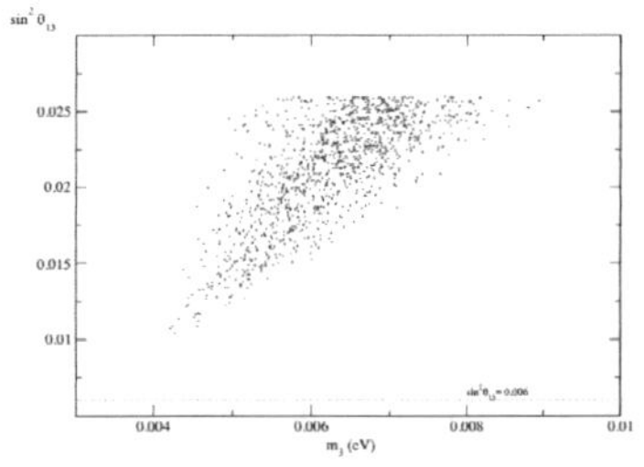

Figure 3. Plot of $\sin^2\theta_{13}$ over m_3 for texture b) and inverted hierarchy.

spectrum. The limits of the lightest neutrino mass for both cases are bounded. We plot in Figures 2 and 3 $\sin^2\theta_{13}$ as a function of the lightest neutrino mass m_1 for NH and m_3 for IH, respectively.

Our results are in agreement with the other constraints considered. We summarise in Table 3 the most important results of this $SU(5) \times Z_4$ GUT model, for more details see Ref. [5].

5. Conclusions

In this work we have studied the minimal deviations from the Hermitic NNI structure, such that the quark mass matrices M_u and M_d could accommodate the experimental quark data. It is shown in Figure 1 the deviations for both sectors as a function of the global deviation, ε_q. One sees that the deviations $\epsilon^{u,d}_{a,b}$ can not be very small, if they are zero for one sector we obtain large deviations in the other sector. It is possible to accommodate the actual quark data on mass matrices in the NNI form with a deviation of order of 20%.

We have shown also that it is possible to obtain the NNI structure for the both up- and down-quark mass matrices through the implementation of Z_4 flavour symmetry at the Lagrangian level in the context of two Higgs doublet model.

Then, with Grand Unification in mind, we extend this Z_4 flavour symmetry in the context of $SU(5)$, where we added right-handed neutrinos and two Higgs quintets. Due to the $SU(5)$ field assignments, the charged lepton mass matrix takes the NNI form while the effective neutrino mass matrix does not, leading to two viable textures a) and b) in Eq. (23). After analising the physical viability of those effective neutrino mass matrix textures we resume in Table 2 the relevant results.

Acknowledgments
We would like to thank the organizers of *DISCRETE'10 - Symposium on Prospects in the Physics of Discrete Symmetries* for the opportunity to present this work. We also thank David Emmanuel-Costa for helping to prepare the talk and the proceeding manuscript. This work was supported by Fundação para a Ciência e a Tecnologia (FCT, Portugal), under the contract SFRH/BD/61623/2009 and the projects CERN/FP/109305/2009 and CFTP-FCT Unit 777 partially funded through POCTI(FEDER) and by the Marie Curie Initial Training Network UNILHC PITN-GA-2009-237920.

References

[1] Fritzsch H 1978 *Phys. Lett. B* **73** 317–322
[2] Branco G C, Lavoura L and Mota F 1989 *Phys. Rev. D* **39** 3443
[3] Branco G, Emmanuel-Costa D and Simões C 2010 *Phys.Lett. B* **690** 62–67 (*Preprint* arXiv:1001.5065)
[4] Fritzsch H, Xing Z z and Zhou Y L 2011 (*Preprint* 1101.4272)
[5] Emmanuel-Costa D and Simões C 2011 * Submited to arXiv * (*Preprint* 1102.xxxx)
[6] Babu K S and Kubo J 2005 *Phys.Rev. D* **71** 056006 (*Preprint* hep-ph/0411226)
[7] Mondragon A 2006 *AIP Conf. Proc.* **857B** 266–282 (*Preprint* hep-ph/0609243)
[8] Ito T, Okamura N and Tanimoto M 1998 *Phys. Rev. D* **58** 077301 (*Preprint* hep-ph/9711454)
[9] Georgi H and Glashow S 1974 *Phys.Rev.Lett.* **32** 438–441
[10] Cabibbo N 1963 *Phys. Rev. Lett.* **10** 531–533
[11] Kobayashi M and Maskawa T 1973 *Prog. Theor. Phys.* **49** 652–657
[12] Nakamura K *et al.* (Particle Data Group) 2010 *J. Phys. G* **37** 075021
[13] Emmanuel-Costa D and Simões C 2009 *Phys. Rev. D* **79** 073006 (*Preprint* 0903.0564)
[14] Rodrigo G and Santamaria A 1993 *Phys.Lett. B* **313** 441–446 (*Preprint* hep-ph/9305305)
[15] Chetyrkin K, Kuhn J H and Kwiatkowski A 1996 *Phys.Rept.* **277** 189–281 revised version
[16] van Ritbergen T, Vermaseren J and Larin S 1997 *Phys.Lett. B* **400** 379–384 (*Preprint* hep-ph/9701390)
[17] Fusaoka H and Koide Y 1998 *Phys.Rev. D* **57** 3986–4001 (*Preprint* hep-ph/9712201)
[18] Xing Z z, Zhang H and Zhou S 2008 *Phys.Rev. D* **77** 113016 (*Preprint* 0712.1419)
[19] Minkowski P 1977 *Phys.Lett. B* **67** 421
[20] Yanagida T 1979
[21] Gell-Mann M, Ramond P and Slansky R To be published in Supergravity, P. van Nieuwenhuizen & D.Z. Freedman (eds.), North Holland Publ. Co., 1979
[22] Mohapatra R N and Senjanovic G 1980 *Phys.Rev.Lett.* **44** 912
[23] Pontecorvo B 1957 *Sov. Phys. JETP* **6** 429
[24] Pontecorvo B 1958 *Sov. Phys. JETP* **7** 172–173
[25] Maki Z, Nakagawa M and Sakata S 1962 *Prog. Theor. Phys.* **28** 870–880
[26] Schwetz T, Tortola M and Valle J W 2008 *New J.Phys.* **10** 113011 (*Preprint* arXiv:0808.2016)
[27] Pascoli S and Petcov S T 2004 *Phys. Lett. B* **580** 280–289 (*Preprint* hep-ph/0310003)
[28] Spergel D N *et al.* (WMAP) 2007 *Astrophys. J. Suppl.* **170** 377 (*Preprint* astro-ph/0603449)

Journal of Physics: Conference Series **335** (2011) 012032 doi:10.1088/1742-6596/335/1/012032

Semileptonic and Leptonic decays in the BABAR experiment

João Firmino da Costa[1]

Laboratoire de l'Acclrateur Linaire Universit Paris-Sud 11, 91898 Orsay Cedex, France

E-mail: joao.costa@cern.ch

Abstract. We describe recent measurements of B and charm semileptonic and leptonic decay branching fractions and the determination of CKM matrix elements.

1. Introduction

The study of semileptonic decays of charm and bottom mesons provides a good testing ground for QCD models. In the following we will summarize new and very competitive results from the BABAR collaboration for semileptonic decays both on the charm and bottom sectors.
In the charm sector the analysis $D^+ \to K^-\pi^+e^+\nu_e$ high statistics analysis provides very competitive results for several physical parameters that can be compared with theoretical predictions and calculations.
In the bottom sector, three analyses are presented which study $B \to X_u l^+\nu_l$ charmless decays of B mesons. The main goal of these analyses is the most accurate determination of the CKM matrix element $|V_{ub}|$. These analysis fall in two different reconstruction approaches whether the final hadronic state is inclusive or exclusive.
Finally a B leptonic decay analysis, $B^+ \to \tau^+\nu_\tau$, is also presented and provides a competitive measurement of the branching ratio for this decay. This last analysis can be used both as a probe for new physics and is also very sensitive to the CKM element $|V_{ub}|$.

2. $D^+ \to K^-\pi^+e^+\nu_e$[1]

A detailed study of the $D^+ \to K^-\pi^+e^+\nu_e$ is of interest for three main reasons :

- It allows measurements of the different $K\pi$ resonant and non-resonant amplitudes that contribute to this decay. In this respect, we have measured the S-wave contribution and searched for radially excited P-wave and D-wave components. Accurate measurements of the various contributions can serve as useful guidelines to B-meson semileptonic decays where there are still missing exclusive final states with mass higher than the D^* mass.
- High statistics in this decay allows accurate measurements of the properties of the $K^{*0}(892)$ meson. Both resonance parameters and hadronic transition form factors can be precisely measured. The latter can be compared with hadronic model expectations and Lattice QCD computations.

[1] Present address: Deutsches Elektronen-Synchrotron, Notkestrasse 85, 22607 Hamburg, Germany

Published under licence by IOP Publishing Ltd

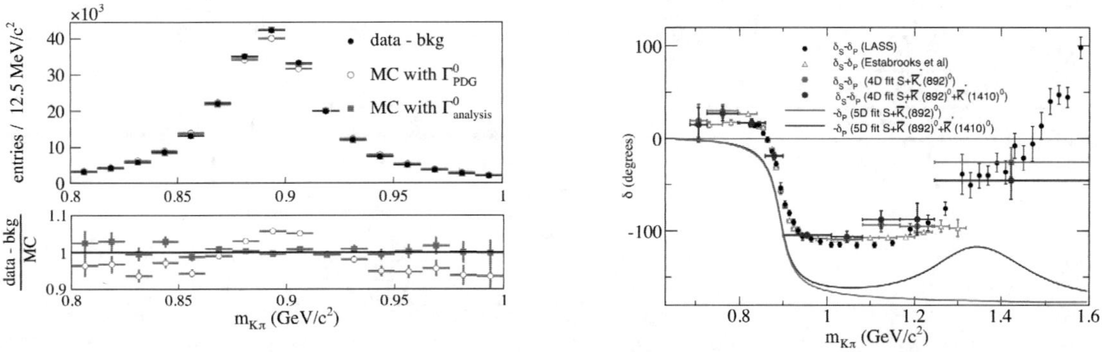

Figure 1. (color online) Left plot: Comparison between measured and fitted $K\pi$ mass distributions in the $\overline{K}^*(892)^0$ region. Results of a fit in which the width of the $\overline{K}^*(892)^0$ meson is fixed to 50.3 MeV/c^2 [2] are also given. Right plot :Difference between the $I = 1/2$ S- and P-wave phase versus the $K\pi$ mass. Measurements are similar whether or not the $\overline{K}^*(1410)^0$ is included in the P-wave parameterization. Results are compared with measurements from $K\pi$ scattering. The red and blue lines give the phase variation with a minus sign ($-\delta_P$) for the $\overline{K}^*(892)^0$ and $\overline{K}^*(892)^0 + \overline{K}^*(1410)^0$, respectively. The difference between these curves and the measured points corresponds to the S-wave contribution.

- Variation of the $K\pi S$ -wave phase versus the $K\pi$ mass can be determined, and compared with other experimental determinations.

Semileptonic decays into a final state with four particles depend on five variables : $q^2, m_{K\pi}$ and three angles ($\cos\theta_K, \cos\theta_e$ and χ)[1]. This allows to separate the different signal components according to angular momentum (S,P,D waves) and gives sensitivity to the interference between these states.
This analysis provides for the first time a full 5D phase space analysis of this decay.
Having reconstructed the charged $K^-\pi^+e^+$ system, the neutrino energy (and D direction) are finally estimated by taking into account the remaining particles in the event. This method provides a high efficiency and results in a non negligible amount of background.
The simulation of signal and background events is improved thanks to extensive studies on the fragmentation of charm mesons using special control samples and comparing data to MC simulation. The background sources ($B\bar{B}$ and $c\bar{c}$) are suppressed using multivariate discriminant variables which separate signal from background based on the topology of the decay ($B\bar{B}$) and the properties of the spectator system ($c\bar{c}$).

The main results are presented in Table 1 and Figure 1. In particular this analysis provides: 1) the most precise results for the hadronic form factor parameters of the $\overline{K}^{*0}$ component, assuming a single pole dominance model and these results serve as reference for LQCD, 2) very precise determinations of the resonance parameters of the $\overline{K}^{*0}$ component, 3) agreement with corresponding S-wave phase measurements done in K^-p interactions producing $K^-\pi^+$ at small transfer. This is a confirmation of these results and illustrates the importance of final state interactions when this phase is obtained from three body Dalitz plot analyses of D-meson hadronic decays and 4) precise determinations of the different branching ratios and very low

[1] For a precise definition of the angles, the reader is refered to [1]

Table 1. Comparison between these measurements and present world average results. Values for $\mathcal{B}(D^+ \to \overline{K}^*(1410)^0/\overline{K}^*_2(1430)^0 e^+\nu_e)$ are corrected for their respective branching fractions into $K^-\pi^+$.

Measured quantity	This analysis	World average [2]
$m_{K^*(892)^0}(\mathrm{MeV}/c^2)$	$895.4 \pm 0.2 \pm 0.2$	896.00 ± 0.25
$\Gamma^0_{K^*(892)^0}(\mathrm{MeV}/c^2)$	$46.5 \pm 0.3 \pm 0.2$	50.3 ± 0.6
$r_{BW}(\mathrm{GeV}/c)^{-1}$	$2.1 \pm 0.5 \pm 0.5$	2.72 ± 0.55
r_V	$1.463 \pm 0.017 \pm 0.031$	1.62 ± 0.08
r_2	$0.801 \pm 0.020 \pm 0.020$	0.83 ± 0.05
$m_A(\mathrm{GeV}/c^2)$	$2.63 \pm 0.10 \pm 0.13$	no result
$\mathcal{B}(D^+ \to K^-\pi^+e^+\nu_e)(\%)$	$4.04 \pm 0.03 \pm 0.04 \pm 0.09$	4.1 ± 0.6
$\mathcal{B}(D^+ \to K^-\pi^+e^+\nu_e)_{\overline{K}^*(892)^0}(\%)$	$3.80 \pm 0.04 \pm 0.05 \pm 0.09$	3.66 ± 0.21
$\mathcal{B}(D^+ \to K^-\pi^+e^+\nu_e)_{S-wave}(\%)$	$0.234 \pm 0.007 \pm 0.007 \pm 0.005$	0.21 ± 0.05
$\mathcal{B}(D^+ \to \overline{K}^*(1410)^0 e^+\nu_e)(\%)$	$0.30 \pm 0.12 \pm 0.18 \pm 0.06$ (< 0.6 at 90% C.L.)	
$\mathcal{B}(D^+ \to \overline{K}^*_2(1430)^0 e^+\nu_e)(\%)$	$0.023 \pm 0.011 \pm 0.011 \pm 0.001$ (< 0.05 at 90% C.L.)	

limits to high mass contributions.

3. $B^0 \to \pi^- l^+ \nu_l, B^+ \to \eta^{(\prime)} l^+ \nu_l$ and $B^{0/+} \to \pi^{-/0} l^+ \nu_l, B^{0/+} \to \rho^{-/0} l^+ \nu_l$

B meson semileptonic exclusive decays are excellent channels for the determination of $|V_{ub}|$ since they provide the best trade-off between statistics and theoretical knowledge of the decay.
The precise determination of $|V_{ub}|$ constrains the description of the Standard Model in the flavour sector. These channels allow also to test the QCD predictions for the form factors.
In the following, two analyses of B decays into semileptonic exclusive modes are presented, both extract $|V_{ub}|$ and test several theoretical models and calculations for the hadronic form factors variation with q^2 (the lepton system mass squared) :

- $\pi - \rho$ analysis[3], which studies four decay modes : $B^{0/+} \to \pi^{-/0} l^+ \nu_l, B^{0/+} \to \rho^{-/0} l^+ \nu_l$
- $\pi - \eta$ analysis[4], which studies three decay modes : $B^0 \to \pi^- l^+ \nu_l, B^+ \to \eta^{(\prime)} l^+ \nu_l$

The $\pi - \rho$ analysis also extracts $|V_{ub}|$ by combining their results on$\Delta\ \mathcal{B}/\Delta q^2$ with the results from the FNAL/MILC collaboration.
The measurement of the missing energy and momentum of the whole event are used to reconstruct the neutrino. This is done without explicitely reconstructing the other B meson, resulting in a high signal sample. The main source of background comes from $B \to X_c l^+ \nu_l$ decays which are much more frequent than $B \to X_u l^+ \nu_l$ decays ($|V_{cb}|^2/|V_{ub}|^2 \approx 50$). The signal is extracted by performing maximum-likelihood fits using the variables m_{ES} (the energy-substituted mass), ΔE and over each of the q^2 bins.

The fitted distributions of$\Delta\ \mathcal{B}/\Delta q^2$ are presented in Figure 2. One finds that the best agreement with data is found for the theoretical prediction coming from HPQCD. However, theory predictions are valid in restricted regions of phase space and their extrapolation on the entire phase space brings large uncertainties. Both analyses find good agreement in$\Delta\ \mathcal{B}/\Delta q^2$ for the common mode $B^0 \to \pi^- l^+ \nu_l$.
A summary of the results on V_{ub} for the two analyses is given in Table 2 using different theoretical models for parameterisation of the form factors. The expression that allows one to get

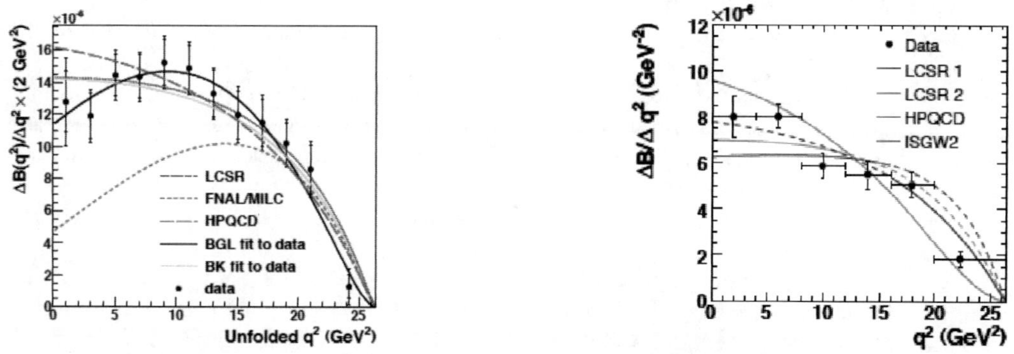

Figure 2. Fitted distributions ofΔ $\mathcal{B}/\Delta q^2$.

Table 2. Results found in $|V_{ub}|$ for both analyses, using different theoretical calculations. Units for $|V_{ub}|$ are (10^{-3})

q^2 (GeV^2)	$\|V_{ub}\|(\pi-\rho)$	$\|V_{ub}\|(\pi-\eta)$
$HPQCD > 16$	$3.21 \pm 0.17^{+0.55}_{-0.36}$	$3.24 \pm 0.13 \pm 0.16^{+0.57}_{-0.37}$
$FNAL > 16$		$3.14 \pm 0.12 \pm 0.16^{+0.35}_{-0.29}$
$LCSR < 12$	$3.78 \pm 0.13^{+0.55}_{-0.40}$	$3.70 \pm 0.07 \pm 0.09^{+0.54}_{-0.39}$

$|V_{ub}|$ fromΔ $\mathcal{B}/\Delta q^2$ is given in Equation 2.

The analysis $\pi-\rho$ makes a combined fit to LQCD and BABAR data. The fit is based on the z-expansion[1] (BGL) and takes advantage of the q^2 variation over the whole q^2 spectrum from the BABAR measurements and the shape plus normalization from LQCD. The determination of the q^2 shape is dominated by the BABAR data.
This procedure allows to reduce the theory error by a factor 2 (8.5% while traditional method gives $^{+17\%}_{-11\%}$). The value extracted for $|V_{ub}| = (2.95 \pm 0.31) \times 10^{-3}$.

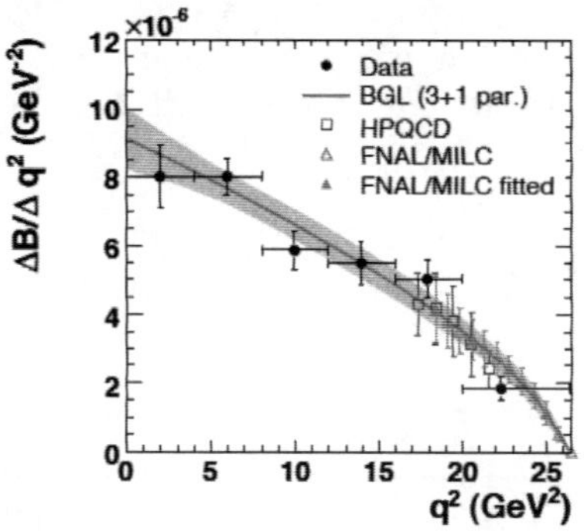

Figure 3. (color online) Simultaneous fits of the BGL parameterization to data (solid points with vertical error bars representing the total experimental uncertainties) and to four of the twelve points of the FNAL/MILC lattice prediction (magenta, closed triangles).The LQCD results are rescaled to the data according to the $|V_{ub}|$ value obtained in the fit. The shaded band illustrates the uncertainty of the fitted function. For comparison, the HPQCD (blue, open squares) lattice results are also shown.

[1] A summary of this expansion may be found in [3]

Journal of Physics: Conference Series **335** (2011) 012032 doi:10.1088/1742-6596/335/1/012032

4. $B \to X_u l^+ \nu_l$ with hadronic tags[5]

The main purpose of this analysis is to determine the channel's branching ratio and from this extract to $|V_{ub}|$.
However, and since there is a huge background from $b \to c$ decays, this type of analysis relies traditionally on hard cuts in the phase space to minimize the background. Therefore one measures partial branching ratios (PBR) and this fact entails that there is a dependence on the theoretical description of these PBR's to extract $|V_{ub}|$. Therefore such studies are done by applying independently various cuts on different phase space regions so to check the consistency of the theoretical description.
With the ever more greater knowledge about B semileptonic decays, the understanding of the background of this channel becomes solid and that permits this analysis to use only mild cuts of the phase space. They are $M_X < 1.55$ GeV/c^2, $M_X < 1.70$ GeV/c^2, $P_+ < 0.66$ GeV/c, $M_X < 1.70$ GeV/c^2 *and* $q^2 > 8$ GeV^2/c^2, $p_l^* > 1.0$ GeV/c, $p_l^* > 1.3$ GeV/c, where M_X stands for the X_u system mass, $P_+ = E_X - P_X$, q^2 the leptonic system mass squared and P_l^* the lepton momentum.
This analysis is an update of an earlier one and uses the entire BABAR $\Upsilon(4S)$ dataset. The event is reconstructed by fully measuring one B meson into hadronic decay modes (this is commonly known as hadronic tag method and this B meson is the B_{reco}). The B_{reco} decays into states DY, where D stands for $D^{(*)0\pm}$ and Y is a system of π and Kaons. The variables$\Delta\ E = E_B - \sqrt{s}/2$ and $m_{ES} = \sqrt{s/4 - p_B^2}$ are used to distinguish events where B_{reco} is well reconstructed from combinatorial background. Furthermore, the amount of combinatorial background is extracted by using an Argus function in the m_{ES} variable.
The remaining particles in the event, not included in B_{reco}, are associated to the X_u candidate system. A lepton of the correct charge tags the semileptonic events.
These events fall into two categories according to whether they pass or not a veto on kaons and D^* candidates. Events that pass the veto are signal enriched and are used in the fit to extract the PBR while the others are signal depleted and this sample is used to cross-check the agreement between data and Monte Carlo.
The PBR is obtained from the observed number of signal events in the kinematic regions considered, normalized to the total number of semileptonic decays, by using Equation 1,

$$\Delta R = \frac{\Delta\mathcal{B}(B \to X_u l^+ \nu_l)}{\mathcal{B}(B \to X l^+ \nu_l)} = \frac{N_u^{true}}{N_{sl}^{true}} \tag{1}$$

where N_u^{true} corresponds to the actual number of signal events, corrected for reconstrucion efficiency and particle identification, and N_{sl}^{true} stands for the total number of B semileptonic decays, also efficiency corrected, determined in a fit to the m_{ES} distribution at the level of the semileptonic tagging of the event.
To extract afterwards the PBF one uses the world average value $\mathcal{B}(B \to X l^+ \nu_l) = (10.75 \pm 0.15)\%$. The result found for the most inclusive fits is given in Table 3 and agreement is found with the Belle analysis.

Table 3. Partial Branching Ratios for most inclusive analyses

Variable, cutΔ	$\mathcal{B}(B \to X_u l^+ \nu_l)_{BABAR} \times 10^{-3}$	$\Delta\mathcal{B}(B \to X_u l^+ \nu_l)_{Belle} \times 10^{-3}$[6]
p_l^*, $p_l^* > 1.0\ GeV/c$	$1.80 \pm 0.13_{stat.} \pm 0.15_{syst.}$	-
(M_X, q^2), $p_l^* > 1.0\ GeV/c$	$1.76 \pm 0.16_{stat.} \pm 0.18_{syst}$	$1.96 \pm 0.17_{stat.} \pm 0.16_{syst.}$

The extraction of $|V_{ub}|$ is made using Equation 2

$$|V_{ub}| = \sqrt{\frac{\Delta\mathcal{B}(B \to X_u l^+ \nu_l)}{\tau_B \Delta\Gamma_{theory}}} \tag{2}$$

where $\Delta\Gamma_{theory}$ is the theoretical $B \to X_u l^+ \nu_l$ width according to the applied cut in phase space. This width depends on the theoretical model used and several models are checked. An estimate of $|V_{ub}|$ using the most inclusive analysis gives $(4.31 \pm 0.35) \times 10^{-3}$ where the signal model systematics are the biggest source of uncertainty.

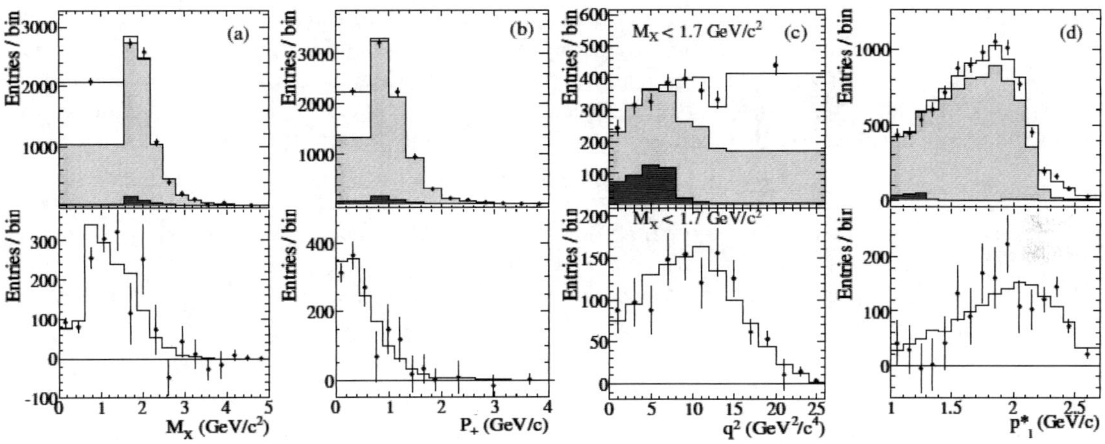

Figure 4. (color online) Upper row (data points): measured M_X (a), P_+ (b), q^2 with $M_X < 1.7GeV/c^2$ (c) and p_l^*(d) spectra. Results from the fit of three MC contributions are shown as histograms : : $B \to X_u l \nu_l$ decays generated inside (no shading) and outside (dark shading) the selected kinematic region, and $B \to X_c l \nu_l$ and other background (light shading). Lower row: corresponding spectra for $B \to X_u l \nu_l$ after $B \to X_c l \nu_l$ and other background subtraction; they have been rebinned in order to show the shape of the kinematic variables. Background-subtracted distributions are not efficiency corrected.

Furthermore, the charged and neutral B-meson decay modes are also studied separately and a search for weak-annihilation effects (isospin breaking effects) is done. It consists of measuring differences in the decay widths. This analysis finds no appreciable differences to the widths and subsequently limits are imposed to the fraction of weak-annihilation effects ($-0.13 < \gamma_{WA}/\Gamma < 0.09$ at $90\%CL$).

5. $B^+ \to \tau^+ \nu_\tau$ with hadronic tags[7]

This decay is, assuming only SM contributions, at tree level, sensitive to the decay constant of the B meson (f_B) and to $|V_{ub}|$. It is an helicity suppressed channel. Its branching ratio expression is given in Equation 3 and the expected value is $(1.2 \pm 0.2) \times 10^{-4}$, using [2] for the input parameters.

$$\mathcal{B}(B^+ \to \tau^+ \nu_\tau) = f_B^2 |V_{ub}|^2 \frac{G_F^2 m_B m_\tau^2}{8\pi} [1 - \frac{m_\tau^2}{m_B^2}]^2 \tau_{B^+} \tag{3}$$

This channel is also sensitive to *possible* extensions of the SM such as two-Higgs doublet models and minimal supersymmetry where the interaction would be mediated by a charged Higgs boson.

In this framework one can constrain the model's parameter space.
This analysis uses the so-called hadronic tags to reconstruct one of the B mesons while the signal decay mode is searched on the recoil of this fully reconstructed meson. This analysis is an update of an older BABAR analysis, using the entire BABAR $\Upsilon(4S)$ dataset.
One B meson is reconstructed in hadronic modes of type $(D^{(*)0}X^-, J/\psi X^-)$ where X^- is a system of light quark mesons (π, K) with total charge -1. Reconstructed events fall in two major categories, those where the former B meson is well reconstructed (peaking events) and those not well reconstructed (combinatorial). The variable $m_{ES} = \sqrt{s/4 - p_B^2}$ is used to evaluate the amount of combinatorial events under the peaking events distribution, a typical procedure in this type of analyses. Furthermore the actual suppression of combinatorial background is done using a likelihood ratio taking variables constructed from the kinematics of the tag B candidate. In the peaking events, one distinguishes between events actually coming from the signal decay channel and the background, using an unbinned maximum likelihood. It is found that the most important discriminating input variable is E_{extra}, the sum of the energies of the neutral clusters not associated with the tag B or with the signal. There is a minimum of 60 MeV for the energy requirement of these clusters for them to be taken into account. One uses this variable to define the probability density functions for the signal and background. Essentially, signal events tend to peak at low E_{extra} while background, with more neutral clusters in average, appear for higher values of this variable.

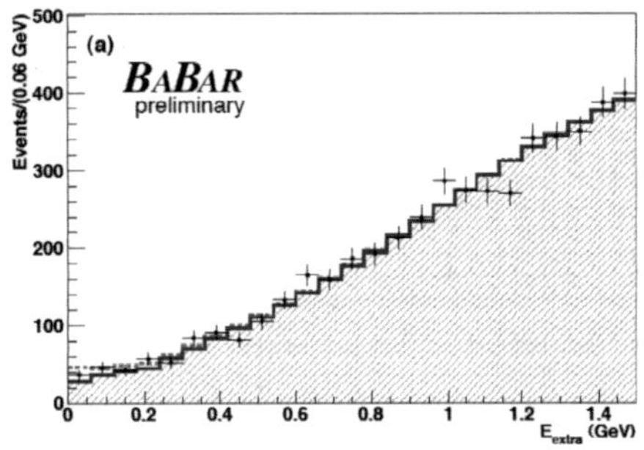

Figure 5. (color online) E_{extra} in data (dots with error bars) with all selection requirements applied and fit results overlaid. The hatched histogram is the background, the red dashed component is the signal.

The τ signal is reconstructed in 4 decay modes : $\tau \rightarrow (e\bar{\nu}_e\nu_\tau, \mu\bar{\nu}_\mu\nu_\tau, \pi\nu_\tau, \rho\nu_\tau)$. The four decay modes are fitted in parallel to extract the branching ratio of the signal decay channel. Both the signal and background PDF are taken from MC simulation. The signal PDF shape is corrected by using specially selected control samples, where both B mesons are reconstructed hadronically. This is done in data and MC. Differences in the shapes of the E_{extra} distributions between data and MC are taken as correction to the signal PDF. The resulting distributions after the fit are presented in Figure 5
This analysis finds that $\mathcal{B}(B^+ \rightarrow \tau^+\nu_\tau) = (1.80^{+0.57}_{-0.54}) \times 10^{-4}$, which corresponds to a statistical significance of 3.6 sigmas.
The dominant source of systematic originates from the background PDF due to the finite statistics of the simulation. The final error budget leads to an overall significance of this decay of 3.3 sigmas.
Combining this result with another BABAR analysis, done with semileptonic tags, leads to a branching ratio of $(1.76 \pm 0.49) \times 10^{-4}$, well in agreement with the $(1.2 \pm 0.2) \times 10^{-4}$ of the SM expectations.

6. Summary

We have presented new results for B and charm semileptonic and leptonic decays from the BABAR collaboration.

7. Acknowledgements

The author thanks the BABAR collaboration for the invitation, the DISCRETE conference organizers , and DESY for the travel arrangements.

References

[1] del Amo Sanchez P. et al. (BABAR Collaboration), *Preprint* arXiv:1012.1810, submitted to Phys. Rev. D
[2] C. Amsler et al. (Particle Data Group), Phys. Lett.B 667, 1 (2008) and 2009 partial update for the 2010 edition.
[3] del Amo Sanchez P. et al. (BABAR Collaboration), *Preprint* arXiv:1005.3288 [hep-ex], accepted by *Phys. Rev. D*
[4] del Amo Sanchez P. et al. (BABAR Collaboration) *Preprint* arXiv:1010.0987, submitted to *Phys. Rev. D*
[5] Shown at *ICHEP2010* in Paris
[6] P. Urquijo et al. [Belle Collaboration], *Phys. Rev. Lett.* 104, 021801 (2010)(*Preprint* arXiv:0907.0379 [hep-ex])
[7] del Amo Sanchez P. et al. (BABAR Collaboration), *Preprint* arXiv:1008.0104, submitted to *Phys. Rev. Lett.*

Journal of Physics: Conference Series **335** (2011) 012033 doi:10.1088/1742-6596/335/1/012033

First ADS analysis of $B^- \to D^0K^-$ decays in hadron collisions

Paola Garosi, on behalf of the CDF Collaboration
INFN Pisa, Largo B. Pontecorvo 3, 56127 Pisa, Italy
University of Siena, via Roma 56, 53100 Siena, Italy

E-mail: paola.garosi@pi.infn.it

Abstract. The CDF experiment reports the first measurement of branching fractions and CP-violating asymmetries of doubly-Cabibbo suppressed $B^- \to D^0K^-$ decays in hadron collisions, using the approach proposed by Atwood, Dunietz and Soni (ADS) to determine the CKM angle γ. Using 5.0 fb^{-1} of data the combined significance of both $B^- \to D^0\pi/K$ signals exceeds 5 sigma. First results in hadron collisions, obtained on 1.0 fb^{-1} of data using the method of Gronau, London and Wyler (GLW) for the Cabibbo suppressed modes are also reported. Both ADS and GLW parameters are determined with accuracy comparable with B factories measurements.

1. Introduction

The measurement of the CKM matrix elements plays a central role both to test the Standard Model consistency and to probe New Physics scenarios. In particular the complex phase of the CKM matrix leads to CP violation in weak processes. Conventionally, CP violating observables are written in terms of the angles α, β and γ of the "Unitarity Triangle", obtained from the unitarity condition of the CKM matrix [1]. While the resolution on α and β reached a good level of precision, the measurement of γ is still limited by the smallness of the branching ratios involved in the processes[2, 3, 4].
Written in terms of the CKM elements, γ is equal to $arg(-V_{ud}V^*_{ub}/V_{cd}V^*_{cb})$ and is related to the element V_{ub} though $V_{ub} = |V_{ub}|e^{-i\gamma}$ [5]. Among the various methods for the γ measurement, those which make use of the tree-level $B^- \to D^0K^-$ decays have the smallest theoretical uncertanties [6, 7, 8]. In fact γ appears as the relative weak phase between two amplitudes, the favored $b \to c\bar{u}s$ transition of the $B^- \to D^0K^-$, whose amplitude is proportional to $V_{cb}V_{us}$, and the color-suppressed $b \to u\bar{c}s$ transition of the $B^- \to \overline{D}^0K^-$, whose amplitude is proportional to $V_{ub}V_{cs}$. A schematic diagram is shown in Fig. 1. The interference between D^0 and $\overline{D}^0$, decaying into the same final state, leads to measurable CP violation effects, from which γ can be extracted. The effects can be also enhanced choosing the interfering amplitudes of the same order of magnitude.

According to the final state of the D^0 we can have the following methods:

- *GLW (Gronau-London-Wyler) method* [6, 9], which uses CP eigenstates of D^0, as $D^0_{CP+} \to K^+K^-, \pi^+\pi^-$ and $D^0_{CP-} \to K^0_s\pi^0, K^0_s\phi, K^0_s\omega$.
- *ADS (Atwood-Dunietz-Soni) method* [7, 10], which uses the doubly Cabibbo suppressed mode $D^0_{DCS} \to K^+\pi^-$.

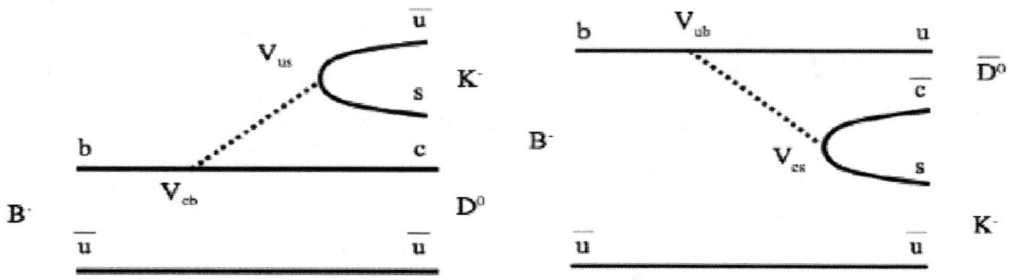

Figure 1. Diagrams contributing to $B \to DK$ modes. On the left the *color favored* transition, on the right the *color suppressed* transition.

- *GGSZ (or Dalitz) method* [8, 10], which uses three body decays of D^0, as $D^0 \to K_s^0\pi^+\pi^-$.

All mentioned methods require no tagging or time-dependent measurements, and many of them only involve charged particles in the final state. They are therefore particularly well-suited to analysis in a hadron collider environment, where the large production of B mesons can be well exploited. The use of specialized trigger based on online detection of secondary vertex (SVT trigger [11]) allows the selection of pure B meson samples.

We will describe in more details the ADS and GLW methods, for which CDF reports the first results in hadron collisions.

2. CDF II detector and trigger

The CDF experiment is located at the Tevatron, a $\sqrt{s} = 1.96$ TeV $p\bar{p}$ collider. The detector [12] is a multipurpose magnetic spectrometer surrounded by calorimeters and muon detectors. The most relevant for B-physics are the tracking, the particle-identification (PID) detectors and the trigger system.

The tracking system provides a determination of the decay point of particles with 15 μm resolution in the transverse plane using six layers of double-sided silicon-microstrip sensors at radii between 2.5 and 22 cm from the beam. A 96-layer drift chamber extending radially from 40 to 140 cm from the beam provides the reconstruction of three-dimensional charged-particles trajectories and excellent transverse momentum resolution, $\sigma_{p_T}/p_T^2 = 0.1\%$ Gev/c^2. Specific ionization measurements in the chamber allow 1.5σ separation between charged kaons and pions, approximately constant at momenta larger than 2 GeV/c.

A three-level trigger system [11] selects events enriched in decays of long-lived particles by exploiting the presence of displaced tracks in the event and measuring their impact parameter with offline-like 30 μm resolution. The trigger requires the presence of two charged particles with transverse momenta greater than 2 GeV/c, impact parameters greater than 100 microns and basic cuts on azimuthal separation and scalar sum of momenta.

3. The Atwood-Dunietz-Soni method

The ADS method [7, 10] takes into account the following decay channels: $B^- \to D^0K^-$ (*color favored*), with $D^0 \to K^+\pi^-$ (*doubly Cabibbo suppressed*) and $B^- \to \overline{D}^0K^-$ (*color suppressed*), with $\overline{D}^0 \to K^+\pi^-$ (*Cabibbo favored*).

The final state $[K^+\pi^-]_DK^-$ is the same and since D^0 and $\overline{D}^0$ are undistinguishable, the direct CP asymmetry can be measured. The interfering amplitudes are of the same order of magnitude, so large asymmetry effects are expected. For simplicity we will call "DCS" the final state $[K^+\pi^-]_D K^-$ and we will use the label $B \to D^0_{DCS}K$ to identify it.

The direct CP asymmetry

$$A_{ADS} = \frac{\mathcal{B}(B^- \to [K^+\pi^-]_D K^-) - \mathcal{B}(B^+ \to [K^-\pi^+]_D K^+)}{\mathcal{B}(B^- \to [K^+\pi^-]_D K^-) + \mathcal{B}(B^+ \to [K^-\pi^+]_D K^+)}$$

can be written in terms of the decay amplitudes and phases:

$$A_{ADS} = \frac{2r_B r_D \sin\gamma \sin(\delta_B + \delta_D)}{r_D^2 + r_B^2 + 2r_D r_B \cos\gamma \cos(\delta_B + \delta_D)},$$

where $r_B = |A(b \to u)/A(b \to c)|$, $\delta_B = arg[A(b \to u)/A(b \to c)]$ and r_D and δ_D are the corresponding amplitude ratio and strong phase difference of the D meson.

The denominator corresponds to another physical observable, the ratio between DCS and Cabibbo favored ("CF") events, the latter coming from the decay channel $B^- \to D^0K^-$ (*color favored*), with $D^0 \to K^-\pi^+$ (*Cabibbo favored*):

$$\begin{aligned} R_{ADS} &= r_D^2 + r_B^2 + 2r_D r_B \cos\gamma \cos(\delta_B + \delta_D) = \\ &= \frac{\mathcal{B}(B^- \to [K^+\pi^-]_D K^-) + \mathcal{B}(B^+ \to [K^-\pi^+]_D K^+)}{\mathcal{B}(B^- \to [K^-\pi^+]_D K^-) + \mathcal{B}(B^+ \to [K^+\pi^-]_D K^+)}. \end{aligned}$$

We can measure the corresponding quantities, A_{ADS} and R_{ADS}, also for the $B^- \to D^0\pi^-$ mode, for which sizeable asymmetries may be found [2].

The invariant mass distributions of CF and DCS modes, using a data sample of 5 fb^{-1} of data, with a nominal pion mass assignment to the track from B, are reported in Fig. 2.

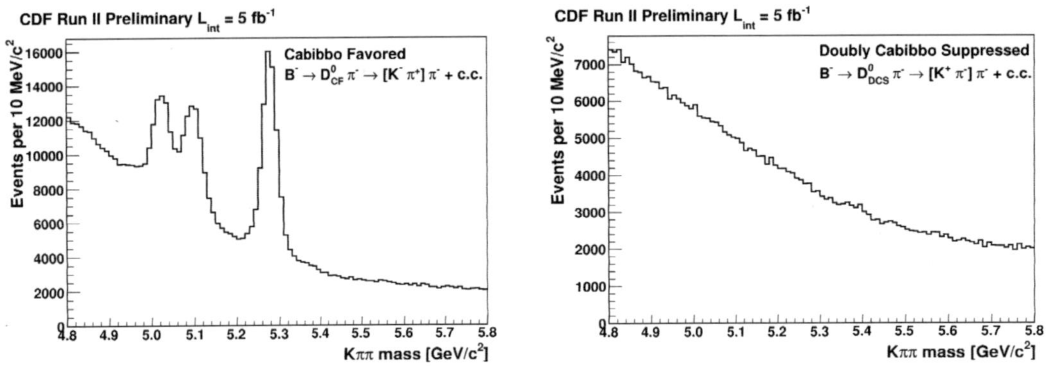

Figure 2. Invariant mass distributions of $B \to D^0h$ candidates for each reconstructed decay mode, Cabibbo favored on the left and doubly Cabibbo suppressed on the right. The pion mass is assigned to the track from the B decay.

A $B \to D^0\pi$ CF signal is visible at the correct mass of about 5.279 GeV/c^2. Events from $B \to D^0K$ decays are expected to form much smaller and wider peak, located about 50 MeV/c^2 below the $B \to D^0\pi$ peak.

The $B \to D^0\pi$ and $B \to D^0K$ DCS signals instead appear to be buried in the combinatorial background. For this reason an important point of this analysis is the suppression of the combinatorial background, obtained through a cuts optimization focused on finding a signal of the $B \to D_{DCS}\pi$ mode. Since the $B \to D_{CF}\pi$ mode has the same topology of the DCS one, but more statistic, we did the optimization using signal (S) and background (B) directly from CF data, choosing a set of cuts which maximize the figure of merit $S/(1.5+\sqrt{B})$ [13].

The *offline cut on the tridimensional vertex quality* χ^2_{3D} and the B *isolation* are powerful handles among the variables used in the optimization. The first exploit the 3D silicon-tracking to resolve multiple vertices along the beam direction and to reject fake tracks. It allows a background reduction by a factor of two and has small inefficiency on signal (less than 10%). The B isolation corresponds to the fraction of momentum carried by the B meson, which is usually greater than the momentum carried by lighter mesons. Another important cut is on the *decay lenght of the* D^0 *with respect to the* B, which allows to reject most of the $B \to hhh$ backgrounds, where h is either π or K. All variables and threshold values applied are described in [14].

The resulting invariant mass distributions of CF and DCS modes are reported in Fig. 3 where the combinatorial background is almost reduced to zero and an excess of events is now visible in the correct DCS signal mass window.

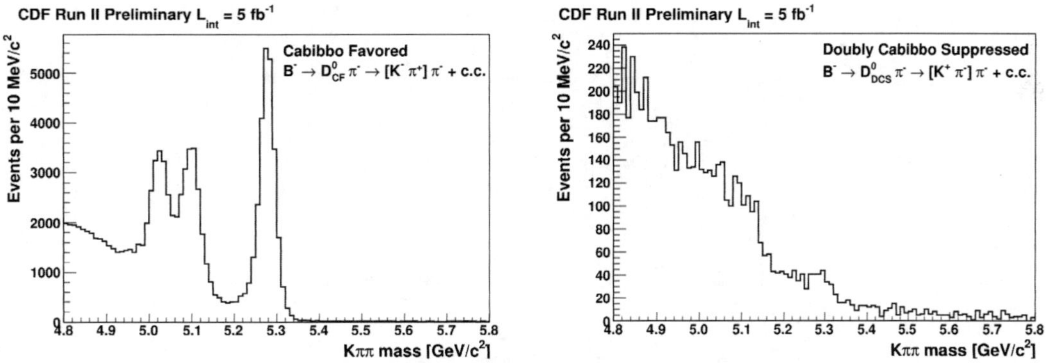

Figure 3. Invariant mass distributions of $B \to D^0 h$ candidates for each reconstructed decay mode, Cabibbo favored on the left and doubly Cabibbo suppressed on the right, after the cuts optimization. The pion mass is assigned to the track from the B decay.

An unbinned likelihood fit, exploiting mass and particle identification information provided by the specific ionization (dE/dx) in the CDF drift chamber, is performed [14] to separate the $B \to DK$ contributions from the $B \to D\pi$ signals and the combinatorial and physics backgrounds. The dominant physics backgrounds for the DCS mode are $B^- \to D^0\pi^-$, with $D^0 \to X$; $B^- \to D^0K^-$, with $D^0 \to X$; $B^- \to D^{0*}\pi^-$, with $D^{0*} \to D^0\pi^0/\gamma$; $B^- \to K^-\pi^+\pi^-$ and $B^0 \to D_0^{*-}e^+\nu_e$.
Fig. 4 shows the DCS invariant mass distributions separated in charge.
We obtained 34 ± 14 $B \to D_{DCS}K$ and 73 ± 16 $B \to D_{DCS}\pi$ signal events.
Since K^+ and K^- have a different probability of interaction in the detector, we evaluated the efficiency using a simulation sample and we corrected the fit results with this value.
The final results for the asymmetries are:

$$A_{ADS}(K) = -0.63 \pm 0.40(\text{stat}) \pm 0.23(\text{syst})$$
$$A_{ADS}(\pi) = 0.22 \pm 0.18(\text{stat}) \pm 0.06(\text{syst})$$

and for the ratios of doubly Cabibbo suppressed mode to flavor eigenstate:

$$R_{ADS}(K) = [22.5 \pm 8.4(\text{stat}) \pm 7.9(\text{syst})] \cdot 10^{-3}$$
$$R_{ADS}(\pi) = [4.1 \pm 0.8(\text{stat}) \pm 0.4(\text{syst})] \cdot 10^{-3}.$$

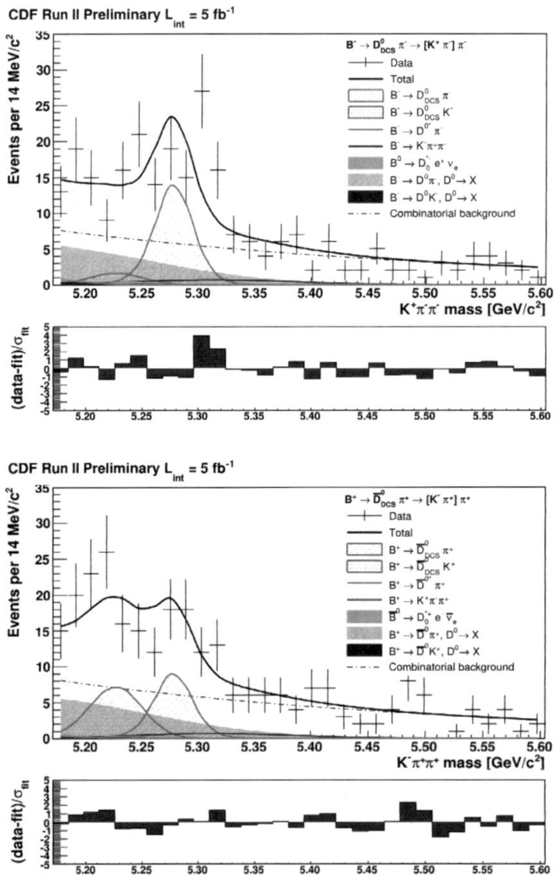

Figure 4. Invariant mass distributions of $B \to D^0_{DCS}h$ candidates for negative (top) and positive (bottom) charges. The pion mass is assigned to the track from the B decay. The projections of the likelihood fit are overlaid.

These quantities are measured for the first time in hadron collisions. The results are in agreement with existing measurements performed at $\Upsilon(4S)$ resonance [2, 4].

4. Gronau-London-Wiler method

In the GLW method [6, 9] the CP asymmetry of $B \to D^0_{CP\pm}K$ is studied, where $CP\pm$ are the CP even and odd eigenstates of the D^0, as $D^0_{CP+} \to K^+K^-, \pi^+\pi^-$ and $D^0_{CP-} \to K^0_s\pi^0, K^0_s\phi, K^0_s\omega$.

We can define four observables:

$$A_{CP\pm} = \frac{\mathcal{B}(B^- \to D^0_{CP\pm}K^-) - \mathcal{B}(B^+ \to D^0_{CP\pm}K^+)}{\mathcal{B}(B^- \to D^0_{CP\pm}K^-) + \mathcal{B}(B^+ \to D^0_{CP\pm}K^+)}$$

$$R_{CP\pm} = 2 \cdot \frac{\mathcal{B}(B^- \to D^0_{CP\pm}K^-) + \mathcal{B}(B^+ \to D^0_{CP\pm}K^+)}{\mathcal{B}(B^- \to D^0_{CF}K^-) + \mathcal{B}(B^+ \to \overline{D}^0_{CF}K^+)},$$

of which only three are indipendent (since $A_{CP+}R_{CP+} = -A_{CP-}R_{CP-}$).

The relations with the amplitude ratios and phases are:

$$A_{CP\pm} = 2r_B \sin\delta_B \sin\gamma / R_{CP\pm}$$

$$R_{CP\pm} \quad = \quad 1 + r_B^2 \pm 2r_B \cos\delta_B \cos\gamma.$$

The GLW method is very clean, in fact for three independent observables we have three unknowns. Unfortunately the sensitivity to γ is proportional to r_B, so we expect to see small asymmetries.

CDF performed the first measurement of branching fraction and CP asymmetry of the $CP+$ modes at a hadron collider, using 1 fb^{-1} of data [15].

The mass distributions obtained for the three modes of interest ($D^0 \to K^+\pi^-$, K^+K^- and $\pi^+\pi^-$) are reported in Fig. 5; a clear $B \to D\pi$ signal can be seen in each plot.

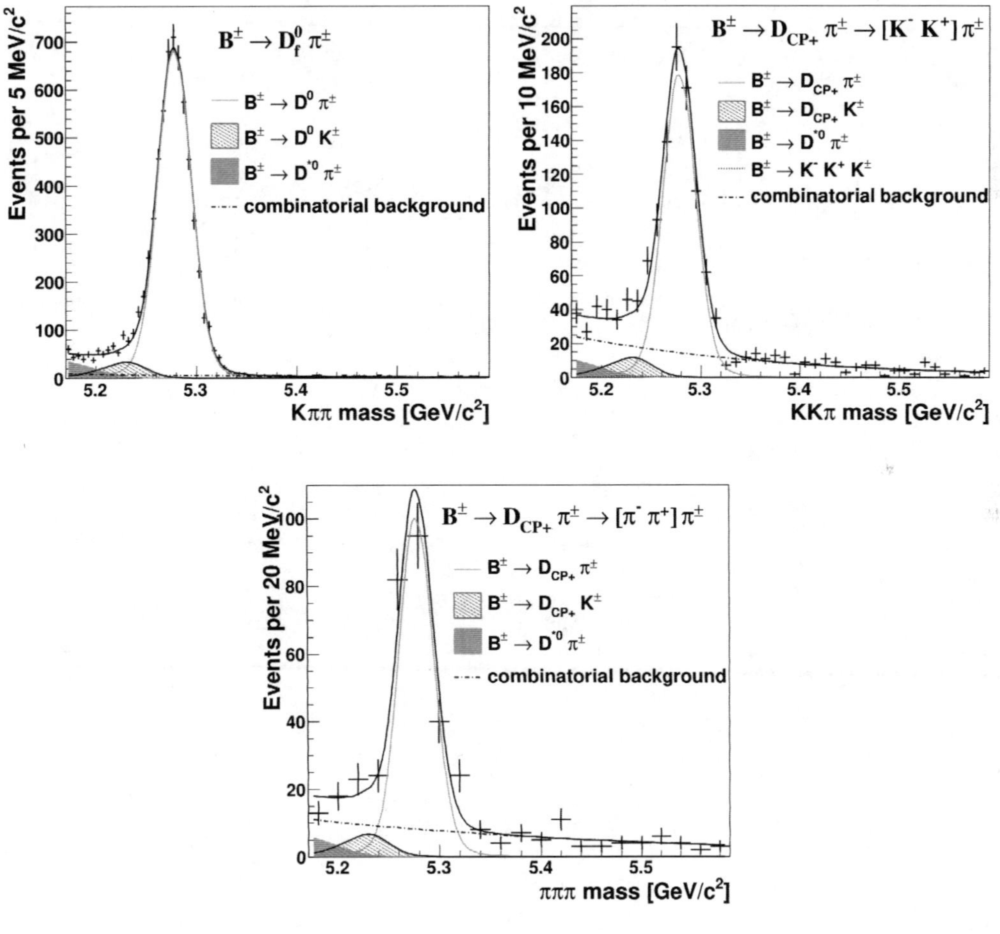

Figure 5. Invariant mass distributions of $B \to D^0 h$ candidates for each reconstructed decay mode, Cabibbo favored on the top left, Cabibbo-suppressed K^+K^- on the top right and Cabibbo-suppressed $\pi^+\pi^-$ on the bottom. The pion mass is assigned to the track from the B decay. The projections of the likelihood fit are overlaid for each mode.

The dominant backgrounds are combinatorial background and mis-reconstructed physics background such as $B^- \to D^{0*}\pi^-$ decay. In the $D^0 \to K^+K^-$ final state also the non resonant $B^- \to K^-K^+K^-$ decay appears, as determined by a study on CDF simulation [16].

An unbinned maximum likelihood fit, exploiting kinematic and particle identification information provided by the dE/dx, is performed to statistically separate the $B \to D^0 K$ contributions from the $B \to D^0 \pi$ signals and from the combinatorial and physics backgrounds.

We obtained about 90 $B \to D^0_{CP+} K$ events and we measured the double ratio of CP-even to flavor eigenstate branching fractions

$$R_{CP+} = 1.30 \pm 0.24(\text{stat}) \pm 0.12(\text{syst})$$

and the direct CP asymmetry

$$A_{CP+} = 0.39 \pm 0.17(\text{stat}) \pm 0.04(\text{syst})$$

These results are in agreement with previous measurements from $\Upsilon(4S)$ decays [2, 4].

5. Conclusions

The CDF experiment is pursuing a global program to measure the γ angle from tree-dominated processes. The published measurement using the GLW method and the preliminary result using the ADS method show competitive results with previous measurements performed at B-factories and demonstrate the feasibility of these kind of measurements also at a hadron collider.

We expect to double the data-set available by the end of the year 2010 and obtain interesting and more competitive results in the near future.

References

[1] Cabibbo N 1963 *Phys. Rev. Lett* **10** 531; Kobayashi M and Maskawa T 1973 *Prog. Theor. Phys.* **49** 652
[2] Asner D *et al.* (The Heavy Flavor Averaging Group), arXiv:1010.1589v1 [hep-ex]
[3] Bona M *et al.* (The UTfit Collaboration), *http://www.utfit.org/*
[4] Charles J *et al.* (CKMfitter Group) 2005 *Eur. Phys. J.* C **41** 1-131, arXiv:0406184 [hep-ex]
[5] Wolfenstein L. 1983 *Phys. Rev. Lett.* **51** 1945
[6] Gronau M Wyler D 1991 *Phys. Lett.* B **265** 172
[7] Atwood D Dunietz I Soni A 1997 *Phys. Rev. Lett.* **78** 3257
[8] Giri A Grossman Y Soffer A Zupan J 2003 *Phys. Rev.* D **68** 054018
[9] Gronau M, arXiv:9802315v1 [hep-ph]
[10] Atwood D Dunietz I Soni A 2001 *Phys. Rev.* D **63** 036005
[11] Ristori L. and Punzi G. 2010 *Annual Review of Nuclear and Particle Science* **60** 595-614
[12] Abe F *et al.* 1988 *Nucl. Instrum. Methods Phys. Res.* A **271** 387; Amidei D *et al.* 1994 *Nucl. Instrum. Methods Phys. Res.* A **350** 73; Abe F *et al.* 1995 *Phys. Rev.* D **52** 4784; Azzi P *et al.* 1995 *Nucl. Instrum. Methods Phys. Res.* A 360 137; Amidei D *et al. The CDF II Detector Technical Design Report*, Fermilab-Pub-96/390-E
[13] Punzi G arXiv:0808063v2 [physics.data-an]
[14] CDF Collaboration CDF Public Note 10309
[15] Aaltonen T *et al.* (CDF Collaboration) 2010 *Phys. Rev.* D **81** 031105
[16] CDF Collaboration CDF Public Note 9109

Journal of Physics: Conference Series **335** (2011) 012034 doi:10.1088/1742-6596/335/1/012034

Studies of Hadronic B Decays with Early LHCb Data

Eduardo Rodrigues[1] **(on behalf of the LHCb Collaboration)**

[1] SUPA, School of Physics and Astronomy, University of Glasgow, Glasgow, G12 8QQ, UK

E-mail: eduardo.rodrigues@glasgow.ac.uk

Abstract. Hadronic B decays offer rich opportunities for CP-violation studies. Decays of the type $B \to DX$, where D represents a charmed meson (D^0, $D^{(*)+}$ or D_s), allow for a theoretically clean determination of the CKM triangle angle γ which will provide a Standard Model benchmark of this quantity, robust against new physics effects. Decays into charmless final states, on the other hand, in general receive significant contributions from loop diagrams and are susceptible to the effects of new heavy particles. The trigger schemes of LHCb, and the particle identification provided by its RICH system, give the experiment high sensitivity to these decays. Signals will be shown from the present LHC run, and the measurement programme with the 2010-11 data set will be discussed.

1. Introduction

The investigation of hadronic B decays offers both a test-bed for the understanding of the (decay) dynamics of heavy flavoured particles and for the study of mixing and CP violation. Hadronic B decays encompass a large range of decays, from two- to multi-body modes, from final states with charged mesons (typically pions and kaons) or baryons (typically protons and Λ resonances) to modes of the form $B \to DX$, where D represents a charmed meson such as D^0, $D^{(*)+}$ or D_s. From an experimental point of view one has at hand a large variety of measurable quantities: lifetimes, branching ratios, time-dependent and time-independent CP asymmetries.

We here focus on decay modes sensitive to the CKM angle $\gamma = arg(-V_{ud}V_{ub}^*/V_{cd}V_{cb}^*)$, the relative weak phase between $b \to u$ and $b \to c$ transitions. The γ angle plays a unique role, being the only CP violating parameter that can be cleanly measured via tree-level decays[1], which are to a large extent insensitive to New Physics (NP). As such, it stands out as a benchmark Standard Model (SM) reference measurement. γ can also be determined with decays involving loop diagrams, potentially providing sensitivity to CP violating effects beyond the SM. A comparison of several independent γ measurements is thus a very powerful tool to probe the SM and search for NP. It is one of the goals of the LHCb experiment to measure γ accurately in tree-level and loop-dominated B decays [1], in particular to improve significantly its present poor knowledge – $\sigma_\gamma \approx 11-20^\circ$ – from direct measurements [2, 3].

In Sec. 2 the aspects of the LHCb experiment most relevant to the physics studies at hand are briefly presented. The following two sections, Secs. 3 and 4, discuss measurements of γ that are insensitive or sensitive to NP, respectively. We conclude with an outlook of the prospects for the near future.

[1] This statement is valid to an excellent approximation, though not strictly true due to (negligible) higher-order effects.

2. The LHCb experiment

The LHCb experiment [4] at CERN's Large Hadron Collider (LHC) is dedicated to (1) the search for NP probing the flavour structure of weak interactions and (2) the study of CP violation and rare b- and c-flavoured hadron decays. The physics goals of LHCb demand excellent tracking and identification of secondary vertices from long-lived particles, and a highly efficient particle identification system capable of separating for example pions from kaons and protons over a large momentum range.

The LHCb detector is a forward spectrometer located around interaction point 8 (IP8) at the LHC. Particles originating from a collision first encounter the vertex locator (VELO), a movable silicon vertex detector designed to measure precisely the production and decay vertices of long-lived particles, thereby separating primary from secondary vertices. It is retracted by 29 mm during LHC beam injection and ramping in energy. The nominal closed position – with the detector at 8 mm from the proton beams – is only achieved once the LHC declares stable beams, *i.e.* when conditions are met for physics data taking.

Downstream of the VELO are two tracking devices and the 4 Tm warm dipole magnet. They provide an accurate determination of the track parameters and an excellent momentum resolution $\delta p/p \approx 0.5\%$.

LHCb comprises a variety of particle identification systems: two Ring Imaging Cherenkov (RICH) detectors for pions, kaons and protons, a set of calorimeters for the identification of photons, electrons and hadrons, and a muon system for triggering on and measuring muons.

2.1. Trigger requirements

Heavy flavour production has a large cross-section at the LHC. But due to an even larger rate of inelastic processes, LHCb requires a fast, flexible and efficient trigger scheme. This is particularly true for triggering on hadronic B decays, which are more challenging than decay modes with muons in their final states.

LHCb opted for a two-level trigger. In the first-level (custom) hardware trigger, high transverse energy E_T and momentum p_T particles are selected based on partial information from the detector thereby providing a fast decision. The highest E_T hadron, electron and photon clusters are selected by the calorimeter system whereas the two highest p_T muons are selected by the muon chambers. The collision rate is brought down to a rate of about 1 MHz. The High Level software Trigger (HLT) runs on a PC farm of several thousand CPU nodes and comprises several steps of increasing complexity in event reconstruction and selection. It has access to the whole detector information for the search of interesting physics signatures. A nominal event rate of 2 kHz is stored for subsequent data analysis.

2.2. The 2010 LHC and LHCb run

The LHC delivered just over 42 pb^{-1} at IP8 during the first $\sqrt{s} = 7$ TeV proton-proton collision run of 2010. The LHCb experiment performed very well in this first data taking period, running stably and at an efficiency around 90%. It recorded close to 38 pb^{-1} of data in spite of the LHC running conditions at times well above nominal design at the LHCb interaction point (the average number of visible proton-proton collisions per LHC bunch crossing attained a value as high as 2.7, to compare with the nominal specification of ≈ 0.4).

3. Measurements of γ from tree-level diagrams

As mentioned in Sec. 1, tree-level decays of the type $B \to DX$ provide a clean determination of γ insensitive to NP. LHCb is at present considering the following modes:

- $B^0 \to D^-\pi^+, DK^{*0}$;
- $B^- \to D^{(*)}K^{(*)-}$;
- $B_s^0 \to D_s^- K^+, D\phi$;

where the charm meson $D_{(s)}$ can either decay via 2-, 3- or 4-body modes[2].

The physics case for this set of modes is rather broad, ranging from the study of branching ratios to mixing and CP violation measurements. The modes $B^- \to DK^-$ are particularly interesting when the final state can be "reached" by both $b \to u$ and $b \to c$ transitions: the quantum interference of the amplitudes gives sensitivity to γ. Several time-independent methods have been developed, exploiting different final states to which both D^0 and $\bar{D}^0$ can decay. Among those methods one can distinguish (1) the GLW method where the D decays to CP eigenstates (e.g. K^+K^-) [5, 6], (2) the ADS method where the D decays to the double Cabibbo-suppressed final state $K^+\pi^-$ [7, 8], and (3) the GGSZ method focusing on the Dalitz plot analysis of three-body D decays (*e.g.* $D \to K_S h^+h^-$, h denoting a pion or a kaon) [9]. The analysis of $B_s \to D_s^- K^+$ provides yet another clean and promising way of determining γ via time-dependent measurements of all decay combinations involved [10].

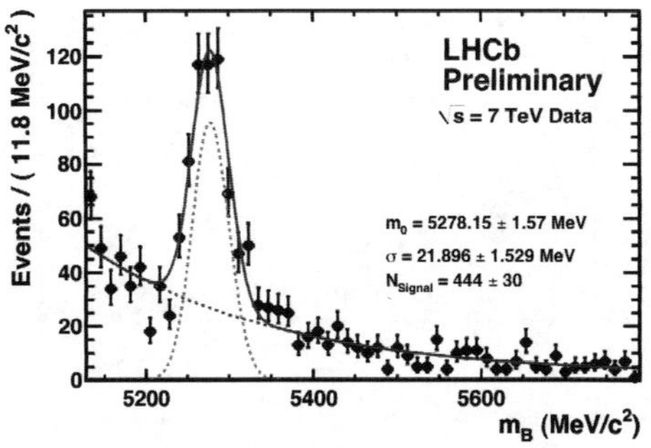

Figure 1. Reconstructed $B^- \to D^0(\to K^-\pi^+)K^-$ signal obtained in ≈ 34 pb^{-1} of the 2010 data taking sample.

Figure 1 shows the $B^- \to D^0K^-$ signal obtained during the LHCb 2010 data run for the Cabibbo-favoured $D^0 \to K^-\pi^+$ mode. Clean signals with large signal-over-background ratios have also been seen in $B^- \to D^0\pi^-$ with $D^0 \to K^-\pi^+, K^+K^-$ and $\pi^+\pi^-$.

Dalitz plot analyses with the 3-body final states $D^0 \to K_S h^+h^-$ are being prepared. Clean signals for $B^- \to D^0\pi^-$ have been seen in the 2010 data sample with $D^0 \to K_S\pi^+\pi^-$ and $K_S K^+K^-$, see Figs. 2 and 3.

Time-dependent CP measurements are foreseen with the $B_s \to D_s^- h^+$ modes. Indeed large CP violation is expected in $B_s \to D_s^{\mp}K^{\pm}$. A fair signal for $B_s \to D_s^-\pi^+$ has been found in the 2010 sample. An example, with $D_s^- \to \phi\pi^-$, is displayed in Fig. 4 (the D_s^- decay is also reconstructed in the modes $K^{*0}K^-$ and non-resonant $K^+K^-\pi^-$).

[2] In the remainder the inclusion of charge conjugate modes will always be implicit.

Journal of Physics: Conference Series **335** (2011) 012034 doi:10.1088/1742-6596/335/1/012034

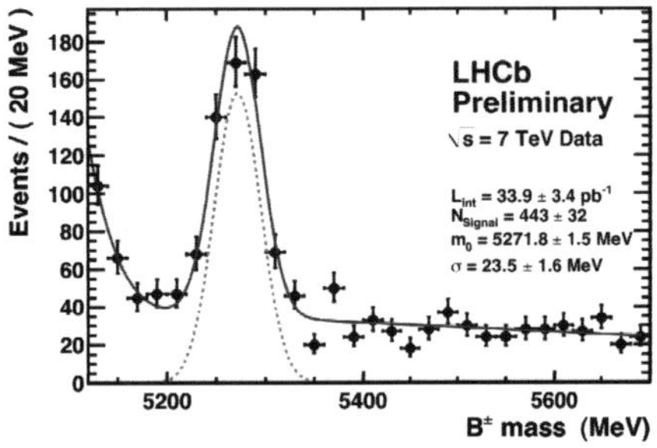

Figure 2. Reconstructed $B^- \to D^0(\to K_S\pi^+\pi^-)\pi^-$ signal obtained in $\approx$ 34 pb^{-1} of the 2010 data taking sample.

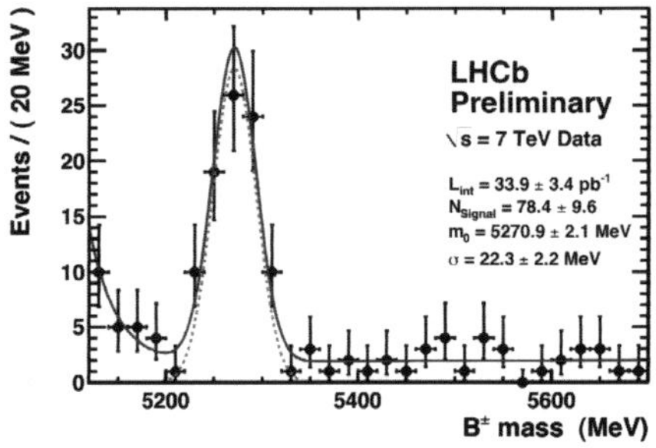

Figure 3. Reconstructed $B^- \to D^0(\to K_SK^+K^-)\pi^-$ signal obtained in $\approx$ 34 pb^{-1} of the 2010 data taking sample.

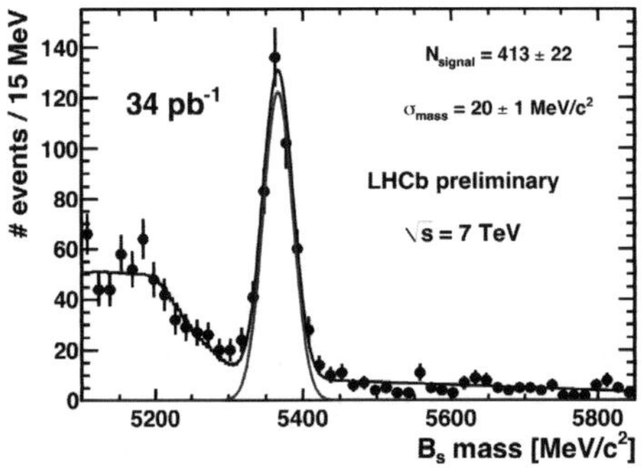

Figure 4. Reconstructed $B_s \to D_s^-(\to \phi\pi^-)\pi^+$ signal obtained in $\approx$ 34 pb^{-1} of the 2010 data taking sample.

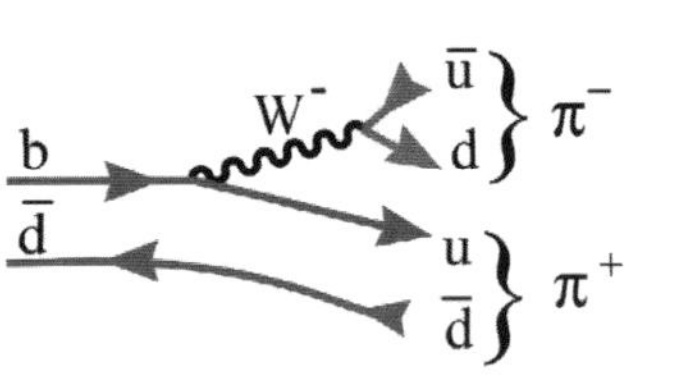

Figure 5. Tree level diagram for the decay $B^0 \to \pi^+\pi^-$.

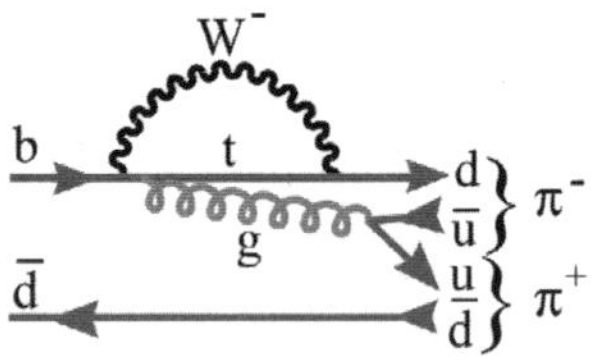

Figure 6. Penguin diagram for the decay $B^0 \to \pi^+\pi^-$.

4. Measurements of γ from loop-level diagrams

LHCb has the possibility to exploit various families of charmless B decays in view of a NP sensitive determination of γ. The families $H_b^0 \to h^+h^{'-}$ (H_b^0 stands for a B^0, B_s orΛ_b) and $B^+ \to h^+h^{'+}h^{''-}$ have been the focus of attentions; in the present paper we will not further discuss the 3-body family.

The $H_b \to h^+h^{'-}$ family allows for a rich list of important measurements such as time-integrated direct CP asymmetries, time-dependent asymmetries, and the determination of the angle γ using U-spin symmetry to relate the K^+K^- and $\pi^+\pi^-$ final states. The U-spin method of extracting γ [11] is sensitive to NP and originates from the interference between the tree-level $b \to u$ transition and the so-called Penguin diagrams ($b \to d, s$ transitions); examples of such diagrams are given in Figs. 5 and 6. In practice the method requires a combined analysis of all $H_b \to h^+h^{'-}$ modes, to better take into account backgrounds and cross-feed from similar final states.

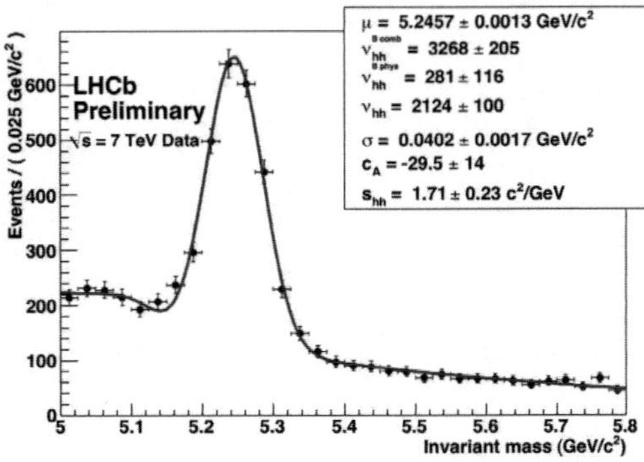

Figure 7. Reconstructed $H_b \to h^+h^{'-}$ signal obtained in ≈ 35 pb^{-1} of the 2010 data taking sample. About 2,100 candidates were selected.

Figure 7 shows the $H_b \to h^+h^{'-}$ spectrum obtained with the LHCb 2010 data sample for all modes combined, assuming the π mass for all daughters; a total of around 2,100 candidates were selected. The various decay modes – *e.g.* $B^0 \to \pi^+\pi^-$, $B_s \to K^+K^-$ – are distinguished after applying particle identification requirements, see Figs. 8 and 9.

The first LHCb (time-integrated) CP analysis with $H_b \to h^+h^{'-}$ is under way. As can be seen comparing Fig. 10 with Fig. 11, the 2010 data already exhibits an expected direct CP asymmetry in both the B^0 and B_s modes decaying into $K\pi$. It is important to note that the visible asymmetry is *raw* and should not be quantitatively interpreted, given that it is not corrected for production and detector asymmetries, which are known to be non-neglible. At present one obtains for the *raw uncorrected* B_s CP asymmetry the value $A_{CP}(B_s \to K\pi) = 0.43 \pm 0.17$ (stat.), see Fig. 12. Being the raw asymmetry, only the magnitude of the statistical error can be compared with the sensitivity obtained in previous measurements. The CDF analysis obtained $A_{CP}(B_s \to K\pi) = 0.39 \pm 0.15\,(\text{stat.}) \pm 0.08\,(\text{syst.})$ with 1 fb^{-1} of proton-antiproton collisions at the TeVatron [12]. A world-best measurement of the $B^0, B_s \to K\pi$ direct CP asymmetries should be achieved with the data to be collected in 2011-12.

Clean signals for the two$\Lambda_b \to ph^-$ modes have also been obtained, see Figs. 13 and 14.

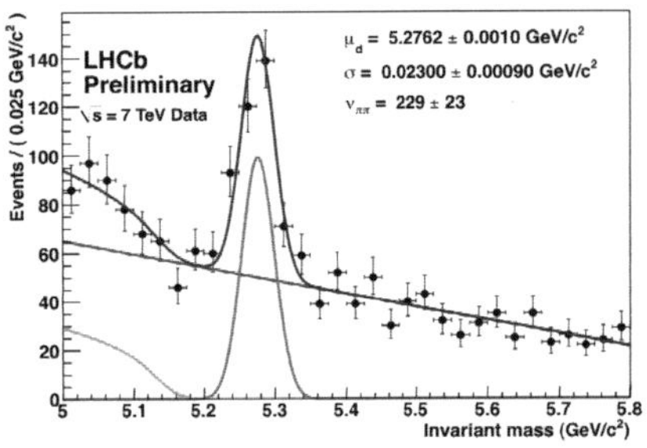

Figure 8. Reconstructed $B^0 \to \pi^+\pi^-$ signal obtained in ≈ 35 pb^{-1} of the 2010 data taking sample.

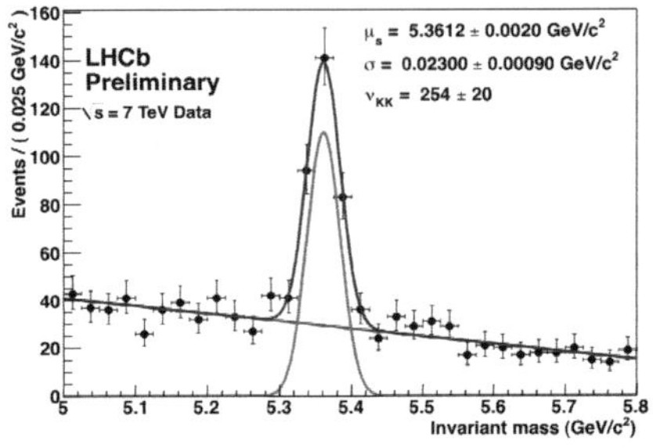

Figure 9. Reconstructed $B_s \to K^+K^-$ signal obtained in ≈ 35 pb^{-1} of the 2010 data taking sample.

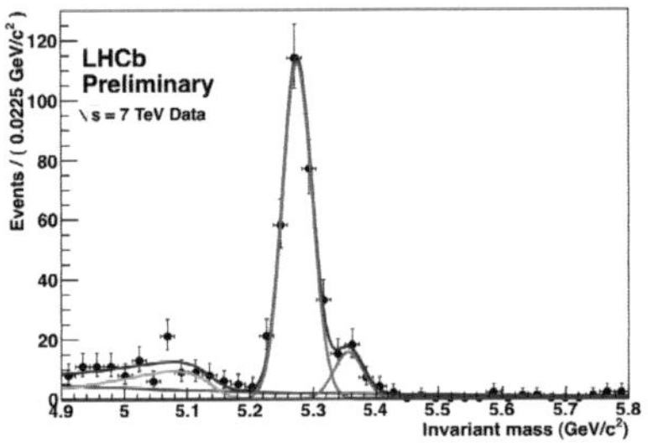

Figure 10. Reconstructed $K^+\pi^-$ invariant mass spectrum obtained with a tight selection in ≈ 35 pb^{-1} of the 2010 data taking sample. Both the B^0 (larger peak) and the $\bar{B}_s$ (smaller peak, right-hand side) resonances are clearly seen.

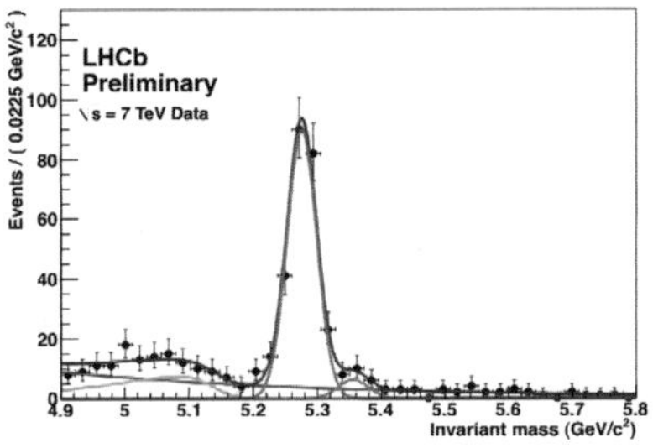

Figure 11. Reconstructed $K^-\pi^+$ invariant mass spectrum obtained with a tight selection in ≈ 35 pb^{-1} of the 2010 data taking sample. Both the $\bar{B^0}$ (larger peak) and the B_s (smaller peak, right-hand side) resonances are clearly seen.

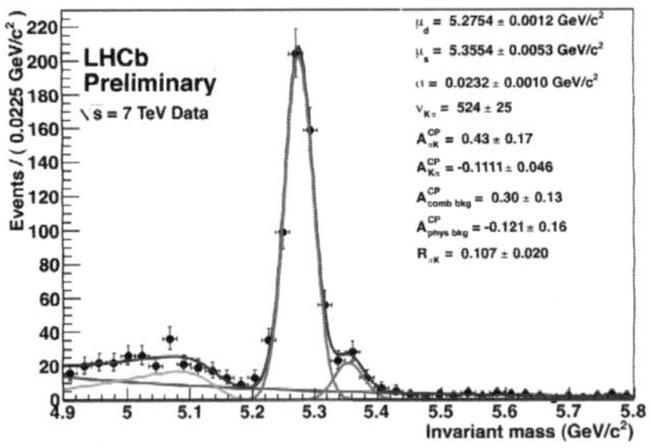

Figure 12. Reconstructed $K\pi$ invariant mass spectrum obtained with a tight selection in ≈ 35 pb^{-1} of the 2010 data taking sample. Both the B^0 (larger peak) and the B_s (smaller peak, right-hand side) resonances are clearly seen.

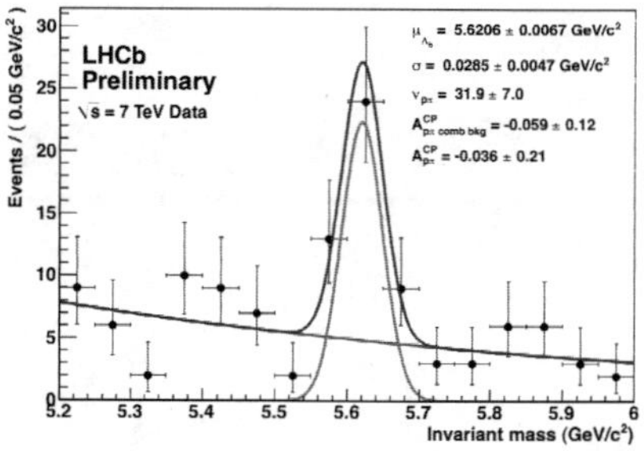

Figure 13. Reconstructed$\Lambda_b \to p\pi^-$ signal obtained in ≈ 35 pb^{-1} of the 2010 data taking sample.

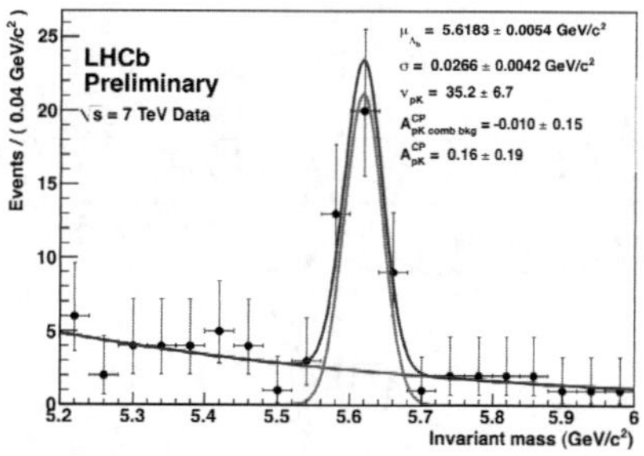

Figure 14. Reconstructed$\Lambda_b \to pK^-$ signal obtained in ≈ 35 pb^{-1} of the 2010 data taking sample.

5. Prospects for the 2011-12 data run

In this first data taking period LHCb has proved its capability as a heavy flavour experiment at a hadron machine. Several (exclusive) B decays have been selected and reconstructed with high efficiency and high signal-to-background significance. Many promising results are to be expected in the near future, with some of them being competitive with previous TeVatron publications or even world-best measurements. In particular, various complementary methods promise a plethora of measurements of the CKM angle γ. Statistical errors on γ of $\sigma_\gamma \approx 7^\circ$ are expected with 1 fb^{-1} of data.

References

[1] The LHCb Collaboration, Roadmap for selected key measurements of LHCb, `arXiv:0912.4179`.
[2] J. Charles *et al.* (CKMfitter Group), *Eur. Phys. J.* C **41** (2005), 1.
Updated results and plots at `http://ckmfitter.in2p3.fr/`.
[3] M. Bona *et al.* (UTfit Collaboration), *JHEP* **0507** (2005) 028.
Updated results and plots at `http://www.utfit.org/`.
[4] The LHCb Collaboration, The LHCb Detector at the LHC, *JINST* **3** (2008) S08005.
[5] M. Gronau, D. London,
How to determine all the angles of the unitarity triangle from $B_d^0 \to DK_S$ and $B_s^0 \to D\phi$,
Phys. Lett. B **253** (1991) 483.
[6] M. Gronau, D. Wyler, On determining a weak phase from charged B decay asymmetries,
Phys. Lett. B **265** (1991) 172.
[7] D. Atwood, I. Dunietz, A. Soni,
Enhanced CP Violation with $B \to KD^0(\bar{D^0})$ Modes and Extraction of the Cabibbo-Kobayashi-Maskawa Angle γ, *Phys. Rev. Lett.* **78** (1997) 3257.
[8] D. Atwood, I. Dunietz, A. Soni,
Improved methods for observing CP violation in $B^\pm \to KD$ and measuring the CKM phase γ,
Phys. Rev. D **63** (2001) 036005.
[9] A. Giri *et al.*, Determining γ using $B^\pm \to DK^\pm$ with multibody D decays, *Phys. Rev. D* **68** (2003) 054018.
[10] R. Fleischer, New strategies to obtain insights into CP violation through $B_s \to D_s^\pm K^\mp, D_s^{*\pm} K^\mp$,... and $B_d \to D^\pm \pi^\mp, D^{*\pm}\pi^\mp$,... decays, *Nucl. Phys. B* **671** (2003) 459.
[11] R. Fleischer, New strategies to extract β and γ from $B_d \to \pi^+\pi^-$ and $B_s \to K^+K^-$,
Phys. Lett. B **459** (1999) 306.
[12] A. Abulencia *et al.* (CDF Collaboration), Observation of $B_s^0 \to K^+K^-$ and Measurements of Branching Fractions of Charmless Two-body Decays of B^0 and B_s^0 Mesons in $p\bar{p}$ Collisions at $\sqrt{s} = 1.96$ TeV,
Phys. Rev. Lett. **97** (2006) 211802.

DIRECT OBSERVATION OF TIME REVERSAL VIOLATION

PABLO VILLANUEVA-PÉREZ

E-mail: pablo.villanueva.perez@ific.uv.es

Abstract. In this presentation a model independent and genuine test of T violation, disentangled from CP and CPT, is built. In the second section, a Monte Carlo simulation is done to show the possibility to perform this analysis at B factories.

1. THEORETICAL MOTIVATION

The violation of CP invariance has been observed in the $K^0 - \bar{K}^0$ and in the $B^0 - \bar{B}^0$ systems (Aubert et al., Abe et al. 2002). Up to now, the experimental results are in agreement with the CKM mechanism in the ElectroWeak Theory. Although all present tests of CPT invariance confirm this symmetry, imposed by any local quantum field theory with Lorentz invariance and Hermicity, it would be of great interest to observe Time-Reversal Violation (TRV) directly in a single experiment. A direct evidence for TRV would mean an experiment that, considered by itself, clearly shows T violation INDEPENDENT and unconnected to the results for CP violation. There is no existing result [1] that clearly demonstrates TRV in this sense. Sometimes the Kabir asymmetry $K^0 \to \bar{K}^0$ vs. $\bar{K}^0 \to K^0$ has been presented [2] as a proof for TRV. This process has, however, besides the drawbacks discussed in [1], the feature that $K^0 \to \bar{K}^0$ is a CPT-even transition, so that it is impossible to separate T violation from CP violation in the Kabir asymmetry: these two transformations are identical experimentally in this case.

There are effects in particle physics that are odd under time $t \to -t$, but they are not genuine T-violating, because do not correspond to an interchange of $in - states$ into $out - states$. These kinds of t-asymmetries, like the macroscopic and the Universe t-asymmetry, can occur in theories which have an exact T-symmetry in fundamental physics [3]. In fact, the t-asymmetry can only be connected [4] to T-asymmetry under the

Journal of Physics: Conference Series **335** (2011) 012035 doi:10.1088/1742-6596/335/1/012035

assumptions of CPT invariance plus the absence of an absorptive part difference between the initial and final states of the transition. As a consequence, we have to disregard these t-asymmetries as direct evidence for T violation.

As shown in [4, 5], B-factories offer the unique opportunity to show SEPARATE evidence for T violation (and CP violation) and measure the corresponding effects. The proposal has been scrutinized by Lincoln Wolfenstein [1] and Helen Quinn [3] with the conclusion that it appears to be a true TRV-effect. The crucial role played by B-factories is the EPR entanglement [6] between the neutral B-mesons produced by the decay of the $\Upsilon(4S)$ resonance. Although this coherence imposed by Bose statistics has only been used for flavor tagging up to now, one has to emphasize, following what quantum mechanics dictates, that the individual state of the neutral meson is not defined before its collapse as a filter imposed by the observation of the decay process of its companion. Similarly to the writing of the physical state of the two particles in terms of Bose-correlated orthogonal B^0 and $\bar{B}^0$, which allows to infer the flavor of the still alive meson by observing the specific flavor decay of the other (and first decaying) meson, one can rewrite the two particle state in terms of any pair of orthogonal states of individual neutral B-mesons. In particular, let us consider the pair of orthogonal states B_+ and B_- of neutral B-mesons, where B_- is the state that decays to $J/\psi K_+$, K_+ being the neutral $K_+ \to \pi\pi$, and B_+ is the orthogonal state to B_-, i.e., not connected to $J/\psi K_+$. We may call the filter imposed by a first observation of one of these decays a "CP-tag" [6], because B_+ and B_- are approximately, up to terms of $Re(\epsilon_K)$ giving the non-orthogonality of K_L and K_S, the neutral B-mesons associated with final states of their decays which are CP-eigenstates, with the identification of $K_+ = K_S$. As we are going to discuss much larger expected effects, one is authorized to use the language of identifying B_- by $J/\psi K_S$, and B_+ by $J/\psi K_L$. To clarify the point, B_- and B_+ should not be associated with CP-eigenstates of the neutral B-mesons themselves.

The theoretical ingredient to be used for this proposal of showing genuine effects for the separate violation of the discrete symmetries T and CP is the EPR entanglement only. The experimental results, and their interpretation, will be thus free of any other theoretical prejudice. Let us consider the two particle state of the neutral B-mesons produced by the decay of $\Upsilon(4S)$,

$$|i>= \frac{1}{\sqrt{2}}[B^0(t_1)\bar{B}^0(t_2) - \bar{B}^0(t_1)B^0(t_2)] = \frac{1}{\sqrt{2}}[B_+(t_1)B_-(t_2) - B_-(t_1)B_+(t_2)], \quad (1)$$

where the states 1 and 2 are defined by the time of their decay with $t_1 < t_2$. We may proceed to a partition of the complete set of events into four categories, defined by the tag in the first decay as B_+ , B_- , B^0 or $\bar{B}^0$.

Let us first take $B^0 \to B_+$ as process I, by observation of l^- (produced by the semileptonic decay of the opposite $\bar{B}^0$ meson) and $J/\psi K_L$ later, denoted as (l^-,$J/\psi K_L$), and consider:

- i) Its CP transformed $\bar{B}^0 \to B_+$ (l^+,$J/\psi K_L$), so that the asymmetry between $B^0 \to B_+$ and $\bar{B}^0 \to B_+$, as a function of $\Delta t = t_2 - t_1$, is a genuine CP-violating effect.
- ii) Its T transformed $B_+ \to B^0$ $(J/\psi K_S, l^+)$, so that the asymmetry between $B^0 \to B_+$ and $B_+ \to B^0$, as a function of $\Delta t = t_2 - t_1$, is a genuine T-violating effect.
- iii) Its CPT transformed $B_+ \to \bar{B}^0$ $(J/\psi K_S, l^-)$, so that the asymmetry between $B^0 \to B_+$and $B_+ \to \bar{B}^0$, as a function of $\Delta t = t_2 - t_1$, is a genuine test of CPT invariance.

Transition	$B^0 \to B_+$	$\bar{B}^0 \to B_+$	$B_+ \to \bar{B}^0$	$B_+ \to B^0$
(X,Y)	$(l^-, J/\psi K_L)$	$(l^+, J/\psi K_L)$	$(J/\psi K_S, l^-)$	$(J/\psi K_S, l^+)$
Transformation	Reference	CP	CPT	T

Table 1: Transitions and symmetry transformations related to process I tag as reference.

One may check, that the events used for the asymmetries i), ii), and iii), sumarized in Table 1, are completely independent. Furthermore, the expectation is that the asymmetry described by ii) will prove and measure, for the first time, T violation with many standard deviations away from zero.

Similarly, one may take as reference process II $B^0 \to B_-$, by observation of l^- first and $J/\psi K_S$ later $(l^-, J/\psi K_S)$. The corresponding transformations are summarized in Table 2.

Transition	$B^0 \to B_-$	$\bar{B}^0 \to B_-$	$B_- \to \bar{B}^0$	$B_- \to B^0$
(X,Y)	$(l^-, J/\psi K_S)$	$(l^+, J/\psi K_S)$	$(J/\psi K_L, l^-)$	$(J/\psi K_L, l^+)$
Transformation	Reference	CP	CPT	T

Table 2: Transitions and symmetry transformations related to process II tag as reference.

Analogously, one may consider the process III $\bar{B}^0 \to B_+$ as the reference, by observation of $(l^+, J/\psi K_L)$. In particular, between $\bar{B}^0 \to B_+$ and $B_+ \to \bar{B}^0$ $(J/\psi K_S, l^-)$ we have again a T reversal transformation as shown in Table 3.

Finally, one may consider the process IV $\bar{B}^0 \to B_-$ as the reference, by observation of $(l^+, J/\psi K_S)$ and the new transformations are summarized in Table 4.

We will develop the results expected for all these genuine asymmetries in the Weisskopf-Wigner effective Hamiltonian approach for the time evolution of the $B^0 - \bar{B}^0$ system [4, 5].

Transition	$\bar{B}^0 \to B_+$	$B^0 \to B_+$	$B_+ \to B^0$	$B_+ \to \bar{B}^0$
(X,Y)	$(l^+, J/\psi K_L,)$	$(l^-, J/\psi K_L,)$	$(J/\psi K_S, l^+)$	$(J/\psi K_S, l^-)$
Transformation	Reference	CP	CPT	T

Table 3: Transitions and symmetry transformations related to process III tag as reference.

Transition	$\bar{B}^0 \to B_-$	$B^0 \to B_-$	$B_- \to B^0$	$B_- \to \bar{B}^0$
(X,Y)	$(l^+, J/\psi K_S,)$	$(l^-, J/\psi K_S,)$	$(J/\psi K_L, l^+)$	$(J/\psi K_L, l^-)$
Transformation	Reference	CP	CPT	T

Table 4: Transitions and symmetry transformations related to process IV tag as reference.

However our goal is to demonstrate and measure the violation of time reversal invariance without using the procedure of fitting parameters in a given theory. The outcome will be highly rewarding as a model-independent observation of T violation.

2. MONTE CARLO STUDY

In this section we describe the procedure we have followed to generate the simulated samples of the required decays, as discussed in the theoretical motivation.On the second part, we estimate the T violation asymmetry and its significance.

We first explain briefly the technique used at B factories to determine the difference of proper-time Δt of the two decaying B mesons, as well as the flavor tagging and the reconstruction. These techniques are described in detail in Ref. [7]. At the asymmetric-energy B factories the $\Upsilon(4S)$ decay products are Lorentz boosted along the longitudinal axis and fly enough for their decay paths to be comparable or larger than the experimental resolution. Since no charged stable particles emerge from the $\Upsilon(4S)$ decay point, the decay length of the individual B mesons from the $\Upsilon(4S)$ decay is unknown and we are left to determine the difference Δz between them, from which Δt can be calculated since the boost of the center-of-mass is well known. One of the two B mesons is fully reconstructed in a CP eigenstate (B_+,B_-), while the other B is partially reconstructed through their decay products (e.g. semileptonic decays, dominant hadronic decays producing charged kaons, etc.) and used for tagging as B^0 or $\bar{B}^0$ and for vertexing. In Figure 1 we sketch the event topology and how the reconstruction of the $\Upsilon(4S)$ decay is performed.

Concerning the Monte Carlo simulation, it has been done using an standard Probability

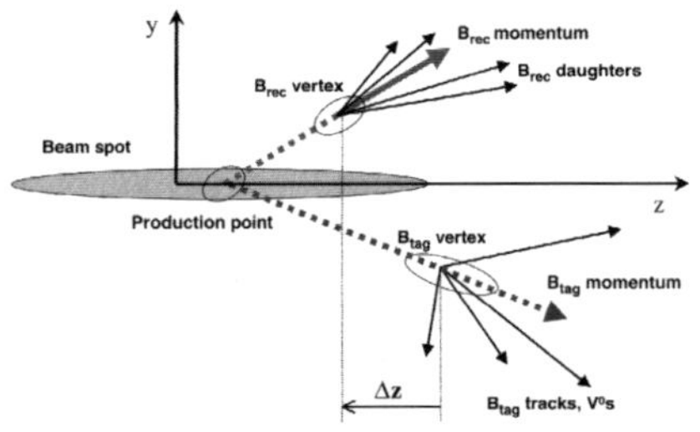

Figure 1: $\Upsilon(4S)$ decay topology and reconstruction. The figure is not drawn to scale.

Density Function (PDF) which contains effects of T, CP and CPT violation and is based on the Wigner-Weiskopff formalism [8]. Neglecting effects from mistags, CPT violation (complex parameter z set to zero), and taking $\Delta\Gamma = 0$, $|q/p| = 1$, the decay rate $g_\pm$ from a neutral B meson to a CP eigenstate (B_+, B_-) is given by equation 2 taken from [8, 9]. To generate our samples we have set the parameter $Im(\lambda_f)$ (i.e. the well known CP-violating parameters $\sin 2\beta$) to 0.672, and $|\lambda_f| = 1$ (i.e. no direct CP violation in the B_+, B_- decays), taken from [9]:

$$\begin{cases} S_f = 0.672 \\ C_f = 0 \end{cases} \Rightarrow \begin{cases} Im(\lambda_f) = 0.672 \\ |\lambda_f| = 1. \end{cases}$$

$$g_\pm(\Delta t) = \frac{e^{\frac{-|\Delta t|}{\tau_{B^0}}}}{4\tau_{B^0}} \{1 \pm [S_f sin(\Delta m_d \Delta t) - C_f cos(\Delta m_d \Delta t)]\} \tag{2}$$

With this PDF the eight samples can be generated, from which we evaluate our asymmetries, as described in section 1.

The T asymmetry is built counting the number of entries in the physics sample (N_a) and its T transformed (N_b). So we subtract bin per bin Δt in both samples, correcting by efficiency effects.

$$A_T = \frac{N_a - cN_b}{N_a + cN_b} = \frac{1 - c\dfrac{N_b}{N_a}}{1 + c\dfrac{N_b}{N_a}} = \frac{1 - \dfrac{\Gamma_b}{\Gamma_a}}{1 + \dfrac{\Gamma_b}{\Gamma_a}}, \tag{3}$$

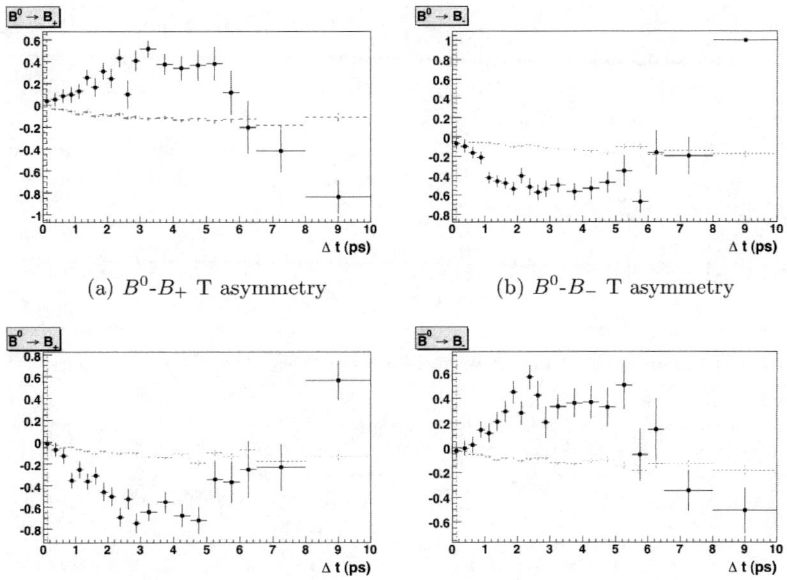

(a) B^0-B_+ T asymmetry

(b) B^0-B_- T asymmetry

(c) $\bar{B}^0$-B_+ T asymmetry

(d) $\bar{B}^0$-B_- T asymmetry

Figure 2: T asymmetries for the four reference processes. The red curve on the plots is the one obtained using the curves without T, CP, and CPT violating effects.

where

$$c = \frac{\varepsilon_{K_L}}{\varepsilon_{K_S}} = \frac{\frac{N^{exp}_{K_L}}{N^{tot}_{K_L}}}{\frac{N^{exp}_{K_S}}{N^{tot}_{K_S}}} = \frac{N^{exp}_{K_L}}{N^{exp}_{K_S}}, \tag{4}$$

or

$$c = \frac{\varepsilon_{K_S}}{\varepsilon_{K_L}} = \frac{\frac{N^{exp}_{K_S}}{N^{tot}_{K_S}}}{\frac{N^{exp}_{K_L}}{N^{tot}_{K_L}}} = \frac{N^{exp}_{K_S}}{N^{exp}_{K_L}}. \tag{5}$$

Eq. (4) has to be used when the a sample is reconstructed as $J/\psi K_L$ and the b sample as $J/\psi K_S$. Analogously, equation 5 will be used when the a sample is reconstructed as $J/\psi K_S$ and the b sample as $J/\psi K_L$.

At BaBar, there are $3255 K_L$ signal events and $7750 K_S$ signal events, as extracted from the experimental analysis. So we obtain the black line in the T asymmetry plots (figure 2). The red dotted curves are those obtained after setting $Im(\lambda_{CP}) = 0$, which in practice correspond to the integration over the bin size of the curve without T violation. These curves have been done using 1 million events (i.e. a virtually infinite statistics). We perform this comparison since our final goal is to compare the asymmetries with T violation and the corresponding ones without T violation with all the experimental effects included. This comparison will be done through a χ^2 test.

We also have included approximate Δt resolution, mistags and reconstruction efficiencies as it is done at the B factories.

We perform a χ^2 test for each of the four statistically independent T asymmetries, using the χ^2 ansatz given by

$$\chi_0^2 = \sum_{\Delta t_i} \frac{(A_T^{exp}(\Delta t_i) - A_T^{NoT-Violation}(\Delta t_i))^2}{\sigma^2_{A_T^{exp}}(\Delta t_i) + \sigma^2_{A_T^{NoT-Violation}}(\Delta t_i)}, \tag{6}$$

from which a confidence level (or equivalently a significance of the T violation effect) can be computed. For this study we assume a χ^2-distributed variable, although in the experimental analysis a careful study of the true distribution will have to be pursued.

Assuming a χ^2 distribution, the results of the χ^2 test for T asymmetry with proper-time resolution, mistags, and efficiency effects for our simulated samples are summarized in table 5. We can compute that $\chi^2/dof \approx 14$ for the TRV test, so we can anticipate to establish directly for the first time T violation with a significance above 10 standard deviations.

T asymmetries Test	$B^0 \to B_+$	$B^0 \to B_-$	$\bar{B}^0 \to B_+$	$\bar{B}^0 \to B_-$
χ_0^2	99.04, 23 bins	159.47, 22 bins	150.05, 22 bins	106.36, 21 bins
Prob($\chi^2 > \chi_0^2$)	$2.06 \cdot 10^{-11}$	$7.69 \cdot 10^{-23}$	$4.69 \cdot 10^{-21}$	$2.13 \cdot 10^{-13}$
Standard Deviations	6.70	9.84	9.42	7.34

Table 5: χ^2 test of the T asymmetries.

3. ACKNOWLEDGEMENTS

This research has been supported by the Grants FPA 2008-02878 from MICINN and PROMETEO 2008/004 from Generalitat Valenciana.

4. REFERENCES

[1] L. Wolfenstein, Int. J. Mod. Phys. E **8**, 501 (1999).
[2] T. Nakada, Discrete'08 Conference, Valencia 2008, J. Phys. Conf. Ser. **171**, 011001 (2009).
[3] H. R. Quinn, Discrete'08 Conference, Valencia 2008, J. Phys. Conf. Ser. **171**, 011001 (2009).
[4] M. C. Bañuls and J. Bernabeu, Phys. Lett. B **464**, 117 (1999), [arXiv:hep-ph/9908353].
[5] M. C. Banñuls, J. Bernabeu, Nucl. Phys. B **590**, 19 (2000), [arXiv:hep-ph/0005323].
[6] M. C. Banñuls, J. Bernabeu, JHEP **032**, 9906 (1999), [arXiv:hep-ph/9807430].
[7] B. Aubert *et al.* (BaBar Collaboration), Phys. Rev. D **66**, 032003 (2002).
[8] B. Aubert *et al.* (BaBar Collaboration), Phys. Rev. Lett. **92**, 181801 (2004); Phys. Rev. D **70**, 012007 (2004).
[9] B. Aubert *et al.* (BaBar Collaboration), Phys. Rev. D **79**, 072009 (2009).

CP Violation Results from *B* Decays at *BaBar*

Pietro Biassoni

Università degli Studi and INFN Milano, via Celoria 16, I-20133 Milano, Italy.
Representing the BaBar Collaboration.

E-mail: pietro.biassoni@mi.infn.it

Abstract. In the present paper we review recent experimental results from the *BABAR* experiment concerning the measurement of the CKM angles. A particular highlight is given to the novel independent determination of the angle α from $B^0 \to a_1(1260)^{\pm}\pi^{\mp}$ and to the recent full-luminosity updates of several angle γ measurements.

1. Introduction

The measurement of *CP* violation (*CPV*) in *B* meson decays provides crucial tests of the Standard Model and of the Cabibbo-Kobayashi-Maskawa (CKM) matrix [1]. The angle β is experimentally measured with a precision of $O(1^\circ)$ in $B^0 \to (c\bar{c})K^0$ [2] and is not covered in this paper. The determination of angles α and γ still suffers from larger experimental uncertainties. We review results from *BaBar*, including the measurement of α in $B^0 \to \rho\rho$ and in the novel decay mode $B^0 \to a_1(1260)^{\pm}\pi^{\mp}$, and full-luminosity updates of several angle γ measurements.

2. Experimental techniques

The results are based on data collected with the *BaBar* detector [3] at the PEP-II asymmetric-energy e^+e^- collider, at a center-of-mass energy near the $\Upsilon(4S)$ resonance.

The B meson is kinematically characterized by the variables$\Delta E \equiv E_B - \frac{1}{2}\sqrt{s}$ and $m_{\rm ES} \equiv \sqrt{s/4 - |\vec{p}_B|^2}$, where $(E_B, \vec{p}_B)$ is the B four-momentum vector expressed in $\Upsilon(4S)$ rest frame. The total integrated luminosity corresponds to about 468×10^6 $B\overline{B}$ pairs.

Background arises primarily from random combinations of particles in $e^+e^- \to q\bar{q}$ events ($q = u, d, s, c$), and is discriminated against $B\overline{B}$ events by using event shape variables, combined into multivariate "shape" classifiers, that are indicated with $\mathcal{SC}$ in the following.

3. Measuring α

The CKM angle α is measured in $b \to u\bar{u}d$ transition, exploiting the interference between the decay of mixed and unmixed B^0 mesons. The signal B meson (B_{CP}) is reconstructed into its decay to a *CP*-eigenstate, accessible from both B^0 and $\overline{B}^0$. From the remaining particles in the event, we reconstruct the decay vertex of the other B meson ($B_{\rm tag}$) and identify its flavor, through the analysis of its decay products [4].

The distribution of the difference$\Delta t \equiv t_{CP} - t_{\rm tag}$ of the proper decay times of B mesons into *CP*-eigenstates, such as $\rho^+\rho^-$, is given by

$$f_q^{\rho\rho}(\Delta t) = \frac{e^{-|\Delta t|/\tau}}{4\tau}\left\{1 - q_{tag}\left[C\cos(\Delta m_d \Delta t) - S\sin(\Delta m_d \Delta t)\right]\right\}, \quad (1)$$

Published under licence by IOP Publishing Ltd

Journal of Physics: Conference Series **335** (2011) 012036 doi:10.1088/1742-6596/335/1/012036

where τ is the mean B lifetime,$\Delta\ m_d$ the $B^0 - \bar{B}^0$ mixing frequency, and $q_{tag} = +1\ (-1)$ if the B_{tag} decays as a B^0 ($\bar{B}^0$). The parameters S and C describe mixing-induced and direct CPV, respectively. Considering just tree level contributions to the process, $S = \sin(2\alpha)$ and $C = 0$. However, non negligible penguin (loop) amplitudes may contribute to $b \to u\bar{u}d$ transitions. The different strong and weak phase of the penguin amplitudes may give rise to direct CPV ($C \neq 0$), and modify S into $S = \sin(2\alpha_{\text{eff}})\sqrt{1-C^2}$, where $\alpha_{\text{eff}} = \alpha - \Delta\alpha$, with $\Delta\alpha \neq 0$. However,$\Delta\ \alpha$ may be extracted via an isospin SU(2) or a flavor SU(3) analysis of the decay.

4. α from $B \to \rho\rho$

4.1. Experimental Inputs

An isospin SU(2) analysis is used to extract α in $B \to \rho\rho$ decays. The $B \to \rho\rho$ decays are $P \to VV$ transitions, where P (V) denotes a pseudoscalar (vector) meson. Hence, the decay is described by three different amplitudes, one for each helicity state, with different CP transformation properties [5]. The analysis of the angular distributions of the B meson decay products allows to extract the fraction f_L of longitudinal polarization. In the helicity formalism, the differential decay rate is

$$\frac{1}{\Gamma}\frac{d^2\Gamma}{d\cos\theta_1 d\cos\theta_2} \propto 4f_L\cos^2\theta_1\cos^2\theta_2 + (1-f_L)\sin^2\theta_1\sin^2\theta_2, \tag{2}$$

where $\theta_{1(2)}$ is the helicity angle between the daughter π and the B recoil direction in the first (second) ρ rest frame. Since experimental measurements have shown the decay to be dominated by the longitudinal, CP-even polarization, a full angular analysis, that allows to separate the definite-CP contributions of the transverse polarization, is not needed.

Several inputs are needed to perform the SU(2) analysis of the $B \to \rho\rho$ decay. They are: the time-dependent (TD) parameters and branching fraction (BF) of $B^0 \to \rho^+\rho^+$ [6, 7], BF and direct CP-asymmetry (A_{CP}) of $B^+ \to \rho^+\rho^0$ [8], and TD parameters and BF of $B^0 \to \rho^0\rho^0$ [9]. In Table 1 we summarize the results of the different $B \to \rho\rho$ analyses, and the number of $B^0\overline{B}^0$ pairs used in each measurement. The $B^+ \to \rho^+\rho^0$ analysis has been updated using the final BABAR dataset.

Mode	BF (10^{-6})	f_L	$C(A_{CP})$	S	$N_{B\overline{B}}$ (10^6)
$B^0 \to \rho^+\rho^-$	$25.5 \pm 2.1^{+3.6}_{-3.9}$	$0.992 \pm 0.024^{+0.026}_{-0.013}$	$0.01 \pm 0.15 \pm 0.06$	$-0.17 \pm 0.20^{+0.05}_{-0.06}$	383
$B^\pm \to \rho^\pm\rho^0$	$23.7 \pm 1.4 \pm 1.4$	$0.950 \pm 0.015 \pm 0.006$	$-0.054 \pm 0.055 \pm 0.010$	-	465
$B^0 \to \rho^0\rho^0$	$0.92 \pm 0.32 \pm 0.14$	$0.75^{+0.11}_{-0.14} \pm 0.04$	$0.2 \pm 0.8 \pm 0.3$	$0.3 \pm 0.7 \pm 0.2$	465

Table 1. Results (BF, f_L, A_{CP}, C, and S) for $B \to \rho\rho$ analyses. The number of $B\overline{B}$ pairs $N_{B\overline{B}}$ used in each analysis is also reported.

Signal yields, A_{CP} and f_L are extracted using a maximum likelihood (ML) fit to m_{ES}, ΔE, SC, and the masses and helicity angles of the ρ mesons. Multidimensional probability density functions are used to properly account for variables correlations in the background.

4.2. Determination of α

Under the isospin SU(2) symmetry, the following relations hold [10]:

$$\frac{1}{\sqrt{2}}A^{+-} = A^{+0} - A^{00}, \quad \frac{1}{\sqrt{2}}\tilde{A}^{+-} = \tilde{A}^{-0} - \tilde{A}^{00}, \tag{3}$$

where A^{ij} ($\tilde{A}^{ij}$) is the amplitude of $B^0(\overline{B}^0) \to \rho^i\rho^j$ process. Neglecting possible electroweak penguins (EWP) amplitudes, which do not obey SU(2) isospin symmetry, $A^{\pm 0}$ receives only tree amplitude contributions. The small A_{CP} value measured in $B^+ \to \rho^+ \rho^0$ decay indicates that the contribution from EWPs is negligible, and the isospin analysis holds within an uncertainty of $1 - 2°$ [11]. Other possible isospin violation effects due to finite ρ width [12] are tested by varying the $\pi\pi$ mass window. Such effects are below the current sensitivity. If A^{+0} and $\tilde{A}^{-0}$ are aligned with a suitable choice of phases, the relations in Eq. (3) can be represented in the complex plane by two triangles, and the phase difference between A^{+-} and $\tilde{A}^{+-}$ is $2\Delta\alpha$. Isospin relations similar to Eq. (3) hold separately for each polarization state. However, since $f_L \sim 1$, only the analysis of *CP*-even longitudinal decay is performed. Constraints on the CKM angle α and on the penguin contribution$\Delta\ \alpha$ are obtained from a confidence level (CL) scan. Assuming the isospin-triangle relations of Eq. (3), a χ^2 for the five amplitudes $(A^{+0}, A^{+-}, A^{00}, \tilde{A}^{+-}, \tilde{A}^{00})$ is calculated from the measurements summarized in Table 1, and minimized with respect to the parameters that don't enter the scan. The $1-CL$ values are then calculated from the probability of the minimized χ^2. Results of such a scan are reported in Fig. 1 Since the measured BFs of

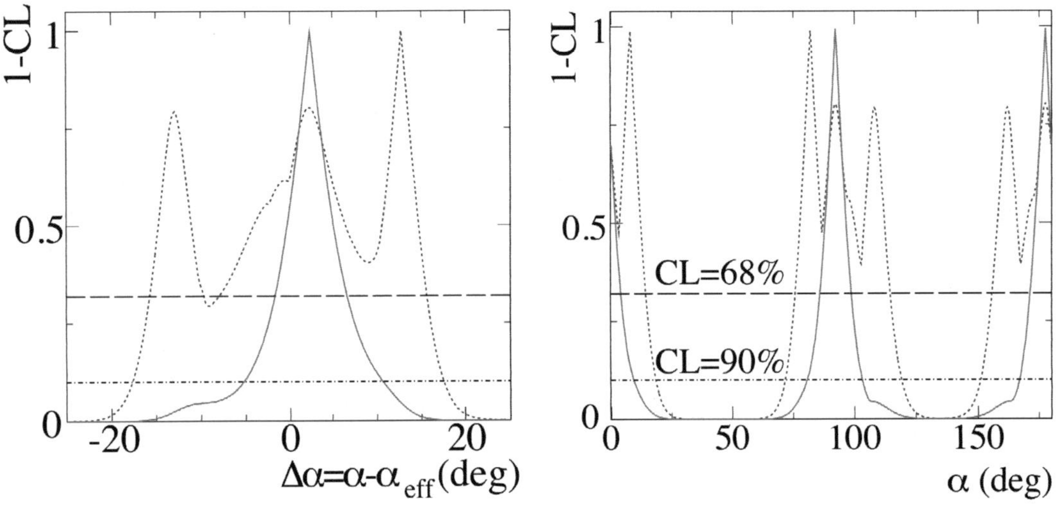

Figure 1. Projection of the $1 - CL$ scan on (left)$\Delta\ \alpha$ and (right) α for the $\rho\rho$ system [8]. The red solid (blue dotted) line represents the result found using the latest [8] (previous [13]) $B \to \rho^+ \rho^0$ BF measurement.

the several reactions satisfy $\mathcal{B}(B^+ \to \rho^+\rho^0) \approx \mathcal{B}(\rho^+\rho^-) \gg \mathcal{B}(B^0 \to \rho^0\rho^0)$, the isospin triangles do not close, *i.e.* $|A^{+-}|/\sqrt{2} + |A^{00}| < |A^{+0}|$. This results in a degeneracy of the eight-fold ambiguity on α into a four-fold ambiguity, corresponding to peaks in the vicinity of 0°, 90° (two degenerate peaks), and 180°. *BABAR* obtains a 68% CL limit $-1.8° < \Delta\alpha < 6.7°$. Taking only the solution consistent with the global CKM fits [14], α is equal to $92.4^{+6.0}_{-6.5}$.

5. α from $B^0 \to a_1(1260)^{\pm}\pi^{\mp}$

A novel independent measurement of α is performed by *BABAR* in the $B \to a_1(1260)^{\pm}\pi^{\mp}$ decay. The TD decay rate of the B meson into the non-*CP* eigenstate $a_1(1260)^{\pm}\pi^{\mp}$ is [15]

$$f_q^{a_1^{\pm}}(\Delta t) \propto \frac{e^{-|\Delta t|/\tau}}{4\tau}(1 \pm A_{CP})\Big\{1 + q_{\text{tag}}\left[(S \pm \Delta S)\sin(\Delta m_d \Delta t) + (C \pm \Delta C)\cos(\Delta m_d \Delta t)\right]\Big\}, \quad (4)$$

and α_{eff} enters this equation via $S \pm \Delta S = \sqrt{1 + (C \pm \Delta C)^2} \times \sin\left(2\alpha_{\text{eff}} \pm \hat{\delta}\right)$, where $\hat{\delta}$ is the strong phase between the tree amplitudes of B^0 decays to $a_1^+\pi^-$ and $a_1^-\pi^+$.

Journal of Physics: Conference Series **335** (2011) 012036 doi:10.1088/1742-6596/335/1/012036

5.1. Experimental Inputs

Since an isospin SU(2) analysis of the $B^0 \to a_1(1260)^\pm \pi^\mp$ decay is not feasible [16], a flavor-SU(3) based approach is used to extract the information on$\Delta\ \alpha$. The experimental inputs needed for the SU(3) analysis are: TD parameters and BF of $B^0 \to a_1(1260)^+ \pi^-$, BFs of $B \to a_1(1260)^\pm K$, and BF of $B \to K_{1A}\pi$. The BFs and TD parameters of $B \to a_1(1260)\pi$ and $B \to a_1(1260)K$ decays have been measured in the last few years [17, 18].

The $B \to K_{1A}\pi$ BF can be extracted using the combined branching fraction of B decays to $K_1(1400)\pi$ and $K_1(1270)\pi$, and the relative magnitude and phase of $B \to K_1(1270)\pi$ and $B \to K_1(1400)\pi$ amplitudes. $K_1(1270)$ and $K_1(1400)$ are both axial vector mesons, they have overlapping mass distributions and both decays to $K\pi\pi$, hence they undergo not-negligible interference effects. In order to include interference effects in the fit, in a recent analysis by *BABAR* [19] the K_1 system is parameterized in terms of a two-resonance, six-channel K-matrix model [20] in the P-vector approach [21]. A partial wave analysis of the diffractively produced $K\pi\pi$ system performed by the ACCMOR collaboration [20] is used to determine the decay couplings and the mass poles of the K-matrix. Signal yields are extracted using a ML fit to $m_{\rm ES}$, ΔE, and $\mathcal{SC}$. The invariant mass of the resonant $K\pi\pi$ system is sensitive to the production parameters of the K_1 system. The combined signal branching fractions are $\mathcal{B}(B^0 \to K_1(1270)^+\pi^- + K_1(1400)^+\pi^-) = 31^{+8}_{-7} \times 10^{-6}$ and $\mathcal{B}(B^+ \to K_1(1270)^0\pi^+ + K_1(1400)^0\pi^+) = 29^{+29}_{-17} \times 10^{-6}$, where the error includes both statistical and systematic contributions. The information about the fraction and phase of the two resonances is used to calculate the contribution of the K_{1A} meson which belongs to the same SU(3) octet as the a_1 meson. The results are $\mathcal{B}(B^0 \to K_{1A}^+\pi^-) = 14^{+9}_{-10} \times 10^{-6}$ and $\mathcal{B}(B^+ \to K_{1A}^0\pi^+) < 36 \times 10^{-6}$, where the latter upper limit is evaluated at the 90% confidence level [19].

5.2. Determination of α

Using a a flavor-SU(3) based approach, the size of penguin amplitudes contributing to the decay is related to the branching fractions of the$\Delta\ S = 1$ partners of the $B^0 \to a_1^\pm \pi^\mp$ decays: $B \to a_1 K$ and $B \to K_{1A}\pi$ [15, 22] . Nonfactorizable contributions to transition amplitudes from exchange and weak annihilation diagrams are neglected [15, 22]. $\hat{\delta}$ is assumed to be negligible. a_1 and K_{1A} form factors, that are needed in the analysis, are obtained from the study of τ decays [23]. A Monte Carlo method is used to derive the 68% and 90% CL upper limits for $\Delta\alpha$ [19]. The result is $|\Delta\alpha| < 11^\circ$ (13°) at the 68% (90%) CL. Combining this bound with the results from $B^0 \to a_1(1260)^\pm \pi^\mp$ TD analysis [18], the final result is $\alpha = (79 \pm 7 \pm 11)^\circ$ for the solution compatible with the CKM global fits, where the first error is statistical and systematic combined and the second is due to penguin pollution.

6. Measuring γ

The CKM angle γ is measured by exploiting the interference between the $b \to c\bar{u}s$ and $b \to u\bar{c}s$ tree amplitudes in CP-violating $B \to D^{(*)}K^{(*)}$ decays. Such amplitudes also depend on the magnitude ratio $r_B \equiv \left|\frac{A(b \to u)}{A(b \to c)}\right|$, and the relative strong phase δ_B between the CKM favored and suppressed amplitudes. These hadronic parameters depend on the B decay and are extracted from data. In the following we report the results of full-luminosity updates of three different analysis [24, 25, 26], based on the GGSZ [27], ADS [28], and GLW [29] methods, respectively.

6.1. GGSZ method: $B^\pm \to D^{(*)}K^{(*)\pm}$, $D \to K_S^0 h^+ h^-$

In the GGSZ method [27], the information on γ is extracted from the Dalitz-plot distribution of the D daughters. The variables sensitive to CP violation are $x_\pm \equiv r_B \cos(\delta_B \pm \gamma)$ and $y_\pm \equiv r_B \sin(\delta_B \pm \gamma)$.

$B^{\pm}\to DK^{\pm}$, $D^*K^{\pm}$ ($D^*\to D\gamma$ and $D\pi^0$), and $DK^{*\pm}$ ($K^{*\pm}\to K^0_S\pi^{\pm}$) decays are studied, with $D\to K^0_S h^+h^-$ ($h = \pi, K$) [24]. Signal yields are extracted using a ML fit to $m_{\rm ES}$, ΔE, and $\mathcal{SC}$. A fit to the Dalitz-plot distribution of the D daughters is used to determine 2D confidence regions for $x_{\pm}$ and $y_{\pm}$, which are shown in Fig. 2. The Dalitz plot model for D^0 and $\overline{D}^0$ decay is studied using the large ($\approx 6.2\times 10^5$) and very pure $D^*\to D\pi$ control sample [30]. The Dalitz model includes a non-resonant part and several intermediate $K^0_S h$ or h^+h^- quasi-two-body decays. The fitted signal yields are about 1000, 500 and 200 events for $B\to DK$, $B\to D^*K$, and $B\to DK^*$, respectively. The fitted $(x_{\pm}, y_{\pm})$ values are reported in Table 2.

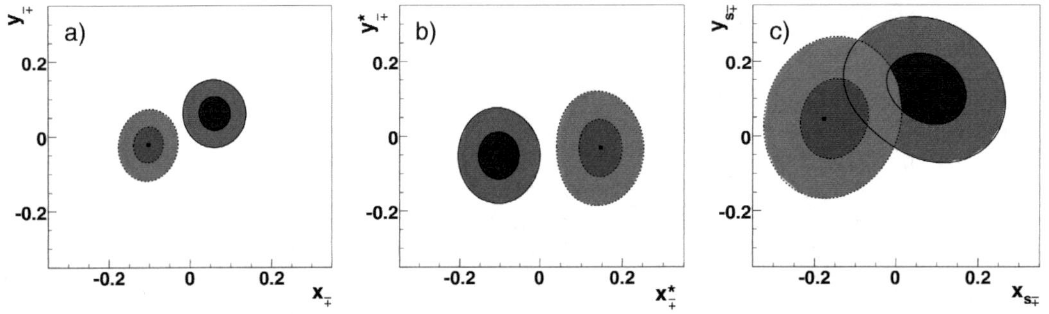

Figure 2. 1σ and 2σ contours in the $(x_{\pm}, y_{\pm})$ planes for (a) $B\to DK$, (b) $B\to D^*K$ and (c) $B\to DK^*$, for B^- (solid lines) and B^+ (dotted lines) decays. $(x^*_{\pm}, y^*_{\pm})$ and $(x_{s\pm}, y_{s\pm})$ denote $(x_{\pm}, y_{\pm})$ in $B\to D^*K$ and $B\to DK^*$ decay, respectively.

	$B^{\pm}\to DK^{\pm}$	$B^{\pm}\to D^*K^{\pm}$	$B^{\pm}\to DK^{*\pm}$
x_+	$-0.103\pm0.037\pm0.006\pm0.007$	$0.147\pm0.053\pm0.017\pm0.003$	$-0.151\pm0.083\pm0.029\pm0.006$
y_+	$-0.021\pm0.048\pm0.004\pm0.009$	$-0.032\pm0.077\pm0.008\pm0.006$	$0.045\pm0.106\pm0.036\pm0.008$
x_-	$0.060\pm0.039\pm0.007\pm0.006$	$-0.104\pm0.051\pm0.019\pm0.002$	$0.075\pm0.096\pm0.029\pm0.007$
y_-	$0.062\pm0.045\pm0.004\pm0.006$	$-0.052\pm0.063\pm0.009\pm0.007$	$0.127\pm0.095\pm0.027\pm0.006$
r_B	0.096 ± 0.029	$0.133^{+0.042}_{-0.039}$	$0.149^{+0.066}_{-0.062}$
δ_B (mod 180°)	$(119^{+19}_{-20})^\circ$	$(-82\pm21)^\circ$	$(111\pm32)^\circ$

Table 2. GGSZ analysis results: $x_{\pm}$, $y_{\pm}$, r_B, and δ_B. For $x_{\pm}$ and $y_{\pm}$ the errors are statistical, systematic and Dalitz-model, respectively. For r_B and δ_B the error is statistical and systematic combined. For $B\pm\to DK^{*\pm}$ decay we report the value of kr_B, where $k = 0.9\pm0.1$ takes into account the K^* finite width.

A frequentist approach is used to obtain r_B, δ_B, and γ for each decay mode. Results of this analysis are reported in Fig. 3. The values of r_B and δ_B for each B decay mode are reported in Table 2. r_B is found to be ≈ 0.1, as expected by the theory. CKM angle γ is found to be equal to $(68\pm14\pm4\pm3)^\circ$ (mod 180°), where the three uncertainties are statistical, systematic and Dalitz-model, respectively. The distance d between (x_+, y_+) and (x_-, y_-) is sensitive to direct CP-violation ($d = 0$ in case of no CPV). Results of the analysis indicate a 3.5σ evidence of direct CPV.

6.2. ADS method: $B^{\pm}\to D^{(*)}K^{\pm}$, $D\to K^{\pm}\pi^{\mp}$

The $B^{\mp}\to DK^{\mp}$ and $D^*K^{\mp}$ ($D^*\to D\gamma$ and $D\pi^0$) decays are studied, where D^0 decays to the $K^+\pi^-$ final state [25]. Events that have same (right) sign kaons are produced in CKM favored decays. Events that have opposite (wrong) sign kaons are produced through a CKM favored

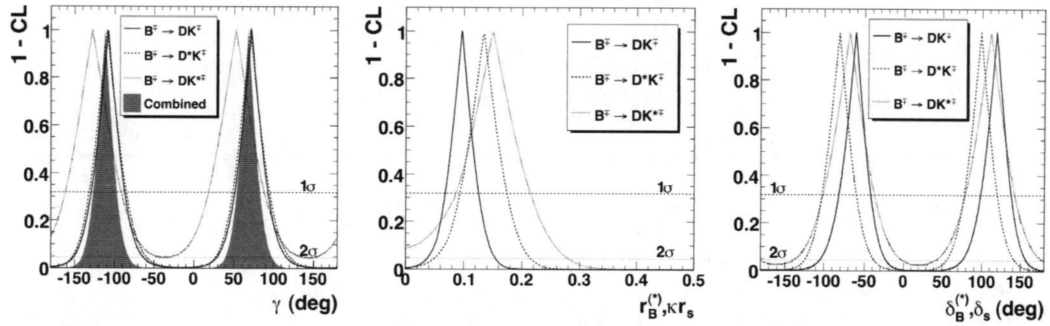

Figure 3. 1-CL as a function of γ (left), r_B (center) and δ_B (right) from the $B \to D^{(*)}K^{(*)}$ Dalitz-plot analysis.

(suppressed) B decay, followed by a CKM suppressed (favored) D decay. The ADS method [28] exploit the interference between these decay chains. Since the total suppression factor is equal for the two decay chains, interfering amplitudes have similar magnitude, thus large interference effects are expected. On the other side, the large suppression implies a $BF \approx O(10^{-7})$. Defining the wrong-to-right sign decay amplitude ratio $R^{\pm} \equiv \frac{\Gamma([K^{\mp}\pi^{\pm}]_D K^{\pm})}{\Gamma([K^{\pm}\pi^{\mp}]_D K^{\pm})}$, the following definitions hold

$$R_{ADS} = \frac{1}{2}(R^+ + R^-) = r_B^2 + r_D^2 \pm 2r_B r_D \cos(\delta_B + \delta_D)\cos\gamma, \quad (5)$$

$$A_{ADS} = \frac{R^- - R^+}{R^- + R^+} = \frac{2r_B r_D \sin\gamma \sin(\delta_B + \delta_D)}{R_{ADS}}, \quad (6)$$

where r_D and δ_D are the ratio and the relative phase between the CKM suppressed and the CKM favored amplitude for D decay, and the $+(-)$ sign in Eq. (5) is used for D and $D^* \to D\pi^0$ ($D^* \to D\gamma$) decays.

To enhance signal purity, a tightΔ E cut is applied. Specialized selection criteria are applied to suppress $K - \pi$ misidentification and $B \to D\pi$, $D \to K^+K^-$ decays, that are the main sources of $m_{\rm ES}$-peaking background. The signal yields, R_{ADS}, and A_{ADS} are determined from fits to $m_{\rm ES}$ and $\mathcal{SC}$. In Table 3 we report the measured values for the signal yield, R_{ADS}, A_{ADS}, and the signal statistical significance (including systematic uncertainties), for each decay mode. A frequentist approach is used to extract the unknowns of Eq. (5)–(6) from the measured

	$B^{\mp} \to DK^{\mp}$	$B^{\mp} \to D^*_{D\pi^0}K^{\mp}$	$B^{\mp} \to D^*_{D\gamma}K^{\mp}$
Wrong-sign Signal Yield	19.4 ± 9.6	10.3 ± 5.5	5.9 ± 6.4
Stat. Sign. (σ)	2.1	2.2	–
R_{ADS} (10^-2)	$1.1 \pm 0.6 \pm 0.2$	$1.8 \pm 0.9 \pm 0.4$	$1.3 \pm 1.4 \pm 0.8$
A_{ADS}	$-0.86 \pm 0.47^{+0.12}_{-0.16}$	$+0.77 \pm 0.35 \pm 0.12$	$+0.36 \pm 0.94^{+0.25}_{-0.41}$

Table 3. Results for $B \to DK$ ADS analysis.

observables. The values of δ_D and r_D are fixed to those reported in [31]. Results are reported in Fig. 4. Despite a poor sensitivity to γ, which is bound to be $54^\circ < \gamma < 83^\circ$, a good determination of r_B is obtained: $r_B^{DK^{\pm}} = 0.095^{+0.051}_{-0.041}$, $r_B^{D^*K^{\pm}} = 0.096^{+0.035}_{-0.051}$.

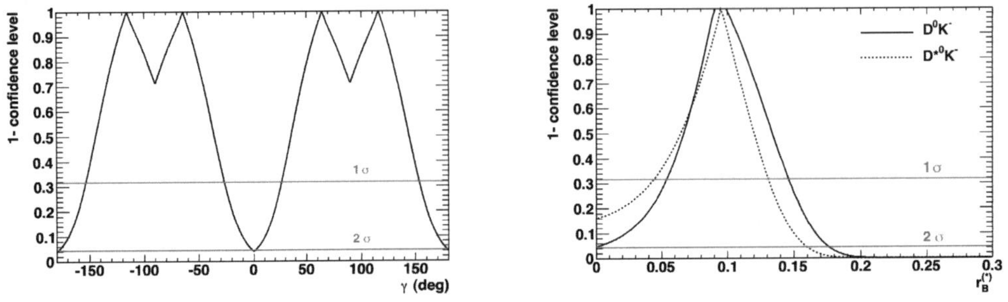

Figure 4. 1-CL as a function of γ (left) and r_B (right) from the $B \to D^{(*)}K$ ADS study.

6.3. GLW method: $B^{\pm} \to DK^{(*)\pm}$, $D \to f_{(\mathrm{CP})}$

In the GLW method [29], the $b \to c\bar{u}s$ and $b \to u\bar{c}s$ amplitude interference is studied via D meson decays into CP eigenstates. The decay amplitudes are used to build the following quantities

$$R_{CP\pm} = \frac{\Gamma(B^- \to D^0_{CP}K^-) + \Gamma\,(B^+ \to D^0_{CP}K^-) +}{\Gamma(B^- \to D^0 K^-) + \Gamma\,(B^+ \to \overline{D}^0 K^-) +} = 1 + r_B^2 + 2\lambda_{CP} r_B \cos\gamma \cos\delta_B, \quad (7)$$

$$A_{CP\pm} = \frac{\Gamma(B^- \to D^0_{CP}K^-) - \Gamma(B^+ \to D^0_{CP}K^-) +}{\Gamma(B^- \to D^0_{CP}K^-) - \Gamma(B^+ \to D^0_{CP}K^-) +} = \frac{2\lambda_{CP} r_B \sin\gamma \sin\delta_B}{R_{CP}}, \quad (8)$$

where D^0_{CP} (D^0) indicates a D decay into a CP (flavor) eigenstate, and λ_{CP} the CP-eigenvalue of the final state. $B^{\pm} \to DK^{\pm}$ decays are reconstructed, with D mesons decaying to CP-even (K^+K^-, $\pi^+\pi^-$), CP-odd ($K^0_S\pi^0$, $K^0_S\phi$, $K^0_S\omega$), and flavor ($K^-\pi^+$) eigenstates [26]. The signal yields are measured, and the partial decay rates determined with a ML fit to m_{ES}, ΔE and $\mathcal{SC}$. The fitted signal yield is about 500 events for both CP-even and CP-odd final states. A_{CP+} (A_{CP-}) and R_{CP+} (R_{CP-}) are extracted from data and are equal to $0.25 \pm 0.06 \pm 0.02$ ($-0.09 \pm 0.07 \pm 0.02$) and $1.18 \pm 0.09 \pm 0.05$ ($1.07 \pm 0.08 \pm 0.04$), respectively. The four observables of Eq. (7)–(8) are used to obtain γ, r_B and δ_B, using a frequentist approach. The results are $0.24 < r_B < 0.45$ ($0.06 < r_B < 0.51$) and, modulo 180°, $11.3° < \gamma < 22.7°$ or $80.9° < \gamma < 99.1°$ or $157.3° < \gamma < 168.7°$ ($7.0° < \gamma < 173.0°$) at the 68% (95%) CL, and are shown in Fig. 5.

In order to compare these results with those from GGSZ method [24], $x_{\pm} = \frac{1}{4}\,[R_{CP+}(1 \mp A_{CP+}) - R_{CP-}(1 \mp A_{CP-})]$ is computed: $x_+ = -0.057 \pm 0.039 \pm 0.015$ and $x_- = 0.132 \pm 0.042 \pm 0.018$. Data from the $D \to K^0_S\phi, \phi \to K^+K^-$ decay are not used to determine $x_{\pm}$, since they are already used in the GGSZ analysis [24].

7. Conclusion

We have reported results for α and γ measurement at *BABAR*. α is measured in $B \to \rho\rho$ decay at 6° level. A novel independent measurement of α in $B \to a_1(1260)\pi$ decay is performed. The measurement of angle γ using the full *BABAR* dataset is performed using three different techniques. Results of the different techniques are consistent inside the experimental uncertainties. The angle γ is measured with a precision of about 15°. The hadronic parameter r_B is found to be ≈ 0.1, as expected by the theory. Finally, evidences of direct CPV at 3.5σ level are reported in $B \to D^{(*)}K$ decays.

Acknowledgments

I would like to thank all my *BABAR* collaborators and especially F. Palombo, A. Gaz, and V. Poireau for their help in preparing this paper.

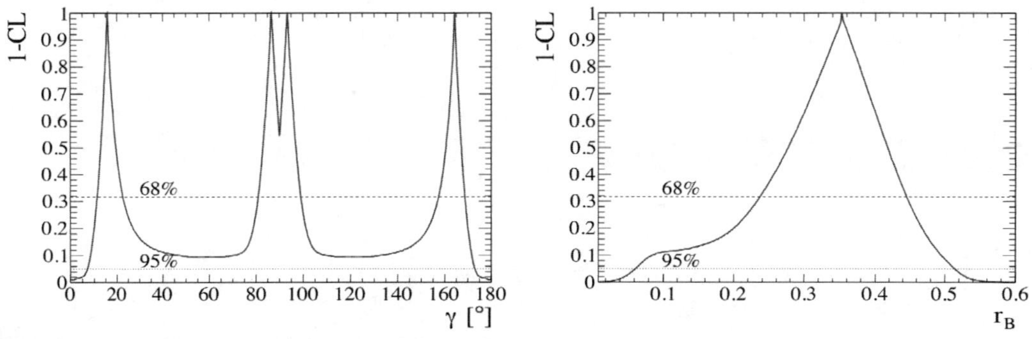

Figure 5. 1-CL as a function of γ mod 180° (left) and r_B (right) from the $B \to DK$ GLW study.

References

[1] N. Cabibbo, Phys. Rev. Lett. **10**, 531 (1963); M. Kobayashi and T. Maskawa, Prog. Theor. Phys. **49**, 652 (1973).

[2] K.-F. Chen *et al.* (Belle Colalboration), Phys. Rev. Lett. **98**, 031802 (2007); B. Aubert *et al.* (*BABAR* Collaboration), Phys. Rev. D **79**, 072009 (2009).

[3] B. Aubert *et al.* (*BABAR* Collaboration), Nucl. Instrum. Methods Phys. Res., Sect. A **479**, 1 (2002).

[4] B. Aubert *et al.* (*BABAR* Collaboration), Phys. Rev. Lett. **94**, 161803 (2005).

[5] A. L. Kagan, Phys. Lett. B **601**, 151 (2004).

[6] Charge conjugation is implied through the paper, unless otherwise specified.

[7] B. Aubert *et al.* (*BABAR* Collaboration), Phys. Rev. D **76**, 052007 (2007).

[8] B. Aubert *et al.* (*BABAR* Collaboration), Phys. Rev. Lett. **102**, 141802 (2009).

[9] B. Aubert *et al.* (*BABAR* Collaboration), Phys. Rev. D **78**, 071104R (2008).

[10] M. Gronau and D. London, Phys. Rev. Lett. **65**, 3381 (1990).

[11] M. Gronau and J. Zupan, Phys. Rev. D **71**, 074017 (2005).

[12] A. F. Falk, Z. Ligeti, Y. Nir, and H. Quinn, Phys. Rev. D **69**, 011502 (2004).

[13] B. Aubert *et al.* (*BABAR* Collaboration), Phys. Rev. Lett. **97**, 261801 (2006).

[14] J. Charles *et al.* (CKMfitter Group), Eur. Phys. J. C **41**, 1-131 (2005), updated results and plots available at `http://ckmfitter.in2p3.fr`; M. Ciuchini *et al.* (UTFIT Group), JHEP **107**, 13 (2001), updated results and plots available at `http://www.utfit.org`.

[15] M. Gronau and J. Zupan, Phys. Rev. D **73**, 057502 (2006);

[16] M. Gronau, Phys. Lett. B **265**, 389 (1991).

[17] B. Aubert *et al.* (*BABAR* Collaboration), Phys. Rev. Lett. **97**, 051802 (2006); Phys. Rev. Lett. **100**, 051803 (2008).

[18] B. Aubert *et al.* (*BABAR* Collaboration), Phys. Rev. Lett. **98**, 181803 (2007).

[19] B. Aubert *et al.* (*BABAR* Collaboration), Phys. Rev. D **81**, 052009 (2010).

[20] C. Daum *et al.* (ACCMOR Collaboration), Nucl. Phys. **B187**, 1 (1981).

[21] I. J. R. Aitchison, Nucl. Phys. A **189**, 417 (1972).

[22] M. Gronau and J. Zupan, Phys. Rev. D **70**, 074031 (2004).

[23] J. C. R. Bloch *et al.*, Phys. Rev. D **60**, 111502(R) (1999); H.-Y. Cheng and K.-C. Yang, Phys. Rev. D **76**, 114020 (2007).

[24] P. del Amo Sanchez *et al.* (*BABAR* Collaboration), Phys. Rev. Lett. **105**, 121801 (2010).

[25] P. del Amo Sanchez *et al.* (*BABAR* Collaboration), Phys. Rev. D **82**, 072006 (2010).

[26] P. del Amo Sanchez *et al.* (*BABAR* Collaboration), Phys. Rev. D **82**, 072004 (2010).

[27] A. Giri, Y. Grossman, A. Soffer and J. Zupan, Phys. Rev. D **68**, 054018 (2003).

[28] D. Atwood, I. Dunietz and A. Soni, Phys. Rev. Lett. **78**, 3257 (1997).

[29] M. Gronau and D. Wyler, Phys. Lett. **B265**, 172; M. Gronau and D. London, Phys. Lett. **B253**, 483 (1991).

[30] P. del Amo Sanchez *et al.* (*BABAR* Collaboration), Phys. Rev. Lett. 105, 081803 (2010).

[31] E. Barbiero *et al.* (HFAG Group), "Avergages of b-hadron and c-hadron Properties at the end of 2007", arXiv:0808.1297v3.

Symposium on Prospects in the Physics of Discrete Symmetries IOP Publishing
Journal of Physics: Conference Series **335** (2011) 012037 doi:10.1088/1742-6596/335/1/012037

CP Violation in B Decays at Belle

Masaya Iwabuchi

Department of Physics, Yonsei University, Seoul 120-749, Korea

E-mail: `iwabuchi@post.kek.jp`

Abstract.
I review the recent Belle measurements on CP violation in B decays. In this review, I cover the experimental constraints on the Cabibbo-Kobayashi-Maskawa (CKM) angles of the Unitarity Triangle by introducing several experimental approaches. Search for new physics beyond the Standard Model in $b \to s\bar{q}q$ transitions is also presented.

1. Introduction

The present status on the measurement of CP violation in the quark sector is mainly based on the results achieved by BaBar and Belle. Those B-factory experiments, already ended their operations, have been spectacularly successful at discovering and investigating CP violation in B-meson decays, thanks to the great success of PEP-II and KEKB asymmetric-energy e^+e^- colliders operating at the center of mass energy corresponding to the mass of $\Upsilon(4S)$. Those experiments have significantly improved our knowledge on CP-violation phenomena over a decade.

In the Standard Model (SM), CP violation arises from a complex phase in the Cabibbo-Kobayashi-Maskawa (CKM) quark mixing matrix [1, 2], where CP violation can be quantified in terms of three parameters, ϕ_1, ϕ_2 and ϕ_3[1], representing the three angles of the so-called Unitarity Triangle. Precise measurements of CKM angles are therefore of fundamental importance for the description of the weak interaction of quarks and the investigation for the new sources of CP violation.

In these proceedings, measurements of CKM angles at the Belle experiment are reviewed. This review also includes the measurement of effective ϕ_1 (ϕ_1^{eff}) in $b \to s\bar{q}q$ transitions for new-physics search.

2. ϕ_1

$\sin 2\phi_1$ can be obtained using $b \to c\bar{c}s$ transitions, where the most precise measurement for ϕ_1 comes from $B^0 \to J/\psi K^0$ decays.

With 535 million $B\overline{B}$ pairs, Belle updated the measurement of $\sin 2\phi_1$ using $B^0 \to J/\psi K^0$ decays [3]. Figure 1 shows the results of a fit to the reconstructed candidates in the fit region. Several kinematic variables are used here; for $B^0 \to J/\psi K^0_S$ decay, the energy difference $\Delta E \equiv E^*_B - E^*_{\text{beam}}$ and the beam-energy constrained mass $M_{\text{bc}} \equiv \sqrt{(E^*_{\text{beam}})^2 - (p^*_B)^2}$ are used in the fit, where E^*_{beam} is the beam energy, and E^*_B and p^*_B are the energy and momentum,

[1] The angles ϕ_1, ϕ_2 and ϕ_3 are also sometimes known as β, α and γ, respectively.

respectively, of the reconstructed B candidate, all measured in the center-of-mass system. p_B^* is used in the fit for $B^0 \to J/\psi K_L^0$ candidates.

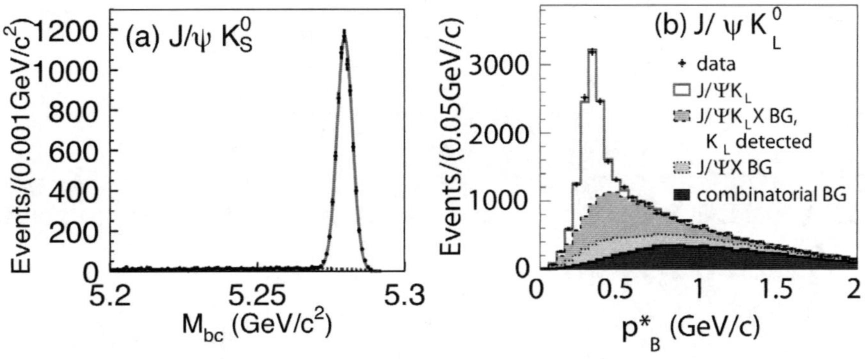

Figure 1. (a) $M_{\rm bc}$ distribution in theΔ E signal region for selected $B^0 \to J/\psi K_S$ candidates and (b) p_B^* distribution for selected $B^0 \to J/\psi K_L$ candidates.

The signal yields are determined to be 7484 ± 87 for $B^0 \to J/\psi K_S$ and 6512 ± 123 for $B^0 \to J/\psi K_L$, where errors are statistical only.

In the decay chain $\Upsilon(4S) \to B^0\overline{B}^0 \to f_{CP} f_{\rm tag}$, where one of the B mesons decays at time t_{CP} to a CP eigenstate f_{CP} and the other decays at time $t_{\rm tag}$ to a final state $f_{\rm tag}$ that distinguishes between B^0 and $\overline{B}^0$, the decay rate has a time dependence given by

$$\mathcal{P}(\Delta t) = \frac{e^{-|\Delta t|/\tau_{B^0}}}{4\tau_{B^0}}\left\{1 + q \cdot \left[\mathcal{S}_{CP}\sin(\Delta m_d \Delta t) + \mathcal{A}_{CP}\cos(\Delta m_d \Delta t)\right]\right\}.$$

Here $\mathcal{S}_{CP}$ and $\mathcal{A}_{CP}$ describe mixing-induced and direct CP-violation, respectively, and τ_{B^0} is the B^0 lifetime,Δ m_d is the mass difference between the two B^0 mass eigenstates,Δ $t = t_{CP} - t_{\rm tag}$, and the b-flavor charge $q = +1$ (-1) when the tagging B meson is a B^0 ($\overline{B}^0$). Here, the SM predicts $S_f = -\xi_f \sin 2\phi_1$ and $\mathcal{A}_{CP} = 0$ for both $b \to c\bar{c}s$ and $b \to s\bar{q}q$ transitions, where $\xi_f = +1$ (-1) corresponds to the CP-even (-odd) final states.Δ t can be determined from the displacement in the electron beamline (z) between the f_{CP} and $f_{\rm tag}$ decay vertices: $\Delta t \simeq (z_{CP} - z_{\rm tag})/(\beta\gamma c) \equiv \Delta z/(\beta\gamma c)$.

An unbinned maximum-likelihood fit to the observedΔ t distribution is performed to determine CP-violation parameters. Figures 2 shows theΔ t distributions and asymmetry. $\sin 2\phi_1$ is measured to be $0.642 \pm 0.031 \pm 0.017$ by the results of the fits toΔ t distributions. Here, the first error is statistical and the second error is systematic.

$\sin 2\phi_1$ will be updated soon at Belle using the full data set (772 million $B\overline{B}$ pairs). The magnitude of the statistical error would be comparable with that of the systematic error, where the statistical error is expected to be around 0.02.

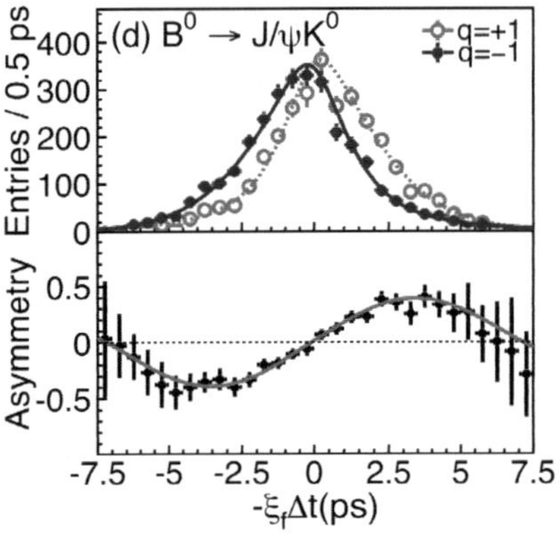

Figure 2. Background subtractedΔ t distributions and asymmetries for events with good tags for $B^0 \to J/\psi K^0$.

3. ϕ_2

$b \to u\overline{u}d$ quark transitions such as $B^0 \to \pi\pi, \rho\pi, \rho\rho$ are sensitive to ϕ_2.

The CP-violation parameters are expected to be $\mathcal{S}_{CP} = \sin 2\phi_2$ and $\mathcal{A}_{CP} = 0$ by the SM with the tree diagram only, however, direct CP violation is possible, $\mathcal{A}_{CP} \neq 0$, due to a large contribution from the electroweak penguin-mediated decay processes. ϕ_2 can be determined by $\mathcal{S}_{CP} = \sqrt{1 - \mathcal{A}^2_{CP}} \sin(2\phi_2 + \kappa)$, where κ is the shift due to the penguin contribution and determined by the SU(2) isospin analysis [4]. For the ϕ_2 measurements through $B^0 \to \pi^+\pi^-$ and $B^0 \to \rho^+\rho^-$ modes, it is possible to determine κ by considering the set of three $B \to hh$ decays where hh is either two pions or two longitudinally polarized ρ mesons. The $B \to hh$ amplitudes obey the complex triangle relations,

$$A_{+0} = \frac{1}{\sqrt{2}} A_{+-} + A_{00}, \quad \overline{A}_{-0} = \frac{1}{\sqrt{2}} \overline{A}_{+-} + \overline{A}_{00}.$$

The electroweak penguins can be neglected in the $B^+ \to h^+h^0$, and thus these triangles share the same base, $A_{+0} = \overline{A}_{-0}$. κ can be determined from the difference between the two triangles. This method has an 8-fold discrete ambiguity in the determination of ϕ_2.

3.1. ϕ_2 from $B^0 \to \pi^+\pi^-$

ϕ_2 measurement is performed using $B^0 \to \pi^+\pi^-$ mode with 535 million $B\overline{B}$ pairs [5].

The signal yield is estimated by a fit to data, making use ofΔ E, $M_{\rm bc}$ and the kaon identification probability. $B^0 \to \pi^+\pi^-$ signal yield is obtained to be 1464 ± 65 events. In the time-dependent analysis, direct CP violation is observed, $A_{CP} = +0.55 \pm 0.08(\text{stat}) \pm 0.05(\text{syst})$ with a significance of 5.5σ, while mixing-induced CP violation is also observed, $S_{CP} = -0.61 \pm 0.10(\text{stat}) \pm 0.04(\text{syst})$ with a significance greater than 5.3σ for any A_{CP} value.

To constrain ϕ_2, the isospin relations are used with measured values of S_{CP} and A_{CP}, the world average (W.A.) branching ratios of $B^0 \to \pi^+\pi^-$, $\pi^0\pi^0$, and $B^+ \to \pi^+\pi^0$, and the W.A. direct CP-asymmetry for $B^0 \to \pi^0\pi^0$ [6]. χ^2 is constructed using the measured values and the six parameters, one of which is ϕ_2. Four different solutions are found after the minimization of the χ^2. With the expectation from other CKM measurements, $(97 \pm 11)^\circ$ is found. The ϕ_2 range $11^\circ < \phi_2 < 79^\circ$ is excluded at the 95% confidence level (C.L.). The resulting two-dimensional confidence regions in the S_{CP} and A_{CP} plane and the difference $1 -$ C.L. plotted for a range of ϕ_2 values are shown in Fig. 3.

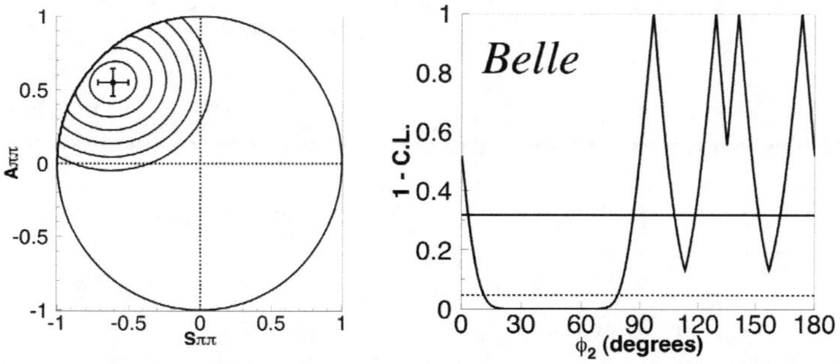

Figure 3. The left plot shows confidence regions for S_{CP} and A_{CP}. The curves show the contour from 1σ to 6σ, from inside to outside. The point with error bars is the S_{CP} and A_{CP} measurement. The right plot shows difference $1 - \text{C.L.}$ plotted for a range of ϕ_2 values obtained with an isospin analysis. The solid and dashed lines indicate C.L. $= 68.3\%$ and 95%, respectively.

3.2. ϕ_2 from $B^0 \to \rho^+\rho^-$

Possible relative angular momentum for the $B^0 \to \rho^+\rho^-$ decay is $L = 0, 1, 2$. The CP-violating parameters receive contributions from a longitudinally polarized state (CP-even) and two transversely polarized states (an admixture of CP-even and CP-odd states). However, measurements of the polarization fraction by Belle and BaBar show that the longitudinal polarization fraction is near unity. There is no need to elaborate the angular analysis, and pure CP-even state (100% longitudinal polarization) is assumed in this analysis.

Time-dependent analysis is performed for $B^0 \to \rho^+\rho^-$ mode with 535 million $B\overline{B}$ pairs [7]. ΔE, $M_{\rm bc}$ and event shape variables for $e^+e^- \to q\overline{q}$ ($q = u, d, s, c$) continuum and $B\overline{B}$ events are used in a fit to estimate the signal yield. $B^0 \to \rho^+\rho^-$ signal yield is obtained to be 576 ± 53 events. Direct CP violation and mixing-induced CP violation are obtained to be $A_{CP} = +0.16 \pm 0.21(\text{stat}) \pm 0.08(\text{syst})$, $S_{CP} = +0.19 \pm 0.30(\text{stat}) \pm 0.08(\text{syst})$, respectively.

In the isospin relations to constrain ϕ_2, the W.A. branching ratios of $B^0 \to \rho^+\rho^-$ and $B^+ \to \rho^+\pi^0$ are used [6], while the branching ratio of $B^0 \to \rho^0\rho^0$ is obtained from the Belle measurement [8] to constrain ϕ_2. ϕ_2 is determined to be $(91.7 \pm 14.9)^\circ$. Figure 4 shows the difference $1 - \text{C.L.}$ plotted for ϕ_2.

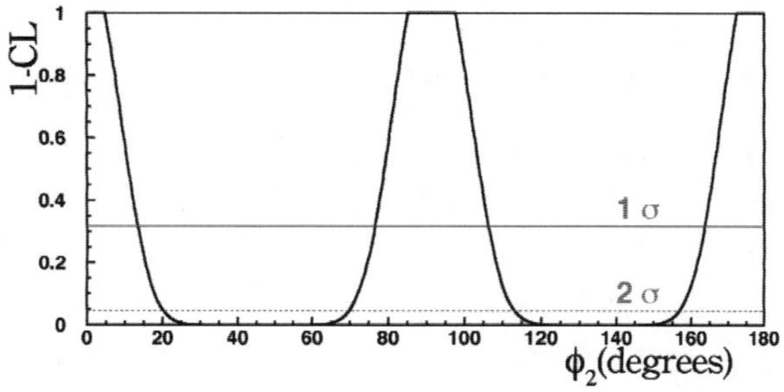

Figure 4. $1 - \text{C.L.}$ vs. ϕ_2 obtained from the isospin analysis of $B \to \rho\rho$ decays.

3.3. ϕ_2 from $B^0 \to \rho\pi$

$B^0 \to \rho\pi$ mode is also sensitive to ϕ_2. Since this mode can be treated as an interference of $B^0 \to \rho^+\pi^-$, $\rho^-\pi^+$ and $\rho^0\pi^0$ modes, a time-dependent Dalitz plot analysis (TDPA) is employed, where measurements of interfering parameters must be included. Measurements for relative sizes and complex phases of the amplitudes are required to extract ϕ_2, however, those amplitudes are related to ϕ_2 through an isospin relation. In this point, $B^0 \to \rho\pi$ channel has an advantage which can reduce discrete ambiguities, in contrast to the previous quasi-two-body modes for the ϕ_2 measurements.

With 449 million $B\overline{B}$ pairs, Belle performed a measurement of ϕ_2 using $B^0 \to \rho\pi$ [9]. The signal yield is determined from a four-dimensional fit to the ΔE-$M_{\rm bc}$ and Dalitz plot distribution. The fit yields 971 ± 42 $B^0 \to \pi^+\pi^-\pi^0$ events in the signal region. To constrain ϕ_2, measurements with results on the branching fractions for $B^0 \to \rho^\pm\pi^\mp$ and $B^+ \to \rho^+\pi^0$, $\rho^0\pi^+$, and flavor asymmetries of the latter two [6] are all combined. χ^2 scans are shown in Fig. 5 where ϕ_2 can be constrained to be $68° < \phi_2 < 95°$ at 68.3% C.L.

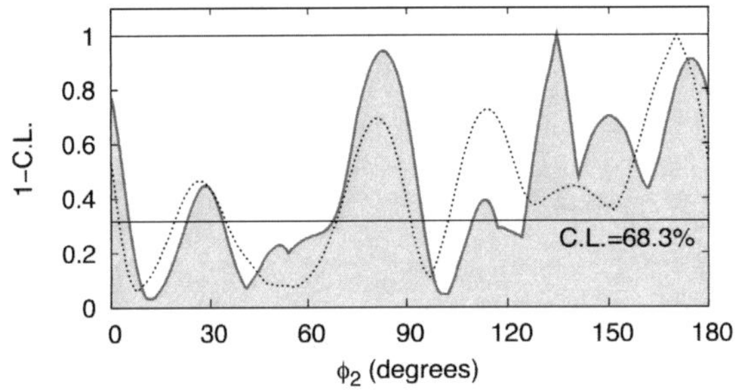

Figure 5. $1 - $ C.L. vs ϕ_2. Dotted and solid curves correspond to the result from the TDPA only and that from the TDPA and an isospin combined analysis, respectively.

4. ϕ_3

$B \to D^{(*)0}K$ modes provide a good probe for the ϕ_3 measurement, where the interference between $B \to D^{(*)0}K$ and $B \to \overline{D}^{(*)0}K$ decays is utilized. The magnitude of the amplitude ratio of these decays (r_B) can be evaluated to be ~ 0.1, using the CKM matrix elements with an additional color suppression factor. The weak phase ϕ_3 arises from the complex phase of V_{ub} included in the $B^+ \to D^{(*)0}K^+$ decay. There is no contribution from the penguin diagram in these decays.

Among the several approaches proposed for the ϕ_3 measurement, the current most precise measurement of ϕ_3 is based on the so-called GGSZ method [10]. In this method, amplitude analysis of the multibody decay of the neutral D meson is required. However, complex phase variations and imperfect assumption of the D-decay model potentially includes the irreducible uncertainties for the ϕ_3 measurement. In other approaches, the so-called ADS method [11] is also considered to be promising. The ADS method can be free from the model uncertainty, while the measurements based on this method have been statistically limited.

4.1. ϕ_3 from the GGSZ method

In the $B^\pm \to DK^\pm$ interference, D can be either D^0 or $\overline{D}^0$, and the amplitude of D can be expressed as

$$|\tilde{D}^0\rangle = r_B e^{\pm i\phi_3 + i\delta_B}|D^0\rangle + |\overline{D}^0\rangle.$$

Here δ_B is the strong phase difference of the B decays. It is possible to distinguish D^0 from $\overline{D}^0$ if the multibody D decays are used.

With 657 million $B\overline{B}$ pairs, Belle updated the ϕ_3 measurement using the GGSZ method with $B^+ \to D^{(*)}K^+$, $D \to K_S\pi^+\pi^-$ decays [12]. This measurement is based on the model-dependent approach, including the model determination of D-meson decay using the $e^+e^- \to q\overline{q}$ continuum processes.

The signal yield is estimated by a two-dimensional fit for the ΔE and $M_{\rm bc}$ distribution for each decay mode. Signal yields of $B^+ \to DK^+$, $B^+ \to D^*K^+$ followed by $D^* \to D\gamma$ or $D^* \to D\pi^0$ are obtained to be 756, 149 and 141 events in each signal region, respectively.

To obtain ϕ_3, Dalitz distributions of the B^+ and B^- are fitted separately, using Cartesian parameters $x_\pm = r_\pm \cos(\pm\phi_3 + \delta)$ and $y_\pm = r_\pm \sin(\pm\phi_3 + \delta)$, where the indices "+" and "−" correspond to B^+ and B^- decays, respectively. Here, r_+ and r_- are not constrained to be equal. The advantage of this approach is for the low bias and simple distribution of the fitted parameters. Several kinematic variables are used in a fit; Dalitz plot distribution,Δ E, $M_{\rm bc}$ and event shape variables that separates the B decays from the $e^+e^- \to q\overline{q}$ continuum processes. The results of the separate B^+ and B^- data fits are shown in Fig. 6. The separation of the B^- and B^+ positions in the (x, y) plane indicates direct CP violation and is measured to be $2r_B|\sin\phi_3|$.

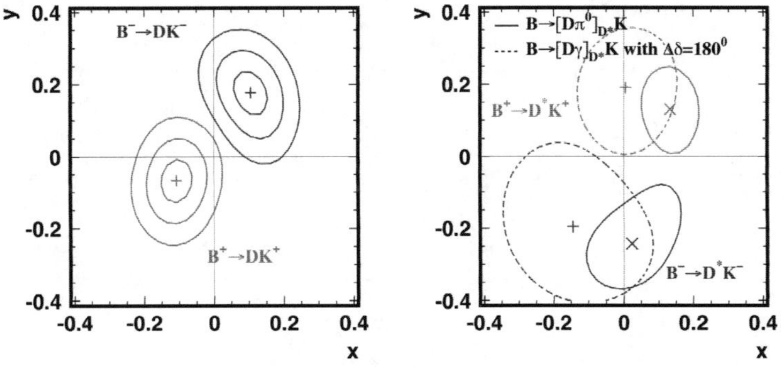

Figure 6. Results of signal fits with free parameters $x = r\cos\theta$ and $y = r\sin\theta$ for $B^+ \to DK^+$ (a) and $B^+ \to D^*K^+$ (b) samples, separately for B^- and B^+ data. Contours indicate 1, 2, and 3 (for $B^\pm \to DK^\pm$) and 1 standard deviation regions (for $B^\pm \to D^*K^\pm$).

A frequentist technique is used to evaluate the statistical significance of the measurements. In order to improve the sensitivity to ϕ_3, event samples of $B^+ \to DK^+$ and $B^+ \to D^*K^+$ are combined. ϕ_3 for the combination of these modes is obtained to be $\phi_3 = 78.4°{}^{+10.8°}_{-11.6°}({\rm stat}) \pm 3.6°({\rm syst}) \pm 8.9°({\rm model})$. CP conservation in this process is ruled out at the confidence level $(1 - {\rm C.L.}) = 5 \times 10^{-4}$, or 3.5 standard deviations, which leads to the evidence for direct CP violation for the $B^+ \to D^{(*)0}K^+$ process.

Model uncertainty is dominated by the uncertainty due to the complex phase. To reduce this model uncertainty, model-independent approach is proposed [10, 13, 14]. Based on this method, CLEO recently updated the results of relative phase between D^0 and $\overline{D}^0$ by studying the $\psi(3770) \to D^0\overline{D}^0$ decays [15]. The model error in the ϕ_3 measurement will be replaced by a statistical error of about a few degrees due to the finite $\psi(3770) \to D^0\overline{D}^0$ sample, while the statistical error associated with the B data sample should increase by 10%-20% due to the binned fit procedure.

4.2. ϕ_3 from the ADS method

In the ADS method, measurements of CP asymmetry ($\mathcal{A}$) and partial rate ($\mathcal{R}$) play an important role for the ϕ_3 measurement. In the decay chain $B^\pm \to DK^\pm$, $D \to K^\pm\pi^\mp$, where D can be either D^0 or $\overline{D}^0$, $\mathcal{A}$ and $\mathcal{R}$ are described by

$$\begin{aligned}
\mathcal{A} &\equiv \frac{\mathcal{B}(B^- \to [K^+\pi^-]_D K^-) - \mathcal{B}(B^+ \to [K^-\pi^+]_D K^+)}{\mathcal{B}(B^- \to [K^+\pi^-]_D K^-) + \mathcal{B}(B^+ \to [K^-\pi^+]_D K^+)} \\
&= 2r_B r_D \sin(\delta_B + \delta_D)\sin(\phi_3)/\mathcal{R}, \\
\mathcal{R} &\equiv \frac{\mathcal{B}(B^- \to [K^+\pi^-]_D K^-) + \mathcal{B}(B^+ \to [K^-\pi^+]_D K^+)}{\mathcal{B}(B^- \to [K^-\pi^+]_D K^-) + \mathcal{B}(B^+ \to [K^+\pi^-]_D K^+)} \\
&= r_B^2 + r_D^2 + 2r_B r_D \cos(\delta_B + \delta_D)\cos(\phi_3).
\end{aligned}$$

Here δ_D is the strong phase difference of the D-meson decays, and r_D is the amplitude ratio of $D^0 \to K^+\pi^-$, doubly Cabibbo suppressed mode, to $\overline{D}^0 \to K^+\pi^-$, Cabibbo favored mode. In the $B^- \to [K^+\pi^-]_D K^-$ interference, either B or D decay is suppressed, while the $B^- \to [K^-\pi^+]_D K^-$ decay process is dominated by the favored modes of B and D decays.

With 772 million $B\overline{B}$ pairs, Belle updated the measurement for the ADS method using the above decay chain, where preliminary results are obtained on $\mathcal{A}$ and $\mathcal{R}$. In this measurement, a tool for multi-variate analysis named NeuroBayes is employed. In order to separate the $B \to DK$ signal from backgrounds, ten kinematic variables are used as inputs for NeuroBayes, such as a likelihood ratio of Fisher discriminant of modified Fox-Wolfram moments, the distance of vertices between the reconstructed B and the other B, and the decay angle of $D \to K\pi$.

The signal yield is extracted by a two dimensional fit onΔ E and NeuroBayes output $\mathcal{NB}$, and determined to be $56.0^{+15.1}_{-14.2}$ for $B^- \to [K^+\pi^-]_D K^-$, where the charge-conjugate mode is included. Figure 7 shows the results of the fit to the reconstructed candidates in the fit region. TheΔ E distributions for $[K^+\pi^-]_D K^-$ and $[K^-\pi^+]_D K^+$ in the $\mathcal{NB}$ signal region are shown.

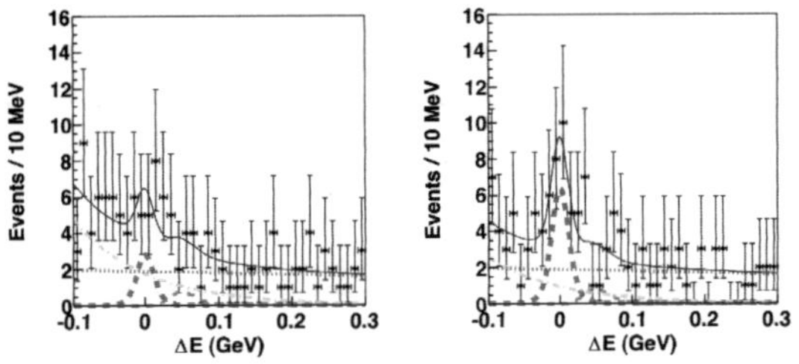

Figure 7. ΔE distributions for $[K^+\pi^-]_D K^-$ (left) and $[K^-\pi^+]_D K^+$ (right). Curves represent the various components from the fit: DK signal (thick dashed line), $D\pi$ backgrounds (dashed line), $B\overline{B}$ backgrounds (dash-dotted line), continuum (dotted line) and sum of all components (solid line).

The first evidence of the DK signal is obtained with a significance of 4.1σ including systematic uncertainties. The obtained results for the CP asymmetry and the ratio of the suppressed to favored decay rate are, $\mathcal{A} = -0.39^{+0.26}_{-0.28}(\mathrm{stat})^{+0.04}_{-0.03}(\mathrm{syst})$, and $\mathcal{R} = [1.63^{+0.44}_{-0.41}(\mathrm{stat})^{+0.07}_{-0.13}(\mathrm{syst})] \times 10^{-2}$. To constrain ϕ_3, r_D, δ_D and other relevant observables will be used.

5. ϕ_1^{eff}

Precise measurement of ϕ_1 also provides a useful comparison between decays mediated by $b \to s\overline{q}q$ and $b \to c\overline{c}s$ transitions. A new physics may enter into the loop diagram, and to search for the new physics effects, various approaches have been tested by investigating deviations of CP violating parameter $\phi_1^{\rm eff}$ from ϕ_1 determined by $b \to c\bar{c}s$ decays.

With 657 million $B\overline{B}$ pairs, Belle performed a measurement of $\phi_1^{\rm eff}$ using $B^0 \to K^+K^-K_S$ decay with a time-dependent Dalitz approach [16]. Several decay modes are included in the signal model: $\phi(1020)K_S$, $f_0(980)K_S$, f_XK_S, $\chi_{c0}K_S$ and nonresonant $K^+K^-K_S$.

The signal yield is estimated by a fit to data, making use ofΔ E, $M_{\rm bc}$ and the flavor-tag quality interval. $B^0 \to K^+K^-K_S$ signal yield is obtained to be 1176 ± 51 events. In the time-dependent analysis, four solutions are found from a fit. A single solution is preferred from all the solutions using the branching fractions of $f_0(980) \to \pi\pi$ and $f_0(1500) \to \pi\pi$ as external information. Figure 8 shows theΔ t distributions and asymmetry of the preferred solution.

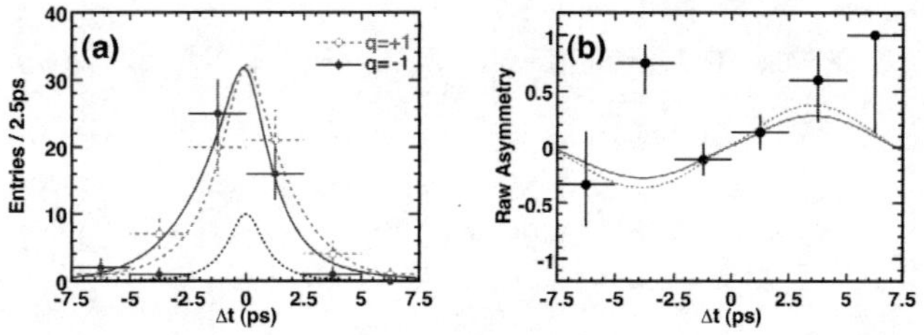

Figure 8. (a)Δ t distribution and (b) raw asymmetry for the $B^0 \to K^+K^-K_S$ candidates in the $\phi(1020)K_S^0$ region, with good tags. In (a), the dotted black curve shows the background component. In (b), the solid curve shows the fit projection, while the dashed curve shows the SM expectation from the time-dependent CP asymmetry measurement in $b \to c\bar{c}s$ decays.

In the preferred solution, $\phi_1^{\rm eff}$ is measured to be $(32.2 \pm 9.0 \pm 2.6 \pm 1.4)^\circ$ for $\phi(1020)K_S$ mode, and $(31.3 \pm 9.0 \pm 3.4 \pm 4.0)^\circ$ for $f_0(980)K_S$ mode. The results are still consistent with the SM expectation like other results from Belle, where measurements of $\phi_1^{\rm eff}$ are all consistent with the value of ϕ_1 obtained from the decay $B^0 \to J/\psi K^0$ within the two standard deviations.

6. Summary

Various studies have been performed to investigate CP violation in the B decays at Belle. The upgraded Belle II experiment will allow us to constrain the CKM angles with much higher statistics, and also help us to explore and clarify the nature of CP violation.

References

[1] N. Cabibbo, Phys. Rev. Lett. **10**, 531 (1963).
[2] M. Kobayashi and T. Maskawa, Prog. Theor. Phys. **49**, 652 (1973).
[3] K.-F. Chen *et al.* (Belle Collaboration), Phys. Rev. Lett. **98**, 031802 (2007).
[4] M. Gronau and D. London, Phys. Rev. Lett. **65**, 3381 (1990).
[5] H. Ishino *et al.* (Belle Collaboration), Phys. Rev. Lett. **98**, 211801 (2007).
[6] Heavy Flavor Averaging Group, `http://www.slac.stanford.edu/xorg/hfag`.
[7] A. Somov *et al.* (Belle collaboration), Phys. Rev. D **76**, 011104 (2007).
[8] C.-C. Chiang *et al.* (Belle collaboration), Phys. Rev. D **78**, 111102(R) (2008).
[9] A. Kusaka *et al.* (Belle collaboration), Phys. Rev. Lett. **98**, 221602 (2007).
[10] A. Giri, Yu. Grossman, A. Soffer and J. Zupan, Phys, Rev. D **68**, 054018 (2003).
[11] D. Atwood, I. Dunietz and A. Soni, Phys. Rev. Lett. **78**, 3357 (1997).
[12] A. Poluektov *et al.* (Belle collaboration), Phys, Rev. D **81**, 112002 (2010).
[13] A. Bondar and A. Poluektov, Eur. Phys. J. C **47**, 347 (2006).
[14] A. Bondar and A. Poluektov, Eur. Phys. J. C **55**, 51 (2008).
[15] J. Libby *et al.* (CLEO collaboration), Phys. Rev. D **82**, 112006 (2010).
[16] Y. Nakahama *et al.* (Belle collaboration), Phys. Rev. D **82**, 073011 (2010).

Journal of Physics: Conference Series **335** (2011) 012038 doi:10.1088/1742-6596/335/1/012038

Angular analysis of $B \to V(\to P_1P_2) + \bar{\ell}\ell$ decays

Christoph Bobeth[1,3], **Gudrun Hiller**[2] **and Danny van Dyk**[2]

[1] Institute of Advanced Study & Excellence Cluster Universe, Technische Universität München, D-85748 Garching, Germany

[2] Institut für Physik, Technische Universität Dortmund, D-44221 Dortmund, Germany

E-mail: `christoph.bobeth@ph.tum.de`

Abstract. The angular analysis of exclusive rare B-meson decays via intermediate vector mesons V into 4-body final states of two pseudo-scalars P_1, P_2 and a pair of light leptons $\ell = e, \mu$ offers a large set of observables. They can be used to test the electroweak short-distance couplings in the Standard Model and to search for New Physics. The two kinematic regions of low and high dilepton mass depend on short-distance physics in complementary ways and can be expanded in powers of $\Lambda_{\rm QCD}/m_b$. These expansions guide towards suitable combinations of observables allowing to *i*) reduce the hadronic uncertainties in the extraction of the short-distance couplings or *ii*) test the lattice QCD $B \to V$ form factors in short-distance independent combinations. Several such possibilities of CP-averaged and CP-asymmetric (T-even and T-odd) quantities are presented for $\bar{B}_d^0 \to \bar{K}^{*0}(\to K^-\pi^+) + \bar{\ell}\ell$ and time-integrated CP-asymmetries without tagging for $\bar{B}_s, B_s \to \phi(\to K^-K^+) + \bar{\ell}\ell$ decays in view of the latest B-factory and CDF results and the forthcoming LHCb measurements.

1. Introduction

The $b \to s + \bar{\ell}\ell$ induced flavour changing neutral current (FCNC) decays of the B-meson are well known to provide sensitive probes of the electroweak short-distance couplings of the Standard Model (SM) and to test scenarios beyond (BSM). Hitherto existing experimental results from the B-factory experiments Babar [1] and Belle [2] as well as the Tevatron experiment CDF [3] are in the ballpark of the SM predictions and provide constraints on the couplings $\mathcal{C}_{9,10}$ related to 4-Fermi operators $(\bar{s}\gamma_\mu P_L b)\,(\bar{\ell}\gamma^\mu\{1,\gamma_5\}\ell)$ [4, 5]. Further experimental progress is expected soon from the final analysis of the full data sets of Babar and Belle, as well as the CDF one with at least doubled statistics. Eventually, LHCb will dominate statistically for final states containing charged particles only, already with a rather low luminosity of a few fb^{-1} [6]. The Super-B factories are expected to contribute to $\ell = e$, just as the B-factories do presently.

Especially the angular analysis of the 4-body final state of $B \to V(\to P_1P_2) + \bar{\ell}\ell$ decays offers a large number of observables in the fully differential distribution [7, 8]. Here the intermediate V is assumed to be on-shell in the narrow-resonance approximation which restricts the number of kinematic variables to four[4]. Using $\bar{B}_d^0 \to \bar{K}^{*0}(\to K^-\pi^+) + \bar{\ell}\ell$ for illustration, they might be chosen as depicted in figure 1.

[3] Speaker
[4] The off-resonance case has been studied in [9].

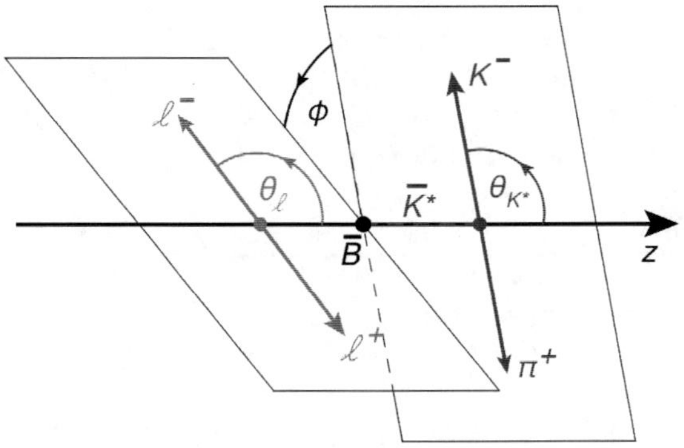

Figure 1. Kinematic variables of $\bar{B}_d^0 \to \bar{K}^{*0}(\to K^-\pi^+) + \bar{\ell}\ell$ decays: *i*) the $(\bar{\ell}\ell)$-invariant mass squared q^2, *ii*) the angle θ_ℓ between $\ell = \ell^-$ and $\bar{B}$ in the $(\bar{\ell}\ell)$ center of mass (c.m.), *iii*) the angle θ_{K^*} between K^- and $\bar{B}$ in the $(K^-\pi^+)$ c.m. and *iv*) the angle ϕ between the two decay planes spanned by the 3-momenta of the $(K\pi)$- and $(\bar{\ell}\ell)$-systems, respectively.

The differential decay rate, after summing over lepton spins, factorises into

$$\begin{aligned}\frac{8\pi}{3}\frac{d^4\Gamma}{dq^2\, d\cos\theta_\ell\, d\cos\theta_{K^*}\, d\phi} &= J_1^s \sin^2\theta_{K^*} + J_1^c \cos^2\theta_{K^*} + (J_2^s \sin^2\theta_{K^*} + J_2^c \cos^2\theta_{K^*})\cos 2\theta_\ell \\ &+ J_3 \sin^2\theta_{K^*} \sin^2\theta_\ell \cos 2\phi + J_4 \sin 2\theta_{K^*} \sin 2\theta_\ell \cos\phi + J_5 \sin 2\theta_{K^*} \sin\theta_\ell \cos\phi \\ &+ (J_6^s \sin^2\theta_{K^*} + J_6^c \cos^2\theta_{K^*})\cos\theta_\ell + J_7 \sin 2\theta_{K^*} \sin\theta_\ell \sin\phi \\ &+ J_8 \sin 2\theta_{K^*} \sin 2\theta_\ell \sin\phi + J_9 \sin^2\theta_{K^*} \sin^2\theta_\ell \sin 2\phi, \quad (1)\end{aligned}$$

that is, into q^2-dependent observables[5] $J_i^j(q^2)$ and the dependence on the angles θ_ℓ, θ_{K^*} and ϕ. No additional angular dependencies can be induced by any extension of the SM operator basis [11] as found by [12, 13]. The following simplifications arise in the limit $m_\ell \to 0$: $J_1^s = 3J_2^s$, $J_1^c = -J_2^c$ and $J_6^c = 0$.

The differential decay rate $d^4\bar{\Gamma}$ of the CP-conjugated decay $B_d^0 \to K^{0*}(\to K^+\pi^-) + \bar{\ell}\ell$ is obtained through the following replacements

$$J_{1,2,3,4,7}^j \to \bar{J}_{1,2,3,4,7}^j[\delta_W \to -\delta_W], \qquad J_{5,6,8,9}^j \to -\bar{J}_{5,6,8,9}^j[\delta_W \to -\delta_W], \quad (2)$$

due to $\ell \leftrightarrow \bar{\ell} \Rightarrow \theta_\ell \to \theta_\ell - \pi$ and $\phi \to -\phi$. The CP-violating (weak) phases δ_W are conjugated.

The angular distribution provides twice as many observables (J_i^j and $\bar{J}_i^j$) when the decay and its CP-conjugate decay are measured separately. This doubles again if the $\ell = e$ and μ lepton flavours are not averaged. Notably, CP-asymmetries can be measured in an untagged sample of B-mesons due to the presence of CP-odd observables ($i = 5, 6, 8, 9$) [7]. Moreover, T-odd observables $\sim \cos\delta_s \sin\delta_W$ ($i = 7, 8, 9$) are especially sensitive to weak BSM phases δ_W [10, 14] contrary to T-even ones $\sim \sin\delta_s \sin\delta_W$ ($i = 1, \ldots, 6$), since the CP-conserved (strong) phase δ_s is often predicted to be small. Note, that in the SM CP-violating effects in $b \to s$ are doubly-suppressed by the Cabibbo angle as $\mathrm{Im}[V_{ub}V_{us}^*/(V_{tb}V_{ts}^*)] \approx \bar{\eta}\lambda \sim 10^{-2}$.

The observables J_i^j are bilinear in the transversity amplitudes $A_a^{L,R}$ ($a = 0, \perp, \|$ in the limit $m_\ell \to 0$[6]) [15]. Here L, R refer to the chirality of the lepton current. The kinematic dependence on the lepton mass is suppressed by m_ℓ^2/q^2. Hence, non-negligible effects from finite lepton masses at $q^2 > 1$ GeV^2 arise only in BSM scenarios, as for example in the MSSM at large $\tan\beta$ due to neutral Higgs boson exchange, see for instance [16].

Depending on q^2, present theoretical predictions suffer from long-distance dominated $(\bar{q}q)$-resonance background ($q = u, d, s, c$) induced by current-current- and QCD-penguin-operators of

[5] Possibilities to extract q^2-integrated J_i^j from single-differential distributions in θ_ℓ, θ_{K^*} or ϕ can be found in [10].
[6] Two more amplitudes contribute for $m_\ell \neq 0$: A_t (time-like) and in the presence of scalar operators, A_S [12].

the $\Delta B = 1$ effective Hamiltonian. Compared to the leading c-current-current contribution, the u-current-current one is suppressed by small CKM elements and the QCD-penguin contribution by small Wilson coefficients. The light (u, d, s)-resonances affect the FCNC $b \to s + \bar{\ell}\ell$ decay at very low $q^2 \lesssim 1$ GeV^2. The narrow $(\bar{c}c)$-resonances J/ψ and ψ' are vetoed in the experimental analysis, while the broader $(\bar{c}c)$-resonances appear at $q^2 > 14$ GeV^2.

This singles out a low-q^2 and a high-q^2 window which are usually chosen in the intervals $q^2 \in [1, 6]$ GeV^2 and $q^2 > 14$ GeV^2, respectively. They correspond to large and low recoil of the K^*-meson, as indicated by the scaling of its energy in the B-meson rest frame: $E_{K^*} \sim m_b/2$ versus $E_{K^*} \sim m_{K^*} + \Lambda_{\rm QCD}$. The heavy quark limit is combined at large recoil with the large energy limit as an application of QCD factorisation (QCDF) [17, 18] whereas at low recoil an operator product expansion (OPE) can be performed [19, 20] in combination with HQET form factor relations [21]. Revealing the symmetries of QCD dynamics in both regions, the expansions reduce the number of hadronic matrix elements. Many works have been inspired in the past decade in order to exploit these symmetries to reduce the hadronic uncertainties in the exclusive decays. Particularly, the angular analysis of the 4-body decay $B \to V(\to P_1 P_2) + \bar{\ell}\ell$ proved most fruitful. For comparison, $B \to P + \bar{\ell}\ell$ decays offer only two additional observables beyond the branching ratio, the lepton forward-backward asymmetry $A_{\rm FB}^\ell$ and the flat term F_H^ℓ [16].

2. $\bar{\mathbf{B}}_{\mathbf{d}}^{\mathbf{0}} \to \bar{\mathbf{K}}^{\mathbf{0}*}(\to \mathbf{K}^-\pi^+) + \bar{\ell}\ell$

The experimental situation in regard to lepton final states is twofold: On the one hand, the (Super-) B-factories are able to measure both $\ell = e$ and μ modes separately. On the other hand, most detectors at hadronic collider machines prefer $\ell = \mu$, however, LHCb might be able to study $\ell = e$, too [22].

2.1. Low-q^2 / Large recoil region

Using QCDF in the low-q^2 region, the seven $B \to V$ QCD form factors $(V, A_{0,1,2}, T_{1,2,3})$ reduce to two universal form factors $\xi_{\perp,\parallel}$ [23]. Furthermore, the $B \to K^* + \bar{\ell}\ell$ amplitudes factorise [17, 18][7], where the numerically leading contributions to the transversity amplitudes are [15]

$$A_{\perp,\parallel}^{L,R} \propto \pm c_{\perp}^{L,R} \times \xi_\perp, \qquad\qquad A_0^{L,R} \propto c_{\parallel}^{L,R} \times \xi_\parallel, \tag{3}$$

with short-distance coefficients $c_{\perp,\parallel}^{L,R}$. The transversity amplitudes have been used to construct the observables $A_T^{(2)}$ [15] and $A_T^{(3,4,5)}$ [26, 27] with reduced hadronic uncertainties and an improved sensitivity to the chirality flipped operators $\mathcal{O}'_{7,9,10}$. Instead of fitting the observables J_i^j from the angular distribution, the authors of reference [26, 27] pursue the possibility to fit directly the transversity amplitudes which necessitates the identification of their rephasing properties. Moreover, the authors investigate the experimental reconstruction uncertainties of their method at LHCb, including among the theoretical uncertainties also lacking subleading QCDF contributions by simple power counting.

The original analysis of the CP asymmetries of the rate $A_{\rm CP}$ and the angular coefficients [7]

$$A_i \equiv \frac{2(J_i^j - \bar{J}_i^j)}{d(\Gamma + \bar{\Gamma})/dq^2} \;\text{ for } i = 3, 6, 9, \qquad A_i^D \equiv \frac{-2(J_i^j - \bar{J}_i^j)}{d(\Gamma + \bar{\Gamma})/dq^2} \;\text{ for } i = 4, 5, 7, 8 \tag{4}$$

based on naive factorisation was extended using QCDF to the low-q^2 region by taking into account the NLO QCD corrections for the strong phase δ_s [10]. Both $A_{3,9}$ vanish in the SM at leading order in QCDF since they are proportional to the interference of SM and

[7] One might use equivalently soft collinear effective theory (SCET) [24, 25].

Table 1. The SM predictions of the q^2-integrated CP-asymmetries of $\bar{B}_d^0 \to \bar{K}^{0*}(\to K^-\pi^+)+\bar{\ell}\ell$ decays ($q^2 \in [1,6]$ GeV^2) and their relative uncertainties from the form factors $\xi_{\perp,\parallel}$ and the renormalisation scale μ_b [10] (see also [12]). Also shown are the ranges of the CP asymmetries after applying the experimental constraints at 90 % C.L. for the generic model-independent BSM scenario [10]. Note that the SM predictions are given in units of 10^{-3}.

		$\langle A_{CP}\rangle$	$\langle A_3\rangle$	$\langle A_4^D\rangle$	$\langle A_5^D\rangle$	$\langle A_6\rangle$	$\langle A_7^D\rangle$	$\langle A_8^D\rangle$	$\langle A_9\rangle$
SM	$\times 10^{-3}$	$4.2^{+1.7}_{-2.5}$	-	$-1.8^{+0.3}_{-0.3}$	$7.6^{+1.5}_{-1.6}$	$-6.4^{+2.2}_{-2.7}$	$-5.1^{+2.4}_{-1.6}$	$3.5^{+1.4}_{-2.0}$	-
	$\xi_{\perp,\parallel}[\%]$	$^{+19}_{-24}$	-	$^{+11}_{-8}$	$^{+10}_{-13}$	$^{+31}_{-39}$	$^{+11}_{-8}$	$^{+7.4}_{-10}$	-
	$\mu_b[\%]$	$^{+33}_{-51}$	-	$^{+2}_{-6}$	$^{+7}_{-8}$	$^{+0}_{-2}$	$^{+42}_{-26}$	$^{+37}_{-53}$	-
BSM	max	+0.10	+0.08	+0.04	+0.07	+0.11	+0.76	+0.48	+0.60
	min	−0.12	−0.08	−0.04	−0.07	−0.13	−0.76	−0.48	−0.62

chirality flipped operators, making them ideal probes of the latter. Furthermore, $A_{7,8}^D$ are subject to cancellations at leading order in QCD, such that the NLO QCD corrections give large corrections. The q^2-integrated SM predictions of the CP asymmetries are summarised in table 1. They are, as expected, tiny and below a percent. A model-independent analysis in the presence of the BSM contributions to the SM and the chirality flipped operators $\mathcal{O}'_{7,9,10}$ shows substantial room for enhancement. The T-odd asymmetries $A_{7,8,9}^D$ can be of order one. However, the experimental uncertainties at LHCb are large for CP-asymmetries, even when considering improved normalisations as discussed for A_{6s}^{V2s} and A_8^V [27].

The rate-normalised CP-averaged quantities $S_i^j \sim (J_i^j + \bar{J}_i^j)$ constitute another important set of observables, of which the CP-odd ones ($i = 5, 6, 8, 9$) require B-tagged samples. The S_i^j have been analysed in the SM, model-independently and model-dependently in great detail [12, 13]. Especially, the lepton forward-backward asymmetry A_{FB} (related to S_6^s) and S_5 could be measured rather precisely with early LHCb data of 2 fb^{-1} [28].

It should be noted that [26, 27] estimate uncertainties due to unknown subleading $\mathcal{O}(\Lambda_{QCD}/E_{K^*})$ contributions to both form factors and $B \to K^* + \bar{\ell}\ell$ amplitudes by power counting. Contrary, [12] does not make use of the large recoil symmetry relations of the form factors in the LO QCD contribution of the said amplitudes. Instead, all seven QCD form factors determined from light cone sum rules (LCSR) [29] are used, thus accounting for the associated unknown subleading $\mathcal{O}(\Lambda_{QCD}/E_{K^*})$ contributions. The direct application of the LCSR results reduces the form factor uncertainties [12] due to more efficient cancellations.

Recently, soft-gluon emission effects due to ($\bar{c}c$)-resonances were found to extend down to the low-q^2 region [30] affecting the q^2-differential rate at most about +15 % at $q^2 = 6$ GeV^2. Due to this relatively large shift such effects should be investigated further and should be considered in future studies of the exclusive rare decays.

2.2. High-q^2 / Low recoil region

Assuming quark-hadron duality, the low recoil region can be studied using an OPE in combination with HQET to treat the long-distance contributions from current-current and QCD-penguin operators [19, 20]. The authors of [20] perform the OPE without subsequent matching to HQET contrary to [19], among further technical details. Duality violating corrections to the OPE are estimated in [20] and found to be small. In [19] it is shown, that at the leading order in the

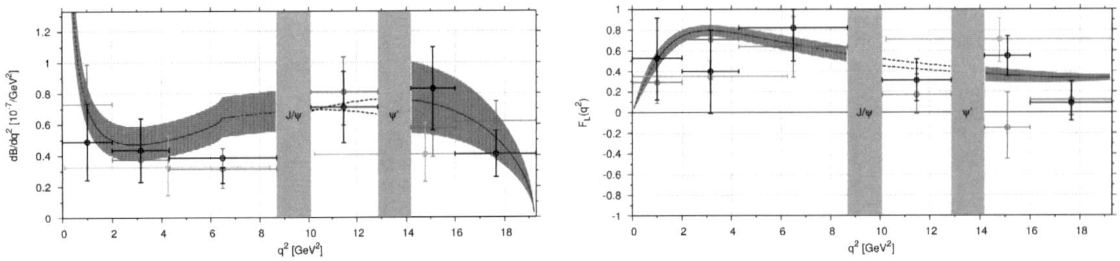

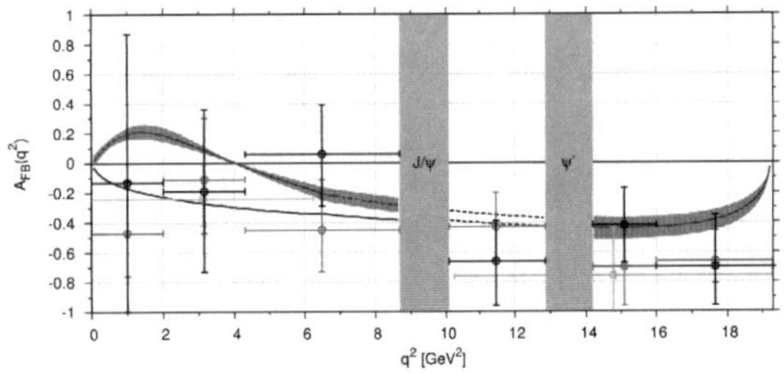

Figure 2. Comparison of BaBar [yellow], Belle [red] and CDF [black] data points with SM predictions (blue) for $\bar{B}_d^0 \to \bar{K}^{*0} + \bar{\ell}\ell$ distributions [4]: $d\mathcal{B}/dq^2$, $A_{\rm FB}$ and F_L. The vertical grey bands are vetoed in experiment.

$\Lambda_{\rm QCD}/Q$, $Q = \{\sqrt{q^2}, m_b\}$ expansion[8] all hadronic matrix elements can be expressed through the 3 QCD form factors $V, A_{1,2}$. As a consequence, a universal short-distance dependence emerges for all transversity amplitudes ($a = 0, \perp, \|$) [4]

$$A_a^{L,R} \propto c_{L,R} \times f_a, \qquad\qquad c_{L,R} = (\mathcal{C}_9^{\rm eff} \mp \mathcal{C}_{10}) + \kappa \frac{2m_b}{q^2} \mathcal{C}_7^{\rm eff}, \tag{5}$$

with the (effective) short-distance couplings $\mathcal{C}_{7,9,10}^{\rm eff}$ and the form factor terms f_a ($a = 0, \perp, \|$) which are linear in $V, A_{1,2}$. The factor $\kappa \sim 1$ is known including NLO in QCD from matching QCD onto HQET [19]. The term $\propto \mathcal{C}_7^{\rm eff}$ is subject to uncertainties of order $\Lambda_{\rm QCD}/m_b$ due to the HQET form factor relations, which are additionally suppressed in the SM by $|\mathcal{C}_7/\mathcal{C}_9| \sim 0.1$. As a consequence of (5) $J_7 = J_8 = J_9 = 0$ and the remaining angular observables

$$\begin{aligned} J_2^c &\sim U_1 = 2\rho_1 f_0^2, & (2J_2^s + J_3) &\sim U_2 = 2\rho_1 f_\perp^2, & (2J_2^s - J_3) &\sim U_3 = 2\rho_1 f_\|^2, \\ J_4 &\sim U_4 = 2\rho_1 f_0 f_\|, & J_5 &\sim U_5 = 4\rho_2 f_0 f_\perp, & J_6^s &\sim U_6 = 4\rho_2 f_\| f_\perp, \end{aligned} \tag{6}$$

depend only on the two short-distance (dominated) combinations $\rho_1 = (|c_R|^2 + |c_L|^2)/2$ and $\rho_2 = (|c_R|^2 - |c_L|^2)/4$ [4]. From (6) one obtains the three long-distance free ratios

$$H_T^{(1)} = \frac{U_4}{\sqrt{U_1 \cdot U_3}} = 1, \qquad H_T^{(2)} = \frac{U_5}{\sqrt{U_1 \cdot U_2}} = 2\,\frac{\rho_2}{\rho_1}, \qquad H_T^{(3)} = \frac{U_6}{\sqrt{U_2 \cdot U_3}} = 2\,\frac{\rho_2}{\rho_1}. \tag{7}$$

After integration over $q^2 \in [14.0, 19.2]$ GeV^2 one obtains in the SM

$$\langle H_T^{(1)} \rangle = +0.997 \pm 0.003, \qquad \langle H_T^{(2)} \rangle = -0.972 \pm 0.010, \qquad \langle H_T^{(3)} \rangle = -0.958 \pm 0.010. \tag{8}$$

Details concerning the definition of $\langle \ldots \rangle$ and the theoretical uncertainties are given in [4] as well as the results for $\mathcal{B}, A_{\rm FB}, F_L$ and $A_T^{(2,3,4)}$. We stress that $\langle H_T^{(1)} \rangle \approx 1$ holds model-independently.

[8] Note, that the subleading order is additionally suppressed by α_s.

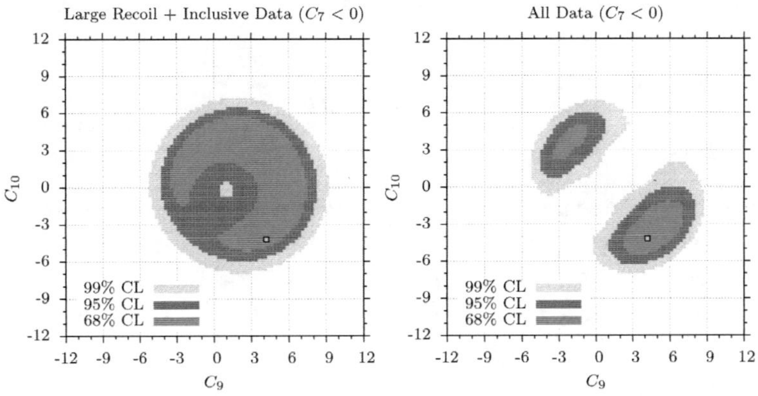

Figure 3. Constraints on real-valued $\mathcal{C}_{9,10}$ at 68%, 95% and 99% CL from low-q^2 bins of $\bar{B}_d^0 \to \bar{K}^{*0}\bar{\ell}\ell$ and $B \to X_s\bar{\ell}\ell$ [left] and when adding high-q^2 bins [right] [4, 5]. Here $\mathcal{C}_7 = +\mathcal{C}_7^{\rm SM}$ is fixed. The square marks the SM value of $(\mathcal{C}_9, \mathcal{C}_{10})$.

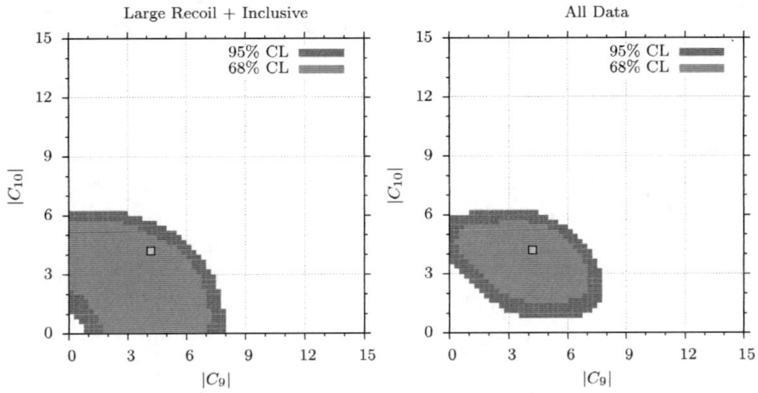

Figure 4. Constraints on $|\mathcal{C}_{9,10}|$ at 68% and 95% CL for complex-valued $\mathcal{C}_{9,10}$ from low-q^2 bins of $\bar{B}_d^0 \to \bar{K}^{*0}\bar{\ell}\ell$ and $B \to X_s\bar{\ell}\ell$ [left] and when adding high-q^2 bins [right] (see [5] for details). The square marks the SM value of $(\mathcal{C}_9, \mathcal{C}_{10})$.

Hence, deviations from this prediction test the validity of the OPE framework. Furthermore, the following short-distance free ratios

$$\frac{f_0}{f_\parallel} = \sqrt{\frac{U_1}{U_3}} = \frac{U_1}{U_4} = \frac{U_4}{U_3} = \frac{U_5}{U_6}, \qquad \frac{f_0}{f_\perp} = \sqrt{\frac{U_1}{U_2}}, \qquad \frac{f_\perp}{f_\parallel} = \sqrt{\frac{U_2}{U_3}} = \frac{\sqrt{U_1 U_2}}{U_4} \tag{9}$$

allow to test the $B \to V$ form factors at high-q^2, such as obtained from lattice QCD, using experimental data [4].

A comparison of the SM predictions and the available data from BaBar, Belle and CDF for low- and high-q^2 regions is shown in figure 2 where the main uncertainties (shaded bands) arise from the form factors and the missing subleading corrections of order $\Lambda_{\rm QCD}/m_b$ as estimated by power counting [4]. The Belle and CDF measurements of the three low- and high-q^2 bins $q^2 \in [1,6], [14.18, 16.0], [16.0, 19.21]$ GeV^2 of $\mathcal{B}$ and $A_{\rm FB}$ (and $q^2 \in [1,6]$ GeV^2 of F_L) yield the first constraints on the short-distance couplings $\mathcal{C}_{9,10}$ of the SM operators $\mathcal{O}_{9,10}$ beyond naive factorisation. The constraints for real-valued $\mathcal{C}_{9,10}$ are shown in figure 3 fixing $\mathcal{C}_7 = +\mathcal{C}_7^{\rm SM}$ (results for $\mathcal{C}_7 = -\mathcal{C}_7^{\rm SM}$ are available in [4]) when employing low-q^2 data only and when adding high-q^2 data[9], respectively. Figure 4 shows $|\mathcal{C}_{10}|$ vs. $|\mathcal{C}_9|$ for complex-valued $\mathcal{C}_{9,10}$ and results of (ϕ_9 vs. $|\mathcal{C}_9|$), (ϕ_{10} vs. $|\mathcal{C}_{10}|$) and (ϕ_{10} vs. ϕ_9) can be found in [5]. Both figures demonstrate the strong constraining power of the high-q^2 region on $\mathcal{C}_{9,10}$, and consistency with low-q^2 data.

CP asymmetries with improved short-distance sensitivity can be designed in analogy to $H_T^{(1,2,3)}$, since $\bar{J}_i^j \sim \bar{\rho}_a$ with $\bar{\rho}_a = \rho_a[\delta_W \to -\delta_W]$. It is convenient to consider the three

[9] Using an extrapolation of the $B \to V$ form factors from low-q^2 QCD LCSR [29].

CP-asymmetries

$$a_{\rm CP}^{(1)} = \frac{\rho_1 - \bar\rho_1}{\rho_1 + \bar\rho_1}, \qquad a_{\rm CP}^{(2)} = \frac{\frac{\rho_2}{\rho_1} - \frac{\bar\rho_2}{\bar\rho_1}}{\frac{\rho_2}{\rho_1} + \frac{\bar\rho_2}{\bar\rho_1}}, \qquad a_{\rm CP}^{(3)} = 2\,\frac{\rho_2 - \bar\rho_2}{\rho_1 + \bar\rho_1}, \tag{10}$$

with various possibilities to measure them [5]. Moreover, $a_{\rm CP}^{(3)}$ can be extracted from an untagged B-meson sample. There are no T-odd CP-asymmetries at low-recoil due to (6): $J_{7,8,9} = 0$. In the SM $|a_{\rm CP}^{(1,2,3)}|_{\rm SM} \lesssim 10^{-4}$ whereas a model-independent analysis for complex-valued $\mathcal{C}_{9,10}$ allows values up to $|\langle a_{\rm CP}^{(1,3)}\rangle| \lesssim 0.2$ and $|\langle a_{\rm CP}^{(2)}\rangle|$ is presently unconstrained (see [5] for details).

3. CP asymmetries without tagging

The decays $\bar B_s, B_s \to \phi(\to K^-K^+) + \bar\ell\ell$ require to account for the mixing of the initial $\bar B_s^0$- and B_s^0-mesons before decaying into the common final state. Consequently, their observables depend additionally on the $\bar B_s - B_s$ mixing phase Φ_s and the lifetime difference $\Delta\Gamma_s$, which are currently being analysed at the Tevatron, for instance in $\bar B_s, B_s \to \phi(\to K^+K^-) + J/\psi(\to \bar\ell\ell)$. In this respect, the CP-odd property of the observables J_i^j with $i = 5, 6, 8, 9$ allows to measure the associated CP-asymmetries from an untagged and time-integrated data set.

The CP-asymmetries A_i^{mix} for $i = 6, 9$ and A_i^{Dmix} for $i = 5, 8$ have been worked out at low-q^2 [10]. They match the corresponding A_i and A_i^D of the flavour-specific unmixed decays for $\Delta\Gamma_s \to 0$. There the CP-averaged rate has been used as normalisation. An improved cancellation of the hadronic uncertainties might be achieved along the lines of the unmixed case by choosing the normalisations proposed for A_{6s}^{V2s} and A_8^V [27].

At high-q^2 the CP-asymmetries associated with J_i^j, $i = 8, 9$ vanish and those with $i = 5, 6$ can be related to the same short-distance dominated $a_{\rm CP}^{mix}$ using the normalisations as indicated in reference [5]. Notably, $a_{\rm CP}^{mix}$ is independent of the sign of $\Delta\Gamma_s$ and the dependence on Φ_s is rather weak for small $\Delta\Gamma_s/\Gamma_s$. Experimentally, $\Delta\Gamma_s/\Gamma_s \sim \mathcal{O}(0.1)$.

4. Conclusion

The angular analysis of the exclusive $\bar B_d^0 \to \bar K^{0*}(\to K^-\pi^+) + \bar\ell\ell$ and $\bar B_s, B_s \to \phi(\to K^-K^+) + \bar\ell\ell$ decays with 4-body final states offers a rich phenomenology to test the short-distance flavour couplings in a large number of observables. The LHCb experiment is expected to improve the experimental situation tremendously in the near future, but also the upcoming final analysis of the complete BaBar, Belle and CDF data sets will contribute to the exploration of the borders of the SM.

The low- and high-q^2 regions are accessible to power expansions which reveal symmetries of QCD and indicate suitable combinations of observables to reduce the hadronic uncertainties. This allows for precise theory predictions for the exclusive decays.

The low-q^2 region is theoretically (QCDF, SCET) well understood and effects of $(\bar cc)$-resonances can be estimated. It offers many interesting observables which wait for confrontation with data.

The high-q^2 region is based on an OPE relying on quark-hadron duality. Violations of the latter could be signaled experimentally by deviations from $H_T^{(1)} = 1$. Currently, the $B \to V$ form factors used at high-q^2 are extrapolated from the low-q^2 region using a dipole formula, but lattice calculations for the high-q^2 regions are in progress [31]. There are two "long-distance free" ratios $H_T^{(2,3)}$ which provide tests of the SM and, moreover, several "short-distance free" ratios which offer direct comparison of the lattice results for ratios of form factors with data.

A new flavour tool "EOS" aimed at the evaluation of the observables covered in this article is developed at TU Dortmund [32] whose first stable release is expected during 2011.

Acknowledgments
C. B. thanks the organisers of the *DISCRETE 2010 - Symposium on Prospects in the Physics of Discrete Symmetries* for the opportunity to present a talk and the kind hospitality in Rome. The work reported here has been supported in part by the Bundesministerium für Bildung und Forschung (BMBF).

References
[1] B. Aubert *et al.* [BABAR Collaboration], Phys. Rev. D **79** (2009) 031102 [arXiv:0804.4412 [hep-ex]].
[2] J. T. Wei *et al.* [BELLE Collaboration], Phys. Rev. Lett. **103** (2009) 171801 [arXiv:0904.0770 [hep-ex]].
[3] T. Aaltonen *et al.* [CDF Collaboration], Phys. Rev. Lett. **106** (2011) 161801 [arXiv:1101.1028 [hep-ex]].
[4] C. Bobeth, G. Hiller and D. van Dyk, JHEP **1007** (2010) 098 [arXiv:1006.5013 [hep-ph]].
[5] C. Bobeth, G. Hiller and D. van Dyk, arXiv:1105.0376 [hep-ph].
[6] B. Adeva *et al.*, [The LHCb Collaboration], LHCb-PUB-2009-029 [arXiv:0912.4179].
[7] F. Kruger, L. M. Sehgal, N. Sinha and R. Sinha, Phys. Rev. D **61** (2000) 114028 [Erratum-ibid. D **63** (2001) 019901] [arXiv:hep-ph/9907386].
[8] C. S. Kim, Y. G. Kim, C. D. Lu and T. Morozumi, Phys. Rev. D **62** (2000) 034013 [arXiv:hep-ph/0001151].
[9] B. Grinstein and D. Pirjol, Phys. Rev. D **73** (2006) 094027 [arXiv:hep-ph/0505155].
[10] C. Bobeth, G. Hiller and G. Piranishvili, JHEP **0807** (2008) 106 [arXiv:0805.2525 [hep-ph]]; PoS **BEAUTY2009** (2009) 047 [arXiv:0911.4054 [hep-ph]].
[11] C. Bobeth, M. Misiak and J. Urban, Nucl. Phys. B **574** (2000) 291 [arXiv:hep-ph/9910220].
[12] W. Altmannshofer, P. Ball, A. Bharucha, A. J. Buras, D. M. Straub and M. Wick, JHEP **0901** (2009) 019 [arXiv:0811.1214 [hep-ph]].
[13] A. K. Alok, A. Datta, A. Dighe, M. Duraisamy, D. Ghosh, D. London and S. U. Sankar, arXiv:1008.2367 [hep-ph].
[14] R. Sinha, arXiv:hep-ph/9608314; C. H. Chen and C. Q. Geng, Nucl. Phys. B **636** (2002) 338 [arXiv:hep-ph/0203003].
[15] F. Kruger and J. Matias, Phys. Rev. D **71** (2005) 094009 [arXiv:hep-ph/0502060].
[16] C. Bobeth, G. Hiller and G. Piranishvili, JHEP **0712** (2007) 040 [arXiv:0709.4174 [hep-ph]].
[17] M. Beneke, T. Feldmann and D. Seidel, Nucl. Phys. B **612** (2001) 25 [arXiv:hep-ph/0106067].
[18] M. Beneke, T. Feldmann and D. Seidel, Eur. Phys. J. C **41** (2005) 173 [arXiv:hep-ph/0412400].
[19] B. Grinstein and D. Pirjol, Phys. Rev. D **70** (2004) 114005 [arXiv:hep-ph/0404250].
[20] M. Beylich, G. Buchalla and T. Feldmann, arXiv:1101.5118 [hep-ph].
[21] B. Grinstein and D. Pirjol, Phys. Lett. B **533** (2002) 8 [arXiv:hep-ph/0201298].
[22] J. Lefrançois and M. H. Schune, LHCb-PUB-2009-008.
[23] M. Beneke and T. Feldmann, Nucl. Phys. B **592** (2001) 3 [arXiv:hep-ph/0008255].
[24] A. Ali, G. Kramer and G. h. Zhu, Eur. Phys. J. C **47** (2006) 625 [arXiv:hep-ph/0601034].
[25] K. S. M. Lee, Z. Ligeti, I. W. Stewart and F. J. Tackmann, Phys. Rev. D **75** (2007) 034016 [arXiv:hep-ph/0612156].
[26] U. Egede, T. Hurth, J. Matias, M. Ramon and W. Reece, JHEP **0811** (2008) 032 [arXiv:0807.2589 [hep-ph]].
[27] U. Egede, T. Hurth, J. Matias, M. Ramon and W. Reece, JHEP **1010** (2010) 056 [arXiv:1005.0571 [hep-ph]].
[28] A. Bharucha and W. Reece, Eur. Phys. J. C **69** (2010) 623 [arXiv:1002.4310 [hep-ph]].
[29] P. Ball and R. Zwicky, Phys. Rev. D **71** (2005) 014029 [arXiv:hep-ph/0412079].
[30] A. Khodjamirian, T. Mannel, A. A. Pivovarov and Y. M. Wang, JHEP **1009** (2010) 089 [arXiv:1006.4945 [hep-ph]].
[31] Z. Liu, S. Meinel, A. Hart, R. R. Horgan, E. H. Muller and M. Wingate, PoS **LAT2009** (2009) 242 [arXiv:0911.2370 [hep-lat]]; arXiv:1101.2726 [hep-ph].
[32] D. van Dyk *et al.*, http://project.het.physik.tu-dortmund.de/eos/.

Journal of Physics: Conference Series **335** (2011) 012039 doi:10.1088/1742-6596/335/1/012039

Probing new physics in electroweak penguins through B_d and B_s decays

Lars Hofer[1,2], **Dominik Scherer**[2], **and Leonardo Vernazza**[3,4]

[1] Institut für Theoretische Teilchenphysik, Karlsruhe Institute of Technology, D–76128 Karlsruhe, Germany

[2] Institut für Theoretische Physik und Astrophysik, Universität Würzburg, D–97074 Würzburg, Germany

[3] Institut für Physik (THEP), Johannes Gutenberg-Universität, D–55099 Mainz, Germany

E-mail: lars.hofer@physik.uni-wuerzburg.de, dominik@particle.uni-karlsruhe.de, vernazza@uni-mainz.de

Abstract. An enhanced electroweak penguin amplitude due to the presence of unknown new physics can explain the discrepancies found between theory and experiment in the $B \to \pi K$ decays, in particular in $A_{\rm CP}(B^- \to \pi^0 K^-) - A_{\rm CP}(\bar{B}^0 \to \pi^+ K^-)$, but the current precision of the theoretical and experimental results does not allow to draw a firm conclusion. We argue that the $\bar{B}_s \to \phi\rho^0$ and $\bar{B}_s \to \phi\pi^0$ decays offer an additional tool to investigate this possibility. These purely isospin-violating decays are dominated by electroweak penguins and we show that in presence of a new physics contribution their branching ratio can be enhanced by about an order of magnitude, without violating any constraints from other hadronic B decays. This makes them very interesting modes for LHCb and future B factories. In [1] we have performed both a model-independent analysis and a study within realistic New Physics models such as a modified-Z^0-penguin scenario, a model with an additional Z' boson and the MSSM. In this article we summarise the most important results of our study.

1. Introduction

The four $B \to \pi K$ decays have been very useful in testing flavour structure and CP violation in the Standard Model (SM) since the late 1990s. These decays have a small branching ratio, $\mathcal{O}(10^{-6})$, and are sensitive to New Physics (NP) contributions. In the past years some discrepancies between $B \to \pi K$ measurements and SM predictions have occurred, provoking speculations on a "$B \to \pi K$ puzzle". To date, the measurements of the branching fractions have fluctuated towards the SM predictions, the latter still suffering from large hadronic uncertainties, and only the following difference of CP asymmetries shows an unexpected behavior [2]:

$$\Delta A_{\rm CP} \equiv A_{\rm CP}(B^- \to \pi^0 K^-) - A_{\rm CP}(\bar{B}^0 \to \pi^+ K^-) \stackrel{\rm SM}{=} 1.9^{+5.8}_{-4.8}\,\% \stackrel{\rm exp.}{=} (14.8 \pm 2.8)\,\%. \quad (1)$$

The theory result, derived within the framework of QCD factorisation (QCDF) [4], and the experimental result, taken from [3], differ by 2.5σ. This result suggests a violation of the isospin symmetry beyond the amount expected in the SM. It can be interpreted as a hint of a NP electroweak (EW) penguin amplitude, or alternatively as an enhanced low-energy hadronic

[4] Alexander-von-Humboldt fellow.

Journal of Physics: Conference Series **335** (2011) 012039 doi:10.1088/1742-6596/335/1/012039

effect which the theory is not able to catch. Present data do not allow to distinguish between these two possibilities.

Our aim is to show that the question can be partially addressed exploiting the large variety of non-leptonic B decays into two charmless mesons. For this purpose, in [1] we have studied the *purely* isospin-violating decays $\bar{B}_s \to \phi\rho^0$ and $\bar{B}_s \to \phi\pi^0$, which are dominated by EW penguins, and we have shown that if NP in this sector exists at a level where it can explain the $\Delta A_{\rm CP}$ puzzle, it could be clearly visible in these purely isospin-violating decays. We present here a summary of the results, referring to [1] for a more detailed discussion.

2. Isospin violation in $B \to \pi K$ and $B_s \to \phi\rho^0$, $B_s \to \phi\pi^0$ decays

The $B \to \pi K$ decays are dominated by the isospin-conserving QCD penguin amplitude. Nevertheless, they receive small contributions from the tree and the EW penguin amplitude, which are isospin-violating. Combining measurements of the four different decay modes $B^- \to \pi^-\bar{K}^0$, $B^- \to \pi^0 K^-$, $\bar{B}^0 \to \pi^+K^-$ and $\bar{B}^0 \to \pi^0\bar{K}^0$, it is possible to construct observables in which the leading contribution from the QCD penguin drops out, so that they are sensitive to isospin violation and therefore to a possible new contribution with the structure of the EW penguins.

The decay amplitudes can be written in terms topological amplitudes r_i which are normalised to the dominant QCD penguin contribution. Introducing as well a NP EW penguin amplitude, such that

$$r_{\rm EW} \to r_{\rm EW} + \tilde{r}_{\rm EW} e^{-i\delta}, \qquad r_{\rm EW}^{\rm C} \to r_{\rm EW}^{\rm C} + \tilde{r}_{\rm EW}^{\rm C} e^{-i\delta}, \qquad r_{\rm EW}^{\rm A} \to r_{\rm EW}^{\rm A} + \tilde{r}_{\rm EW}^{\rm A} e^{-i\delta}, \tag{2}$$

one obtains

$$\Delta A_{\rm CP} \simeq -2\,{\rm Im}\,(r_{\rm C})\sin\gamma + 2\,{\rm Im}\left(\tilde{r}_{\rm EW} + \tilde{r}_{\rm EW}^{\rm A}\right)\sin\delta\,. \tag{3}$$

Here $r_{\rm EW}$, $r_{\rm EW}^{\rm (C,A)}$ represent respectively the SM colour-allowed, colour-suppressed and annihilation electroweak penguin amplitudes, while the $\tilde{r}_{\rm EW}^i$ are the corresponding NP contributions. It is clear from (3) that a large $\Delta A_{\rm CP}$ can be obtained both through enhanced low-energy effects in ${\rm Im}\,(r_{\rm C})$, as well as by means of a NP contribution. An ${\rm Im}\,(r_{\rm C})$ large enough to solve the $\Delta A_{\rm CP}$ discrepancy, however, would break QCDF, so that within this framework a NP solution is favoured. Although one can construct many other observables which are sensitive to isospin violation, the colour-allowed EW penguin contribution and the colour-suppressed tree contribution cannot be disentangled within the $B \to \pi K$ decays alone since they enter the amplitudes exclusively in the combination

$$r_{\rm EW} - r_{\rm C}\, e^{-i\gamma}. \tag{4}$$

Our proposal is now to get a more complete picture by considering as well the decays

$$\boxed{\bar{B}_s \to \phi\rho^0 \qquad \text{and} \qquad \bar{B}_s \to \phi\pi^0.}$$

These decays are purely isospin-violating and their structure in the SM is very simple. Even though we are facing the same type of EW penguin vs colour-suppressed tree amplitude pollution as in the $B \to \pi K$ decays, various elements suggest that the analysis of these decays may be interesting:

- these decays are dominated by the EW penguin amplitude such that a NP effect coming from a new EW penguin amplitude should have more spectacular effects in these decays than in the QCD penguin dominated $B \to \pi K$: A new EW amplitude of the same order as the SM one can enhance the branching ratios by about an order of magnitude.

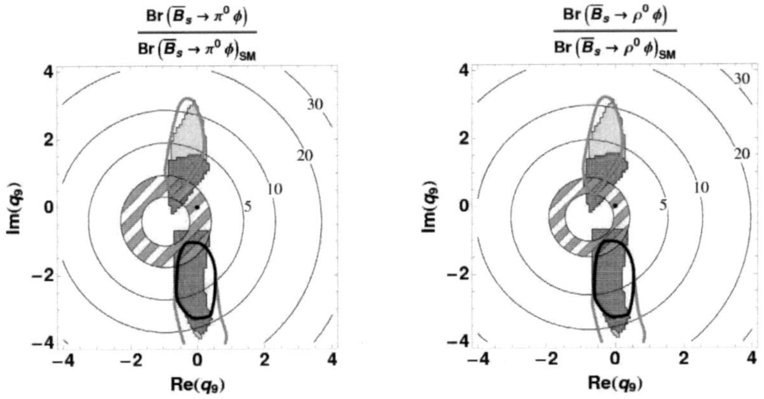

Figure 1: Enhancement factors of the $\bar{B}_s \to \phi\rho^0, \phi\pi^0$ branching ratios with respect to their SM values. The black dot represents the SM result while the red striped region shows the theoretical uncertainty in the SM. The dark green area is the region allowed by the $2\,\sigma$ constraints from $\bar{B} \to \pi K^{(*)}, \rho K^{(*)}, \phi K^{(*)}$ and $\bar{B}_s \to \phi\phi, \bar{K}K$ decays; for comparison, the light green area represents the area allowed by constraints from isospin-sensitive observables only, considering only $\bar{B} \to \pi K, \pi K^{(*)}, \rho K$ decays. The solid black line represents the 1σ CL of the fit with $S_{CP}(\bar{B}^0 \to \pi \bar{K}^0)$, while the solid grey line represents the 1σ CL of the fit without it. Here the scenario $q_9 \neq 0$ is shown.

- the nature of these B_s decays is different from the $B \to \pi K$ decays, the first being PV, VV decays, respectively, while the latter are PP decays. The non-perturbative low-energy QCD dynamics is expected not to be correlated among the two classes of decays, and the exact relation cannot be determined within QCDF. Phenomenological comparison with other decays such as $B \to \pi\pi$, $B \to \pi\rho$, $B \to \rho\rho$ suggests a decreasing of non-perturbative effects when vector mesons appear in the final state. A NP contribution is instead of high-energy origin, and its effects can be reliably studied within perturbation theory, allowing for a correlation among the $B \to \pi K$ and the B_s decays of interest.

3. Model-independent analysis

We first perform a model-independent analysis. We parameterise NP in EW penguins by adding corresponding terms $C_{7,9}^{(\prime)\mathrm{NP}}$ to the Wilson coefficients of the effective weak Hamiltonian,

$$C_{7,9}^{(\prime)\mathrm{NP}}(m_W) = C_9^{\mathrm{LO}}(m_W)\, q_{7,9}^{(\prime)}, \qquad q_{7,9}^{(\prime)} = |q_{7,9}^{(\prime)}| e^{i\phi_{7,9}^{(\prime)}}. \tag{5}$$

Here C_9^{LO} denotes the leading-order SM coefficient, the primed operators are obtained from the SM $Q_{7,9}$ by flipping the chiralities of the quark fields. The purpose is to perform a χ^2-fit in order to determine the NP parameters in such a way that they describe better $B \to \pi K$ data. In particular we look for a solution of the ΔA_{CP} discrepancy. Further hadronic decays like $B \to \rho K, \pi K^*, \rho K^*$ are used to impose additional constraints at the $2\,\sigma$ level. The result of the fit is used to study the decays $\bar{B}_s \to \phi\pi^0, \phi\rho^0$ and quantify a potential enhancement of their branching fractions. Such an analysis, correlating different hadronic decay modes, is only possible if hadronic matrix elements are calculated from first principles like in the framework of QCDF. A method based on flavour symmetries, as it has been used in most studies of $B \to \pi K$ decays so far, could not achieve this. In particular, the decays $\bar{B}_s \to \phi\pi^0, \phi\rho^0$ are not related to any other decay via $SU(3)_F$ so their branching fractions cannot be predicted in this way.

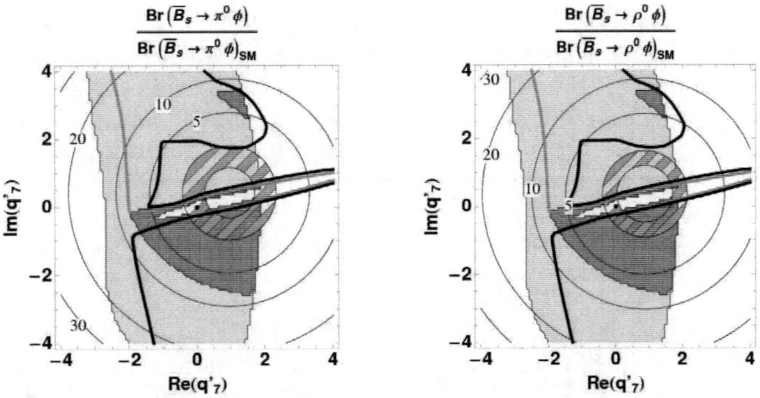

Figure 2: Enhancement factors of the $\bar{B}_s \to \phi\rho^0, \phi\pi^0$ branching ratios with respect to their SM values. The meaning of the contours and regions is the same as in fig. 1. Here the scenario $q_7' \neq 0$ Is displayed. The 1σ region of the fit is the region to the right of the black (grey) curve.

As far as the $B \to \pi K$ decays are concerned, the new contributions to the EW penguin amplitudes are given by

$$\sum_{i=7,9,7',9'} \tilde{r}_{\mathrm{EW},\,i}\, e^{-i\delta_i} = (q_7 - q_7') \left[(-0.12)^{+0.04}_{-0.05} + (-0.02)^{+0.07}_{-0.02}\, i \right] + (q_9 - q_9') \left[0.12^{+0.05}_{-0.04} + 0.02^{+0.02}_{-0.07}\, i \right],$$

$$\sum_{i=7,9,7',9'} \tilde{r}^{\mathrm{C}}_{\mathrm{EW},\,i}\, e^{-i\delta_i} = (q_7 - q_7') \left[0.10^{+0.03}_{-0.02} + 0.01^{+0.01}_{-0.06}\, i \right] + (q_9 - q_9') \left[0.04^{+0.02}_{-0.03} + (-0.005)^{+0.016}_{-0.026}\, i \right],$$

$$\sum_{i=7,9,7',9'} \tilde{r}^{\mathrm{A}}_{\mathrm{EW},\,i}\, e^{-i\delta_i} = (q_7 - q_7') \left[0.03^{+0.04}_{-0.07} + (-0.06)^{+0.12}_{-0.01}\, i \right] + (q_9 - q_9') \left[0.007^{+0.003}_{-0.010} + (-0.006)^{+0.012}_{-0.003}\, i \right]. \quad (6)$$

From these expressions one draws e.g. the following conclusions:

- parity-symmetric models with $q_7 = q_7'$ and $q_9 = q_9'$ do not contribute to $B \to \pi K$ at all. Therefore such a scenario cannot solve the ΔA_{CP} discrepancy.
- The contributions $\tilde{r}_{\mathrm{EW},\,7^{(\prime)}}$ and $\tilde{r}_{\mathrm{EW},\,9^{(\prime)}}$ tend to cancel each other. Hence in the scenarios with $q_7 = q_9$ and $q_7' = q_9'$ only a negligible new colour-allowed EW penguin contribution is generated.
- Whereas $\mathrm{Re}(\tilde{r}^{\mathrm{C}}_{\mathrm{EW},\,9^{(\prime)}})$ features the typical colour-suppression with respect to $\mathrm{Re}(\tilde{r}_{\mathrm{EW},\,9^{(\prime)}})$, this pattern is not obeyed by the $q_7^{(\prime)}$ terms.
- The annihilation coefficient $\tilde{r}^{\mathrm{A}}_{\mathrm{EW},\,7^{(\prime)}}$ develops a large imaginary part. In scenarios with non-vanishing $q_7^{(\prime)}$ this term gives the dominant contribution to ΔA_{CP}.

From eq. (3) we see that the ΔA_{CP} discrepancy can be solved either through $\tilde{r}_{\mathrm{EW}}$ or through $\tilde{r}^{\mathrm{A}}_{\mathrm{EW}}$. Except for parity-symmetric models, any scenario with at least one of $q_{7,9}^{(\prime)}$ different from zero can achieve such a solution. The minimal $|q|$ - value needed to reduce the ΔA_{CP} tension below the $1\,\sigma$ level varies among different scenarios. One finds e.g. $|q_7| \gtrsim 0.3$ for a model with only $q_7 \neq 0$, $|q_9| \gtrsim 0.8$ for a model with only $q_9 \neq 0$ and $|q_7| = |q_9| \gtrsim 0.4$ for a model with $q_7 = q_9$ and the primed coefficients being zero. The fact that in the $q_7 = q_9$ case only a small NP contribution is needed, in spite of the absence of $\tilde{r}_{\rm EW}$, demonstrates the importance of the annihilation term $\tilde{r}^{\rm A}_{\rm EW}$. Finally, we like to stress that the solution of the $\Delta A_{\rm CP}$ discrepancy via a minimal $|q|$ - value requires the adjustment of the phase ϕ to a certain value. Realistic scenarios avoiding such a fine-tuning have larger $|q|$ - values, typically $|q| \sim 1$.

For the q_7-only and the q_9-only scenarios, the result of the fit to $B \to \pi K$ data and the consequences for the B_s decays of interest are shown in figs. figs. 1,2 For details about the fit and the observables used we refer to [1]. We find that the $B \to \pi K$ and related decays set quite strong constraints on the parameter space, especially in scenarios where $q_9 \neq 0$ or $q_9' \neq 0$. This basically rules out the possibility of having $|q_i| \gtrsim 5$, i.e. NP corrections cannot be much larger than the EW penguins of the SM. The SM point is always excluded at the $2\,\sigma$ level as a direct consequence of the ΔA_{CP} data. According to the sign pattern in eq. (6), the $B \to \pi K$ fits of the primed and unprimed scenarios in figs. 1,2 are related to each other through rotation by 180°. The fit works best in the $q_9^{(\prime)}$ scenario where the best fit point is given by

$$|\hat{q}_9^{(\prime)}| = 1.9 \qquad \hat{\varphi}_9^{(\prime)} = -100^\circ\,(+180^\circ). \tag{7}$$

This parameter point yields a full agreement of all the $B \to \pi K$ observables with the experimental mean values. In the $q_7^{(\prime)} = q_9^{(\prime)}$ case a plateau of $\chi^2 = 0$ points arises due to the large theoretical errors. It turns out that the $B \to \pi K$ observables are not very sensitive to the $q_7^{(\prime)}$-only scenarios and so the fit does not work well here.

From figs. 1,2 the enhancement $\mathrm{Br}^{\rm SM+NP}/\mathrm{Br}^{\rm SM}$ of the B_s branching fractions can be read off with respect to the different constraint- and fit-regions. In most scenarios an enhancement of an order of magnitude or more is possible, especially in those involving $q_7^{(\prime)} \neq 0$. The fact that large parts of the allowed regions do not overlap with the SM uncertainty regions is encouraging. It means that, if such NP is realised in nature, it could be possible to probe it easily.

Exceptions are $\bar{B}_s \to \phi\pi$ for $q_7^{(\prime)} = q_9^{(\prime)}$ and $\bar{B}_s \to \phi_L \rho_L$ (with the subscript L denoting longitudinally polarised vector-mesons) for parity-symmetric models, where no enhancement is expected. Furthermore, effects in the q_9' and the $q_7' = q_9'$ scenarios are limited by the small allowed region resulting from the $B \to \pi K$ fit. Largest effects occur as expected in the scenarios which are least constrained by $B \to \pi K$, i.e. the single $q_7^{(\prime)}$ and the parity-symmetric models. Especially in these cases a $\bar{B}_s \to \phi\pi$ measurement would complement $B \to \pi K$ data and, while the parity-symmetric models lack the motivation via the $\Delta A_{\rm CP}$ discrepancy, the q_7' setting resolves it with ease. Moreover, we like to stress that $B \to \pi K$ data alone cannot distinguish among opposite-parity scenarios because such scenarios generate equal results for the $B \to \pi K$ observables (for 180°-rotated parameter points). Therefore an analysis of $B \to \pi K$ should for example be supported by the analysis of a PV decay, suggesting $\bar{B}_s \to \phi\pi^0$ as an ideal candidate.

4. Analysis of specific NP models

The results of the previous section raise the question: which concrete models for NP can provide a large new EW penguin amplitude without being excluded by present data? In this section we provide a survey of results obtained analysing a set of well-motivated NP models, referring to [1] for more details. The main difference with respect to the model-independent analysis is the possibility of adding constraints from other flavour processes beyond the hadronic B modes, e.g. the semileptonic decay $\bar{B} \to X_s e^+ e^-$, the radiative decay $\bar{B} \to X_s \gamma$ and B_s-$\bar{B}_s$ mixing. These processes usually yield tight constraints on new flavour structures and it has to be investigated if the effects in $B \to \pi K$ and $\bar{B}_s \to \phi\rho^0, \phi\pi^0$ survive these constraints.

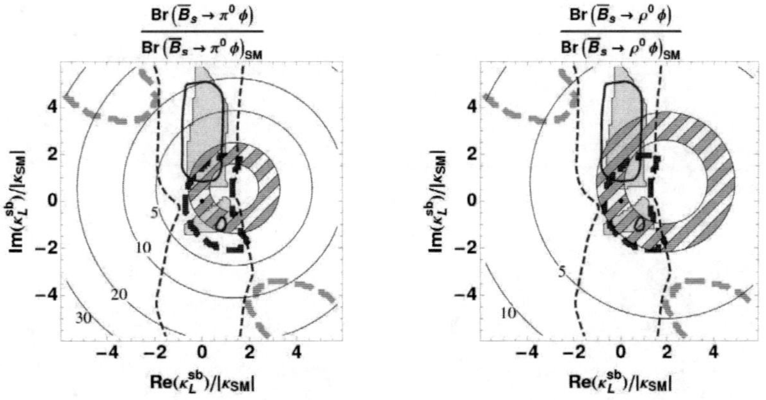

Figure 3: Enhancement factor for the $\bar{B}_s \to \phi\rho^0, \phi\pi^0$ branching ratios with respect to their SM values in the modified-Z^0-penguin scenario. The green area represents the region allowed by the 2σ constraints from all the considered hadronic decays, while the area inside the dashed blue line represents the region allowed by the 2σ constraint from semi-leptonic decays. The areas inside the dashed orange line represent the parameter values for which the modified-Z^0-penguin would explain the B_s-$\bar{B}_s$ mixing phase

.

4.1. The modified Z^0-penguin scenario

The simplest class of models with large new contributions to EW penguins comprises models with a modified $Z\bar{s}b$ coupling. Such a FCNC coupling can either be generated by integrating out new heavy particles, e.g. in supersymmetric models or fourth-generation models, or it can exist at tree-level in more exotic scenarios like models with non-sequential quarks.

The model is described in terms of two parameters, $\kappa^{sb}_{L,R}$, which parameterise the effective coupling of the Z^0 to left- and right-handed quarks, respectively. A scenario with only $\kappa^{sb}_L \neq 0$ scenario shares its most important features with the q_9-only setup of the model-independent study, while the case with only $\kappa^{sb}_R \neq 0$ is similar to the q_7-only case. This expectation is confirmed by the graphs in fig. 3. We only note that we get a 180° rotation due to the signs of δC_9 and C'_7. The main difference is that we now face additional constraints from semileptonic decays and B_s-$\bar{B}_s$ mixing. The allowed region for the former is given by the interior of the blue dashed curve, the allowed region for the latter by the orange areas outside the zone preferred by the $B \to \pi K$ fit. We see that the Δ_s anomaly of B_s-$\bar{B}_s$ mixing cannot be resolved in a modified Z scenario when fulfilling at the same time the semileptonic constraints. However, in models with loop-induced modified Z couplings the Z exchange contribution to B_s-$\bar{B}_s$ mixing has to be regarded as a subleading effect and therefore the corresponding constraint does not apply in this case. Focussing on the semileptonic constraints, we find that they are compatible with the 1σ region of the $B \to \pi K$ fit for all three cases, but draw the FCNC couplings $\kappa^{sb}_{L,R}$ to very small values. As a consequence, we expect no significant effects in $\bar{B}_s \to \phi\pi^0, \phi\rho^0$.

4.2. Models with an additional $U(1)'$ gauge symmetry

The presence of a heavy Z' boson associated with an additional $U(1)'$ gauge symmetry is a well-motivated extension of the SM. This additional symmetry has not been invented to solve a particular problem of the SM, but rather occurs as a byproduct in many models like e.g. Grand Unified Theories, various models of dynamical symmetry breaking and Little-Higgs models.

The case of interest, in which the flavour effect is assumed to contribute mainly to the

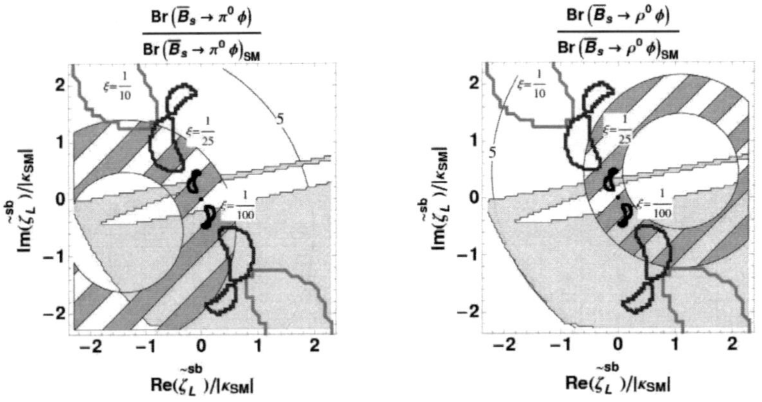

Figure 4: Enhancement factors of $\mathrm{Br}(\bar{B}_s \to \phi\pi^0)$ and $\mathrm{Br}(\bar{B}_s \to \phi\rho^0)$ for $\tilde{\zeta}_L^{sb} \neq 0$. The red-hatched ring corresponds to the SM uncertainty. The green area is allowed by the 2σ constraints from all hadronic B decays while the regions inside the blue lines are compatible with the constraint from B_s-$\bar{B}_s$ mixing. From the biggest to the smallest region they stand for $\xi = 1/10$, $\xi = 1/25$ and $\xi = 1/100$, respectively.

electroweak penguin, can be described in terms of two parameters, $\tilde{\zeta}_{L,R}^{sb}$, corresponding to the properly normalised couplings of the Z'-boson to left- and right-handed quarks, respectively. The three scenarios with only $\tilde{\zeta}_L^{sb} \neq 0$, only $\tilde{\zeta}_R^{sb} \neq 0$ and with $\tilde{\zeta}_L^{sb} = \tilde{\zeta}_R^{sb} \neq 0$ exactly correspond to the q_7-only, q_9'-only and the $q_7 = q_9'$ scenarios in our model-independent analysis, except for a normalisation factor. The exclusion regions from the $2\,\sigma$ constraints and the confidence levels from the fit can be immediately read off from figs. 1 and 2, provided one rescales the axes by an appropriate normalisation factor and rotates the pictures by 180° to take into account a minus sign in the Wilson coefficients. In this scenario no direct constraints arise from semileptonic decays, assuming leptophobic Z' couplings. The relation to B_s-$\bar{B}_s$ mixing moreover depends on an additional parameter $\xi \equiv \frac{g_{U(1)'}^2}{g^2}\frac{M_W^2}{M_{Z'}^2}$, which is related to the overall strength of the Z' coupling to fermions, and to its mass. From the diagrams in fig. 4 we see that the B_s-$\bar{B}_s$ mixing constraint is in general very tight. It prohibits large effects in $\bar{B}_s \to \phi\pi^0, \phi\rho^0$ for realistic values of the parameter $\xi \lesssim 1/25$, which would correspond for example to $g_{U(1)'} \sim g$ and $M_{Z'} \sim 400\mathrm{GeV}$. We find that enhancement of a factor ~ 5 is possible in the ζ_L^{sb} and ζ_R^{sb} scenarios whereas no effect can occur in the $\zeta_L^{sb} = \zeta_R^{sb}$. For $\xi = 1/100$ the constraints from B_s-$\bar{B}_s$ mixing become so strong that no effect in $\bar{B}_s \to \phi\pi^0, \phi\rho^0$ would be detectable. A measurement of a significant enhancement would therefore set a lower limit on ξ, equivalent to an upper limit on the Z' mass.

4.3. MSSM

The main contribution to the electroweak penguin in supersymmetry comes from Z penguins induced by chargino-squark loops. This scenario is equivalent to a modified Z^0 penguin model with only κ_L^{sb} different from zero. Constraints from flavour observables and squark masses, however, strongly constrain the parameter space available. In Fig. 5 we provide a zoom of the κ_L^{sb} scenario with a superposition of points obtained from a scan of the MSSM parameter space. The constraints from squark and chargino masses and from $\mathrm{Br}(\bar{B} \to X_s\gamma)$ are taken into account. One sees that these points are not able to decrease the ΔA_{CP} discrepancy below the $2\,\sigma$ level, nor to create a large enhancement in the B_s decays branching ratio.

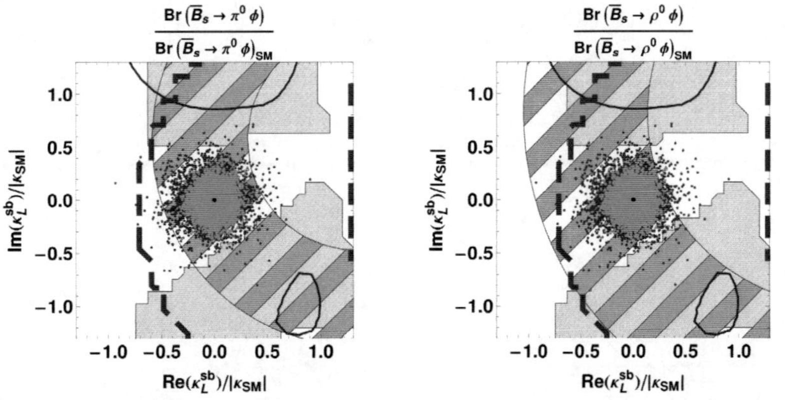

Figure 5: Zoom of the upper plots in fig. 3. On top we add the $\kappa_L^{sb}/|\kappa_{\rm SM}|$ values resulting from chargino-induced flavour-violating Z couplings in a parameter scan.

5. Conclusion

We have studied the possibility of probing isospin-violating NP in hadronic B decays. Our analysis is motivated by discrepancies found in $B \to \pi K$ decays. In particular, the $2.5\,\sigma$ discrepancy found in the observable $\Delta A_{\rm CP}$ can be interpreted as a sign of NP in the EW penguin sector of the theory. Since this discrepancy alone is not able to provide evidence of NP, we have proposed here to consider the decays $\bar{B}_s \to \phi\pi^0$ and $\bar{B}_s \to \phi\rho^0$ as well, which are extremely sensitive to EW penguins. First, we have demonstrated in a model-independent analysis that the solution of the $B \to \pi K$ discrepancy is obtained with an additional NP contribution to the EW penguin operators $Q_7^{(\prime)}, ..., Q_{10}^{(\prime)}$ of the same order of magnitude as the leading SM coefficient $C_9^{\rm SM}$. Then, we have studied the corresponding enhancement of the $\bar{B}_s \to \phi\pi^0, \phi\rho^0$ branching ratios for various scenarios, taking into account constraints from other hadronic B decays. The results show that in many cases a large enhancement of about an order of magnitude is possible. Exceptions are parity-symmetric models, which have no impact on the VV decay $\bar{B}_s \to \phi\rho^0$, and scenarios with (approximately) equal contributions to $C_7^{(\prime)}$ and $C_9^{(\prime)}$, which cancel in $\bar{B}_s \to \phi\pi^0$.

The last part of the analysis concerns a survey of concrete NP models such as modified Z^0 penguin, a model with an additional $U(1)'$ gauge symmetry and the MSSM. In such models additional constraints arise from the semileptonic decays $\bar{B} \to X_s e^+e^-$ and $\bar{B} \to K^* l^+ l^-$ and from B_s-$\bar{B}_s$ mixing. While the solution of the $\Delta A_{\rm CP}$ discrepancy is in some cases still possible, like in the modified-Z^0-penguin scenario or in a model with an additional Z' boson, the enhancement of the $\bar{B}_s \to \phi\pi^0, \phi\rho^0$ decays is usually strongly constrained by either the constraint from semileptonic decays or from B_s-$\bar{B}_s$ mixing. Last, we find that the new contribution to the EW penguins in the MSSM is always marginal and cannot solve the $\Delta A_{\rm CP}$ discrepancy, nor create a large enhancement of the B_s decays considered here.

Acknowledgements

L.V. would like to thank the organizers of Discrete 2010 for putting together an exciting conference in a very pleasant environment.

References

[1] L. Hofer, D. Scherer, L. Vernazza, [arXiv:1011.6319 [hep-ph]].

[2] M. Gronau, J. L. Rosner, Phys. Rev. **D59** (1999) 113002. [hep-ph/9809384]; S. Baek, D. London, Phys. Lett. **B653** (2007) 249-253. [hep-ph/0701181]; R. Fleischer, S. Recksiegel, F. Schwab, Eur. Phys. J. **C51** (2007) 55-61. [hep-ph/0702275 [HEP-PH]].
[3] E. Barberio *et al.* [Heavy Flavor Averaging Group Collaboration], [arXiv:0808.1297 [hep-ex]].
[4] M. Beneke, G. Buchalla, M. Neubert *et al.*, Nucl. Phys. **B606** (2001) 245-321. [hep-ph/0104110]; M. Beneke, M. Neubert, Nucl. Phys. **B675**, 333-415 (2003). [hep-ph/0308039].

Journal of Physics: Conference Series **335** (2011) 012040 doi:10.1088/1742-6596/335/1/012040

B_s^0 Decays at Belle

Remi Louvot

(On behalf of the Belle collaboration)
École Polytechnique Fédérale de Lausanne (EPFL), Lausanne, Switzerland

E-mail: remi.louvot@epfl.ch

Abstract. The large data sample recorded with the Belle detector at the $\Upsilon(5S)$ energy provides a unique opportunity to study the poorly-known B_s^0 meson. Several analyses, made with a data sample representing an integrated luminosity of 23.6 fb^{-1}, are presented. We report the study of the large-signal $B_s^0 \to D_s^{(*)-}h^+$ ($h^+ = \pi^+, \rho^+$) decays. In addition, several results on B_s^0 decays related to CP-violation studies are described. Beside the non-flavor specific $B_s^0 \to D_s^{\mp}K^{\pm}$ decay, the following CP-eigenstate decays are studied: $B_s^0 \to J/\psi\,\eta^{(\prime)}$, $B_s^0 \to J/\psi\, f_0(980)$, the charmless $B_s^0 \to h\bar{h}$ ($h = \pi^+, K^+, K_S^0$) and the three $B_s^0 \to D_s^{(*)+}D_s^{(*)-}$ modes from which $\Delta\Gamma_s^{CP}/\Gamma_s$ is extracted.

1. Introduction

The Belle experiment [1], located at the interaction point of the KEK B factory [2], was designed for the study of B mesons[1] produced in e^+e^- annihilation at the $\Upsilon(4S)$ resonance ($\sqrt{s} \approx 10.58$ GeV). After having recorded an unprecedented sample of ~ 550 millions of $B\bar{B}$ pairs, the Belle collaboration started to record collisions at higher energies, opening the possibility to study the B_s^0 meson. A significant theoretical effort on the B_s^0 potential for studying CP violation in the Standard Model (SM) and beyond has taken place from the 90's [3]. The non-flavor specific decays such as $B_s^0 \to D_s^{\mp}K^{\pm}$ are expected to be insensitive to new physics (NP); they can be used to measure the CKM angle γ [4] and help to resolve the ambiguity in the width difference in the $B_s^0 - \bar{B}_s^0$ system, $\Delta\Gamma_s$ [5]. The penguin-dominated decays $B_s^0 \to K^+K^-$ are expected to be sensitive to NP [6] and the CKM angles β and γ can be extracted by comparing its branching fraction (BF) with that of $B^0 \to \pi^+\pi^-$ [7]. Finally, the decays involving a $b \to c\bar{c}s$ transition (the "golden mode" $B_s^0 \to J/\psi\phi$, $B_s^0 \to J/\psi\,\eta^{(\prime)}$, $B_s^0 \to J/\psi\, f_0(980)$, $B_s^0 \to D_s^{(*)+}D_s^{(*)-}$, etc.) are probably the most promising, as the CP-violation effects in these modes are expected to be tiny in the SM [8]. A particular attention will be given to the $B_s^0 \to D_s^{(*)+}D_s^{(*)-}$ modes because the width difference $\Delta\Gamma_s^{CP}$ can be extracted from their total BF [9].

Up to now, a data sample of integrated luminosity of $L_{\text{int}} = (23.6 \pm 0.3)\,\text{fb}^{-1}$ (out of a total of 120 fb^{-1}) has been analyzed at the energy of the $\Upsilon(5S)$ resonance ($\sqrt{s} \approx 10.87$ GeV). Since the $\Upsilon(5S)$ resonance is above the $B_s^0\bar{B}_s^0$ threshold, it was naturally expected that the B_s^0 meson could be studied with $\Upsilon(5S)$ data as well as the B mesons are with $\Upsilon(4S)$ data. The main advantage with respect to the hadron colliders is the better prospects of absolute BF measurements. However, the abundance of B_s^0 mesons in $\Upsilon(5S)$ hadronic events has to be precisely determined. Above the $e^+e^- \to u\bar{u}, d\bar{d}, s\bar{s}, c\bar{c}$ continuum events, the $e^+e^- \to b\bar{b}$

[1] The notation "B" refers either to a B^0 or a B^+. Moreover, charge-conjugated states are implied everywhere.

process can produce different kinds of final states involving a pair of non-strange B mesons [10] ($B^*\bar{B}^*$, $B^*\bar{B}$, $B\bar{B}$, $B^*\bar{B}^*\pi$, $B^*\bar{B}\pi$, $B\bar{B}\pi$, $B\bar{B}\pi\pi$ and $B\bar{B}\gamma$), a pair of B_s^0 mesons ($B_s^*\bar{B}_s^*$, $B_s^*\bar{B}_s^0$ and $B_s^0\bar{B}_s^0$), or final states involving a lighter bottomonium resonance below the open-beauty threshold [11]. The B^* and B_s^* mesons always decay by emission of a photon. The total $e^+e^- \to b\bar{b}$ cross section at the $\Upsilon(5S)$ energy was measured to be $\sigma_{b\bar{b}} = (302 \pm 14)$ pb [12, 13] and the fraction of B_s^0 events to be[2] $f_s = \sigma(e^+e^- \to B_s^{(*)}\bar{B}_s^{(*)})/\sigma_{b\bar{b}} = (20.2 \pm 3.6)$ % [15]. The dominant B_s^0 production mode, $b\bar{b} \to B_s^*\bar{B}_s^*$, represents a fraction $f_{B_s^*\bar{B}_s^*} = (90.1^{+3.8}_{-4.0} \pm 0.2)$ % of the $b\bar{b} \to B_s^{(*)}\bar{B}_s^{(*)}$ events, as measured with $B_s^0 \to D_s^-\pi^+$ events (see next section and Ref. [16]).

For all the exclusive modes presented here, the B_s^0 candidates are fully reconstructed from the final-state particles. From the reconstructed four-momentum in the CM, ($E^*_{B_s^0}$,$\boldsymbol{p}^*_{B_s^0}$), two variables are formed: the energy difference $\Delta E = E^*_{B_s^0} - \sqrt{s}/2$ and the beam-constrained mass $M_{\rm bc} = \sqrt{s/4 - \boldsymbol{p}^{*2}_{B_s^0}}$. The signal coming from the dominant $e^+e^- \to B_s^*\bar{B}_s^*$ production mode is extracted from a two-dimensional fit performed on the distribution of these two variables. The corresponding BF is then extracted using the total efficiency (including sub-decay BFs) determined with Monte-Carlo (MC) simulations, $\sum \varepsilon\mathcal{B}$, and the number of B_s^0 mesons produced via the $e^+e^- \to B_s^*\bar{B}_s^*$ process, $N_{B_s^0} = 2 \times L_{\rm int} \times \sigma_{b\bar{b}} \times f_s \times f_{B_s^*\bar{B}_s^*} \sim 2.5 \times 10^6$.

2. Dominant CKM-favored B_s^0 Decays

We report the measurement of exclusive $B_s^0 \to D_s^{(*)-}h^+$ ($h^+ = \pi^+$ or ρ^+) decays [16]. These modes are expected to produce an abundant signal because of their relatively large predicted BFs [17] and their clean signatures: four charged tracks and up to two photons. The leading amplitude for the four $B_s^0 \to D_s^{(*)-}\pi^+$ and $B_s^0 \to D_s^{(*)-}\rho^+$ modes is a $b \to c$ tree diagram of order λ^2 (in the Wolfenstein parametrization [18] of the CKM quark-mixing matrix [19]) with a spectator s quark. Besides being interesting in their own right, such measurements, if precise enough, can be of high importance for the current and forthcoming hadron collider experiments. It was for example pointed out [20] that the search for the very rare decay $B_s^0 \to \mu^+\mu^-$ will be systematically limited by the poor knowledge of B_s^0 production, in case NP will enhance the decay probability by no more than a factor 3 above the SM expectation.

In addition, polarization measurements of B decays have become of high interest since the observation of a surprisingly large transverse polarization in $B \to \phi K^*$ decays by BaBar [21] and Belle [22]. The relative strengths of the longitudinal and transverse states can be measured with an angular analysis of the decay products. In the helicity basis, the expected $B_s^0 \to D_s^{*-}\rho^+$ differential decay width is proportional to $\frac{{\rm d}^2\Gamma(B_s^0 \to D_s^{*-}\rho^+)}{{\rm d}\cos\theta_{D_s^{*-}}\,{\rm d}\cos\theta_{\rho^+}} \propto 4f_L \sin^2\theta_{D_s^{*-}} \cos^2\theta_{\rho^+} + (1-f_L)(1+\cos^2\theta_{D_s^{*-}})\sin^2\theta_{\rho^+}$, where f_L is the longitudinal polarization fraction, and $\theta_{D_s^{*-}}$ (θ_{ρ^+}) is the helicity angle of the D_s^{*-} (ρ^+) defined as the supplement of the angle between the B_s^0 and the D_s^- (π^+) momenta in the D_s^{*-} (ρ^+) frame.

The D_s^- mesons are reconstructed via three modes : $D_s^- \to \phi(\to K^+K^-)\pi^-$, $D_s^- \to K^{*0}(\to K^+\pi^-)K^-$ and $D_s^- \to K_S^0(\to \pi^+\pi^-)K^-$. Based on the ratio of the second and the zeroth Fox-Wolfram moments [23], R_2, the continuum events are efficiently rejected by taking advantage of the difference between their event geometry (jet like, high R_2) and the signal event shape (spherical, low R_2). The $B_s^0 \to D_s^-\pi^+$ and $B_s^0 \to D_s^{*-}\pi^+$ ($B_s^0 \to D_s^-\rho^+$ and $B_s^0 \to D_s^{*-}\rho^+$) candidates with R_2 smaller than 0.5 (0.35) are kept for further analysis. A best candidate selection, based on the intermediate-particle reconstructed masses, is then implemented in order to keep only one B_s^0 candidate per event per mode. The $M_{\rm bc}$ and ΔE distributions of the

[2] The BF values presented here are calculated with $f_s = (19.3 \pm 2.9)\%$ [14]. The BFs of $B_s^0 \to D_s^-\pi^+$, $B_s^0 \to D_s^{\mp}K^{\pm}$ and those in Sections 4 and 5 are calculated with $f_s = (19.5^{+3.0}_{-2.3})\%$, also provided in Ref. [14].

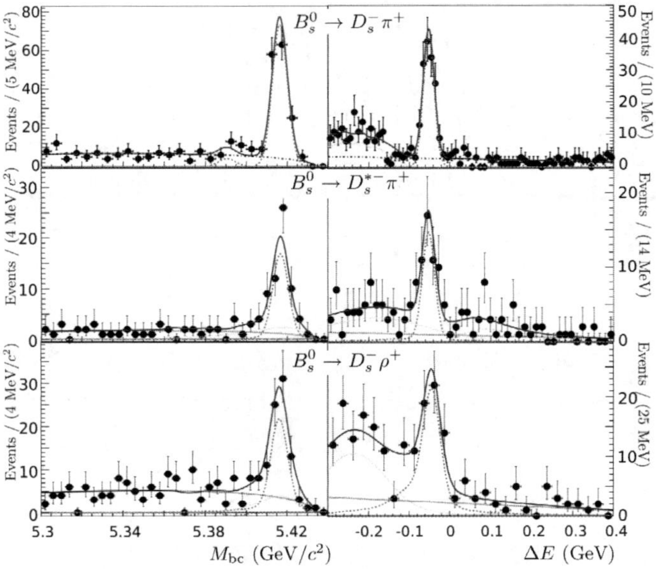

Figure 1. Left: M_{bc} distributions for the $B_s^0 \to D_s^- \pi^+$ (top) $B_s^0 \to D_s^{*-}\pi^+$ (middle) and $B_s^0 \to D_s^- \rho^+$ (bottom) candidates with ΔE restricted to the $B_s^* \bar{B}_s^*$ signal region. Right: ΔE distributions with M_{bc} restricted to the $B_s^* \bar{B}_s^*$ signal region. The black- (green-) dotted curve represents the continuum (peaking) background, while the red-dashed curves are the signal shapes. The larger one is the signal in the $B_s^* \bar{B}_s^*$ kinematic region and the two others, which are very close to 0, are the signals in the two other B_s^0 production modes ($B_s^* \bar{B}_s^0$ and $B_s^0 \bar{B}_s^0$).

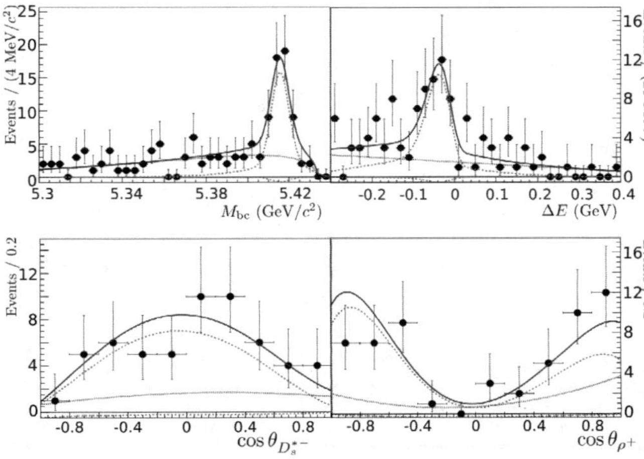

Figure 2. Fit of the $B_s^0 \to D_s^{*-}\rho^+$ candidates. Top: M_{bc} and ΔE distributions, similarly to Fig. 1. Bottom: helicity distributions of the D_s^{*-} (left) and ρ^+ (right) with M_{bc} and ΔE restricted to the $B_s^* \bar{B}_s^*$ kinematic region. The black-dotted curve represents the background, while the two red-dashed curves are the signal. The large (small) signal shape corresponds to the longitudinal (transverse) component.

selected B_s^0 candidates for the three D_s^- modes are shown in Figs. 1 and 2, where the various components of the probability density function (PDF) used for the fit are described. The $B_s^0 \to D_s^{*-}\rho^+$ candidates are observed with two additional variables, $\cos\theta_{D_s^{*-}}$ and $\cos\theta_{\rho^+}$, which are the cosines of the helicity angles defined above. They are needed for the measurement of f_L.

Table 1 presents a summary of the numerical results obtained for the $B_s^0 \to D_s^{(*)-}\pi^+$ and $B_s^0 \to D_s^{(*)-}\rho^+$ modes. The different sources of systematic uncertainties affecting the measurements are identified and quoted as a second error. Our results on the B_s^0 decays are consistent with theoretical predictions [17] and with existing measurements (Table 1).

3. Evidence for $B_s^0 \to D_s^{\mp} K^{\pm}$

We searched for the Cabibbo-suppressed counterpart of $B_s^0 \to D_s^- \pi^+$, $B_s^0 \to D_s^{\mp} K^{\pm}$, which is not flavour specific [16]. The analysis is performed in the same way as that of $B_s^0 \to D_s^- \pi^+$, by replacing the π^+ by a K^+ candidate. The fit also includes the $B_s^0 \to D_s^- \pi^+$ contamination (when the pion is misidentified as a kaon). A 3.5σ evidence with $6.7^{+3.4}_{-2.7}$ events is obtained

Table 1. Summary of the results for the four $B_s^0 \to D_s^{(*)-}\pi^+$ and $B_s^0 \to D_s^{(*)-}\rho^+$ modes [16]. Top: signal yields in the $B_s^*\bar{B}_s^*$ production mode ($N_{B_s^*\bar{B}_s^*}$), significances (S) including systematics, total signal efficiencies (ε) including all sub-decay BFs, and BFs ($\mathcal{B}$), where the uncertainty due to f_s (third error) is separated from the others systematics (second error). The first error represents the statistical uncertainties. Bottom: other measurements, obtained with the $B_s^0 \to D_s^-\pi^+$ and $B_s^0 \to D_s^{*-}\rho^+$ analyses.

Mode	$N_{B_s^*\bar{B}_s^*}$	S	ε (10^{-3})	$\mathcal{B}$ (10^{-3})
$B_s^0 \to D_s^-\pi^+$	145^{+14}_{-13}	21σ	15.8	$3.7^{+0.4}_{-0.3} \pm 0.4 \pm 0.5$
$B_s^0 \to D_s^{*-}\pi^+$	$53.4^{+10.3}_{-9.4}$	7.1σ	9.13	$2.4^{+0.5}_{-0.4} \pm 0.3 \pm 0.4$
$B_s^0 \to D_s^-\rho^+$	$92.2^{+14.2}_{-13.2}$	8.2σ	4.40	$8.5^{+1.3}_{-1.2} \pm 1.1 \pm 1.3$
$B_s^0 \to D_s^{*-}\rho^+$	$77.8^{+14.5}_{-13.4}$	7.4σ	2.67	$11.9^{+2.2}_{-2.0} \pm 1.7 \pm 1.8$

Observable	Our measurement
Masses (MeV/c^2)	$m(B_s^0) = 5364.4 \pm 1.3 \pm 0.7$; $m(B_s^*) = 5416.4 \pm 0.4 \pm 0.5$
Prod. mode (%)	$f_{B_s^*\bar{B}_s^*} = 90.1^{+3.8}_{-4.0} \pm 0.2$; $f_{B_s^*\bar{B}_s^0} = 7.3^{+3.3}_{-3.0} \pm 0.1$; $f_{B_s^0\bar{B}_s^0} = 2.6^{+2.6}_{-2.5}$
$f_L(B_s^0 \to D_s^{*-}\rho^+)$	$1.05^{+0.08+0.03}_{-0.10-0.04}$

in the $B_s^*\bar{B}_s^*$ signal region (Fig. 3), leading to the branching fraction $\mathcal{B}(B_s^0 \to D_s^{\mp}K^{\pm}) = (2.4^{+1.2}_{-1.0}(\text{stat.}) \pm 0.3(\text{syst.}) \pm 0.3(f_s)) \times 10^{-4}$, in agreement with the CDF result [24].

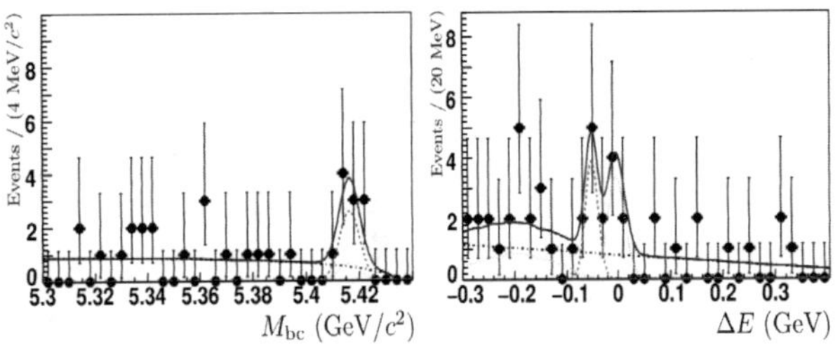

Figure 3. Distributions, similarly to Fig. 1, of the $B_s^0 \to D_s^{\mp}K^{\pm}$ candidates and the fitted PDF as solid blue curve. The red, black and green dotted curves represent the signal, the continuum and the $B_s^0 \to D_s^-\pi^+$ component of the PDF, respectively.

4. Observation of $B_s^0 \to K^+K^-$ and Searches for $B_s^0 \to \pi^+\pi^-$, $B_s^0 \to K^-\pi^+$ and $B_s^0 \to K_S^0K_S^0$

We present our results for the $B_s^0 \to K^+K^-$, $B_s^0 \to K^-\pi^+$, $B_s^0 \to \pi^+\pi^-$ and $B_s^0 \to K_S^0K_S^0$ charmless decays [25]. The charged pion and kaon candidates are selected using charged tracks and identified with energy deposition, momentum and time-of-flight measurements. The K_S^0 candidates are reconstructed via the $K_S^0 \to \pi^+\pi^-$ decay. A likelihood based on 16 modified Fox-Wolfram moments [26] is implemented to reduce the continuum, which is the main source of background.

We do observe a 5.8σ excess of 24 ± 6 events in the $B_s^*\bar{B}_s^*$ region for the $B_s^0 \to K^+K^-$ mode (Fig. 4). The BF, $\mathcal{B}(B_s^0 \to K^+K^-) = (3.8^{+1.0}_{-0.9}(\text{stat.}) \pm 0.5(\text{syst.}) \pm 0.5(f_s)) \times 10^{-5}$, is derived. However, no significant signal is seen for the other modes. Including the systematic uncertainties, we set the following upper limits at 90% confidence level: $\mathcal{B}(B_s^0 \to \pi^+\pi^-) < 1.2 \times 10^{-5}$, $\mathcal{B}(B_s^0 \to K^-\pi^+) < 2.6 \times 10^{-5}$ and, assuming $\mathcal{B}(B_s^0 \to K^0\bar{K}^0) = 2 \times \mathcal{B}(B_s^0 \to K_S^0K_S^0)$,

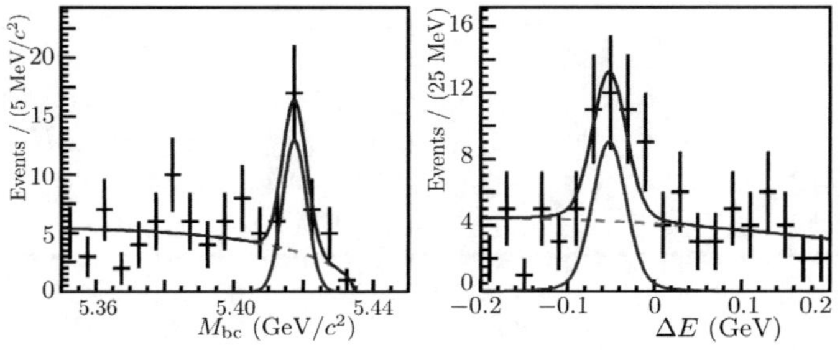

Figure 4. Distributions, similarly to Fig. 1, of the $B_s^0 \to K^+K^-$ candidates and the fitted PDF as a solid blue curve. The solid-red and the dotted-grey curves represent the signal and the continuum component of the PDF, respectively.

$\mathcal{B}(B_s^0 \to K^0\bar{K}^0) < 6.6 \times 10^{-5}$. The later is the first limit set for the $B_s^0 \to K^0\bar{K}^0$ mode. All the other values are compatible with the CDF results [27, 28].

5. Study of $B_s^0 \to J/\psi\,\eta^{(\prime)}$ and Search for $B_s^0 \to J/\psi\, f_0(980)$

Results about the first observation of $B_s^0 \to J/\psi\,\eta$ and the first evidence for $B_s^0 \to J/\psi\,\eta^{\prime}$ are reported [29]. The J/ψ candidates are formed with oppositely-charged electron or muon pairs, while η candidates are reconstructed via the $\eta \to \gamma\gamma$ and $\eta \to \pi^+\pi^-\pi^0$ modes. A mass (mass and vertex) constrained fit is then applied to the η (J/ψ) candidates. The $\eta^{\prime}$ candidates are reconstructed via the $\eta^{\prime} \to \eta\pi^+\pi^-$ and $\eta^{\prime} \to \rho^0\gamma$ modes, while the ρ^0 candidates are selected from $\pi^+\pi^-$ pairs. If more than one candidate per event satisfies all the selection criteria, the one with the smallest fit residual is selected. The main background is the continuum, which is reduced by requiring $R_2 < 0.4$. The combined $M_{\rm bc}$ and ΔE distributions are presented in Figs. 5 ($B_s^0 \to J/\psi\,\eta$) and 6 ($B_s^0 \to J/\psi\,\eta^{\prime}$). We obtain $\mathcal{B}(B_s^0 \to J/\psi\,\eta) = (3.32 \pm 0.87(\text{stat.})^{+0.32}_{-0.28}(\text{syst.}) \pm 0.42(f_s)) \times 10^{-4}$ and $\mathcal{B}(B_s^0 \to J/\psi\,\eta^{\prime}) = (3.1 \pm 1.2(\text{stat.})^{+0.5}_{-0.6}(\text{syst.}) \pm 0.4(f_s)) \times 10^{-4}$. This is, respectively, the first observation (7.3σ) and the first evidence (3.8σ) for these modes.

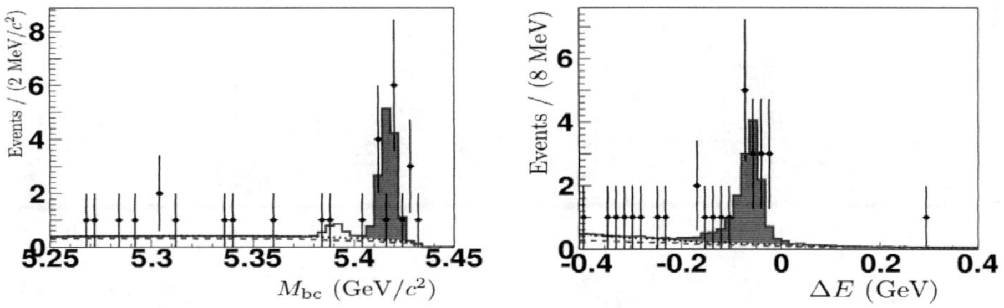

Figure 5. $M_{\rm bc}$ (left) and ΔE (right) distributions, similarly to Fig. 1, of the $B_s^0 \to J/\psi\,\eta$ candidates (points with error bars) and the fitted PDF as a solid curve. The sub-modes $\eta \to \gamma\gamma$ and $\eta \to \pi^+\pi^-\pi^0$, which are fitted separately, are summed in these plots. The green-dotted curve (plain red region) represents the continuum (signal) component of the PDF. The small peak in the $M_{\rm bc}$ plot is the signal contribution in the $B_s^*B_s^0$ region.

The $B_s^0 \to J/\psi\, f_0(980)$ mode is especially interesting for the hadron-collider experiments because it has only four charged tracks in its final state. Recent calculations and measurements predict the ratio $R_{f/\phi} = \frac{\mathcal{B}(B_s^0 \to J/\psi\, f_0(980)) \times \mathcal{B}(f_0(980) \to \pi^+\pi^-)}{\mathcal{B}(B_s^0 \to J/\psi\, \phi) \times \mathcal{B}(\phi \to K^+K^-)}$ to be ≈ 0.2 [30], 0.42 ± 0.11 [31] or ≈ 0.24 [32], in agreement with other predictions from generalized QCD factorization [33].

With the same selection for the J/ψ as described above, and the reconstruction of $f_0(980) \to \pi^+\pi^-$ candidates, the $B_s^0 \to J/\psi\, f_0(980)$ signal is fitted using the distribution of the energy

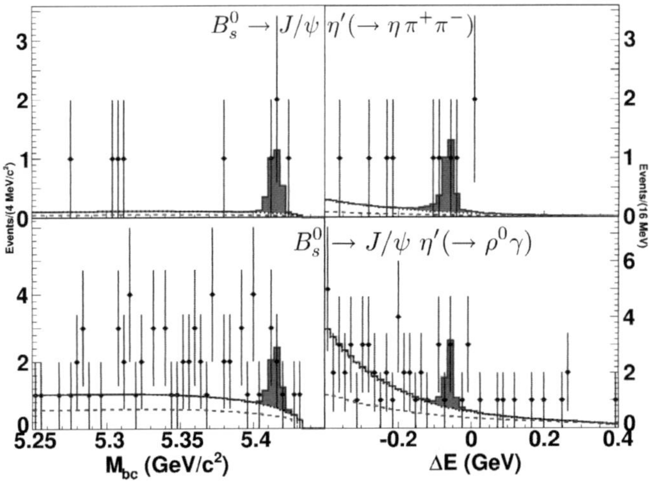

Figure 6. $M_{\rm bc}$ (left) and ΔE (right) distributions, similarly to Fig. 1, of the $B_s^0 \to J/\psi\, \eta'$ candidates (points with error bars) and the fitted PDF as a solid curve. The green-dotted curve represents the continuum component of the PDF. The red region represents the signal component of the PDF.

difference, ΔE, and the $f_0(980)$ mass, $M_{\pi^+\pi^-}$ (Fig. 7). No significant signal (6.0 ± 4.4 events, 1.7σ) is seen and we set the upper limit [34]

$$\mathcal{B}(B_s^0 \to J/\psi\, f_0(980)) \times \mathcal{B}(f_0(980) \to \pi^+\pi^-) < 1.63 \times 10^{-4} \text{ (at 90\% C.L.)},$$

or, similarly, $R_{f/\phi} < 0.275$ (at 90% C.L.) using our preliminary result of $\mathcal{B}(B_s^0 \to J/\psi\, \phi)$ [35]. These limits are clearly in the region of interest and an update using our full data sample ($120\,\text{fb}^{-1}$) and an improved selection is being performed.

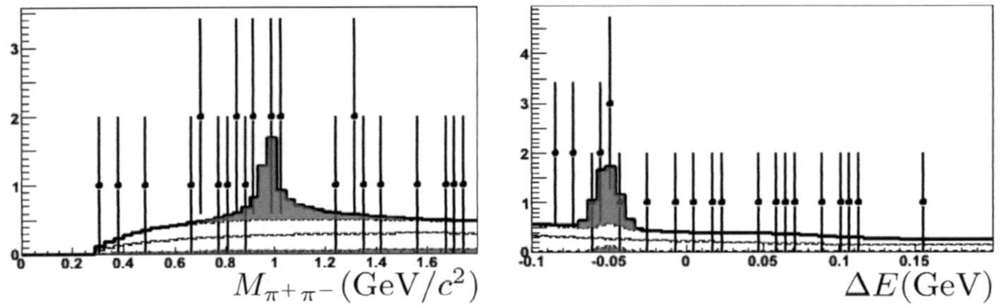

Figure 7. $f_0(980)$ mass (left) and ΔE (right) distributions of the $B_s^0 \to J/\psi\, f_0(980)$ candidates. The solid-black curve is the total fitted PDF. The plain green region represents the contribution of the non-resonant $B_s^0 \to J/\psi\, \pi^+\pi^-$, while the plain red region is the signal. The dotted-black curve is the contribution of the other $B_s^0 \to J/\psi\, X$ modes.

6. Study of $B_s^0 \to D_s^{(*)+} D_s^{(*)-}$ and Measurement of $\Delta\Gamma_s^{CP}/\Gamma_s$

We finally report the results from our analysis of the $B_s^0 \to D_s^{(*)+} D_s^{(*)-}$ decays [36]. These modes are CP eigenstates *and* CKM favored ($b \to c\bar{c}s$ transition of order λ^2). In the heavy-quark limit, they are CP even and dominate $\Delta\Gamma$ [9]. The relative width difference of the $B_s^0 - \bar{B}_s^0$ system can be obtained from the relation $\Delta\Gamma_s^{CP}/\Gamma_s = \frac{2\times\mathcal{B}(B_s^0 \to D_s^{(*)+} D_s^{(*)-})}{1-\mathcal{B}(B_s^0 \to D_s^{(*)+} D_s^{(*)-})}$. In order to reconstruct the $B_s^0 \to D_s^{(*)+} D_s^{(*)-}$ candidates, we form D_s^- candidates from 6 modes: $D_s^- \to \phi\pi^-$, $D_s^- \to K^{*0}K^-$, $D_s^- \to K_S^0 K^-$, $D_s^- \to \phi\rho^-$, $D_s^- \to K^{*0}K^{*-}$ and $D_s^- \to K_S^0 K^{*-}$. Only one

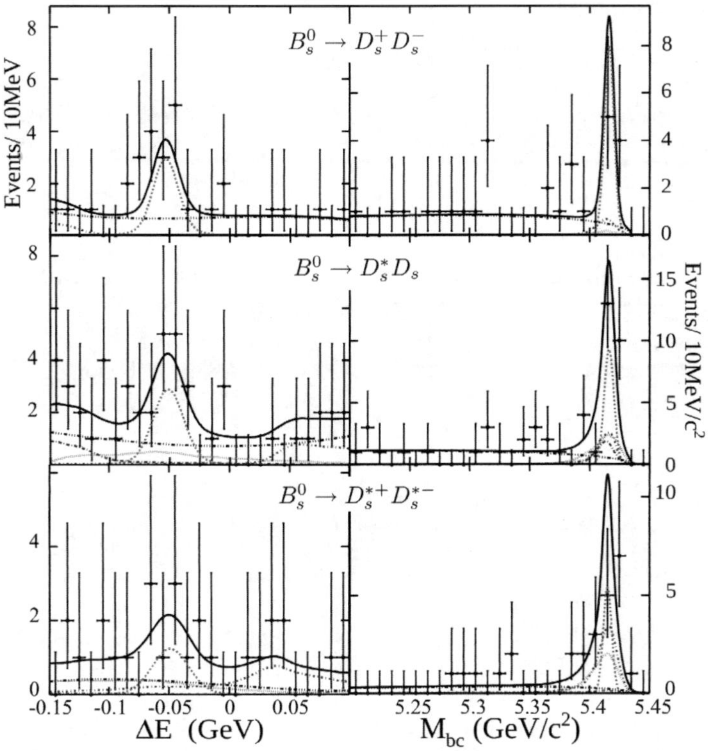

Figure 8. ΔE (left) and $M_{\rm bc}$ (right) distributions, similarly to Fig. 1, of the $B_s^0 \to D_s^+ D_s^-$ (top), $B_s^0 \to D_s^{*\pm} D_s^{\mp}$ (middle) and $B_s^0 \to D_s^{*+} D_s^{*-}$ (bottom) candidates, together with the fitted PDF. Except the continuum background component, which is shown by the black dashed-dotted curve, all the other contributions are peaking in $M_{\rm bc}$. The correct (wrong) combination signal, shown by the peaking (smooth) red dashed curve and the cross-feed components, shown by the blue dashed-dotted curve are well separated in ΔE.

Table 2. Signal event yields, $N_{\rm sig.}$, significances, S, including systematics and BFs, $\mathcal{B}$, for the three $B_s^0 \to D_s^{(*)+} D_s^{(*)-}$ modes and their sum.

Mode	$N_{\rm sig.}$	S	$\mathcal{B}$	$\mathcal{B}$ World Average [14]
$B_s^0 \to D_s^{*+} D_s^{*-}$	$4.9^{+1.9}_{-1.7}$	3.2σ	$(3.1^{+1.2}_{-1.0}(\text{stat.}) \pm 0.8(\text{syst.}))\%$	First evidence
$B_s^0 \to D_s^{*\pm} D_s^{\mp}$	$9.2^{+2.8}_{-2.4}$	6.6σ	$(2.8^{+0.8}_{-0.7}(\text{stat.}) \pm 0.7(\text{syst.}))\%$	First observation
$B_s^0 \to D_s^+ D_s^-$	$8.5^{+3.2}_{-2.6}$	6.2σ	$(1.0^{+0.4}_{-0.3}(\text{stat.})^{+0.3}_{-0.2}(\text{syst.}))\%$	$(1.04^{+0.37}_{-0.34})\%$
$B_s^0 \to D_s^{(*)+} D_s^{(*)-}$	$22.6^{+4.7}_{-3.9}$		$(6.9^{+1.5}_{-1.3}(\text{stat.}) \pm 1.9(\text{syst.}))\%$	$(4.0 \pm 1.5)\%$

candidate per event is selected based on the values of $M(D_s^-)$ and $M(D_s^{*-}) - M(D_s^-)$. The same likelihood as in Sec. 4 is used to reject 80% of the continuum events, while 95% of the signal is kept. The ΔE and $M_{\rm bc}$ distributions for each of the three $B_s^0 \to D_s^{(*)+} D_s^{(*)-}$ modes are fitted simultaneously. The signal PDF is made of two components studied with signal MC simulations: the correctly reconstructed candidates and the wrong combinations in which a non-signal track (photon) is included in place of a true daughter track (photon). In addition the so-called cross-feed contributions are included: a $D_s^{*\pm} D_s^{\mp}$ ($D_s^{*+} D_s^{*-}$) event can be selected as a $D_s^+ D_s^-$ ($D_s^{*\pm} D_s^{\mp}$) candidate with a lower energy because one photon is missing; conversely, a $D_s^+ D_s^-$ ($D_s^{*\pm} D_s^{\mp}$) candidate can be reconstructed as a $D_s^{*\pm} D_s^{\mp}$ ($D_s^{*+} D_s^{*-}$) candidate with an additional photon, hence its energy larger than expected.

The fit results can be seen in Fig. 8 while the numerical values are reported in Table 2. With the relation above, we extract

$$\Delta\Gamma_s^{CP}/\Gamma_s = (14.7^{+3.6}_{-3.0}(\text{stat.})^{+4.4}_{-4.2}(\text{syst.})) \times 10^{-2}\,.$$

This value is in agreement with the SM expectations [37] and with the results from ALEPH, $(25^{+21}_{-14})\%$ [38], DØ, $(7.2\pm3.0)\%$ [39], and CDF, $(12^{+9}_{-10})\%$ [40]. With only 23 fully-reconstructed signal events, our measurement is already competitive with the Tevatron values.

7. Conclusion

We presented recent results on B_s^0 decays obtained from 23.6 fb^{-1} of $\Upsilon(5S)$ data recorded by the Belle detector. While modes with large statistics can provide precise measurements of BFs and $B_s^{(*)}$ properties, first observations of several CP-eigenstate B_s^0 decays are a confirmation of the large potential of our $120\,\text{fb}^{-1}$ $e^+e^- \to \Upsilon(5S)$ data sample and advocate an ambitious B_s^0 program at super-B factories.

References

[1] A. Abashian *et al.* (Belle Collaboration) 2002 *Nucl. Instrum. Methods Phys. Res., Sect. A* **479** 117.
[2] S. Kurokawa and E. Kikutani 2003 *Nucl. Instrum. Methods Phys. Res., Sect. A* **499** 1.
[3] I. Dunietz 1995 *Phys. Rev. D* **52** 3048; I. Dunietz, R. Fleischer and U. Nierste 2001 *Phys. Rev. D* **63** 114015.
[4] R. Aleksan, I. Dunietz and B. Kayser 1992 *Z. Phys. C* **54** 653; R. Fleicher 2003 *Nucl. Phys. B* **671** 459.
[5] S. Nandi and U. Nierste 2008 *Phys. Rev. D* **77** 054010.
[6] D. London and J. Matias 2004 *Phys. Rev. D* **70** 031502.
[7] R. Fleischer 1999 *Phys. Lett. B* **459** 306.
[8] P. Ball and R. Fleischer 2000 *Phys. Lett. B* **475** 111.
[9] R. Aleksan *et al.* 1993 *Phys. Lett. B* **316** 567.
[10] A. Drutskoy *et al.* (Belle Collaboration) 2010 *Phys. Rev. D* **81** 112003.
[11] K.F. Chen *et al.* (Belle Collaboration) 2008 *Phys. Rev. Lett.* **100** 112001.
[12] A. Drutskoy *et al.* (Belle Collaboration) 2007 *Phys. Rev. Lett.* **98** 052001.
[13] G.S. Huang *et al.* (CLEO Collaboration) 2007 *Phys. Rev. D* **75** 012002.
[14] K. Nakamura *et al.* (Particle Data Group) 2010 *J. Phys. G* **37** 075021.
[15] D. Asner *et al.* (Heavy Flavor Averaging Group) 2010 arXiv:1010.1589 [hep-ex].
[16] R. Louvot *et al.* (Belle Collaboration) 2009 *Phys. Rev. Lett.* **102** 021801; 2010 *Phys. Rev. Lett.* **104** 231801.
[17] A. Deandrea *et al.* 1993 *Phys. Lett. B* **318** 549; R.H. Li, C.D. Lü and H. Zou 2008 *Phys. Rev. D* **78** 014018.
[18] L. Wolfenstein 1983 *Phys. Rev. Lett.* **51** 1945.
[19] N. Cabibbo 1963 *Phys. Rev. Lett.* **10** 531; M. Kobayashi and T. Maskawa 1973 *Prog. Theor. Phys.* **49** 652.
[20] B. Adeva *et al.* (LHCb Collaboration) 2009 arXiv:0912.4179v1 [hep-ex].
[21] B. Aubert *et al.* (BaBar Collaboration) 2003 *Phys. Rev. Lett.* **91** 171802.
[22] K.F. Chen *et al.* (Belle Collaboration) 2003 *Phys. Rev. Lett.* **91** 201801.
[23] G.C. Fox and S. Wolfram 1978 *Phys. Rev. Lett.* **41** 1581.
[24] T. Aaltonen *et al.* (CDF Collaboration) 2009 *Phys. Rev. Lett.* **103** 191802.
[25] C.C. Peng *et al.* (Belle Collaboration) 2010 *Phys. Rev. D* **82** 072007.
[26] S.H. Lee *et al.* (Belle Collaboration) 2003 *Phys. Rev. Lett.* **91** 261801.
[27] A. Abulencia *et al.* (CDF collaboration) 2006 *Phys. Rev. Lett.* **97** 211802.
[28] T. Aaltonen *et al.* (CDF Collaboration) 2009 *Phys. Rev. Lett.* **103** 031801.
[29] I. Adachi *et al.* (Belle Collaboration) 2009 arXiv:0912.1434 [hep-ex].
[30] S. Stone and L. Zhang 2009 *Phys. Rev. D* **79** 074024.
[31] K.M. Ecklund *et al.* (CLEO Collaboration) 2009 *Phys. Rev. D* **80** 052009.
[32] P. Colangelo, F. De Fazio and W. Wang 2010 *Phys. Rev. D* **81** 074001.
[33] O. Leitner *et al.* 2010 *Phys. Rev. D* **82** 076006.
[34] R. Louvot 2010 *Proceedings of Science* PoS(FPCP 2010)015.
[35] R. Louvot 2009 arXiv:0905.4345v2 [hep-ex].
[36] S. Esen *et al.* (Belle Collaboration) 2010 *Phys. Rev. Lett.* **105** 201802.
[37] A. Lenz and U. Nierste 2007 *J. High Energy Phys.* JHEP06(2007)072.
[38] R. Barate *et al.* (ALEPH Collaboration) 2000 *Phys. Lett. B* **486** 286.
[39] V.M. Abazov *et al.* (D0 Collaboration) 2009 *Phys. Rev. Lett.* **102** 091801.
[40] T. Aaltonen *et al.* (CDF Collaboration) 2008 *Phys. Rev. Lett.* **100** 121803.

New measurement of the B_s^0 mixing phase and observation of suppressed B_s^0 decays at CDF

Louise Oakes, for the CDF Collaboration

Technische Universität München, Excellence Cluster Universe, Boltzmannstraße 2, D-85748 Garching, Germany

E-mail: loakes@fnal.gov

Abstract. Recent CDF measurements of B_s^0 decay parameters are presented. The *CP* violating phase β_s has been measured in $B_s^0 \to J/\psi\phi$ decays using 5.2 fb^{-1} integrated luminosity of CDF data. This updated β_s measurement includes the contribution of $B_s^0 \to J/\psi K^+K^-$ or $B_s^0 \to J/\psi f_0$ events to the signal sample, where the f_0 and non-resonant K^+K^- are *S*-wave states. Additional improvements for this update include more than doubling the signal sample, improved selection and particle ID, and fully calibrated flavour tagging for the complete dataset. This measurement shows good agreement with the Standard Model expectation for β_s, with a significance of $< 1\sigma$ for deviation from the expected value. In a related analysis, two suppressed B_s^0 decay channels, $B_s^0 \to J/\psi K^{*0}$ and $B_s^0 \to J/\psi K_s$, have been observed for the first time, in 5.9 fb^{-1} CDF data. Their branching fractions and relative branching ratios are reported. These newly observed channels have the potential for extraction of important CKM parameters including the phase β_s.

1. Introduction

The study of neutral B meson properties can provide important tests of the Standard Model (SM) including constraints on parameters of the Cabibbo-Kobayashi-Maskawa (CKM) matrix. While the B^0 system has been thoroughly investigated by B factories, precision measurements in the B_s^0 system are a more recent development, driven largely by the Tevatron experiments. The $B_s^0 - \bar{B}_s^0$ system has the potential to yield indirect observations of New Physics (NP), through the presence of non-SM particles in second order weak interaction processes such as flavour mixing. The golden mode for this measurement is $B_s^0 \to J/\psi\phi$. The $J/\psi\phi$ final state is common to B_s^0 and $\bar{B}_s^0$ decays; *CP* violation occurs in the mixing through interference between decays with and without B_s^0 mixing. The phase, β_s, between these two amplitudes is predicted to be close to zero in the SM [1], so a significant excess would be a clear indication of evidence for NP in this channel.

The updated measurement is of particular interest as a published CDF $B_s \to J/\psi\phi$ analysis and a recent update showed deviations from the Standard Model expected value of approximately 2σ [2, 3], and a similar effect was seen in a recent combined Tevatron result [4].

The observation of previously unseen decays $B_s^0 \to J/\psi K^{*0}$ and $B_s^0 \to J/\psi K_s$ [5] is also of interest for flavour physics. The $J/\psi K_s$ final state is a *CP* odd eigenstate, which, with sufficient statistics, could yield a measurement of the pure B_s^H mass eigenstate lifetime. It has been suggested [6] that this channel could also be used to extract the unitarity angle γ. The decay $B_s^0 \to J/\psi K^{*0}$ has an admixture of *CP* states for the final state, and could give a measurement

Published under licence by IOP Publishing Ltd

of the CP violating phase β_s analagously to the $B_s^0 \to J/\psi\phi$ channel. Branching ratios have been measured relative to the equivalent B^0 decays to the same final states.

2. New measurement of B_s^0 mixing phase

2.1. *CP violation in $B_s \to J/\psi\phi$*

The flavour eigenstates of B_s^0 mesons in the SM are not the same as the mass eigenstates, leading to oscillations between $|B_s^0\rangle = (\bar{b}s)$ and $|\bar{B}_s^0\rangle = (b\bar{s})$ via the second order weak interactions. The phenomenology of this weak mixing is described by the CKM matrix. The time evolution of the $B_s^0 - \bar{B}_s^0$ system is approximated by the Schrödinger equation

$$i\frac{d}{dt}\begin{pmatrix} B_s^0(t) \\ \bar{B}_s^0(t) \end{pmatrix} = \mathcal{H}\begin{pmatrix} B_s^0(t) \\ \bar{B}_s^0(t) \end{pmatrix} \equiv \left[\begin{pmatrix} M_0 & M_{12} \\ M_{12}^* & M_0 \end{pmatrix} - \frac{i}{2}\begin{pmatrix} \Gamma_0 & \Gamma_{12} \\ \Gamma_{12}^* & \Gamma_0 \end{pmatrix}\right]\begin{pmatrix} B_s^0(t) \\ \bar{B}_s^0(t) \end{pmatrix} \quad (1)$$

where the M and Γ matrices describe the mass and decays of the system. The mass eigenstates can be obtained by diagonalising $\mathcal{H}$, giving:

$$|B_s^H\rangle = p|B_s^0\rangle - q|\bar{B}_s^0\rangle \quad (2)$$

and

$$|B_s^L\rangle = p|B_s^0\rangle + q|\bar{B}_s^0\rangle \quad (3)$$

where $|q/p| = 1$ in the case of no CP violation in mixing, as predicted in the $J/\psi\phi$ channel. The indices H and L label the heavy and light eigenstates respectively. The mass difference, Δm_s, between the heavy and light states is proportional to the frequency of B_s^0 mixing and is approximately equal to $2|M_{12}|$. The mass eigenstates have a small but non negligible lifetime difference, which can be described in terms of the decay width difference

$$\Delta\Gamma_s = \Gamma_L - \Gamma_H \approx 2|\Gamma_{12}|\cos(2\phi_s) \quad (4)$$

where the CP violating phase is defined as $\phi_s = arg(-M_{12}/\Gamma_{12})$ and the mean decay width $\Gamma_s = 1/\tau_s$. The SM predicts ϕ_s^{SM} to be of order 0.004, and it would be expected to increase in the presence of NP.

The full angular and time dependent equations for the measurement of $B_s^0 \to J/\psi\phi$, including the addition of the S-wave KK component are detailed in [8].

The relative phase, β_s, between decays of a B_s^0 meson to $J/\psi\phi$ directly, and after mixing to $\bar{B}_s^0$, is defined in the SM as

$$\beta_s^{SM} = arg\left(\frac{-V_{ts}V_{tb}^*}{V_{cs}V_{cb}^*}\right) \approx 0.02 \quad (5)$$

A New Physics phase, contributing to the weak mixing diagrams in the neutral B_s^0 system would introduce a new physics phase ϕ_s^{NP} to β_s such that the measured value would be $2\beta_s = 2\beta_s^{SM} - \phi_s^{NP}$. The same NP phase would enhance ϕ_s, giving $\phi_s = \phi_s^{SM} + \phi_s^{NP}$. As both β_s^{SM} and ϕ_s^{SM} are predicted to be close to zero, the NP phase would dominate, and the measured phase would be $2\beta_s \approx -\phi_s \approx \phi_s^{NP}$. This approximation is valid given the current experimental resolution, but future high precision measurements may be able to distinguish between these quantities.

2.2. *Experimental strategy*

The decay $B_s^0 \to J/\psi(\to \mu^+\mu^-)\phi(\to K^+K^-)$ is fully reconstructed from events which pass the di-muon trigger. A Neural Network (NN) selection procedure [9] is used to reconstruct $\sim$ 6500 signal events in 5.2 $^{-1}$ of data. The NN is trained on background events from the B_s^0 mass side bands, and MC signal events. The NN output for these training samples is shown in Figure 1.

Journal of Physics: Conference Series **335** (2011) 012041 doi:10.1088/1742-6596/335/1/012041

For this updated analysis, rather than using a standard $S/\sqrt(S+B)$ figure of merit, the NN cut is optimised by selecting the cut value which minimises the statistical error on β_s in pseudo experiments. This, along with a full re-calibration of particle identification information from dE/dx and Time of Flight (TOF), has lead to a $B_s^0 \to J/\psi\phi$ signal yield of more than twice that in the previous 2.8 fb^{-1} data sample [3]. Figure 1 shows the $B_s^0 \to J/\psi\phi$ mass distribution after selection.

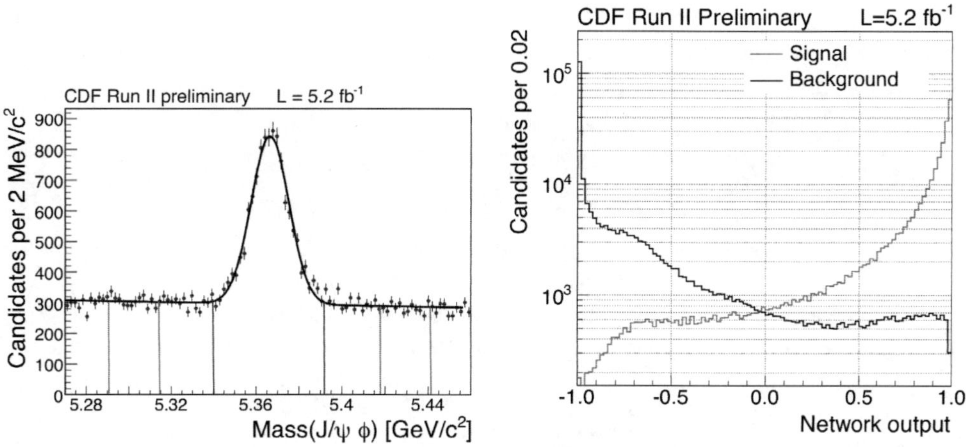

Figure 1. [left] B_s^0 mass distribution for CDF 5.2 fb $^-1$ data sample, fitted with a single Gaussian for the signal and a line to describe the background component. [right] NN training: output for signal and background events.

The final state is an admixture of CP odd and even states, which can be separated according to their angular momentum. The total angular momentum of the $J/\psi\phi$ state can be $L = 0$, 1, or 2, and the CP of the state is $(-1)^L$, so the $L = 0, 2$ states are CP even, and the $L = 1$ state is CP odd. These CP states can be separated using the angular distribution of the four final state particles, the muons and kaons from the decay of the J/ψ and ϕ. The transversity basis [10] is used to define the angular dependence of the final state, where the relative directions of the four particles can be described in terms of three transversity angles, $\{\cos\theta_T, \phi_T, \cos\psi_T\}$ which are defined by the direction of the decaying J/ψ and ϕ mesons. In the transversity basis, the decay amplitude can be separated into three components which represent different linear polarisation states. The CP even states correspond to the vector mesons either being longitudinally polarised, or transverse to their direction of motion and parallel to each other (0, $\|$), for the CP odd state the mesons are transversely polarised with respect to their direction of motion, and perpendicular to each other ($\perp$). The amplitudes of these states are $A_0, A_\|$ and $A_\perp$ respectively. The use of the transversity basis to separate the CP odd and even states in this way means that the measurement is sensitive to the CP violating phase with or without flavour tagging information for the initial state.

The angular analysis is combined with time development and mass dependence in a multivariate likelihood fit. In the simplest case, the fit without flavour tagging information, the likelihood function has a four fold ambiguity under the transformations $\{\beta_s, \Delta\Gamma, \phi_\|, \phi_\perp\} \Leftrightarrow \{\pi/2 - \beta_s, -\Delta\Gamma, 2\pi - \phi_\|, \pi - \phi_\perp\}$ and $\beta_s \Leftrightarrow -\beta_s$, where the strong phases are defined in terms of the transversity amplitudes, $\phi_\| \equiv arg(A_\| A_0^*)$ and $\phi_\perp \equiv arg(A_\perp A_0^*)$.

2.3. Flavour tagging

By flavour tagging the initial B_s^0 meson, the time development of B_s^0 and $\bar{B}_s^0$ states can be followed separately, which removes the insensitivity to the sign of β_s and $\Delta\Gamma$. This reduces the ambiguity to two points. The flavour of the decaying B meson is tagged using a combination of opposite side (OST) and same side (SST) tagging algorithms. The OST tags on the b quark content of a B meson from the same production vertex as the candidate B_s^0, the SST tags according to the s quark content of a kaon produced with the candidate.

For this updated analysis, the SST has been re-calibrated for the full sample on data through a B_s^0 mixing measurement [11]. This technique uses the fact that a measured mixing amplitude of ≈ 1 means that the tagger accurately assesses its performance, and an amplitude > 1 or < 1 implies an under or over estimation of its power, respectively.

This calibration uses the modes:

$$\begin{aligned}
B_s^0 &\rightarrow D_s^-\pi^+, D_s^- \rightarrow \phi\pi^-, \phi \rightarrow K^+K^- \\
B_s^0 &\rightarrow D_s^-\pi^+, D_s^- \rightarrow K^*K^-, K^* \rightarrow K^+\pi^- \\
B_s^0 &\rightarrow D_s^-\pi^+, D_s^- \rightarrow \pi^-\pi^-\pi^+ \\
B_s^0 &\rightarrow D_s^-\pi^+\pi^+\pi^-, D_s^- \rightarrow \phi\pi^-, \phi \rightarrow K^+K^-
\end{aligned}$$

The first channel accounts for about 50% of the statistics, the yield for this channel is shown in Figure 2.

The amplitude measured for this calibration is $\mathcal{A} = 0.94 \pm 0.15$ (stat.) $\pm$ 0.13(syst.), shown in Figure 2. The mixing frequency, $\Delta m_s = 17.79 \pm 0.07$ ps^{-1}, with statistical errors only, is in good agreement with the CDF published measurement.

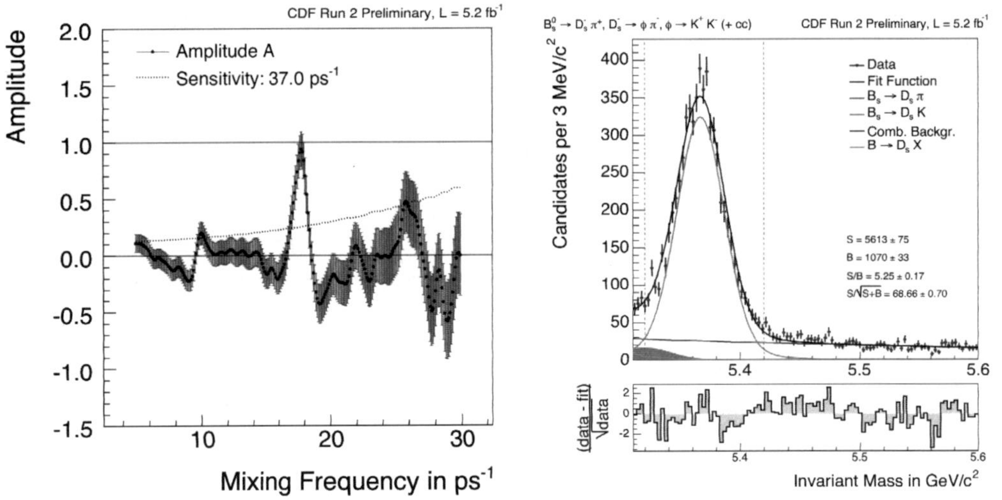

Figure 2. [left] B_s^0 mixing amplitude scan, for calibration of SST. [right] Invariant $D_s^-\pi^+$ mass showing yield for the "golden mode" used in the B_s^0 mixing amplitude measurement.

2.4. S-wave KK component

It has been suggested [12] that a potential contamination of the signal ϕ meson by S-wave f_0 or non-resonant KK of $\sim$ 10% could bias the measurement of β_s towards the SM value. This latest CDF analysis includes the S-wave component in the full angular and time-dependent analysis. Both the f_0 and non-resonant KK components are considered flat in mass within the small selection window, and the ϕ meson mass is modelled by an asymmetric relativistic Breit Wigner; however this mass is integrated over in the fit function and is not used in the fit. The $J/\psi K^+K^-$ or $J/\psi f_0$ final state is a pure CP odd state, and thus follows the time dependence of the CP odd component of the $B_s^0 \to J/\psi\phi$ decay.

A preliminary study of the S-wave contamination of the ϕ meson signal was carried out by studying the invariant KK mass distribution, giving no strong indication of a large additional component, as shown in Figure 3.

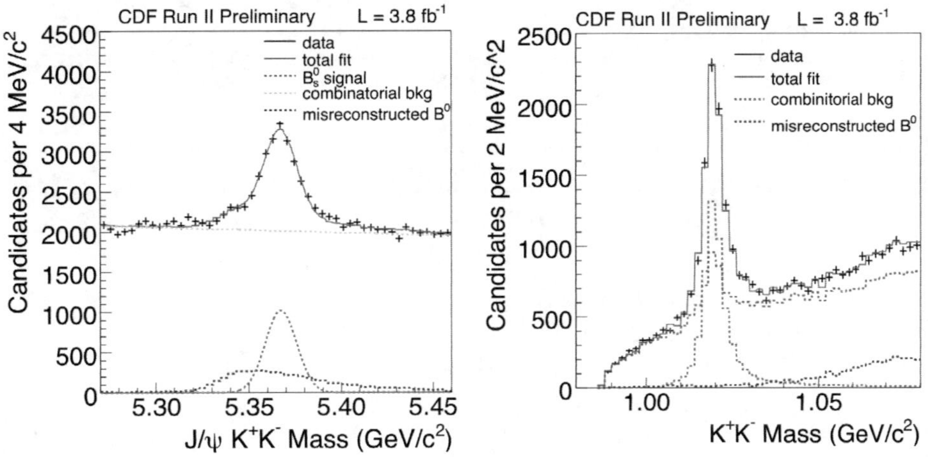

Figure 3. The B_s^0 mass is plotted (left) with a loose ϕ mass cut window, which allows contamination from $B^0 \to J/\psi K^*$ misreconstructed as $B_s^0 \to J/\psi\phi$, this reflection component is fitted with a MC template, the signal B_s^0 mass is fitted with a Gaussian and the combinatorial background with a line.

2.5. Results

It is not currently possible to quote a point value for the phase β_s, without adding external constraints, due to the symmetries in the likelihood function and the non-Gaussian error distribution for β_s. Instead the results are presented as frequentist likelihood contours; a profile-likelihood ratio ordering technique is used to ensure full coverage. Figure 4 shows the likelihood contours in the $\beta_s - \Delta\Gamma$ plane. The confidence interval for $\beta_s^{J/\psi\phi}$ at the 68 % confidence level is $[0.02, 0.52] \cup [1.08, 1.55]$ and at the 95 % confidence level, $[-\pi/2, -1.44] \cup [-0.13, 0.68] \cup [0.89, \pi/2]$.

The upper limit on the fraction of the S-wave KK (f_0) component in the $B_s^0 \to J/\psi\phi$ signal was measured to be $< 6.7\%$ at the 95% confidence level.

In the hypothesis of no CP violation ($\beta_s = 0.0$), the values of the B_s^0 lifetime, $\tau_s = 1.53 \pm 0.025$ (stat.)± 0.01 (syst.) ps, decay width difference $\Delta\Gamma_s = 0.075 \pm 0.035$ (stat.)± 0.01 (syst.) ps^{-1} the transversity amplitudes, $|A_\parallel|^2 = 0.231 \pm 0.014$ (stat.) $\pm$ 0.015 (syst.) and $|A_0|^2 = 0.524 \pm 0.013$ (stat.) $\pm$ 0.015 (syst.) and the strong phase $\phi_\perp = 2.95 \pm 0.64$ (stat.) $\pm$ 0.07 (syst.) are determined [13] from the flavour tagged fit.

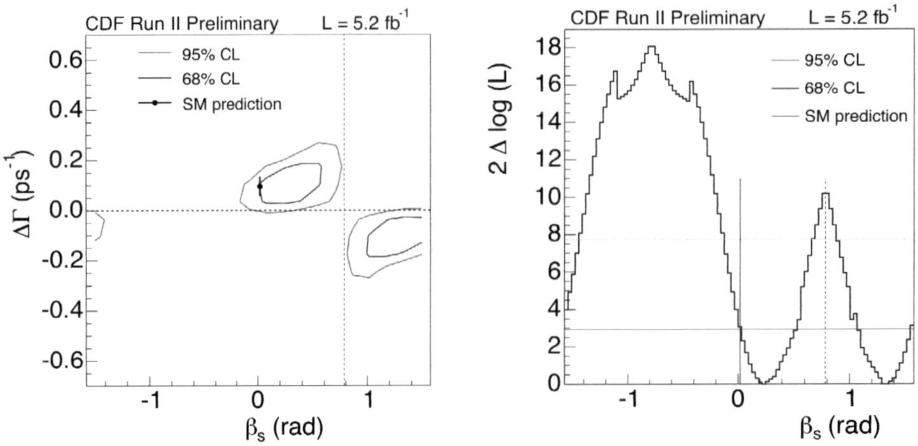

Figure 4. Confidence regions in 2-dimensions in the $\beta_s^{J/\psi\phi} - \Delta\Gamma_s$ plane, including systematics effects on the statistical uncertainties (left) and in 1-dimension for $\beta_s^{J/\psi\phi}$ (right). The SM predicted value is marked with a black point or line.

3. Observation of suppressed B_s^0 decays

The branching ratios for the suppressed decays $B_s^0 \to J/\psi K^{*0}$ and $B_s^0 \to J/\psi K_s$ are measured relative to the more common B^0 decays in the same channels, using the relation [5]:

$$\mathcal{B}(B_s^0 \to J/\psi K)/\mathcal{B}(B^0 \to J/\psi K) = A_{rel} \times f_d/f_s \times N(B_s^0 \to J/\psi K)/N(B^0 \to J/\psi K) \quad (6)$$

where K is K_s^0 or K^{*0}, A_{rel} is the relative acceptance measured from MC and f_d/f_s is the ratio of fragmentation fractions, taken from a combination of CDF measurements and PDG values.

3.1. Experimental technique

This analysis uses 5.9 fb^{-1} integrated luminosity of CDF data collected with the di-muon trigger. From the enhanced sample of $J/\psi \to \mu\mu$ events, $B^0 \to J/\psi K^{*0}$ and $B^0 \to J/\psi K_s^0$ decays are reconstructed. The signal selection is optimised for these B^0 decays then applied to the B_s^0 channel. Selection for the $J/\psi K^{*0}$ channel is carried out using rectangular cuts, optimised on the quantity $S/(1.5+\sqrt{B})$. The $J/\psi K_s^0$ channel is selected using a Neural Network (NN) as the B_s^0 contribution is expected to be smaller in this decay, making good combinatorial background suppression essential. In a binned likelihood fit, the B^0 mass distributions are modelled with a three Gaussian template taken from Monte Carlo, and an identical template is used for the B_s^0 mass with the mean shifted appropriately.

3.2. Results

Figure 5 shows the invariant mass distributions for the $B_s^0 \to J/\psi K_s$ and $B_s^0 \to J/\psi K^{*0}$ channels. The $B_s^0 \to J/\psi K_s$ channel is observed with a significance of 7.2σ over the null hypothesis, the $B_s^0 \to J/\psi K^{*0}$ channel with 8.0σ.

The ratios of branching fractions for the B_s^0 channels with respect to the reference B^0 decays are:

$$\frac{\mathcal{B}(B_s^0 \to J/\psi K^{*0})}{\mathcal{B}(B^0 \to J/\psi K^{*0})} = (0.062 \pm 0.009 \text{ (stat.) } \pm 0.025 \text{ (syst.) } \pm 0.008 \text{ (frag.)})$$

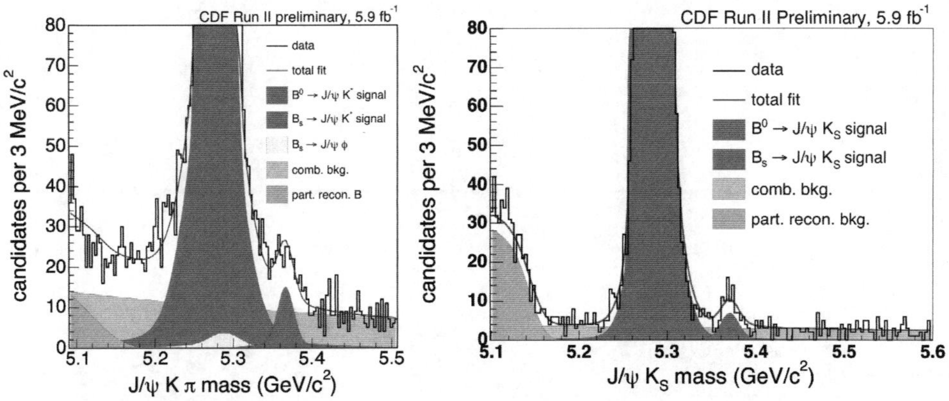

Figure 5. Invariant mass distribution, zoomed in signal region, for (left) $B_s^0 \to J/\psi K^{*0}$ and (right) $B_s^0 \to J/\psi K_s$

$$\frac{\mathcal{B}(B_s^0 \to J/\psi K_s^0)}{\mathcal{B}(B^0 \to J/\psi K_s^0)} = (0.041 \pm 0.007 \text{ (stat.) } \pm 0.004 \text{ (syst.) } \pm 0.005 \text{ (frag.)}) \quad (7)$$

From these, using the PDG values for the branching fractions of the reference B^0 decays, the absolute branching fractions for the B_s^0 decays can be calculated as:

$$\begin{aligned} \mathcal{B}(B_s^0 \to J/\psi K^0) &= (3.5 \pm 0.6 \text{ (stat.) } \pm 0.4 \text{ (syst.) } \pm 0.4 \text{ (frag.) } \pm 0.1 \text{ (norm)}) \times 10^{-5} \\ \mathcal{B}(B_s^0 \to J/\psi K^{*0}) &= (8.3 \pm 1.2 \text{ (stat.) } \pm 3.4 \text{ (syst.) } \pm 1.0 \text{ (frag.) } \pm 0.4 \text{ (norm)}) \times 10^{-5} \quad (8) \end{aligned}$$

4. Conclusions

Both the CDF [3] and DØ [14] experiments have previously produced measurements of the B_s^0 mixing phase β_s, independently and in a combined Tevatron result [4], which indicated a shift of about 2 σ from the Standard Model expectation. The new $\beta_s^{J/\psi\phi}$ measurement from the CDF experiment shows good agreement with the Standard Model expected value, with a p-value of 44%, equivalent to a deviation of 0.8 σ.

These latest results benefit from an improved selection and particle ID to take advantage of the full available statistics, updated flavour tagging, and the inclusion of the S-wave KK component of the signal fraction in the fit. Additionally, results from the β_s measurement include the world's most precise single measurements of the B_s^0 lifetime and decay width difference in the hypothesis of no CP violation.

The obervation of two suppressed B_s^0 decay channels opens up the possibility (in the case of sufficient statistics) of further measurements of CKM parameters, in particular the study of β_s in the decay $B_s^0 \to J/\psi K^{*0}$, and the potential to measure individually the CP odd $B_s^{0,H}$ lifetime in the channel $B^0 \to J/\psi K_s^0$.

In the near future, each of the discussed measurements at CDF can be improved by doubling the data samples used, as the integrated luminosity delivered by the Tevatron has already surpassed 10 fb^{-1}.

References

[1] Bona, M. et al (UTfit Collaboration) arXiv:hep-ph.0803.0659
[2] T.Aaltonen et al (CDF collaboration) *Phys. Rev. Lett.* **100**, 161802 (2008)
[3] The CDF collaboration, Public Note 9458

[4] http://tevbwg.fnal.gov/results/Summer2009_betas/
[5] The CDF collaboration, Public Note 10708 (2010)
[6] R. Fleischer, Eur. Phys. J. C 10, 299-306, (1999)
[7] A.Abulencia et al (CDF collaboration) *Phys. Rev. Lett.* **97**, 242003 (2006)
[8] F.Azfar et al, *JHEP* **1011**, 158 (2010), arXiv:1008.4283 (2010)
[9] M. Feindt, A Neural Bayesian Estimator for Conditional Probability Densities, arXiv:physics/0402093.
[10] A. S. Dighe, I. Dunietz and R. Fleischer, *Eur. Phys. J. C* **6**, 647 (1999)
[11] The CDF collaboration, Public Note 10108
[12] S.Stone, L.Zhang, *Phys. Rev.* D **79**, 074024 (2009) , arXiv:0812.2832 (2009)
[13] The CDF collaboration, Public Note 10206 (2010)
[14] The DØ collaboration, DØ Conference Note 5933-conf

Symposium on Prospects in the Physics of Discrete Symmetries IOP Publishing
Journal of Physics: Conference Series **335** (2011) 012042 doi:10.1088/1742-6596/335/1/012042

Prospects for CP violation in $B_s^0 \to J/\psi\phi$ from first LHCb data

Olivier Leroy, on behalf of the LHCb collaboration
CPPM, 163 avenue de Luminy Case 907, 13288 Marseille cedex 9, France
E-mail: Olivier.Leroy@in2p3.fr

Abstract.
The determination of the CP-violating phase in $B_s^0 \to J/\psi\phi$ decays is one of the key goals of the LHCb experiment. Its value is predicted to be very small in the Standard Model but can be significantly enhanced in many models of New Physics. The steps towards a precise determination of this phase with a flavour-tagged, time-dependent, angular analysis of the decay $B_s^0 \to J/\psi\phi$ will be reviewed and first studies performed with data collected at LHC in pp collisions at 7 TeV center-of-mass energy will be presented. Prospects are also given for the measurement of the flavour specific asymmetry using semileptonic $B_{d,s}^0$ decays.

1. Introduction

The phenomenology of CP violation in B_s^0 decays is described in many articles [1]. Here we just recall the main aspects. The time evolution of the $B_s^0(\sim \bar{b}s)$ and $\overline{B}_s^0(\sim b\bar{s})$ mass eigenstates is described by a Schrödinger equation:

$$i\frac{\partial}{\partial t}\begin{pmatrix} |B_s^0(t)\rangle \\ |\overline{B}_s^0(t)\rangle \end{pmatrix} = \left(\mathbf{M} - i\frac{\mathbf{\Gamma}}{2}\right)\begin{pmatrix} |B_s^0(t)\rangle \\ |\overline{B}_s^0(t)\rangle \end{pmatrix},$$

where $\mathbf{M}$ and $\mathbf{\Gamma}$ are the mass and decay matrices. We can define three parameters describing the phenomenon of B_s^0–$\overline{B}_s^0$ mixing:

$$\Delta m_s = M_H - M_L, \qquad \Delta\Gamma_s = \Gamma_L - \Gamma_H, \qquad \phi_s = \arg\left(-\frac{M_{12,s}}{\Gamma_{12,s}}\right), \tag{1}$$

where the average B_s^0 mass and width (indices L and H for the two mass eigenstates) are:

$$M_{B_s^0} = \frac{M_H + M_L}{2}, \qquad \Gamma_s = \frac{\Gamma_L + \Gamma_H}{2}. \tag{2}$$

There are several ways of probing New Physics using B_s^0–$\overline{B}_s^0$ mixing. In this document, we describe the golden method using $B_s^0 \to J/\psi\phi$ decays. We also present an alternative method using semileptonic decays.

Within the Standard Model, the decay $B_s^0 \to J/\psi\phi$ is dominated by $\bar{b} \to \bar{c}c\bar{s}$ quark level transitions, as represented in Figure 1 (left), with a weak phase $\Phi_D = \arg(V_{cb}V_{cs}^*)$[1]. The B_s^0

[1] The possible small penguin contribution (same figure, right) is neglected here and discussed further in [2]

meson can also oscillate to a $\overline{B}^0_s$ before decaying to $J/\psi\phi$, with a weak phase $\Phi_M = 2\arg(V^*_{ts}V_{tb})$ (Figure 2). The interference between both processes gives rise to a CP violating weak phase $\phi_s^{J/\psi\phi} = \Phi_M - 2\Phi_D = -2\beta_s$, where $\beta_s = \arg(-V_{ts}V^*_{tb}/V_{cs}V^*_{cb})$ is the smallest angle of the "b−s unitarity triangle". The indirect determination via global fits to experimental data gives $2\beta_s = (0.0363 \pm 0.0017)$ rad [3]. The direct measurement of this phase is one of the key goals of the LHCb experiment. Indeed, $\phi_s^{J/\psi\phi}$ is one of the CP observables with the smallest theoretical uncertainty in the Standard Model, and New Physics could significantly modify this prediction. if new particles contribute with a new phase to the B^0_s–$\overline{B}^0_s$ box diagram. Both CDF and DØ have reported constraints on $\phi_s^{J/\psi\phi}$ with large uncertainties [4].

An alternative way to search for New Physics is to measure the flavour specific asymmetry, which is linked to the quantity

$$a^s_{fs} = \frac{\Delta\Gamma_s}{\Delta m_s}\tan\phi_s\,, \tag{3}$$

as explained in [5] (see also Eq. 7 in Section 4). Within the Standard Model, $\phi_s^{SM} = (3.40^{+1.32}_{-0.77}) \times 10^{-3}$ rad [3]. DØ reported in 2010 a 3.2σ deviation with respect to the Standard Model, measuring the the like-sign dimuon charge asymmetry in semileptonic b-hadron decays [6]. The quantity $\phi_s^{J/\psi\phi}$ and ϕ_s are two different observables. However, assuming new physics affects only the mixing box diagram, they would be both shifted by the same New Physics phase, ϕ_s^Δ:

$$\phi_s^{J/\psi\phi} = -2\beta_s + \phi_s^\Delta\,, \tag{4}$$

$$\phi_s = \phi_s^{SM} + \phi_s^\Delta\,. \tag{5}$$

where we have used the notation of [7].

In Section 2, we present the method to measure $\phi_s^{J/\psi\phi}$ at LHCb and the expected sensitivity, based on Monte Carlo estimates. In Section 3, the first preliminary results obtained with the 2010 real data are given. Eventually in Section 4, we present a way to constrain ϕ_s via semileptonic decays.

2. $B^0_s \to J/\psi\phi$ studies based on Monte Carlo

$B^0_s \to J/\psi\phi$ is a pseudo-scalar to vector-vector decays. The final state is a superposition of three polarizations amplitudes (A_0, $A_\parallel$ and $A_\perp$), and an angular analysis is required to disentangle statistically the CP-odd and CP-even components. The argument of the polarization amplitudes are strong phases denoted δ_0, $\delta_\parallel$ and $\delta_\perp$. The three decay product angles are shown in Figure 3, in the transversity basis, for $B^0_s \to J/\psi\phi$.

The differential decay rate of $B^0_s \to J/\psi\phi$ is a function of six observables (the proper time, the $J/\psi\phi$ invariant mass, the initial B^0_s flavour and the three decay angles) and nine physics parameters: the weak phase ($\phi_s^{J/\psi\phi}$), the mass and width difference between B^0_s mass eigenstates (Δm_s and $\Delta\Gamma_s$), the mass and width of the B^0_s, two amplitudes and two strong phases [8]. The possibility to account for a KK S-wave in the $B^0_s \to J/\psi K^+K^-$ decay is discussed in [9].

In the rest of this section, we present the trigger and selection, the measurement of proper time and decay angles, the flavour tagging, the fit and systematics studies, based on Monte Carlo [10].

2.1. Trigger and selections

The $B^0_s \to J/\psi\phi$ channel is triggered and selected in a uniform way together with two control channels ($B^0 \to J/\psi K^{*0}$ and $B^+ \to J/\psi K^+$), in order to extract the mistag rate (defined in Section 2.2) without applying large corrections. The trigger and selections are designed to maximize the sensitivity to $\phi_s^{J/\psi\phi}$ while avoiding large proper time and angular acceptance

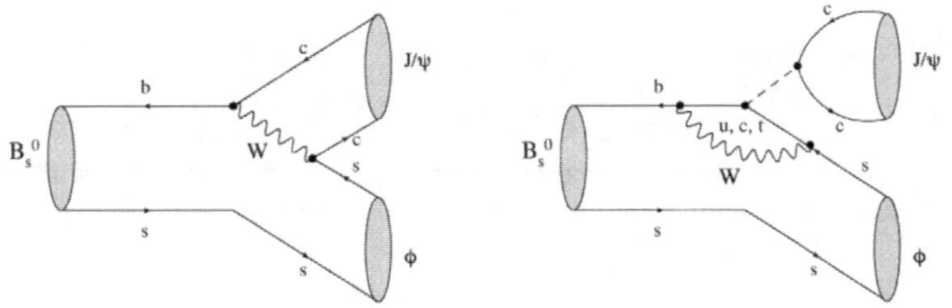

Figure 1. Feynman diagrams contributing to the decay $B_s^0 \to J/\psi\phi$, within the Standard Model. Left: tree; right: penguins.

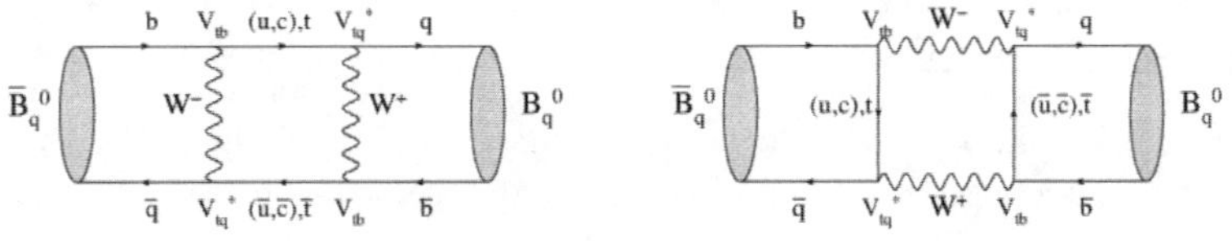

Figure 2. Feynman diagrams responsible for B_q–$\overline{B}_q$ mixing, within the Standard Model (q=s,d).

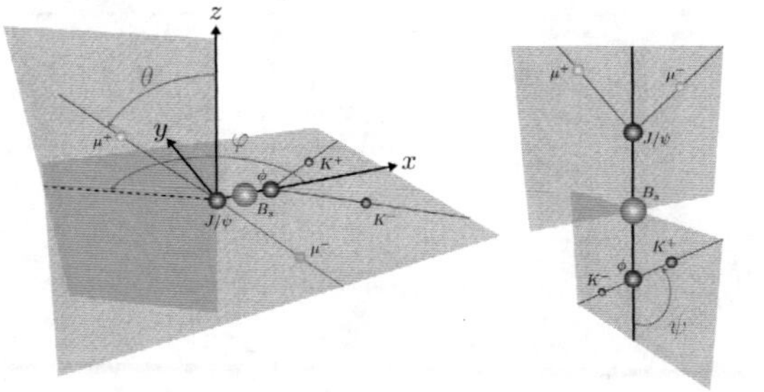

Figure 3. Angle definition: θ is the angle formed by the positive lepton (ℓ^+) and the z axis, in the J/ψ rest frame. The angle φ is the azimuthal angle of ℓ^+ in the same frame. In the ϕ meson rest frame, ψ is the angle between $\vec{p}(K^+)$ and $-\vec{p}(J/\psi)$.

corrections [11]. The J/ψ is reconstructed from its decay to two muons, while the ϕ is reconstructed from two kaons. Based on a Monte Carlo simulation, with a center-of-mass energy of 14 TeV, we expect 117 000 $B_s^0 \to J/\psi\phi$ per 2 fb^{-1}. The total trigger efficiency is $\sim 70\%$. The background is dominated by prompt events, since no cut which can bias the lifetime is used. The angular acceptances for the three transversity angles show distortions below 8%. These are mainly from the LHCb forward geometrical acceptance.

2.2. Flavour tagging

The flavour tagging algorithm and its calibration in LHCb are described in [12, 13]. It is characterized by a mistag rate ($\omega = W/(W+R)$), a tagging efficiency ($\varepsilon_{\rm tag} = (W+R)/(W+R+U)$) and an effective tagging efficiency $\varepsilon_{\rm eff} = \varepsilon_{\rm tag}(1-2\omega)^2$; where W, R and U denotes the number of events wrongly tagged, correctly tagged and untagged. Within each event, all available information is used and combined using a neural network. The combined effective tagging efficiency measured on simulated $B_s^0 \to J/\psi\phi$ events is 5.3%.

2.3. Fits and systematics

The fit procedure is described in [10]. The eight physics parameters $\{\phi_s^{J/\psi\phi}, \Gamma_s, \Delta\Gamma_s, R_\perp, R_0, \delta_\perp, \delta_\parallel, \Delta m_s\}$ are determined by an unbinned likelihood fit to six physical observables (mass, proper time, initial flavour and three transversity angles), taking into account 18 detector parameters. The total PDF is the sum of the signal PDF plus two PDFs for modelling the prompt and the long-lived background. The proper time resolution, angular acceptance and mistag fraction are taken into account. The projections of data and fitted probability density function on the transversity angles and on the proper time can be found in Figure 4. The fitted PDFs and Monte Carlo data show very good agreement. Using hundreds of toy Monte Carlo experiments, the statistical uncertainty on $\phi_s^{J/\psi\phi}$ is estimated to be 0.07 rad, with 1 fb^{-1} of data taken at a LHC centre-of-mass energy of 7 TeV[2]. A summary of the systematic effects is given in Table 1. No irreducible systematic uncertainty has been identified. Figure 5 shows the statistical uncertainty on $\phi_s^{J/\psi\phi}$ versus the integrated luminosity. The latest Tevatron results are also indicated.

Parameter	Variation	$\lvert\phi_s^{\rm wrong} - \phi_s^{\rm true}\rvert/\phi_s^{\rm true}$
Angular distortions	$\pm5\%$	7%
Proper time resolution	(38 ± 5) fs	6%
Mistag	$(34 \pm 1)\%$	7%

Table 1. Relative systematic variation on $\phi_s^{J/\psi\phi}$ (column 3), due to parameter variations (columns 1 and 2).

3. First $B_s^0 \to J/\psi\phi$ results based on 2010 real data

In 2010, the LHCb detector recorded 37 pb^{-1} of data. The detector worked remarkably well, despite being exposed to a pile-up $\sim$ 5 times higher than the design. The invariant mass of $B^+ \to J/\psi K^+$, $B^0 \to J/\psi K^{*0}$ and $B_s^0 \to J/\psi\phi$, triggered and selected in $\sim$ 33 pb^{-1} are given in Figure 6. 877 ± 32 $B_s^0 \to J/\psi\phi$ are reconstructed with a proper time above 0.3 ps. Note that this cut on proper time is just used for illustration purpose, to remove the huge prompt background. It will not be used in the final fit. The proper time resolution of the B_s^0 candidates is estimated to be 50 fs, roughly 30% worse than Monte Carlo expectation, but still twice better than the one reported by Tevatron experiments and easily sufficient to resolve the fast B_s^0–$\overline{B}_s^0$ oscillations, which have a period of 352 fs. First results on opposite side flavour tagging using $B^0 \to D^{*-}\mu^+\nu_\mu$ event gives 60% of the performance expected from Monte Carlo. Since the DISCRETE symposium, much progress in this area has been achieved and new results including same-side tagging will soon be reported.

[2] Results given in [10] were obtained assuming 14 TeV and 2 fb^{-1}. They have simply been rescaled here to 7 TeV.

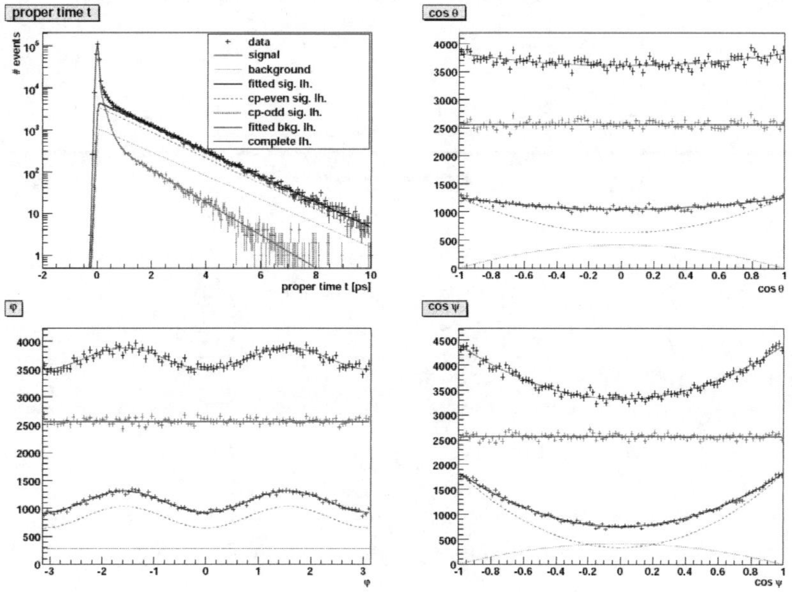

Figure 4. The projections of data and fitted signal PDF including both the angular and proper time acceptance effects, in a sample of toy MC events, including prompt and long-lived background. Also shown are the CP-even (dashed) and the CP-odd (dotted) components.

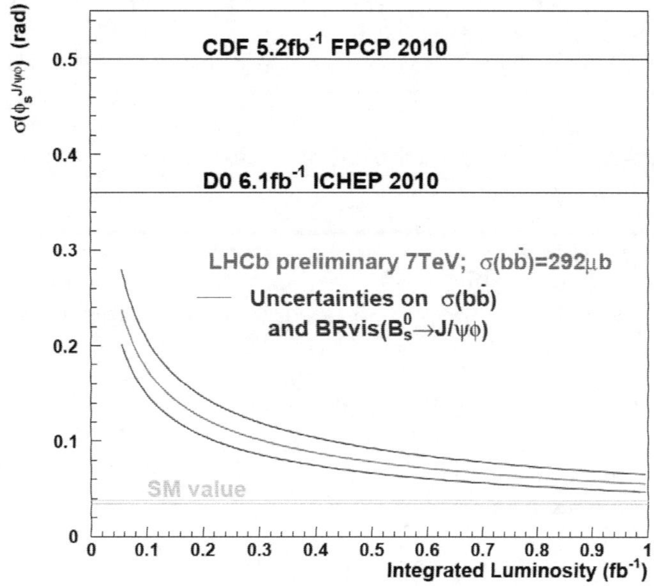

Figure 5. Monte Carlo expected uncertainty on $\phi_s^{J/\psi\phi}$ at LHCb versus integrated luminosity, for a LHC centre-of-mass energy of 7 TeV.

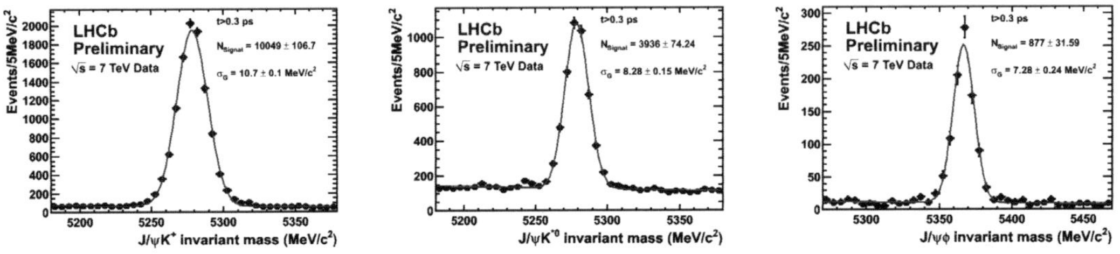

Figure 6. Invariant mass of $B^+ \to J/\psi K^+$, $B^0 \to J/\psi K^{*0}$ et $B^0_s \to J/\psi\phi$ candidates reconstructed in the first data ($33\,\mathrm{pb}^{-1}$) collected by the LHCb experiment in 2010.

4. Flavor-specific asymmetry in $B^0_{d,s}$ decays

An alternative way to look for New Physics in the mixing box diagram is to measure the flavour specific asymmetry:

$$A^q_{\mathrm{fs}} = \frac{\Gamma(f) - \Gamma(\overline{f})}{\Gamma(f) + \Gamma(\overline{f})} \qquad \mathrm{q} = \mathrm{d}, \mathrm{s}\,. \tag{6}$$

The index $\mathrm{q} = \mathrm{d}, \mathrm{s}$ stands for B^0 and B^0_s mesons. This observable is related to a_{fs} (Eq. 3) via:

$$A^{\mathrm{q}}_{\mathrm{fs}}(t) = \frac{a^{\mathrm{q}}_{\mathrm{fs}}}{2} - \frac{\delta^{\mathrm{q}}_c}{2} - (\frac{a^{\mathrm{q}}_{\mathrm{fs}}}{2} + \frac{\delta^{\mathrm{q}}_p}{2})\frac{\cos(\Delta m_{\mathrm{q}} t)}{\cosh(\Delta\Gamma_{\mathrm{q}} t/2)} + \frac{\delta^{\mathrm{q}}_b}{2}\left(\frac{B}{S}\right)^{\mathrm{q}} \qquad \mathrm{q} = \mathrm{d}, \mathrm{s} \tag{7}$$

where δ_c is the detector asymmetry ($\sim 10^{-2}$), δ_p the production asymmetry ($\sim 10^{-2}$), δ_b the background asymmetry ($\sim 10^{-3}$) and B/S the background over signal ratio. Given the order of magnitude of these polluting contributions, the direct extraction of a_{fs} is extremely challenging at LHCb. The Standard Model predictions are: $a^{\mathrm{d}}_{\mathrm{fs}}(SM) = (-6.4^{+1.6}_{-1.8}) \times 10^{-4}$, $a^{\mathrm{s}}_{\mathrm{fs}}(SM) = (3.0^{+1.2}_{-1.3}) \times 10^{-5}$[14]. That is why we propose to use $B^0_s \to D^-_s \mu^+ \nu$ and $B^0 \to D^- \mu^+ \nu$ with the same final state $K^+K^-\pi^-\mu^+$, so that the detector asymmetry cancels in the difference $A^{\mathrm{d}}_{\mathrm{fs}} - A^{\mathrm{s}}_{\mathrm{fs}}$. With the huge number of semileptonic decays that will be collected in 2011 at LHCb, we expect a tiny statistical uncertainty on $a^{\mathrm{d}}_{\mathrm{fs}} - a^{\mathrm{s}}_{\mathrm{fs}}$. However, the control of the systematic uncertainties will still be a demanding task.

5. Conclusions and Prospects

The $B^0_s \to J/\psi\phi$ channel will allow LHCb to probe possible New Physics effects in the B^0_s–$\overline{B}^0_s$ box diagram. We have presented the work-plan to perform the measurement of the CP-violating weak phases in this channel, based on Monte Carlo, together with first results obtained with the real data taken in 2010. 877 $B^0_s \to J/\psi\phi$ candidates have been fully reconstructed with $33\,\mathrm{pb}^{-1}$. With $1\,\mathrm{fb}^{-1}$ of data taken at a center-of-mass energy of 7 TeV, as expected by the end of 2011, we anticipate a statistical uncertainty on $\phi_s^{J/\psi\phi}$ of 0.07 rad. We have also presented prospects to measure the semileptonic asymmetry using the $B^0_s \to D^-_s(K^+K^-\pi^-)\mu^+\nu_\mu$ and $B^0 \to D^-(K^+K^-\pi^-)\mu^+\nu_\mu$ channels.

References

[1] See for example: R. Fleischer, "Flavour physics and CP violation", hep-ph/0608010, (2006).
[2] H. Boos, T. Mannel and J. Reuter, "The gold-plated mode revisited: $\sin(2\beta)$ and $B^0 \to J/\psi K^0_S$ in the Standard Model", Phys. Rev. D 70, 036006 (2004);
S. Mishima, "Study of penguin pollution in the $B^0 \to J/\psi K^0_S$ decay", arXiv:hep-ph/0703193v1 (2007);
H. N. Li and S. Mishima, "Penguin pollution in the $B^0 \to J/\psi K^0_S$ decay", JHEP 0703, 009 (2007).
S. Faller, R. Fleischer and T. Mannel, "Precision physics with $B^0_s \to J/\psi\phi$ at the LHC: The quest for New

Physics", arXiv:0810.4248v1 [hep-ph] (2008).
M. Gronau and J. L. Rosner, "Doubly CKM-suppressed corrections to CP asymmetries in $B^0 \to J/\psi K^0_S$", arXiv:0812.4796 [hep-ph] (2008).
M. Ciuchini et al. proceedings of the CKM workshop, Warwick, UK (2010).

[3] J. Charles *et al.* (CKMfitter group), Eur. Phys. J. C41, 1-131 (2005), hep-ph/0406184, updated results and plots available at http://ckmfitter.in2p3.fr/

[4] CDF collaboration, CDF/ANAL/BOTTOM/PUBLIC/10206, 2 November 2010.
DØ collaboration, Note 6098-CONF, 22 July 2010.
CDF colaboration, "First flavor-tagged determination of bounds on mixing-induced CP violation in $B^0_s \to J/\psi\phi$ decays", Phys. Rev. Lett. 100, 161802 (2008).
CDF collaboration, "An updated measurement of the CP violating phase $\phi_s^{J/\psi\phi}$", CDF/ANAL/BOTTOM/PUBLIC/9458 (2008).
DØ collaboration, "Measurement of B^0_s mixing parameters from the flavor-tagged decay $B^0_s \to J/\psi\phi$", Phys. Rev. Lett. 101, 241801 (2008).
CDF and DØ collaborations, "Combination of DØ and CDF results on $\Delta\Gamma_s$ and the CP-violationg phase $\beta_s^{J/\psi\phi}$", CDF/PHYS/BOTTOM/CDFR 9787, DØ note 59-28-CONF (2009).

[5] N. Brook *et al.* "LHCb's potential to measure flavour-specific CP-asymmetry in semileptonic and hadronic B^0_s decays", LHCb note 2007-054 (2007).

[6] DØ collaboration, "Evidence for an anomalous like-sign dimuon charge asymmetry", Fermilab-PUB-10/114-E, arXiv:1005.2757 [hep-ex] (2010).

[7] A. Lenz, "Theoretical status of B^0_s-mixing and lifetimes of heavy hadrons", arXiv:0705.3802v2 [hep-ph] (2007).

[8] A. Dighe *et al.*, "Angular distribution and lifetime difference in $B^0_s \to J/\psi\phi$ decays", arXiv:hep-ph/951136v1 (1995).

[9] P. Clarke *et al.*, "Determination of β_s in $B^0_s \to J/\psi KK$ decays in the presence of a KK S-wave contribution", JHEP09 074 (2009) and LHCb-PUB-2009-007 (2009).

[10] The LHCb Collaboration,"Roadmap for selected key measurements of LHCb.", LHCB-PUB-2009-029, arXiv:0912.4179.

[11] M. Calvi *et al.*, "Lifetime unbiased selection of $B^0_s \to J/\psi\phi$ and related control channels: $B^0 \to J/\psi K^{*0}$ and $B^+ \to J/\psi K^+$", LHCb-PUB-2009-025 (2009).

[12] M. Calvi *et al.*, "Flavour tagging algorithms and performances in LHCb", LHCb note 2007-058 (2007).

[13] M. Calvi *et al.*, "Calibration of flavour tagging with $B^+ \to J/\psi K^+$ and $B^0 \to J/\psi K^{*0}$ control channels at LHCb", LHCb-PUB-2009-020 (2009).

[14] A. Lenz et al, "Anatomy of New Physics in B-Bbar mixing", arXiv:1008.1593 [hep-ph] (2010).

Journal of Physics: Conference Series **335** (2011) 012043 doi:10.1088/1742-6596/335/1/012043

Studies of CP Violation and Mixing in the D Mesons decays from *BABAR*

Giulia Casarosa for the *BABAR* Collaboration
Università di Pisa and INFN - Sezione di Pisa
Largo B. Pontecorvo 3, 56127 Pisa, Italy
E-mail: giulia.casarosa@pi.infn.it

Abstract. After an introduction on the mixing and CP violation (CPV) in the neutral charmed mesons system, we present the current status of the measurements. We discuss some recent analyses performed on the data sample collected by the *BABAR* detector at the PEPII e^+e^- asymmetric collider.

1. Introduction

Mixing occurs in neutral mesons systems in which the flavour eigenstates differ from the mass eigenstates. The $D^0 - \overline{D}^0$ system is one of the four systems in which this phenomenon is predicted by the Standard Model (SM) and observed experimentally, the other systems are $B^0 - \overline{B}^0$, $K^0 - \overline{K}^0$ and $B_s - \overline{B}_s$. The oscillation rate is different in each system, the D system is the one that shows the smallest mixing, this is the reason why it's the last one in which evidence of mixing has been found experimentally.

2. Theory and Formalism

The following is a brief overview on mixing and CP violation, for more details see PDG [1].

The time evolution in the D^0-$\overline{D}^0$ system is described by a Schrödinger equation:

$$i\frac{\partial}{\partial t}\begin{pmatrix} D^0(t) \\ \overline{D}^0(t) \end{pmatrix} = \left(\mathbf{M} - \frac{i}{2}\mathbf{\Gamma}\right)\begin{pmatrix} D^0(t) \\ \overline{D}^0(t) \end{pmatrix} \tag{1}$$

where $\mathbf{M}$ and $\mathbf{\Gamma}$ are two Hermitian matrices. C, P and T conservations impose constraints on the elements of $\mathbf{M}$ and $\mathbf{\Gamma}$. In the following we will assume CPT conservation, which implies $M_{11} = M_{22}$ and$\Gamma_{11} = \Gamma_{22}$.

The two D^0 mass eigenstates are obtained diagonalizing the equation above and can be represented as:

$$|D_1\rangle = p|D^0\rangle + q|\overline{D}^0\rangle \qquad |D_2\rangle = p|D^0\rangle - q|\overline{D}^0\rangle \tag{2}$$

where

$$\left(\frac{q}{p}\right)^2 = \frac{M_{12}^* - \frac{i}{2}\Gamma_{12}^*}{M_{12} - \frac{i}{2}\Gamma_{12}}. \tag{3}$$

The normalization condition is $|p|^2 + |q|^2 = 1$. The mass eigenstate $|D_i\rangle$ has a mass M_i and widthΓ_i. Assuming $CP|D^0\rangle = +|\overline{D}^0\rangle$, in case of CP conservation the $|D_1\rangle$ is the CP-even state.

Published under licence by IOP Publishing Ltd

Journal of Physics: Conference Series **335** (2011) 012043 doi:10.1088/1742-6596/335/1/012043

2.1. CP Violation

CP Violation is an important ingredient when neutral meson mixing is involved. The only source of CP Violation (CPV) in the Standard Model is the complex phase in the CKM matrix.

There are three possible manifestations of CPV:

(i) in the decay (*direct* CPV);

(ii) in mixing (*indirect* CPV);

(iii) in the interference between mixing and decay (*indirect* CPV).

There is direct CPV when the probability of a meson D to decay into a final state f differs from the probability of the anti-meson to decay into the CP conjugate state $\bar{f}$. In other words when:

$$\frac{|A_f|}{|\bar{A}_{\bar{f}}|} \neq 1 \tag{4}$$

where $A_f = \langle f|H_d|D\rangle$ and $\bar{A}_{\bar{f}} = \langle \bar{f}|H_d|\bar{D}\rangle$, H_d is the Hamiltonian responsible of the decay. It's important to underline that the moduli $|\bar{A}_{\bar{f}}|$ and $|A_f|$ can be different only if the decay proceeds through the sum of two different contributions, with different phases. If the phase is the same for the two contributions, then the asymmetry will be zero.

This kind of CPV is independent of mixing and it is expected to be negligible in the decays of D mesons [2].

When CPV in mixing occurs then the probability of the transition $D^0 \to \overline{D}^0$ is different from the probability of the inverted process $\overline{D}^0 \to D^0$. The signature for CPV in mixing is:

$$R_{\mathrm{M}} \equiv \left|\frac{q}{p}\right| \neq 1. \tag{5}$$

The CPV in the interference between mixing and decay can occur when both the D^0 and the $\overline{D}^0$ have access to a common final state f. In this case f can be reached in two possible ways:

(i) with a direct decay: $D^0 \to f$;

(ii) with a decay after the mixing: $D^0 \to \overline{D}^0 \to f$.

CPV occurs in the interference between these two paths and it is possible also when CP is conserved in the decay and in the mixing separately. In this case we define λ_f:

$$\lambda_f = \frac{q}{p}\frac{\bar{A}_f}{A_f} = R_{\mathrm{M}}\left|\frac{\bar{A}_f}{A_f}\right| e^{i(\delta_f+\phi_f)} \tag{6}$$

where δ_f and ϕ_f are, respectively, the strong and the weak phase differences between the two paths. There is CPV when the weak phase difference is different from zero:

$$\phi_f \neq 0 \tag{7}$$

3. Mixing and CPV in the charm sector

The mixing in the D system is described by two real parameters, the *mixing parameters* that are proportional to the difference of masses and widths of the mass eigenstates:

$$x_{\mathrm{D}} = \frac{M_1 - M_2}{\Gamma} \quad , \quad y_{\mathrm{D}} = \frac{\Gamma_1 - \Gamma_2}{2\Gamma} \ . \tag{8}$$

with$\Gamma \equiv (\Gamma_1 + \Gamma_2)/2$. If either x_{D} or y_{D} is non-zero, mixing will occur. If D_1 and D_2 have different lifetimes then $y_{\mathrm{D}} \neq 0$.

The mixing parameters defined above are the ones directly related to the physical quantities that drive mixing. Other parameters not directly sensitive to these quantities may be defined. The parameter $y_{\rm D}^{CP}$ [3] measures how much the lifetime of the CP-even eigenstate (τ_{CP+}) differs from the D^0 lifetime:

$$y_{\rm D}^{CP} = \frac{\tau_{D^0}}{\tau_{CP+}} - 1. \tag{9}$$

In case of no CPV then $y_{\rm D}^{CP} = y_{\rm D}$.

The CPV is also quantified in the parameter$\Delta\ Y$:

$$\Delta Y = \frac{\tau_{D^0}}{\langle \tau_{CP+} \rangle} A_\tau \quad \text{with} \quad A_\tau = \frac{\Gamma(D^0 \to f_+) - \Gamma(\overline{D}^0 \to f_+)}{\Gamma(D^0 \to f_+) + \Gamma(\overline{D}^0 \to f_+)} \tag{10}$$

In case of no mixing$\Delta\ Y$= 0. In the limit of CP conservation $y_{\rm D}^{CP} = y_{\rm D}$ and$\Delta\ Y$=0.$\Delta\ Y \neq 0$ is the signature for indirect CPV.

3.1. Standard Model Predictions and Measurements

Standard Model predictions for $x_{\rm D}$ e $y_{\rm D}$ are of the order of 10^{-2} or smaller for both, as showed in Fig. 1.

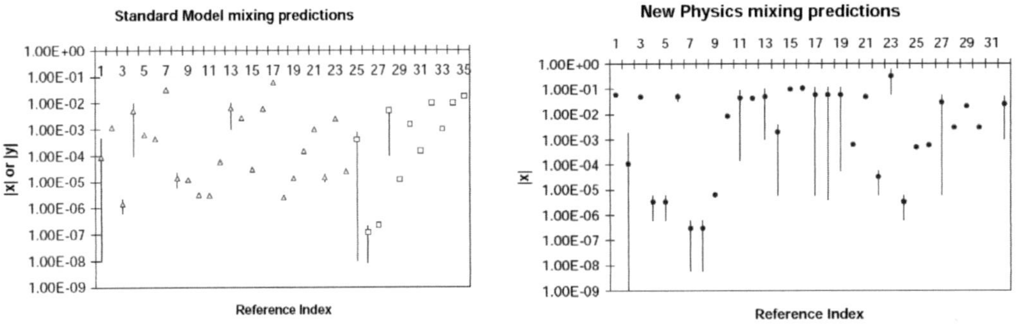

Figure 1. left: SM predictions for $|x_{\rm D}|$ (triangles) and $|y_{\rm D}|$ (squares) published by different authors; right: predictions for $x_{\rm D}$ coming from theories beyond the SM. See [4] for the reference index.

The suppression of mixing in the D system with respect to the other systems has been understood within the Standard Model. Since the mixing process changes the flavour of two units, the transition must proceed through an intermediate state that can be either real or virtual. One possible transition proceeding through a virtual intermediate state is represented in Fig. 2 (left). The contribution of these diagrams is small [5] due to the CKM factors associated to the vertices: the contribution of the b quark is doubly Cabibbo suppressed (DCS) while the contributions of the d and s quark are GIM suppressed. The biggest uncertainty related to the computation of these diagrams is due to the poor knowledge of the hadronization processes. It's important to notice that the D system is the only one in which the virtual quarks are down-type since the top quark is too heavy to form long-living bound states and the π^0 coincides with its anti-particle. From this point of view the D system is unique.

Some examples of real diagrams are reported in Fig. 2 (right), the intermediate states are all the possible final states common to the D^0 and the $\overline{D}^0$ decays, such as K^+K^- and $\pi^+\pi^-$. This second class of diagrams is expected to be dominant in the mixing process though the uncertainty associated to their computation is even larger [5].

Figure 2. left: Feynman diagrams of D^0 mixing through virtual intermediate states; right: mixing processes involving real intermediate states.

The first evidence of D^0 mixing was found by the Belle experiment [6] and the *BABAR* experiment [7] in 2007, and it was confirmed a year latex by the CDF experiment [8]. The measurements are all in agreement with the SM predictions and show no evidence of CPV. Fig. 3 shows the mean world values for x_D e y_D, extracted from all measurements, computed by the HFAG group [9].

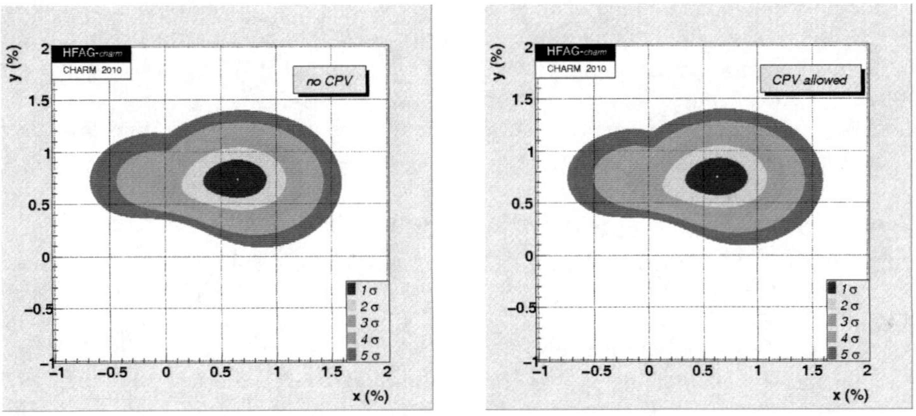

Figure 3. World mean values computed by the HFAG group under the hypothesis of CP conservation (left) and allowing for CPV (right).

3.2. Impact of the measurement

Historically mixing measurements have played an important role as SM tests and in building New Physics (NP) theories. The D system is the last one in which mixing has been observed, still not yet with a precision that can bring information on a possible or impossible NP structure and on CPV.

Measuring mixing in the D system completes the picture of neutral meson mixing within the SM bringing new information since this is the only system in which the virtual quark are down-type. Moreover, since the mixing parameters values are quite small, NP could manifest itself with a higher relative value (v. Fig. 1). Unfortunately, except for big values of x_D and y_D (hypothesis currently not favored by the experimental results), the interpretation of the results is not straightforward, it is limited by the uncertainties in the SM predictions.

The search for CPV is probably the most effective probe for NP: an evidence of CPV with the current experimental sensitivity would be a clear sign of NP.

4. The *BABAR* Experiment

The *BABAR* detector [10] has been running at the B-Factory PEP II, the e^+e^- asymmetric collider located at the SLAC National Accelerator Laboratory. The data taking started in 1999 and ended in April 2008, resulting in a total integrated luminosity of 530 fb^{-1}. The collider operated mostly at a center of mass energy of 10.58 GeV, corresponding to the mass of the $\Upsilon(4S)$ resonance (Run1 to Run6). During the last period of data taking it also operated at lower energies, corresponding to the masses of the $\Upsilon(3S)$ and $\Upsilon(2S)$ (Run7). The asymmetry in the beam energies led to a boost $\beta\gamma = 0.56$. The internal part of the detector consists of the Silicon Vertex Tracker (SVT), the drift chamber (DCH), the Cherenkov light detector (DIRC) and the electromagnetic calorimeter (EMC). Outside there is the instrumented flux return (IFR) and a superconductive solenoid which produces a uniform magnetic field of 1.5 T parallel to the direction of the electron beam. The B Factory produced around 690 millions of $e^+e^- \to c\bar{c}$ events, working also as a Charm Factory.

4.1. The D^0 mesons tagging

In CPV measurements it is of crucial importance to identify the flavour of the neutral D meson at production. In order to do so the decay chain $D^{*+} \to D^0\pi^+$ is reconstructed and the charge of the pion tag the flavour of the D meson. Samples of D mesons reconstructed in this way are referred to as tagged samples. In mixing analyses the knowledge of the flavour of the D is not always needed and in this cases we talk of untagged samples.

The tagged samples contain a much lower level of background but are around four times less populated than the correspondent untagged samples.

5. A time-dependent analysis on the Dalitz Plot: $D^0 \to K^0_S\pi^+\pi^+, K^0_S K^+K^-$

A direct measurement of the mixing parameters is possible through a full time-dependent analysis of the Dalitz plot (DP) for the three-body final states $K^0_S h^+h^-$ ($h = \pi, K$) with the K^0_S reconstructed in the $\pi^+\pi^-$ final state.

The three-body D^0 decay is assumed to proceed through two-body intermediate states where one is a resonance, and it is described by two independent Dalitz variables, m^2_+ and m^2_- which are the reconstructed invariant mass squared of $K^0_S h^+$ and $K^0_S h^-$. In the following we assume CP conservation in the decay, *i.e.* $A(m^2_-, m^2_+) = \bar{A}(m^2_+, m^2_-)$. The total amplitude $A(m^2_-, m^2_+)$ can be written as the superimposition of the amplitudes of each resonance (r) plus a non-resonant (NR) term:

$$A(m^2_-, m^2_+) = \sum_r a_r e^{i\phi_r} A_r(m^2_-, m^2_+) + a_{NR} r^{i\phi_{NR}} \tag{11}$$

where a_r and ϕ_r are the modulus and the phase of the amplitudes and $A_r(m^2_-, m^2_+)$ reproduces the dependence on the Dalitz variables. According to [11] the modulus and the phases are extracted from data while a phenomenological model for each $A_r(m^2_-, m^2_+)$ is assumed. After selecting a sample of D^0 of purity greater than 98% for both the D^0 decay modes, we discriminate signal from background events fitting the distribution in the ($\Delta m = m_{D^*} - m_{D^0}, m_{D^0}$) plane for the signal and background categories. An extended maximum likelihood fit is then performed on the combined samples of $D^0 \to K^0_S\pi^+\pi^-$ and $D^0 \to K^0_S K^+K^-$ decays in order to extract the Dalitz model and the resolution function parameters, yielding [12]:

$$\begin{aligned} x_{\rm D} &= [1.6 \pm 2.3({\rm stat}) \pm 1.2({\rm syst}) \pm 0.8({\rm model})] \times 10^{-3} \\ y_{\rm D} &= [5.7 \pm 2.0({\rm stat}) \pm 1.3({\rm syst}) \pm 0.7({\rm model})] \times 10^{-3} \end{aligned} \tag{12}$$

This measurement, done on 485 fb^{-1}, is the most precise direct measurement of $x_{\rm D}$ and $y_{\rm D}$. The contour plot for the 1 to 3 standard deviations regions is reported in Fig. 4.

The most important systematic errors are related to the Dalitz model assumption (referred to as "model" in the error), the fit bias and the detector misalignment.

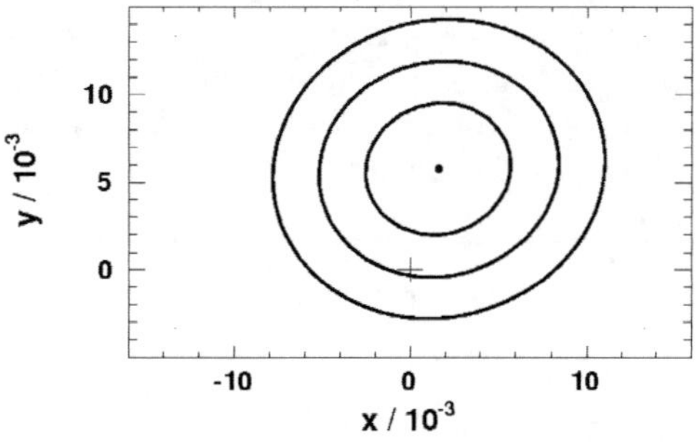

Figure 4. Fit results for $K_S^0\pi^+\pi^- - K_S^0K^+K^-$ combined fit with the confidence-level (CL) contours for $1-\mathrm{CL} = 0.317\ (1\sigma)$, $4.55\times10^{-2}\ (2\sigma)$ and $2.70\times10^{-3}\ (3\sigma)$. Systematic uncertainties are included. The no-mixing point is shown as a plus sign (+).

6. Lifetime Ratio analyses: $D^0 \to K^-\pi^+, K^+K^-, \pi^+\pi^+$

The *BABAR* Collaboration has presented two analyses on the measurement of the ratio between the lifetimes of D^0 reconstructed in CP eigenstates and in CP mixed states, y_{D}^{CP}. One was performed on an untagged sample, the other on a statistically independent sample of tagged events. In both analyses τ_{D^0} is extracted from the proper time distribution of D^0 reconstructed in the final state $K^-\pi^+$ assuming an exponential decay[1]. The same hypothesis is under the extraction of the value of τ_{CP+} from K^+K^- final state and, in the tagged analysis only, the $\pi^+\pi^-$ final state. The value of y_{D}^{CP} extracted from the untagged sample of D^0 is [13]:

$$y_{\mathrm{D}}^{CP} = (1.12 \pm 0.26 \pm 0.22)\%. \tag{13}$$

In the case of the tagged sample a measurement of CPV is also possible through the parameter ΔY. The extracted values for mixing and CPV are [7] :

$$\begin{aligned} y_{\mathrm{D}}^{CP} &= (1.24 \pm 0.39 \pm 0.13)\% \\ \Delta Y &= (-0.26 \pm 0.36 \pm 0.08)\% \end{aligned} \tag{14}$$

By combining the tagged and the untagged samples, *BABAR* obtained a value of

$$y_{\mathrm{D}}^{CP} = (1.16 \pm 0.22 \pm 0.18)\%, \tag{15}$$

with a significance for mixing of 4.1 standard deviations, resulting the most significant mixing measurement.

7. T violation with $D^0 \to K^+K^-\pi^+\pi^+$

Considering the four-body decay $D^0 \to K^+K^-\pi^+\pi^+$, the T odd observable $C_T \equiv \vec{p}_{K^+}\cdot(\vec{p}_{\pi^+}\times \vec{p}_{\pi^-})$ built with the momenta of the tracks, is used to measure a T asymmetry:

$$A_T = \frac{\Gamma(D^0, C_T > 0) - \Gamma(D^0, C_T < 0)}{\Gamma(D^0, C_T > 0) + \Gamma(D^0, C_T < 0)} \tag{16}$$

Since final state interactions can produce a non zero value of A_T, the relevant parameter is $\mathcal{A}_T = \frac{A_T - \bar{A}_T}{2}$, where $\bar{A}_T$ is extracted from a $\bar{D}^0$ sample, and therefore a tagged sample of D^0 is

[1] Since the mixing effect is smaller than 1%, the assumption is experimentally acceptable.

needed. In order to extract the value of $\mathcal{A}_T$ from a sample of 470 fb^{-1} a two-steps analysis is performed. First, the resolution function parameters for signal and background components are extracted from data in the (Δm,m_{D^0}) plane. Then the sample is divided in four sub-samples (depending on the flavour of the D and the sign of C_T) and, after fixing the resolution function, the number of events in each sub-sample is extracted and $\mathcal{A}_T$ is computed.

The measurement is still statistically limited [15]:

$$\mathcal{A}_T = [1.0 \pm 5.1(\text{stat}) \pm 4.4(\text{syst})] \times 10^{-3} \tag{17}$$

but is represents an improvement of one order of magnitude with respect to the previous FOCUS measurement. The most important systematic error is associated to the particle identification, other less important contributions come from the cut on the D^0 momentum inn the center of mass, the detector misalignment and the fit bias.

8. CP violation with $D^+ \to K^0_S\pi^+$

A measurement of the direct CP asymmetry is possible studying the decay $D^+ \to K^0_S\pi^+$ and and computing:

$$A_{\text{CP}} = \frac{\Gamma(D^+ \to K^0_S\pi^+) - \Gamma(D^- \to K^0_S\pi^-)}{\Gamma(D^+ \to K^0_S\pi^+) + \Gamma(D^- \to K^0_S\pi^-)}. \tag{18}$$

A CP violation contribution from K^0–$\bar{K}^0$ mixing must also be taken into account. The expected value of A_{CP} due to the CP violation in the K^0 system is $A^K_{\text{CP}} = [-0.332 \pm 0.006]\%$ [14], any deviation from it would be evidence of CP violation.

The measured asymmetry $A = \frac{N_{D^+} - N_{D^-}}{N_{D^+} + N_{D^-}}$ will have two contributions besides the CP asymmetry: the forward-backward asymmetry and the detector-induced asymmetry in the reconstruction of the charged particles. The former is extracted directly from data exploiting its oddness in the center of mass polar angle. In order to remove the latter a new data-driven technique has been developed.

8.1. *The track reconstruction efficiency*

If the detector is not completely symmetric in charge reconstruction, the measurement of CP violation will be affected by a detector-induced asymmetry. In the signal channel the only track affected by this asymmetry is the charged pion from the D decay. In order to estimate this effect and subtract it from the final result a high populated control sample with no asymmetry from physics is selected: if no detector asymmetry is present then the number of reconstructed π^+ ($N^{\pi^+}_{\text{rec}}$) should be equal to the number of reconstructed π^- ($N^{\pi^-}_{\text{rec}}$). Using the $\Upsilon(4S)$ decays[2] the ratio of the charged pion detection efficiency as a function of the D momentum was computed as:

$$R(\vec{p}) = \frac{\epsilon^+(\vec{p})}{\epsilon^-(\vec{p})} = \frac{N^{\pi^+}_{\text{rec}}(\vec{p})}{N^{\pi^-}_{\text{rec}}(\vec{p})} \tag{19}$$

and this correction is applied to the negative D candidates. In Fig. 5 the value of the ratio of the detection efficiency and its error are reported: the plots shows that the correction is small and dominated by statistical fluctuations.

8.2. *Results*

After the subtraction of the forward-backward asymmetry (v. fig 6), the measured CP asymmetry is:

$$A_{\text{CP}} = [-4.4 \pm 1.3(\text{stat}) \pm 1.0(\text{syst})] \times 10^{-3}. \tag{20}$$

[2] these events are produced at rest and decay isotropically, continuum subtraction was applied.

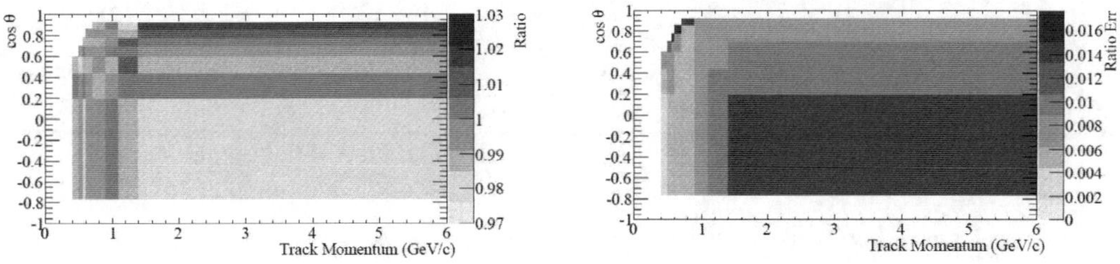

Figure 5. Value of ratio of detection efficiency as a function of the D momentum (left) and its error (right).

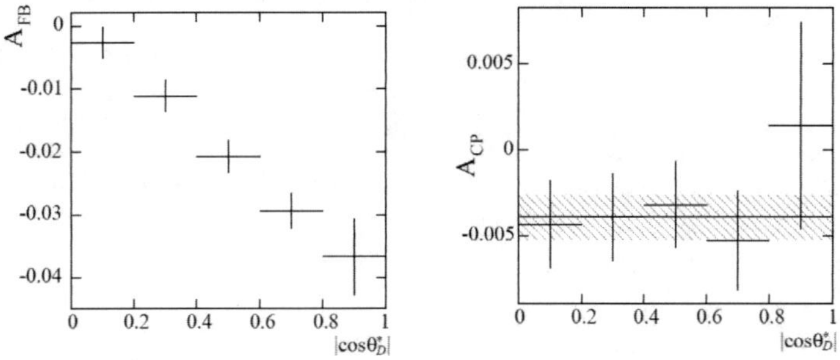

Figure 6. Forward-backward asymmetry (left) and CP asymmetry (right) in bins of cosine of the polar angle in the center of mass.

it is consistent with $A^K_{\rm CP}$ and therefore with no CP violation in the charm sector, as predicted by the Standard Model at these levels of error. This measurement is the most precise CP asymmetry measurement at B-Factories. The most important systematic errors are relative to the charge asymmetry correction and the K^0-$\overline{K}^0$ regeneration in the material.

References

[1] Nakamura, K. and others (Particle Data Group Collaboration), J. Phys., **G37** (2010) 075021;
[2] Buccella, F. and Lusignoli, M. and Miele, G. and Pugliese, A. and Santorelli, P., Phys. Rev. D **51** (1995) 3478;
[3] Bergmann, S. and Grossman, Y. and Ligeti, Z. and Nir, Y. and A.A. Petrov, Phys. Lett. B **486** (2000) 418;
[4] Petrov, Alexey A, Int. J. Mod. Phys. **A21** (2006) 5686-5693;
[5] Golowich, E. and Hewett, J. and Pakvasa, S. and Petrov, A. A., Phys. Rev. D **76** (2007) 095009;
[6] Staric, M. and others (Belle collaboration), Phys. Rev. Lett. **98** (2007) 211803;
[7] Aubert, B. and others (*BABAR* Collaboration), Phys. Rev. D **78** (2008) 011105;
[8] Aaltonen, T. and others (CDF Collaboration), Phys. Rev. Lett. **100** (2008) 121802;
[9] Asner, D. and others (Heavy Flavor Averaging Group), http://www.slac.stanford.edu/xorg/hfag/;
[10] Aubert, B. and others (*BABAR* Collaboration) Nucl. Instrum. Meth. **A479** (2002) 1-116;
[11] Aubert B. and others (*BABAR* collaboration), Phys. Rev. D **78** (2008) 034023
[12] del Amo Sanchez P. and others (*BABAR* collaboration), Phys. Rev. Lett. **105** (2010) 081803
[13] Aubert, B. and others (*BABAR* Collaboration), Phys. Rev. D **80(7)**, (2009) 071103;
[14] H. J. Lipkin and Z. z. Xing, Phys. Lett. B **450**, 405,(1999)
[15] del Amo Sanchez P. and others (*BABAR* collaboration), Phys. Rev. D **81**, (2010) 111103

Journal of Physics: Conference Series **335** (2011) 012044 doi:10.1088/1742-6596/335/1/012044

Precision measurement of CP Violation in $D^0 \to \pi^+\pi^-$

Fabrizio Ruffini, on behalf of the CDF Collaboration.
INFN of Pisa, Largo B. Pontecorvo, 56127 Pisa - Italy
University of Siena, Via Roma 56, 53100 Siena - Italy
E-mail: fabrizio.ruffini@pi.infn.it

Abstract. We report a measurement of the CP violating asymmetry in $D^0 \to \pi^+\pi^-$ decays using approximately 215,000 decays reconstructed in about 5.94 fb^{-1} of CDF data. We use the strong $D^{\star +} \to D^0\pi^+$ decay ("$D^\star$ tag") to identify the flavor of the charmed meson at production time and exploit CP-conserving strong $c\bar{c}$ pair-production in $p\bar{p}$ collisions. Higher statistic samples of Cabibbo-favored $D^0 \to K^-\pi^+$ decays with and without $D^\star$ tag are used to highly suppress systematic uncertainties due to detector effects. The result is the world's most precise measurement to date and it is fully consistent with no CP violation.

1. Introduction and motivation

Time integrated CP-violating asymmetries of singly-Cabibbo suppressed transitions as $D^0 \to \pi^+\pi^-$ and $D^0 \to K^+K^-$ are powerful probes of new physics (NP). Contribution to these decays from "penguin" amplitudes are negligible in the Standard Model (SM), but presence of NP particles could enhance the size of CP-violation with respect to the SM expectation. Any asymmetry significantly larger than few 0.1%, as expected in the CKM hierarchy, may unambiguously indicate new physics contributions [1]. We present a measurement of time-integrated CP violating asymmetry in the Cabibbo-suppressed $D^0 \to \pi^+\pi^-$ decay:

$$A_{\text{CP}}(\pi^+\pi^-) = \frac{\Gamma(D^0 \to \pi^+\pi^-) - \Gamma(\overline{D}^0 \to \pi^-\pi^+)}{\Gamma(D^0 \to \pi^+\pi^-) + \Gamma(\overline{D}^0 \to \pi^-\pi^+)}. \tag{1}$$

Both direct and mixing-induced CP violation contribute to the asymmetry. The latter source produces a time-dependent asymmetry, whose expression when neutral charmed mesons decay into CP eigenstates is [1]

$$A_{\text{CP}}(t) \approx \frac{\eta_{CP}}{2}\frac{t}{\tau}\left[y\left(\left|\frac{p}{q}\right| - \left|\frac{q}{p}\right|\right)\cos(\varphi) + x\left(\left|\frac{p}{q}\right| + \left|\frac{q}{p}\right|\right)\sin(\varphi)\right], \tag{2}$$

that persists when integrated over time. In eq. (2) η_{CP} is the CP-parity of the decay final state (+1 for $\pi^+\pi^-$), x, y, p and q are the usual parameters used to describe flavored mesons mixing, φ is the weak CP violating phase and t/τ the proper decay time in unit of D^0 lifetime ($\tau \approx 0.5$ ps). The measured integrated asymmetry, owing to the slow mixing rate of charm mesons, reduces at first order to a sum of two terms:

$$A_{\text{CP}}(\pi^+\pi^-) = a_{\text{CP}}^{\text{dir}} + \int_0^\infty A_{\text{CP}}(t)\text{D}(t)dt \approx a_{\text{CP}}^{\text{dir}} + \frac{\langle t\rangle}{\tau}a_{\text{CP}}^{\text{ind}}. \tag{3}$$

The first term arises from direct and the second one from mixing-induced CP violation. The integration in eq. (3) is performed over the observed distribution of proper decay time, $D(t)$. Since the value of $\langle t \rangle$ depends strongly on $D(t)$, different values of $A_{\rm CP}$ could be observed in different experimental environments because of different sensitivities to $a_{\rm CP}^{\rm dir}$ or $a_{\rm CP}^{\rm ind}$. Since the trigger used in this analysis imposes requirements on minimum impact parameters of the D^0 decay particles, our sample is enriched of higher-valued proper decay time candidates with respect to B-factory experiments. This makes this analysis more sensitive to mixing-induced CP violation.

The measurement, described with further details in [2], has been performed on about 5.94 fb^{-1} of $p\bar{p}$ collisions at $\sqrt{s} = 1.96$ TeV recorded by the CDF II detector at Fermilab's Tevatron collider.

2. Detector and trigger

The CDF II detector [3] is a magnetic spectrometer surrounded by calorimeters and muon detectors. It provides a determination of the decay point of particles with 15 μm resolution in the transverse plane using six layers of double-sided silicon-microstrip sensors at radii between 2.5 and 22 cm from the beam. A 96-layer drift chamber extending radially from 40 to 140 cm from the beam provides excellent momentum resolution, yielding approximately 8 MeV/c^2 mass resolution for two body charm decays. A three-level trigger system selects events enriched in decays of long-lived particles by exploiting the presence of displaced tracks in the event and measuring their impact parameter with offline-like 30 μm resolution. The trigger requires presence of two charged particles with transverse momenta greater than 2 GeV/c, impact parameters greater than 100 microns and basic cuts on azimuthal separation and scalar sum of momenta.

3. Analysis overview

We measure the asymmetry using $D^0 \to \pi^+\pi^-$ decays from charged $D^\star$ mesons through fits of the $D^0\pi$ mass distributions. The observed asymmetry includes a possible contribution from actual CP violation, diluted in much larger effects from instrumental charge-asymmetries. We exploit a fully data-driven method that uses higher statistic samples of $D^\star$-tagged (indicated with an asterisk) and untagged Cabibbo-favored $D^0 \to K^-\pi^+$ decays to correct for all detector effects thus suppressing systematic uncertainties to below the statistical ones. We define "raw" asymmetries the observed asymmetries in signal yields,

$$A_{\rm CP}^{\rm raw}(D^0 \to f) = \frac{N_{\rm obs}(D^0 \to f) - N_{\rm obs}(\overline{D}^0 \to \bar{f})}{N_{\rm obs}(D^0 \to f) + N_{\rm obs}(\overline{D}^0 \to \bar{f})},$$

before any correction for instrumental effects has been applied. Then the uncorrected "raw" asymmetries in the three samples can be written as a sum of several contributions:

$$\begin{aligned}
A_{\rm CP}^{\rm raw}(\pi\pi^\star) &= A_{\rm CP}(\pi\pi) + \delta(\pi_s)^{\pi\pi^\star} \\
A_{\rm CP}^{\rm raw}(K\pi^\star) &= A_{\rm CP}(K\pi) + \delta(\pi_s)^{K\pi^\star} + \delta(K\pi)^{K\pi^\star} \\
A_{\rm CP}^{\rm raw}(K\pi) &= A_{\rm CP}(K\pi) + \delta(K\pi)^{K\pi},
\end{aligned}$$

Journal of Physics: Conference Series **335** (2011) 012044 doi:10.1088/1742-6596/335/1/012044

where

- $A_{\rm CP}(\pi\pi)$ and $A_{\rm CP}(K\pi)$ are the actual physical asymmetries;
- $\delta(\pi_s)^{\pi\pi^\star}$ and $\delta(\pi_s)^{K\pi^\star}$ are the instrumental asymmetries in reconstructing a positive or negative soft pion associated to a $\pi^+\pi^-$ and a $K^+\pi^-$ (or $K^-\pi^+$) charm decay. This is mainly induced by charge-asymmetric track-reconstruction efficiency at low transverse momentum.
- $\delta(K\pi)^{K\pi}$ and $\delta(K\pi)^{K\pi^\star}$ are the instrumental asymmetries in reconstructing a $K^+\pi^-$ or a $K^-\pi^+$ charm decay respectively for the untagged and the $D^\star$-tagged case. These are mainly due to the difference in interaction cross-section with matter between positive and negative kaons. Smaller effect are due to charge-curvature asymmetries in track triggering and reconstruction.

The physical asymmetry is extracted by subtracting the instrumental effects through the combination

$$A_{\rm CP}(\pi\pi) = A_{\rm CP}^{\rm raw}(\pi\pi^\star) - A_{\rm CP}^{\rm raw}(K\pi^\star) + A_{\rm CP}^{\rm raw}(K\pi), \tag{4}$$

that is valid if kinematics distributions are equal across samples. Any instrumental effect can vary as a function of a number of kinematic variables or environmental conditions in the detector, but if the kinematic distributions of soft pions are consistent in $K\pi^\star$ and $\pi\pi^\star$ samples, and the distributions of D^0 decay products are consistent in $K\pi^\star$ and $K\pi$ samples, then $\delta(\pi_s)^{\pi\pi^\star} \approx \delta(\pi_s)^{K\pi^\star}$ and $\delta(K\pi)^{K\pi^\star} \approx \delta(K\pi)^{K\pi}$. This condition was verified in the analysis by inspecting a large set of kinematic distributions and applying small corrections (reweight) when needed.

4. Measurement

The trigger selects pair of tracks from oppositely charged particles that are consistent with originating from a secondary decay vertex separated from the beamline, requiring an impact parameter greater than 100 μm. Using these tracks we reconstruct signals consistent with the desired two-body decays ($\pi^+\pi^-$ or $K^-\pi^+$ or $K^+\pi^-$) of a neutral charmed meson (D^0 or $\overline{D}^0$). To remove most part of non-promptly produced charmed mesons we also require the impact parameter of the D^0 candidate not to exceed 100 μm. Then we associate a low-momentum charged particle to the meson candidate to construct a $D^{\star+}$ (or $D^{\star-}$) candidate. The flavor of the charmed meson is determined from the charge of the pion in the strong $D^{\star+} \to D^0\pi^+$ (or $D^{\star-} \to \overline{D}^0\pi^-$) decay. Sample-specific mass requirements are used for the two tagged samples: we ask the two-body invariant mass ($M(K\pi)$ for the $D^0 \to K\pi$ case and $M(\pi\pi)$, for the $D^0 \to \pi\pi$ case) to lie within 24 MeV/c^2 of the nominal D^0 mass.

We reconstruct approximately 215,000 $D^\star$-tagged $D^0 \to \pi^+\pi^-$ decays, 5 million $D^\star$-tagged $D^0 \to \pi^+K^-$ decays and 29 million $D^0 \to \pi^+K^-$ decays where no tag was required. The much larger statistics of $D^0 \to \pi^+K^-$ channels, with respect to the signal sample, is used for correction of instrumental asymmetries and ensures smaller systematic uncertainties than statistical ones on the final result.

We extract independent signal yields for D^0 and $\overline{D}^0$ candidates without using particle identification in the analysis. In the two $D^\star$-tagged samples this is done using the charge of the soft pion. In the untagged $D^0 \to K^-\pi^+$ sample we randomly divided the sample in two indipendent subsamples similar in size. In each subsample we calculate the mass of each candidate with a specific mass assignments: $K^-\pi^+$ in the first subsample and $K^+\pi^-$ in the second one. In one sample the $D^0 \to K^-\pi^+$ signal is correctly reconstructed and appears as a narrow peak (about 8 MeV/c^2 wide), overlapping a $\sim$ 10 times broader peak of the misreconstructed $\overline{D}^0 \to K^+\pi^-$ component (red and green curves in figs. 1 (e)-(f)). The viceversa applies the other sample. The yield asymmetry is extracted by fitting the number of candidates populating the two narrow peaks.

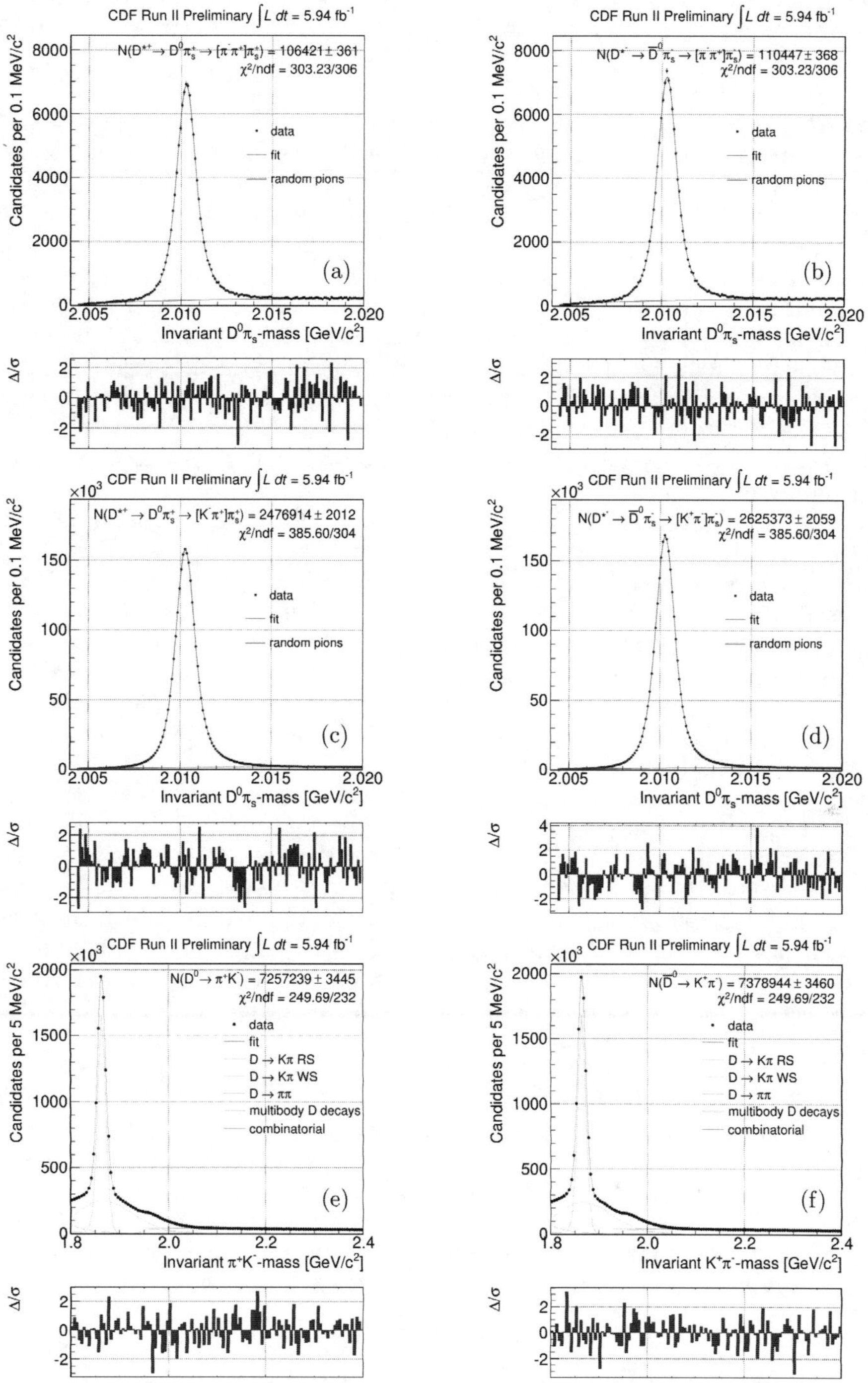

Figure 1. Projections of the combined fit on data for tagged $D^0 \to \pi^+\pi^-$ (a)-(b), tagged $D^0 \to K^-\pi^+$ (c)-(d) and untagged $D^0 \to K^-\pi^+$ (e)-(f) decays.

We determine the yields by performing a binned χ^2 fit to the $D^0\pi_s$-mass ($K\pi$-mass) distribution combining positive and negative decays of both tagged (untagged) samples. The fits projections are shown in fig. 1, the resulting raw asymmetries are: $A_{\rm CP}^{\rm raw}(\pi\pi^\star) = (-1.86 \pm 0.23)\%$, $A_{\rm CP}^{\rm raw}(K\pi^\star) = (-2.91 \pm 0.05)\%$, $A_{\rm CP}^{\rm raw}(K\pi) = (-0.83 \pm 0.03)\%$.

5. Systematic uncertainties

The analysis technique has been extensively tested on Monte Carlo simulation using samples simulated with a wide range of physical and detector asymmetries to verify that the cancellation works regardless of the specific configuration. These studies confirm the validity of our approach and provide a quantitative estimate of possible asymmetries induced by higher order detector effect that may not get fully cancelled or effects of not factorization of $K\pi$ and π_s reconstruction efficiencies. This upper limit is used as systematic uncertainty and amount to 0.009%.

We evaluate all other systematic uncertainties from data. In most cases, this implied varying slightly the shape of the functional forms used in fits, repeating the fit on data, and using the difference between the results of these and the central fit as a systematic uncertainty. This overestimates the size of the systematic effects because it introduces an additional statistical source of fluctuation in the results. But we can comfortably afford that given the large event samples size involved.

Small differences between $D^0\pi_s$-mass distributions of positive and negative $D^\star$ candidates selected in their $D^0(\to K\pi)\pi$ decay are present. This may be due to possible small differences in tracking resolutions between positive and negative tracks at low momentum. These effects impact at first order the observed asymmetry. Insignificant differencies are observed in the $K\pi$-mass distributions of the untagged $D^0 \to K\pi$ sample. To evaluate an associated systematic uncertainty we repeated the fits after fixing signal shapes to be the same and/or leaving background shapes to vary independently for positive and negative $D^\star$ candidates. The maximum observed variations (0.088% for the tagged case and 0.044% for the untagged one) are used as systematic uncertainties.

A contamination by charm mesons produced in b–hadron decays could affect the asymmetry measurement in case CP–violating asymmetries in B decays induce an asymmetric source of charm and anti-charm mesons. These effect may be sizable for a single exclusive mode, but are expected to average to very small values for inclusive $B \to D^0/D^\star X$ decays. In the analysis we exclude the majority of non-primary charm contributions by applying an upper threshold on the D^0 candidate impact parameter. However, a fit to the impact parameter distribution determines a residual 16.6% fraction of charm from B in our sample. To assess the effect of these events we repeat the measurement using only charm mesons with large impact parameters, enriched in b–hadron decays. The observed asymmetry is

$$A_{\rm CP}(B \to D^0/D^\star X) = (-0.21 \pm 0.20)\%.$$

The uncertainty on this number, and the fraction of the non-prompt contribution that survives the D^0 impact parameter cut, is used to asses a conservative estimate of the systematic uncertainty caused by the non-prompt contamination in our samples.

Tab. 1 summarizes the set of all systematic uncertainties considered in the measurement. Assuming they are independent and summing in quadrature we obtain a total systematic uncertainty on our final $A_{\rm CP}(\pi\pi)$ measurement of 0.11%, approximately half of the statistical uncertainty.

6. Final result and conclusions

We report the measurement of the CP asymmetry in the decay $D^0 \to \pi^+\pi^-$ using 5.94 fb^{-1} of data collected by the CDF displaced track trigger. The final result is

$$A_{\rm CP}(D^0 \to \pi^+\pi^-) = \left[+0.22 \pm 0.24\ (stat.) \pm 0.11\ (syst.)\right]\%,$$

Source of systematic uncertainty	Variation on $A_{CP}(\pi\pi)$
Approximations in the method	0.009%
Beam drag effects	0.004%
Contamination of non-prompt D^0s	0.034%
Templates used in fits	0.010%
Templates charge differences	0.098%
Asymmetries from non-subtracted backgrounds	0.018%
Imperfect sample reweighing	0.0005%
Sum in quadrature	0.105%

Table 1. Summary of systematic uncertainties.

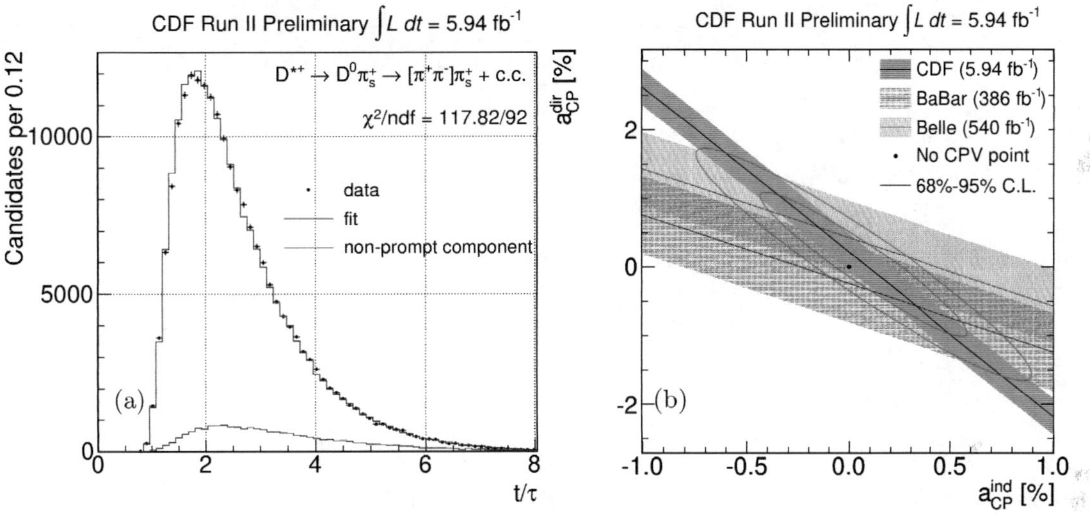

Figure 2. Fit to the proper decay time (in units of D^0 lifetime) distribution of sideband-subtracted tagged $D^0 \to \pi\pi$ data (a). Comparison of our measurement with current best results from B-factories in the parameter space $(a_{\rm CP}^{\rm ind}, a_{\rm CP}^{\rm dir})$ (b).

which is consistent with CP conservation and also with the SM predictions.

To disentangle the independent contributions of direct and indirect CP violation in $D^0 \to \pi^+\pi^-$ decays, an analysis where the time evolution of charm decays is studied is needed. Nevertheless some interesting conclusions could be derived either comparing our result with B-factories measurements or making some theoretical assumptions.

The observed asymmetry is at first order the linear combination of a direct, $a_{\rm CP}^{\rm dir}$, and an indirect, $a_{\rm CP}^{\rm ind}$, CP violating asymmetry through a coefficient that is the mean proper decay time of D^0 candidates in the data sample (see eq. (3),). Fig. 2 (a) shows a fit to the mean proper decay time distribution of our tagged $D^0 \to \pi^+\pi^-$ sample, the resulting mean value is 2.40 ± 0.03 $(stat. + syst.)$ times the D^0 lifetime. Our measurement therefore describes a straight band in the plane $(a_{\rm CP}^{\rm ind}, a_{\rm CP}^{\rm dir})$ with angular coefficient -2.4. The same holds for B-factories' measurements, with angular coefficient -1 [4], due to their unbiased acceptance in charm decay time. The three measurements in the plane $(a_{\rm CP}^{\rm ind}, a_{\rm CP}^{\rm dir})$ are shown in fig. 2 (b), where the bands are 1σ wide and the red curves represent the 68% and 95% CL limits of the combined result assuming Gaussian uncertainties.

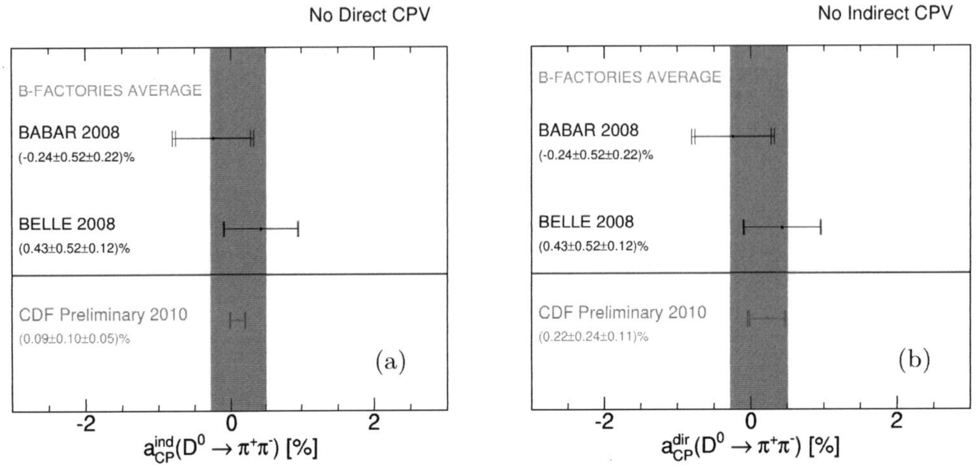

Figure 3. Comparison of our measurement with B-factories experiments [4] assuming no direct (a) or indirect (b) CP violation.

If we assume no direct CP violation in the charm sector eq. (3) simplifies to

$$A_{\rm CP}(\pi^+\pi^-) \approx \frac{\langle t \rangle}{\tau} a_{\rm CP}^{\rm ind}$$

so this measurement implies

$$a_{\rm CP}^{\rm ind} = \left[+0.09 \pm 0.10\ (stat.) \pm 0.05\ (syst.)\right]\%,$$

that means the range $[-0.124, 0.307]\%$ covers $a_{\rm CP}^{\rm ind}$ at the 95% CL. Note that, since $\langle t \rangle/\tau$ in our sample is greater than in B-factories ones, this range is more than five times tighter than the ones obtained using B-factories measurements, as shown in fig. 3 (a).

Conversely, assuming $a_{\rm CP}^{\rm ind} = 0$, our number is directly comparable to other measurements in different experimental configurations. In this case, fig. 3 (b), our statistical uncertainties are half those from the best B-factories measurements, and also systematic uncertainties are smaller.

We have measured the CP asymmetry in the $D^0 \to \pi^+\pi^-$ decay with unprecedented precision, and find a result compatible with zero. A new measurement is expected from the channel $D^0 \to K^+K^-$, which is more abundant, although the higher level of background requires additional care in the analysis. It is expected that these high precision measurements will allow to put tight constraints on NP in the up-quark sector.

Acknowledgments
I would like to thank the Discrete 2010 organizers for the opportunity of speak at this conference and the colleagues of the CDF Collaboration for assisting me while preparing the talk and this document.

References

[1] D.-S. Du, *Eur. Phys. J. C* **50**, 579 (2007), hep-ph/0608313; Y. Grossman, A. Kagan, and Y. Nir, *Phys. Rev D* **75**, 036008 (2006), hep-ph/0609178; I. I. Y. Bigi, arXiv:0902.3048 [hep-ph]; S. Bianco *et al.*, *Riv. Nuovo Cim.* 26N7, 1 (2003), hep-ex/0309021; Z.-Z. Xing, *Phys. Rev. D* **55**, 196 (1997), hep-ph/9606422.

[2] Public CDF note 10296, `http://www-cdf.fnal.gov/physics/new/bottom/100916.blessed-Dpipi6.0/cdf10296_cpv_D0pipi.pdf`.

[3] CDF Collaboration (D. Acosta *et al.*), *Phys. Rev. D* **71**, 032001 (2005), hep-ex/0412071.
[4] BaBar Collaboration (B. Aubert *et al.*), *Phys. Rev. Lett.* **100**, 061803 (2008), arXiv:0709.2715 [hep-ex]; Belle Collaboration (M. Staric *et al.*), *Phys. Lett. B* **670**, 190 (2008), arXiv:0807.0148 [hep-ex].

Charm Physics at LHCb

Jonas Rademacker on behalf of LHCb
H H Wills Physics Laboratory, University of Bristol, UK
E-mail: Jonas.Rademacker@bristol.ac.uk

Abstract. LHCb is set to significantly increase the world charm sample in the next data taking period, leading charm physics to a new level of precision. In these proceedings we present recent LHCb results and prospects for the 2011/2012 data taking period.

1. Introduction

LHCb [1], the dedicated flavour physics experiment at the LHC [2], will exploit the huge $b\bar{b}$ and $c\bar{c}$ cross section of $> 7\,\text{TeV}$ proton proton collisions to explore the precision frontier at the LHC. The same characteristics that optimise LHCb for B physics, also optimise it for charm physics. The charm cross section is about 20 times larger than the b cross section. LHCb is expected to select huge charm samples, leading to new opportunities of high-precision measurements in charm, with unique sensitivity to physics beyond the Standard Model. In these proceedings we review the current status of the LHCb charm physics programme and prospects for the future, especially for the 2011/2012 run, which is also discussed in [3].

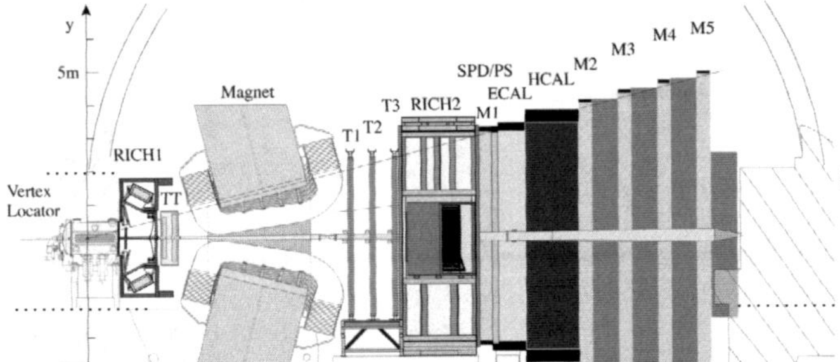

Figure 1. Schematic of the LHCb detector inside the cavern. The interaction point is on the left hand side of the diagram, inside the VELO detector.

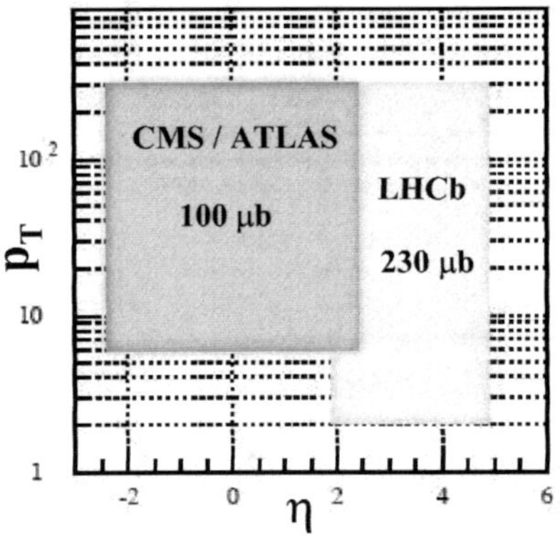

Figure 2. Pseudo rapidity (η) and transverse momentum (p_T) coverage of different LHC experiments. (Also shown: approximate $b\bar{b}$ cross section integrated over the corresponding η, p_T region.)

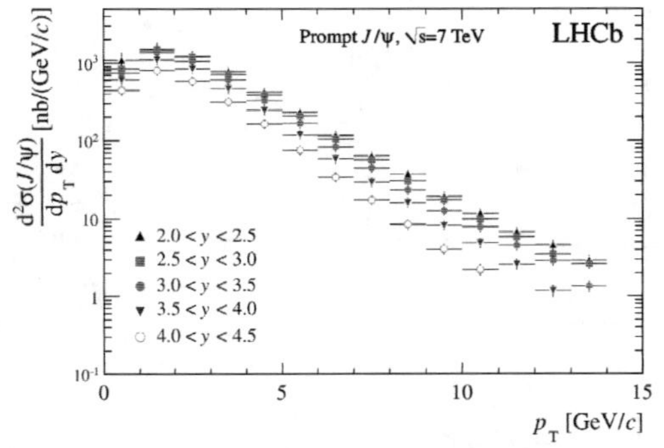

(a) Differential production cross-section for prompt J/ψ in as a function of p_T in bins rapidity y.

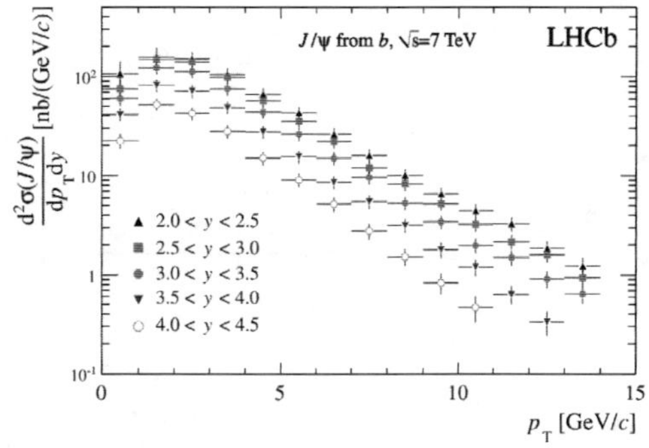

(b) Differential production cross-section for J/ψ from b as a function of p_T in bins rapidity y.

Figure 3. J/ψ cross section measurements at LHCb, based on $5.2\,\text{pb}^{-1}$ [14]. The results shown assume unpolarised J/ψ - a discussion of this assumption, and results assuming polarised J/ψ, and can be found in the publication [14].

2. Charm Cross Sections

Heavy flavour hadro-production is an experiment-led field which has continued to defy theoretical predictions ever since the heavy flavour production cross sections was found to be much larger than predicted by Next-to-Leading-Order (NLO) QCD calculations (see for example [4] and [5]). The discrepancy between data and theory was particularly dramatic in the quarkonium production, where the "Colour Singlet Model" leading-order QCD calculation underestimates the measured cross section by more than an order of magnitude (e.g.[6]).

Since these measurements were made about a decade ago, there has been a number of theoretical advances. Fixed-Order Next to Leading Logarithm (FONLL) calculations [7] describe the open charm and b production cross sections well. Competing models have been put forward that describe the observed quarkonium production rates and p_T spectra well. None of these predicted however the J/ψ polarisation as a function of p_T measured by CDF [8]. More recent models devised since this measurement provide a better, but still not perfect description of the observed J/ψ polarisation [9, 10, 11, 12, 13].

It is clear that LHCb, the dedicated flavour physics experiment at the LHC, covering a unique kinematic range, will play a crucial role in further investigating the hadro-production mechanisms of heavy flavour.

2.1. Hidden Charm Cross Section Results

The differential J/ψ cross section as a function of transverse momentum for both prompt and secondary J/ψ is shown in Fig. 3. Prompt production is distinguished from J/ψ resulting from b decays using the characteristically long lifetime of b hadrons, which leads to displaced vertices. Figure 3(a) shows the measured prompt cross section, Fig. 3(b) the cross section for J/ψ from b decays.

This measurement is in good agreement with, and considerably extends the kinematic range of the recent measurement by CMS [15]. The authors of the LHCb paper [14] compare the result to various theoretical models and find acceptable agreement. The integrated prompt J/ψ production cross section over the fiducial region $p_T < 14\,\text{GeV}$ and $2.0 < y < 4.5$ is found to be [14]

$$\sigma_{J/\psi\,\text{prompt}}\left(p_T < 14\,\text{GeV}, 2.0 < y < 4.5\right) = \left(10.52 \pm 0.04 \pm 1.40^{+1.64}_{-2.20}\right)\mu b$$

Journal of Physics: Conference Series **335** (2011) 012045 doi:10.1088/1742-6596/335/1/012045

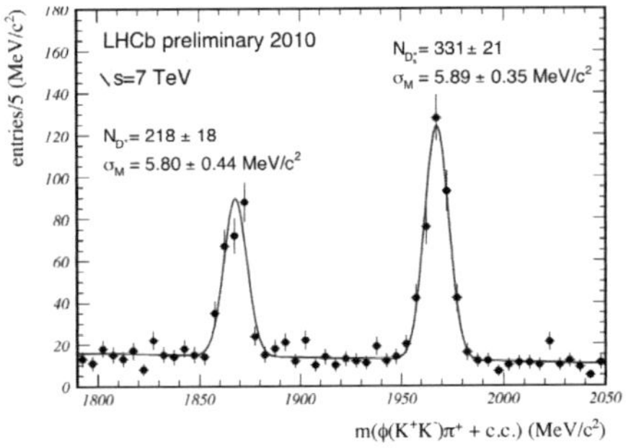

Figure 4. Reconstructed mass of $D^{\pm}, D_s \to KK\pi$ candidates for the D_s cross section measurement in $1.8\,\text{nb}^{-1}$ of LHCb data. (The D^+ cross section has been measured in $D^{\pm} \to K\pi\pi$).

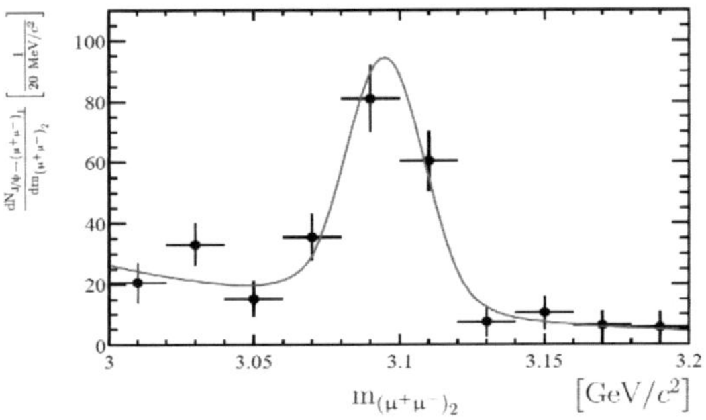

Figure 5. The invariant mass of two muon pairs, $(\mu^+\mu^-)_1$ and $(\mu^+\mu^-)_2$ is reconstructed. The plot shows the number of J/ψ found in $(\mu^+\mu^-)_1$ as a function of the reconstructed mass in $(\mu^+\mu^-)_2$.

where the first uncertainty is statistical, the second systematic, and the third due to the unknown polarisation of the J/ψ. For J/ψ from b decays, this is [14]

$$\sigma_{J/\psi\text{ from }b}\,(p_T < 14\,\text{GeV}, 2.0 < y < 4.5) = (1.14 \pm 0.01 \pm 0.16)\,\mu b$$

where the first uncertainty is statistical, the second systematic (for secondary J/ψ the polarisation is not an additional source of uncertainty). This measurement can be used to calculate the $b\bar{b}$ cross section, and the result, $\sigma\,(pp \to b\bar{b}X) = (288 \pm 4 \pm 48)\ \mu\text{b}$, is in excellent agreement with a previous measurement by LHCb, based on semileptonic b decays to charm, of $\sigma\,(pp \to b\bar{b}X) = (288 \pm 20 \pm 49)\ \mu\text{b}$ [16].

2.2. Double J/ψ Production
In 2011, LHCb performed the first observation of double J/ψ production in proton proton collisions. For this measurement, the invariant mass of two muon pairs, $(\mu^+\mu^-)_1$ and $(\mu^+\mu^-)_2$ is reconstructed. Figure 5 shows the number of J/ψ found in the $(\mu^+\mu^-)_1$ distribution, in bins of of the reconstructed $(\mu^+\mu^-)_2$ mass. The preliminary result for the production cross section of $J/\psi\,J/\psi$ pairs in $\sqrt{s} = 7\,\text{TeV}$ proton proton collisions, in the rapidity range $2 < y^{J\psi} < 4.5$, for transverse momenta $p_T^{J/\psi} < 10\,\text{GeV}$, is [17]:

$$\sigma^{J/\psi\,J/\psi} = (5.6 \pm 1.1 \pm 1.2)\ \text{nb}$$

where the first error is statistical and the second systematic.

2.3. Open Charm Cross Section Results
LHCb have measured the open charm production cross section in various channels. Figure 6 shows the differential cross sections as function of p_T in different y bins for D^0, D^*, $D^{\pm}$ and D_s production, which are in good agreement with calculations by M. Cacciari, S. Frixione, M. Mangano, P. Nason, G. Ridolfi [20]; and B.A. Kniehl, G. Kramer, I. Schienbein, H. Spiesberger [18]; both groups have updated earlier calculations [21, 22, 19] to account for LHC running conditions and LHCb's unique kinematic range. We use the measured differential cross sections to extrapolate to a total $c\bar{c}$ cross section for pp collisions at $\sqrt{s} = 7\,\text{TeV}$ of:

$$\sigma(pp \to c\bar{c}) = (6.10 \pm 0.93)\ \text{mb}$$

approximately $20\times$ the $b\bar{b}$ cross section.

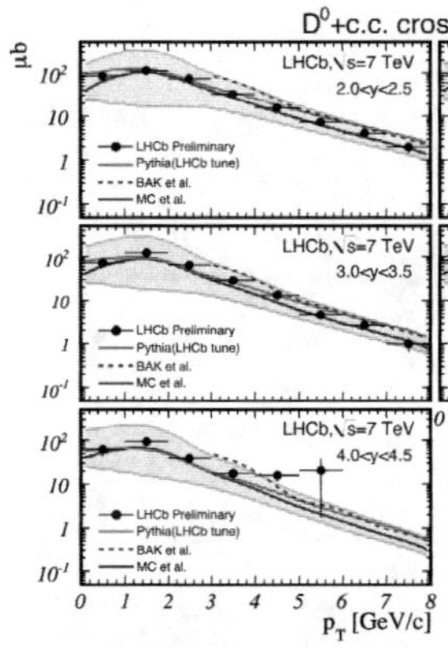

(a) $D^0 + \overline{D}^0$ cross section measured using the decay $D^0 \to K\pi$.

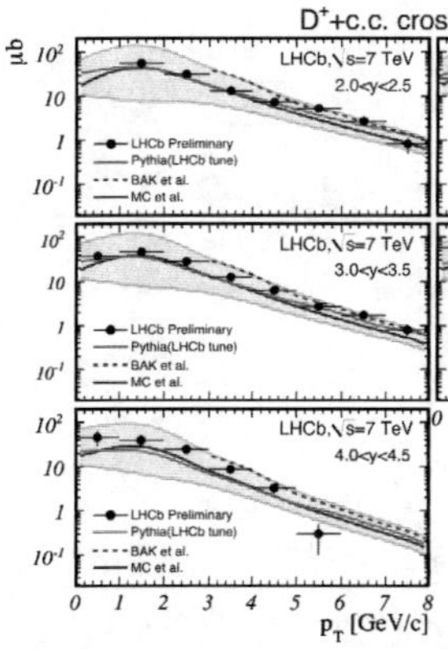

(b) $D^+ + D^-$ cross section measured using the decay $D^+ \to K\pi\pi$.

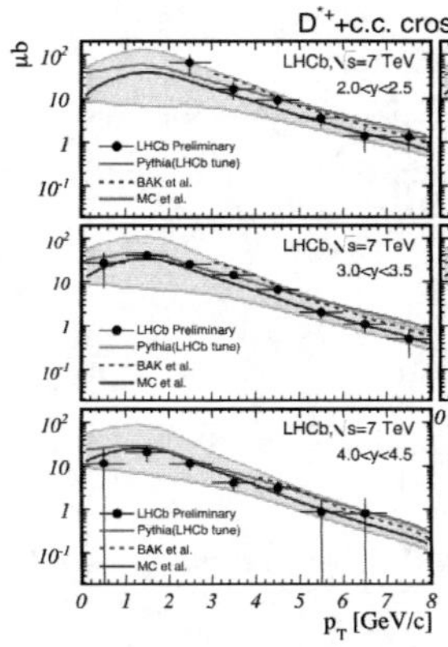

(c) $D^{*+} + D^{*-}$ cross section measured using the decay $D^* \to D\pi$, $D^0 \to K\pi$.

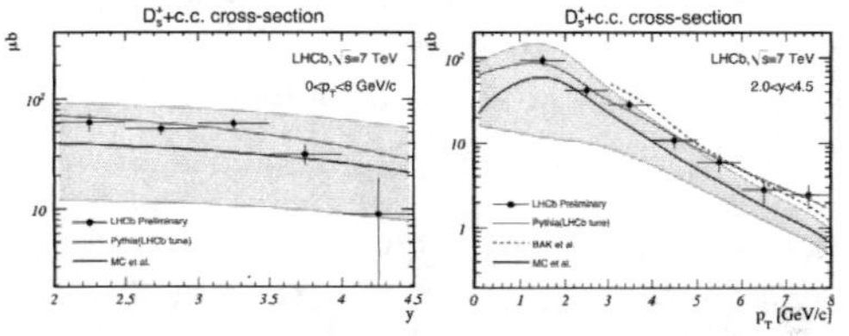

(d) $D_s^+ + D_s^-$ cross section measured using the decay $D_s \to \phi\pi$.

Figure 6. Open charm production cross sections as a function of transverse momentum p_T for different rapidity (y) ranges using $1.8\,\mathrm{nb}^{-1}$ of LHCb data. Also shown: The cross section from Pythia (LHCb tune, in red) and calculations by B.A. Kniehl, G. Kramer, I. Schienbein, and H. Spiesberger (broken line) [18, 19]; and M. Cacciari, S. Frixione, M. Mangano, P. Nason, G. Ridolfi (solid line), [20, 21, 22]. All experimental results are preliminary.

2.4. LHCb's Charm Zoo

Already with as little as $0.5 - 35\,\mathrm{pb}^{-1}$, LHCb has started to observe many rare open and hidden charm states and decays including the X(3872) that was only recently discovered [23]. A set of examples is given in Fig. 7.

3. Charm mixing and CP violation

3.1. The neutral D^0 system

The neutral D system is the only neutral meson system that mixes and consists of up-type quarks. It therefore provides a unique window on Flavour Changing Neutral Currents (FCNC's) between up-type quarks, which can be affected by new physics in a very different way than those in down-type quarks, such as investigated in the $B^0_{s,d}$ and K^0 systems. Particularly interesting is CP violation in the D system, which is essentially zero in the Standard Model, but could

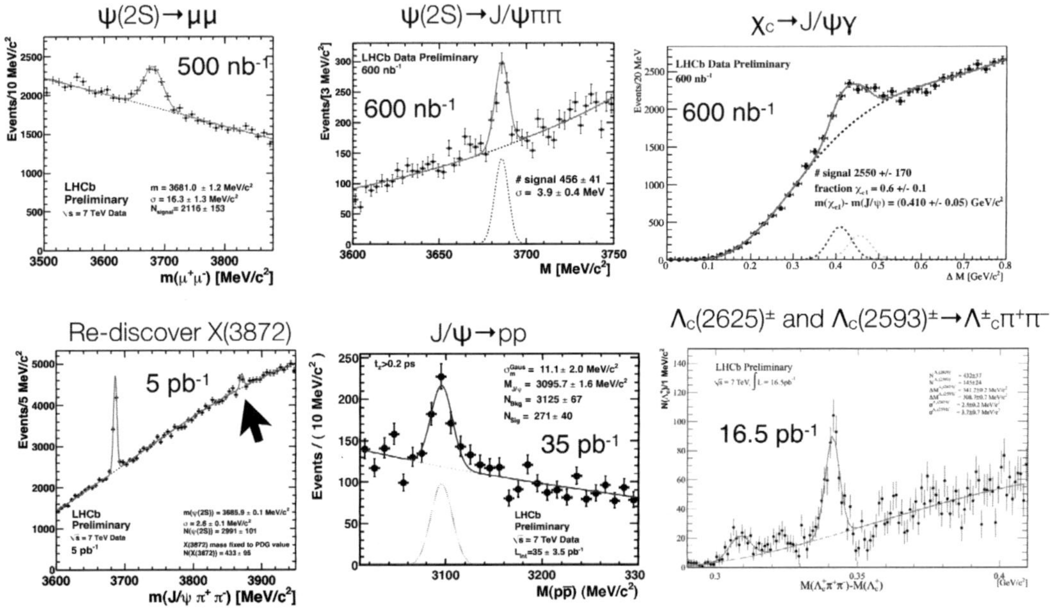

Figure 7. Re-discovering charm decays, including (left-to-right, top to bottom) $\psi(2\mathrm{S}) \to \mu\mu$ in $0.6\,\mathrm{pb}^{-1}$, $\psi(2\mathrm{S}) \to \mathrm{J}/\psi$ in $0.6\,\mathrm{pb}^{-1}$, $\chi_\mathrm{c} \to \mathrm{J}/\psi\gamma$, the "re-discovery" of the X(3872) in $5\,\mathrm{pb}^{-1}$, a baryonic decay of the J/ψ, in charmed baryons.

be significant in many New Physics scenarios. The discovery of CP violation in charm decays would be a clear signal for physics beyond the Standard Model.

Mixing Parameters The physical mass eigenstates D_1, D_2 of the neutral D meson are superpositions of the flavour-specific states D^0 and $\bar{\mathrm{D}}^0$:

$$\mathrm{D}_1 = p\mathrm{D}^0 + q\bar{\mathrm{D}}^0, \qquad \mathrm{D}_2 = p\mathrm{D}^0 - q\bar{\mathrm{D}}^0 \tag{1}$$

where p and q are complex numbers satisfying $|p|^2 + |q|^2 = 1$. The mass and the width difference between D_1 and D_2 are$\Delta\, m$ and $\Delta\Gamma$. Mixing in the neutral D system is conventionally parametrised by the parameters x and y given by

$$x = \frac{\Delta m}{\bar{\Gamma}}, \qquad y = \frac{\Delta\Gamma}{2\bar{\Gamma}}, \tag{2}$$

where $\bar{\Gamma}$ is the average width of the D system. Evidence for mixing has been found by BaBar, BELLE and CDF [24, 25, 26, 27, 28, 29, 30, 31]. While no single experiment has found 5σ evidence for mixing, combining these measurements provides evidence for mixing beyond 5σ [32]. With $x = 0.65^{+0.18}_{-0.19}$ and $y = 0.74 \pm 0.12$, values that are consistent with the (rather broad) Standard Model expectation [33, 34, 35, 36, 37, 38].

CP Violation in Charm CP violation in charm can manifest itself as CP violation in mixing, CP violation in the interference between mixing and decay and as direct CP violation. In the Standard Model, CP violation in charm decays is expected to be small, at a level $< 10^{-3}$ [39, 40, 41]. Observing it would be a clear sign of physics beyond the Standard Model.

CP violation in the mixing occurs if $\left|\frac{q}{p}\right| \neq 1$. CP violation in the interference between mixing and decay is parametrised by the phase ϕ. The phase ϕ is the exact equivalent in the neutral D^0 system of the parameter -2β in the neutral B^0_d system or $\phi_s = -2\beta_1$ in the $\mathrm{B_s}$ system. The phase of the ratio $\frac{q}{p}$ is convention dependent; usually, a phase convention is chosen where $\frac{q}{p} = \left|\frac{q}{p}\right| e^{i\phi}$, so that both of the aforementioned types of CP violation are encoded in the same complex ratio q/p. Current world-averages for $|q/p|$ and ϕ are consistent with the no CP violation hypothesis [32]. Finally, CP violation in charm decays can also manifest itself as direct CP violation in time-integrated decay rate asymmetries:

$$A_{CP}(D \to f) \equiv \frac{\Gamma(D \to f) - \Gamma\left(\overline{D} \to \overline{f}\right)}{\Gamma(D \to f) + \Gamma\left(\overline{D} \to \overline{f}\right)} \tag{3}$$

Of particular interest are singly Cabibbo suppressed (SCS) decays which are sensitive to new contributions from QCD penguin operators. This could yield direct CP violating effects of $\mathcal{O}(10^{-2})$ [42, 43]. Current limits are already approaching this level of precision. Particularly encouraging are the recent CDF results [44, 45] $A_{CP}(D \to \pi\pi) = (0.22 \pm 0.24(\text{stat}) \pm 0.11(\text{sys}))\,\%$ and $A_{CP}(D \to KK) = (-0.24 \pm 0.22(\text{stat}) \pm 0.10(\text{sys}))\,\%$ which constitute the single most precise measurements in the respective channels, with approximately the same precision as the previous world-averages [32]. This demonstrates that such difficult measurements are possible at hadron colliders. Of course, LHCb will have to deal with the additional complication of the non CP symmetric proton-proton initial state, in contrast to CDF's $p\bar{p}$, but on the other hand will have far greater statistical power. At least initially, rather than measuring $A_{CP}(\mathrm{D} \to \pi\pi)$ and $A_{CP}(\mathrm{D} \to \mathrm{KK})$ individually, LHCb will measure the difference $A_{CP}(\mathrm{D} \to \mathrm{KK}) - A_{CP}(\mathrm{D} \to \pi\pi)$ in order to control systematic uncertainties.

3.2. Mixing and CP violation through lifetime ratios

The parameter y_{CP} is the ratio of D^0 lifetimes measured in decays to a (nearly) flavour specific and a CP-specific states:

$$y_{CP} \equiv \frac{\tau\left(D^0 \to K^-\pi^+\right)}{\tau\left(D^0 \to KK\right)} - 1 = y\cos\phi - \frac{1}{2}\left(\left|\frac{q}{p}\right| - \left|\frac{p}{q}\right|\right) x\sin\phi \tag{4}$$

In the absence of CP violation, $y_{CP} = y$. Even in the presence of CP violation, the deviation of y_{CP} from y will be only be of second order in terms of the CP violating parameters. In contrast, the lifetime asymmetry between CP-conjugate decays to CP eigenstates, is linearly sensitive to CP violation [46]:

$$A_\Gamma \equiv \frac{\tau\left(\overline{D}^0 \to K^+K^-\right) - \tau\left(D^0 \to K^+K^-\right)}{\tau\left(\overline{D}^0 \to K^+K^-\right) + \tau\left(D^0 \to K^+K^-\right)} = \frac{1}{2}\left(\left|\frac{q}{p}\right| - \left|\frac{p}{q}\right|\right) y\cos\phi - x\sin\phi \tag{5}$$

As well as $\mathrm{D} \to \mathrm{KK}$, any other CP-specific final state such as $\pi\pi$ is suitable. Both measurements, y_{CP} and A_Γ, have the appealing feature that many detector effects cancel in the lifetime ratio, which will help control systematics - crucial given the excellent statistical precision we expect at LHCb. The current world average for y_{CP} is $(1.107 \pm 0.217)\,\%$ [32], dominated by BaBar's measurement [47], based on 2.7 M Kπ and 260 k KK events in $0.38\,\mathrm{ab}^{-1}$. The world average of A_Γ is $(0.123 \pm 0.248)\,\%$ [32] which is based on measurements by BaBar [48] and BELLE [49], with a combined event sample of approximately 2 M Kπ pairs, 180 k KK and 80 k $\pi\pi$ events. At LHCb, this measurement will be performed using $\mathrm{D}^{*+} \to \mathrm{D}^0\pi^+$ and $\mathrm{D}^{*-} \to \overline{\mathrm{D}}^0\pi^-$ decays. The

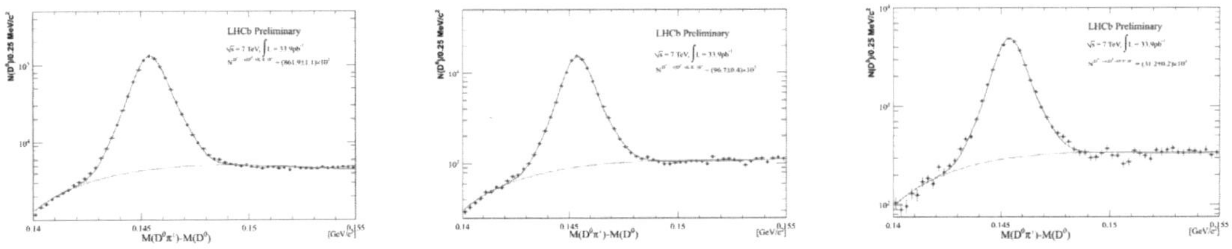

(a) $\sim 862,000$ tagged $D \to K\pi$ events. (b) $\sim 97,000$ tagged $D \to KK$ events. (c) $\sim 34,000$ tagged $D \to \pi\pi$ events.

Figure 8. The plots show the mass difference$\Delta\ m = m_{D\pi} - m_D$ between reconstructed mass of $D^* \to D\pi$ and D candidates, for different decay modes of the D, in $34\,\mathrm{pb}^{-1}$ of LHCb data. Note the logarithmic scale - the background levels are small.

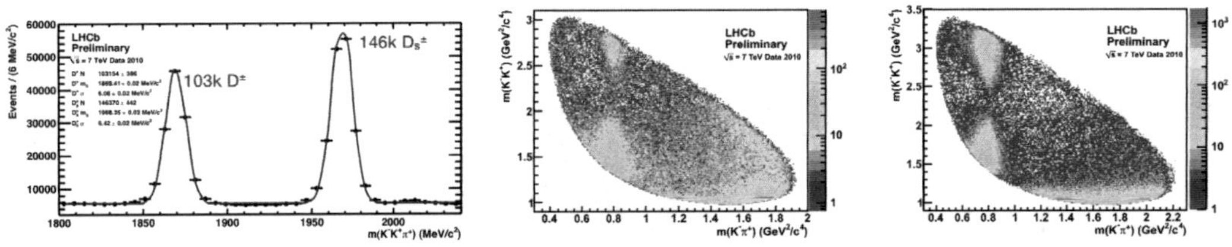

(a) $K^+K^-\pi^+$ mass distribution (b) $D^+ \to K^+K^-\pi^+$ (c) $D_s^+ \to K^+K^-\pi^+$

Figure 9. Mass distribution and Dalitz plots for $D_{(s)} \to KK\pi$

charge of the D^* tags the flavour of the D^0. An additional benefit is that the characteristically small mass difference$\Delta\ m$ between the D^* and the D^0 provides a very powerful discriminating variable and thus very clean data samples. Figure 8 shows$\Delta\ m$ distributions for the relevant decay modes in $34\,\mathrm{pb}^{-1}$ of LHCb data, which corresponds to about half the number $D \to KK$ of decays the current world average on y_{CP} is based on. While it is difficult to make precise extrapolations due to varying run conditions, these numbers suggest that LHCb will very soon be in a position to significantly improve the precision on y_{CP} and A_Γ.

4. Dalitz Analyses

Dalitz plot [50] analyses and their 4-body generalisations are of huge significance for the extraction of γ from $B^\pm \to DK^\pm$ decays [51, 52, 53, 54, 55, 56], and for charm mixing and CP violation measurements [24, 27, 57, 58], and form a central element of LHCb's charm programme. To illustrate LHCb's progress in this area, one of the first Dalitz plots at LHCb is shown in Fig. 9.

5. Summary

With a charm cross section that is $20\times$ the $b\bar{b}$ cross section, the LHC is set to produce vast quantities of charm, and LHCb, optimised for flavour physics, is the experiment in the best position to exploit this bonanza. With these data, LHCb will soon be able to turn charm mixing from a discovery into a precision measurement, and probe for CP violation in charm in a variety of ways, providing a unique window on possible new physics contributions. Charm physics at LHCb is an extremely rich and varied field, and in these proceedings we had to leave out many important topics such as rare decays, charm spectroscopy, amplitude analyses, charm's impact on γ and many more [3]. The step-change in precision offers the potential for new and surprising result, and will truly over-constrain the description of charm in the Standard Model, and in fact any other model that might be a candidate to take its place.

[1] Alves Jr A A *et al.* (LHCb) 2008 *Journal of Instrumentation* **3** S08005
[2] Voss R and Breskin A (eds) 2009 vol 1, 2 ISBN 978-92-9083-336-9 and 978-92-9083-337-6
[3] Adeva B *et al.* (LHCb) 2009 (*Preprint* 0912.4179)
[4] Acosta D E *et al.* (CDF) 2002 *Phys. Rev.* **D65** 052005 (*Preprint* hep-ph/0111359)
[5] Abbott B *et al.* (D0) 2000 *Phys. Rev. Lett.* **85** 5068–5073 (*Preprint* hep-ex/0008021)
[6] Abe F *et al.* (CDF) 1997 *Phys. Rev. Lett.* **79** 572–577
[7] Cacciari M, Greco M and Nason P 1998 *JHEP* **05** 007 (*Preprint* hep-ph/9803400)
[8] Abulencia A *et al.* (CDF Collaboration) 2007 *Phys. Rev. Lett.* **99** 132001
[9] Zhang Y J, Ma Y Q, Wang K and Chao K T 2010 *Phys. Rev. D* **81** 034015
[10] Baranov S P 2006 *Phys. Rev. D* **73** 074021
[11] Campbell J, Maltoni F and Tramontano F 2007 *Phys. Rev. Lett.* **98** 252002
[12] Artoisenet P 2008 *AIP Conf. Proc.* **1038** 55–62
[13] Artoisenet P, Campbell J, Lansberg J P, Maltoni F and Tramontano F 2008 *Phys. Rev. Lett.* **101** 152001
[14] Aaij R *et al.* (LHCb) 2011 submitted to Eur. Phys. J. C (*Preprint* 1103.0423)
[15] Khachatryan V *et al.* (CMS) 2010 Submitted to Eur. Phys. J. C. (*Preprint* 1011.4193)
[16] Aaij R *et al.* (LHCb) 2010 *Physics Letters B* **694** 209 – 216 ISSN 0370-2693
[17] Collaboration L 2011 LHCb-CONF-2011-009
[18] Kniehl B A, Kramer G, Schienbein I and Spiesberger H , as in [19] (private communications)
[19] Kniehl B A, Kramer G, Schienbein I and Spiesberger H 2009 *Phys. Rev. D* **79** 094009
[20] Cacciari M, Frixione S, Mangano M L, Nason P and Ridolfi G , as in [21], see also [22] (priv. comm.)
[21] Cacciari M and Nason P 2003 *JHEP* **09** 006 (*Preprint* hep-ph/0306212)
[22] Cacciari M, Frixione S, Mangano M L, Nason P and Ridolfi G 2004 *JHEP* **07** 033 (*Preprint* hep-ph/0312132)
[23] Choi S K *et al.* (Belle Collaboration) 2003 *Phys. Rev. Lett.* **91** 262001
[24] Abe K *et al.* (BELLE) 2007 *Phys. Rev. Lett.* **99** 131803 (*Preprint* 0704.1000)
[25] Aubert B *et al.* (BABAR) 2007 *Phys. Rev. Lett.* **98** 211802 (*Preprint* hep-ex/0703020)
[26] Aaltonen T *et al.* (CDF) 2008 *Phys. Rev. Lett.* **100** 121802 (*Preprint* 0712.1567)
[27] del Amo Sanchez P *et al.* (BABAR Collaboration) 2010 *Phys. Rev. Lett.* **105** 081803
[28] Aubert B *et al.* (BABAR) 2009 *Phys. Rev. Lett.* **103** 211801 (*Preprint* 0807.4544)
[29] Aubert B *et al.* (BABAR) 2008 *Phys. Rev.* **D78** 011105 (*Preprint* 0712.2249)
[30] Aubert B *et al.* (BABAR) 2009 *Phys. Rev.* **D80** 071103 (*Preprint* 0908.0761)
[31] del Amo Sanchez P *et al.* (The BABAR) 2010 *Phys. Rev. Lett.* **105** 081803 (*Preprint* 1004.5053)
[32] Asner D *et al.* (Heavy Flavor Averaging Group) 2010 And online update at http://www.slac.stanford.edu/xorg/hfag (*Preprint* 1010.1589)
[33] Petrov A A 2006 *Int. J. Mod. Phys.* **A21** 5686–5693 (*Preprint* hep-ph/0611361)
[34] Falk A F, Grossman Y, Ligeti Z, Nir Y and Petrov A A 2004 *Phys. Rev. D* **69** 114021
[35] Bigi I I and Uraltsev N G 2001 *Nuclear Physics B* **592** 92 – 106 ISSN 0550-3213
[36] Falk A F, Grossman Y, Ligeti Z and Petrov A A 2002 *Phys. Rev.* **D65** 054034 (*Preprint* hep-ph/0110317)
[37] Donoghue J F, Golowich E, Holstein B R and Trampetić J 1986 *Phys. Rev. D* **33** 179–183
[38] Wolfenstein L 1985 *Phys. Lett.* **B164** 170
[39] Buccella F *et al.* 1993 *Phys. Lett.* **B302** 319–325 (*Preprint* hep-ph/9212253)
[40] Buccella F *et al.* 1995 *Phys. Rev.* **D51** 3478–3486 (*Preprint* hep-ph/9411286)
[41] Golden M and Grinstein B 1989 *Phys. Lett.* **B222** 501
[42] Grossman Y, Kagan A L and Nir Y 2007 *Phys. Rev. D* **75** 036008
[43] Bianco S, Fabbri F L, Benson D and Bigi I 2003 *Riv. Nuovo Cim.* **26N7** 1–200 (*Preprint* hep-ex/0309021)
[44] CDF Collaboration 2010 CDF note 10296
[45] Acosta D *et al.* (CDF Collaboration) 2005 *Phys. Rev. Lett.* **94** 122001
[46] Bergmann S, Grossman Y, Ligeti Z, Nir Y and Petrov A A 2000 *Physics Letters B* **486** 418 – 425
[47] Aubert B *et al.* (BABAR) 2009 *Phys. Rev. D* **80** 071103
[48] Aubert B *et al.* (BABAR) 2008 *Phys. Rev. D* **78** 011105
[49] Staric M *et al.* (Belle) 2007 *Phys. Rev. Lett.* **98** 211803 (*Preprint* hep-ex/0703036)
[50] Dalitz R H 1953 *Phil. Mag.* **44** 1068
[51] Gronau M and Wyler D 1991 *Phys. Lett. B* **265** 172
[52] Gronau M and London D 1991 *Phys. Lett. B* **253** 483
[53] Atwood D, Dunietz I and Soni A 1997 *Phys. Rev. Lett.* **78** 3257 (*Preprint* hep-ph/9612433)
[54] Giri A, Grossman Y, Soffer A and Zupan J 2003 *Phys. Rev. D* **68** 054018
[55] Bondar A and Poluektov A 2007 (*Preprint* hep-ph/0703267)
[56] Rademacker J and Wilkinson G 2007 *Phys. Lett. B* **647** 400 (*Preprint* hep-ph/0611272)
[57] Asner D M *et al.* (CLEO) 2005 *Phys. Rev.* **D72** 012001 (*Preprint* hep-ex/0503045)
[58] Zhang L M *et al.* (Belle Collaboration) 2007 *Phys. Rev. Lett.* **99** 131803

The search for neutrinoless double beta decay with Cuoricino and Cuore

Elena Guardincerri for the CUORE collaboration

Laboratori Nazionali del Gran Sasso, S.S. 17 bis km 18.910, 67010 Assergi, L'Aquila, Italy

E-mail: elena.guardincerri@lngs.infn.it

Abstract. Cuoricino was a cryogenic bolometric detector operating in Gran Sasso Underground Laboratory, Italy, from March 2003 to June 2008. With its 40.7 kg of $^{130}TeO_2$ in the form of an array of 62 crystals it has set the currently best lower limit on the halflife of 130Te neutrinoless double beta decay, $T_{1/2} > 2.94 \cdot 10^{24}$ y at 90 % CL. It has moreover proven the feasibility of the CUORE experiment, whose aim is to be sensitive to values of the effective neutrino mass as low as few tens of meV. We will report on the latest results from Cuoricino and on the status of the CUORE project.

1. Introduction

The recent observations of neutrino oscillations by atmospheric[1], solar[2], and reactor[3] neutrino experiments established that neutrinos are massive particles.

The oscillation experiments have measured or constrained the elements of the neutrino mixing matrix and the values of the differences between the squared mass eigenvalues participating in the oscillations.

The question whether neutrinos are Dirac or Majorana fermions is however still open and neutrinoless double beta decay ($0\nu\beta\beta$) is currently the only experimentally viable way to answer it.

Besides this, the same process can probe the absolute neutrino mass scale by measuring the effective Majorana mass of the electron neutrino m_{ee}.

The goal of the CUORE experiment is to measure m_{ee} with a sensitivity in the 10-100 meV range, where the spread is due to the uncertainty on the nuclear matrix element (see [4], [5] and the references listed) involved in the determination of m_{ee} from the $^{130}Te_{0\nu\beta\beta}$ lifetime measurement.

The feasibility of CUORE has been proved by its prototype, CUORICINO, which has been the most massive $0\nu\beta\beta$ experiment currently running and whose performance will be discussed below.

2. Principles of operation of CUORE and Cuoricino

The search for $0\nu\beta\beta$ is pursued by a bolometric technique: the detectors consist of TeO_2 crystals operated at a temperature of ~8 mK inside a $^3He/^4He$ dilution refrigerator.

The heat capacity of dielectric materials at this temperature is very low according to the Debye law, so that even tiny energy deposits in the crystals cause an appreciable rise in their temperature.

Journal of Physics: Conference Series **335** (2011) 012046 doi:10.1088/1742-6596/335/1/012046

Thermal pulses are recorded by neutron transmutated doped Ge thermistors glued on each crystals [6].

Among the few even-even nuclei candidates for $0\nu\beta\beta$ decay, ^{130}Te has been chosen for its high transition energy (2527.518 ± 0.013 keV) and natural isotopic abundance (33.87%).

Besides achieving high energy resolutions, close to those obtained by Ge detectors, the bolometric technique provides the possibility to choose among different nuclei as sources for $0\nu\beta\beta$ decay, thus allowing for a cross check in case of discovery.

3. Cuoricino experiment

3.1. CUORICINO setup

The CUORICINO detector[7] was an array of 62 $^{130}TeO_2$ bolometers operated in the Hall A of Laboratori Nazionali del Gran Sasso, Italy.

The total sensitive mass was 40.7 kg and the mass of ^{130}Te was 11.3 kg. The crystals were arranged in 13 planes: 11 of them consisted of four $5 \times 5 \times 5$ cm^3 crystals with a mass of 790 g each, 2 of them were made of nine $3 \times 3 \times 6$ cm^3 crystals whose mass was 330 g.

All crystals were made of natural tellurium except for four $3 \times 3 \times 6$ cm^3 cm^3 ones: two of these are enriched in ^{128}Te with an isotopic abundance of 82.3% and two in ^{130}Te with an isotopic abundance of 75%.

Great care was taken to reduce radioactive contaminations and background sources at all stages of the detector construction and assembly: the crystals were grown from low radioactive materials at the Shangai Institute of Ceramics and shipped to Italy by sea to minimize cosmic activation.

Once in Italy their surface was lapped with radiopure abrasives.

To avoid external vibrations reaching the detectors, the tower was mechanically decoupled from the cryostat through a steel spring.

A 1.2 cm shield of Roman lead with ^{210}Pb activity of 4 mBq/kg was framed around the array to reduce the backgrounds induced by contaminants on the thermal shields of the cryostat.

The refrigerator itself was externally shielded by two layers of lead of 10 cm minimal thickness each.

The background due to environmental neutrons was reduced by a layer of borated polyethylene of 10 cm minimum thickness.

The refrigerator sits inside a Plexiglass anti-radon box flushed with clean N_2 and a Faraday cage was framed around the whole setup to reduce electromagnetic interferences.

The detector was calibrated every month by inserting two thoriated tungsten wires between the refrigerator and the external lead shield.

3.2. Cuoricino results

CUORICINO took data from February 2003 to June 2008. The background spectrum collected up to June 2008, corresponding to an exposure of 19.75 kg ^{130}Te·y, is shown in figure 2.

The background level in the $0\nu\beta\beta$ region is 0.153 ± 0.006 counts/(keV·kg·y) and the average energy resolution (FWHM), calculated on the ^{208}Tl gamma line at 2615 keV is (6.3 ± 2.5) keV for the large crystals and (9.9 ± 4.2) keV for the small crystals.

Apart from the sum of the ^{60}Co gamma lines and the aforementioned ^{208}Tl line (not visible in figure 2), no peak is found near the 2527 keV ^{130}Te $0\nu\beta\beta$ Q value.

This allowed us to set a lower limit on the $0\nu\beta\beta$ ^{130}Te hal-flife of 2.8×10^{24} y at 90 % CL.

Depending on the nuclear matrix element calculation adopted ([8][9][10][11]), this can be translated into an upper limit on m_{ee} in the range $m_{ee} < (0.3 - 0.70)$ eV.

This constraint is currently the most restrictive for ^{130}Te and is comparable with the values obtained with Ge diodes.

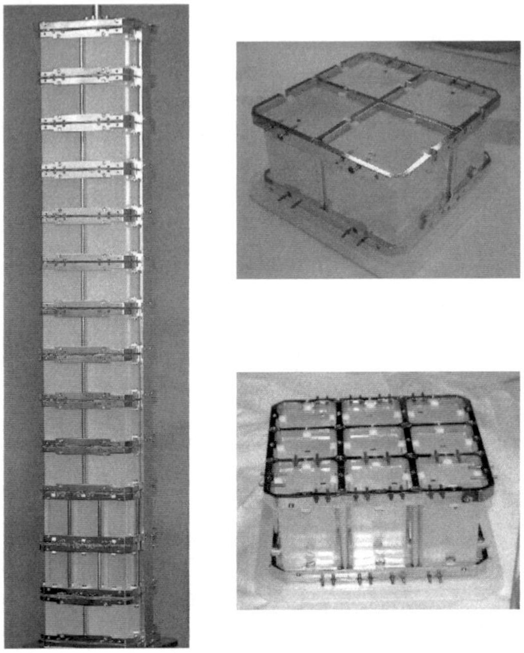

Figure 1. CUORICINO array (left) and details of the planes hosting the $5 \times 5 \times 5$ cm^3 crystals (top right) and $3 \times 3 \times 6$ cm^3 crystals (bottom right)

The current limit is less stringent than the previously published one [7] due to the recent filling of the dip which was present in the spectrum around the ^{130}Te Q value.

4. From CUORICINO to CUORE

The CUORE detector[12] will consist of a cylindrical array of 988 TeO_2 bolometers arranged in 19 towers of 52 crystals each (figure 3), for a total mass of $\sim$ 741 kg, 203 kg of which will be ^{130}Te.

The detector and its inner 6 cm of lead will be cooled to $T \simeq 8$ K inside a dilution refrigerator in Hall A of Laboratori Nazionali del Gran Sasso, next to CUORICINO.

A further layer of 30 cm of low activity lead will shield the array from the dilution unit of the refrigerator and from environmental activity.

A borated polyethylene shield and an air-tight cage will enclose the cryostat.

The detector structure will be more compact than in Cuoricino: the $^{130}TeO_2$ crystals will be closely packed and this will increase the effectiveness of the anticoincidence techniques used in the data analysis.

The detector will furthermore benefit from the improved mechanical decoupling between the experimental and the cryogenic set-ups and from a better crystal holding structure specifically designed to reduce vibrational noise responsible of worsening the energy resolution.

In 5 years of running CUORE sensitivity to the $0\nu\beta\beta$ half-life of ^{130}Te will be $S_{0\nu} \simeq 2.1 \times 10^{26}$ years: this will provide an upper limit on m_{ee} in the range $0.035 - 0.082$ eV.

This sensitivity will result from reducing the background in the $0\nu\beta\beta$ region to $B \simeq 0.01$ counts/(keV·kg·y) and by improving the energy resolution to the level of $\Gamma(2.5 \text{ MeV}) = 5$ keV: both goals will be accomplished thanks to the improvements described above and to the recent development of new, better material cleaning techniques.

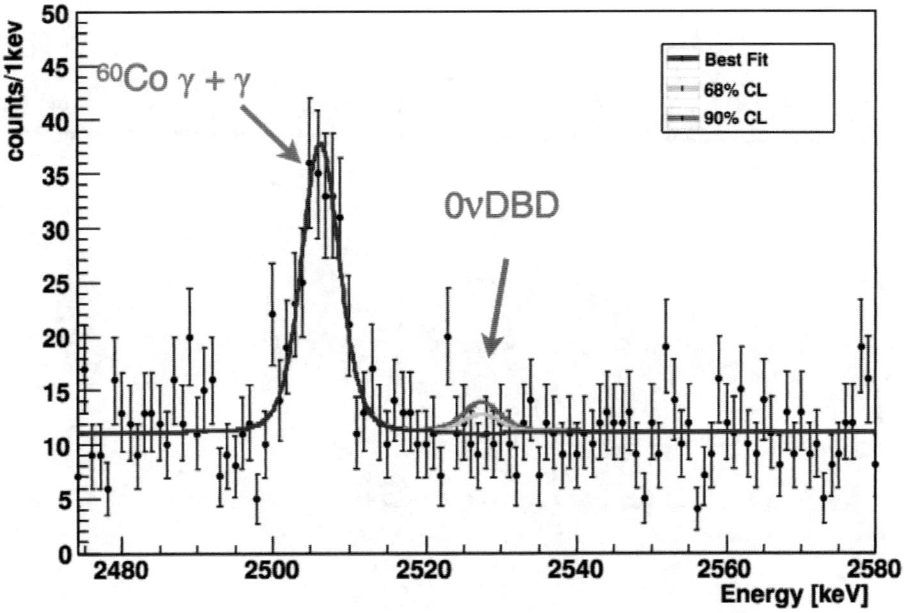

Figure 2. CUORICINO spectrum in the $0\nu\beta\beta$ region

5. CUORE prospects and current status

Cuore construction is now proceeding in the Hall A of Laboratori Nazionali del Gran Sasso: the hut where the detector will be placed has been built as well as the clean room for the detector assembly.

In parallel, the collaboration is also preparing Cuore 0, a prototype of Cuore consisting a single tower made of 52 already available Cuore crystals, prepared and assembled according to the Cuore standards and operated within Cuoricino cryostat.

This detector will be the ultimate test of Cuore concept before Cuore start, and it is scheduled to take data in 2011-2012.

As mentioned in the previous section, the ultimate goal of CUORE is to measure or constrain m_{ee} in the $\sim 0.035 - 0.082$ eV range.

This will allow to probe part of the inverse and degenerate neutrino mass hierarchy pattern scenarios envisioned by neutrino oscillations experiments[13].

6. Acknowledgements

This work was performed under the auspices of the Italian Istituto Nazionale di Fisica Nucleare and the U. S. Department of Energy by the University of California, Lawrence Berkeley National Laboratory under Contract number KB0401022.

References

[1] Hosaka J *et al.* (Super-Kamkiokande) 2006 *Phys. Rev.* **D73** 112001 (*Preprint* hep-ex/0508053)
[2] Ahmad Q R *et al.* (SNO) 2002 *Phys. Rev. Lett.* **89** 011301 (*Preprint* nucl-ex/0204008)
[3] Abe S *et al.* (KamLAND) 2008 *Phys. Rev. Lett.* **100** 221803 (*Preprint* 0801.4589)
[4] Avignone F T 2007 *AIP Conf. Proc.* **942** 1–7
[5] Rodin V 2009 (*Preprint* 0910.5866)

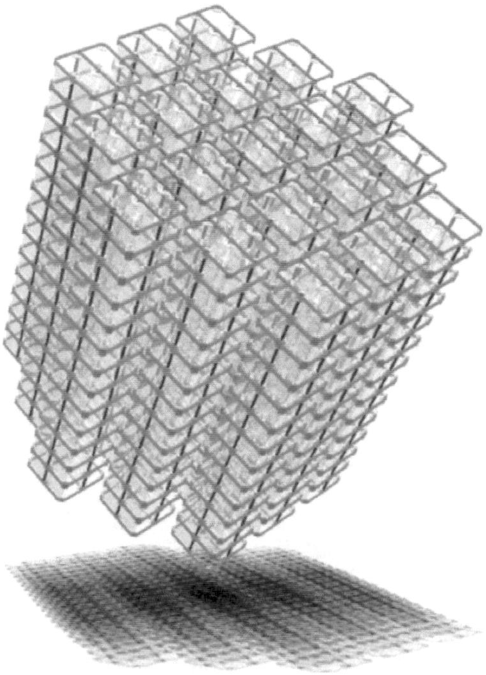

Figure 3. CUORE detector: the bolometers array is made of 19 CUORICINO like towers

[6] Haller E E 1995 *J. Appl. Phys.* **77** 2857
[7] Arnaboldi C *et al.* (CUORICINO) 2008 *Phys. Rev.* **C78** 035502 (*Preprint* 0802.3439)
[8] Simkovic F, Faessler A, Rodin V, Vogel P and Engel J 2008 *Phys. Rev.* **C77** 045503 (*Preprint* 0710.2055)
[9] Civitarese O and Suhonen J 2009 *J. Phys. Conf. Ser.* **173** 012012
[10] Menendez J, Poves A, Caurier E and Nowacki F 2009 *Nucl. Phys.* **A818** 139–151 (*Preprint* 0801.3760)
[11] Barea J and Iachello F 2009 *Phys. Rev.* **C79** 044301
[12] Beeman J *et al.* 2006 *AIP Conf. Proc.* **850** 1623–1626
[13] Strumia A and Vissani F 2006 (*Preprint* hep-ph/0606054)

Journal of Physics: Conference Series **335** (2011) 012047 doi:10.1088/1742-6596/335/1/012047

LUCIFER: Neutrinoless Double Beta decay search with scintillating bolometers

Luca Pattavina on behalf of the LUCIFER collaboration
INFN Milano-Bicocca, Piazza della Scienza 3, 20126 Milano, ITALY
Università degli Studi di Milano-Bicocca, Dipartimento di Fisica "G. Occhialini", Piazza della Scienza 3, 20126 Milano, ITALY
E-mail: luca.pattavina@mib.infn.it

Abstract. One of the fundamental open questions in elementary particle physics is the value of the neutrino mass and its nature of Dirac or Majorana particle. Neutrinoless double beta decay (DBD0ν) is a key tool for investigating these neutrino properties and for finding answers to the open questions concerning mass hierarchy and absolute scale.
Experimental techniques based on the calorimetric approach with cryogenic particle detectors are proved to be suitable for the search of this rare decay, thanks to high energy resolution and large mass of the detectors. One of the main issues to access an increase of the experimental sensitivity is strictly related to background reduction, trying to perform possibly a zero background experiment.
The LUCIFER (Low-background Underground Cryogenic Installation For Elusive Rates) project, funded by the European Research Council, aims at building a background-free DBD0ν experiment, with a discovery potential comparable with the present generation experiments. The idea of LUCIFER is to measure, simultaneously, heat and scintillation light with ZnSe bolometers.
Detector features and operational procedures are reviewed. The expected performances and sensitivity are also discussed.

1. Introduction

Interest in double-beta decay (DBD) has seen a significant renewal in recent years after evidences for neutrino oscillations were obtained from the results of solar, atmospheric, nuclear reactor and accelerator neutrino experiments [1, 2, 3, 4, 5, 6]. These results are an impressive proof that neutrinos have a non-zero mass. However, the experiments studying neutrino oscillations are not sensitive to the nature of the neutrino (Dirac or Majorana) and provide no information on the absolute scale of the neutrino masses, since they are sensitive only to the difference of the masses, Δm^2. The search for DBD0ν is an important experimental technique for understanding the inclusion of the massive neutrino in the Standard Model of Particle Physics.

Double beta decay is a rare nuclear process that can occur in certain even-even nuclei. It is the dominant decay mode when the single beta decay is energetically forbidden. Two-neutrino double beta decay (DBD2ν) is allowed by the Standard Model and has been observed in the laboratory for several isotopes. In the Standard Model, which does not include a neutrino mass term, DBD0ν is a forbidden decay. However, for a massive Majorana neutrino, the decay is allowed.

The double beta decay without emission of neutrinos violates the lepton number by two units, and has never been observed. The observation of this nuclear decay would imply that neutrinos are Majorana particles.

2. Double beta decay

Double beta decay is a second order nuclear weak process and it corresponds to the transition from a nucleus (A, Z) to its isobar $(A, Z+2)$ with the emission of two electrons. The transition may occur via a Standard Model allowed process in which two electron antineutrinos are emitted along with the electrons: $(A, Z) \to (A, Z+2) + 2e^- + 2\overline{\nu}_e$. This decay mode, known as two-neutrino double beta decay, can be thought of as two simultaneous beta decays. The expected rate for DBD2ν decay was first calculated by Goeppert-Mayer in 1935 [7], and the decay has now been observed in various nuclides. Double beta decay half-lives for nuclei that undergo the process are very long, on the order of 10^{19} y, since the decay is second order in the weak interaction. In principle, a nucleus (A, Z) can decay by double beta decay as long as the nucleus $(A, Z+2)$ is lighter. However, if the nucleus can also decay by single beta decay, $(A, Z+1)$, the branching ratio for the DBD will be too difficult to be observed due to the overwhelming background rate from the single beta decay. Therefore, candidate isotopes for detecting double beta decay are even-even nuclei that, due to the nuclear pairing force, are lighter than the odd-odd $(A, Z+1)$ nucleus, making single beta decay kinematically forbidden (see Figure 1).

If neutrinos are Majorana fermions, there is an additional double beta decay mode,

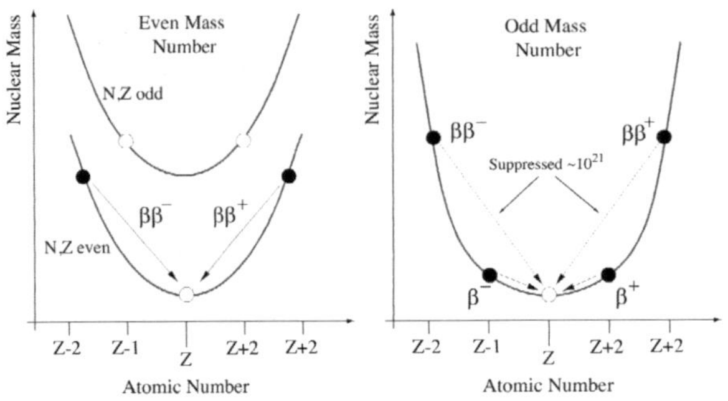

Figure 1. Atomic mass as a function of the atomic number Z in the case of an isobar candidate with A even (left) and A odd (right).

$(A, Z) \to (A, Z+2) + 2e^-$, with no (anti)neutrinos in the final state. This decay mode is known as neutrinoless double beta decay and has never been observed (except for one controversial claim [9]). The first calculations of the rate for DBD0ν, performed by Furry [10], yielded a much faster rate than for DBD2ν decay, which prompted initial interest in experimental detection of the decay. However, at that time the chiral nature of the weak interaction was not yet known so a severe suppression of the DBD0ν rate was not incorporated in the calculations.

Neutrinoless double beta decay is forbidden in the Standard Model since it breaks lepton number conservation. Of course lepton number conservation is broken anyway if neutrinos are Majorana particles. Feynman diagrams for DBD2ν and DBD0ν modes are shown in Figure 2. The DBD2ν diagram contains only Standard Model interactions, while the DBD0ν diagram requires only a massive Majorana neutrino. One can think of the virtual neutrino in the Feynman diagram as an antineutrino (equal to a neutrino since it is Majorana particle) at one vertex which is absorbed as a neutrino at the other vertex. In addition to the Majorana equivalence of neutrino

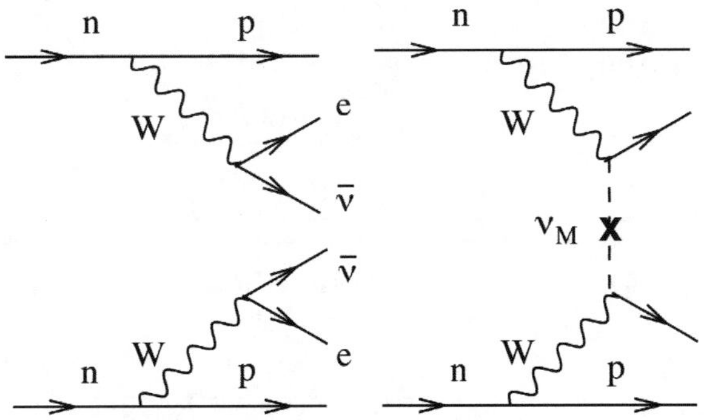

Figure 2. Double beta decay diagrams for the DBD2ν (left) and the DBD0ν (right) decay modes. The DBD0ν diagram assumes that the process is mediated by the exchange of a Majorana neutrino.

and antineutrino, a non-zero neutrino mass is required to flip the helicity since antineutrinos are right-handed and neutrinos are left-handed. The helicity flip and the smallness of the neutrino mass cause the rate of DBD0ν decay to have a smaller rate compare to the DBD2ν.
The DBD0ν rate is given, to a good approximation [11] by:

$$\frac{1}{T_{1/2}^{DBD0\nu}} = G_{0\nu}(Q,Z)\left|M^{0\nu}\right|^2 \langle m_{\beta\beta}\rangle^2 \tag{1}$$

where $G_{0\nu}$(Q,Z) is the phase-space factor including coupling constants, $|M^{0\nu}|$ is the nuclear matrix element of the transition and $\langle m_{\beta\beta}\rangle$ is the effective Majorana mass, defined as:

$$\langle m_{\beta\beta}\rangle \equiv \left|\sum_{i=1}^{3} U_{ei}^2\, m_i\right| \tag{2}$$

where U_{ei} represents the Pontecorvo-Maki-Nakagawa-Sakata mixing matrix that describes the mixing between the neutrino mass eigenstates and the neutrino flavor eigenstates, and m_i are the different mass eigenvalues. The information concerning the neutrino mass is contained in $m_{\beta\beta}$. The phase-space factor $G_{0\nu}$(Q,Z) is calculable but the evaluation of the nuclear matrix element is a challenging problem in nuclear theory, and it is the main source of uncertainty in the evaluation of the neutrino mass.

3. Sensitivity for DBD0ν experiments

It is now analyzed the concept of sensitivity in the case of an experiment with background. Defining the sensitivity as the average upper limit that would be obtained by an ensemble of identical replicas of such an experiment, each one with the same mean expected background and no true signal. Translating in mathematical formulas this concept [12], we define the number of background events, B, can be written as:

$$B = b \cdot \Delta E \cdot M \cdot t, \tag{3}$$

where b is the background rate in counts·keV^{-1}·kg^{-1}·y^{-1}, M is the mass of the $\beta\beta$ emitter in kg, t is the experimental running time in years and ΔE is the width of the region of interest (ROI) in keV and is proportional to the energy resolution of the detector. In this case, we assume that

Journal of Physics: Conference Series **335** (2011) 012047 doi:10.1088/1742-6596/335/1/012047

the background in our ROI is approximately constant over the interval ΔE. In the case that ΔE is very large, this approximation is less valid because events from DBD2ν decay can fall into the interval ΔE.
The number of DBD0ν events ($N_{\beta\beta}$) can be expressed in terms of the decay rate ($\Gamma_{\beta\beta}$) as follows:

$$N_{\beta\beta} = \Gamma_{\beta\beta} \cdot n \cdot \epsilon \cdot t \;\propto\; \Gamma_{\beta\beta} \cdot a \cdot M \cdot \epsilon \cdot t, \tag{4}$$

where n is the total number of nuclide that can DBD0ν decay, a the isotopic abundance of the nuclide candidate and ϵ the detection efficiency for DBD0ν events.
It is convenient to define a general criterion for discovery potential for a DBD0ν experiment in terms of σ of the Poisson distribution. If we require $N_{\beta\beta}$ to be equal to $\sqrt{B}$ (68% C.L.), then we will have:

$$k \cdot \Gamma_{\beta\beta} \cdot a \cdot \epsilon \cdot M \cdot t = \sqrt{b \cdot \Delta E \cdot M \cdot t} \tag{5}$$

k is a constant that includes the Avogadro's number and the atomic mass of the nuclide candidate. The equation 5 results in the following expression for the sensitivity:

$$T_{1/2}^{DBD0\nu} = k \cdot a \cdot \epsilon \cdot \sqrt{\frac{M \cdot t}{b \cdot \Delta E}}. \tag{6}$$

Once the technique for detecting the decay products and the isotopic abundance of the candidates are chosen, the only parameters on which experimentalists can play are: b, M and t. They are related to the neutrino mass value by the fourth root law, therefore big efforts are required to improve the sensitivity. Since the variable *time* will never improve fast enough, the search for the DBD0ν must be carried out increasing the source mass and reducing the background as much as possible.
A perfect tool, for the search of this rare decay, able to scale to big masses (tens of kg) [13, 14] and potentially have an extremely small background are the scintillating bolometer detectors.

4. Scintillating bolometer detectors

Bolometers are calorimeters operating at low temperature, in which the energy deposited by a particle is converted into phonons and is detected as a temperature variation. Almost the total released energy can be detected in this device.
The energy deposited by a single particle into an energy absorber (connected to a heat sink) produces an increase of the temperature T of the absorber. This variation corresponds to the ratio between the energy released by the interacting particle and the heat capacity of the absorber. The only requirements needed for running this type of devices are therefore to operate the detector at low temperatures (mK scales), in order to reduce the value of the heat capacity (large temperature variations), and to have a sensitive thermometer coupled to the absorber. The thermometer is usually a thermistor, a resistive device with a strong dependence of the resistance to the temperature.
When the bolometer crystal is an efficient scintillator at low temperatures, a small but significant fraction of the deposited energy (up to a few percentage) is converted into scintillation while the remaining part is detected in the form of heat. The emitted light can be measured by a light detector facing the scintillating bolometer [15] (see Figure 3). The idea to use a bolometer as light detector was first developed by Bobin [16] and then optimized for Dark Matter (DM) searches. Now, it is evident that the best way to detect the scintillating light is to developed an independent detector opaque to the emitted light and equiped with its own thermal sensor. The size and the relative position to the scintillating bolometer must be chosen in a way to maximize the light collection.
The idea to use these detectors for DBD0ν purposes was born in 1998 [17].

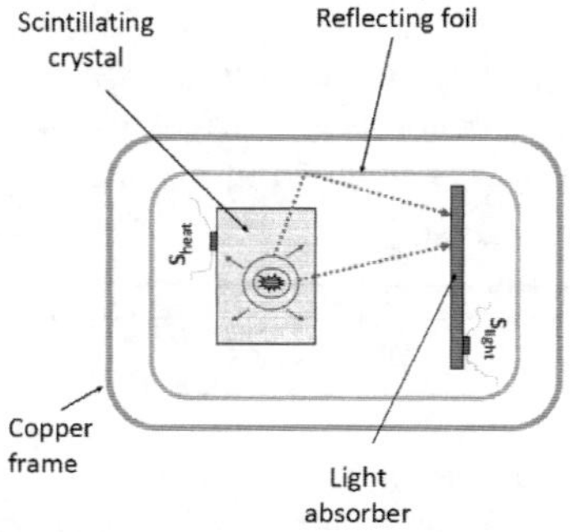

Figure 3. Pictorial view of a scintillating bolometer.

5. Background sources

The experimental signature of the DBD0ν is in principle very clear: a peak (at the $Q_{\beta\beta}$ value) in the two-electron summed energy spectrum. In spite of such characteristics, the rarity of the process makes the identification very difficult. Such a signal has to be disentangled from a background produced by natural radioactive decay chains, cosmogenic-induced activity, and artificial radioactivity. All of these background sources can release energy in the region of interest for the DBD, unfortunately at a higher rate. Thus, the main task in DBD0ν searches is to suppress the natural background using state-of-the-art ultralow background techniques and, hopefully, identifying the signal.

Among the various background sources (external γ background, neutrons, surface contaminations and ^{238}U and ^{232}Th internal contaminations) surface contaminations represent the main source of background for bolometric detectors, such as CUORICINO [18]. This source of background plays a role in almost all detectors but turns out to be crucial for fully active detectors, as in the case of bolometers: a radioactive nucleus located within a few μm of a surface facing the detector can emit an α particle (whose energy is roughly between 4 and 9 MeV). Those particles can lose part, or even all, of their energy in the few microns of dead layer before reaching the bolometer. The energy spectra read by the bolometer, therefore, will result in a continuum between 0 and 9 MeV, covering, unfortunately all the possible $Q_{\beta\beta}$ values (see Figure 4). Furthermore, the same mechanism holds in the case of surface contaminations on the bolometer itself. The use of scintillating crystals as absorbers can overcome these background limitations,

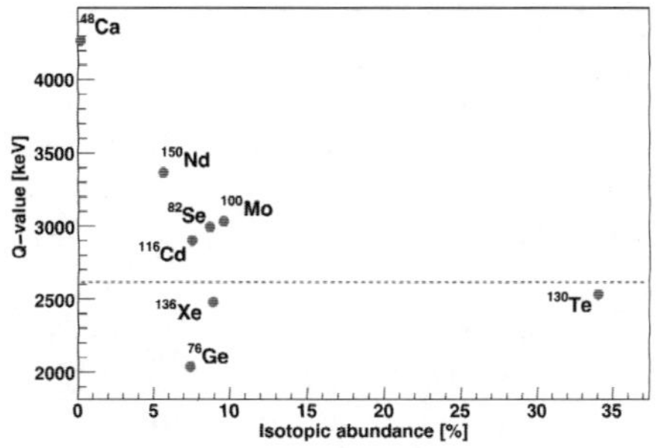

Figure 4. Natural abundances and Q-value for several DBD0ν isotopes. The dashed lines represents the 2615 keV γ line, one of the highest natural γ emission.

Table 1. Sensitivity of three experimental options for a scintillating bolometer experiment on neutrinoless double beta decay (isotope candidate enriched at 98% level). The value $m_{\beta\beta}$ is the 1σ sensitivity calculated with the Feldman Cousin approach for 5 years running, considering a background level of 0.001 counts·keV^{-1}·kg^{-1}·y^{-1}, an energy resolution of 5 keV and the matrix elements described in [22, 23, 24, 25].

Crystal	Mass	Useful material	Sensitivity to $m_{\beta\beta}$ [meV]
$CdWO_4$	^{116}**Cd** 15.1 kg	32 %	65 –80
$ZnMoO_4$	^{110}**Mo** 11.3 kg	44 %	67 –73
ZnSe	^{82}**Se** 17.6 kg	56 %	52 –65

providing a powerful tool to improve the experimental DBD0ν sensitivities. In fact, by means of the various scintillation yield of various particle interactions (namely β/γ, α and neutrons) they can be very different.

6. The LUCIFER experiment

LUCIFER (Low-background Underground Cryogenic Installation For Elusive Rates) is a Double Beta Decay pilot project based on scintillating bolometers and it is funded by the European Community.

6.1. Isotope candidate

Practical scintillators for DBD searches can be built out Mo, Cd and Se [19, 20]. These three nuclides contain isotopes that can potentially decay through DBD0ν channel: ^{100}Mo, ^{116}Cd and ^{82}Se, with a $Q_{\beta\beta}$ value higher than the characteristic gamma emission of ^{208}Tl (2.6 MeV) (see Figure 4), which is one of the most energetic natural gamma line. This feature implies that the gamma background will be significantly reduced.
If we still look at Figure 4, the three mentioned candidates have a small isotopic abundance, compare to ^{130}Te, so if these isotopes are used for a DBD0ν experiment they will require a large mass (difficult to handle) or an isotopic enrichment.
The candidates crystals are $CdWO_4$, $ZnMoO_4$, ZnSe. In Table 1, we compare these nuclei with respect to physics sensitivity for a given detector mass, relative amount of DBD emitter in the crystal (48 crystals of 125 cm^3), background rejection factor (Quenching Factor [21]). The Quenching Factor is usually defined as the ratio between the amplitude of light pulses corresponding to alpha and β/γ particles of the same energies.

Looking at Table 1 the ZnSe choice seems to be the best for what concerns the mass source and the useful material. But if we consider the quenching factor (Q.F.) values for these scintillating crystals (Table 2), ZnSe has a non-standard value: it is larger than 1. This implies that alpha particles scintillate more compare to the β/γ in the crystal (see Figure 5). This behavior is supposed to be caused by the energy distribution in radiative (detectable) and non-radiative or long-living metastable (non-detectable) channels.

Discrimination problems may arise in these crystals, in fact if there is an inefficient light collection, we can have a leak of alpha events in the β/γ region, producing a background in our region of interest. Anyway it should be stressed that the Q.F. for ZnSe crystals is quite large, and the particle discrimination is still efficient (see Figure 6). In the specific, preliminary studies reveal the capability of an α power rejection larger than 99% [26].

Table 2. Alpha quenching factor for three different scintillating crystals [19, 29]

Crystal	Q.F. (α/β)
$CdWO_4$	0.19
$ZnMoO_4$	0.16
ZnSe	4.2

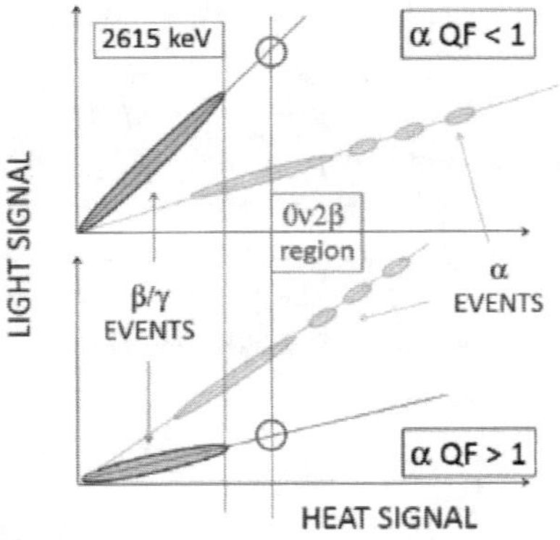

Figure 5. Scatter plot for heat-light events, for a scintillating crystal with Q.F. smaller (top) and greater (bottom) than 1.

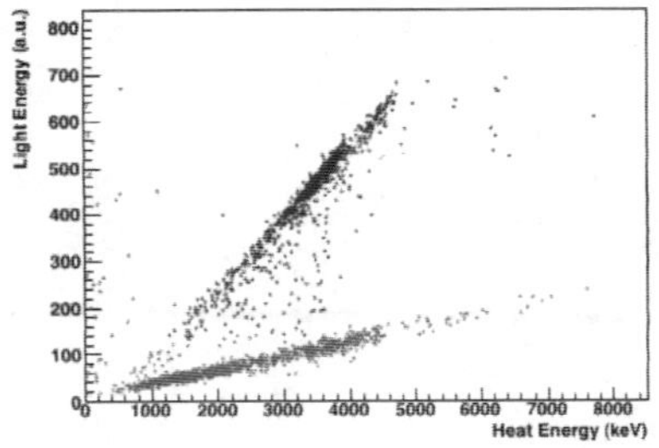

Figure 6. Comparison between α events from ^{234}U and ^{238}U sources (in blue) and γ events generated by neutrons from AmBe source (in red).

6.2. The detector structure

As every large mass bolometric detector, LUCIFER will consist of an array of tens of individual bolometers arranged in a close-packed configuration and housed by the same cryogenic infrastructure. This approach allows to achieve large masses ($\sim$ 50 kg) keeping the single crystal mass reasonable (hundreds of grams) and providing a rough granularity, very helpful for background identification and rejection. The array will be composed by a set of identical elementary modules (see Figure 7).

The proposed structure of each single module exploits the experience gathered during the development of past and present DBD bolometric experiments. The configuration resembles

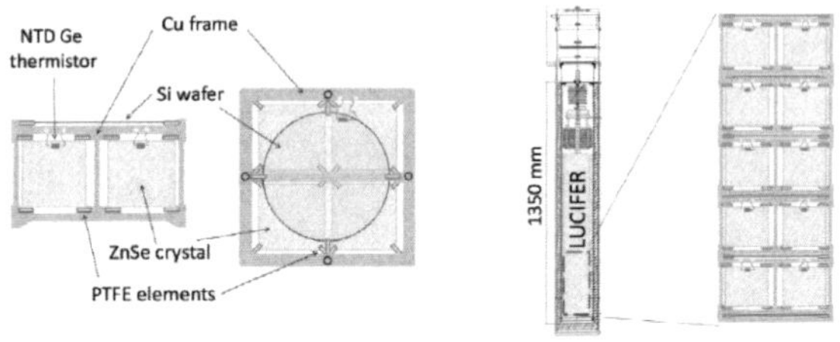

Figure 7. Elementary module equipped with four ZnSe scintillating crystals read out by a single light detector (left), and the whole LUCIFER detector inside the cryostat (right).

closely the one selected and extensively tested for CUORE [27], with an additional light detector, designed and developed during the scintillating-bolometer R&D [28]. 12 single modules will be assembled one above the other. The total tower height will be about 85 cm, exploiting the available space in the CUORICINO cryostat in the present internal shielding arrangement. The total bolometric mass will be 31.7 kg, and the corresponding mass of the enriched material 17.6 kg.

7. Conclusions

The project LUCIFER aims to unveil the nature of the neutrino mass. A positive result of the experiment would give an answer to one of the fundamental open questions in Nature, that neutrino is a Majorana particle. Beside this fundamental assessment, it would constitute a proof of existence of New Physics and it would give a scale to the absolute neutrino mass.

This project will have the merit of indicating the way to go for next generation experiments, by setting a limit (or hopefully a value) on the DBD0ν half-life, using such a detector mass 30 kg w. The technology demonstrated by this project is indeed fully scalable.

8. Aknowledgments

The project LUCIFER is funded from the European Research Council under the European Unions Seventh Framework Program (FP7/2007-2013) / ERC grant.

References

[1] SNO Collaboration, Ahmad Q R et al. 2002 *Phys. Rev. Lett.* **89**
[2] SNO Collaboration, Ahmad Q R et al. 2004 *Phys. Rev. Lett.* **92**
[3] Super-Kamiokande Collaboration, Ashie Y et al. 2005 *Phys. Rev. D* **71**
[4] KamLAND Collaboration, Abe S. et al. 2008 *Phys. Rev. Lett.* **100**
[5] K2K Collaboration, Ahn M H et al. 2006 *Phys. Rev. D* **74**
[6] MINOS Collaboration, Adamson P et al. 2008 *Phys. Rev. Lett.* **101**
[7] Goeppert-Mayer M 1935 *Phys. Rev.* **48**, 512
[8] Cremonesi O 2010 *Journal of Physics: Conference Series* **202**
[9] Klapdor-Kleingrothaus H V et al. 2001 *Mod. Phys. Lett. A* **16** 2049
[10] Furry W H 1939 *Phys. Rev.* **56**
[11] Avignone III F T, Elliott S R, and Engel J 2008 *Rev. Mod. Phys.* **80** 481
[12] Avignone III F T , King III G S, and Zdesenko Yu G 2005 *New J. Phys.* **7** 6
[13] Angloher G et al. and CRESST Collaboration 2009 *Astropart. Phys.* **31** 270
[14] Andreotti E et al. 2010 130Te Neutrinoless Double-Beta Decay with CUORICINO (*Arxiv preprint* arXiv:1012.3266v1)
[15] Gonzalez-Mestre L and Perret-Gallix D 1989 *Nucl. Instrum. Meth. A* **279** 382
[16] Bobin C et al. 1994 *Nucl. Instr. Meth. A* **386** 453
[17] Alessandrello A. et al. 1998 *Phys. Lett. B* **420** 109
[18] Arnaboldi C et al. 2005 *Phys. Rev. Lett.* **95** 142501

[19] Pirro S et al. 2006 *Phys. Atom. Nucl.* **69** 2109
[20] Gironi L 2010 *Nucl. Instrum. Meth. A* **617** 478-481
[21] Tretyak V I 2010 *Astropart. Phys.* **33** 40
[22] Menendez J et al. 2009 *NPA* **818** 139
[23] Civitarese O et al. 2009 *JoP:Conference series* **173** 012012
[24] Simkovic F et al. 2008 *Phys. Rev. C* **77** 045503
[25] Barea J and Iachello F *Phys. Rev. C* **79** 044301
[26] Arnaboldi C et al. 2011 *Astropart. Phys.* **34** 344-353
[27] Fiorini E 1998 CUORE: a cryogenic underground observatory for rare events *Physics Reports*
[28] Pirro S et al. 2006 *Nucl. Instrum. Meth. A* **559** 361-363
[29] Gironi L et al., 2009 CdWO4 bolometers for Double Beta Decay search (arXiv:0809.5126)

Journal of Physics: Conference Series **335** (2011) 012048 doi:10.1088/1742-6596/335/1/012048

Lepton Number Violation in TeV Scale See-Saw Extensions of the Standard Model

A Ibarra[1], E Molinaro[2,5] and S T Petcov[3,4,6]

[1] Physik-Department T30d, Technische Universität München, James-Franck-Straße, 85748 Garching, Germany

[2] Centro de Física Teórica de Partículas (CFTP), Instituto Superior Técnico, Avenida Rovisco Pais 1, 1049-001, Lisboa, Portugal

[3] SISSA and INFN-Sezione di Trieste, Via Bonomea 265, 34136 Trieste, Italy

[4] Institute for the Physics and Mathematics of the Universe, University of Tokyo, Kashiwa, 277-8583, Japan

Abstract.
The low-energy neutrino physics constraints on the TeV scale type I see-saw scenarios of neutrino mass generation are revisited. It is shown that lepton charge (L) violation, associated to the production and decays of heavy Majorana neutrinos N_j having masses in the range of $M_j \sim (100 \div 1000)$ GeV and present in such scenarios, is hardly to be observed at ongoing and future particle accelerator experiments, LHC included, because of very strong constraints on the parameters and couplings responsible for the corresponding $|\Delta L| = 2$ processes. If the heavy Majorana neutrinos N_j are observed and they are associated only with the type I mechanism, they will behave effectively like pseudo-Dirac fermions. Conversely, the observation of effects proving the Majorana nature of N_j would imply that these heavy neutrinos have additional relatively strong couplings to the Standard Model particles or that light neutrino masses compatible with the observations are generated by a mechanism other than see-saw (e.g., radiatively at one or two loop level) in which the heavy Majorana neutrinos N_j are nevertheless involved.

1. Introduction

Neutrino oscillation experiments [1] have revealed undeniable evidence for flavour non-conservation in the lepton sector. Flavour neutrino eigenstates, $\nu_{\ell L}$ ($\ell = e, \mu, \tau$), oscillate, changing their lepton flavour quantum number [2, 3], due to non-zero neutrino masses and neutrino mixing. The latter is parametrized by the well known Pontecorvo-Maki-Nakagawa-Sakata (PMNS) neutrino mixing matrix [4].

All present neutrino oscillation data can be described assuming 3-flavour neutrino mixing in vacuum. The number of massive neutrinos can, in general, be bigger than 3, if, for instance, there exist right-handed (RH) sterile neutrinos [5] and they mix with the left-handed (LH) flavour neutrinos. It follows from the existing data that at least 3 of the neutrinos must be light, with masses $m_1 \neq m_2 \neq m_3$, $m_{1,2,3} \lesssim 1$ eV and at least two of the m_j different from zero. At present there are no compelling experimental evidences for the existence of more than 3 light neutrinos.

[5] Talk given by the author at DISCRETE 2010, 6 December 2010, Rome, Italy.

[6] Also at: Institute of Nuclear Research and Nuclear Energy, Bulgarian Academy of Sciences 1784 Sofia, Bulgaria.

The smallness of neutrino masses, $m_j/m_e \lesssim 10^{-6}$, where m_e is the electron mass, suggests the presence of a new fundamental scale in particle physics and thus to new physics beyond the Standard Model (SM). Among the possible extensions of the SM which can explain neutrino mass generation, a well known proposal is the see-saw mechanism [6]. In its simplest formulation, the type I see-saw scenario, at least two "heavy" $SU(2)_L \times U(1)_Y$ singlet RH neutrinos ν_{aR} are introduced in the theory. Within the see-saw framework, the latter are assumed to possess a Majorana mass term as well as Yukawa type couplings with the Standard Model lepton and Higgs doublets.

Below the electroweak (EW) symmetry breaking scale, the light neutrino masses arise from the Lagrangian:

$$\mathcal{L}_\nu = -\overline{\nu_{\ell L}}\,(M_D)_{\ell a}\,\nu_{aR} - \frac{1}{2}\,\overline{\nu^C_{aL}}\,(M_N)_{ab}\,\nu_{bR} + \text{h.c.}\,, \tag{1}$$

where $\nu^C_{aL} \equiv C\overline{\nu_{aR}}^T$ $(a = 1,\ldots,k)$, C being the charge conjugation matrix, $M_N = (M_N)^T$ is the $k \times k$ Majorana mass matrix of the RH neutrinos, and M_D is the $3 \times k$ neutrino Dirac mass matrix, which is generated by the matrix of neutrino Yukawa couplings after the EW symmetry breaking: $M^D = v\lambda$, $\lambda_{\ell a}$ being the matrix of neutrino Yukawa couplings and $v = 174$ GeV being the v.e.v. of the SM Higgs doublet. The matrices M_N and M_D are complex, in general. A Majorana mass term m_ν for the active left-handed neutrinos is indeed generated by the interplay between the Dirac mass term and the Majorana mass term of the heavy Majorana neutrinos. The well know see-saw mass relation is in this case: $m_\nu \cong -M_D M_N^{-1}(M_D)^T$.

In grand unified theories, M_D is typically of the order of the charged fermion masses. In $SO(10)$ theories, for instance, M_D coincides with the up-quark mass matrix. Taking indicatively $m_\nu \sim 0.05$ eV, $M_D \sim 100$ GeV, one finds $M_N \sim 2 \times 10^{14}$ GeV, which is close to the scale of unification of the electroweak and strong interactions, $M_{GUT} \cong 2 \times 10^{16}$ GeV. In GUT theories with RH neutrinos one finds that indeed the heavy Majorana neutrinos naturally obtain masses which are by few to several orders of magnitude smaller than M_{GUT}. The estimate of M_N thus obtained is effectively based on the assumption that the neutrino Yukawa couplings are large: $|\lambda_{\ell a}| \sim 1$. The alternative possibility is to have heavy Majorana neutrino masses in the range of $\sim (100 \div 1000)$ GeV, i.e. TeV scale see-saw generation of neutrino masses. This possibility has been recently revisited by several authors (see e.g. [7] and references quoted therein).

Following the discussion reported in [7], the low-energy neutrino physics constraints on the TeV scale type I see-saw scenarios are reviewed. In particular, it is analyzed in detail, within three different frameworks relying on the type I see-saw extension of the SM, the possibility of observing $|\Delta L| = 2$ processes at current and future particle accelerators, LHC included, due to the production and decays of the heavy Majorana neutrinos.

2. Non-unitarity effects, see-saw mechanism and $(\beta\beta)_{0\nu}$-decay

Following the formalism developed in [7], after the diagonalization of the full neutrino mass matrix in (1), the charged current (CC) and neutral current (NC) weak interactions involving the light Majorana neutrinos χ_j with definite mass m_j are

$$\mathcal{L}^\nu_{CC} = -\frac{g}{\sqrt{2}}\,\bar{\ell}\,\gamma_\alpha\,\nu_{\ell L}\,W^\alpha + \text{h.c.} = -\frac{g}{\sqrt{2}}\,\bar{\ell}\,\gamma_\alpha\,((1+\eta)U)_{\ell i}\,\chi_{iL}\,W^\alpha + \text{h.c.}\,, \tag{2}$$

$$\mathcal{L}^\nu_{NC} = -\frac{g}{2c_w}\,\overline{\nu_{\ell L}}\,\gamma_\alpha\,\nu_{\ell L}\,Z^\alpha = -\frac{g}{2c_w}\,\overline{\chi_{iL}}\,\gamma_\alpha\,\left(U^\dagger(1+\eta+\eta^\dagger)U\right)_{ij}\,\chi_{jL}\,Z^\alpha\,. \tag{3}$$

The unitary matrix U diagonalizes the light Majorana mass matrix, $m_\nu = U^*\text{diag}(m_1, m_2, m_3)U^\dagger$. In the basis in which the three charged lepton fields are equal to the respective mass eigenstates, U corresponds to the PMNS neutrino mixing matrix. The hermitian matrix η, therefore, parametrizes the deviation from unitarity of the neutrino mixing matrix due to the mixing of

Journal of Physics: Conference Series **335** (2011) 012048 doi:10.1088/1742-6596/335/1/012048

the flavour neutrino fields $\nu_{\ell L}$ with the heavy RH fields in the neutrino mass Lagrangian (1). As shown below, η depends on a combination of the see-saw parameters responsible for lepton flavour violating processes. Notice that in deriving the expressions of the CC and NC weak interactions it is implicitly assumed that $\eta_{ij} \ll 1$. Indeed, for SM singlet fermions with masses above the EW symmetry breaking scale, i.e. bigger than $\sim$ 100 GeV, the combined data on neutrino oscillation experiments and EW interaction processes provide the upper bound [8, 9]: $|\eta_{ij}| < 6 \times 10^{-3}$.

The charged current and the neutral current interactions of the heavy Majorana neutrino mass eigenstates N_j with $W^{\pm}$ and Z^0 formally read [7]:

$$\mathcal{L}_{CC}^{N} = -\frac{g}{2\sqrt{2}}\,\bar{\ell}\,\gamma_\alpha\,(RV)_{\ell k}(1-\gamma_5)\,N_k\,W^\alpha \,+\, \text{h.c.}\,, \tag{4}$$

$$\mathcal{L}_{NC}^{N} = -\frac{g}{2c_w}\,\overline{\nu_{\ell L}}\,\gamma_\alpha\,(RV)_{\ell k}\,N_{kL}\,Z^\alpha \,+\, \text{h.c.}\,, \tag{5}$$

where the unitary matrix V diagonalizes the RH neutrino mass term in (1), $M_N \simeq V^*\text{diag}(M_1,\ldots,M_k)V^\dagger$, and $R \equiv (M_D M_N^{-1})^*$. The experimental limits on the deviations from unitarity in the neutrino mixing matrix affect the couplings $(RV)_{\ell j}$. Indeed, in this class of models [7]: $\eta = -\frac{1}{2}RR^\dagger$.

It is important to remark that, apart from non-unitarity effects, further constraints on $(RV)_{\ell j}$ and/or a particular combination of them can be obtained at low energy considering $i)$ the scale of the symmetric matrix $m_\nu = -M_D M_N^{-1}(M_D)^T \simeq -R^* M_N R^\dagger \lesssim 1$ eV, resulting from the see-saw mechanism of generation of neutrino masses, and $ii)$ from the experimental upper bound on the effective Majorana mass, $|(m_\nu)_{ee}|$ [10], which enters in the expression of the $(\beta\beta)_{0\nu}$-decay rate of even-even nuclei (see e.g. [11, 12]). Indeed, for the heavy Majorana neutrinos N_k with masses $M_k \sim M_R \geq 100$ GeV, barring "accidental" cancellations or extreme fine-tuning (at the level of $\sim 10^9$), the constraint $i)$ implies:

$$|(RV)_{\ell j}| \lesssim 3 \times 10^{-6}\left(\frac{100\text{ GeV}}{M_R}\right)^{1/2}. \tag{6}$$

The estimate of $|(RV)_{\ell j}|$ thus obtained makes the heavy Majorana neutrinos N_j practically unobservable at particle accelerators (see e.g. [13]). In order for the CC and NC couplings of the heavy Majorana neutrinos N_j to $W^{\pm}$ and Z^0, eqs (4) and (5), to be sufficiently large, that is $|(RV)_{\ell j}| \approx 10^{-2}$, so that the see-saw mechanism could be partially or completely tested in experiments at the currently operating and planned future accelerators (LHC included), the small size of the m_ν matrix elements should be due to strong mutual compensation between the parameters that enter in the see-saw mass relation. In general, such cancellations arise naturally from symmetries in the lepton sector, corresponding, for instance, to the conservation of some additive lepton charge $\hat{L}$ (see, e.g. [11, 14, 15, 16]). In the limit in which $\hat{L}$ is exact, the RH neutrinos ν_{aR} $(a = 1,\ldots,k)$ should be Dirac particles, which is possible for all heavy neutrinos only if the number of the RH singlet neutrino fields is even: $k = 2q$, $q = 1, 2, \ldots$. If their number is odd, barring again "accidental" cancellations, some (odd number) of the discussed heavy Majorana neutrinos will have strongly suppressed couplings to the $W^{\pm}$ and charged leptons and will be practically unobservable in the current and the future planned accelerator experiments.

The resulting mass spectrum will depend on the assumed total lepton charge $\hat{L}$. Indeed, if L_i $(i = e, \mu, \tau, \ldots, k)$ are the ordinary lepton charges of the fermion fields, in the basis in which the charged lepton mass matrix is diagonal, the conserved lepton charge $\hat{L}$ takes the general form:

$$\hat{L} = \sum_{i=e,\mu,\tau,\ldots,k} (-1)^{n_i}\,a_i\,L_i\,, \tag{7}$$

where $n_i = 0, 1$ and $a_i = 0, 1$ for $i = e, \mu, \tau, \dots, k$, with at least one $a_i \neq 0$. In this case, it can be shown [14] that the number of massive Dirac neutrinos and massless neutrinos in the theory is, respectively, min $\left(n_+(\hat{L}), n_-(\hat{L})\right)$ and $\left|n_+(\hat{L}) - n_-(\hat{L})\right|$, where $n_+(\hat{L})$ ($n_-(\hat{L})$) is the number of ordinary lepton charges in (7) which enter with the plus (minus) sign.

The correct light Majorana neutrino mass spectrum can be generated by small perturbations that violate the corresponding symmetry $\hat{L}$. In this case, each massive Dirac neutrino is split into two Majorana neutrinos with close but different masses, i.e. the Dirac neutrinos become pseudo-Dirac fermions [17, 18]. Typically, for $|(RV)_{\ell j}| \sim 10^{-3}$ (10^{-4}), the expected splitting between the masses of the two heavy Majorana neutrinos forming a pseudo-Dirac pair, is roughly 1 (100) MeV for masses of the order of 100 (1000) GeV. Such small mass difference will not be observable at LHC and planned future accelerator experiments, thus preventing any possible experimental test at collider of the Majorana nature of the heavy Majorana neutrinos of the type I see-saw mechanism.

The other low energy constraint to the see-saw parameter space, strictly related to the Majorana nature of the heavy and light Majorana neutrinos, is provided by $(\beta\beta)_{0\nu}$-decay experiments [10]. In this case, in addition to the standard contribution due to the light Majorana neutrino exchange, the $(\beta\beta)_{0\nu}$-decay effective Majorana mass $|(m_\nu)_{ee}|$ receives a contribution from the exchange of the heavy Majorana neutrinos N_k:

$$|(m_\nu)_{ee}| \cong \left|\sum_i (U_{\rm PMNS})^2_{ei}\, m_i - \sum_k F(A, M_k)\, (RV)^2_{ek}\, M_k\right| , \qquad (8)$$

where $F(A, M_k)$ is a known real (positive) function of the atomic number A of the decaying nucleus and of the mass M_k of N_k [19, 20, 21]. For the range $M_k \sim (100 \div 1000)$ GeV of interest, one has with good approximation (see, e.g. [20, 21, 22, 23]) $F(A, M_k) \simeq (M_a/M_k)^2 f(A)$, with $M_a \simeq 0.9$ GeV and $f(A) \simeq \mathcal{O}(10^{-2})$.

3. The case of a broken lepton charge $\hat{L}$

It is shown in the following that, within the type I see-saw scenario of generation of light neutrino masses, if $m_\nu \neq 0$ arises as result of the breaking of a global symmetry corresponding to some conserved lepton charge $\hat{L}$, the Majorana nature of the heavy Majorana neutrinos could be hardly tested at collider physics. As stated before, the reason is that if the heavy Majorana fields involved in the see-saw mechanism of LH neutrino mass generation, have couplings to charged leptons and weak gauge bosons sufficiently large to produce and observe them at collider, they would behave effectively as a pseudo-Dirac fermion and all lepton number violating processes, e.g. same sign di-muon production in proton-proton collisions, would be strongly suppressed.

Consider for simplicity the case of three RH neutrinos ν_{aR} ($a = 1, 2, 3$) and a conserved lepton charged $\hat{L} = L_e + L_\mu + L_\tau + L_3 - L_1$ (see e.g. [7]). A possible choice of the Dirac and Majorana mass matrix which preserves $\hat{L}$ is:

$$M_D = \begin{pmatrix} 0 & 0 & m^D_{e3} \\ 0 & 0 & m^D_{\mu 3} \\ 0 & 0 & m^D_{\tau 3} \end{pmatrix} , \qquad M_N = \begin{pmatrix} M_{11} & 0 & 0 \\ 0 & 0 & M_{23} \\ 0 & M_{23} & 0 \end{pmatrix} , \qquad (9)$$

where, without loss of generality, all the parameters in M_D and M_N are taken real and positive. As a consequence of the simplifying choice made, $m^D_{\ell 1} = 0$, $l = e, \mu, \tau$, the Majorana field $N_1 = (\nu_{1R} + \nu^C_{1L})/\sqrt{2}$ is decoupled and does not take part in the see-saw mechanism of generation of LH neutrino masses. In this case $n_+(\hat{L}) = 4$ and $n_-(\hat{L}) = 1$ and the spectrum of the theory contains 3 massless neutrinos and 1 massive Dirac fermion. The latter is given by the combination $N_D = (N_2+N_3)/\sqrt{2}$ and has a mass $M = M_{23} > 0$, where the two heavy Majorana neutrinos N_2 and N_3 have the same mass $M_2 = M_3 = M_{23}$ and satisfy the Majorana conditions $C\overline{N_k}^T = \rho_k N_k$, $k = 2, 3$, with $\rho_2 = -1$ and $\rho_3 = +1$.

In order to have a light neutrino mass spectrum compatible with the observations, it is necessary to softly break $\hat{L}$. Taking $(M_D)_{\ell 2} \equiv m^D_{\ell 2} \neq 0$, it is easy to show [7] that two of the light neutrinos acquire a finite mass, $0 < m_2 < m_3$, while the Dirac fermion N_D is split into two different Majorana fields N_2 and N_3 with$\Delta M \equiv M_3 - M_2 = m_3 - m_2 = 2A/M_{23}$, where A depends explicitly on the lepton charge breaking parameters $m^D_{\ell 2}$: $A = m^D_{e2}m^D_{e3} + m^D_{\mu 2}m^D_{\mu 3} + m^D_{\tau 2}m^D_{\tau 3}$. The light neutrino mass scale, $m_3 - m_2 \lesssim 1$ eV, imposes strong constraints on the symmetry breaking parameters. More explicitly, for $M_{23} \approx 100$ GeV one has: $|m^D_{\ell 2}m^D_{\ell' 3}| \lesssim 10^{-7}\text{GeV}^2$ $(\ell, \ell' = e, \mu, \tau)$. Because of the small mass splittingΔM, the heavy Majorana neutrinos $N_{2,3}$ effectively behave in the collider physics as a pseudo-Dirac fermion $N_{PD} = (N_2 + N_3)/\sqrt{2}$.

Consider now the process of same sign di-muon production in proton-proton collisions, assuming that one of the muons, say μ^-, is produced together with real or virtual $N_{2,3}$ in the decay of a virtual W^-, while the second μ^- originates from the decay $N_{2,3} \to W^+\mu^-$, with virtual or real W^+ which decays further into, e.g. two hadronic jets. In order to collect a sufficiently large number of events, it is necessary to have sizable couplings of the heavy Majorana neutrinos to the muon field. The latter can be easily computed [7]: $(RV)_{\mu 2} = m^D_{\mu 3}\cos\theta/M_{23}$ and $(RV)_{\mu 3} = m^D_{\mu 3}\sin\theta/M_{23}$, with $\tan^2\theta = M_2/M_3$. In the present scenario $\theta \simeq \pi/4$ because N_2 and N_3 form a pseudo-Dirac fermion. If, for instance, $M_{23} \approx 100$ GeV and $|(RV)|_{\mu 2,3} \approx 10^{-2}$, one obtains $m^D_{\mu 3} \approx 1$ GeV and $\hat{L}$ charge breaking parameter $m^D_{\mu 2} \approx 100$ eV, due to the constraint which arises from the small neutrino masse scale (see the discussion above). The relevant part for the amplitude of the process under discussion is [7]:

$$A(p\,p\,\to\,\mu^-\,\mu^-\,2\text{jets}\,X) \propto \frac{(m^D_{\mu 3})^2}{M^2_{23}}\,\frac{M_2M_3}{M_3+M_2}\,\frac{M_3^2 - M_2^2 - i(\Gamma_3M_3 - \Gamma_2M_2)}{(p^2 - M_3^2 + i\Gamma_3M_3)(p^2 - M_2^2 + i\Gamma_2M_2)}\,. \quad (10)$$

Taking into account that $M_2M_3 \simeq M^2_{23}$ and$\Gamma_{2(3)} \propto \sum_\ell |(RV)_{\ell 2,3}|^2 G_F\, M^3_{2(3)}$, it is easy to show that the amplitude for this lepton number violating process is proportional to the factor $\Delta M/M_{23}$. As discussed previously, in the pseudo-Dirac scenario of interest, which is consequence of an almost conserved lepton charge present in the theory, the heavy Majorana neutrino mass splitting is strongly bounded by the light neutrino mass scale,$\Delta M \lesssim 1$ eV, thus making the same sign di-muon production at LHC unobservable.

4. $\hat{L}$ is not conserved but $m_\nu = 0$ at leading order

A simple realization of this scenario is obtained from the mass textures reported in eq. (9), in which the Majorana mass matrix, M_N, is conveniently modified with the $\hat{L}$-breaking mass term: $(M_N)_{33} = M_{33} > 0$. In this case, even if lepton number is not conserved, at leading order one has $m_\nu = M_D\,M_N^{-1}\,M_D^T = 0$ and, therefore, the tree-level constraints on the see-saw parameter space from the light neutrino mass scale do not apply here: light neutrino masses, compatible with neutrino data, can be generated at higher order (one or two loops). This in turn could lead to sufficiently large $N_{2,3}$ production rates at LHC to make the observation of the two heavy Majorana neutrinos possible. Indeed, in such scenario$\Delta M = M_3 - M_2 = M_{33}$ and, in principle, one can observe the same sign di-muon production at LHC for a sufficiently large mass splitting (see eq. (10)). Anyway, it should be clear from the preceding discussion that all $|\Delta L| = 2$ Majorana type effects should vanish in the limit of $M_{33} = 0$.

As discussed before, in such class of models, the heavy Majorana neutrinos might provide a sizable contribution to the $(\beta\beta)_{0\nu}$-decay rate. Indeed, from eq. (8), neglecting the standard contribution due to the tree-level exchange of light Majorana neutrino mass eigenstates, the effective Majorana mass $|(m_\nu)_{ee}|$ results proportional to the mass splitting$\Delta\ M$ [7]:

$$|(m_\nu)_{ee}| \;\cong\; \left|\frac{(m^D_{e3})^2}{M^2_{23}}\, f(A)\, M^2_a\, \frac{M_3 - M_2}{M_2\, M_3}\right| \cong \left|\frac{(m^D_{e3})^2}{M^2_{23}}\, f(A)\, \frac{M^2_a}{M^2_{23}}\, M_{33}\right| . \tag{11}$$

If $(m^D_{e3})^2/M^2_{23}$ will be determined from an independent measurement, the bound on $|(m_\nu)_{ee}|$ will lead to a strong constraint on M_{33}/M_{23}. Taking e.g. $|(m_\nu)_{ee}| \lesssim 1$ eV, $f(A) = 0.078$ (corresponding to ^{76}Ge) and the maximal value of $(m^D_{e3})^2/M^2_{23}$ allowed by the non-unitarity constraints (see [8]), 8×10^{-3}, one finds: $M_{33} \lesssim 1.8 \times 10^{-5}\, M_{23}(M_{23}/M_a)$. For $M_{23} = 100$ GeV this implies $M_{33} \lesssim 2 \times 10^{-3} M_{23} \cong 0.2$ GeV $\ll M_{23}$. Such a small mass difference would render the Majorana-type effects associated with $N_{2,3}$ hardly observable. If, however, $|(m^D_{e3}/M_{23})^2| \lesssim 1.6 \times 10^{-6}$, $M_{33} \lesssim M_{23}$ and a signature of the process $pp \to \mu^- \mu^-\, 2$ jets$\,X$, generated by the production and decay of real or virtual $N_{2,3}$, is detectable at LHC.

5. The extreme fine-tuning case

Consider now the type I see-saw Lagrangian in which the RH neutrino masses are given between $(100 \div 1000)$ GeV and the see-saw mechanism provides the observed LH neutrino mass scale and mixing without assuming any softly broken conserved lepton charge in the theory. This possibility requires, in general, a huge tuning of parameters. Namely, demanding $(M_D)_{ij} \sim \mathcal{O}(1\,\text{GeV})$ and $M_j \sim \mathcal{O}(100\,\text{GeV})$ requires a tuning of one part in 10^9 in order to produce a neutrino mass $m_i \sim \mathcal{O}(10^{-2}\,\text{eV})$ [7].

Unlike the case discussed earlier, now the see-saw parameter space is constrained by both the small light neutrino mass scale, $m_\nu \lesssim 1$ eV, and the upper bound on the effective Majorana mass in $(\beta\beta)_{0\nu}$-decay experiments. Consider, for instance, the two heavy Majorana neutrinos mass eigenstates N_2 and N_3 with masses $M_2 > 0$ and $M_3 \equiv M_2(1+z)$, $z > 0$, and satisfying standard Majorana conditions, $C\overline{N}^T_{2,3} = N_{2,3}$. For large neutrino Yukawa couplings, the CC and NC weak interactions of N_2 and N_3 with charged leptons and gauge bosons, eqs (4) and (5), satisfy, with a good approximation, the relation [7]: $(RV)_{\ell 3} \simeq i(RV)_{\ell 2}/\sqrt{1+z}\ (\ell = e, \mu, \tau\,)$, with $|(RV)_{\mu 2}| \approx |(RV)_{e2}|$. From eq. (8), neglecting again the contribution due to the exchange of the light Majorana neutrino mass eigenstates, the effective Majorana mass is given by

$$|(m_\nu)_{ee}| \cong 2z\, |(RV)_{e2}|^2\, \frac{M^2_a}{M_2}\, f(A)\,. \tag{12}$$

Assuming production and detection of the RH Majorana fields at collider, i.e. $|(RV)_{\ell 2}| \approx 10^{-2}$ $(\ell = e, \mu)$, one has that $|(m_\nu)_{ee}| \lesssim 0.2$ eV for $z \lesssim 10^{-3}\,(10^{-2})$ and $M_1 \approx 100\,(1000)$ GeV. Therefore, the non-observation of the $(\beta\beta)_{0\nu}$-decay requires in this scenario a degeneracy in the right-handed neutrino masses of at least one per cent. As discussed above, the cross section for same sign di-muon production in proton-proton collisions is proportional to the mass difference of the right-handed neutrinos. Thus, even in this extremely fined-tuned scenario, the Majorana nature of the right-handed neutrinos will be difficult to probe at colliders.

6. Conclusions

The see-saw mechanism provides a natural explanation for the smallness of neutrino masses. In such scenario, a new mass scale is introduced in the theory, which is linked to the light neutrino mass scale and can in principle be accessible to present and forthcoming particle physics accelerators, LHC included. In its simplest formulation, the type I see-saw scenario, the Standard Model (SM) is extended with at least two "heavy" Majorana neutrinos, which are singlets of $SU(2)_L \times U(1)_Y$ and are coupled to the SM lepton and Higgs doublets. It is shown that, if the type I see-saw scenario provides the correct mechanism of generation of light neutrino masses and the Majorana fields N_j have masses in the range $M_j \sim (100 \div 1000)$ GeV, the physical effects associated with the Majorana nature of these heavy neutrinos N_j, are so small that they are unlikely to be observable in the currently operating and future planned accelerator experiments (including LHC). This is a consequence of the existence of very strong constraints on the parameters and couplings, responsible for the corresponding $|\Delta L| = 2$ processes in which N_j are involved, and/or on the couplings of N_j to the weak $W^{\pm}$ and Z^0 bosons. The strongest constraints are provided by the experimental upper limit on *i*) the light neutrino mass matrix, $|(m_\nu)_{ll'}| \lesssim 1$ eV, $l, l' = e, \mu, \tau$ and *ii*) the effective Majorana mass $|(m_\nu)_{ee}|$ obtained in the $(\beta\beta)_{0\nu}$-decay experiments. The latter, is sensitive to the heavy Majorana neutrino mass scale and splitting(s).

As consequence, it is shown that charged and neutral current interactions of the heavy Majorana fields N_j with the Standard Model charged leptons and neutrinos are extremely suppressed, unless the heavy Majorana neutrinos form a pseudo-Dirac pair. Therefore, the Majorana nature of the heavy neutrinos will not be detected in collider experiments: either the production cross section is highly suppressed or the heavy neutrinos behave to a high level of precision as Dirac fermions.

The suppression of the $|\Delta L| = 2$ processes can be avoided only if there exist additional TeV scale interaction terms in the Lagrangian between the heavy Majorana neutrinos and the Standard Model particles. If this is the case, the production cross section of heavy neutrinos will not necessarily be suppressed, while their charged and neutral current interactions with the Standard Model charged leptons and neutrinos can still be tiny.

Acknowledgements

Emiliano Molinaro would like to thank the organizers of the conference DISCRETE 2010, Symposium on Prospects in the Physics of Discrete Symmetries, for the opportunity of presenting the topics discussed in these Proceedings. The work of Emiliano Molinaro is supported by the Fundação para a Ciência e a Tecnologia (FCT, Portugal) through the project PTDC/FIS/098188/2008 and CFTP-FCT Unit 777 which are partially funded through POCTI (FEDER).

References

[1] B.T. Cleveland *et al.*, Astrophys. J. **496**, 505 (1998); Y. Fukuda *et al.* [Kamiokande Collaboration], Phys. Rev. Lett. **77**, 1683 (1996); J.N. Abdurashitov *et al.*, Phys. Rev. C **80**, 015807 (2009); P. Anselmann *et al.*, Phys. Lett. B **285** (1992) 376; W. Hampel *et al.*, Phys. Lett. B **447** (1999) 127; M. Altmann *et al.*, Phys. Lett. B **616** (2005) 174; S. Fukuda *et al.* [Super-Kamiokande Collaboration], Phys. Lett. **B539** (2002) 179; Q.R. Ahmad *et al.* [SNO Collaboration], Phys. Rev. Lett. **87** (2001) 071301 and **89** (2002) 011301; Y. Fukuda *et al.* [Super-Kamiokande Collaboration], Phys. Rev. Lett. **81** (1998) 1562; Y. Ashie *et al.* [Super-Kamiokande Collaboration], Phys. Rev. Lett. **93** (2004) 101801; K. Eguchi *et al.* [KamLAND Collaboration], Phys. Rev. Lett. **90** (2003) 021802; T. Araki *et al.*, Phys. Rev. Lett. **94** (2005) 081801; C. Arpesella *et al.*, Phys. Lett. B **658** (2008) 101; Phys. Rev. Lett. **101** (2008) 091302; M. H. Ahn *et al.* [K2K Collaboration], Phys. Rev. D **74** (2006) 072003; D.G. Michael *et al.* [MINOS Collaboration], Phys. Rev. Lett. **97** (2006) 191801; P. Adamson *et al.*, Phys. Rev. Lett. **101** (2008) 131802.
[2] B. Pontecorvo, Zh. Eksp. Teor. Fiz. (JETP) **33** (1957) 549 and **34** (1958) 247.
[3] Z. Maki, M. Nakagawa and S. Sakata, Prog. Theor. Phys. **28** (1962) 870.
[4] C. Amsler *et al.* [Particle Data Group], Phys. Lett. B **667** (2008) 1.
[5] B. Pontecorvo, Sov. Phys. JETP **26** (1968) 984. [Zh. Eksp. Teor. Fiz. **53**, 1717 (1967)].
[6] P. Minkowski, Phys. Lett. B **67** (1977) 421; M. Gell-Mann, P. Ramond and R. Slansky, *Proceedings of the Supergravity Stony Brook Workshop*, New York 1979, eds. P. Van Nieuwenhuizen and D. Freedman; T. Yanagida, *Proceedinds of the Workshop on Unified Theories and Baryon Number in theUniverse*, Tsukuba, Japan 1979, ed.s A. Sawada and A. Sugamoto; R. N. Mohapatra and G. Senjanovic, Phys. Rev. Lett. **44** (1980) 912.
[7] A. Ibarra, E. Molinaro and S. T. Petcov, JHEP **1009** (2010) 108 [arXiv:1007.2378 [hep-ph]].

[8] S. Antusch, J. P. Baumann and E. Fernandez-Martinez, Nucl. Phys. B **810** (2009) 369.
[9] S. Antusch *et al.*, JHEP **0610** (2006) 084.
[10] See, e.g. C. Aalseth *et al.*, arXiv:hep-ph/0412300; C.E. Aalseth *et al.* [IGEX Collaboration], Phys. Atomic Nuclei **63** (2000) 1225; H. V. Klapdor-Kleingrothaus *et al.*, Phys. Lett. B **586** (2004) 198; A. S. Barabash [NEMO Collaboration], arXiv:0807.2336 [nucl-ex]; C. Arnaboldi *et al.* [CUORICINO Collaboration], Phys. Rev. C **78** (2008) 035502.
[11] S. M. Bilenky and S. T. Petcov, Rev. Mod. Phys. **59** (1987) 671.
[12] S.M. Bilenky, S. Pascoli and S.T. Petcov, Phys. Rev. D **64** (2001) 053010; S.T. Petcov, Phys. Scripta T **121** (2005) 94; S. Pascoli and S. T. Petcov, Phys. Rev. D **77** (2008) 113003.
[13] A. Atre *et al.*, JHEP **0905** (2009) 030.
[14] C.N. Leung and S.T. Petcov, Phys. Lett. B **125** (1983) 461.
[15] D. Wyler and L. Wolfenstein, Nucl. Phys. B **218** (1983) 205.
[16] G. C. Branco, W. Grimus and L. Lavoura, Nucl. Phys. B **312** (1989) 492.
[17] L. Wolfenstein, Nucl. Phys. B **186** (1981) 147.
[18] S.T. Petcov, Phys. Lett. B **110** (1982) 245.
[19] A. Halprin, S.T. Petcov and S.P. Rosen, Phys. Lett. B **125** (1983) 335.
[20] J. Vergados, Nucl. Phys. B **218** (1983) 109.
[21] W.C. Haxton and J. Stephenson, Prog. Part. Nucl. Phys. **12** (1984) 409.
[22] H. Paes *et al.*, Phys. Lett. B **498** (2001) 35.
[23] M. Blennow et *al.*, JHEP **1007** (2010) 096 [arXiv:1005.3240 [hep-ph]].

Journal of Physics: Conference Series **335** (2011) 012049 doi:10.1088/1742-6596/335/1/012049

Bimaximal Neutrino Mixing with Discrete Flavour Symmetries

Luca Merlo

Physik-Department, Technische Universität München
James-Franck-Str. 1, D-85748 Garching, Germany
and
TUM Institute of Advanced Study
Lichtenbergstr. 2a, D-85748 Garching, Germany

E-mail: merlo@tum.de

Abstract. In view of the fact that the data on neutrino mixing are still compatible with a situation where Bimaximal mixing is valid in first approximation and it is then corrected by terms of order of the Cabibbo angle, we present examples where these properties are naturally realized. The models are supersymmetric in 4-dimensions and based on the discrete non-Abelian flavour symmetry S_4.

1. The flavour puzzle and the neutrino mixing patterns

The information on masses and mixings are encoded in the Yukawa terms, which are theoretically unpredicted. On the contrary, from the experiments we know that the charged fermions have strong hierarchical masses, spanning from tenths of MeV up to few hundreds GeV, while the neutrinos do not show a strong hierarchy and only upper bounds of few eV have been put on their masses. Furthermore, the solar and the atmospheric anomalies find an elegant explanation in the oscillation of three active massive neutrinos and the corresponding frequencies have been measured in terms of mass squared difference of neutrinos eigenvalues. In table 1, we can see the result of two independent global fits on the neutrino oscillation data. An uncertainty is still present on the sign of the atmospheric mass squared difference and this reflects our ignorance on the type of the neutrino spectrum: normal hierarchy, inverse hierarchy or quasi degeneracy.

When looking at the mixings, other differences are underlined: the quark mixing matrix, V_{CKM}, has only small angles which can be expressed in terms of powers of the Cabibbo angle, $\lambda \approx 0.23$, through the Wolfenstein parametrisation, while the lepton mixing matrix, U_{PMNS}, presents two very large angles and the third compatible to be vanishing: in particular the atmospheric angle, in the 23 sector, is compatible with the maximal value well inside the 1σ region; the solar angle, in the 12 sector, is large, but about 5σ's from being maximal; the reactor angle, in the 13 sector, is compatible to be vanishing at about 1σ level as a result of global fits on the neutrino oscillation data, but only an upper bound of about λ has been put on $\sin\theta_{13}$ by reactor experiments. A very good approximation to the lepton mixing matrix is provided by the so-called Tri-Bimaximal (TB) patter [3, 4],

$$\sin^2\theta_{12} = 1/3\;, \quad \sin^2\theta_{23} = 1/2\;, \quad \sin\theta_{13} = 0\;, \tag{1}$$

Table 1. Neutrino oscillation parameters from independent global fits [1, 2].

parameter	Ref. [1] best fit(1σ)	Ref. [1] 3σ-interval	Ref. [2] best fit(1σ)	Ref. [2] 3σ-interval
$\Delta m^2_{21}\ [10^{-5}\mathrm{eV}^2]$	$7.67^{+0.16}_{-0.19}$	$7.14-8.19$	$7.65^{+0.23}_{-0.20}$	$7.05-8.34$
$\lvert\Delta m^2_{31}\rvert\ [10^{-3}\mathrm{eV}^2]$	$2.39^{+0.11}_{-0.8}$	$2.06-2.81$	$2.40^{+0.12}_{-0.11}$	$2.07-2.75$
$\sin^2\theta_{12}$	$0.312^{+0.019}_{-0.018}$	0.26-0.37	$0.304^{+0.022}_{-0.016}$	0.25-0.37
$\sin^2\theta_{23}$	$0.466^{+0.073}_{-0.058}$	0.331-0.644	$0.50^{+0.07}_{-0.06}$	0.36-0.67
$\sin^2\theta_{13}$	$0.016^{+0.010}_{-0.010}$	≤ 0.046	$0.010^{+0.016}_{-0.011}$	≤ 0.056

which agrees at the 1σ level with the data. Note that U^{TB} does not depend on the mass eigenvalues, in contrast with the quark sector, where the entries of the CKM matrix can be written in terms of the ratio of the quark masses. Moreover it is a completely real matrix, since the factors with the Dirac phase vanish (the Majorana phases can be factorized outside). The best measured neutrino mixing angle θ_{12} is just about 1σ below the TB value, while the other two angles are well inside the 1σ interval.

In a series of papers [5, 6, 7, 8, 9] it has been pointed out that a broken flavour symmetry based on the discrete group A_4 appears to be particularly suitable to reproduce this specific mixing pattern as a first approximation. Other solutions based on alternative discrete or continuous flavour groups have also been considered [10, 11, 12, 13, 14], but the A_4 models have a very economical and attractive structure, e.g. in terms of group representations and of field content. In all these models, when the symmetry is broken, some corrections to the mixing angles are introduced: in general all of them are of the order of λ_C^2 and therefore these models indicate a value for the reactor angle which is well compatible with zero (for a different approach see [15]).

There is an experimental hint for a non-vanishing reactor angle [1, 2] and, if a value close to the present upper bound is found in the future experiments, this could be interpreted as an indication that the agreement with the TB mixing is only accidental. Looking for an alternative leading principle, it is interesting to note that the data suggest a numerical relationship between the lepton and the quark sectors, known as the complementarity relation, for which $\theta_{12}+\lambda_C\simeq\pi/4$ [16, 17, 18]. However, there is no compelling model which manages to get this nice feature, without parameter fixing. Our proposal is to relax this relationship. Noting that $\sqrt{m_\mu/m_\tau}\simeq\lambda_C$, we can write the following expression, which we call *weak complementarity* relation

$$\theta_{12}\simeq\frac{\pi}{4}-\mathcal{O}\left(\sqrt{\frac{m_\mu}{m_\tau}}\right)\,. \tag{2}$$

The idea is first to get a maximal value both for the solar and the atmospheric angles and then to correct θ_{12} with relatively large terms. To reach this task, the bimaximal (BM) pattern [19] can be extremely useful: it corresponds to the requirement that $\theta_{13}=0$ and $\theta_{23}=\theta_{12}=\pi/4$. The unitary matrix which corresponds to this patter is given by

$$U^{BM}=\begin{pmatrix}1/\sqrt{2} & -1/\sqrt{2} & 0\\ 1/2 & 1/2 & -1/\sqrt{2}\\ 1/2 & 1/2 & +1/\sqrt{2}\end{pmatrix}\,. \tag{3}$$

Note that U^{BM} does not depend on the mass eigenvalues and is completely real, like the TB pattern. The most general mass matrix, m_ν^{BM}, diagonalized by this mixing is $\mu-\tau$ symmetric and satisfies to an additional symmetry for which $(m_\nu^{BM})_{1,1}=(m_\nu^{BM})_{2,2}+(m_\nu^{BM})_{2,3}$:

$$m_\nu^{BM}=\begin{pmatrix} x & y & y \\ y & z & x-z \\ y & x-z & z \end{pmatrix}. \tag{4}$$

Starting from the BM scheme, the corrections introduced from the symmetry breaking must have a precise pattern: $\delta\sin^2\theta_{12}\simeq\lambda_C$, while $\delta\sin^2\theta_{23}\leq\lambda_C^2$ and $\delta\sin\theta_{13}\leq\lambda_C$ in order to be in agreement with the experimental data. This feature is not trivially achievable.

2. The model building

In this part we present two flavour models [20, 21] in which the neutrino mixing matrix at the leading order (LO) is the BM scheme, while the charged lepton mass matrix is diagonal with hierarchical entries; moreover the models allow for corrections that bring the mixing angles in agreement with the data. The strategy is to use the flavour group $G_f=S_4\times Z_4\times U(1)_{FN}$, where S_4 is the group of the permutations of four objects, and to let the SM fields transform non-trivially under G_f; moreover some new fields, the flavons, are introduced which are scalars under the SM gauge symmetry, but transform under G_f; these flavons, getting non-vanishing vacuum expectation values (VEVs), spontaneously break the symmetry in such a way that two subgroups are preserved, $G_\ell=Z_4$ in the charged lepton sector and $G_\nu=Z_2\times Z_2$ in the neutrino sector. It is this breaking chain of S_4 which assures that the LO neutrino mixing matrix, U_ν, is the BM pattern in the basis of diagonal charged lepton mass matrix. The additional terms $Z_4\times U(1)_{FN}$ forbid dangerous operators and allow for the correct charged lepton mass hierarchy.

The first model [20] deals only with the lepton sector in a supersymmetric SM scenario, implementing the weak complementarity. In the second model [21], we extend also to the quark sector in a Pati-Salam GUT context.

2.1. *The lepton model*

We formulate our model in the framework of the See-Saw mechanism. For this we choose the 3 generations of left-handed lepton doublets ℓ and of RH neutrinos ν^c transforming as $(\mathbf{3},1)$ under $S_4\times Z_4$, while the RH charged leptons e^c, μ^c and τ^c transform as $(\mathbf{1},-1)$, $(\mathbf{1}',-i)$ and $(\mathbf{1},-i)$, respectively. The S_4 symmetry is then broken by suitable flavons, φ_ℓ and χ_ℓ, transforming as $(\mathbf{3},i)$ and $(\mathbf{3}',i)$, and φ_ν and ξ_ν, transforming as $(\mathbf{3},1)$ and $(\mathbf{1},1)$. A flavon θ, carrying only a negative unit of the $U(1)_{FN}$ charge, acquires a VEV and breaks $U(1)_{FN}$. In view of a possible GUT extension of the model at a later stage, we adopt a supersymmetric context, so that two Higgs doublets $H_{u,d}$, invariant under $S_4\times Z_4$, are present in the model. The usual continuous $U(1)_R$ symmetry, related to R-parity, is implemented in the model. Supersymmetry also helps producing and maintaining the hierarchy $\langle H_{u,d}\rangle=v_{u,d}\ll\Lambda_f$ whereΛ $_f$ is the cutoff scale of the theory.

The complete superpotential can be written as

$$w=w_e+w_\nu+w_d\,, \tag{5}$$

where w_d is responsible for the flavon VEV alignment as discussed in [20], while w_e and w_ν refer to the charged lepton and neutrino sectors and can be written as

$$\begin{aligned} w_e &= \frac{y_e^{(1)}}{\Lambda_f^2}\frac{\theta^2}{\Lambda_f^2}e^c(\ell\varphi_\ell\varphi_\ell)H_d+\frac{y_e^{(2)}}{\Lambda_f^2}\frac{\theta^2}{\Lambda_f^2}e^c(\ell\chi_\ell\chi_\ell)H_d+\frac{y_e^{(3)}}{\Lambda_f^2}\frac{\theta^2}{\Lambda_f^2}e^c(\ell\varphi_\ell\chi_\ell)H_d+ \\ &\quad +\frac{y_\mu}{\Lambda_f}\frac{\theta}{\Lambda_f}\mu^c(\ell\chi_\ell)'H_d+\frac{y_\tau}{\Lambda_f}\tau^c(\ell\varphi_\ell)H_d+\ldots \\ w_\nu &= y(\nu^c\ell)H_u+M\Lambda_f(\nu^c\nu^c)+a(\nu^c\nu^c\xi_\nu)+b(\nu^c\nu^c\varphi_\nu)+\ldots \end{aligned} \tag{6}$$

indicating with $(\ldots)$ the singlet $\mathbf{1}$, with $(\ldots)'$ the singlet $\mathbf{1}'$ and with $(\ldots)_R$ the representation R ($R = \mathbf{2},\ \mathbf{3},\ \mathbf{3}'$). Note that the parameter M defined above is dimensionless. In the above expression for the superpotential w, only the lowest order operators in an expansion in powers of $1/\Lambda_f$ are explicitly shown. Dots stand for higher dimensional operators. The stated symmetries ensure that, for the leading terms, the flavons that appear in w_e cannot contribute to w_ν and viceversa.

We showed in [20] that the potential corresponding to w_d possesses an isolated minimum for the following VEV configuration:

$$\langle\varphi_\ell\rangle/\Lambda_f \sim \begin{pmatrix} 0 \\ 1 \\ 0 \end{pmatrix} v\,, \quad \langle\chi_\ell\rangle/\Lambda_f \sim \begin{pmatrix} 0 \\ 0 \\ 1 \end{pmatrix} cv\,, \quad \langle\varphi_\nu\rangle/\Lambda_f \sim \begin{pmatrix} 0 \\ 1 \\ -1 \end{pmatrix} v'\,, \quad \langle\xi_\nu\rangle/\Lambda_f \sim c'v'\,, \tag{7}$$

where c and c' are order one coefficients which parametrize the ratio among the different VEVs. Similarly, the Froggatt-Nielsen flavon θ gets a VEV, determined by the D-term associated to the local $U(1)_{FN}$ symmetry, and it is denoted by $\langle\theta\rangle/\Lambda_f = t$.

With this VEVs configuration, the charged lepton mass matrix is diagonal

$$M_e \sim \mathrm{diag}\left(t^2,\, t,\, 1\right) v v_d/\sqrt{2}\,, \tag{8}$$

where for simplicity we omit to indicate all the coupling constants and the coefficient c. As a result, at the LO, there is no contribution to the lepton mixing matrix from the diagonalisation of charged lepton masses. It is possible to estimate the values of v and t by looking at the mass ratios of charged leptons:

$$m_\mu/m_\tau \sim t\,, \qquad\qquad m_e/m_\mu \sim vt\,. \tag{9}$$

In order to fit these relations with the data, we must have approximately $t \sim 0.06$ and $v \sim 0.08$ (modulo coefficients of $\mathcal{O}(1)$).

The light neutrino Majorana mass matrix, given by the See-Saw relation, is diagonalised by the BM matrix and the eigenvalues are

$$U_\nu^T m_\nu U_\nu = \mathrm{diag}(m_1,\, m_2,\, m_2) \qquad \text{with} \quad m_i \equiv \frac{|y^2| v_u^2}{2M_i}\,, \tag{10}$$

where M_i are the masses of the heavy RH neutrinos

$$M_1 = 2\left|M + v'\left(a - \sqrt{2}bc'\right)\right|\Lambda_f\,, \quad M_2 = 2\left|M + v'\left(a + \sqrt{2}bc'\right)\right|\Lambda_f\,, \quad M_3 = 2\left|M + av'\right|\Lambda_f\,. \tag{11}$$

Notice that since the neutrino sector is not charged under the Z_4 symmetry, we have operators of dimension 5 which contribute to the neutrino masses and may correspond to some heavy exchange other than the right-handed neutrinos ν^c. When considering the interesting domain of parameters, we find that this effective contribution is subdominant.

At the LO, the light neutrino mass matrix depends on only 2 effective parameters indeed the terms M and av' enter the mass matrix in the combination $F \equiv M + av'$. The coefficients $y_e^{(i)}$, y_μ, y_τ, y, a and b are all expected to be of $\mathcal{O}(1)$. A priori M could be of $\mathcal{O}(1)$, corresponding to a RH neutrino Majorana mass of $\mathcal{O}(\Lambda_f)$, but, actually, we saw that it must be of the same order as v'.

To summarise, at the LO we have diagonal and hierarchical charged leptons together with the exact BM mixing for neutrinos. It is clear that substantial the next-to-leading order (NLO) corrections are needed to bring the model to agree with the data on θ_{12}. A crucial feature of our model is that the neutrino sector flavons φ_ν and ξ_ν are invariant under Z_4 which is not the case for the charged lepton sector flavons φ_ℓ and χ_ℓ. The consequence is that φ_ν and ξ_ν can contribute at the NLO to the corrections in the charged lepton sector, while at the NLO φ_ℓ and χ_ℓ cannot modify the neutrino sector couplings. As a results the dominant genuine corrections to the BM mixing only occur at the NLO through the diagonalisation of the charged leptons. Without entering in the details of the NLO discussion (see [20]), we find that the NLO corrections, coming from the higher order terms, are not democratic and the final corrected mixing angles are

$$\sin^2\theta_{12} = \frac{1}{2} - \frac{1}{\sqrt{2}}(c_1 + c_2)v' \,, \qquad \sin^2\theta_{23} = \frac{1}{2} \,, \qquad \sin\theta_{13} = \frac{1}{\sqrt{2}}(c_1 - c_2)v' \,, \qquad (12)$$

where c_1 and c_2 are coefficients of order one. When v' is of the order of the Cabibbo angle, then θ_{12} is brought in agreement with the experimental data; in the meantime the reactor angle is corrected of the same amount, suggesting a value for θ_{13} close to its present upper bound. Note that the atmospheric angle remains uncorrected at this order. Any quantitative estimates are clearly affected by large uncertainties due to the presence of unknown parameters of order one, as we can see in figure 1, but in our model a value of θ_{13} much smaller than the present upper bound would be unnatural.

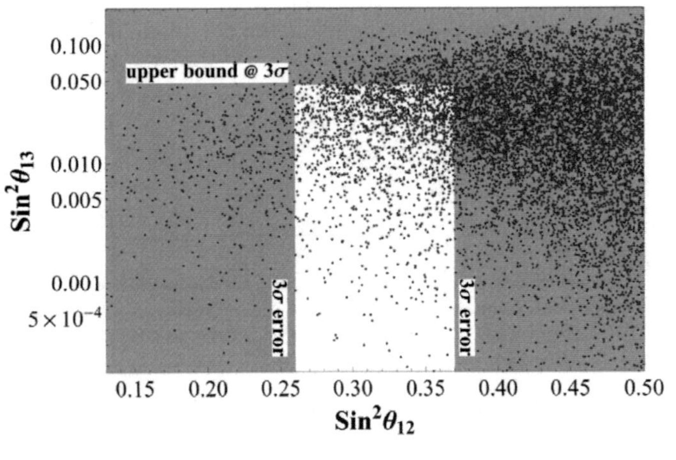

Figure 1. $\sin^2\theta_{13}$ as a function of $\sin^2\theta_{12}$ is plotted, following eqs. (12). The parameters c_1 and c_2 are treated as random complex numbers of absolute value between 0 and 0.30. The gray bands represents the regions excluded by the experimental data [1, 2]: the horizontal one corresponds to the 3σ-upper bound for $\sin^2\theta_{13}$ of 0.46 and the vertical ones to the region outside the 3σ error range $[0.26 - 0.37]$ for $\sin^2\theta_{12}$.

It is then interesting to verify the agreement of the model with other sectors of the neutrino physics, such as the $0\nu2\beta$-decay and the leptogenesis (See [22, 23] for a general approach). The result of the analysis is that the model presents a normal ordered – moderate hierarchical or quasi degenerate – spectrum with a suggested lower bound for the lightest neutrino mass and for the effective $0\nu2\beta$-mass parameter $|m_{ee}|$ of about 0.1 meV. On the other hand it is compatible with the constraints from leptogenesis as an explanation of the baryon asymmetry in the Universe.

2.2. The full model

In this section we are interested in the extension to the quark sector. A first attempt is to adopt for quarks the same representations under S_4 that have been used for leptons: the left-handed quark doublets q transform as a triplet $\mathbf{3}$, while the right-handed quarks (u^c, d^c), (c^c, s^c) and (t^c, b^c) transform as $\mathbf{1}$, $\mathbf{1}'$ and $\mathbf{1}$, respectively. We can similarly extend to quarks the transformations of Z_4 (and $U(1)_R$) given for leptons. As a result, it is easy to see that the quark mass matrices are diagonal, at the LO on the expansion parameters, exactly as for the charged leptons and to account for the correct mass hierarchies the $U(1)_{FN}$ has to be suitably implemented. At this level the CKM matrix is the unity matrix and to get realistic mixings, the higher-order corrections should switch on off-diagonal entries with a well-defined pattern: $(12) \sim \lambda$, $(23) \sim \lambda^2$ and $(13) \sim \lambda^3$. By an explicit computation we find the following result for the quark mass matrices: in terms of order of magnitude

$$M_d = \begin{pmatrix} v\,t^2 & v\,v'\,t^2 & v\,v'\,t^2 \\ v'\,t & t & v'^2\,t \\ v' & v'^2 & 1 \end{pmatrix} \frac{v\,v_d}{\sqrt{2}}\,, \qquad M_u = \begin{pmatrix} v\,t^3 & v\,v'\,t^3 & v\,v'\,t^3 \\ v'\,t^2 & t^2 & v'^2\,t^2 \\ v' & v'^2 & 1 \end{pmatrix} \frac{v\,v_u}{\sqrt{2}}\,. \tag{13}$$

Calculating now the unitary matrices which diagonalise $M_d^\dagger M_d$ and $M_u^\dagger M_u$ we find

$$V_d \sim V_u = \begin{pmatrix} 1 & v' & v' \\ -v' & 1 & v'^2 \\ -v' & -v'^2 & 1 \end{pmatrix}\,. \tag{14}$$

At a first sight, barring cancellations among the single entries, we can see that the CKM matrix $V = V_u^\dagger V_d$ should be similar to V_u and V_d and as a consequence it cannot correctly describe the quark mixings: while the entries (12) and (23) well reproduce the measured values, the large values in the (13) entry would require a large fine-tuning of order λ^2.

An alternative possibility is to further investigate on the complementarity relations:

$$\theta_{12} + \lambda \simeq \pi/4\,, \qquad \theta_{23} + \lambda^2 \simeq -\pi/4\,. \tag{15}$$

These equations suggest that the angles in the CKM and PMNS matrices may have a common origin which can be motivated for example in Pati-Salam models, where the following relation holds,

$$U_e \sim V_d\,. \tag{16}$$

We can use this to write the CKM and PMNS matrices as

$$\begin{aligned} U &= R_{23}\left(-\tfrac{\pi}{4}\right) R_{13}(\lambda) R_{12}\left(\tfrac{\pi}{4} - \lambda\right) = \underbrace{\left(R_{23}\left(\frac{\pi}{4}\right) R_{13}(\lambda) R_{12}(\lambda)\right)^\dagger}_{U_e} \underbrace{R_{12}\left(\frac{\pi}{4}\right)}_{U_\nu} \\ V &= R_{12}(\lambda) = \underbrace{\left(R_{23}\left(\frac{\pi}{4}\right) R_{13}(\lambda) R_{12}(\lambda)\right)^\dagger}_{V_u} \underbrace{R_{23}\left(\frac{\pi}{4}\right) R_{13}(\lambda) R_{12}(\lambda))}_{V_d}\,. \end{aligned} \tag{17}$$

Here $R_{ij}(\alpha)$ stand for rotations in the (ij) plane of the angle α (apart from coefficients $\mathcal{O}(1)$ in front of each angles). The coefficients of the angles in the rotations in $V_u^\dagger$ and V_d should be such that the rotations cancel each other in the (13) sector, but not in the (12) sector. Thus we should introduce terms which distinguish between the up- and down-quark sectors, indeed possible within the Pati-Salam context.

Moving to the explicit form of the mass matrices, the generic Majorana neutrino mass matrix m_ν which is diagonalised by $U_\nu = R_{12}\left(\frac{\pi}{4}\right)$,

$$m_\nu^{diag} = R_{12}\left(\frac{\pi}{4}\right)^T m_\nu\, R_{12}\left(\frac{\pi}{4}\right)\,, \tag{18}$$

is given by

$$m_\nu \sim \begin{pmatrix} a & b & 0 \\ b & a & 0 \\ 0 & 0 & c \end{pmatrix}\,. \tag{19}$$

Considering the charged lepton mass matrix M_e, the product $M_e^\dagger M_e$ should be diagonalised by the action of V_d as in eq. 17,

$$R_{12}(-\lambda)R_{13}(-\lambda)R_{23}\left(-\frac{\pi}{4}\right) M_e^\dagger M_e R_{23}\left(\frac{\pi}{4}\right) R_{13}(\lambda)R_{12}(\lambda)\,. \qquad (20)$$

In the limit $m_e \to 0$ we find the generic structure for the product $M_e^\dagger M_e$:

$$M_e^\dagger M_e \sim \frac{m_\tau^2}{2}\begin{pmatrix} 0 & \lambda & \lambda \\ \lambda & 1 & 1 \\ \lambda & 1 & 1 \end{pmatrix} + \frac{m_\mu^2}{2}\begin{pmatrix} 0 & \lambda & -\lambda \\ \lambda & 1 & -1 \\ -\lambda & -1 & 1 \end{pmatrix} + \mathcal{O}(\lambda^2)\,, \qquad (21)$$

that can be obtained if M_e is given by

$$M_e \sim \frac{m_\tau}{\sqrt{2}}\begin{pmatrix} 0 & 0 & 0 \\ 0 & 0 & 0 \\ \lambda & 1 & 1 \end{pmatrix} + \frac{m_\mu}{\sqrt{2}}\begin{pmatrix} 0 & 0 & 0 \\ \lambda & 1 & -1 \\ 0 & 0 & 0 \end{pmatrix} + \mathcal{O}(\lambda^2)\,. \qquad (22)$$

It is interesting to note that, moving to the basis of diagonal charged leptons and considering only the LO terms, the neutrino mass matrix results to be of the classical BM type.

From eq. 16, the relation $M_e \sim M_d$ follows and therefore the down-quark matrix has a similar structure as in eq. 22. Looking at eq. (17) we find that also M_u should have a similar structure. We find that we can satisfy the constraint on the eqs. (12) and (13) rotations if the third columns of M_u and M_d are proportional to each other, but the second columns are not.

So far we have not given yet any explanation of the origin of these mass matrices and mixings and we leave this analysis to [21]. Here we only say that such a construction is possible in a Pati-Salam realisation where the flavour symmetry is $S_4 \times Z_4 \times U(1)_{FN}$: as for the non-GUT model described above, key points are a suitable choice of the group representations for the flavons and the particular VEV misalignment whose effects are a reactor angle and a deviation from $\pi/4$ of the solar angle of the order of λ, while introducing small deviations of the order λ^2 from the maximal value of the atmospheric angle.

A difficulty with respect the non-GUT model refers to the study of the gauge coupling running and of the Higgs potential: while in a general Pati-Salam model, in particular without any flavour symmetry implementation, it is possible to reproduce a realistic sequential symmetry breaking chain, with the usual SM or MSSM Higgs fields at the electroweak scale, the introduction of a flavour symmetry puts strong constraints. Indeed we need additional scalars which transform under the gauge group and the effect is to sandwich the energy scales of the different symmetry breakings, lowering to 10^{14} GeV the energy scale of the (almost) unification.

3. Conclusions

We have illustrated two models based on the flavour symmetry $S_4 \times Z_4 \times U(1)_{FN}$ where the BM mixing is realised at the LO in a natural way. The hierarchy of charged lepton masses is obtained as a combined effect of the $U(1)_{FN}$ and of $S_4 \times Z_4$ symmetry breaking.

Since exact BM mixing implies a value of $\tan\theta_{12}$ which is excluded by the data, large corrections are needed. The dominant corrections to the BM mixing arise at the NLO. The shifts of the quantities $\sin^2\theta_{12}$ and $\sin\theta_{13}$ from the BM values are linear in the parameter v', which is expected to be of the same order as v, but not necessarily too close, as v and v' are determined by two different sets of minimisation equations. From the experimental value $\tan^2\theta_{12} = 0.45 \pm 0.04$, which is sizably different than the BM value $\tan^2\theta_{12} = 1$, we need $v' \sim \mathcal{O}(\lambda)$. As in most models where the BM mixing is only corrected by the effect of charged lepton diagonalisation, one also expects $\theta_{13} \sim \mathcal{O}(\lambda)$. A value of θ_{13} near the present bound would be a strong indication in favour of this mechanism and a hint that the closeness of the measured values of the mixing angles to the TB values may be purely an accident. In addition, a very important feature of our models is that the shift of $\sin^2\theta_{23}$ from the maximal mixing value of 1/2 vanishes at the NLO and is expected to be of $\mathcal{O}(\lambda^2)$ at most. In our S_4 models, this property is obtained by only allowing the breaking of S_4 in the neutrino sector via flavons transforming as **1** and **3** (in particular with no doublets).

The quark sector is discussed only in the second model and realistic fermion mass hierarchies and a correct CKM matrix are found. This model is constructed in a supersymmetric Pati-Salam model, where a combined study of the flavour and Higgs sectors is performed.

Studies on flavour changing neutral currents as in [24, 25, 26] and on the Higgs phenomenology as in [27, 28, 29] will follow.

Acknowledgments

We thank the organizers of *DISCRETE'10 - Symposium on Prospects in the Physics of Discrete Symmetries* for giving the opportunity to present my talk and for the kind hospitality in Rome.

References

[1] G. L. Fogli, E. Lisi, A. Marrone, A. Palazzo, and A. M. Rotunno, in the proceedings of 4th International Workshop on Neutrino Oscillations in Venice: Ten Years after the Neutrino Oscillations, Venice, Italy, 15-18 Apr 2008, pp. 21–28 [arXiv: 0809.2936].
[2] T. Schwetz, M. A. Tortola, and J. W. F. Valle, New J. Phys. **10** (2008) 113011 [arXiv: 0808.2016].
[3] P. F. Harrison, D. H. Perkins, and W. G. Scott, Phys. Lett. **B530** (2002) 167 [arXiv: hep-ph/0202074].
[4] Z.-z. Xing, Phys. Lett. **B533** (2002) 85–93 [arXiv: hep-ph/0204049].
[5] E. Ma and G. Rajasekaran, Phys. Rev. **D64** (2001) 113012 [arXiv: hep-ph/0106291].
[6] K. S. Babu, E. Ma, and J. W. F. Valle, Phys. Lett. **B552** (2003) 207–213 [arXiv: hep-ph/0206292].
[7] G. Altarelli and F. Feruglio, Nucl. Phys. **B720** (2005) 64–88 [arXiv: hep-ph/0504165].
[8] G. Altarelli and F. Feruglio, Nucl. Phys. **B741** (2006) 215–235 [arXiv: hep-ph/0512103].
[9] A. Y. Smirnov, Talk given at DISCRETE'10 - Symposium on Prospects in the Physics of Discrete Symmetries.
[10] F. Feruglio, C. Hagedorn, Y. Lin, and L. Merlo, Nucl. Phys. **B775** (2007) 120–142 [arXiv: hep-ph/0702194].
[11] F. Bazzocchi, L. Merlo, and S. Morisi, Nucl. Phys. **B816** (2009) 204–226 [arXiv: 0901.2086].
[12] F. Bazzocchi, L. Merlo, and S. Morisi, Phys. Rev. **D80** (2009) 053003 [arXiv: 0902.2849].
[13] I. de Medeiros Varzielas and G. G. Ross, Nucl. Phys. **B733** (2006) 31–47 [arXiv: hep-ph/0507176].
[14] I. de Medeiros Varzielas, S. F. King, and G. G. Ross, Phys. Lett. **B644** (2007) 153–157 [arXiv: hep-ph/0512313].
[15] Y. Lin, Nucl. Phys. **B813** (2009) 91–105 [arXiv: 0804.2867].
[16] A. Y. Smirnov, Phys. Scripta **T121** (2005) 57–64 [arXiv: hep-ph/0412391].
[17] M. Raidal, Phys. Rev. Lett. **93** (2004) 161801 [arXiv: hep-ph/0404046].
[18] H. Minakata and A. Y. Smirnov, Phys. Rev. **D70** (2004) 073009 [arXiv: hep-ph/0405088].
[19] F. Vissani, arXiv: hep-ph/9708483.
[20] G. Altarelli, F. Feruglio, and L. Merlo, JHEP **05** (2009) 020 [arXiv: 0903.1940].
[21] R. de Adelhart Toorop, F. Bazzocchi, and L. Merlo, JHEP **08** (2010) 001 [arXiv: 1003.4502].
[22] E. Bertuzzo, P. Di Bari, F. Feruglio, and E. Nardi, JHEP **11** (2009) 036 [arXiv: 0908.0161].
[23] D. Aristizabal Sierra, F. Bazzocchi, I. de Medeiros Varzielas, L. Merlo, and S. Morisi, Nucl. Phys. **B827** (2010) 34–58 [arXiv: 0908.0907].
[24] F. Feruglio, C. Hagedorn, Y. Lin, and L. Merlo, Nucl. Phys. **B809** (2009) 218–243 [arXiv: 0807.3160].
[25] F. Feruglio, C. Hagedorn, Y. Lin, and L. Merlo, Nucl. Phys. **B832** (2009) 251–288 [arXiv: 0911.3874].
[26] F. Feruglio, C. Hagedorn, and L. Merlo, JHEP **03** (2010) 084 [arXiv: 0910.4058].
[27] R. de Adelhart Toorop, F. Bazzocchi, L. Merlo, and A. Paris, arxiv: 1012.1791.
[28] R. de Adelhart Toorop, F. Bazzocchi, L. Merlo, and A. Paris, arxiv: 1012.2091.
[29] R. de Adelhart Toorop, Talk given at DISCRETE'10 - Symposium on Prospects in the Physics of Discrete Symmetries.

Journal of Physics: Conference Series **335** (2011) 012050 doi:10.1088/1742-6596/335/1/012050

A model for neutrino masses and mixing based on the non-abelian discrete symmetry A_4

Davide Meloni

Institut fuer Theoretische Physik und Astrophysik Universitaet Wuerzburg Am Hubland 97074 Wuerzburg

E-mail: davide.meloni@physik.uni-wuerzburg.de

Abstract. In this talk I discuss a see-saw A_4 model [1] for Tri-Bimaximal mixing which is based on a very economical flavour symmetry and field content and still possesses all the good features of A_4 models. In particular the charged lepton mass hierarchies are determined by the $A_4 \times Z_4$ flavour symmetry itself without invoking a Froggatt-Nielsen U(1) symmetry. Tri-Bimaximal mixing is exact in leading order while all the mixing angles receive corrections of the same order in next-to-the-leading approximation. As a consequence the predicted value of θ_{13} is within the sensitivity of the experiments which will take data in the near future. The light neutrino spectrum with its phenomenological implications is discussed in detail.

1. Introduction

It is an experimental fact [2] that within measurement errors the observed neutrino mixing matrix [3] is compatible with the so called Tri-Bimaximal (TB) form [4]. The best measured neutrino mixing angle θ_{12} is just about 1σ below the TB value $\tan^2\theta_{12} = 1/2$, while the other two angles are well inside the 1σ interval. It has been pointed out [5] that a broken flavour symmetry based on the discrete group A_4 appears to be particularly suitable to reproduce this specific mixing pattern in leading order (LO). Other solutions based on alternative discrete or continuous flavour groups have also been considered, but the A_4 models have a very economical and attractive structure, e.g. in terms of group representations and of field content. In most of the models A_4 is accompanied by additional flavour symmetries, either discrete like Z_N or continuous like U(1), which are necessary to eliminate unwanted couplings, to ensure the needed vacuum alignment and to reproduce the observed charged lepton mass hierarchies. Given the set of flavour symmetries and having specified the field content, the non leading corrections to the TB mixing arising from loop effects and higher dimensional operators can be evaluated in a well defined expansion. In this talk I present an A_4 model for TB mixing which is based on a most economical flavour symmetry and field content and still possesses all the features described above. In particular TB mixing is exact in LO while all mixing angles receive corrections at higher orders. The charged lepton mass hierarchies are determined by the $A_4 \times Z_4$ flavour symmetry itself without invoking a Froggatt-Nielsen $U(1)$ symmetry, as a consequence of a particular alignment as proposed in refs. [6]. Our model, which is of the see-saw type, differs from those in ref.[6] because the flavour symmetry is smaller and the pattern of corrections to TB mixing is more general and flexible.

Journal of Physics: Conference Series **335** (2011) 012050 doi:10.1088/1742-6596/335/1/012050

2. The structure of the model

The model is formulated in terms of the A_4 realization in the T diagonal basis (see [1] for more details). Since we want to work in the framework of the see-saw mechanism, we assign the 3 generations of left-handed (LH) lepton doublets l and of right-handed (RH) neutrinos ν^c to two triplets 3, while the RH charged leptons e^c, μ^c and τ^c all transform as 1 (while the most usual classification in A_4 models is as 1, 1'' and 1'). The A_4 symmetry is then broken by suitable flavons. All the flavon fields are singlets under the Standard Model gauge group. The complete flavour symmetry is $A_4 \times Z_4$. We adopt a supersymmetric context, so that two Higgs doublets $h_{u,d}$, invariant under A_4, are present in the model. A $U(1)_R$ symmetry related to R-parity and the presence of driving fields in the flavon superpotential are common features of supersymmetric formulations. The field content and the symmetry assignments are as in Tab.1. For the class of

Field	ν^c	ℓ	e^c	μ^c	τ^c	h_d	h_u	φ_T	ξ'	φ_S	ξ	φ_0^T	φ_0^S	ξ_0
A_4	3	3	1	1	1	1	1	3	1'	3	1	3	3	1
Z_4	-1	i	1	i	-1	1	i	i	i	1	1	-1	1	1
$U(1)_R$	1	1	1	1	1	0	0	0	0	0	0	2	2	2

Table 1. *Transformation properties of leptons, electroweak Higgs doublets and flavons under $A_4 \times Z_4$ and $U(1)_R$.*

models of ref.[6] the crucial feature is the alignment

$$\begin{aligned} \langle \varphi_S \rangle &= (v_S, v_S, v_S) \\ \langle \xi \rangle &= u \end{aligned} \tag{1}$$

$$\langle \xi' \rangle = u' \neq 0 \; , \qquad \langle \varphi_T \rangle = (0, v_T, 0) \; , \qquad v_T = -\frac{h_1 u'}{2h_2} \; . \tag{2}$$

Note that this differs from the usual A_4 alignment in that $\langle \varphi_T \rangle = (0, v_T, 0)$ replaces $\langle \varphi_T \rangle = (v_T, 0, 0)$. The difference is that, while $(1,0,0)^n = (1,0,0)$ (i.e. all positive powers are aligned in the same direction), for $(0,1,0)$ we have $(0,1,0)^2 = (0,0,1)$ and $(0,1,0)^3 = (1,0,0)$. These 3 directions are important in order to obtain the observed hierarchy of charged lepton masses: the electron, muon and tauon masses arise at order $(\langle \varphi_T \rangle/\Lambda)^3$, $(\langle \varphi_T \rangle/\Lambda)^2$, and $\langle \varphi_T \rangle/\Lambda$, respectively, where Λ is the cutoff. For $\langle \varphi_T \rangle/\Lambda \sim \mathcal{O}(\lambda_C^2)$, with λ_C being the Cabibbo angle, the correct hierarchy is reproduced. In the following we first assume that the stated alignment actually occurs and describe the LO structure of the model. Then in subsect. 2.3 we will show that the alignment is indeed naturally realized at LO from the most general superpotential allowed by the symmetry of the model.

2.1. Charged leptons

The leading order structure of the vacua in eqs.(1,2) automatically generates a diagonal charged lepton matrix, through the following superpotential terms:

$$\begin{aligned} w_l \;=\; & \frac{y_\tau}{\Lambda}\tau^c(\ell\varphi_T)\,h_d + \\ & \frac{y_\mu}{\Lambda^2}\mu^c(\ell\varphi_T\varphi_T)\,h_d + \frac{y'_\mu}{\Lambda^2}\mu^c(\ell\varphi_T)''\xi'\,h_d + \\ & \frac{y_e}{\Lambda^3}e^c(\ell\varphi_T\varphi_T)''\xi'\,h_d + \frac{y'_e}{\Lambda^3}e^c(\ell\varphi_T)'\xi'^2\,h_d + \frac{y''_e}{\Lambda^3}e^c(\ell\varphi_T)'(\varphi_T\varphi_T)''\,h_d + \\ & \frac{y'''_e}{\Lambda^3}e^c(\ell\varphi_T)''(\varphi_T\varphi_T)'\,h_d + \frac{y^{\rm iv}_e}{\Lambda^3}e^c(\ell\varphi_T)_1(\varphi_T\varphi_T)_1\,h_d + ... \end{aligned} \tag{3}$$

In the above expression for the superpotential w_l, for each charged lepton flavour, only the lowest order operators in an expansion in powers of $1/\Lambda$ are explicitly shown. Dots stand for higher dimensional operators that will be discussed later on. After symmetry breaking, the mass matrix has the form:

$$m_\ell = \begin{pmatrix} \frac{v_T v_d}{\Lambda^3}\left(2y_e v_T u' + y'_e u'^2 + y''_e v_T^2\right) & 0 & 0 \\ 0 & \frac{v_T v_d}{\Lambda^2}\left(2y_\mu v_T + y'_\mu u'\right) & 0 \\ 0 & 0 & \frac{y_\tau v_d v_T}{\Lambda} \end{pmatrix} , \quad (4)$$

where $v_d = \langle h_d \rangle$. As a result, the charged lepton mass matrix is diagonal and with hierarchical entries. To estimate the order of magnitude of v_T and u', we can use the experimental information on the ratio of lepton masses. Assuming that all the y coefficients are of $\mathcal{O}(1)$, one obtains:

$$\left(\frac{m_\mu}{m_\tau}\right) \sim 2\varepsilon + \varepsilon_u \simeq 0.06 \qquad \left(\frac{m_e}{m_\tau}\right) \sim 2\,\varepsilon\,\varepsilon_u + \varepsilon^2 + \varepsilon_u^2 \simeq 0.0003$$

where we introduced the small quantities $\varepsilon = v_T/\Lambda$ and $\varepsilon_u = u'/\Lambda$. These relations are satisfied for both sets of values $(\varepsilon, \varepsilon_u) = (0.043, -0.025)$ and $(\varepsilon, \varepsilon_u) = (0.077, -0.094)$. As we see, we can roughly assume that both ε and ε_u are of the same order of magnitude, $\mathcal{O}(\lambda_C^2)$.

2.2. Neutrinos

In the neutrino sector the superpotential is given by:

$$w_\nu = y_\nu (\nu^c \ell)\, h_u + (M + a\,\xi)\, \nu^c \nu^c + b\, \nu^c \nu^c\, \varphi_S \quad (5)$$

where a and b are generic coefficients and M is a constant with dimension of mass. Note that we also included in the LO neutrino superpotential linear terms in φ_S and ξ.

The Dirac mass matrix is obtained from the first term in eq.(5) and it is given by:

$$m_D = y_\nu\, v_u \begin{pmatrix} 1 & 0 & 0 \\ 0 & 0 & 1 \\ 0 & 1 & 0 \end{pmatrix} = v_u\, Y_\nu\ . \quad (6)$$

The other terms lead to the Majorana mass matrix:

$$m_M = \begin{pmatrix} M + a\,u + 2\,b\,v_S & -b\,v_S & -b\,v_S \\ -b\,v_S & 2\,b\,v_S & M + a\,u - b\,v_S \\ -b\,v_S & M + a\,u - b\,v_S & 2\,b\,v_S \end{pmatrix} . \quad (7)$$

The light neutrino mass matrix is then given by the see-saw formula $m_{light} = -m_D^T\, m_M^{-1}\, m_D$. Note that all matrices m_D, m_M, m_M^{-1} and m_{light} are of the general form

$$m = \begin{pmatrix} x & y & y \\ y & x+v & y-v \\ y & y-v & x+v \end{pmatrix} \quad (8)$$

and therefore are diagonalized by $U_{\rm TB}$:

$$U_{\rm TB} = \begin{pmatrix} \sqrt{2/3} & 1/\sqrt{3} & 0 \\ -1/\sqrt{6} & 1/\sqrt{3} & -1/\sqrt{2} \\ -1/\sqrt{6} & 1/\sqrt{3} & +1/\sqrt{2} \end{pmatrix} . \quad (9)$$

Journal of Physics: Conference Series **335** (2011) 012050 doi:10.1088/1742-6596/335/1/012050

The m_{light} eigenvalues are:

$$\begin{aligned} m_1 &= -\left(\frac{v_u^2\, y_\nu^2}{M + a\,u + 3\,b\,v_S}\right) \\ m_2 &= -\left(\frac{v_u^2\, y_\nu^2}{M + a\,u}\right) \\ m_3 &= \left(\frac{v_u^2\, y_\nu^2}{M + a\,u - 3\,b\,v_S}\right). \end{aligned} \tag{10}$$

2.3. Alignment
At LO the most general driving superpotential w_d invariant under $A_4 \times Z_4$ with $R = 2$ is given by

$$\begin{aligned} w_d &= M(\varphi_0^S \varphi_S) + g_1(\varphi_0^S \varphi_S \varphi_S) + g_2\xi(\varphi_0^S \varphi_S) + g_3\xi_0(\varphi_S\varphi_S) + g_4\xi_0\xi^2 + M_\xi \xi_0 \xi \\ &+ M_0^2\, \xi_0 + h_1\xi'(\varphi_0^T \varphi_T)'' + h_2(\varphi_0^T \varphi_T \varphi_T)\ . \end{aligned} \tag{11}$$

The equations giving the vacuum structure for the fields φ_T and ξ' are:

$$\begin{aligned} \frac{\partial w}{\partial \varphi_{01}^T} &= 2h_2(\varphi_{T1}^2 - \varphi_{T2}\,\varphi_{T3}) + h_1\,\xi'\,\varphi_{T3} = 0 \\ \frac{\partial w}{\partial \varphi_{02}^T} &= 2h_2(\varphi_{T2}^2 - \varphi_{T1}\,\varphi_{T3}) + h_1\,\xi'\,\varphi_{T2} = 0 \\ \frac{\partial w}{\partial \varphi_{01}^T} &= 2h_2(\varphi_{T3}^2 - \varphi_{T1}\,\varphi_{T2}) + h_1\,\xi'\,\varphi_{T3} = 0 \end{aligned} \tag{12}$$

whose solutions are:

$$\langle \xi' \rangle = u' \neq 0\ , \qquad \langle \varphi_T \rangle = (0, v_T, 0)\ , \qquad v_T = -\frac{h_1 u'}{2h_2}\ , \tag{13}$$

with u' undetermined.

With the same procedure, from Eq. (11) we can obtain the equations from which to extract the vacuum expectation values for φ_S and ξ; an extremum solution is of the form:

$$\begin{aligned} \langle \varphi_S \rangle &= (v_S, v_S, v_S) \\ \langle \xi \rangle &= u\ . \end{aligned} \tag{14}$$

Note that we expect a common order of magnitude for the VEV's (scaled by the cutoff Λ):

$$\frac{v_T}{\Lambda} \sim \frac{u'}{\Lambda} \sim \varepsilon, \qquad \frac{v_S}{\Lambda} \sim \frac{u}{\Lambda} \sim \varepsilon'\ . \tag{15}$$

However, as the minimization equations for the two sets are separate, we can tolerate a moderate hierarchy between ε and ε'.

3. Beyond the leading order

At the next level of approximation each term w_l, w_ν and w_d of the superpotential is corrected by operators of higher dimension whose contributions are suppressed by at least one power of VEV's/Λ. The corrections to w_d determine small deviations from the LO VEV alignment configuration. The next to the leading order (NLO) corrections to mass and mixing matrices are obtained by inserting the corrected VEV alignment in the LO operators plus the contribution of the new operators evaluated with the unperturbed VEV's.

The VEV configuration obtained from $w_d + \Delta w_d$, where Δw_d is the most general set of terms suppressed by one power of the cutoff is given by:

$$\begin{aligned} \langle \varphi_S \rangle &= (v_S + \delta v_S, v_S + \delta v_S, v_S + \delta v_S) \\ \langle \varphi_T \rangle &= (\delta v_{T_1}, v_T + \delta v_{T_2}, \delta v_{T_3}) \\ \langle \xi \rangle &= u + \delta u \end{aligned} \tag{16}$$

and u' is still undetermined. Thus $\langle \varphi_S \rangle$ acquires $\mathcal{O}(1/\Lambda)$ corrections in the same direction, whereas all components of $\langle \varphi_T \rangle$ acquire different corrections so that its alignment is tilted.

In the charged lepton sector, the correction$\Delta\ w_l$ is obtained by adding to each term of w_l one factor of φ_S/Λ or ξ/Λ in all possible ways with arbitrary coefficients. As a result, each diagonal entry gets a small correction, while all non diagonal entries become non vanishing and of the order of the diagonal term in each row multiplied by ε':

$$m_\ell = \varepsilon\, v_d \begin{pmatrix} a_1\, \varepsilon^2 & a_2\, \varepsilon^2\, \varepsilon' & a_3\, \varepsilon^2\, \varepsilon' \\ b_1\, \varepsilon\, \varepsilon' & b_2\, \varepsilon & b_3\, \varepsilon\, \varepsilon' \\ c_1\, \varepsilon' & c_2\, \varepsilon' & c_3 \end{pmatrix} \tag{17}$$

where the coefficients a_i, b_i and c_i are $\mathcal{O}(1)$ unspecified constants. This pattern is not altered when one adds the corrections from inserting the shifted VEV's in the LO expression of w_l. The matrix $m_\ell^\dagger\, m_\ell$ can be diagonalized to Diag$[|a_1^2\, \varepsilon^4|, |b_2^2\, \varepsilon^2|, |c_3^2|]$ by the unitary transformation

$$U_\ell = \begin{pmatrix} 1 & (\frac{b_1}{b_2}\, \varepsilon')^* & (\frac{c_1}{c_3}\, \varepsilon')^* \\ -\frac{b_1}{b_2}\, \varepsilon' & 1 & (\frac{c_2}{c_3}\, \varepsilon')^* \\ -\frac{c_1}{c_3}\, \varepsilon' & -\frac{c_2}{c_3}\, \varepsilon' & 1 \end{pmatrix} . \tag{18}$$

Note that, at this order, the coefficients of the electron row in eq.(17) do not enter in U_ℓ. From the matrix in eq.(18), we can compute the corresponding corrections to the TB mixing matrix according to $U_{\rm PMNS} = U_\ell^\dagger\, U_{\rm TB}$ and all entries of $U_{\rm TB}$ get corrected to $\mathcal{O}(1/\Lambda)$.

In the neutrino sector the corrections due to inserting the VEV shifts in the LO operators do not affect the Dirac mass at all while the changes in the Majorana mass are still of the form that is diagonalized by the TB mixing matrix. However the corrections from operators of higher dimension obtained by inserting one extra power of φ_S/Λ or ξ/Λ, of order ε', alter both the Dirac and the Majorana mass in such a way that the TB mixing pattern is completely violated by small corrective terms. As a result, the overall correction to TB mixing arises from the most general symmetric matrix of order $v_S\varepsilon'$:

$$m_\nu = -m_D^T\, m_M^{-1}\, m_D = (m_\nu)_{TB} + v_S\varepsilon' \begin{pmatrix} A & B & C \\ B & D & F \\ C & F & E \end{pmatrix} \tag{19}$$

where by $(m_\nu)_{TB}$ we denote the matrix, diagonalizable by U_{TB}, which is obtained from the LO term plus the corrective terms that can be reabsorbed in the LO coefficients as they preserve TB mixing. In conclusion when the NLO corrections are included TB mixing is violated by small terms and one expects:

$$\begin{aligned} \sin^2\theta_{12} &= \frac{1}{3} + \mathcal{O}(\varepsilon') \\ \sin^2\theta_{23} &= \frac{1}{2} + \mathcal{O}(\varepsilon') \\ \sin\theta_{13} &= \mathcal{O}(\varepsilon')\ . \end{aligned} \tag{20}$$

We have already noted that the data require that $\varepsilon' \leq \mathcal{O}(\lambda_C^2)$.

Journal of Physics: Conference Series **335** (2011) 012050 doi:10.1088/1742-6596/335/1/012050

4. Light neutrino spectrum and constraints from r

It is useful to study the constraints on the model imposed by the observed values of$\Delta\ m^2_{atm}$ and of the ratio $r = \Delta m^2_{sol}/|\Delta m^2_{atm}|$. Here $\Delta m^2_{sol} = |m_2|^2 - |m_1|^2 > 0$,$\Delta\ m^2_{atm} = |m_3|^2 - |m_1|^2$. We do this in the LO approximation. In fact the results of the previous section indicate that the corrections to the spectrum are sufficiently small to be neglected for a first orientation. For this discussion we adopt the following parameterization:

$$\begin{aligned} A &= M + a\,u = |A|\,e^{i\,\phi_A} \\ B &= 3\,b\,v_S = |B|\,e^{i\,\phi_B} \\ \alpha &= \frac{|B|}{|A|} \\ \phi &= \phi_B - \phi_A\ . \end{aligned} \tag{21}$$

In LO the masses can then be written as:

$$\begin{aligned} m_1 &= -\frac{v_u^2\,y_\nu^2}{|A|\,e^{i\,\phi_A}}\left(\frac{1}{1+\alpha\,e^{i\,\phi}}\right) \\ m_2 &= -\frac{v_u^2\,y_\nu^2}{|A|\,e^{i\,\phi_A}} \\ m_3 &= \frac{v_u^2\,y_\nu^2}{|A|\,e^{i\,\phi_A}}\left(\frac{1}{1-\alpha\,e^{i\,\phi}}\right)\ . \end{aligned} \tag{22}$$

In terms of the parameters α (which is real and positive) and ϕ, the ratio r is:

$$r = \frac{\Delta m^2_{sol}}{|\Delta m^2_{atm}|} = \frac{(1+\alpha^2 - 2\,\alpha\,\cos\phi)(\alpha + 2\cos\phi)}{4\,|\cos\phi|}\ . \tag{23}$$

The limit $\alpha \to 0$ gives $r = 1/2$, which is too large compared to the experimental value. Thus in order to accomodate $r \sim 1/30$ one needs a value of α of $\mathcal{O}(1)$, which implies that (see eq.(21)) $|A| = |M + a\,u| \sim |B| = 3\,|b\,v_S|$. With $|a|, |b| \sim \mathcal{O}(1)$ this is obtained if $|M| \sim u, v_S$, or, in other words, $|M|$ must be sizeably smaller than the cutoff Λ. We interpret this result as related to the fact that the RH neutrino Majorana mass M must empirically be close to M_{GUT}. This means that in the context of a grand unified theory M must be of $\mathcal{O}(M_{GUT})$ rather than of $\mathcal{O}(M_{Planck})$. More precisely, from Eq. (23) one recognizes that a small r can be reproduced if

$$\begin{aligned} &a)\ \cos\phi \sim \alpha \sim 1 \\ &b)\ \cos\phi = -\frac{\alpha}{2} + \delta\alpha \qquad \delta\alpha \sim \mathcal{O}(r)\ . \end{aligned} \tag{24}$$

The first condition $a)$ corresponds to a normal hierarchy spectrum whereas the second condition $b)$ leads to an inverted hierarchy scheme.

In the normal hierarchy case, $a)$, by expanding around $\cos\phi \sim \alpha \sim 1$, we obtain:

$$\begin{aligned} |m_1|^2 &= \frac{1}{3}\Delta m^2_{atm}\ r + ... \\ |m_2|^2 &= \frac{4}{3}\Delta m^2_{atm}\ r + ... \\ |m_3|^2 &= \left(1 + \frac{r}{3}\right)\Delta m^2_{atm} + ... \\ |m_{ee}|^2 &= \frac{16}{27}\Delta m^2_{atm}\ r + ... \sim (0.007\ \text{eV})^2\ , \end{aligned} \tag{25}$$

where we have expressed the parameters in terms of$\Delta\ m^2_{atm}$ and r. Dots denote terms of order r^2 as well as corrections beyond the LO. Note that in this model $|m_1|$ cannot vanish. In the last line $|m_{ee}|$ is the effective mass combination controlling the violation of the total lepton number in neutrinoless double beta decay.

In the inverse hierarchy case b), we set $2\cos\phi = -\alpha + \delta$ with δ positive and small and vary α in the range between 0.07 and 2 (the lower limit comes from the absolute bound on the squared masses at fixed$\Delta\ m^2_{atm}$ and r, taken indicatively at $|m_i| \leq 0.5\ eV$). The quantity δ is determined in terms of r:

$$\delta = \frac{2\alpha r}{1 + 2\alpha^2} \tag{26}$$

while the scale of the squared masses is fixed by $|\Delta m^2_{atm}|$. In terms of α one obtains:

$$\begin{aligned} |m_2|^2 &= |m_1|^2 + |\Delta m^2_{atm}|\, r = |\Delta m^2_{atm}| \left(\frac{1+2\alpha^2}{2\alpha^2}\right)\left[1 + r\left(1 + \frac{1}{(1+2\alpha^2)^2}\right)\right] + \ldots \\ |m_3|^2 &= |m_1|^2 - |\Delta m^2_{atm}|\,. \end{aligned} \tag{27}$$

Note that the r terms are sufficiently small that the corresponding contributions could be overshadowed by the neglected NLO corrections. Omitting these additional linear terms in r, m_{ee} is given by:

$$|m_{ee}|^2 = |\Delta m^2_{atm}| \left(\frac{1+2\alpha^2}{2\alpha^2}\right)\left(1 - \frac{2}{9}\alpha^2\right) + \ldots \tag{28}$$

$|m_{ee}|^2$ is a decreasing function of α in the physical range and is close to $|\Delta m^2_{atm}| \sim (0.05\ \text{eV})^2$ for $\alpha = 1$ and to $|\Delta m^2_{atm}|/8$ for $\alpha = 2$. The behaviour of $|m_{ee}|$ as a function of α, including the neglected terms in r, is shown in the left panel of Fig.(1). The most pronounced inverse hierarchy is realized for $\alpha = 2$ where $|m_3| \sim |m_1|/3$, as it is seen from the right panel of Fig.(1).

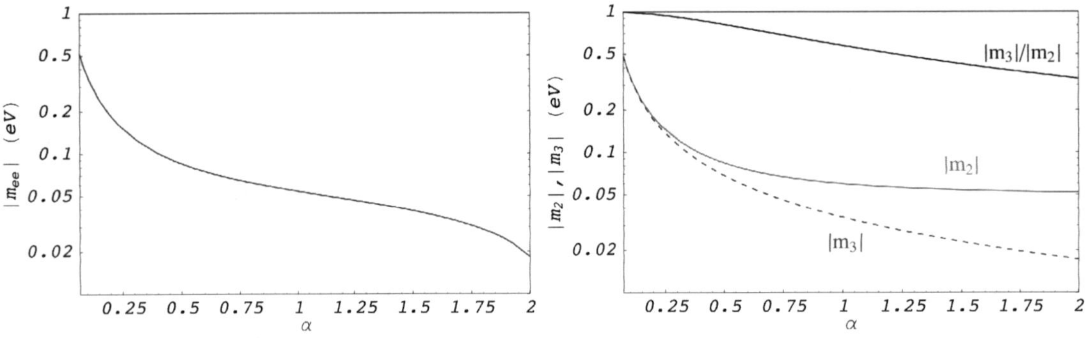

Figure 1. *Behaviour of neutrino masses in the inverted hierarchy case (at fixed Δm^2_{atm} and r) as a function of α in the range between 0.07 and 2 (the lower bound on α corresponds to an upper bound on $|m_i|$). Left panel: $|m_{ee}|$. Right panel: $|m_2|$, $|m_3|$ and the ratio $|m_3|/|m_2|$.*

Using reasonable values for the parameters, like $v_u = 170$ GeV and $y_\nu = 1$, we estimate $|A| \sim \mathcal{O}(10^{14} - 10^{15})$ GeV (see eq. (21)), with a significant preference of the normal (inverted) hierarchy toward larger (smaller) values of $|A|$, as we can see in Fig.(2), where we show the behaviour of $|A|$ as a function of $\cos\phi$ (left panel) and α (right panel) for both neutrino spectra. The allowed parameter space is built from the 3-σ constraints from [2].

In conclusion the observed value of r requires that M is at or just below M_{GUT}. Once this is realized the observed spectrum can be reproduced.

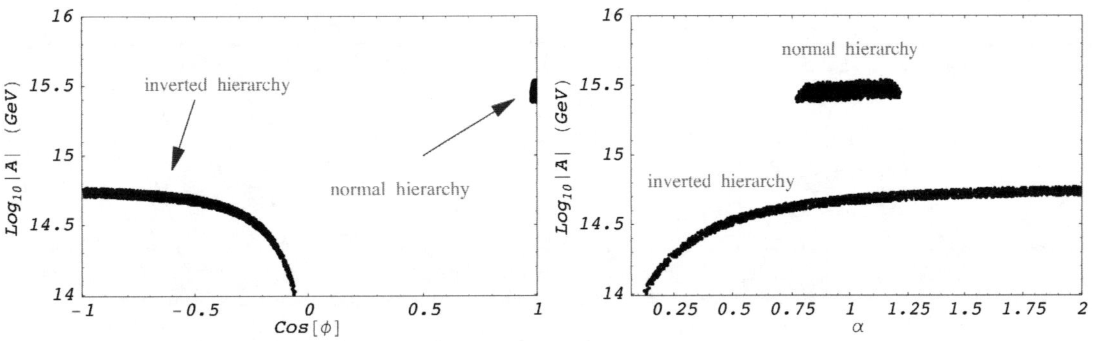

Figure 2. *Values of* $|A|$ *(in log-scale) as a function of* $\cos\phi$ *(left panel) and* α *(right panel) obtained by imposing the constraints from [2].*

5. Conclusion

We have presented and discussed an A_4 model for TB mixing of the see-saw type which, in spite of being based on a most economical flavour symmetry and field content, still it is phenomenologically viable. In particular TB mixing is exact in LO while all mixing angles receive corrections at higher orders. The charged lepton mass hierarchy is determined by the $A_4 \times Z_4$ flavour symmetry itself without invoking a Froggatt-Nielsen $U(1)$ symmetry. A value of $\theta_{13} \sim \mathcal{O}(\lambda_C^2)$ is indicated which is within the sensitivity of the experiments which are now in preparation and will take data in the near future. This example shows once more that the results derived from A_4 are robust and, in particular, do not depend on the detailed mechanism that produces the hierarchy of charged lepton masses. The model is compatible with either a normal hierarchy or an inverse hierarchy spectrum. We have studied the spectrum in detail in these different cases and discussed the predictions for the mass eigenvalues and the angle θ_{13}.

References

[1] G. Altarelli, D. Meloni, J. Phys. G **G36**, 085005 (2009). [arXiv:0905.0620 [hep-ph]].

[2] T. Schwetz, M. Tortola and J. W. F. Valle, New J. Phys. **10** (2008) 113011 [arXiv:0808.2016 [hep-ph]]; M. Maltoni and T. Schwetz, arXiv:0812.3161 [hep-ph].

[3] G. Altarelli and F. Feruglio, New J. Phys. **6** (2004) 106 [arXiv:hep-ph/0405048].

[4] P. F. Harrison, D. H. Perkins and W. G. Scott, Phys. Lett. B **530** (2002) 167 [arXiv:hep-ph/0202074]; P. F. Harrison and W. G. Scott, Phys. Lett. B **535** (2002) 163 [arXiv:hep-ph/0203209]; Z. z. Xing, Phys. Lett. B **533** (2002) 85 [arXiv:hep-ph/0204049]; P. F. Harrison and W. G. Scott, Phys. Lett. B **547** (2002) 219 [arXiv:hep-ph/0210197]; P. F. Harrison and W. G. Scott, Phys. Lett. B **557** (2003) 76 [arXiv:hep-ph/0302025]; P. F. Harrison and W. G. Scott, arXiv:hep-ph/0402006; P. F. Harrison and W. G. Scott, Phys. Lett. B **594**, 324 (2004) [arXiv:hep-ph/0403278].

[5] G. Altarelli, F. Feruglio, Rev. Mod. Phys. **82**, 2701-2729 (2010). [arXiv:1002.0211 [hep-ph]]; H. Ishimori, T. Kobayashi, H. Ohki *et al.*, Prog. Theor. Phys. Suppl. **183**, 1-163 (2010). [arXiv:1003.3552 [hep-th]].

[6] Y. Lin, Nucl. Phys. B **813**, 91 (2009) [arXiv:0804.2867 [hep-ph]].

Journal of Physics: Conference Series **335** (2011) 012051 doi:10.1088/1742-6596/335/1/012051

OPERA: A first tau-neutrino appearance candidate

Björn Wonsak for the OPERA Collaboration

Hamburg University, D-22761 Hamburg, Germany

E-mail: bwonsak@mail.desy.de

Abstract. OPERA is a long-baseline neutrino experiment dedicated to the study of muon-neutrino to tau-neutrino oscillation. Using the high-energy CERN neutrinos to Gran Sasso beam (CNGS), it is the first experiment directly searching for tau-neutrino appearance from oscillation of muon-neutrinos. Since 2008 runs with CNGS neutrinos have been successfully accomplished. After a brief introduction on the OPERA hybrid detector and the main parameters of the experiment, recent results are presented. A first candidate for a tau-neutrino charged-current event is described in detail. The background and the corresponding significance of the event is evaluated.

1. Introduction

In the past two decades, experimental results from various sources have established the picture of a three-neutrino oscillation scenario with two large mixing angles. Most of these results are based solely on dissappearance analyses, with the lack of observed neutrinos explained by oscillation. However, from the SNO results, it is clear that the total flux of active neutrinos remains constant for solar neutrinos [1], while the flavour of the appearing oscillation partners still has not been directly observed. In addition, indirect methods on a statistical basis have been applied by Super-Kamiokande, strongly favouring the hypothesis that the oscillation partner of the atmospheric ν_μ is the ν_τ [2]. On the other hand, recent MiniBooNE data [3], as well as the preceding LSND data [4], seem to include an excess of $\bar{\nu}_e$, that, if interpreted as an oscillation effect, is incompatible with the three-flavour solution. Furthermore, the first anti-neutrino data of the MINOS experiment suggest some slight tension with respect to the oscillation parameters measured with neutrinos [5]. However, these last observations require confirmation with significantly more statistic.

Thus, it is of the utmost importance to unambiguously answer as many questions as possible. The OPERA experiment [6] is specially designed to address one of these questions. Its goal is to directly observe the appearance of ν_τ in a pure ν_μ beam by identifying the τ leptons produced in their CC interaction.

The OPERA detector is located in the Gran Sasso underground laboratory (LNGS) in Italy, 732 km away from its neutrino source at CERN. The CNGS [7, 8] is a high-energy ν_μ-beam with an average neutrino energy ($\bar{\mathrm{E}}_{\nu_\mu} \approx 17$ GeV) well above the production threshold for τ leptons. The $\bar{\nu}_\mu$ beam contamination in terms of interactions is about 2.1 %, while ν_e and $\bar{\nu}_e$ together contribute less than 1 %, and the prompt ν_τ production is negligible. With a nominal CNGS beam intensity of 4.5×10^{19} protons on target (p.o.t.) per year, and assuming $\Delta m^2_{23} = 2.5 \times 10^{-3}$ eV^2 and full mixing, about 10 ν_τ events are expected to be observed in OPERA in 5 years of data taking, with selection criteria reducing the background to 0.75 events.

Journal of Physics: Conference Series **335** (2011) 012051 doi:10.1088/1742-6596/335/1/012051

The ν_τ signature is given by the decay topology of the short-lived τ lepton produced in ν_τ Charged Current (CC) interactions, decaying after typically 1 mm (at CNGS energies) to one charged particle (electron, muon or hadron) or three charged hadrons. In order to resolve this topology, a very fine spatial resolution is needed. In addition, the detector has to provide the interaction target for the neutrinos, and thus has to be massive. The technology chosen for this challenge are emulsion films interleaved with lead plates, historically called Emulsion Cloud Chamber (ECC).

2. The OPERA detector

The OPERA ECC are arranged into small light-tight packages called bricks. Each brick consists of 57 thin nuclear emulsion films, interspaced with 56 lead plates, 1 mm in thickness, with the dimensions of $10.2 \times 12.7 \times 7.5$ cm^3. The processing of events in these bricks is a complex and time-consuming task. Therefore further information is needed to reduce the scanning area as much as possible. For this reason, the OPERA detector has a hybrid structure, combining ECC-bricks with electronic detectors. The latter act as a kind of trigger for the bricks in the sense that they allow to locate the interaction vertex. Furthermore, they provide kinematical information and identify muons. Two additional emulsion films, packed in a removable box, are attached to the downstream face of each brick. These so-called Changeable Sheets (CS) are used for a first cross-check between the electronic detector data and the brick and help to reduce the scanning area in the brick films.

The detector consists of two super modules (SM1 and SM2), each of them made up of a target section and a muon spectrometer. The target section contains the bricks. In total, around 150000 bricks are used, amounting to a target mass of 1.25 kton. The bricks are arranged in walls, separated by planes of horizontal and vertical scintillator strips composing the target tracker (TT). The TT strips have a width, perpendicular to the beam direction, of 2.63 cm, allowing the identification of bricks containing an interaction vertex with a high probability.

The main component of the muon spectrometers, one behind each target, is a dipole magnet, operated at 1.52 T. The iron slabs of its core are interleaved with resistive plate chambers (RPC), used as inner trackers. In addition, the spectrometer is equipped with high-resolution drift tube chambers, the precision tracker (PT), allowing the measurement of muon momenta up to 50 GeV/c with a resolution of better than 20 % and the charge determination with an efficiency of better than 98 %. In order to remove ambiguities in the reconstruction of particle trajectories, in particular in multi-track events, each spectrometer is further instrumented with additional RPC planes in front of the magnet, with two crossed strip planes rotated by $\pm 42.6°$ with respect to the horizontal (XPC). Together with the TT the spectrometer allows the identification of μ leptons with an efficiency of 95 %.

Another two RPC planes (VETO) are placed in front of the detector, acting as a veto for charged particles originating from interactions in the upstream material (mainly muons from neutrino interactions in the rock).

A detailed description of the complete detector can be found in [9]. The event processing and analysis procedures are described in [10] and [11].

3. A ν_τ appearance candidate

OPERA has been collecting data from beam neutrinos since several years now. After a short commissioning run in 2007, with 38 beam-induced events inside the detector target, three years of physics runs have been successfully accomplished until 2010. Within these three years, neutrinos produced by a total of 9.34×10^{19} p.o.t. have been sent in the CNGS beam towards the Gran Sasso, and 9637 beam events have been collected within the OPERA target until November 23rd, 2010. The processing of these events, particularly the scanning of emulsion films, is continuously going on. Even though the event processing is very time-consuming and complex,

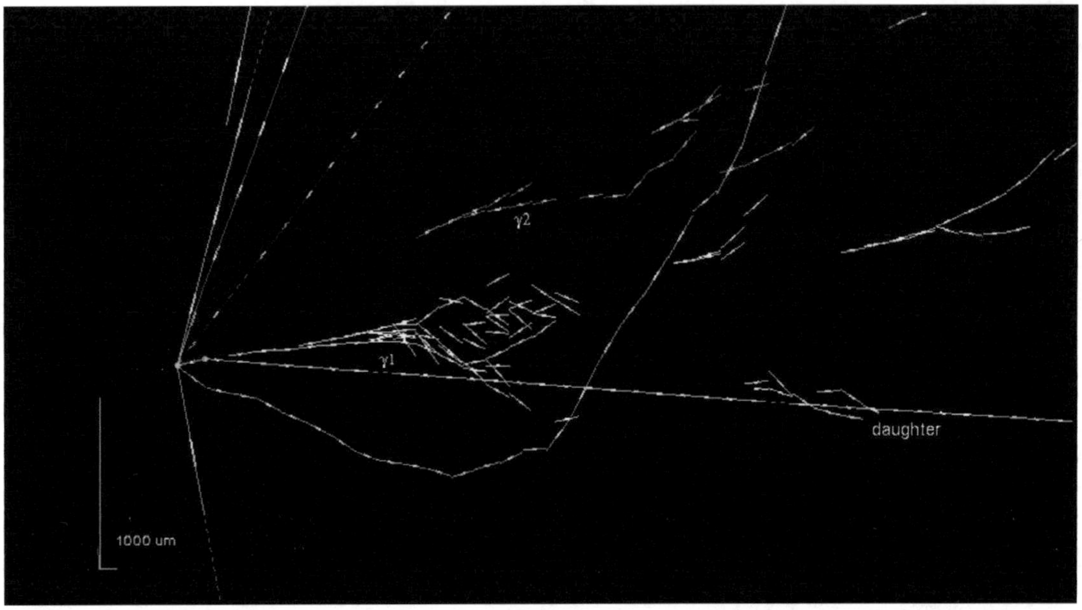

Figure 1. Side view of the ν_τ candidate event. The short yellow markings indicate the measured tracks in the emulsions. All other lines are extrapolations. The red track is the τ candidate, with its decay daughter in light blue. Two electromagnetic showers $\gamma 1$ and $\gamma 2$ are visible, pointing in the direction of the primary and secondary vertex.

at the beginning of 2010 the analysis of an unbiased sub-sample of 1088 events had been finalised, out of which 901 have been classified as CC interactions. Taking into account a combined brick finding and vertex location efficiency of about 60%, the total sub-sample corresponds to 1813 interactions occurring in the target. Considering the energy spectrum of the beam, this number of interactions is expected for an integrated beam intensity of 1.89×10^{19} p.o.t., which is about 35% of the 2008 and 2009 statistics. Assuming $\Delta m^2_{23} = 2.5 \times 10^{-3}\text{eV}^2$ and full mixing, the expected number of observed ν_τ events with these statistics is 0.54 ± 0.13 (syst.). This subsample, including 2008 and 2009 data, has been fully analysed in order to be able to extract first results.

One ν_τ candidate event has been found, fulfilling all the selection criteria determined in the experiment proposal [6]. To reject background, and for the right choice of cuts, it is very important to verify whether there are any muons attached to the primary neutrino vertex or the secondary vertex, being the possible decay vertex. Considering the range of the tracks involved as well as scattering and ionisation in the emulsions, no muon track could be found in this event. However, to verify the muonless nature of the event, three of the tracks that do not end in the interaction brick have been followed through the emulsions of the downstream bricks, up to their endpoint. For two of those, an interaction at the endpoint could be confirmed, ruling out the possibility of muons. The last track has only a short range, resulting in a probability of about 10^{-3} to be left by a muon.

The procedure to follow down tracks to check their possibility of being muons is an upgrade with respect to the procedures taken into account within the proposal. With this new procedure, the residual probability of an event for being a ν_μ CC interaction with an undiscovered large angle low momentum muon can be reduced from ~5 %, as was assumed in the proposal, to ~1 %. However, for the calculations of background and sensitivity given below, the nominal value of the proposal has been used.

Another distinctive feature of this event are the two γ showers that have been found to belong to it, clearly visible in figure 1. One of the selection criteria for the decay of a τ to a single charged hadron is that the total transverse momentum (P_T) of the decay products is larger than 0.6 GeV/c if there are no photons emitted at the decay vertex, and 0.3 GeV/c otherwise. Since P_T has a relative small expectation value of $\lesssim$100 GeV in pion interactions, this cut rejects most of the hadronic reinteractions mimicking the decay topology. The cut is lower if at least one gamma shower could be attached to the decay vertex, because MC studies have shown that the expected background is reduced if a π_0 is emitted at the decay vertex [12]. The probability of the gamma showers to point to the secondary vertex has been calculated to be 32 % for the first shower, and 82 % for the second. Their probability to be attached to the primary vertex is less than 0.1 %, and 10 %, respectively. For the most probable hypothesis of both showers being attached to the secondary vertex, the total transverse momentum P_T of the daughter particles is $(0.47^{+0.24}_{-0.12})$ GeV/c. If the second γ shower is instead attached to the primary vertex, this number is changed by no more than 50 MeV/c. Thus, in both cases the P_T cut criteria mentioned above is satisfied.

The invariant mass of both γ is $(120 \pm 20\,(\text{stat.}) \pm 35\,(\text{syst.}))$ MeV/c^2, supporting the hypothesis that they originate from a π^0 decay. Similarly, the invariant mass of the charged decay product, assumed to be a π, and of the two γ amounts to $(640^{+125}_{-80}\,(\text{stat.})^{+100}_{-90}\,(\text{syst.}))$ MeV/c^2, being compatible with the $\rho(770)$ mass. The branching ratio of the decay mode $\tau \rightarrow \rho\pi^0$ is about 25 % [13].

The two dominant sources of background for a 1-prong hadronic τ decay are decays of charmed particles, produced at the primary vertex, and 1-prong inelastic interactions of primary hadrons. In both cases, ν_μ NC and CC interactions contribute. The latter can efficiently be rejected if the primary muon is identified. Following the experiment proposal [6], $0.007 \pm 0.004\,(\text{syst.})$ charm background events are expected in the analysed sample and $0.011 \pm 0.006\,(\text{syst.})$ background events from hadronic reinteractions. However, the current estimations of charm background are conservative, since they are based on the original scanning strategy. If all decay channels of the τ are considered, the total background, including large-angle muon scattering, amounts to $0.045 \pm 0.02\,(\text{syst.})$ events.

The probability that the total expected background of $0.018 \pm 0.007\,(\text{syst.})$ events to the 1-prong hadronic decay channel of the τ fluctuates to 1 event is 1.8 % ($2.36\,\sigma$) while for all decay channels the probability is 4.5 % ($2.01\,\sigma$).

A detailed description of the event can be found in [14]. Further information regarding the analysis procedures and the background can be found in [11].

4. Conclusions

The OPERA experiment searches for ν_τ appearing at a distance of 732 km in the CNGS ν_μ beam. A complete analysis of a subsample corresponding to 1.89×10^{19} p.o.t. has been performed. One muonless event with a $\tau \rightarrow$ 1-prong hadron decay topology has been detected that passes the selection criteria. It is the first candidate event for $\nu_\mu \rightarrow \nu_\tau$ appearance.

The observation of one possible ν_τ candidate in the 1-prong hadronic decay channel has a significance of 2.36 σ of not being a background fluctuation. If one considers all decay modes included in the search, the significance of the observation is 2.01 σ.

References

[1] Q. R. Ahmad *et al.* [SNO Collaboration], Phys. Rev. Lett. **89** (2002) 011301
[2] K. Abe *et al.* [Super-Kamiokande Collaboration], Phys. Rev. Lett. **97** (2006) 171801
[3] A. A. Aguilar-Arevalo *et al.* [The MiniBooNE Collaboration], Phys. Rev. Lett. **105** (2010) 181801
[4] A. Aguilar *et al.* [LSND Collaboration], Phys. Rev. D **64** (2001) 112007
[5] Patricia Vahle [for the Minos Collaboration], presentation at the Neutrino 2010 in Athens
[6] M. Guler *et al.* [OPERA Collaboration], LNGS-P25-00

[7] Ed. K. Elsener, CERN 98-02, INFN/AE-98/05
[8] R. Bailey et al., (Addendum to CERN 98-02, INFN/AE-98/05), CERN-SL/99- 034(DI), INFN/AE-99/05
[9] R. Acquafredda *et al.*, JINST **4** (2009) P04018
[10] N. Agafonova *et al.*, JINST **4** (2009) P06020
[11] C. Lazzaro, "Summary of the OPERA analysis", this conference proceedings
[12] M. Guler *et al.* [OPERA Collaboration], LNGS-EXP-30-2001-ADD-1
[13] K. Nakamura et al. (Particle Data Group), J. Phys. G 37, 075021 (2010)
[14] N. Agafonova *et al.* [OPERA Collaboration], Phys. Lett. B **691** (2010) 138

Journal of Physics: Conference Series **335** (2011) 012052 doi:10.1088/1742-6596/335/1/012052

Summary of the OPERA analysis

C. Lazzaro on behalf of the OPERA collaboration

E-mail: claudia@lazzaro.ch

Abstract. The OPERA experiment is located in the Gran Sasso underground laboratory and designed to directly detect the appearance of ν_τ in a pure ν_μ beam by observing the decay of charged τ leptons produced in ν_τ charged current interactions. The OPERA detector is a hybrid detector, containing electronic detectors (ED) and high resolution nuclear emulsions. A short introduction to the detector is presented. The analysis procedures will be described, including the analysis flow chain and the vertex decay search procedure. First results will be presented, including the main kinematic measurements obtained until now and the cross checks made on the background expectation.

1. Introduction

In the past two decades several experiments have provided strong evidence supporting the neutrino oscillation hypothesis. The ν oscillation was first established by the SuperKamiokande experiment [1] on atmospheric neutrinos and afterwards confirmed by the K2K [2] and MINOS [3] accelerator experiments. These experiments observed the disappearance of ν_μ neutrinos; the unambiguous, direct observation of the appearance of a ν_τ from an oscillated ν_μ neutrino is still missing. The OPERA experiment [4] has been designed to directly observe the ν_τ appearance in the CNGS ν_μ beam [5] from CERN. The ν_τ signature is given by the observation of the decay topology of the charged τ leptons produced in ν_τ Charged Current(CC) interactions.

2. The OPERA experiment

2.1. The CNGS beam

The CNGS beam [5] is a wide band neutrino beam ($\langle E_{\nu_\mu}\rangle = 17\ GeV$), which was optimized for the tau production and decay observation at the Gran Sasso laboratory. It is produced by a 400 GeV/c proton beam extracted from the SPS accelerator and that hits a carbon target producing mainly pions and kaons. Positively charged π and K are energy selected and focussed with the horn and the reflector into a 1 km long decay tunnel in the direction of the Gran Sasso laboratory. The mesons decay mainly into ν_μ and μ. The nominal integrated beam intensity is 4.5×10^{19} protons on target ($p.o.t.$), for 200 days of beam operation per year. The time structure of the beam is characterized by a CNGS cycle of 6 s duration with two extraction spills separated by 50 ms and lasting 10.5 μs each, as shown in Fig. 1.

2.2. The OPERA detector

The decay length of the τ particles at CNGS energies is below 1 mm on average, thus, to reconstruct the decays, a spatial resolution of the order of about a micron is required. Also, due to the low neutrino cross section, a large mass experiment is mandatory. To fulfil these

Published under licence by IOP Publishing Ltd

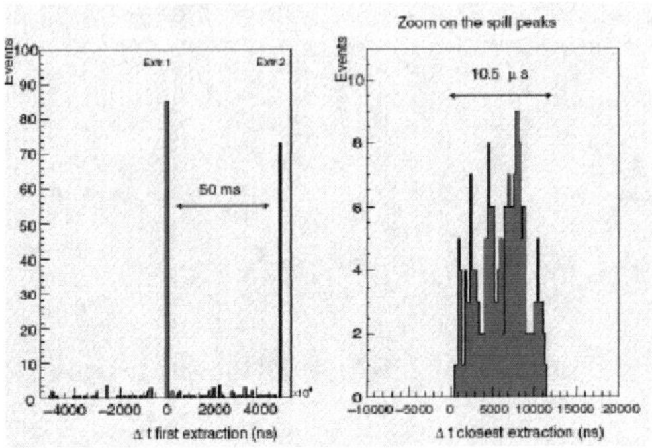

Figure 1. Time structure of the OPERA events following the CNGS beam cycle. On the left the time separation of the two extractions, and on the right a zoomed view of one extraction is shown.

requirements, OPERA was built as a hybrid detector consisting of a massive lead/photographic emulsion target to reach the micron resolution at the event vertex, complemented by electronic detectors, see [4] for details.
The detector is made of two identical super-modules (SM), as shown in Fig. 2. Each one contains 31 vertical target walls (perpendicular to the beam), able to hold the basic units of OPERA: the so-called emulsion cloud chamber (ECC) bricks. Each brick is a sandwich of 56 lead plates, 1 mm thick, interleaved by 57 emulsion films; it weighs about 8.3 kg. The transverse dimensions of the brick are 12.5×10 cm^2, the thickness along the beam direction is 7.9 cm, corresponding to about 10 radiation lengths. The two target modules contain in total about $150,000$ bricks for a mass of 1.25 kton.
Behind each target wall there are two planes of horizontal and vertical scintillator strips (TargetTracker, TT); the strips are 1 cm thick, 2.6 cm wide and 6.9 m long. The scintillators are mainly used to predict the brick in which a neutrino interaction occurred, to measure the hadronic energy and the precise time of the event. Attached to the downstream side of each brick are two additional films, called Changeable Sheets (CS). They are used to validate the brick, which was predicted to contain the interaction by the electronic detector.
A magnetic spectrometer is placed behind the target section in each super module: each arm of the dipole magnet is composed of 12 iron slabs, interleaved by 11 Resistive Plate Chambers (RPC). Before, after and between each arm a station of vertical drift tubes (Precision Trackers, PT) permit an accurate measurement of the curvature of the charged particles in the magnetic field region, yielding a measurement of the charge and momentum of particles leaving the target, mainly muons. Two additional glass RPC planes (VETO) are mounted in front of the detector to select the charged particles originating from outside the target, due to neutrino interactions in the surrounding rock material.
Events generated by the CNGS neutrino beam are selected using a delayed coincidence between the proton extraction spill from the SPS and events in the detector. The synchronization between the SPS at CERN and the OPERA detector at LNGS is based on a GPS system, with a precision of 100 ns.

2.3. The τ detection principle
The τ events are identified selecting the decay topology of the short-lived τ leptons produced in the ν_τ Charged Current (CC) interactions, decaying to one prong (electron, muon or hadron) or three prongs (hadrons). The one-prong events are characterized by a kink at the tau decay

Figure 2. The OPERA detector at the Gran Sasso laboratory.

Table 1. Expected number of τ and background events in OPERA for 5 nominal years of data taking.

decay channel	signal ($\Delta m^2 = 2.5 \times 10^{-3} eV^2$)	background
$\tau \to \mu$	2.9	0.17
$\tau \to e$	3.5	0.17
$\tau \to h$	3.1	0.24
$\tau \to 3h$	0.9	0.17
ALL	10.4	0.75

vertex or by a track with a large impact parameter with respect to the primary vertex. The reconstruction of the event topology is possible due to the high resolution of the emulsions in the brick.

The brick analysis and the event reconstruction in the brick include the momentum measurement of charged particles by the multiple Coulomb scattering (MCS) in the lead plates, a measurement of electromagnetic showers and muon-pion separation.

Assuming 5 years of data taking with nominal beam intensity, the total number of ν_μ interactions expected in the detector is about 24,000, and about 170 ν_e and $\overline{\nu}_e$ CC interactions. The number of ν_τ charged current interactions is about 115 (for $\Delta m^2 = 2.5 \times 10^{-3} eV^2$ and full mixing); including decay branching ratios and experimental efficiencies, about 10 tau decays from ν_τ CC events are expected to be observed, with tight selection conditions keeping the background below one event. The main background sources are:

- events with hadronic re-interactions mimicking tau decay topologies,
- the production of (short-lived) charmed particles in charged current ν_μ interactions where the primary lepton is not identified,

The expected number of τ and background events for each channel are summarized in Table 1.

3. The analysis chain

The selection of neutrino events is performed off-line. Every time a trigger of the electronic detectors is compatible with an interaction inside the target and in time with the beam, an algorithm is applied to the electronic detector data to select the most probable brick in which the interaction occurred. The analysis of the electronic detector includes the reconstruction of the event, that permits first of all a selection between NC-like (Neutral Current) and CC-like events. In the case of a CC event, the muon track is reconstructed with PT, RPC and TT data. Different procedures, including a Hough transformation and a Kalman filter, are applied to estimate the position of the muon on the CS of a brick, to select in this way the brick containing the interaction. A measurement of the muon momentum is obtained using the information of the drift tubes (PT) in the spectrometer with a resolution of 20% up to a momentum of 30 GeV/c [6]. If no muon track is reconstructed, the TT data of the hadronic shower are used to estimate the position of the interaction vertex. In both cases an estimation of the hadronic energy of the event is obtained by the analysis of electronic detector data.

For each event a probability map for the closest bricks to contain the neutrino interaction is calculated, and the brick with the highest probability is extracted from the target, the CS films are developed and analyzed to validate the brick prediction. If the event is found in the CS, the brick will be exposed to X-rays and cosmic rays for the alignment of the emulsion films. The brick is then disassembled, the films developed and sent to the different scanning laboratories to perform the complete analysis; there are ten scanning laboratories in Europe and two in Japan.

The first scanning procedure in the brick (Scan Back phase) is focused on the neutrino interaction vertex location: the tracks found in the CS are followed from the most downstream film of the bricks to their stopping point (a track is considered as stopped, if it is not found in three consecutive films along its upstream extrapolation). Next, a volume around the stopping point is scanned (Total Scan): an area of 1 cm^2 is analysed around the stopping point in 10 films downstream and 5 upstream of the stopping plate. With the tracks found in the volume, the vertex is constructed, using the impact parameter (IP) of the tracks to the vertex; the IP of a track is defined as the shortest distance between the reconstructed candidate vertex and the track. As can be seen in Fig. 6, a precise measurement of the IP is fundamental to have a good τ decay detection capability. In addition, tracks attached to the vertex can be followed in the downstream direction (Scan-Forth) for kinematical measurements.
In order to detect all possible decay topologies near the primary vertex, a systematic decay search procedure is applied to the events; this includes three main steps:

- analysis of the impact parameter distribution,
- detection of possible decay topologies on tracks attached to the primary vertex,
- search for extra tracks from neutral decays or gamma conversion not attached to the primary vertex.

The measurement of a charged particle's momentum in a brick is possible due to the precise measurements of the angular deviations produced by multiple Coulomb scattering in the lead plates; for charged particles up to 6 GeV/c a resolution better than 22% [7] is obtained. For higher momentum particles (passing through an entire brick), the measurement is based on position deviations and the resolution is better than 33% up to 12 GeV/c.

For the detection of the decay topology it is especially important to reconstruct the gamma rays in the bricks, since 70% of 1-prong hadronic τ decays include one or more π^0. The γ reconstruction is performed by the analysis of the electromagnetic shower; its energy is determined with a Neural Network algorithm that uses as input mainly the number of tracks found in the shower and its shape and size.

With the decay search procedure neutrino induced charm events are also found. Charmed particles are produced in about 4% of CC ν_μ interactions at the CNGS energies, and they have

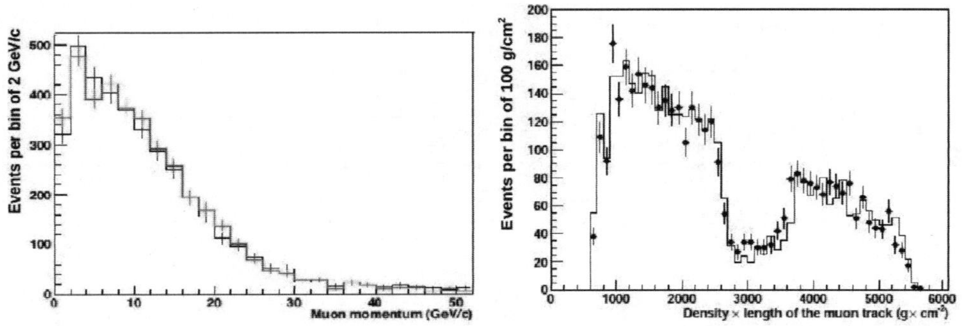

Figure 3. On the left: data-MC comparison for the reconstructed muon momentum and on the right data-MC comparison of muon track lengths in the detector in g/cm^2. MC distributions have been normalized to data

similar lifetime and decay topologies as the τ particle; this also means that they can mimic a τ decay in case the primary muon is not identified. Hence, the study of charmed particle decays in the OPERA experiment is necessary to validate the procedures used for the selection and identification of short-lived particles and their efficiencies estimated by the Monte-Carlo, and to measure the background from charm decays.

4. Physics results

The results presented here were obtained from electronic detector data collected in 2008-2009, and the decay search analysis applied to about 35% of the 2008/-009 emulsion data sample, this corresponds to an integrated beam intensity of $1.89 \times 10^{19} p.o.t$ and 1088 ν events of which 901 were classified as CC interactions. The analysis of this sample was reported together with the first τ decay candidate in [8][1]. In this section the data-MC comparison of some kinematical variables and the background estimation used to validate the analysis and to cross-check the MC is presented.

4.1. Eletronic detector data analysis

Many kinematical variables were studied, mostly related to the muon track, such as the reconstructed muon momentum in the CC events and the length of the muon track times the density of the crossed material, which is the main variable used for the muon identification. These are shown in Fig. 3.

An important cross check is the comparison of the measured and expected ratio between neutral current and charged current interactions. The measured (visible) NC/CC ratio is taking into account the true NC/CC ratio (= 0.3), the muon identification efficiency, the fraction of the CC misidentified as NC events, and the fiducial volume selection efficiency. The fiducial volume selection is necessary to reduce the background accumulation at target borders due to particles produced in beam-neutrino interactions outside the target in the rock and surrounding material. Very good agreement is found for this ratio between the MC-expectation and the measurement in the preliminary analysis of 2008 and 2009 data, not including systematic errors under study, as shown in Table 2.

A good agreement between electronic detector data and MC expectation for the other investigated variables was observed.

[1] The tau decay candidate is not described here, since it is covered in the contribution of Bjoern Wonsak [9].

Journal of Physics: Conference Series **335** (2011) 012052 doi:10.1088/1742-6596/335/1/012052

Table 2. Measured (visible) and expected MC ration NC/CC.

MC $NC/CC = 0.236 \pm 0.005(stat)$
Data 2008-2009 $NC/CC = 0.230 \pm 0.008(stat)$

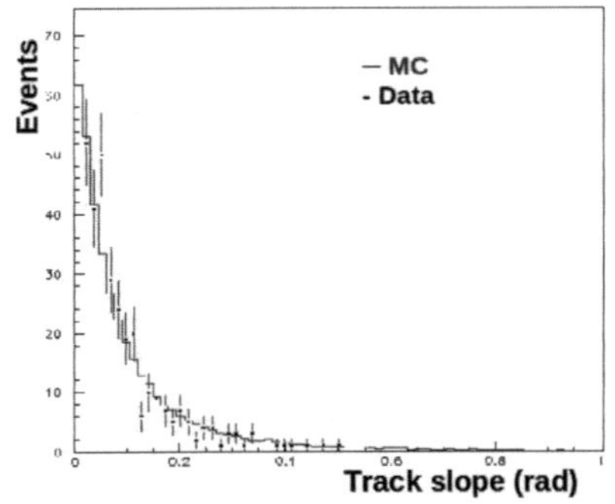

Figure 4. Distribution of muon slopes for ν_μ CC events measured at the primary vertex and comparison with the MC expectation.

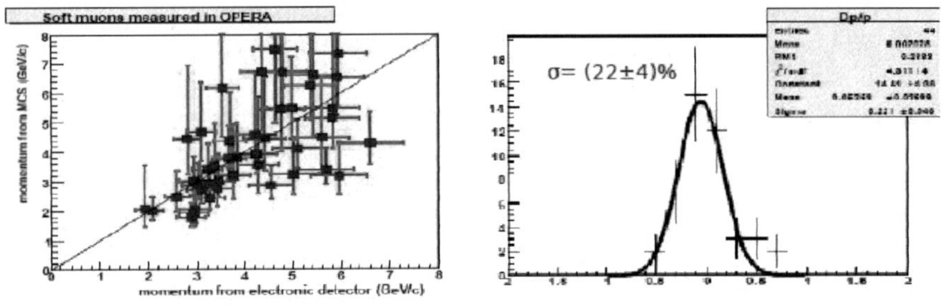

Figure 5. Correlation between muon momentum measurements in the brick (using MCS) and by the electronic detector.

4.2. Analysis of emulsions

The vertex position in the plane transverse to the beam (x-y coordinate distribution) is observed to be uniform, indicating that the transverse vertex location procedure is not biased. In Fig. 4 the measured muon slopes are compared with the MC expectation.

The muon momentum measured in the emulsions (by multiple Coulomb scattering) for the soft muons is compared with the measurement obtained with the ED to validate the procedure. They are in good agreement, as shown in Fig. 5, with a resolution of about 20%.

The IP distribution at the ν_μ vertex is shown in Fig. 6; it is well contained within 10 microns; the IP distribution for the daughter tracks from τ decays form a Monte Carlo simulation is also shown, typical τ decay IP values are well beyond the region dominated by the experimental resolution.

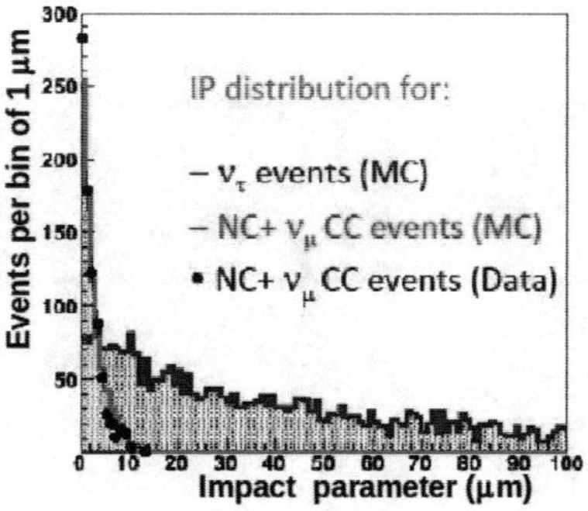

Figure 6. Simulated tau daughters impact parameter distribution is shown by the blue shaded histogram, the impact parameter distribution measured for tracks emitted at the primary vertex in MC ν_μ interaction events is shown by the red shaded histogram. The corresponding observed distribution (black dots) does not include the tau and charm decay candidates.

4.3. Background
As discussed in the previous section, the two most important background sources for the OPERA experiment are hadron re-interactions and charm decays.

4.3.1. Hadronic re-interaction To study the hadronic re-interaction background, first of all a MC analysis has been done, after simulation of a very large number of hadron interactions in lead The kink probabilities are evaluated by applying the same cuts as for the tau decay analysis; integration over the ν_μ NC hadronic spectrum yields a background probability of $(1.9\pm0.1)\times10^{-4}$ kinks/NC. This probability decreases to $(3.8\pm0.2)\times10^{-5}$ per NC event when taking into account the cuts on the event global kinematics. With the collection of a sufficient number of events a search for hadronic re-interactions in the data was also started. So far, a total of 9 m of hadrons tracks have been followed for the presence of re-interactions far from the vertex region. No events were found that satisfy the selection criteria for ν_τ candidates, corresponding to an upper limit of 1.54×10^{-3} kinks/NC event ($90\% C.L.$). This search will be extended to about 100 m of tracks in order to validate and eventually replace the MC estimation.

4.3.2. Charm decay events To limit this source of background it is extremely important to have a good identification of the primary muon; in OPERA the muon identification reaches $\sim 95\%$. The charm background expectation for the 1-prong hadronic tau decay channel (the found tau candidate is in this channel) for the statistics collected until now is 0.007 ± 0.004 (syst) events.

An example of a neutral charm candidate decaying in 4 prongs is shown in Fig. 7: 3 tracks have been reconstructed at the primary vertex. All the measured impact parameters at the two vertices are of the order of a micron. The minimum value of the invariant mass was estimated to be 1.74 GeV/c^2; the angle ϕ between the directions of the parent track and of the primary hadron shower momentum in the transverse plane, is found to be close to $180°$ which corresponds to a back-to-back emission of the decaying particle with respect to the muon, strengthening the charm hypothesis.

Improvement of the charm background reduction is under evaluation by implementing the systematic follow-down of low energy tracks in the bricks and the inspection of their end-range for a better hadron-muon separation.

For the statistics collected in 2008-2009 , the charm decays found and expected from MC

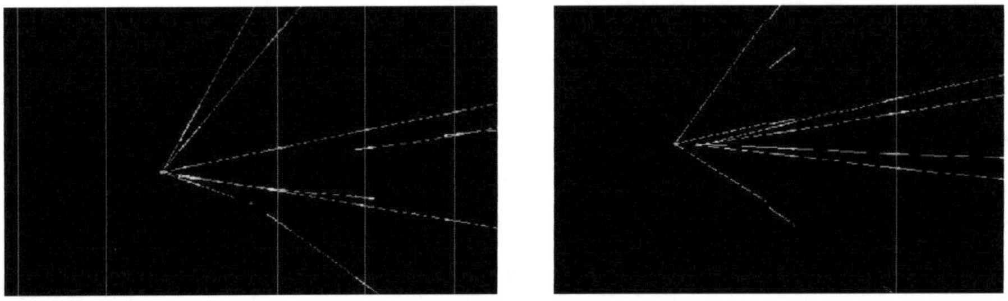

Figure 7. Charm events: charged charm with a clear kink shown by the red dot on the left and D^0 decay into 4 prongs on the right.

simulations are shown in Table 3.

Table 3. Charm decays found for the statistics collected in 2008-2009, and MC expectation.

20 charm candidates
Expected: 16 $\pm$ 2.9

5. Conclusions

OPERA is collecting neutrino interactions in its lead-emulsion targets since 2008. Up to now, 1921 neutrino vertices have been precisely located and the decay search analysis has been completed for a sub-sample of 1088 events. One tau decay candidate was found [8], and the full analysis chain was tested. The electronic detectors are running reliably and the performance is well understood. A systematic decay search was started in 2008 and 2009 located neutrino interactions in order to find all possible decay topologies. The good agreement between the observed number of charm decays and the MC expectation confirms the capability to find and reconstruct decays of short-lived particles with the OPERA detector. The background sources are under control, and the data analysis of the 2009-2010 run is in progress. The experiment is expected to continue data taking for two more years in CNGS beam.

References

[1] Y. Fukuda, et al., Phys. Rev. Lett. 81 (1998) 1562.
[2] M.H. Ahn, et al., Phys. Rev. D 74 (2006) 072003.
[3] D.G. Michael, et al., Phys. Rev. Lett. 101 (2008) 131802
[4] R. Acquafredda et al. [OPERA Collaboration], JINST 4 (P04018) (2009).
[5] CNGS project. http://proj-cngs.web.cern.ch/proj-cngs/.
[6] R. Zimmermann et al., Nucl. Instrum. Meth. A 555 (2005) 435; R. Zimmermann et al., erratum, Nucl. Instrum. Meth. A 557 (2006) 690.
[7] M. De Serio at al., Nucl. Instrum. Meth. 512 (2003) 539; M. Besnier, PhD thesis (2008) Université de Savoie, LAPP-T-2008-02
[8] OPERA Collaboration, Phys.Lett.B691:138-145,2010.
[9] B. Wonsak, *OPERA: a first tau-neutrino appearance candidate*, this conference proceedings.

Journal of Physics: Conference Series **335** (2011) 012053 doi:10.1088/1742-6596/335/1/012053

T2K experiment: Status and first results

F. Dufour, for the T2K collaboration

University of Geneva, 24 Quai Ernest-Ansermet, 1205 Geneva, Switzerland

E-mail: fanny.dufour@unige.ch

Abstract. The long baseline neutrino oscillation experiment T2K, which uses a high intensity off-axis muon neutrino beam produced at JPARC, aims at a discovery of oscillation of muon neutrinos into electron neutrinos and measurement of the mixing angle θ_{13}, a key parameter for further search for CP violation in the lepton sector. T2K is also designed to precisely measure the oscillation parameters Δm^2_{23} and θ_{23} in a disappearance oscillation mode. The major components of T2K include a neutrino beam line, muon monitors, a near detector complex ND280 located at 280 m from the proton interaction target, and the far detector Super-Kamiokande (295 km from the neutrino source), at a 2.5 degree off-axis angle from the beam. The measured neutrino spectrum at ND280 and measurements of the primary proton beam combined with the NA61 hadron production data will be used to predict the unoscillated spectrum at Super-Kamiokande. The oscillation analysis will be done by comparing the observed and predicted spectra at the far detector. The construction of the T2K neutrino beam line and the near neutrino detectors is completed. For the first data taking period with the neutrino beam, about 3×10^{19} 30 GeV protons on target were accumulated. The design, performance, current status of the experiment and near future perspectives will be discussed. The results of the first physics run will be presented.

1. Introduction

The Tokai to Kamioka (T2K) experiment is a long baseline muon neutrino beam experiment located in Japan. The main components of T2K include a neutrino beam line, muon monitors, a near detector complex ND280 located at 280 m from the proton interaction target, and the far detector Super-Kamiokande (295 km from the neutrino source), at a 2.5 degree off-axis angle from the beam. The main goals of the experiment are to measure the mixing angle θ_{13} for the first time, and to improve the measurement of θ_{23} and Δm^2_{23}. In addition, several neutrino cross-section studies will be conducted with the near detector. In these proceedings, I will describe the current status of the experiment and the data collected in the first run which started in January 2010 and ended in June 2010.

2. Overview of the T2K experiment

A schematic view of the experiment is present in Fig. 1. Protons are accelerated to 30 GeV throughout the Tokai accelerator complex, and collide on a graphite target. From these collisions pions are created. Positive pions are selected and focused with a series of three magnetic horns. These pions then decay into muon and muon neutrinos inside a 94 m long decay tunnel. The muons are observed with the muons monitors, 120 m away from the target. The neutrino near detector is located 280 m away from the target (which explains its name: ND280). It consists of two parts: the on-axis and the off-axis detector. As their names suggest, the on-axis near

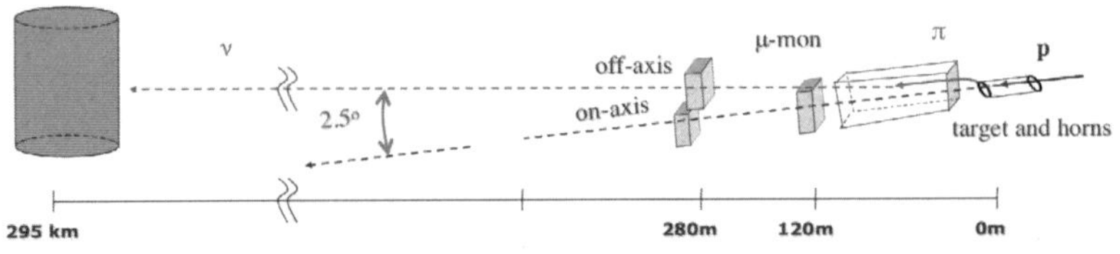

Figure 1. Overview of the T2K experiment.

detector (INGRID) is on axis with the neutrino beam, while the off-axis near detector is at a 2.5° off-axis angle from the neutrino beam. A more detailed description of the near detector is given in Section 5 and in the ND280 technical design report [1]. Finally, the far detector (Super-Kamiokande) is located 295 km away and at a 2.5° off-axis angle from the target.

3. Physics goals

The physics goals of T2K are to improve the measurement of Δm^2_{23} and $\sin^2 2\theta_{23}$ using the ν_μ disappearance analysis, and to measure for the first time the θ_{13} oscillation parameter through the ν_e appearance analysis. The sensitivity of the T2K experiment for the measurement of θ_{13} is presented in Fig. 2 and Fig. 3. In addition the near detector will be able to improve neutrino cross-section measurements around 750 MeV.

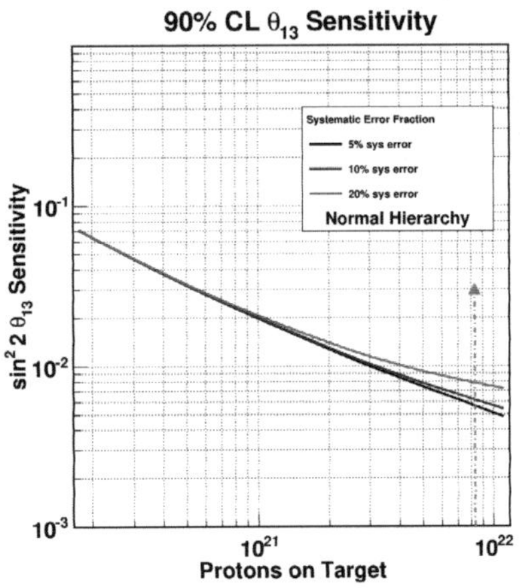

Figure 2. T2K sensitivity to θ_{13} at the 90% confidence level as a function of protons on target. 5%, 10% and 20% systematic error fractions are plotted. The dashed arrow indicates a 5-year run at 750 kW. The following oscillation parameters are assumed: $\sin^2 2\theta_{12} = 0.8704$, $\sin^2 2\theta_{23} = 1.0$, $\delta m^2_{12} = 7.6 \times 10^{-5}\text{eV}^2$, $\delta m^2_{23} = 2.4 \times 10^{-3}\text{eV}^2$, $\delta_{CP} = 0$, normal hierarchy.

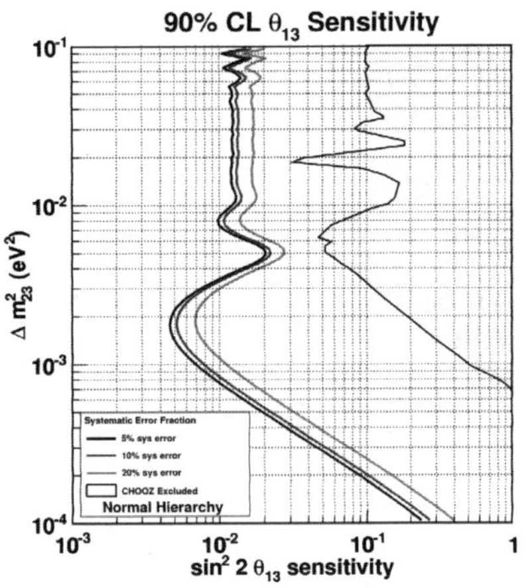

Figure 3. T2K sensitivity to θ_{13} at the 90% confidence level as a function of Δm^2_{23}. Beam is assumed to be running at 750kW for 5 years, using the 22.5 kton fiducial volume SK detector. 5%, 10% and 20% systematic error fractions are plotted. The yellow region has already been excluded to 90% confidence level by the Chooz reactor experiment. The oscillation parameters are the same as in Fig. 2.

3.1. Oscillation analysis strategy

Both oscillation analyses rely on the comparison of the data collected at the near detector and the data collected at the far detector. The number of events at the far detector can be estimated using the following equation:

$$N(E_\nu^{rec}) = \Phi_{SK}^{exp}(E_\nu^{true}) \times \sigma(E_\nu^{true}) \times P_{osc}(E_\nu^{true}) \times \epsilon_{SK}(E_\nu^{true}) \times f(E_\nu^{rec}, E_\nu^{true}), \qquad (1)$$

where Φ_{SK}^{exp} is the expected flux at SK, $\sigma(E_\nu^{true})$ the neutrino cross-section, $P_{osc}(E_\nu^{true})$ the oscillation probability, $\epsilon_{SK}(E_\nu^{true})$ the efficiency to detect a neutrino at SK and $f(E_\nu^{rec}, E_\nu^{true})$ is the energy response function. The expected flux at SK, Φ_{SK}^{exp}, is given by $\Phi_{SK}^{exp} = R_{SK/ND} \times \Phi_{ND}$ and the near detector flux, Φ_{ND}, is given by $\Phi_{ND} = N_{ND}^{obs}/(\sigma_{ND} \times \epsilon_{ND})$. In order to know Φ_{ND} reliably, we need inputs from the hadron production experiment NA61, and we need to know neutrino cross-sections (σ_{ND}). We will use neutrino cross-sections measured by other experiments, like SciBooNE, but also by the T2K near detector. The flux at SK, Φ_{SK}^{exp}, relies on a good understanding of the normalization and of the neutrino energy spectrum of the near detector data, and a good understanding of the far/near ratio ($R_{SK/ND}$).

4. Beam commissioning for the 2010a run

The T2K experiment started to take data in January 2010. The first run lasted from January to June 2010 and accumulated 3.23×10^{19} protons on target (POT) at 30 GeV. The beam was run stably at 50 kW and we also achieved powers of 100 kW. During this running period, the Super-Kamiokande detector and the INGRID detector were 99% efficient and the off-axis near detector was very stable after it started running in February 2010. An overview of the 2010a run is presented in Fig. 4.

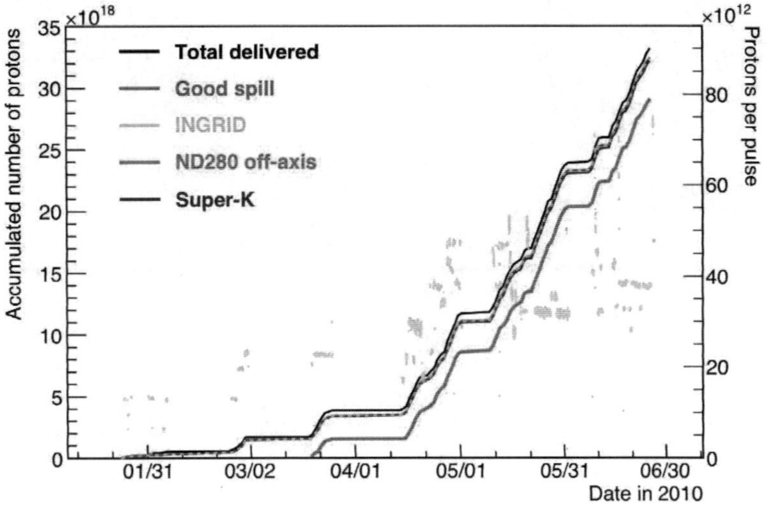

Figure 4. Number of protons on target during the 2010a data-taking period. Yellow points are the number of proton per pulse and should be read with the scale on the right side of the plot.

5. Current status of the near detector (ND280)

An overview of the near detector is presented in Fig. 5 and a detailed view of the off-axis near detector is presented in Fig. 6. The purpose of the on-axis INGRID detector is to monitor the direction of the beam by looking at neutrinos themselves. The off-axis near detector is located inside 0.2 Tesla magnetic field produced by the UA1/NOMAD magnet. It consists of several fine grained detectors and trackers. All sub-detectors were installed in 2009 and ready for data-taking in 2010 apart from the barrel ECAL which was installed in summer 2010. The purpose of the off-axis near detector is to:

- measure charge-current interactions from beam muon neutrinos in order to do the normalization with the far detector.
- measure charge-current interactions from beam electron neutrinos in order to predict the background contamination for the ν_e appearance analysis.
- measure neutrino cross-sections.

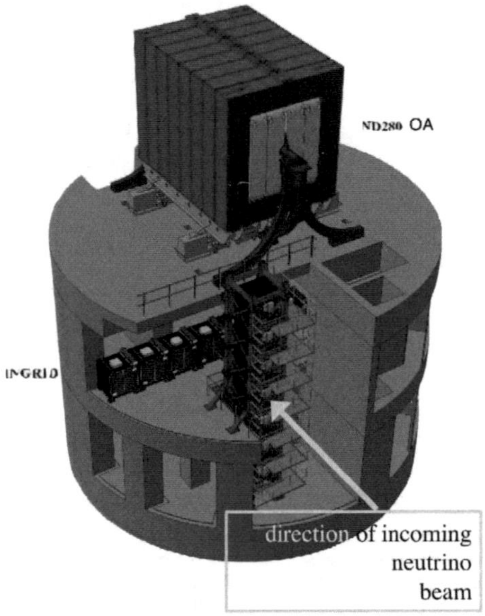

Figure 5. General view of the near detector.

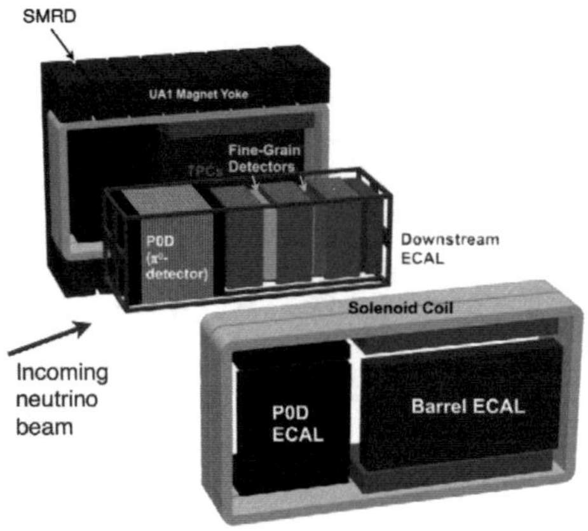

Figure 6. General view of the off-axis near detector.

5.1. INGRID

As stated before, the main purpose of the INGRID detector is to monitor the beam with neutrinos themselves and at off-axis angles ranging from 0 to 2.5 degrees. Data collected with the INGRID detector tell us that the neutrino beam for the 2010a run was where we expected it to be. Figure 7 and Fig. 8 show the horizontal and vertical beam profile and that the neutrino beam is properly centered.

5.2. Time projection chambers (TPC) and Fine-grained detector (FGD)

The time projection chambers (TPC) and the fine grained detector (FGD) form the tracker part of the off-axis near detector. The 2 FGDs act as an active target and give timing information while the TPCs allow for particle identification, charge, and momentum determination. There are 3 TPC modules and they have a sensitive area of $1.8 \times 2 \times 0.70$ m^3. The readout is handled through MicroMeGas modules with about 124,000 channels. The TPCs particle identification

Journal of Physics: Conference Series **335** (2011) 012053 doi:10.1088/1742-6596/335/1/012053

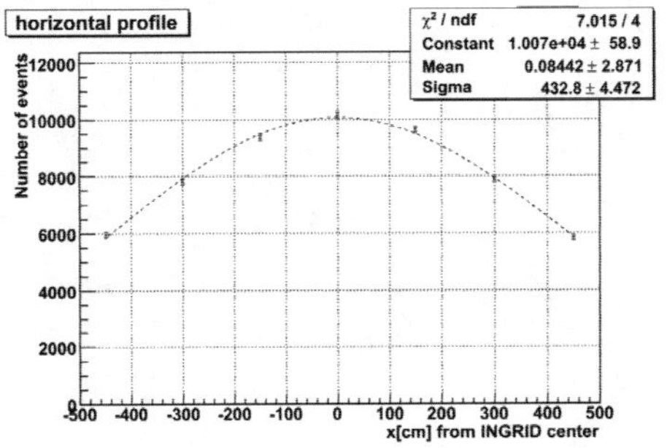

Figure 7. Horizontal beam profile recorded by INGRID.

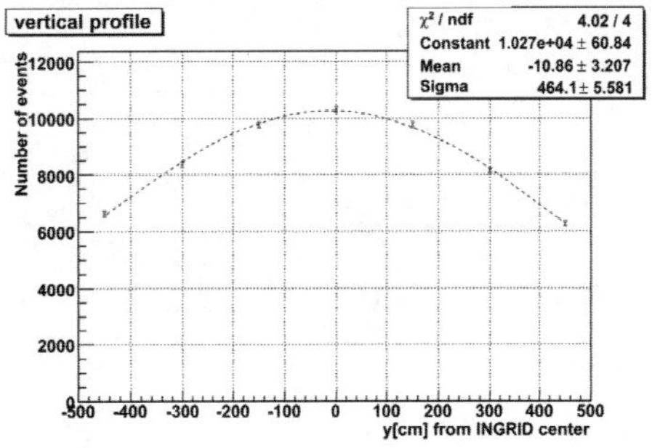

Figure 8. Vertical beam profile recorded by INGRID.

performance is presented in Fig. 9 for negative particles and Fig. 10 for positive particles. The TPCs have been described in great detail in a NIM paper submitted in December 2010 [2]. There are 2 modules of fine-grained detector with 2×1.3 tons of active target. The first FGD is only made of plastic, while the second has plastic and water. The scintillaton light is detected by Geiger-mode avalanche photo-diode multi-pixel photon counter (MPPC). There are about 9500 channels.

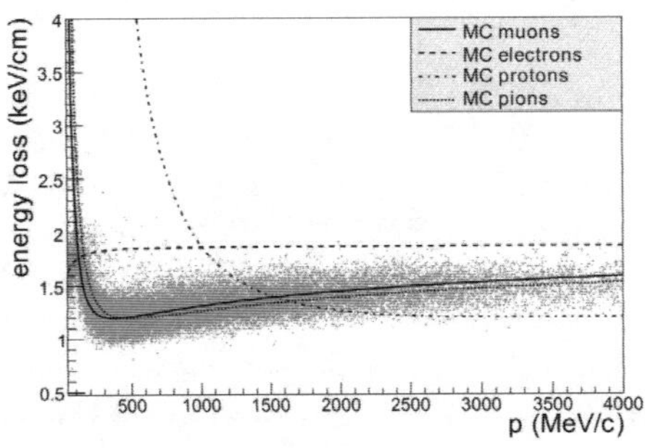

Figure 9. TPC PID performance for negative particles. Energy loss versus momentum.

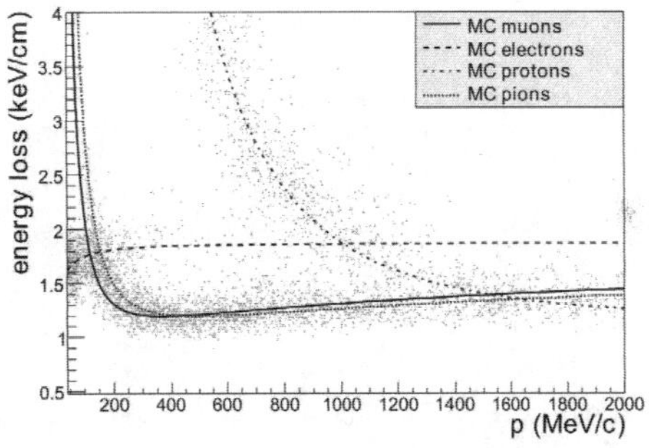

Figure 10. TPC PID performance for positive particles. Energy loss versus momentum.

5.3. The π^0 detector (P0D), electromagnetic calorimeter (ECAL and DSECAL), Side muon range detector (SMRD) performance

The π^0 detector, the downstream electromagnetic calorimeter (DSECAL) and the side muon range detectors have all been installed prior to the 2010a data taking run and were operating smoothly during the run. The remaining part of the electromagnetic calorimeter, the barrel ECAL, was installed over the summer of 2010 and is functioning well since we restarted taking data in November 2010.

5.4. Selection of event displays from the 2010a run

In Fig. 11, we present a selection of events collected with the off-axis near detector during the 2010a run. It can be seen that all sub-detectors of the off-axis near detector are functioning properly.

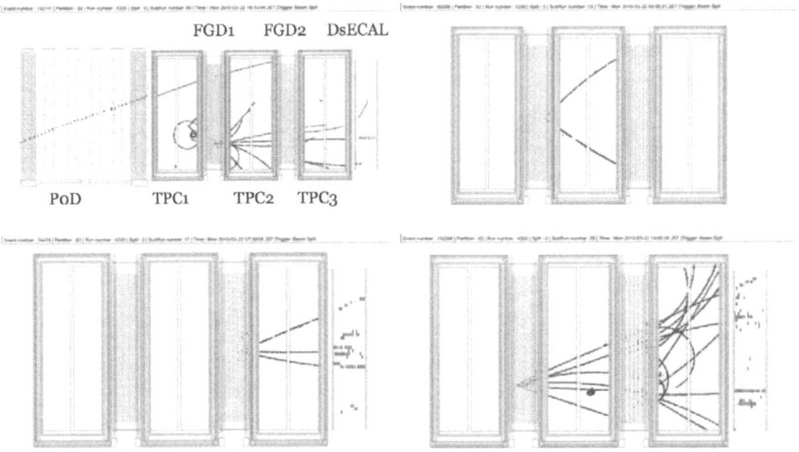

Figure 11. ND280 event displays with possible interactions types. Top left: deep inelastic scattering event in FGD1 + cosmic muon. Top right: CCQE event in FGD1. Bottom left: CC single π event in FGD2. Bottom right: deep inelastic scattering event in FGD1.

6. Current status of the far detector (Super-Kamiokande)

The Super-Kamiokande detector is located close to the town of Kamioka in the Gifu prefecture in Japan. It is a well-known water Cherenkov neutrino detector which consists of a 50 kton water tank with a 22.5 kton fiducial volume. There are about 11,000 PMT's facing inward and looking for neutrino interactions that happen in the inner detector. There are also about 2000 PMT's facing outward to look for light coming from interactions which happen in the outer detector (a 2.5 m wide shell around the inner detector). The outer detector data is used as a veto for cosmic ray muons. The Super-Kamiokande detector has been described in great detail in a NIM paper in 2003 [3].

6.1. Performance of the Super-Kamiokande detector

After about 15 years of running, the Super-Kamiokande detector is now very well understood. During the 2010a data taking period, it was running very reliably. Among others, the water transparency, the electronics, the energy scale calibration and the water scattering parameters were closely monitored and considered perfectly adequate for normal data taking. The event rate and timing distributions of the events collected by the Super-Kamiokande detector during the 2010a run are presented in Fig. 12 and Fig. 13. Fig. 12 shows the data collected within 1 millisecond of the T2K beam spill. We can see that the low energy data (LE) is flat, and so are the outer detector data (OD). For the T2K analyses, we will use only events that are fully-contained in the inner detector (FC) and we can see that such events happen only during the 3 μs T2K beam spill and we therefore have no atmospheric neutrino contamination. Fig. 13 shows the bunch structure of the spill and the distribution of the FC events among the bunches.

6.2. Summary of analysis cuts for the far detector dataset

To perform the ν_μ disappearance analysis and the ν_e appearance analysis, a set of cuts was fixed before data taking began and are summarized in Table 1. The first four cuts are common to

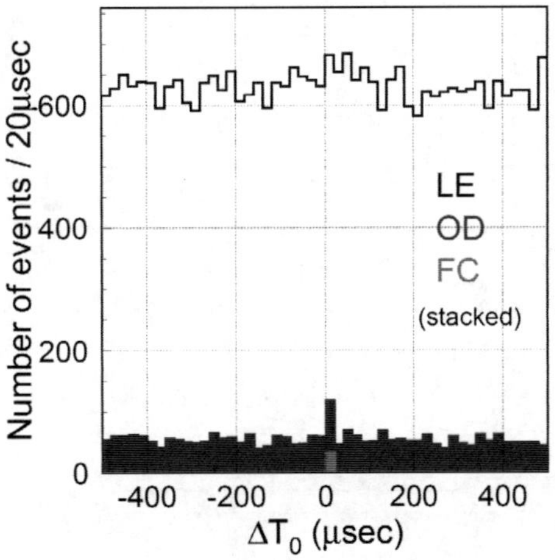

Figure 12. Relative event time to the spill time. LE are low energy events, OD are events with a vertex in the outer detector, FC are fully-contained events. Only FC events are considered for T2K analyses.

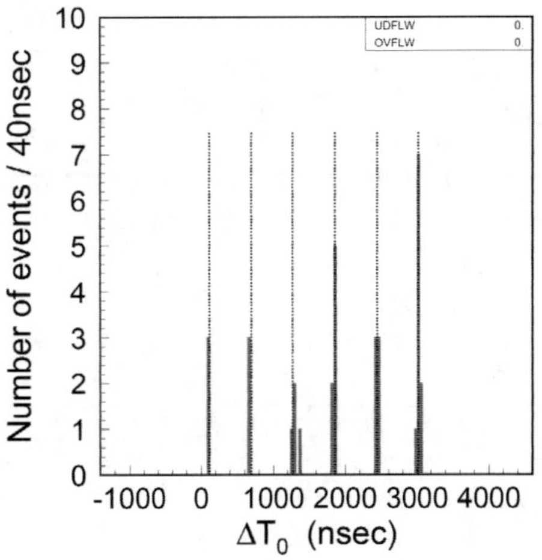

Figure 13. Relative event time to the spill, zoomed in to see the bunch structure of the beam.

Table 1. Summary of cuts applied for both appearance and disappearance analyses.

For ν_μ disappearance	For ν_e appearance
Timing coincident with beam time + (TOF)	
Fully contained (no OD activity)	
Vertex in fiducial volume (2 m from the wall)	
Single ring event	
Visible energy (E_{vis}) >30 Mev	Visible energy (E_{vis}) >100 Mev
μ-like ring	e-like ring
	No decay electron
	Invariant mass of forced 2nd ring < 105 MeV
	E_ν^{rec} < 1250 MeV

both analyses. We select events that are within the timing window of the T2K beam spill, that are fully contained inside the Super-Kamiokande inner detector (ie. we allow less than 16 hits in the outer detector). We also require that the vertex is reconstructed more than 2 m away from the inner wall of the detector to ensure proper reconstruction of the events. Finally we require single ring events. For the ν_μ disappearance analysis, we then require a visible energy greater than 30 MeV and that the ring found is μ-like. For the ν_e appearance measurement, we require a visible energy greater than 100 MeV, and that the ring found is e-like. In addition we ask for no decay electron, and if we force our algorithm to look for a second ring, we ask that the invariant mass of the found ring and the forced ring be below 105 MeV to reject neutral pions decaying to two photons. Finally we ask that the reconstructed neutrino energy be smaller than 1250 MeV to limit background contamination from higher energy.

6.3. Overview of the 2010a dataset collected by the Super-Kamiokande detector

During the 2010a run, 33 events in time with the T2K beam and fully-contained (FC) inside the inner detector were found. Out of these 33 events, 23 events pass the fiducial volume cut and the requirement that the visible energy be greater than 30 MeV. Fig. 14 and Fig. 15 show the distributions of these 33 events inside the Super-Kamiokande detector.

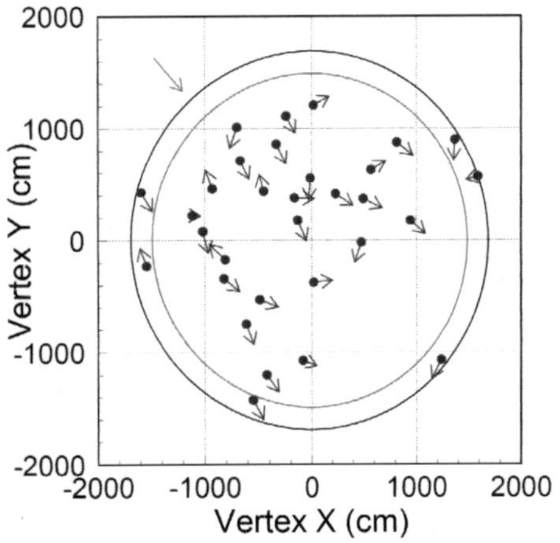

Figure 14. Vertex distributions (XY projection) of FC events. The dots are the reconstructed vertices, the blue arrows are the reconstructed direction from the ring. The red arrow is the direction of the T2K beam.

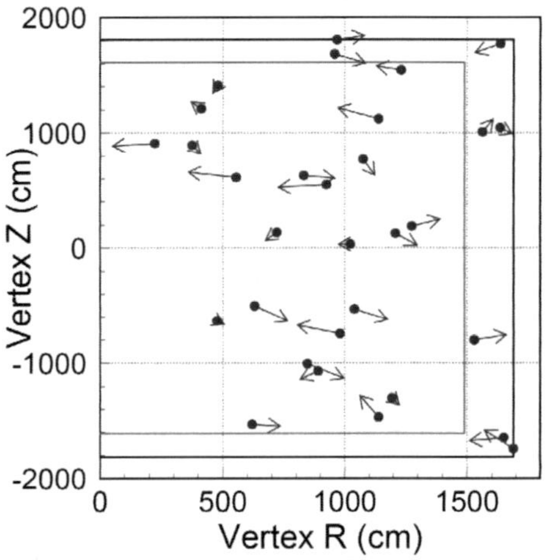

Figure 15. Vertex distributions (RZ projection) of FC events. The dots are the reconstructed vertices, the arrows are the reconstructed direction from the ring

7. Conclusions

The first data-taking run of the T2K experiment has been very successful. We accumulated 3.23×10^{19} proton on target at 30 GeV, and the proton intensity is increasing steadily. The performances of both the near detector and the far detector have been superb during the entire data-taking period. After the 2010a run, the barrel ECAL installation was completed and we started a new data-taking run in November 2010, at around 100 kW. The T2K experiment aims to accumulate an integrated proton power on target of 150kW $\times 10^7$ seconds by July 2011.

8. References

[1] URL `http://www.nd280.org/info`
[2] T2K ND280 TPC collaboration (*Preprint* `arXiv.1012.0865 (2010)`)
[3] Fukuda Y *et al.* 2003 *Nucl. Instrum. Meth.* **A501** 418–462

Journal of Physics: Conference Series **335** (2011) 012054 doi:10.1088/1742-6596/335/1/012054

Hint of CPT Violation in Short-Baseline Electron Neutrino Disappearance

Carlo Giunti

INFN, Sezione di Torino, and Department of Theoretical Physics, University of Torino, Via P. Giuria 1, I–10125 Torino, Italy

Marco Laveder

Dipartimento di Fisica "G. Galilei", Università di Padova, and INFN, Sezione di Padova, Via F. Marzolo 8, I–35131 Padova, Italy

Abstract. We analyzed the electron neutrino data of the Gallium radioactive source experiments and the electron antineutrino data of the reactor Bugey and Chooz experiments in terms of neutrino oscillations. We found a hint of a CPT-violating asymmetry of the effective neutrino and antineutrino mixing angles.

The GALLEX [1–3] and SAGE [4–7] Gallium solar neutrino experiments have been tested with intense artificial ^{51}Cr and ^{37}Ar radioactive sources placed inside the detectors. The results of these "Gallium radioactive source experiments" indicate a ratio R of measured and predicted ^{71}Ge event rates which is smaller than unity:

$$R_{\text{B}}^{\text{GALLEX-Cr1}} = 0.953 \pm 0.11 \,, \tag{1}$$

$$, R_{\text{B}}^{\text{GALLEX-Cr2}} = 0.812^{+0.10}_{-0.11} \,, \tag{2}$$

$$R_{\text{B}}^{\text{SAGE-Cr}} = 0.95 \pm 0.12 \,, \tag{3}$$

$$R_{\text{B}}^{\text{SAGE-Ar}} = 0.791^{+0.084}_{-0.078} \,. \tag{4}$$

The average ratio is [8]

$$R_{\text{B}}^{\text{Ga}} = 0.86^{+0.05}_{-0.05} \,. \tag{5}$$

Thus, the number of measured events is about 2.8σ smaller than the prediction. This is the "Gallium anomaly" [8–16].

As indicated by the "B" subscript, the ratios in Eqs. (1)–(5) have been calculated with respect to the rate estimated using the best-fit values of the cross sections of the detection process

$$\nu_e + {}^{71}\text{Ga} \to {}^{71}\text{Ge} + e^- \tag{6}$$

calculated by Bahcall [17] for ^{51}Cr and ^{37}Ar neutrinos. The uncertainties of these cross sections are not taken into account in the experimental ratios in Eqs. (1)–(4). These uncertainties are large [17–19], because only the cross section of the transition from the ground state of ^{71}Ga to the ground state of ^{71}Ge is known with precision from the measured rate of electron capture decay of ^{71}Ge to ^{71}Ga. Electron neutrinos produced by ^{51}Cr and ^{37}Ar radioactive sources can be absorbed also through transitions from the ground state of ^{71}Ga to two excited states of ^{71}Ge at 175 keV and 500 keV, with cross sections which are inferred using a nuclear model from $p + {}^{71}\mathrm{Ga} \to {}^{71}\mathrm{Ge} + n$ measurements [20].

Hence, at least part of the deficit of measured events with respect to the prediction could be explained by an overestimation of the transitions to the two excited states of ^{71}Ge [6, 7, 21]. However, since the contribution of the transitions to the two excited states of ^{71}Ge is only 5% [17], even the complete absence of such transitions would reduce the ratio of measured and predicted ^{71}Ge event rates to about 0.91 ± 0.05, leaving an anomaly of about 1.7σ [14].

We think that for a correct assessment of the statistical significance of the Gallium anomaly simple approaches based on either accepting the Bahcall cross section without taking into account its uncertainty or suppressing without theoretical motivations the transitions to the two excited states of ^{71}Ge are insufficient. A correct assessment of the statistical significance of the Gallium anomaly can be done by taking into account the large uncertainties of the transitions to the two excited states of ^{71}Ge [17–19]. The most reliable estimate of these transitions and their uncertainties have been done by Haxton in Ref. [19], leading to cross sections for ^{51}Cr and ^{37}Ar neutrinos which are larger than the Bahcall cross sections [8]. This leads to an enhancement of the Gallium anomaly. However, since uncertainties of the cross sections are large, a correct assessment of the statistical significance of the Gallium anomaly requires an accurate statistical treatment.

In Ref. [8] we calculated the average ratio R^{Ga} of measured and predicted ^{71}Ge event rates using the Haxton cross sections and their uncertainties:

$$R^{\mathrm{Ga}} = 0.76^{+0.09}_{-0.08} \,. \tag{7}$$

We found that the probability of $R^{\mathrm{Ga}} < 1$ is 99.86% (3.0σ anomaly), slightly larger than the probability of $R^{\mathrm{Ga}}_{\mathrm{B}} < 1$, which is 99.75% ($2.8\sigma$ anomaly).

For the four individual Gallium radioactive source experiments, using the same method, we obtained [8]

$$R^{\text{GALLEX-Cr1}} = 0.84^{+0.13}_{-0.12} \,, \tag{8}$$

$$R^{\text{GALLEX-Cr2}} = 0.71^{+0.12}_{-0.11} \,, \tag{9}$$

$$R^{\text{SAGE-Cr}} = 0.84^{+0.14}_{-0.13} \,, \tag{10}$$

$$R^{\text{SAGE-Ar}} = 0.70^{+0.10}_{-0.09} \,. \tag{11}$$

Since the Gallium anomaly is confirmed by the new statistical analysis which takes into account the uncertainty of the detection cross section, it is plausible that it is due to a physical mechanism. In the following, we consider the possibility of electron neutrino disappearance due to short-baseline oscillations [9–15, 22] (another explanation based on quantum decoherence in neutrino oscillations has been proposed in Ref. [23]).

We consider the electron neutrino survival probability

$$P^{\mathrm{SBL}}_{\nu_e \to \nu_e}(L, E) = 1 - \sin^2 2\vartheta_\nu \sin^2\left(\frac{\Delta m^2_\nu L}{4E}\right) \,, \tag{12}$$

where ϑ_ν is the mixing angle, Δm^2_ν is the squared-mass difference, L is the neutrino path length and E is the neutrino energy. This survival probability is effective in short-baseline (SBL) experiments in the framework of four-neutrino mixing schemes (see Refs. [24–27]), which are the simplest extensions of three-neutrino mixing schemes which can accommodate the two measured small solar and atmospheric squared-mass differences $\Delta m^2_{\mathrm{SOL}} \simeq 8 \times 10^{-5}\,\mathrm{eV}^2$ and $\Delta m^2_{\mathrm{ATM}} \simeq 2 \times 10^{-3}\,\mathrm{eV}^2$ and one larger squared-mass difference for short-baseline neutrino

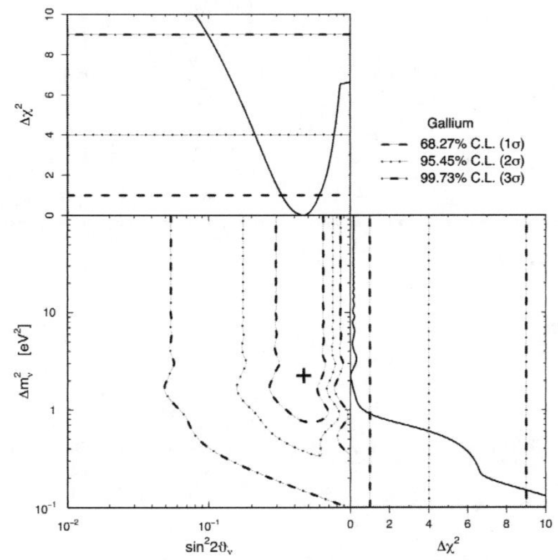

Figure 1. Results of the fit of the data of the Gallium radioactive source experiments [8]. The best-fit point is indicated by a cross.

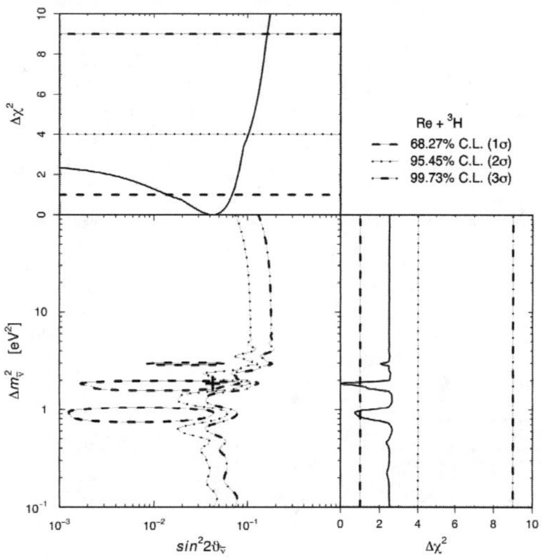

Figure 2. Results of the fit of the data of reactor and Tritium β-decay experiments [14]. The best-fit point is indicated by a cross.

oscillations, $\Delta m^2 \gtrsim 0.1\,\text{eV}^2$. The existence of a fourth massive neutrino corresponds, in the flavor basis, to the existence of a sterile neutrino ν_s.

Figure 1 shows the allowed regions in the $\sin^2 2\vartheta_\nu$–Δm^2 plane and the marginal $\Delta\chi^2 = \chi^2 - \chi^2_{\text{min}}$'s for $\sin^2 2\vartheta_\nu$ and Δm^2, from which one can infer the corresponding uncorrelated allowed intervals. The best-fit values of the oscillation parameters are

$$\sin^2 2\vartheta_{\nu,\text{bf}} = 0.46\,, \quad \Delta m^2_{\nu,\text{bf}} = 2.24\,\text{eV}^2\,. \tag{13}$$

The null hypothesis of no oscillations is disfavored at 99.95% C.L. (3.5σ) by $\Delta\chi^2 = 15.4$. This indication in favor of neutrino oscillations is slightly stronger than the indication in favor of the Gallium anomaly obtained from Eq. (7) (3.0σ), because neutrino oscillations give different values for the ratio R in GALLEX and in the two SAGE experiments, which allow for a better fit of the four data in Eqs. (8)–(11). In fact, the best-fit values of the oscillation parameters in Eq. (13) give $R^{\text{GALLEX-Cr1}} = R^{\text{GALLEX-Cr2}} = 0.77$, $R^{\text{SAGE-Cr}} = 0.76$ and $R^{\text{SAGE-Ar}} = 0.75$.

From Fig. 1 one can see that the marginal distributions of $\sin^2 2\vartheta_\nu$ and Δm^2 indicate that, at 3σ,

$$\sin^2 2\vartheta_\nu \gtrsim 0.1\,, \quad \Delta m^2 \gtrsim 0.2\text{eV}^2\,. \tag{14}$$

These bounds indicate that the short-baseline disappearance of electron neutrinos may be larger than that of electron antineutrinos, which is bounded by the results of reactor neutrino experiments [12–14,28,29]. Our analysis of the data of the Bugey and Chooz reactor antineutrino experiments in terms of the effective short-baseline electron antineutrino survival probability

$$P^{\text{SBL}}_{\bar\nu_e\to\bar\nu_e}(L,E) = 1 - \sin^2 2\vartheta_{\bar\nu} \sin^2\!\left(\frac{\Delta m^2_{\bar\nu} L}{4E}\right)\,, \tag{15}$$

where $\vartheta_{\bar\nu}$ is the effective antineutrino mixing angle and $\Delta m^2_{\bar\nu}$ is the effective antineutrino squared-mass difference, gives the best-fit values [14]

$$\sin^2 2\vartheta_{\bar\nu,\text{bf}} = 0.042\,, \quad \Delta m^2_{\bar\nu,\text{bf}} = 1.85\,\text{eV}^2\,. \tag{16}$$

Journal of Physics: Conference Series **335** (2011) 012054 doi:10.1088/1742-6596/335/1/012054

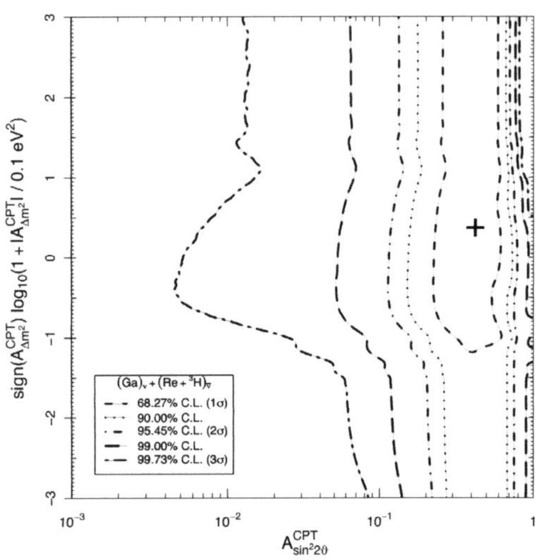

Figure 3. Allowed regions in the $A^{\rm CPT}_{\sin^2 2\vartheta}$–$A^{\rm CPT}_{\Delta m^2}$ plane [16]. The best-fit point is shown by a cross.

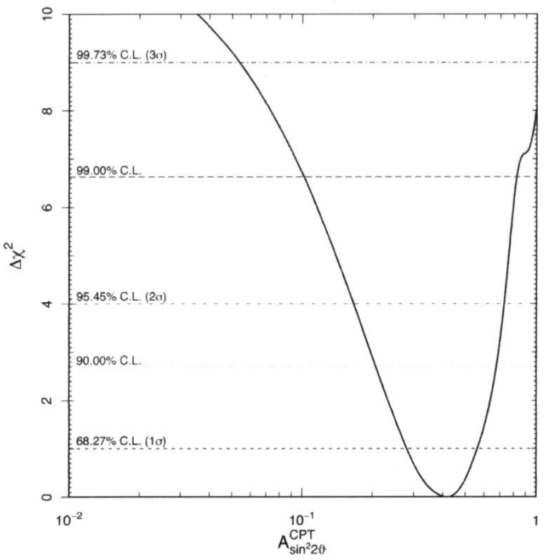

Figure 4. Marginal $\Delta\chi^2$ for $A^{\rm CPT}_{\sin^2 2\vartheta}$ [16].

Figure 2 shows the allowed regions in the $\sin^2 2\vartheta_{\bar{\nu}}$–$\Delta m^2_{\bar{\nu}}$ plane and the marginal $\Delta\chi^2$'s, obtained taking into account also the constraints on the mixing given by the results of the Mainz and Troitsk Tritium β-decay experiments [14].

CPT symmetry implies that the survival probabilities of neutrinos and antineutrinos are equal (see Ref. [30]), i.e. $\sin^2 2\vartheta_{\nu} = \sin^2 2\vartheta_{\bar{\nu}}$ and $\Delta m^2_{\nu} = \Delta m^2_{\bar{\nu}}$. Figs. 1 and 2 show that $\sin^2 2\vartheta_{\nu}$ is likely to be larger than about 0.1, whereas $\sin^2 2\vartheta_{\bar{\nu}}$ is likely to be smaller than about 0.1. The incompatibility of neutrino and antineutrino data in the case of CPT symmetry is quantified by a 0.2% parameter goodness-of-fit [31]. Hence, we have a hint of CPT violation in short-baseline ν_e and $\bar{\nu}_e$ disappearance [16] which could be complementary to that found recently in the MINOS long-baseline ν_μ and $\bar{\nu}_\mu$ disappearance experiment [32].

Analyzing the Gallium data and the reactor plus Tritium data in terms of the CPT mass and mixing asymmetries

$$A^{\rm CPT}_{\Delta m^2} = \Delta m^2_{\nu} - \Delta m^2_{\bar{\nu}} \,, \tag{17}$$

$$A^{\rm CPT}_{\sin^2 2\vartheta} = \sin^2 2\vartheta_{\nu} - \sin^2 2\vartheta_{\bar{\nu}} \,, \tag{18}$$

we obtained the best-fit values [16]

$$(A^{\rm CPT}_{\sin^2 2\vartheta})_{\rm bf} = 0.42 \,, \quad (A^{\rm CPT}_{\Delta m^2})_{\rm bf} = 0.37\,{\rm eV}^2 \,. \tag{19}$$

The allowed regions in the $A^{\rm CPT}_{\sin^2 2\vartheta}$–$A^{\rm CPT}_{\Delta m^2}$ plane are shown in Fig. 3. We used a logarithmic scale for $A^{\rm CPT}_{\sin^2 2\vartheta}$, considering only the interval $10^{-3} \leq A^{\rm CPT}_{\sin^2 2\vartheta} \leq 1$ which contains all the allowed regions. For $A^{\rm CPT}_{\Delta m^2}$ we used an antisymmetric logarithmic scale, which allows us to show both positive and negative values of $A^{\rm CPT}_{\Delta m^2}$, enlarging the region of small values of $A^{\rm CPT}_{\Delta m^2}$ between 0.1 and 1 eV^2.

The best-fit value $(A^{\rm CPT}_{\Delta m^2})_{\rm bf}$ of the mass asymmetry is small, but Fig. 3 shows that in practice any value of the mass asymmetry is allowed, with a slight preference for positive values. On the other hand, we obtain a very interesting result for the mixing asymmetry: the best-fit value $(A^{\rm CPT}_{\sin^2 2\vartheta})_{\rm bf}$ is large and positive and Fig. 3 shows that zero or negative values are disfavored.

From Fig. 3 one can see that the smallest value of $A^{\text{CPT}}_{\sin^2 2\vartheta}$ included in the 3σ allowed region is about 0.005 at $A^{\text{CPT}}_{\Delta m^2} \simeq -0.15\,\text{eV}^2$. However, since in practice $A^{\text{CPT}}_{\Delta m^2}$ is not bounded, the statistically reliable limits on $A^{\text{CPT}}_{\sin^2 2\vartheta}$ are given by the marginal $\Delta\chi^2 = \chi^2 - \chi^2_{\text{min}}$ function for $A^{\text{CPT}}_{\sin^2 2\vartheta}$ depicted in Fig. 4. One can see that $A^{\text{CPT}}_{\sin^2 2\vartheta} > 0.055$ at 3σ.

The marginal $\Delta\chi^2$ of a null asymmetry ($A^{\text{CPT}}_{\sin^2 2\vartheta} = 0$) is 12.0, with an associated p-value of 0.05%. Hence, there is an indication of a positive asymmetry $A^{\text{CPT}}_{\sin^2 2\vartheta}$ at a level of about 3.5σ.

The indication in favor of a CPT asymmetry that we have found is robust, because it is obtained by confronting the observations on the disappearance of electron neutrino and antineutrino, which should be equal if the CPT symmetry is not violated. We considered the simplest case of a difference of the effective squared-masses and mixings of neutrinos and antineutrinos. The analysis of the data in the framework of other, more complicated, models would lead to a similar indication of a CPT asymmetry in the space of the parameters of the specific model under consideration.

The short-baseline disappearance of electron neutrinos can be tested in the future with new Gallium radioactive source experiments, as that proposed in Ref. [15]. The Borexino collaboration is studying the possibility of a radioactive source experiment [33] which could provide a "smoking gun" signal by measuring the oscillation pattern inside the detector. By using radioactive sources of electron neutrinos and antineutrinos it may be possible to test the CPT asymmetry. Some possible measurements with radioactive sources and different detector types has been recently discussed in Ref. [34].

Other experiments under study which can test the CPT asymmetry by measuring the disappearance of both electron neutrinos and antineutrinos with sources which emit well-known neutrino and antineutrino fluxes and detection processes with well-known cross sections are near-detector beta-beam [35] and neutrino factory [36, 37] experiments.

In conclusion, we have found an indication of a CPT-violating asymmetry in the short-baseline disappearance of electron neutrinos and antineutrinos by confronting the neutrino data of the Gallium radioactive source experiments and the antineutrino data of the reactor Bugey and Chooz experiments. Considering the simplest case of a difference of squared-masses and mixings of neutrinos and antineutrinos, we found that the squared-mass asymmetry is practically not bounded, whereas the mixing asymmetry is positive with a statistical significance of about 3.5σ.

Acknowledgments
We would like to thank E. Bellotti, A. Ianni, G. Ranucci and S. Schoenert for interesting discussions on the possibility of a radioactive source experiment with the Borexino detector.

References

[1] GALLEX, P. Anselmann et al., Phys. Lett. B342 (1995) 440.
[2] GALLEX, W. Hampel et al., Phys. Lett. B420 (1998) 114.
[3] F. Kaether et al., Phys. Lett. B685 (2010) 47, 1001.2731.
[4] SAGE, J.N. Abdurashitov et al., Phys. Rev. Lett. 77 (1996) 4708.
[5] SAGE, J.N. Abdurashitov et al., Phys. Rev. C59 (1999) 2246, hep-ph/9803418.
[6] SAGE, J.N. Abdurashitov et al., Phys. Rev. C73 (2006) 045805, nucl-ex/0512041.
[7] SAGE, J.N. Abdurashitov et al., Phys. Rev. C80 (2009) 015807, 0901.2200.
[8] C. Giunti and M. Laveder, (2010), 1006.3244.
[9] M. Laveder, Nucl. Phys. Proc. Suppl. 168 (2007) 344, Talk presented at the Workshop on Neutrino Oscillation Physics (NOW 2006), Otranto, Lecce, Italy, 9-16 Sep 2006.
[10] C. Giunti and M. Laveder, Mod. Phys. Lett. A22 (2007) 2499, hep-ph/0610352.
[11] C. Giunti and M. Laveder, Phys. Rev. D77 (2008) 093002, 0707.4593.
[12] M.A. Acero, C. Giunti and M. Laveder, Phys. Rev. D78 (2008) 073009, 0711.4222.
[13] C. Giunti and M. Laveder, Phys. Rev. D80 (2009) 013005, 0902.1992.
[14] C. Giunti and M. Laveder, Phys. Rev. D82 (2010) 053005, 1005.4599.
[15] V.N. Gavrin et al., (2010), 1006.2103.

[16] C. Giunti and M. Laveder, Phys. Rev. D82 (2010) 113009, 1008.4750.
[17] J.N. Bahcall, Phys. Rev. C56 (1997) 3391, hep-ph/9710491.
[18] N. Hata and W. Haxton, Phys. Lett. B353 (1995) 422, nucl-th/9503017.
[19] W.C. Haxton, Phys. Lett. B431 (1998) 110, nucl-th/9804011.
[20] D. Krofcheck et al., Phys. Rev. Lett. 55 (1985) 1051.
[21] G. Fogli et al., (2006), hep-ph/0605186, 3rd International Workshop on NO-VE: Neutrino Oscillations in Venice: 50 Years after the Neutrino Experimental Discovery, Venice, Italy, 7–10 Feb 2006.
[22] J.N. Bahcall, P.I. Krastev and E. Lisi, Phys. Lett. B348 (1995) 121, hep-ph/9411414.
[23] Y. Farzan, T. Schwetz and A.Y. Smirnov, JHEP 07 (2008) 067, 0805.2098.
[24] S.M. Bilenky, C. Giunti and W. Grimus, Prog. Part. Nucl. Phys. 43 (1999) 1, hep-ph/9812360.
[25] M. Maltoni et al., New J. Phys. 6 (2004) 122, hep-ph/0405172.
[26] A. Strumia and F. Vissani, (2006), hep-ph/0606054.
[27] M.C. Gonzalez-Garcia and M. Maltoni, Phys. Rept. 460 (2008) 1, 0704.1800.
[28] Bugey, B. Achkar et al., Nucl. Phys. B434 (1995) 503.
[29] CHOOZ, M. Apollonio et al., Eur. Phys. J. C27 (2003) 331, hep-ex/0301017.
[30] C. Giunti and C.W. Kim, Fundamentals of Neutrino Physics and Astrophysics (Oxford University Press, Oxford, UK, 2007).
[31] M. Maltoni and T. Schwetz, Phys. Rev. D68 (2003) 033020, hep-ph/0304176.
[32] P. Vahle, (2010), Neutrino 2010, 14-19 June 2010, Athens, Greece.
[33] A. Ianni, D. Montanino and G. Scioscia, Eur. Phys. J. C8 (1999) 609, hep-ex/9901012.
[34] S.K. Agarwalla and R.S. Raghavan, (2010), 1011.4509.
[35] S.K. Agarwalla, P. Huber and J.M. Link, JHEP 01 (2010) 071, 0907.3145.
[36] C. Giunti, M. Laveder and W. Winter, Phys. Rev. D80 (2009) 073005, 0907.5487.
[37] J. Bernabeu et al., (2010), 1005.3146.

Double-Chooz Neutrino Experiment

Carmen Palomares for the Double Chooz Collaboration
CIEMAT, Av. Complutense 22, Madrid, Spain
E-mail: mc.palomares@ciemat.es

Abstract. The Double Chooz experiment will use the electron anti-neutrinos produced by the Chooz nuclear power station to search for a non-vanishing value of the θ_{13} neutrino mixing angle. Double Chooz will be the first of a new generation of neutrino experiments using identical detectors at different distances from the neutrino source to reduce the systematic errors due to the uncertainties on the neutrino flux and to the detector acceptance. The far detector will be operative by the beginning of 2011. Installation of the near detector will occur in 2012.

1. Physics Motivation

Experiments with solar, atmospheric, reactor and beam neutrinos have provided compelling evidence for the existence of neutrino oscillations driven by non-zero neutrino masses and neutrino mixings [1]. The mixing in the leptonic sector can be parametrized in terms of 3 mixing angles (θ_{12}, θ_{23}, θ_{13}) and the CP violating phase δ_{CP}. θ_{12} and θ_{23} have been measured to be large (almost maximal) while the value of the third mixing angle, θ_{13}, is not known, and is constrained to be small compared to the other two angles. δ_{CP} is totally unknown. The three-flavor neutrino oscillation framework also includes two mass-squared differences: Δm^2_{21} and Δm^2_{31}.

The best limit of θ_{13} was established by the CHOOZ experiment [2] from the non-observation of disappearance of reactor electron anti-neutrino at the scale of Δm^2_{32}. However, the current bound is obtained from a combination of global existing neutrino oscillation data [3], being $\sin^2 2\theta_{13} < 0.135$ at 90% C.L. The determination of θ_{13} is one of the primary goals in neutrino physics, in fact, genuine three flavor oscillation effects occur only for a finite value of θ_{13}. First, δ_{CP} becomes unphysical if θ_{13} is zero because it is also a three flavor effect. Hence, leptonic CP violation can only be tested by forthcoming experiments if θ_{13} is non-zero. In addition, any realistic possibility to determine the sign of Δm_{31} relies on a not-too-small θ_{13}. Finally, the determination of θ_{13} will provide important information on the mechanism of neutrino mass generation and the flavor structure in the leptonic sector [4].

Two different classes of experiments are able to measure small values of θ_{13}: accelerator neutrino beam experiments and nuclear reactor experiments. Accelerator experiments look for the appearance of ν_e in an almost pure ν_μ beam, due to oscillations. The appearance probability depends on all six oscillation parameters, being especially important the correlation of θ_{13} value with δ_{CP} and the sign of Δm_{31}, indeed, the correlation between δ_{CP} and θ_{13} can be only resolved if the accelerator operates also with anti-neutrinos.

Reactor neutrino experiments are able to provide a clean measurement of θ_{13} as they do not suffer from degeneracies and correlations between different oscillation parameters. However, reactor experiments are not sensitive to the value of δ_{CP}.

2. Nuclear reactor experiments

The nuclear reactor experiments look for the disappearance of electron anti-neutrinos produced in reactor cores with an average energy of 2 MeV. Nuclear reactors are very intense sources of $\bar{\nu}_e$ coming from the β-decay of the fission products of fuel elements (^{235}U, ^{238}U, ^{239}Pu, ^{241}Pu). The survival probability is given by

$$P_{ee} \approx 1 - \sin^2 2\theta_{13} \sin^2 \frac{\Delta m^2_{31} L}{4E_\nu} + O(\alpha^2)$$

with $\alpha = \Delta m^2_{21}/\Delta m^2_{31}$. In order to measure θ_{13}, a distance to the detector of about 1 km is chosen to maximize the disappearance probability at the scale of Δm^2_{32}. This short baseline also prevents reactor experiment measurements to be affected by matter effects.

The $\bar{\nu}_e$ are detected via the inverse β-decay reaction: $\bar{\nu}_e + \mathrm{p} \rightarrow \mathrm{e}^+ + \mathrm{n}$. The signature is a delayed coincidence between the prompt e^+ signal and the signal from the neutron capture. The neutron capture signal delay is about 200 μs in liquid scintillator and is reduced to about 30 μs in gadolinium loaded liquid scintillator. The positron energy is related to the $\bar{\nu}_e$ one by: $E_{\mathrm{e}^+} = E_{\bar{\nu}_e} - (M_\mathrm{n} - M_\mathrm{p})$ and the γ-ray energy released after the n-capture goes from 2 MeV (capture on H) to 8 MeV (capture on Gd). This characteristic signature is powerful enough to provide a $\bar{\nu}_e$ sample with a moderate level of background just applying a few selection cuts.

Nevertheless, the signal may be mimicked by background events which can be divided into two classes: accidental and correlated. The single events due to radioactivity from detector materials and surrounding rock is the responsible of the accidental background. The random coincidence between these events and a cosmogenic neutron would be drastically reduced by a careful selection of the materials used to build the detector. On the other hand, the two main contributions to the correlated background are: fast external neutrons and unstable isotopes, such as ^{9}Li and ^{8}He, with a β-neutron cascade. Fast neutrons may enter the detector and create recoil protons, faking the e^+ signal, and be captured after thermalization. Both types of events are coming from spallation processes of high energy muons. However, in the case of ^{9}Li and ^{8}He, the correlation with a muon signal is very difficult due to the relatively long-life of these isotopes (about 0.1 s).

3. Improving CHOOZ: The Double Chooz concept

CHOOZ experiment measured the electron anti-neutrino rate with a 2.8% statistical and a 2.7% systematic error [2]. The aim of upcoming experiments is to improve CHOOZ sensitivity by a factor 5 at least. This could be obtained by improving a factor five both statistics and systematics. To gain a factor 25 in the number of neutrinos with a detector twice bigger than CHOOZ, the detector must take data at least for 3 years. The stability of Gd doped liquid scintillator is the main limitation factor for long data taking. More difficult is the reduction of systematics. The dominant systematic error from the absolute measurement of the reactor neutrino flux (2%) and spectral shape will be removed by using an additional detector located a few hundred meters from the nuclear core to monitor the unoscillated neutrino flux. Figure 1 shows the neutrino oscillation probability at different reactor distances for a non-zero value of θ_{13}. In addition, if the detectors are identical, the uncertainty related to detector acceptance will be largely reduced.

The Double Chooz experiment [5] is located close to the twin reactor cores of the Chooz nuclear power station in the Ardennes (France). The far detector is placed at 1050 m distance from the cores in the same laboratory used by the CHOOZ experiment. It provides a quickly prepared and well-shielded (300 m.w.e.) site with near-maximal oscillation effect. The second identical detector (near detector) will be installed 410 m away from the nuclear cores under a hill of 115 m.w.e.

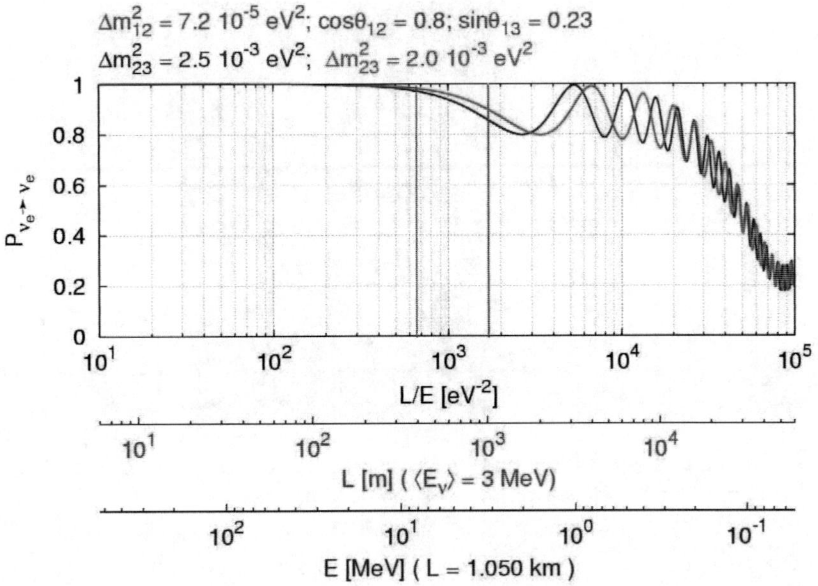

Figure 1. Neutrino oscillation probability as a function of L/E for $\sin\theta_{13} = 0.23$. The vertical blue line indicates the position foreseen for the near detector ($\sim$ 400m) while the red one shows the position of the far detector ($\sim$ 1km).

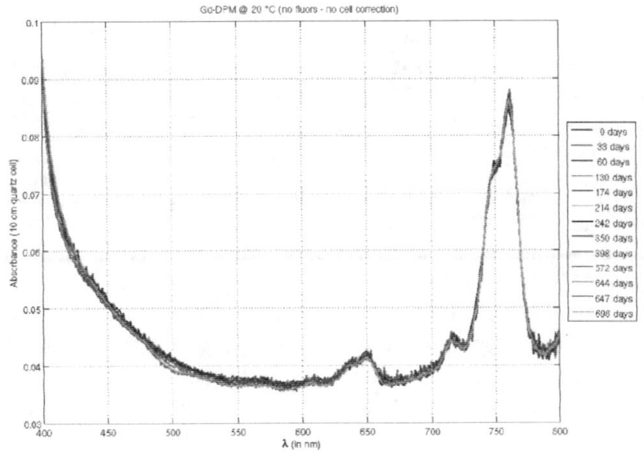

Figure 2. Optical properties stability of Double Chooz Gd loaded liquid scintillator. The absorbance for wavelength from 400 to 800 nm has been monitored for more than 2 years.

The detector has been designed to double the fiducial volume of the previous CHOOZ detector and the stability of a newly developed liquid scintillator has been successfully tested over 3 years, as it is shown in figure 2. The data taking period will extend to 3-5 years, in order to achieve 0.5% statistical error.

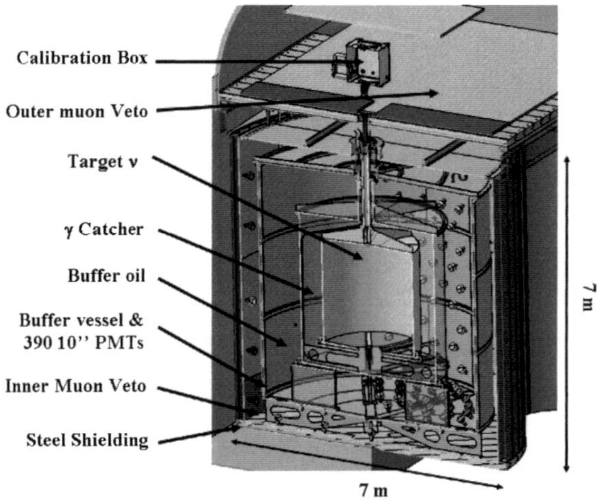

Figure 3. The Double Chooz detector

4. Double Chooz detector

The Double Chooz detectors (see figure 3) consist of concentric cylinders. The innermost volume is the neutrino target, a 10.3 m^3 acrylic cylinder, filled with 0.1% Gd loaded liquid scintillator. The target is surrounded by a 55 cm thick layer of unloaded scintillator (γ-catcher) contained in a second acrylic vessel. The function of this region is to get the full positron energy, as well as most of the neutron energy from the neutron capture on Gd. In addition, there is a 105 cm buffer of non scintillating oil to decrease the rate of single events from photomultiplier tubes (PMT) radioactivity. The buffer vessel (3 mm thick stainless steel cylinder) serves at the same time as support structure for PMTs. The detection system consists of 390 10 inches PMTs, providing about 13% photocathode coverage. The detector is encapsulated within an active cosmic-ray muon veto: a 50 cm thick region filled with liquid scintillator and instrumented with 78 8-inch PMTs. Outside this vessel, a 15 cm thick low activity steel shielding will protect the detector from the natural radioactivity of the rocks around the pit. Together with the buffer, it allows to reduce by two orders of magnitude the single event rate with respect to previous CHOOZ experiment. Finally, the upper part of the detector will be covered by plastic scintillator detector as an additional *outer muon veto*. The detector has been designed to minimize background, achieving in the far detector an expected background to signal ratio lower than 5%.

The detector is calibrated by a permanent non-intrusive tool of optical fibers that inject blue light to monitor the stability of liquids and PMT response. Other calibration systems will be deployed periodically into the detector allowing to measure the detection efficiency and the detector energy response for gammas and neutrons. The deployment of radioactive sources will be performed in a clean environment under a dry nitrogen atmosphere.

The far detector is in commissioning period, all PMT channels are operational, the read-out electronics is ready and the background rate is at the expected level.

5. Expected detector performance-sensitivity

In Double Chooz, the error due to the uncertainty on the anti-neutrino flux will be replaced by the relative normalization between detectors. The goal of the experiment is to reduce the systematic uncertainty to 0.6%. The expected systematic errors are shown in Table 1 together with the CHOOZ results. An effort is made for a precise measurement of the number of free

	CHOOZ	Double Chooz
Neutrino flux and reactor power	2.1%	-
Number of protons	0.8%	0.2%
H/C ratio and Gd concentration	1.2%	<0.2%
Detection efficiency	1.5%	0.5%

Table 1. Contributions to the overall systematic uncertainty on the absolute normalization factor in CHOOZ [2] and the expectations for Double Chooz.

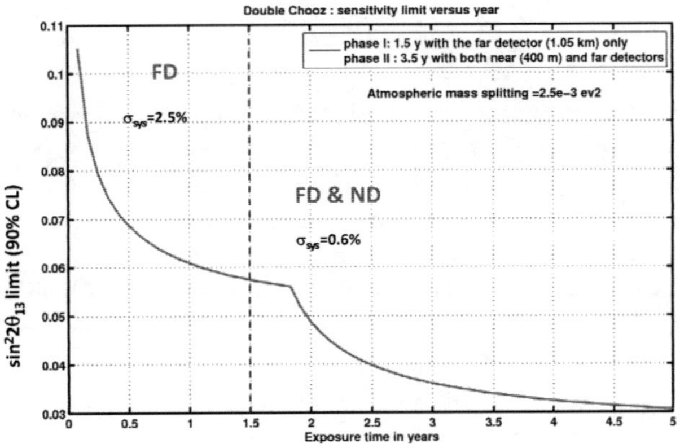

Figure 4. Double Chooz expected sensitivity limit (90% C.L.) to $\sin^2 2\theta_{13}$ as a function of time.

protons inside the target volumes, since the anti-neutrino rate is proportional to it. To do that, the target mass has been weighed with a precision of 0.2%. Since only one batch of liquid scintillator will be used to fill both detectors, the uncertainty in the molecular composition will be smaller than 0.2%. In addition, the optimization of the detector design allows to simplify the analysis and to reduce the detection efficiency systematic error up to 0.5%.

As we reported in the previous section, the signal to background ratio will be kept above 20 for the far detector and above 100 for the near one. The error originated by the background subtraction in both detectors is expected to be less than 1%.

The data taking of Double Chooz will be divided in two phases. During the first one (starting at the beginning of 2011) only the far detector will be present during 1.5 years. The expected systematic error is 2.5%, where the main contribution (about 2%) comes from reactor neutrino flux uncertainty. In this first period the sensitivity will be $\sin^2 2\theta_{13} < 0.055$ (at 90% C.L., for $\Delta m^2_{atm} = 2.5 \times 10^{-3}$ eV^2), a factor 2 better than current limit. The limit from CHOOZ will be reached in just a few weeks of operation, as it shown in figure 4. In the second phase both detectors will take data for 3 years. The full experiment will then achieve a final sensitivity of $\sin^2 2\theta_{13} < 0.03$ in the case of non-oscillation, assuming a relative normalization error of 0.6%.

6. Conclusions

The Double Chooz experiment is accomplishing the commissioning period of the far detector and will start the data taking in the following weeks. The near lab site will be available at the beginning of 2012 to accommodate the second detector. Both, near and far detectors, will be operational by mid 2012. After 3 years of operation of both detectors Double Chooz will be able to measure θ_{13} with 3σ effect if $\sin^2 2\theta_{13} > 0.05$. An intense R & D work has been carried out by the Double Chooz collaboration to validate the robustness of the double detector concept.

References

[1] M.C. González-García and M. Maltoni. *Phenomenology with massive neutrinos.* Phys.Rept.460 (2008) 1-129.
[2] M. Apollonio (CHOOZ Collaboration), et al. *Search for neutrino oscillations on a long base-line at the CHOOZ nuclear power station* Eur. Phys. J. C27 (2003) 331.
[3] T. Schwetz. *Three-flavour neutrino oscillation update.* 0808.2016 [hep-ph] (2008).
[4] M. Mezzetto,T. Schwetz. θ_{13}*: phenomenology, present status and prospect.* J. Phys. G 37:103001 (2010).
[5] F. Ardellier, et al. (Double Chooz Collaboration). *Double Chooz: A search for the neutrino mixing angle theta(13).* hep-ph/0606025 (2006).

Journal of Physics: Conference Series **335** (2011) 012056 doi:10.1088/1742-6596/335/1/012056

Foreseeing Neutrino spectra in Deep Core

Daniele Fargion and Daniele D'Armiento,
Paola Di Giacomo, Paolo Paggi

Physics Department, Rome University 1, INFN 1 Roma, Ple. A. Moro 2, Italy

E-mail: daniele.fargion@roma1.infn.it

Abstract. In the present paper we evaluate the effect of neutrino flavor mixing and possible CPT violation (MINOS like) for up-going atmospheric muon neutrinos measured with the Deep Core detector of the IceCube Neutrino Observatory at South Pole. Our evaluation of rate and spectrum is based on the known Cosmic Ray flux and its consequent atmospheric neutrino spectrum derived at Super-Kamiokande, and it takes into account propagation and flavor mixing inside the Earth, interaction into the Deep Core (variable) volume, and projection along the physical Deep Core detector. Our results differ from other ones; however when the Deep Core results will be available our predictions can be compared directly with them. Interestingly, the Minos CPT violation hint might be confirmed or not.

The observed spectrum should also exhibit an anomalous minimum, due to neutrino oscillation, offering a "silent" narrow detection window for neutrinos with energies around 20 GeV, which could be very useful for a new Neutrino Astronomy.

1. Introduction

Mixing among muon and tau neutrino flavors has been observed since nearly 13 years, and it has been already considered in several recent high energy astrophysical frameworks [1],[2],[3]. Recently MINOS results [4] hint for a CPT violation in neutrino sector. In a previous paper [5] we studied the potentialities of a hypothetical long base experiment with a neutrino beam from CERN (OPERA like experiment) or Fermilab (FNL) across the Earth to the Deep Core detector of the IceCube Neutrino Observatory at South Pole, in order to clarify such a puzzle, i.e. to confirm or not the CPT violation claim. This experiment might provide a novel signature for muon neutrino and anti-neutrino mixing, while it could more generally focus on the study of the neutrino flavor mixing [5].

In the present paper, starting from a previous work [6], we address a different scenario, i.e. the impact of neutrino flavor mixing and possible CPT violation (MINOS like) directly on the signal of up-going atmospheric muon neutrinos in the Deep Core detector. The accurate evaluation of the spectrum, elaborated here in detail, is based on the known Super-Kamiokande (SK) data [7, 8], scaled to Deep Core volumes (and environment); the production of atmospheric neutrinos up to their interactions, the corresponding muon event rate and tracks projected along each channel string[1] of the Deep Core detector are considered. Results are somehow surprising and not completely in agreement with other independent evaluations [9].

[1] The detector is composed by an array of "strings", each holding several Digital Optical Modules (DOMs) deployed below the surface of the South Pole ice.

2. From the atmospheric cosmic ray to the event rate in Deep Core

Let us remind the neutrino energy flux defined as

$$\phi_{\nu_\mu}(E_{\nu_\mu}) = E^2_{\nu_\mu} \cdot \frac{dN_{\nu_\mu}}{dE_{\nu_\mu}\, dt\, dA\, d\Omega} \tag{1}$$

that we parameterized, here for the first time, fitting the well known curve from ref.[7] for muon neutrinos with the following analytical formula:

$$\phi_\nu(E_\nu) = 10^2 \left(\frac{1}{9.5\, E_\nu^{-1.12}} + \frac{1}{4} + \frac{1}{27\, E_\nu} \right)^{-1} \mathrm{m}^{-2}\, \mathrm{s}^{-1}\, \mathrm{sr}^{-1}\, \mathrm{GeV}, \tag{2}$$

where the energy E_ν is expressed in GeV units; the energy range of this formula is extended from $\sim$ 100 MeV up to 1 TeV. This averaged isotropic spectrum must be enhanced (empirically) at horizons (with mild energy dependence) because of longer pion, kaon and muon decay flights. We disregarded for Deep Core the geomagnetic cut off influence; (its role at lowest energy has been taken into account at SK for azimuthal directions). To take into account the enhancement (of horizontal with respect to vertical rate) we derived (from a fit to the SK data) a correction factor to be multiplied to the previous flux, eq.2:

$$F(E_\nu, \theta) = f(E_\nu, \theta) + \left(1 - \frac{\int_0^{\pi/2} d\theta\, f(E_\nu, \theta)}{\pi/2} \right) \tag{3}$$

with

$$f(E_\nu, \theta) = \left[1 + \left(\frac{E_\nu}{50} \right)^{\frac{1}{5}} (\sin\theta)^{0.7} \right]^{-1}, \tag{4}$$

where the energy is in GeV units, and the horizontal angle θ (complemental to the zenith angle $\hat{z}$) ranges from 0 (horizon) to $\pi/2$ (upward vertical direction). The total mass target in Deep Core for incoming neutrino is taken from the effective volume estimated (for each energy) in [9], without online veto. The ratio between all events (fully contained (FC), partially contained (PC), and upward stopping and through going, i.e. without on line veto, events) and only FC+PC, i.e. with on-line veto, events is unity up to $\sim$ 100 GeV, reaching a ratio $\sim$ 1.5 at 400 GeV, or $\sim$ 2 at 1 TeV. At the highest energies the event rate is low and it is not so relevant in our neutrino oscillation study, mostly focused below 200 GeV and signals below 50 channels, with the channel number defined as the number of illuminated DOMs along the muon track. Therefore we considered here all the events without veto, with less than 50 channels, and within the Deep Core energy dependent volume (following Ref. [9]):

$$V(E_\nu) = 1.17 \cdot 10^6 \cdot (5.5\, E_\nu^{0.2} - 5.5) \cdot \left[3.2 + 0.8 \cdot \ln\left(\frac{E_\nu}{80} \right) \right] \mathrm{m}^3. \tag{5}$$

with E_ν in GeV units. Neutrino Charge Current (CC) cross section for muon production is considered (see ref. [10]) to be linear increasing with energy, in the energy range considered, while antineutrino cross section is taken about a half of the neutrino one. Electron neutrino and tau neutrinos CC and all flavor neutral current (NC) events are source of spherical showers whose role may influence (and pollute) only few channels (3-5); they will be discussed in detail elsewhere.

The present calculation includes neutrino oscillation probability $P_{\nu_\mu \to \nu_i}(E_\nu, \theta), i = 1, 2, 3$ as a function both of energy and arrival observation angle, given the Earth chord distance for each angle below the horizon; we take into account three neutrino flavor oscillations (see [5]), discussing the out-coming (observable) muon tracks. We also take into account the matter

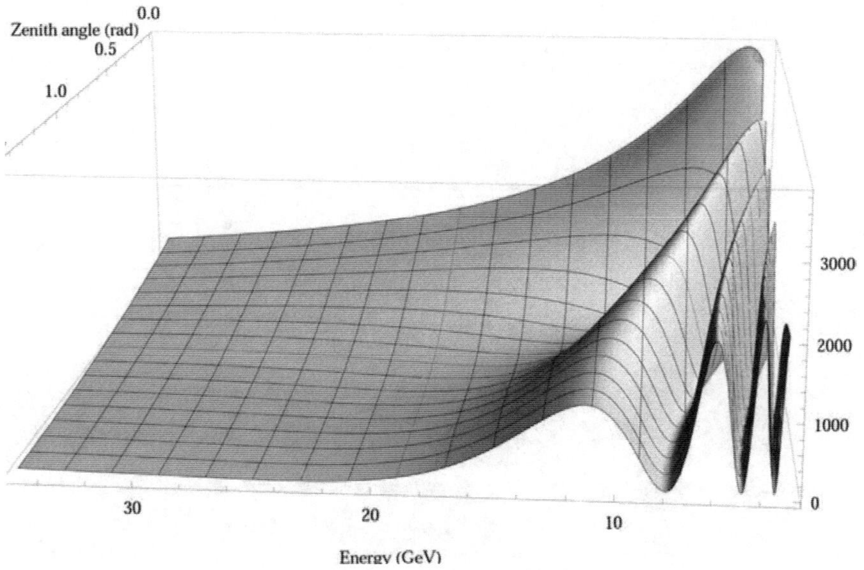

Figure 1. The differential number of event counts per year in Deep Core as a function of the muon neutrino energy (3–35 GeV) and angle, assuming vacuum along the neutrino flight.The horizontal axis is the energy in GeV, the vertical one is the number of event counts per year in Deep Core, the depth axis is the zenith angle variable in radiant.

influence on the mixing inside the Earth assuming at each zenith angle the corresponding average Earth density. We will show elsewhere that the exact integral in the variable Earth density do not change our present results in the 10–100 GeV range, but mostly induces changes at lower energies.

Thus we obtain the differential number of events (per unit energy, solid angle, and unit time) as a function of neutrino energy and the horizontal angle θ (in our definition θ ranges from $0°$ at horizon to $90°$ at the vertical upward direction):

$$\frac{dN_\mu}{dE_\nu\, dt\, d\theta}(E_\nu, \theta) \cdot \left(\frac{\Delta t}{\mathrm{yr}}\right) = 2\pi\, \frac{\phi_\nu(E_\nu)}{E_\nu^2}\, f(E_\nu, \theta)\, \sigma_{CC}(E_\nu)\, N_A\, V(E_\nu) \left(\frac{\Delta t}{\mathrm{yr}}\right) P_{\nu_\mu \to \nu_\mu}(E_\nu, \theta). \quad (6)$$

See Fig.1 for the energy range 3–35 GeV, and Fig.2 for the energy range 3–65 GeV. The event rate in one year of observation is obtained integrating over θ from $37°$ below the horizon (equal to zenith angle $\cos(\hat{z}) = \cos(\pi/2 - \theta) = \sin(\theta) = 0.6$ in Deep Core zenith angle) to $90°$, and integrating over an appropriate energy bin:

$$\Delta N_\mu(E_\nu) = \frac{dN_\mu}{dt}(E_\nu) \left(\frac{\Delta t}{\mathrm{yr}}\right) = \int_{E_i}^{E_i+\Delta E} \int_{\theta=37°}^{\theta=90°} \frac{dN_\mu}{dE_\nu\, dt\, d\theta}(E_\nu, \theta) \cdot \left(\frac{\Delta t}{\mathrm{yr}}\right) d\theta\, dE_\nu, \quad (7)$$

where N_A is the Avogadro number (multiplied by 10^6, because ice density $\rho_{ICE} \approx 10^3$ kg m^{-3}) and $V(E_\nu)$ is the effective interaction volume in Deep Core (expressed in m^3, like eq. (5)). The resulting event distributions are shown in figs. 3, 4, 5, and 6 for several energy resolutions (20 MeV with neutrino propagation in vacuum, and 20 Mev, 1 GeV, 2.8 GeV with neutrino propagation in matter). Then the same spectra of muon tracks has to be projected along the vertical axis of Deep Core, as shown finally in fig. 7.

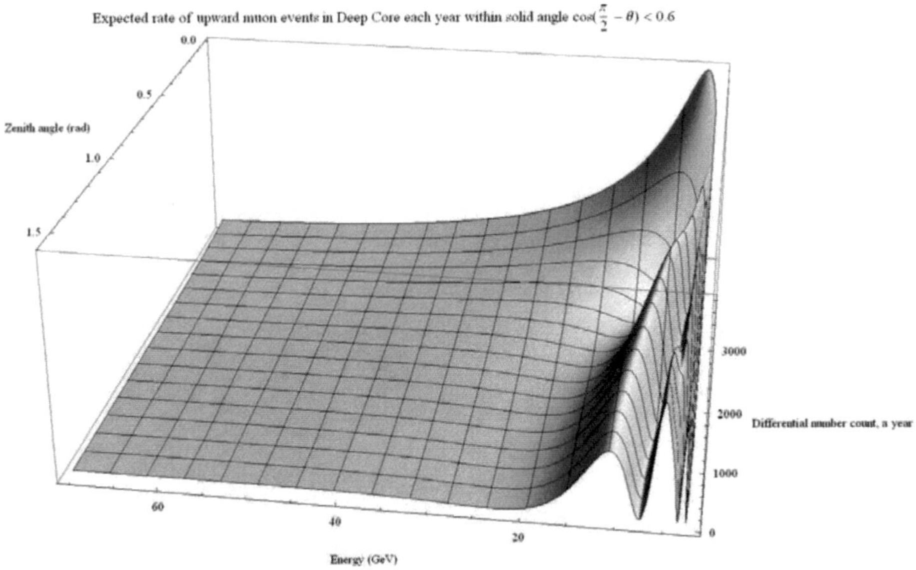

Figure 2. As the previous figure but in a wider energy range (3–65 GeV), and assuming an average Earth density $\rho = 7$ during up-going neutrino flight within Earth. Most of the inner oscillating structure is lost because of the Deep Core un-ability to reveal a few GeVs muon track. The flux suppression at 20 GeV survives and is responsible of the spectrum minimum in Channel 6–9

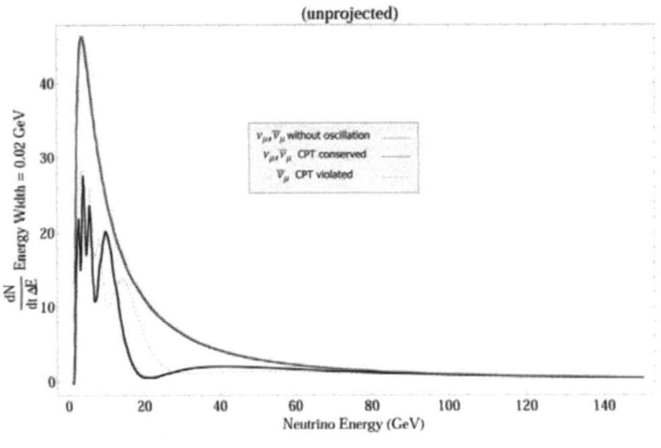

Figure 3. The ideal count rate that Deep Core would be able to observe with an energy resolution of 20 MeV. It is evident the difference between non oscillated case (red non oscillating curve), the mixed CPT conserved case (oscillating blue curve), and the thin-dashed CPT violated (green) alternative signal. The propagation in vacuum is assumed.

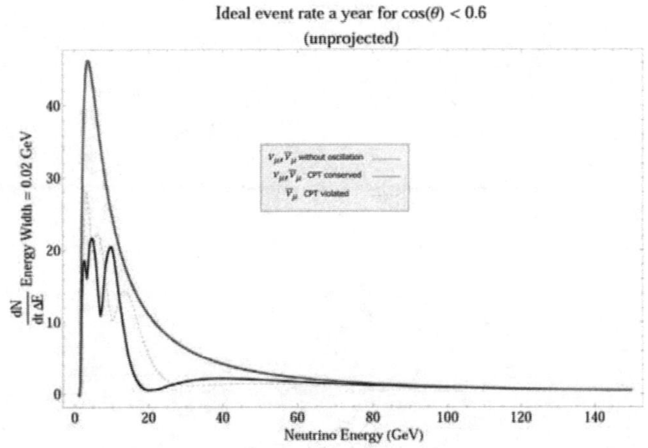

Figure 4. As the previous figure, where neutrino mixing occurs in matter. Note the little difference in spectra at very low energies.

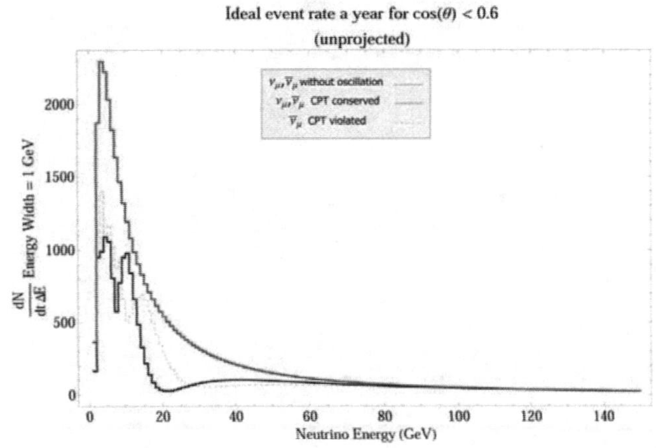

Figure 5. As the previous figure, but with an energy resolution of 1 GeV. Most, but not all, of the low energy oscillations are lost.

The muon produced in a Charged Current interaction has nearly half the neutrino energy, and undergoes an average energy loss of $dE_\mu/dX \simeq 200$ MeV m^{-1} or $1/5$ GeV m^{-1}. Therefore the muon length (keeping in mind that $E_\mu \sim E_\nu/2$) in ice is approximately: $L_\mu = 5\,E_\nu/2$ m, with energy as usual expressed in GeV. We used indeed the exact muon lenght in ice:

$$L_\mu = 3.9^{-1} \cdot 10^4 \ln \left(1 + \frac{3.9}{4.0} \cdot 10^{-3}\, E_{\nu_\mu}\right) \text{ m.} \tag{8}$$

To obtain a comparable plot of events distribution versus channel number we assumed an average projection factor, $\langle\cos(\pi/2 - \theta)\rangle = \langle\sin\theta\rangle$ of a muon track along the DOM strings due to the

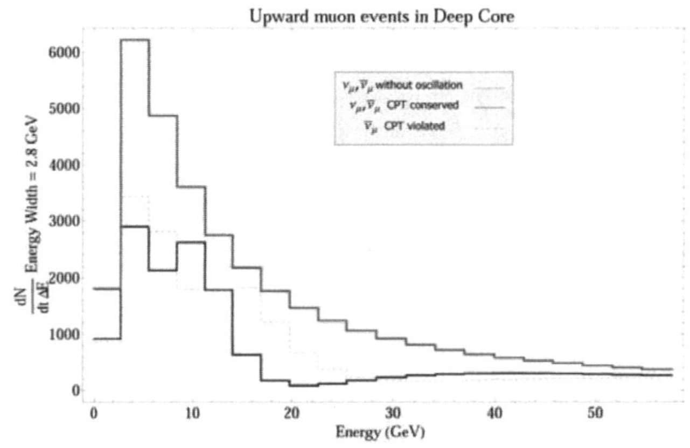

Figure 6. As the previous figure, but with an energy resolution of 2.8 GeV. The smearing of the low energy oscillations is complete.

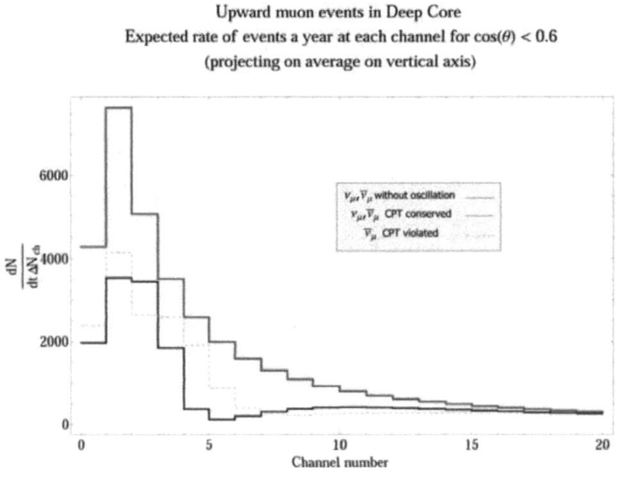

Figure 7. The most realistic count rate per year as a function of the channel with an energy resolution of 3.88 GeV. As above the red curve marks the un-oscillated case, the blue one the CPT conservation case and the thin-dashed green line the CPT violation one.

spread of the muon track direction (in the range $0.6 < \sin\theta < 1$). We have two projections: the first one has to be $\langle \sin\theta \rangle = 0.8$ at the same plane containing the fired string; an additional projection factor 0.8 is needed because muons tracks may be often skewed with respect to the string line. The overall factor ranges (in average between 0.8^2 and 0.8). We assumed the medium projection factor equal to 0.72. Finally the number of events for each channel is:

$$\Delta N_\mu(N_{ch}) = \frac{dN_\mu}{dt}(E_\nu)\left(\frac{\Delta t}{\text{yr}}\right) = \int_{E_i}^{E_i+\Delta E_{ch}/\langle\cos\theta\rangle}\int_{\theta=37^\circ}^{\theta=90^\circ}\frac{dN_\mu}{dE_\nu\,dt\,d\theta}(E_\nu,\theta)\cdot\left(\frac{\Delta t}{\text{yr}}\right)d\theta\,dE_\nu. \tag{9}$$

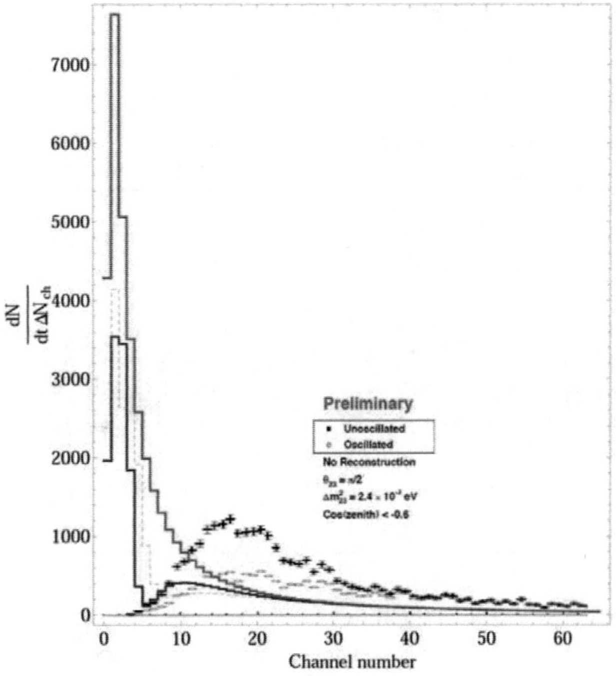

Figure 8. Comparison of our expected count rate (histograms) for upgoing muon tracks versus the most recent Deep Core expectations (dots) [9]. As above the red curve marks the un–oscillated case, the blue one the CPT conservation case and the thin-dashed green line the CPT violation one. The black (un–oscillated case) and red (oscillated case - CPT conserved) dots are the spectrum predictions from Ref.[9]. Our blu line should be compared to the red dots, and our red line to black dots.

The distance between contiguous DOMs corresponds to a minimum muon energy of 2.8 GeV (as for a vertical muon), but because of the inclined muon track projected along the vertical string $< \sin\theta > \simeq 0.72$, on average, to a greater energy value $2.8/0.72 = 3.88$ GeV. This projecting factor smears the event rate spectrum, and it masks most of the low energy neutrino oscillations. However the main deep (minimum) in the channel spectrum (see Fig. 7) at about 20 GeV survives the smearing and it characterizes our expected count rate.

Our results differ from an independent evaluation of the channel spectra [9], see Fig.8, especially in the lowest channel region (3-4), where we expect an excess of events. Moreover in the case of neutrino oscillations (CPT conserved case) the maximum of our spectrum is around channel 10, while it is around channel 18-20 for the spectrum from Ref.[9]. When the Deep Core telescope rate spectrum will be published our predictions can be compared with it. If at some channels (mostly 3-4-5) there will be an over production of signals, than the Minos CPT violation hint may be confirmed. It should be taken into account that NC (neutral current) events at tens GeV may spread light in spherical shape increasing and polluting the signal with thousands of events ($\simeq 3000 - 4000$) in channels 2-5. However their timing and intensity structure may help in the rejection of this background. In any case channels from 6 to 8 present low atmospheric background offering a rare window for 20 GeVs neutrino astronomy. To improve the observations for this astronomy in the whole arrival directions, we suggest to double the present Deep Core array strings, allowing a simultaneous double string detection. The highest Solar Flare neutrinos

[3] may be among the first interesting observations.

3. Conclusions

In the present paper we show step by step the procedure to estimate the Deep Core rate and spectrum based on the known Cosmic Ray rate, on the derived muon neutrino flux at SK, taking into account flavor mixing during propagation (see Appendix) and interaction in Deep Core (variable) volume, leading to lepton track signals in the Deep Core detector. Our results differ from other ones [9], see Fig 8. When the Deep Core telescope rate spectrum will be published our predictions can be compared with it, and the Minos CPT violation hint might be confirmed or not.

The minimum in the rate spectrum for atmospheric neutrinos around channel 6 offers a narrow window around twenty GeV that could be very useful for Neutrino Astronomy. By doubling the string array we may foresee a higher rate, a more complete (zenith and azimuth) atmospheric neutrino detection, and a new exciting neutrino Astronomy. The highest Solar Flare neutrinos [3] may be among the first interesting signals.

Acknowledgments
The authors wish to thank A. Di Domenico for useful suggestions during the paper preparation.

4. Appendix: 3 flavor neutrino mixing

The evolution equation for three neutrino mixing in matter reads :

$$i\frac{d}{dx}\Psi_\alpha = H\Psi_\alpha, \quad \text{with } H = \frac{1}{2E}\left(UMU^\dagger + A\right), \ \alpha = 1,2,3 \tag{10}$$

which is a Schrodinger equation with a matter potential A,

$$A = \begin{pmatrix} A_{\rm CC} & 0 & 0 \\ 0 & 0 & 0 \\ 0 & 0 & 0 \end{pmatrix}, \quad M = \begin{pmatrix} 0 & 0 & 0 \\ 0 & \Delta \mathrm{m}^2_{21} & 0 \\ 0 & 0 & \Delta \mathrm{m}^2_{31} \end{pmatrix}, \quad \Psi_\alpha = \begin{pmatrix} \psi_{\alpha e} \\ \psi_{\alpha\mu} \\ \psi_{\alpha\tau} \end{pmatrix}$$

where $A_{CC} = 2\sqrt{2}EG_F N_e = 0.76 \cdot 10^{-4} eV^2 (\frac{E}{GeV})(\frac{\rho}{g/cm^3})$ is the potential only due to ν_e charged current interaction.

This equation is a set of 9 differential equations whose solutions for probability conversion $P_{\alpha\beta} = \Psi^2_{\alpha\beta}$, with $\alpha, \beta = e, \mu, \tau$, has boundaries conditions $P_{\alpha\beta}(0)$ equal to 1 for the starting flavor μ, 0 for the others (electron conversion into muon is negligible). This system of equations is symmetric since $P_{\alpha\beta} = P_{\beta\alpha}$.

Among the simplest numerical solutions to show, we found a compact survival probability, for instance in vacuum for muon to muon, at a distance equal to the Earth diameter, corresponding to $\theta = 90°$, as a function of neutrino energy:

$$P_{\mu\mu} = \left|0.240128 + 0.304714\ e^{-2.5859i/En} + 0.455158\ e^{-75.9607i/En}\right|^2$$

The more general case taken into account in the present study, with earth matter density included, is too complex to be reported here.

References

[1] Fargion D., astro-ph/9704205, ApJ, v.570,909-925, 2002; Fargion D. et all, Mod. Phys. Lett. A, Vol. 11, No. 17, pp. 1363-1369, (1996);
[2] Belotsky K., Fargion D., Khlopov M, Konoplich R.V., Phys.Atom.Nucl.71:147-161,2008.
[3] Fargion D., Di Giacomo P., Nucl.Phys.Proc.Suppl.188,(2009)142-145; Fargion D., Phys.Scripta T127 (2006) 22-24 .

[4] MINOS Collaboration, http://www-numi.fnal.gov/PublicInfo/forscientists.html
[5] Fargion D., *et al.*, arXiv:1012.3245v1;
[6] Fargion D., D'Armiento D. *Prog. Part. Nucl. Phys.* **66**: 473 (2011) arXiv:1012.5271; arXiv:1101.1991.
[7] Honda M. *et al.*, 2004, *Phys. Rev.* **D 70**, 043008
[8] SuperKamiokande Collaboration, Phys.Rev. **D71** (2005) 112005; arXiv:hep-ex/0501064v2
[9] Montaruli T., for the IceCube Collaboration, *Proc. of CRIS 2010 Conference*, Catania, Sep 2010, arXiv:1012.0881v1 [astro-ph.HE]; D. Gratt et al. (IceCube collab.) Procs. 31st ICRC (Poland), 2009
[10] Nakamura K. *et al.* (Particle Data Group), J. Phys. G 37 (2010) 075021

Journal of Physics: Conference Series **335** (2011) 012057 doi:10.1088/1742-6596/335/1/012057

The Entropic Uncertainty Principle for Decaying Systems and $\mathcal{CP}$ violation

Beatrix C. Hiesmayr

University of Vienna, Faculty of Physics, Boltzmanngasse 5, 1090 Vienna, Austria; *and* Research Center for Quantum Information, Institute of Physics, Slovak Academy of Sciences, Dúbravská cesta 9, 84511 Bratislava, Slovakia

E-mail: Beatrix.Hiesmayr@univie.ac.at

Abstract. Employing an effective formalism for decaying system we are able to investigate Heisenberg's uncertainty relation for observables measured at accelerator facilities. In particular we investigate the neutral K–meson system and show that, firstly, due to the time evolution an uncertainty between strangeness measurements at different times is introduced and, secondly, due to the imbalance of matter and antimatter ($\mathcal{CP}$ violation) an uncertainty in the evolution of the eigenstates of the effective Hamiltonian of the system. Consequently, the existence of $\mathcal{CP}$ violation is linked to uncertainties of observables, i.e. the outcomes cannot be predicted even *in principle* to arbitrary precisions.

1. Introduction

Heisenberg's uncertainty relation is one of the most well known phenomenon in popular science, it captures one of the key features of quantum theory and essentially describes the stability of atoms. In detail it bounds the uncertainties of the outcomes of any two observables on a system in terms of their commutativity or differently stated uncertainty relations constrain the potential knowledge an observer can have about the physical properties of a system. In high energy physics one usually observes different decay events at different times corresponding to different distances from the point of particle creation. In this note we enlighten the uncertainties in the time evolution of unstable two–state systems with two different decay constants.

We start by repeating Heisenberg's uncertainty relation and presenting an entropic version of it. Considering typical observables in the neutral K-meson system including $\mathcal{CP}$ violation ($\mathcal{C}\ldots$ charge conjugation; $\mathcal{P}\ldots$ parity) we show the uncertainties in this system. This symmetry violation, which captures the difference between a world of matter and a world of antimatter, has been proven in many precision experiments but its origin is still an open puzzle in Particle Physics. In this note we show that it adds uncertainty to the system. Deriving the uncertainty relation independent of the initial state of the system is only possible because we use an effective formalism which enables us to rewrite any observable in the Heisenberg picture, i.e. time–dependent, and effectively living in the surviving part of the in general infinite dimensional Hilbert space which was recently introduced in Ref. [1].

Meson–antimeson systems, as e.g. the neutral K–meson or B–meson system, have turned out to be very suitable to discuss various quantum foundation issues as e.g. quantum erasing and marking procedures [2, 3, 4, 5] or Bell inequalities [6, 7, 8, 9, 10, 11, 12] or for testing decoherence or entanglement in the system [13, 14, 15, 16, 17, 18, 19, 20, 21] or Bohr's complementary relation [22] or possible $\mathcal{CPT}$ violation effects [23, 24, 25].

2. The Uncertainty Principle For Two Observables

Let us repeat the derivation of the Robertson version [26] of the uncertainty principle of Heisenberg for two arbitrary observables O_n, O_m. Consider an arbitrary operator C and note that

$$\langle CC^\dagger\rangle_\psi \;=\; \langle C^\dagger C\rangle_\psi \geq 0 \tag{1}$$

as $CC^\dagger$ is Hermitian. Consider two Hermitian operators O_n, O_m with $(\Delta O)^2_\psi := \langle O^2\rangle_\psi - \langle O\rangle^2_\psi$ being the standard mean value and choose $C = \frac{O_n - \langle O_n\rangle}{\Delta O_n} + i\frac{O_m-\langle O_m\rangle}{\Delta O_m}$ then it follows that the inequality

$$(\Delta O_n)_\psi \cdot (\Delta O_m)_\psi \geq \frac{1}{2}\,|\langle[O_n, O_m]\rangle_\psi| \tag{2}$$

has to be satisfied for any state ψ. In case of $O_n = \hat{x}$ and $O_m = \hat{p}$, being the position and momentum operator, we obtain Heisenberg's well known uncertainty relation $\Delta\hat{x}\Delta\hat{p} \geq \frac{\hbar}{2}$.

It was criticized that the version of the uncertainty principle by Robertson (2) has a lower bound which is depending in general on the state. It was shown that there exists an information theoretic formulation of uncertainty principle which lacks this problem and has in general stronger bounds [27], hence puts a stronger limit on the extent to which the two observables can be simultaneously peaked. The entropic uncertainty relation of two non-degenerate observables is given by (introduced by D. Deutsch [28], improved in Ref. [29] and proven in Ref. [30])

$$H(O_n) + H(O_m) \;\geq\; -2\log_2\left(\max_{i,j}\{|\langle\chi_n^i|\chi_m^j\rangle|\}\right) \tag{3}$$

where

$$H(O_n) = -p(n)\log_2 p(n) - (1-p(n))\log_2(1-p(n)) \tag{4}$$

is the binary entropy for a certain prepared pure state ψ and the $p(n)$'s are the probability distribution associated with the measurement of O_n for ψ, hence $p(n) = |\langle\chi_n|\psi\rangle|^2$. The maximal value of the right hand side of the entropic uncertainty relation is obtained for

$$|\langle\chi_n|\chi_m\rangle| \;=\; \frac{1}{\sqrt{2}}\,, \tag{5}$$

in this case the two observables are commonly called complementary to each other (their eigenvalues have to be nondegenerate), for example if the operators are σ_x and σ_z. In general a non-zero value of the right hand side of inequality (3) means that the two observables do not commute, i.e. it quantifies the complementarity of the observables. The binary entropies on the left hand side quantify the gain of information on average when we learn about the value of the random variable associated to O_n. Alternatively, one can interpret the entropy as the uncertainty *before* we obtain the result of the random variable. Remarkably, the right hand side of the entropic uncertainty relation also does not depend on the eigenvalues (except to test the non-degeneracy), this means that if the state is prepared in an eigenstate say of O_n then the two eigenvalues of O_m are equally probable as measured values, i.e. the exact knowledge of the measured value of one observable implies maximal uncertainty of the measured value of the other, independent of the eigenvalues.

3. Observables at Accelerator Facilities in the Heisenberg Picture

Particle detectors at accelerator facilities detect or reconstruct different decay products at different distances from the point of creation, usually by a *passive* measurement procedure, i.e. observing a certain decay channel at a certain position in the detector without having control over the decay channel nor the time point (distance from point of creation). More rarely by an *active* measurement procedure, e.g. placing a block of matter in the beam and forcing the incoming neutral meson to react with the material. In such a way e.g. the strangeness content of a neutral kaon at a certain time point (position of the block of matter) can be measured *actively*, i.e. the experimenter decides what physical property (strangeness) and when. For measurements of the lifetime of neutral kaons also an *active* measurement procedure was found in Refs. [2, 3], one waits for $4.8\tau_S$ (τ_S lifetime of the short lived state) and if one observes a decay (any decay channel) within this time duration one denotes the event as a short lifetime measurement otherwise as a long lifetime measurement. In this case we have a conceptual error which is of the amount of $\mathcal{CP}$ violation (for details consult Ref. [3])

In principle one can generally ask the following questions to the mesonic quantum systems

- Are you a certain quasispin $|k_n\rangle$ at a certain time t_n or not?
- Or: Are you a certain quasispin $|k_n\rangle$ or its orthogonal state $|k_n^\perp\rangle$ ($\langle k_n^\perp|k_n\rangle = 0$) at a certain time t_n?

where we denote by a quasispin k_n any superposition of the mass eigenstates $|K_{S/L}\rangle$ which are the solutions of the effective Schrödinger equation, i.e. $|k_n\rangle = \cos\frac{\alpha}{2}|K_S\rangle + \sin\frac{\alpha}{2}\cdot e^{i\phi_n}\,|K_L\rangle$. We will consider the first question because it includes all information that is *in principle* obtainable in these decaying systems.

Let us now discuss what is learnt by finding a certain quasispin $|k_n\rangle$ at a certain time t_n or not, compared to the situation to find a quasispin k_m or not at the point of creation t_m. Certainly, this result also quantifies our uncertainty *before* we learn the result (Yes, No) at t_n and (Yes, No) at t_m. In particular, if we compare observables of same quasispin at different times, we obtain the uncertainty due to the time evolution.

Differently stated, we can view it in the following way, two experimenters, Alice and Bob, choose two different measurements corresponding to the observables O_n and O_m. Alice prepares a certain state ψ and sends it to Bob. Bob carries out one of the two measurements O_n or O_m and announces his choice n or m to Alice. She wants to minimize her uncertainty about Bob's measurement result. Alice's result is consequently bounded by the inequality (3).

Usually, the effect of the time evolution is asserted to the state vector describing the quantum system to a given time point. The difficulty with unstable systems is that then the state vector is no longer normalized and a simple re-normalizing by dividing by the norm of the state vector neglects some information before that particular time point that is *in principle* obtainable. A way out is to consider decaying systems as open quantum system and enlarging there Hilbert space by at least twice, see Ref. [31]. Herewith, the effect of the decay of these unstable systems can be effectively described as a certain "*kind of decoherence*" as it is effectively described by a Markovian master equation. The price to pay is that for $t > 0$ the state is no longer pure and the time evolution of multipartite meson systems is complicated to compute. Another way out was recently presented in Ref. [1] by using the Heisenberg picture, i.e. making the observables time dependent, and by that being able to effectively describe the system by a Hilbert space that has the same dimension as number of degrees of freedom.

Journal of Physics: Conference Series **335** (2011) 012057 doi:10.1088/1742-6596/335/1/012057

4. An Effective Formalism For Decaying Systems

For two-state decaying systems any expectation value for finding a certain quasispin $|k_n\rangle$ at a certain time t_n or not for given initial state ρ (not necessarily pure) is given by

$$\begin{aligned} E(k_n,t_n) &= P(\text{Yes}:k_n,t_n) - P(\text{No}:k_n,t_n) \\ &\overset{P(\text{No}:k_n,t_n)+P(\text{Yes}:k_n,t_n)=1}{=} 2\,P(\text{Yes}:k_n,t_n) - 1 \\ &= Tr(O_n^{eff}\,\rho) \end{aligned} \tag{6}$$

with the effective 2×2 operator

$$O_n^{eff} := O^{eff}(\theta_n,\phi_n,t_n) = \lambda_1(\theta_n,t_n,\delta)\,|\chi_n^1\rangle\langle\chi_n^1| + \lambda_2\,|\chi_n^2\rangle\langle\chi_n^2| \tag{7}$$

with the eigenvectors

$$\begin{aligned} |\chi_n^1\rangle &= \frac{1}{\sqrt{N(t_n)}}\Big\{\langle K_S|k_n\rangle\cdot e^{i\lambda_S^* t_n}\,|K_1\rangle + \langle K_L|k_n\rangle\cdot e^{i\lambda_L^* t_n}\,|K_2\rangle\Big\} \\ |\chi_n^2\rangle &= \frac{1}{\sqrt{N(-t_n)}}\Big\{-\langle K_L|k_n\rangle^*\cdot e^{i\lambda_S t_n}\,|K_1\rangle + \langle K_S|k_n\rangle^*\cdot e^{i\lambda_L t_n}\,|K_2\rangle\Big\} \\ \text{with}\quad N(t_n) &= e^{-\Gamma_S t_n}\,|\langle K_S|k_n\rangle|^2 + e^{-\Gamma_L t_n}\,|\langle K_L|k_n\rangle|^2 \end{aligned} \tag{8}$$

and the eigenvalues

$$\begin{aligned} \lambda_1(\theta_n,t_n,\delta) &= -1 + (e^{-\Gamma_S t_n} - e^{-\Gamma_L t_n})(1-\delta^2)\cos\theta_n \\ &\quad + (e^{-\Gamma_S t_n} + e^{-\Gamma_L t_n})(1+\delta^2+2\delta\cos\phi_n\sin\theta_n) \\ \lambda_2 &= -1\,. \end{aligned} \tag{9}$$

Here $\Gamma_{S,L}$ are the decay constants of the eigenvectors of the Hamiltonian, i.e. the short lived state $|K_S\rangle$ and the long lived state $|K_L\rangle$. $|K_{1/2}\rangle$ denote the $\mathcal{CP}$ eigenstates and δ is the $\mathcal{CP}$ violating parameter. Note that due to the decaying property of the system the first eigenvalue λ_1 changes in time and approaches for $t_n \longrightarrow \infty$ the value of the second eigenvalue λ_2 which is independent of any choice of the observer.

For spin–$\frac{1}{2}$ systems, the most general observable is given by $\vec{n}\vec{\sigma}$ (σ_i Pauli operators) where any normalized quantization direction ($|\vec{n}|=1$) can be parameterized by the azimuth angle θ_n and the polar angle ϕ_n. In case of unstable systems the effective observable is given by the set of operators $O_n^{eff} = (|\vec{n}(\theta_n,\phi_n,t_n)|-1)\mathbb{1} + \vec{n}(\theta_n,\phi_n,t_n)\vec{\sigma}$ for which the "*quantization direction*" is no longer normalized and its loss results in an additional contribution in form of "white noise", i.e. the expectation value gets a contribution independent of the initial state for $t_n > 0$.

5. The Entropic Uncertainty Principle in the Neutral Kaon Systems

Equipped with this effective observable formalism in the Heisenberg picture we can derive the lower bound on the entropic uncertainty relation (3), for which we have to find the maximum of the normalized eigenvectors of the effective operators O_n^{eff} and O_m^{eff}

$$\max\Big\{|\langle\chi_m^1|\chi_n^1\rangle|, |\langle\chi_m^1|\chi_n^2\rangle|, |\langle\chi_m^2|\chi_n^1\rangle|, |\langle\chi_m^2|\chi_n^2\rangle|\Big\}\,. \tag{10}$$

To get a geometrical visualization of the eigenvectors corresponding to the question "*Is the neutral kaon meson system in the state* $|k_n\rangle$ *or not at time* t_n" we can represent them by three dimensional real Bloch points

$$\vec{b}_n^i = \langle\chi_n^i|\vec{\sigma}|\chi_n^i\rangle\,. \tag{11}$$

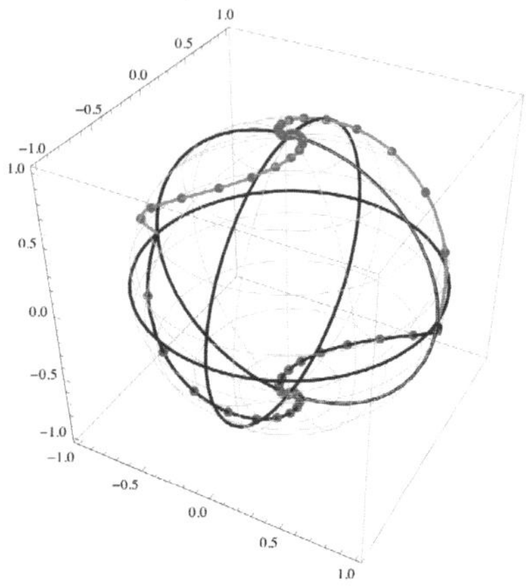

Figure 1. (Colour online) This figure depicts the Bloch sphere for the eigenvectors of observable $O^{eff}(\frac{\pi}{2}, 0, tn)$ $(\equiv |K^0\rangle)$ or $O^{eff}(\frac{\pi}{2}, \pi, tn)$ $(\equiv |\bar{K}^0\rangle)$ respectively. The eigenvectors χ^1 represented by the corresponding Bloch vectors $\vec{b}^i_n$, Eq. (11), start at the equatorial plane and move with time (red dots) to the south pole $(\equiv |K_L\rangle)$ while the eigenvectors χ^2 being orthogonal to χ^1 consequently move to the north pole $(\equiv |K_S\rangle)$. We see that when crossing the longitude at $\phi = \pm\pi/2$ the uncertainty in time for strangeness measurements become ma ximal compared to $t = 0$ measurements $(\equiv \chi^{1/2}$ in the equatorial plan) and again when reaching the south/north pole.

on a so called Bloch sphere. The Bloch sphere is an abstract sphere with extreme opposite points (or radial vectors) representing different orthogonal states of a qubit system, i.e. a system with two degrees of freedom. We choose for our representation that the north pole $(\theta = 0)$ corresponds to the state $|K_S\rangle$ and the state $|K_L\rangle$ $(\theta = \pi)$ is presented by the north pole, in case of neglecting $\mathcal{CP}$ violation. Any other pure state of the two–state system corresponds to a point on the surface of the Bloch sphere, in particular the strangeness eigenstates $(|K^0\rangle, (\theta = \frac{\pi}{2}, \phi = 0)$; $|\bar{K}^0\rangle, (\theta = \frac{\pi}{2}, \phi = \pi)$ are located at the equatorial plane.

In Fig. 1 we chose $\theta_n = \pi/2$ and $\phi = 0(\pi)$, i.e. corresponding to the question "*Is the neutral kaon meson system in the state* $|K^0\rangle$ $(|\bar{K}^0\rangle)$ *or not at time* t_n". We draw the eigenvectors of this observable for different time point t_n and observe a spiral movement to the south pole of the Bloch sphere for χ^1 and, consequently, a mirrored movement to the north pole for χ^2 (remember that these two eigenvectors have to be orthogonal which means that $\vec{b}^1 = -\vec{b}^2$).

If we are interested in the complementarity for the observable asking the question "*Is the neutral kaon system in the state* $|K^0\rangle$ *or not at time* $t = 0$" compared to the question "*Is the neutral kaon system in the state* $|K^0\rangle$ *or not at time* t", i.e. comparing the complementary introduced by the time evolution in case of strangeness measurements, we have to consider the scalar product between the radial Bloch vector for $t = 0$ (equatorial plane) to a radial Bloch vector corresponding to a later time. We see that the angle given by the scalar product $|\langle\chi^1_n|\chi^1_m\rangle| = |\langle\chi^2_n|\chi^2_m\rangle|$ increases, i.e. the complementary relation increases until the maximal value, then the scalar product between $|\langle\chi^1_n|\chi^2_m\rangle| = |\langle\chi^2_n|\chi^1_m\rangle|$ becomes greater but due to the fact that the short lived state dies out, the complementarity does not vanish anymore.

In Fig. 1 we plotted the eigenvectors of different choices of the azimuth angle θ_n in case of conservation of the $\mathcal{CP}$ symmetry $(a), (b)$ and in case of violation $(c), (d)$. We observe that the $|K_L\rangle$ (south pole) and the $|K_S\rangle$ (north pole) are the attractors of the time evolution of the open quantum system for the eigenvectors, respectively, and that the life time observables are

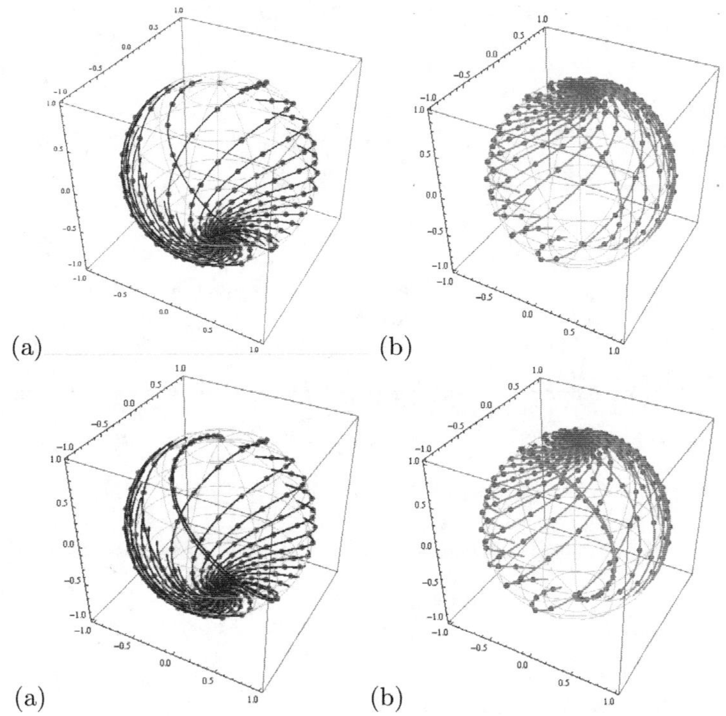

Figure 2. (Colour online) This picture depicts the time evolution of the eigenstates for different choices of quasispins with $\phi = 0, \pi$: (a),(c) of the χ^1's and (b),(d) of the χ^2's. We observe that the $|K_L\rangle$ (south pole) and the $|K_S\rangle$ (north pole) are the attractors of the time evolution of the open quantum system for the eigenvectors, respectively. Fig.(c),(d) include also $\mathcal{CP}$ violation. We observe that the eigenvectors of the observable corresponding to the question about a short lived state K_S or not get then also time dependent. Indeed any nonzero value of $\mathcal{CP}$ violating parameter causes a time evolution of the eigenstates. Consequently, a maximal uncertainty is always obtained at a time dependent on the value of the $\mathcal{CP}$ violating parameter. The maximal uncertainty for the $\mathcal{CP}$ violation parameter $\delta = 3.322 \cdot 10^{-3}$ (world average [32]) is reached for $t_n = 11.4\tau_S$ (see Ref. [1]).

persisting in time. Including $\mathcal{CP}$ violation(Figs. (c),(d)) we observe that the lifetime observables are no longer persisting in time, since $\langle K_S|K_L\rangle \neq 1$ is no longer orthogonal. Note that, for deriving the physical measurable quantities, i.e. the expectation values we trace over the operator times the initial state which is as well affected by $\mathcal{CP}$ violation. Thus the presented uncertainty is only due to the observable.

6. Summary

We discussed the entropic uncertainty principle for observables in High Energy Physics. We show that strangeness measurements at some time t_n compared to strangeness measurements at a later time $t_m \neq t_n$ introduce always a nonzero uncertainty, i.e a lack of knowledge about the physical properties of this quantum system. Then we discuss the effect of imbalance of matter and antimatter (CP violation) and show that it constrains the potential knowledge an observer can have about the eigenstates of the effective Hamiltonian at a certain time. Given the amount of the CP violation parameter we always find a "*complementary*" time, i.e. a time point when the uncertainty reaches the maximal value, i.e. the two observables become complementary.

Acknowledgments
I would like to thank the organizers of the DISCETE 2010 conference in Rome, in particular Antonio Di Domenico for inviting and discussions and the Austrian Fund project FWF-P21947N16 and EU project QESSENCE.

References

[1] Di Domenico A, Gabriel A, Hiesmayr B C, Hipp F, Huber M, Krizek G, Mhlbacher K, Radic S, Spengler C and Theussl L 2011 Heisenberg's Uncertainty Relation and Bell Inequalities in High Energy Physics *Preprint* arXiv:1101.4517
[2] Bramon A, Garbarino G and Hiesmayr B C 2004 Quantum marking and quantum erasure for neutral kaons *Phys. Rev. Lett.* **92** 020405
[3] Bramon A, Garbarino G and Hiesmayr B C 2004 Active and passive quantum eraser for neutral kaons *Phys. Rev.* A **69** 062111
[4] Bramon A, Garbarino G and Hiesmayr B C 2004 Quantitative complementarity in two-path interferometry *Phys. Rev.* A **69** 022112
[5] Amelino-Camelia G et.al. 2010 Physics with the KLOE-2 experiment at the upgraded DAPHNE *European Physical Journal* C **68**, Number 3-4 619.
[6] Hiesmayr B C 2007 Nonlocality and entanglement in a strange system *European Physical Journal* C **50**, 73
[7] Hiesmayr B C 2001 A generalized Bell inequality and decoherence for the K0 anti-K0 system *Found. of Phys. Lett.* 14, 231
[8] Bertlmann R A and Hiesmayr B C 2001 Bell inequalities for entangled kaons and their unitary time evolution *Phys. Rev.* A **63** 062112
[9] Bertlmann R A, Bramon A, Garbarino G and Hiesmayr B C 2004 A Bell inequality in high energy physics really experimentally violated? *Phys. Lett.* A **332** 355
[10] Bramon A, Escribano R and Garbarino G 2006 Bell's inequality tests with meson-antimeson pairs *Found. Phys.* **26** 563
[11] Bramon A, Escribano R and Garbarino G 2005 Bell's inequality tests: from photons to B-mesons *J. Mod. Opt.* **52** 1681
[12] Li J and C.F. Qiao CF 2009 New Possibilities for Testing Local Realism in High Energy Physics *Phys. Lett.* A **373** 4311
[13] Bertlmann R A, Grimus W and Hiesmayr B C 1999 Quantum mechanics, Furry's hypothesis and a measure of decoherence *Phys. Rev.* D **60** 114032
[14] Mavromatos NE 2009 Decoherence and CPT Violation in a Singy Model of Space-Time Foam *Preprint* arXiv:0906.2712
[15] Mavromatos NE and Sarka S 2005 Liouville Decoherence in a Model of Flavour Oscillations in the Presence of Dark Energy *Phys. Rev.* D **72** 065016
[16] Blasone M, Capolupo A and Vitiello G 2002 Quantum Field Theory of three flavor neutrino mixing and oscillations with CP violation *Phys.Rev.* D **66** 025033
[17] Blasone M, Dell'Anno F, De Siena S, Di Mauro M and Illuminati F 2008 Multipartite entangled states in particle mixing *Phys. Rev.* D **77** 096002
[18] Capolupo A, Capozziello S and Vitiello G 2009 Dark energy and particle mixing *Phys.Lett.* A **373** 601
[19] Bertlmann R A, Durstberger K and Hiesmayr B C 2003 Decoherence of entangled kaons and its connection to entanglement measures *Phys. Rev.* A **68** 012111.
[20] Caban P, Rembielinski J, Smolinski KA, Walczak Z and Wlodarczyk M 2006 An open quantum system approach to EPR correlations in K0-K0 systems *Phys. Lett.* A **357** 6
[21] Yabsley BD 2008 Quantum entanglement at the psi(3770) and Upsilon (4S), *Flavor Physics & CP Violation Conference*, Taipei
[22] Hiesmayr B C and Huber M 2008 Bohr's complementarity relation and the violation of the CP symmetry in high energy physics *Physics Letters* A **372** 3608
[23] Bernabeu J, Ellis J, Mavormatos NE, Nanopoulos DV and Pavassiliou J 2006 CPT and Quantum Mechanics Tests with Kaons *Preprint* arXive:hep-ph/0607322
[24] Di Domenico A and KLOE coll. 2010 CPT symmetry and quantum mechanics tests in the neutral kaon system at KLOE *Found. Phys.* **40** 852
[25] Capolupo A, Ji CR, Mishchenko Y and Vitiello G 2004 Phenomenology of flavor oscillations with non-perturbative effects from quantum field theory *Physics Letters* B **594** 135
[26] Robertson HP 1929 The Uncertainty Principle *Phys. Rev.* **34** 163

[27] Bialynici-Birula I and Rudnicki L 2010 Entropic Relation in Quantum Physics *Preprint* arXiv:1001.4668
[28] Deutsch D 1983 Uncertainty in quantum measurements *Phys. Rev. Lett.* **50** 631
[29] Kraus K 1987 Complementary observables and uncertainty relations *Phys. Rev.* D **35** 3070
[30] Maassen H and Uffink JBM 1988 Generalized Entropy Uncertainty Relation *Phys. Rev. Lett.* **60** 1103
[31] Bertlmann R A, Grimus W and Hiesmayr B C2006 An open–quantum–system formulation of particle decay *Phys. Rev.* A **73** 054101
[32] Nakamura K et al. (Particle Data Group) 2010 *J. Phys.* G **37** 075021
[33] Spengler C, Huber M and Hiesmayr B C 2011 A geometric comparison of entanglement and quantum nonlocality in discrete system *J. Phys.* A: Math. Theor. **44** 065304
[34] Spengler C, Huber M and Hiesmayr BC 2010 A composite parameterization of unitary groups, density matrices and subspaces, *J. Phys.* A: Math. Theor. **43** 385306

Symposium on Prospects in the Physics of Discrete Symmetries IOP Publishing
Journal of Physics: Conference Series **335** (2011) 012058 doi:10.1088/1742-6596/335/1/012058

STATUS AND PROSPECTS FOR LORENTZ AND CPT VIOLATION TESTS AT KLOE AND KLOE-2

ANTONIO DE SANTIS
Dipartimento di Fisica, Università di Roma 'La Sapienza'
and I.N.F.N. Sezione di Roma
P.le A. Moro, 2
I-00185 Rome, Italy
E-mail: antonio.desantis@roma1.infn.it

on behalf of the KLOE/KLOE-2 collaborations

Abstract. The neutral kaon system offers a unique possibility to perform fundamental tests of CPT invariance. In this contribution the KLOE prospects for the measurements of CPT violation in the context of the Standard-Model Extension are presented. Preliminary results on the CPT violating parameters already obtained by KLOE analyzing only half of the available data with a simplified analysis scheme will be reported: $\Delta a_X = (-6.3 \pm 6.0) \times 10^{-18}$ GeV, $\Delta a_Y = (2.8 \pm 5.9) \times 10^{-18}$ GeV, $\Delta a_Z = (2.4 \pm 9.7) \times 10^{-18}$ GeV. In this contribution we will discuss a new analysis method that will allow to measure the complete set of four parameters (Δa_μ) also with the perspective given by the KLOE-2 data-taking campaign.

1. The KLOE experiment

The KLOE experiment operates at DAΦNE, the Frascati ϕ–factory. DAΦNE is an e^+e^- collider running at a center of mass energy of ∼1020 MeV, the mass of the ϕ meson. Positron and electron beams collide at an angle of π-25 mrad, producing ϕ mesons with small boost in the orbit plane ($p_x(\phi) \sim -15$ MeV).

The KLOE detector consists of a large cylindrical drift chamber (DC) surrounded by a lead-scintillating fiber electromagnetic calorimeter (EMC). A superconducting coil around the EMC provides a 0.52 T axial field. The DC[1] is 4 m in diameter and 3.3 m long and has 12,582 all-stereo tungsten sense wires. The chamber shell is made of carbon fiber-epoxy composite and the gas used is a 90% helium, 10% isobutane mixture. These features maximize transparency to photons and reduce $K_L \rightarrow K_S$ regeneration and multiple scattering. The position resolutions are $\sigma_{r,\phi} \sim 150$ μm and $\sigma_z \sim 2$ mm. The momentum resolution is $\sigma(p_\perp)/p_\perp \sim 0.4\%$. The calorimeter[2] is divided into a barrel and two end-caps, and covers 98% of the solid angle. The modules are read-out at both ends by photo-multipliers, both in amplitude and time for a total of 2440 cells per side arranged in five layers. Cells close in time and space are grouped into calorimeter clusters. The cluster energy E is the sum of the cell energies. The cluster time T and position $\vec{R}$ are energy-weighed averages. Energy and time resolutions are $\sigma_E/E = 5.7\%/\sqrt{E(\mathrm{GeV})}$ and $\sigma_t = 57\mathrm{ps}/\sqrt{E(\mathrm{GeV})} \oplus 100$ ps, respectively.

At present KLOE has already acquired 2.5 fb^{-1} of data and a new extensive campaign of data taking is starting aiming at an integrated luminosity of 25 fb^{-1}.

The production cross section for the process $e^+e^- \to \phi$ is 3.3 μb and the ϕ meson decays into $K_0\overline{K}_0$ with a branching fraction of $\sim$34%. The initial state of the kaon pair is produced via strong interaction with quantum numbers $J^{PC} = 1^{--}$:

$$|i\rangle = \frac{|K_0\rangle|\overline{K}_0\rangle - |\overline{K}_0\rangle|K_0\rangle}{\sqrt{2}} = \mathcal{N}(|K_S\rangle|K_L\rangle - |K_S\rangle|K_L\rangle), \quad (1)$$

where $|K_S/K_L\rangle$ are the kaon mass eigenstate and $\mathcal{N}$ is a normalization factor. The two kaons are produced in an antisymmetric correlated state. The time evolution of the system that decay in the pair of final states f_1 and f_2 can be expressed as a function of the difference of decay time ($\Delta t = t_2 - t_1$) as:

$$I_{f_1 f_2}(\Delta t) \propto e^{-\Gamma|\Delta t|}\Big[|\eta_1|^2 e^{\frac{\Delta\Gamma}{2}\Delta t} + |\eta_2|^2 e^{\frac{\Delta\Gamma}{2}\Delta t} - 2\Re e\Big(\eta_1\eta_2 e^{-i\Delta m \Delta t}\Big)\Big] \quad (2)$$

where $\eta_j = \langle f_j|K_L\rangle/\langle f_j|K_S\rangle$, $\Gamma = \Gamma_S + \Gamma_L$ and $\Delta\Gamma = \Gamma_S - \Gamma_L$.

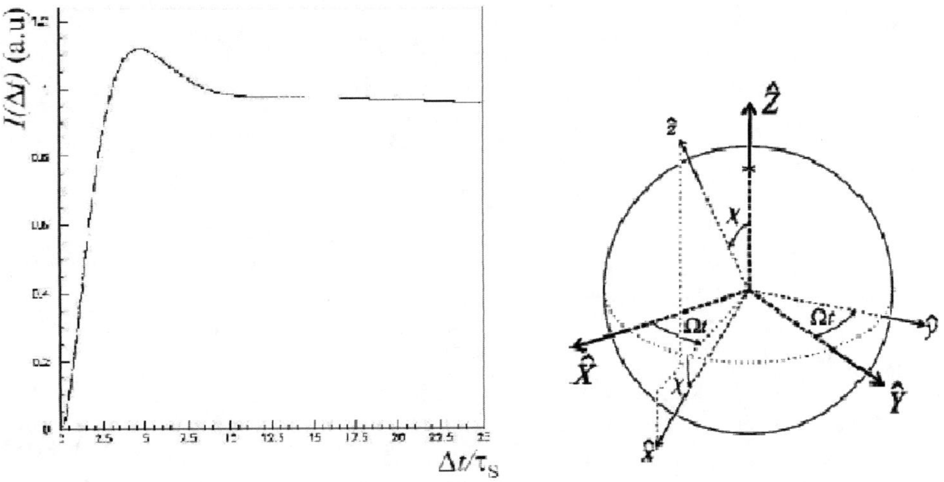

Figure 1. Left: distribution of eq. 2 for $f_1 = f_2 = f$. Right: description of the sidereal reference frame and relative coordinate transformation.

If the final states considered are the same for the two kaons ($f_1 = f_2 = f$) it is not possible to distinguish between them. In the standard Quantum Mechanics description of the time evolution of the system, a fully destructive interference is expected for equal decay time ($|\Delta t| = 0$) as shown if fig. 1(left). In the case $f = \pi^+\pi^-$ we have $\eta_f = \eta_\pm \simeq \varepsilon_K - \delta_K$, where ε_K and δ_K are the CP and CPT violation parameters in the mixing, respectively (violations in the decay have been neglected being irrelevant here). In the following we will use the notation $I_\pm(\Delta t) = I_{f_1,f_2}(\Delta t)$ with $f_1 = f_2 = \pi^+\pi^-$.

2. CPT and Lorentz symmetry breaking

A general theoretical possibility for CPT violation is based on spontaneous breaking of Lorentz symmetry, as developed by Kostelecky[3, 4, 5]: the Standard-Model Extension (SME). Following this theoretical approach, for neutral kaons, CPT violation manifests to lowest order only in the parameter δ_K and is related to Lorentz Invariance breaking. This results in a dependence of δ_K on the 4-momentum of the kaon:

$$\delta_K \approx i \sin\phi_{SW} e^{i\phi_{SW}} \gamma_K (\Delta a_0 - \vec{\beta}_K \cdot \Delta\vec{a})/\Delta m, \tag{3}$$

where γ_K and $\vec{\beta}_K$ are the kaon boost factor and velocity in the observer frame, and Δa_μ are the CPT violation coefficients for the two valence quarks in the kaon.

The simple expression in (3) holds if the observer reference frame is at rest with respect to fixed stars (sidereal reference frame). To take into account that our laboratory on the Earth is moving we have to consider a coordinates transformation. The rotation of the Earth will results in a dependence of the δ_K parameter on the sidereal time. Choosing a three-dimensional basis $(\hat{X}, \hat{Y}, \hat{Z})$ in a non-rotating frame, with the $\hat{Z}$ axis along the Earth's rotation axis, and a basis $(\hat{x}, \hat{y}, \hat{z})$ for the laboratory frame[4] (see fig. 1) eq. 3 will transform as:

$$\begin{aligned} \delta_K(\vec{p}_K, t) = \frac{i \sin\phi_{SW} e^{i\phi_{SW}}}{\Delta m} \gamma_K \Big[& \Delta a_0 + \beta_K \Delta a_Z (\cos\vartheta \cos\chi - \sin\vartheta \cos\phi \sin\chi) \\ & - \beta_K \Delta a_X \sin\vartheta \sin\phi \sin\Omega T \\ & + \beta_K \Delta a_X (\cos\vartheta \sin\chi + \sin\vartheta \cos\phi \sin\chi) \cos\Omega T \\ & + \beta_K \Delta a_Y (\cos\vartheta \sin\chi + \sin\vartheta \cos\phi \sin\chi) \sin\Omega T \\ & - \beta_K \Delta a_Y \sin\vartheta \sin\phi \cos\Omega T \Big], \end{aligned} \tag{4}$$

where T is the sidereal time, Ω is the Earth's sidereal frequency and χ is the angle between the Earth rotation axis and the $\hat{z}$ direction in the laboratory frame; ϑ and ϕ are the polar and azimuthal angles in the laboratory frame.

Every Δa_μ component in eq. 4 has his own dependence with respect to sidereal time and kaon flight direction. This kind of dependence will be exploited in our analysis of the KLOE data to access the complete set of SME parameters.

3. KLOE approach

Using KLOE data, it is possible to perform different analysis to determine the SME parameters for the kaon system, extensively discussed in Ref. [6]. In this contribution we will discuss the possibility offered by kaon interferometry when using the same final state, i.e. $f_1 = f_2 = \pi^+\pi^-$.

In KLOE the two kaons decay with opposite momenta[1] thus experiencing, on a event by event basis, two different modulation for the δ_K parameter appearing in $\eta_\pm$ coefficients as described in eq. 3.

Ordering the two decay times according to the kaon direction in the lab frame, it is possible to construct the decay amplitude in eq. 2 directly as function of Δt even in this case in which is not possible to distinguish the two decays. Because of eq. 4 we expect small asymmetries in the two branches of the function ($\Delta t < 0$ vs $\Delta t > 0$) that should exhibit the expected modulation as a function of sidereal time and kaon directions. This will allow to define the following quantity:

$$A(\Delta t) = \frac{I_\pm(\Delta t > 0) - I_\pm(\Delta t < 0)}{I_\pm(\Delta t > 0) + I_\pm(\Delta t < 0)}. \tag{5}$$

The above asymmetry for $\Delta t \gg \tau_S$ tends to zero, because ε_K and δ_K are 90° out of phase [7]:

$$A(\Delta t \gg \tau_S) \simeq -2\Re e\left(\frac{\delta_K}{\varepsilon_K}\right) \sim 0, \tag{6}$$

[1] Two kaons are not produced exactly back to back because of the ϕ boost in the lab frame. This small effect will be fundamental to access all the four Δa_μ parameters as discussed in the text.

while for $|\Delta t| \leq 5\tau_S$

$$A(|\Delta t| < 5\tau_S) \propto -2\Im m\left(\frac{\delta_K}{\varepsilon_K}\right) . \tag{7}$$

Eq. 7 implies that one of the most critical point in the analysis will be the resolution on the difference of decay times. To improve as much as possible on this respect we have to apply severe selection criteria to guarantee the high quality and low contamination of the gathered dataset. At the end of the selection chain the background contamination in the sample is of the order of 1.7% and is manly due to the kaon regeneration on the beam pipe.

Since the signal has a very clean topology, two decay vertex reconstructed from two pairs of tracks, we exploit the possibility to use global event information in the signal hypothesis to improve the reconstruction. This is done by performing a global fit of the event to redetermine the decay length of the two kaons leaving free the position of the primary vertex. In the standard reconstruction the position of the primary vertex is determined by using a special sample of bhabha events for the whole run. This method will provide, for a standard run, a resolution of the order of centimeters, while the global fit can provide results up to few millimeters, as shown in fig. 2-left.

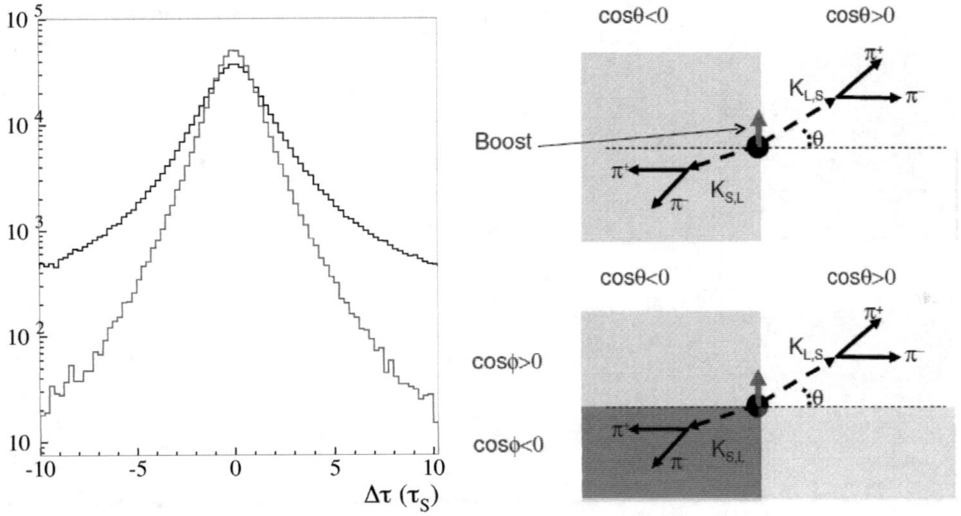

Figure 2. Left: Resolution on the proper decay time difference. The black is the output of the standard reconstruction, the red one is the output of the global fit. The events are the same in both cases. Right: pictorial description of the methods used in the preliminary analysis (top) and current approach (bottom). The red arrow represents the ϕ boost.

With KLOE data we already performed a preliminary analysis [8] using only half of the total available statistics (1 fb^{-1}). In this analysis we used the forward-backward asymmetry along z direction as a function of sidereal time only (e.g. integrating over azimuthal angle see fig. 2-right). With this approach only the vector part of the Δa_μ four-vector can be observed[2]. Our preliminary results on $\Delta a_{X,Y,Z}$ are:

$$\Delta a_X = (-6.3 \pm 6.0) \times 10^{-18} \text{ GeV},$$

[2] The integration over azimuthal angle does not allow to make observations as a function of the kaon momenta value (e.g. γ_K). According to eq. 4 this will wash out the Δa_0 contribution to the asymmetry.

$$\Delta a_Y = (2.8 \pm 5.9) \times 10^{-18} \text{ GeV},$$
$$\Delta a_Z = (2.4 \pm 9.7) \times 10^{-18} \text{ GeV}. \qquad (8)$$

The Δa_0 parameter is coupled only with the γ_K factor without any sidereal or spatial modulation, as shown in eq. 3. Since at DAΦNE the ϕ is produced with a small boost in the horizontal plane the kaons have different values for γ_K as a function of the azimuthal angle. This will make possible a different approach with respect to the one used for the preliminary results. The kaons can be ordered according to their z component of the momentum, as in the previous case (forward-backward). Then we can split the event sample according to the sign of the x component for the kaon going in the forward direction. As shown in fig. 2-right, this choice corresponds to divide the data sample in two subset for which the kaons going in the forward direction has higher (lower) value for γ_K since it is emitted along (opposite) to the ϕ boost. In this way we can enhance the effect of the SME parameter Δa_0, as shown in fig. 3-left. In the plots a simulation with all parameters but Δa_0 set to zero is shown. For values of Δa_0 of the order $\mathcal{O}(10^{-18})$ GeVwe expect an effect on the $I_{\pm}(\Delta t)$ up to 1%-2% in the region $|\Delta t| < 5\tau_S$ from which we expect to be able to put limits of the order $\mathcal{O}(10^{-17})$-$\mathcal{O}(10^{-18})$ GeV.

The Δa_μ parameters can be all simultaneously measured by performing a proper sidereal time dependent analysis of asymmetries of $I_{\pm}(\Delta t)$, eq. 2 and eq. 4. An accuracy $\mathcal{O}(10^{-18})$ GeV could be reached with the analysis of the full KLOE data sample.

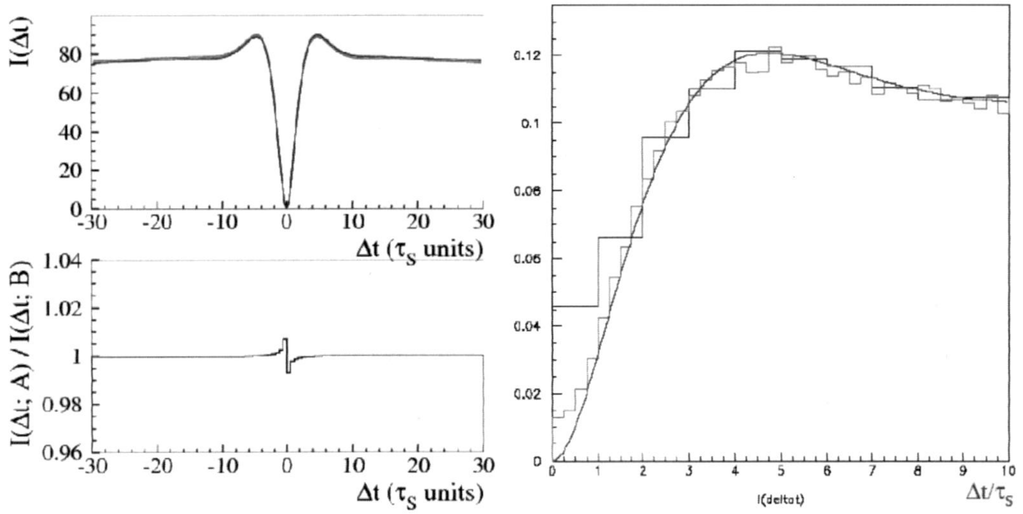

Figure 3. Left: effect of Δa_0 parameters on the decay rate distribution. Red and blue in the top plot refers to the different pair of sectors chosen, while bottom plot is the ratio between the two. Right: effect on the resolution with the upgrades proposed for KLOE-2.

4. Conclusions and future plans

All four Δa_μ parameters of the SME can be independently measured at KLOE, completing results obtained by fixed beam experiments.

The continuation of the KLOE physics program with KLOE-2 [9] at an improved DAΦNE machine is currently starting. The data taking campaign will be organized in two different steps. During the first one we will double the statistics already taken by KLOE with a new beam interaction scheme [10] and with the inclusion of two pairs of electron-positron taggers [11] for the study of the gamma-gamma physics. In this first phase we will benefit of the increased statistics and also we can expect improvements due to the new interaction region geometry both for regeneration, because the interior part of the beam pipe has been redesigned, and for the increased crossing angle, that will results in a increased range for γ_K coefficient.

The second phase aims to an integrated luminosity of $\sim$25 fb^{-1} including several upgrades for the KLOE detector:

- a pair of crystal calorimeters (CCALT[12]) near the interaction region to improve the angular acceptance for low-θ particles;
- a pair of tile calorimeters (QCALT[13]) covering the quadrupoles along the beam pipe made of tungsten foil and singly read-out scintillator tiles to improve the angular coverage for particles coming from the active volume of the DC;
- a small and light tracker (IT[14]) made of four planes of cylindrical GEM to improve the resolution of the vertex reconstruction around the interaction point and to increase the low-ϑ charged particles acceptance.

In this second phase we expect an improvement of the decay vertex resolution, because of the Inner Tracker. As shown in fig. 3, we should obtain an increase in the resolution of Δt of a factor of three/four. This, together with the increased statistics, should allow us to improve the results on CPT violation parameters by more than one order of magnitude.

References

[1] M. Adinolfi *et al.*, KLOE Collaboration, Nucl. Inst. Meth. A 488 (2002) 51.
[2] M. Adinolfi *et al.*, KLOE Collaboration Nucl. Inst. Meth. A 482 (2002) 363.
[3] V.A. Kostelecký, Phys. Rev. Lett. **80**, 1818 (1998).
[4] V.A. Kostelecký, Phys. Rev. D **61**, 016002 (1999).
[5] V.A. Kostelecký, Phys. Rev. D **64**, 076001 (2001).
[6] A. Di Domenico, in V.A. Kostelecký, ed., *Fourth Meeting on CPT and Lorentz Symmetry*, World Scientific, Singapore, 2008; A. Di Domenico, in A. Di Domenico ed., *Handbook on neutral kaon interferometry at a Φ-factory*, Frascati Physics Series, Vol. 43.
[7] C. D. Buchanan *et al.*, Phys. Rev. D **45**, 4088 (1992).
[8] A. Di Domenico, J. Phys. Conf. Ser. **171**, 012008 (2009).
[9] G. Amelino-Camelia *et al.*, arXiv:1003.3868.
[10] C. Milardi *et al.*, DAFNE Collaboration, arXiv:1006.1487.
[11] D. Babusci *et al.*, Nucl. Instrum. Meth. A **617**, 81 (2010).
[12] F. Happacher *et al.*, Nucl. Phys. Proc. Suppl. **197**, 215 (2009).
[13] M. Cordelli *et al.*, Nucl. Instrum. Meth. A **617**, 105 (2010).
[14] F. Archilli *et al.*, KLOE-2 Collaboration, arXiv:1002.2572.

Journal of Physics: Conference Series **335** (2011) 012059 doi:10.1088/1742-6596/335/1/012059

Measurement of the ground-state hyperfine splitting of antihydrogen

B Juhász[1], E Widmann[1] and S Federmann[1,2]

[1]Stefan Meyer Institute for Subatomic Physics, Boltzmanngasse 3, 1090 Wien, Austria
[2]CERN, 1211 Genève 23, Switzerland

E-mail: bertalan.juhasz@oeaw.ac.at

Abstract. The ASACUSA collaboration at the Antiproton Decelerator of CERN is planning to measure the ground-state hyperfine splitting of antihydrogen using an atomic beam line, consisting of a cusp trap as a source of partially polarized antihydrogen atoms, a radiofrequency spin-flip cavity, a superconducting sextupole magnet as spin analyser, and an antihydrogen detector. This will be a measurement of the antiproton magnetic moment, and also a test of the CPT invariance. Monte Carlo simulations predict that the antihydrogen ground-state hyperfine splitting can be determined with a relative precision of $\sim 10^{-7}$. The first preliminary measurements of the hyperfine transitions will start in 2011.

1. Motivation

The validity of the Charge-Parity-Time reversal (CPT) invariance is assumed – either implicitly or explicitly – in all our mainstream quantum field theories and in the Standard Model of particle physics. This assumption is rooted in a mathematical theorem, which states that any quantum field theory which respects Lorentz invariance, locality and unitarity (i.e. conservation of probability) must be symmetric under the CPT transformations. However, some of the three conditions of this CPT theorem might be violated in certain scenarios, for example around microscopic black holes at the Planck scale, or in string theory. Therefore it is important to test experimentally the validity of CPT invariance.

CPT symmetry implies that the properties of a particle and its antiparticle, or more generally, a system and its anti-system, should be the same or exactly opposite. These properties, among others, are the mass, charge, spin, magnetic moment, or various energy levels. Thus one way to test the CPT invariance is to study antimatter systems and compare them to their corresponding matter counterpart. However, even without such comparisons, the apparent abundance of matter in the Universe is a hint that the matter-antimatter symmetry might not be perfect. One possible explanation to the lack of antimatter in the Universe is either CP or CPT violation.

2. Theoretical background

The $1s$ ground state of hydrogen is split due to the interaction between the electron spin and the proton spin, and can have quantum numbers $(F, M) = (0,0)$, $(1,-1)$, $(1,0)$ and $(1,1)$, where F is the total spin, and M is the projection of F. The states with quantum numbers $F = 0$ (singlet state) and $F = 1$ (triplet state) have different energies even at zero external magnetic

Journal of Physics: Conference Series **335** (2011) 012059 doi:10.1088/1742-6596/335/1/012059

field. This ground-state hyperfine splitting (GS-HFS) frequency has been measured to a very high precision of 10^{-12} [1]:

$$\nu_{\mathrm{HFS}} = 1420.405\ 751\ 766\ 7(9)\ \mathrm{MHz}. \tag{1}$$

Thus measuring the ν_{HFS} of the antimatter counterpart of hydrogen, the antihydrogen atom ($\overline{\mathrm{H}}$), which consists of an antiproton ($\overline{\mathrm{p}}$) and a positron (e^+), would allow us to test the CPT invariance to a similarly high precision.

The ground-state hyperfine splitting frequency in hydrogen (antihydrogen) is proportional in the leading order to the proton (antiproton) spin magnetic moment μ_p:

$$\nu_{\mathrm{HFS}} = \frac{16}{3}\left(\frac{m_p}{m_p+m_e}\right)^3 \frac{m_e}{m_p}\frac{\mu_p}{\mu_N}\alpha^2 c R_\infty(1+\delta), \tag{2}$$

where m_p and m_e are the proton (antiproton) and the electron (positron) masses, respectively, μ_N is the nuclear magneton, and R_∞ is the Rydberg constant. The correction term δ is ~1100 ppm, and it is mainly due to higher-order QED corrections. It is theoretically calculated with a precision of slightly under 1 ppm; this uncertainty comes from the proton structure-dependent corrections [2]. The antiproton spin magnetic moment $\mu_{\overline{\mathrm{p}}}$, on the other hand, is known to a precision of only 0.3% from experiments with antiprotonic lead [3] and later from experiments with antiprotonic helium [4]. Therefore a measurement of ν_{HFS} of antihydrogen (which has never been done before) with a precision of only 10^{-6} can already lead to an improvement of the value of $\mu_{\overline{\mathrm{p}}}$ by three orders of magnitude.

The group of V.A. Kostelecký has developed an extension of the standard model of elementary particles that includes both CPT violating and Lorentz invariance violating terms in the Lagrangian of the quantum field theory [5]. These terms result in a correction to the energies of the $1s$ and $2s$ hyperfine sublevels:

$$\begin{aligned} \Delta E^{\mathrm{H}}(m_J, m_I) = {} & a_0^e + a_0^p - c_{00}^e m_e - c_{00}^p m_p \\ & + (-b_3^e + d_{30}^e m_e + H_{12}^e) m_J/|m_J| \\ & + (-b_3^p + d_{30}^p m_p + H_{12}^p) m_I/|m_I|, \end{aligned} \tag{3}$$

where m_J and m_I are the characteristic spin quantum numbers of the electron (positron) and the proton (antiproton), respectively, at large external magnetic fields. The parameters a, d and H change sign for antihydrogen, therefore the GS-HFS transition frequencies in hydrogen and antihydrogen can be different.

An important feature of this theory is that the parameters a and b are not dimensionless; instead, they have a dimension of energy (or frequency). From this it follows that it is not the relative but the *absolute* precision of an experiment that matters when one is comparing matter and antimatter systems for CPT tests. Therefore the $\overline{\mathrm{H}}$ GS-HFS measurement can be competitive with the oft-quoted 'most sensitive CPT limit' of $|m_{K^0} - m_{\overline{K}^0}|/m_{K^0} < 10^{-18}$ [6]. This is because the K^0-$\overline{K}^0$ measurement corresponds to only ~100 kHz on the absolute scale, while the $\overline{\mathrm{H}}$ GS-HFS measurement with a relative precision of only 10^{-6} would have an absolute precision of 1.42 GHz $\times 10^{-6}$ = 1.42 kHz.

3. Proposed experimental method

The experimental resolution of the ground-state hyperfine transitions is inversely proportional to the interaction time of the (anti)hydrogen atoms with the radiofrequency (RF) field. Thus the most precise measurements of the transition frequencies of hydrogen have been done with hydrogen masers, where the atoms can be kept in the RF field for several minutes. Unfortunately,

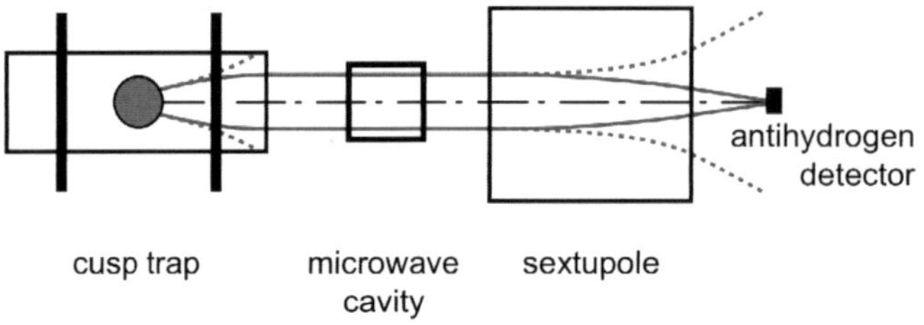

Figure 1. Schematic view of the proposed atomic beam spectrometer with the cusp trap as the antihydrogen source, a radiofrequency cavity for spin flip, and a sextupole magnet for spin analysis. The trajectories drawn with solid red lines represent $\overline{\text{H}}$ atoms in low-field seeker states, while the dashed green lines represent $\overline{\text{H}}$ atoms in high-field seeker states.

antimatter atoms cannot be kept in a maser cavity, since they would annihilate when colliding with the cavity walls.

The neutral antihydrogen atoms could be kept in the RF field by a confining magnetic field gradient. However, the required fields would be very strong and would shift and broaden the energy levels too much, which would make high-precision spectroscopy measurements very difficult.

Thus the ASACUSA collaboration at CERN's Antiproton Accelerator (AD) plans to use an atomic beam line [7, 8] to measure the $\overline{\text{H}}$ GS-HFS. This method is based on the same Stern-Gerlach principle as the classic experiments which first measured the hyperfine splitting of hydrogen. It has the advantage that it can work with relatively high-temperature (50-100 K) atoms, while trapping would require ultra-cold (< 1 K) atoms. A schematic view of the proposed atomic spectrometer line is shown in Fig. 1.

The antihydrogen atoms will be produced in a cusp trap i.e. a pair of anti-Helmholtz coils [9]. The currents running in the opposite direction in the two coils produce zero field in the centre of the trap, but the magnetic field strength increases linearly with the distance from the trap centre. Such an inhomogeneous magnetic field can confine not only charged but also neutral particles. Thus the produced antihydrogen atoms, which are most likely created in higher-lying Rydberg states, can be trapped long enough to de-excite to the ground state before they eventually leave the trap and reach the radiofrequency resonator [10].

Figure 2 shows the energy levels of the four hyperfine states of antihydrogen as a function of the external magnetic field. Two hyperfine states are so-called low-field seekers, while the other two are high-field seekers. When moving in the inhomogeneous magnetic field of the cusp trap, these two types will have different trajectories: the low-field seekers will be focussed, while the high-field seekers will be defocussed as they leave the trap (see Fig. 1). Thus the emerging $\overline{\text{H}}$ beam will be at least partially polarized: more low-field seeker atoms will reach the radiofrequency cavity as high-field seekers.

Figure 2 also shows the three allowed hyperfine transitions which can convert one of the two low-field seeker states into one of the two high-field seeker states (or vice versa). Two of these transitions, π_1 and σ_1, lie close to each other in weak magnetic fields, therefore a single radiofrequency resonator operating around 1.42 GHz can be used to flip the spin of the $\overline{\text{H}}$ atoms as they pass through it. A superconducting sextupole magnet will be placed after this cavity, which will focus the low-field seeker atoms onto an antihydrogen detector, while it will defocus the high-field seekers. This forms the basis of the spectroscopy method: when the radiofrequency field is on resonance with one of the hyperfine transitions, fewer atoms will reach the antihydrogen detector. The advantage of the sextupole magnet is that its magnetic field gradient increases linearly with the distance from the magnet axis, therefore all low-field seeker

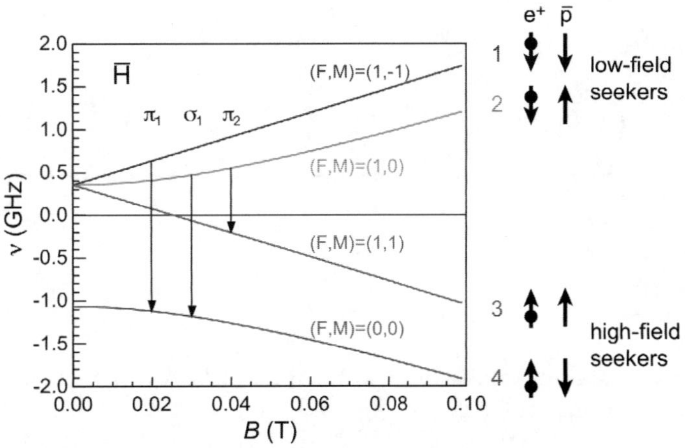

Figure 2. Frequencies (i.e. energies) of the four hyperfine states of antihydrogen as a function of the external static magnetic field B. The allowed transitions which convert low-field seekers into high-field seekers (π_1, σ_1, π_2) are also drawn.

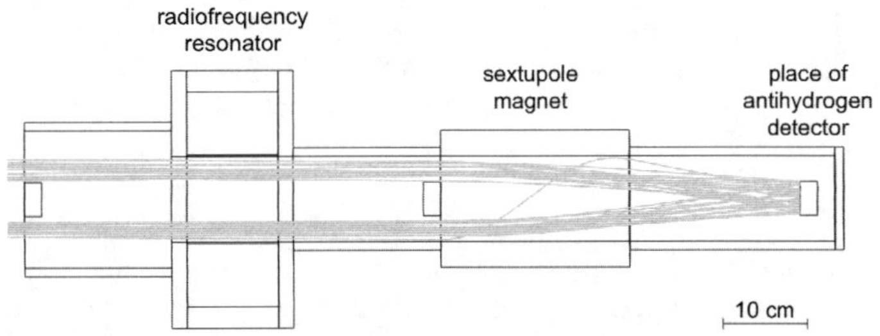

Figure 3. Simulated trajectories of low-field seeker antihydrogen atoms emerging from the cusp trap, then passing through the radiofrequency cavity and the sextupole magnet, and finally hitting an antihydrogen detector (not shown). The internal diameter of the sextupole magnet is 10 cm.

atoms with the same velocity will be focussed into a single focal point, independently of where they enter the magnet.

4. Monte Carlo simulations

Simulations using the GEANT4 toolkit [11] have been carried out to determine the necessary size and field strength of the sextupole magnets, and to estimate the expected count rates and experimental resolution. It was assumed that the cusp trap can provide an antihydrogen beam with a polarization ratio of 4:1 1 m downstream of the cusp trap, i.e. where the RF resonator will be connected. Such a simulation with the cusp trap is shown in Fig. 3. The sextupole magnet needs to have a large internal diameter of 10 cm in order to increase the transmission efficiency for the $\overline{\mathrm{H}}$ atoms. The simulation results showed that if the antihydrogen atoms have a temperature of 5 K, then the sextupole magnet has to have a pole tip field of 0.35 T and an effective length (field FWHM) of 22 cm. The necessary pole tip field strength increases linearly with the temperature of the atoms. Since the superconducting magnet can reach a maximum pole tip field of 3 T, atoms of up to 50 K can be used in the experiment.

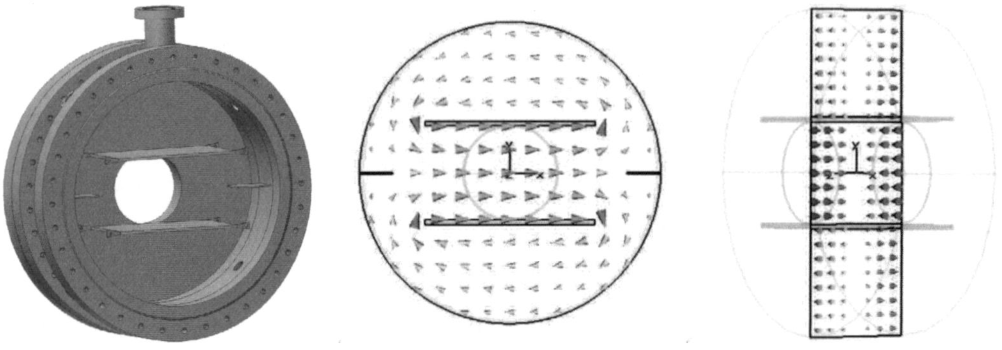

Figure 4. Left [13]: Schematic view of the cylindrical radiofrequency resonator with the double stripline and two auxiliary plates. Of the two vacuum flanges which close the chamber, only one is shown. Center [13]: Oscillating magnetic field of the double stripline cavity in the X-Y plane (perpendicular to the beam axis). Right [13]: Oscillating magnetic field in the Y-Z plane.

5. Radiofrequency resonator

As explained above, the purpose of the radiofrequency resonator is to induce a spin flip of the antihydrogen atoms. Ideally, the spin flip should be a 'π-pulse' i.e. when the spin of of the atoms makes exactly one half of a Rabi oscillation. This way the width$\Delta\ \nu$ of the GS-HFS line in the RF spectrum can reach the theoretical minimum, which can be calculated for monoenergetic (monovelocitic) atoms as [12]:

$$\Delta\nu = \frac{0.799}{T}, \tag{4}$$

where T is the observation time of the transition i.e. the time it takes for one atom to pass through the RF resonator and interact with the oscillating magnetic field. It can be readily seen that the longer the observation time, the narrower the transition line.

The resonator has to be tunable within the range of 1420–1425 MHz. This will be achieved by using a low-Q cavity in which the frequency of the oscillating field can be changed simply by changing the frequency of the external frequency source.

5.1. Oscillating magnetic field

The spin flip of the antihydrogen atom has to be induced by a magnetic field oscillating at 1.42 GHz. With such an oscillating field, the π_1 and the σ_1 transitions in Fig. 2 can be observed. Ideally, the oscillating magnetic field should be perfectly homogenous in all directions. However, Maxwell's equations forbid to generate a perfectly homogenous oscillating magnetic (or electric) field in a volume whose dimensions are comparable or larger than the half-wavelength of the field. This is exactly the case in this setup, since the size of the volume in which the oscillating field has to be present is at least $\sim 10\times10\times10$ cm^3, while the half-wavelength of the radiofrequency field is $\lambda_{\mathrm{HFS}}/2 \simeq 10.5$ cm. Therefore at least in one direction the magnetic field cannot be homogenous.

To generate the required oscillating magnetic field, a cylindrical cavity with a double stripline, i.e two parallel conducting plates, has been chosen [13, 14] (see Fig. 4). The length of the plates (measured along the beam) has to be an integer multiple of the desired half-wavelength, while the width and the distance between the plates can be arbitrary. The plates are placed inside a cylindrical vacuum chamber, which has two openings on the front and back plates. These openings are covered with fine metallic meshes, which allow the $\overline{\mathrm{H}}$ atoms to enter and leave the cavity, but keep the RF field from leaking out.

Figure 4 shows that the oscillating magnetic field of the double stripline is very homogenous perpendicularly to the beam (X-Y plane): the field inhomogeneity is $\pm 2\%$. Along the beam (Z direction), however, the field strength follows a $\sin(Z)$ distribution: the field is zero in the center plane, and has maxima at the front and back walls of the resonator. Thus the magnetic field points at opposite directions in the two half-volumes. When the frequency of the oscillating field is on resonance with one of the GS-HFS transitions, the effects of the two half-volumes cancel each other. Consequently, one gets zero signal on resonance (independently of the velocity of the atoms), and gets a non-zero signal away from the resonance: the peak of each transition in the GS-HFS spectrum is split into two, as it can be seen in Figs. 5 and 6. This can be thought of as a Doppler splitting, since the standing wave in the cavity is a superposition of two travelling waves, one moving in the direction of the $\overline{\mathrm{H}}$ beam, and the other moving the opposite way. The splitting is thus proportional to the velocity (i.e. energy) of the antihydrogen atoms.

5.2. Static magnetic field

In order to prevent the spontaneous Majorana spin flips of the antihydrogen atoms inside the cavity, a homogeneous static magnetic field has to be created there using one or more pairs of Helmholtz coils. Thus the GS-HFS frequency at zero external magnetic field needs to be obtained from the π_1 and σ_1 transition frequencies (see Fig. 2) measured at non-zero magnetic field(s). To excite the σ_1 transition, the static magnetic field and the oscillating magnetic field have to be parallel. To excite the π_1 transition, however, they have to be perpendicular to each other. Therefore in order to excite both transitions while keeping the static field unchanged, the static and the oscillating fields have to be at an angle of 45° with respect to each other.

6. Determination of the GS-HFS frequency

Of the two above-mentioned GS-HFS transitions, σ_1 has only a weak, approximately second-order dependence on the external magnetic field strength:

$$\Delta\nu_{\sigma_1} \simeq 2.76 \text{ kHz/G}^2 \tag{5}$$

for small external fields. Therefore this transition is not significantly broadened by the inhomogeneities of the external magnetic field. Figure 5 illustrates this with simulated resonance profiles at 1 G external magnetic field with various field inhomogeneities. It can be seen that even a 10% inhomogeneity hardly distorts the line shape.

The π_1 transition line, on the other hand, has a very strong, approximately linear dependence on the external magnetic field:

$$\Delta\nu_{\pi_1} \simeq 1400 \text{ kHz/G} + 1.38 \text{ kHz/G}^2 \tag{6}$$

for small external fields. Therefore this transition is severely broadened by the magnetic field inhomogeneities. Figure 6 shows that at an inhomogeneity of 0.5%, the double peak structure is already washed out, and an inhomogeneity of 2% already makes the transition line too broad to reliably detect it.

Knowing the above, the zero-field GS-HFS frequency ν_{HFS} can be determined in two ways. If one can create a static magnetic field in the RF resonator with an inhomogeneity of not more than 1%, then both the σ_1 and the π_1 transition frequencies can be measured at the same external magnetic field of e.g. 1 G, and from these two frequencies the zero-field GS-HFS frequency can be calculated. The Monte Carlo simulations showed that with an external magnetic field of 1 G $\pm$ 5 mG (i.e. magnetic field inhomogeneity of 1%), ν_{HFS} can be obtained with an accuracy of $\sim$100 Hz, which corresponds to a relative precision of $\sim 10^{-7}$.

A static magnetic field with such a good homogeneity, however, is not easily achievable, since the cusp trap and the sextupole magnet produce a strong ($\sim$100 G) inhomogeneous stray field at

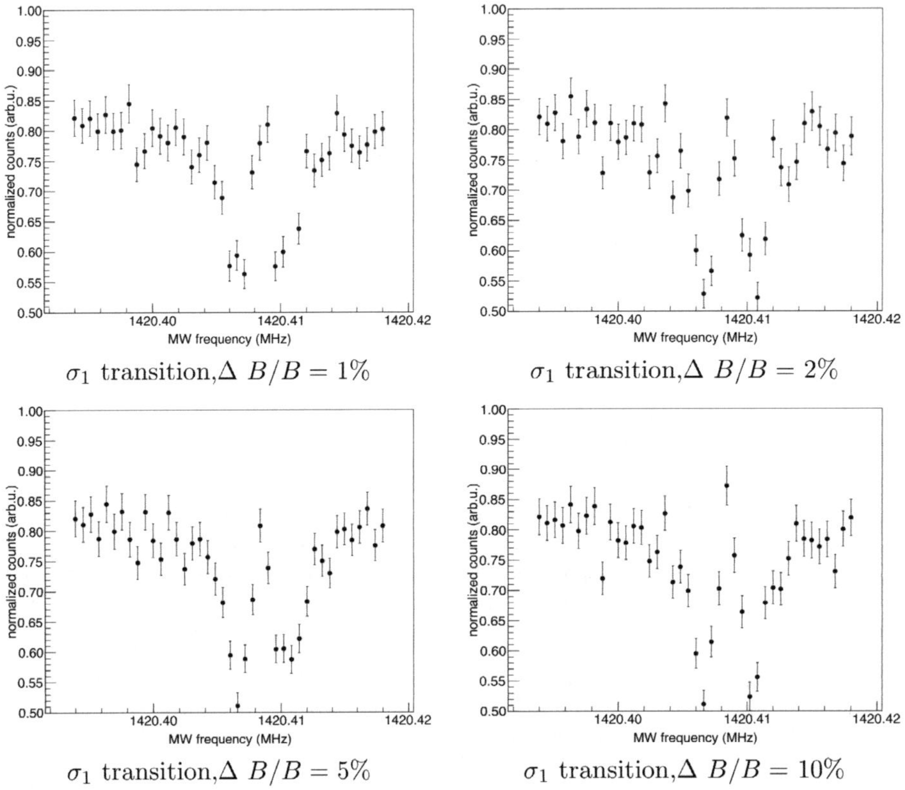

Figure 5. Simulated resonance profiles of the σ_1 transition at an external magnetic field B of 1 G but at different magnetic field inhomogeneities$\Delta\ B/B$.

the position of the RF resonator. Therefore in the first stage of the experiment, only the 'easy' σ_1 transition will be measured at different external magnetic field strengths, and ν_{HFS} will be obtained by extrapolating the measured frequencies to zero. This method has the disadvantage that the achievable precision is lower: according to the Monte Carlo simulations, by measuring ν_{σ_1} at three different magnetic fields between 1–10 G with a constant absolute inhomogeneity of 0.1 G, ν_{HFS} can be determined with a a relative precision of $\sim 5 \times 10^{-7}$.

7. Project status

In 2010, antihydrogen atoms could be produced for the first time in the cusp trap. The average production rate was $\sim$30 $\overline{\mathrm{H}}$/s in Rydberg states [15]. This was an important step towards the measurement of the ground-state hyperfine splitting of antihydrogen. In 2011, the $\overline{\mathrm{H}}$ synthesis process will be further optimized to increase the number of atoms produced. The RF resonator, the sextupole magnet and the antihydrogen detector will be installed, but the magnetic shielding and the Helmholtz coils around the resonator will not yet be in their final form. Nevertheless, this setup will be enough to attempt the first preliminary antihydrogen spin flip experiment. If successful, it will be the first ever spectroscopy measurement of an atomic system made entirely of antimatter. Afterwards the final, high-resolution scans of the transition lines will commence from 2012.

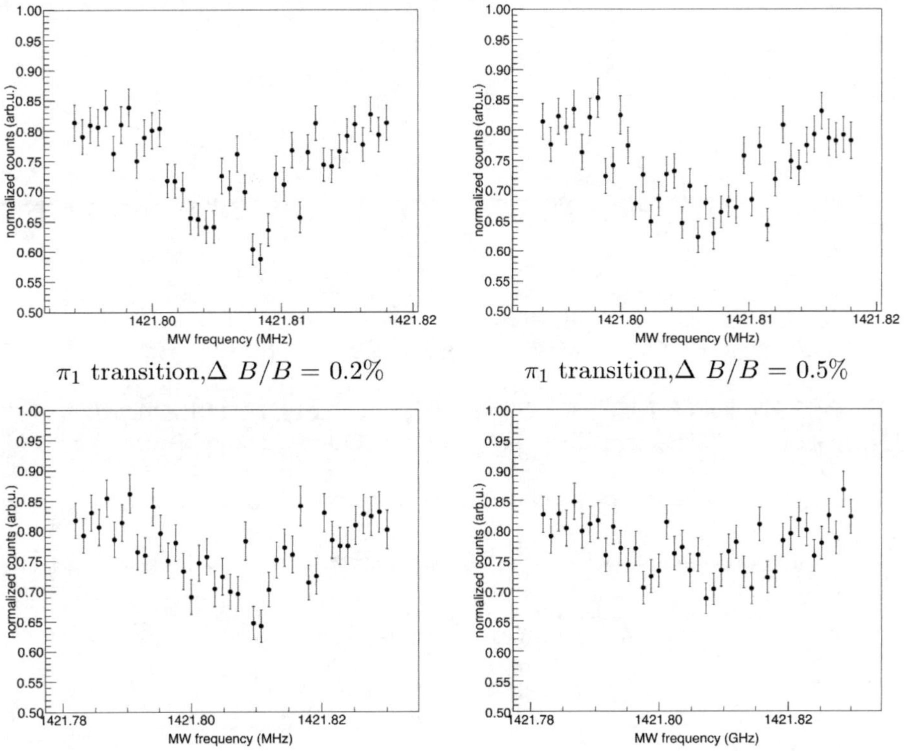

π_1 transition, $\Delta B/B = 0.2\%$ — π_1 transition, $\Delta B/B = 0.5\%$ — π_1 transition, $\Delta B/B = 1\%$ — π_1 transition, $\Delta B/B = 2\%$

Figure 6. Simulated resonance profiles of the π_1 transition at an external magnetic field B of 1 G but at different magnetic field inhomogeneities $\Delta B/B$.

Acknowledgments
We are greatly indebted to F. Caspers, T. Kroyer, E. Mahner and B. Salvant for the fruitful discussions on the design of the resonance cavity. The authors would like to thank E. Ciapala and E. Chapochnikova as well as the CERN ABP group for their support. This work was supported by the Austrian Federal Ministry of Science and Research.

References

[1] Ramsey N F 1990 *Rev. Mod. Phys.* **62** 541
[2] Carlson C E, Nazaryan V and Griffioen K 2008 *Phys. Rev.* A **78** 022517
[3] Kreissl A *et al* 1998 *Z. Phys.* C **37** 557 (1988)
[4] Pask T *et al* 2009 *Phys. Lett.* B **678** 55–59
[5] Bluhm R, Kostelecký V A and Russell N 1999 *Phys. Rev. Lett.* **82** 2254
[6] Amsler C *et al* (Particle Data Group) 2008 *Phys. Lett.* B **667** 1
[7] *ASACUSA proposal addendum* 2005 CERN/SPSC 2005-002, SPSC P-307 Add.1
[8] Juhász B and Widmann E 2009 *Hyp. Int.* **193** 305
[9] Mohri A and Yamazaki Y 2003 *Europhys. Lett.* **63** 207
[10] Pohl T, Sadeghpour H R, Nagata Y and Yamazaki Y 2006 *Phys. Rev. Lett.* **97** 213001
[11] Agostinelli S *et al* 2003 *Nucl. Instrum. Methods* A **506** 250
[12] Kusch P and Hughes V W 1959 *Encyclopedia of Physics* vol XXXVII/1, ed S Flügge (Berlin: Springer) p 56
[13] Kroyer T 2008 *CERN Report* CERN-AB-Note-2008-016
[14] Federmann S, Caspers F, Mahner E, Juhász B and Widmann E 2010 *1st Int. Particle Accelerator Conf. (Kyoto)* (Kyoto: IPAC'10 OC/ACFA) p 1095
[15] Enomoto Y *et al* 2010 *Phys. Rev. Lett.* **105** 243401

Journal of Physics: Conference Series **335** (2011) 012060 doi:10.1088/1742-6596/335/1/012060

Testing the Pauli Exclusion Principle for electrons

J. Marton[1], S. Bartalucci[2], S. Bertolucci[4], M. Bragadireanu[2,3], M. Cargnelli[1], C. Curceanu (Petrascu)[2,3], S. Di Matteo[5], J.-P. Egger[6], C. Guaraldo[2], M. Iliescu[2], T. Ishiwatari[1], M. Laubenstein[7], E. Milotti[8], D. Pietreanu[2], T. Ponta[2,3], A. Rizzo[2], A. Romero Vidal[2], E. Sbardella[2], A. Scordo[2], D.L. Sirghi[2,3], F. Sirghi[2,3], L. Sperandio[2], O. Vazquez Doce[2], E. Widmann[1] and J. Zmeskal[1]

[1] The Stefan Meyer Institute for Subatomic Physics, Boltzmanngasse 3, A-1090 Vienna, Austria
[2] INFN, Laboratori Nazionali di Frascati, CP 13, Via E. Fermi 40, I-00044, Frascati (Roma), Italy
[3] "Horia Hulubei" National Institute of Physics and Nuclear Engineering, Str. Atomistilor no. 407, P.O. Box MG-6, Bucharest - Magurele, Romania
[4] CERN, CH-1211, Geneva 23, Switzerland
[5] Institut de Physique UMR CNRS-UR1 6251, Université de Rennes1, F-35042 Rennes, France
[6] Institut de Physique, Université de Neuchâtel, 1 rue A.-L. Breguet, CH-2000 Neuchâtel, Switzerland
[7] Laboratori Nazionali del Gran Sasso, S.S. 17/bis, I-67010 Assergi (AQ), Italy
[8] Dipartimento di Fisica, Università di Trieste and INFN– Sezione di Trieste, Via Valerio, 2, I-34127 Trieste, Italy

E-mail: johann.marton@oeaw.ac.at

Abstract.
The Pauli Principle represents one of the most important rules in physics and explains numerous phenomena as well as characteristic properties of matter, like its stability. Testing the validity of this principle at the highest possible sensitivity is a challenging experimental task - the VIP experiment at the Gran Sasso underground laboratory aims at a limit for the violation probability of the order of 10^{-29} to 10^{-30}. The method is based on the search for Pauli-forbidden x-ray transitions in a pure copper conductor using silicon x-ray detectors with high resolution in energy. The experimental setup, results obtained so far and new ideas to further enhance the sensitivity will be presented.

1. Introduction

Wolfgang Pauli formulated a rule in 1925 which became the Pauli Exclusion Principle (PEP)[1]. PEP was remarkably successful and could easily explain the shell structure of the atoms. It was found that PEP is valid for all spin half particles, i.e. particles obeying the Fermi-Dirac statistics called fermions. In this way many features of nature, from stability of matter and neutron stars to the quark picture of hadrons found a clear explanation using PEP. On the other hand only with a complicated argumentation an explanation for PEP could be found by Pauli [2]. The Pauli principle has manifested its role as a pillar in quantum theory and should hold true to an extremely high level. The question about a tiny violation of PEP came up with the availability

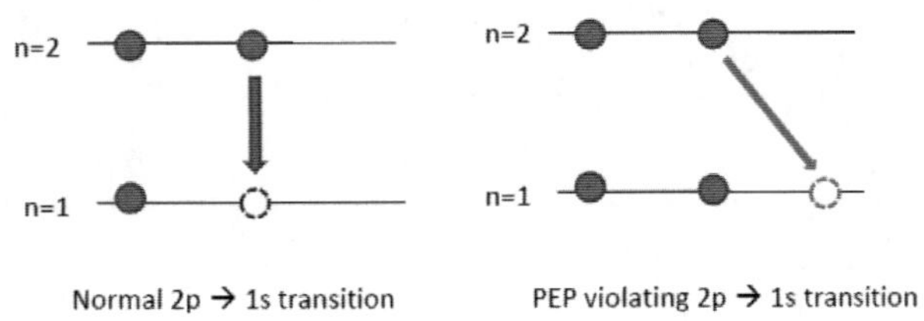

Figure 1. VIP is searching for anomalous, Pauli-forbidden x-ray transitions in copper. The non-Paulian K_α transition energy would be shifted by about 300 eV from 8.040 to 7.729 keV.

of precision methods. In the last 20 years, several experiments have been performed to search for possible small PEP violations using different experimental approaches [3, 4, 5, 6, 7, 8].

For electrons Ramberg and Snow [9] performed a dedicated experiment at Fermilab, searching for PEP forbidden x-ray transitions giving a limit for the probability of PEP violation

$$\frac{\beta^2}{2} < 1.7 \times 10^{-26}. \tag{1}$$

The VIP (<u>VI</u>olation of the <u>P</u>auli Exclusion Principle) Collaboration used a much improved version of the Ramberg and Snow experimental setup yielding a higher sensitivity [10]. The goal of VIP is an improvement of the PEP violation limit for electrons by several orders of magnitude. For the x-ray detectors an array of charged coupled devices (CCDs) [11, 12, 13, 14, 15] were used which provide excellent energy resolution and background reduction by selection of x-ray pixel hits. After first measurements at LNF-INFN the VIP apparatus was installed in the underground laboratory of LNGS (Laboratori Nazionali di Gran Sasso) of the Italian Institute for Nuclear Physics (INFN) which has the advantage of a highly suppressed background from cosmic rays.

2. Experimental Method

The VIP experiment is searching for Pauli-forbidden transitions in copper (see fig.1), i.e. the transition of an electron to the already fully occupied 1s state. The energy of the Pauli-forbidden K_α transition is shifted by about 300 eV, i.e. from 8.040 to 7.729 keV according to calculations [16]. The experiment uses the same experimental approach of the Ramberg-Snow experiment [9] in which the x-ray spectrum of a copper conductor with and without circulating current was compared.

2.1. Experimental setup

The VIP setup employs a thin cylinder of ultrapure (99.995%) copper (radius 45 mm, 88 mm height, thickness 50 μ) through which current is flowing (40 A). The circulating current of 40 A corresponds to $\sim 2.4\times10^{20}$/s test electrons probing the PEP validity. For the x-ray detection an array of 16 CCDs surround the copper target. These CCDs were successfully used for x-ray spectroscopy of kaonic atoms in the DEAR experiment [17]. The arrangement is installed in an insulation vacuum container in order to cool the CCD x-ray detectors to 170 K for obtaining the required high energy resolution. CCDs allow to single out x-ray events by analysis of the hit pattern thus suppressing background [18].

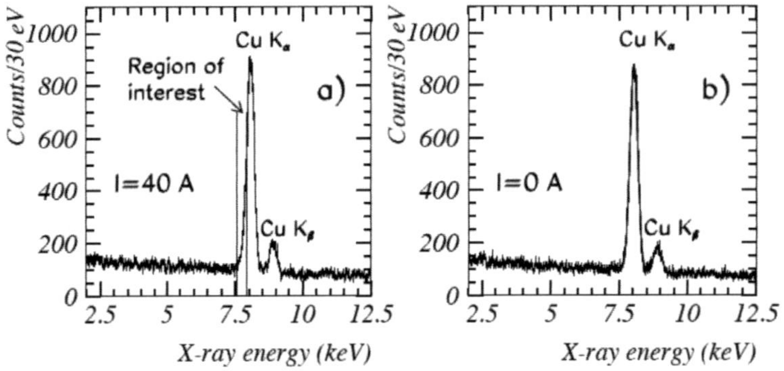

Figure 2.
Calibrated energy spectra with current a) and without current b) flowing through the copper cell. The data a) and b) were measured within the same time of 10 days.

The experimental setup of VIP improves the experiment of Ramberg-Snow to a large extent (see below)

- X-ray detection: CCDs instead of gaseous detectors, improving the times the energy resolution and providing higher stability;
- Underground site: Clean and low-background environment of the underground LNGS Laboratory;
- Higher statistics: longer data acquisition periods (due to the stability of CCDs)

2.2. Results
Prior to the installation in the final destination in the Gran Sasso laboratory, the VIP apparatus was set up and tested at the LNF-INFN laboratory. First measurements were performed with current circulating in the copper target (about 10 days with 40A current) and a background measurement without current with the same measurement time. The obtained energy-calibrated x-ray spectra are shown in figure 2. The spectra show the copper x-ray lines sitting on the continuous background but no other x-ray lines proving the high purity of the structure materials used in the VIP setup.

From the x-ray spectrum with current the spectrum without current was subtracted and a value for the PEP violation factor $\beta^2/2$ was determined applying the Ramberg-Snow method [9]. Compared with the result of Ramberg-Snow the limit for PEP violation for electrons was improved by a factor of 40 [19].

$$\frac{\beta^2}{2} < 4.5 \times 10^{-28} \tag{2}$$

Vip was transferred to LNGS and continued the measurements with and without current. From the VIP data collected at LNGS a further improved limit (preliminary) for $\beta^2/2$ could be deduced:

$$\frac{\beta^2}{2} < 4.7 \times 10^{-29} \tag{3}$$

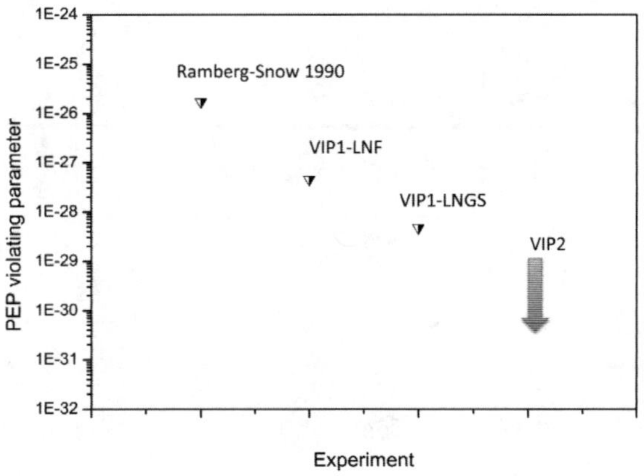

Figure 3. Progress in lowering the limit for PEP violation for electrons. The follow-up experiment VIP2 is aiming at a limit below 10^{-30}.

3. Future Opportunities

The VIP setup uses CCD detectors which are excellent x-ray detectors in respect to energy resolution in the soft x-ray domain and background rejection based on pattern analysis. However, CCDs have no timing capability and therefore no active shielding of the apparatus against background is possible. To overcome this limitation new x-ray detectors, Silicon Drift Detectors (SSDs), are in consideration for the next VIP setup (VIP2). They provide excellent energy resolution comparable with CCDs, but have timing capability (Δ(FWHM) $\simeq 1\mu$s) and large detection area (100 mm^2) [21]. SDDs were successfully used in the SIDDHARTA experiment [22] for measurements of the x-ray spectra of kaonic hydrogen and kaonic helium isotopes at the DAΦNE accelerator of LNF-INFN. A compact design consisting of 6 SDDs looking at a cooled thin planar copper foil could be surrounded by anti-coincidence scintillation detectors. The setup will be additionally passively shielded by lead. With larger solid angle and possibly higher current - demanding high power cooling - a significantly higher sensitivity to non-Paulian transition in copper can be expected in the future.

Acknowledgement

The very important assistance of the INFN-LNGS laboratory staff during all phases of preparation, installation and data taking and the support from the HadronPhysics FP6 (506078) and HadronPhysics2 FP7 (227431) projects are gratefully acknowledged.

References

[1] W. Pauli, *Z. Physik* **31** 765 (1925).
[2] W. Pauli, *Phys. Rev.* **58** 716 (1940).
[3] Bernabei R *et al.* 1997 *Phys. Lett. B* **408** 439
[4] Borexino Colaboration, Back H O 2005 *et al. Eur. Phys. J.* **C37** 421
[5] Hilborn R C and Yuca C L 1996 *Phys. Rev. Lett.* **76** 2844
[6] Nemo Colaboration 2000 *Nucl. Phys. B (Proc. Suppl.)* **87** 510
[7] Nolte E *et al.* 1991 *J. Phys. G: Nucl. Part. Phys.* **17** S355

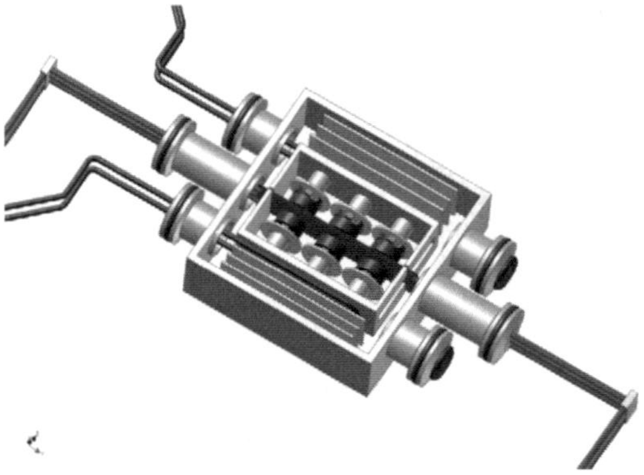

Figure 4. Design study of the inner part of a future setup employing 6 SDDs.

Figure 5. Schematic layout of the future apparatus. The inner part of the future setup will installed inside a lead housing thus active and passive shielding by lead is provided.

[8] Tsipenyuk Y, Barabash A, Kornoukhov V and Chapyzhnikov B, 1998 *Radiat. Phys. Chem.* **51** 507
[9] Ramberg E and Snow G A 1990 *Phys. Lett. B* **238** 438
[10] The VIP proposal, LNF-LNGS Proposal, September, 2004, http://www.lnf.infn.it/esperimenti/vip.
[11] Culhane J L 1990 *Nucl. Instrum. Methods A* **310** 1
[12] Egger J P, Chatellard D and Jeannet E 1993 *Particle World* **3** 139
[13] Fiorucci G *et al.* 1990 *Nucl. Instrum. Methods A* **292** 141
[14] Varidel D *et al.* 1990 *Nucl. Instrum. Methods A* **292** 147
[15] Kraft R P *et al.* 1995 *Nucl. Instrum. Methods A* **372** 372
[16] Di Matteo S, Sperandio L, 2006 *VIP Note, IR-04, 26 April 2006; The energy shift has been computed by P. Indelicato, private communication*
[17] Ishiwatari T *et al.* 2004 *Phys. Lett. B* **593** 48; Beer G *et al.* 2005 *Phys. Rev.Lett.* **94** 212302

[18] Ishiwatari T *et al.* 2006 *Nucl. Instrum. Methods Phys. Res. A* **556** 509
[19] Bartalucci S *et al.* (VIP Collaboration) 2006 *Phys. Lett. B* **641** 18
[20] Sperandio L, 2008 *Ph D thesis "New experimental limit on the Pauli Exclusion Principle violation by electrons from the VIP experiment" at University "Tor Vergata", Roma, 5 March 2008*
[21] J. Marton *et al.*, 2009 *Trans. Nucl. Sci.* **56** 1400.
[22] Curceanu C *et al.* 2007 *Eur Phys* **J A31** 537-539; Bazzi M *et al.* 2009 *Phys. Lett.* **B681** 310

Journal of Physics: Conference Series **335** (2011) 012061 doi:10.1088/1742-6596/335/1/012061

Deviating from the canonical: induced noncommutativity

P A Aguilar Nuñez, C Chryssomalakos, H Hernandez-Coronado and E Okon

Instituto de Ciencias Nucleares
Universidad Nacional Autónoma de México
Apdo. Postal 70-543, 04510 México, D.F., MÉXICO

E-mail: pedro.aguilar,chryss,hcoronado,eliokon@nucleares.unam.mx

Abstract. The motion of a quantum particle, constrained to move along a space curve via a confining potential, is governed by a hamiltonian that depends on the geometry of the curve. Adiabatic changes in the shape of the curve are shown to give rise to geometric phases. The effect of the latter on the quantization of the motion of the curve itself can be naturally described by a deformation of the canonical commutation relations, providing an example of a noncommutative effective quantum field theory.

1. Introduction

At the heart of quantum field theory lie the canonical commutation relations among the dynamical variables of the theory, which, in particular, imply discrete symmetries that experiments have amply verified. For the last two decades or so, several attempts have been made to develop a noncommutative version of the theory, *i.e.*, one where the above commutation relations are somehow deformed. To our knowledge, none of these attempts derives the particular deformation scheme employed from some basic principle, be it physical in nature or mathematical, a state of affairs that leaves the questions of mathematical self-consistency and physical relevance wide open. In the present short communication we study a physical system consisting of a "slow" and a "fast" part in the adiabatic regime, and find that the effective field theory describing the slow part may be cast in noncommutative form, in the sense that a natural set of dynamical variables emerges the commutation relations of which deviate from the canonical. The system consists of a quantum particle constrained to move along a closed space curve by a confining potential with a sharp minimum centered on the curve. For a particular classical configuration of the curve, the 1D effective hamiltonian for the particle's motion along the curve is found, and its wavefunction is determined. Then the focus shifts on the dynamics of the curve itself, and the effect the particle's presence has on it. It is found that geometrical phases that accumulate as the curve is deformed, influence its own effective dynamics in that both positions and momenta end up covariantized, satisfying deformed commutation relations. What might be considered an appealing feature of the theory is that the exact form of noncommutativity is unambiguously determined by standard quantum mechanics, the deformed commutators being given by Berry's curvature.

Regarding the structure of this note, section 2 reviews the basics of constrained quantum mechanics, following [1, 2], in the example of a particle bound by a 2D harmonic oscillator potential in the plane normal to a closed space curve. It is explained furthermore, how cyclic deformations of the curve give rise to geometric phases, as shown in [3]. Section 3 illustrates the mechanism by which the effective theory for the slow part of a "fast-slow" composite system becomes noncommutative, drawing on an example discussed in [4]. The dynamics of the curve itself is discussed in section 4, while the final synthesis of all the ingredients is outlined in section 5.

2. Particles constrained on a curve and geometric phases

Consider a quantum particle that is forced to live on a closed space curve via a confining potential V, that has a sharp minimum on the curve. For concreteness' shake, take V to be a 2D harmonic oscillator potential, in the plane normal to the curve. Calling s a parameter along the curve, and a, b, coordinates in the normal plane, along the normal $\mathbf{n}$ and binormal $\mathbf{b}$ of the curve, respectively, we take $V(s,a,b) = (a^2+b^2)/\eta^4$, where η is a small number, $\eta \ll 1$, that controls the penetration depth of the particle in the normal plane. Switching to rescaled normal coordinates $\alpha = a/\eta$, $\beta = b/\eta$, so that the particle's wavefunction takes significant values where α, β are of order 1, one obtains for the 3D hamiltonian H of the particle

$$H = -\frac{1}{2|G|^{1/2}}\partial_A G^{AB}|G|^{1/2}\partial_B + \frac{(\alpha^2+\beta^2)}{2\eta^2}\,, \tag{1}$$

where G_{AB} is the euclidean metric in the adapted coordinates (s,α,β). The normalization condition for the 3D wavefunction Φ of the particle is $\int ds d\alpha d\beta G^{\,1/2}|\Phi|^2 = 1$, where $G = \det(G_{AB})$. In order to derive an effective 1D hamiltonian for the motion of the particle along the curve, we rescale the wavefunction,$\Phi \;\to \Psi \equiv e^{-h}\,\Phi$, with $e^{-h} = (G/g)^{1/4}$ and $g = \rho^2$ the metric of the curve, so that the normalization condition becomes $\int ds d\alpha d\beta g^{\,1/2}|\Psi|^2 = 1$, and $\int d\alpha d\beta\,|\Psi|^2$ can be interpreted as the effective 1D probability density for the particle at the position s along the curve, using the conventional measure $g^{1/2}ds$. At the same time, we have to transform the hamiltonian,

$$H \to \tilde{H} = e^{-h}He^{h}\,, \quad \text{where} \quad \tilde{H} = \frac{1}{\eta^2}\tilde{H}_{-2} + \tilde{H}_0 + O(\eta)\,, \tag{2}$$

and

$$\tilde{H}_{-2} = -\frac{1}{2}\left(\partial_\alpha^2 + \partial_\beta^2\right) + \frac{1}{2}(\alpha^2+\beta^2)\,, \qquad \tilde{H}_0 = -\frac{1}{2}\left(\rho^{-1}\partial_s - \tau L\right)^2 - \frac{1}{8}\kappa^2\,. \tag{3}$$

The operator L above is the (antihermitean) generator of rotations in the normal plane, $L = \alpha\partial_\beta - \beta\partial_\alpha$. Factorizing Ψ as $\Psi(s,\alpha,\beta) = \chi(\alpha,\beta)\psi(s)$, and assuming the normal wavefunction χ to satisfy the eigenfunction equations $\tilde{H}_{-2}\chi_{\sigma n} = (n+1)\chi_{\sigma n}$, and $L\chi_{\sigma n} = i\sigma\chi_{\sigma n}$, converts the 3D relation $\tilde{H}\Psi = E\Psi$ to the following equation for ψ,

$$\begin{aligned} H_0\psi_{\sigma m} &\equiv -\frac{1}{2}\left(\rho^{-1}\partial_s - i\sigma\tau\right)^2\psi_{\sigma m} - \frac{1}{8}\kappa^2\psi_{\sigma m} \\ &= -\frac{1}{2\rho^2}\psi''_{\sigma m} + \left(i\sigma\rho^{-1}\tau + \frac{\rho'}{2\rho^3}\right)\psi'_{\sigma m} + \frac{1}{2}\left(i\sigma\rho^{-1}\tau' + \sigma^2\tau^2 - \frac{1}{4}\kappa^2\right)\psi_{\sigma m} = E_{\sigma m}\psi_{\sigma m}\,, \end{aligned} \tag{4}$$

where κ and τ are the curvature and torsion of the curve, respectively. Notice that the index m in the above equation is discrete, owing to the periodic boundary conditions appropriate for a closed curve. The $\tilde{H}$ eigensystem is then given by

$$\Psi_{\sigma nm} = \chi_{\sigma n}\psi_{\sigma m}\,, \qquad E_{\sigma nm} = \frac{(n+1)}{\eta^2} + E_{\sigma m}\,. \tag{5}$$

A look at (4) reveals that the shape of the curve, codified in the two functions κ, τ, plays the role of an infinite set of external parameters for H_0. Cyclic deformations of the curve then, may lead to the appearance of geometric phases in the particle's wavefunction, and this is indeed what happens, as was shown in [3]. To calculate the corresponding Berry's curvature, we consider two infinitesimal deformations with parameters $\xi(t)$, $\zeta(t)$, respectively, and calculate the perturbed hamiltonian $\mathcal{H}$,

$$\mathcal{H} = H_0 + \xi H^\xi + \zeta H^\zeta \,, \quad \text{where} \quad H^\xi = i\sigma\tau_\xi \partial_s + \frac{i}{2}\sigma\tau'_\xi + \sigma^2\tau\tau_\xi - \frac{1}{4}\kappa\kappa_\xi \,. \tag{6}$$

κ_ξ, τ_ξ above are the derivatives of κ, τ, w.r.t. ξ, given by

$$\kappa_\xi = \mathbf{v}''^{\,n} \tag{7}$$

$$\tau_\xi = \kappa^{-1}\mathbf{v}'''^{\,b} - \kappa^{-1}\tau\mathbf{v}''^{\,n} - \kappa^{-2}\kappa'\mathbf{v}''^{\,b} + \kappa\mathbf{v}'^{\,b} + \tau\mathbf{v}'^{\,t} \,, \tag{8}$$

where $\dot\xi(0)\mathbf{v}$ is the deformation velocity field along the curve, at $t = 0$ — similar expressions hold for H^ζ. Next, one computes the eigenfunction $\psi(s;\xi,\zeta)$ of $\mathcal{H}$ and, from it, the rescaled wavefunction $\Psi(s,\alpha,\beta;\xi,\zeta)$ and, finally, Φ. Notice that although Φ is eigenfunction of the cartesian coordinate hamiltonian (1), it is nevertheless expressed in the adapted coordinates. To properly apply Berry's formula for the curvature,

$$K^{(n)}_{\xi\zeta} \equiv -2\,\mathrm{Im}(\partial_\xi\langle n,\xi\,|)(\partial_\zeta|n,\xi\,\rangle) \,, \tag{9}$$

one has to express $\Phi(s,\alpha,\beta;\xi,\zeta) = \langle s,\alpha,\beta\,|n;\xi,\zeta\rangle$ in the cartesian coordinates, then take the partials w.r.t. the parameters, and reexpress the result in the adapted coordinates, this latter step facilitating the evaluation of the relevant integrals. The same result is obtained if one uses total, rather than partial, derivatives in (9),

$$K^{(n)}_{\xi\zeta} = i\int ds\, d^2y\sqrt{G}\left\{\left[(\partial_\xi - \mathbf{V}^\xi)\Phi_n\right]^*\left[(\partial_\zeta - \mathbf{V}^\zeta)\Phi_n\right] - \text{c.c.}\right\} \,, \tag{10}$$

where $\dot\xi(0)\mathbf{V}^\xi$ is the initial velocity field, under the ξ-deformation, of points with fixed adapted coordinates, observed in the ambient $\mathbb{R}^3$, as they are dragged along by the moving curve (similarly for $\mathbf{V}^\zeta$), and we have reverted to pre-Dirac notation to show explicitly the integration measure used.

Particularly interesting, for our purposes, is the hamiltonian in the adapted coordinate frame, namely, the generator of time evolution for the adapted frame wavefunction. This can be obtained by transforming the cartesian coordinate time-dependent Schroedinger equation to the adapted frame. Under the change of coordinates $(t,x^1,x^2,x^3) \to (t',y^1,y^2,y^3)$, with $t' = t$ and $y^i = y^i(x;\xi(t))$, the time derivative in the time dependent Schroedinger equation transforms as

$$\partial_t = \partial_{t'} + \dot\xi(\partial y^i/\partial\xi)\partial_{y^i} \equiv \partial_{t'} - \dot\xi\mathbf{V} \,, \tag{11}$$

where the assumption was made that the time dependence of the coordinate transformation is entirely through $\xi(t)$. The term involving $\dot\xi$ is to be absorbed in the adapted frame hamiltonian $\mathcal{H}$,

$$\mathcal{H} = H + i\dot\xi\,\mathbf{V} \,. \tag{12}$$

When, subsequently, the wavefunction is rescaled, and all operators undergo a similarity transformation, as in (2), the time derivative of the rescaling factor also contributes to the hamiltonian,

$$\tilde{\mathcal{H}} = \tilde H + i\dot\xi(\tilde{\mathbf{V}} - \partial_\xi h) = \tilde H + i\dot\xi(\mathbf{V} - d_\xi h) \equiv \tilde H + \dot\xi\mathcal{H}^{\dot\xi} \,, \tag{13}$$

where $d_\xi h \equiv \partial_\xi h - \mathbf{V}^i\partial_i h$, $\tilde H$ is as in (2), and the last equation defines $\mathcal{H}^{\dot\xi}$. In the case that two parameters, ξ and ζ, are present, the adapted frame hamiltonian takes the form

$$\tilde{\mathcal{H}} = \tilde H + \xi\tilde H^\xi + \zeta\tilde H^\zeta + i\dot\xi(\mathbf{V} - d_\xi h) + i\dot\zeta(\mathbf{W} - d_\zeta h) \,, \tag{14}$$

where $\mathbf{W}$ is the 3D vector field corresponding to the ζ-deformation. Notice that $\tilde{\mathcal{H}}$ depends on both the parameters *and* their time derivatives.

There is considerably more that can be said about how the two hamiltonians in (6), (14), predict the same geometric phase, and this despite the fact that the curvatures computed from each differ — we defer a detailed treatment of these matters to a forthcoming publication, currently in progress [5].

Journal of Physics: Conference Series **335** (2011) 012061 doi:10.1088/1742-6596/335/1/012061

3. A toy example of induced noncommutativity

The core idea behind what we call induced noncommutativity is best appreciated in a simplified context, involving a spin 1/2 coupled to the position of an otherwise free particle, a system that has been studied before (see [4] and references therein). The relevant hamiltonian is

$$H_{\rm tot} = \frac{P^2 + X^2}{2} + \lambda \mathbf{X} \cdot \boldsymbol{\sigma} \, . \tag{15}$$

Assume the initial state of the spin, at $t = 0$, to be $|\mathbf{X_0}+\rangle$ and that, furthermore, that its evolution can be approximated adiabatically, namely, the motion of the particle is slow enough for the spin to "lock" onto the instantaneous eigenstate $|\mathbf{X_t}+\rangle = \cos(\Theta/2) \mathrm{e}^{-i\Phi/2}|+\rangle + \sin(\Theta/2)\mathrm{e}^{i\Phi/2}|-\rangle$, where $|\pm\rangle$ are the σ_z eigenstates, and (Θ, Φ) are the particle's standard spherical angular coordinates. Up to the this point, we have treated the particle as classical, with a well defined position $\mathbf{X}$, which shows up as external parameter in the spin's wavefunction. Now the operator nature of $\mathbf{X}$ is invoked, and, in particular, its noncommutativity with the momentum $\mathbf{P}$ is taken into account. We calculate the effective hamiltonian for the particle by taking the expectation value of the full hamiltonian in the spin's assumed quantum state, thus obtaining,

$$H_{\rm eff} = \langle \mathbf{X}+|H_{\rm tot}|\mathbf{X}+\rangle = \frac{1}{2}\left(\mathbf{P} - \mathbf{A}(\mathbf{X})\right)^2 + \Phi\left(\mathbf{X}\right) + E^{+}(\mathbf{X}) \, , \tag{16}$$

where $A_i(\mathbf{X}) = i\langle \mathbf{X}+|\frac{\partial}{\partial \mathbf{X}_i}|\mathbf{X}+\rangle$ is Berry's connection and

$$\Phi(\mathbf{X}) = \frac{1}{4}(X^2+Y^2+Z^2)^{-1} + \frac{1}{2}(X^2+Y^2+Z^2) \, , \qquad E^{+}(\mathbf{X}) = \lambda(X^2+Y^2+Z^2)^{1/2} \, . \tag{17}$$

What most interests us here is the natural appearance of the "covariantized" momentum $\mathcal{P} \equiv \mathbf{P} - \mathbf{A}(\mathbf{X})$, the components of which fail to commute,

$$[\mathcal{P}_i, \mathcal{P}_j] = \frac{\partial \mathbf{A}_j}{\partial \mathbf{X}_i} - \frac{\partial \mathbf{A}_i}{\partial \mathbf{X}_j} = -iK_{ij} \, , \tag{18}$$

where in the second equation, Berry's curvature has made a suggestive appearance. Starting from this well known example, we propose to dualize the situation, in that we now consider a spin 1/2 coupled to the *momentum* of an otherwise free particle,

$$\tilde{H}_{\rm tot} = \frac{P^2 + X^2}{2} + \lambda \mathbf{P} \cdot \boldsymbol{\sigma} \, . \tag{19}$$

Mirroring the above manipulations, the resulting effective hamiltonian is now

$$\tilde{H}_{\rm eff} = \langle \mathbf{P}+|\tilde{H}_{\rm tot}|\mathbf{P}+\rangle = \frac{1}{2}\left(\mathbf{X} - \tilde{\mathbf{A}}(\mathbf{P})\right)^2 + \tilde{\Phi}(\mathbf{P}) + \tilde{E}^{+}(\mathbf{P}) \, , \tag{20}$$

where $\tilde{A}_i(\mathbf{P}) = i\langle \mathbf{P}+|\frac{\partial}{\partial \mathbf{P}_i}|\mathbf{P}+\rangle$, and $\tilde{\Phi}(\mathbf{P})$, $\tilde{E}^{+}(\mathbf{P})$ are given by (17), with $\mathbf{X} \to \mathbf{P}$. As before, our interest is aroused primarily by the appearance of the "covariantized" position operator $\mathcal{X} \equiv \mathbf{X} - \tilde{\mathbf{A}}(\mathbf{P})$, the components of which, similarly to what transpired above, fail to commute,

$$[\mathcal{X}_i, \mathcal{X}_j] = \frac{\partial \tilde{\mathbf{A}}_j}{\partial \mathbf{P}_i} - \frac{\partial \tilde{\mathbf{A}}_i}{\partial \mathbf{P}_j} \equiv -i\tilde{K}_{ij} \, . \tag{21}$$

Clearly, when the spin's hamiltonian depends on both the position and the momentum of the particle, the effective hamiltonian for the particle will be naturally expressible in terms of covariantized position and momentum operators, with modified, in general, commutation relations, compared to the canonical ones. We return now to the problem of the particle living on a curve, but focus, this time, on the curve.

4. The dynamics of the curve

We assume the curve is actually a wire, made of an elastic material, naturally shaped into a circle. Small deformations are penalized as in Bernoulli's elastica — the corresponding Lagrangian density is taken to be, in schematic form,

$$\mathcal{L} = \frac{1}{2}\rho \dot{\mathbf{X}} \cdot \dot{\mathbf{X}} - \frac{1}{2}\mu(\delta\kappa)^2 - \frac{1}{2}\frac{\mu}{\lambda}(\delta s)^2 \,, \tag{22}$$

where ρ is a uniform linear mass density, and $\delta\kappa$, δs, are local infinitesimal changes in curvature and arclength, resulting from the above mentioned deformations. Rather than imposing *locally arclength preserving* (LAP) deformations, we let sufficiently small values of λ suppress stretching and compression of the wire. We describe the deviation of the wire from the circular shape by an infinitesimal 3D vector field $\epsilon\boldsymbol{\varphi}(s)$, $\epsilon \ll 1$, the Frenet components $\epsilon(\varphi^t, \varphi^n, \varphi^b)$ of which are the fields $\mathcal{L}$ depends on,

$$\epsilon^{-2}\mathcal{L} = \frac{1}{2}\rho\left((\dot{\varphi}^t)^2 + (\dot{\varphi}^n)^2 + \left(\dot{\varphi}^b\right)^2 \right) - \frac{1}{2}\mu\left(\varphi_n + \partial_s^2\varphi^n\right)^2 - \frac{1}{2}\frac{\mu}{\lambda}\left(\partial_s\varphi^t - \varphi^n\right)^2 . \tag{23}$$

The fields are now expanded in Fourier modes,

$$\varphi^i(s,t) = \frac{1}{\sqrt{2\pi}}\sum_k e^{iks}\varphi_k^i(t)\,, \tag{24}$$

the equations of motion of which are diagonalized by the normal mode amplitudes

$$\phi_k^1 = \varphi_k^t - ik\varphi_k^n\,, \qquad \phi_k^2 = -ik\varphi_k^t + \varphi_k^n\,, \qquad \phi_k^3 = \varphi_k^b\,, \tag{25}$$

with corresponding frequencies

$$\omega_{1\,k}^2 = \frac{\mu}{\rho}\frac{k^2\left(1-k^2\right)^2}{1+k^2}\,, \qquad \omega_{2\,k}^2 = \frac{\mu}{\rho}\left(\frac{\left(1-k^2\right)^2}{1+k^2} + \frac{1+k^2}{\lambda}\right), \qquad \omega_{3\,k}^2 = 0\,. \tag{26}$$

The free motion of ϕ_k^3, implied by the vanishing of $\omega_{3\,k}^2$, is an artifact of our perturbative treatment — higher order terms, in ϵ, eventually tame deformations along the binormal. The Lagrangian L may now be obtained, as a function of the normal mode amplitudes, by integrating $\mathcal{L}$ along the circle,

$$\begin{aligned} \epsilon^{-2}L(\boldsymbol{\phi}_k, \dot{\boldsymbol{\phi}}_k) = \frac{1}{2}\sum_k & \frac{\rho}{1+k^2}(\dot{\phi}_k^{1*}\dot{\phi}_k^1 + \dot{\phi}_k^{2*}\dot{\phi}_k^2) + \rho\dot{\phi}_k^{3*}\dot{\phi}_k^3 \\ & - \frac{\mu k^2(1-k^2)^2}{(1+k^2)^2}\phi_k^{1*}\phi_k^1 - \mu\left(\frac{(1-k^2)^2(1-2k^2)}{(1+k^2)^2} + \frac{1}{\lambda}\right)\phi_k^{2*}\phi_k^2 + \mathcal{O}(\lambda)\,. \end{aligned} \tag{27}$$

Finally, we arrive at the hamiltonian of the curve, expressed in terms of the ϕ_k^i and their conjugate momenta,

$$\begin{aligned} \epsilon^{-2}H(\boldsymbol{\phi}_k, \boldsymbol{\pi}_k) = \frac{1}{2}\sum_k & \frac{1+k^2}{\rho}\pi_k^{1*}\pi_k^1 + \frac{1+k^2}{\rho}\pi_k^{2*}\pi_k^2 + \frac{1}{\rho}\pi_k^{3*}\pi_k^3 \\ & + \frac{\rho\,\omega_{1\,k}^2}{1+k^2}\phi_k^{1*}\phi_k^1 + \frac{\rho\,\omega_{2\,k}^2}{1+k^2}\phi_k^{2*}\phi_k^2 + \rho\,\omega_{3\,k}^2\phi_k^{3*}\phi_k^3 + \mathcal{O}(\lambda)\,, \end{aligned} \tag{28}$$

where $\pi_k^i \equiv \epsilon^{-2}\partial L/\partial\dot{\phi}_k^i$ are given by

$$\pi_k^1 = \frac{\rho}{1+k^2}(\dot{\phi}_k^1)^*\,, \qquad \pi_k^2 = \frac{\rho}{1+k^2}(\dot{\phi}_k^2)^*\,, \qquad \pi_k^3 = \rho(\dot{\phi}_k^3)^*\,. \tag{29}$$

5. Backreaction and induced noncommutativity

We are finally ready to consider the backreaction of the particle on the curve. As the technical aspects of the discussion quickly degenerate into long formulas, we concentrate in this section on the conceptual side of the problem, leaving a detailed analysis to a lengthier publication, currently in progress. We begin by examining the cartesian coordinate description, in which the particle's hamiltonian is given by (6) while the curve's by (28). It is crucial, for our discussion, to realize that the parameter ξ in (6) can be made to coincide with one of the Fourier modes φ^i_k of (24) if the corresponding deformation is taken to be an exponential. Needless to say, a precise statement of the above idea would necessitate working with real deformations, as those assumed to accompany ξ, but this can of course be easily arranged by working with a Fourier expansion in sines and cosines, instead of complex exponentials — in the loose spirit of this section advertised above, we bypass details of this nature. Then, the particle hamiltonian depends on the curve's Fourier amplitudes φ^i_k, or, which is the same, the normal mode amplitudes ϕ^i_k, and we are essentially in the same situation as with the spin-coupled-to- position example of section 3. When the effective hamiltonian for the curve is computed, by taking the expectation value of the full hamiltonian in the particle's assumed quantum state, the normal mode amplitudes that appear as external parameters in the particle's wavefunction, will be promoted to their full quantum operator status, and will give rise to covariantized conjugate momenta, the commutator of which will involve Berry's curvature, as given in (10). Thus, the effective hamiltonian for the curve, in the presence of the particle, can be naturally expressed in terms of the normal modes ϕ^i_k and covariantized conjugate momenta $\tilde{\pi}^i_k$, which do not commute among themselves. Of course, one may as well choose to work with the original canonical variables instead, so that the induced noncommutativity, although quite natural, is entirely optional. The physical system under study then, provides us with a laboratory where noncommutative field theory can be explored, and its techniques tried and refined, while an alternative commutative treatment is available all along, so that all results can be checked against those obtained by standard techniques.

The situation is even richer, when the analysis is carried out in the frame adapted to the curve. Then the particle's hamiltonian is $\tilde{\mathcal{H}}$, as given in (14), and clearly depends not only on $\xi \sim \varphi^i_k$ but also on $\dot{\xi} \sim \pi^i_k$. Accordingly, noncommuting normal mode amplitudes $\tilde{\phi}^i_k$ will also naturally appear in the curve's effective hamiltonian, and a fully noncommutative field theory will be available to explore. In this case, all canonical commutation relations among the dynamical variables, are, in principle, modified, providing a concrete deformation scheme with guaranteed self consistency, originating, as it is, in standard quantum mechanics.

6. Concluding remarks

We have shown, in this short note, that the presence of a quantum particle, constrained to move along a closed space curve, induces, in the adiabatic limit, magnetic potential-like terms in the curve's effective hamiltonian, which are naturally absorbed in covariantized, and hence noncommuting, in general, dynamical variables. The availability, in parallel, of the commutative treatment, allows constant control over results obtained by novel techniques. The fact that the resulting deformation scheme originates in standard quantum mechanics resolves from the very beginning possible consistency doubts, and provides a paradigm of sensible noncommutativity. Given the modified commutation relations among the covariantized normal mode amplitudes $\tilde{\phi}^i_k$, it is particularly interesting to determine the corresponding modifications in the commutators among the deformation fields $\tilde{\varphi}^i(s) = \sum_k \tilde{\varphi}^i_k e^{iks}/\sqrt{2\pi}$ (compare with (24)).

The nature of the results obtained here is such that one can reasonably expect them to be true generically: in the quantum world, physical systems are invariably accompanied by "parasites", like the cloud of virtual particles hovering over, say, a soccer ball in flight. What is to be inferred from the above analysis, is that the effective description of a great number of physical systems might well involve a natural noncommutativity, so that noncommutative field theory, instead of a passing two-decade fad, might be here to stay.

Acknowledgments
C.C. would like to acknowledge partial financial support from DGAPA-UNAM project IN 118309 and CONACyT project 103486.

References
[1] Maraner P and Destri C 1993 *Mod. Phys. Lett. A* **8** 861
[2] Schuster P C and Jaffe R L 2003 *Ann. Phys.* **307** 132
[3] Chryssomalakos C, Hernandez-Coronado H, Gelbwaser-Klimovsky D and Okon E 2008 *Mod. Phys. Lett. A* **23/36** 3087–3094
[4] Berry M 1989 *Geometric Phases in Physics* ed Shapere A and Wilczek F (World Scientific) pp 7–28
[5] Aguilar-Nunez P A, Chryssomalakos C, Hernandez-Coronado H and Okon E Work in progress.

Strong Interactions, (De)coherence and Quarkonia

Stefano Bellucci[1]

[1]INFN-Laboratori Nazionali di Frascati, Via E. Fermi 40, 00044 Frascati, Italy.

E-mail: `bellucci@lnf.infn.it`

Vinod Chandra[2]

[2]Department of Theoretical Physics, Tata Institute of Fundamental Research, Homi Bhabha Road Mumbai-400005, India.

E-mail: `joshi.vinod@gmail.com`

Bhupendra Nath Tiwari[1]

[1]INFN-Laboratori Nazionali di Frascati, Via E. Fermi 40, 00044 Frascati, Italy.

E-mail: `tiwari@lnf.infn.it`

Abstract. Quarkonia are the central objects to explore the non-perturbative nature of non-abelian gauge theories. We describe the confinement-deconfinement phases for heavy quarkonia in a hot QCD medium and thereby the statistical nature of the inter-quark forces. In the sense of one-loop quantum effects, we propose that the "quantum" nature of quark matters follows directly from the thermodynamic consideration of Richardson potential. Thereby we gain an understanding of the formation of hot and dense states of quark gluon plasma matter in heavy ion collisions and the early universe. In the case of the non-abelian theory, the consideration of the Sudhakov form factor turns out to be an efficient tool for soft gluons. In the limit of the Block-Nordsieck resummation, the strong coupling obtained from the Sudhakov form factor yields the statistical nature of hadronic bound states, e.g. kaons and Ds particles.

Keywords: CPT symmetry, decoherence, Lorentz symmetry breaking.
PACS: 02.40.-k; 14.40.Pq; 12.40.Nn; 14.70.Dj

1. Introduction

In this work, we study the geometric nature of the quark matter formation. Specifically, we shall illustrate that the components of the vacuum fluctuations define a set of local pair correlations against the vacuum parameters, e.g. charge, mass and angular momentum. Our consideration follows from the notion of the thermodynamic geometry [1–3]. Importantly, this framework provides a mathematical platform to exactly understand the nature of the pair local correlations and underlying geometric structures pertaining to the global phase transitions in quarkonium systems. This perspective yields a well-known understanding for the phase structures of mixtures of gases, black holes in string theory [4–7] and in other diverse contexts, as well.

The main purpose of the present investigation is to determine the thermodynamic properties of the quarkonium configurations, in general. Quantum chromodynamics (QCD), as the theory of strong interactions, celebrates physics [8–10] at both the high and low temperature domains. Thereby, our consideration plays a crucial role in understanding the phases and stability of the matter formation. In the soft gluon limit, viz., for a small transverse momentum $k_\perp$, the behavior of the abelian theory follows directly from the Poisson distribution and its numerical counterparts. On the other hand, the statistical nature of the non-abelian soft gluons is understood in terms of the Sudhakov form factor [12]. Let us recall that the QCD coupling $\alpha_s(k_\perp)$ never lies near the limit $k_\perp \to 0$, and so the QCD effects are limited for the bound state thus formed after the resummation. As mentioned in the Ref. [15], an integration over $k_\perp$ requires the Sudhakov form factor

$$h(b) = \int \frac{dk_l}{2k}\frac{dk_\perp}{k_\perp^2}(1 - \exp(-i\vec{k_\perp}.\vec{b})), \tag{1}$$

where $k_\perp \in (0, m_{P/2})$ is due to a physical reason, i.e. that on average each soft gluon can take as much as half of the initial center of mass energy. Towards the determination of the index p, an interesting argument follows from the work of Polyakov [11] which we shall explore further from the perspective of thermodynamic geometry in the subsequent consideration. Before doing so, let us consider the joint effects of the (i) confinement and (ii) rotation, and thus make a platform to describe the thermodynamic geometry of rotating quarkonia. Such a simplest configuration is described by Regge trajectories with the leading order effective potential

$$V(r, J) = \frac{J(J+1)}{r^2} + Cr^{2p-1}. \tag{2}$$

In this case, the effective theory [13, 14] is inspired from the limiting QCD strong coupling

$$\alpha_s(Q^2) = b^{-1}\frac{p}{\ln(1 + p(\frac{Q^2}{\Lambda_{QCD}^2})^p)}, \tag{3}$$

where $b := (33 - 2N_f)/12\pi$. In the sense of one-loop quantum effects, we propose that the "quantum" nature of quark matters follows directly from the thermodynamic intrinsic geometry of the Richardson type potentials. In the momentum space, the net effective potential offers the right quarkonium bound states, after taking account of the one-loop exchange terms.

Let us now focus our attention on the thermodynamic geometry of quarkonium bound states with finitely many parameters of the effective field theory. We consider here a framework of the intrinsic Riemannian geometry whose covariant metric tensor is defined as the Hessian matrix of the QCD coupling, with respect to a finite number of arbitrary parameters carried by the soft gluons and quarks. Such a consideration yields the space spanned by n parameters of the strong QCD coupling α_s, which, in the present treatment, exhibits a n-dimensional intrinsic Riemannian manifold M_n. As per the definition of the thermodynamic geometry [1–7], the components of the covariant metric tensor are given by

$$g_{ij} := \frac{\partial^2 \alpha_s(\vec{x})}{\partial x^j \partial x^i}, \tag{4}$$

where the vector $\vec{x} \in M_n$. In the strongly coupled quarkonium effective configuration, there are only a few physical parameters, what makes the analysis fairly simple. As mentioned in the Ref. [15], the variables of the interest of the above quarkonium configurations are the momentum scale parameter, $Q^2 := q$, the mass M, and the angular momentum J, if any.

2. Massless Quarkonia

Let us first examine the massless non-rotating quarkonia and study the thermodynamic stability properties as emphasized earlier, and subsequently include the rotation. Considering Eqn.(3), one obtains the following expression

$$A(q, p) := \frac{p}{b\ln(1 + p(q/L)^p)}. \tag{5}$$

for the strong QCD coupling, where $L := \Lambda_{QCD}^2$. To compute the thermodynamic metric tensor in the space of the parameters $\{q, p\}$, we employ the Eqn.(4), which leads to the following expression for the components of the metric tensor

$$g_{qq} = \frac{p^3}{bq^2}\frac{n_{11}^Q}{r_{11}^Q}, \quad g_{qp} = \frac{p^2}{bq}\frac{n_{12}^Q}{r_{12}^Q}, \quad g_{pp} = \frac{1}{b}\frac{n_{22}^Q}{r_{22}^Q}. \tag{6}$$

In order to simplify the subsequent notations, let us define the logarithmic factor as

$$l(p) := \ln(1 + p(q/L)^p). \tag{7}$$

Thus, the factors in the numerator of the local pair correlations can be expressed as

$$\begin{aligned} n_{11}^Q &:= 2p^2(q/L)^{2p} + l(p)((q/L)^p - p(q/L)^p + (q/L)^{2p}p), \\ n_{12}^Q &:= 2(q/L)^{2p}p + l(p)(2(q/L)^{2p}p^2\ln(q/L) - 3(q/L)^p - 2(q/L)^{2p}p - (q/L)^p p\ln(q/L)), \\ n_{22}^Q &:= 2(q/L)^{2p}(p + 2p^2\ln(q/L) + p^3\ln(q/L)^2) - l(p)((q/L)^p + (q/L)^{2p}p \\ &\quad +4(q/L)^p p\ln(q/L) + 2(q/L)^{2p}p^2\ln(q/L) + (q/L)^p p^2\ln(q/L)^2). \end{aligned} \tag{8}$$

In the case of the massless non-rotating quarkonia, we find that the factors $\{r_{ii}^Q | \ i = 1, 2, 3\}$ take the uniform value $l(p)^3 \exp(2l(p))$ in all denominators of the local pair correlation functions. In a given QCD phase, this happens when the parameters $\{q, p\}$ are confined in the domain

$$\{\mathcal{D} := (q, p) \in \ M_2 | \ n_{11}^Q > 0, \ n_{22}^Q > 0\}. \tag{9}$$

Over the above domain of $\{q, p\}$, the massless non-rotating quarkonia is well-behaved and locally stable. From the definition of the thermodynamic geometry, we find further for the generic value of the parameters that the Gaussian fluctuations form a stable set of correlations over $\{q, p\}$, if the determinant of the metric tensor

$$\|g\| = \frac{p^3}{b^2q^2l(p)^5\exp(3l(p))} n_g^Q \tag{10}$$

remains a positive function on the intrinsic qp-surface $(M_2(R), g)$. Explicitly, we obtain that the numerator of the determinant of the metric tensor can be expressed as

$$\begin{aligned} n_g^Q(q, p) : \ = \ & 2(q/L)^{3p}(p + 3p^2 + 2p^2ln(q/L) + 2p^3ln(q/L) + p^3ln(q/L)^2) \\ & -l(p)(2(q/L)^{2p} + p(q/L)^{3p} + 2p^2(q/L)^{3p}ln(q/L) + 4p(q/L)^{2p}ln(q/L) \\ & +p^2(q/L)^{2p}ln(q/L)^2 + 2p^2(q/L)^{2p}ln(q/L) + 4p^2(q/L)^{3p} + 7p(q/L)^{2p}). \end{aligned} \tag{11}$$

In this case, we find that the scalar curvature reduces to the following specific form

$$R(q, p) = \frac{bl(p)}{2p^2(n_g^Q)^2}(n_R^{(0)Q} + n_R^{(1)Q}l(p) + n_R^{(2)Q}l(p)^2 + n_R^{(3)Q}l(p)^3). \tag{12}$$

We find that the factors of the numerator of the scalar curvature take a set of interesting expressions [15]. In the case when the Ricci scalar curvature $R(q, p)$ vanishes, the underlying quarkonium system is found to be in equilibrium. Such a state of the configuration can arise with $\{n^{(i)Q} = 0, i = 0, 1, 2, 3\}$, if the other factors of the scalar curvature remain non-zero. In the other case, when the scalar curvature $R(q, p)$ diverges, the configuration goes over a transition. Such an extreme behavior of the quarkonia is expected to happen, when either the index p or the numerator of the determinant of the metric tensor n_g^Q vanish. We further observe that the stability of massless non-rotating quarkonia exists in certain bands. A closer view shows that the quarkonia are decaying and interacting particles in the Coulombic limit. For the regime of the rising potential, the respective determinant of the metric tensor and scalar curvature shows that the limiting rising potential quarkonia are stable and non-interacting particles. In the Regge trajectory model [15], we find interestingly that all possible local and global thermodynamic stability behavior of the three parameter $\{q, p, J\}$ massless quarkonia remains the same up to the sign of b_1, as if there were no effects of the rotation in the underlying configuration. The fact that the efficiency of the rotation induces a mass to the quarkonium is analyzed by considering the Bloch-Nordsieck resummation of the angular phases.

3. Massive Quarkonia

Let us firstly illustrate the cases for the two parameter configurations with either $\{q, J\}$ or $\{q, m\}$ fluctuating and then systematically extend the consideration for the generic three parameter quarkonia.

3.1. Two Parameter Quarkonia

For the of massive quarkonia, the resummed strong QCD coupling takes the following form

$$A(q,J) := \frac{1}{b}\frac{p(1-J(0,a\sqrt{q}))}{\ln(1+p(q/L)^p)}\ln(\frac{\sqrt{J}+\sqrt{J-q}}{\sqrt{J}-\sqrt{J-q}}), \tag{13}$$

where $J(\nu,x)$ is the Bessel function of the first kind of the order ν. In order to describe fluctuations in the QJ-plane, let the logarithmic factor concerning the Bloch-Nordsieck rotation be defined as

$$f(q,J) := \ln(\frac{\sqrt{J}+\sqrt{J-q}}{\sqrt{J}-\sqrt{J-q}}). \tag{14}$$

From the Eqn.(4), we find that the components of the metric tensor are

$$g_{qq} = -\frac{p(n_{11}^{(0)J}+n_{11}^{(1)J}l(p)+n_{11}^{(2)J}l(p)^2)}{4bl(p)^3\exp(2l(p))q^{5/2}(J-q)^{3/2}},\ g_{qp} = \frac{p(n_{12}^{(0)J}+n_{12}^{(1)J}l(p))}{2bl(p)^2\exp(l(p))q^{3/2}\sqrt{J}(J-q)^{3/2}},\ g_{pp} = -\frac{p(2J-q)(1-J(0,a\sqrt{q}))}{2bl(p)J^{3/2}(J-q)^{3/2}}. \tag{15}$$

Without any approximation, the factors in the numerator of the pure qq and qJ-components can be expressed as linear combinations of the integer powers of the scaling $(q/L)^{np}$, where $n \in Z$. As a result, we see that the geometric nature of the parametric pair correlation functions turns out to be remarkably interesting, viz., the domains of the local stability of the fluctuating quarkonia may be easily described in terms of the parameter q and J. Under the Gaussian fluctuations of $\{q,J\}$, the local stability of the system requires that (i) qq- fluctuations satisfy the constraint

$$n_{11}^{(0)J}+n_{11}^{(1)J}l(p)+n_{11}^{(2)J}l(p)^2 < 0 \tag{16}$$

and (ii) JJ- fluctuations be constrained to the following limiting values of the Bessel function

$$\begin{aligned} J(0,a\sqrt{q}) &< 1,\ q>2J, \\ &> 1,\ q<2J. \end{aligned} \tag{17}$$

In this case, we find that the determinant of the metric tensor reduces to the following expression

$$\|g\| = -\frac{p^2(n_g^{(0)J}+n_g^{(1)J}l(p)+n_g^{(2)J}l(p)^2)}{8b^2l(p)^4\exp(2l(p))q^{5/2}J^{3/2}(J-q)^{3/2}}, \tag{18}$$

where $n_g^{(1)J} = 4p^2(q/L)^p(n_g^{(12)J}+n_g^{(13)J}p(q/L)^p)$ and $n_g^{(2)J} = n_g^{(20)J}+2n_g^{(21)J}p(q/L)^p+n_g^{(23)J}(q/L)^{2p}p^2$. Furthermore, it turns out that all factors appearing in the numerator of the determinant of the metric tensor, e.g., $\{n_g^{(0)J}, n_g^{(12)J}, n_g^{(13)J}, n_g^{(20)J}, n_g^{(21)J}, n_g^{(23)J}\}$, can be presented as linear combinations of the scaling $(q/L)^{np}$, where $n \in Z$. Combining the effects of all fluctuations of the $\{q,J\}$, we observe that the quarkonia are stable for $q := Q^2 \in (1,4)$. In general, the global stability requires that the determinant of the metric tensor must be positive definite, which in the present case transform as

$$n_g^J := n_g^{(0)J}+n_g^{(1)J}l(p)+n_g^{(2)J}l(p)^2 \quad < \quad 0. \tag{19}$$

It turns out that the thermodynamic curvature may be written as the series of the charmonium logarithmic factor $l(p)$ and Bloch-Nordsieck logarithmic factor $f(q,J)$ of rotation as the coefficient of the expansion. Systematically, the exact expression for the scalar curvature takes the form

$$R(q,J) = \frac{bl(p)}{2p^2(n_g^J)^2}\sum_n B_n \times (l(p))^n, \tag{20}$$

where the B_n in the numerator of the scalar curvature are polynomials in p, whose coefficients are the functions of the Bloch-Nordsieck logarithmic factor $f(q,J)$. In sequel, we find that the quantitative properties of the scalar curvature and the Riemann curvature tensor remain similar [15], when we consider the variable Q as the transverse momentum with the understanding that $k = k_\perp$ and the mass as the other variable of the arbitrary two parameter real quarkonium system. The present consideration explicates the regions of the thermodynamic (in)stability for the massive quarkonia and the (un)stable phases of the one loop QCD. As the non-linear effects become stronger and stronger, it turns out that the thermodynamic instability and correlations grow further. This motivates us to extend our analysis to the case of more general quarkonium configurations.

3.2. Three Parameter Quarkonia

Finally, let us analyze the general massive rotating quarkonia, when all parameters, viz., the scale q, index p and angular momentum J of the theory are allowed to fluctuate. In the framework of the Bloch-Nordsieck resummation, the strong QCD coupling takes the following form

$$A(q,p,J) = \frac{1}{b}\frac{p(1-J(0,a\sqrt{q}))}{\ln(1+p(q/L)^p)}\ln(\frac{\sqrt{J}+\sqrt{J-q}}{\sqrt{J}+\sqrt{J-q}}). \quad (21)$$

After some simplification, we find that the components of the metric tensor are

$$\begin{aligned} g_{qq} &= -\frac{p(n_{11}^{(0)G}+n_{11}^{(1)G}l(p)+n_{11}^{(2)G}l(p)^2)}{4bl(p)^3\exp(2l(p))q^{5/2}(J-q)^{3/2}},\ g_{qp} = \frac{(n_{12}^{(0)G}+n_{12}^{(1)G}l(p)+n_{12}^{(2)G}l(p)^2)}{2bl(p)^3\exp(2l(p))q^{3/2}(J-q)^{1/2}}, \\ g_{qJ} &= \frac{p(n_{13}^{(0)G}+n_{13}^{(1)G}l(p))}{2bl(p)^2\exp(l(p))q^{3/2}(J-q)^{3/2}J^{1/2}},\ g_{pp} = \frac{f(-1+J(0,a\sqrt{q}))(n_{22}^{(0)G}+n_{22}^{(1)G}l(p))}{bl(p)^3\exp(2l(p))}, \\ g_{pJ} &= -\frac{(-1+J(0,a\sqrt{q}))(n_{23}^{(0)G}+l(p)n_{23}^{(1)G})}{bl(p)^2\exp(l(p))(J-q)^{1/2}J^{1/2}},\ g_{JJ} = \frac{p(-1+J(0,a\sqrt{(}q)))(2J-q)}{2bl(p)(J-q)^{3/2}J^{3/2}}, \end{aligned} \quad (22)$$

where the coefficients $\{n_{11}^{(1)G}, n_{11}^{(2)G}, n_{12}^{(1)G}, n_{12}^{(2)G}\}$ appearing in the components of the metric tensor are shown to factorize as before. In the powers of p, the exact factorizations are given as follows:
(i) the qq-component

$$\begin{aligned} n_{11}^{(1)G} &= 4p^2(q/L)^p n_{11}^{(12)G} + 4p^3(q/L)^{2p} n_{11}^{(12)G}, \\ n_{11}^{(2)G} &= n_{11}^{(20)G} + 2p(q/L)^p n_{11}^{(21)G} + p^2(q/L)^{2p}, \end{aligned} \quad (23)$$

(ii) the qp-component

$$\begin{aligned} n_{12}^{(1)G} &= p(q/L)^p n_{12}^{(11)G} + p^2(q/L)^{2p} n_{12}^{(12)G}, \\ n_{12}^{(2)G} &= n_{12}^{(20)G} + 2p(q/L)^p n_{11}^{(21)G} + p^2(q/L)^{2p} n_{12}^{(22)G}. \end{aligned} \quad (24)$$

Interestingly, the factors appearing in the numerator of the qq, qp and qJ-components can be cascaded as a linear combination in the integer powers of the scaling $(q/L)^{pn}$, where $n \in Z$. Similarly, the corresponding factors of numerator of the pp-components take the following expressions

$$\begin{aligned} n_{22}^{(0)G} &= -2p(q/L)^{2p}(1+2p\ln(q/L)+p^2\ln(q/L)^2), \\ n_{22}^{(1)G} &= (q/L)^p(2+4p\ln(q/L)+p^2\ln(q/L)^2)+p(q/L)^{2p}(1+2p\ln(q/L)). \end{aligned} \quad (25)$$

Finally, the factors in the numerator of the pJ-components are

$$\begin{aligned} n_{23}^{(0)G} &= -p(q/L)^p - p^2\ln(q/L)(q/L)^p, \\ n_{23}^{(1)G} &= 1+p(q/L)^p. \end{aligned} \quad (26)$$

The local stability of the configuration requires a set of constraints on the domains of the parameters. Specifically, for the same sign of $\{p,b\}$, we find that the local stability of the fluctuating quarkonia enforces the following simultaneous requirements
(i) the qq- fluctuations satisfy

$$n_{11}^{(0)G} + n_{11}^{(1)G}l(p) + n_{11}^{(2)G}l(p)^2 < 0, \quad (27)$$

(ii) the JJ- fluctuations remain within the limiting values of the Bessel function

$$\begin{aligned} J(0,a\sqrt{q}) &> 1,\ 2J < q, \\ &< 1,\ 2J > q, \end{aligned} \quad (28)$$

(iii) the pp- fluctuations satisfy

$$\begin{aligned} n_{22}^{(0)G} + n_{22}^{(1)G}l(p) &> 0,\ J(0,a\sqrt{q}) > 1, \\ &< 0,\ J(0,a\sqrt{q}) > 1 \end{aligned} \quad (29)$$

for the same sign of $\{f, b\}$. The silent feature of the JJ-fluctuations is that we find the two distinct local behaviors for $J > 0$ and $J < 0$. Subsequently, we find that the qp-surface is stable, if there exits a positive surface minor

$$p_S^G := -\frac{(n_S^{(0)G} + n_g^{(1)G} l(p) + n_S^{(2)G} l(p)^2 + n_S^{(3)G} l(p)^3)}{4b^2 l(p)^5 \exp(3l(p)) q^{5/2} (J-q)^{3/2}}. \tag{30}$$

It is worth mentioning that $\{n_S^{(0)G}, n_S^{(1)G}, n_S^{(2)G}, n_S^{(3)G}\}$ have the following factorizations

$$\begin{aligned} n_S^{(0)G} &= (q/L)^{3p} n_S^{(03)G}, \\ n_S^{(1)G} &= (q/L)^{2p} n_S^{(12)G} + (q/L)^{3p} n_S^{(13)G}, \\ n_S^{(2)G} &= (q/L)^{p} n_S^{(21)G} + (q/L)^{2p} n_S^{(22)G} + (q/L)^{3p} n_S^{(23)G}, \\ n_S^{(3)G} &= n_S^{(30)G} + (q/L)^{p} n_S^{(31)G} + (q/L)^{2p} n_S^{(32)G} + (q/L)^{3p} n_S^{(33)G}. \end{aligned} \tag{31}$$

In this identification, the factors appearing in the powers of $(q/L)^p$ have the following structures

$$\begin{aligned} n_S^{(03)G} &= 8f^2 n_S^{(032)G}, \\ n_S^{(1i)G} &= n_S^{(1i0)G} + n_S^{(1i1)G} f + n_S^{(122)G} f^2, i = 2, 3, \\ n_S^{(2i)G} &= n_S^{(2i0)G} + n_S^{(2i1)G} f + n_S^{(2i2)G} f^2, i = 1, 2, 3, \\ n_S^{(3i)G} &= n_S^{(3i0)G} + n_S^{(3i1)G} f + n_S^{(3i2)G} f^2, i = 0, 1, 2, 3. \end{aligned} \tag{32}$$

Further, the factors 032, $12i$, $13i$, $21i$, $22i$ and $23i$ take the following explicit expressions

$$\begin{aligned} n_S^{(032)G} &= p^4 n_S^{(0321)G} + p^5 n_S^{(0322)G} + p^6 n_S^{(0323)G}, \\ n_S^{(12i)G} &= p^2 n_S^{(12i2)G} + 2p^3 n_S^{(21i3)G} + p^4 n_S^{(12i4)G}, i = 0, 1 \\ n_S^{(122)G} &= p^2 n_S^{(2122)G} + 2p^3 n_S^{(2123)G} + p^4 n_S^{(2124)G} + 4p^5 n_S^{(2125)G}, \\ n_S^{(13i)G} &= 4p^3 n_S^{(13i3)G} + 8p^4 n_S^{(13i4)G} + 4p^5 n_S^{(13i5)G}, i = 1, 2, \\ n_S^{(21i)G} &= 2pn_S^{(21i1)G} + 2p^2 n_S^{(21i2)G} + p^3 n_S^{(21i3)G}, i = 1, 2, \\ n_S^{(22i)G} &= p^2 n_S^{(22i1)G} + 2p^3 n_S^{(22i2)G} + p^4 n_S^{(22i3)G}, i = 1, 2, \\ n_S^{(23i)G} &= p^3 n_S^{(23i3)G} + 2p^4 n_S^{(23i4)G}, i = 1, 2. \end{aligned} \tag{33}$$

Interestingly, the sub-factors of the factors 032, 120, 121, 122, 130, 131, 211, 212, 220, 221, 222, 230, 231 and 232 can be expressed as the linear combinations of the integral powers of the scaling $\ln(q/L)$ and the integral powers of the zeroth order Bessel function. As per the expectation, we observe that the $3ij$-factors that appear are relatively simpler than their foregoing counterparts. In fact, we find that the 30-factors can be seen as a two variable polynomial in J, q. For the general quarkonia with $\{q, p, J\}$ fluctuating, the stability of the qp-surface requires that the principle minor p_S^G remains positive on (M_3, g). Correspondingly, this leads to the constraint that the $\{q, p, J\}$ satisfy

$$n_S^{(0)G} + n_g^{(1)G} l(p) + n_S^{(2)G} l(p)^2 + n_S^{(3)G} l(p)^3 \quad < \quad 0. \tag{34}$$

As per the generic fluctuations of the strongly coupled massive rotating quarkonia, we obtain the following general expressions for the determinant of the metric tensor

$$\|g\| \quad = \quad -\frac{pf(-1 + J(0, a\sqrt{q}))}{8b^3 l(p)^6 \exp(3l(p)) q^{5/2} J^{3/2} (J-q)^{3/2}} (n_g^{(0)G} + n_g^{(1)G} l(p) + n_g^{(2)G} l(p)^2 + n_g^{(3)G} l(p)^3). \tag{35}$$

As mentioned before, we find that the $\{n_g^{(0)G}, n_g^{(1)G}, n_g^{(2)G}, n_g^{(3)G}\}$-terms factorize as follows

$$\begin{aligned} n_g^{(0)G} &= (q/L)^{3p} (8p^4 n_g^{(04)G} + 8p^5 n_g^{(05)G} + 8p^6 n_g^{(06)G}), \\ n_g^{(1)G} &= (q/L)^{2p} n_g^{(12)G} + (q/L)^{3p} n_g^{(13)G}, \\ n_g^{(2)G} &= (q/L)^{p} n_g^{(21)G} + ((q/L)^{2p} n_g^{(22)G} + (q/L)^{3p} n_g^{(23)G}, \\ n_g^{(3)G} &= n_g^{(30)G} + (q/L)^{p} n_g^{(31)G} + (q/L)^{2p} n_g^{(32)G} + (q/L)^{3p} n_g^{(33)G}, \end{aligned} \tag{36}$$

Journal of Physics: Conference Series **335** (2011) 012062 doi:10.1088/1742-6596/335/1/012062

After a direct simplification, we obtain the sub-factorizations

$$\begin{aligned}
n_g^{(0i)G} &= n_g^{(0i0)G} + fn_g^{(0i1)G}, i = 4, 5, 6, \\
n_g^{(1i)G} &= n_g^{(1i0)G} + fn_g^{(1i1)G}, i = 2, 3, \\
n_g^{(2i)G} &= n_g^{(2i0)G} + fn_g^{(2i1)G}, i = 1, 2, 3, \\
n_g^{(3i)G} &= n_g^{(3i0)G} + fn_g^{(3i1)G}, i = 0, 1, 2, 3.
\end{aligned} \tag{37}$$

As per our computation, we observe that the terms appearing in the various powers of $l(p)$ have a sub-factorization in the index p. Specifically, we find that the factors of 12, 13, 21, 22 and 23-components are given by

$$\begin{aligned}
n_g^{(120)G} &= n_g^{(1202)G}p^2 + n_g^{(1203)G}p^3 + n_g^{(1204)G}p^4, \\
n_g^{(121)G} &= n_g^{(1212)G}p^2 + n_g^{(1213)G}p^3 + n_g^{(1214)G}p^4 + n_g^{(1215)G}p^5, \\
n_g^{(130)G} &= n_g^{(1303)G}p^3 + n_g^{(1304)G}p^4 + n_g^{(1305)G}p^5, \\
n_g^{(131)G} &= n_g^{(1313)G}p^3 + n_g^{(1314)G}p^4 + n_g^{(1315)G}p^5 + n_g^{(1316)G}p^6, \\
n_g^{(21i)G} &= n_g^{(21i1)G}p + n_g^{(21i2)G}p^2 + n_g^{(21i3)G}p^3, i = 0, 1, \\
n_g^{(22i)G} &= n_g^{(22i2)G}p^2 + n_g^{(22i3)G}p^3 + n_g^{(22i4)G}p^4, i = 0, 1, \\
n_g^{(23i)G} &= n_g^{(23i3)G}p^3 + n_g^{(23i4)G}p^4, i = 0, 1.
\end{aligned} \tag{38}$$

As in the case of the surface minor, we observe that all the individual sub-factorizations, (e.g. 040, 041, 050, 051, 060, 061, $120i$, $121j$, $130i$, $130j$, 2303, 2304, 2313, 2314), can be expressed as the linear combination over the integral powers of the zeroth order Bessel function and the scaling $\ln(q/L)^n$. In the case of the $l(p)^3$ terms, we find that the sub-factors pertaining to 30, 31 and 32-factors have no dependence on $\ln(q/L)^n$ and thus they are expressible as polynomial expressions in the Bessel function only. From this observation, we predict that the regions of the thermodynamic stability are present for $q, J \in (1, 4)$. Globally, the stability of (M_3, g) constrains the principle minors $\{g_{ii}, p_S^G, \|g\|\}$ to remain positive. Specifically, for the same sign of $\{b, p, f\}$, the volume stability of the (M_3, g) imposes the following constraint

$$\begin{aligned}
\sum_{i=0}^{3} n_g^{(i)G} l(p)^i &> 0, \ J(0, a\sqrt{q}) < 1, \\
&< 0, \ J(0, a\sqrt{q}) < 1.
\end{aligned} \tag{39}$$

Importantly, it is worth mentioning that both the limiting configurations with $J = q$ and $J(0, a\sqrt{q}) = 1$ are abided from the thermodynamic stability constraints. To summarize the phases of generic quarkonia, the exact formula for the scalar curvature may analogously be deduced, as the one we have offered for the fluctuations in the qJ-plane. In this sense, we find that the summation over $l(p)$ naturally arises with the B_n as the polynomials in p, whose coefficients can be expressed as the functions of the Bloch-Nordsieck logarithmic factor $f(q, J)$. Based on the analysis of the present paper, the thermodynamically stable index of the Bloch-Nordsieck rotating massive quarkonia is constrained by the following set

$$\mathcal{P}_{m\neq 0} := \{p \mid n_{ii}^G > 0, \ n_S^G > 0, \ n_g^G > 0\}. \tag{40}$$

As the gluons become softer and softer, we find, in the limit of Bloch-Nordsieck resummation, that the underlying Sudhakov form factor offers all possible thermodynamically stable phases of the strongly coupled quarkonia. As examined for the QJ-plane, we would like to explicitly understand properties of the set $\mathcal{P}_{m\neq 0}$ and the associated Ricci scalar curvature. Up to a phase of QCD, the global properties of the three parameter quarkonia remain the same as we have exactly indicated for the two parameter quarkonia. As mentioned in the Ref. [15], we have computed the intrinsic geometric properties of the Bessel function of the first kind convoluted with two logarithmic functions of the respective weights $(0, 2p)$. Further analysis of the geometric features of these exploitations is left for the future.

4. Conclusion and Outlook

We have examined the role of the thermodynamic intrinsic geometry for a class of quarkonium configurations. We have offered a geometric perspective to the confinement- deconfinement phase of (heavy) quarkonia in a hot QCD medium and thereby described the statistical nature of the inter-quark forces. Specifically, the intrinsic geometric analysis provides a set of physical indications encoded in the geometric quantities, e.g., the scalar curvature and possible geometrically non-trivial invariants offer the global correlation properties of an ensemble or subensemble of equilibrium configurations. In the sense of statistical mechanics, our analysis involving a Gaussian distribution of the particles ensures the thermodynamical properties of the underlying quarkonia in the late time limit. From the perspective of one-loop quantum effects, the nature of quark matter is shown to follow directly from the thermodynamic consideration of the Richardson potential.

Our study of the quarkonia could further be explored towards other configurations concerning the non-perturbative and non-abelian nature of the gauge theories. Such a consideration provides a unified description, encompassing all the regimes of QCD at finite temperature, i.e. the Coulombic, the linear rising and the Regge rotating regimes, for both massless and massive quarkonia. Phenomenologically, our results can be thus be used to investigate the statistical nature of soft gluons and the associated phenomenon at the LHC.

Acknowledgments

The work of S.B. has been supported in part by the European Research Council grant n. 226455, *"SUPERSYMMETRY, QUANTUM GRAVITY AND GAUGE FIELDS (SUPERFIELDS)"*.

B.N.T. would like to thank organizers of the *"DISCRETE 2010, Rome, Italy"* for providing stimulating environment towards this work. B.N.T. further acknowledges postdoctoral research fellowship of the *"INFN-Laboratori Nazionali di Frascati, Roma, Italy"*.

References

[1] F. Weinhold, J.Chem. Phys. **63**, 2479 (1975), *ibid* J. Chem. Phys **63**, 2484 (1975).
[2] G. Ruppeiner, Rev. Mod. Phys **67** 605 (1995), Erratum **68**,313 (1996); M. Santoro, A. S. Benight, math-ph/0507026.
[3] G. Ruppeiner, Phys. Rev. **A 20**, 1608 (1979); Phys. Rev. Lett **50**, 287 (1983); Phys. Rev. **A 27**, 1116 (1983); G. Ruppeiner, C. Davis, Phys. Rev. **A 41**, 2200 (1990).
[4] S. Bellucci, V. Chandra, B. N. Tiwari, To Appear in Int. J. Mod. Phys. A. arXiv:0812.3792; S. Bellucci, V. Chandra, B. N. Tiwari, arXiv:1010.4405v1 [hep-th]; S. Bellucci, V. Chandra, B. N. Tiwari, arXiv:1010.4225v1 [hep-th].
[5] B. N. Tiwari, arXiv:0801.4087v1 [hep-th]; T. Sarkar, G. Sengupta, B. N. Tiwari, JHEP **10**, 076 (2008) arXiv:0806.3513v1 [hep-th]; S. Bellucci, B. N. Tiwari, Entropy 2010, **12**, 2097-2143; S. Bellucci, B. N. Tiwari, JHEP **05** 023 (2010), S. Bellucci, B. N. Tiwari, Phys. Rev. D **82** (2010) 084008.
[6] G. Ruppeiner, Phys. Rev. **D 75**, 024037 (2007).
[7] J. E. Aman, I. Bengtsson, N. Pidokrajt, Gen. Rel. Grav. **35**, 1733 (2003) gr-qc/0304015; G. Arcioni, E. L. Tellechea, Phys. Rev. **D 72**, 104021 (2005) hep-th/ 0412118; J. Y. Shen, R. G. Cai, B. Wang, R. K. Su, gr-qc/0512035; J. E. Aman, N. Pidokrajt, Phys. Rev. **D73**, 024017 (2006) hep-th/0510139; J. E. Aman, I. Bengtsson, N. Pidokrajt, gr-qc/0601119.
[8] J. P. Blaizot, E. Iancu, A. Rebhan, Phys. Rev. **D 63**, 065003 (2001); J. P. Blaizot, E. Iancu, A. Rebhan, Phys. Rev. **D 63**, 065003 (2001); J. P. Blaizot, E. Iancu, hep-ph/0101103 (2001).
[9] F. Karsch, Nucl. Phys. **A698**, 199 (2002) arXiv:hep-ph/0103314; C. R. Allton, S. Ejiri, S. J. Hands, O. Kaczmarek, F. Karsch, E. Laermann, C. Schmidt, L. Scorzato, Phys. Rev. **D 66**, 074507 (2002); F. Karsch, C. R. Allton, S. Ejiri, S. J. Hands, O. Kaczmarek, E. Laermann, C. Schmidt, L. Scorzato, Nucl. Phys. Proc. Suppl. **129**, 614 (2004), hep-lat/0309116.
[10] V. Chandra, R. Kumar, V. Ravishankar, Phys. Rev. **C 76**, 054909 (2007); V. Chandra, A. Ranjan, V. Ravishankar, arXiv:0801.1286; V. Chandra, V. Ravishankar, arXiv:0805.4820; arXiv:0812.1430.
[11] A. M. Polyakov, JETP Lett. **20** 194 (1974).
[12] J. C. Collins, Adv. Ser. Direct. High Energy Phys. **5** 573 (1989).
[13] L. Durand, Phys. Rev. D **32** 1257 (1985).
[14] J. H. Kuhn, XXIV Cracow School of Theoretical Physics, Zakopane, Poland, CERN-TH/ **4083** (1984).
[15] S. Bellucci, V. Chandra, B. N. Tiwari, arXiv:1010.4225.

Axial charge segregation during a first order phase transition in the presence of hypermagnetic fields

Gabriella Piccinelli

Centro Tecnológico, FES Aragón, Universidad Nacional Autónoma de México, Av. Rancho Seco S/N, Bosques de Aragón, Nezahualcóyotl, Estado de México 57130, México.

E-mail: itzamna@unam.mx

Abstract.
We study the scattering of fermions off a finite width bubble wall during a first order phase transition in the presence of a background hypermagnetic field. We show that the chiral nature of the fermion coupling with the background field in the symmetric phase generates an axial asymmetry in the scattering processes. We briefly discuss possible implications of such axial charge segregation as a source of CP violation during a process of non-local baryogenesis.

1. Introduction

The problem of baryogenesis is still one of the open questions in cosmology, despite the large amount of work devoted to find a viable explanation. The conditions for developing a baryon asymmetry in an initially symmetric universe were laid down by Sakharov in 1967 [1] and we are still looking for a scenario that effectively encompasses them. These three well-known ingredients are: (1) existence of interactions that violate baryon number, (2) C and CP violation and (3) departure from thermal equilibrium.

1.1. *Electroweak baryogenesis*

The standard model (SM) of electroweak interactions meets all these requirements, provided that the electroweak phase transition (EWPT) is first order since, at this stage of the universe evolution, it is the only possible source of departure from thermal equilibrium. Nonetheless, it is well known that neither the amount of CP violation within the minimal SM, nor the strength of the EWPT are enough to generate a sizable baryon number [2, 3]. Regarding the existence of processes violating baryon number, the SM provides us with a valuable mechanism, tied to the vacuum structure of a non-Abelian theory. The *Sphaleron* [4] is a static and unstable solution of the field equations of the EW model, corresponding to the top of the energy barrier between topologically distinct vacua. Each minimum corresponds to configurations with zero gauge field energy but different baryon number. In such a way, topological transitions between different minima are associated to baryon number ($n_B - n_{\bar{B}}$) violation [5] and can either induce baryogenesis or wash out the previously generated asymmetry. That is why one of the difficult steps along the baryogenesis process is to preserve the baryon asymmetry. This requires that the transition between different topological vacua is supressed in the broken phase, when the universe recovers thermal equilibrium, leading to a rather difficult bound to meet for the parameters involved in the PT [6, 3].

1.2. Non-local baryogenesis

Among the different scenarios for baryogenesis proposed in the framework of the EWPT, non-local baryogenesis offers an interesting possibility: it does not require that the three Sakharov conditions occur in a single process at a single space-time point, making them easier to be fulfilled (see [7] for a review). In particular, it takes advantage of *CP* violation and out-of-equilibrium conditions that can develop during the EWPT and of anomalous baryon violation processes occuring with a large rate in the symmetric phase. In this scenario, the *CP* violating charge asymmetry generated in the bubble wall can be transported back in the symmetric region by particle reflection off the bubble wall and be converted to baryon number outside the bubble, in the symmetric region. Moreover, in [8, 9], it was shown that the required *CP* violation can be obtained through the generation of an asymmetry in another quantum number, i.e. in any non conserved charge which is not orthogonal to baryon number. These authors proposed an extension of the SM, where a *CP* violating phase is introduced, at tree level, in the scalar potential of a two doublets Higgs model, acting as an effective chemical potential for fermionic hypercharge. As a consequence, an axial charge segregation is generated between the two phases, providing a bias for baryon over anti-baryon production in the sphaleron solution.

1.3. Electroweak baryogenesis in the presence of (hyper)magnetic fields

An interestig attempt to link the baryogenesis process to the EWPT results from taking into account the effect of another possible ingredient: primordial magnetic fields. Before the EWPT, these fields would in fact couple through hypercharge and recieve the name of *hypermagnetic fields*, belonging to the $U(1)_Y$ group. Magnetic fields seem to be pervading the entire universe and their generation may be either primordial or associated to the process of structure formation. They have been observed in galaxies, clusters, intracluster medium and high redshift objects [10]. Although there is at present no conclusive evidence about the origin of magnetic fields, their existence prior to the EWPT cannot certainly be ruled out and it is important to investigate their possible influence on different cosmological events and in particular on the baryogenesis process (see [11], for a review). In fact, it has been shown that they act on each of the three Sakharov ingredients.

The first mechanism that was largely explored was the effect of the presence of a large scale magnetic field on the dynamics of the phase transition. This process has been analytically studied classically [12], to one-loop order [13] and up to the contribution of ring diagrams [14, 15], that have prooved to be crucial for the description of the long wavelength properties of the theory [16]. It has also been adressed by means of lattice simulations [17]. These calculations all agree that the stength of the PT is enhanced in presence of hypermagnetic fields, although not enough to satisfy the requirements imposed by baryon number preservation. However, other authors [18] reach the conclusion that these fields inhibit the first order phase transition.

Regarding the sphaleron transitions, contrary to the previous encouraging results, it has been found [19] that the presence of the magnetic field is counter-productive for the desired result, since the coupling of the sphaleron dipole moment to the external field lowers the potential barrier, making easier the erasure of the asymmetry.

On another hand, the influence of magnetic fields on the enhancement of *CP* violation has received much less attention and this is the aspect that we want to examine here. In [20] and [21], it has been shown that the external field is able to produce an axially asymmetric scattering of fermions off first order phase transition bubble walls. This *CP* violating reflection is due to the chiral nature of the couplings of right- and left-handed modes with the external field in the symmetric phase. This mechanism produces an axial charge segregation which can then be transported in the broken symmetry phase where sphaleron induced transitions - taking place with a large rate - can convert it into baryon number generation. In this way, this mechanism is providing an additional source of *CP* violation within the SM.

The work is organized as follows: in Sect. 2 we describe the Higgs field vacuum expectation value evolution through a first-order PT and the consequent behavior of the masses of the particles involved in the PT. In Sect. 3 we present the movement equations for fermions in the symmetric and the broken symmetry phases in the presence of a background (hyper)magnetic field. We devote Sect. 4 to the study of the scattering of fermions off the bubble walls: using analytical and numerical solutions of the above equations in both phases, we compute the transmission and reflection coefficients of fermions through the wall and we show that these coefficients differ for distinct helicity modes, leading to an axial asymmmetry on both sides of the wall. Finally, in Sect. 5, we discuss the possible implications of such axial charge segregation as a source of *CP* violation during a process of non-local baryogenesis.

2. Kink solution in a first order phase transition

A first order PT implies the coexistence of two phases and an order parameter that undergoes a discontinuous change. In the case of the EWPT, the two phases are the symmetric one, built on the false vacuum of the effective potential, and the broken symmetry phase, characterized by the true vacuuum; the order parameter is the expectation value of the Higgs field. The conversion from one phase to the other happens through the nucleation and propagation of true vacuum bubbles. The region separating both phases is called the wall and its properties depend on the effective, finite temperature, Higgs potential. Under the assumption that the wall is thin and that the PT happens when the energy densities of both phases are degenerate, it is possible to find a one-dimensional analytical solution for the (dimensionless) Higgs field strength $\varphi(z)$, called the *kink* [22]. This is given by

$$\varphi(z) = 1 + \tanh\left(\frac{z}{\lambda}\right), \qquad (1)$$

where z is the coordinate along the direction of the phase change and λ is the width of the wall. It can be expressed in terms of the effective potential parameters and the temperature at wich the transition takes place. (See for instance Ref. [13] for explicit expressions of the effective potential and the parameters involved).

If we are interested in describing the evolution of particles from the symmetric to the broken phase, one of the aspects that we must take care of is their interaction with the bubble walls. In particular, we are interested in the scattering from the symmetric to the broken phase since particles in the high temeprature phase are reflected (or transmitted) off the advancing bubble walls, while most particles in the low temperature phase are unable to catch up with the receeding walls. During this processes, since particles acquire mass through the Higgs boson, masses cannot be treated as constant quantities and some assumption must be adopted for their evolution. If we consider the above behavior of the Higgs field vacuum expectation vaue, the profile for the mass will be $m(z) = m_0(1 + \tanh(z/\lambda))$. In terms of the kink solution we can see that $z = -\infty$ represents the region outside the bubble, that is the region in the symmetric phase, where the particles are massless. Conversely, for $z = +\infty$, the system is inside the bubble, that is in the broken phase, and the particles have acquired a finite mass m_0.

The problem can be further simplified by considering the limit when the width of the wall approaches zero. In such a case, the kink solution becomes a step function,$\Theta(\ z)$, and consequently the expression for the particles' mass becomes $m(z) = m_0\Theta(z)$.

3. Dirac equation for fermions in the symmetric and broken symmetry phases in a background (hyper)magnetic field

When scattering is not affected by diffusion, the problem of fermion reflection and transmission through the wall can be casted in terms of solving the Dirac equation, with a position dependent fermion mass proporional to the Higgs field [23].

In the presence of an external magnetic field, we need to consider that fermion modes couple differently to the field in the symmetric and in the broken symmetry phases. In [20] and [21] we have derived the Dirac equation for such fermions and explicitly constructed the solutions in both phases, with a constant magnetic field, requiring that these match at the interface $z = 0$. In [20], working with a step function for the mass profile, we obtained rather simple analytical solutions but we also encountered some problems derived from the discontinuos behavior of the Higgs field value. In [21], we used the more realistic kink solution for the mass evolution and had to resort to numerical methods for obtainig the solution in the symmetric phase. The results obtained with both mass profiles essentially coincide. We will report here the results of this last more relistic case. After exhibiting the Dirac equation in both phases, we only present graphs for the transmission and reflection coefficients for fermions interacting with the wall. These coefficients were built from the spinor wave functions through the incident, reflected and transmitted currents. We refer the interested reader to the original papers for the details on the methods we employed for solving these equations and the matching procedure at the interface. For definitness, we have worked with fermions mooving in the normal direction to the bubble wall, *i.e.* along the $\hat{z}$ axis, with a constant (hyper)magnetic field in this same direction. In another work [24], this restriction has been lifted, solving the Dirac equation in three spatial dimensions, and obtaining essentially the same result.

3.1. Symmetric phase

In the symmetric phase, for $z \leq 0$, the coupling of fermions with the hypermagnetic field is chiral. Let

$$\begin{aligned} \Psi_R &= \frac{1}{2}(1+\gamma_5)\,\Psi \\ \Psi_L &= \frac{1}{2}(1-\gamma_5)\,\Psi \end{aligned} \tag{2}$$

represent, as usual, the right and left-handed chirality modes for the spinorΨ , respectively. Then, the equations of motion for these modes, as derived from the electroweak interaction Lagrangian, are

$$\begin{aligned} (i\partial\!\!\!/ - \frac{y_L}{2}g' A\!\!\!/)\Psi_L - m(z)\Psi_R &= 0 \\ (i\partial\!\!\!/ - \frac{y_R}{2}g' A\!\!\!/)\Psi_R - m(z)\Psi_L &= 0\,, \end{aligned} \tag{3}$$

where $y_{R,L}$ are the right and left-handed hypercharges corresponding to the given fermion, respectively, g' the $U(1)_Y$ coupling constant and we take $A^\mu = (0, \mathbf{A})$ representing a four-vector potential having non-zero components only for its spatial part, in the rest frame of the wall. These equations can be written as a couple of sets of equations, that describe the independent motion of the spin component parallel and antiparallel to the magnetic field respectively. The second is obtained from the first one by changing B to $-B$. Both set of equations involve modes coupled with y_R and y_L and in the limit when $y_R = y_L = e$, they decouple as in the case when describing the interaction of fermions with the magnetic field through their electric charge.

3.2. Broken symmetry phase

We now look at the corresponding equation in the broken symmetry phase. For $z \geq 0$ the coupling of the fermion with the external field is through the electric charge e and thus, the equation of motion is simply the Dirac equation describing an electrically charged fermion in a background magnetic field, namely,

$$\left\{i\partial\!\!\!/ - eA_\mu\gamma^\mu - m(z)\right\}\Psi = 0\,. \tag{4}$$

4. Fermion scattering off the bubble walls and axial charge segregation

We study here the scattering of left and right-handed fermions off a finite width kink wall during the EWPT in the presence of a background hypermagnetic field. Using analytical and numerical solutions of the above equations in both phases, we compute the transmission and reflection coefficients of fermions through the wall, for the case when they move from the symmetric to the broken symmetry phase. Since these are related to the corresponding coefficients for fermions incident from the broken symmetry phase, it is easy to obtain the total flux of fermions for the two distinct chirality modes in the symmetric phase. We work in particular with top quarks, since all other fermions have small Yukawa couplings and therefore penetrate the bubble wall with little reflection.

Once we have found the solutions for the above equations, for modes with spin components parallel and antiparallel to the magnetic field, prepared as incident waves coupled with y_L or with y_R, the problem is completely defined by matching the solutions and their derivatives across the interface.

Since in the broken symmetry phase there should not be a propagating component corresponding to the Z^0 field [12, 13], the magnetic field strength B in the symmetric phase is related to B' in the broken one and Weinberg's angle θ_W by

$$B' = \frac{B}{\cos\theta_W}\,, \tag{5}$$

which in turn implies that the coupling of the fermion with the magnetic field is given by

$$eB' = g'B\,. \tag{6}$$

We find that the amplitudes for the distinct axial modes in the symmetric phase are not the same, meaning that there is the possibility of building an axial asymmetry during the scattering of fermions off the wall. To quantify the asymmetry, we need to compute the corresponding reflection and transmission coefficients. These are built from the reflected, transmitted and incident currents of each type. Recall that for a given spinor wave functionΨ , the current normal to the wall is given by

$$J = \Psi^\dagger\gamma^0\gamma^3\Psi\,. \tag{7}$$

The currents need to be computed in the asymptotic regions far away from the wall where the amplitudes represent plane waves with well defined direction of motion.

For a left-handed incident particle, the reflection and transmission coefficients are given as the ratios of the corresponding reflected and transmitted currents, to the incident one, respectively, projected along a unit vector normal to the wall. These are

$$\begin{aligned} R_{l\to r} &= -J^r_{\rm ref}/J^l_{\rm inc} \\ T_{l\to l} &= J^l_{\rm tra}/J^l_{\rm inc}\,, \end{aligned} \tag{8}$$

where indexes l and r denote helicity modes. The corresponding coefficients for the axially conjugate process are

$$\begin{aligned} R_{r\to l} &= -J^l_{\rm ref}/J^r_{\rm inc} \\ T_{r\to r} &= J^r_{\rm tra}/J^r_{\rm inc}\,. \end{aligned} \tag{9}$$

For an incident wave coupled with y_L (y_R), the fact that the differential equations mix up the solutions means that the reflected wave will also include a component coupled with y_R

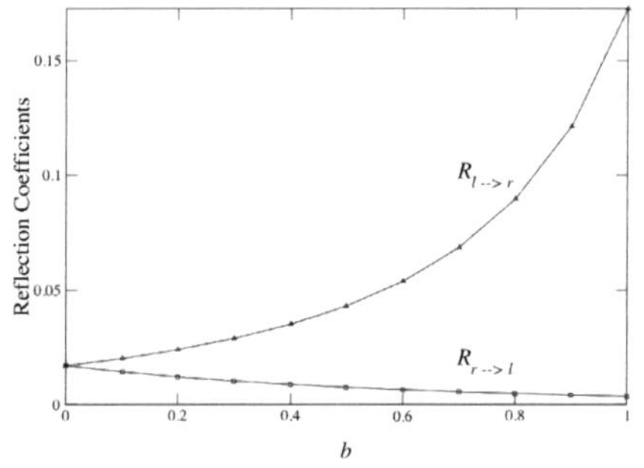

Figure 1. Coefficients $R_{l\to r}$ and $R_{r\to l}$ as a function of the magnetic field parameter b for $\xi = 3.5$, $\epsilon = 7.03$, $y_R = 4/3$, $y_L = 1/3$. The value for the $U(1)_Y$ coupling constant is taken as $g' = 0.344$, corresponding to the EWPT epoch. The dots represent the computed values.

(y_L). We have prepared the incident fermion from the symmetric phase in such a way that when coupled with a given *chirality*, it corresponds to the same *helicity*. Since in the symmetric phase the fermion mass is asymptotically zero, both the chirality and helicity operators can be simultaneously defined and their eigenvalues coincide. This is no longer the case when the fermion moves in the broken symmetry phase where its mass is different from zero. Nevertheless, since scattering off the wall does not change the direction of the fermion spin, the fermion helicity is preserved during transmission and reversed upon reflection.

Our results are displayed in Figure 1 and Figure 2, where we are working with magnitudes scaled by the width of the domain wall, λ. b is the parameter representing the magnetic field strength given by $b = \lambda^2 B$; ϵ is the energy parameter, $\epsilon = \lambda E$, where E is the energy of the incident particle, and ξ represents the heigth of the barrier, being twice the ratio of the fermion mass to the Higgs mass. Figure 1 shows the coefficients $R_{l\to r}$ and $R_{r\to l}$ as a function of the magnetic field parameter b for a value of $\xi = 3.5$, an energy parameter $\epsilon = 7.03$, hypercharge values $y_R = 4/3$, $y_L = 1/3$ and for a value of $g' = 0.344$, as appropriate for the EWPT epoch. Notice that when $b \to 0$, these coefficients approach each other and that the difference grows with increasing field strength.

Figure 2 shows the reflection and transmission coefficients as a function of the energy parameter ϵ scaled by twice the height of the barrier 2ξ. Figure 2a shows the coefficients $R_{l\to r}$ and $T_{l\to l}$ and Fig. 2b the coefficients $R_{r\to l}$ and $T_{r\to r}$ for $b = 0.5$.

Since the solutions are computed assuming that the transmitted waves are not exponentially damped, their energies have to be taken larger then the heigth of the barrier, $\epsilon \geq \sqrt{4\xi^2 - g'b}$, for waves with spin components parallel to the magnetic field, whereas for antiparallel spin components, $\epsilon \geq \sqrt{4\xi^2 + g'b}$. These requirements basically mean that, by energy conservation, the energy of the incident particle must be larger then the mass the particle will aquire while crossing the phase boundary. It can be checked that $R_{r\to l} + T_{r\to r} = 1$ and $R_{l\to r} + T_{l\to l} = 1$ within the numerical precision of the calculation, which means that the analysis respects unitarity.

5. Discussion and conclusions

We have studied the scattering of fermions off a first order EWPT bubble wall with a finite width in the presence of a background magnetic field directed along the fermion direction of motion. In the symmetric phase, the chiral nature of the coupling of fermions to the hypermagnetic field implies that it is possible to build an axial asymmetry during the scattering of fermions off the wall. We have computed reflection and transmission coefficients showing explicitly that they differ for left and right-handed incident particles from the symmetric phase.

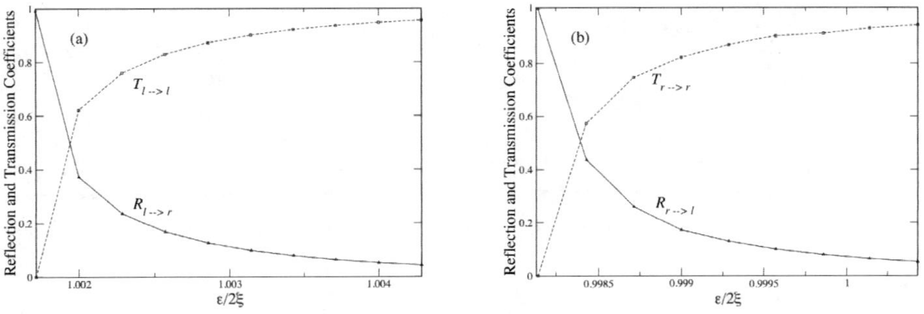

Figure 2. Reflection and transmission coefficients as a function of the energy parameter ϵ scaled by twice the height of the barrier 2ξ for $b = 0.5$ and $\xi = 3.5$, $y_R = 4/3$, $y_L = 1/3$, $g' = 0.344$. Figure 2a (left panel) shows the coefficients for incident, left-handed helicity modes and Fig. 2b (right panel) for incident, right-handed helicity modes. In both figures, the dots represent the computed values.

The results of this more realistic, albeit numerical calculation where we allow for a finite wall width are in qualitative and quantitative agreement with those previously found in Ref. [20], where the wall was modeled as a step function.

Some remarks about the particular physical situation we have considered are in order here. Let us first look at the situation in which the direction of the magnetic field is reversed with respect to the case studied here, changing B to $-B$ in the two sets of equations we had to solve. The only effect of this switch will be the interchange of one set of equations with the other, leaving intact the coupling. Physically this is also easy to understand since the fermion coupling with the external field is through its spin. Changing the direction of the field exchanges the role of each spin component but since each chirality mode contains both spin orientations, it does not affect the final probabilities. Now suppose that the original direction of motion of the fermion is not parallel to the direction of the magnetic field. The effect of this situation is also negligible since a boost can be done to the frame where the transverse momentum vanishes [9]. The case has been explicitly addressed in [24], obtaining a similar result to the present one. Finally, under the assumptions of CPT invariance and unitarity, the total axial asymmetry - which includes contributions both from particles and antiparticles- can be quantified in terms of the particle asymmetry. Let ρ_i represent the number density for species i. The net densities in left-handed and right-handed axial charges are obtained by taking the differences $\rho_L - \rho_{\bar{L}}$ and $\rho_R - \rho_{\bar{R}}$, respectively. It is straightforward to show [9] that CPT invariance and unitarity imply that the above net densities are given by

$$\begin{aligned} \rho_L - \rho_{\bar{L}} &= (f^s - f^b)(R_{r\to l} - R_{l\to r}) \\ \rho_R - \rho_{\bar{R}} &= (f^s - f^b)(R_{l\to r} - R_{r\to l}), \end{aligned} \tag{10}$$

where f^s and f^b are the statistical distributions for particles or antiparticles (since the chemical potentials are assumed to be zero or small compared to the temperature, these distributions are the same for particles or antiparticles) in the symmetric and the broken symmetry phases, respectively. Thus, the asymmetry in the axial charge density is finally given by

$$(\rho_L - \rho_{\bar{L}}) - (\rho_R - \rho_{\bar{R}}) = 2(f^s - f^b)(R_{r\to l} - R_{l\to r}). \tag{11}$$

This asymmetry, built on either side of the wall, is dissociated from non-conserving baryon number processes and can subsequently be converted to baryon number in the unbroken symmetry phase where sphaleron induced transitions are taking place with a large rate. It is interesting to notice that, with this mechanism, we are not generating a net excess of one type of particle (left or right-handed) over the other; it is merely a segregation between the two sides of the bubble wall.

The relation of this axial asymmetry to *CP* violation is to be found in the dynamics of the scattering process, that can be thought of as describing the mixing of two levels: right and left-handed quarks coupled to an external magnetic field. When the two chirality modes interact only with the external field, they evolve separately. It is only the scattering with the bubble wall what allows a finite transition probability for one mode to become the other. Since the modes are coupled differently to the external field, these probabilities are different. *CP* is violated in the process because, though *C* is conserved, *P* is violated and thus is *CP*.

Acknowledgments
Support for this work has been received in part by DGAPA-UNAM under PAPIIT grant number IN112308.

References

[1] Sakharov A D 1967 *Pis'ma Zh. Eksp. Teor. Fiz.* **5** 32 (1967 *JETP Lett.* **5** 24)
[2] Gavela M B, Hernandez P, Orloff J and Pene O 1994 *Mod. Phys. Lett.* A **9** 795
[3] Kajantie K, Laine M, Rummukainen K and Shaposhnikov M 1996 *Nucl. Phys.* B **466** 189
[4] Klinkhamer F R and Manton N S 1984 *Phys. Rev.* D **30** 2212
[5] t' Hooft G 1976 *Phys. Rev.* D **37** 8; *Phys. Rev.* D **14** 3432
[6] Shaposhnikov M E 1987 *Nucl. Phys.* B **287** 757
[7] Trodden M 1999*Rev. Mod. Phys.* **71** 1463
[8] Cohen A G, Kaplan D B and Nelson A E 1991 *Phys. Lett.* B **263** 86
[9] Nelson A E, Kaplan D B and Cohen A G 1992 *Nucl. Phys.* B **373** 453
[10] For comprehensive reviews see Kronberg P P 1994 *Rep. Prog. Phys.* **57** 325
Beck R, Brandenburg A, Moss D, Shukurov A and Sokoloff D 1996 *Annu. Rev. Astron. Astrophysics* **34** 155
Carilli C L and Taylor G B 2002 *Annu. Rev. Astron. Astrophys.* **40** 319
Govoni F and Feretti L 2004 *Int. J. Mod. Phys.* D **13** 1549
[11] Piccinelli G and Ayala A 2004 *Lecture Notes on Physics* **646** 293
[12] Giovannini M and Shaposhnikov M E 1998 *Phys. Rev.* D **57** 2186
[13] Elmfors P, Enqvist K and Kainulainen K 1998 *Phys. Lett.* B **440** 269
[14] Sánchez A, Ayala A and Piccinelli G 2007 *Phys. Rev.* D **75** 043004
[15] Navarro J, Sánchez A, Tejeda-Yeomans M E, Ayala A and Piccinelli G 2010 *Phys. Rev.* D **82** 123007
[16] Carrington M E 1992 *Phys. Rev.* D **45** 2933
[17] Kajantie K, Laine M, Peisa J, Rummukainen K and Shaposhnikov M 1999 Nucl. Phys. B **544** 357
[18] Skalozub V and Demichik V 1999 "Can baryogenesis survive in the standard model due to strong hypermagnetic field?" *Preprint* hep-ph/9909550
[19] Comelli D, Grasso D, Pietroni M and Riotto A 1999 *Phys. Lett.* B **B458** 304
[20] Ayala A, J. Besprosvany J, Pallares G and Piccinelli G 2001 *Phys. Rev.* D **64** 123529
[21] Ayala A, Piccinelli G and Pallares G 2002 *Phys. Rev.* D **66** 103503
[22] Liu B H, McLerran L and Turok N 1992 *Phys. Rev.* D **46** 2668
[23] Ayala A, Jalilian-Marian J, McLerran L and Vischer A P 1994 *Phys. Rev.* D **49** 5559
[24] Ayala A and Besprosvany J 2003 *Nucl. Phys.* B **651** 211

Direct neutralino searches in the NMSSM with gravitino LSP in the degenerate scenario

Grigoris Panotopoulos

Departament de Fisica Teorica, Universitat de Valencia, E-46100, Burjassot, Spain, and Instituto de Fisica Corpuscular (IFIC), Universitat de Valencia-CSIC, Edificio de Institutos de Paterna, Apt. 22085, E-46071, Valencia, Spain

E-mail: grigoris.panotopoulos@uv.es

Abstract. In the present work a two-component dark matter model is studied adopting the degenerate scenario in the R-parity conserving NMSSM. The gravitino LSP and the neutralino NLSP are extremely degenerate in mass, avoiding the BBN bounds and obtaining a high reheating temperature for thermal leptogenesis. In this model both gravitino (absolutely stable) and neutralino (quasi-stable) contribute to dark matter, and direct detection searches for neutralino are discussed. Points that survive all the constraints correspond to a singlino-like neutralino.

1. Introduction

There is accumulated evidence both from astrophysics and cosmology that about 1/4 of the energy budget of the universe consists of so called dark matter, namely a component which is non-relativistic and neither feels the electromagnetic nor the strong interaction. For a review on dark matter see e.g. [1]. Although the list of possible dark matter candidates is long (for a nice list see [2]), it is fair to say that the most popular dark matter candidate is the lightest supersymmetric particle (LSP) in supersymmetric models with R-parity conservation [3]. For supersymmetry and supergravity see [4]. The simplest supersymmetric extension of the standard model that solves the mu problem [5] is the next-to-minimal supersymmetric standard model (NMSSM) [6]. If we do not consider the axion [7] and the axino [8], the superpartners that have the right properties for playing the role of cold dark matter in the universe are the gravitino and the lightest neutralino. By far the most discussed case in the literature is the case of the neutralino (see the classic review [9]), probably because of the prospects of detection. However, in the case is which neutralino is assumed to be the only dark matter component, one has to face the fine-tuning problem and the gravitino problem [10]. In most of the parameter space the neutralino relic density turns out to be either too small or too large [11]. Furthermore, unstable gravitinos will undergo late-time cascade decays to a neutralino LSP. These decays will destroy the light element abundances built up in BBN, unless $T_R < 10^5$ GeV [12], which poses serious difficulties to the thermal leptogenesis scenario [13]. If, on the other hand, gravitino is the LSP and therefore stable, playing the role of cold dark matter in the universe, it is then the neutralino that will undergo late time decays into gravitino and hadrons, and the gravitino problem is re-introduced [14].

It has been shown that in the degenerate scenario [15] the BBN and CMB constraints are

Journal of Physics: Conference Series **335** (2011) 012064 doi:10.1088/1742-6596/335/1/012064

avoided, and high values of the reheating temperature are obtained compatible with thermal leptogenesis. Here we focus on the scenario in which the masses of the gravitino LSP and neutralino NLSP are extremely degenerate in mass. Under this assumption neutralino becomes quasi-stable participating to the cold dark matter of the universe together with gravitino, and it is still around and it can be seen in direct detection searches experiments.

This article is organized as follows. In the next section we present the theoretical framework. In section 3 we discuss all the relevant constraints from colliders and from cosmology, and we show our results. Finally, we conclude.

2. Theoretical Framework

2.1. The NMSSM

The particle physics model is defined by the superpotential

$$W = \epsilon_{ij}\left(Y_u\,H_2^j\,Q^i\,u + Y_d\,H_1^i\,Q^j\,d + Y_e\,H_1^i\,L^j\,e\right) - \epsilon_{ij}\lambda\,S\,H_1^iH_2^j + \frac{1}{3}\kappa S^3 \tag{1}$$

as well as the soft breaking masses and couplings

$$\begin{aligned}-\mathcal{L}_{\rm soft} = \quad & m_{\tilde Q}^2\,\tilde Q^*\,\tilde Q + m_{\tilde U}^2\,\tilde u^*\,\tilde u + m_{\tilde D}^2\,\tilde d^*\,\tilde d + m_{\tilde L}^2\,\tilde L^*\,\tilde L + m_{\tilde E}^2\,\tilde e^*\,\tilde e \\ & + m_{H_1}^2\,H_1^*\,H_1 + m_{H_2}^2\,H_2^*H_2 + m_S^2\,S^*S \\ & + \epsilon_{ij}\left(A_u\,Y_u\,H_2^j\,\tilde Q^i\,\tilde u + A_d\,Y_d\,H_1^i\,\tilde Q^j\,\tilde d + A_e\,Y_e\,H_1^i\,\tilde L^j\,\tilde e + \text{H.c.}\right) \\ & + \left(-\epsilon_{ij}\lambda\,A_\lambda S H_1^i H_2^j + \tfrac{1}{3}\kappa\,A_\kappa\,S^3 + \text{H.c.}\right) \\ & - \tfrac{1}{2}\left(M_3\,\lambda_3\,\lambda_3 + M_2\,\lambda_2\,\lambda_2 + M_1\,\lambda_1\,\lambda_1 + \text{H.c.}\right) \end{aligned} \tag{2}$$

When the singlet acquires a vaccum expectation value, S, we obtain an effective μ parameter, $\mu_{eff} = \lambda S$. Imposing universality at the GUT scale, a small controllable number of free parameters remains, namely

$$tan\beta = v_u/v_d, m_0, A_0, m_{1/2}, \lambda, A_k$$

and the sign of the effective μ parameter can be chosen at will.

Because of the extra singlet superfield, in the NMSSM there is a larger higgs sector and a larger neutralino sector. The neutralino mass matrix is characterized by the appearence of a fifth neutralino state, meaning that the composition of the lightest neutralino has an extra singlino contribution

$$\tilde\chi_1^0 = N_{11}\tilde B^0 + N_{12}\tilde W_3^0 + N_{13}\tilde H_1^0 + N_{14}\tilde H_2^0 + N_{15}\tilde S \tag{3}$$

In the following, neutralinos with $N_{11}^2 > 0.9$, or $N_{15}^2 > 0.9$, will be referred to as bino- or singlino-like, respectively.

Furthermore, in the Higgs sector we have now two CP-odd neutral, and three CP-even neutral Higgses. We make the assumption that there is no CP-violation in the Higgs sector, and therefore the CP-even and CP-odd states do not mix. We are not interested in the CP-odd states, while the CP-even Higgs interaction and physical eigenstates are related by the transformation

$$h_a^0 = S_{ab}H_b^0 \tag{4}$$

where S is the unitary matrix that diagonalises the CP-even symmetric mass matrix, $a, b = 1, 2, 3$, and the physical eigenstates are ordered as $m_{h_1^0} < m_{h_2^0} < m_{h_3^0}$.

Journal of Physics: Conference Series **335** (2011) 012064 doi:10.1088/1742-6596/335/1/012064

2.2. Production of gravitinos

In the usual case (not in the degenerate scenario) gravitinos can be produced after inflation in two ways. One way to produce gravitinos is with scatterings from the thermal bath, and another is from the out-of-equillibrium decays of the NLSP, which decouple from the thermal bath before primordial Big-Bang Nucleosynthesis and decay after the BBN time. Thus, imposing the WMAP bounds [16] we can write for the gravitino abundance

$$0.1097 < \Omega_{3/2}h^2 = \Omega_{3/2}^{TP}h^2 + Y_{3/2}^{NLSP}h^2 < 0.1165 \tag{5}$$

where

$$\Omega_{3/2}^{NLSP}h^2 = \frac{m_{\tilde{G}}}{m_{NLSP}}\,\Omega_{NLSP}h^2 \tag{6}$$

with m_{NLSP} the mass of the NLSP, and $\Omega_{NLSP}h^2$ the abundance the NLSP would have, had it not decayed into the gravitino. The thermal contribution is given by (approximately for a light gravitino, $m_{\tilde{G}} \ll m_{\tilde{g}}$) [17]

$$\Omega_{3/2}^{TP} \simeq 0.27\left(\frac{T_R}{10^{10}\ GeV}\right)\left(\frac{m_{\tilde{g}}}{TeV}\right)^{-2}\left(\frac{m_{\tilde{G}}}{100\ GeV}\right) \tag{7}$$

In the limit where $m_{NLSP} \to m_{\tilde{G}}$ and $\tau_{NLSP} \gg 10^{17}\ sec$ the scenario looks as if one would have a two-component dark matter with the NLSP contribution $\Omega_{NLSP}h^2$, and a gravitino contribution from thermal production only, $Y_{3/2}^{TP}$. Therefore, in the degenerate scenario with $m_{NLSP} \simeq m_{\tilde{G}}$ the WMAP bound becomes

$$0.1097 < \Omega_{cdm}h^2 = \Omega_{NLSP}h^2 + \Omega_{3/2}^{TP}h^2 < 0.1165 \tag{8}$$

where from now on the NLSP is the lightest neutralino, $\chi = NLSP$.

3. Constraints and results

- Spectrum and collider constraints: We have used the computer software NMSSMTools [18], we have performed a random scan over the whole parameter space (with fixed $\mu > 0$ motivated by the muon anomalous magnetic moment), and we have selected only those points that satisfy all the theoretical requirements, as well as the LEP bounds on the Higgs mass, collider bounds on SUSY particle masses, and experimental data from B-physics [19, 20]. For all these good points the lightest neutralino is either a bino or a singlino, and contrary to the case where neutralino is the dark matter particle, here we do not require that the neutralino relic density falls within the allowed WMAP range.

- As we have already mentioned, the total dark matter abundance, and not the neutralino one, should satisfy the cold dark matter constraint [16]

$$0.1097 < \Omega_{cdm}h^2 = \Omega_\chi h^2 + \Omega_{3/2}^{TP}h^2 < 0.1165 \tag{9}$$

that relates the reheating temperature after inflation to the gravitino mass as follows

$$0.11 = A(m_{\tilde{G}}, m_{\tilde{g}})T_R + \Omega_\chi h^2 \tag{10}$$

For a given point in the cNMSSM parameter space, the complete spectrum and couplings have been computed, and we are left with two more free parameters, namely the gravitino mass and the reheating temperature after inflation. The gravitino mass is equal essentially to the neutralino mass, and the precise value can be determined if we specify the neutralino lifetime. In the discussion to follow we have used a neutralino lifetime $\tau = 10^{26}\ sec$, although the results are not sensitive to it, and the figures we have produced for different values of the lifetime cannot be distinguished. Finally, the reheating temperature after inflation is obtained from the cold dark matter constraint. The thermal production contribution cannot be larger than the total dark matter abundance, and for this we can already obtain an upper bound on the reheating temperature

$$T_R \leq 4.1 \times 10^9 \left(\frac{m_{\tilde{G}}}{100\ \text{GeV}}\right) \left(\frac{\text{TeV}}{m_{\tilde{g}}}\right)^2 GeV \tag{11}$$

Assuming a gluino mass $m_{\tilde{g}} \sim 1$ TeV, we can see that for a heavy gravitino, $m_{\tilde{G}} \sim 100$ GeV, it is possible to obtain a reheating temperature large enough for thermal leptogenesis.

- For neutralino NLSP in the degenerate scenario, the only decay mode is $\chi \to \gamma\tilde{G}$, for which the decay width can be computed once the supergravity Largrangian is known [21], and it is given by [14, 22]

$$\Gamma(\chi \to \gamma\tilde{G}) = \frac{|N_{11}\cos\theta_W + N_{12}\sin\theta_W|^2}{48\pi M_*^2} \frac{m_\chi^5}{m_{\tilde{G}}^2} \left[1 - \frac{m_{\tilde{G}}^2}{m_\chi^2}\right]^3 \left[1 + 3\frac{m_{\tilde{G}}^2}{m_\chi^2}\right] \tag{12}$$

where M_* is the Planck mass, m_χ is the neutralino mass, and θ_W is the weak angle. In the limit where the mass difference $\Delta m \equiv m_\chi - m_{\tilde{G}}$ is much lower than the masses themselves, $\Delta m \ll m_\chi, m_{\tilde{G}}$, the neutralino lifetime becomes

$$\tau = \frac{1.78 \times 10^{13}\ sec}{|N_{11}\cos\theta_W + N_{12}\sin\theta_W|^2} \left(\frac{GeV}{\Delta m}\right)^3 \tag{13}$$

From this formula one can see that for a mostly bino-neutralino a mass difference of 1 MeV is already enough to give a neutralino lifetime larger than the age of the universe.

- Neutralino-Nucleon spin-independent cross-section: LHC is now running and collecting data. Although LHC is a powerful machine to look for physics beyond the standard model, it is known that other facilities are also needed to offer complementary information towards the direction of searching for supersymmetry and identifying dark matter. The gravitino interactions are suppressed by the Planck mass, and therefore direct production of gravitinos at colliders and/or direct detection prospects seem to be hopeless. On the other hand, for a weakly interacting neutralino there are existing as well as future experiments that put experimental limits on the nucleon-neutralino cross-section. The spin-independent cross-section is given by

$$\sigma_{\chi-N} = \frac{4m_r^2}{\pi} f_N^2 \tag{14}$$

where m_r is the Nucleon-neutralino reduced mass, $m_r = m_N m_\chi/(m_N + m_\chi)$, and

$$\frac{f_N}{m_N} = \sum_{q=u,d,s} f_{Tq}^{(N)} \frac{\alpha_q}{m_q} + \frac{2}{27} f_{TG}^{(N)} \sum_{q=c,b,t} \frac{\alpha_{3q}}{m_q} \tag{15}$$

In the above, $f_{TG}^{(N)} = 1 - \sum_{q=u,d,s} f_{Tq}^{(N)}$, we have taken the following values for the hadronic matrix elements [23]:

$$\begin{aligned} f_{Tu}^{(p)} &= 0.020 \pm 0.004\,, & f_{Td}^{(p)} &= 0.026 \pm 0.005\,, & f_{Ts}^{(p)} &= 0.118 \pm 0.062\,, \\ f_{Tu}^{(n)} &= 0.014 \pm 0.003\,, & f_{Td}^{(n)} &= 0.036 \pm 0.008\,, & f_{Ts}^{(n)} &= 0.118 \pm 0.062\,. \end{aligned} \tag{16}$$

and α_q is the coupling in the effective Lagrangian

$$\mathcal{L}_{\text{eff}} = \alpha_i\, \bar{\chi}\, \chi\, \bar{q}_i\, q_i \tag{17}$$

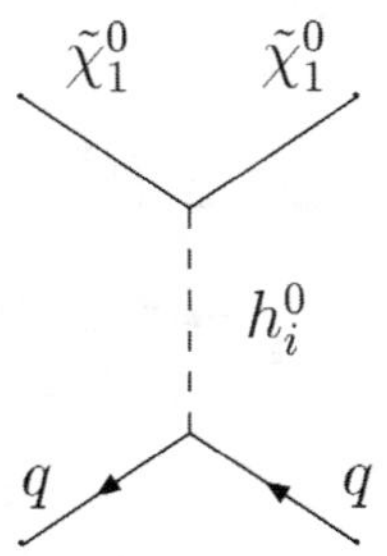

Figure 1. Neutralino scattering off a nucleon by a neutral Higgs boson exchange.

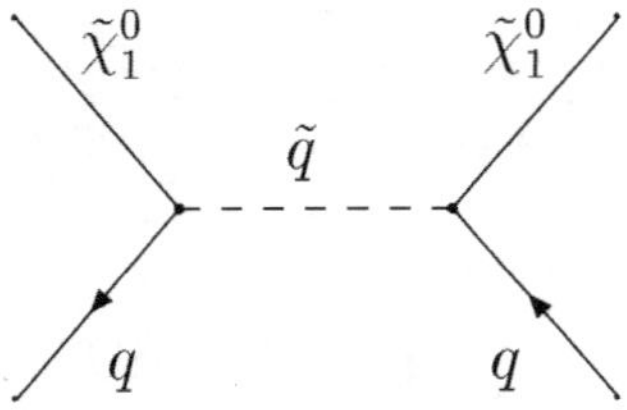

Figure 2. Neutralino scattering off a nucleon by a squark exchange.

where $i = 1, 2$ denotes up- and down-type quarks, and the Lagrangian is summed over the three quark generations. The coupling α_q can be decomposed into two parts, $\alpha_q = \alpha_q^h + \alpha_q^{\tilde{q}}$, where the first term is the t-channel exchange of a neutral Higgs (Fig. 1), while the second term is the s-channel exchange of a squark (Fig. 2). The expressions for α_q is terms of the masses and couplings of the model can be found in [24].

Our main results are summarized in the figures below. In Fig. 3 and Fig. 4 we show the Nucleon-neutralino spin-independent cross section (in cm^2) versus neutralino mass and lightest Higgs boson (in GeV) respectively. The blue region corresponds to a bino neutralino, while the green region corresponds to a singlino neutralino, and the curves are the current experimental limits from CDMS [25]. According to our results the bino scenario is already ruled out, while in the singlino case the upper region can be probed by future experiments. In Fig. 5 we show the reheating temperature after inflation as a function of the neutralino/gravitino mass. The blue region corresponds to a bino, the blue points correspond to singlino, and finally the red points correspond to singlino with relatively high values of the cross-section, namely $\sigma_{\chi-N} > 10^{-47}\ cm^2$. The largest values of T_R correspond to a bino, which is ruled out, and for the singlino with relatively high values of cross-section we obtain a reheating temperature $T_R \simeq 5 \times 10^9\ GeV$ for a neutralino/gravitino mass $m_\chi \simeq m_{\tilde{G}} \simeq 200\ GeV$. In the last figure we show the (m_0-$m_{1/2}$) plane (m_0 and $m_{1/2}$ in GeV) for singlino points with a cross-section larger than $10^{-47}\ cm^2$, or lower than $10^{-47}\ cm^2$. We see that m_0 is not larger than 600 GeV, and therefore future direct detection experiments cannot probe a region of the parameter space which can neither be probed by LHC.

4. Conclusion

In the framework of NMSSM, which solves the mu problem, we have assumed that the gravitino LSP and the lightest neutralino NLSP are degenerate in mass. Under this assumption the neutralino becomes extremely long-lived avoiding the BBN bounds. In this scenario we have a two component dark matter made out of the absolutely stable gravitino and the quasi-stable neutralino. We have performed a random scan over the whole parameter space keeping the

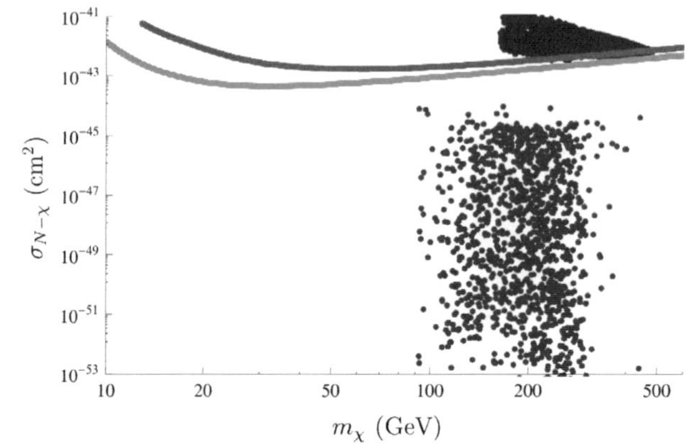

Figure 3. Spin-independent neutralino-nucleon (proton) cross-section versus neutralino mass. Shown are the available experimental bounds from CDMS, and the predictions of the theoretical model. The blue region corresponds to a bino-like neutralino, while the green points correspond to a singlino-like neutralino.

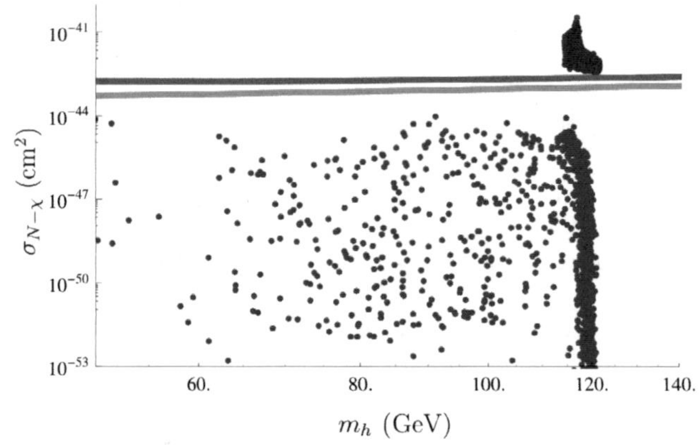

Figure 4. Spin-independent neutralino-nucleon (proton) cross-section versus the lightest Higgs mass. Shown are the available experimental bounds CDMS, and the predictions of the theoretical model. The blue region corresponds to a bino-like neutralino, while the green points correspond to a singlino-like neutralino.

points that satisfy the available collider constraints plus the WMAP bound for dark matter. These points correspond to either a bino or a singlino neutralino. We have computed the neutralino-nucleon spin-independent cross section as a function of the neutralino mass and the lightest Higgs mass, and we find that the bino is ruled out. Then we explored the $(m_0 - m_{1/2})$ parameter space, and the reheating temperature dependence of the neutralino/gravitino mass for the singlino points that correspond to cross section values to be probed by future experiments.

Acknowledgments
We acknowledge financial support from FPA2008-02878, and Generalitat Valenciana under the grant PROMETEO/2008/004.

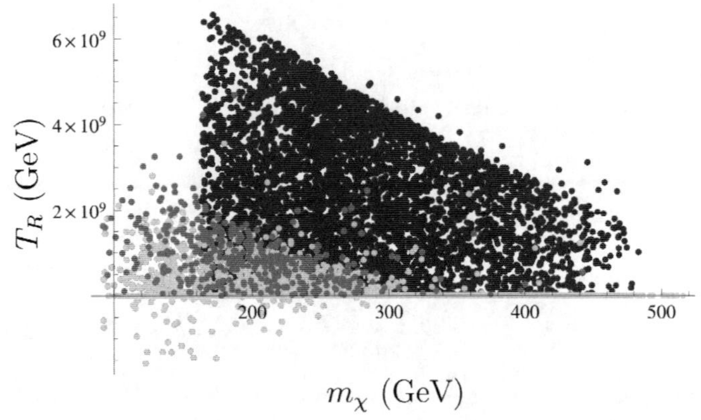

Figure 5. Reheating temperature versus neutralino (or gravitino mass). Blue points correspond to bino, green points correspond to singlino, and red points correspond to singlino with a cross-section larger than 10^{-47} cm^2.

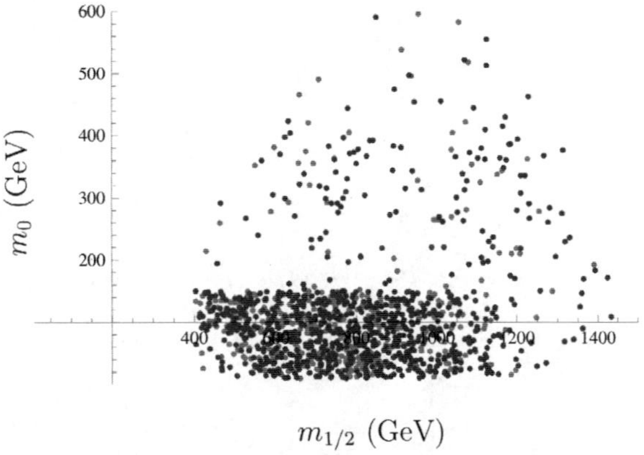

Figure 6. The (m_0-$m_{1/2}$) plane for the singlino points. One color corresponds to a cross-section larger than 10^{-47} cm^2, and the other color corresponds to a cross-section lower than 10^{-47} cm^2.

References

[1] K. A. Olive, arXiv:astro-ph/0301505;
C. Munoz, Int. J. Mod. Phys. A **19** (2004) 3093 [arXiv:hep-ph/0309346].
[2] M. Taoso, G. Bertone and A. Masiero, JCAP **0803** (2008) 022 [arXiv:0711.4996 [astro-ph]].
[3] J. L. Feng, *In the Proceedings of 31st SLAC Summer Institute on Particle Physics: Cosmic Connection to Particle Physics (SSI 2003), Menlo Park, California, 28 Jul - 8 Aug 2003, pp L11* [arXiv:hep-ph/0405215].
[4] H. P. Nilles, Phys. Rept. **110** (1984) 1;
H. E. Haber and G. L. Kane, Phys. Rept. **117** (1985) 75;
J. Wess and J. Bagger, *Supersymmetry and supergravity* (Princeton University Press, Princeton, 1992);
S. P. Martin, arXiv:hep-ph/9709356.
[5] J. E. Kim and H. P. Nilles, Phys. Lett. B **138** (1984) 150.
[6] H. P. Nilles, M. Srednicki and D. Wyler, Phys. Lett. B **120** (1983) 346;
J. R. Ellis, J. F. Gunion, H. E. Haber, L. Roszkowski and F. Zwirner, Phys. Rev. D **39** (1989) 844;
M. Drees, Int. J. Mod. Phys. A **4** (1989) 3635;
U. Ellwanger, M. Rausch de Traubenberg and C. A. Savoy, Phys. Lett. B **315** (1993) 331 [arXiv:hep-ph/9307322];

U. Ellwanger, M. Rausch de Traubenberg and C. A. Savoy, Z. Phys. C **67** (1995) 665 [arXiv:hep-ph/9502206];
U. Ellwanger, M. Rausch de Traubenberg and C. A. Savoy, Nucl. Phys. B **492** (1997) 21 [arXiv:hep-ph/9611251];
T. Elliott, S. F. King and P. L. White, Phys. Lett. B **351** (1995) 213 [arXiv:hep-ph/9406303];
S. F. King and P. L. White, Phys. Rev. D **52** (1995) 4183 [arXiv:hep-ph/9505326].
[7] R.D.Peccei and H.R.Quinn, Phys. Rev. Lett. 38 (1977) 1440;
R.D.Peccei and H.R.Quinn, Phys. Rev. D 16 (1977) 1791;
S.Weinberg, Phys. Rev. Lett. 40 (1978) 223;
F.Wilczek, Phys. Rev. Lett. 40 (1978) 279.
[8] L. Covi, J. E. Kim and L. Roszkowski, Phys. Rev. Lett. **82** (1999) 4180 [arXiv:hep-ph/9905212];
L. Covi, H. B. Kim, J. E. Kim and L. Roszkowski, JHEP **0105** (2001) 033 [arXiv:hep-ph/0101009].
[9] G. Jungman, M. Kamionkowski and K. Griest, Phys. Rept. **267** (1996) 195 [arXiv:hep-ph/9506380].
[10] J. R. Ellis, A. D. Linde and D. V. Nanopoulos, Phys. Lett. B **118** (1982) 59;
M. Y. Khlopov and A. D. Linde, Phys. Lett. B **138** (1984) 265.
[11] J. Ellis and K. Olive, Phys. Lett. B 514 (2001) 114;
H. Baer and A. Box, Eur. Phys. J. C 68 (2010) 523;
H. Baer, A. Box and H. Summy, arXiv:1005.2215 (2010).
[12] M. Kawasaki, K. Kohri and T. Moroi, Phys. Lett. B **625** (2005) 7 [arXiv:astro-ph/0402490].
[13] W. Buchmuller, P. Di Bari and M. Plumacher, Nucl. Phys. B **643** (2002) 367 [Erratum-ibid. B **793** (2008) 362] [arXiv:hep-ph/0205349];
G. F. Giudice, A. Notari, M. Raidal, A. Riotto and A. Strumia, Nucl. Phys. B **685** (2004) 89 [arXiv:hep-ph/0310123].
[14] J. L. Feng, S. Su and F. Takayama, Phys. Rev. D **70** (2004) 075019 [arXiv:hep-ph/0404231];
L. Roszkowski, R. Ruiz de Austri and K. Y. Choi, JHEP **0508** (2005) 080 [arXiv:hep-ph/0408227];
D. G. Cerdeno, K. Y. Choi, K. Jedamzik, L. Roszkowski and R. Ruiz de Austri, JCAP **0606** (2006) 005 [arXiv:hep-ph/0509275].
[15] L. Boubekeur, K. Y. Choi, R. Ruiz de Austri and O. Vives, JCAP **1004** (2010) 005 [arXiv:1002.0340 [hep-ph]].
[16] E. Komatsu *et al.* [WMAP Collaboration], Astrophys. J. Suppl. **180** (2009) 330 [arXiv:0803.0547 [astro-ph]].
[17] M. Bolz, A. Brandenburg and W. Buchmuller, Nucl. Phys. B **606** (2001) 518 [arXiv:hep-ph/0012052].
[18] U. Ellwanger, J. F. Gunion and C. Hugonie, JHEP **0502** (2005) 066 [arXiv:hep-ph/0406215];
U. Ellwanger and C. Hugonie, Comput. Phys. Commun. **175** (2006) 290 [arXiv:hep-ph/0508022];
G. Belanger, F. Boudjema, C. Hugonie, A. Pukhov and A. Semenov, JCAP **0509** (2005) 001 [arXiv:hep-ph/0505142].
[19] S. Heinemeyer, "Electroweak precision data and gravitino dark matter," arXiv:hep-ph/0611372.
[20] W. M. Yao *et al.* [Particle Data Group], "Review of particle physics," J. Phys. G **33** (2006) 1.
[21] E. Cremmer, S. Ferrara, L. Girardello and A. Van Proeyen, Nucl. Phys. B **212** (1983) 413.
[22] J. R. Ellis, K. A. Olive, Y. Santoso and V. C. Spanos, Phys. Lett. B **588** (2004) 7 [arXiv:hep-ph/0312262].
[23] J. R. Ellis, A. Ferstl and K. A. Olive, Phys. Lett. B **481** (2000) 304 [arXiv:hep-ph/0001005].
[24] D. G. Cerdeno, C. Hugonie, D. E. Lopez-Fogliani, C. Munoz and A. M. Teixeira, JHEP **0412** (2004) 048 [arXiv:hep-ph/0408102];
D. G. Cerdeno, E. Gabrielli, D. E. Lopez-Fogliani, C. Munoz and A. M. Teixeira, JCAP **0706** (2007) 008 [arXiv:hep-ph/0701271].
[25] Z. Ahmed *et al.* [The CDMS-II Collaboration], Science **327** (2010) 1619 [arXiv:0912.3592 [astro-ph.CO]].

Journal of Physics: Conference Series **335** (2011) 012065 doi:10.1088/1742-6596/335/1/012065

Looking for a charge asymmetry in cosmic rays

I Masina[1,2]

[1] Dip. di Fisica dell'Università di Ferrara and INFN Sez. di Ferrara, Via Saragat, Ferrara, Italy
[2] CP3-Origins, University of Southern Denmark, Campusvej, Odense, Denmark

E-mail: masina@fe.infn.it

Abstract. We combine the data from PAMELA and FERMI-LAT cosmic ray experiments by introducing a simple sum rule. This allows to investigate whether the lepton excess observed by these experiments is charge symmetric or not. We also show how the data can be used to predict the positron fraction at energies yet to be explored by the AMS-02 experiment.

1. Introduction

The data collected by PAMELA [1] indicate that there is a positron excess in the cosmic ray (CR) energy spectrum above 10 GeV. The rising behavior observed by PAMELA does not fit previous estimates of the CR formation and propagation implying the possible existence of a direct excess of CR positrons of unknown origins. Interestingly PAMELA's data show no excess in the anti-protons.

While ATIC [2] and PPB-BETS [3] reported unexpected structure in the all-electron spectrum in the range 100 GeV- 1 TeV, the picture has changed with the higher-statistics measurements by FERMI-LAT [4] and HESS [5], leading to a possible slight additional unknown component in the CR $e^{\pm}$ flux over and above the specific Moskalenko and Strong model prediction [6, 7] which assumes a single-power-law injection spectrum.

These interesting features have drawn much attention, and many explanations have been proposed: For example, these excesses could be due to an inadequate account of the CR background in previous modeling; The presence of new astrophysical sources; They could also originate from annihilations and/or decays of dark matter. We refer to [8] for a recent review.

Whatever the origin of these excesses might be, following Ref. [9], we show how to derive constraints about their charge asymmetry. Clearly, this will eventually shed some light on the physical nature of their source.

2. A useful sum rule

We start by writing the observed flux of electrons and positrons as the sum of two contributions: A background component, $\phi^B_{\pm}$, due to known astrophysical sources (which actually are better known for the electrons rather than the positrons), and an unknown component, $\phi^U_{\pm}$, in formulae:

$$\phi_{\pm} = \phi^U_{\pm} + \phi^B_{\pm} \ . \qquad (1)$$

The component $\phi^U_{\pm}$ is the one needed to explain the features in the spectra observed by PAMELA and FERMI-LAT.

These experiments measure respectively the positron fraction and the total electron and positron fluxes as a function of the energy E of the detected $e^{\pm}$, i.e.:

$$P(E) = \frac{\phi_+(E)}{\phi_+(E) + \phi_-(E)} , \qquad F(E) = \phi_+(E) + \phi_-(E) . \tag{2}$$

The left-hand side of the equations above refer to the experimental measures. The contribution from the unknown source is then:

$$\phi_+^U(E) = P(E)\, F(E) - \phi_+^B(E) , \tag{3}$$
$$\phi_-^U(E) = F(E)\, (1 - P(E)) - \phi_-^B(E) . \tag{4}$$

In terms of their difference and sum:

$$\begin{aligned} \phi_+^U(E) - \phi_-^U(E) &= F(E)\, (2P(E) - 1) + (\phi_-^B(E) - \phi_+^B(E)) , \\ \phi_+^U(E) + \phi_-^U(E) &= F(E) - (\phi_-^B(E) + \phi_+^B(E)) . \end{aligned} \tag{5}$$

The latter equation implies $F(E) \geq \phi_-^B(E) + \phi_+^B(E)$. We model the background spectrum using $\phi_\pm^B(E) = N_B B^\pm(E)$, where N_B is a normalization coefficient such that $F(E)/(B^-(E) + B^+(E)) \geq N_B$ and $B^\pm(E)$ are provided using specific astrophysical models. In this paper we adopt the popular Moskalenko and Strong model [6, 7], for which N_B is less than 0.75 and $B^\pm(E)$ are given by:

$$B^+ = \frac{4.5E^{0.7}}{1 + 650E^{2.3} + 1500E^{4.2}} , \tag{6}$$
$$B^- = B_1^- + B_2^- , \tag{7}$$
$$B_1^- = \frac{0.16E^{-1.1}}{1 + 11E^{0.9} + 3.2E^{2.15}} , \tag{8}$$
$$B_2^- = \frac{0.70E^{0.7}}{1 + 110E^{1.5} + 600E^{2.9} + 580E^{4.2}} , \tag{9}$$

where E is measured in GeV and the Bs in GeV^{-1}cm^{-2}sec^{-1}sr^{-1} units. We checked that our results remain unchanged when replacing the parameterization above with the one adopted by the Fermi Collaboration (model zero) [10, 11].

It is convenient to introduce the following parameter:

$$r_U(E) \equiv \frac{\phi_-^U(E)}{\phi_+^U(E)} = \frac{F(E)\, (1 - P(E)) - \phi_-^B(E)}{P(E)\, F(E) - \phi_+^B(E)} , \tag{10}$$

which quantifies the level of charge asymmetry in the unknown component of electrons and positrons. The equation above can be rewritten as

$$R(E) \equiv \frac{F(E)}{B^-(E)} \frac{1 - (1 + r_U(E))P(E)}{1 - r_U(E)\frac{\phi_+^B(E)}{\phi_-^B(E)}} = N_B . \tag{11}$$

Although the sum rule $R(E)$ seems to depend on the energy it should, in fact, be a constant as is clear from the right hand side of the previous equation. *This leads to a nontrivial constraint linking together in an explicit form the experimental results, the model of the backgrounds and the dependence on the energy of the unknown components.*

We now turn to the actual data and show in which way the sum rule provides essential information on the unknown components of the CRs.

Since we use simultaneously the results of FERMI-LAT and PAMELA we are obliged to consider only the common energy range. Note that the CRs energy range below $10-20$ GeVs, where the spectrum is affected by the Sun, is outside the common range. Within the relevant but limited range of energies we will consider here it is therefore sensible to assume r_U to be nearly constant.

We find useful to plot the function $R(E)$ for different values of r_U in order to test the sum rule. This would imply that this function is independent of the energy. The associated constant value would then be identified with the background CRs normalization factor N_B. We report the results in Fig. 1. The straight (red) line is the $N_B = 0.75$ value which is the largest one can assume for the background not to be larger than the FERMI-LAT results. The shaded region is obtained by using the PAMELA and FERMI-LAT data released in 2010, assuming a 1σ error band for both. We observe that there is a clear tendency for the combined data to predict a lower value of the constant N_B, say about $0.66, 0.64, 0.62, 0.55$ for increasing value of the ratio $r_U = 0, 1, 2, 4$. This is clear when looking, from top to bottom, at the different panels of Fig. 1. It is interesting to note that we find a plateau, in the relevant energy range, up to r_U near the value of 4 when $R(E)$ starts showing some deviation.

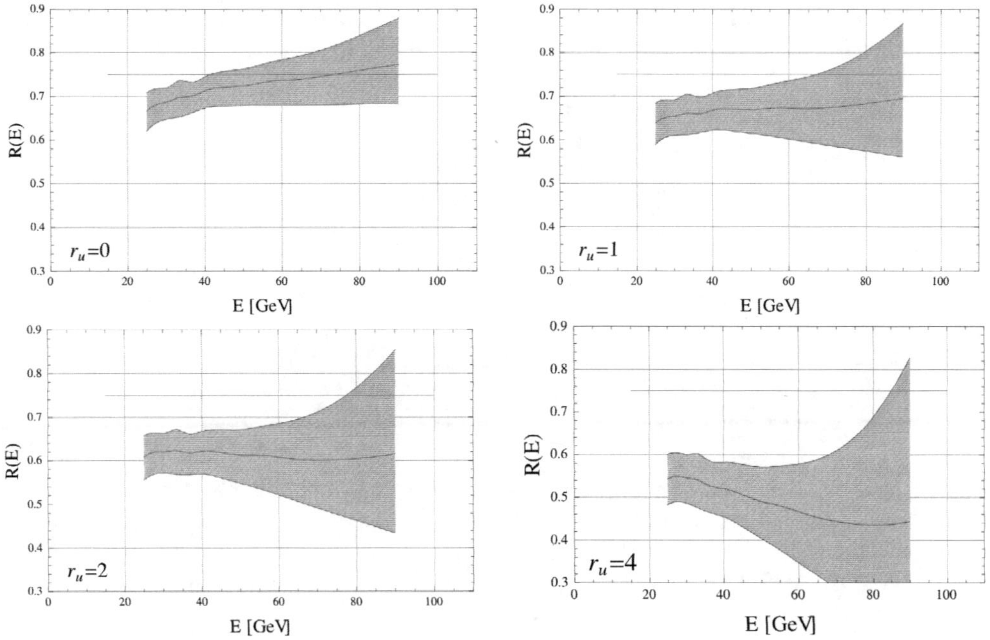

Figure 1. Ratio $R(E)$ as a function of the energy E of electrons and positrons and for values of $r_U = 0, 1, 2, 4$. The shaded region accounts for the 1σ error in PAMELA and FERMI-LAT data (2010 release). Secondaries are estimated according to the expressions in [6, 7]. The value of $R(E)$ thus corresponds to the normalization factor for the background, called N_B in the text.

Given the large uncertainties (for PAMELA, the 2010 1σ error bands are larger than in 2008) we cannot yet provide a more solid conclusion, however we can use the derived normalization N_B

for each different ratios of the unknown components to *predict* the positron fraction at energies higher than the ones provided so far by PAMELA.

3. PAMELA and the charge asymmetry of the unknown component

In order to be able to make such a prediction we first rewrite eq. (11) as follows:

$$P(E) = \frac{1}{1+r_U}\left(1 - \frac{\phi_-^B(E)}{F(E)}(1 - r_U\frac{\phi_+^B(E)}{\phi_-^B(E)})\right) , \tag{12}$$

where we use for each r_U the estimated associated N_B (referring to Fig. 1, we consider the range of N_B values allowed for $E \approx 25$ GeV). The different predictions for the positron fraction, assuming that r_U remains constant over the entire energy range, up to 1000 GeV are shown in Fig. 2. The resulting picture disfavours both very small and very large values of r_U which, in

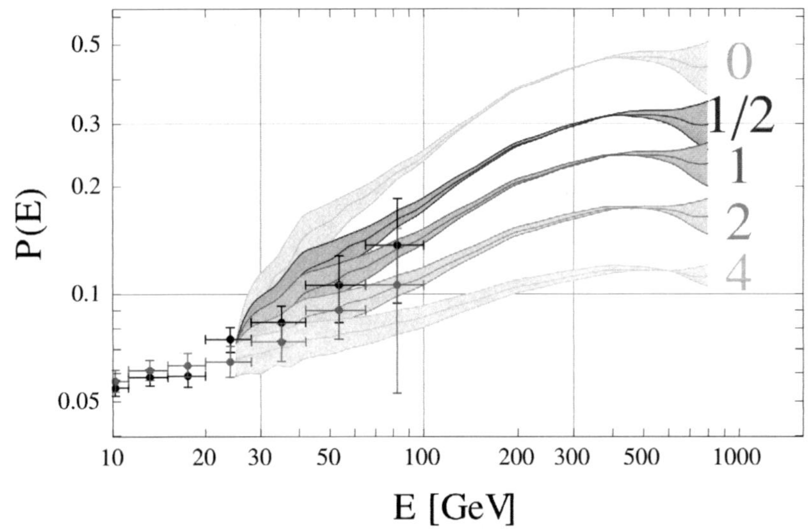

Figure 2. Model independent prediction for the positron fraction $P(E)$ as a function of the energy E of electrons and positrons and for vales of $r_U = 0, 1/2, 1, 2, 4$, from top to bottom. Secondaries are estimated according to the expressions in [6, 7] reported in the main text. The shaded regions account for the range of the allowed N_B values (see Fig. 1), obtained in turn by considering the 2010 PAMELA and FERMI-LAT data at 1σ. We also display the 2010 (red) and 2008 (black) PAMELA data, with 1σ error bars.

turn, means that one expects the electron fraction to be neither be small nor too large with respect to the positron fraction.

As the experimental data on the positron fraction should hopefully soon become more accurate [12], the method outlined above will allow to extract crucial information about the charge asymmetry of the unknown component of the electrons and positrons in CRs.

4. Conclusions

The model independent analysis of the combined PAMELA and FERMI-LAT data proposed in ref. [9] shows that current data still allow for approximately equal contributions of the electrons and positrons from the unknown components of the associated CRs, but disfavor electron to positron fractions much smaller than 1/2 or larger than 4. Hence, the typical oversimplifying model assumption of charge symmetry ($r_U = 1$) used so far, actually constitutes just a small portion of the allowed models still left unconstrained by the present data.

Interestingly enough, we emphasize that the planned improvements in the experimental sensitivity to the positron fraction, as well as further data at energies higher than the one explored so far by PAMELA, could reveal whether the unknown component of the CR electrons and positrons is charge asymmetric or not. If it turned out to be charge asymmetric, an explanation in terms of new physics beyond the Standard Model would clearly become more compelling with respect to an astrophysical one.

Acknowledgments
We thank F.Palmonari and R.Sparvoli for useful comments and suggestions.

References
[1] Adriani O *et al.* 2009 *Nature* **458** 607 (*Preprint* arXiv:0810.4995 [astro-ph]). Adriani O *et al.* 2010 *Astropart. Phys.* **34** 1-11 (*Preprint* arXiv:1001.3522 [astro-ph.HE]). Adriani O *et al.* 2010 *Phys. Rev. Lett.* **105** 121101 (*Preprint* arXiv:1007.0821 [astro-ph.HE]).
[2] Chang J *et al.* 2008 *Nature* **456** 362.
[3] Yoshida K *et al.* 2008 *Adv. Space Res.* **42** 1670. Torii S *et al. Preprint* arXiv:0809.0760 [astro-ph].
[4] Abdo A *et al.* 2009 *Phys. Rev. Lett.* **102** 181101 (*Preprint* arXiv:0905.0025 [astro-ph.HE]).
[5] Aharonian F *et al.* 2008 *Phys. Rev. Lett.* **101** 261104 (*Preprint* arXiv:0811.3894 [astro-ph]). Aharonian F *et al.* 2009 *Astron. Astrophys.* **508** 561 (*Preprint* arXiv:0905.0105 [astro-ph.HE]).
[6] Strong A W and Moskalenko I V 1998 *Astrophys. J.* **509** 212 (*Preprint* arXiv:astro-ph/9807150).
[7] Baltz E A and Edsjo E 1998 *Phys. Rev.* D **59** 023511 (*Preprint* arXiv:astro-ph/9808243).
[8] Fan Y Z Zhang B and Chang J 2010 *Int. J. Mod. Phys.* **D19** 2011-2058 (*Preprint* arXiv:1008.4646 [astro-ph.HE]).
[9] Frandsen M T Masina I and Sannino F *Preprint* arXiv:1011.0013 [hep-ph].
[10] Grasso D *et al.* 2009 *Astropart. Phys.* **32** 140 (*Preprint* arXiv:0905.0636 [astro-ph.HE]).
[11] Ibarra A Tran D and Weniger C 2010 *J. High Energy Phys.* JCAP01(2010)009 (*Preprint* arXiv:0906.1571 [hep-ph]).
[12] The AMS-02 webpage: http://ams.cern.ch/

Journal of Physics: Conference Series **335** (2011) 012066 doi:10.1088/1742-6596/335/1/012066

Search for Dark Matter in Cosmic Rays with the AMS-02 space spectrometer

F Palmonari, V Bindi, A Contin, N Masi and L Quadrani
On behalf of the AMS International Collaboration
INFN & University of Bologna, Italy
E-mail: palmonari@bo.infn.it

Abstract.
The cosmic rays spectrometer AMS-02 will be placed on the ISS Space Station in April 2011 and should take data for at least 10 years. After a short review of the impressive observational data collected in the last few years by satellites and underground experiments in the quest for Dark Matter (DM), we describe the AMS-02 space spectrometer with its powerful particle identification capabilities. The performances of the experimental apparatus and the perspectives opened by this new powerful observational instrument for the DM search will be briefly reviewed. In particular the extension at the TeV scale of the positron and electron spectra with high statistical significance well beyond the PAMELA results will hopefully clarify the quest for Dark Matter evidence in cosmic rays. Finally the present status of the AMS-02 payload for the Shuttle transportation and installation on the International Space Station , and the implementation of the Control Center for data collection and analysis is reviewed.

1 Introduction

The astrophysical growing observational evidence for Dark Matter (DM) gives a solid base to the Cosmological Standard Model. The particles and interactions Standard Model, however, has no answer on DM nature. Experimental evidence of DM from underground experiments is under scrutiny while recent results from space experiments have no clear interpretation.

The last NASA Shuttle Mission STS-134, scheduled in 2011, will bring and install on the International Space Station (ISS) the Cosmic Ray Spectrometer AMS-02, a powerful upgrade of AMS-01 which was flown for 10 days in June 1998.

The ISS scientific program has been extended up to at least 2020, so that the AMS-02 spectrometer will be a permanent Cosmic Ray observatory looking for DM, AntiMatter, primary Cosmic Ray Nuclei and gamma rays.

2 Review of recent results of DM searches

The claim for a WIMP Cold DM particle interaction signal comes from the DAMA/LIBRA experiment at Gran Sasso [1]. After 7 years of observations they find an annual modulation effect of the inelastic DM scattering in the scintillator, in the 2 – 6 keV energy range of amplitude 0.0114 ± 0.0016 counts/day/kg/keV . The DAMA/LIBRA result, a model independent piece of evidence, has not yet been confirmed by other experiments of the same kind, like CDMS, whose 2 candidate events (with an expected background of 0.9 ± 0.2) have been up to now interpreted as zero signal [2]. In conclusion, concerning the inelastic WIMP collisions in matter we must wait for a confirmation or final rejection of the DAMA effect.

Interesting data have shown up recently from satellite experiments detecting cosmic rays, like PAMELA, or gamma rays, like FERMI. Both experiments can detect the annihilation products of Dark Matter candidate particles, like cold or hot relics in Supersymmetric theories. PAMELA data confirm with high statistics and extend to above 100 GeV the earlier findings of a spectacular rise in the positron fraction from 0.05 around 10 GeV to 0.15 at 100 GeV [3]. FERMI and HESS [4] observe the same rise in the $e^+ + e^-$ cumulative spectra with respect to the expected power law E^{-3} decrease.

Finally experimental evidence has been claimed [5], analysing the FERMI gamma ray spectra from a small region of our galaxy center, for a component not seen in the emission spectrum of point gamma sources. The mean energy of the anomalous gamma rays is of the order of few GeV. A possible interpretation put forward is the annihilation, in the galaxy nucleus where the DM density should be much higher than in the galaxy arms, of about 8 GeV/c^2 mass DM particles in tau leptons. However the FERMI collaboration has analysed the gamma ray spectra in terms of possible gamma peaks and no evidence of spectral lines has been seen in the present data [6].

To conclude this short review, thanks to its large acceptance and unprecedented particle identification capability AMS-02 will be able to answer to the present day questions both on positron-electron spectrum anomalies and the gamma observations from the galactic centre. In our hopes in 10 years of observations on board of the ISS AMS-02 will contribute to cosmic ray physics like the Hubble telescope is contributing to observational astrophysics.

3 The AMS-02 Cosmic Rays spectrometer on the ISS

The success of the AMS-01 10 days mission in 1998 [7], gave a strong boost to the approval of the AMS-02 design. The permanent magnet was replaced by a 5 times stronger magnetic field generated by an array of superconducting coils working in a liquid He bath around 1.8 K to keep the He superfluid with an high thermal conductivity.

The AMS-02 project was financed and started in year 2001, the most challenging project being the superconducting magnet(SM), which should have survived the Shuttle flight and the severe thermal conditions in space. Despite technical and financial difficulties the SM was finally completed in September 2009 and tested extensively. The design was for a 3 years continuous operation with 2500 l of liquid He. The final test performed in the ESA space simulator chamber at ESTEC, Holland, showed that the operational time of the SM was around 28 months.

3.1 *The spectrometer magnet*

A superconducting magnet was ideal for a three year stay on ISS, from 2005 to 2008, as originally planned for AMS. After the 2003 Shuttle accident the AMS program was slowing down while the ISS operation was scheduled to end in 2010. Last year the ISS scientific program has been strengthened and its lifetime has been extended to 2020 (2028). The Shuttle flights, however will go to end in 2011, thus eliminating any possibility of returning and refilling the SM.

A superconducting magnet was therefore no longer the ideal choice. Most importantly, the permanent magnet option will have 10(18) years time to collect data, providing much more sensitivity to search for new phenomena. The replacement of the SM with the old AMS-01

Journal of Physics: Conference Series **335** (2011) 012066 doi:10.1088/1742-6596/335/1/012066

permanent magnet (PM) was performed in less than 5 months, thanks to the fact that the SM had been designed with the same aperture and the same tracker planes as those equipping the PM.

Another important challenge for the experiment was to loose not too much analyzing power at high particles momenta with a 5 times lower magnetic field. This was achieved, at least for high rigidities, with a modified set of tracker planes, exploiting a characteristic property of a magnetic spectrometer. In fact the analyzing power of a dipole field, i.e. the momentum resolution (Δp/p) is the sum of two contributions:

1. Measurement inside the magnet with an effective length L

$$(Z/p)\cdot(\Delta p/p) \propto 1/BL^2$$

2. Measurement of the incident (θ1) and exit (θ2) angles which depend on the lever arm L1

$$(Z/p)\cdot(\Delta p/p) \propto 1/BLL1$$

where L1 is the distance first - second, and second last - last planes of the tracker. For both magnets, L ~ 80 cm, but in the permanent magnet B is 5 times smaller. To maintain the same Δp/p the lever arm L1 has been increased from ~15 cm (SM) to ~125 cm (PM).

In figure 1 the comparison of the resolution in rigidity for the SM and the PM is shown. The two lines refer to the calculated rigidity resolution, whereas the points show the results obtained in the 400 GeV proton test beam at CERN for the two configurations of the spectrometer. As one can see, above 1 TV the difference in the resolution is of the order of 12% and is decreasing as the rigidity becomes higher. This is of particular importance for the measurements of electron and positron spectra at high momenta.

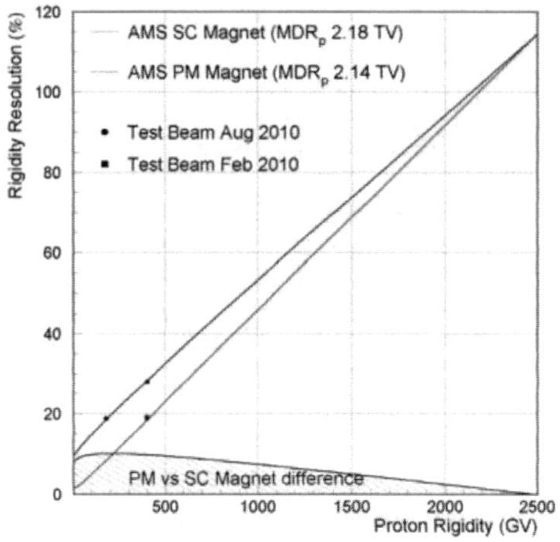

Figure 1 - Comparison of the rigidity % resolution of SM and PM.

3.2 *Particle identification*

The cut-out view of the AMS-02 spectrometer is shown in figure 2, where the different subdetectors are labeled. A cosmic ray particle traversing the spectrometer downward encounters layer 1 of the silicon strip tracker detector (TRACKER), a straw tubes gas transition detector (TRD), the upper two scintillator hodoscope planes of the time of flight detector (UTOF), then inside the cylindrical dipole permanent magnet (PM) the 7 central layers of the TRACKER and, covering the internal magnet surface, the scintillation counters of the cylindrical anticoincidence detector (ACC).

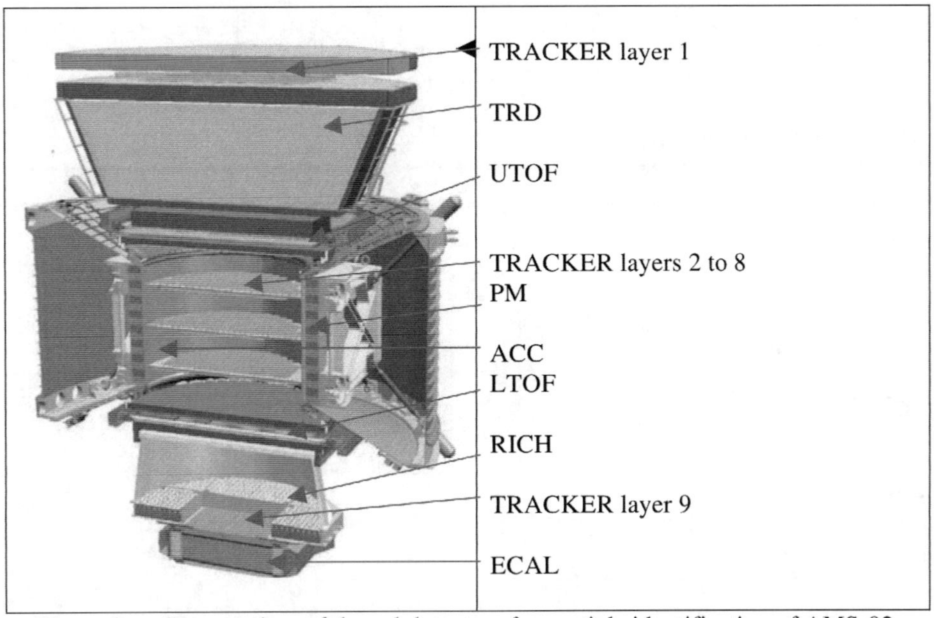

Figure 2 - Cut-out view of the subdetectors for particle identification of AMS-02.

Below the magnet there are the lower two planes (LTOF) of the TOF detector, the ring imaging Cerenkov detector (RICH), the last layer 9 of the TRACKER and the Pb-scintillating fibre electromagnetic calorimeter (ECAL).

3.2.1. TRACKER **.** The TRACKER detector is made of 9 planes of Silicon crystals arranged in arrays of parallel lecture strips, which in the collector side are orthogonal to the bending plane and on the other side inclined at 45°. Figure 2 shows the actual disposition of the planes. The residuals of a track in the curvature plane of the magnet are of the order of 10 micrometers (see figure 3 below).

3.2.2 TRD . The TRD detector is made of 20 layers of radiator and straw tubes filled with a Xe/Co2 mixture. The TRD is placed at the entrance of the spectrometer and is of great importance in distinguishing positrons from protons up to around 300 GeV and together with the electromagnetic calorimeter it will ensure the separation power between leptons and protons needed for the DM search. The TRD alone at 290 GeV/c beam momentum can detect electrons with a 90% efficiency and reject protons at the level of better than a factor 100 (figure 3).

3.2.3. TOF. The two scintillator [8] hodoscopes UTOF and LTOF, consisting each of two planes of 12 cm wide paddles, respectively in the x and y orthogonal directions, define the spectrometer acceptance for charged particles and provide the fast trigger to start the event data acquisition of all detectors. The scintillator counters, seen at both ends by 2 photomultipliers, can measure on a flight path of 120 cm the particle velocity with a resolution of 4%. The time of flight gives a 10^{10} rejection for upgoing particles excluding to this level the confusion of e.g. an He nucleus going upward with an AntiHelium going down from space. In figure 4 the charge resolution and the time resolution as function of charge are shown.

3.2.4. RICH. Before the calorimeter there is a ring imaging Cerenkov counter consisting of a 2 cm thick Aerogel radiator, a 45 cm expansion space for the ring with a conical mirror and a planar array of PMTs forming a 9 x 9 mm pixel grid. The velocity resolution (figure 3) is of the order of 0.1% , ensuring a good separation of isotopes in the study of the light nuclei.

3.2.5. ECAL **.** The main detector for an efficient separation of electrons/positrons from protons is the 17 X_0 Pb-scintillator electromagnetic calorimeter made of 9 superlayers of

Journal of Physics: Conference Series **335** (2011) 012066 doi:10.1088/1742-6596/335/1/012066

orthogonal fibres. It allows excellent shower shape reconstruction and has an energy resolution of the order of 4 % (figure 3) . The calorimeter rejection factor for protons is standalone of the order of 10^4 . Together with the TRD the proton rejection amounts to a factor 10^6 .

4 Overall performances of AMS-02

In figure 3, 4 the overall performances of the subdetectors relevant for particle identification, as measured on test beams of protons, selected secondary particle momenta and charge, heavy Ions, are shown.

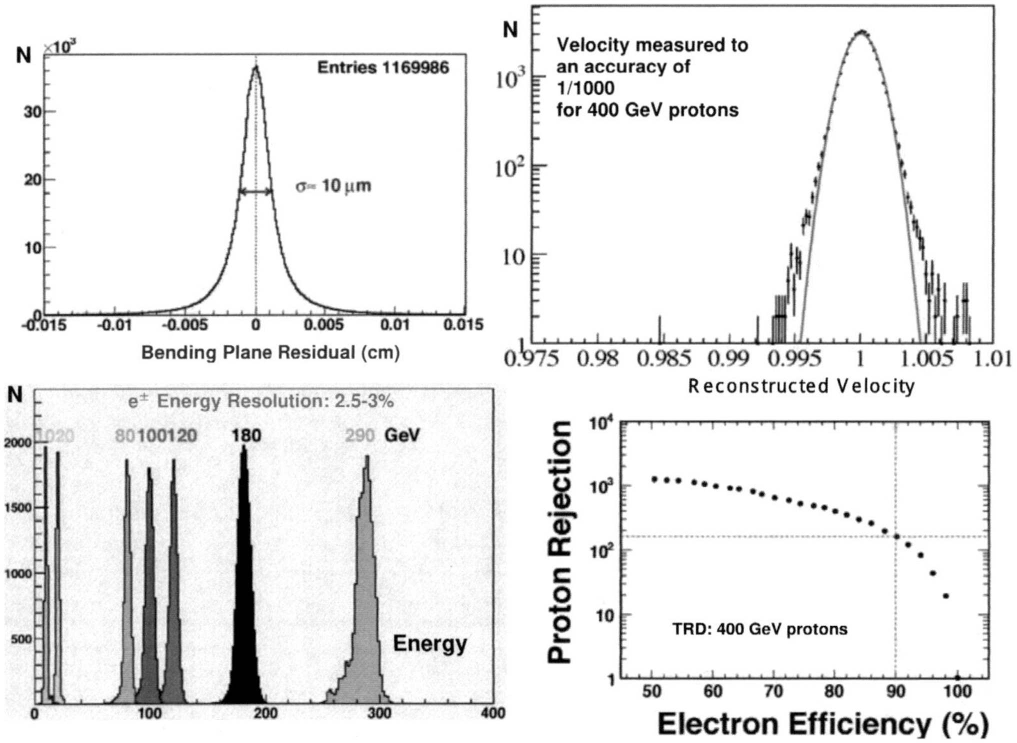

Figure 3 - Test Beam Results with permanent magnet – 8-20 Aug 2010. From left clockwise: The measured residuals of the TRACKER fitted tracks in the curvature plane of the magnet. The measured RICH velocity resolution, about 0,1 %. The measured TRD separation between electrons and protons. The measured ECAL energy spectra of electrons from 10 to 290 GeV.

The performances of the AMS-02 spectrometer are also monitored periodically running a cosmic ray trigger on ground since the first assembly. All detectors are found to behave according to the design characteristics. No degradation of the detectors has been shown from CERN, where it was assembled first, to ESTEC in Holland where underwent the space qualification tests, then at CERN to change from the SM to the PM, and finally to the NASA laboratory KSC in Florida, where is undergoing integration in the Shuttle and the ISS.

Finally we must mention that AMS-02 with the unique features of its electromagnetic calorimeter, a thickness of 17 X_0, and granularity permitting 3D reconstruction of showers, will be also an excellent gamma ray spectrometer. ECAL will measure γ to 1 TeV, with an angular resolution $\sigma(\theta)$ of 2 arc-sec.

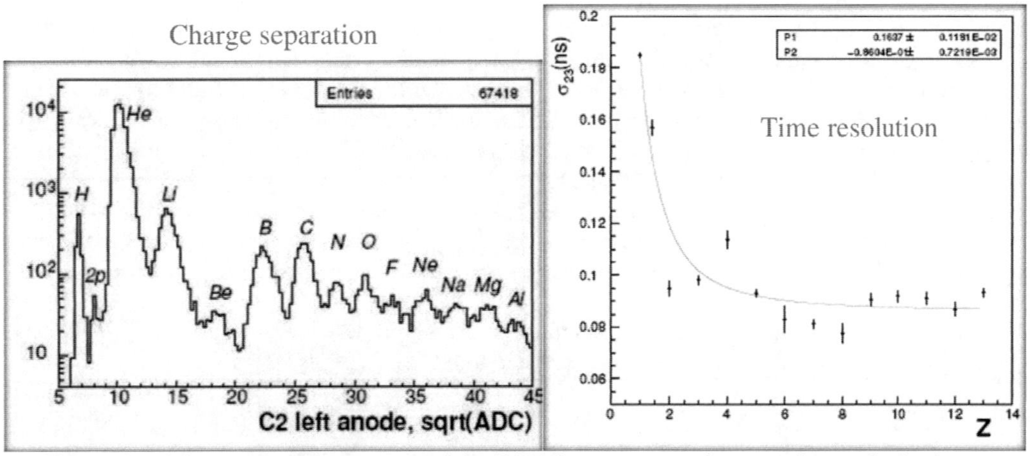

Figure 4 - Measured performances of the TOF detector on an Ions test beam. Left: charge resolution of beam Ions. Right: time of flight resolution as function of Ion charge.

5 AMS-02 and the search for DM

Given the excellent particles identification power and the large acceptance of AMS-02, we expect to improve in few months the present day measurements of electrons, positrons and antiprotons, in the quest for DM signals.

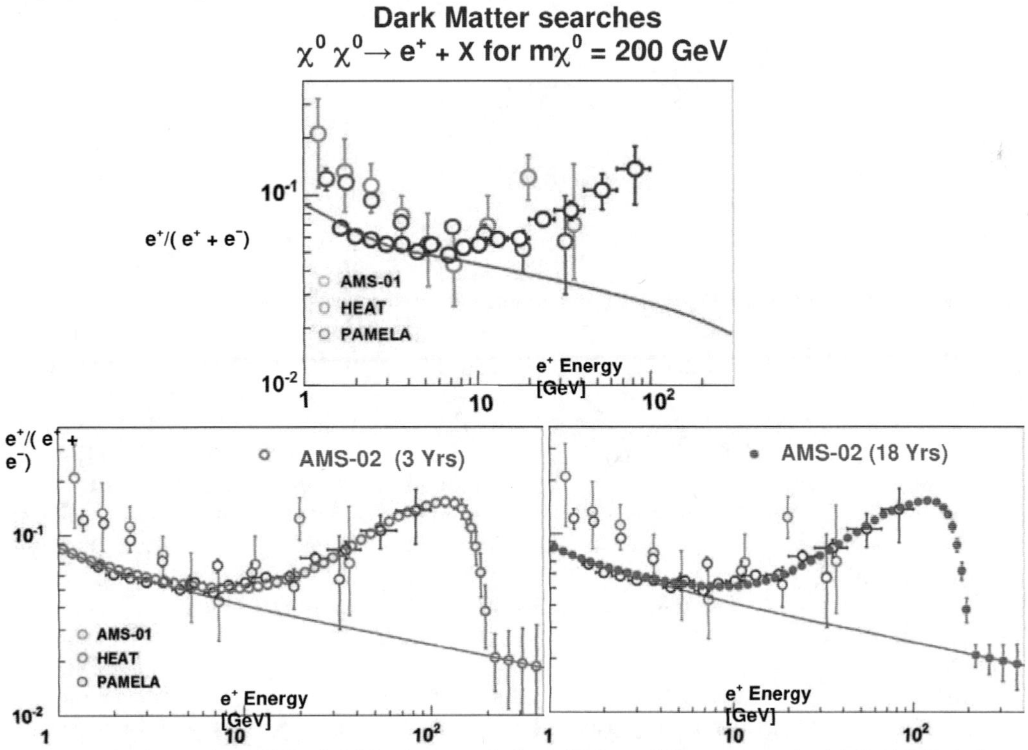

Figure 5 - Potential of AMS-02 in the search for a DM particle compared with PAMELA data.

The excellent rejection at the 10^{-6} level of protons ensures a clean reconstruction of the electron and positron fluxes up to 300 GeV and above. As examples of that figure 5 shows how the anomalous rise in the $e^+ / (e^+ + e^-)$ ratio would appear in the AMS-02 data, if due to the annihilation of supersymmetric WIMP neutralino DM particles with a mass of 200 GeV.

As another example, Kaluza-Klein Bosons are also DM candidates. In fig. 6 the expected signal in the positron fraction for a TeV Scale Singlet DM candidate is shown. The measured positron fraction rise and fall requires in this case a much longer observation time, to have a statistically significant signal at such high energy.

The lifetime of the ISS in fact could be extended well behind year 2020 and the PM AMS-02 spectrometer has been proved to be capable of taking data for such a long time. The gas system of the TRD straw tubes, for example, has been tested for gas leakages and proven to have a potential lifetime of 30 years.

In any case in less than one year, due to an order of magnitude wider acceptance of AMS-02, the PAMELA results will be recovered and extended to higher energies.

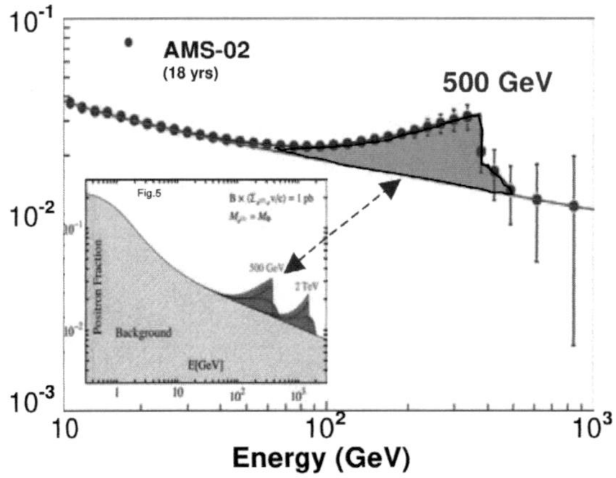

Figura 6 - AMS-02 potential for the search of a 500 GeV/c^2 mass Kaluza-Klein type DM.

6 The preparation of the AMS Payload and science data flow

Figure 7 shows various stages of the AMS-02 spectrometer preparation before launch. The Payload Operations Control Center (POCC), which will monitor the spectrometer and the data taking of AMS-02 before and after the installation on the ISS is still at KSC, Florida, where it is directly connected to the Payload (figure 7d). At the same time a backup POCC has been established at JSC, Houston Tx, to test communications and prepare the final POCC at CERN.

Once installed and completely tested on the ISS in fact, from 15 June 2011, the AMS-02 data, received at ground at the Marshall laboratory in Alabama via the NASA transmission channels, will be transmitted directly via Internet to the CERN POCC. All operations of monitoring, run control and data storage will be then made there. The expected dataflow has already been tested. From JSC to CERN the network bandwidth is 34Mbit/s (AMS requires 10 Mbit/s).

7 Conclusion

In summer 2011 CERN will host the AMS-02 Control Center. An international collaboration of 60 scientific institutes and 500 physicists will have for the next decade the most advanced instrument in space to study CR, from protons to heavy nuclei, from antiparticles to γ rays.

The impact of AMS-02 for astroparticle physics could be comparable to the impact of the Hubble telescope to astronomy. The most exciting objective of AMS is to probe the unknown; to search in nature for phenomena that we have not yet imagined nor had the tools to discover.

The origin and propagation of CR, the AntiMatter mystery, the nature of Dark Matter are the astroparticle physics research themes attached by AMS-02 in the same place, the CERN laboratories, where the LHC experiments try to unveil the cosmological mysteries from the microscopic point of observation, looking beyond the Standard Model.

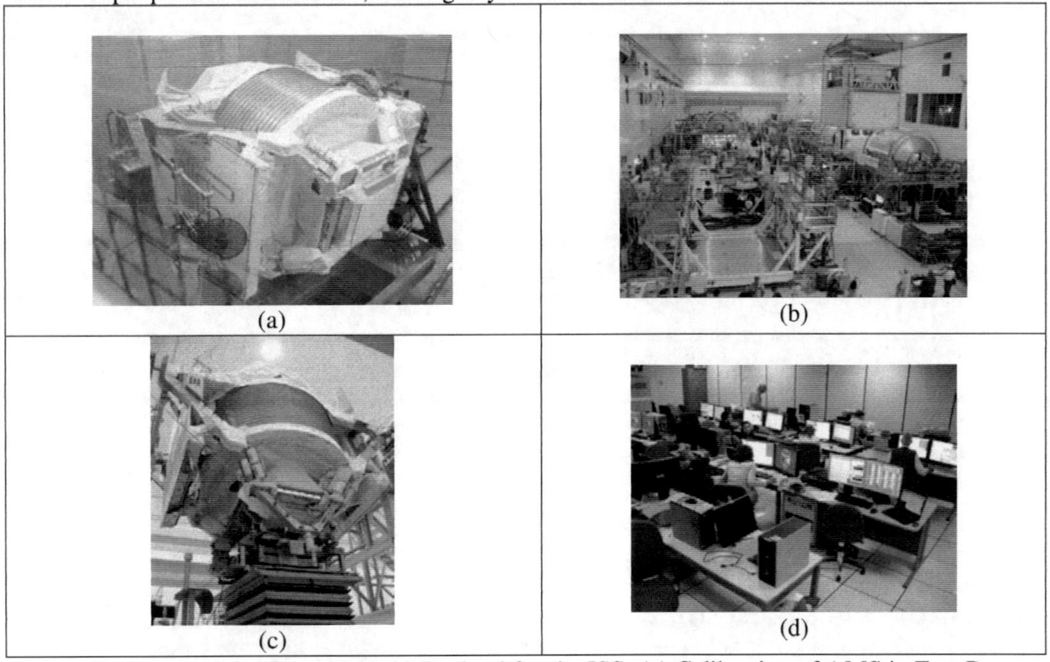

Figure 7 - Preparation of the AMS-02 Payload for the ISS: (a) Calibration of AMS in Test Beam with permanent magnet – 8-20 Aug 2010. (b) AMS being installed in Space Station Processing Facility – 26 Aug 2010. (c) AMS mated with a Payload Attach System simulator - 22 October 2010. (d) Since 28 August, AMS has been undergoing tests at the KSC control POCC.

A very exciting decade of experimental results is at our hands and we look forward for discoveries opening new windows in the understanding of our Universe.

References

[1] DAMA Collab., R. Bernabei *et al.*, Eur. Phys. J. C Phys. Lett. **56**, (2008) 333-355.
[2] CDMS Collab., Z. Ahmed *et al.*, Phys. Rev. Lett. **102**, 011301 (2009).
[3] PAMELA Collab., O. Adriani *et al.*, Nature **458**, 607 (2009).
[4] FERMI/LAT Collab., A.A.Abdo *et al.*, Phys. Rev. Lett. **102**, 181101 (2009).
HESS Collab., F. Aharonian *et al.*, Eur. Phys. Rev. Lett. **101**, 261104 (2008).
[5] Dan Hooper and Lisa Goodenough, FERMILAB-PUB-10-414-A,
[6] FERMI/LAT Collab., A.A. Abdo *et al.*, Phys. Rev. Lett. **104**, 091302 (2010).
[7] AMS Collab., M. Aguilar *et al.*, Phys. Rep. **366**, (2002) 331-405.
[8] V. Bindi *et al.*, NIM A. **623**, (2010) 968.

Acknowledgements

Many thanks to the DISCRETE 2010 organizers for inviting. I would like to acknowledge all colleagues, students and PhD of the INFN Bologna group, who in the last 10 years have accomplished the TOF detector, with the invaluable help of all the INFN technical staff, especially C. Guandalini, M. Lolli, F. Massera and R. Pilastrini.

Journal of Physics: Conference Series **335** (2011) 012067 doi:10.1088/1742-6596/335/1/012067

U boson searches at KLOE

Simona Giovannella

Laboratori Nazionali di Frascati dell'INFN, via Enrico Fermi 40, I-00044 Frascati (Rome) Italy

E-mail: simona.giovannella@lnf.infn.it

on behalf of the KLOE-2 Collaboration

Abstract. The existence of a secluded gauge sector could explain several puzzling astrophysical observations. This hypothesis can be tested at low energy e^+e^- colliders such as DAΦNE. Preliminary results obtained with KLOE data and perpectives for the KLOE-2 run, where a larger data sample is expected, are discussed.

1. Introduction

In recent years, several astrophysical observations have failed to find easy interpretations in terms of standard astrophysical and/or particle physics sources. A non exhaustive list of these observations includes the 511 keV gamma-ray signal from the galactic center observed by the INTEGRAL satellite [1], the excess in the cosmic ray positrons reported by PAMELA [2], the total electron and positron flux measured by ATIC [3], Fermi [4], and HESS [5, 6], and the annual modulation of the DAMA/LIBRA signal [7, 8].

An intriguing feature of these observations is that they suggest the existence of a WIMP dark matter particle belonging to a secluded gauge sector under which the Standard Model (SM) particles are uncharged [9, 10, 11, 12, 13, 14, 15, 16, 17, 18]. An abelian gauge field, the U boson with mass near the GeV scale, couples the secluded sector to the SM through its kinetic mixing with the SM hypercharge gauge field. The kinetic mixing parameter, ϵ, can naturally be of the order 10^{-4}–10^{-2}. In a very minimal scenario, in addition to the U, it is natural to have a secluded Higgs boson, the h', which spontaneously breaks the secluded gauge symmetry. A consequence of the above hypotheses is that observable effects can be induced in $\mathcal{O}$(GeV)–energy e^+e^- colliders [19, 20, 21, 22, 23, 24] and fixed target experiments [25, 26, 27, 28].

2. Searches for dark forces at KLOE

The KLOE experiment operates at DAΦNE, the e^+e^- Frascati ϕ-factory. From 2000 to 2006, KLOE collected 2.5 fb^{-1} of collisions at the ϕ meson peak and about 240 pb^{-1} below the ϕ resonance ($\sqrt{s} = 1$ GeV). The ϕ meson predominantly decays into charged and neutral kaons, thus allowing KLOE to make precision studies in the fields of flavor physics, low energy QCD and test of discrete symmetries [29].

A new beam crossing scheme allowing a reduced beam size and increased luminosity is operating at DAΦNE [30]. The KLOE-2 detector was successfully installed in this new interaction region and has been upgraded with small angle tagging devices to detect both high and low energy electrons or positrons in $e^+e^- \to e^+e^-X$ events. About 5 fb^{-1} are expected in the first year of running. An inner tracker and small angle calorimeters are scheduled to be installed in a subsequent step, providing larger acceptance both for charged particles and photons. A detailed description of the KLOE-2 physics program can be found in Ref. [31].

The KLOE detector consists of a large cylindrical drift chamber, DC, surrounded by a lead-scintillating fiber electromagnetic calorimeter, EMC. A superconducting coil around the EMC provides a 0.52 T magnetic field. The all-stereo drift chamber [32], 4 m in diameter and 3.3 m long, is made of carbon fiber-epoxy composite and operates with a light gas mixture (90% helium, 10% isobutane). The position resolutions are $\sigma_{xy} \sim 150\ \mu$m and $\sigma_z \sim 2$ mm. The momentum resolution is $\sigma(p_\perp)/p_\perp \approx 0.4\%$. Vertexes are reconstructed with a spatial resolution of ~ 3 mm. The calorimeter [33] is divided into a barrel and two endcaps and covers 98% of the solid angle. The modules are read out at both ends by photomultipliers with a readout granularity of $\sim (4.4 \times 4.4)$ cm^2. The arrival times of particles and the positions in three dimensions of the energy deposits are obtained from the signals collected at the two ends. Cells close in time and space are grouped into a calorimeter cluster. The cluster energy E is the sum of the cell energies. The cluster time T and position $\vec{R}$ are energy weighted averages. Energy and time resolutions are $\sigma_E/E = 5.7\%/\sqrt{E\ (\text{GeV})}$ and $\sigma_t = 57\ \text{ps}/\sqrt{E\ (\text{GeV})} \oplus 50$ ps, respectively.

The U boson can be produced at DAΦNE through radiative decays of neutral mesons, such as $\phi \to \eta U$. With the statistics already collected at KLOE, this decay can potentially probe couplings down to $\epsilon \sim 10^{-3}$ [21], covering most of the parameter's range of interest for the theory. The U boson can be observed by its decay into a lepton pair, while the η can be tagged by one of its not-rare decays.

Assuming also the existence of the h', both the U and the h' can be produced at DAΦNE if their masses are smaller than M_ϕ. The mass of the U and h' are both free parameters, and the possible decay channels can be very different depending on which particle is heavier. In both cases, an interesting production channel is the h'-strahlung, $e^+e^- \to Uh'$ [19]. Assuming the h' to be lighter than the U boson, it turns out to be very long-lived, so that the signature process will be a lepton pair, generated by the U boson decay, plus missing energy. In the case $m_{h'} > m_U$, the dark higgs frequently decays to a pair of real or virtual U's. In this case one can observe events with 6 leptons in the final state, due to the h'-strahlung process, or 4 leptons and a photon, due to the $e^+e^- \to h'\gamma$ reaction.

Another possible channel to look for the existence of the U boson is the $e^+e^- \to U\gamma$ process [19]. The expected cross section can be as high as $\mathcal{O}$(pb) at DAΦNE energies. The on-shell boson can decay into a lepton pair, giving rise to a $l^+l^-\gamma$ signal.

In the following, progresses on the analyes of $\phi \to \eta U$ and $e^+e^- \to Uh'$ channels are reported, together with perspectives for the new KLOE-2 run.

3. The $\phi \to \eta U$ decay

As discussed above, the search of the U boson can be performed at KLOE using the decay chain $\phi \to \eta U$, $U \to l^+l^-$. An irreducible background due to the Dalitz decay of the ϕ meson, $\phi \to \eta l^+l^-$, is present. This decay has been studied by SND and CMD-2 experiments, which measured a branching fraction of $BR(\phi \to \eta e^+e^-) = (1.19 \pm 0.19 \pm 0.07) \times 10^{-4}$ and $BR(\phi \to \eta e^+e^-) = (1.14 \pm 0.10 \pm 0.06) \times 10^{-4}$ respectively [34, 35]. This corresponds to a cross section of $\sigma(\phi \to \eta l^+l^-) \sim 0.7$ nb, while for the signal

$$\sigma(\phi \to \eta U) = \epsilon^2\, |F_{\phi\eta}(m_U^2)|^2\, \frac{\lambda^{3/2}(m_\phi^2, m_\eta^2, m_U^2)}{\lambda^{3/2}(m_\phi^2, m_\eta^2, 0)}\, \sigma(\phi \to \eta\gamma) \sim 40\ \text{fb} \quad (1)$$

using $\epsilon = 10^{-3}$, $|F_{\phi\eta}(m_U^2)|^2 = 1$. In Eq. (1), $|F_{\phi\eta}(m_U^2)|^2$ is the form factor of the $\phi \to \eta\gamma$ decay evaluated at the U mass while the following term represent the ratio of the kinematic functions

Journal of Physics: Conference Series **335** (2011) 012067 doi:10.1088/1742-6596/335/1/012067

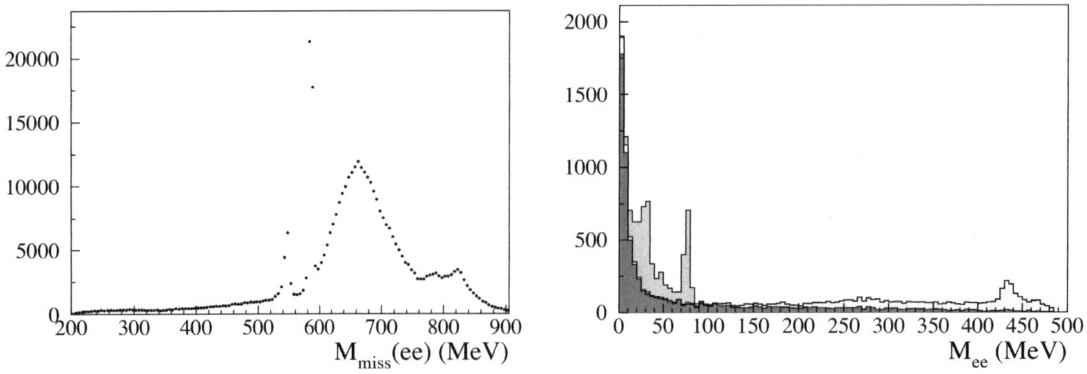

Figure 1. Left: missing mass of the e^+e^- pair for data sample after preselection cuts. The $\phi \to \eta e^+e^-$ signal is clearly visible in the peak corresponding to η mass. The second peak at ~ 590 MeV is due to $K_S \to \pi^+\pi^-$ events with wrong mass assignment. Right: M_{ee} distribution for data at different analysis steps: preselection (green), conversion (yellow) and ToF (purple) cuts.

of the involved decays. Despite the values of the cross section, the different shapes of the e^+e^- invariant mass for the two processes allow to test the ϵ parameter down to 10^{-3} with the whole KLOE data set [21].

The best channel to search the for the $\phi \to \eta U$ process at KLOE is the $U \to e^+e^-$ decay for two reasons: (i) a wider range of U boson mass can be tested; (ii) $e^\pm$ are easily identified using the time-of-flight (ToF) measurement. The η can be tagged by the three-pion or two-photon final state, which represent $\sim 85\%$ of the total decay rate. We have performed a preliminary analysis using the $\eta \to \pi^+\pi^-\pi^0$ channel, which provide a clean signal with four charged tracks and two photon in the final state. Studies are under way also for the $\eta \to \gamma\gamma$ sample.

3.1. The $\eta \to \pi^+\pi^-\pi^0$ final state

The preliminary analysis of the $\eta \to \pi^+\pi^-\pi^0$ final state has been performed on 739 pb^{-1}, corresponding to 40% of the total sample. Preselection cuts require: (i) four tracks in a cylinder around the interaction point (IP) plus two photon candidates; (ii) best $\pi^+\pi^-\gamma\gamma$ match to the η mass using the pion hypothesis for tracks; (iii) other two tracks assigned to e^+e^-; (iv) loose cuts on η and π^0 invariant masses. These simple cuts allow to clearly see the peak due to $\phi \to \eta e^+e^-$ events in the distribution of the missing mass to the e^+e^- pair, $M_{\rm miss}(ee)$ (see Fig. 1-left). A cut $535 < M_{\rm miss}(ee) < 560$ MeV is then applied.

A residual background contamination, due to $\phi \to \eta\gamma$ events with photon conversion on beam pipe (BP) or drift chamber walls (DCW), is rejected by tracking back to BP/DCW surfaces the two e^+, e^- candidates and then reconstructing the electron-positron invariant mass (M_{ee}) and the distance between the two particles (D_{ee}). Both quantities are small if coming from photon conversion. $\phi \to K\bar{K}$ and $\phi \to \pi^+\pi^-\pi^0$ events surviving analysis cuts have more than two pions in the final state. They are rejected using time-of-flight to the calorimeter. When an EMC cluster is connected to a track, the arrival time to the calorimeter is evaluated both with calorimeter ($T_{\rm cluster}$) and drift chamber ($T_{\rm track}$) information. Events with an e^+, e^- candidate outside a $3\,\sigma$'s window on the $DT = T_{\rm track} - T_{\rm cluster}$ variables are rejected. In Fig. 1-right the M_{ee} distribution evaluated at IP for data at different steps of the analysis is shown. The conversion and ToF cuts remove events at low and high invariant mass values respectively.

In Fig. 2 the comparison between data and Monte Carlo (MC) events for M_{ee} and $\cos\psi_{ee}$ variables is shown. The second variable is the three-dimensional opening angle of the e^+e^- pair.

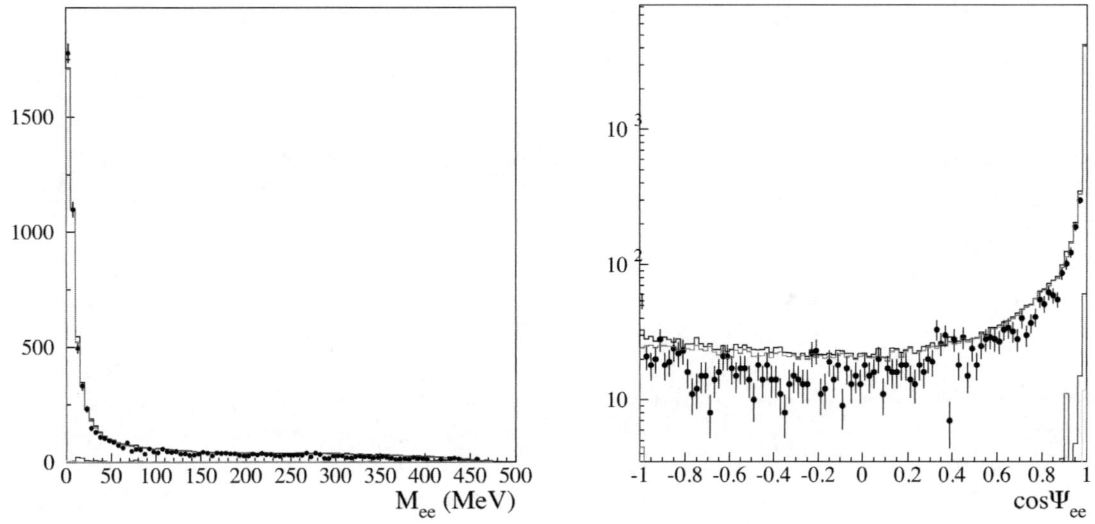

Figure 2. Invariant mass (left) and opening angle (right) of the e^+e^- pair for $\phi \to \eta e^+e^-$, $\eta \to \pi^+\pi^-\pi^0$ events. Dots are data, the red line represents the expected MC shape for signal while the residual background contamination from ϕ decays is shown in blue. The black solid line is the sum of all MC contributions.

About 7000 $\phi \to \eta e^+e^-$, $\eta \to \pi^+\pi^-\pi^0$ candidates are present in the analysed data set, with a negligible residual background contamination. The MC description of these kind of events is done using Vector Meson Dominance, and the form factor slope parameter is taken from the measurement of SND experiment, obtained with 213 events [34]. Since an accurate description of the irreducible background is important for the search of the $\phi \to \eta U$ signal, we extract it directly from our data. The decay parametrization is taken from Ref. [36], with form factor approximation according to [37]:

$$\frac{d}{dq^2}\frac{\Gamma(\phi \to \eta e^+e^-)}{\Gamma(\phi \to \eta\gamma)} = \frac{\alpha}{3\pi}\frac{|F_{\phi\eta}(q^2)|^2}{q^2}\sqrt{1-\frac{4m^2}{q^2}}\left(1+\frac{2m^2}{q^2}\right)\left[\left(1+\frac{q^2}{m_\phi^2-m_\eta^2}\right)^2 - \frac{4m_\phi^2 q^2}{(m_\phi^2-m_\eta^2)^2}\right]^{3/2} \quad (2)$$

$$F(q^2_{\phi\eta}) = \frac{1}{1-q^2/\Lambda^2_{\phi\eta}} \quad (3)$$

The preliminary fit to the M_{ee} shape, reported in Fig. 3, has $\chi^2/\mathrm{ndf} = 1.4\%$. The resulting accuracy on form factor slope, $b = \Lambda^{-2}_{\phi\eta}$, is 4.6%.

The $\phi \to \eta U$ MC signal has been produced according to Ref. [21], with a flat distribution of the e^+e^- invariant mass. Events are then divided in sub-samples of 1 MeV width. For each M_{ee} value, signal hypothesis has been excluded at 95% C.L. using two different methods, CLS technique [38] and Bayesian approach [39], obtaining consistent results. The expected shape for the irreducible background $\phi \to \eta e^+e^-$ is taken from our fit to the M_{ee} distribution. In Fig. 4-top the preliminary exclusion plot on number of events obtained with CLS technique is shown. In the region $25 < M_{ee} < 425$ MeV, an upper limit of $\sim$ 20 events is obtained. Systematics are not yet included. The exclusion plot for ϵ parameter is obtained using Eq. (1), taking into account analysis efficiency and including the opening of the $U \to \mu^+\mu^-$ threshold. Results are reported in Fig. 4-bottom. The mean value in the $25 < M_{ee} < 425$ MeV region is $\epsilon \sim 3 \times 10^{-3}$.

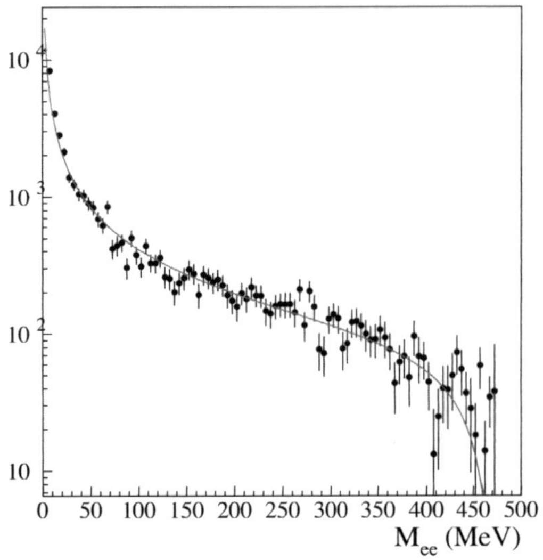

Figure 3. Fit to the M_{ee} spectrum for the Dalitz decays $\phi \to \eta e^+e^-$, using the $\eta \to \pi^+\pi^-\pi^0$ final state.

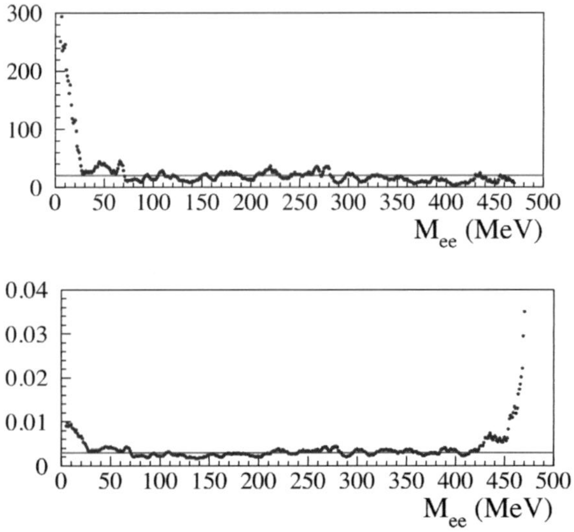

Figure 4. Top: exclusion plots on number of events for the decay chain $\phi \to \eta U$, $\eta \to \pi^+\pi^-\pi^0$, $U \to e^+e^-$. Bottom: exclusion plot on ϵ parameter, obtained from the previous one in the hypothesis that the U boson decays only to lepton pairs.

3.2. The $\eta \to \gamma\gamma$ final state

A similar analysis strategy has been developed for the decay chain $\phi \to \eta U$, $U \to e^+e^-$, $\eta \to \gamma\gamma$. The preselection requires: (i) two tracks coming from a cylinder around IP, classified as $e^\pm$ using ToF information; (ii) two photon candidates; (iii) the cosine of the angle between photons in η rest frame to be ~ 1; (iv) a loose cut on the total invariant mass of the system, $950 < M_{\rm tot} < 1150$ MeV.

The most severe background is generated by double radiative Bhabha scattering events and it is strongly reduced by cutting on the opening angle between the charged tracks and the photons. Residual non-Bhabha background is rejected by using further electron identification, based on the E/p ratio for the e^+e^- candidates. The resulting background reduction is still not enough for the search of $\phi \to \eta U$ events. The M_{ee} spectrum obtaind with 1.7 fb^{-1} shows a clear evidence of $\phi \to \eta e^+e^-$ Dalitz decays at low values and some residual background contamination at high M_{ee} due to Bhabha events. Work is in progress to further improve signal to background ratio.

4. The higgs$'$-strahlung channel

The feasibility of the search for the process $e^+e^- \to Uh'$ has been done considering the $m_{h'} < m_U$ case. At DAΦNE energies, for $\epsilon \sim 10^{-3}$, a production cross section of ≈ 20 fb is expected and the h' has $\tau_{h'} > 10$ μs, escaping the detection. The signature is therefore a lepton pair from the U boson plus missing energy.

The selection strategy has been optimized using Monte Carlo events. The signal has been generated according to Ref. [19] in a discrete set of mass values in the range $m_U \leq 900$ MeV, $m_{h'} \leq 400$ MeV. The $U \to e^+e^-$ events are not selected by any official KLOE event classification (ECL) algorithms, which divide the events on the basis of topological information and provide reconstructed data to be used for different analyses. On the contrary, ECL is fully efficient for $U \to \mu^+\mu^-$ events when $m_{h'} < 300$ MeV. We therefore considered the $\mu^-\mu^-$ final state only.

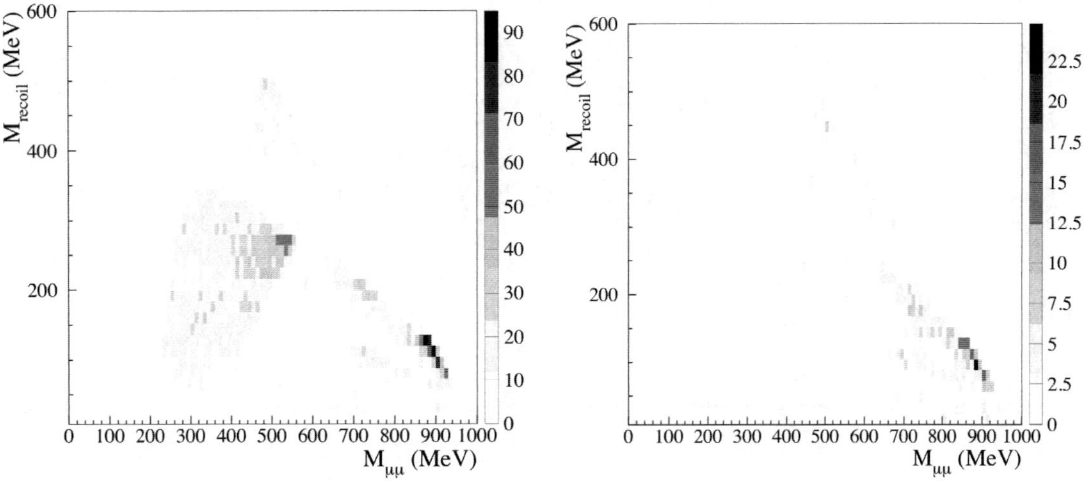

Figure 5. Search for $e^+e^- \to h'U$, $U \to \mu^+\mu^-$, $h' \to$ "*invisible*" events: recoil mass to the $\mu^+\mu^-$ pair as a function of the di-muon invariant mass for data taken at the ϕ mass (left) and at $\sqrt{s} = 1$ GeV (right).

Muons are identified and separated from electrons and pions using a neural network algorithm based on energy depositions along the shower depth in the calorimeter and E/p, β variables. The other relevant cuts to reduce background contamination are: (i) missing momentum direction in the barrel calorimeter; (ii) a tight cut on vertex-IP distance and (iii) no clusters in the calorimeter, with the exception of the two associated to tracks. The residual background contamination is due to $e^+e^- \to \pi^+\pi^-\gamma/\mu^-\mu^-\gamma$ continuum events with an undetected photon, and to $\phi \to K^+K^- \to \mu + \mu^-\nu\bar{\nu}$ with early decaying kaons.

In Fig. 5-left the distribution of the recoil mass to the $\mu^+\mu^-$ pair (M_{recoil}) as a function of the di-muon invariant mass obtained with 1.65 fb^{-1} is reported. M_{recoil} is evaluated using the center of mass energy of each run measured with Bhabha scattering events and the momenta of the muons. Continuum background, which can be further reduced tuning the π/μ identification algorithm, is concentrated in the band at $M_{\mu^+\mu^-} > 700$ MeV. The $\phi \to K^+K^-$ channel covers a wider region of the plane ($M_{\mu^+\mu^-} < 600$ MeV, $M_{\text{recoil}} < 300$ MeV). This background, having only two muons in the final state and missing energy due to neutrinos, has the same signature of the signal. The efficiency for $e^+e^- \to Uh'$ events is 15–40%, depending on m_U, $m_{h'}$ masses. Taking into account the total integrated luminosity, a signal would show up as a sharp peak with ≤ 10 events in the M_{recoil}-$M_{\mu\mu}$ plane for $\epsilon \sim 10^{-3}$.

Being the $\phi \to K^+K^-$ background a nasty background source, we repeated the analysis using the off-peak sample, 0.2 fb^{-1} taken at center of mass energy of 1 GeV. As can be seen in Fig. 5-right, the contribution from resonant background is not present anymore, providing a much cleaner sample for the search of $e^+e^- \to Uh'$ candidates.

5. Summary and perspectives for KLOE-2

The search for $\phi \to \eta U$ with $\eta \to \pi^+\pi^-\pi^0$, using 739 pb^{-1} of KLOE data, results in a preliminary upper limit on the ϵ parameter: $\epsilon < 3 \times 10^{-3}$ @ 95% C.L. in the $25 < M_{ee} < 425$ MeV range. With a sample of $\mathcal{O}(20 \text{ fb}^{-1})$ expected at KLOE-2, this value can be improved to 1.3×10^{-3} using the same η decay channel. The inclusion of other final states, such as $\eta \to \gamma\gamma$ and $\eta \to \pi^0\pi^0\pi^0$, will further improve this result.

The search of the higgs$'$-strahlung channel, $e^+e^- \to Uh'$ with $U \to \mu^+\mu^-$ plus missing energy,

Journal of Physics: Conference Series **335** (2011) 012067 doi:10.1088/1742-6596/335/1/012067

is limited by a non negligible $\phi \to K^+K^-$ background in a wide region of the $M_{\mu^+\mu^-}$, M_{recoil} plane. Work is in progress to reduce this contribution on the KLOE data sample. At KLOE-2, the improvement on the vertex resolution, achievable with the insertion of the inner tracker, will provide a higher rejection factor. The feasibility of a high statistics run at 1 GeV, where the resonant background contribution is naturally reduced, is also under discussion.

References

[1] P. Jean et al., Astron. Astrophys. 407, L55 (2003).
[2] O. Adriani *et al* , Nature 458, 607 (2009).
[3] J. Chang *et al* , Nature 456, 362 (2008).
[4] A.A. Abdo *et al* , Phys. Rev. Lett. 102, 181101 (2009).
[5] F. Aharonian *et al* , Phys. Rev. Lett. 101, 261104 (2008).
[6] F. Aharonian *et al* , Astron. Astrophys. 508, 561 (2009).
[7] R. Bernabei *et al* , Int. J. Mod. Phys. D 13, 2127 (2004).
[8] R. Bernabei *et al* , Eur. Phys. J. C 56, 333 (2008).
[9] M. Pospelov, A. Ritz, M.B. Voloshin, Phys. Lett. B 662, 53 (2008).
[10] N. Arkani-Hamed, D.P. Finkbeiner, T.R. Slatyer, N. Weiner, Phys. Rev. D 79, 015014 (2009).
[11] D.S.M. Alves, S.R. Behbahani, P. Schuster, J.G. Wacker, Phys. Lett. B 692, 323 (2010).
[12] M. Pospelov, A. Ritz, Phys. Lett. B 671, 391 (2009).
[13] J. Hisano, S. Matsumoto, M.M. Nojiri, Phys. Rev. Lett. 92, 031303 (2004).
[14] M. Cirelli, M. Kadastik, M. Raidal, A. Strumia, Nucl. Phys. B 813, 1 (2009).
[15] J. March-Russell, S.M. West, D. Cumberbatch, D. Hooper, JHEP 07, 058 (2008).
[16] I. Cholis, G. Dobler, D.P. Finkbeiner, L. Goodenough, N. Weiner, Phys. Rev. D 80, 123518 (2009).
[17] I. Cholis, D.P. Finkbeiner, L. Goodenough, N. Weiner, JCAP 0912, 007 (2009).
[18] N. Arkani-Hamed, N. Weiner, JHEP 12, 104 (2008).
[19] B. Batell, M. Pospelov, A. Ritz, Phys. Rev. D 79, 115008 (2009).
[20] R. Essig, P. Schuster, N. Toro, Phys. Rev. D 80, 015003 (2009).
[21] M. Reece, L.T. Wang, JHEP 07, 051 (2009).
[22] F. Bossi (2009), arXiv:0904.3815.
[23] N. Borodatchenkova, D. Choudhury, M. Drees, Phys. Rev. Lett. 96, 141802 (2006).
[24] P.F. Yin, J. Liu, S.h. Zhu, Phys. Lett. B 679, 362 (2009).
[25] J.D. Bjorken, R. Essig, P. Schuster, N. Toro, Phys. Rev. D 80, 075018 (2009).
[26] B. Batell, M. Pospelov, A. Ritz, Phys. Rev. D 80, 095024 (2009).
[27] R. Essig, P. Schuster, N. Toro, B. Wojtsekhowski, JHEP 1102, 009 (2011).
[28] M. Freytsis, G. Ovanesyan, J. Thaler, JHEP 01, 111 (2010).
[29] F. Bossi *et al* , Rivista del Nuovo Cimento, Vol. 031, 531 (2008).
[30] C. Milardi *et al* , ICFA Beam Dyn. Newslett. 48, 23 (2009).
[31] G. Amelino-Camelia *et al* , Eur. Phys. J. C 68, 619 (2010).
[32] M. Adinolfi *et al* , Nucl. Inst. and Meth. A 488, 51 (2002).
[33] M. Adinolfi *et al* , Nucl. Inst. and Meth. A 482, 363 (2002).
[34] M. N. Achasov *et al* , Phys. Lett. B 504, 275 (2001).
[35] R. R. Akhmetshin *et al* , Phys. Lett. B 501, 191 (2001).
[36] L.G. Landsberg, Phys. Rep. 128, 301 (1985).
[37] N.N. Achasov, A.A. Kozhevnikov, Sov. J. Nucl. Phys. 55, 449 (1992).
[38] T. Junk, Nucl. Instr. Meth. A 434, 435 (1999).
[39] J.M. Bernardo and A.F.M.Smith, *Bayesian theory,* Wiley (2000).

Hidden Valley Search at ATLAS

Monica Verducci

On behalf of the ATLAS Collaboration

Department of Physics, University of Washington, PO Box 351560, Seattle, WA 98195, USA

monica.verducci@cern.ch

Abstract. A number of extensions of the Standard Model result in neutral and weakly-coupled particles that decay with macroscopic decay lengths to multiple hadrons or leptons. These particles with decay paths that can be comparable with ATLAS detector dimensions represent, from an experimental point of view, a challenge both for the trigger and for the reconstruction capabilities of the ATLAS detector. We will present a set of signature driven triggers for the ATLAS detector that target such displaced decays and evaluate their performances for some benchmark models and describe analysis strategies and limits on the production of such long-lived particles. A first estimation of the Hidden Valley trigger rates has been evaluated with 6pb-1 of data collected at ATLAS during the data taking of 2010.

1. The Hidden Valley Scenario

A number of extensions of the Standard Model (SM) result in particles that are neutral, weakly-coupled and have macroscopic decay lengths that can be comparable with LHC (Large Hadron Collider) detector dimensions.

The Hidden Valley (HV) scenario represents a possible extension of the SM where a new sector, neutral under SM forces and only weakly coupled to the SM sector, results in neutral long-lived HV particles (π_v) that decay to heavy quark pairs and tau pairs. These particles can be produced in Higgs boson decays, SUSY processes or Z' decays, which allow the interactions between it and the SM sector.

A Hidden Valley model is structured as illustrated in figures 1 and 2. To the SM is appended a hidden sector, the "v-sector" for short, and a communicator (or communicators) which interacts with both sectors. A barrier (perhaps the communicator's high mass, weak couplings or small mixing angles) weakens the interactions between the two sectors, making production even of light v-sector particles ("v-particles") rare at low energy. At the LHC, by contrast, production of v-particles, through various possible channels, may be observable [1]. We present the results of a first study of the ATLAS Detector performance for the Higgs decay $h^0 \rightarrow \pi_v \pi_v$, where the π_v is neutral and has a displaced decay mainly to bottom quarks, see figure 2.

Here we present results based on a simulation [2,3] of the process $h^0 \rightarrow \pi_v \pi_v$ with the following parameters: $m_{h0} = 140$ GeV, $m_{\pi v} = 40$ GeV and $c\tau_{\pi v} = 1500$ mm.

2. The ATLAS Detector

ATLAS is one of the four large detectors at the proton-proton collider LHC, currently running at CERN, see figure 3.

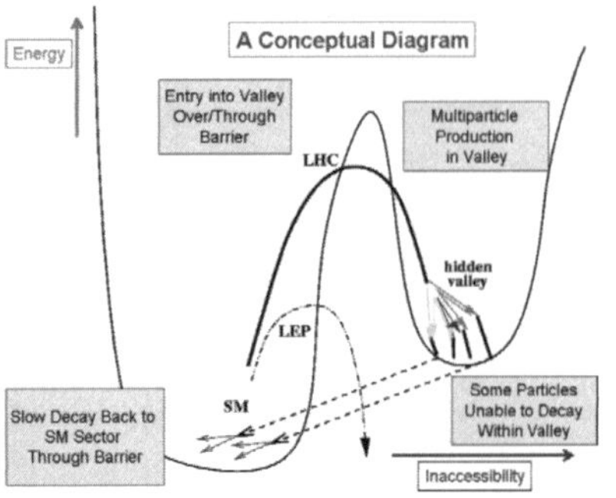

Figure 1. A hidden sector, separated from the Standard Model sector by a barrier, becomes accessible at LHC energies. The mass gap stabilizes some v-particles, allowing them to decay slowly back through the barrier to Standard Model particles.

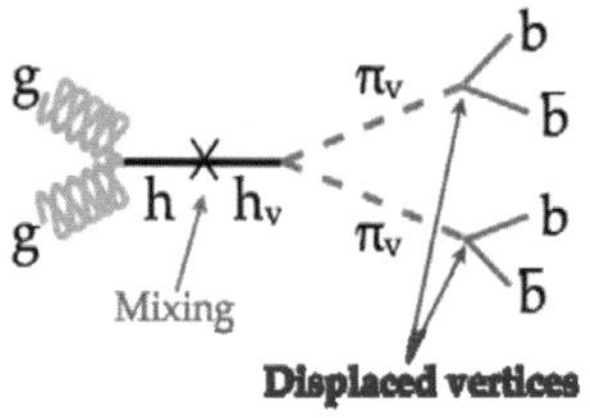

Figure 2. Neutral long lived HV particles (π_v) model that decay to bb quark pairs (90%) and tau pairs (10%). Higgs bosons mediate the interaction between the Standard Model and the Hidden Valley sector.

ATLAS is a general-purpose detector, designed to explore a wide energy region and to probe further into the fundamental forces of nature, those currently known but even those unexpected.

The detector design is therefore guided by Standard Model issue among which the Higgs search, the electroweak symmetry breaking and the search of super-symmetric particles. The ATLAS detector must be able to measure the energy-momentum four-vector with great precision and high efficiency for all particles.

The overall layout consists of different specialized sub-detectors: the innermost detector around the

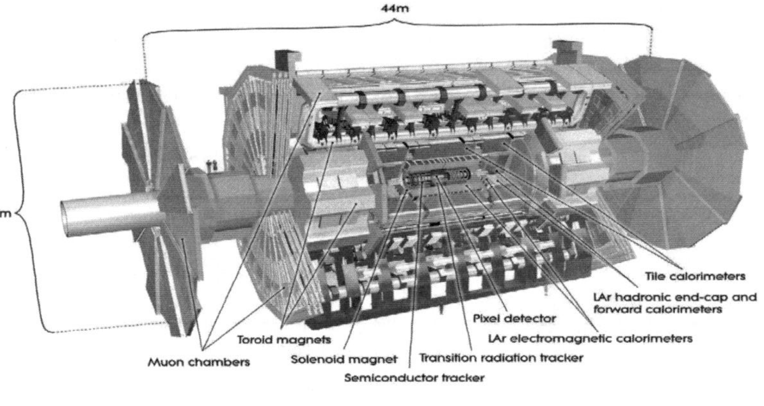

Figure 3. The ATLAS Detector. The position in ATLAS is defined by the cylindrical coordinate system (z,R,ϕ), where z is the position in the direction along the beam axis, R the radius, ϕ the azimuth angle, in addition we use η the pseudo-rapidity defined as $\eta=-\ln(\tan(\theta/2))$, with θ the polar angle.

interaction region is the Inner Detector (ID), followed by a calorimeter system (Electromagnetic and Hadronic calorimeter) and, in the outer region, by the Muon Spectrometer.

The ID provides pattern recognition and precise measurements of the decay vertex and momentum of charged particles. The calorimeter system measures with great accuracy the energy of the jet showers and the Muon Spectrometer, mounted around a large superconducting air-core toroid, provides a good momentum resolution of about 10% for a muon momentum of 1TeV.

The enormous data flow arising from the high LHC design collision rate of 10^9 Hz must be reduced to about 300 MB/s, corresponding to a maximum trigger of about 200 Hz out of the ATLAS Trigger system.

In fact, the ATLAS trigger system has been designed to bridge this gap via a complex system of three different levels of selection algorithms optimized to efficiently select the interested events with a short latency time.

The first level trigger (LVL1) is designed to operate at a maximum pass rate of about 75 kHz, with a decision time (latency) less than 2.5 μs. The LVL1 decision is based on information with a coarse granularity of two sub-detector systems: the muon trigger chambers and the calorimeters. For events accepted by LVL1, the information of all sub-detector systems is pre-processed and stored in the Read-Out Buffers (ROBs).

The second level trigger (LVL2) uses both the LVL1 output and the data stored in the ROBs to further reduce the data rate to a maximum of 3 kHz. Even though the LVL2 has access to the full data, the selection is generally restricted to 'Regions of Interest' (ROI) flagged by the output of the LVL1. The LVL2 mean target processing time is about 40 ms (the time depends on the complexity of the event processed). For events accepted by the LVL2, the data fragments stored in the ROBs are collected by the so-called Event Builder and written into the Full Event Buffers.

The third trigger stage is called Event Filter and uses the information stored in the Full Event Buffers to keep the event rate sent to mass storage below 200 Hz, or about 300 MB/s. The Event Filter accesses the complete event information using complex offline selection algorithms.

3. Detector Signatures and Trigger Strategies for Displaced Decays

Hidden Valley events are characterized by highly displaced decays, leading to jets appearing throughout the volume of ATLAS. Due to the low Higgs mass resulting in soft jets, which are highly pre-scaled at the trigger level, and tracks that do not point to the IP (Interaction Point) coming from the highly displaced vertex, the standard ATLAS triggers select only a small fraction of these unique events.

However, it is possible to use the displaced vertex signature as a trigger object in order to increase the fraction of events accepted. In this section we discuss the b-jet signatures that are observed in different parts of the ATLAS detector from decays of π_v's to bb pairs. We consider four detector regions:

a. Decays from end of the Hadron Calorimeter (HCal) to the first Muon Spectrometer trigger plane;
b. Decays in the calorimeters;
c. Decays in the Inner Detector beyond the pixel layers to the end of the Transition Radiation Tracker (TRT);
d. Decays in the beam pipe and pixel layers.

3.1. Decays in the Muon Spectrometer

Decays occurring near the end of the HCal and before the first muon trigger plane are characterized by a large number of charged hadrons traversing a narrow (η;ϕ) region of the Muon Spectrometer and result in a large number of clustered LVL1 "muon" RoIs (ATLAS LVL1 trigger object which describes a Region of Interest in ($\eta;\phi$) to be examined at LVL2) . The cluster of LVL1 RoIs will be accompanied by a lack of activity in the ID and Calorimeters.

These events are rejected by the current ATLAS LVL2 trigger which requires a matching ID track for the LVL1 muons.

The algorithm implemented for a stand-alone HV LVL2 trigger requires that the cluster contains at least three muon RoIs in a cone of $\Delta R = 0.4$, no jets in a cone of $\Delta R = 0.7$ centered on muon RoI cluster and no inner detector track in ($\Delta\eta,\Delta\varphi$) region of (0.2x0.2).

The efficiency obtained is about 70% in barrel and 25% in the endcap of the muon system using this new LVL2 algorithm, see figure 4.

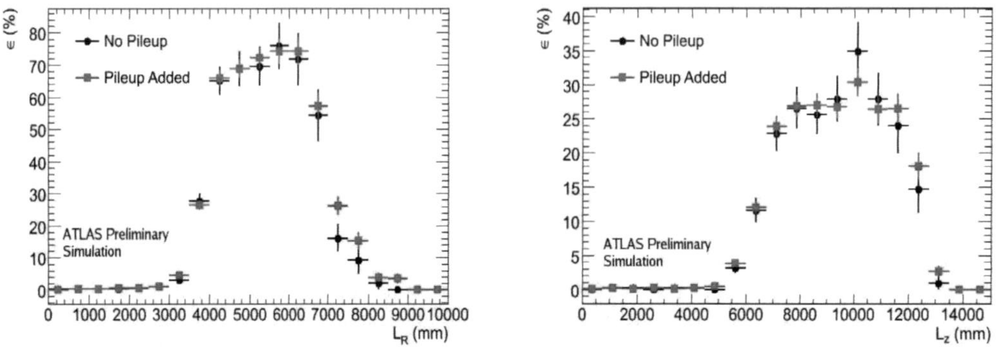

Figure 4. Efficiency distribution obtained in the barrel (left plot) and in the endcap region (right plot) by the Hidden Valley LVL2 trigger based on the Muon Spectrometer.

3.2. Decays in Calorimeters

If the π_v decays beyond the Electromagnetic calorimeter (ECal) the ratio of energy deposited in the HCal to that in ECal will be larger than that which is normally observed for jets originating at the IP.

This suggests to use the ratio between the energy deposited in the HCal and ECal, $\log(E_{HAD}/E_{EM})$, as a trigger object. Figure 5 shows the energy ratio as a function of the π_v decay point; making very clear that for this type of event the ECal records little or no energy, while the HCal has a large energy deposition. No visible track connects this jet with the IP.

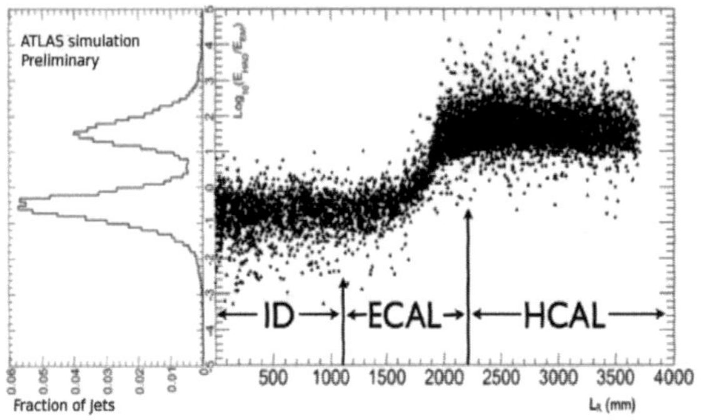

Figure 5. Distribution of the $\log_{10}(E_{HAD}/E_{EM})$ variable. Jets from the decay of π_v's in the HCal have $\log_{10}(E_{HAD}/E_{EM}) \sim 1.5$, while decays in the ID and ECal have a value less than 1.

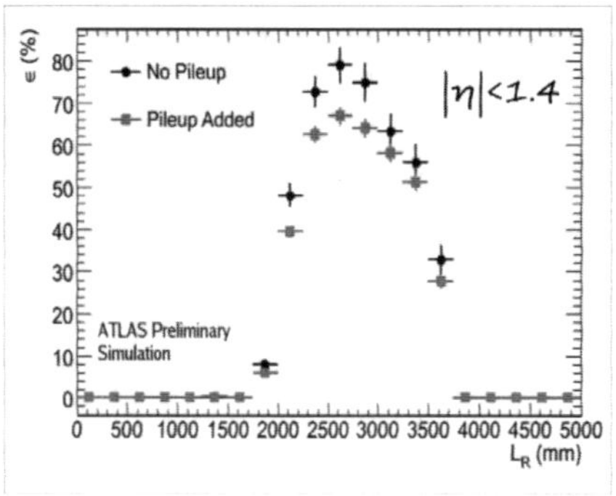

Figure 6. Efficiency distribution obtained in the barrel region by the "Calo Ratio" Hidden Valley LVL2 trigger.

The HV trigger algorithm requires:

- $\log_{10}(E_{HAD}/E_{EM}) > 1$
- No ID tracks in region around jet direction

The efficiency obtained in barrel is about 60% (see figure 6) and about 35% in endcap regions.

3.3. Decays in TRT

Decays in the outer Silicon layers and TRT (Transition Radiation Tracker) lead to jets with no connecting track to the IP (trackless jets). Selecting such events at LVL2 requires a robust LVL2 TRT vertex finding algorithm, which unfortunately is not currently available. Thus we examine trackless jets as trigger objects for this event signature and we added the new trigger chain. We require a jet of energy greater than 35 GeV and no tracks matching it in the ID. The efficiency obtained by this selection algorithm is still not optimized and requires further improvements in the future.

3.4. Decays in the Beam Pipe and Pixels

Decays in this region will look very much like decays of heavy flavor particles produced in Standard Model processes. The b-tagging algorithms should find these events with good efficiency and consequently special LVL2 triggers are not needed (not used).

3.5. Dependence of Trigger Efficiencies on π_v Decay Length

The relative efficiency of the new signature driven LVL2 triggers as a function of the π_v decay length is shown in figure 7. From the efficiency for triggering on a π_v as a function of its decay position, we predict the fraction of Higgs decays that would be accepted by our triggers as a function of the π_v proper lifetime $c\tau$, assuming a 100% branching fraction for $h^0 \to \pi_v \pi_v$.

Figure 7 shows the expected fraction of events vs the π_v lifetime for each of our three triggers, and the total fraction accepted. The right scale gives the number of expected events for an integrated luminosity of $100 pb^{-1}$.

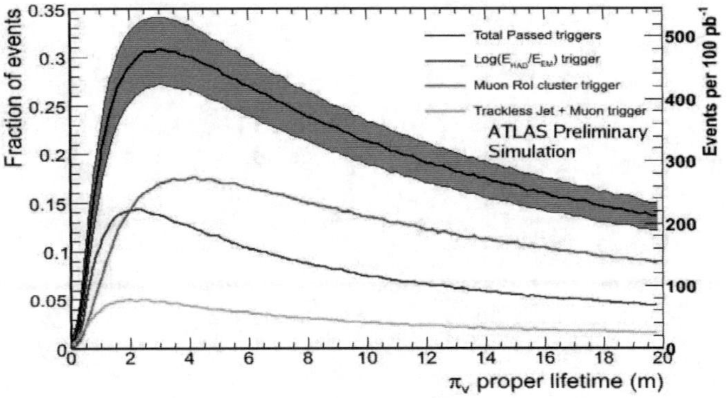

Figure 7. $h^0 \to \pi_v \pi_v$ events accepted by our long-lived particle triggers vs the π_v lifetime for m_{h0} =140 GeV and $m_{\pi v}$ = 40 GeV. Left axis: fraction of events accepted. Right axis: number of events for an integrated luminosity of 100 pb^{-1} assuming Br ($h^0 \to \pi_v \pi_v$)=100%.

4. Trigger Performance with Early Data

The new HV algorithms have been tested during the first period of ATLAS Data taking. We measured the final trigger rate for all the three trigger algorithms; as results of this measurement, we obtained trigger rates well inside the experimental data taking limits at the luminosity of 10^{32}.

Figure 8 shows the luminosity achieved in the run analyzed while in the figures 9, 10 and 11 the rates obtained for each trigger menu are shown.

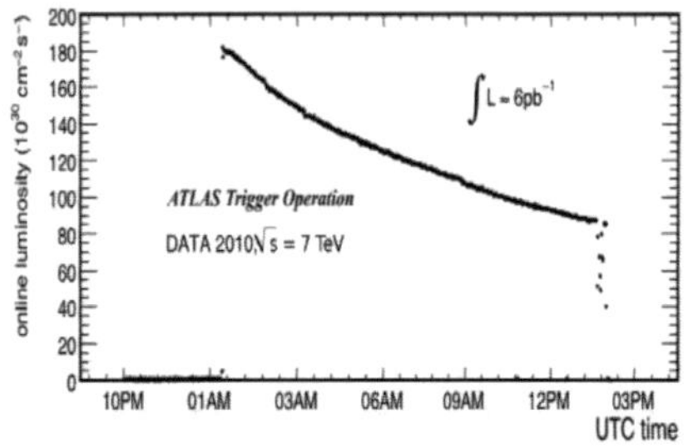

Figure 8. Online luminosity for 6pb^{-1} of data analyzed.

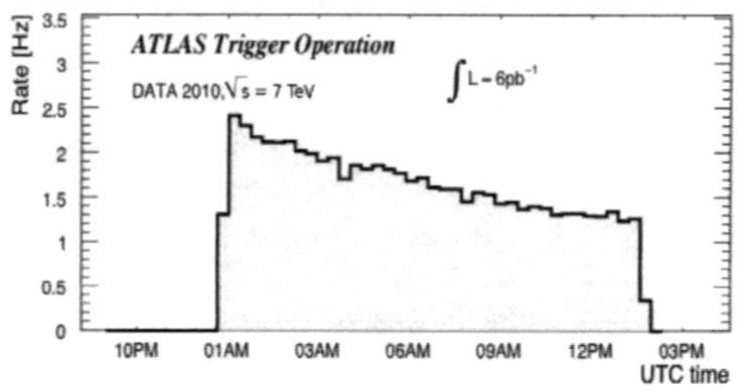

Figure 9. LVL2 Trigger Rate for the HV trigger algorithm Muon Spectrometer based.

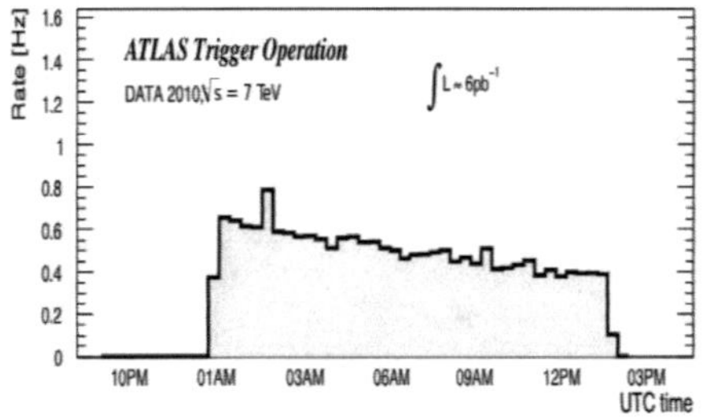

Figure 10. LVL2 Trigger Rate for the HV trigger algorithm based on the Calorimeter energy.

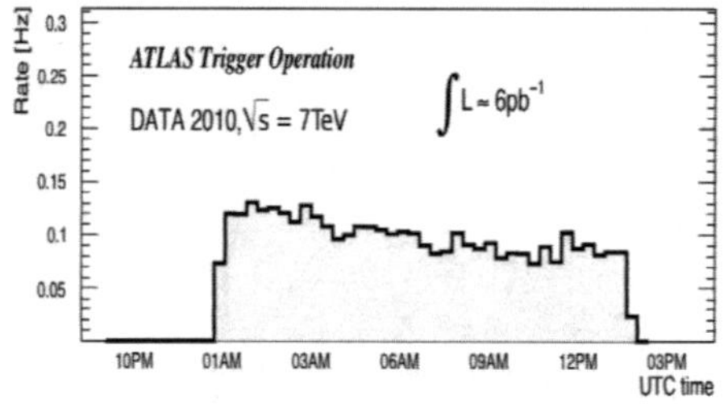

Figure 11. LVL2 Trigger Rate for the HV trigger algorithm based on the TRT tracks.

5. Conclusions

Different models predict extensions of the Standard Model with new particles that are neutral and have lifetimes comparable with ATLAS detector dimensions. In this work we have shown that it is necessary to define a set of signature-driven triggers in order to select these types of events with a reasonable efficiency. Algorithms for these triggers have been implemented in the ATLAS High Level Trigger framework. Rate studies of these new triggers using 6pb^{-1} of data have been presented. Studies on QCD di-jets and cosmic backgrounds are ongoing.

References

[1] M. J. Strassler and K. M. Zurek, Phys. Lett. B 651 (2007) 374-379; M. J. Strassler and K. M.Zurek,Phys. Lett. B 661 (2008) 263-267; M. J. Strassler,arXiv:hep- ph/0607160.
[2] PYTHIA 6.301 manual, T. Sjöstrand et al., LU TP 03-38, hep-ph/0308153.
[3] GEANT4, J. Sulkimo et al., Nucl. Instr. and Meth. in Phys. Research A 506 (2003) 250-303.
[4] The ATLAS Collaboration, G. Aad et al., 2008 JINST 3 S08003.
[5] The ATLAS Collaboration, G. Aad et al., 2009, "Triggering on Long-Lived Neutral Particles in the ATLAS Detector", ATL-PHYS-PUB-2009-082.

Journal of Physics: Conference Series **335** (2011) 012069 doi:10.1088/1742-6596/335/1/012069

Searches for supersymmetry with the CMS detector at the LHC

Edmund Widl on behalf of the CMS collaboration

Institute of High Energy Physics, Nikolsdorfer Gasse 18, A-1050 Wien, Austria

E-mail: edmund.widl@oeaw.ac.at

Abstract. Supersymmetry may give rise to striking events that could be discovered early in LHC running. Search strategies based on the generic event signatures of high jet multiplicity and large missing transverse momentum, optionally including leptons and photons in the final state, are discussed. An important aspect of such searches is the commissioning of search variables with LHC data, and demonstrating a good understanding of the detector, which is covered in detail. Techniques for estimating the contributions from Standard Model background processes using data are presented. Finally prospects for a discovery are reviewed.

1. Introduction

The CMS experiment [1] will perform extensive studies to find proof of the existence of supersymmetric (SUSY) particles [2], the predicted partners of the known standard model (SM) particles. However, searches for SUSY particles are a highly non-trivial task, mostly due to the fact that the theoretical framework of SUSY allows for a broad range of different phenomenological features. These include different mass spectra with distinct individual decay channels or different cross sections and branching ratios. In spite of these difficulties the CMS collaboration decided to base the strategies for early searches [3] on generic event signatures comprised of high transverse momentum jets and large amounts of missing transverse energy (MET). Optionally, the signatures may also contain leptons and photons in the final state.

This strategy is motivated by several facts. First of all, the LHC is a proton-proton collision machine, for which the strong production of gluinos and squarks would be the dominant processes for SUSY production. Hence high transverse momentum jets originating from the decay of the colored SUSY particles in the hard scattering process are expected. Secondly, motivated by mSUGRA scenarios, the existence of an uncharged and weakly interacting *lightest supersymmetric particle* (LSP) is assumed. Under the assumption of R-parity conservation all decay chains containing SUSY particles should end with an LSP, thus being a potential source of a high excess of MET.

The strategy is supplemented with a set of mSUGRA scenarios [4] that are used as benchmark points for analyses (see Figure 1). These very well understood and not yet ruled out scenarios cover a large variety of distinct signatures. For this reason it is assumed that model depended prejudices due to restrictions within the framework of mSUGRA are avoided as far as possible.

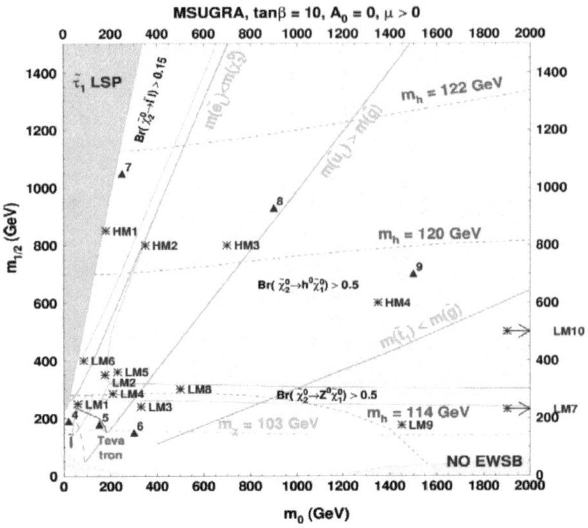

Figure 1. The low mass (LM) and high mass (HM) mSUGRA benchmark points used by CMS, depicted in the m_0-$m_{1/2}$-plane.

2. Comissioning of search variables with data

Backgrounds are expected to be orders of magnitude larger than any SUSY signal. A good understanding of the complex physics observables and control over backgrounds from SM processes hence are an inevitable experimental requirement. Careful commissioning of the search variables is therefore the key to success, especially the performance of the reconstruction of jets and MET.

2.1. Performance of jet reconstruction

For SUSY analyses jets are experimental signatures of the strong production of squarks and gluinos, which are produced in highly energetic hard scattering of partons in p-p collisions. A detailed understanding of the energy calibration and resolution of jets is of crucial importance and is a leading source of systematic uncertainty for any analyses with jets in the final state. The instrumentation of CMS allows the reconstruction of jets in four different ways [5], which combine information from individual sub-detectors to form the inputs to the jet clustering algorithm in

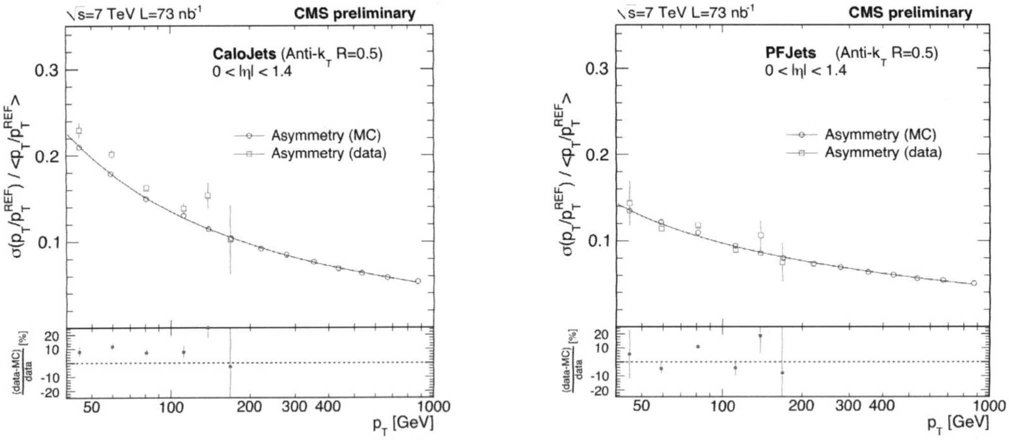

Figure 2. Calorimeter (left) and PF (right) jet resolution for $0 \leq |\eta| \leq 1.4$ from QCD simulation and compared with the results from data.

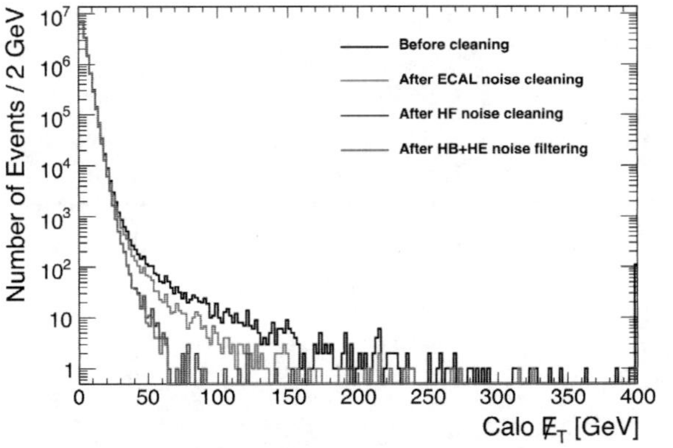

Figure 3. Distribution of the calorimeter MET before and after noise cleaning and filtering.

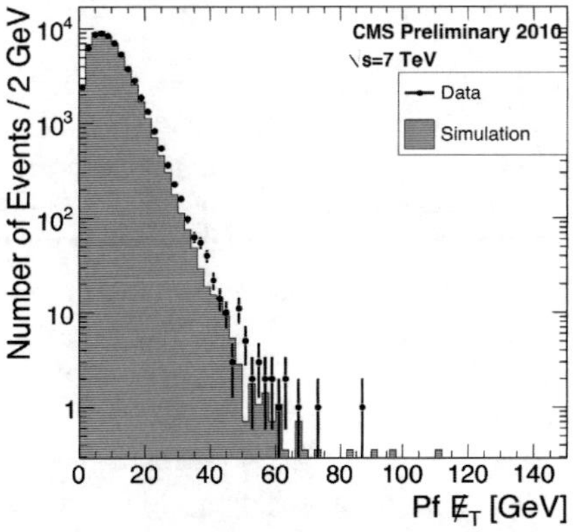

Figure 4. Distribution of the PF MET for dijet events from simulation and compared with data.

different ways: calorimeter jets, Particle-Flow (PF) jets, Jet-Plus-Track jets, and track jets.

Calorimeter jets are reconstructed using information from the calorimeters only, by combining energy deposits in the electromagnetic and hadronic calorimeter cells into so-called calorimeter towers. The Particle Flow algorithm [6, 7] combines the information from all CMS sub-detectors to identify and reconstruct all particles in the event, namely muons, electrons, photons, charged hadrons and neutral hadrons. PF jets are then reconstructed from the resulting list of particles.

CMS has developed a factorized multi-step procedure for the jet energy calibration [8], comprising so-called offset, relative and absolute corrections. The offset correction aims to correct the jet energy for the excess unwanted energy due to electronics noise and pile-up. The relative correction removes variations in jet response due to the uniformity in η of the detector. The absolute correction removes variations in jet response versus jet p_T. Figure 2 shows the jet energy resolution for calorimeter and PF jets after applying all correction steps. The good agreement between data and simulation confirms that the jet reconstruction is working properly.

2.2. Performance of MET reconstruction

MET is generally calculated as the magnitude of the negative vector sum of the momentum transverse to the beam axis of all final-state particles reconstructed in the detector. The instrumentation of CMS allows the reconstruction of MET in four different ways [9, 10, 11], which combine the information from the individual sub-detectors in different ways: calorimeter MET,

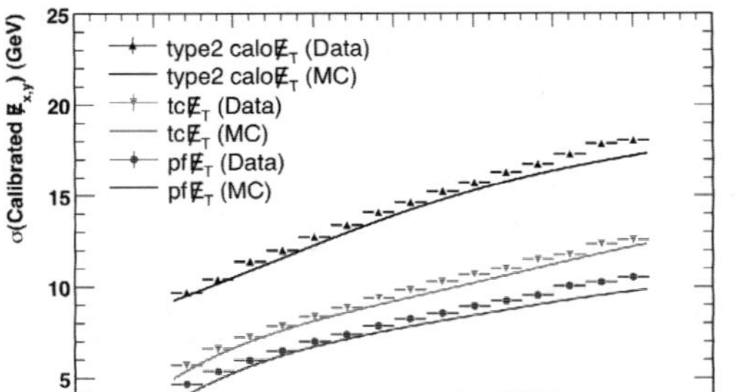

Figure 5. Calorimeter (black), PF (blue) and TC (red) MET resolution from simulation and compared with the results from data.

Particle-Flow (PF) MET, track-corrected (TC) MET, and MET calculated using reconstructed jets only.

While MET is an important observable not only for SUSY searches, it is also one of the most complex observables. Its reconstruction is very sensitive to various detector malfunctions, particles impinging poorly-instrumented regions of the detector, cosmic-ray particles, and beam-halo particles. Especially in the extreme tails of the distribution these effects may result in an artificial MET. This is demonstrated by Figure 3 that shows for the case of the calorimeter MET the importance of noise cleaning, i.e. the removal of noisy channels from the computation of the MET, and filtering, i.e. the removal of a complete event. Thus, great care is required to understand the MET distribution as measured by the detector (see Figure 4). Figure 5 shows the resulting Gaussian core resolution of the calorimeter MET, the PF MET and the TC MET, after correcting for all before mentioned unwanted sources of artificial MET. Both TC MET and PF MET show improvements in the resolution compared to the calorimeter only MET, and the PF MET yields the smallest MET resolution.

3. Treatment of backgrounds in SUSY searches

Searches for SUSY with MET depend on a good understanding of the extreme tails of the MET distribution, since the cross sections of the background channels are assumed to be orders of magnitude larger than any SUSY signal. Especially the cross section of QCD is not predicted with high enough precision and key topological and kinematical features such as number of jets, angular distributions or momentum spectra are difficult to model in simulations. Electroweak backgrounds (e.g. W+jets, Z+jets, $t\bar{t}$) present similar challenges, even though they are not as harsh as those for QCD. Therefore backgrounds will be determined using data-driven methods whenever possible, with multiple methods for cross-checks.

3.1. Suppression of backgrounds

QCD events in which the hadronic activity is mis-measured are a significant source of artificial MET. Such events, where the observed MET is primarily due to jet resolution effects rather than to real missing momentum carried by unobserved particles, have to be carefully suppressed. To this end a variety of kinematical variables or information from complementary detector systems can be used.

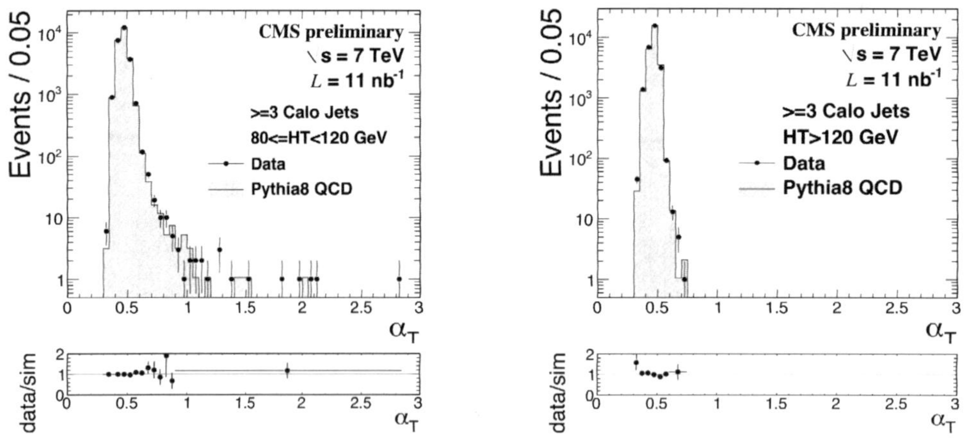

Figure 6. Comparison of the distribution of α_T between data (black) and simulation (yellow) for different values of H_T for multi-jet events ($\geq$3 jets).

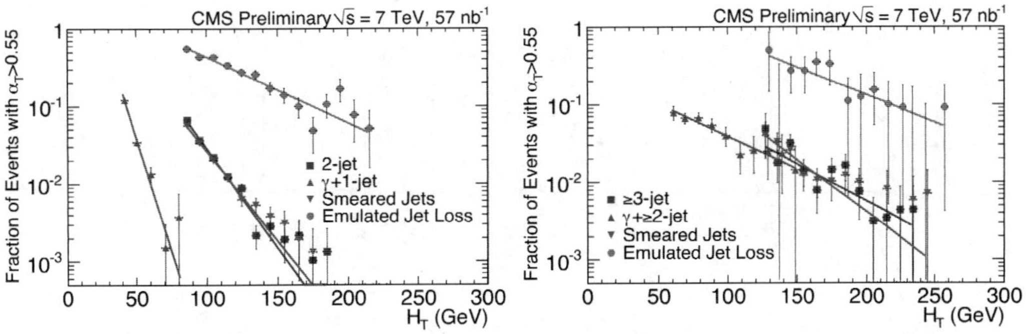

Figure 7. Fraction of events with $\alpha_{\mathrm{T}} \geq 0.55$ in dependence on H_{T} as extracted from data.

As an example [3] for the suppression of background, the variable α_{T} [12, 13] is discussed here. It is a dimensionless quantity that characterizes the overall transverse momentum balance of the event by using the ratios of magnitudes of transverse momenta:

$$\alpha_{\mathrm{T}} = \frac{1}{2} \frac{H_{\mathrm{T}} - \Delta H_{\mathrm{T}}}{\sqrt{H_{\mathrm{T}}^2 - \mathrm{MHT}^2}}$$

where

$$H_{\mathrm{T}} = \sum_{\mathrm{jets}\, j} p_{\mathrm{T},j}, \qquad \Delta H_{\mathrm{T}} = \min_{\mathrm{jets}\, j} \left(p_{\mathrm{T,\,pseudojet\,1}} - p_{\mathrm{T,\,pseudojet\,2}}\right), \qquad \mathrm{MHT} = \Big|\sum_{\mathrm{jets}\, j} -\vec{p}_{\mathrm{T},j}\Big|.$$

The partition of the multi-jet system into two pseudo-jets can be performed in multiple ways. For the computation of α_{T} typically the configuration with the minimal ΔH_{T} of all possible partitions is chosen.

Like the MET, α_{T} is a powerful discriminator against QCD background. While QCD background is not the only or even the dominant background in the jets+MET search, it is perhaps the most challenging to control due to the above mentioned problems with predicting QCD precisely. For the suppression of other backgrounds, such as W+jets or $\mathrm{t\bar{t}}$, other methods have to be applied.

Figure 6 shows that in practice, QCD background is largely confined to the region with $\alpha_{\mathrm{T}} \leq 0.55$. Figure 7 demonatrates that the suppression, i.e. the fraction of events with $\alpha_{\mathrm{T}} \leq 0.55$, even improves with increasing values of H_{T}. This is not only true for carefully reconstructed data from di-jet events (blue), but also for artificially degraded data using photon-triggered events dominated by mis-identified jets (green), events with emulated jet loss (red) or momentum-smeared jets (violet). In all these cases an exponential trend is confirmed. Thanks to the reliability and robustness of the method, this behavior can even be used to calculate limits on the suppression rate for higher H_{T}-bins.

3.2. Background estimation

Even though variables like α_{T} provide a good handle on the suppression of background channels, the remaining background still has to be quantified using supplementary measurements and methods. In this section two examples [3] are presented.

Both examples make use of control samples in which no genuine MET is present to make predictions for the background at high values of MET, where SUSY signals might be observed. The control samples are divided into bins of number of jets and H_{T} and for each bin the shape of the MET spectrum – the so-called *MET template* – is reconstructed. Figure 8 shows such a set of

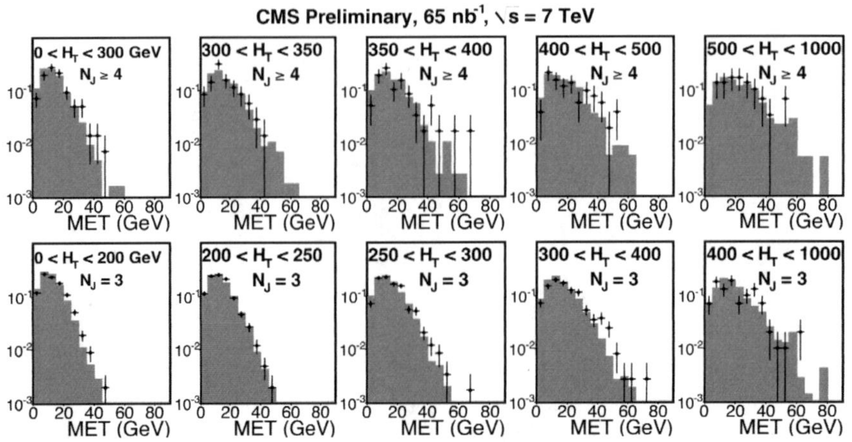

Figure 8. MET templates reconstructed from a multi-jet sample.

templates extracted from a multi-jet QCD sample. For a given signal channel the corresponding distributions of the number jets and H_T are measured. To obtain a background estimate the control sample's MET templates are weighted accordingly to these distributions.

The before mentioned MET templates extracted from the multi-jet QCD sample were used to perform a consistency test in the γ+jets+MET channel. Since the kinematical effects are expected to get diluted, the templates from the multi-jet sample should give a good estimate for artificial MET. Figure 9 confirms this assumption. For the region with MET$\geq$15 GeV the prediction of 12.5 events and the observation of 11 events is statistically consistent.

A similar example is shown in Figure 10, where the template method was used to estimate the background MET distribution in the di-photon+MET channel, coming from electrons and jets misidentified as photons. A kinematically similar QCD control sample was used, in which fake photon candidates were selected by inverting the isolation criteria. The templates were then re-weighted according to the photon p_T-spectrum. For the region with MET$\geq$20 GeV the prediction of 4.2$\pm$1.5 events and the observation of 4 events is again statistically consistent.

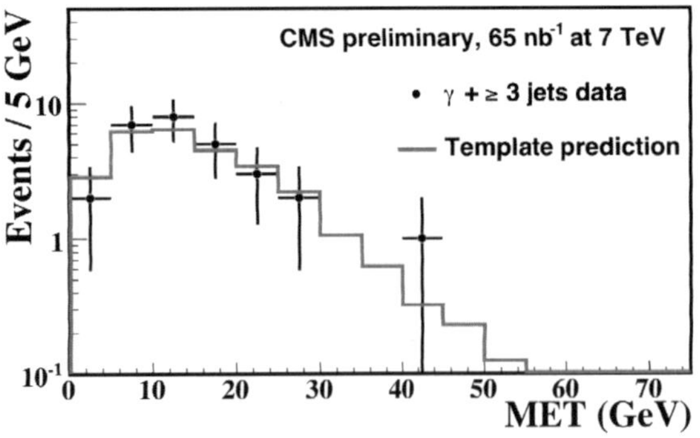

Figure 9. Consistency test for MET templates in the γ+jets channel.

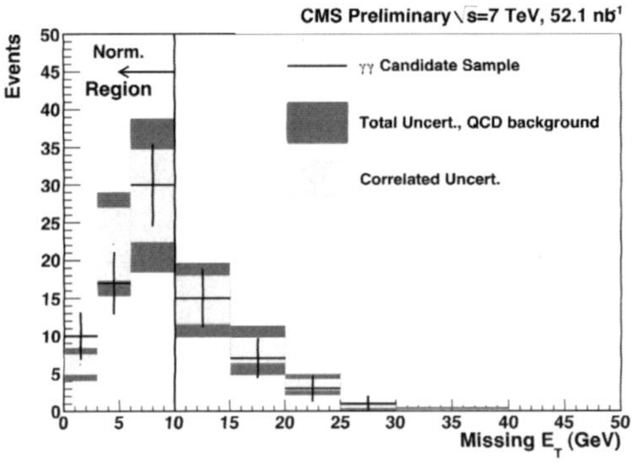

Figure 10. Consistency test for MET templates in the di-photon channel.

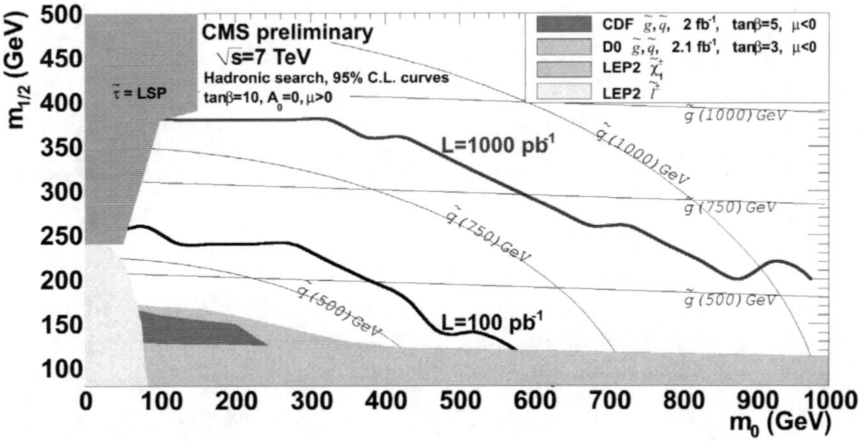

Figure 11. Estimated 95% confidence level exclusion limits for the all-hadronic SUSY search,

4. Early data SUSY searches

The two searches discussed here are representative of the CMS SUSY search program [12, 13, 14, 15]. Here we give a brief sketch of how these analyses are performed.

In the all-hadronic search, events with isolated muons or electrons above a certain momentum threshold (10 GeV/c for muons and 15 GeV/c for electrons) are vetoed, so the search is nearly statistically independent from the leptonic searches. At least three jets with a $p_{\mathrm{T}} \geq 50$ GeV/c are required. The angle of the missing momentum vector is required to point away from the leading jets, since fluctuations in the jet-energy measurement can lead to artificial MHT. For the sensitivity plots at 100 pb^{-1} values of $H_{\mathrm{T}} \geq 400$ GeV and MHT ≥ 225 GeV/c are demanded, whereas for 1 fb^{-1}, the cuts are tightened to $H_{\mathrm{T}} \geq 500$ GeV and MHT ≥ 250 GeV/c. Other cuts are used to help ensure the reliability of the quantities used in these selection procedures. Due to the generic nature of these requirements, the all-hadronic search typically has the highest efficiency of all early CMS searches over the mSUGRA parameter space. Figure 11 shows the

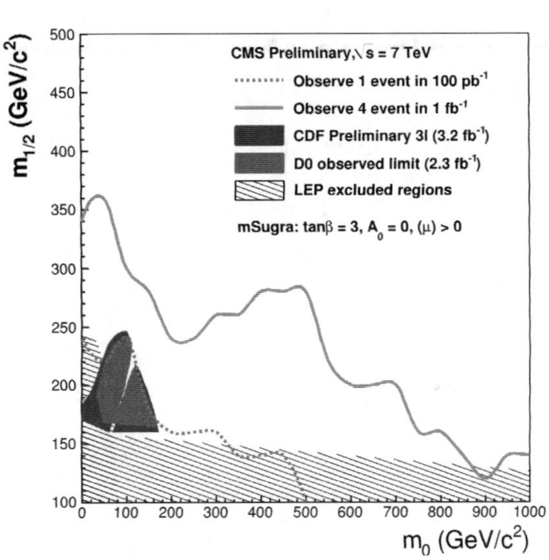

Figure 12. Estimated 95% confidence level exclusion limits for the like-sign di-lepton SUSY search. The expected SM background at 100 pb^{-1} (1 fb^{-1}) is 0.4 (4.0) events; an observed yield of 1 event (4 events) for setting these exclusion limits is assumed.

corresponding 95% confidence level contour lines.

The like-sign di-lepton analysis is performed in the three channels $\mu^{\pm}\mu^{\pm}$, $\mu^{\pm}e^{\pm}$ and $e^{\pm}e^{\pm}$, with the requirement of two like-sign, isolated leptons above minimum $p_{\rm T}$-thresholds (both leptons with $p_{\rm T}$ ≥10 GeV/c and at least one with $p_{\rm T}$ ≥20 GeV). A minimum of three jets above $p_{\rm T}$ ≥30 GeV/c is required, and the jets must satisfy $H_{\rm T}$ ≥200 GeV. Additionally, MHT ≥80 GeV/c is demanded. Figure 11 shows the corresponding 95% confidence level contour lines.

These results indicate that in the 7 TeV run, CMS should have sensitivity to regions of SUSY (mSUGRA) parameter space beyond the currently existing limits. Both of the channels discussed here should be able to yield interesting sensitivities well before 1 fb^{-1}.

5. Conclusion and Outlook

Jets and missing transverse energy play a key role for early SUSY searches at the LHC. The CMS Collaboration has put a lot of effort into the accurate reconstruction of these observables. Both are by now well understood, as confirmed by the good agreement seen in the comparison between data and simulation. A good understanding of backgrounds from the SM is another prerequisite for SUSY searches. Methods for suppressing these backgrounds and data-driven techniques for estimating what remains have been developed. By now, these methods have been successfully tested and validated with data.

Due to the excellent performance of CMS and the LHC, the data collected so far (≈42 pb^{-1}) will allow to draw new conclusions about the existence of SUSY very soon. Also, as more data will arrive in the near future, plenty of new results are expected.

References

[1] CMS Collaboration *JINST* **3** (2008) S08004 The CMS Experiment at the CERN LHC
[2] Martin S 2008 *A Supersymmetry Primer*, arXiv:hep-ph/9709356v5
[3] CMS Collaboration 2010 *CMS Physics Analysis Summary* **SUS-10-001** Performance of Methods for Data-Driven Background Estimation in SUSY Searches
[4] CMS Collaboration 2007 *J. Phys. G* **34** 995 CMS physics Technical Design Report, Volume II : Physics Performance
[5] CMS Collaboration 2010 *CMS Physics Analysis Summary* **JME-10-003** CMS Jet Performance in pp Collisions at $\sqrt{s}$ = 7 TeV
[6] CMS Collaboration 2009 *CMS Physics Analysis Summary* **PFT-09-001** Particle-Flow Event Reconstruction in CMS and Performance for Jets, Taus, and E_T^{miss}
[7] CMS Collaboration 2010 *CMS Physics Analysis Summary* **PFT-10-002** Commissioning of the Particle-Flow Reconstruction in Minimum-Bias and Jet Events from pp Collisions at 7 TeV
[8] CMS Collaboration 2007 *CMS Physics Analysis Summary* **JME-07-002** Plans for Jet Energy Corrections at CMS
[9] CMS Collaboration 2009 *CMS Physics Analysis Summary* **JME-09-006** Performance of Missing Transverse Energy Using Calorimeter and Tracks in CMS
[10] CMS Collaboration 2009 *CMS Physics Analysis Summary* **JME-09-010** Performance of Track-Corrected Missing ET in CMS
[11] CMS Collaboration 2010 *CMS Physics Analysis Summary* **JME-10-004** CMS MET Performance in Jet Events from pp Collisions at $\sqrt{s}$ = 7 TeV
[12] CMS Collaboration 2008 *CMS Physics Analysis Summary* **SUS-08-005** SUSY searches with dijet events
[13] CMS Collaboration 2009*CMS Physics Analysis Summary* **SUS-09-001** Search for SUSY with exclusive hadronic N-Jet Events
[14] CMS Collaboration 2009 *CMS Physics Analysis Summary* **SUS-09-004** Data-Driven Background Estimates for SUSY Di-Photon Searches
[15] CMS Collaboration 2010 *CMS-NOTE* **2010-008** The CMS physics reach for searches at 7 TeV

Status and prospects for the KLOE-2 experiment

Stefano Miscetti
on behalf of the KLOE-2 collaboration
Laboratori Nazionali di Frascati dell'INFN, via Enrico Fermi 40, I-00044 Frascati (Rome) Italy
E-mail: stefano.miscetti@lnf.infn.it

Abstract. The KLOE-2 experiment is starting a data taking campaign at DAϕNE, the e^+e^- collider of the Frascati Laboratories of INFN, with the aim of integrating O(20 fb^{-1}) in few years of running. The KLOE-2 detector has been upgraded with the insertion of small angle taggers for $\gamma\gamma$ physics. The construction of larger upgrades such as an Inner Tracker and two small angle calorimeters is underway.

1. From KLOE to KLOE-2

The KLOE experiment is starting a new data taking campaign (as KLOE-2) at DAϕNE, the e^+e^- collider of the Frascati Laboratories of INFN. The machine has been recently modified by using a Crabbed Waist scheme [1] in the interaction region. An increase of $\sim$ a factor of 3 in luminosity is expected to get O(20 fb^{-1}) in few years of running. The KLOE-2 [2] detector has also been upgraded with the insertion of small angle taggers for $\gamma\gamma$ physics. An integrated luminosity of 5 fb^{-1} should be collected in the first year of running, phase-1. The construction of larger upgrades (see Fig. 1): an Inner Tracker, IT, and two small angle calorimeters is underway with the aim of completing their insertion at the beginning of 2012, phase-2.

The KLOE-2 roll-in in DAϕNE was succesfully completed in January 2010 while the insertion of the new IP region, surrounded by our background screens, was done in June 2010. The KLOE magnet was cooled down in June and turned on at the nominal value of 5.2 kGauss for a month in July. This allowed us to perform all final calibrations for the calorimeter, EMC, and the tracking, DCH, systems. After the August shutdown, we got troubles in cooling the magnet which needed to be warmed up to room temperature before succeeding to reach a stable cool down. In November, first beams with normal optics were circulating in DAϕNE. Currents up to 1 A were reached both for the e^- and e^+ beams and many improvements found (vertical beam dimension down to 150 μm achieved). First collisions were tried at low currents (up to 300 mA/beam) without turning on the Crab sextupoles and delivering instantaneous luminosity of few 10^{31} $cm^{-2}sec^{-1}$. However the machine background in the detector was too high to collect data in a reasonable way. On January 2011, the septum magnet for the e^+ injection line accidentaly broke and we keep running e^- beam for few weeks to improve the background conditions. Apart few changes in the optics, we succeeded to run with 700 mA of electrons with all detector "on" after having hardened the tripping condition of the DCH. The machine background was not yet optimal since we got $\sim$ 10 MHz of single EMC clusters due to machine background. In February, we made a 5 days access on the KLOE-2 detector to add some additional lead screens (1 cm thick) and a special lead plug around the inner quadrupoles area.

At the moment of writing, we have completed our fixing and a new septum magnet has been inserted. We expect to start the run in the next weeks.

2. The physics program

The KLOE-2 physics program is very wide and aims to improve the experimental studies on hadronic physics at low energy, kaons, $\gamma\gamma$ physics, the contribution of hadron vacuum polarization to the muon anomalous magnet, CKM unitarity and Lepton Universality, CPT symmetry and quantum interferometry, QI, studies. A full description of this program can be found elsewhere [3]. Detailed analysis of QI studies and searches for a "U" boson from a dark gauge sector were presented at this meeting and can be found in the same proceedings [4, 5]. The relevance of $\gamma\gamma$ physics is reported in section 3 where we describe also the implication of the new tagger system. Regarding the quantum interferometry, we recall only that, in the phase-2 of the experiment, the presence of the IT will improve by a factor between 3 to 4 the vertex resolution and consequently the reconstruction of the relevant QI parameters. As an example, assuming that the K_S, K_L pair decays to the same final state $\pi^+\pi^-$, the distribution of $\Delta T = |t1-t2|$, where t1 and t2 are the decay times of the K_S-K_L, shows a deep around zero since no simultaneous decays can occur due to destructive quantum interference. The reconstruction of the deep gets better increasing the vertex resolution reaching almost an "ideal" curve with the insertion of the IT in the tracking. This resolution improvement acts on the determination of the QI parameters as a fourfold increase on luminosity.

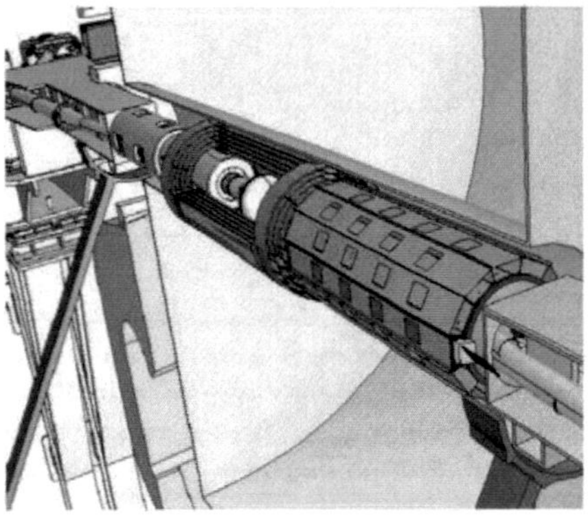

Figure 1. Expanded view of the inner KLOE-2 region. The IT and QCALT detectors are shown close to the IP.

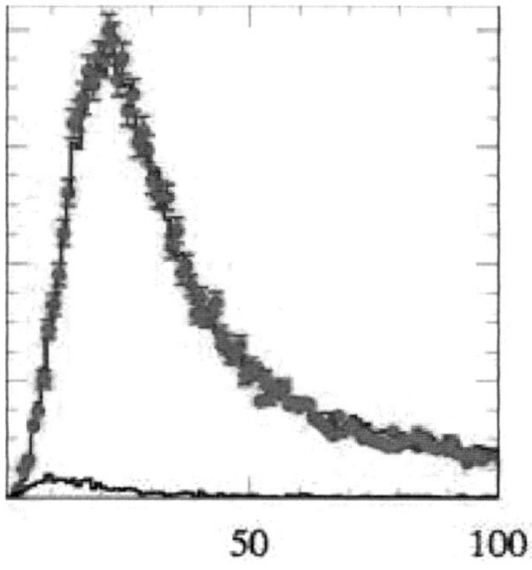

Figure 2. Distribution of χ^2 for events with six photons and a K_L-crash. Red points (data), blue line (MC bkg), black line (MC signal).

In this section, we describe in detail only two analysis examples which can be carried out with the first 5 fb^{-1}: the search for $\phi \to KK\gamma$ and $K_S \to 3\pi^0$ decay chains.

The $\phi \to K\bar{K}\gamma$ process should proceed through the $\phi \to [a_0(980) + f_0(980)]\gamma$ intermediate state. In KLOE, we searched for this channel by looking at the decay chain $\phi \to K_S K_S \gamma \to \pi^+\pi^-\pi^+\pi^-\gamma$. The main backgrounds being the resonant $e^+e^- \to \phi\gamma \to K_S K_L$ with an IRS photon and the continuum $e^+e^- \to \pi^+\pi^-\pi^+\pi^-\gamma$ processes. From an integrated luminosity of 2.2 fb^{-1}, 5 candidates are found in data, while 3.2 ± 0.7 background events are expected from MC . This leads to BR($\phi \to K\bar{K}\gamma$) $< 1.9 \times 10^{-8}$ at 90% CL. Scaling these numbers to the KLOE-2 statistics a sensitivity of $BR(\phi \to K^0\bar{K}^0\gamma) < 1 \times 10^{-8}$ is expected.

The best limit on the CP–violating decay $K_S \rightarrow 3\pi^0$ comes from the analysis of $\sim$ 450 pb^{-1} collisions taken during 2001-2002. In this search, we tagged the K_S by means of a K_L interaction in the calorimeter, K_L-crash, and looked for events with six prompt photons. Discriminating variables based on kinematic fit and photon pairing were used to elimate the obvious background of $K_S \rightarrow 2\pi^0$ + 2 additional photons. We got 2 candidates with an expected background of 3.13 $\pm$ 0.82 events and a reconstruction efficiency of 24.4 %. We set the limits to be 1.2×10^{-7} at 90 % C.L.

We have lately started a production campain with a larger sample (1.7 fb^{-1} data, 3.4 fb^{-1} MC) aiming to a much stronger background reduction. Driven by MC studies of signal and background, we add two cuts: (i) splitting recovery, by cutting on the minimal 3-D distance between clusters, R_{min}, and (ii) fake-rejection, by hardening the K_L-crash selection (E > 150 MeV and $0.200 < \beta_K < 0.225$). In Fig. 2, we show a good data/MC agreement in the χ^2 distribution of the kinematic fit for the six photon sample. The new cuts show their rejection power in MC. At the end of a preliminary optimization, we find 0 candidates in the simulated sample while keeping a reasonable overall efficiency of $\sim$ 22 %. We expect to reach a limit which is not anymore background limited and so will scale with increasing luminosity at values of $\sim$ 3 (1) $\times 10^{-8}$ for the KLOE (2.5 fb^{-1}) or KLOE-2 phase-1 (5 fb^{-1}) data sets. The addition of the upgrades and a larger data set will lead to observe for the first time such a decay.

3. $\gamma\gamma$ physics and status of the $\gamma\gamma$ taggers

The $e^+e^- \rightarrow e^+e^-\gamma^*\gamma^* \rightarrow e^+e^-X$ process, where X is the final state resulting from the fusion of two photons, gives access to states with $J^{PC} = 0^{\pm+}$, $2^{\pm+}$. In the low-energy region covered by the KLOE detector, $m_{\pi^0} < W_{\gamma\gamma} < 700$ MeV, existing measurements are affected by large statistical and systematic uncertainties. At KLOE, where there is no tagging of the outgoing e^+e^-, $\gamma\gamma$ interactions have been studied using 240 pb^{-1} collected at $\sqrt{s} = 1$ GeV, as the main source of background comes from ϕ decays. We started looking for the $e^+e^- \rightarrow e^+e^-\pi^0\pi^0$ process. A large excess with respect to known background is evident at low $M_{4\gamma}$ values. The η radiative width,$\Gamma_{\eta\gamma\gamma}$, is usually extracted from the measurement of the $e^+e^- \rightarrow e^+e^-\eta$ cross section. The KLOE analysis is performed using the $\eta \rightarrow \pi^+\pi^-\pi^0$ decay channel. After background rejection, distributions of longitudinal momentum (p_L) and recoil missing mass (M^2_{miss}) are independently fitted with the superposition of MC shapes for signal and background More than 600 $e^+e^- \rightarrow e^+e^-\eta \rightarrow e^+e^-\pi^+\pi^-\pi^0$ events are obtained.

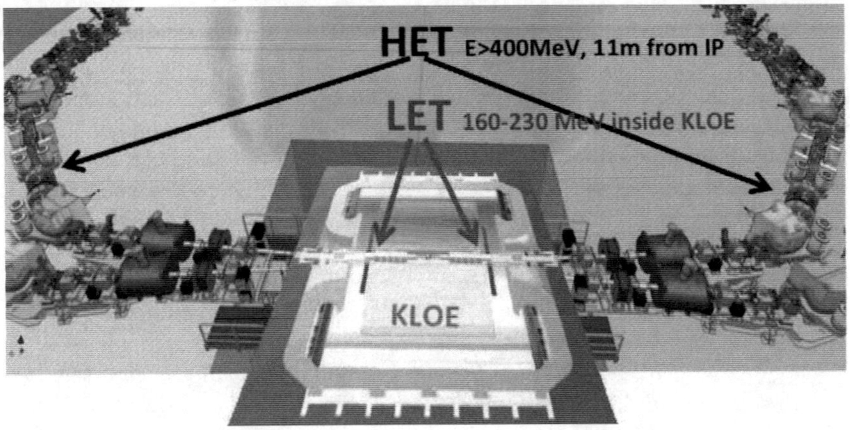

Figure 3. Scheme of the $\gamma\gamma$ taggers in DAΦNE and KLOE.

These KLOE analyses represent also a feasibility test for the study of $\gamma\gamma$ processes in KLOE-

2, that will take advantage of a tagging system for the detection of scattered e^+/e^- in the final state thus allowing to perform these measurements also at the ϕ peak. Moreover, with the tagger, it would be possible to measure the transition form factors $\mathcal{F}_{X\gamma^*\gamma^*}(q_1^2, q_2^2)$ as a function of the momentum of the virtual photons, q_1^2 and q_2^2, relevant for the evaluation of the hadronic light-by-light contribution to the muon magnetic anomaly. The installation of $\gamma\gamma$ taggers [6] is almost completed. In Fig. 3, the taggers system is shown. It consists of a low energy tagger, LET, inside KLOE and a high energy tagger, HET, after the first bending magnet 11 m far away from the IP. In year 2010 the detectors have been realized and the LET calorimeters have been installed on the final focusing region of DAϕNE. The LET is composed by two stations of 20 LYSO crystals, each readout by 3×3 mm^2 SIPM's from Hamamatsu. The expected energy resolution is of ~ 7 % at 100 MeV. All channels have been acquired by the KLOE DAQ system and test with Cosmic rays are in progress. The HET consists instead of two stations of position detectors composed of 32 scintillator counters of $5 \times 6 \times 3$ mm^3 readout by small size PMTs. The HET optical system is ready, including shaped optical guides and scintillators wrapped with aluminized mylar. The three main components of the HET mechanics are the insertion and vacuum systems, the structural support for the phototubes and the external support. The insertion and vacuum systems have been installed inside the beam pipe. The structural support for the phototubes and the external support have been designed and are under construction. Two out of four boards for the acquisition chain, i.e., the Front End Buffer board (FEBb), and the Fast Discriminator Shaper board (FDSb), are being tested, while the others, i.e. the Acquisition V5 board (ACQ-V5b) and the Slow Environment Control System board (SECSb), are being constructed.

4. The Inner Tracker

The Inner Tracker[7] (IT) is composed by four cylindrical layers of triple-GEM (CGEM) at radii from 13 to 23 cm, and, as shown in Fig. 1, will get installed between the spherical beam pipe and the drift chamber. The total active length for all layers is 70 cm. The anode readout of each CGEM is segmented with 650–μm–pitch, XV–patterned strips at a stereo angle of $\sim 40°$, for a total of 30,000 FEE channels based on GASTONE ASIC [8], a dedicated 64–channels chip developed for the KLOE-2 experiment.

The construction and extensive test of a full-size CGEM prototype has demonstrated the feasibility of such a novel low–mass and dead–zone–free vertex detector. The final readout configuration has been successfully validated testing small planar prototypes operated in magnetic field. During 2010, to conclude the R&D phase, a large area planar GEM prototype of 300×700 mm^2 area has been built. This detector has been equipped with the GASTONE 64–channels final release, readout by the Off Gastone Electronic (OGE) board and tested at CERN-PS T9 beam-line in October, 2010.

The detector mechanics, assembly and quality control tools have been prepared and are now ready for the construction of the Inner Tracker. The Vertical Insertion System (VIS) has been built. This system will be used to assembly each IT layer by inserting the cylindrical electrodes one inside the other with high precision. In 2010 the activity has been mainly focused on the design of the detector components and tools necessary for the construction of the Inner Tracker: (i) **Detector Components**, the design of the detector components and construction tools originates from previous construction experience and, more specifically, for the construction of the full scale prototype of the CGEM [9]. In particular, the materials used for the CGEM were largely tested and validated for high-rate environments and for different gas mixtures (Ar, i-C4H10, CF4, CO2). The CGEM has the typical triple-structure of such a kind of micro-pattern gas detector: the gaps among the different electrodes of the detector (cathode-G1,G1-G2,G2-G3 and G3-anode) define the various regions of the detector itself: drift (3 mm wide), transfers and induction (2 mm wide). The frames are realized in EPCG203 fiberglass, a composite material

Figure 4. (Top) the aluminum moulds used for the construction of the cylindical electrodes, (bottom) Visual Inspection tool.

Figure 5. The assembly machine (VIS) used for the insertion of the electrodes one inside the other.

that allows precise machining, and are glued at the edge of the cylindrical electrodes outside their active area, define the various gaps of the detector. The most relevant modification with respect to the prototype design is represented by the embedded-anode, consisting of a very light honeycombed carbon fiber cylinder (CFC) on which the anode readout circuit is glued. The CFC acts as a rigid support for the whole detector layer. The 700 mm long GEMs are realized with the new single mask procedure that allows large area foils to be built[10]. In order to overcome the limit on the width of the raw material used for the GEM foils (600 mm), all the large electrodes of each layer are realized by splicing three smaller foils. The GEM foils, as well as the XV strip-pad patterned anode readout circuits and the cathode foils, are realized on the basis of our design by the CERN EST-DEM Printed Circuit Board Workshop.
(ii) **Construction and tooling**, the construction of the CGEM layers composing the IT are the same used for the prototype. The main construction steps can be summarized as follows: 1. the three GEMs, as well as the anode and cathode foils are glued together in order to obtain a single large foil. For this operation we exploit a precise Alcoa plane and a vacuum bag technique. 2. the large foil is then rolled on a very precise aluminum cylindrical mould covered with a 0.4 mm machined Teflon film for easy and safe extraction of the cylindrical electrode (fig. 4-top.left). The mould is then enveloped with the vacuum bag, and vacuum is applied for the glue curing time (about 12 hours). 3. the final assembling of a CGEM layer is performed by means of the Vertical Insertion System (VIS), fig. 5-right, a quite complex device that allows a precise and safe insertion of the cylindrical electrodes one inside the other.
(iii)**Material preparation and Quality Controls**, before the final assembly of the different detector parts, each component follows a well defined preparation procedure that generally includes a global optical inspection (see Fig. 4.bottom-left), a cleaning and an HV test. In

particular for GEMs the HV test is repeated at each construction step, in order to avoid the assembly of damaged GEM and to minimize the losses of precious components.
(iv)**Final readout**, the IT will be inserted between the beam pipe and the Drift Chamber inner wall. This location has very poor heat-exchange capability, it is not accessible after system installation and there is a reduced space for routing the cables. In addition, due to the GEM moderate gas gain, the front-end electronics must be installed on the detector to maximize S/N ratio. Therefore low power consumption and low number of connections are required together with high reliability of the entire system. To fulfill these requirements the GASTONE chip has been developed. This chip, built with the 0.35 CMOS technology, implements a mixed analog-digital circuit with 64 channels. Each channel is made of 4 blocks: a charge preamplifier with a sensitivity of about 20 mV/fC, a shaper, a leading-edge discriminator with programmable threshold and a monostable circuit to stretch the discriminated signals then allowing their synchronization with the KLOE trigger. Up to 180 chips assembled on 90 boards are required to fully instrument the detector. Digital output signals are serially readout by means of dedicated boards (then minimizing I/O cables) located about 4 m from the detector. Readout boards provide also LV and L0 to the front-end boards and ECS connection.

We have now the material to build two layers and we are proceeding with the construction of the cathods. We are still waiting the completion of the cleaning room to start assembly the layers which is foreseen to start in spring 2011.

5. The QCALT: a Quadrupole Calorimeter with Tiles

To increase the photon detection efficiency for $K_L \rightarrow 3\pi^0$ events we are developing a new Quadrupole Calorimeter to cover the area where the inner focalizing quadrupoles are located. There will be [11] two detectors, one per side with respect to the IP, named QCALT, (see Fig. 6), each one consisting of a dodecagonal structure, $\sim$ 0.9 m long, composed by a sampling of five layers of 5 mm thick scintillator plates alternated with 3.5 mm thick W/Cu (90/10) plates. The total absorption depth will be of 4.75 cm (5.5 X_0). The active part of each plane is divided into sixteen tiles of $\sim$5$\times$5 cm^2 area with 1 mm diameter WLS fibers embedded in circular grooves. Each fiber is then optically connected to a silicon photomultiplier of 1 mm^2 area, SiPM, for 80 channels/module and 1920 channels in total. The "T" in QCALT stays for tile, since we will read-out each tile singularly to get O(2 cm) resolution along z and be able to substain large rates O(MHz/module) thanks to the high granularity. Relying to the new quality SIPM of Hamamatsu, we have reached [12] a light yield of $\sim$ 30 p.e/MIP and a time resolution of $\sim$ 1 ns enough to obtain a reconstruction efficiency above 95% for photons with energy > 20 MeV.

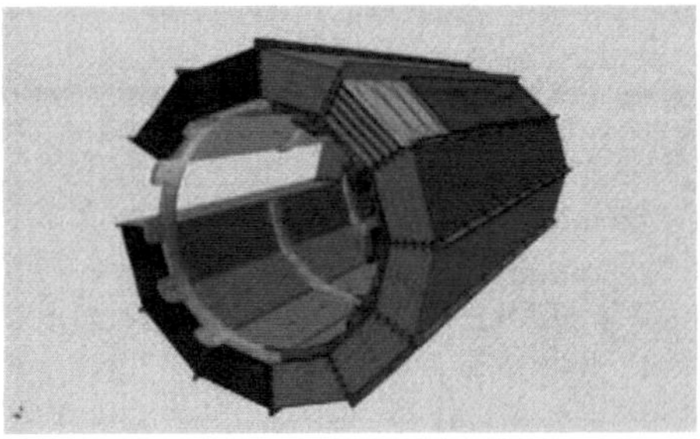

Figure 6. CAD drawings of the QCALT calorimeter.

Figure 7. Picture of the QCALT module-0.

During 2010, we have realized the first full size "module-0", (see Fig. 7), in order to understand the assembly techniques and the problems of integrating the readout in the reduced space available inside KLOE. During this construction we have learnt: (i) to survey the plastic scintillator thickness and how to prepare the grooves; (ii) to get well machined W/Cu plates; (iii) to "wrap" the tiles, easyness of construction and uniformity indicates that the best method to follow is to spray the tile with reflective paint; (iv) to couple fibers to SIPM's, a first set of fiber-SIPM PVC connector has been realized to check the optical contact reproducibility. We discovered that the mechanical precision in positioning should be better than 0.1 mm. To improve on this, we need for each module, to position the fibers in a PVC holder and match them with a PCB where 80 SIPM's will be bond; (v) to calibrate after assembly, this has been done acquiring cosmic rays,CR, in most of the columns (i.e. five planes at a time) by triggering with an external scintillator; (vi) to test the FEE; a set of twenty single channel amplifiers (dimension 10×20 mm^2) has been produced together with an HV driver done in NIM standard.

We have also studied the long-term stability of the tiles by firing a reference tile with a UV LED in a temperature controlled environment. We observed no hints of signal deterioration along an observation period of few months. We have then carried out a complete study on how to safely handle the calorimeter both during assembly and installation over the beam pipe support. During assembly, we will leave the fibers passing through the fiber holder and connect them to reference SIPM's by means of a special aluminum connector. This will allow to check the construction of each plane by firing with a UV LED along the different tiles. We plan to assembly one plane/day, one module/week for each assembly station. We will then bring the module to the SPCM of LNF for final fiber milling over the PVC holder.

Due to the difficulty of positioning 80 SIPM's in the front calorimeter face, we are working to integrate them over a PCB. First PCB prototypes have been realized and the IRST/FBK firm of Trento has built for us the first 160 SMD SIPMs of 1.2×1.2 mm^2 dimension to minimize the problem of alignment to the fiber. The IRST will also bond the SIPM's on the PCB. We expect to have a final working PCB for spring 2011.

The FEE, on detector, will consists of four boards/module of 20 channels/each where the preamplifiers and the HV regulator will be located. The OFF-detector electronics will consist of a 40 channel transition board that will do both the discrimination and the HV control. The readout will be organized by a TDC multihit with a range between 0.6 – 0.8 μsec and 1 ns resolution. Design of the final FEE boards is underway. We expect to have the first prototypes ready for spring 2011. At the end of 2010, we have ordered the material to start constructing the first six modules. We expect to start construction on spring 2011.

6. The CCALT: a Crystal Calorimeter with Timing

The CCALT is a crystal calorimeter which covers the solid angle between 10 and 18 degrees in order to extend our acceptance to low energy photons. The necessity of a calorimeter with high photon reconstruction efficiency (ϵ_γ >99% for $E_\gamma > 20$ MeV) and good time resolution (σ_t 200-500 ps) in this angular region has been proposed [11] with the double task of (i) vetoing the presence of unwanted photons in rare η and K_S channels ($K_S \to 2\gamma$ or $\eta \to \pi^0\gamma\gamma$) (ii) increasing the acceptance for very rare decays such as $K_S \to 3\pi^0$.

A good timing resolution and a fast response, ($\tau_{scint} < 50$ ns), are needed since the CCALT will be positioned (see Fig. 8) in between the interaction point and the first inner quadrupole, QD0, where an intense flux of photons of O(MHz)/side, produced from Tousheck particle, is expected. The detector is composed by two calorimeters, one per side with respect to the IP, each subdivided in twelve modules. Each module is constituted by four crystals of special trapezoidal shape with $\sim$ 9 cm length and a transversal area of $\sim 25 \times 25$ mm^2 for the largest face. Given the reduced space available, we need fast and high density crystals with a large light yield. The crystal matching these requirements is the LYSO which has a density of 7.5 g/cm^3, a X_0 of 0.9 cm, a τ of 40 ns and a light yield of 500 p.e./ MeV when readout by a Bialcali PM. In 2009, we

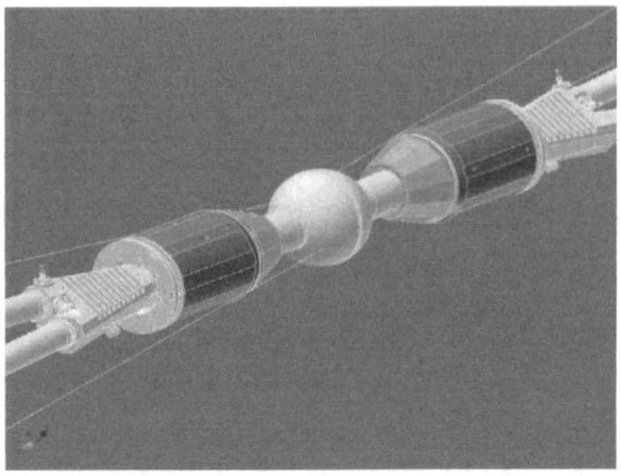

Figure 8. CAD drawings of the CCALT calorimeters (in blue) positioned over the IR.

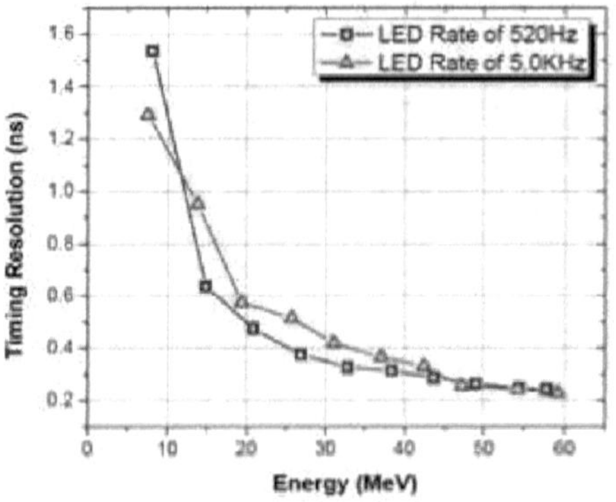

Figure 9. Dependence of the timing resolution of LYSO readout by SIPM's as a function of energy.

have built a first prototype done by an inner matrix of ten LYSO crystals readout by means of 5×5 mm^2 APD's. We have tested this prototype with 100 to 500 MeV electrons at the Beam Test Facility (BTF) [13] of LNF obtaining an energy resolution of 5% (13%) at 500 (100) MeV and a timing resolution between 100 to 200 ps. The energy resolution term proportional to 1/E resulted too large with respect to the measured noise term of 500 keV/channel. In order to understand that, we have constructed a better prototype readout by larger area APD's (10×10 mm^2) achieving a noise term of 150 keV/channel in laboratory tests. We plan to study the components of the energy resolution at the tagged photon beam facility of Mainz Microtron (MAMI) in spring 2011.

The integration with the other detectors, IT and QCALT, asks for tight requirements such as reducing as much as possible the thermal dissipation while respecting the space constraints. We have therefore investigated if we can succesfully readout the CCALT by means of large area SIPM's thus avoiding the usage of preamplifiers which have a typical power consumption of 100 mV/channel and a minimal transverse area of 10×20 mm^2. Two crystals have been optically connected with optical grease to two different SIPM's, a 4×4 (6×6) mm^2 from IRST/FBK (Hamamatsu) and their response to a UV LED has been tested both in charge and in time. In Fig. 9, we report the timing resolution of the crystal readout by IRST SIPM'a as a function of the energy for low rate pulsing. We see no deterioration of the timing response up to 50 kHz and a worsening up to a factor of 2 for rate close to the MHz. The timing resolution remains between 200 to 450 ps also at low energies, ($E_\gamma > 50$ MeV), showing that this technique will not be worse than a factor of two with respect to the APD readout.

References

[1] C. Milardi *et al.*, ICFA Beam Dyn. Newslett. 48 (2009) 23.
[2] http://www.lnf.infn.it/lnfadmin/direzione/KLOE2-LoI.pdf.
[3] G. Amelino-Camelia *et al.*, Eur. Phys. J. C 68 (2010) 619
[4] A. DeSantis "Status and prospects for Lorents and CPT violation tests at KLOE and KLOE-2 ", on these proceedings.
[5] S. Giovannella "U boson searches at KLOE-2 ", on these proceedings.
[6] D. Babusci *et al.*, arXiv:0906.0875 and LNF-10/17(P) INFN-LNF, Frascati, 2009.
[7] F. Archilli et al., arXiv:1002.2572v1 and LNF-10/3(P) INFN-LNF, Frascati,2010.
[8] A. Balla et al., Nucl. Inst. & Meth. **A 604** (2009) 23.
[9] G. Bencivenni et al., NSS Conference Record, 2007 IEEE, Volume 6 pp. 4666-4670.
[10] M. Alfonsi et al., Nucl. Inst. & Meth. **A 617** (2010) 151.

[11] F. Bossi et al., LNF-07/19(IR) INFN-LNF, Frascati, 2007.
[12] M. Cordelli et al., Nucl. Inst. & Meth. **A 617** (2010) 105.
[13] M. Cordelli et al., Nucl. Inst. & Meth. **A 617** (2010) 109.

Journal of Physics: Conference Series **335** (2011) 012071 doi:10.1088/1742-6596/335/1/012071

The NA62 experiment at CERN

Gianluca Lamanna
CERN, CH-1211 Geneve 23, Switzerland
E-mail: gianluca.lamanna@cern.ch

Abstract. The Branching Ratio (BR) of the rare decay $K^+ \to \pi^+\nu\bar{\nu}$ is predicted with a small error in the framework of the standard model (SM). The contribution of long distance terms is suppressed thanks to the GIM mechanism and the short distance dynamics is almost completely described by well measured semileptonic operators. The cleanness of the SM prediction together with the small BR ($\sim 10^{-10}$) makes this decay an interesting opportunity to describe the presence and the contribution of the new physics using relatively low energy processes. The NA62 experiment at CERN aims at measuring O(100) events in two years of data taking with 10% background. From an experimental point of view this measurement is very challenging because the weakness of the signal signature and the huge amount of background processes. In this paper the key points of the experiment will be described after a brief introduction on theoretical aspects and present results.

1. Introduction

Among the GIM suppressed rare decays in B and K meson, the $K \to \pi\nu\bar{\nu}$ processes play an important role in search for new physics. These decays are completely dominated by short distance dynamics and the amplitude is governed by just one single semileptonic operator, defined using the well measured semileptonic kaon decays. The relevant diagrams are boxes and penguins where the top contribution is dominant. The contribution from the CKM dynamics comes from the top coupling with d and s quark. For that reason the matrix element is proportional to λ^5 and the BR is very small. For what concerns the $K^+ \to \pi^+\nu\bar{\nu}$ a small contribution from the c quark coupling has to be taken into account. This contribution gives an irreducible theoretical error of 3% in the BR prediction, while all the other contributions in the total error are parametric, depending on measured quantities. The present SM prediction [1], including NNLO QCD corrections [2], electroweak corrections [3] and long distance contribution [4] is

$$BR(K^+ \to \pi^+\nu\bar{\nu}) = (8.5 \pm 0.7) \cdot 10^{-11}$$

The $K_L \to \pi^0\nu\bar{\nu}$ is also cleaner from a theoretical point of view, with an irreducible error of $\sim 1\%$. The precision in the prediction for these decays allows a stringent test for the SM and, at the same time, a very powerful research for new physics beyond the SM. Several extensions, including models with 4th quarks generation [5], Littlest Higgs [6], Randall and Sundrum mechanism [7] and supersymmetric flavour models [8], give sizable differences in both $K^+ \to \pi^+\nu\bar{\nu}$ and $K_L \to \pi^0\nu\bar{\nu}$ branching ratios. For that reason the measurement of these decays will be very important not only to understand the magnitude of contribution of new physics, but also to distinguish among different models. Nevertheless the very good precision in theoretical prediction needs an experimental measurement with compatible error. At the

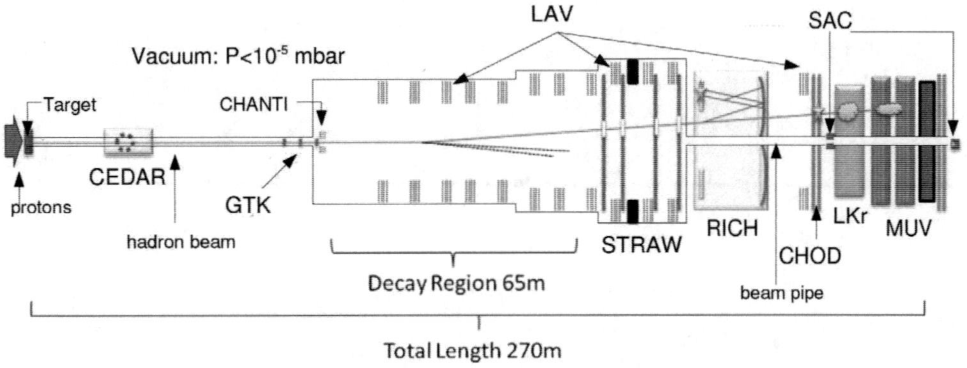

Figure 1. NA62 experiment layout.

moment the experimental precision is quite far from the goal. In particular in the $K^+ \to \pi^+\nu\bar{\nu}$ the most precise results have being obtained by the E787 and E949 experiments at BNL, by studying stopped kaon [9]. The final BR result

$$BR(K^+ \to \pi^+\nu\bar{\nu}) = (17.3^{+11.5}_{-10.5}) \cdot 10^{-11}$$

is based on 7 measured events.

The NA62 experiment aims at measuring O(100) events in about 2 years of data taking, with a small systematic uncertainty. To this purpose at least 10^{13} K^+ decays are required, assuming an acceptance of 10%. To obtain a contribution of the background below 10% a factor of 10^{12} in rejection the other kaon decay mode is needed. Signal acceptance, high intensity beam and rejection factor for the background drive the NA62 design, as will be described in the following.

2. NA62: general overview

In figure 1 the experimental layout is shown. The hadronic beam, composed by $\sim 6\%$ of kaons, is produced by using the 400 GeV/c proton beams coming from the SPS accelerator, impinging in a 40 cm long beryllium target. The beam transport system allow to select an unseparated hadron beam with a momentum of 75 GeV/c and spread of 1%. Differently with respect to the previous experiments, the NA62 technique is based on such high energy beam, instead of using stopped kaons. The advantage of this is both in the possibility to have an intense secondary beam and a higher energy in the decay products of the kaons: thanks to that the requirements to the photon veto system, a very essential part of the experiment as we will see, can be easily fulfilled. On the other hand the use of a high energy hadron beam imply a long decay region and the impossibility to separate the secondary beam components. In any case the large fraction of the decay products of pions present in the beam remains along the beam trajectory, thanks to the large longitudinal momentum with respect the transverse momentum of the decay. Nevertheless the kaon is positively identified in the beam. The decay region begins $\sim$ 100 m after the target, and is housed in a $\sim$ 117 m long and 2.4 m diameter (in average) vacuum tube. The vacuum inside the decay region is maintained at the level of 10^{-6} mbar using diffusion cryogenic pumps, in order to reduce the secondary interaction of the K decay products and to avoid, as much as possible, inelastic scattering of the pions present in the beam. Along the decay region a system is placed in order to veto events not fully contained in the acceptance of the downstream detectors. The momentum of the K decay products is measured inside the vacuum by a spectrometer based on straws chambers and other detectors, described in details in the following, are employed for particle identification and vetoing. The detector

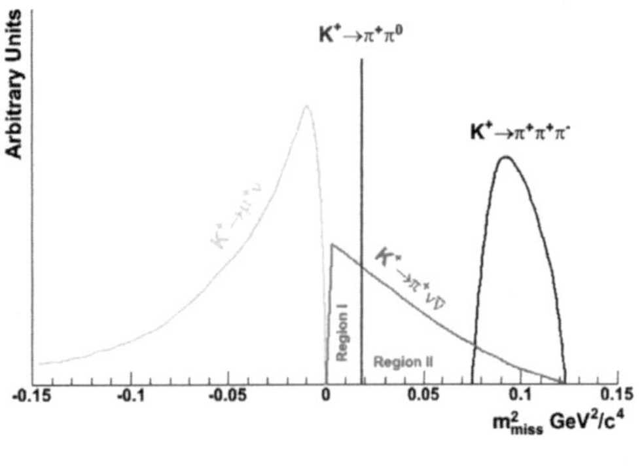

Figure 2. Distribution of kinematically constrained background.

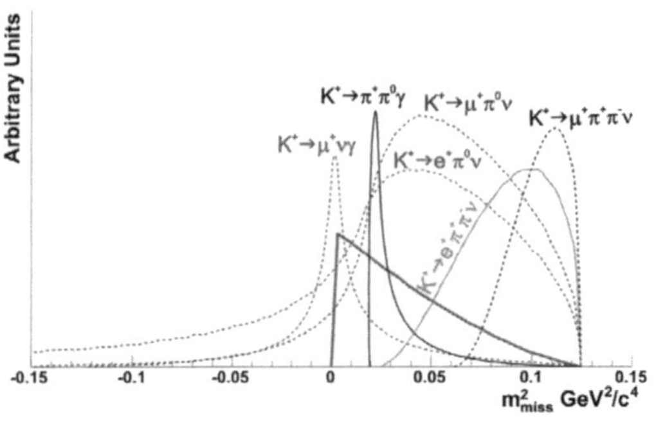

Figure 3. Distribution of not kinematically constrained background.

is designed to have a complete coverage for the K decay products: the very feeble branching ratio and the weak experimental signature, impose an identification of the decay channel on event-by-event base. In table 1 the main K decay modes are listed. The $\pi^+\pi^0$ and the $\mu\nu$ are

Table 1. Branching ratio of the main K decay modes.

Decay mode	Branching Ratio
$\mu^+\nu$	63.5%
$\pi^+\pi^0$	20.7%
$\pi^+\pi^+\pi-$	5.6%
$e^+\pi^0\nu$	5.1%
$\mu^+\pi^0\nu$	3.3%
$\pi^+\pi^0\pi^0$	1.8%

clearly the most dangerous, but almost any decay mode could be a sources of background. In order to reduce the contribution of the spurious decay modes, the background is studied as a function of the squared missing mass, defined below . In the Fig.2 is shown that two regions with potentially good signal to noise ratio can be identified in the "ideal" missing mass distribution. In the real world a good kinematic reconstruction has to be applied in order to have negligible contribution of resolution tails in the signal regions. We refer to the background that can be separated using the signal regions definition as "kinematically constrained". This amounts to $\sim$ 92% of the total background. The remaining $\sim$ 8% is "not kinematically constrained" and spans over the whole range of the missing mass distribution, as shown in Fig.3. The rejection of this component of the background is completely based on the efficiency of particle rejection and identification. The tolerable level of background is around 10% of the signal events. In order to collect O(100) events in a reasonable data taking period, the intense kaon beam produces about 10 MHz of events in the downstream detectors. Particular care has been devoted in designing an efficient data acquisition and trigger system to cope with a big data flux without introducing any kind of additional inefficiency. Summarizing, the key components of the experiment are: a

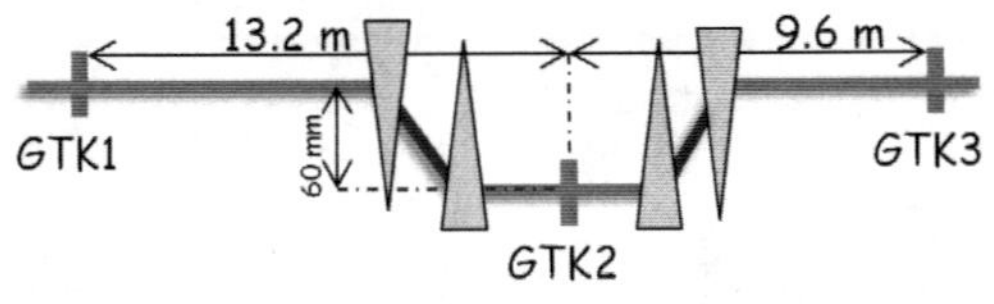

Figure 4. The GTK momentum measurement principle.

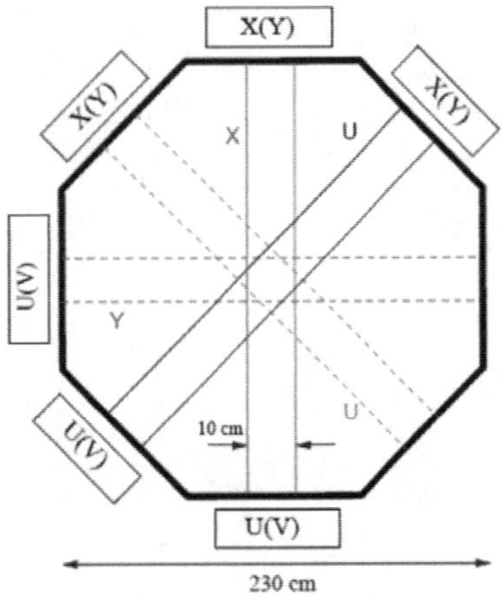

Figure 5. Straws disposition in a chamber.

good kinematic rejection in order to reduce the kinematically constrained background, a high efficiency veto system and particle identification to recognize the not kinematically constrained kaon decays and the resolution tails of the main decay modes, and a trigger system to allow a reliable, lossless and efficient data acquisition at high rate. In the next sections these points will be presented in details.

3. Kinematics rejection

The signal region, as already mentioned above, is defined using the kinematics of $K^+ \to \pi^+ \nu\bar{\nu}$. In particular the squared missing mass (m^2_{miss}) defined as the squared difference of K^+ and downstream single track in the π^+ hypothesis 4-momenta. This variable separates the signal from most of the kaon decays, as already shown in Fig.2. Although clean regions can be defined exploiting the kinematic reconstruction of the background events (mainly $\pi^+\pi^0$ and $\mu^+\nu$), resolution effects can bring non-signal events in the signal regions. To increase the resolution in the m^2_{miss} variable, both kaon and pion momenta have to be measured with high precision and with small non-gaussian tails. For that reasons the multiple scattering has to be reduced as much as possible by using light mass detectors placed in vacuum. In addition, the beam spectrometer (for the kaons) and the downstream spectrometer (for the pions) information has to be associated with a good time resolution, due to the high intensity beam.

The designed beam spectrometer, called GigaTracker (GTK), consists of 3 Silicon pixel stations. Each station is 6.3x2.9 cm^2 to match the beam dimension. A system of analyzing dipole magnets, as shown in figure 4, allow to measure momentum and direction of any particle in the beam (about 800 MHz). The detector has to be very thin in order to reduce the multiple scattering: 200 μm for the Si sensor and 100 μm for the bump bonded readout chip. The achievable time resolution has been measured at a test beam to be better than $200ps$. The whole detector will count about 18000 pixels, with a maximum rate of 150 kHz per pixel in the central part. Particular care has been devoted to the cooling system design in order to decrease the contribution of the leakage current in a radiation hard environment. The resolution in the momentum measurement will be in the order of $\sim$ 0.2% for 75 GeV/c particles, while the direction will be measured with a precision of $\sim$ 16 μrad.

The downstream spectrometer is formed by four chambers made with 9.6 mm in diameter

Journal of Physics: Conference Series **335** (2011) 012071 doi:10.1088/1742-6596/335/1/012071

straw tubes and one dipole magnet. To reduce the multiple scattering the spectrometer is placed in the same decay region vacuum. Each chamber is equipped with 16 planes of 2.1 m long tubes oriented in the x, y and ±45 degrees directions. In each plane a strip in the central part is keept free from tubes in order to create an octagonal hole around the center of the chamber where the beam of undecayed particles passes through (see figure 5). The resolution of this system is (in GeV/c):

$$\frac{\sigma(P_\pi)}{P_\pi} = 0.3\% \oplus 0.007\% \cdot P_\pi$$

$$\frac{\sigma(dX/dZ)}{dX/dz} = 15 - 45\mu rad$$

The STRAWS and the GigaTracker allow to have a 10^4 and 10^5 rejection factor on $K^+ \to \pi^+\pi^0$ and $K^+ \to \mu^+\nu$, with a definition of the signal region sufficient to have 10% acceptance.

4. Veto system

The kinematic constrains are not enough to reject all the $K^+ \to \pi^+\pi^0$ and $K^+ \to \mu^+\nu$. The additional rejection factor, independent from the kinematics, will be given by the direct detection of γ and μ. Moreover the detection of the not constrained background is completely based on the particle detection. The muon detection will be discussed in the next session. The photon veto system design is driven by the necessity to detect the $K^+ \to \pi^+\pi^0$ with an inefficiency at the level of 10^{-8} for the π^0. In order to cover the whole acceptance for the γ's, the veto system is subdivided into different detector: the large angle veto (LAV) in the region between 8.5 and 50 mrad, the LKr calorimeter in forward region between 1 and 8.5 mrad, and the small angle vetoes (SAV), for gammas with angle less than 1 mrad.

The requirement for the LAV is to have an inefficiency on single photon detection at level of 10^{-4}, in the energy range from 200 MeV up to several tens of GeV. The LAV is made of 12 stations placed along the decay region, inside the spectrometer and after the RICH; three possible options have been tested: scintillating fibers and lead (KLOE calorimeter like), scintillator and lead tiles or lead-glass crystals (from the OPAL calorimeter). Details of the prototypes studies could be found in [10]. All the options match the requirement of 10^{-4} inefficiency, the lead glass solution has been selected. In each station the lead glass blocks are disposed radially in five rings in order to assure at least $20X_0$ (three blocks) hit by any gamma impinging on the detector (figure 6). The whole system uses more than 2500 lead glass crystals. The time resolution has been measured at a test beam to be better than 700 ps.

For the forward gammas the requirements in terms of photon veto inefficiency depend on the energy, going from 10^{-3} for 1 GeV up to 10^{-6} for energy greater than 10 GeV. The LKr calorimeter is a quasi-homogeneous electromagnetic calorimeter $27X_0$ in depth, with a good energy, position and time resolutions, built for the NA48 experiment [11]. It's composed by more than 13000 $2 \times 2cm^2$ cells, continuously digitized at 40 MHz. The performance as photon veto detector has been measured in a special run in 2006 on a 75 GeV/c kaon beam. The readout electronics will be completely rebuilt in order to cope with the higher rate with respect to the NA48 runs.

The region below 1 mrad is covered by two detectors: the IRC and the SAC. Both are Shashlyk calorimeter. The first is placed in front of the LKr to cover the region around the beam pipe where the calorimeter is not efficient enough, the second is placed at the very end of the experiment, in the beam dump, to catch the gammas from the π^0 remaining inside the beam pipe. The undecayed charged beam is deflected before this detector. Before the decay region, an additional charged particles veto, the CHANTI, is placed around the third GTK station to detect the hadronic interactions in the beam spectrometer.

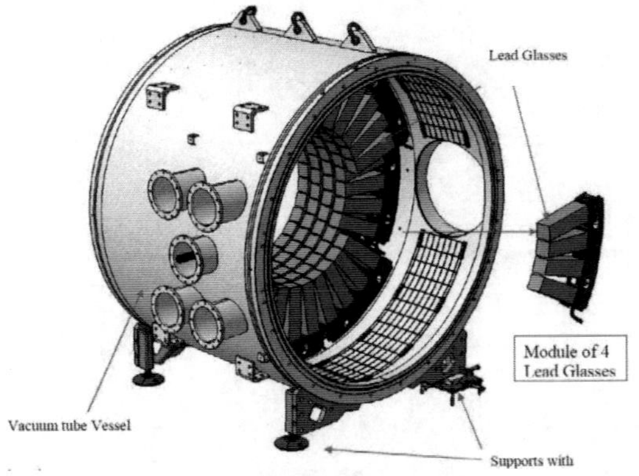

Figure 6. Lead-glass blocks in a LAV station.

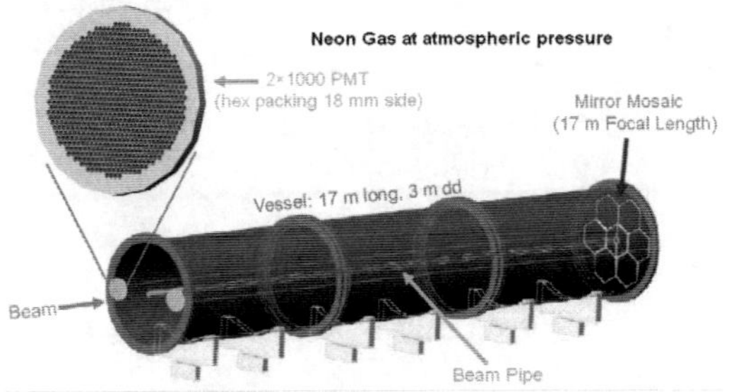

Figure 7. The RICH detector.

5. Particle identification

Mainly to reject the dominant $K^+ \to \mu^+\nu$, decay a redundant particle identification system has been designed, to provide the additional 10^4 rejection factor. The PID system is also employed (in some case together with the veto system) to reach the rejection power for not constrained decay with muons or electrons (as, for instance, $K^+ \to \mu^+\pi^0\nu$ and $K^+ \to e^+\pi^0\nu$).

The RICH (figure 7) is built in a 17 m long and 2 m in diameter vessel, with 1 atm Neon as radiator. The Cherenkov light is focused by two mirrors of 17 m of focal length in two spots covered by about 1000 phototubes each. The phototubes are 1.8 cm in diameter with a quantum efficiency sufficient to have about 18 photo-electrons per rings generated by 35 GeV/c pions. The separation inefficiency between pion and muon is below 1% in the range 15-35 GeV/c . The very good time resolution of 70 ps per track is exploited to match the decay information with the kaon passed inside the GTK.

The MUV, placed after the LKr calorimeter, is subdivided in three subsystems: the MUV1 and MUV2 are composed by iron and scintillators to measure the shape and the energy deposition of the impinging particle to distinguish between muons and pions. The MUV3, placed after a further 80 cm iron wall, is used as fast muon detector for trigger purposes.

For what concerns the electrons identification the RICH and LKr, comparing the electromagnetic energy and the momentum measured by the STRAWS, will be used.

The CEDAR is a differential Cherenkov detector made with H_2 at 1 atm and placed before the GTK, directly on the unseparated hadron beam. Through an optical system made with mirrors and slits, the Cherenkov ring, coming from kaons, is observed in 8 spots, using photomultipliers. In order to decrease the rate per phototubes, the Cherenkov light in each spot is spread in a large surface covered by several phototubes. The main purpose of this detector is to positive identify the kaon to reduce the contribution of the pion from the beam scattering on the residual gas in the decay region.

6. Trigger system

The rate of events in the decay region is strongly dominated by background. According to preliminary simulations the rate on the main detectors is around 10 MHz (table 2). An additional rate of at least ~ 1 MHz of muons coming from the beam production target, must be taken into account. In this environment the requests to the DAQ and trigger systems are:

- Very low DAQ inefficiency ($< 10^{-8}$);
- High trigger efficiency ($> 95\%$);

Table 2. Rates on principal detectors.

Detector	Rate (MHz)
CEDAR	50
GTK	800
LAV	9.5
STRAWS	8
RICH	8.6
LKR	10.5
MUV	9.2
SAC	1.5
Muon Halo	1

- Fully monitored systems;
- Readout without zero suppression for candidates;
- Low random veto probability at trigger level;
- Scalability in terms of bandwidth.

The first request is uncommon in other DAQ systems, but it's crucial for the NA62 experiment, where the full reconstruction of the background is an important issue. For the same reason zero suppression, mainly in the veto detectors, must be avoided as much as possible during the acquisition process. A good trigger efficiency can be obtained using information coming from several detectors with an excellent time resolution, in order to reduce the random veto probability.

A fully digital and integrated DAQ and trigger system has been designed: the digitization in the early stage of the readout system allows efficient monitoring of each stage of the chain, in order to detect any possible source of losses. The trigger system will be split into different levels: the first stage (L0), implemented in hardware using FPGAs, will be used to reduce the total rate to $\sim$ 1 MHz, while the second and third stages (L1 and L2) will be completely software based exploiting powerful PC-farms with large input bandwidth. The data accepted by the L2 will be directly transmitted to the CERN computer center, to be permanently recorded. The factor $\sim$ 10 in rate reduction at the first stage, will be obtained by a L0 trigger processor (L0TP) using information coming from RICH, LAV, LKr calorimeter and MUV detectors. The trigger primitives from each detector involved in the L0 trigger decision, will be built directly in the same data acquisition board devoted to digitization and monitoring: the TEL62 mother board, developed starting from the TELL1 board designed for the LHCb experiment [12]. The TEL62 board (9U format) houses 5 Altera Stratix III FPGAs allowing a fully customizable configuration. A total RAM memory of 8 GB gives the possibility to store the data in a first buffer stage, waiting for the trigger decision delivered to the board through the TTC [13] interface. A credit card PC (CCPC) allows to control all the functionality of the board. The output stage uses a quad GigaBit Ethernet card (total output bandwidth of 4 GB/s). The input stage can be adapted to different purposes using 4 custom daughter boards. For instance, the time of the data coming from the detectors front end could be digitized using the appositely developed TDCB daughter board[14]. The use of uniform system for all the subdetectors allows to have a common fully integrated trigger and readout architecture, exploiting the possibility to use the same data chain to monitor the whole system and avoiding the complications due to independent trigger and readout chains. The L1 and L2 are completely in software. In the former the primitives are constructed using "single detector" information, while in the L2 the decision is taken on the information coming from all the detectors. Thanks to the trigger system the final acquisition rate will be of the order of tens of kHz.

A pilot project in the NA62 TDAQ, aims at investigating the use of the video card processors (GPU) for trigger purposes. The advantage of this approach is given by the huge computing power offered by the modern commercial GPUs at very competitive costs. The use of the video cards at L0, where the input rate is 10 MHz and the maximum latency is 1 ms, is very challenging, but is very promising to produce high quality primitives to build selective trigger conditions [15].

7. Conclusions

The ultra-rare $K^+ \rightarrow \pi^+\nu\bar{\nu}$ is an unique instrument to test the SM and to study possible contribution of new physics, thanks to clearness of the theoretical predictions. The possibility to distinguish between new physics model and to measure the dynamic structure of the coupling of the new physics particles, is complementary to the possible discover at hadron colliders. From an experimental point of view the measurement of this decay is very challenging due to the weakness of the signal signature and the presence of very huge background. The NA62 experiment aims at measuring O(100) events with 10% error. After three years of successful R&D program, the NA62 detector is now under construction and will be ready for commissioning and early physics runs in 2012.

References

[1] G. Buchalla and A. J. Buras, Nucl. Phys. B **548** (1999) 309 [arXiv:hep-ph/9901288].
[2] A. J. Buras, M. Gorbahn, U. Haisch and U. Nierste, JHEP **0611** (2006) 002 [arXiv:hep-ph/0603079].
[3] J. Brod and M. Gorbahn, Phys. Rev. D **78** (2008) 034006 [arXiv:0805.4119 [hep-ph]].
[4] G. Isidori, F. Mescia and C. Smith, Nucl. Phys. B **718** (2005) 319 [arXiv:hep-ph/0503107].
[5] A. J. Buras, B. Duling, T. Feldmann, T. Heidsieck and C. Promberger, JHEP **1009** (2010) 104 [arXiv:1006.5356 [hep-ph]].
[6] M. Blanke, A. J. Buras, B. Duling, S. Recksiegel and C. Tarantino, Acta Phys. Polon. B **41** (2010) 657 [arXiv:0906.5454 [hep-ph]].
[7] M. Blanke, PoS E **PS-HEP2009** (2009) 196 [arXiv:0908.2716 [hep-ph]].
[8] W. Altmannshofer, A. J. Buras, S. Gori, P. Paradisi and D. M. Straub, Nucl. Phys. B **830** (2010) 17 [arXiv:0909.1333 [hep-ph]].
[9] A. V. Artamonov *et al.* [BNL-E949 Collaboration], Phys. Rev. D **79** (2009) 092004 [arXiv:0903.0030 [hep-ex]].
[10] A. Antonelli *et al.*, J. Phys. Conf. Ser. **160** (2009) 012020 [Nucl. Phys. Proc. Suppl. **197** (2009) 224].
[11] V. Fanti *et al.* [NA48 Collaboration], Nucl. Instrum. Meth. A **574** (2007) 433.
[12] G. Haefeli, A. Bay, A. Gong, H. Gong, M. Muecke, N. Neufeld and O. Schneider, Nucl. Instrum. Meth. A **560** (2006) 494.
[13] B. G. Taylor [RD12 Project Collaboration], IEEE Trans. Nucl. Sci. **45** (1998) 821.
[14] C. Avanzini *et al.* [NA62 Collaboration], Nucl. Instrum. Meth. A **623** (2010) 543.
[15] G. Collazuol, G. Lamanna, M. Sozzi Nucl. Instrum. Meth. A **628** (2011) 457.

Future Neutrino Beam Facilities

Pasquale Migliozzi
Istituto Nazionale Fisica Nucleare, Napoli
E-mail: migliozzi@na.infn.it

Abstract. After a brief summary of the status of neutrino oscillation searches, we discuss the proposed Neutrino Beam Facilities and compare their sensitivity in the $\theta_{13} - \delta_{\rm CP}$ plane. Finally, we also recall the detector technologies that have been proposed to exploit neutrino beams for future facilities.

1. Introduction

The experimental evidences for neutrino oscillations collected in the last fifteen years represent a major discovery in modern particle physics. The oscillation phenomenon allows the measurement of fundamental parameters of the Standard Model and provides the first insight beyond the electroweak scale [1]. Moreover, neutrino oscillations are important for many fields of astrophysics and cosmology and open the possibility to study CP violation in the leptonic sector [2].

Neutrino flavor oscillations can be described in terms of three mass eigenstates ν_1, ν_2, ν_3 with mass values m_1, m_2 and m_3 that are connected to the flavor eigenstates ν_e, ν_μ and ν_τ by a mixing matrix U (Pontecorvo, Maki, Nakagawa and Sakata (PMNS) matrix), usually parameterized as

$$U(\theta_{12},\theta_{23},\theta_{13},\delta_{\rm CP}) = \begin{pmatrix} c_{13}c_{12} & c_{13}s_{12} & s_{13}e^{-i\delta_{\rm CP}} \\ -c_{23}s_{12} - s_{13}s_{23}c_{12}e^{i\delta_{\rm CP}} & c_{23}c_{12} - s_{13}s_{23}s_{12}e^{i\delta_{\rm CP}} & c_{13}s_{23} \\ s_{23}s_{12} - s_{13}c_{23}c_{12}e^{i\delta_{\rm CP}} & -s_{23}c_{12} - s_{13}c_{23}s_{12}e^{i\delta_{\rm CP}} & c_{13}c_{13} \end{pmatrix} \tag{1}$$

where the short-form notation $s_{ij} \equiv \sin\theta_{ij}, c_{ij} \equiv \cos\theta_{ij}$ is used. As a result, the neutrino oscillation probability depends on 3 mixing angles, $\theta_{12}, \theta_{23}, \theta_{13}$, 2 mass differences, $\Delta m^2_{12} = m^2_2 - m^2_1$, $\Delta m^2_{23} = m^2_3 - m^2_2$, and a CP phase $\delta_{\rm CP}$. Additional phases are present in case neutrinos are Majorana particles, but they do not influence at all neutrino flavor oscillations. Furthermore, the neutrino mass hierarchy, the ordering with which mass eigenstates are coupled to flavor eigenstates, can be fixed by measuring the sign of Δm^2_{23}. In vacuum the oscillation probability between two neutrino flavors α, β is:

$$P(\nu_\alpha \to \nu_\beta) = -4\sum_{k>j} Re[W^{jk}_{\alpha\beta}]\sin^2\frac{\Delta m^2_{jk}L}{4E_\nu} \pm 2\sum_{k>j} Im[W^{jk}_{\alpha\beta}]\sin^2\frac{\Delta m^2_{jk}L}{2E_\nu} \tag{2}$$

where $\alpha = e, \mu, \tau$, $j = 1, 2, 3$, $W^{jk}_{\alpha\beta} = U_{\alpha j}U^*_{\beta j}U^*_{\alpha k}U_{\beta k}$. In the case of only two neutrino flavor oscillation it can be written as:

Journal of Physics: Conference Series **335** (2011) 012072 doi:10.1088/1742-6596/335/1/012072

$$P(\nu_\alpha \to \nu_\beta) = \sin^2 2\theta \cdot \sin^2 \frac{1.27\ \Delta m^2(eV^2)\cdot L(km)}{E_\nu(GeV)}. \quad (3)$$

Therefore, two experimental quantities are relevant for neutrino oscillations: the neutrino energy E_ν and the baseline L (distance of the neutrino source from the detector). They can be varied in order to explore different values of the mixing parameters. When neutrinos pass through matter, the oscillation probability is perturbed (the so-called MSW effect [3]) depending on sign(Δm^2_{23}) [4].
The present knowledge of the PMNS is summarized in Ref. [5]. The numerical values of the matrix elements are:

GS98 with Gallium cross-section	AGSS09 with modified Gallium cross-section
$\Delta m^2_{21} = 7.59 \pm 0.20\ (^{+0.61}_{-0.69}) \times 10^{-5}\ \mathrm{eV}^2$	Same
$\Delta m^2_{31} = \begin{cases} -2.36 \pm 0.11\ (\pm 0.37) \times 10^{-3}\ \mathrm{eV}^2 \\ +2.46 \pm 0.12\ (\pm 0.37) \times 10^{-3}\ \mathrm{eV}^2 \end{cases}$	Same
$\theta_{12} = 34.4 \pm 1.0\ (^{+3.2}_{-2.9})^\circ$	$34.5 \pm 1.0\ (^{+3.2}_{-2.8})^\circ$
$\theta_{23} = 42.8\ ^{+4.7}_{-2.9}\ (^{+10.7}_{-7.3})^\circ$	Same
$\theta_{13} = 5.6\ ^{+3.0}_{-2.7}\ (\leq 12.5)^\circ$	$5.1\ ^{+3.0}_{-3.3}\ (\leq 12.0)^\circ$
$[\sin^2\theta_{13} = 0.0095\ ^{+0.013}_{-0.007}\ (\leq 0.047)]$	$[0.008\ ^{+0.012}_{-0.007}\ (\leq 0.043)]$
$\delta_{CP} \in [0,\ 360]$	Same

Although gigantic steps forward have been achieved in the past fifteen years and with the recent observation of a τ candidate from the appearance of a ν_τ in a pure ν_μ beam [6], there are still two unknown angles (θ_{13} and δ_{CP}), the mass hierarchy is still to be determined, the evidence for neutrino oscillations in the $\bar{\nu}_\mu \to \bar{\nu}_e$ channel at a scale of $\Delta m^2 \simeq 0.1\mathrm{eV}^2$ is still debated and needs more solid confirmation [7]. Furthermore, the absolute neutrino mass is still unknown [8]. In the coming year the T2K experiment [9] should provide an unambiguous evidence for a non-zero θ_{13} if above 3°. The discovery of a non-zero value for θ_{13} will disclose the possibility to search for CP violation in the leptonic sector. Many ideas have been put forward to measure a non-zero value of δ_{CP}: Super-Beams, Beta-Beams and Neutrino Factory and will be reviewed in the next Sections. For details on the future facility sensitivities we refer to [10].

2. Super-Beams

In a Super-Beam facility neutrinos are produced in the standard manner, by bombarding a target with a proton beam and focusing the produced pions. However, the Super-Beam projects are different in two respects from the long-baseline facilities currently under construction: (1) the proton beam power is higher, around 4 MW, and (2) the neutrino detector is much bigger, typically around 1 Mt.
An interesting option for the Super-Beams is the possibility to tilt the beam axis a few degrees with respect to the position of the far detector (Off-Axis beams). According to the two body π-decay kinematics, all the pions above a given momentum produce neutrinos of similar energy at a given angle $\theta \neq 0$ with respect to the direction of parent pion (contrary to the $\theta = 0$ case where the neutrino energy is proportional to the pion momentum). These neutrino beams have several advantages with respect to the corresponding on-axis ones: they are narrower, lower energy and with a smaller ν_e contamination (since ν_e's mainly come from three body decays) although the neutrino flux can be significantly smaller.

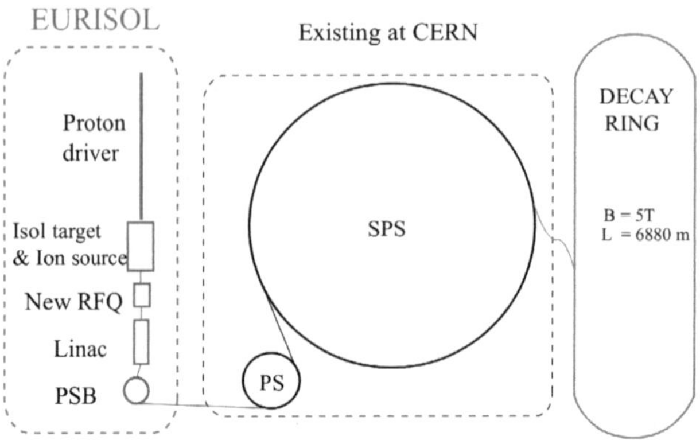

Figure 1. A schematic layout of the BetaBeam complex. At left, the low energy part is largely similar to the EURISOL project. The central part (PS and SPS) uses existing facilities. At right, the decay ring has to be built.

The technology of the Super-Beam is considered the most straightforward way to explore leptonic mixing in case of positive result from T2K and/or Double-Chooz [11] and, therefore, it is viewed as a solid option in Japan (Tokai to HyperKamiokande) and US (Fermilab/Project X to DUSEL) [12]. Unfortunately, it also shows evident limitations:

- It is not a "pure" source of neutrinos of a given flavor, being plagued by the ν_e produced by the decay-in-flight of the kaons and of the muons. When seeking for subdominant $\nu_\mu \to \nu_e$ transitions, the systematics on the knowledge of the ν_e contamination will likely be the main limitation for a precise determination of CP violation in the leptonic sector [13];
- It produces mainly ν_μ and, therefore, the leptonic mixing is studied through $\nu_\mu \to \nu_e$ and its CP conjugate $\overline{\nu}_\mu \to \overline{\nu}_e$, i.e. looking for *electrons* in the final state instead of muons. It thus requires huge, low density detectors to be hosted in underground sites (see below);
- in a Super-Beam, CP violation appears as an asymmetry between the probability of transition $\nu_\mu \to \nu_e$ and its CP conjugate $\overline{\nu}_\mu \to \overline{\nu}_e$. At low proton energies, the production yield of π^- is suppressed with respect to π^+. This suppression, together with the suppression due to the cross section ($\sigma_{\bar{\nu}}/\sigma_\nu \simeq 1/2$), makes the antineutrino run much more time-consuming than the neutrino run.

A multi-MW proton accelerator and the corresponding neutrino beamline is considered feasible in less than a decade [14] but it requires a substantial effort in accelerators and targetry, a major resource investment and feedback from the operation of neutrino facilities in the sub-MW range (firstly T2K).

3. Beta-Beams

The enormous progress in the technology of Radioactive Ion Beams has led P. Zucchelli [15] to the proposal of a neutrino facility based on the decay in flight of β-unstable ions (for a full review see [16]). "Beta-Beams" (BB) are pure sources of $\overline{\nu}_e$ or, in the occurrence of β^+ decays, of ν_e. Hence, they are ideal tools to study $\nu_e \to \nu_\mu$ transitions and their CP-conjugate. They share with the Neutrino Factory (NF) the nearly complete absence of systematics in the knowledge of the source with the bonus of no "right sign muon" background (no ν_μ in the initial state). On the other hand, due to the very different mass-to-charge ratio between muons and β-unstable ions, the energy of the neutrinos are typically much smaller than what can be obtained at the

NF. The original proposal of [15] was tuned to leverage at most the present facilities of CERN - the PS and the SPS - and it was based on ^{6}He and ^{18}Ne as $\overline{\nu}_e$, and ν_e sources respectively, see Figure 1. Beta-Beam triggered the interest of nuclear physics community, which was offered a stimulating synergy with the neutrino programme at CERN. As a result, such proposal [17, 18] was studied in a systematic manner within the framework of the EURISOL Design Study[1] . The study aimed at 2.9×10^{18} antineutrinos per year from ^{6}He and 1.1×10^{18} neutrinos per year from ^{18}Ne. The outcome was extremely encouraging, except for the production of ^{18}Ne, which cannot attain the needed rate using standard methods and medium-intensity proton accelerators (200 kW). Along this line, the most straightforward alternative would be direct production on MgO based on a 2 MW, few MeV, proton accelerators, which are quite similar to the linacs that have to be built for the International Fusion Materials Irradiation Facility [19]. In this case, the BB would partially miss the advantage of a low-power front-end compared with the multi-MW accelerators needed for the Superbeams and for the NF, although a few tens of MeV MW accelerator is anyway a much simpler machine than a few GeV MW Linac. From the point of view of the physics performance, an additional weakness stems from the fact that the SPS is able to accelerate these ions up to relatively low energies, so that the corresponding emitted neutrinos are just in the sub-GeV range. This impacts on the choice of the detectors and on the cross-sections, which are highly depleted.

To improve the performance of the Eurisol Beta-Beam several alternatives to the SPS have been considered: a high energy SPS ("SuperSPS" [20]) accelerating protons up to 1 TeV (a machine originally envisaged for the energy and luminosity upgrade of the LHC) or even the LHC itself [21, 22]. These configurations improve the sensitivities to CP violation and the mass hierarchy at the expense of a large increase of costs: large investments are needed especially for the construction of the decay ring since the length of the ring depends from the magnetic rigidity of the circulating ions, which is proportional to their Lorentz γ factor, and for the compensation of potential flux reduction due to the longer lifetime of the ion in the laboratory frame.

In 2006, C. Rubbia et al. [23] proposed the use of ^{8}Li and ^{8}B as neutrino sources noting that these isotopes could be produced in a multiturn passage of a low-energy ion beam through a low-Z target. In this case, ionization cooling techniques could increase the circulating beam lifetime and thus enhance the ion production to a level suitable for the BB. This option has the advantage of employing isotopes with higher Q-value than ^{18}Ne and ^{6}He, increasing correspondingly the neutrino energy (from ~ 0.5 to ~ 1.5 GeV for the SPS-based BB). This alternative approach will be at focus in the framework of the EUROν Design Study[2]. A drawback with respect to the use of low-Q ions is that the flux at the far location is smaller due to the larger beam divergence and a larger amount of ions stacked in the decay ring is needed. Although the BB optimization is a complex task [25, 26, 27], some simple scaling laws can be used as reference. For a given number of decays per year N_β, if we label with γ the Lorentz factor of the ion (which depends on the machine employed to accelerate the ion), with Q the Q-value of the isotope and with L the source-detector distance, the events at the far location are proportional to the convolution of the flux ($\phi \sim \gamma^2/L^2$), of the cross section ($\sigma \sim E_\nu \sim Q\gamma$) and of the oscillation probability, times N_β. If the facility is operated at the maximum of the oscillation probability, then $1.27\Delta m^2 L/E_\nu = \pi/2$; therefore, $L \sim Q\gamma$. As a result the number of events are proportional to $N_\beta \gamma/Q$. Note, however, that also N_β has a dependence on γ^{-1} due to the increase of the ion lifetime in the lab frame at larger γ.

[1] The EURISOL Design Study was a Project funded by the European Community within the 6th Framework Programme as a Research Infrastructures Action under the "Structuring the European Research Area Specific Programme". The Project started officially on Feb 2005, and has been completed on spring 2009.

[2] EUROν [24] is a FP7 Design Study which started on Sept 2008 and will run for 4 years. The primary aims are to study three possible future neutrino oscillation facilities for Europe (a Superbeam from CERN-to-Frejus, a RAL or CERN based NF and high-Q BB) and do a cost and performance comparison.

Summarizing, a high-Q BB needs a smaller γ for the same neutrino energy, with the advantage that an accelerator of larger energy than the SPS would not be needed and that the length of the decay ring could be shortened. On the other hand, given the Beta Beam kinematics, for the same baseline L an high-Q BB needs an order of magnitude more ions at the source to match the performance of a high-γ BB.
In general, the clarification of the issue of the ion production yield is considered a crucial milestone for the BB. Given an appropriate yield, the acceleration and stacking is viewed as less demanding than what is needed for a NF both from the point of view of R&D and cost. Clearly, the possibility of employing existing facilities (e.g. the CERN PS-SPS complex or its upgrades) might substantially strengthen this option.

4. Neutrino Factory

Production, acceleration and stacking of high intensity muon beams for muon colliders have been envisaged since the 60's and it has been noted very early that their decays might produce useful beams of ν_μ and $\overline{\nu}_e$ (exploiting μ^- decays into $e^-\overline{\nu}_e\nu_\mu$) or $\overline{\nu}_\mu$ and ν_e (μ^+ decays into $e^+\nu_e\overline{\nu}_\mu$). However, realistic layouts to get intense neutrino sources have become available only in recent times [28]. In the modern formulation of the "Neutrino Factory" concept, muons are created from an intense pion source at low energies, their phase space compressed to produce a bright beam, which is then accelerated to the desired energy and injected into a storage ring with long straight sections pointing in the desired direction. In 1997 S. Geer [29] noted that this source could be ideal to study $\nu_e \to \nu_\mu$ oscillations at the atmospheric scale, i.e. the T-conjugate of the channel observed in Superbeams ($\nu_\mu \to \nu_e$). Since μ^+ decays into $e^+\nu_e\overline{\nu}_\mu$, it is possible to investigate $\nu_e \to \nu_\mu$ oscillations seeking for the appearance of μ^- from ν_μ CC events ("wrong sign muons"), provided that we are able to separate these events from the bulk of μ^+ ("right sign muons") coming from unoscillated $\overline{\nu}_\mu$. A. De Rujula et al. [30] underlined that the simultaneous exploitation of μ^- and μ^+ decays would be an ideal tool to address CP violation in the leptonic sector, with outstanding performance compared with pion-based sources. The only condition to be fulfilled was a large ratio $\Delta m^2_{12}/\Delta m^2_{23}$ and, of course, the finite size of θ_{13}. The results from KAMLAND and SNO [31] confirming $\Delta m^2_{12}/\Delta m^2_{23} \simeq 1/30$ boosted enormously Geer's proposal, together with the above-mentioned proposals for Superbeams. The realization of the NF represents a major accelerator challenge compared with Super-Beams. It is met through a world-wide R&D programme [32]; in Europe this programme is especially fostered by UK. Among the NF-oriented projects we recall MICE [33] at the Rutherford Appleton Laboratories (ionization cooling), HARP [34] at CERN (hadroproduction for the front-end proton accelerator), MERIT [35] at CERN (targetry), EMMA [36] at Daresbury (fixed-field alternating-gradient accelerators) and the MUCOOL R&D [37] at Fermilab (radio-frequency and absorbers). Moreover, the NF has to be seeded by a very powerful low-energy proton accelerator (4 MW); its realization requires similar R&D as for the Superbeams, although its optimal energy lays in the few-GeV range (e.g. the aforementioned SPL). Current designs aim at 10^{21} muon decays per year running with a muon energy of 50 GeV, although more realistic scenarios suggest decay rates in the ballpark of $2-5\times 10^{20}$ decays/y and with energies in the 20-50 GeV range (corresponding to neutrinos in the 10-30 GeV range) [38]. After the work of the International Scoping Study (ISS) [12, 39, 40], there is a rather widespread consensus on the fact that the Neutrino Factory can be considered the most performing facility for the determination of θ_{13}, CP violation and the mass hierarchy. With respect to Super-Beams, neutrino factories profit of much smaller systematics in the knowledge of the source and much higher energies (i.e. statistics, due to the linear rise of the deep-inelastic ν_μ cross section with energy). In fact, the energy is so high that for any realistic baseline (< 7000 km) the ratio L/E will be off the peak of the oscillation maximum at the atmospheric scale. This condition is the main cause of the occurrence of multiple solutions when the mixing parameters are extracted from the physics observables, i.e. the rates of appearance of wrong sign muons. In jargon, this issue is dubbed the "degeneracy problem" [12, 41]. It also affects other facilities than NF but it is particularly severe for experiments running off the peak of the oscillation probability. The ISS suggests as an ideal solution the positioning of two detectors at baseline around 3000 and 7000 km. An alternative to the second 7000 km detector could be the detection of $\nu_e \to \nu_\tau$ at baseline around 1000 km ("silver channel") [42] (see below). The exploitation of the silver channel, moreover, is useful to investigate the occurrence of non-standard interactions in the neutrino sector [43]. Although the superior physics reach of the Neutrino Factory is nearly undisputed and no evident showstoppers have been identified, the R&D needed to build this facility remains impressive. In turn, the time schedule for its realization and the cost estimate are vague ($\sim$ 2020 after an investment of 1-2 Billion$). On the other hand, a clear indication on the size of θ_{13} will enormously boost the interest of particle physics on this technology. Neutrino Factories are virtually capable of performing real precision physics on the leptonic mixing in a way that resembles the former physics potential of the b-factories on quark mixing.

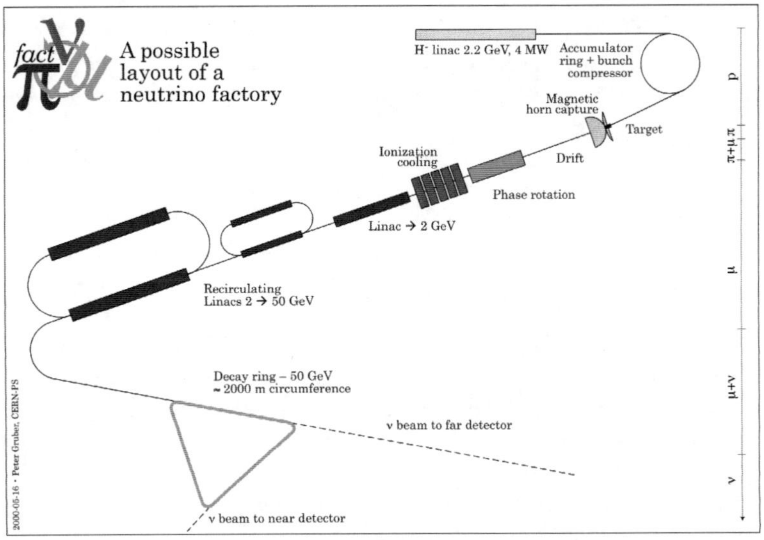

Figure 2. Expected layout for a neutrino factory at CERN.

5. Detectors

A discussion of the possible detector technologies that will exploit future neutrino facilities is not the goal of this paper. Therefore, we only briefly recall the optimal technology for each facility and we refer to the detector group of the International Scoping Study for details [40]. The baseline detector options for each possible neutrino beam are defined as follows:

- a very massive (Megaton) water Cherenkov detector is the baseline option for a sub- GeV Beta-Beam and Super-Beam facility;
- there are a number of possibilities for either a Beta-Beam or Super-Beam medium energy facility between 1-5 GeV. These include a totally active scintillating detector (TASD), a liquid argon TPC or a water Cherenkov detector;
- A 100 kton magnetized iron neutrino detector (MIND) is the baseline to detect the wrong sign muon final states (golden channel) at a high energy (20-50 GeV) neutrino factory from

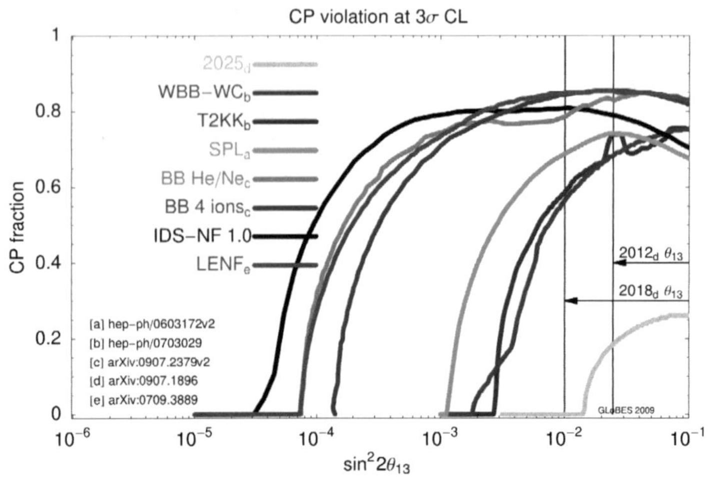

Figure 3. Comparison of the physics reach of different future facilities for the CP violation discovery reach (3σ). Figure taken from Ref. [44].

muon decay. A 10 kton hybrid neutrino magnetic emulsion cloud chamber detector for wrong sign tau detection (silver channel) is a possible complement to MIND, if one needs to resolve degeneracies that appear in the $\delta_{\text{CP}} - \theta_{13}$ parameter space.

6. Conclusion

We want to conclude this paper by comparing the CP violation discovery reach of the different experiments, see Fig. 3. From this figure it is clear that the current generation of experiments will explore a very small region of the CP phase and most likely will not be able to establish CP violation. The Super-Beam upgrades discussed before can measure CP violation for most of the parameter space if $\sin^2 2\theta_{13} > 0.01$. It is worth noting that values of $\sin^2 2\theta_{13}$ larger that 0.01 will be in the reach of the current generation of accelerator and reactor experiments. For smaller values of $\sin^2 2\theta_{13}$ ($\ll 0.01$) the Super-Beams are useless and only different versions of the Neutrino Factory or of the Beta-Beam can measure CP violation. For details on the physics reach of future neutrino facilities we refer to [10].

References

[1] A. De Gouvea, Mod. Phys. Lett. A **19** (2004) 2799 [arXiv:hep-ph/0503086].
[2] A. Strumia, F. Vissani, [hep-ph/0606054].
[3] L. Wolfenstein, Phys. Rev. D17 (1978) 2369. S.P. Mikheev and A.Y. Smirnov, Nuovo Cim. C9 (1986) 17.
[4] K. Kimura, A. Takamura and H. Yokomakura, Phys. Rev. D **66** (2002) 073005 [arXiv:hep-ph/0205295]. E. K. Akhmedov, R. Johansson, M. Lindner, T. Ohlsson and T. Schwetz, JHEP **0404** (2004) 078 [arXiv:hep-ph/0402175]. M. Freund, Phys. Rev. D **64** (2001) 053003 [arXiv:hep-ph/0103300].
[5] M. C. Gonzalez-Garcia, M. Maltoni, J. Salvado, JHEP **1004** (2010) 056. [arXiv:1001.4524 [hep-ph]].
[6] N. Agafonova *et al.* [OPERA Collaboration], Phys. Lett. **B691** (2010) 138-145. [arXiv:1006.1623 [hep-ex]].
[7] A. A. Aguilar-Arevalo *et al.* [The MiniBooNE Collaboration], Phys. Rev. Lett. **105** (2010) 181801 [arXiv:1007.1150 [hep-ex]].
[8] O. Cremonesi, these proceeedings.
[9] Y. Itow *et al.* [The T2K Collaboration], arXiv:hep-ex/0106019.
[10] M. Mezzetto, these proceeedings.
[11] F. Ardellier *et al.* [Double Chooz Collaboration], arXiv:hep-ex/0606025.
[12] A. Bandyopadhyay *et al.* [ISS Physics Working Group], arXiv:0710.4947 [hep-ph].
[13] P. Huber, M. Mezzetto and T. Schwetz, JHEP **0803** (2008) 021.

[14] J. Hylen, talk at 11th International Workshop on Neutrino Factories, Superbeams and Beta Beams (Nufact09), Chicago, IL, July 20-25, 2009.
[15] P. Zucchelli, Phys. Lett. B **532** (2002) 166.
[16] M. Lindroos and M. Mezzetto, "Artificial Neutrino Beams: Beta Beams", Imperial College Press, Aug. 2009.
[17] J. E. Campagne, M. Maltoni, M. Mezzetto and T. Schwetz, JHEP **0704** (2007) 003.
[18] J. Bouchez, M. Lindroos and M. Mezzetto, AIP Conf. Proc. **721** (2004) 37.
[19] M. Martone et al., "International Fusion Materials Irradiation Facility Conceptual Design Activity", RT/ERG/FUS/96-11.
[20] O. Bruning *et al.*, "LHC luminosity and energy upgrade: A feasibility study," CERN-LHC-PROJECT-REPORT-626.
[21] J. Burguet-Castell, D. Casper, J. J. Gomez-Cadenas, P. Hernandez and F. Sanchez, Nucl. Phys. B **695** (2004) 217; F. Terranova, A. Marotta, P. Migliozzi and M. Spinetti, Eur. Phys. J. C **38** (2004) 69; S. K. Agarwalla, A. Raychaudhuri and A. Samanta, Phys. Lett. B **629** (2005) 33.
[22] A. Donini, E. Fernandez-Martinez, P. Migliozzi, S. Rigolin, L. Scotto Lavina, T. Tabarelli de Fatis and F. Terranova, Eur. Phys. J. C **48** (2006) 787; A. Donini *et al.*, Eur. Phys. J. C **53** (2008) 599.
[23] C. Rubbia, A. Ferrari, Y. Kadi and V. Vlachoudis, Nucl. Instrum. Meth. A **568** (2006) 475.
[24] Information and documentation available at `http://www.euronu.org/`
[25] A. Donini and M. Lindroos, "Optimisation of a beta beam," Talk at 10th International Workshop on Neutrino Factories, Superbeams and Betabeams: Nufact08, Valencia, Spain, 30 Jun - 5 Jul 2008.
[26] S. K. Agarwalla, S. Choubey, A. Raychaudhuri and W. Winter, JHEP **0806** (2008) 090.
[27] W. Winter, Talk given at 10th International Workshop on Neutrino Factories, Superbeams and Betabeams: Nufact08, Valencia, Spain, 30 Jun - 5 Jul 2008, arXiv:0809.3890 [hep-ph].
[28] S. Geer, Lecture at the 1st International Neutrino Factory Summer Institute, Abingdom, UK, June 24-29, 2002.
[29] S. Geer, Phys. Rev. D **57** (1998) 6989 [Erratum-ibid. D **59** (1999) 039903].
[30] A. De Rujula, M. B. Gavela and P. Hernandez, Nucl. Phys. B **547** (1999) 21.
[31] Q. R. Ahmad *et al.* [SNO Collaboration], Phys. Rev. Lett. **87** (2001) 071301.
[32] A. Bross, "The Neutrino Factory: The Final Frontier in Neutrino Physics?" FERMILAB-CONF-09-169-APC, Presented at Particle Accelerator Conference (PAC 09), Vancouver, BC, Canada, 4-8 May 2009.
[33] G. Gregoire et al. [MICE Collaboration] "MICE: Muon Ionisation Cooling Experiment, Technical Reference Document,", MICE-TRD-2005 (2005).
[34] M.G. Catanesi et al., [HARP Collaboration] "Proposal to study hadron production for the neutrino factory and for the atmospheric neutrino flux" , CERN-SPSC/99-35 (1999).
[35] H.G. Kirk et al., "The MERIT High-Power Target Experiment at the CERN PS", EPAC08-WEPP169, FERMILAB-CONF-08-224-APC, presented at the European Particle Accelerator Conference (EPAC 08), Genova, Italy, 23-27 Jun 2008.
[36] R. Edgecock et al., "EMMA - the World's First Non-scaling FFAG", EPAC08-THPP004 (2008), presented at the European Particle Accelerator Conference (EPAC 08), Genova, Italy, 23-27 Jun 2008.
[37] J. Norem et al. "The MUCOOL RF Program.", FERMILAB-CONF-06-387-AD, JLAB-ACC-06-481; presented at the European Particle Accelerator Conference (EPAC 06), Edinburgh, Scotland, 26-30 Jun 2006.
[38] S. Geer and M. S. Zisman, Prog. Part. Nucl. Phys. **59** (2007) 631.
[39] J. S. Berg *et al.* [ISS Accelerator Working Group], JINST **4** (2009) P07001.
[40] T. Abe *et al.* [ISS Detector Working Group], JINST **4** (2009) T05001.
[41] V. Barger, D. Marfatia and K. Whisnant, Phys. Rev. D **65** (2002) 073023.
[42] A. Donini, D. Meloni and P. Migliozzi, Nucl. Phys. B **646** (2002) 321. D. Autiero *et al.*, Eur. Phys. J. C **33** (2004) 243.
[43] J. Kopp, T. Ota and W. Winter, Phys. Rev. D **78** (2008) 053007.
[44] J. Bernabeu, M. Blennow, P. Coloma *et al.*, "EURONU WP6 2009 yearly report: Update of the physics potential of Nufact, superbeams and betabeams," [arXiv:1005.3146 [hep-ph]].

Journal of Physics: Conference Series **335** (2011) 012073 doi:10.1088/1742-6596/335/1/012073

The Mu2e Experiment at Fermilab

F. CERVELLI ***for the Mu2e collaboration***

INFN Pisa, Italy

E-mail: franco.cervelli@pi.infn.it

Abstract. The Mu2e Collaboration has proposed an experiment at Fermilab to search for the coherent decay of a muon to an electron in the Coulomb field of a nucleus with an expected sensitivity less than 10^{-16}. In this article the physics motivations and status of the experiment are described.

1. Introduction

The recently proposed Mu2e experiment at Fermilab [1] is a search for the Charged Lepton Flavor Violation (CLFV) process $\mu^- N \rightarrow e^- N$, which is the coherent conversion of a muon into an electron in the field of a nucleus. In the process of muon to electron conversion the initial state is a muonic atom and the final state consists of a mono-energetic electron recoiling against an intact atomic nucleus. There are no neutrinos in the final state. The recoiling nucleus is not observed, leaving an observed final state of just the electron, which has the energy of the muon rest mass less corrections for the nuclear recoil and the K-shell binding energy of the muon.

The result of the experiment will be expressed by the ratio:

$$R_{\mu e} = \frac{\Gamma(\mu^- N(A,Z) \rightarrow e^- N(A,Z))}{\Gamma(\mu^- N \rightarrow \nu_\mu N^*)}$$

where N(A,Z) denotes a nucleus with mass number A and atomic number Z. The numerator is the rate for the conversion process and the denominator is the rate for muon capture on the same nucleus (N* is the nucleus after the capture).

In the Standard Model, the predicted ratio, $R_{\mu e}$, for the $\mu^- N \rightarrow e^- N$ CLFV process is less than 10^{-50} [2]. Though many experimental observations suggest the existence of New Physics (NP) beyond the Standard Model and most new physics scenarios predict that this CLFV process will occur at some level. In many NP models that include a description of neutrino mass, the rates for these processes are enormously enhanced. In particular scenarios predicting that SUSY is within the reach of the LHC also foresee $R_{\mu e} \sim 10^{-15}$, a rate for which Mu2e expects to observe, after 2 years of running, about 40 events on a background of fewer than 0.5 events. An excellent review of CLFV and the flavor physics of leptons can be found in reference [3]. Two classes of diagrams can contribute to conversion. The first class includes magnetic moment loop diagrams with a photon exchanged between the loop and the nucleus; these diagrams can proceed with many different sorts of particles in the loop, including, but not limited to, SUSY particles, heavy neutrinos and a second Higgs Doublet. This class of diagrams also produces non-zero rates for the process $\mu \rightarrow e\gamma$.

Published under licence by IOP Publishing Ltd

In the second class, the $\mu^- N \rightarrow e^- N$ process is sensitive to NP contributions via contact interactions, such as those expected in Compositeness, Leptoquark, and GUT models with additional gauge bosons and/or anomalous couplings.

TABLE 1. Estimated background contributions for the Mu2e experiment. These numbers assume 10^{18} stopped muons are delivered in 2 x 10^7 seconds of run time, an inter-pulse extinction ratio of 10^{-9}, and a cosmic ray veto efficiency of 99.99%.

Category	Source	Events
Intrinsic	μ Decay in Orbit	0.225
	Radiative μ Capture	<0.002
Late Arriving	Radiative π Capture	0.072
	Beam Electrons	0.036
	μ Decay in Flight	<0.063
	π Decay in Flight	<0.001
Miscellaneous	Cosmic Ray	0.016
	Long Transit	0.006
	Pat. Recognition Errors	<0.002
TOTAL		**0.42**

2. The experimental technique and background

The Mu2e apparatus is described in detail in the Mu2e proposal [1]. The basic idea behind Mu2e is motivated by the MECO experiment, which in turn derives from the MELC proposal [4].

Using 8 GeV protons from the Fermilab accelerator complex, a beam of low momentum negative muons (<60 MeV/c) is produced via upstream charged pion decays. The beam is stopped on a set of thin Al target foils, where the muons drop to the K-shell, forming a muonic atom. The Bohr radius of the K-shell of muonic Al is about 20 fm and the nuclear radius of Al is about 4 fm, which yields a large overlap between the muon wavefunction and that of the nucleus. The stopped muons have a characteristic decay time, $\tau_\mu^N = 864$ ns.

Genuine $\mu^- N \rightarrow e^- N$ events produce an isolated electron with energy of 104.96 MeV. Energy loss and scattering in the target and detector material result in an observed energy spectrum with a mean of about 104 MeV and an estimated FWHM of 0.9 MeV [5]. About 62% of the signal electrons have energies in the range 103.6-105.0 MeV, which is defined as the signal energy window. Therefore the objective of the Mu2e experiment is to search for an excess in this endpoint energy region.

Muons which slow down and stop in the stopping target can either undergo beta decay in the field of the nucleus (DIO, decay in orbit), be captured (OMC, ordinary muon capture) or convert to electrons. The first is a potential source of background, the second provides the denominator to $R_{\mu e}$ and the third is our signal. Decay in orbit occurs about 40% of the time [6] and the capture about 60% of the time [7].

DIO produces electrons with a continuous energy spectrum: the Michel spectrum of free muon decay is distorted because the outgoing electron can recoil against the atomic nucleus. Near the endpoint the spectrum falls roughly as $(E_{max}-E)^5$ [8] and has a long radiative tail that extends exactly to the muon mass, assuming zero neutrino mass. In one extreme configuration, both neutrinos are at rest and the electron recoils against the intact Al nucleus: this is the configuration where the electron has the maximum energy in the lab frame, 104.96 MeV for muonic Al [9].

The ordinary muon capture, $\mu^- N_Z \rightarrow \nu N_{Z-1}$, is not a source of background electrons, apart a very small contribution from photon produced in radiative capture (RMC), as reported in Table 1.

So the technique to identify the background from DIO electrons is to carefully measure the energy spectrum from electrons emitted from the target foils. The spectrometer in the Mu2e proposal has a target resolution of <1 MeV at FWHM for 105 MeV electrons. As an example, the fraction of DIO electrons which are expected inside the signal energy window is estimated to be about 2.5 x 10^{-18}.

3. Other background sources

Backgrounds arising from interactions that occur at the production target are overwhelmingly prompt and arrive at the stopping target in time with the muon beam. These backgrounds are eliminated by only accepting electrons appearing 700 ns after the main beam pulse. However, out-of-time protons impinging on the production target will produce background electrons that fall into this delayed signal-timing window. To suppress these Late Arriving Backgrounds (LAB) is necessary to minimize the ratio of out-of-time to in-time protons impinging on the production target. An extinction ratio of 10^{-10} was determined to be required [10] to achieve the background levels listed in Table 1.The largest contributions to LAB come from pion capture on the nucleus and from muon decays in flight (i.e, when the electron scatters into the spectrometer acceptance due to interactions in the stopping target material). Electrons, mostly from upstream π° decays, can also scatter in the target to produce background. The Late Arriving Backgrounds all scale linearly with the extinction factor and taken together account for about 40% of the total background. The required extinction factor is obtained by means of a pair of AC dipoles, synchronized with the beam so to have the in-time beam passing through a narrow channel in a collimator, designed to absorb all out-of-time protons.
The remaining background sources are dominated by contributions from cosmic rays. Cosmic ray muons can produce background electrons either by decay or by interactions in the material. A combination of passive and active shielding is required to suppress the cosmic ray background to sufficiently small levels. As this background scales linearly with the inefficiency of the veto system, a veto efficiency of 99,99%c is necessary to achieve the cosmic ray background estimate listed in Table 1.

4. Beamline

The 8 GeV bunches of protons with a full width of about 100 ns is steered onto a pencil shaped Au (or W) target located in the middle of a solenoid, the Production Solenoid (PS), shown in Figure 1. The PS has a graded magnetic field, with a field of 5 T at the proton-downstream end, falling to about 2.2 T at the proton-upstream end. In the production target, p-Au (W) interactions produce pions that are captured into helical trajectories in the field of the solenoid; these pions decay into muons that are also captured by the field of the solenoid. Mu2e is interested in the backscattered muon beam. The PS graded fiels also forms a magnetic mirror and reflects a portion of the forward going pions and muons, thereby increasing the yield of captured muons.
From the PS the muons enter the S-bend graded field Transport Solenoid (TS), also shown in Figure 1.The bend in the TS induces a dipole term which allows, by appropriate placement of absorbers and collimators, the sign selection of the muon beam and the stopping of any anti-protons accompanying the muon beam. Furthermore the S-shape eliminates a line-of-sight between the PS and the detector to reduce background, rate, and heat load at the detector. The TS has an axial length of 13 meters and a 0.5 meter bore diameter. The field is graded and decreases from 2.5 Tesla near the production target to 2.0 Tesla near the Detector Solenoid (DS).
Inside the DS the muon beam encounters the stopping target. This target is composed of 17 aluminum foils each 200 μm thick with radii that taper from 10.0 to 6.5 cm in the direction of the incoming muon beam. The tapering radii reduce the number of scatters for a signal electron originating in one of the upstream foils. About half the incoming muons will stop in the foils. Downstream of the target there is the Mu2e Detector. Also the DS magnetic field is graded so to form a magnetic mirror that reflects

some backwards going electrons towards the tracker. In the volume occupied by the detector the DS magnetic field is highly uniform at 1.0 T.

The Mu2e experiment will receive an average of about $2x10^{13}$ protons/s in micro-bunches of about 3-6 x 10^7protons each with a bunch spacing of 1.7 μs. Several schemes are being explored with duty factors in the range of 50-90%. The number of stopped muons per proton on target is estimated to be 0.0025 with a 20% uncertainty, dominated by modeling uncertainties in the π^- production at the PS. This modeling is in the process of being tuned to recent data [11].

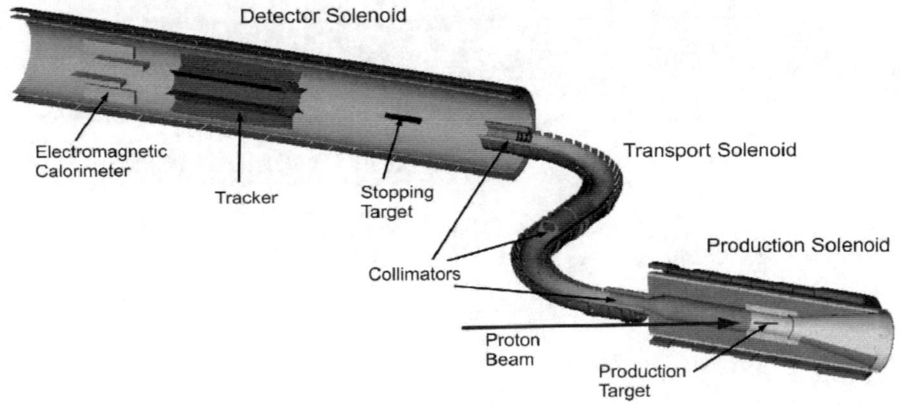

FIGURE 1. Diagram of the Mu2e muon beam-line and detector. The proton beam enters from the left. A back-scattered muon beam is captured by the Production Solenoid and transported through the S-bend Transport Solenoid to the stopping targets. Conversion electrons, produced in the stopping target are captured by the magnetic field in the Detector Solenoid and transported through the Tracker, which makes a precision measurement of the momentum. The conversion electrons then strike the Electromagnetic Calorimeter, which produces an event trigger.

5. Detector

The Mu2e detector (see Fig.2) has two main components, the tracker and the calorimeter. In the present design the tracker consists of 2.8k 2.4 meter long straw tubes oriented parallel with the muon beamline. The inner radius of the tracker is approximately 38 centimeters. When a conversion or DIO electron is emitted from the stopping target, it travels in a helical trajectory, and, if it has sufficient transverse momentum (P_T), its trajectory will be measured by the tracking system. With an inner radius of 38 cm, only electrons with $P_T > 55$ MeV/c will reach the detector and only those with $P_T >$ 90 MeV/c will intersect enough of the tracker to form a reconstructible track. Because almost all tracks from DIO have $P_T < m_\mu /2$ they will never reach the tracker. This is the key to making a measurement of $R_{\mu e}$ with a sensitivity of $0(10^{-17})$, as the apparatus is only sensitive to the tail of the DIO energy distribution. The momentum resolution of the tracker is dominated by scattering (120 keV/c) and by energy loss across the material: the resulting FWHM is estimated to be 900 keV/c for signal electrons. Electrons that pass through the tracker will eventually intersect the electromagnetic calorimeter where they will provide an event trigger plus an energy measurement and a position measurement, both of which can be used to confirm track candidates.

In the present design electromagnetic calorimeter consists of 1.2k Lead Tungstate crystals. They ensure an energy measurement with a resolution of about 5% at 100 MeV and a position resolution of about 1 cm.

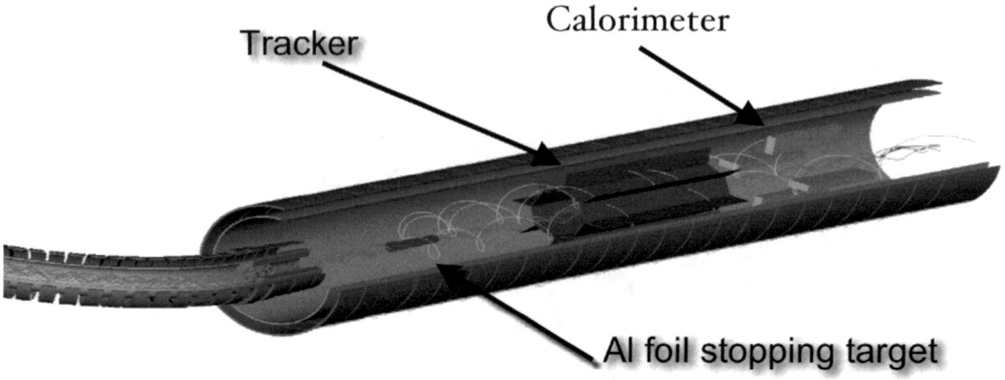

FIGURE 2. Layout of the Mu2e detector.

6. $R_{\mu e}$ calculation

The Mu2e Experiment will measure the ratio of the muon-to-electron conversion rate to the muon capture rate in Al, i.e:

$$R_{\mu e} = \frac{\Gamma(\mu N \to eN)}{\Gamma(\mu N \to \nu_\mu N^*)} = \frac{N_{\mu e}/N_s \cdot 1/\epsilon_{\mu e}}{\Lambda_{\mu\nu}/\Lambda_{\rm tot}\,(= 0.609)},$$

where: $N_{\mu e}$ is the number of events in the signal window E > 103.6 MeV: N_s is the total number of muon stops; $\varepsilon_{\mu e}$ is the overall detector acceptance (0.076), which is the product of the calorimeter trigger efficiency (0.80). the fraction of muon captures in the time window (0.51), and the reconstruction and selection efficiency (0.19): $\Lambda_{\mu e}$ is the muon capture rate: and Λ_{tot} is the total decay rate (in Al). For a correct evaluation of $R_{\mu e}$ it is evident that some means of confirming the rate and integral number of negative muons which stop on the target foils is crucial. An effective and reliable Muon Stopping Target Monitor will be established by observing the prompt production of muonic x-rays which signal the formation of muonic atoms in the target foils. This objective will be achieved locating a germanium detector where it can view, without serious deadtime. photons coming from the target foils. Such x-rays are unambiguously characteristic of a muonic atom's atomic number Z.

7. Status

The Mu2e experiment has received CD-0 approval from the U.S. Department of Energy. At present, a broad R&D program is underway to re-optimize the design of the experiment and to demonstrate the feasibility of the most technically challenging parts of the proposal. The CD-1 approval from the U.S. DOE is expected in Summer of 2011. The critical path for the Mu2e apparatus is the design and construction of the solenoid system. If all resources are made available as required, the solenoids could be installed by 2016.

8. Conclusions

The goal of the Mu2e experiment is to observe μ to e conversion or to set an upper limit of $< 6 \times 10^{-17}$ at the 90% CL and to do so in two years of running. This is 10,000 times better than the previous best limit [12] and mass scales up to 0(10,000 TeV) are within reach. For $R_{\mu e} = 10^{-15}$ the detector would see about 40 events on a background of less than 0.5 events.

ACKNOWLEDGMENTS

The author would like to thank the Fermilab management and staff for their strong support of the Mu2e experiment.

REFERENCES

1. R. M. Carey, et al. (Mu2e Collaboration), Fermilab Proposal 0973 (2008)
2. See for example, Y. Kuno and Y. Okada, Rev. Mod. Phys. 73, 151 (2001)
3. M. Raidal et al., Eur.Phys. J. C57:13-182, 2008 (arXiv:0801.1826)
4. The MECO homepage: http://www.bnl.gov/rsvp/MECO.htm . V.S. Abdjev, et al, "MELC Experiment to search for μ-A-> e-A Process", INR Preprint 786/92 (1992)
5. D.F. Measday, Physics Report 354(2001), 243-409
6. The DIO fraction is f_{DIO}=(1 -f_{OMC}), where f_{OMC} is the ordinary muon capture fraction taken from Ref [7].
7. T. Suzuki et al., Phys. Rev. C 35, 2212 (1977)
8. O. Shanker, Phys. Rev. D55(1977) 7307-7308
9. P.Bergbusch et al., Phys. Rev. C 59, 2853 (1999).
10. Mu2e Document, doc-db 1087
11. HARP Collaboration, arXiv:0909.03 37 [hep-ex] (2009).
12. W. Bertl et al. (SINDRUM II Collab.) Eur Phys. J. C47:337, 2006.

Journal of Physics: Conference Series **335** (2011) 012074 doi:10.1088/1742-6596/335/1/012074

ICARUS-T600 and beyond: LAr-TPC for neutrino physics and proton decay search

Christian Farnese, for the *ICARUS* collaboration

Istituto Nazionale di Fisica Nucleare and Dep. of Physics G. Galilei, via Marzolo 8, 35131 Padova, Italy

E-mail: `christian.farnese@pd.infn.it`

Abstract. Liquid Argon TPCs are performing detectors for neutrino and astroparticle physics thanks to their high granularity, good energy resolution and 3D imaging, allowing a precise event reconstruction. At the end of May 2010 the ICARUS-T600, the first underground large mass LAr-TPC devoted to the experimental study of neutrinos, matter stability and, more generally, rare phenomena, recorded the first CNGS neutrino interaction and is presently taking data at INFN Gran Sasso Laboratory. ICARUS-T600 represents also an important step of the Liquid Argon technique towards the realization of much larger LAr-TPC detectors for future neutrino and matter stability projects.

1. Introduction

The Liquid Argon Time Projection Chamber (LAr TPC), first proposed by C. Rubbia in 1977 [1], is a powerful detection technique that can provide a 3D imaging of any ionizing event. This continuously sensitive and self triggering detector is characterized by an high granularity and spatial resolution, similar to a bubble chambers. Moreover it is an excellent calorimeter that allow also an efficient particle identification based on the energy deposition vs range measurements.

The operating principle of the Liquid argon TPC is rather simple: a ionizing event taking place in highly purified LAr can be transported practically undistorted by an appropriate uniform electric field over macroscopic distances. Three parallel wires planes, oriented with different directions and placed at the end of the drift path record the signals induced by the drifting electrons providing simultaneous different projections of the same event. The informations from these three projection allow a precise reconstruction of the recorded particle trajectory and a precise calorimetric measurement.

The ICARUS T600 LAr-TPC detector presently taking data in Hall B of LNGS underground laboratory is the largest liquid Argon TPC ever built, with a size of about 600 t of imaging mass. Its detection technique offers the possibility to detect "bubble chamber like" neutrino events due first of all to the CNGS neutrino beam from the CERN-SPS to Gran Sasso to look for the $\nu_\mu \rightarrow \nu_\tau$ oscillations. At the same time ICARUS can study events due to solar and atmospheric neutrino and finally it can be used to search for proton decay in a background free mode.

In the following the T600 detector and its physics potentials are presented.
Finally a new proposal based on LAr-TPC detectors is described.

2. The ICARUS T600 detector

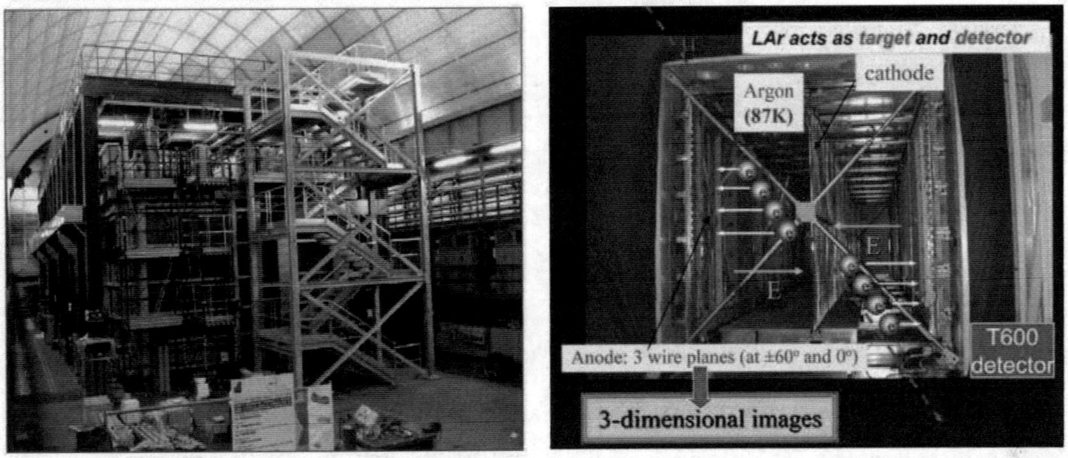

Figure 1. The ICARUS T600 detector in Hall B at the LNGS underground laboratory (left) and a simple scheme of the LAr-TPC working principle for one of the two semi-module of the T600.

The ICARUS-T600 detector [2] (see Fig.1) consists of a large cryostat split into two identical, adjacent half-modules, with internal dimensions $3.6 \times 3.9 \times 19.9$ m^3 each. Each half-module houses two Time Projection Chambers (TPC) separated by a common cathode, a field shaping system, monitors and probes. Each TPC is made of three parallel wire planes, 3 mm apart, the first with horizontal wires and the other two at $\pm 60^0$ from the horizontal direction. The distance between the cathode and the wire planes, corresponding to the maximal drift length, is 1.5 m equivalent to 1 ms drift time and the nominal drift field 500 V/cm. By appropriate voltage biasing, the first two planes faced to the drift region (Induction planes) provide signals in non-destructive way: the charge is collected in the last wire plane called Collection View. The signals coming from each wire are continuously read and digitized every 400 ns and recorded into a circular buffer. The total number of wires in the T600 detector is about 54000.

2.1. ICARUS Trigger

The passage of charged particles through the LAr ionizes it and generates a copious scintillation signal, with emission of 128 nm photons. This signal can be collected by an array of PMTs [3] coated with TPB wavelenght-shifter and located behind the wire planes and used in the trigger of this detector.

The trigger rate is dominated by cosmic rays muons and it is $\sim$ 40 mHz. In addition, using the information about the absolute time provided by an LNGS atomic clock, the early warning signals provided by CERN before each CNGS proton extraction is used to trigger in coincidence with PMT signals the CNGS neutrino events with a 10^5 suppression of cosmic rays. The analysis of the time distributions of the neutrino interactions, recorded during the CNGS spill gate, allows to reconstruct the 10.5 μs spill width of the two CNGS proton spill and at the same time the $\sim$ 2.4 ms delay is in agreement with the neutrino travel time from CERN to Gran Sasso (fig 2).

Moreover the PMT signals are used to establish the "T_0 time", i.e. the time of the ionizing event: this information, combined with the knowledge of the electron drift velocity, provides the absolute position of the ionizing event along the drift coordinate.

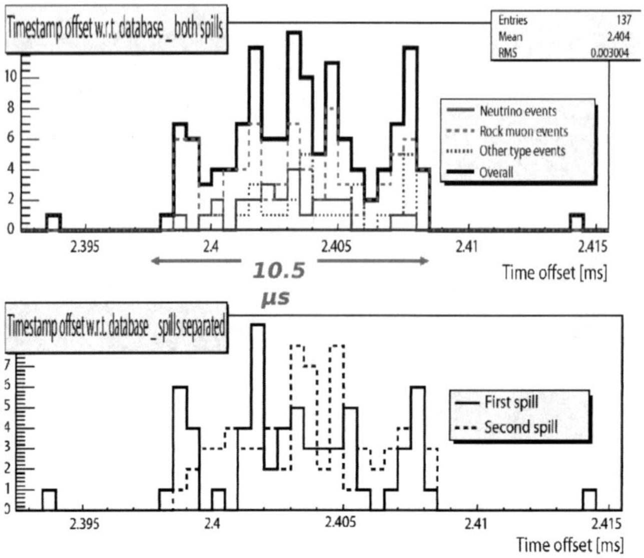

Figure 2. Distributions of the difference between the neutrino interaction ICARUS timestamp and the time of the corresponding CNGS proton extraction separated for different event type (top) or different spill (bottom).

2.2. Electron Lifetime in ICARUS

The main technological challenge of the development of the LAr-TPC is the capability to ensure and maintain a sufficiently long free electron lifetime. The ICARUS detector is equipped with both a gas and liquid recirculation system to reduce and keep at an exceptionally low level the electro-negative impurities, especially Oxygen and Nitrogen, in order to obtain free electron lifetime of several milliseconds. The electron lifetime is continuously measured in real time studying the charge signal attenuation vs drift time along through-going muon tracks without evident associated δ-rays and γ's in Collection view (Fig. 3). A precise and systematic measurement of the electron lifetime in LAr was initiated soon after the liquid recirculation in the two half-modules started. Thanks to the liquid recirculation system in the first 5 months of data taking the LAr purity reaches $\sim$ 7 ms in the West module, corresponding to a free electron charge attenuations less then 13% in the maximal drift lenght, and 2.8 ms in the East module, corresponding to a signal attenuation $\sim$ 30%.

3. Physics potentials and performances of ICARUS T600

ICARUS T600 is the major milestone towards the realization of a much more massive multikiloton LAr-TPC detector. Thanks to its high resolution and $\sim$ mm granularity, the information redundancy and the particle identification capability, ICARUS can address some interesting physics in itself, both in underground physics (proton decay search and cosmic neutrino measurements) and long base-line and high precision neutrino physics [6]. The particle ID capabilities allow to identify the interacting neutrino flavour and provide a powerful discrimination of the ν_e CC from the ν_μ NC interactions with high efficiency, by identifying and measuring the electrons emitted in the primary vertex. Overall, exploiting the event topology study, the dE/dx signal

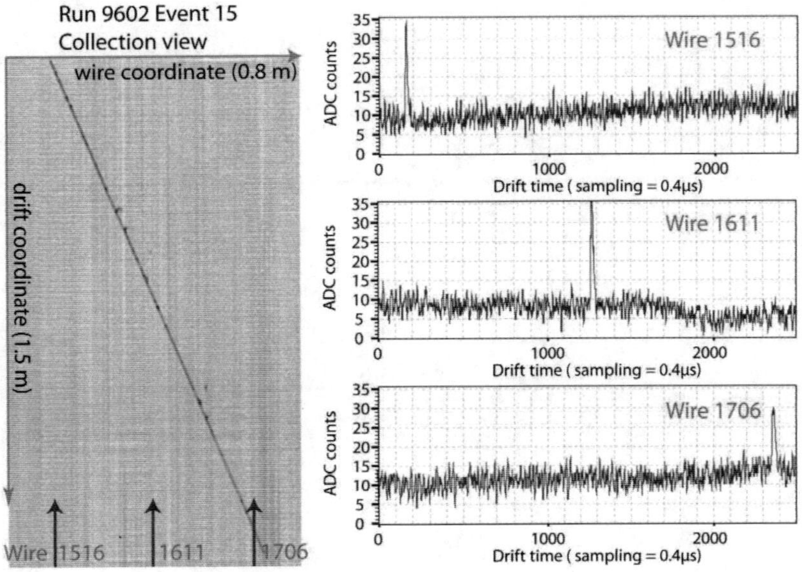

Figure 3. A muon track candidate for the purity measurement and three different wire signals.

comparison and the π^0 invariant mass measurement, it is possible to obtain a 10^{-3} electron - π^0 separation with a 90% electron identification efficiency [4].
ICARUS will observe $\sim$ 1200 CNGS neutrino CC interaction per year ($4.5 \cdot 10^{19}$ pot) with an efficiency of 90% inside the fiducial volume that means $\sim$ 2800 CNGS neutrino events for 10^{20} pot will be collected. It is possible to measure the muon momentum by the study of the multiple scattering within $\sim 10 \div 15\%$ accuracy for the ν_μ CC from CNGS [5]. The $\nu_\mu \rightarrow \nu_\tau$ oscillation will be searched in the CNGS neutrino beam recognizing the τ decay on the basis of kinematical criteria; in addition ICARUS can also search for sterile neutrino in LNSD parameter space looking at a deep e-like inelastic CC events excess. ICARUS can also study cosmic neutrino events: at least 100 CC atmospheric neutrino interaction per year are expected and also solar electron neutrino interaction with an energy greater than 8 MeV will be studied. Finally, thanks to the powerful background rejection, ICARUS can perform exclusive nucleon decay modes measurements in a background free mode, in particular for interesting channels not accessible to Čerenkov detectors due to the complicated events topology or because the emitted particles are below the Čerenkov threshold. For this reasons with an exposure of a few years ICARUS can improve the knows limits on some "super-symmetric favored" nucleon decay channel (fig. 4).

An example of CNGS ν_μ CC event detected in ICARUS T600 is shown in fig. 5: in this interaction a long muon track, $\sim$4 meters, is clearly visible. In addition, looking near the vertex in the Collection and Induction II views, the two photons emitted in a π^0 decay are visible. Another example of low energy neutrino interaction is shown in figure 6: the total visible energy is $\sim$ 890 MeV.

4. ICARUS T600 after CNGS-2: a new proposal for a future neutrino experiment at CERN-PS

The operation of ICARUS T600 demonstrate the important milestones achieved, opening the way to the development and the realization of much larger LAr-TPC detectors, for a next

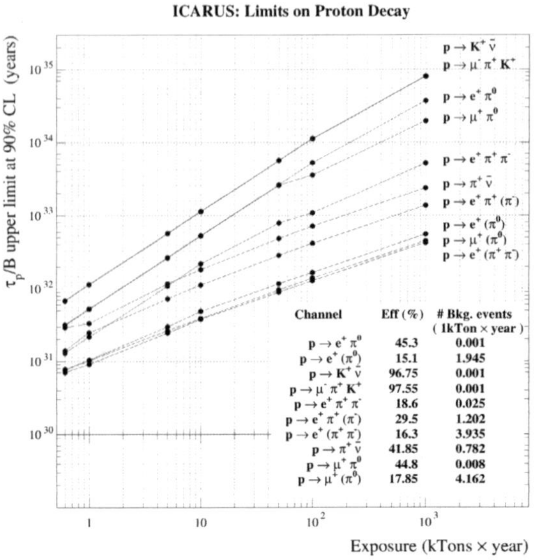

Figure 4. The proton decay limit as a function of the exposure

generation of experiments.
A detector of the same size as the T600, i.e. requiring in practice no R.&D. to become operative soon, would be an ideal device to solve the anomalies that have been found by the LNSD and MiniBooNE experiments thanks to its detection capability of ν_e. For these reasons a new experimental search based on two strictly identical LAr-TPC detector and on the observation of the ν_e signal using a refurbished ν beam at CERN-PS has been proposed [7]. The neutrino beam is a low energy ν_μ beam produced by 19.2 GeV protons of intensity 1.25 10^{20} pot/yr. The far detector, located at $\sim$ 850 m from the target, will have the same size of the ICARUS T600 while the near detector, at $\sim$ 127 m, will have a active mass of $\sim$ 150 ton (fig. 7). A possible solution for these two detector is to transport, at the end of the CNGS neutrino beam, the ICARUS-T600 and to use it as Far detector. At the same time to ensure the maximum similarity between Near and Far detectors, the Near one will be a clone of a module of T600 with length reduced by a factor 2, keeping untouched the inner detector layout. The energy resolution and granularity of these detectors, largely adequate for the study low energy neutrino events (below 3 GeV), allow to obtain a very high ν_e detection efficiency maintaining at the same time an extremely high level of rejection of associated NC background events, in particular due to π^0.

In absence of oscillations, apart some small beam related spatial corrections, the two experimentally observed event distributions in the Near and Far detector must be a precise copy of each other and can be compared without any need of Monte Carlo corrections. Therefore an exact observed proportionality between the two ν_e spectra implies directly the absence of neutrino oscillations over the measured interval of L/E. At the same time any difference between the Near and Far observed event spectra would be a direct proof of neutrino oscillations allowing to determinate separately both the mixing angle and the mass difference. As shown in fig. 8, very different and distinguishable pattern can be obtained for different values in the (Δm^2,sen^2 2θ) plane: in any case, the magnitude of the LSND expected oscillatory behavior, for the moment completely unknown, is well above the background, also considering the very high statistical impact and resolution of the experimental measurement. The expected sensitivity in a two-

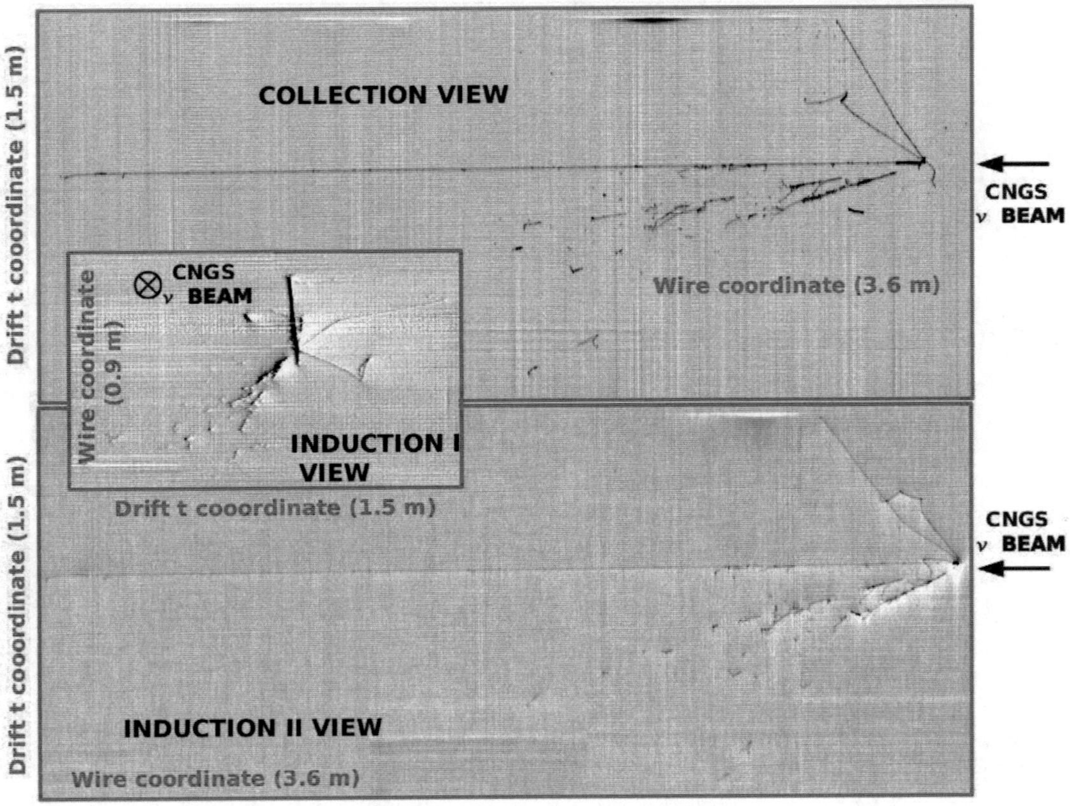

Figure 5. Example of CNGS ν_μ CC event.

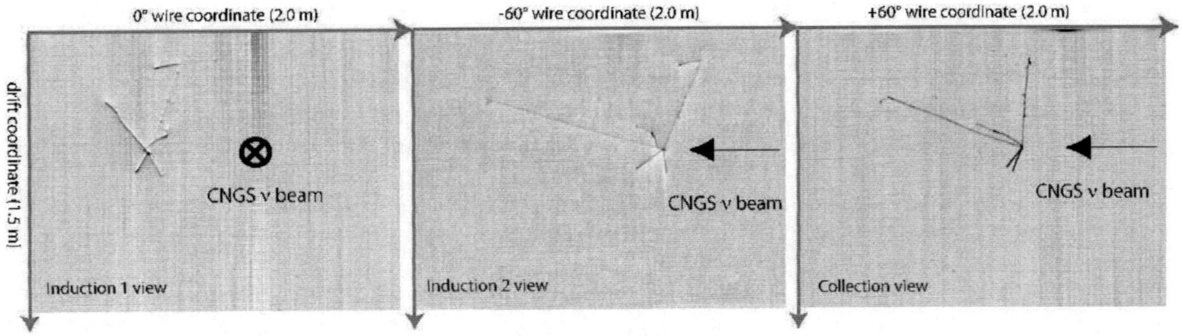

Figure 6. Example of low energy neutrino interaction.

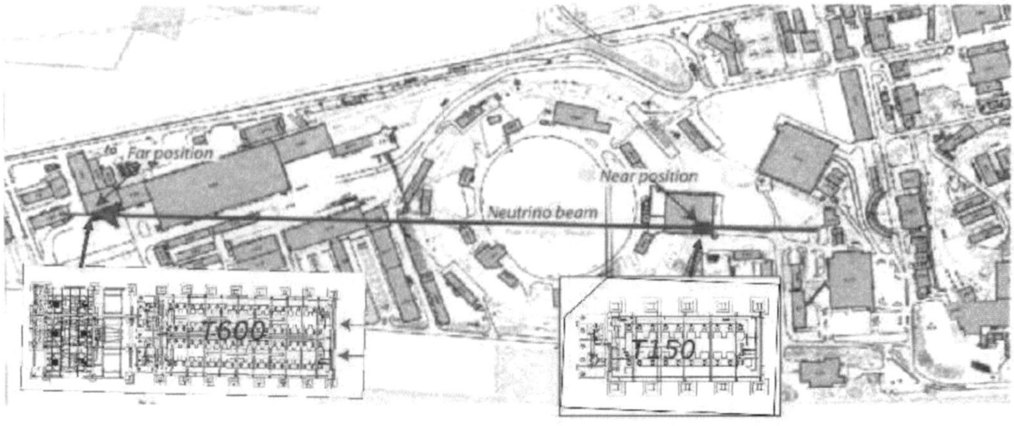

Figure 7. Neutrino beam from the CERN-PS and positions of the two LAr-TPC detectors.

flavour oscillation scheme for the neutrino beam considering 2.5 10^{20} pot, shown in fig. 8, extends well over the region explored by LSND and MiniBoone. Even in anti-neutrino mode, 2 years of data taking would be sufficient to cover the whole LNSD region.

5. Conclusions

The ICARUS-T600 detector is presently taking data at the LNGS underground laboratory. Neutrino interactions have been observed and collected for 6 months. The successful assembly and operation of this LAr-TPC is the experimental proof that this technology is mature: this experiment is so far the most important milestone for the LAr-TPC technology and acts as a full-scale test-bed located in a underground environment. The unique imaging capability of this detector, its spatial and calorimetric resolutions and the powerful separation between the electron and π^0 signals allow to reconstruct and identify events in a new way with respect to the other neutrino experiments. Its high imaging granularity and resolution allow to address a wide physics programme. ICARUS-T600 is collecting events from the CNGS neutrinos beam from CERN-SPS to search the $\nu_\mu \rightarrow \nu_\tau$ oscillations. In addition ICARUS-T600 will study with few year data taking the solar and atmospheric neutrino and explore in a new way the nucleon stability in particular channels beyond the present limits. Finally the LAr-TPC technology can be employed to solve the LSND/MiniBooNe anomalies : for this reason a novel search with a refurbished ν beam at the CERN-PS has been proposed after the ICARUS T600 exploitation at LNGS.

6. Acknowledgements

I thank all the ICARUS Collaboration for the contributions in the preparation of the talk and of this proceeding. In particular I wish to thank Prof. Milla Baldo Ceolin who introduces me in the neutrino physics and provides me precious advices for this talk. I would like also to thank the Organizing Committee for the possibility to present this talk during the DISCRETE 2010.

References

[1] C. Rubbia, The Liquid-Argon Time Projection Chamber: A new Concept for Neutrino Detector, CERN-EP/77-08 (1977).
[2] S. Amerio *et al.* [ICARUS Collab.], Nucl. Instr. and Methods A527 (2004) 329.

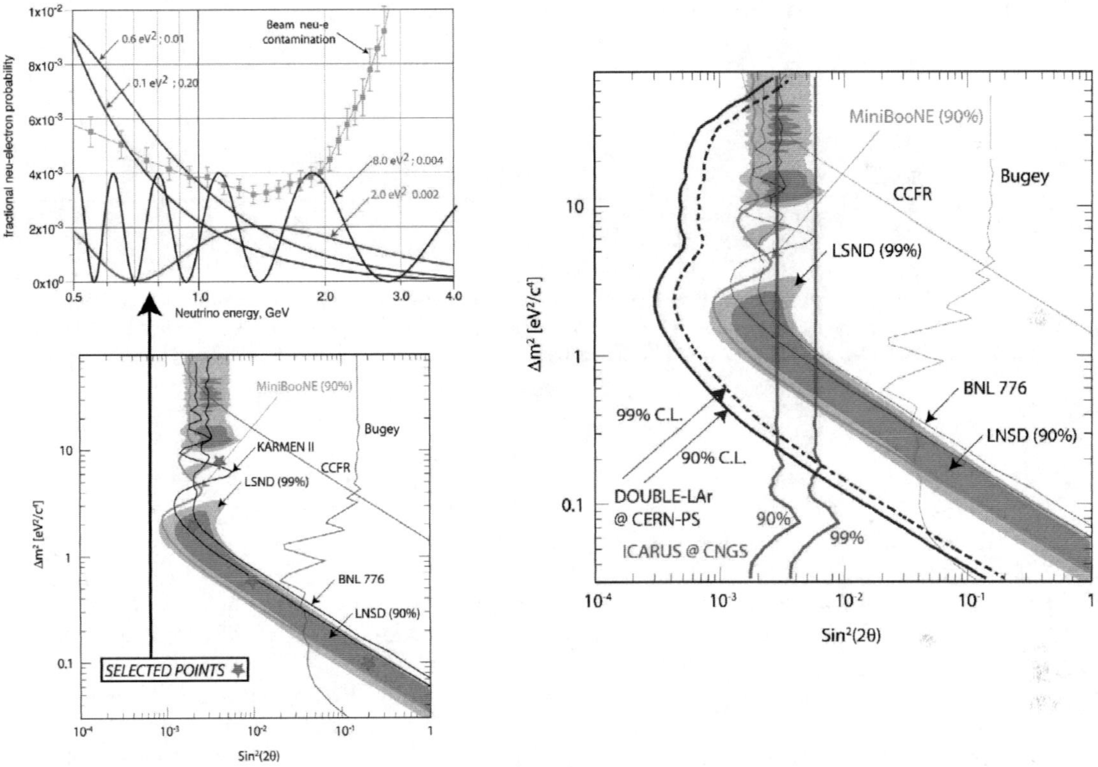

Figure 8. Left: the experimentally expected oscillation patterns at 850 m are shown for some particular points of the LSND allowed region; the expected background is also shown. Right: expected sensitivity for the proposed experiment exposed at the CERN-PS neutrino beam considering 2.5 10^{20} pot; the expectations from the ICARUS-T600 at LNGS are also shown.

[3] A. Ankowski *et al.* [ICARUS Collab.] Nucl. Instr. and Meth. A556 (2005) 146.
[4] A. Ankowski *et al.* [ICARUS Collab.], Acta Phys. Polon. B41 (2010) 103.
[5] A. Ankowski *et al.* [ICARUS Collab.], E. Phys. J. C48, (2006) 667.
[6] F. Arneodo *et al.*[ICARUS Collab.], ICARUS initial physics program, LNGS P28/01, LNGS-EXP 13/89 add. 1/01; Cloning of T600 modules to reach the design sensitive mass, LNGS-EXP 13/89 add. 2/01.
[7] B.Baibussionov *et al.*, A new search for anomalous neutrino oscillations at the CERN-PS Preprint 0909.0355

Journal of Physics: Conference Series **335** (2011) 012075 doi:10.1088/1742-6596/335/1/012075

First Results of Searches for New Physics at $\sqrt{s}$=7 TeV with the CMS detector

Paolo Azzurri

Scuola Normale Superiore, Piazza dei Cavalieri 7, 56126 Pisa, Italy

E-mail: paolo.azzurri@cern.ch

Abstract. First searches for new physics phenomena using the LHC 7 TeV proton-proton collision data collected by the CMS detector in 2010 are reviewed. Results are presented of searches for new physics in events with hadronic jet pairs, and for heavy stable charged particles, including a dedicated search for long-lived particles that stop in the detector and decay in periods between beam crossings.

1. Introduction

During the 2010 LHC running, the CMS experiment [1] recorded over 40 pb^{-1} of pp collisions at 7 TeV center-of-mass energy. The large momentum transfers accessible at these collision energies allow to test for new short range physics scenarios beyond the predictions of Quantum Chromodynamics (QCD) in dijet events, and to search for the production of new heavy quasi-stable particles [2].

In the following, results for the search of new physics in dijet events are presented for resonances in the dijet mass spectrum in Section 2, and for new interactions affecting dijet angular distributions in Sections 3 and 4. The outcome of searches for generic heavy stable charged particles are given in Section 5, while the pursuit of long-lived gluinos that stop and decay in the CMS detector is detailed in Section 6.

2. Narrow dijet resonances

In proton-proton collisions QCD predicts that the invariant mass spectrum of the two jets with largest p_{T} falls steeply and smoothly, while new massive objects that couple to quarks (q) and gluons (g) would show up as resonant structures in the dijet mass. A data sample of 2.9$\pm$0.3 pb^{-1} is employed for this analysis [3], using a single-jet trigger with a transverse energy threshold of 50 GeV, that is measured to be over 99.5% efficient for dijet masses above 220 GeV.

Jets are reconstructed with the anti-k_T algorithm [4] and $R = 0.7$. The reconstructed jet energy E is corrected as a function of p_{T} and η for the non-linearity and inhomogeneity of the calorimeter response [5]. The two jets with the leading p_{T} define the dijet system. The inclusive dijet mass is shown in Figure 1, where the data points are compared with a prediction from PYTHIA [6], that is in agreement within the 10% jet energy scale uncertainty. The data is also compared with a smooth fit to a functional form with four free parameters, and there is no indication of narrow resonances.

The dijet mass data points, the fitted background parameterization, and the expected dijet resonance shapes are used to set specific Bayesian limits on new particles decaying to parton

Published under licence by IOP Publishing Ltd

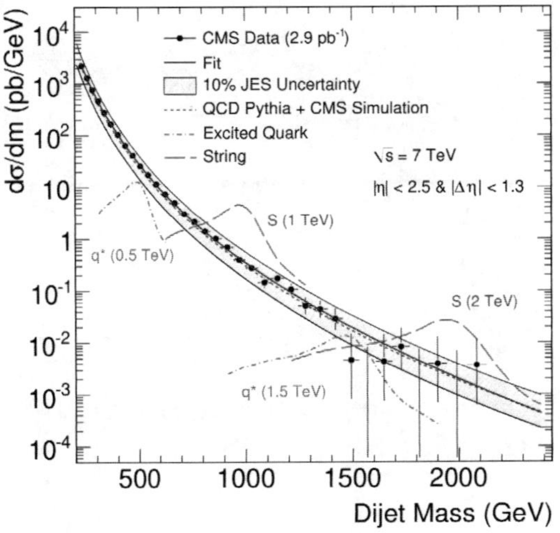

Figure 1. Dijet mass spectrum. The data points are superimposed with a smooth fit (solid line) and to QCD predictions [6] (short-dashed line). The shaded band shows the jet energy scale uncertainty. Possible excited quark signals and string resonance signals are also shown.

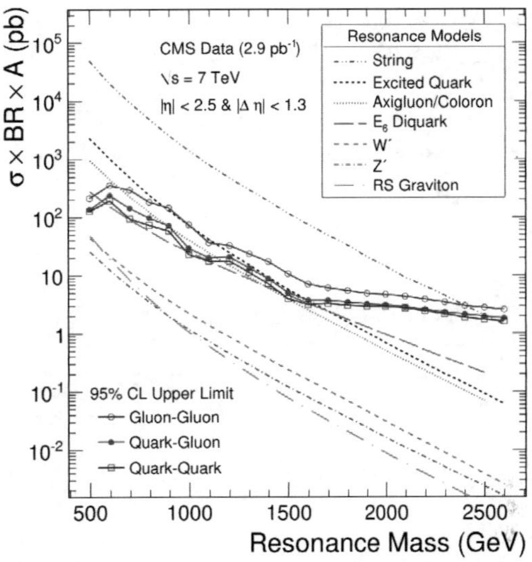

Figure 2. Production upper limits for resonances decaying to gluon-gluon, quark-gluon, and quark-quark dijets, compared to theoretical predictions from different models.

pairs. The obtained 95% confidence level (CL) upper limits on the resonances production are shown in Figure 2 and compared to the expected productions in different models. The resulting excluded mass ranges for string resonances [7], excited quarks [8], axigluons [9], colorons [10] and E_6 diquarks [11] all extend previous exclusions [3], while for new gauge bosons W' and Z' [12], and Randall-Sundrum gravitons [13] no mass interval is excluded with the present data.

3. Dijet centrality ratio

In proton-proton collisions the QCD dijet production is dominated by t-channel processes and peaks in the forward direction (along the beam axis) while new physics scenarios, including models of quark compositeness, predict more isotropic angular distribution, leading to enhanced jet production at smaller pseudorapidity values.

The dijet centrality ratio R_η between the number of events with the two leading jets in the region $|\eta| < 0.7$ and the number of events with the two jets in the region $0.7 < |\eta| < 1.3$, is defined to use less angular information than a full dijet angular distribution analysis, but allows for fine binning in the dijet mass and cancels many sources of systematic uncertainty, providing an accurate test of QCD and probe to new physics. The analysis [14] is based on the same data sample, events selection and jet reconstruction used for the previously described dijet resonances search [3]. Figure 3 shows the predicted and observed values of R_η a a function of the dijet mass m_{jj}. The measured R_η is nearly flat with a value of about 0.5, as predicted by QCD, while the presence of quark contact interactions would lead to an increase of R_η above a value of m_{jj} that depends on the compositeness energy scale Λ.

A log-likelihood-ratio $\mathcal{R}_{LL}$ is used to test for the presence of quark compositeness, and, given the consistency of the data with the QCD hypothesis, a modified frequentist method with the CL_s approach [15], is employed to derive the 95% CL limit $\Lambda > 4.0$ TeV , as displayed in Figure 4

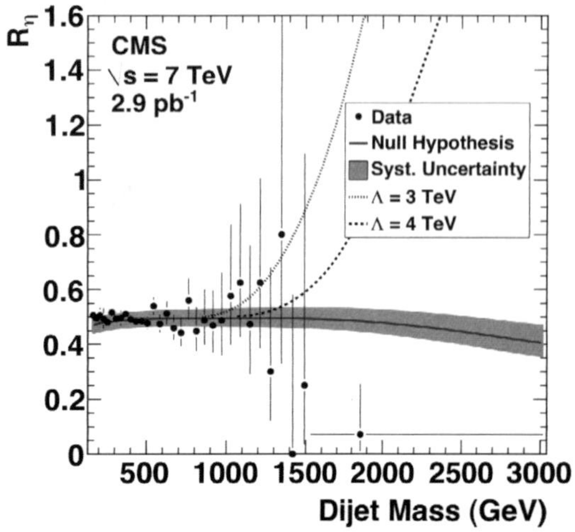

Figure 3. The observed dijet centrality ratio as a function of m_{jj} compared with QCD predictions (solid line), with the systematic uncertainty band, and to two hypotheses of quark contact interactions with $\Lambda = 3, 4$ TeV.

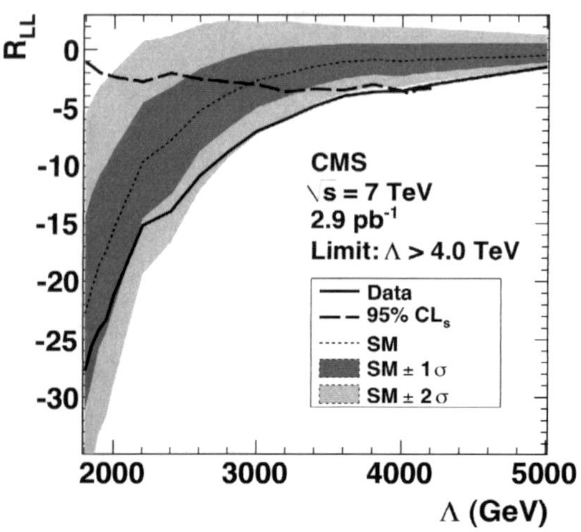

Figure 4. Exclusions for the contact interaction scale Λ. We show $\mathcal{R}_{LL}$ versus Λ for the data (solid line), the 95% CL_s (dashed line), and the SM expectation (dotted line) with 1σ (dark) and 2σ (light) bands.

4. Dijet angular distributions

The complete dijet angular distribution is measured though the differential cross section $(1/\sigma_{\rm dijet})(d\sigma_{\rm dijet}/d\chi_{\rm dijet})$ where $\chi_{\rm dijet} = (1+|\cos\theta^*|)/(1-|\cos\theta^*|)$, and θ^* is the polar scattering angle in the dijet center-of-mass (CM) frame [16]. This choice is motivated by the fact that $d\sigma_{\rm dijet}/d\chi_{\rm dijet}$ is rather flat for QCD (Rutherford) parton scattering.

A data sample of up 36 ± 4 pb^{-1} is employed for this analysis, and events are selected with five inclusive single-jet triggers, with different jet transverse momentum $p_{\rm T}$ thresholds. Offline jets are reconstructed with the anti-k_T algorithm [4] with $R = 0.5$, and the reconstructed jet energy is corrected [5]. The differential dijet angular distributions for different m_{jj} ranges, normalized to their respective integrals, are shown in Fig. 5. The data are compared with perturbative QCD (pQCD) predictions at next-to-leading order (NLO), for all m_{jj} ranges.

Given the good agreement with pQCD, the data are used to to set limits on quark compositeness represented by a four-fermion contact interaction term in addition to the QCD Lagrangian. As in the previous section, a statistical test based on the log-likelihood-ratio $\mathcal{R}_{LL}$ is used to derive limits on Λ, using the modified frequentist $\mathrm{CL_s}$ approach [15]. The results are shown in Figure 6, and a 95% confidence level bound Λ >5.6 TeV is observed, the currently most stringent limit on the contact interaction scale of left-handed quarks.

5. Heavy Stable Charged Particles

Many extensions of the standard model predict the existence of heavy (quasi-)stable charged particles (HSCPs), that would travel across the size of typical particle detector [2], and could be observable as high momentum (p) particles with exceptional ionization energy loss (dE/dx).

The CMS search [17] is based on an integrated luminosity of 3.1 pb^{-1}, and has two approaches. A "*tracker-only*" selection looks for charged particles in the inner tracker with large dE/dx and $p_{\rm T}$. A "*tracker-plus-muon*" selection additionally requires associated hits in the outer muon

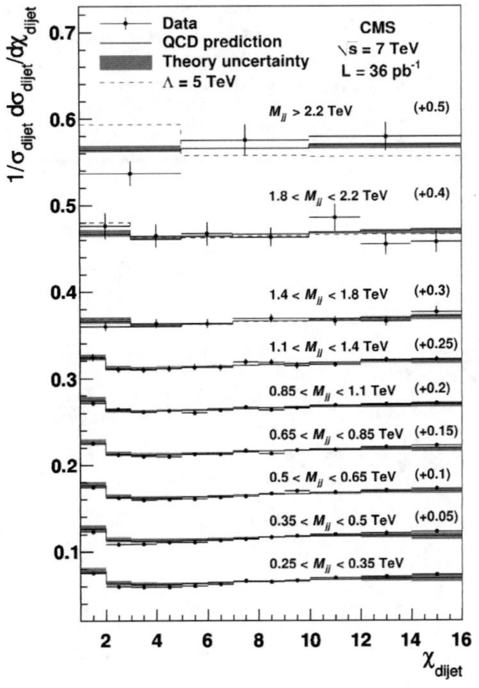

Figure 5. Normalized dijet angular distributions in m_{jj} ranges, shifted vertically (as given in parentheses) for clarity. The data points include full uncertainty bars and are compared with the predictions of pQCD at NLO (solid histogram) and with the predictions including a contact interaction term of compositeness scale $\Lambda = 5$ TeV (dashed histogram). The shaded band shows the effect on the NLO pQCD predictions due to factorization and renormalization scale variations and parton distribution functions uncertainties, as well as the uncertainties from the non-perturbative corrections added in quadrature.

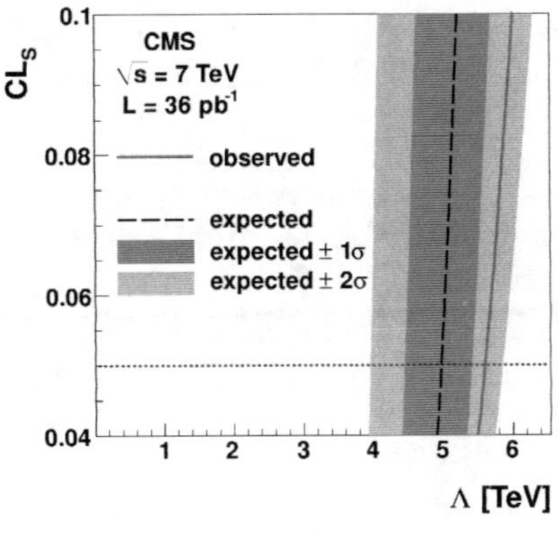

Figure 6. The observed $\mathrm{CL_s}$ (solid) and expected $\mathrm{CL_s}$ (dashed) with one (two) standard deviation indicated by the dark (light) band as a function of the contact interaction scale Λ, from the dijet angular distributions. The 95% confidence level limits on Λ are extracted from the intersections of the $\mathrm{CL_s}$ lines with the horizontal line at $\mathrm{CL_s}$=0.05.

detectors. For both selections, the mass of the candidate is estimated from the measured p and dE/dx. For candidate tracks with $|\eta| < 2.5$ and $p_T > 15$ GeV/c, an estimator I_{as} of the degree of compatibility of the observed charge measurements in the silicon detector with the minimum-ionizing particle (MIP) hypothesis is used to point out non-relativistic HSCP candidates. The data and Monte Carlo (MC) normalized distributions of p_{T} and I_{as} for pre-selected tracker-only candidates are shown in Figure 7. One MC sample contains events from standard QCD processes, while the second MC sample contains signal events from pair-production of stable $\widetilde{\mathrm{g}}$ with a mass of 200 GeV/c^2. Both samples are generated with PYTHIA [6].

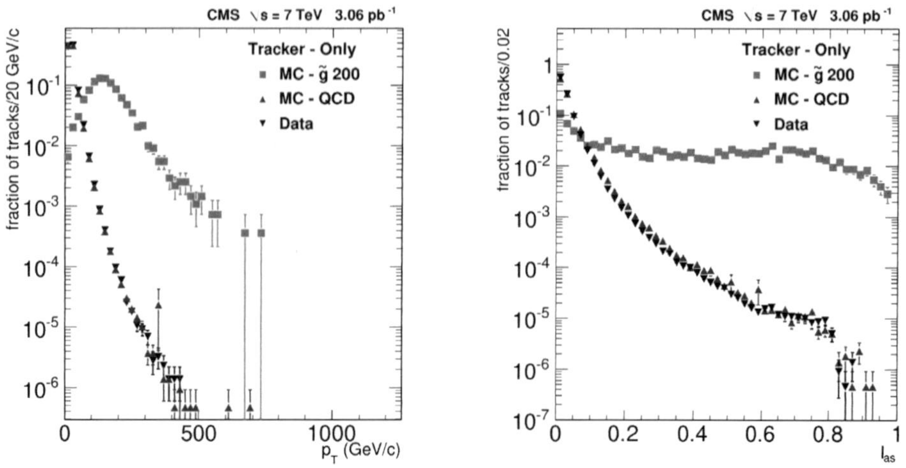

Figure 7. Data normalized distributions of p_{T} (left) and I_{as} (right), compared to simulations from standard QCD processes and from pair-production of $\tilde{g}$ with a mass of 200 GeV$/c^2$.

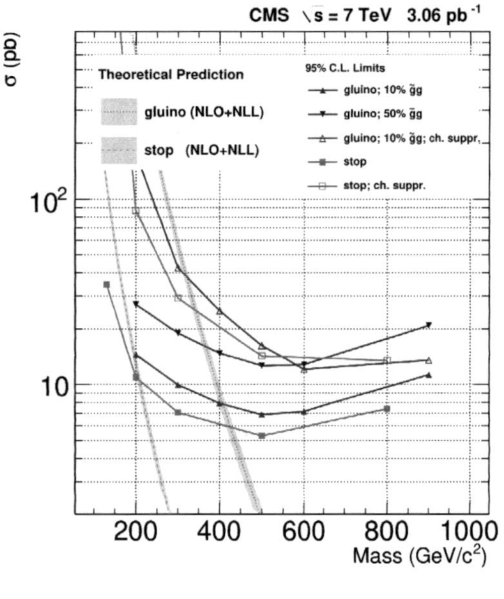

Figure 8. HSCP theoretical production cross section and observed 95% CL upper limits for different combinations of models and scenarios: pair production of supersymmetric stop and gluinos; different fractions, f, of R-gluonball hadronization states and charge suppression scenarios [17]. Only the results obtained with the most sensitive selection are reported: tracker-only for the charge suppression scenarios and tracker-plus-muon for all other cases. The bands represent the theoretical uncertainties on the cross section values.

The mass of the candidates is extracted from the measured dE/dx and p values, inverting a parametrized Bethe-Bloch fitted from data. The search is performed as a counting experiment with candidates required to have I_{as} and p_T greater than threshold values, and the candidates mass to be in the range of 75 to 2000 GeV$/c^2$.

No signal candidates are observed in the data, with 0.074 ± 0.011 expected form standard processes for the tracker-only selection, and 0.025 ± 0.004 for the tracker-plus-muon selection. Therefore cross section upper limits at the 95% CL are set on the HSCP production for two benchmark scenarios: direct production of $\tilde{g}$ pairs and $\tilde{t}_1$ pairs. The signal efficiencies depend on the hadronization fractions and interactions of the resulting signal R-hadrons. A summary of cross section upper limits, set with a Bayesian method, are shown in Figure 8. The tracker-plus-muon selection set a 95% CL lower mass limit of 398 GeV$/c^2$ for long-lived $\tilde{g}$ with a $f = 0.1$

fraction of R-gluonball states, and of 202 GeV/c^2 for long-lived $\tilde{t}_1$. In a pessimistic scenario of complete charge suppression the tracker-only selection yields a $\tilde{g}$ mass limit of 311 GeV/c^2.

6. Search for stopped gluinos

A significant fraction of slow charged R-hadrons, resulting from the production of long-lived gluinos, can lose sufficient energy to be stopped inside the CMS detector and may decay at later times (with $\tilde{g} \rightarrow g\tilde{\chi}_1^0$) resulting in jet-like signals in the calorimeters, out-of-time with respect to LHC collisions. The presence of a stopped-particle signal is searched for in 62 hours of trigger live-time, following collision data corresponding to 10 pb^{-1} recorded with 1×10^{32} cm^{-2}s^{-1} peak luminosity. A control sample of 95 hours of trigger live-time collected during fills with $2 - 7 \times 10^{27}$ cm^{-2}s^{-1} luminosity is used to estimate the background rates [18].

A dedicated trigger is used that requires a jet-like signal when no beam is crossing CMS, and outside a 75 ns window around beam crossings. Care is taken to reject beam-halo events, cosmic rays and calorimeter instrumental noise. Both a counting experiment and a time-profile analysis are performed for candidate events, building a probability density function (PDF) for the gluino decay signal. Figure 9 shows an example of such a PDF for a 1 μs lifetime, together with the in-orbit positions of two candidate events.

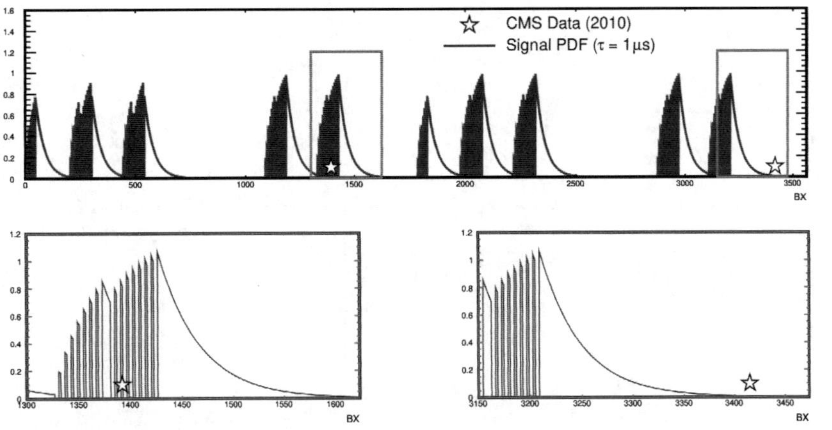

Figure 9. In-orbit positions of two observed events recorded during an LHC fill with 140 colliding bunches. The decay profile for a 1 μs lifetime hypothesis is overlaid. The bottom panels are zoomed views of the boxed regions around the two events.

Finally no excess of signal events over the expected background rates is observed in the search sample, for any lifetime hypothesis, so that 95% CL limits are set on $\sigma(pp \rightarrow \tilde{g}\tilde{g}) \times BR(\tilde{g} \rightarrow g\tilde{\chi}_1^0)$ for a given mass difference $m_{\tilde{g}} - m_{\tilde{\chi}_1^0} > 100$ GeV/c^2, and are summarized in Figure 10. For $m_{\tilde{g}} = 300$ GeV/c^2 lifetimes from 75 ns to 3×10^5 s are excluded with the counting experiment.

The excluded regions as a function of the gluino mass are plotted in Figure 11. For lifetimes between 10 μs and 1000 s $m_{\tilde{g}} < 370$ GeV/c^2 is excluded, and if only electromagnetic (EM) interactions are considered the limit becomes $m_{\tilde{g}} < 302$ GeV/c^2.

References

[1] R. Adolphi *et al.* [CMS Collaboration], JINST **3** (2008) S08004.
[2] M. Fairbairn et al., "Stable massive particles at colliders", *Phys. Rept.* **438** (2007) 1,
[3] V. Khachatryan *et al.* [CMS Collaboration], "Search for Dijet Resonances in 7 TeV pp Collisions at CMS," Phys. Rev. Lett. **105** (2010) 211801 [arXiv:1010.0203 [hep-ex]].
[4] M. Cacciari, G. P. Salam, and G. Soyez, "The anti-k(t) jet clustering algorithm", *JHEP* **04** (2008) 063.
[5] CMS Collaboration, "Jet Performance in pp Collisions at $\sqrt{s}$=7 TeV", *CMS Physics Analysis Summary* **CMS-PAS-JME-10-003** (2010).
[6] T. Sjostrand et al., "High-energy-physics event generation with PYTHIA 6.1", *Computer Physics Commun.* **135** (2001) 238.

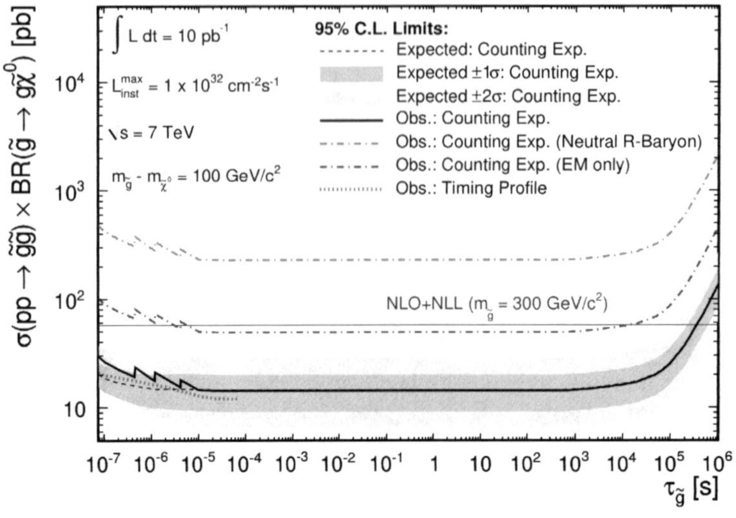

Figure 10. Expected and observed 95% CL limits on gluino pair production cross section times branching fraction as a function of the gluino lifetime from the counting experiment and the time-profile analysis. Alternative R-hadron interaction scenarios are also presented. The NLO+NLL production expectation for $m_{\tilde{g}} = 300$ GeV/c^2 is also indicated.

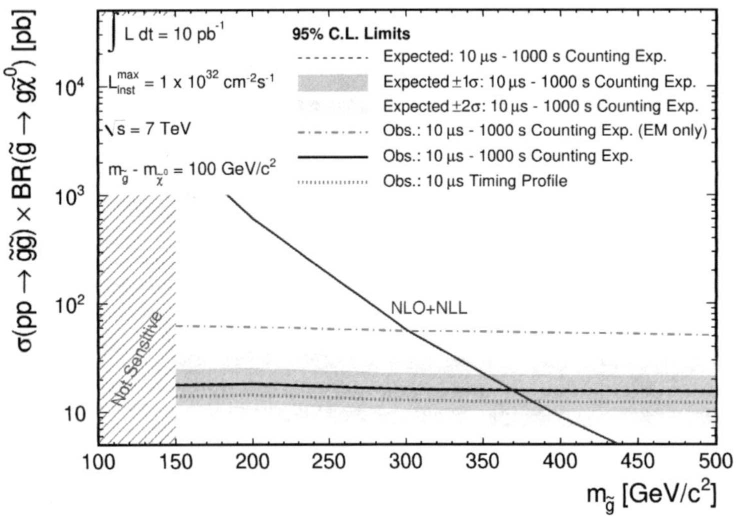

Figure 11. Upper 95% CL limits on gluino pair production cross section times branching fraction as a function of the gluino mass with strong or EM only R-hadron interactions. The $m_{\tilde{g}} - m_{\tilde{\chi}_1^0}$ difference is set at 100 GeV/c^2 and $m_{\tilde{\chi}_1^0} > 50$ GeV/c^2 is assumed. The NLO+NLL expected production are also shown. The indicated lifetimes maximize the counting experiment and time-profile analysis sensitivity.

[7] L. A. Anchordoqui, H. Goldberg, S. Nawata et al., "Jet signals for low mass strings at the Large Hadron Collider", *Phys. Rev. Lett.* **101** (2008) 241803.
[8] U. Baur, I. Hinchliffe, and D. Zeppenfeld, "Excited Quark Production at Hadron Colliders", *Int. J. Mod. Phys.* **A2** (1987) 1285.
[9] P. H. Frampton and S. L. Glashow, "Chiral color: An alternative to the standard model", *Phys. Lett.* **B190** (1987) 157.
[10] E. H. Simmons, "Coloron phenomenology", *Phys. Rev.* **D55** (1997) 1678.
[11] J. L. Hewett and T. G. Rizzo, "Low-energy phenomenology of superstring-inspired E_6 models", *Phys. Rept.* **183** (1989) 193.
[12] E. Eichten, I. Hinchliffe, K. Lane et al., "Supercollider physics", *Rev. Mod. Phys.* **56** (1984) 579.
[13] L. Randall and R. Sundrum, "An alternative to compactification", *Phys. Rev. Lett.* **83** (1999) 4690.
[14] V. Khachatryan *et al.* [CMS Collaboration], "Search for Quark Compositeness with the Dijet Centrality

Ratio in pp Collisions at $\sqrt{s}$=7 TeV," Phys. Rev. Lett. **105** (2010) 262001 [arXiv:1010.4439 [hep-ex]].
[15] A. L. Read, "Modified frequentist analysis of search results (The CL(s) method)", in *First Workshop on Confidence Limits, CERN, Geneva, Switzerland*, pp. 81–101. Jan, 2000.
[16] V. Khachatryan *et al.* [CMS Collaboration], "Measurement of Dijet Angular Distributions and Search for Quark Compositeness in pp Collisions at 7 TeV," arXiv:1102.2020 [hep-ex].
[17] V. Khachatryan *et al.* [CMS Collaboration], "Search for Heavy Stable Charged Particles in pp collisions at $\sqrt{s}$=7 TeV," arXiv:1101.1645 [hep-ex].
[18] V. Khachatryan *et al.* [CMS Collaboration], "Search for Stopped Gluinos in pp collisions at $\sqrt{s}$ = 7 TeV," Phys. Rev. Lett. **106** (2011) 011801 [arXiv:1011.5861 [hep-ex]].
[19] W. Beenakker et al., "Soft-gluon resummation for squark and gluino hadroproduction", *JHEP* **12** (2009) 041,

Journal of Physics: Conference Series **335** (2011) 012076 doi:10.1088/1742-6596/335/1/012076

New Physics with the ATLAS detector: experimental prospects

Giovanni Siragusa *on behalf of the ATLAS collaboration*
Institut für Physik, Staudingerweg, 7, 55128 Mainz, Germany.
E-mail: siragusa@uni-mainz.de

Abstract. During 2010 the ATLAS detector has collected 45 pb^{-1} of *proton-proton* collisions at $\sqrt{s} = 7$ TeV. These data have been used for a wide range of searches such as high-mass final states and contact interactions. Early inclusive SUSY searches have been also performed for a wide range of final states. The most recent results of searches of physics beyond the Standard Model with the ATLAS detector are presented. Prospects for physics searches with ~ 1 fb^{-1} of data will be discussed together with the most relevant performance results.

1. Introduction

During the data taking period of 2010, the LHC has delivered *proton-proton* (*pp*) collisions at $\sqrt{s} = 7$ TeV for a total integrated luminosity of 48 pb^{-1}, reaching the peak istantaneous luminosity of 2.1×10^{32} $cm^{-2}s^{-1}$. Further details on the luminosity measurement and beam conditions can be found in Ref. [1].
The ATLAS detector [2], a general purpose detector optimized for a wide range of physics processes involving high-p_T leptons, jets and missing energy, has been used to search for evidence of new physics beyond the Standard Model (SM). During the 7 TeV run in 2010, it has collected 45 pb^{-1} of data, which correspond to a data-taking efficiency of 93.6%.
Recent experimental results from ATLAS will be presented, discussing their implications in the context of several theoretical models, which predict new phenomena at the TeV scale.

2. Search for new physics in final states containing two jets

The production and properties of dijet events in high energy collisions at the LHC have been measured in ATLAS [3] showing good agreement between data and QCD expectations. Detailed analyses have been performed to search for new physics in the dijet final states. In particular, in the following sections, we report on results for the search of resonances and for deviations in several sensitive observables, which should account for quark contact-interactions.
In the following, jets are reconstructed by using the anti-kT jet clustering algorithm [4] with a radius parameter $R = 0.6$. Detailed studies on the jet reconstruction and calibration have been performed using Monte Carlo (MC) simulated events, test beam and collision data [5, 6].

2.1. Dijet resonances

Several extensions beyond the SM predict the presence of new particle states with heavy masses accessible at LHC energies and decaying into two energetic partons. Such states include an excited composite quark q^*, in models postulating quark substructure [7, 8, 9]. This is the benchmark model that will be used for the present study.

Particularly sensitive to such new objects is the dijet invariant mass observable, defined as $m_{jj} = \sqrt{(E^{j_1} + E^{j_2})^2 - (\vec{p}^{j_1} + \vec{p}^{j_2})^2}$, where E and $\vec{p}$ are the jet energy and three-momentum, respectively. Several experiments have examined m_{jj} distributions in search of new resonances. Recently, 1.13 fb^{-1} of $p\bar{p}$ collision data at $\sqrt{s} = 1.96$ TeV collected at the Fermilab Tevatron collider, excited quarks q^* with a mass of $260 < m_{q^*} < 870$ GeV have been excluded [10].

In ATLAS, the analysis technique [11] consisted of a model-independent search for a narrow dijet mass resonance on top of a smooth and rapidly falling spectrum, relying on the measured m_{jj} distribution to estimate the background level to this new possible signal. In the absence of an observed new physics signal, upper limits were determined on products of cross section (σ) times signal acceptance ($\mathcal{A}$). The analysis is based on a data sample corresponding to an integrated luminosity of 3.1 pb^{-1}.

The lowest unprescaled jet-trigger was used, which retains events with a jet satisfying the requirement $p_T^{jet} > 150$ GeV. Events were required to have a primary vertex reconstructed from at least three tracks with p_T above 150 MeV and be compatible with the average beam spot position. Events with at least two jets were retained if the highest p_T jet (the "leading" jet) satisfied $p_T^{j1} > 150$ GeV and the next-to-leading jet satisfied $p_T^{j2} > 30$ GeV. The two leading jets were required to satisfy several quality criteria [12] and to lie outside detector regions where the jet energy was not yet measured in an optimal way, such as the interval $1.3 < |\eta^{jet}| < 1.8$. Finally, both jets were required to be in the pseudorapidity region $|\eta^{jet}| < 2.5$, and their pseudorapidity difference was required to satisfy $|\eta^{j_1} - \eta^{j_2}| < 1.3$. The final event sample was selected by requiring the dijet invariant mass to satisfy $m_{jj} > 200$ GeV in order to eliminate any potential kinematic bias in the m_{jj} distributions from the selection requirements on the jet candidates.

The invariant mass distribution is shown in Fig. 1 together with the estimated QCD background, which is obtained by fitting the data with a function of the form $f(x) = p_1(1-x)p_2 \cdot x^{p_3 + p_4 \ln x}$, where $x \equiv m_{jj}/\sqrt{s}$ and p_i are free parameters [10]. The presence or absence of detectable m_{jj} resonances in this distribution was determined by performing several statistical tests of the background-only hypothesis [11].

In the absence of any observed discrepancy with the zero-signal hypothesis, a Bayesian approach was used to set 95% confidence level (CL) upper limits on $\sigma \cdot \mathcal{A}$ for hypothetical new particles decaying into dijets with $|\eta^{jet}| < 2.5$. For more details on the statistical treatment see Ref. [11]. The effects of the systematic uncertainties (jet energy scale, background fit, integrated luminosity, and jet energy resolution), were incorporated as nuisance parameters into the likelihood function used to extract the limit, as reported in in Ref. [11]. Figure 2 depicts the resulting 95% CL upper limits on $\sigma \cdot \mathcal{A}$ as a function of the q^* resonance mass after incorporation of systematic uncertainties. A 95% CL q^* mass exclusion region was determined to be $0.50 < m_{q^*} < 1.53$ TeV.

2.2. Quark Contact Interactions

This analysis [13] focuses on dijet angular distributions, which have been shown by previous experiments [14, 15, 16, 17] to be sensitive measures for testing the predictions of QCD and for searching for new processes. Dijet angular distributions are well suited to the analysis of early LHC data, since they are little affected by the main systematic uncertainties associated with the jet energy scale (JES) and the luminosity. QCD calculations predict that high-p_T dijet production is dominated by t-channel gluon exchange, leading to angular distributions that are peaked at $|\cos\theta^*|$ close to 1, where θ^* is the polar scattering angle in the two-parton center-of-mass (CM) frame. By contrast, models of new processes characteristically predict angular distributions that would be more isotropic than those of QCD.

The highest exclusion limits on quark contact interactions set by any previous experiment [17],

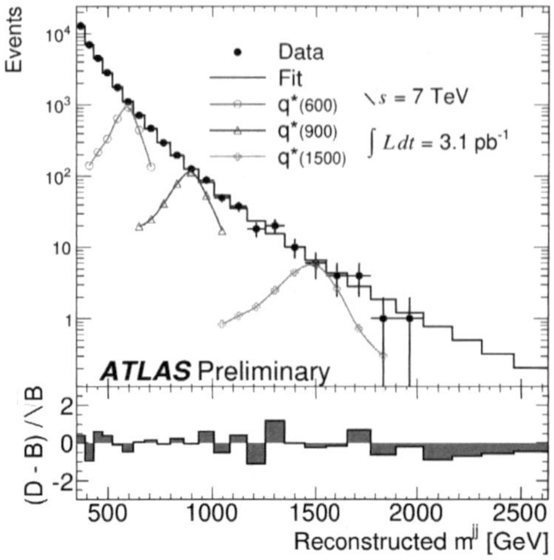

Figure 1. The data (D) dijet mass distribution (filled points) fitted using a binned background (B) distribution (histogram). The predicted q^* signals for excited-quark masses of 500, 800, and 1200 GeV are overlaid, and the bin-by-bin significance of the data-background difference is shown.

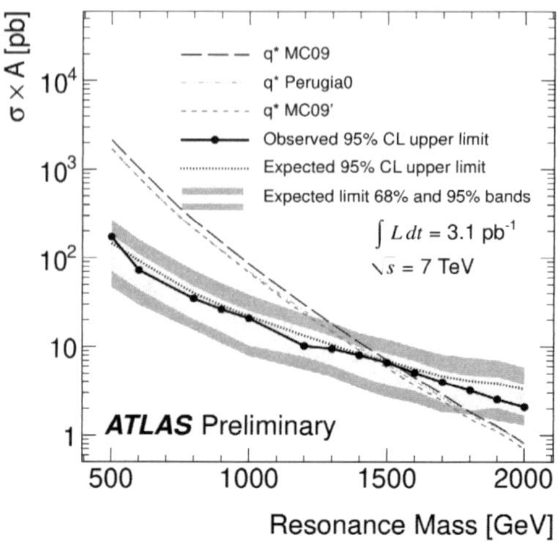

Figure 2. The 95% CL upper limit on $\sigma \cdot \mathcal{A}$ as a function of dijet resonance mass (black filled circles). The black dotted curve shows the expected 95% CL upper limit and the light and dark yellow shaded bands represent the 68% and 95% credibility intervals of the expected limit, respectively.

for several statistical analyses, ranges from 2.8 to 3.1 TeV at 95% CL for the compositeness scale Λ. The reference model assumed here is one with isoscalar left-left interaction with positive interference.

The variable χ, used in the first angular distributions considered in this study, is derived from the rapidities of the two jets defining the dijet topology (y_1 and y_2). For a given scattering angle θ^*, the corresponding rapidity in the CM frame (in the massless particle limit) is $y^* = \frac{1}{2}(y_1 - y_2)$. Defining $\chi \equiv \exp|y_1 - y_2| = \exp 2|y^*|$, the evaluation of $\mathrm{d}N/\mathrm{d}\chi$ in QCD events shows that this distribution is almost constant in χ. By contrast, the angular distributions characteristic of new processes are more isotropic in θ^*, leading to additional dijet events at low χ.

The second angular distribution considered is the dijet centrality ratio, R_C. For this analysis, the detector is divided into two pseudorapidity regions: central and non-central. R_C is defined as the ratio of the number of events in which the two highest p_T jets both fall into the central region ($|\eta_{1,2}| < 0.7$) to the number of events in which the two highest p_T jets both fall into the non-central region ($0.7 < |\eta_{1,2}| < 1.3$). Since new processes are expected to produce more central activity than QCD, their signal would appear as an increase in R_C above some m_{jj} threshold, with the increase being directly related to the cross section of the new signal.

Vertex and trigger requirements are as those discussed in Sec. 2.1. Events with at least two jets are retained if the highest p_T jet (the "leading" jet) satisfies $p_T > 60$ GeV and the next-to-leading jet satisfies $p_T > 30$ GeV. The two selected jets are required to satisfy quality criteria [12]. They are also required to be found within the pseudorapidity region $|\eta| < 2.8$, where the jet energy scale is known to highest precision.

The benchmark process considered in this analysis is a quark contact interaction, which may be used to model the onset of kinematic properties that would characterize quark compositeness: the hypothesis that quarks are composed of more fundamental particles. The model Lagrangian

for this benchmark process is a four-fermion contact interaction, the analog of the Fermi four-fermion interaction used to describe effects of the weak interaction. The effects of the contact interaction would be expected to appear below or near a characteristic energy scale Λ.
The JES represents the dominant uncertainty for this analysis. Typical values of the JES uncertainty in the considered phase space are between 5% and 7%. The resulting bin-wise uncertainties are up to 9% for the χ observable, and up to 7% for the R_C observable. The dominant sources of theoretical uncertainty on the QCD cross section and the PDF uncertainties. Since no signal of new physics processes is observed in the χ (Fig. 3) and R_C distributions, limits have been obtained on the compositeness scale Λ of quark contact interactions, based on analyses of the χ distributions. The contact term hypothesis is tested in the highest dijet mass bin in Fig. 3, which begins at $m_{jj} = 1200$ GeV. For the χ distribution in this mass bin, the parameter F_χ is defined as the ratio of the number of events in the first four χ bins to the number in all χ bins. The upper boundary of the fourth bin is at $\chi = 3.32$. A frequentist analysis is employed, as reported in Ref. [13].
The observed limit on Λ is 3.4 TeV. This limit is found from the point where the F_χ 95% CL contour crosses the measured F_χ value, which is shown in Fig. 4. All values of Λ less than this value are excluded with 95% CL. The expected limit, found from the crossing at the QCD prediction, is 3.5 TeV. This corresponds to a distance scale of $\sim 6 \cdot 10^{-5}$ fm.

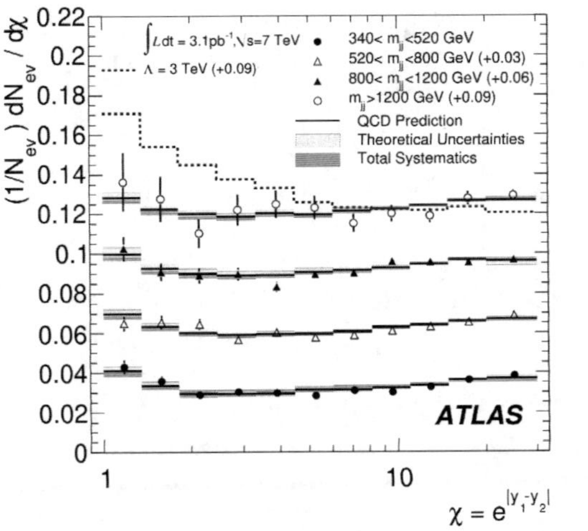

Figure 3. Normalized χ distributions for different mass intervals. Shown are the QCD predictions with systematic uncertainties (bands), and data points with statistical uncertainties. The prediction for QCD with an added quark contact term with $\Lambda = 3.0$ TeV is shown for the highest mass bin.

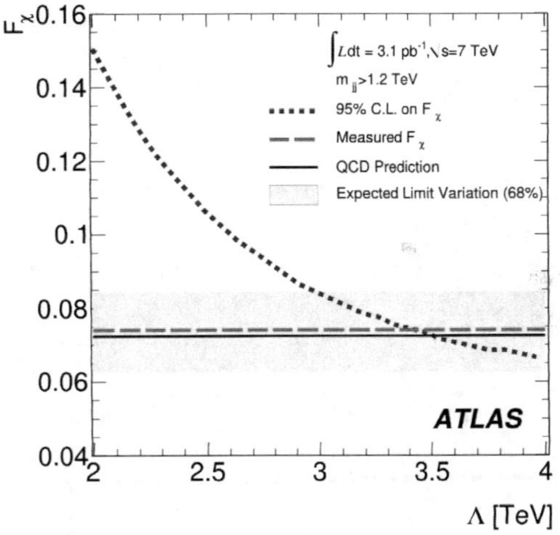

Figure 4. The dashed horizontal line is the measured F_χ and the solid horizontal line is the QCD prediction, with a band to illustrate a 1-σ variation of the expected limit. The dotted curve is the 95% CL exclusion contour for F_χ with quark contact interactions, used to set the exclusion limit on Λ.

3. High-mass states with electron plus missing transverse energy

One extension to the SM common to many models is the existence of additional, heavy gauge bosons [18]. It is common to use W' to denote any charge ± 1, spin 1 particle outside the SM and here that convention is used. The analysis has been performed on the search for a W' boson

decaying to an electron and a neutrino, whose production is inferred from Missing Transverse Energy (E_T^{miss}) [19]. Measurements at the Fermilab Tevatron experiments [20, 21] rule out the existence of a W' with massless than 1 TeV assuming the Sequential SM, i.e. the same couplings as those for the SM W boson.
Although ATLAS has not yet recorded enough luminosity to improve on this value, a limit on $\sigma \cdot B$ (the product of the production cross section and the branching fraction) is set in the electron decay channel over the mass range between 150 and 600 GeV. This limit is based on collision data equivalent to an integrated luminosity of only 317 nb^{-1}.
The kinematic variable used to identify the W' is the transverse mass $m_T = \sqrt{p_T E_T^{miss}(1 - \cos\varphi)}$ which has a Jacobian peak which falls sharply above the boson mass. Here p_T is the electron transverse momentum and φ is the angle between the transverse components of the electron momentum and the missing momentum.
The main background to the W' signal comes from the SM W boson. Other backgrounds are Z bosons decaying into two electrons where one electron is not reconstructed, W or Z decaying to tau leptons where the tau subsequently decays to an electron, and QCD and $t\bar{t}$ production where a light or heavy hadron decays into an electron or a jet is misidentified as an electron.
Events are required to be triggered by an electron with $p_T > 20$ GeV and to have a primary vertex reconstructed from at least three tracks with p_T above 150 MeV and longitudinal distance less than 15 cm from the center of the collision region. Spurious tails in E_T^{miss} arising from calorimeter noise are suppressed by checking the quality of each reconstructed jet and discarding events with any jet which has a shape indicating possible noise contamination following standard ATLAS criteria for jet cleaning [12]. Events are required to have a candidate electron defined as follows. A candidate electron is one reconstructed with $E_T > 20$ GeV, $|\eta| < 1.37$ or $1.52 < |\eta| < 2.47$ and satisfying the ATLAS "medium" electron requirement, which is defined in Ref. [2]. In addition, a fiducial cut is made to remove events with electrons near problematic regions of the electromagnetic calorimeter. In addition, the inner detector track associated with the electron is required to be close to the primary vertex, specifically with transverse distance of approach satisfying $|r_0^{PV}| < 1$ mm and longitudinal distance at this point $|z_0^{PV}| < 5$ mm. Events are required to have exactly one candidate electron.
To suppress QCD sources of background, the electron is required to be isolated, defining the isolation by $R_{isol} = \frac{\sum p_T^{trk}}{p_T^e}$ and requiring $R_{isol} < 0.05$. Here p_T^e is the electron transverse momentum and the sum in the numerator is over the transverse momenta of the inner detector tracks with $p_T^{trk} > 1.0$ GeV in a cone $\Delta R < 0.30$ ($\Delta R = \sqrt{(\Delta\eta)^2 + (\Delta\varphi)^2}$) around the direction of the electron, excluding the track associated to the electron itself. For the final selection, additionally E_T^{miss} is required to exceed 25 GeV.
The m_T spectrum (Fig. 5) shows no evidence for the existence of a W' and the data are used to set limits on $\sigma \cdot B$ for a series of W' masses ranging from 150 to 600 GeV. Limits are obtained by counting the number of events with $m_T > 0.7 \cdot m_{W'}$.
A single-bin likelihood analysis is performed to set a limit at each mass using the observed number of events. Limits for 95% CL exclusion are calculated as described in Ref. [19]. Inputs to this calculation include the observed number of events, the signal efficiencies and uncertainties. The systematic uncertainties used in the limit calculations include the uncertainty on the number of background events, the event selection efficiency uncertainties, and the uncertainty on the integrated luminosity. These are assumed to be uncorrelated. Results are shown in Fig. 6. The intersection between the limits and these values provides a 95% CL estimate upper limit on the W' mass in the SSM model. Using linear interpolation between the points, a limit of 465 GeV is obtained at 95% CL.

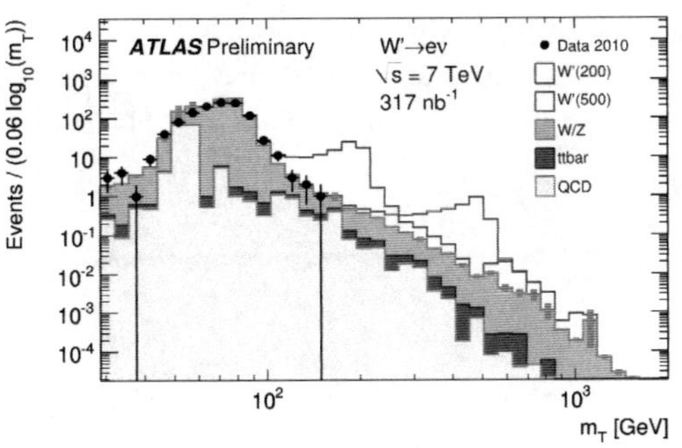

Figure 5. Transverse mass spectra after the final selection. Points are ATLAS data and the filled histograms show the Monte Carlo background from QCD, $t\bar{t}$, W and Z boson contributions. Open histograms are W' signals added to the background.

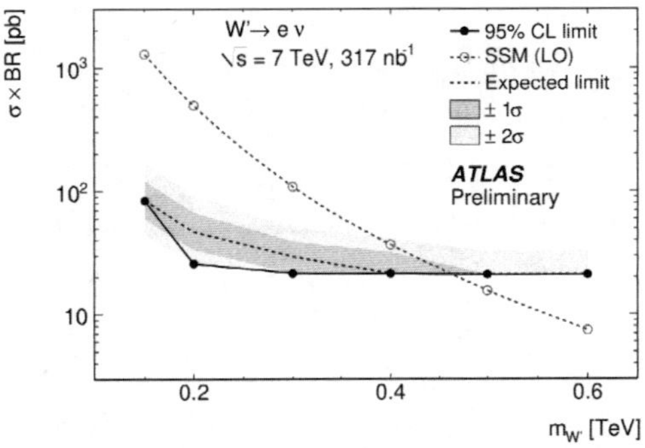

Figure 6. Limits on W' production. Limits and the Pythia SSM predictions are shown on log scale.

4. Diphoton events with large missing E_T

In the SM, the production in pp collisions of diphoton ($\gamma\gamma$) events with large E_T^{miss} is mainly due to $W/Z + \gamma\gamma$ processes. Taking into account the branching ratios of W/Z decays involving at least one neutrino, the cross sections are only a few femtobarns for 7 TeV pp collisions. In contrast, some new physics models predict much larger $\gamma\gamma$+E_T^{miss}rates.

Universal Extra Dimension (UED) models [22] postulate the existence of additional spatial dimensions in which all SM particles can propagate, leading to the existence for each SM particle of a series of excitations, known as a Kaluza-Klein (KK) tower. The same signature exists in gauge mediated supersymmetry with a gravitino as a LSP and Bino as NLSP. This analysis considers the case of a single TeV^{-1} sized UED, with compactification radius R. The masses of the states of successive levels in the tower are separated by $\sim 1/R$.

The UED model considered is defined by specifying R and Λ, the ultraviolet cut-off used in the calculation of radiative corrections to the KK masses. This analysis treats R as a free parameter and sets Λ such that $\Lambda \cdot R = 20$.

The reconstruction of photons is described in detail in Ref. [23]. To select photon candidates, EM calorimeter clusters were required to pass several quality criteria and to lie outside problematic calorimeter regions. Photon candidates were required to have $|\eta| < 1.81$ and to be outside the transition region $1.37 < |\eta| < 1.52$ between the barrel and the end-cap calorimeters. The analysis uses a "loose" photon selection [2]. The loose selection provides a high photon efficiency with modest rejection against the background from jets.

The data sample analysed corresponds to an integrated luminosity of 3.1 pb^{-1}. The events selected had to satisfy a trigger requiring at least one loose photon candidate with $E_T > 20$ GeV, and had to contain at least one reconstructed primary vertex consistent with the average beam spot position and with at least three associated tracks. Events were retained if they had at least two photon candidates, each with $E_T > 25$ GeV. In addition, a photon isolation cut was applied, where in the E_T in a radius of 0.2 in the $\eta - \varphi$ space around the center of the cluster (excluding the cells belonging to the photon cluster) had to be less than 35 GeV. The background was evaluated entirely using data.

Given the good agreement between the measured E_T^{miss} spectrum and the expected background (Fig. 7), a limit was set on 1/R in the specific UED model considered here. A Bayesian approach was used to calculate a limit based on the number of observed and expected events with $E_T^{miss} > 75$ GeV, as explained in detail in Ref. [24]. A Poisson distribution was used as the likelihood function for the expected number of signal events, and a flat prior was used for the signal cross section [24]. The observed 95% CL exclusion region is $1/R < 728$ GeV, as shown in Fig. 8.

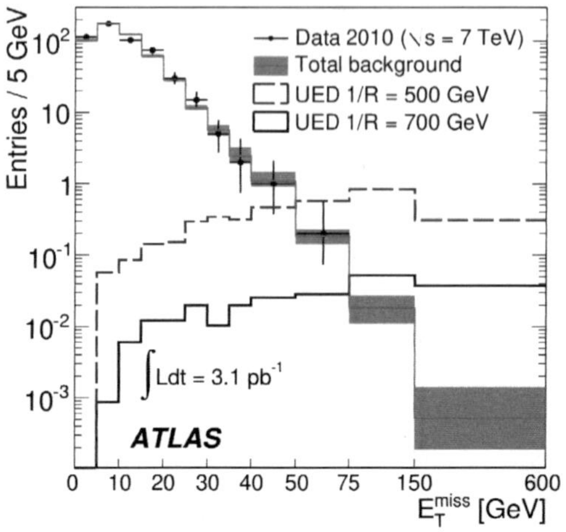

Figure 7. E_T^{miss} spectrum for the $\gamma\gamma$ candidates, compared to the total SM background as estimated from data. Also shown are the expected UED signals for $1/R = 500$ GeV and 700 GeV.

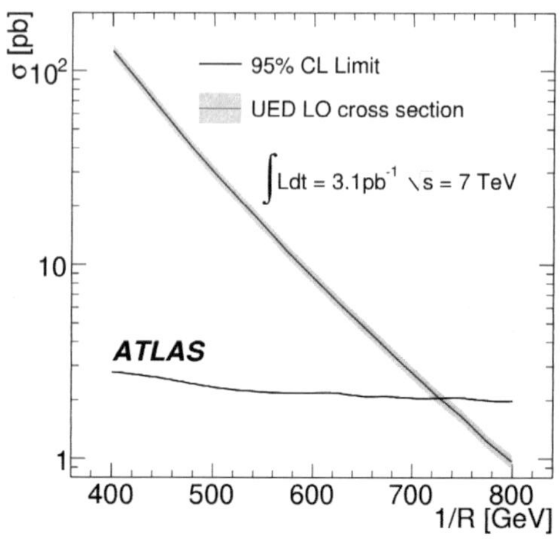

Figure 8. 95% CL upper limits on the UED production cross section, and the LO theory cross section prediction, as a function of 1/R. The shaded band shows the pdf uncertainty.

5. SUSY searches

A wide range of SUSY searches have been performed with the first LHC data. The available data sample has been used to make a first study of the backgrounds and to optimize the analysis. Results from inclusive searches of SUSY signals are to be published. The expected sensitivity with an integrated luminosity of around 500 pb^{-1} has also been estimated [25]. With such a luminosity, the most promising final states for the search of SUSY are the one including 4 jets and 0 leptons (Ref. [26]), for which discovery of squarks up to a mass of $\sim$ 700 GeV and gluinos to a mass of $\sim$ 600 GeV is possible in the scenario of minimal Supergravity with R-parity conservation [27]. With increasing luminosity, starting from $\sim$ 1 fb^{-1}, will become important final states including leptons (Ref. [28]). Also final states with b-jets are expected to be really sensitive to SUSY scenarion (Ref. [29]).

6. Conclusion

The most recent results for the search of new physics with the ATLAS experiment have been shown after the detector has collected a total integrated luminosity of 40 pb^{-1} at $\sqrt{s} = 7$ TeV. The detector is working nicely and the physics performance are under control and are being improved quickly. With this first round of analysis, several limits on Exotics physics processes have been extended beyond the results of previous experiments.
Several benchmark searches are being refined and are expected to give interesting results once the full available luminosity will be analyzed. The optimization of the analysis strategies and

the improving of the understanding of the detector will allow for several sensitive searches with the data expected in the next year of LHC running.

References

[1] ATLAS Collaboration 2010 ATLAS-CONF-2010-060
[2] Aad G *et al.* (ATLAS Collaboration) 2008 *J. Instrum.* **3** S08003
[3] ATLAS Collaboration 2010 ATLAS-CONF-2010-085
[4] Cacciari M, Salam G P and Soyez G 2008 *Journal of High Energy Physics* **2008** 063
[5] Abat E *et al.* 2010 *Nucl. Instrum. Meth. A* **621** 134 – 150
[6] ATLAS Collaboration 2010 ATLAS-CONF-2010-050
[7] Eichten E J, Lane K D and Peskin M E 1983 *Phys. Rev. Lett.* **50** 811–814
[8] Baur U, Hinchliffe I and Zeppenfeld D 1987 *Int. J. Mod. Phys. A* **2** 1285
[9] Baur U, Spira M and Zerwas P M 1990 *Phys. Rev. D* **42** 815–824
[10] Aaltonen T *et al.* (CDF Collaboration) 2009 *Phys. Rev. D* **79** 112002
[11] ATLAS Collaboration 2010 ATLAS-CONF-2010-093
[12] ATLAS Collaboration 2010 ATLAS-CONF-2010-038
[13] Aad G *et al.* (ATLAS Collaboration) 2011 *Physics Letters B* **694** 327 – 345
[14] Arnison G *et al.* (UA1 Collaboration) 1984 *Physics Letters B* **136** 294 – 300
[15] Bagnaia P *et al.* (UA2 Collaboration) 1984 *Physics Letters B* **144** 283 – 290
[16] Abe F *et al.* ((CDF Collaboration)) 1996 *Phys. Rev. Lett.* **77** 5336–5341
[17] Abazov V M *et al.* (The D0 Collaboration) 2009 *Phys. Rev. Lett.* **103** 191803
[18] Amsler C *et al.* 2008 *Physics Letters B* **667** review of Particle Physics
[19] ATLAS Collaboration 2010 ATLAS-CONF-2010-089
[20] Abazov V M *et al.* (D0 Collaboration) 2008 *Phys. Rev. Lett.* **100** 031804
[21] Abulencia A *et al.* (CDF Collaboration) 2007 *Phys. Rev. D* **75** 091101
[22] Appelquist T, Cheng H C and Dobrescu B A 2001 *Phys. Rev. D* **64** 035002
[23] ATLAS Collaboration 2010 ATLAS-CONF-2010-077
[24] 2010 Search for diphoton events with large missing transverse energy in 7 TeV proton-proton collisions with the ATLAS detector. to be published in Phys. Rev. Lett. (*Preprint* `arXiv:1012.4272`)
[25] ATLAS Collaboration 2010 ATL-PHYS-PUB-2010-010
[26] ATLAS Collaboration 2010 ATLAS-CONF-2010-065
[27] Chamseddine A H, Arnowitt R and Nath P 1982 *Phys. Rev. Lett.* **49** 970–974
[28] ATLAS Collaboration 2010 ATLAS-CONF-2010-066
[29] ATLAS Collaboration 2010 ATLAS-CONF-2010-079

Journal of Physics: Conference Series **335** (2011) 012077 doi:10.1088/1742-6596/335/1/012077

Light dark forces at flavor factories

L Barzè[1,2]

[1] Dipartimento di Fisica Nucleare e Teorica, Università di Pavia, via A Bassi, 6, 27100, Pavia, Italy

[2] INFN, Sezione di Pavia, via A Bassi, 6, 27100, Pavia, Italy

E-mail: luca.barze@pv.infn.it

Abstract. SuperB experiment could represent an ideal environment to test a new $U(1)$ symmetry related to light dark forces candidates. A promising discovery channel is represented by the resonant production of a boson U, followed by its decay into lepton pairs. Beyond approximations adopted in the literature, an exact tree level calculation of the radiative processes $e^+e^- \to \gamma, U \to \mu^+\mu^-\gamma, e^+e^-\gamma$ and corresponding QED backgrounds is performed, including also the most important higher-order corrections. The calculation is implemented in a release of the generator BabaYaga@NLO useful for data analysis and interpretation. The distinct features of U boson production are shown and the statistical significance is analysed.

1. Introduction and theoretical framework

The aim of this paper, which is based on [1], is to illustrate Dark Matter models which imply the existence of a new vector boson, which carries a new *dark* force and which is lightly coupled to the photon. In particular a Monte Carlo event generator useful to describe this kind of light dark forces signatures at leptonic colliders will be presented and potentiality of a SuperB experiment for this kind of studies will be shown.

The existence of an abelian gauge simmetry with an associated light U boson which can interact with a really small coupling with Standard Model (SM) particles has been proposed by a wide class of new physics models[2–6].

An important phenomenological support to this class of models comes in recent years, when standard astrophysics and particle physics seem fail to explain striking experimental signals in terms of known sources.

These signals are represented by the 511 KeV gamma-ray signal from the galactic center observed by the INTEGRAL satellite[7], the excess in the cosmic ray positrons reported by PAMELA[8], the total electron and positron flux measured by ATIC[9], Fermi[10], and HESS[11], the annual modulation of the DAMA/LIBRA signal[12] and the features of the low-energy spectrum of rare events reported by the CoGeNT collaboration[13].

If a new secluded gauge sector under which the SM particles remain uncharged is included into the theoretical description, these evidences can be comprehensively interpreted in terms of WIMP Dark Matter (DM) particles connected to SM ones with interaction terms varying from model to model.

The simplest assumption[6] is to add an extra $U(1)$ symmetry to SM symmetry group which describe a new *dark* force and suppose this force as carried by a new vector boson (U). It is

Journal of Physics: Conference Series **335** (2011) 012077 doi:10.1088/1742-6596/335/1/012077

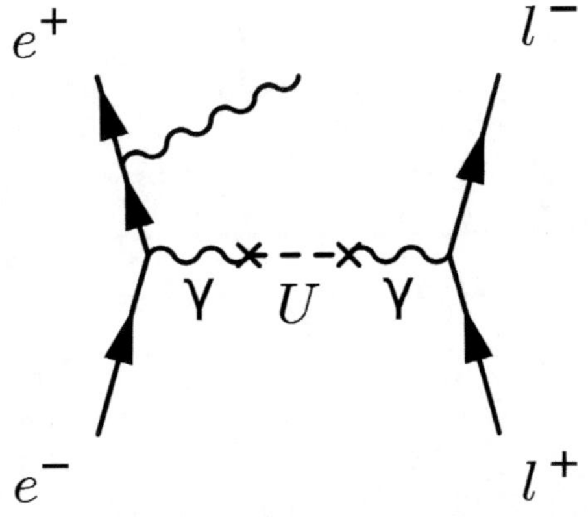

Figure 1. Typical reaction useful at leptonic colliders to study U boson properties.

possible to suppose also that U boson can communicate with the SM through a kinetic mixing term of the form

$$\mathcal{L}_{mix} = \frac{\varepsilon}{2} F^{\rm em}_{\mu\nu} F^{\mu\nu}_{\rm dark}, \tag{1}$$

describing the interaction of the U boson with SM photon. This mixing term could occour through loop effects due to really massive WIMPs both coupled to ordinary photon and U boson. In this case the ε parameter should be lower than about $10^{-2} - 10^{-3}$.

With reactions involving WIMPs in the initial state and standard particles as positrons in the final state and supposing these reactions mediated by the U boson it would be possible to explain the experimental signals described above. Since no astrophysical data involves anomalous production of antiproton, it is necessary to require the U boson mass (M_U) to be lower than the mass of two protons.

An interesting consequence of the existence of such a light U boson is that it can be directly produced in a controlled environment, such as fixed target experiments [14–16] or high-luminosity e^+ e^- colliders at the GeV scale (flavor factories)[17–19].

At flavor factories, e.g. at DAΦNE, BESIII and present and future B-factories, a particularly clean and simple channel, insensitive to the details of how the U boson takes a mass, is surely represented by the resonant radiative production of a U boson, followed by its decay into a lepton - antilepton pair (see Fig. 1).

A distinctive feature of the expected signal is the appearance of a Breit-Wigner peak in the shape of the invariant mass distribution of the lepton pairs induced by the mechanism of photon radiative return and corresponding to U boson resonant production. The drawback of this channel is the fairly small value, over a wide range of parameters, of the signal cross section in comparison with the large rate of the backgrounds given by the QED radiative processes $e^+e^- \to l^+l^-\gamma$, $l = e, \mu$. The signal over background ratio can be enhanced by cutting on the invariant mass of the lepton pair.

The conclusion presented so far in the literature is that the $U\gamma$ production process allows to identify a U boson at present flavor factories if its mass is in the range $0.1 - 1$ GeV and the kinetic mixing parameter ε is greater than $\sim 10^{-3}$.

These analyses are generally based on the evaluation of the number of signal events through the calculation of the differential cross section of the $2 \to 2$ process $e^+e^- \to U\gamma$, including the decay of the on-shell U boson into lepton pairs by means of branching ratios[17, 20], and/or an approximate estimate of the backgrounds. More importantly, all the studies so far performed neglect the contribution of higher-order initial and final state QED corrections, which are known to be a phenomenologically relevant effect at GeV-scale e^+e^- colliders[21].

To investigate whether and how the above approximations affect the physical observables of experimental interest, a Monte Carlo event generator, which is still missing for such studies, has been made available[1] by improving the existng BabaYaga event generator for QED processes at flavor factories.

The generator can be useful for data analysis at flavor factories to study every physical observable, i.e. invariant mass distributions but also angular distributions or correlations and so on.

It includes:

- an exact tree-level calculation of the signal and background processes contributing to the signatures $e^+e^- \to \gamma, U \to \mu^+\mu^-\gamma, e^+e^-\gamma$;
- the effects of the most important higher-order corrections induced by multiple photon radiation and vacuum polarization.

We used our calculation in order to assess its impact on the experimental sensitivity as evaluated in the literature, and show how this can be enhanced by means of event selections not considered so far.

The calculation has been done by means of the ALPHA[22] algorithm, a tool to compute tree-level matrix elements. We implemented the calculation in the BabaYaga@NLO event generator[21, 23, 24], to simulate distributions of experimental interest and account for realistic event selection criteria. In the U boson propagator we include the total dark photon width using the formulae of Ref. [18] for the partial widths into leptons and hadrons, i.e.

$$\Gamma_{U\to f^+f^-} = \frac{1}{3}\alpha\varepsilon^2 M_U \beta_f \left(1 + \frac{2m_f^2}{M_U^2}\right) R(S = M_U^2) \tag{2}$$

where $\beta_f = \sqrt{1 - \frac{4m_f^2}{M_U^2}}$, $R = 1$ for leptonic decays of the U boson and $R = \sigma_{e^+e^-\to\text{hadrons}}/\sigma_{e^+e^-\to\mu^+\mu^-}$ for hadronic decays[25]. It is worth to note that the U boson is a dramatically tiny resonance, being is width well under any experimental sensitivity, varying from 10^{-7} to 10^{-2} for reasonable values of M_U and ε.

It is well known that at flavor factories multiple soft and collinear radiation emitted by the colliding beams may have a strong impact on the measured cross section and on the shape of the distributions. The effect of higher-order corrections in our Monte Carlo generator is taken into account using the popular QED structure function approach[26]. Initial state radiation modifies the tree-level cross section as follows

$$\mathrm{d}\sigma(s,t) = \int_0^1 \mathrm{d}x_1\mathrm{d}x_2\mathrm{d}y_1\mathrm{d}y_2\mathrm{d}\sigma_0 D^e(x_1,s)D^e(x_2,s)D^l(y_1,\hat{s})D^l(y_2,\hat{s}) \tag{3}$$

where $\hat{s}, \hat{t}$ are the Mandelstam invariants after photon radiation and $D(x,s)$ is the electron structure function, which describes the probability to find an electron with a momentum fraction x inside an electron of a given momentum s. If we take into account only the most important terms, due to resummed multiple soft and hard photon emission in the collinear approximation $(\mathcal{O}(\alpha))$, the structure function has the form:

[1] you can find the code on our website (`http://www.pv.infn.it/~hepcomplex`).

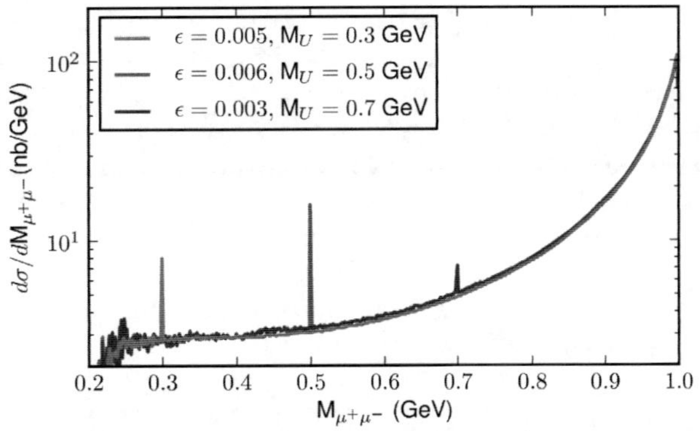

Figure 2. Invariant mass distribution of the muon pairs for different values of ε and M_U with $s = 1.02$ GeV

$$D(x,s) = \frac{e^{\frac{\beta}{2}\left(\frac{3}{4}-\gamma_E\right)}}{\Gamma\left(1+\frac{\beta}{2}\right)}\frac{\beta}{2}(1-x)^{\frac{\beta}{2}-1}-\frac{\beta}{4}(1+x)+\frac{\beta^2}{32}\left[(1+x)(-4\ln(1-x)+3\ln x)-4\frac{\ln x}{1-x}-5-x\right] \tag{4}$$

where $\beta = 2\alpha/\pi(\ln(s/m_f^2)-1)$, Γ is the Euler gamma function and γ_E is the Euler - Mascheroni constant.

In Eq. 3 $D^e(x_i, s)$ refers to the structure functions of the initial state electron and $D^l(y_i, \hat{s})$ refers to final state fermions.

Being an effect of size comparable to that of photon radiation, we also consider in our calculation the running of the electromagnetic coupling constant according to (see Ref. [21] for a recent review)

$$\alpha(q^2) = \frac{\alpha}{1-\Delta\alpha(q^2)}$$
$$\Delta\alpha(q^2) = \Delta\alpha_l(q^2) + \Delta\alpha_h(q^2) \tag{5}$$

where $\Delta\alpha_l$ is the contribution due to leptons, which is analitically known, and $\Delta\alpha_h$ is the non-perturbative hadronic contribution, included according to Ref. [25] or [27].

Fig. 2 illustrates a typical signal, the invariant mass distribution of the muon pairs for three values of ε and M_U at DAΦNE energies ($s = 1.02$ GeV), as obtained with our calculation in the lowest order approximation. It is possible to see a peak in the distribution due to the U boson effects.

2. Experimental sensitivity

If we saw a peak in a distribution, for example the one related to the invariant mass of final state fermions, we would be able to describe the U boson properties like its coupling and its mass. Instead if we couldn't see nothing but the SM background, we would exclude a certain region in the M_U - ε parameter space. So we need to calculate which are the U boson properties that made possible to confuse the U boson's peak with a statistical fluctuation of the background for a certain experiment with its experimental resolution and its luminosity (*experimental sensitivity*).

Journal of Physics: Conference Series **335** (2011) 012077 doi:10.1088/1742-6596/335/1/012077

On the grounds of the calculation described in Section 1, we revisited the experimental sensitivity to a dark force signal evaluated in the literature according to the approximations previously discussed. For concreteness, we consider the case of the KLOE/KLOE-2 experiment at the upgraded DAΦNE [28] and of future experiments as SuperB factories[29, 30]. We also made a comparison with BaBar's analysis made by [20] and the results totally agree, which means that other calculations already present in the literature are a really good estimate of the exact calculation and that the effects due to the finite width of the U boson are negligible for the invariant mass distribution. Neverthless the Monte Carlo event generator can be useful for the experimental analysis and for studies of backward-forward asymmetry (which is a really smaller signal with respect to the presence of the peak in the invariant mass distribution).

We define the statistical significance (S) as

$$S = \frac{N_S}{\sqrt{N_B}} = \sqrt{L}\frac{(\sigma_F - \sigma_B)}{\sqrt{\sigma_B}} \tag{6}$$

where σ_F is the full cross section including the exchange of a virtual photon and U boson, σ_B (N_B) is the SM cross section (number of events) background, N_S is the expected number of events due to the presence of U boson and L is the integrated luminosity. In the above equation we require S to be greater than 5 to claim a discovery.

We assume L = 5 fb^{-1} for KLOE/KLOE-2 and L = 100 ab^{-1} for a B SuperB factory, respectively. To simulate detector acceptances, we also impose the following energy and angular cuts:

KLOE $35^\circ \leq \theta_{l^\pm,\gamma} \leq 145^\circ$, $E^{min}_{l^\pm,\gamma} = 10$ MeV

SuperB $30^\circ \leq \theta_{l^\pm,\gamma} \leq 150^\circ$, $E^{min}_{l^\pm,(\gamma)} = 30, (20)$ MeV

using as c.m. energies $\sqrt{s} = 1.02, 10.56$ GeV, respectively. To improve S, signal events can be detected as peaks in the lepton pair invariant mass close to the value M_U, in a window $M_U \pm \delta_M$. We want δ_M to be as small as possible, optimally coinciding with the detector resolution, a crucial parameter for these studies.

As detector resolutions we use $\delta_M = \pm 1$ MeV for KLOE/KLOE-2 and the values obtained according to the empirical relations of Ref. [20] for B/SuperB factories, giving $\delta_M \sim \pm[1-10]$ MeV for a mass M_U in the range 0.1 - 5 GeV.

In Fig. 3 we show the reach potential of KLOE/KLOE-2 experiment, where the sensitivity to the kinetic mixing parameter ε is shown as a function of M_U. Actually, we observed that corrections affect with about the same amount both the signal and the background cross sections; hence the systematics induced by photon radiation largely cancel in the experimental sensitivity and the conclusions in the presence of QED radiative corrections confirm the ones obtained in the lowest order approximation.

Exclusion limits imposed by the precision measurements of the anomalous magnetic moment of the leptons and of α_{QED}[31] are indicated as gray areas.

The most striking features which could be seen in Fig. 3 are the following:

- the muon channel has a better reach than the e^+e^- channel, obviously because of the smaller background. The e^+e^- channel could instead be interesting for low values of M_U, but an experimental difficulty, due to the fact that it is really hard to distinguish a low energy electron with a photon, exists and limits this kind of analysis. If we want to study a U boson lighter than the $\mu^+\mu^-$ threshold we have to use some different techniques;
- the sensitivity is significantly degraded if M_U is around the ϱ resonance because the branching fraction $U \to l^+l^-$ is suppressed by the dominant decay mode $U \to \pi^+\pi^-$, which would be a more convenient channel in this region.

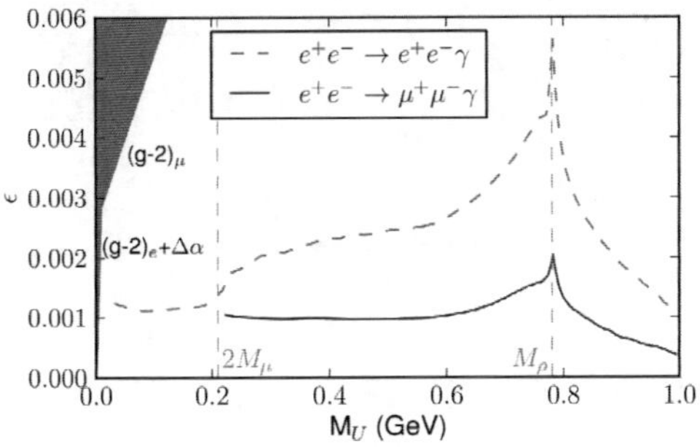

Figure 3. Experimental sensivity for KLOE/KLOE-2

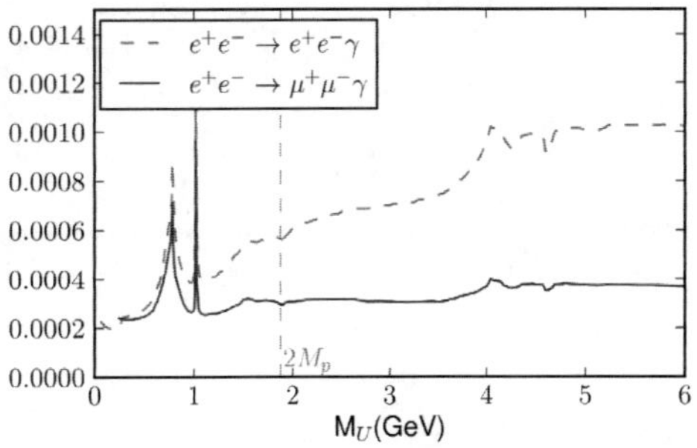

Figure 4. Experimental sensivity for SuperB

We have noticed that the maximum sensitivity achievable at the upgraded DAΦNE with a luminosity $L \simeq 5$ fb^{-1} , i.e. $\varepsilon \sim 0.001 - 0.002$, is equivalent to the one of the present B-factory experiments BaBar/Belle with $L \simeq 500$ fb^{-1}[20]. This fact could be easily understood since the reach on ε follows the rule $\varepsilon^2 \propto E/\sqrt{L}$.

In our analysis we realized that radiative corrections do not significatively alter the experimental sensitivity because both the background and the signal are modified by the same amount.

Fig. 4 presents the experimental sensitivity at a SuperB factory, where it could be possible to exclude an epsilon greater than few per 10^{-4} for all mass values up to 2 proton mass. Hence the really high statistics of a SuperB collider will allow to probe values of ϵ about an order of magnitude smaller than those reachable by present flavor factories.

Finally, Figs. 5 and 6 show the sensitivity for the small angle selection, defined in Eq. 7, for the KLOE/KLOE2 experiment.

$$35^o \leq \theta_{l^\pm} \leq 145^o, E_{l^\pm}^{min} = 10\text{MeV} \\ |\cos(\theta_\gamma)| \geq \cos(15) \tag{7}$$

The small angle selection is also compared with the large angle selection previously discussed. It is possible to see that there is a relevant gain in sensitivity for M_U greater than ~ 0.5 GeV.

3. Conclusion

With existing experiments and analysis of existing data we can exclude a U boson with an ϵ coupling of about 10^{-3} for a wide range of mass (flavor factories[1]) and some particular area in the ϵ - M_U parameter space for a really light U boson (beam dump experiment[4]). With a SuperB factory we would be able to see a U boson if its kinetic mixing is above few 10^{-4}. Anyway a really big area in the $\varepsilon - M_U$ space would be allowed also after the analysis of a SuperB experiment. A U boson with such properties would be able to explain the astrophysical data. To cover this area it would be useful a new fixed target experiment[16] in which a really high luminosity could be reached.

References

[1] Barzè L, Balossini G, Bignamini C, Calame C M C, Montagna G, Nicrosini O and Piccinini F 2010 6 URL http://arxiv.org/abs/1007.4984

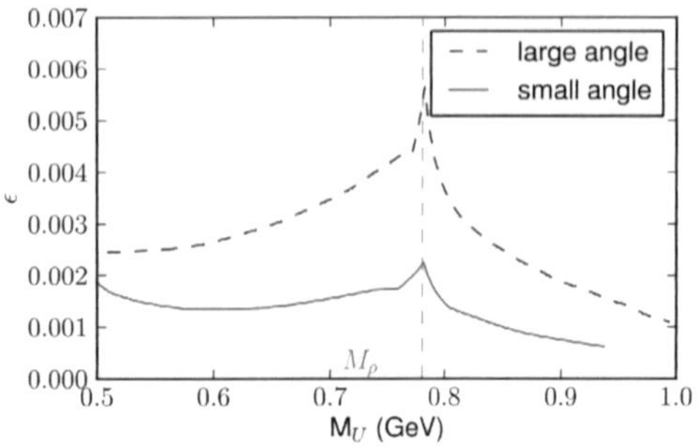

Figure 5. Experimental sensivity for KLOE/KLOE-2 - *Small Angle* selection - μ channel

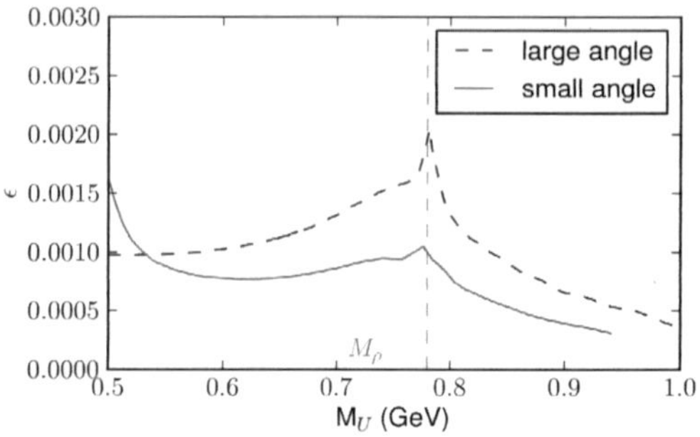

Figure 6. Experimental sensivity for KLOE/KLOE-2 - *Small Angle* selection - e channel

[2] Boehm C and Fayet P 2004 *Nuclear Physics B* **683** 219–263 ISSN 05503213 URL http://arxiv.org/abs/hep-ph/0305261

[3] Pospelov M, Ritz A and Voloshin M 2008 *Physics Letters B* **662** 53–61 ISSN 03702693 URL http://arxiv.org/abs/0711.4866

[4] Pospelov M 2009 *Physical Review D* **80** 14 ISSN 1550-7998 URL http://arxiv.org/abs/0811.1030

[5] Cholis I, Dobler G, Finkbeiner D P, Goodenough L and Weiner N 2009 *Physical Review D* **80** 28 ISSN 1550-7998 URL http://arxiv.org/abs/0811.3641

[6] Arkani-Hamed N, Finkbeiner D P, Slatyer T R and Weiner N 2009 *Physical Review D* **79** 22 ISSN 1550-7998 URL http://arxiv.org/abs/0810.0713

[7] Jean P, Knodlseder J, Lonjou V, Allain M, Roques J P, Skinner G K, Teegarden B J, Vedrenne G *et al.* 2003 *Astronomy and Astrophysics* **407** L55–L58 ISSN 0004-6361 URL http://arxiv.org/abs/astro-ph/0309484

[8] Adriani O, Barbarino G C, Bazilevskaya G A, Bellotti R, Boezio M, Bogomolov E A, Bonechi L, Bongi M *et al.* 2009 *Nature* **458** 607–9 ISSN 1476-4687 URL http://arxiv.org/abs/0810.4995

[9] Chang J, Adams J H, Ahn H S, Bashindzhagyan G L, Christl M, Ganel O, Guzik T G, Isbert J *et al.* 2008 *Nature* **456** 362–5 ISSN 1476-4687 URL http://dx.doi.org/10.1038/nature07477

[10] Abdo A, Ackermann M, Ajello M, Atwood W, Axelsson M, Baldini L, Ballet J, Barbiellini G *et al.* 2009 *Physical Review Letters* **102** 6 ISSN 0031-9007 URL http://arxiv.org/abs/0905.0025

[11] Aharonian F, Akhperjanian A, Barres de Almeida U, Bazer-Bachi A, Becherini Y, Behera B, Benbow W, Bernlöhr K *et al.* 2008 *Physical Review Letters* **101** 5 ISSN 0031-9007 URL http://arxiv.org/abs/0811.3894

[12] Bernabei R, Belli P, Cappella F, Cerulli R, Dai C J, DAngelo A, He H L, Incicchitti A *et al.* 2008 *The European Physical Journal C* **56** 333–355 ISSN 1434-6044 URL http://arxiv.org/abs/0804.2741

[13] Aalseth C, Barbeau P, Cerdeño D, Colaresi J, Collar J, de Lurgio P, Drake G, Fast J *et al.* 2008 *Physical Review Letters* **101** ISSN 0031-9007 URL http://arxiv.org/abs/0807.0879

[14] Bjorken J, Essig R, Schuster P and Toro N 2009 *Physical Review D* **80** 14 ISSN 1550-7998 URL http://arxiv.org/abs/0906.0580

[15] Freytsis M, Ovanesyan G and Thaler J 2009 38 URL http://arxiv.org/abs/0909.2862

[16] Essig R, Schuster P, Toro N and Wojtsekhowski B 2010 *Physical Review D* **80** URL http://arxiv.org/abs/0903.3941

[17] Borodatchenkova N, Choudhury D and Drees M 2006 *Physical Review Letters* **96** 4 ISSN 0031-9007 URL http://arxiv.org/abs/hep-ph/0510147

[18] Batell B, Pospelov M and Ritz A 2009 *Physical Review D* **79** 14 ISSN 1550-7998 URL http://arxiv.org/abs/0903.0363

[19] Bossi F 2009 URL http://arxiv.org/abs/0904.3815

[20] Reece M and Wang L 2009 *Journal of High Energy Physics* **2009** 051–051 ISSN 1029-8479 URL http://arxiv.org/abs/0904.1743

[21] Actis S, Arbuzov A, Balossini G, Beltrame P, Bignamini C, Bonciani R, Carloni Calame C M, Cherepanov V *et al.* 2010 *The European Physical Journal C* **66** 585–686 ISSN 1434-6044 URL http://arxiv.org/abs/0912.0749

[22] Caravaglios F and Moretti M 1995 *Physics Letters B* **358** 332–338 ISSN 03702693 URL http://arxiv.org/abs/hep-ph/9507237

[23] Carloni Calame C M 2000 *Nuclear Physics B* **584** 459–479 ISSN 05503213 URL http://arxiv.org/abs/hep-ph/0003268

[24] Balossini G, Carloni Calame C M, Montagna G, Nicrosini O and Piccinini F 2006 *Nuclear Physics B* **758** 227–253 ISSN 05503213 URL http://arxiv.org/abs/hep-ph/0607181

[25] Teubner T, Hagiwara K, Liao R, Martin A D and Nomura D 2010 6 URL http://arxiv.org/abs/1001.5401

[26] Nicrosini O and Trentadue L 1987 *Physics Letters B* **196** 551–556 ISSN 03702693 URL http://dx.doi.org/10.1016/0370-2693(87)90819-7

[27] Jegerlehner F 2006 *Nuclear Physics B - Proceedings Supplements* **162** 22–32 ISSN 09205632 URL http://arxiv.org/abs/hep-ph/0608329

[28] Amelino-Camelia G, Archilli F, Babusci D, Badoni D, Bencivenni G, Bernabeu J, Bertlmann R A, Boito D R *et al.* 2010 60 URL http://arxiv.org/abs/1003.3868

[29] Collaboration S 2007 524 URL http://arxiv.org/abs/0709.0451

[30] Golob B 2010 URL http://arxiv.org/abs/1006.4208

[31] Pospelov M and Ritz A 2009 *Physics Letters B* **671** 391–397 ISSN 03702693 URL http://arxiv.org/abs/0810.1502

Physics potential of SuperB project

Alberto Cervelli on behalf of SuperB collaboration
Universitá and INFN Pisa, Largo Bruno Pontecorvo 3, Pisa 56124, Italy
E-mail: alberto.cervelli@pi.infn.it

Abstract. In the following we present the physics prospects of SuperB project, a proposed asymmetric e^+e^- collider, projected to have a mean luminosity of $\mathcal{L} = 10^{36}\text{cm}^{-2}\text{s}^{-1}$, and the option for an 80% electron beam polarization, which would provide, with a five-year data taking period, a data sample consisting of an integrated luminosity of 75 ab^{-1}. Such a large data sample will allow to perform precision measurement in many different sectors. In this paper we will focus on three main topics: the searches for τ lepton flavor violating decays, with in depth discussion about $\tau \to \ell\gamma$ and $\tau \to \ell_1\ell_2\ell_3$ (where $\ell_i = e, \mu$); search for CP violation in the charm sector, which has never been observed and if observed would provide a clear hint for new physics; and precision measurements in the Electro-Weak sector, both measuring $\sin^2\theta_W$ in an energy region previously unexplored, and testing neural current universality in the γ-Z interference and Z-b$\bar{\text{b}}$ coupling. The Physics program of SuperB project will be vast and the potential reach for new physics discovery great thanks to the many topics that will be studied with the large statistic available and the novel machine techniques expected for SuperB.

1. Introduction

The Standard Model of fundamental interactions (SM) has been tested in the last decades to great extent, without any finding of large discrepancies from theoretical expectations. The great success of SM was confirmed by the measurement on Electroweak sector at hadron and e^+e^- machines, such as LEP results, discovery of the top quark, measurement of CP violation in K and B sectors observed by experiment at CERN, Fermilab and at the B-Factories KEKB and PEP-II. On the other hand SM is not able to explain some physical observations and some crucial questions are still left unanswered as the lack of original antimatter in the Universe or the nature of dark matter. It is a general opinion that physics processes not expected by the SM need to be looked for, and further experimentation is needed to look for new physics (NP). Results from neutrino oscillations [1] from dedicated experiments suggest that NP could be within experimental sensitivities and Lepton Flavor Violation (LFV) would be one of the most clear signals of it. A direct way to look for NP is now pursued by LHC experiments, where the available energy for the interaction is enough to produce particles not predicted by the SM. The other way is to look at the indirect effects of NP in interference processes (as CP violation and Electroweak effects) and in rare or forbidden decays (like in LFV processes in μ and τ decays).

The complementarity between the two approaches is based on the fact that NP processes are mainly governed by two parameters: the NP energy scale Λ and the effective coupling of the processes C, which would have different intensities and different patterns, depending on the particular NP flavor model. Super Flavor Factories, producing 7-9 $\times 10^{10}$ pairs of B and D mesons and of τ leptons, will provide enough statistics to look for NP in both cases whether LHC will produce evidence or not. If NP is found at LCH, such high statistics would provide

Published under licence by IOP Publishing Ltd

Journal of Physics: Conference Series **335** (2011) 012078 doi:10.1088/1742-6596/335/1/012078

precision measurements on the flavor structure of NP, through CPV phases, LFV processes, and electroweak parameters, along with possible hints of heavier states not yet found at LHC. If the NP is to be expected at largerΛ , the Super Flavor Factories will be able look for indirect searches for NP and constrain the phase space of NP models up to the order of several tens or even hundred of TeV.

In the following we present the physics potential of the SuperB factory, with particular attention to CP violation in the charm sector, searches for LFV in the τ decays, and precision measurements in the Electroweak sector.

2. Search for Lepton Flavor Violation in τ decays

The τ physics will assume great importance to probe new physics beyond the SM. The heavy lepton sector, with the use of larger luminosities available at SuperB, will provide precise measurement of both direct effects, via LFV processes, and indirect effects, visible in g-2 and electron dipole moment (EDM) [2] of τs.

The SuperB machine is expected to deliver mean luminosities of $10^{36}\text{cm}^{-2}\text{s}^{-1}$, achieving an integrated luminosity ($\mathcal{L}$) of about 75 ab^{-1} during a five year long period of data-taking. Differently from other proposed super flavor factories, SuperB machine design provides the option of a polarized e^- beam, with mean polarization of $\sim 80\%$, which may be used to both reduce backgrounds in LFV searches and shed light on the chiral coupling, and flavor structure of the NP if LFV signal is observed.

LFV is indeed one of the clearest signature of NP if detected, in fact SM predicts LFV in the charged sector only at very small rates (less than $\mathcal{O}(10^{-40})$) [3], which are well beyond present (and future) flavor factory sensitivities, on the other hand many NP models predict enhanced branching fractions (BF) for LFV processes as large as $\mathcal{O}(10^{-8})$ [4]. An observation of LFV processes would be a clear signature for NP, and if LFV is observed in more than one channel the ratios between different processes BFs would identify the NP flavor structure [5].

2.1. Search for $\tau^- \to \ell^-\gamma, \ell = e, \mu$

Most NP model predict $\tau^- \to \ell^-\gamma, \ell = e, \mu$ to be the LFV process with the largest BF, so the search for this particular decay is of uttermost importance for NP searches since the sensitivity needed for the discovery is smaller than the one needed for other channels. Searches at the B-factories were limited by irreducible backgrounds coming from radiative τ decays such as $\tau \to \mu\nu\nu\gamma$. Radiative τ decays present the same topology, reconstructed neutral energy, and lepton momenta as a signal event when the νs are produced almost at rest. BaBar, in its analysis [6], performed a blind search for $\tau^- \to \ell^-\gamma, \ell = e, \mu$ decays, and 5.1 background events were expected in the 2σ blinded signal region, among them 1.7 came from lepton tags, 2.0 expected in single hadron tags (π and ρ), and the rest coming from 3 hadron tags. Most of the background is produced by real τ decays, and 86% of the total is due to $\tau \to \mu\nu\nu\gamma$ decays: if no further backgrounds become relevant with the increase in statistics expected at SuperB, the backgrounds predicted at higher luminosities could be extrapolated by a simple scaling. Considering a sample of 75 ab^{-1}, SuperB would expect about 300 background events in the signal region, leading to an expected Upper Limit (UL) of $\sim 6.4 \times 10^{-9}$, which would scale as $\sqrt{\mathcal{L}}$ from BaBar ULs. The $\sqrt{\mathcal{L}}$ scaling is expected if background can not be reduced more efficiently than in previous experiment, so if no further handles for signal selection are available the UL improvement would be strongly suppressed.

SuperB machine is expected to have an 80% polarized e^- beam, this would allow to have information on the dynamical properties for the decay, especially if single hadron modes ($\tau \to \pi\nu$ and $\tau \to \rho\nu$) are used in the tagging,with the presence of only one ν in the final state for signal events. The presence of only one ν has two main consequences: first of all the event kinematics is fully reconstructed by identifying the missing momentum of the event to be the ν momentum,

Journal of Physics: Conference Series **335** (2011) 012078 doi:10.1088/1742-6596/335/1/012078

and secondly and more importantly the event helicity is fixed, since the ν may be considered massless and hence chiral. Angular momentum conservation can be used to reduce backgrounds by looking at the correlation between the signal and tracks helicity angles: the helicity angle is defined as the angle between the track momentum in the parent τ rest frame and the momentum of the other τ in the laboratory frame. The correlation is shown for both signal and background in Fig. 1.

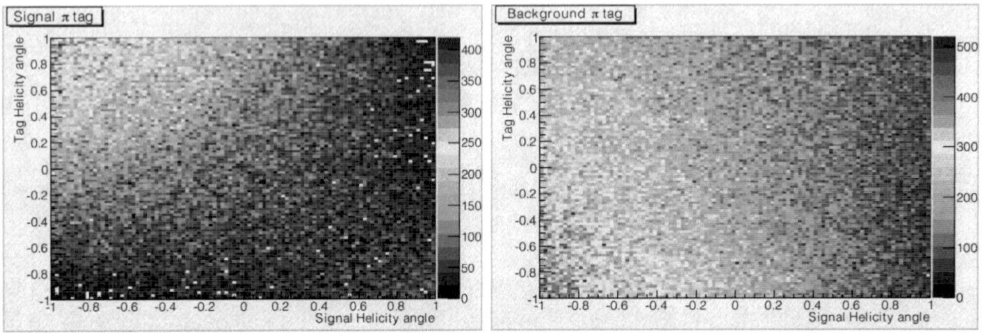

Figure 1. Correlation between signal track and tag track helicity angles for signal events (left) and background events (right) for $\tau \to \pi\nu$ tag. On the horizontal axis the signal helicity angle is plotted while on vertical axis tag helicity angle is shown.

Applying a simple selection on the helicity angle correlation it is possible to strongly reduce backgrounds coming from radiative τ decays, even by a factor 10, while retaining almost 50% of the signal. Using only single hadron tags, along with angular momentum conservation the background could be reduced to $\sim$ 15 events, making it possible for the UL to scale almost linearly with $\mathcal{L}$. The improvement expected with the use of polarized beams would make it possible to reach UL of 3.9×10^{-9}, corresponding to an increase of $\mathcal{L}$ (in the background dominated case, using all tags) of a factor 2.6, albeit using only 25% of the statistics. Further improvements are expected using the same method on lepton tags, even if the effect is going to be diluted due to the presence of two νs in the event.

2.2. Search for $\tau^- \to \ell_1^- \ell_2^+ \ell_3^-, \ell_i = e, \mu$

The search for $\tau^- \to \ell_1^- \ell_2^+ \ell_3^-$ decays allows the simultaneous study of all the six channels compatible with charge conservation. Theoretical models generally predict smaller BF for these channels with respect to $\tau^- \to \ell^- \gamma$ decays, making them an useful, but less important way to discover LFV processes, on the other hand being able to simultaneously measure six channels would make it possible to study the flavor structure of NP and the chiral coupling of the NP particles. Background expectation in previous BaBar analysis [7] were small for all the six channels, so even considering a background increase proportional to $\mathcal{L}$, and without considering any improvement in both selection and detection, the ULs are expected to scale almost $\propto \mathcal{L}$ as shown in Fig.2. SuperB thanks to its large statistics would reach ULs of $\mathcal{O}(10^{-10})$ for some channels, with an improvement of almost 2 orders of magnitude for all the channels.

3. CP Violation in charm sector

The recent observation of large $D^0\bar{D}^0$ oscillation at the B-Factories [8] rise the exciting possibility of finding CP violation in the charm decay, which would be a major hint for physics beyond the SM, in fact no CP violation is expected in the up sector due to the presence of only one phase

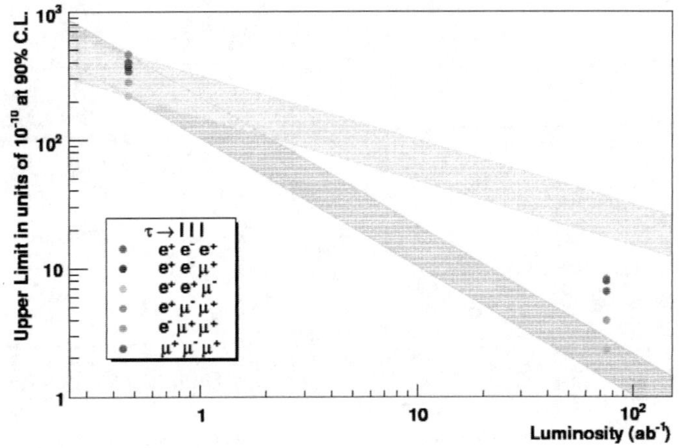

Figure 2. UL measured at the B-factories and extrapolation at 75 ab^{-1}. The yellow shading presents the background dominated expectation ($\propto \sqrt{\mathcal{L}}$) and the green shading the background free expectation ($\propto \mathcal{L}$) for the ULs

in the CKM matrix. SuperB will be able to make comprehensive studies in the charm sector, with the large amount of data taken on the $\Upsilon(4S)$ resonance, SuperB will also be able to run at$\Psi\,'(3770)$ resonance, featuring a collider designed to run at lower center of mass energies with reduced luminosity of $10^{35}\mathrm{cm}^{-2}\mathrm{s}^{-1}$.

SuperB has many distinctive features useful to study CPV in the charm sector. The experimental environment is very clean, with small background contribution with respect to the great rate at which D's are produced, at $\Upsilon(4S)$ energy the mesons can be tagged through $D^\star \rightarrow D\pi$ decay, making it possible to have a flavor tag on the produced Ds. On the other hand D production at$\Psi\,'(3770)$ energy would allow a coherent production of the $D^0\bar{D}^0$ pair, opening novel ways to study CPV in the up sector. While running at threshold would offer lower backgrounds and access to both direct and indirect CPV it comes at the expense of statistics. SuperB great luminosity at $\Upsilon(4S)$ energy make it possible to obtain great results in time dependent measurements: the vertexing detector of SuperB should be able to achieve precisions of few tens of microns, which would make it possible to study the decay lengths of the D's with the beam position known with nm precision allowing time dependent measurements. In Fig. 3 present and future precision for CPV parameters in the D sector are shown.

4. Testing the Electro-Weak sector

The combination of high luminosity and polarized electrons present at SuperB provides a unique opportunity to measure, with precision comparable or better than SLC and LEP, a number of electroweak neutral current parameters at transfered momenta q^2 of (10.58 GeV)^2. In particular use of polarized beams allows the measurement of $\sin^2\theta_W$, using μ and τ pair production differential cross sections, with precision comparable to SLD but at much lower energies. Another important aspect for Electroweak precision tests is the measurement of the neutral current vector coupling to both μs and b quarks, such measurements are sensitive to a Z' and can prove neutral current universality at high precision.

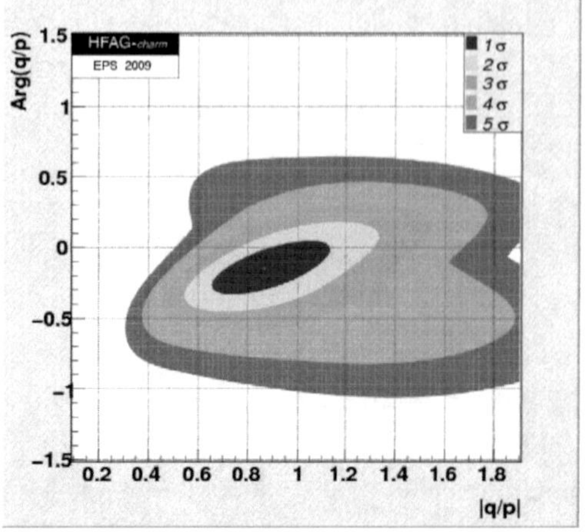

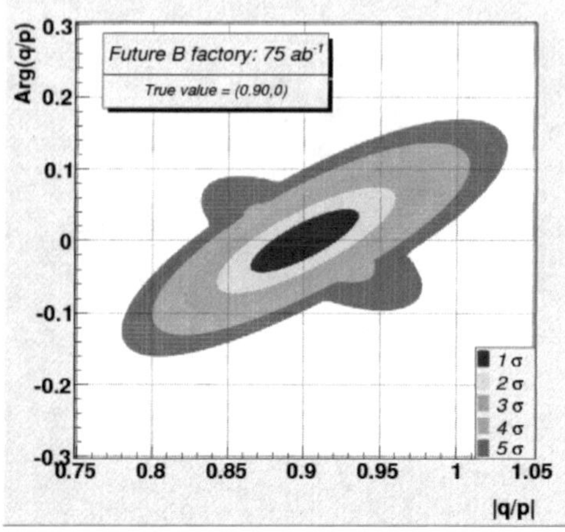

Figure 3. Two dimensional contours for Charm mixing and CP violation as a function of q/p and a strong phase δ, at present (left) and after 5 years of data-taking of SuperB (right). Note the different scales on the axis.

4.1. Measure of $\sin^2(\theta_W)$ *with* $e^+e^- \to \mu^+\mu^-$ *processes*

With polarization, SuperB will be able to make a relatively straightforward measurement of the left-right asymmetry in the $e^+e^- \to \mu^+\mu^-$ production in a manner identical to the one used by the SLC collaboration [9], which operated at the Z pole. SuperB will measure the A_{LR} asymmetry defined as $A_{LR} = \frac{\sigma_L-\sigma_R}{\sigma_L+\sigma_R}\frac{1}{<P>}$, where $< P >$ is the mean polarization of the beams, $\sin^2(\theta_W)$ can be extracted from the A_{LR} parameter since $A_{LR} \propto (T_3^f - 2Q_f \sin^2(\theta_W))$. SLC measured $\sin^2(\theta_W) = 0.23098 \pm 0.00026$, where the systematic uncertainty (accounting for almost 50% of the error) is due to the 0.5% uncertainty on the polarization.

An analysis made using BaBar sample, and then scaled to the SuperB luminosity, revealed that selection efficiencies of 53% can be achieved with 99.6% sample purity. Considering a sample of 46 billion μ-pairs available after 5 year of data taking at SuperB, and an 80% mean polarization, the statistical error on A_{LR} is expected to be of $\mathcal{O}(10^{-2})$. If the systematic error on polarization measurement can be kept under 1% level the uncertainty on $\sin^2(\theta_W)$ may be as low as 0.0002, competitive to the SLC measurement, but at a different q^2. The same A_{LR} can be measured in τ and charm production, with smaller efficiencies and purities, providing the most stringent test to neutral current universality. Fig.4 shows the present results and the planned measurement that will be made by SuperB.

4.2. Z-$b\bar{b}$ *coupling and* g_V^b *measure*

As SuperB will be running mostly at Υ(4S) energy, the left right asymmetry for B mesons will be sensitive to the product of the electron neutral current axial coupling and b quark neutral current vector coupling g_V^b, the same formalism adopted for measuring g_V^s at Φ factories can be used and is described elsewhere [10]. Using 10^9 $B\bar{B}$ pairs produced at the Υ(4S) energy, with an 80% e^- polarized beam, SuperB will provide a measurement of g_V^b competitive with the measurement from LEP and SLC, but at smaller q^2 values. Present measurement have a relative error of about 2.4% with the statistics available at SuperB a 0.3% error could be achieved if the systematics error on polarization are kept under control. The main concern comes from the non resonant production of B-mesons, however the problem has been studied and results prove that

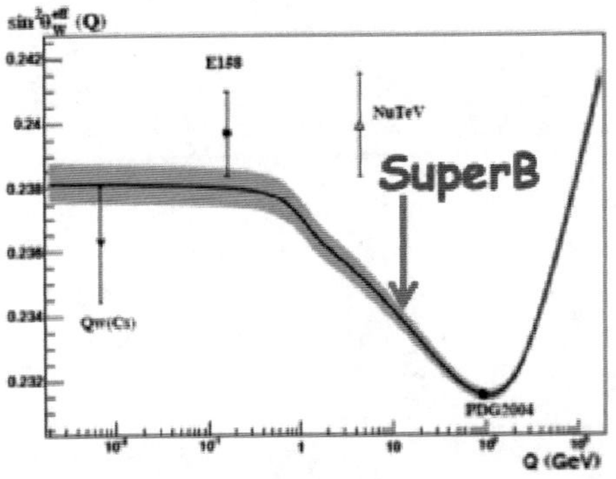

Figure 4. Summary of experiments that have measured $\sin^2(\theta_W)$. SM predictions and uncertainty are presented as the solid line and gray area respectively. SuperB will provide a point at $q = 10.58$ GeV, a region previously unexplored.

A_{LR} from $B\bar{B}$ at $\Upsilon(4S)$ give access to the $Z - b\bar{b}$ coupling in a manner independent from the production process. Such high precision measurement can be obtained at SuperB thanks to the fact that no QCD corrections are needed at such low energy, and that the coupling between the final states and the parent pseudo scalars couple directly without any hadronization processes.

It is also possible to measure the ratio between the left right asymmetry measured in muon and b-quark production, testing the universality of neutral vector coupling with great precision, since polarization systematics cancel out in the ratio, giving another robust way to test neutral current universality. In addition to probing NP, this measurement will shed light on the long standing 3σ difference between the measurement of $\sin^2(\theta_W)$ obtained from the forward backward asymmetry of b-quarks and the leptons. In Fig.5 the present status of g_V^b and g_A^b measurement is shown along with the expected precision of the measurement made at SuperB.

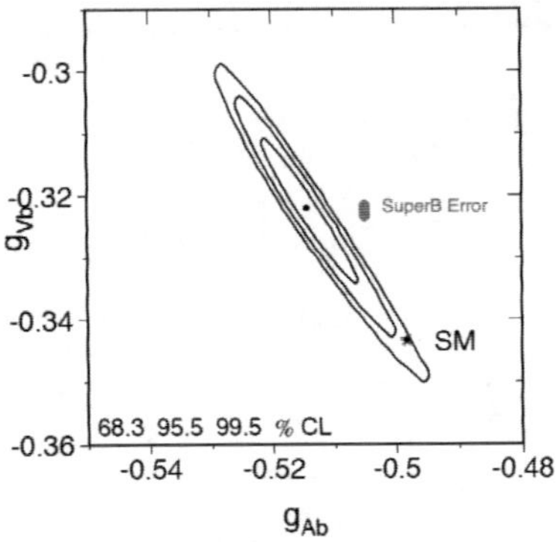

Figure 5. Present status of measurement of g_V^b and g_A^b, contours represent experimental uncertainties, the red cross represent SM prediction, the pink shaded area is the expected error obtained by SuperB measurement.

4.3. Open issues in Electroweak test at SuperB

The main issues for Electroweak testing at SuperB come from the systematic errror on polarization, which are dominated by the luminosity dependent polarization: it is of great importance to know $< P >$ at a level better than 1% in order to reach sensitivities competitive to those of previous experiment. SuperB with its great luminosity could in principle have a great knowledge of the polarization value for each bunch, and new ways to measure $< P >$ with great precision are under study. On the other hand in order to compare the SuperB results with LEP and SLC results g_V^b should be run to larger q^2 values, which is a huge theoretical effort. The last issues comes from the possibility to use$\Psi\,'(3770)$ running to make the same measurement for charm as it was done for strange and bottom coupling, however large polarization at lower energy is needed along with sufficient luminosity, the reach of charm running and the systematics associated are still under study by SuperB collaboration.

5. Conclusions

SuperB machine with its large luminosity and the possibility to have a polarized electron beam will have clear edges in the search for NP in the near future. Its design and physics case is complimentary to LHC experiments, and SuperB seeks NP in an indirect way, looking for discrepancies from SM predictions and rare decays. SuperB has a strong physics case in the search for LFV in the τ sector being able with its polarized beams to reduce UL of $\tau \to \ell\gamma$ of almost two order of magnitude, albeit the present searches are background dominated, and the same will happen for $\tau \to \ell\ell\ell$ LFV decays. The nanometric beams and the large statistics, along with the possibility to run at$\Psi\,'(3770)$ will allow to look for CPV in the charm sector in both direct CPV and in the mixing, or even with time dependent measurement. On the precision frontier, SuperB will be able to measure Electroweak parameters in q^2 regions previously unexplored and probably shedding light on discrepancies observed by previous experiments such as LEP and SLC. SuperB has a vast physics case ranging from B-physics to electroweak precision test and its sensitivity to such processes is a great tool for NP discovery and for NP flavor study, if NP is going to be observed at LHC.

References

[1] K2K Collaboration, M. H. Ahn *et al.*, Phys. Rev. Lett. **90**, 041801 (2003); KamLAND Collaboration, K. Eguchi *et al.*, Phys. Rev. Lett. **90**, 021802 (2003); SNO Collaboration, Q. R. Ahmad *et al.*, Phys. Rev. Lett. **89**, 011301 (2002).
[2] J. Bernabéu, G. A. Gonzalez-Sprinberg, and J. Vidal, Nucl. Phys. **B763**, 283 (2007), hep-ph/0610135.
[3] W.J. Marciano and A.I. Sanda, Phys. Lett. B **67** 303 (1977).
[4] P. Paradisi, JHEP **0510**, 006 (2005); K. S. Babu and C. Kolda, Phys. Rev. Lett. **89**, 241802 (2002); A. Brignole and A. Rossi, Phys. Lett. B **566**, 217 (2003).
[5] A.G. Akeroyd, M. Aoki, and Y. Okada, Phys. Rev. D **76**, 013004 (2007).
[6] B. Aubert *et al.*, Phys. Rev. Lett. **104**, 021802 (2010).
[7] J.P. Lees *et al.*, Phys. Rev. D **81**, 111101(R) (2010).
[8] B. Aubert *et al.*, Phys. Rev. Lett. **98**, 211802 (2007).
[9] K. Abe *et al.*, Phys. Rev. Lett. **73**, 25 (1994); S. Schael *et al.*, Phys. Rept. **427**, 257 (2006).
[10] J. Bernabeu, F.J. Botella, O. Vives, Nucl.Phys. **B472** (1996) 659-682.